Graphene-Based Terahertz Electronics and Plasmonics

Graphene-Based Terahertz Electronics and Plasmonics

Detector and Emitter Concepts

edited by

Vladimir Mitin | Taiichi Otsuji | Victor Ryzhii

Published by

Jenny Stanford Publishing Pte. Ltd.
Level 34, Centennial Tower
3 Temasek Avenue
Singapore 039190

Email: editorial@jennystanford.com
Web: www.jennystanford.com

British Library Cataloguing-in-Publication Data
A catalogue record for this book is available from the British Library.

Graphene-Based Terahertz Electronics and Plasmonics: Detector and Emitter Concepts

ISBN 978-981-4800-75-4 (Hardcover)
ISBN 978-0-429-32839-8 (eBook)

Contents

PART 4 Stimulated Emission and Lasing in Graphene and Graphene Structures

Preface

The emergence and realization of new materials always strongly affected humankind.

Iron and aluminum as well as their various compounds are the main materials used in almost all machines: ships, automobiles, trains, air- and spacecraft, and others. Could we fly over oceans in liners taking hundreds of passengers from one point in the globe to another if Mother Nature did not supply us with aluminum. Humankind should be also happy that the Earth provided a huge amount of silicon and that its dioxide is a very stable substance. Would we have supercomputers, cell phones, personal computers, the internet, and many other things without silicon-based micro- and nanoelectronics? The utilization of other semiconductor materials, such as gallium arsenide and its relatives, also greatly contributes to modern engineering. Some materials, known for decades, found their applications only in recent years. This is true for graphite and, particularly, its stack of a few monolayers. The unique electrical properties of the graphite monolayer have been known for a rather long time. The band structure of graphite and its monolayer and their consequences on electric characteristics were first described by P. R. Wallace in his paper published in *Physical Review* in 1947 [1]. The first graphite monolayers were produced in 1962 by H.-P. Boehm and his colleagues (see, e.g., [2]). Such monolayers, which are allotropes of carbon consisting of a single layer of carbon atoms forming a hexagonal lattice, is now widely called *graphene*. The name *graphene* is a combination of *graphite* and the suffix *-ene* [3].

However, thanks to pioneering works by A. Geim, K. Novoselov, W. de Heer, Ph. Kim, and their coworkers, only in the beginning of this century did it become understood that the unique properties of graphene can provide it wide application in different areas, in particular, in electronics and optoelectronics (see, e.g., [4]).

This discovery caused a huge wave of theoretical, technological, and experimental works and, consequently, publications in the world scientific literature. There is also a large number of real demonstrations of novel graphene-based devices: transistors, optical modulators, sources of electromagnetic radiation, and so forth. In 2010 A. Geim and K. Novoselov received the Nobel Prize for their pathbreaking work, which stressed the importance and prospects of graphene. The prize undoubtedly stimulated the pertinent research and application activities.

Among the unique characteristics of graphene, one can indicate the virtually linear dispersion law (of the Dirac type) for electrons and holes accompanied with their gapless energy spectrum and very high electron and hole velocities (about 10^8 cm/s). Due to such features, graphene is fairly prospective for detectors and sources of electromagnetic radiation in the ranges of frequencies from terahertz to infrared and higher. The high electron and hole characteristic velocity promotes their very high mobilities, particularly at room temperature (much higher than in all other semiconductor materials). Therefore, it is possible to realize the so-called ballistic (collisionless) electron and hole transport across samples (and devices) with micrometer sizes and larger. The latter is very promising for high-speed electronic devices operating in the terahertz frequency range. Excellent dynamic properties of electrons and holes in graphene open prospects in realization of novel plasmonic terahertz devices.

In spite of enormous research activity in the area of devices based on graphene, the amount of extensive review material on terahertz devices based on graphene is small. We hope that this review volume, which comprises journal articles by V. Mitin, T, Otsuji, V. Ryzhii, and their collaborators published from 2007 to present, would fill the gap and can be used by researchers and engineers working in the fields of electronics, plasmonics, and influence of plasmonics on device performance. Moreover, we think that the book can be used as a required text for advanced courses for PhD students and as a supplementary material for master-level courses. We used it successfully for term paper assignment as well as for projects of master-level students. For example, several teams of master's students can be assigned to deliver consecutive chapters of a selected part for presentation in the class. We also assigned students to work on details of derivation, as in many sections the details are omitted. Moreover, the book will come handy while writing small programs to plot results of the final equations first for the same set of parameters, as given in the sections, to be sure that the program is working properly and then to expand the calculation to a new set of parameters to better understand the limits of the results.

Apart from the above introductory remarks, the collection comprises six parts separated into chapters. Each chapter presents an original journal article.

Part 1 comprises papers devoted to the studies of general electronic and plasmonic phenomena in graphene and some graphene-based structures. It starts from the chapter in which the spectrum of plasmonic waves (the spatio-temporal variations of carrier density accompanied with the ac self-consistent electric field) was calculated for graphene covered by a highly conducting electrode separated by an insulating layer (gated graphene). The main result of this chapter is that the wave velocity always exceeds the characteristic electron and hole velocity and, therefore, the Fermi velocity in graphene. This implies an important limitation on plasmon Landau damping and related processes. Several chapters in this part deal with the studies of the features of various diode and transistors using the graphene lateral channel, the graphene bilayer, lateral and vertical multiple graphene layer structures, and structures with two graphene layers separated by a tunneling barrier.

In Part 2, results related to radiation detectors, in particular, terahertz detectors based on lateral graphene-layer structures, are presented. As demonstrated, graphene-based terahertz detectors can use interband transitions exploiting the absence of the energy gap as well the heating of the carriers by absorbing radiation (hot-carrier bolometric detectors). The excitation of the plasmonic oscillations in such detectors can result in their resonant response. The latter can enable very high detector responsivity at resonant conditions.

Part 3 demonstrates that optical and injection pumping creating interband population inversion leads to negative dynamic conductivity in different graphene-layer structures. The negative dynamic conductivity in the pumped graphene arises as a result of competition between the interband electron transitions from the conduction band to the valence band and the intraband processes of the photon absorption by the carriers (Drude absorption).

Due to this, negative dynamic conductivity at the condition of population inversion appears in an intermediate spectral range corresponding to terahertz frequencies.

In particular, it is shown that the generation of hot electrons and holes in graphene by optical pumping can result in elevated effective temperatures of the whole electron–hole plasma, adding complexity to the realization of interband population inversion.

In Part 4, methods of forming interband population inversion in graphene structures are proposed and analyzed. Also, application for efficient terahertz sources of radiation, in particular, terahertz-emitting diodes and lasers, are demonstrated. Different versions of graphene-based sources are considered, for example, with dielectric and slot-line waveguides, including that with tooth-like metal strips working for a distributed feedback cavity.

Part 5 deals with detectors and emitters based on photon/plasmon-assisted tunneling between graphene layers separated with an atomically thin insulating/semiconducting layer. If an n-doped layer of graphene is separated from a p-doped layer of graphene by a thin dielectric, electron resonant tunneling would take place from the n-layer into the p-layer if the Dirac points of both layers are aligned—that is, if energy and momentum are conserved during the tunneling. The shift of the alignment results in photon-assisted tunneling. Depending on the direction of the shift, a photon is either absorbed or emitted during the tunneling. That allows to have well-controlled emitters or detectors of terahertz radiation.

Part 6 expands the concepts of devices considered in Part 5 to a wider group of van der Waals heterostructures based on graphene. In the case of wider barriers, there is no tunneling between graphene layers, but there is a possibility for electrons from the filled valence band go to the wider-bandgap material absorbing photons. In the case of transport perpendicular to the layers, the detector is analogous to the conventional quantum well photodetector, but its performance is superior to the latter as its operation is caused by interband absorption that has higher probability and does not depend on polarization. In the case of in-plane transport, the conductivity depends on the difference in conductivity of graphene layers and barrier layers. Graphene–phosphorene heterostructures are of special interest for in-plane transport as mobility in phosphorene is strongly anisotropic and can be substantially lower than the mobility of graphene. This leads to a well-pronounced negative differential conductivity because with an increase in the electric field, electrons can transfer from graphene into phosphorene (real space transfer). As the result of real space transfer, the photoconductivity in graphene-phosphorene heterostructures can be positive or negative.

We would like to acknowledge our professional colleagues and collaborators with whom we published the papers included in this volume. Also, we wish to thank our loved ones for their support and for forgiving us for not devoting more time to them while working on this volume.

Vladimir Mitin
Taiichi Otsuji
Victor Ryzhii

References

1. P. R. Wallace, The band theory of graphite. *Phys. Rev.*, **71**, 622–634 (1947).
2. H. P. Boehm, A. Clauss, G. O. Fischer, U. Hofmann, Das adsorptionsverhalten sehr dünner kohlenstoff-folien. *Z. Anorg. Allg. Chem.*, **316**, 119–127 (1962), doi:10.1002/zaac.19623160303.
3. H. P. Boehm, R. Setton, and E. Stumpp, Nomenclature and terminology of graphite intercalation compounds. *Pure Appl. Chem.*, **66**(9), 1893–1901 (1994).
4. K. S. Novoselov, A. K. Geim, S. V. Morozov, et. al., Electric field effect in atomically thin carbon films. *Science*, **306,** 666–669 (2004).

Part 1

Electronic and Plasmonic Properties of Graphene and Graphene Structures

Chapter 1

Plasma Waves in Two-Dimensional Electron-Hole System in Gated Graphene Heterostructures*

V. Ryzhii,[a] A. Satou,[a] and T. Otsuji[b]
[a]*Computer Solid State Physics Laboratory, University of Aizu, Aizu-Wakamatsu 965-8580, Japan*
[b]*Research Institute of Electrical Communication, Tohoku University, Sendai 980-8577, Japan*
v-ryzhii@u-aizu.ac.jp

Plasma waves in the two-dimensional electron-hole system in a graphene-based heterostructure controlled by a highly conducting gate are studied theoretically. The energy spectra of two-dimensional electrons and holes are assumed to be conical (neutrinolike), i.e., corresponding to their zero effective masses. Using the developed model, we calculate the spectrum of plasma waves (spatio-temporal variations of the electron and hole densities and the self-consistent

---

Graphene-Based Terahertz Electronics and Plasmonics: Detector and Emitter Concepts
Edited by Vladimir Mitin, Taiichi Otsuji, and Victor Ryzhii

ISBN 978-981-4800-75-4 (Hardcover), 978-0-429-32839-8 (eBook)
www.jennystanford.com

electric potential). We find that the sufficiently long plasma waves exhibit a linear (soundlike) dispersion, with the wave velocity determined by the gate layer thickness, the gate voltage, and the temperature. The plasma wave velocity in graphene heterostructures can significantly exceed the plasma wave velocity in the commonly employed semiconductor gated heterostructures. The gated graphene heterostructures can be used in different voltage tunable terahertz devices which utilize the plasma waves.

1.1 Introduction

Recent progress in fabrication of *graphene,* a monolayer of carbon atoms forming a dense honeycomb two-dimensional (2D) crystal structure (which can also be considered as an unrolled single-wall carbon nanotube or as a giant flat fullerene molecule), and the demonstration of its exceptional electronic properties have attracted much interest [1]. The possibility of creating novel electronic devices on the basis of graphene heterostructures (see, for instance, Refs. [2] and [3]) has caused a surge of experimental and theoretical publications. The massless neutrinolike energy spectrum of electrons and holes in graphene can lead to specific features of the transport properties [4, 5], photon-assisted transport [6], and quantum Hall effect [7–9], as well as the unusual high-frequency properties and collective behavior of the 2D electron or hole (2DEG or 2DHG) systems in graphene-based heterostructures [10–13]. The plasma waves in graphene-based heterostructures, in which the 2DEG or 2DHG system can serve as a resonant cavity or as a voltage-controlled waveguide, can be used in different devices operating in the terahertz (THz) range of frequencies [14, 15]. Since the spectrum of plasma waves is sensitive to the electron (hole) mass, the plasma waves in graphene heterostructures can exhibit specific features associated with the zero electron and hole masses. The electron and hole sheet densities in graphene heterostructures with the conducting gate can be effectively varied by the gate voltage [1]. Thus, the massless electron and hole spectra and voltage tunability of the electron and hole densities open up new opportunities to create novel THz plasma wave devices based on graphene heterostructures on silicon substrates.

In this work, the spectrum of plasma waves (spatio-temporal variations of the electron and hole densities and self-consistent electric potential) in a gated graphene heterostructure in wide ranges of the gate voltages and the temperatures is studied. A gated graphene heterostructure (with n^+-Si substrate serving as the gate and the gate layer made of SiO_2) shown in Fig. 1.1 is considered. A limiting case of relatively large (positive) gate voltages at sufficiently low temperatures was considered recently by one of the present authors [13]. In this case, the plasma waves are associated with a degenerate 2DEG. Here, we generalize the study of plasma waves in gated graphene heterostructures for the situation when the gate voltage and temperature vary in wide ranges, so that the contribution of both electrons and holes and the features of their energy distributions can be essential.

1.2 Equations of the Model

We shall consider "classical" plasma waves with the wavelength λ markedly exceeding the characteristic length of the electron de Broglie wave λ_F. In this case, we can use the following kinetic equations coupled with the equation governing the self-consistent electric potential:

$$\frac{\partial f_e}{\partial t} + \mathbf{v_p}\frac{\partial f_e}{\partial \mathbf{r}} + e\frac{\partial f_e}{\partial \mathbf{p}}\frac{\partial \varphi}{\partial \mathbf{r}}\bigg|_{z=0} = I_e, \tag{1.1}$$

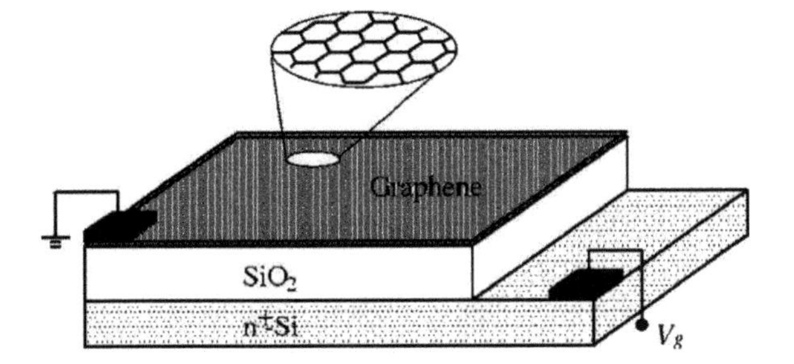

Figure 1.1 Schematic view of a gated graphene heterostructure.

$$\frac{\partial f_h}{\partial t} + \mathbf{v_p}\frac{\partial f_h}{\partial \mathbf{r}} - e\frac{\partial f_h}{\partial \mathbf{p}}\frac{\partial \varphi}{\partial \mathbf{r}}\bigg|_{z=0} = I_h, \tag{1.2}$$

Here, $f_e = f_e(\mathbf{p}, \mathbf{r}, t)$ and $f_h = f_h(\mathbf{p}, \mathbf{r}, t)$ are the electron and hole distribution functions, respectively, $\varphi = \varphi(\mathbf{r}, z, t)$ is the electric potential, $\mathbf{p} = (p_x, p_y)$ is the electron (or hole) in-plane momentum, $\mathbf{r} = (\mathbf{x}, \mathbf{y})$ (x and y correspond to the directions in the graphene layer plane, while axis z is directed perpendicular to this plane), and $e = |e|$ is the electron charge. The quantity $\mathbf{v_p}$ is the velocity of an electron and a hole with momentum $\mathbf{p}$: $\mathbf{v_p} = \partial\varepsilon_\mathbf{p}/\partial\mathbf{p}$, where $\varepsilon_\mathbf{p} = \upsilon_F|\mathbf{p}|$ corresponds to the electron and hole massless spectra and υ_F is the characteristic (Fermi) velocity. As a result, $\mathbf{v_p} = \upsilon_F\mathbf{p}/p$, where $p = |\mathbf{p}|$. The terms I_e and I_h on the right-hand sides of Eqs. (1.1) and (1.2) govern the processes of electron and hole scattering and recombination. For the sake of simplicity, we disregard the electron and hole scattering processes and their recombination, and therefore set $I_e = 0$ and $I_h = 0$. This is justified when the characteristic frequency of plasma waves significantly exceeds the pertinent collision frequencies.

The Poisson equation which supplements Eqs. (1.1) and (1.2) is presented as

$$\frac{\partial^2\varphi}{\partial x^2} + \frac{\partial^2\varphi}{\partial y^2} + \frac{\partial^2\varphi}{\partial z^2} = \frac{4\pi e}{æ}(\Sigma_e - \Sigma_h)\delta(z), \tag{1.3}$$

where æ is the dielectric constant of the surrounding space and $\delta(z)$ is the Dirac delta function playing the role of the form factor of the 2DEG and 2DHG localization in the direction perpendicular to the graphene plane. Here, we have neglected the finiteness of the localization region thickness. The electron and hole sheet densities $\Sigma_e = \Sigma_e(\mathbf{r}, t)$ and $\Sigma_h = \Sigma_h(\mathbf{r}, t)$ and the distribution functions $f_e = f_e(\mathbf{p}, \mathbf{r}, t)$ and $f_h = f_h(\mathbf{p}, \mathbf{r}, t)$ are related by the following equations:

$$\Sigma_e = \frac{2}{(2\pi\hbar)^2}\int d^2\mathbf{p} f_e, \quad \Sigma_h = \frac{2}{(2\pi\hbar)^2}\int d^2\mathbf{p} f_h, \tag{1.4}$$

where $\hbar$ is the reduced Planck constant.

In equilibrium, the electron and hole distribution functions are given by

$$f_{e0} = \left[1 + \exp\left(\frac{v_F p - \varepsilon_F}{k_B T}\right)\right]^{-1},$$

$$f_{h0} = \left[1 + \exp\left(\frac{v_F p - \varepsilon_F}{k_B T}\right)\right]^{-1}, \tag{1.5}$$

where ε_F is the Fermi energy, T is the temperature, and k_B is the Boltzmann constant. The Fermi energy is determined by the condition of neutrality $æV_g/W_g = 4\pi e(\Sigma_{e0} - \Sigma_{h0})$, where V_g is the gate voltage, W_g is the gate layer thickness, and Σ_{e0} and Σ_{h0} are the equilibrium electron and hole densities, respectively. Considering Eq. (1.4) the latter condition can be presented as

$$\frac{\pi æ\hbar^2 V_g}{2eW_g} = \int d^2\mathbf{p}(f_{e0} - f_{h0}). \tag{1.6}$$

Taking into account Eq. (1.5), Eq. (1.6) can be rewritten as

$$\frac{æ\hbar^2 V_g}{4eW_g} = \int_0^\infty \mathrm{d}p\, p \left\{ \left[1+\exp\left(\frac{v_F p - \varepsilon_F}{k_B T}\right)\right]^{-1} - \left[1+\exp\left(\frac{v_F p + \varepsilon_F}{k_B T}\right)\right]^{-1} \right\},$$

or, introducing the variable $\xi = \upsilon_F p/k_B T$, as

$$\frac{æ\hbar^2 v_F^2 V_g}{4eW_g (k_B T)^2} = F\left(\frac{\varepsilon_F}{k_B T}\right), \tag{1.7}$$

where

$$F\left(\frac{\varepsilon_F}{k_B T}\right) = \int_0^\infty \mathrm{d}\xi\, \xi \left[\frac{1}{1+\exp(\xi - \varepsilon_F/k_B T)} - \frac{1}{1+\exp(\xi + \varepsilon_F/k_B T)}\right]. \tag{1.8}$$

At $V_g = 0$, Eqs. (1.7) and (1.8) naturally yield $\varepsilon_F = 0$ (which corresponds to $\Sigma_{e0} = \Sigma_{h0}$).

1.3 Dispersion Equation for Plasma Waves

Using the standard small-signal analysis, i.e., assuming that $f_{e,h}(\mathbf{p}, \mathbf{r}, t) = f_{e0,h0}(p) + \delta f_{e,h}(p)\exp[i(kx-\omega t)]$ and $\varphi(\mathbf{r}, \mathbf{z}, t) = \delta\varphi(z)\exp[i(kx-\omega t)]$, where $|\delta f_{e,h}| \ll f_{e0,h0}$ are small perturbations, k and ω are the wavenumber and frequency of the plasma wave, respectively, from Eqs. (1.1) and (1.2) we arrive at the following equations:

$$(kv_{\mathrm{F}}p_x/p-\omega)\delta f_{\mathrm{e}} = -ek\frac{\partial f_{\mathrm{e0}}}{\partial p_x}\delta\phi\bigg|_{z=0}, \tag{1.9}$$

$$(kv_{\mathrm{F}}p_x/p-\omega)\delta f_{\mathrm{h}} = ek\frac{\partial f_{\mathrm{h0}}}{\partial p_x}\delta\varphi\bigg|_{z=0}. \tag{1.10}$$

Assuming that the gate is equipotential (its potential is equal to V_{g}), Eq. (1.3) results in the following:

$$\delta\varphi\big|_{z=0} = -\frac{4\pi e(\delta\Sigma_{\mathrm{e}}-\delta\Sigma_{\mathrm{h}})}{æ\hbar^2|k|\left[\coth(|k|W_{\mathrm{g}})+1\right]}. \tag{1.11}$$

Using Eqs. (1.4) and (1.11), we obtain

$$\delta\varphi\big|_{z=0} = -\frac{2e}{\pi æ\,\hbar^2|k|[\coth(|k|W_{\mathrm{g}})+1]}\int d^2\mathbf{p}(\delta f_{\mathrm{e}}-\delta f_{\mathrm{h}}). \tag{1.12}$$

Substituting δf_{e} and δf_{h} from Eqs. (1.9) and (1.10) into Eq. (1.12), we arrive at the dispersion equation relating the wavenumber and frequency of the plasma wave,

$$\frac{\pi æ\hbar^2}{2e^2}\frac{[\coth(|k|W_{\mathrm{g}})+1]}{(k/|k|)} = \int\frac{d^2\mathbf{p}}{(\omega-kv_{\mathrm{F}}p_{\mathrm{x}}/p)}\left[-\frac{\partial(f_{\mathrm{e0}}+f_{\mathrm{h0}})}{\partial p_{\mathrm{x}}}\right]. \tag{1.13}$$

Taking into account Eqs. (1.5) and (1.6), Eq. (1.13) can be transformed as

$$\frac{\pi æ\,\hbar^2 v_{\mathrm{F}}^2|k|[\coth(|k|W_{\mathrm{g}})+1]}{4e^2k_{\mathrm{B}}T} = \int_0^{\pi}\frac{d\Theta\cos\Theta}{(\omega/kv_{\mathrm{F}}-\cos\Theta)}$$
$$\times\int_0^{\infty}d\xi\left[\frac{1}{1+\exp(\xi-\varepsilon_{\mathrm{F}}/k_{\mathrm{B}}T)}+\frac{1}{1+\exp(\xi+\varepsilon_{\mathrm{F}}/k_{\mathrm{B}}T)}\right]. \tag{1.14}$$

Here, we have used the variables $\Theta = \cos^{-1}(p_x/p)$ and $\xi = v_{\mathrm{F}}p/k_{\mathrm{B}}T$.

Integrating in Eq. (1.14) over $d\Theta$ and $d\xi$, we arrive at the following dispersion equation for plasma waves:

$$\frac{(\omega/kv_{\mathrm{F}})}{\sqrt{(\omega/kv_{\mathrm{F}})^2-1}} = 1+\left(\frac{æ\hbar^2v_{\mathrm{F}}^2}{4e^2W_{\mathrm{g}}k_{\mathrm{B}}T}\right)$$
$$\times\frac{|k|W_{\mathrm{g}}[\coth(|k|W_{\mathrm{g}})+1]}{\ln[2+\exp(\varepsilon_{\mathrm{F}}/k_{\mathrm{B}}T)+\exp(-\varepsilon_{\mathrm{F}}/k_{\mathrm{B}}T)]}. \tag{1.15}$$

Equation (1.15) yields

$$\omega = \frac{v_F k}{\sqrt{1-(\alpha DK)^2/(\alpha DK+1)^2}}, \quad (1.16)$$

where

$$\alpha = \frac{4e^2 W_g \varepsilon_F}{æ\hbar^2 v_F^2}, \quad (1.17)$$

$$D = \frac{k_B T}{\varepsilon_F} \ln\left[2 + \exp\left(\frac{\varepsilon_F}{k_B T}\right) + \exp\left(-\frac{\varepsilon_F}{k_B T}\right)\right], \quad (1.18)$$

and

$$K^{-1} = |k| W_g[\coth(|k| W_g) + 1]. \quad (1.19)$$

1.4 Limiting Cases

1.4.1 Low Temperatures or High Gate Voltages

At sufficiently low temperatures or at sufficiently high (positive) gate voltages, the electron density substantially exceeds the density of holes and the electron Fermi energy is much larger than the thermal energy ($\varepsilon_F = \hbar v_F \sqrt{æV_g/2eW_g} \gg k_B T$), so that 2DEG is degenerate and $D \simeq 1$. In this case, Eq. (1.14) and (1.15) yield

$$\omega = \frac{v_F k}{\sqrt{1-(\alpha K)^2/(\alpha K+1)^2}}, \quad (1.20)$$

with

$$\alpha = \frac{4e^2 W_g \varepsilon_F}{æ\hbar^2 v_F^2} \simeq \sqrt{\frac{8e^3 W_g V_g}{æ\hbar^2 v_F^2}}. \quad (1.21)$$

At sufficiently high gate voltages, $\alpha \gg 1$. In this case, from Eq. (1.16) we obtain

$$\omega \simeq v_F k \sqrt{\frac{\alpha}{2|k| W_g[\coth(|k| W_g) + 1]}}. \quad (1.22)$$

For long plasma waves ($|k| W_g \ll 1$), Eq. (1.22) yields (compare with Ref. [13])

$$\omega \simeq v_F k \left(1 - \frac{|k| W_g}{2}\right)\sqrt{\frac{\alpha}{2}}. \tag{1.23}$$

One can see that relatively long plasma waves (longer than the gate layer thickness) in the gated graphene exhibit a linear (soundlike) spectrum with the wave phase and group velocities, which depend on the gate voltage (via the voltage dependence of parameter α). In this case, the wave phase and group velocities, ω/k and $\delta\omega/\delta k$, are equal to

$$s \simeq v_F \sqrt{\frac{\alpha}{2}} = \sqrt{\frac{2e^2 W_g \varepsilon_F}{æ\hbar^2}} \propto W_g^{1/4} V_g^{1/4}. \tag{1.24}$$

These velocities depend on the gate voltage (via the voltage dependence of parameter α) as $V_g^{1/4}$. If æ = 3.8, $v_F = 10^8$ cm/s, $W_g = 3 \times 10^{-5}$ cm (as in Ref. [1]), and $V_g = 1 - 10$ V, one obtains $\alpha \simeq 48 - 152$ and $s \simeq (4.9\text{–}8.7) \times 10^8$ cm/s. In this case, the electron sheet density $\Sigma_0 \simeq 7 \times (10^{10}\text{–}10^{11})$ cm^{-2} corresponds to the Fermi energy $\varepsilon_F \simeq 40\text{–}126$ meV. Since $\alpha \gg 1$ in a wide range of parameters, one can assume that $s > v_F$ (or even $s \gg v_F$).

For short plasma waves ($|k| W_g > 1$), Eq. (1.20) yields

$$\omega \simeq v_F \sqrt{|k|}\sqrt{\frac{\alpha}{4W_g}} = v_F \sqrt{|k|}\sqrt{\frac{e^2 \varepsilon_F}{æ\hbar^2 v_F^2}}. \tag{1.25}$$

1.4.2 Elevated Temperatures and Low Gate Voltages

At low gate voltages, the Fermi energy is small. In particular, when $V_g = 0$, one has $\varepsilon_F = 0$. Therefore, at low gate voltages one can assume that $\varepsilon_F < k_B T$. In this case, $D \simeq (k_B T/\varepsilon_F)\ln 4$. As a result, from Eq. (1.16) one can obtain

$$\omega = \frac{v_F k}{\sqrt{1 - (\beta K)^2/(\beta K + 1)^2}}, \tag{1.26}$$

where

$$\beta = \frac{8 \ln 2e^2 W_g k_B T}{æ\hbar^2 v_F^2}. \tag{1.27}$$

At elevated temperatures $T \gg T_g = (æ\hbar^2 v_F^2/4 \ln 4k_B e^2 W_g)$, parameter $\beta \gg 1$, and

$$\omega \simeq v_F k \sqrt{\frac{\beta}{2|k|W_g[\coth(|k|W_g)+1]}}. \quad (1.28)$$

If æ = 3.8, $\upsilon_F = 10^8$ cm/s, and $W_g = 3 \times 10^{-5}$ cm, one obtains the following estimate: $T_g \simeq 7$ K. For long plasma waves ($|k|W_g \ll 1$) one obtains

$$\omega \simeq k\left(1-\frac{|k|W_g}{2}\right)\sqrt{\frac{4\ln 2e^2 W_g k_B T}{æ\hbar^2}}, \quad (1.29)$$

so that the phase and group velocities can be presented as

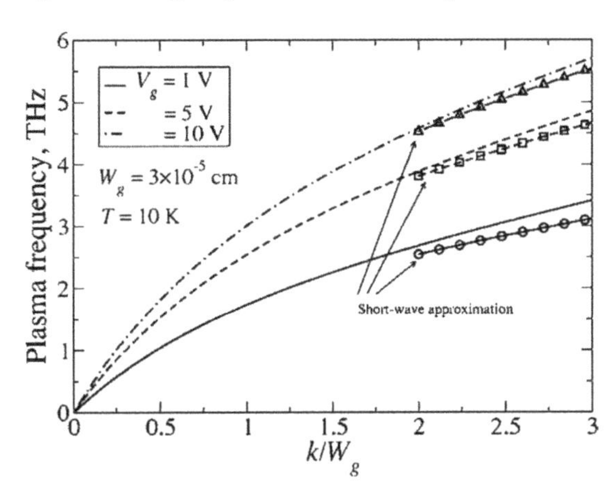

Figure 1.2 Plasma wave frequency versus wavenumber calculated for a graphene heterostructure ($W_g = 3 \times 10^{-5}$ cm and T = 10 K) at different gate voltages.

$$s \simeq \sqrt{\frac{4\ln 2e^2 W_g k_B T}{æ\hbar^2}} \propto W_g^{1/2} T^{1/2}. \quad (1.30)$$

As follows from Eq. (1.24), the frequency and the phase and group velocities of long plasma waves increase with the temperature as $T^{1/2}$. This is due to the pertinent increase in the electron and hole densities.

For relatively short plasma waves ($|k|W_g \gg 1$) we arrive at [11]

$$\omega \simeq \sqrt{|k|}\sqrt{\frac{2\ln 2e^2 k_B T}{æ\hbar^2}}. \quad (1.31)$$

1.5 General Results and Discussion

The spectra of plasma waves in a graphene heterostructure with $v_F = 10^8$ cm/s at $T = 10$ K calculated using Eq. (1.15) invoking Eq. (1.7) for different gate voltages are shown in Fig. 1.2. The plasma frequency versus wavenumber dependences are close to linear dependences at small wavenumbers ($\omega \propto k$). At large wavenumbers (short plasma waves), the plasma frequency increases with the wavenumber as $\omega \propto \sqrt{k}$. As seen, the plasma frequency pronouncedly increases with increasing gate voltage. Figure 1.3 shows the dependence of the plasma wave phase velocity s (at small k) as a function of the gate voltage and the temperature. As follows from Fig. 1.3, the phase velocity is sensitive to the temperature only at low gate voltages when 2DEG and 2DHG systems are not degenerate. This is confirmed by the dependences shown in Fig. 1.4.

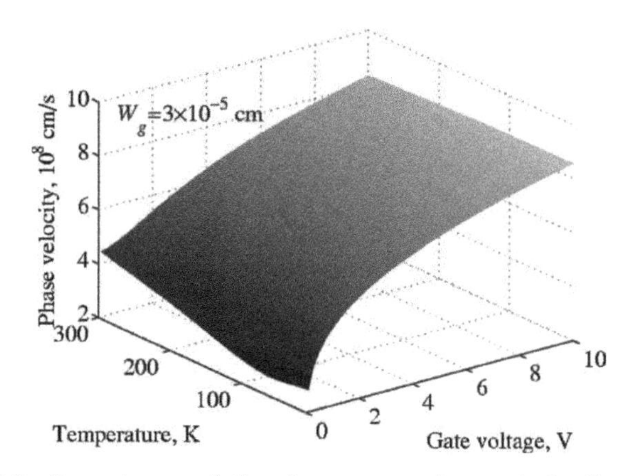

Figure 1.3 Dependences of the plasma wave phase velocity (long plasma waves) on the gate voltage and temperature.

Taking into account that for a degenerate 2DEG with massless electrons [as it follows from Eq. (1.4)] the dc electron density [16] is

$$\Sigma_{e0} = \frac{\varepsilon_F^2}{2\pi\hbar^2 v_F^2}, \tag{1.32}$$

and introducing the fictitious electron mass

$$m = \frac{\varepsilon_F}{v_F^2} = \frac{\hbar}{v_F}\sqrt{\frac{æV_g}{2eW_g}} \propto W_g^{-1/2}V_g^{1/2}, \tag{1.33}$$

the wave velocity s can be presented in the form

$$s = \sqrt{\frac{4\pi e^2 W_g \Sigma_{e0}}{æm}} = \sqrt{\frac{eV_g}{m}}, \qquad (1.34)$$

which formally coincides with that obtained for the "normal" 2DEG,

$$s^* = \sqrt{\frac{4\pi e^2 W_g \Sigma_{e0}}{æ^* m^*}} = \sqrt{\frac{eV_g}{m^*}}, \qquad (1.35)$$

where m^* is the electron (hole) "normal" effective mass and $æ^*$ is the dielectric constant of the pertinent material (see, for instance, Refs. [14] and [15]). The fictitious mass m determines the relationship between the electron mobility and electron transport time [16]. However, the voltage dependences of s in the 2DEG of massless and normal electrons are different. The point is that in both cases $\Sigma_e \propto V_g$. But, the effective mass of the massless electrons increases with V_g as $m \propto \varepsilon_F \propto V_g^{1/2}$, whereas an increase in the effective mass of normal electrons is weak because it is associated with a modest nonparabolicity of their energy spectrum. One can see from Eqs. (1.34) and (1.35)

$$\frac{s}{s^*} = \sqrt{\frac{æ^* m^*}{æm}} \propto W_g^{1/4} V_g^{-1/4}. \qquad (1.36)$$

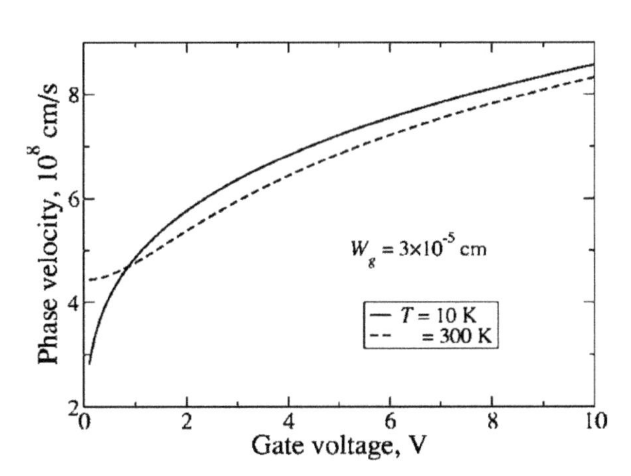

Figure 1.4 Dependences of the plasma wave phase velocity (long plasma waves) on the gate voltage in graphene heterostructures at different temperatures.

Since $m < m^*$ in certain ranges of thicknesses and voltages and usually æ < æ*, the plasma wave velocity in graphene heterostructures can pronouncedly exceed that in normal heterostructures. For example, assuming that $\upsilon_F = 10^8$ cm/s, æ = 3.8, æ* = 12, $W_g = (1\text{–}3) \times 10^{-5}$ cm, and $V_g = 1\text{–}3$ V, the ratio of the wave velocity in graphene heterostructures to that in AlGaAf/GaAs heterostructures with $m^* = 6 \times 10^{-29}$ g can be estimated as $s/s^* \simeq 3\text{–}5$.

For a nondegenerate 2DEG-2DHG system ($V_g = 0$ and $\varepsilon_F = 0$), Eq. (1.4) yields

$$\Sigma_{e0} = \Sigma_{h0} = \frac{1}{\pi}\left(\frac{k_B T}{\hbar v_F}\right)^2 \int_0^\infty \frac{d\xi\,\xi}{1+\exp\xi} = \frac{\pi}{12}\left(\frac{k_B T}{\hbar v_F}\right)^2. \quad (1.37)$$

In this case, taking into account that the net carrier density is equal to $\Sigma_{e0} + \Sigma_{h0}$, the wave velocity can be presented in the form

$$s = \sqrt{\frac{4\pi e^2 W_g (\Sigma_{e0} + \Sigma_{h0})}{æ m}}, \quad (1.38)$$

in which $m = (\pi^2 k_B T / 12\upsilon_F^2)$.

As follows from Eq. (1.16), the plasma wave phase velocity always exceeds the electron (hole) velocity, i.e., $\omega/k > \upsilon_F$, disregarding the shape of the electron (hole) distribution. This implies that the Landau damping of the plasma waves in the system under consideration is absent. The plasma wave damping is associated with the electron (hole) scattering. The damping of the plasma waves associated with the scattering can be estimated as $\gamma = 1/2\tau$, where τ is the electron transport time. This time and the electron mobility μ in the massless 2DEG are related by an equation which is different from that for the customary 2DEG. Taking into account that for the massless 2DEG the dc conductivity is proportional to $\varepsilon_F \tau$ [16], one can express τ via μ: $\tau = \mu m/e$, where, in contrast to the semiconductors with the customary electron spectrum, $m = \varepsilon_F/\upsilon^2_F$, i.e., the fictitious mass introduced above.

Acknowledgment

This work was financially supported by the Grant-in-Aid for Scientific Research (S) from the Japan Society for Promotion of Science, Japan.

References

1. K. S. Novoselov, A. K. Geim, S. V. Morozov, D. Jiang, M. I. Katsnelson, I. V. Grigorieva, S. V. Dubonos, and A. A. Firsov, *Nature*, **438**, 197 (2005).
2. C. Berger, Z. Song, T. Li, X. Li, A. Y. Ogbazhi, R. Feng, Z. Dai, A. N. Marchenkov, E. H. Conrad, P. N. First, and W. A. de Heer, *J. Phys. Chem.*, **108**, 19912 (2004).
3. B. Obradovic, R. Kotlyar, F. Heinz, P. Matagne, T. Rakshit, M. S. Giles, M. A. Stettler, and D. E. Nikonov, *Appl. Phys. Lett.*, **88**, 142102 (2006).
4. K. Nomura and A. H. MacDonald, e-print cond-mat/0606589 (2006).
5. I. L. Aleiner and K. B. Efetov, *Phys. Rev. Lett.*, **97**, 236801 (2006).
6. B. Trauzettel, Ya. M. Blanter, and A. F. Morpurgo, e-print cond-mat/0606505 (2006).
7. V. P. Gusynin and S. G. Sharapov, *Phys. Rev. Lett.*, **95**, 146801 (2005).
8. Y. Zhang, Y.-W. Tan, H. Stormer, and P. Kim, *Nature*, **438**, 201 (2005).
9. D. Abanin, P. A. Lee, and L. S. Levitov, *Phys. Rev. Lett.*, **96**, 176803 (2006).
10. V. P. Gusynin, S. G. Sharapov, and J. P. Carbotte, *Phys. Rev. Lett.*, **96**, 256802 (2006).
11. L. A. Falkovsky and A. A. Varlamov, e-print cond-mat/0606800 (2006).
12. O. Vafek, *Phys. Rev. Lett.*, **97**, 266406 (2006).
13. V. Ryzhii, Jpn. *J. Appl. Phys.*, Part 2, **45**, L923 (2006).
14. M. Dyakonov and M. Shur, *IEEE Trans. Electron Devices*, **43**, 1640 (1996).
15. M. S. Shur and V. Ryzhii, Int. J. High Speed Electron. Syst., **13**, 7 (2003).
16. V. P. Gusynin and S. G. Sharapov, *Phys. Rev. B*, **73**, 245411 (2006).

Chapter 2

Device Model for Graphene Bilayer Field-Effect Transistor*

V. Ryzhii,[a,b] M. Ryzhii,[a,b] A. Satou,[a,b]
T. Otsuji,[b,c] and N. Kirova[d]
[a]*Computational Nanoelectronics Laboratory, University of Aizu, Aizu-Wakamatsu 965-8580, Japan*
[b]*Japan Science and Technology Agency, CREST, Tokyo 107-0075, Japan*
[c]*Research Institute for Electrical Communication, Tohoku University, Sendai 980-8577, Japan*
[d]*Laboratoire de Physique des Solides, Univ. Paris-Sud, CNRS, UMR 8502, F-91405 Orsay Cedex, France*
v-ryzhii@u-aizu.ac.jp

We present an analytical device model for a graphene bilayer field-effect transistor (GBL-FET) with a graphene bilayer as a channel, and with back and top gates. The model accounts for the dependences of the electron and hole Fermi energies as well as energy gap in different sections of the channel on the bias back-gate and top-gate

Graphene-Based Terahertz Electronics and Plasmonics: Detector and Emitter Concepts
Edited by Vladimir Mitin, Taiichi Otsuji, and Victor Ryzhii

ISBN 978-981-4800-75-4 (Hardcover), 978-0-429-32839-8 (eBook)
www.jennystanford.com

voltages. Using this model, we calculate the dc and ac source-drain currents and the transconductance of GBL-FETs with both ballistic and collision dominated electron transport as functions of structural parameters, the bias back-gate and top-gate voltages, and the signal frequency. It is shown that there are two threshold voltages, $V_{th,1}$ and $V_{th,2}$, so that the dc current versus the top-gate voltage relation markedly changes depending on whether the section of the channel beneath the top gate (gated section) is filled with electrons, depleted, or filled with holes. The electron scattering leads to a decrease in the dc and ac currents and transconductances, whereas it weakly affects the threshold frequency. As demonstrated, the transient recharging of the gated section by holes can pronouncedly influence the ac transconductance resulting in its nonmonotonic frequency dependence with a maximum at fairly high frequencies.

2.1 Introduction

The features of the electron and hole energy spectra in graphene provide the exceptional properties of graphene-based heterostructures and devices [1–6]. However, due to the gapless energy spectrum, the interband tunneling [7] can substantially deteriorate the performance of graphene field-effect transistors (G-FETs) with realistic device structures [8–11]. To avoid drawbacks of the characteristics of G-FETs based on graphene monolayer with zero energy gap, the patterned graphene (with an array of graphene nanoribbons) and the graphene bilayers can be used in graphene nanoribbon FETs (GNR-FETs) and in graphene bilayer FETs (GBL-FETs), respectively. The source-drain current in GNR-FETs and GBL-FETs, as in the standard FETs, depends on the gate voltages. The positively biased back gate provides the formation of the electron channels, whereas the negative bias voltage between the top gate and the channels results in forming a potential barrier for electrons which controls the current. By properly choosing the width of the nanoribbons, one can fabricate graphene structures with a relatively wide band gap [12] (see also Refs. [13–16]). Recently, the device dc and ac characteristics of GNR-FETs were assessed using both numerical [14] and analytical [17–19] models. The effect of the transverse electric field (to the GBL plane) on the energy spectrum of GBLs [20–22] can also be used to manipulate and optimize the GBL-

FET characteristics. A significant feature of GBL-FETs is that under the effect of the transverse electric field not only the density of the two-dimensional electron gas in the GBL varies, but the energy gap between the GBL valence and conduction bands appears. This effect can markedly influence the GBL-FET characteristics. The structure of a GBL-FET is shown in Fig. 2.1. In this chapter, we present a simple analytical device model for a GBL-FET, obtain the device dc and ac characteristics, and compare these characteristics with those of GNR-FETs.

The chapter organized as follows. In Section 2.2, we consider the GBL-FET band diagrams at different bias voltages and estimate the energy gaps and the Fermi energy in different sections of the device. Section 2.3 deals with the Boltzmann kinetic equation, which governs the electron transport at dc and ac voltages and the solutions of this equation. The cases of the ballistic and collision dominated electron transport are considered. In Sections 2.4 and 2.5, the dc transconductance and the ac frequency-dependent transconductance are calculated using the results of Section 2.3. Section 2.6 deals with the demonstration and analysis of the main obtained results, numerical estimates, and comparison of the GBL-FET properties with those of GNR-FETs. In Section 2.7, we draw the main conclusions. In Appendix, some intermediate calculations related to the dynamic recharging of the gated section by holes due to the interband tunneling are singled out.

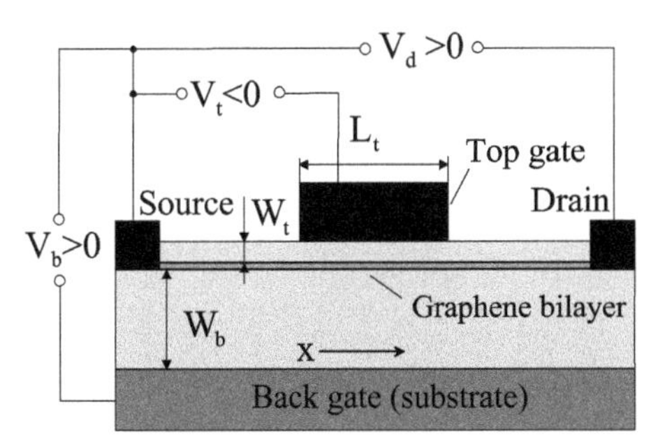

Figure 2.1 Schematic view of the GBL-FET structure.

2.2 GBL-FET Energy Band Diagrams

We assume that the bias back-gate voltage $V_b > 0$, while the bias top-gate voltage $V_t < 0$. The electric potential of the channel at the source and drain contacts are $\varphi = 0$ and $\varphi = V_d$, respectively, where V_d is the bias drain voltage. The former results in the formation of a 2DEG in the GBL. The distribution of the electron density Σ along the GBL is generally nonuniform due to the negatively biased top gate forming the barrier region beneath this gate. Simultaneously, the energy gap E_g is also a function of the coordinate x (its axis is directed in the GBL plane from the source contact to the drain contact) being different in the source, top-gate, and drain sections of the channel (see Fig. 2.2). Since the net top-gate voltage apart from the bias component V_t comprises the ac signal component $\delta V(t)$, the height of the barrier for electrons entering the section of the channel under the top gate (gated section) from the source side can be presented as

$$\Delta(t) = \Delta_0 + \delta\Delta(t). \tag{2.1}$$

Depending on the Fermi energy in the extreme sections of the channel, in particular, on its value, ε_F in the source section and on the height of the barrier in this section Δ_0, there are three situations. The pertinent the GBL-FET energy band diagrams are demonstrated in Fig. 2.2. The spatial distributions of electrons and holes in the GBL channel are different depending on the relationship between the top-gate voltage V_t and two threshold voltages, $V_{th,1}$ and $V_{th,2}$. These threshold voltages are determined in the following.

When $V_{th,2} < V_{th,1} < V_t$, the top of the conduction band in the gated section is below the Fermi level (Fig. 2.2a). In this case, an n^+-n-n^+ structure is formed in the GBL channel. At $V_{th,2} < V_t < V_{th,1}$, the Fermi level is between the top of the conduction band and the bottom of the valence band in this section (Fig. 2.2b). This top-gate voltage range corresponds to the formation of an n^+-i-n^+ structure. If $V_t < V_{th,2} < V_{th,1}$, both band edges are above the Fermi level (Fig. 2.2c), so that n^+-p and p-n^+ junctions are formed beneath the edges of the top gate. In the first and third ranges of the top gate voltage (a and b ranges), the electron and hole populations of the gated section are essential. In the second range (range b), the gated section is depleted. In the voltage range a, the source-drain current is associated with a hydrodynamical electron flow (due to effective electron-electron scattering) in the gated section. In this case, the source-drain current and GBL-FET characteristics are determined by the conductivity of the gated section, which, in turn, is determined by the electron density and scattering mechanisms including the electron-

electron scattering mechanism, and by the self-consistent electric field directed in the channel plane. In such a situation, different hydrodynamical models of the electron transport (including the drift-diffusion model) can be applied (see, for instance, Refs. [23–27]).

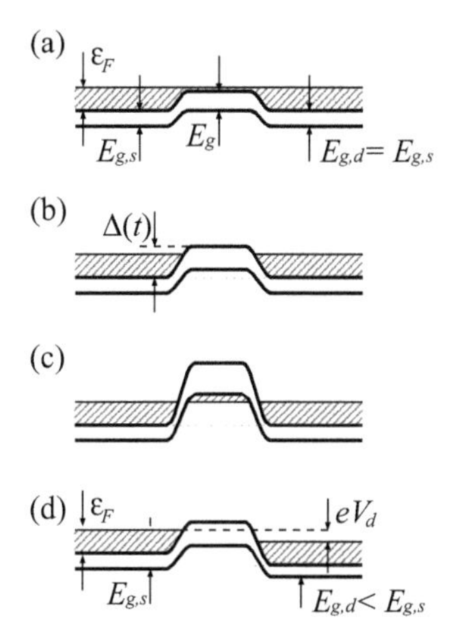

Figure 2.2 Band diagrams at different top gate bias voltages ($V_b > 0$, $V_d = 0$): (a) $V_{th,2} < V_{th,1} < V_t$, (b) $V_{th,2} < V_t < V_{th,1}$ (depleted gated section), and (c) $V_t < V_{th,2} < V_{th,1}$ (gated section filled with holes), Panel (d) corresponds to $V_{th,2} < V_t < V_{th,1}$ but with $V_d > 0$.

If $V_{th,2} < V_t < V_{th,1}$, considering the potential distribution in the direction perpendicular to the GBL plane invoking the gradual channel approximation [28, 29] and assuming for simplicity that the thicknesses of the gate layers separating the channel and the pertinent gates, W_b and W_t, are equal to each other $W_b = W_t = W$, we obtain

$$\Delta_0 = -e\frac{(V_b + V_t)}{2}, \quad \delta\Delta(t) = -\frac{1}{2}e\delta V(t), \tag{2.2}$$

where $e = |e|$ is the electron charge. In the voltage range in question, the electron system in the gated section is not degenerate. This voltage range, as well as the range $V_t < V_{th,2} < V_{th,1}$, correspond to the GBL-FET "off-state." Similar formulas take place for the barrier height from the drain side (with the replacement of Δ_0 by $\Delta_0 + eV_d$).

In the cases when $V_{th,2} < V_{th,1} < V_t$ or $V_t < V_{th,2} < V_{th,1}$,

$$\Delta_0 = -e\frac{(V_b + V_t)}{2} \pm \frac{2\pi eW}{k}\Sigma_0^{\mp},$$
$$\delta\Delta(t) = -\frac{1}{2}e\delta V(t) \pm \frac{2\pi eW}{k}\delta\Sigma^{\mp}(t). \tag{2.3}$$

Here $\Sigma_0^{\mp} + \delta\Sigma^{\mp}(t)$ are the electron and hole densities in the gated section and κ is the dielectric constant of the gate layers. In the most interesting case when the electron densities in the source and drain sections are sufficiently large, so that the electron systems in these sections are degenerate. Considering this, the height of the barrier Δ_0 is given by

$$\Delta_0 = -e\frac{V_t(a_B/8W)}{[1+(a_B/4W)]} \simeq -eV_t\left(\frac{a_B}{8W}\right), \tag{2.4}$$

$$\Delta_0 = -e\frac{V_t(a_B/8W) + (V_t - V_b)(d/2W)}{[1+(a_B/4W)]}, \tag{2.5}$$

when $V_{th,2} < V_{th,1} < V_t$ and $V_t < V_{th,2} < V_{th,1}$, respectively. Here $a_B = \kappa\hbar^2/me^2$ is the Bohr radius, d is the effective spacing between the graphene layers in the GBL which accounts for the screening of the electric field between these layers [20, 21]. This quantity is smaller than the real spacing between the graphene layers in the GBL $d_0 \simeq 0.36$ nm. The Bohr radius a_B can be rather different in different materials of the gate layers. In the cases of SiO_2 and HfO_2 (with $k \simeq 20$ [30, 31]) gate gate layers, $a_B \simeq 4$ nm and $a_B \simeq 20$ nm, respectively. In deriving Eqs. (2.4) and (2.5), we have taken into account that in real GBL-FETs, $(a_B/8W) \ll 1$.

For the normal GBL-FET operation, the electron densities, Σ_s^- and Σ_d^-, induced by the back-gate voltage in the source and drain sections, respectively, should be sufficiently high markedly exceeding their thermal value $\Sigma_T = 2 \ln 2mk_BT/\pi\hbar^2$, where k_B is the Boltzmann constant and T is the temperature. This occurs when $V_b > \ln 2(8W/a_B) \times (k_BT/e) + V_T$. In such a case, i.e., at sufficiently high back-gate voltages, the Fermi energy in the source section is given by

$$\varepsilon_F = \frac{k_BT}{[1+(a_B/8W)]}\ln\left[\exp\left(\frac{a_B}{8W}\frac{eV_b}{k_BT}\right) - 1\right]. \tag{2.6}$$

Here we considered that the electron density in the source section $\Sigma_s^- = \kappa V_b/4\pi eW$ (the electron density in the drain section of the channel is approximately equal to $\Sigma_d^- = \kappa(V_b - V_d)/4\pi eW$). If $a_B = 4 - 20$ nm at T = 300 K, $V_{\rm T} \simeq 0.035 - 0.173$ V. As follows from Eq. (2.6), the Fermi energy in the source section at $V_{\rm b} \gtrsim V_{\rm T}$ is a linear function of $V_{\rm b}$, practically independent of the temperature, and it can be presented as

$$\varepsilon_{\rm F} \simeq eV_{\rm b}\frac{(a_{\rm B}/8W)}{[1+(a_{\rm B}/8W)]} \simeq eV_{\rm b}\left(\frac{a_{\rm B}}{8W}\right) \gg k_{\rm B}T. \tag{2.7}$$

Setting W = 5 nm, $a_{\rm B}$ = 4 – 20 nm, and $V_{\rm b}$ = 1 V, we obtain $\varepsilon_{\rm F}$ = 48 – 91 meV. This case corresponds to the electron density in the source section is $\Sigma_{\rm s}^{-} \simeq 4\times10^{12} - 2\times10^{13}$ cm^{-2}. At $V_{\rm d} \lesssim V_{\rm b}$, the electron density in the drain section is somewhat smaller but of the same order of magnitude.

Comparing Eqs. (2.2), (2.4), and (2.5), one can see that the height of the barrier Δ_0 increases with increasing absolute value of the top-gate voltage rather slow in the voltage ranges *a* and "*c*" in contrast with its steep increase in the voltage range *b*. Since the energy gaps in GBLs $E_{\rm g,s}$, $E_{\rm g}$, and $E_{\rm g,d}$ depend on the local transverse electric field [20–22], they are different in different sections of the channel depending on the bias voltages,

$$E_{\rm g,s} = \frac{edV_{\rm b}}{2W}, E_{\rm g} = \frac{ed_0(V_{\rm b}-V_{\rm t})}{2W}, E_{\rm g,d} = \frac{ed(V_{\rm b}-V_{\rm d})}{2W}. \tag{2.8}$$

One can see that at $V_{\rm t} < 0$ and $V_{\rm d} > 0$, one obtains $E_{\rm g} > E_{\rm g,s} \geq E_{\rm g,d}$. Since $d \ll a_{\rm B}$, the energy gap in the source section is much smaller than the Fermi energy in this section. Indeed, assuming $d \simeq d_0 = 0.36$ nm and W = 5 nm for $V_{\rm b}$ = 1 V from Eq. (2.8), we obtain $E_{\rm g,s} \simeq 36$ meV. However, the energy gap in the gated section at sufficiently high top-gate voltages can be relatively large (see below). Naturally, an increase in the top-gate voltage leads to an increase in the Fermi energy in the source (drain) section as well to an increase in the energy gaps in all sections.

The threshold voltages $V_{\rm th,1}$ and $V_{\rm th,2}$ are determined by the conditions $\Delta_0 = \varepsilon_{\rm F}$ and $\Delta_0 = \varepsilon_{\rm F} + E_{\rm g}$, respectively. The latter implies that the Fermi energy of holes in the gated section $\varepsilon_{\rm F}^{\rm (hole)} = \Delta_0 - \varepsilon_{\rm F} - E_{\rm g} = 0$. As a result, the threshold voltages are given by

$$V_{\rm th,1} \simeq -V_{\rm b}\left(1+\frac{a_{\rm B}}{4W}\right), \quad V_{\rm th,2} \simeq -V_{\rm b}\left(1+\frac{a_{\rm B}}{4W}+\frac{d_0}{W}\right). \tag{2.9}$$

Since one can assume that $d \ll W$, the threshold voltages are close to each other, $|V_{th,1}| \lesssim |V_{th,2}|$ with $|V_{th,2} - V_{th,1}| \simeq (2d_0/W)V_b \gtrsim 4E_{g,s}/e$. The values of the energy gap in the gated section at the threshold top gate voltages are given by

$$E_g\Big|_{V_t=V_{th,1}} \lesssim E_g\Big|_{V_t=V_{th,2}} \simeq \frac{ed_0V_b}{W} \simeq 2E_{g,s}\left(\frac{d_0}{d}\right). \quad (2.10)$$

Using the same parameters as in the above estimate of the energy gap in the source section, for the energy gap in the gated section at $V_t \simeq V_{th,1} \simeq -1$ V, we obtain $E_g \simeq 72$ meV. In the following we restrict our consideration by the situations when the height of the barrier for electrons in the gated section is sufficiently large (so that $\Delta_0 > \varepsilon_F$), which corresponds to the band diagrams shown in Figs. 2.2b and c.

2.3 Boltzmann Kinetic Equation and Its Solutions

The quasiclassical Boltzmann kinetic equation governing the electron distribution function $f_\mathbf{p} = f_\mathbf{p}(x, t)$ in the section of the channel covered by the top gate (gated section) can be presented as

$$\frac{\partial f_\mathbf{p}}{\partial t} + v_x\frac{\partial f_\mathbf{p}}{\partial x} = \int d^2\mathbf{q}\, w(q)(f_{\mathbf{p}+\mathbf{q}} - f_\mathbf{p})\delta(\varepsilon_{\mathbf{p}+\mathbf{q}} - \varepsilon_\mathbf{p}). \quad (2.11)$$

Here, taking into account that the electron (and hole) dispersion relation at the energies close to the bottom of the conduction band is virtually parabolic with the effective mass m ($m \sim 0.04\, m_0$, where m_0 is the bare electron mass), for the energy of electron with momentum $\mathbf{p} = (p_x, p_y)$ we put $\varepsilon_\mathbf{p} = p^2/2m = \varepsilon$, $v_x = p_x/m = p\cos\theta/m$, where $\cos\theta = p_x/p$ (the x-axis and the y-axis are directed in the GBL plane) and $w(q)$ is the probability of the electron scattering on disorder and acoustic phonons with the variation of the electron momentum by quantity q. The density of the electron (thermionic) current, $J = J(x, t)$, in the gated section of the channel (per unit length in the y-direction) can be calculated using the following formula:

$$J = \frac{4e}{(2\pi\hbar)^2}\int d^2\mathbf{p}\, v_x f_\mathbf{p}, \quad (2.12)$$

where $\hbar$ is the reduced Planck constant. Disregarding the electron-electron collisions in the gated section of the channel (due a low electron density in this section in contrast to the source and drain sections where the electron-electron collisions are essential), we consider two limiting cases: ballistic transport of electrons across the gated section and strongly collisional electron transport.

2.3.1 Ballistic Electron Transport

If $\delta V(t) = \delta V_{\omega} e^{-i\omega t}$, where $\delta V_{\omega} \ll |V_{\mathrm{t}}|$ and ω are the amplitude and frequency of the ac signal, then the electron distribution function can be searched as $f_{\mathbf{p}} = F_0 + \delta F_{\omega}(x)\, e^{-i\omega t}$ and $\Delta = \Delta_0 + \delta\Delta_{\omega} e^{-i\omega t}$. Assuming that $eV_{\mathrm{d}} \gg k_{\mathrm{B}}T$ and solving Eq. (2.11) with the boundary conditions,

$$f_{\mathbf{p}}\Big|_{p_x \geq 0, x=0} = \exp\left[\frac{\varepsilon_{\mathrm{F}} - \Delta(t) - \varepsilon}{k_{\mathrm{B}}T}\right], \quad f_{\mathbf{p}}\Big|_{p_x \leq 0, x=L_t} \simeq 0, \qquad (2.13)$$

where L_{t} is the top gate length, we obtain

$$F_0 \simeq \exp\left(\frac{\varepsilon_{\mathrm{F}} - \Delta_0 - \varepsilon}{k_{\mathrm{B}}T}\right)\Theta(p_x), \qquad (2.14)$$

$$\delta F_{\omega}(x) = \exp\left(\frac{\varepsilon_{\mathrm{F}} - \Delta_0 - \varepsilon}{k_{\mathrm{B}}T} + i\omega\sqrt{\frac{m}{2\varepsilon}}\frac{x}{\cos\Theta}\right)$$

$$\times\left(-\frac{\delta\,\Delta_{\omega}}{k_{\mathrm{B}}T}\right)\Theta(p_x). \qquad (2.15)$$

Here, $\Theta(p_x)$ is the unity step function. The first boundary condition given by Eq. (2.14) corresponds to quasi equilibrium electron distribution in the source section of the channel and the injection of electrons with the kinetic energy exceeding the barrier height $\Delta(t)$ from the source section to the gated section (at $x = 0$). The injection of electrons from the drain source to the gated section (at $x = L_{\mathrm{t}}$) is negleted due to $eV_{\mathrm{d}} \gg k_{\mathrm{B}}T$; this inequality leads to rather high barrier near the drain edge of the gated section. The presence of the unity step fuction $\Theta(p_x)$ in Eqs. (2.16) and (2.17) reflects the fact that there are no electrons propagating backwards due to the absence of the electron scattering in the gated section.

Using Eqs. (2.12), (2.14), and (2.15), we arrive at the following formulas for the dc and ac components, J_0 and δJ_ω, of the current at the drain edge of the gated section (i.e., at $x = L_t$),

$$J_0 = e\frac{\sqrt{2m}(k_B T)^{3/2}}{\pi^{3/2}\hbar^2}\exp\left(\frac{\varepsilon_F - \Delta_0}{k_B T}\right) = J_0^B, \tag{2.16}$$

$$\frac{\delta J_\omega}{J_0} = \left(-\frac{\delta\Delta_\omega}{k_B T}\right)\int_0^\infty d\xi\sqrt{\xi}e^{-\xi}\mathcal{F}_\omega(\xi). \tag{2.17}$$

Here

$$\mathcal{F}_\omega(\xi) = \frac{2}{\sqrt{\pi}}\int_0^1 dy\,\exp\left(i\frac{\omega\tau}{\sqrt{\xi}\sqrt{1-y^2}}\right)$$

$$\simeq 2\frac{\xi^{1/4}}{\sqrt{\omega\tau}}\exp\left(i\frac{\omega\tau}{\sqrt{\xi}}\right)\left[C\left(\sqrt{\frac{\omega\tau}{2\sqrt{\xi}}}\right) + iS\left(\sqrt{\frac{\omega\tau}{2\sqrt{\xi}}}\right)\right],$$

where $\tau = L_t\sqrt{m/2k_B T}$ is the effective ballistic transit time across the gated section of electrons with the thermal velocity $\upsilon_T = \sqrt{2k_B T/m}$ and $C(x)$ and $S(x)$ are Frenel's cosine and sine functions. At $\omega\tau \gg 1$,

$$\mathcal{F}_\omega(\xi) \simeq \frac{\sqrt{2}\xi^{1/4}}{\sqrt{\omega\tau}}\exp\left(i\frac{\omega\tau}{\sqrt{\xi}} + i\frac{\pi}{4}\right).$$

At $\omega\tau \ll 1$, $F_\omega(\xi)$ tends to unity.

2.3.2 Collisional Electron Transport

In the case of strongly collisional electron transport, the distribution function in the gated section is close to isotropic and it can be searched in the form

$$f_P = F + g\cos\theta. \tag{2.18}$$

Here $F = F(\varepsilon, x, t)$ is the symmetrical part of the electron distribution function (which is generally not the equilibrium function). The second term in Eq. (2.18) presents the asymmetric part of the distribution function with $g = g(\varepsilon, x, t)$. Similar approach was used for the calculation of characteristics of heterojunction bipolar transistors [32] (see also Refs. [19, 33]). As a result, after the averaging of Eq. (2.11) over the angle θ, one can arrive at the following coupled equations:

$$\frac{\partial F}{\partial t} = -\sqrt{\frac{\varepsilon}{2m}}\frac{\partial g}{\partial x}, \quad \frac{\partial gF}{\partial t} + vg = -\sqrt{\frac{2\varepsilon}{m}}\frac{\partial F}{\partial x}. \tag{2.19}$$

Here in the case of $\omega(q) = \omega =$ const, which corresponds to the scattering of electrons on short-range defects, $v = m\omega/2$. Equation (2.19) are reduced to the following equation for function F:

$$\frac{\partial}{\partial t}\left(\frac{\partial F_\varepsilon}{\partial t} + vF\right) = \frac{\varepsilon}{m}\frac{\partial^2 F}{\partial x^2}. \tag{2.20}$$

In the most interesting case when $eV_{\rm d} \gg k_{\rm B}T$, the boundary conditions for Eq. (2.20) at $x = 0$ and $x = L_{\rm t}$ can be adopted in the following form:

$$F\big|_{x=0} = \exp\left[\frac{\varepsilon_F - \Delta(t) - \varepsilon}{k_{\rm B}T}\right], \quad F\big|_{x=L_{\rm t}} \simeq 0. \tag{2.21}$$

The boundary condition under consideration imply that at $x = 0$ there is the electron injection from the source section of the channel, whereas at $x = L_{\rm t}$ an effective extraction of the electrons into the drain section occurs due to a strong pulling dc electric field. Due to a strong electron scattering a significant portion of the injected electrons returns back to the source section.

Setting as above $\delta V(t) = \delta V_\omega e^{-i\omega t}$ and, hence, $F = F_0 + \delta F_\omega e^{-i\omega t}$ and $g = F_0 + \delta g_\omega e^{-i\omega t}$, we obtain

$$\frac{d^2 F_0}{dx^2} = 0, \tag{2.22}$$

$$\frac{d^2 \delta F_\omega}{dx^2} - \frac{m\omega(\omega + iv_p)}{\varepsilon}\delta F_\omega = 0, \tag{2.23}$$

and arrive at

$$F_0 = \exp\left[\frac{\varepsilon_F - \Delta_0 - \varepsilon}{k_{\rm B}T}\right]\left(1 - \frac{x}{L_t}\right), \tag{2.24}$$

$$\delta F_{\omega} = \exp\left[\frac{\varepsilon_F - \Delta_0 - \varepsilon}{k_B T}\right]\frac{\sinh[\alpha_{\omega}(x - L_t)]}{\sinh(\alpha_{\omega} L_t)}\left(\frac{\delta\Delta_{\omega}}{k_B T}\right), \quad (2.25)$$

where $\alpha_{\omega} = \sqrt{m\omega(\omega + i\nu)/\varepsilon}$. Considering Eqs. (2.19), (2.24), and (2.25), we obtain

$$g_0 = -\frac{1}{\nu}\sqrt{\frac{2\varepsilon}{m}}\frac{\partial F_0}{\partial x}$$

$$= \frac{1}{\nu L_t}\sqrt{\frac{2\varepsilon}{m}}\exp\left[\frac{\varepsilon_F - \Delta_0 - \varepsilon}{k_B T}\right], \quad (2.26)$$

$$\delta g_{\omega} = -\frac{i}{(\omega + i\nu)}\sqrt{\frac{2\varepsilon}{m}}\frac{\partial \delta F_{\omega}}{\partial x}$$

$$= -i\exp\left[\frac{\varepsilon_F - \Delta_0 - \varepsilon}{k_B T}\right]\sqrt{\frac{2\omega}{(\omega + i\nu)}}$$

$$\times\frac{\cosh[\alpha_{\omega}(x - L_t)]}{\sinh(\alpha_{\omega} L_t)}\left(\frac{\delta\,\Delta_{\omega}}{k_B T}\right). \quad (2.27)$$

After that, using Eqs. (2.12), (2.26), and (2.27), we arrive at the following formulas for J_0 and δJ_{ω} (at $x = L_t$):

$$J_0 = e\frac{2(k_B T)^2}{\pi\hbar^2 L_t \nu}\exp\left(\frac{\varepsilon_F - \Delta_0}{k_B T}\right) = J_0^C, \quad (2.28)$$

$$\frac{\delta J_{\omega}}{J_0} = \left(-\frac{\delta\Delta_{\omega}}{k_B T}\right)\int_0^{\infty} d\xi\sqrt{\xi}e^{-\xi}\mathcal{H}_{\omega}(\xi). \quad (2.29)$$

Here

$$\mathcal{H}_{\omega}(\xi) = \frac{i\tau\nu}{\sinh\sqrt{[2\omega(\omega + i\nu)\tau^2/\xi]}}\sqrt{\frac{2\omega}{\omega + i\nu}}.$$

According to Eq. (2.28), $J_0^C \propto 1/L_t\nu$. One needs to stress that the collisional case under consideration corresponds actually to $\nu\tau \gg 1$. In the frequency range $\omega \ll \nu/\tau^2$, $\mathcal{H}_{\omega}(\xi) \simeq \sqrt{\xi}$. At $\omega \gg \nu/\tau^2$, one obtains

$$\mathcal{H}_{\omega}(\xi) \simeq 2(1 + i)\sqrt{\omega\tau^2\nu}\exp\left[-\frac{(1 + i)\sqrt{\omega\tau^2\nu}}{\sqrt{\xi}}\right].$$

2.4 GBL-FET DC Transconductance

Equations (2.16) and (2.28) provide the dependences of the source-drain dc current J_0 as a function of the device structural parameters, temperature, and back- and top-gate voltages for GBL-FETs with ballistic and collisional electron transport, respectively (in the limit $eV_{\rm d} \gg k_{\rm B}T$). Using Eq. (2.16), one can find the dc transconductance $G_0 = (\partial J_0/\partial V_{\rm t})|_{V_{\rm b}}$ of a GBL-FET with the ballistic electron transport:

$$G_0^B = e^2 \frac{\sqrt{2mk_{\rm B}T}}{\pi^{3/2}\hbar^2} \exp\left(\frac{\varepsilon_F - \Delta_0}{k_{\rm B}T}\right) = \frac{eJ_0^B}{2k_{\rm B}T}, \tag{2.30}$$

when $V_{\rm th,2} < V_{\rm t} < V_{\rm th,1}$, and

$$G_0^B = e^2 \frac{\sqrt{2mk_{\rm B}T}}{\pi^{3/2}\hbar^2} \exp\left(\frac{\varepsilon_F - \Delta_0}{k_{\rm B}T}\right) \mathcal{R}_0 = \frac{eJ_0^B}{2k_{\rm B}T} \mathcal{R}_0, \tag{2.31}$$

when $V_{\rm t} < V_{\rm th,2} < V_{\rm th,1}$. Here

$$\mathcal{R}_0 \simeq \left(\frac{a_{\rm B}}{4W}\right). \tag{2.32}$$

Similarly, using Eq. (2.28), we obtain the following formulas for the GBL-FET transconductance in the case of collisional electron transport:

$$G_0^C = e^2 \frac{k_{\rm B}T}{\pi\hbar^2 L_{\rm t} v} \exp\left(\frac{\varepsilon_F - \Delta_0}{k_{\rm B}T}\right) = \frac{eJ_0^C}{2k_{\rm B}T}, \tag{2.33}$$

when $V_{\rm th,2} < V_{\rm t} < V_{\rm th,1}$, and

$$G_0^C = e^2 \frac{k_{\rm B}T}{\pi\hbar^2 L_{\rm t} v} \exp\left(\frac{\varepsilon_F - \Delta_0}{k_{\rm B}T}\right) \mathcal{R}_0 = \frac{eJ_0^C}{2k_{\rm B}T} \mathcal{R}_0, \tag{2.34}$$

when $V_{\rm t} < V_{\rm th,2} < V_{\rm th,1}$.

As follows from the comparison of Eq. (2.30) with Eq. (2.31) and Eq. (2.33) with Eq. (2.34), the GBL-FET dc transconductance in the top gate voltage range $V_{\rm th,2} < V_{\rm t} < V_{\rm th,1}$ (in the range *b*) might be much larger than that when $V_{\rm t} < V_{\rm th,2} < V_{\rm th,1}$ (in the range *c*) since $a_{\rm B} \ll 8W$. This is due to relatively slow increase in Δ_0 with increasing $|V_{\rm t}|$ when the hole density in the gated section becomes essential.

The voltage dependences of the dc transconductance can be obtained using Eqs. (2.30), (2.31), (2.33), and (2.34) and invoking Eqs. (2.2), (2.4), and (2.6). In particular, in a rather narrow voltage range $V_{\rm th,2} < V_{\rm t} < V_{\rm th,1}$, one obtains

$$G_0^C = \frac{\sqrt{\pi}}{v\tau} G_0^B \propto \exp\left[\frac{eV_\mathrm{b}}{k_\mathrm{B}T}\left(\frac{a_\mathrm{B}}{8W}\right)\right]\exp\left[\frac{e(V_\mathrm{b}+V_\mathrm{t})}{2k_\mathrm{B}T}\right]$$
$$= \exp\left[\frac{e(V_\mathrm{t}-V_{\mathrm{th},1})}{2k_\mathrm{B}T}\right]. \quad (2.35)$$

At sufficiently large absolute values of the top-gate voltage when $V_\mathrm{t} < V_{\mathrm{th},2} < V_{\mathrm{th},1}$, the transconductance vs voltage dependence is given by

$$G_0^C = \frac{\sqrt{\pi}}{v\tau} G_0^B \propto \left(\frac{a_\mathrm{B}}{4W}\right)\exp\left[\frac{e(V_\mathrm{t}-V_\mathrm{b})}{k_\mathrm{B}T}\left(\frac{d_0}{2W}\right)\right]$$
$$\times \exp\left[\frac{e(V_\mathrm{t}-V_{\mathrm{th},2})}{k_\mathrm{B}T}\left(\frac{a_\mathrm{B}}{8W}\right)\right]. \quad (2.36)$$

2.5 GBL-FET AC Transconductance

According to the Shockley–Ramo theorem [34, 35], the source-drain ac current is equal to the ac current induced in the highly conducting quasi-neutral portion of the drain section of the channel and in the drain contact by the electrons injected from the gated section. This current is determined by the injected ac current given by Eq. (2.16) or Eq. (2.28), as well as the electron transit-time effects in the depleted portion of the drain section. However, if $\omega\tau_\mathrm{d} \ll 1$, where the τ_d is the electron transit time in depleted region in question, the induced ac current is very close to the injected ac current [19, 36, 37]. Since, at moderate drain voltages, the length of depleted portion of the drain section L_d can usually be shorter than the top gate length L_t, in the most practical range of the signal frequencies $\omega \lesssim \tau^{-1}$, one can assume that $\tau_\mathrm{d} \ll \tau$ and, hence, $\omega\tau_\mathrm{d} \ll 1$. Considering this and using Eqs. (2.17) and (2.28), the GBL-FET ac transconductance $G_\omega = (\partial\delta J_\omega/\partial\delta V_\omega)|_{V_\mathrm{b}}$ at different electron transport conditions can be presented as

$$G_\omega^B = \frac{J_0^B}{k_\mathrm{B}T}\left(-\frac{\partial\delta\Delta_\omega}{\partial\delta V_\omega}\bigg|_{V_\mathrm{b}}\right)\int_0^\infty d\xi\sqrt{\xi}e^{-\xi}\mathcal{F}_\omega(\xi), \quad (2.37)$$

$$G_\omega^C = \frac{J_0^C}{k_\mathrm{B}T}\left(-\frac{\partial\delta\Delta_\omega}{\partial\delta V_\omega}\bigg|_{V_\mathrm{b}}\right)\int_0^\infty d\xi\sqrt{\xi}e^{-\xi}\mathcal{H}_\omega(\xi), \quad (2.38)$$

respectively.

In the range of gate voltages $V_{\text{th},2} < V_{\text{t}} < V_{\text{th},1}$ (range *b*), Eq. (2.2) yields

$$\left.\frac{\partial \delta\Delta_\omega}{\partial \delta V_\omega}\right|_{V_{\text{d}}} = -\frac{e}{2}. \tag{2.39}$$

In this case, Eqs. (2.37) and (2.38) result in

$$G_\omega^B = \frac{eJ_0^B}{2k_{\text{B}}T}\int_0^\infty d\xi\,\sqrt{\xi}e^{-\xi}\mathcal{F}_\omega(\xi), \tag{2.40}$$

$$G_\omega^C = \frac{eJ_0^C}{2k_{\text{B}}T}\int_0^\infty d\xi\,\sqrt{\xi}e^{-\xi}\mathcal{H}_\omega(\xi). \tag{2.41}$$

As follows from Eqs. (2.40) and (2.41), the characteristic frequencies of the ac transconductance roll-off are $1/\tau$ and ν/τ^2 in the case of the ballistic and collisional electron transport, respectively, i.e., the inverse times of the ballistic and diffusive transit across the gated section of the channel. Indeed, the quantity ν/τ^2 can be presented as D/L_t^2, where D is the electron diffusion coefficient.

The situation becomes more complex in the range of the top gate bias voltages $V_{\text{t}} < V_{\text{th},2} < V_{\text{th},1}$ (range *c*). As follows from Eq. (2.4) in this voltage range, the quantity $\delta\Delta_\omega$ is determined not only by the ac voltage δV_ω but also by the ac component of the hole density in the gated section $\delta\Sigma_\omega^+$. Moreover, at sufficiently high signal frequencies, the hole system in the gated section can not manage to follow the variation of the ac voltage. Taking into account the dynamic response of the hole system (see Appendix), instead of Eq. (2.39) one can obtain

$$\left.\frac{\partial \delta\Delta_\omega}{\partial \delta V_\omega}\right|_{V_{\text{d}}} = -\frac{e}{2}\mathcal{R}_\omega. \tag{2.42}$$

Here (see Appendix)

$$\mathcal{R}_\omega = \left(\frac{a_{\text{B}}}{4W}\right)\left\{\left(\frac{a_{\text{B}}}{4W}\right) + \frac{1}{1 - i\omega\tau_r}\right\}^{-1}$$

$$= \mathcal{R}_0\left(\frac{1 - i\omega\tau_r}{1 - i\omega\tau_r\mathcal{R}_0}\right), \tag{2.43}$$

where τ_{r} is the time of the gated section recharging associated with changing of the hole density due to the tunneling or/and generation-recombination processes. Generally, τ_r depends on the top gate length L_t.

Accounting for Eq. (2.42), we arrive at the following formulas for the GBL-FET ac transconductance when $V_t < V_{th,2} < V_{th,1}$:

$$G_\omega^B = \frac{eJ_0^B}{2k_BT}\mathcal{R}_\omega\int_0^\infty d\xi\,\sqrt{\xi}e^{-\xi}\mathcal{F}_\omega(\xi), \tag{2.44}$$

$$G_\omega^C = \frac{eJ_0^C}{2k_BT}\mathcal{R}_\omega\int_0^\infty d\xi\,\sqrt{\xi}e^{-\xi}\mathcal{H}_\omega(\xi). \tag{2.45}$$

If $\omega \gg \tau_r^{-1}$, one obtains $R_\omega \simeq 1$ and the ac transconductances in both b and c ranges of the top gate voltage are close to each other (compare Eqs. (2.40) and (2.41) with Eqs. (2.42) and (2.44)). However, at low signal frequencies ($\omega \gg \tau_r^{-1}$), the ac transconductance given by Eqs. (2.44) or (2.45) for the voltage range c are markedly smaller than those given by Eqs. (2.40) and (2.41) valid in the voltage range b.

2.6 Analysis of the Results and Discussion

Comparing G_0^B and G_0^C given by Eqs. (2.30) and (2.35), we obtain $G_0^C/G_0^B = \sqrt{\pi}/\nu\tau$. This implies that the above ratio markedly decreases with increasing collision frequency (with decreasing electron mobility) and the top gate length, i.e., with the departure from the ballistic transport.

As shown above, the dc current steeply drops in a narrow top-gate voltage range $V_{th,2} < V_t < V_{th,1}$. Indeed, the ratio $J_0^B|_{V_t = V_{th,2}}/J_0^B|_{V_t = V_{th,1}} \simeq \exp[-(ed_0V_b/Wk_BT)]$. Setting $W = 5$ nm, $T = 300$ K, and $V_b = 1–2$ V, we find $J_0^B|_{V_t = V_{th,2}}/J_0^B|_{V_t = V_{th,1}} \simeq 3\times10^{-3} - 6\times10^{-2}$. The estimate of the dc current at $V_t = V_{th,2} \simeq -V_b$ (which might be interesting for the GBL-FET applications in digital large scale circuits) with $W = 5$ nm, $T = 300$ K, and $V_b = 1 - 2$ V yields $J_0^B \simeq 1\times10^{-3} - 2\times10^{-2}$ A/cm. At $T = 77$ K and $V_b = 1$ V, one obtains $J_0^B \simeq 7\times10^{-7}$ A/cm. In the case of a GBL-FET with the width $H = 1$ μm, the latter corresponds to the characteristic value of the "off-current" $J_0^B H \simeq 70$ pA. Similar values can be obtained at $T = 300$ K when $-V_t = V_b \simeq 4.6$ V.

As follows from Eq. (2.35), the GBL dc transconductance in the range of the top gate voltages $V_{th,2} < V_t < V_{th,1}$ is particularly large when $V_t \lesssim V_{th,1}$. This is due to a sharp voltage-sensitivity of the dc current and its relatively high values at such voltages. Indeed, using Eq. (2.28) at $T = 300$ K and $V_t \lesssim V_{th,1}$, we obtain $G_0^B \lesssim 2500$ mS/mm. In GBL-FETs with $W_t \ll W_b$, the dc transconductance can be even larger.

The pre-exponential factor in the right-hand side of Eq. (2.36) is proportional to a small parameter $(a_B/8W)$. The argument of the exponential function in this equation comprises small parameters $(a_B/8W)$ and $(d_0/2W)$. This implies that the dc transconductance in the voltage range $V_t < V_{th,2} < V_{th,1}$ described by Eq. (2.36) is relatively small and is a faily weak function of the top-gate voltage. As follows from Eqs. (2.35) and (2.36), the ratio of the dc transconductance at $V_t \lesssim V_{th,2}$ to that at $V_t \lesssim V_{th,1}$ is equal approximately to the following small value: $(a_B/4W)\exp[ed_0(V_{th,2} - V_b)/2Wk_BT] = (a_B/4W)\exp(-2eE_{g,S}d_0/dk_BT)$; it is smaller then the ratio of the dc currents by parameter $(a_B/4W)$.

Figure 2.3 shows the ac transconductance $|G_\omega|$ normalized by the dc transconductance at the ballistic transport G_0^B as a function of the normalized signal frequency $\omega\tau$ calculated for GBL-FETs with both ballistic and collisional electron transport. It is assumed that the top gate voltage is in the range $V_{th,2} < V_t < V_{th,1}$. The inset in Fig. 2.3 shows the dependence of the normalized threshold frequency $\omega_t\tau$ on $\nu\tau \propto \nu L_t$. The threshold frequency is defined as that at which $|G_\omega|/G_0 = 1/\sqrt{2}$. One can see that $|G_\omega|$ pronouncedly decreases with increasing collision frequency. However, as seen from the inset in Fig. 2.3, the decrease in ω_t with increasing ν is markedly slower: the ratio of ω_t at $\nu\tau = 2$ and ω_t at $\nu\tau = 6$ (i.e., three times larger) is approximately equal to 1.42. Setting $L_t = 100 - 500$ nm and $T = 300$ K, for the threshold frequency $f_t = \omega_t/2\pi$ at the ballistic transport, we obtain $f_t^B \simeq 0.485 - 0.97$ THz. To realize the near ballistic regime of the electron transport ($\nu\tau \ll 1$) in GBL-FETs with such gate lengths, the electron mobility $\mu > (1 - 5) \times 10^4$ cm^2/V s is required. The possibility of the latter mobilities at room temperatures was discussed recently (see, for instance, Ref. [6]). At a shorter top gate, $L_t = 75$ nm, one obtains $f_t^B \simeq 1.29$ THz. In the case of a GBL-FET with relatively long top gate and moderate mobility ($L_t = 500$ nm and $\mu = 2 \times 10^4$ cm^2/V s) when the effect of scattering is strong ($\nu \simeq 2 \times 10^{12}$ s^{-1} and $\nu\tau \simeq 2.7$), we obtain $f_t^C \simeq 94$ GHz.

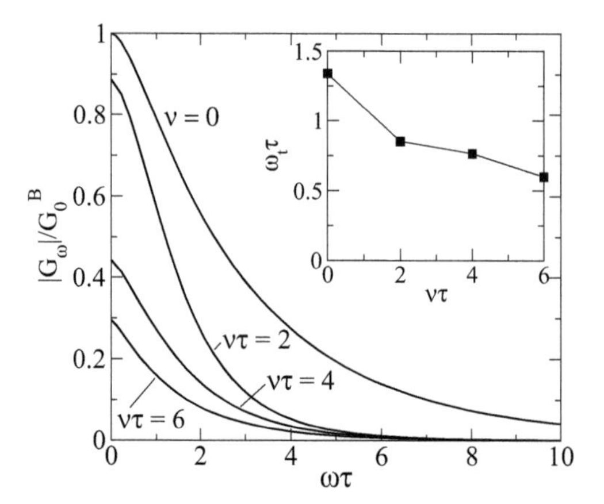

Figure 2.3 The ac transconductance (normalized by G_0^B) versus $\omega\tau$ calculated for GBL-FETs with ballistic ($\nu = 0$) and collisional electron transport ($\nu\tau = 2$, 4, and 6) at the top gate voltage in range *b*, i.e., $V_{th,2} < V_t < V_{th,1}$. For $L_t = 100$ nm and $T = 300$ K, $\omega\tau = 10$ corresponds to $f = \omega/2\pi \simeq 3.3$ THz. The inset shows the normalized threshold frequency $\omega_t\tau$ as a function of parameter $\nu\tau$.

In Fig. 2.4, the similar dependences calculated for a GBL-FET with collisional electron transport at $V_{th,2} < V_t < V_{th,1}$ and at $V_t < V_{th,2} < V_{th,1}$ are demonstrated. Since in the latter top-gate voltage range the electron scattering on holes accumulated in the gated section can be strong, so that the realization of the ballistic transport at such top-gate voltages might be problematic, only the dependences corresponding to the collisional electron transport are shown. As follows from Eqs. (2.44) and (2.45) and seen from Fig. 2.4, the ac transconductance at the top gate voltages $V_t < V_{th,2} < V_{th,1}$ is fairly small at low frequencies $\omega \lesssim \tau_r^{-1}$ being close to the dc transconductance (due to a smallness of parameter $R \simeq a_B/4W$), whereas it becomes much larger in the intermediate frequency range $\tau_r^{-1} \ll \omega \lesssim \tau^{-1}$ (or $\tau_r^{-1} \ll \omega \lesssim \nu\tau^{-2}$). This is due to the effect of holes in the gated section. Owing to this effect, the low frequency noises can be effectively suppressed.

Using the above results for GBL-FETs and those obtained previously [19] for GNR-FETs, one can compare the GBL and GNR characteristics. In particular, considering the expressions for J_0^B found for GBL-FETs and GNR-FETs, we obtain $J_0^B/J_0^{B,GNR} = 1/2$. For the case of collisional transport, one obtains $J_0^C/J_0^{C,GNR} \sim 1$. As a result, the GBL-FET and GNR-FET dc transconductances are close to each other. The ratio of the GBL-FET and GNR-FET ac transconductances at high signal frequencies is $G_\omega{}^{B}/G_\omega{}^{B,GNR} \propto 1/\sqrt{\omega\tau}$, i.e., the GBL-FET ac transconductance falls more steeply with increasing ω than the GNR-FET ac transconductance.

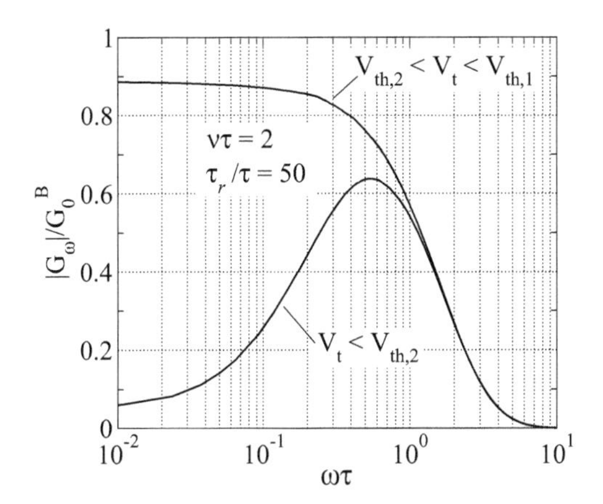

Figure 2.4 The ac transconductance (normalized by G_0^B) versus $\omega\tau$ calculated for GBL-FETs with collisional electron transport ($\nu\tau$ = 2) at the top voltage in range *c* ($V_t < V_{th,2} < V_{th,1}$ at $\tau_r/\tau = 50$.

The GBL-FET dc and ac characteristics obtained are valid if the interband tunneling source-drain current through the $n^+ - p$ and $p - n^+$ junctions beneath the edges of the top gate are small in comparison with the thermionic current created by the electrons overcoming the potential barrier in the gated section. Such a tunneling current can be essential in the voltage range *c* ($V_t < V_{th,2} < V_{th,1}$) depending on the energy gap near the top gates and the length of the $n^+ - p$ and $p - n^+$ junctions in question. This implies that there is a limitation when the top-gate voltage is not too high V_t in comparison with the threshold voltage $V_{th,2}$ (i.e., when $V_t \lesssim V_{th,2}$), so that the calculated characteristics correspond to the most interesting voltage range where the ac transconductance can be rather large.

2.7 Conclusions

We have presented an analytical device model for a GBL-FET. Using this model, we have calculated the GBL-FET dc and ac characteristics and shown that:

(1) The dependence of the dc current on the top gate voltage is characterized by the existence of three voltage ranges, corresponding to (a) the population of the gated section by electrons, (b) the depletion of this section, and (c) its essential filling with holes, and determined by the top-gate threshold voltages $V_{th,1}$ and $V_{th,2}$.
(2) The ac current is most sensitive to the top-gate voltage V_t and the dc and ac transconductances are large when $V_{th,2} < V_t \lesssim V_{th,1}$.
(3) The electron scattering in the gated section results in a marked reduction in the dc and ac transconductances. However, the threshold frequency corresponding to $|G_\omega|/G_0 = 1/\sqrt{2}$ decreases with increasing collision frequency relatively smoothly.
(4) The transient recharging of the gated section by holes (at $V_t < V_{th,2}$) leads to a nonmonotonic frequency dependence of the ac transconductance with a pronounced maxima in the range of fairly high frequencies. This effect might be used for the optimization of GBL-FETs with reduced sensitivity to low frequency noises.
(5) The fabrication of GBLs with high electron mobility at elevated temperatures opens up the prospects of realization of terahertz GBL-FETs with ballistic electron transport operating at room temperatures and surpassing FETs based on A_3B_5 compounds.

Acknowledgments

The authors are grateful to Professor S. Brazovskii (University Paris-Sud) and Professor E. Sano (Hokkaido University) for fruitful discussions and valuable information. The work was supported by the Japan Science and Technology Agency, CREST, Japan. One of the authors (N.K.) acknowledges the support by the INTAS grant 7972, and by the ANR program in France (the project BLAN07-3-192276).

Appendix: Dynamic Response of the Hole System in the Gated Section

At sufficiently large values $|V_t|$, the gated section is essentially populated with the holes. As follows from Eq. (2.4),

$$\delta\Delta_\omega = -\frac{1}{2}e\delta V_\omega - \frac{2\pi e^2 W}{\kappa}\delta\Sigma_\omega^+. \tag{2.A1}$$

The ac component of the hole density $\delta\Sigma_\omega^+$ obeys the continuity equation, which can be presented in the following form:

$$-i\omega\delta\Sigma_\omega^+ = \delta G_\omega + \frac{\delta J_\omega^T\big|_{x=0} - \delta J_\omega^T\big|_{x=L_t}}{eL_t}. \tag{2.A2}$$

Here δG_ω is the variation of the generation of holes in the gated section (associated, say, with the generation of holes by the thermal radiation [38]) and $\delta J_\omega^T\big|_{x=0}$ and $\delta J_\omega^T\big|_{x=L_t}$ are the interband tunneling ac currents near the source and drain edges of the top gate, respectively. For normal operation of GBL-FETs, these tunneling current should relatively small. This is achieved in GNL-FETs by proper choice of the energy gap in the different sections of the channel ($E_{g,s}$, E_g, and $E_{g,d}$, which should be not too small. The terms in the right-hand side of Eq. (2.A2) can be presented as

$$\delta G_\omega = K_g(\delta\Delta_\omega - \delta\varepsilon_F^+), \tag{2.A3}$$

$$\delta J_\omega^T\big|_{x=0} - \delta J_\omega^T\big|_{x=L_t} = \frac{2(\delta\Delta_\omega - \delta\varepsilon_F^+)}{eR_t}. \tag{2.A4}$$

Here $\delta\varepsilon_F^+ \simeq (\pi\hbar^2/2m)\,\delta\Sigma_\omega^+$, $K_g = 2m/\pi\hbar^2\tau_g$, where τ_g is the characteristic time of the generation-recombination processes in the gated section, R_t is the tunneling resistance of the *p-n* junctions induced by the negative top gate voltage near the edges of the top gate: $\ln R_t \propto E_g^{3/2}/\varepsilon_{||} \propto (V_b - V_t)^{3/2}/V_t$. From Eqs. (2.A1)–(2.A4), taking into account the limit $\omega \to 0$ (see Eqs. (2.5) and (2.30)), we find

$$\delta\Delta_\omega = -\frac{e}{2}\delta V_\omega$$

$$\times\left(\frac{a_{\rm B}}{4W}\right)\left\{\left(\frac{a_{\rm B}}{4W}\right)+\frac{\dfrac{1}{\tau_g}+\left(\dfrac{a_{\rm B}}{4W}\right)\dfrac{1}{\tau_{RC}}}{\dfrac{1}{\tau_g}+\left(\dfrac{a_{\rm B}}{4W}\right)\dfrac{1}{\tau_{RC}}-i\omega}\right\}^{-1}$$

$$=-\frac{e}{2}\delta V_\omega\left(\frac{a_B}{4W}\right)\left\{\left(\frac{a_{\rm B}}{4W}\right)+\frac{1}{1-i\omega\tau_r}\right\}^{-1}, \qquad (2.A5)$$

where $\tau_{\rm RC} = R_{\rm t}C_{\rm t}$ is the time of the gated section recharging by the tunneling currents, $R_{\rm t}$ is the tunneling resistance of the *p-n* junctions induced by the negative top gate voltage near the edges of the top gate, the quantity $\tau_{\rm r} = \tau_{\rm RC}\tau_{\rm g}/[\tau_{\rm RC} + (a_{\rm B}/4W)\tau_{\rm g}]$ is the characteristic time of the gated section recharging by holes, and $C_{\rm t} = \kappa L_{\rm t}/2\pi W$ is the capacitance of the gated section. At $\omega \to 0$, Eq. (2.A5) leads to

$$\delta\Delta_\omega \simeq -\frac{1}{2}e\delta V_\omega\mathcal{R}_0, \quad \left.\frac{\partial\delta\Delta_\omega}{\partial\delta V_\omega}\right|_{V_{\rm d}} \simeq -\frac{e}{2}\mathcal{R}_0, \qquad (2.A6)$$

where

$$\mathcal{R}_0 \simeq \frac{(a_{\rm B}/4W)}{1+(a_{\rm B}/4W)} \simeq (a_{\rm B}/4W),$$

i.e., coincides with the value given by Eq. (2.28). When $\omega \gg \tau_{\rm r}^{-1}$, Eq. (2.A5) yields

$$\delta\Delta_\omega \simeq -\frac{1}{2}e\delta V_\omega, \quad \left.\frac{\partial\delta\Delta_\omega}{\partial\delta V_\omega}\right|_{V_{\rm d}} \simeq -\frac{e}{2}. \qquad (2.A7)$$

References

1. C. Berger, Z. Song, T. Li, X. Li, A.Y. Ogbazhi, R. Feng, Z. Dai, A. N. Marchenkov, E. H. Conrad, P. N. First, and W. A. de Heer, *J. Phys. Chem.*, **108**, 19912 (2004).
2. K. S. Novoselov, A. K. Geim, S. V. Morozov, D. Jiang, M. I. Katsnelson, I. V. Grigorieva, S. V. Dubonos, and A. A. Firsov, *Nature*, **438**, 197 (2005).
3. B. Obradovic, R. Kotlyar, F. Heinz, P. Matagne, T. Rakshit, M. D. Giles, M. A. Stettler, and D. E. Nikonov, *Appl. Phys. Lett.*, **88**, 142102 (2006).
4. J. Hass, R. Feng, T. Li, X. Li, Z. Zong, W. A. de Heer, P. N. First, E. H. Conrad, C. A. Jeffrey, and C. Berger, *Appl. Phys. Lett.*, **89**, 143106 (2006).

5. A. K. Geim and K. S. Novoselov, *Nat. Mater.*, **6**, 183 (2007).
6. S. V.Morozov, K. S. Novoselov, M. I. Katsnelson, F. Schedin, D. C. Elias, J. A. Jaszczak, and A. K. Geim, *Phys. Rev. Lett.*, **100**, 016602 (2008)
7. V. V. Cheianov and V. I. Fal'ko, *Phys. Rev. B*, **74**, 041403 (2006).
8. B. Huard, J. A. Sulpizio, N. Stander, K. Todd, B. Yang, and D. Goldhaber-Gordon, *Phys. Rev. Lett.*, **98**, 236803 (2007).
9. M. C. Lemme, T. J. Echtermeyer, M. Baus, and H. Kurz, *IEEE Electron Device Lett.*, **28**, 282 (2007).
10. V. Ryzhii, M. Ryzhii, and T. Otsuji, *Appl. Phys. Express*, **1**, 013001 (2008),
11. V. Ryzhii, M. Ryzhii, and T. Otsuji, Phys. *Stat. Sol. A*, **205**, 1527 (2008).
12. Z. Chen, Y.-M. Lin, M. J. Rooks, and P. Avouris, *Physica E*, **40**, 228 (2007).
13. K. Wakabayshi, Y. Takane, and M. Sigrist, *Phys. Rev. Lett.*, **99**, 036601 (2007).
14. Y. Quyang, Y. Yoon, J. K. Fodor, J. Guo, *Appl. Phys. Lett.*, **89**, 203107 (2006).
15. G. Liang, N. Neophytou, D. E. Nikonov, and M. S. Lundstrom, *IEEE Tran. Electron Devices*, **54**, 677 (2007).
16. G. Fiori and G. Iannaccone, *IEEE Electron Device Lett.*, **28**, 760 (2007).
17. V. Ryzhii, M. Ryzhii, A. Satou, and T. Otsuji, *J. Appl. Phys.*, **103**, 094510 (2008).
18. V. Ryzhii, V. Mitin, M. Ryzhii, N. Ryabova, and T. Otsuji, *Appl. Phys. Express*, **1**, 063002 (2008).
19. M. Ryzhii, A. Satou, V. Ryzhii, and T. Otsuji, *J. Appl. Phys.*, **104**, 114505 (2008).
20. T. Ohta, A. Bostwick, T. Seyel, K. Horn, E. Rotenberg, *Science*, **313**, 951 (2006).
21. E. McCann, *Phys. Rev. B*, **74**, 161403 (2006).
22. E. V. Castro, K. S. Novoselov, S. V. Morozov, N. M. R. Peres, J. M. L. dos Santos, J. Nilsson, F. Guinea, A. K. Geim, and A. H. Castro Neto, *Phys. Rev. Lett.*, **99**, 216802 (2007)
23. L. A. Falkovski, *Phys. Rev. B*, **75**, 033409 (2007).
24. F. T. Vasko and V. Ryzhii, *Phys. Rev. B*, **76**, 233404 (2007).
25. V. Vyurkov and V. Ryzhii, *JETP Lett.*, **88**, 322 (2008).
26. M. G. Ancona, 2008 International Conference on Semiconductor Processes and Devices (SISPAD 2008), Sept. 9–11, 2008, Hakone, Japan, p. 169.

27. V. Vyurkov, I. Seminikhin, M.Ryzhii, T. Otsuji, and V. Ryzhii, Techn. Digest, International Symposium on Graphene Devices: Technology, Physics, and Modeling (ISGD2008), Nov. 17–19, 2008, Aizu-Wakamatsu, Japan, p. 32.

28. M. Shur, *Physics of Semiconductor Devices* (Prentice Hall, New Jersey, 1990).

29. S. M. Sze, *Physics of Semiconductor Devices* (Wiley, New York, 1981).

30. E. P. Gusev, C. Carbal Jr., M. Copel, C. D'Emic, and M. Gribelenuk, *Microelectron. Eng.*, **63**, 145 (2003).

31. Y. Xuan, Y. Q. Wu, T. Shen, M. Qi, M. A. Capano, J. A. Cooper, and P. D. Ye, *App. Phys. Lett.*, **92**, 013101 (2008).

32. V. I. Ryzhii, A. A. Zakharova, and S. N. Panasov, *Sov. Phys. Semicond.*, **19**, 298 (1985).

33. E. Gnani, A. Gnudi, S. Reggiani, and G. Baccarani, *IEEE Trans. Electron Devices*, **55**, 2918 (2008)

34. W. Shockley, *J. Appl. Phys.*, **9**, 635 (1938)

35. S. Ramo, *Proc. IRE*, **27**, 584 (1939).

36. V. Ryzhii and G. Khrenov, *IEEE Trans. Electron Devices*, **42**, 166 (1995).

37. V. Ryzhii, A. Satou, M. Ryzhii, T. Otsuji, and M. S. Shur, *J. Phys.: Condens. Matter*, **20**, 384207 (2008).

38. F. T. Vasko and V. Ryzhii, *Phys. Rev. B*, **77**, 195433 (2008).

Chapter 3

Electrically-Induced n-i-p Junctions in Multiple Graphene Layer Structures*

M. Ryzhii,[a] V. Ryzhii,[a] T. Otsuji,[b]
V. Mitin,[c] and M. S. Shur[d]
[a]*Computational Nanoelectronics Laboratory, University of Aizu, Aizu-Wakamatsu 965-8580 and Japan Science and Technology Agency, CREST, Tokyo 107-0075, Japan*
[b]*Research Institute of Electrical Communication, Tohoku University, Sendai 980-8577 and Japan Science and Technology Agency, CREST, Tokyo 107-0075, Japan*
[c]*Department of Electrical Engineering, University at Buffalo, State University of New York, NY 14260, USA*
[d]*Department of Electrical, Computer, and Systems Engineering, Rensselaer Polytechnic Institute, Troy, New York 12180, USA*
m-ryzhii@u-aizu.ac.jp

The Fermi energies of electrons and holes and their densities in different graphene layers (GLs) in the n and p regions of the electrically induced n-i-p junctions formed in multiple-GL structures

*Reprinted with permission from M. Ryzhii, V. Ryzhii, T. Otsuji, V. Mitin, and M. S. Shur (2010). Electrically-induced n-i-p junctions in multiple graphene layer structures, *Phys. Rev. B*, **82**, 075419. Copyright © 2010 The American Physical Society.

Graphene-Based Terahertz Electronics and Plasmonics: Detector and Emitter Concepts
Edited by Vladimir Mitin, Taiichi Otsuji, and Victor Ryzhii

ISBN 978-981-4800-75-4 (Hardcover), 978-0-429-32839-8 (eBook)
www.jennystanford.com

are calculated both numerically and using a simplified analytical model. The reverse current associated with the injection of minority carriers through the *n* and *p* regions in the electrically-induced n-i-p junctions under the reverse bias is calculated as well. It is shown that in the electrically-induced n-i-p junctions with moderate numbers of GLs the reverse current can be substantially suppressed. Hence, multiple-GL structures with such n-i-p junctions can be used in different electron and optoelectron devices.

3.1 Introduction

The possibility to form electrically-induced n-p and n-i-p junctions [1–3] in gated graphene layers (GLs), as well as lateral arrays of graphene nanoribbons and graphene bilayer, opens up prospects of creation novel electronic and optoelectronic devices [4–8]. In contrast to GL structures with chemically doped n- and p-region in the GL structures with electrically-induced n-p and n-i-p junctions, there is a possibility of their voltage control. Recent success in fabricating high-quality multiple GLs Refs. [9–11] (see also the recent review article [12] and references therein) stimulates an interest in different prospective devices using multiple-GL structures. These structures constitute the stacks of disoriented GLs (with the non-Bernal stacking). Each GL exhibits linear dispersion relation for electrons and holes similar to that in a single GL. This is a well-established experimental fact (see, for instance, Refs. [11–14]. Using the multiple-GL structures instead of single-GL structures can provide a significant improvement in device performance of terahertz tunneling transit-time oscillators (similar to that considered in Ref. [4]), lasers with optical and electrical pumping, and high performance interband photodetectors [15–17]. Gated multiple-GL structures can be also used in high-frequency field-effect transistors [18] and other devices (such as terahertz frequency multipliers, and plasmonic devices). However the penetration of the electric field (transverse to the GL plane) beyond the topmost GL as well as its sceening by electron or hole charges in GLs can substantially limit the influence of the gates (the effect of the quantum capacitance [19]). In this chapter, we study the influence of screening in gated multiple-GL structures on the formation and characteristics of *n* or *p* regions and n-i-p junctions in these structures. We calculate the electron and hole Fermi energies and densities in GLs in the *n* and *p* regions as functions of the GL index, gate voltage, and temperature. Using these data, we find the voltage and temperature dependences of the reverse current in the n-i-p junctions with different structural parameters. The multiple-GL structures are often obtained by thermal decomposition from 4H-SiC substrate (see, for instance, Refs. [9] and [10]) on the C-terminated surface. The majority of the top GLs is practically neutral, whereas a buffer bottom GL (BGL) or a few BGLs in the very close vicinity of the interface with SiC is highly conducting (doped) as a result of the charge transfer from SiC [9]. For the device applications akin those considered in Refs. [15–17], such a highly conducting BGL plays a negative role. Due to this, we focus on the multiple-GL structures in which the BGL (or BGLs) and, hence, the effect of SiC substrate are eliminated. The multiple-GL structures without the BGL (BGLs) can be fabricated using chemical/mechanical reactions and transferred substrate techniques, which include chemically etching the substrate and the highly conducting GL (or GLs) and transferring the upper portion of the multiple-GL structure on a Si or equivalent substrate. The mechanical peeling of the upper portion of GL with the subsequent placement on a substrate is another option. Possible modifications of the effects under consideration in the multiple-GL structures with the highly conducting BGL are briefly discussed below. The structures with few non-Bernal stacked GLs are also fabricated on 3C-SiC(110) on Si(110) substrate [20].

3.2 Equations of the Model

Let us consider a multiple-GL structure with the side Ohmic contacts to all GLs and two split gates (isolated from GLs) on the top of this structure as shown in Fig. 3.1a. Applying the positive ($V_n = V_g > 0$) or negative ($V_p = -V_g < 0$) voltage between the gate and the adjacent contact (gate voltage), one can obtain the electrically-induced *n* or *p* region. In the single- and multiple-GL structures with two split top gates under the voltages of different polarity, one can create lateral n-p or n-i-p junctions. Generally, the source-drain voltage V can be applied between the side Ohmic contacts to GLs. Depending on the polarity of this voltage, the n-p and n-i-p junctions can be either direct or reverse biased. We assume that the potentials of the first (source) contact and the pertinent gate are $\varphi_s = 0$ and $\varphi_g = V_g > 0$, respectively, and the potentials of another gate and contact (drain) are $\varphi_g = -V_g < 0$ and $\varphi_d = V = 0$ (or $\varphi_d = V \neq 0$. If the slot between the gates $2Lg$ is sufficiently wide (markedly exceeds the thickness of the gate layer W_g separating the gates and the topmost GL), there are intrinsic *i* regions in each GL under the slot. Thus the n-i-p junction is formed. The pertinent band diagrams are shown in Figs. 3.1b and 3.1c. Since the side contacts are the Ohmic contacts, the electron Fermi energy in the *k*-th GL sufficiently far from the contacts are given by $\mu_k = \pm e\varphi_k$. Here *e* is the electron charge and $\varphi_k = \varphi|_{z\,=\,kd}$ is the potential of the *k*-th GL, $k = 1, 2, ..., K$, where *K* is the number of GLs in the structure, *d* is the spacing between GLs, and the axis *z* is directed perpendicular to the GL plane with $z = 0$ corresponding to the topmost GL and $z = z_K = Kd$ to the lowest one.

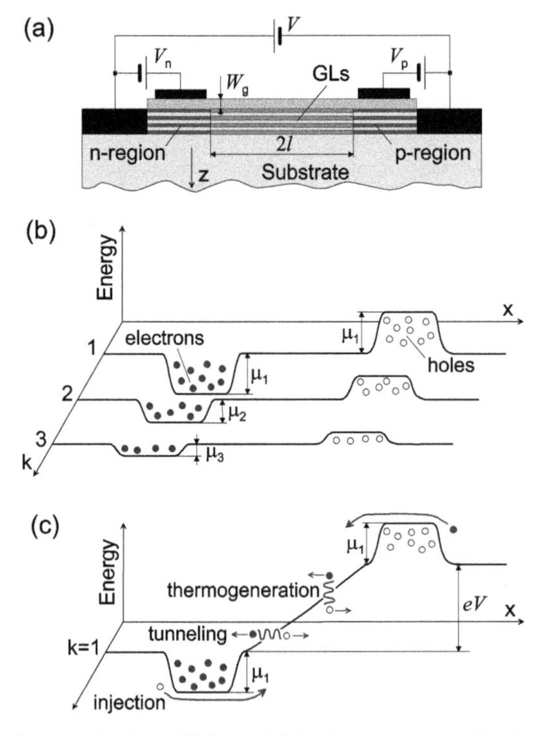

Figure 3.1 Schematic view of (a) a multiple-GL structure and its band diagrams at (b) zero bias voltage ($V = 0$) and (c) at reverse bias (for GL with $k = 1$). Arrows indicate the directions of propagation of injected electrons and holes as well as those thermogenerated and generated due to interband tunneling.

Focusing on the n region (the p region can be considered in a similar way) and introducing the dimensionless potential $\psi = 2\varphi/V_g$, one can arrive at the following one-dimensional Poisson equation governing the potential distribution in the z direction (in the n region):

$$\frac{d^2\psi}{dz^2} = \frac{8\pi e}{æ V_g} \sum_{k=1}^{K} \left(\Sigma_k^- - \Sigma_k^+\right) \cdot \delta(z - kd + d). \tag{3.1}$$

Here æ is the dielectric constant, $\delta(z)$ and Σ_k^- and Σ_k^+ are the equilibrium sheet densities in the k-th GL of electrons and holes, respectively. These densities, taking into account the linear dispersion low for electrons and holes in graphene, are expressed via the electron Fermi energy as

$$\begin{aligned}\Sigma_k^{\mp} &= \frac{2}{\pi}\left(\frac{k_B T}{\hbar v_F}\right)^2 \int_0^{\infty} \frac{d\xi\xi}{1 + \exp(\xi \mp \mu_k/k_B T)} \\ &= \frac{12\Sigma_T}{\pi^2} \int_0^{\infty} \frac{d\xi\xi}{1 + \exp(\xi \mp \mu_k/k_B T)},\end{aligned} \tag{3.2}$$

where $\Sigma_T = (\pi/6)(k_B T/\hbar v_F)^2$ is the electron anwd hole density in the intrinsic graphene at the temperature T, $v_F \simeq 10^8$ cm/s is the characteristic velocity of electrons and holes in graphene, and $\hbar$ and k_B are the Planck and Boltzmann constants, respectively. Here it is assumed that the electron (hole) energy spectrum is $\varepsilon = v_F p$, where p is the absolute value of the electron momentum. The boundary conditions are assumed to be as follows:

$$\psi|_{z=-0} = 2 + W_g \left.\frac{d\psi}{dz}\right|_{z=-0}, \quad \left.\frac{d\psi}{dz}\right|_{z=z_K+0} = 0. \tag{3.3}$$

Equations (3.1)–(3.3) yield

$$2 - \psi_1 = \Gamma\Phi(\psi_1) \tag{3.4}$$

for $K = 1$,

$$\frac{d}{W_g}(2 - \psi_1) - \psi_1 + \psi_2 = \frac{d}{W_g}\Gamma\Phi(\psi_1),$$

$$\psi_1 - \psi_2 = \Gamma\Phi(\psi_2) \tag{3.5}$$

for $K = 2$, and

$$\frac{d}{W_g}(2 - \psi_1) - \psi_1 + \psi_2 = \frac{d}{W_g}\Gamma\Phi(\psi_1),$$

$$\psi_{k-1} - 2\psi_k + \psi_{k+1} = \frac{d}{W_g}\Gamma\Phi(\psi_k),\ \ (2 \le k \le K-1),$$

$$\psi_{K-1} - \psi_K = \frac{d}{W_g}\Gamma\Phi(\psi_K) \tag{3.6}$$

for $K > 2$. Here

$$\Phi(\psi) = \frac{12}{\pi^2}\left[\int_0^\infty \frac{d\xi\xi}{1+\exp(\xi - U_g\psi)} - \int_0^\infty \frac{d\xi\xi}{1+\exp(\xi + U_g\psi)}\right], \tag{3.7}$$

where $\Gamma = (8\pi/æ)(eW_g\Sigma_T/V_g) \propto T^2/V_g$ and $U_g = eV_g/2k_BT$.

3.3 Numerical Results

Equations (3.4)–(3.7) were solved numerically. The results of the calculations are shown in Figs. 3.2–3.5. In these calculations, we assumed that $æ = 4$, $d = 0.35$ nm, and $W_g = 10$ nm.

Figure 3.2 shows the dependences of the electron Fermi energy

$$\mu_k = \frac{eV_g}{2}\psi_k \tag{3.8}$$

as a function of the GL index k calculated for multiple-GL structures with different number of GLs K at different gate voltages and temperatures. One can see that the Fermi energy steeply decreases with increasing GL index. However, in GLs with not too large k, the Fermi energy is larger or about of the thermal energy. As one might expect, the electron Fermi energies in all GLs at $T = 77$ K are somewhat larger than at $T = 300$ K (see also Fig. 3.5). The obtained values of the electron Fermi energies in topmost GLs are $\mu_1 \simeq 92$ meV and $\mu_1 \simeq 77$ meV for $V_g = 1000$ mV at $T = 77$ K and $T = 300$ K, respectively.

Figure 3.3 shows the voltage dependences of the electron Fermi energies in some GLs in multiple-GL structure with $K = 2$, $K = 10$, and $K = 50$ at $T = 300$ K.

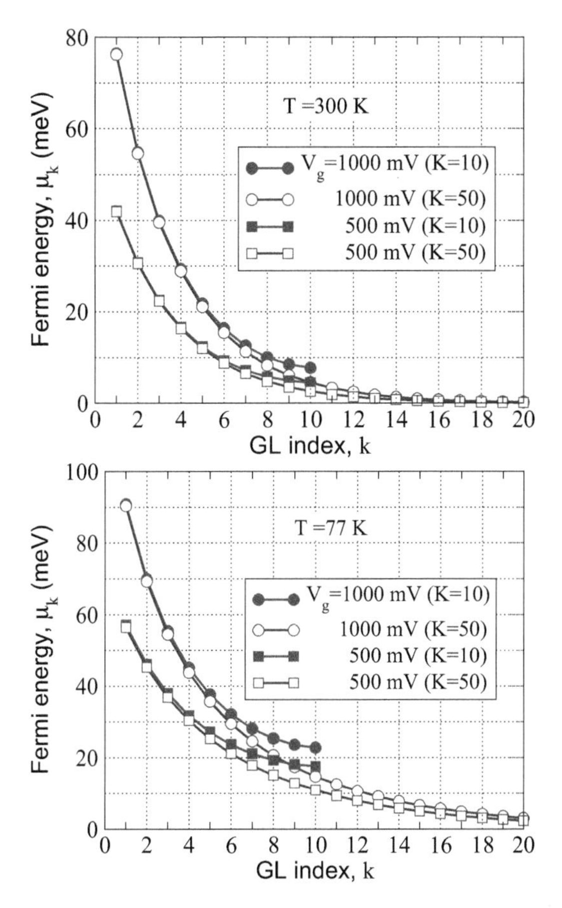

Figure 3.2 The electron Fermi energy μ_k as a function of the GL index k calculated for multiple-GL structures with different number of GLs K for different gate voltages V_g at $T = 300$ K (upper panel) and $T = 77$ K (lower panel).

Figure 3.4 shows the electron densities Σ_k^- in the structures with different number of GLs K at different temperatures. One can see that the calculated electron densities in GLs with sufficiently large indices ($k > 15$ at $T = 77$ K and $T = 300$ K) approach to their values in the intrinsic graphene ($\Sigma_T = 0.59 \times 10^{10}$ cm^{-2} and 8.97×10^{10} cm^{-2}). The electron densities in GLs in the structures with different K are rather close to each other, particularly, in GLs with small and moderate indices.

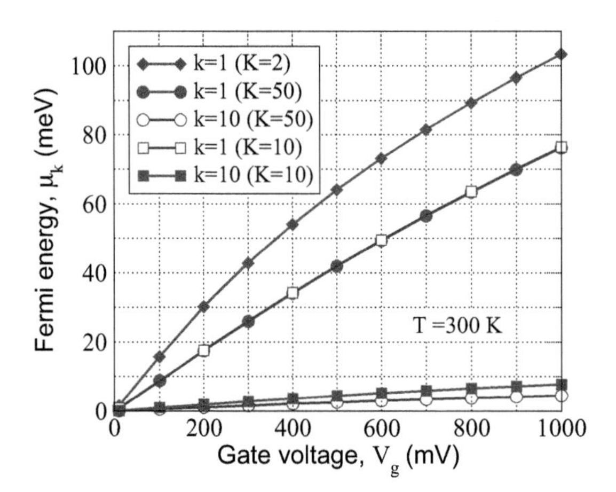

Figure 3.3 Voltage dependences of electron Fermi energies in some GLs in multiple-GL structure with different K at T = 300 K.

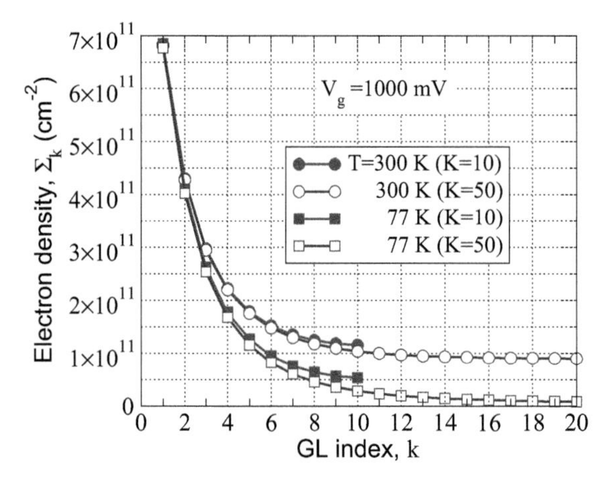

Figure 3.4 Electron density vs GL index in multiple GL-structures with different number of GLs (K = 10 and K = 50) at different temperatures and V_g = 1000 mV.

Figure 3.5 presents the Fermi energies in GLs with different indices at different temperatures. One can see from Fig. 3.5 (as well as from Fig. 3.2) that the higher T corresponds to lower μ_k. This is due to an increasing dependence of the density of states on the energy and the thermal spread in the electron energies.

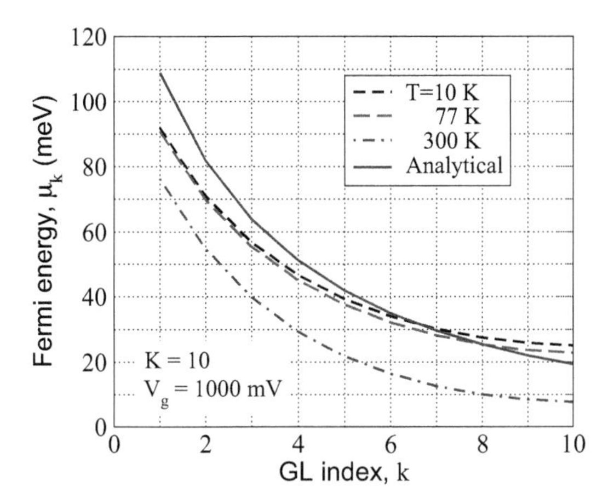

Figure 3.5 Comparison of the μ_k vs k dependences calculated for different temperatures using numerical and simplified analytical (solid line) models.

3.4 Analytical Model

At not too low gate voltages when $U_g \gg 1$, one can assume that in a number of GLs the electrons under the gate are degenerate, i.e., $\mu_k \gg k_B T$, and the contribution of holes (nondegenerate) can be disregarded, hence, from Eq. (3.7) we obtain

$$\Phi(\psi) \simeq \frac{6}{\pi^2} U_g^2 \psi^2. \tag{3.9}$$

In this case, for a single-GL structure ($K = 1$), Eq. (3.4) yields

$$2 - \psi_1 \simeq \frac{6}{\pi^2} U_g^2 \psi_1^2. \tag{3.10}$$

Solving Eq. (3.10) and considering Eq. (3.8), for the electron Fermi energy in a GL (in a single-GL structure), we obtain

$$\mu_1 \simeq \mu_g \sqrt{1 - \frac{\mu_g}{eV_g}}, \tag{3.11}$$

where $\mu_g = \hbar v_F \sqrt{æV_g / 4eW_g}$. For the same parameters as those used in Figs. 3.2–3.4, $\mu_1 \simeq 150$ meV. Such a value is markedly larger than those calculated for the topmost GLs in multiple-GL structures (although it is somewhat exaggerated because the temperature spread in the electron energies is disregarded). This can be attributed to the fact that in multiple-GL structures the electron density is shared between the topmost GL and underlying GLs resulting in lower Fermi energies in all of them.

Considering multiple-GL structure with large K, one can neglect the discreteness of the structure and replace the summation in Eq. (3.1) by the integration. As a result, following Ref. [17], one can arrive at

$$\frac{d^2\psi}{dz^2} = 0 \quad (-W_g < z < 0, z > z_K), \tag{3.12}$$

$$\frac{d^2\psi}{dz^2} = \frac{\Gamma}{dW_g}\Phi(\psi) \quad (0 < z < z_K). \tag{3.13}$$

In this case, considering Eq. (3.7), we arrive at

$$\frac{d^2\psi}{dz^2} = \frac{\psi^2}{L_s^2} \tag{3.14}$$

with the characteristic screening length

$$L_s = \frac{\pi\sqrt{dW_g}}{\sqrt{6\Gamma U_g}} = \hbar v_F\sqrt{\frac{æd}{2e^3V_g}} \propto V_g^{-1/2}. \tag{3.15}$$

The boundary conditions for Eq. (3.14) are given by Eq. (3.3)

In multiple-GL structures with a large number of GLs ($K \gg 1$), one can extend the coordinate of the lowest GL to infinity and set $d\psi/dz|_{z=\infty} = 0$ with $\psi|_\infty = 0$. Solving Eq. (3.14) with the latter boundary conditions, we arrive at

$$\psi = \frac{1}{(C + z/\sqrt{6}L_s)^2}, \tag{3.16}$$

where C satisfies the following equation:

$$C^3 - C/2 = (W_g/\sqrt{6}L_s). \tag{3.17}$$

Since in reality $W_g \gg \sqrt{6}L_s$, one obtains $C \simeq (W_g/\sqrt{6}L_s)^{1/3} \propto V_g^{1/6}$.

Taking into account Eq. (3.8), Eq. (3.16) yields

$$\mu_k \simeq \frac{eV_g}{2[C + (k-1)d/\sqrt{6}L_s]^2} = \mu_1 a_k. \tag{3.18}$$

Here

$$\mu_1 = \frac{eV_g}{2C^2} \propto \left(\frac{V_g}{W_g}\right)^{2/3} \tag{3.19}$$

is the Fermi energy of electrons in the topmost GL in the n section (holes in the p section),

$$a_k = [1 + (k - 1)\gamma]^{-2}, \tag{3.20}$$

and

$$\gamma = \frac{d}{\sqrt{6}L_s C} \propto \frac{d}{W_g^{1/3} L_s^{2/3}} \propto \left(\frac{V_g}{W_g}\right)^{1/3}. \tag{3.21}$$

Setting d = 0.35 nm, W_g = 10 nm, æ = 4, and V_g = 1 V, one can obtain $L_s \simeq 0.44$ nm. $C \simeq 2.14$, $\mu_1 \simeq 109$ meV, and $\gamma \simeq 0.153$. The μ_k versus k dependence obtained using our simplified analytical model, i.e., given by Eqs. (3.15) and (3.17) is shown by a solid line in Fig. 3.5.

Equations (3.9)–(3.21) are valid when $\mu_K > k_B T$, i.e., at sufficiently large V_g or/and sufficiently small T. Since the Fermi energy (at fixed electron density) decreases with increasing temperature, the above formulas of our simplified (idealistic) model yield somewhat exaggerated values of this energy [compare the dependences in Fig. 3.5 obtained numerically using Eqs. (3.4)–(3.7) and that found analytically using Eqs. (3.18)–(3.21)].

3.5 The Reverse Current

The current across the n-i-p junctions under their reverse bias ($V < 0$, see Fig. 3.2c) is an important characteristic of such junctions [21]. In particular, this current can substantially affect the performance of the terahertz tunneling transit-time oscillators and interband photodetectors [5, 17]. This current is associated with the thermogeneration and tunneling generation of the electron-hole pairs in the i region. A significant contribution to this current can be provided by the injection of minority carriers (holes in the n region and electrons in the p region). Such an injection current in the k-th GL is determined by the height of the barrier for the minority carrier which, in turn, is determined by the Fermi energy μ_k of the majority carrier. The latter, as shown above, depends on the gate voltage and the GL index. As a result, the reverse current can be presented as

$$J = J_{\rm i} + J_{\rm th} + J_{\rm tunn}, \tag{3.22}$$

where the injection current (which is assumed to be of the thermionic origin) is given by

$$J_{\rm i} = \frac{2ev_{\rm F}}{\pi^2}\left(\frac{k_{\rm B}T}{\hbar v_{\rm F}}\right)^2 \int_{-\pi/2}^{\pi/2} d\theta\cos\theta$$

$$\times\sum_{\rm k=1}^{\rm K}\int_0^{\infty}\frac{d\xi\xi}{1+\exp(\xi+\mu_{\rm k}/k_{\rm B}T)}$$

$$= \frac{24ev_{\rm F}\Sigma_{\rm T}}{\pi^3}\sum_{\rm k=1}^{\rm K}\int_0^{\infty}\frac{d\xi\xi}{1+\exp(\xi+\mu_{\rm k}/k_{\rm B}T)},$$

$$= \frac{12J_{\rm T}\Sigma_{\rm T}}{\pi^2}\sum_{\rm k=1}^{\rm K}\int_0^{\infty}\frac{d\xi\xi}{1+\exp(\xi+\mu_{\rm k}/k_{\rm B}T)} \tag{3.23}$$

with $J_{\rm T} = 2ev_{\rm F}\Sigma_{\rm T}/\pi$. At $T = 77 - 300$ K, $J_{\rm T} \simeq 0.06 - 0.9$ A/cm. Deriving Eq. (3.24), we have taken into account that the distribution function of holes which enter the n region overcoming the barrier in the k-th GL with the height $\mu_{\rm k}$ is $f_{\rm k}^{\rm pn} \simeq \{1 + \exp[(\mu_{\rm k} + v_{\rm F}p)/k_{\rm B}T]\}^{-1}$. Similar formula is valid for electrons in the p region. The factor 2 appears in Eq. (3.24) due to the contribution of both holes and electrons. Equation (3.24) is valid when the bias voltage is not too small: $eV > k_{\rm B}T$. At $V \to 0$ one has $J_{\rm i} \to 0$ (as well as $J_{\rm th}$ and $J_{\rm tunn}$). Scattering of holes in the n region and electrons in the p region resulting in returning of portions of them back to the contacts leads to some decrease in $J_{\rm i}$.

The temperature dependences of the reverse current associated with the injection from the n and p region calculated using Eq. (3.23) (with the quantities $\mu_{\rm k}$ shown in Fig. 3.5) are presented in Fig. 3.6. As one might expect, the reverse current sharply increases with the temperature and the number of GL. The latter is due to relatively low energy barriers for minority carriers in the n and p regions in GLs with large indices. For comparison, the injection current in a single-GL structure calculated using Eq. (3.23) with Eq. (3.8), is given by

$$J_{\rm i}^{\rm S} = \frac{12J_{\rm T}}{\pi^2}\int_0^\infty \frac{d\xi\xi}{1+\exp(\xi+\mu_1/k_{\rm B}T)}$$

$$\simeq \frac{12J_T}{\pi^2}\exp\left(-\frac{\hbar v_F\sqrt{æV_g/8eW_g}}{k_BT}\right). \qquad (3.24)$$

At the same parameters as above and T = 300 K, Eq. (3.24) yields $J_{\rm i}^{\rm S} \simeq 0.01$ A/cm.

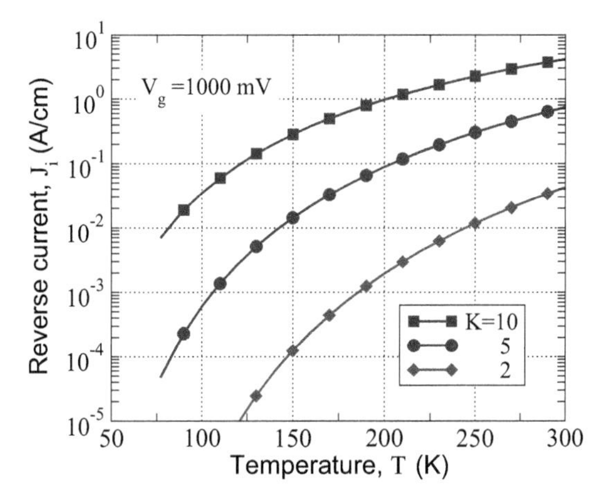

Figure 3.6 Temperature dependence of reverse current (injection component, $J_{\rm i}$) for multiple-GL structures with different number of GLs.

When $K \gg \gamma^{-1}\sqrt{\mu_1/k_{\rm B}T}$, the main contributions to the reverse current is associated with GLs with large indices (in which the barriers are very low), so that one obtains

$$J_{\rm i} \lesssim KJ_{\rm T}. \qquad (3.25)$$

As follows from Eq. (3.25), the injection current (and, therefore, the net reverse current) can be fairly large due to the "shortcut" by GLs with large indices (placed deep below the gate).

Since the thermogeneration is associated primarily with the absorption of optical phonons [21], the pertinent rate $g_{\rm th}$ is independent of the electric field in the *i* region, but it is proportional to the *i* section length $2l$ ($l \lesssim L_{\rm g}$). The contribution of the thermogeneration to the reverse current can be presented as

$$J_{\rm th} = 4Kelg_{\rm th}. \qquad (3.26)$$

The quantity g_{th} as a function of the temperature was calculated in Ref. [20]. Equations (3.23)–(3.26) are valid if $2l < l_R$, where l_R is the recombination length. In the situations when the bias voltage between the side contacts is not too small (as it should be, for instance, in GL-based interband photodetectors), the recombination length is fairly long. Indeed, assuming that the recombination time (at $T = 300$ K) and the drift velocity are $\tau_R = 5 \times 10^{-10}$ s [22] and $< v > \sim v_F/2 = 5 \times 10^7$ cm/s [18], respectively, one obtains $l_R \simeq 250$ μm.

Assuming that $2l = 10$ μm with $g_{th} = 10^{13}$ cm^{-2}s^{-1} and $g_{th} = 10^{21}$ cm^{-2}s^{-1} at $T = 77$ K and $T = 300$ K, respectively [22], we obtain $J_{th} \simeq 3.2 \times (10^{-9}–10^{-1})$ A/cm. One can see that the thermogeneration contribution to the reverse current is much smaller than the injection contribution at lower temperatures, while it can be substantial at $T = 300$ K in the n-i-p structures with long i region.

The tunneling generation can significantly contribute to the reverse current at elevated electric fields in the i region, i.e., in relatively short GL structures at elevated bias voltages [1, 5]. This current can be calculated using the following formula which follows from the expression for the tunneling probability in GLs [1, 2] (see, for instance [5]):

$$J_{tunn} = \frac{ev_F}{\pi^2 l^2}\left(\frac{eV}{2l\hbar v_F}\right)^{3/2} \propto \frac{V^{3/2}}{l^{1/2}}. \tag{3.27}$$

Depending on the n-i-p junction applications, the quantities V and l should be chosen to provide either domination of tunneling current (as in tunneling transit-time oscillators [4]) or its suppression (as in the interband photodetectors [17]).

3.6 Multiple-GL Structures with Highly Conductive BGL

As it was mentioned in the introduction, the multiple-GL structures obtained by the thermal decomposition from 4H-SiC substrate, include a highly conductive BGL. Its presence can strongly affect the potential and electron density distributions across the multiple-GL structure. However, the BGL influence depends on the connection between the BGL and the side contacts. The following options can be considered: (a) *all the GLs are directly connected with the side contacts.*

If the bias voltage between the side contacts $V = 0$, the potentials of GLs with different indices $1 \leq k \leq K$ (the BGL index is equal to $K+1$) can be found using the same equations as above, i.e., Eqs. (3.4)–(3.7) or Eq. (3.14) but with modified boundary conditions. Considering high conductivity of BGL, its potential can be assumed to be constant and equal to the side contact potential, the second condition in Eq. (3.6) should be replaced by $\psi|_{z=z_{K+1}} = 0$. In the case of multiple-GL structures with sufficiently large K, both $d\psi/dz$ and ψ steeply tend to zero with increasing z. This directly follows from Fig. 3.2 (see the curves calculated for $K = 50$), as well as from Eqs. (3.16)–(3.21). Thus, the difference between the potential distributions calculated using the boundary conditions $d\psi/dz|_{z=z_K} = 0$ and $\psi|_{z=z_{K+1}} = 0$ is fairly small. Hence, the results obtained above are also applicable for the multiple-GL structures with the highly conductive BGL if $K \gg 1$. Since the highly conductive BGL effectively shorts out the side contacts, for the devices whose operation requires $V \neq 0$ the GBL should be cut between the side contacts. (b) *All the GLs, except the BGL, are connected with the side contacts and a bias voltage* V_b is applied between the GBL and one of the side contacts. In this case, the GBL can be considered as the back gate so that the potential distribution is determined by both V_g and V_b. However, in the absence of the top gates, the obtained results can still be used if W_g and V_g in all equations substituted with d and V_b, respectively. (c) *All the GLs, except the BGL, are connected with the side contacts, but the GBL is isolated from the side contacts.* In such a case, the GBL plays the role of a floating gate. This case needs a separate analysis and will be considered elsewhere.

3.7 Conclusions

We calculated the Fermi energies and electron and hole densities in the electrically induced n-i-p junctions formed in multiple-GL structures as functions of the GL indices, gate voltage, temperature, and structural parameters. Using the obtained values of the Fermi energies and the heights of potential barriers for minority carriers in the n and p regions, we found the temperature dependences of the reverse injection current for multiple-GL structures with different numbers of GLs. This analysis shows the possibility of the formation of effective electrically induced *n* and *p* regions and n-i-p junctions, i.e., the n-i-p junctions with suppressed reverse currents in multiple-GL structures. These electrically induced n-i-p junctions in multiple-GL structures can be used in different devices, such as terahertz tunneling transittime oscillators, lasers, and high-performance interband photodetectors. They could enhance the device performance (increasing output power and responsivity) and expand their functions by allowing for the gate-voltage control.

Acknowledgments

This work was supported by the Japan Society for Promotion of Science and by the Japan Science and Technology Agency, CREST, Japan.

References

1. V. V. Cheianov and V. I. Fal'ko, *Phys. Rev. B*, **74**, 041403(R) (2006).
2. L. M. Zhang and M. M. Fogler, *Phys. Rev. Lett.*, **100**, 116804 (2008).
3. B. Huard, J. A. Sulpizio, N. Stander, K. Todd, B. Yang, and D. Goldhaber-Gordon, *Phys. Rev. Lett.*, **98**, 236803 (2007).
4. B. Özyilmaz, P. Jarillo-Herrero, D. Efetov, D. Abanin, L. S. Levitov, and P. Kim, *Phys. Rev. Lett.*, **99**, 166804 (2007).
5. V. Ryzhii, M. Ryzhii, V. Mitin, and M. S. Shur, *Appl. Phys. Express*, **2**, 034503 (2009).
6. M. Ryzhii and V. Ryzhii, *Jpn. J. Appl. Phys. Part 2*, **46**, L151 (2007).
7. F. Xia, T. Murller, Y-M. Lin, A. Valdes-Garsia, and F. Avouris, *Nat. Nanotechnol.*, **4**, 839 (2009).
8. M. Ryzhii and V. Ryzhii, *Phys. Rev. B*, **79**, 245311 (2009).
9. F. Varchon, R. Feng, J. Hass, X. Li, B. Ngoc Nguyen, C. Naud, P. Mallet, J.-Y. Veuillen, C. Berger, E. H. Conrad, and L. Magaud, *Phys. Rev. Lett.*, **99**, 126805 (2007).
10. M. Orlita, C. Faugeras, P. Plochocka, P. Neugebauer, G. Martinez, D. K. Maude, A.-L. Barra, M. Sprinkle, C. Berger, W. A. de Heer, and M. Potemski, *Phys. Rev. Lett.*, **101**, 267601 (2008).
11. P. Neugebauer, M. Orlita, C. Faugeras, A.-L. Barra, and M. Potemski, *Phys. Rev. Lett.*, **103**, 136403 (2009).

12. M. Orlita and M. Potemski, *Semicond. Sci. Technol.*, **25**, 063001 (2010).

13. D. L. Miller, K. D. Kubista, G. M. Rutter, M. Ruan, W. A. de Heer, P. N. First, and J. A. Stroscio, *Science*, **324**, 924 (2009).

14. M. Sprinkle, D. Siegel, Y. Hu, J. Hicks, A. Tejeda, A. Taleb- Ibrahimi, P. Le Fèvre, F. Bertran, S. Vizzini, H. Enriquez, S. Chiang, P. Soukiassian, C. Berger, W. A. de Heer, A. Lanzara, and E. H. Conrad, *Phys. Rev. Lett.*, **103**, 226803 (2009).

15. V. Ryzhii, M. Ryzhii, A. Satou, T. Otsuji, A. A. Dubinov, and V. Ya. Aleshkin, *J. Appl. Phys.*, **106**, 084507 (2009).

16. V. Ryzhii, A. A. Dubinov, T. Otsuji, V. Mitin, and M. S. Shur, *J. Appl. Phys.*, **107**, 054505 (2010).

17. V. Ryzhii, M. Ryzhii, V. Mitin, and T. Otsuji, *J. Appl. Phys.*, **107**, 054512 (2010).

18. V. Ryzhii, M. Ryzhii, and T. Otsuji, *Phys. Status Solidi A*, **205**, 1527 (2008).

19. S. Luryi, *Appl. Phys. Lett.*, **52**, 501 (1988); see also S. Luryi, in *High-Speed Semiconductor Devices*, by S. M. Sze (John Wiley, New York, 1990), p. 57.

20. H. Fukidome, Y. Miyamoto, H. Handa, E. Saito, and M. Suemitsu, *Jpn. J. Appl. Phys.*, **49**, 01AH03 (2010).

21. M. Shur, *Physics of Semiconductor Devices* (Prentice-Hall, New Jersey, 1990).

22. F. Rana, P. A. George, J. H. Strait, S. Shivaraman, M. Chanrashekhar, and M. G. Spencer, *Phys. Rev. B*, **79**, 115447 (2009).

23. R. S. Shishir, D. K. Ferry, and S. M. Goodnick, *J. Phys.: Conf. Ser.*, **193**, 012118 (2009).

Chapter 4

Tunneling Recombination in Optically Pumped Graphene with Electron-Hole Puddles*

V. Ryzhii,[a,c] **M. Ryzhii,**[a,c] **and T. Otsuji**[b,c]
[a]*Computational Nanoelectronics Laboratory, University of Aizu, Aizu-Wakamatsu 965-8580, Japan*
[b]*Research Institute for Electrical Communication, Tohoku University, Sendai 980-8577, Japan*
[c]*Japan Science and Technology Agency, CREST, Tokyo 107-0075, Japan*
v-ryzhii@u-aizu.ac.jp

We evaluate recombination of electrons and holes in optically pumped graphene associated with the interband tunneling between electron-hole puddles and calculate the recombination rate and time. It is demonstrated that this mechanism can be dominant in a wide range of pumping intensities. We show that the tunneling recombination rate and time are nonmonotonic functions of the

*Reprinted with permission from V. Ryzhii, M. Ryzhii, and T. Otsuji (2011). Tunneling recombination in optically pumped graphene with electron-hole puddles, *Appl. Phys. Lett.*, **99**, 173504. Copyright © 2011 American Institute of Physics.

Graphene-Based Terahertz Electronics and Plasmonics: Detector and Emitter Concepts
Edited by Vladimir Mitin, Taiichi Otsuji, and Victor Ryzhii

ISBN 978-981-4800-75-4 (Hardcover), 978-0-429-32839-8 (eBook)
www.jennystanford.com

quasi-Fermi energies of electrons and holes and optical pumping intensity. This can result in hysteresis phenomena.

The gapless energy spectrum of electrons and holes in graphene layers (GLs) and non-Bernal stacked multiple graphene layers (MGLs) [1–3], which leads to specific features of its electrical and optical processes, opens up prospects of building of novel devices exploiting the interband absorption and emission of terahertz (THz) and infrared (IR) photons. Apart from an obvious possibility to use of graphene in THz and IR photodetectors [4–6], graphene layers (GLs) and multiple graphene layers (MGLs) can serve as active media in THz or IR lasers with optical [7–12] and injection [13] pumping. The achievement of sufficiently strong population inversion, which is necessary for negative dynamic conductivity in the THz or IR ranges of frequencies, can be complicated by the recombination processes. The recombination of electrons and holes in GLs at not too low temperatures is mainly determined by the emission of optical photons [14]. Due to a relatively high energy of optical phonons ($\hbar\omega_0 \simeq 0.2$ eV), the recombination rate in GLs associated with this mechanism can be acceptable even at the room temperatures [10–12]. The radiative recombination in the practically interesting temperatures is weaker than that due to the optical phonon emission [15]. The acoustic phonon recombination mechanism as well as the Auger mechanism are forbidden due to the linearity of the electron and hole spectra even in the case their modifications associated with the inter-carrier interaction [16, 17]. More complex processes involving the electron and hole scattering on impurities assisted by acoustic phonons also provide rather long recombination times which are longer than the radiative recombination time [18]. However in GLs with a disorder caused, for instance, by fluctuation of the surface charges resulting in the formation of the electron and hole puddles [19] (see also, for instance, Refs. [20–22]), the recombination can also be associated with the interband tunneling in the spots where the build-in fluctuation field is sufficiently strong. In this chapter, we find the dependences of the recombination rate and time in optically pumped graphene with electron-hole puddles on the quasi-Fermi energy of electrons and holes (pumping intensity). The obtained results can be used for the interpretation of experimental observations and to promote the realization of graphene-based THz and IR lasers.

The electric potential $\varphi = \varphi\ (x, y, z)$, where x and y are the coordinates in the GL plane and the coordinate z is directed perpendicular to this plane, is governed by the Poisson equation presented in the following form:

$$\Delta\varphi = \frac{4\pi e}{æ}(\Sigma_{\rm e} - \Sigma_{\rm h} - \Sigma_{\rm i})\cdot\delta(z). \tag{4.1}$$

Here,

$$\Sigma_{\rm e} = \frac{2}{\pi\hbar^2}\int_0^\infty \frac{dp\,p}{1+\exp\left(\dfrac{v_{\rm W}p - \mu_{\rm e} + e\varphi}{k_{\rm B}T}\right)},$$

$$\Sigma_{\rm h} = \frac{2}{\pi\hbar^2}\int_0^\infty \frac{dp\,p}{1+\exp\left(\dfrac{v_{\rm W}p + \mu_{\rm h} - e\varphi}{k_{\rm B}T}\right)},$$

and $\Sigma_{\rm i}$ are the electron, hole, and charged impurity sheet densities, respectively, $e = |e|$, æ, $\hbar$, and $k_{\rm B}$ are the electron charge, dielectric constant, reduced Planck's constant, and Boltzmann's constant, $v_W \simeq 10^8$ cm/s is the characteristic velocity, p is the electron and hole momentum, $\mu_{\rm e}$ and $\mu_{\rm h}$ are the electron and hole quasi-Fermi energies counted from the Dirac point, T is the temperature, and Δ is the three dimensional Laplace operator. The delta function $\delta(z)$ reflexes the localization of electrons, holes, and impurities in the GL plane $z = 0$. It is assumed that the electron spectra or electrons and holes are linear: $\varepsilon_{\rm e} = v_{\rm W}p$ and $\varepsilon_{\rm h} = -v_{\rm W}p$. In the equilibrium, i.e., in the absence of pumping, $\mu_{\rm e} = \mu_{\rm h} = \mu_{\rm i}$, where $\mu_{\rm i}$ is determined by the spatially averaged impurity density $\langle\Sigma_{\rm i}(x, y)\rangle$ or by the gate voltage (in gated GL structures). In the case of pumping of intrinsic GLs, $\mu_{\rm e} \simeq -\mu_{\rm h} = \mu \neq 0$. In the latter case, T can differ from the lattice temperature $T_{\rm l}$ being both higher than $T_{\rm l}$ or lower [23].

Considering in the following the case $\langle\Sigma_{\rm d}(x, y)\rangle = 0$ and assuming that the fluctuations are not too strong ($e|\varphi| < k_{\rm B}T$), Eq. (4.1) can be linearized

$$\Delta\varphi = \left[\frac{\varphi}{r_{\rm s}}\ln\left(1+e^{\mu/k_{\rm B}T}\right) - \frac{4\pi e}{æ}\Sigma_{\rm i}\right]\cdot\delta(z), \tag{4.2}$$

where $r_{\rm s} = (æ\hbar^2 v_{\rm W}^2/16e^2k_{\rm B}T)$ is the screening length.

Solving Eq. (4.2) with the boundary conditions $|\varphi|_{z=\pm\infty} = 0$ assuming that

$$\Sigma_{\rm i} = \sum_{q_x, q_y} \Sigma^{(i)}_{q_x, q_y} \exp[i(q_x x + q_y y)],$$

$$\varphi(x,y,z) = \sum_{q_x, q_y} \Phi_{q_x, q_y} \exp(-q|z|) \exp[i(q_x x + q_y y)],$$

where $\Sigma^{(i)}_{q_x, q_y}$ and Φ_{q_x, q_y} are the pertinent amplitudes and $q = \sqrt{q_x^2 + q_y^2}$, for the amplitude of the potential at the GL plane we obtain

$$\psi_{q_x, q_y} = \frac{2\pi e}{æ\left[q + \dfrac{1}{2r_s}\ln(1 + e^{\mu/k_BT})\right]} \Sigma^{(i)}_{q_x, q_y}. \tag{4.3}$$

Setting $q \sim \pi/\bar{a}$, where $\bar{a}$ is the characteristic size of the puddles, and using Eq. (4.3), one can express $\psi = \max|\psi_{q_x, q_y}|$ via its value in the absence of screening $\bar{\psi} = 2e\bar{a}\bar{\Sigma}/æ$

$$\psi = \frac{\bar{\psi}}{\left[1 + \dfrac{\bar{a}}{2\pi r_s}\ln(1 + e^{\mu/k_BT})\right]}. \tag{4.4}$$

According to Eq. (4.4), the fluctuation electric field between positively and negatively charged puddles can be estimated as $\varepsilon \sim \psi/\bar{a}$.

Using the general formulas for the probability of the interband tunneling in GLs [24–26] and considering Eq. (4.4), the rate of the tunneling recombination can be presented as

$$R \sim \frac{2\sqrt{2}e^{3/2}}{\pi^2\hbar^{3/2}u_W^{1/2}}\sqrt{\frac{\bar{\psi}}{\bar{a}}}\left(\frac{\mu}{e\bar{a}}\right) = \frac{\bar{R}(\mu/e\bar{\psi})}{\sqrt{1 + \eta_s\ln(1 + e^{\mu/k_BT})}} \tag{4.5}$$

$$R \sim \frac{2\sqrt{2}e^{3/2}}{\pi^2\hbar^{3/2}u_W^{1/2}}\left(\frac{\psi}{\bar{a}}\right)^{3/2} = \frac{\bar{R}}{[1 + \eta_s\ln(1 + e^{\mu/k_BT})]^{3/2}} \tag{4.6}$$

at relatively weak ($\mu < e\psi$) and strong ($\mu > e\psi$) optical pumping, respectively. Here,

$$\bar{R} \sim \frac{2\sqrt{2}}{\pi^2 v_W^{1/2}}\left(\frac{e\bar{\psi}}{\bar{a}\hbar}\right)^{3/2}, \quad \eta_s = \frac{\bar{a}}{2\pi r_s} = \frac{8}{\pi æ}\frac{e^2\bar{a}k_BT}{\hbar^2 v_W^2}.$$

Assuming that æ = 4, $\bar{a}$ = 30 nm, and $\bar{\Sigma}$ = 4 × 10^{10} cm^{-2} [19], for the amplitude of the potential profile fluctuations $e\bar{\psi}$ and for the characteristic rate of the tunneling recombination $\bar{R}$, we obtain the following estimates: $e\bar{\psi}$ = 8.5 meV and $\bar{R} \simeq 2.76 \times 10^{23}$ $cm^{-2}s^{-1}$, respectively. At T = 77 – 300 K, one obtains $r_s \simeq 0.67 - 2.62$ nm and η_s = 0.47 – 1.82. These values are used in the following.

In the case of strong pumping, the effective temperature of the electron-hole plasma can be much higher than the lattice temperature [23]. As a result, the electron-hole plasma can become nondegenerate, so that $\mu_e = \mu < 0$ while $\mu_h = -\mu > 0$. This is possible when the plasma density increases with increasing pumping intensity slower that the effective temperature rises. In such a case, the screening vanishes ($\psi \simeq \bar{\psi}$), and $R < 0$. This implies that when μ changes its sign, the tunneling recombination of electron-hole pairs turns to their tunneling generation. This is because when the plasma density is lower than it would be in equilibrium but at the effective temperature T, the tunneling generation tends to establish an equilibrium density corresponding to T.

The dependences of the tunneling recombination rate R versus the quasi-Fermi energy μ for different temperatures T are shown in Fig. 4.1. For the calculations we used a formula for R which interpolates the dependences given by Eqs. (4.5) and (4.6). Figure 4.2 demonstrates also the dependences of the recombination rate R + R_0 associated with both tunneling and optical phonon mechanisms (dashed lines). The contribution of the optical phonon mechanism R_0 is taken into account using a simplified analytical formula [10, 23] derived from more rigorous one [14]:

$$R_0 \simeq \bar{R}_0 \exp\left(-\frac{\hbar\omega_0}{k_B T}\right)\left[\exp\left(\frac{2\mu}{k_B T}\right) - 1\right]. \qquad (4.7)$$

Here $\bar{R}_0 \simeq 3 \times 10^{24}$ $cm^{-2}s^{-1}$ [14] is the pertinent characteristic recombination rate. Disregarding the effects of electron-hole heating or cooling and the effect of optical phonon heating [23], we have neglected the difference in the effective temperature T and the lattice temperature T_l. One can see that R is a nonmonotonic function of μ with a maximum at a certain value of μ. This is attributed to an increase in R at small μ due to an increase in the electro and hole densities, i.e., in the number of carriers participating in the tunneling processes. However, at higher values of μ, the screening of the potential fluctuations leads to a decrease in the electric field between the puddles and, consequently, in a decrease in the tunneling probability. When μ increases further, the tunneling mechanism gives way to the optical phonon mechanism which becomes dominating. This results in a dramatic rise of $R + R_0$ in the range of large μ. The quantity μ is governed by an equation equalizing the rate of recombination $R + R_0$ and the rate of carrier photogeneration $G = \beta I$, where $\beta = 0.023$ is the GL absorption coefficient and I is the photon flux. Hence, the $\mu - I$ dependence should exhibit the S-shape (see the dashed line in Fig. 4.1 corresponding to $T = 200$ K) leading to a hysteresis which might be pronounced at lowered temperatures.

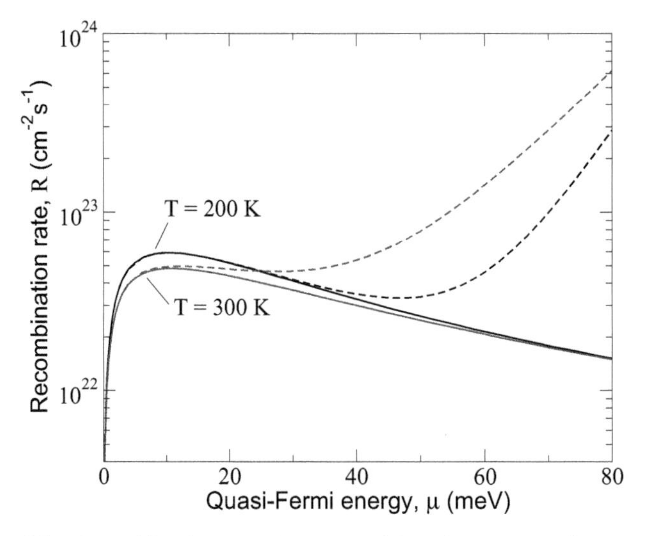

Figure 4.1 Recombination rate R vs quasi-Fermi energy μ for tunneling mechanism (solid lines) and for combination of tunneling and optical phonon mechanism (dashed lines) in a GL at different temperatures T.

The recombination time is given by

$$\tau_{\mathrm{R}} = \frac{\langle(\Sigma_{\mathrm{e}} + \Sigma_{\mathrm{h}})\rangle}{2(R + R_0)}$$

$$= \frac{2}{\pi(R+R_0)}\left(\frac{k_B T}{\hbar v_W}\right)^2 \int_0^\infty \frac{d\varepsilon\,\varepsilon}{1+\exp(\varepsilon-\mu/k_B T)}. \tag{4.8}$$

The recombination time calculated without considering optical phonon mechanisms (solid lines) and with the latter (dashed lines) is shown in Fig. 4.2. As follows from Fig. 4.2, the τ_R – μ dependence is also nonmonotonic with a minimum at moderate μ (where $\tau_R \sim 2$ ps) and a maximum at rather high μ (where $\tau_R \sim 5$ – 20 ps).

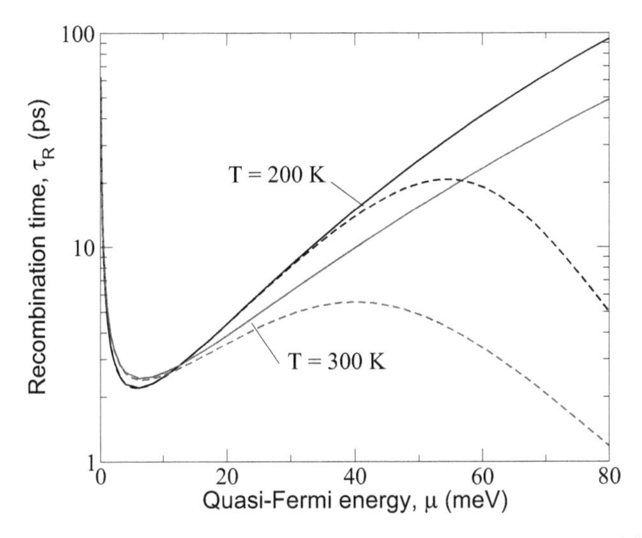

Figure 4.2 Recombination time τ_R vs quasi-Fermi energy μ at different temperatures T. Solid lines correspond to tunneling recombination, while dashed lines correspond to recombination associated with both tunneling and phonon mechanisms.

In optically pumped MGL structures, the electron and hole densities in each GL are close to each other provided the number of GLs K is not too large [27]. If the fluctuations are associated with the charges located near the lowermost GL, say, near the interface between this GL and the substrate, these fluctuations are screened by all GLs. In cases of such MGL structures, the first two terms in the right-hand side of Eq. (4.1) can be multiplied by a factor K, so that the screening length becomes much shorter: $r_s^{(K)} = (æ\hbar^2 v_W^2 / 16Ke^2 k_B T)$. Consequently, parameter η_s should be substituted by $K\eta_s$. Naturally, this leads to a stronger suppression of the fluctuations (by a stronger screening) and hence to weakening of the tunneling recombination under consideration as shown in Fig. 4.3. If each GLs contains its own fluctuation charges, the averaging over a number of GL leads to a significant smoothening of the potential relief in all GL and, hence, to the suppression of the tunneling recombination is such MGL structures.

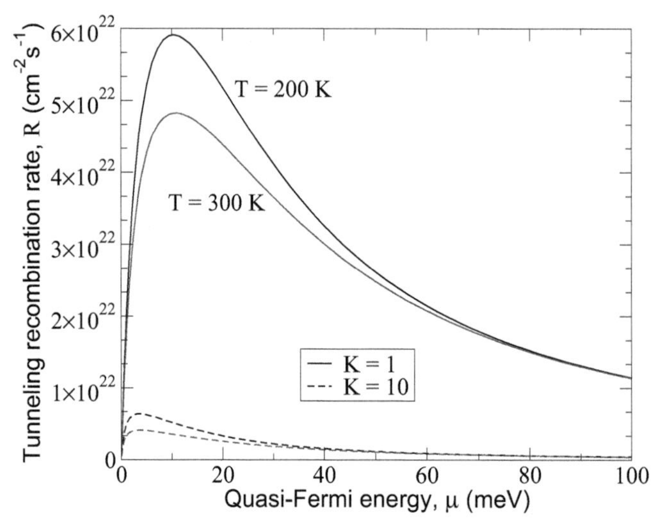

Figure 4.3 Tunneling recombination time τ_R vs quasi-Fermi energy μ at different temperatures T in a GL (solid lines) and in MGL structure with ten GLs, i.e., $K = 10$ (dashed lines).

In conclusion, we calculated the recombination and time as functions of the quasi-Fermi energy in optically pumped graphene with electron-hole puddles. It was shown that the tunneling recombination can be dominant recombination mechanism at low and moderate values of quasi-Fermi energy and, hence, at low and moderate pumping intensities. The dependences of the recombination and time on the quasi-Fermi energy and pumping intensity can be nonmonotonic resulting in hysteresis phenomena. The role of tunneling recombination can be diminished in MGL structures.

The authors are grateful to N. Ryabova for comments on the manuscript and S. Boubanga Tombet for useful information. This work was supported by the Japan Science and Technology Agency, CREST and by the Japan Society for Promotion of Science, Japan.

References

1. A. H. Castro Neto, F. Guinea, N. M. R. Peres, K. S. Novoselov, and A. K. Geim, *Rev. Mod. Phys.*, **81**, 109 (2009).
2. M. Sprinkle, D. Suegel, Y. Hu, J. Hicks, A. Tejeda, A. Taleb-Ibrahimi, P. Le Fevre, F. Bertran, S. Vizzini, H. Enriquez, S. Chiang, P. Soukiassian, C. Berger, W. A. de Heer, A. Lanzara, and E. H. Conrad, *Phys. Rev. Lett.*, **103**, 226803 (2009).
3. M. Orlita and M. Potemski, *Semicond. Sci. Technol.*, **25**, 063001 (2010).
4. V. Ryzhii and M. Ryzhii, *Phys. Rev. B*, **79**, 245311 (2009).
5. F. Xia, T. Murller, Y.-M. Lin, A. Valdes-Garsia, and F. Avouris, *Nat. Nanotechnol.*, **4**, 839 (2009).
6. V. Ryzhii, M. Ryzhii, V. Mitin, and T. Otsuji, *J. Appl. Phys.*, **106**, 084512 (2009).
7. V. Ryzhii, A. Satou, and T. Otsuji, *J. Appl. Phys.*, **101**, 024509 (2007).
8. F. Rana, *IEEE Trans. Nanotechnol.*, **7**, 91 (2008).
9. A. Dubinov, V. Ya. Aleskin, M. Ryzhii, and V. Ryzhii, *Appl. Phys. Express*, **2**, 092301 (2009).
10. V. Ryzhii, M. Ryzhii, A. Satou, T. Otsuji, A. A. Dubinov, and V. Ya. Aleshkin, *J. Appl. Phys.*, **106**, 084507 (2009).
11. V. Ryzhii, A. A. Dubinov, T. Otsuji, V. Mitin, and M. S. Shur, *J. Appl. Phys.*, **107**, 054505 (2010).
12. A. A. Dubinov, V. Ya. Aleshkin, V. Mitin, T. Otsuji, and V. Ryzhii, *J. Phys.: Condens. Matter*, **23**, 145302 (2011).
13. M. Ryzhii and V. Ryzhii, *Jpn. J. Appl. Phys.*, **46**, L151 (2007).
14. F. Rana, P. A. George, J. H. Strait, S. Shivaraman, M. Chanrashekhar, and M. G. Spencer, *Phys. Rev. B*, **79**, 115447 (2009).
15. A. Satou, F. T. Vasko, and V. Ryzhii, *Phys. Rev. B*, **78**, 115431 (2008).
16. M. S. Foster and I. L. Aleiner, *Phys. Rev. B*, **79**, 085415 (2009).
17. D. M. Basko, S. Piscanec, and A. C. Ferrari, *Phys. Rev. B*, **80**, 165413 (2009).
18. F. T. Vasko and V. V. Mitin, *Phys. Rev. B*, **84**, 155445 (2011).
19. J. Martin, N. Akerman, G. Ulbricht, T. Lohmann, J. H. Smet, K. von Klitzing, and A. Yacoby, *Nat. Phys.*, **4**, 144 (2008).

20. Y. Zhang, V. W. Brar, C. Girit, A. Zett, and M. F. Cromme, *Nat. Phys.*, **5**, 722 (2009).
21. J. M. Poumirol, W. Escoffer, A. Kumar, M. Goiran, R. Raquet, and J. M. Broto, *New. J. Phys.*, **12**, 083006 (2010).
22. P. Parovi-Azar, N. Nafari, and M. Reza Rahimi Tabat, *Phys. Rev. B*, **83**, 165434 (2011).
23. V. Ryzhii, M. Ryzhii, V. Mitin, A. Satou, and T. Otsuji, *Jpn. J. Appl. Phys.*, **50**, 9 (2011).
24. V. V. Cheianov and V. I. Fal'ko, *Phys. Rev. B*, **74**, 041403(R) (2006).
25. A. Ossipov, M. Titov, and C. W. J. Beenakker, *Phys. Rev. B*, **75**, 251401(R) (2007).
26. V. Ryzhii, M. Ryzhii, and T. Otsuji. *Phys. Status Solidi A*, **205**, 1527 (2008).
27. M. Ryzhii, V. Ryzhii, T. Otsuji, V. Mitin, and M. S. Shur, *Phys. Rev. B*, **82**, 075419 (2010).

Chapter 5

Analytical Device Model for Graphene Bilayer Field-Effect Transistors Using Weak Nonlocality Approximation*

V. Ryzhii,[a,d] M. Ryzhii,[a,d] A. Satou,[b,d] T. Otsuji,[b,d] and V. Mitin[c]

[a]*Computational Nanoelectronics Laboratory, University of Aizu, Aizu-Wakamatsu 965-8580, Japan*
[b]*Research Institute for Electrical Communication, Tohoku University, Sendai 980-8577, Japan*
[c]*Department of Electrical Engineering, University at Buffalo, State University of New York, NY 14260, USA*
[d]*Japan Science and Technology Agency, CREST, Tokyo 107-0075, Japan*
v-ryzhii@u-aizu.ac.jp

We develop an analytical device model for graphene bilayer field-effect transistors (GBL-FETs) with the back and top gates. The model is based on the Boltzmann equation for the electron transport and

Graphene-Based Terahertz Electronics and Plasmonics: Detector and Emitter Concepts
Edited by Vladimir Mitin, Taiichi Otsuji, and Victor Ryzhii

ISBN 978-981-4800-75-4 (Hardcover), 978-0-429-32839-8 (eBook)
www.jennystanford.com

the Poisson equation in the weak nonlocality approximation for the potential in the GBL-FET channel. The potential distributions in the GBL-FET channel are found analytically. The source-drain current in GBL-FETs and their transconductance are expressed in terms of the geometrical parameters and applied voltages by analytical formulas in the most important limiting cases. These formulas explicitly account for the short-gate effect and the effect of drain-induced barrier lowering. The parameters characterizing the strength of these effects are derived. It is shown that the GBL-FET transconductance exhibits a pronounced maximum as a function of the top-gate voltage swing. The interplay of the short-gate effect and the electron collisions results in a nonmonotonic dependence of the transconductance on the top-gate length.

5.1 Introduction

The unique properties of graphene layers, graphene nanoribbon arrays, and graphene bilayers [1–3] along with graphene nanomeshs [4] make them promising for different nanoelectronic device applications. The gapless energy spectrum of graphene layers allows to use them in terahertz and mid-infrared detectors and lasers [5–9]. However, the gapless energy spectrum of graphene layers (GLs) is an obstacle for creating transistor digital circuits based on graphene field-effect transistors (G-FETs) due to relatively strong interband tunneling in the FET off-state [10, 11]. The reinstatement of the energy gap in graphene-based structures like graphene nanoribbons, graphene nanomeshs, and graphene bilayers appears to be unavoidable to fabricate FETs with a sufficiently large on/off ratio. Recently, the device dc and ac characteristics of graphene nanoribbon and graphene bilayer FETs (which are referred to as GNR-FETs and GBL-FETs, respectively) were both numerically and analytically assessed [12–19]. The device characteristics of GNR-FETs operating in near ballistic and drift-diffusion regimes can be calculated analogously with those of nanowire- and carbon nanotube-FETs (see, for instance [20–22] and references therein). The GBL-FET characteristics can, in principle, be found using the same approaches as those realized previously for more customary FETs with a two-dimensional electron system in the channel [23–31].

However, some important features of GBL-FETs, in particular, the dependence of both the electron density and the energy gap in different sections of the GBL-FET channel on the gate and drain voltages should be considered [32–34], along with the "short-gate" effect and the drain-induced barrier lowering [29].

In this chapter, we use a substantially generalized version of the GBL-FET analytical device model [17, 18] to calculate the characteristics of GBL-FETs (the threshold voltages, current-voltage characteristics, and transconductance) in different regimes and analyze the possibility of a significant improvement of the ultimate performance of these FETs by shortening of the gate and decreasing of the gate layer thickness. The device model under consideration which presents the GBL-FET characteristics in closed analytical form allows a simple and clear evaluation of the ultimate performance of GBL-FETs and their comparison. The following effects are considered: (a) dependences of the electron density and energy gap in different sections of the channel on the applied voltages and the inversion of the gated section charge; (b) degeneracy of the electron system, particularly in the source and drain sections of the channel; (c) the short-gate effect and the effect of drain-induced barrier lowering; (d) electron scattering in the channel.

Our model is based on the Boltzmann kinetic equation for the electron system in the GBL-FET and the Poisson equation in the weak nonlocality approximation [35, 36]. The use of the latter allows us to find the potential distributions along the channel in the most interesting cases and obtain the GBL-FET characteristics analytically.

The chapter is organized as follows. In Section 5.2, the GBL-FET device model under consideration is presented and the features of GBL-FET operation are discussed. In Section 5.3, the main equations of the model are cited. The general formulas for the source-drain current and the GBL-FET transconductance simplified for the limiting cases (far below the threshold, near threshold, and at low top-gate voltages corresponding to the on-state) are also presented. In this section, the source-drain current and the GBL-FET transconductance are expressed in terms of the Fermi energy in the source and drain contacts and the height of the potential barrier in the channel. To find the barrier height, the Poisson equation is solved for different limiting cases in Section 5.4. The obtained potential distributions are used for the derivation of the explicit formulas for source-drain current and the transconductance as functions of the applied voltages and geometrical parameters. Section 5.5 deals with a brief discussion of some effects (role of the device geometry, electron scattering, charge inversion in the channel, and interband tunneling), which influence the GBL-FET characteristics. In Section 5.6, we draw the main conclusions. Some reference data related to the voltage dependences of the Fermi energy and the energy gap in different sections of the GBL-FET channel are contained in the Appendix.

5.2 Device Model and Features of Operation

We consider a GBL-FET with the structure shown in Fig. 5.1. It is assumed that the back gate, which is positively biased by the pertinent voltage $V_b > 0$, provides the formation of the electron channel in the GBL between the Ohmic source and drain contacts. A relatively short top gate serves to control the source-drain current by forming the potential barrier (its height Δ_m depends on the top gate voltage V_t and other voltages) for the electrons propagating between the contacts.

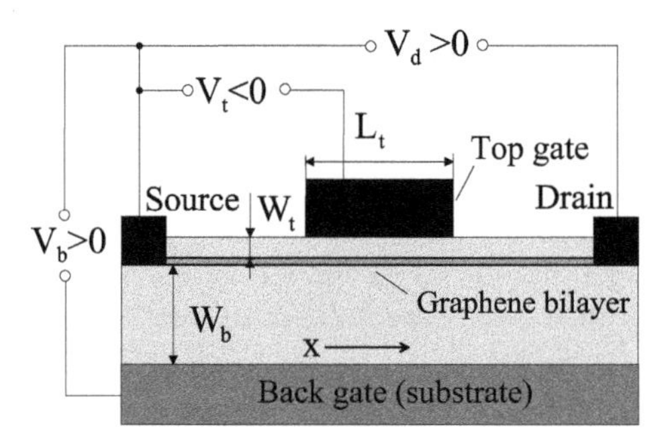

Figure 5.1 Schematic view of the GBL-FET structure.

We shall assume that the GBL-FETs under the conditions when the electron systems in the source and drain sections are degenerate, i.e., $\varepsilon_F \gg k_B T$. This implies that the back gate voltage is sufficiently high to induce necessary electron density in the source and drain sections.

In the GBL-FET the energy gap is electrically induced by the back gate voltage [32–34] (see also [18]). Thus in GBL-FETs, the back gate plays the dual role: it provides the formation of the electron channel and the energy gap. Since the electric field component directed perpendicular to the GBL plane in the channel section below the top gate (gated section) is determined by both V_b and V_t, the energy gap can be different in different sections of the GBL channel: $E_{g,s}$ (source section), E_g (gated section), and $E_{g,d}$ (drain section) [17, 18]. At sufficiently strong top-gate voltage ($V_t < V_{th} < 0$, where V_{th} is the threshold voltage), the gated section becomes depleted. Since the energy gaps in GBLs are in reality not particularly wide, at further moderate increase in $|V_t|$, the gated section of the channel becomes filled with holes (inversion of the charge in the gated section) if $V_t < V_{in} < V_{th}$, where V_{in} is the inversion voltage. As a result, the GBL-FETs with moderate energy gap are characterized by the threshold and inversion voltages: $V_{th} < 0$ and $V_{in} < 0$. The explicit formulas for V_{th} and V_{in} shall be given in the following. The cases $V_t = V_{th}$ and $V_t = V_{in}$ correspond to the alignment of the Fermi level in the source section of the channel with the conduction band bottom and the valence band top, respectively, in the gated section.

5.3 Main Equations of the Model

Due to relatively high energy of optical phonons in graphene, the electron scattering in the GBL-FET channel is primarily due to disorder and acoustic phonons. Considering such quasielastic scattering, the quasiclassical Boltzmann kinetic equation governing the steady state electron distribution function $f_{\mathbf{p}} = f_{\mathbf{p}}(x)$ in the gated section of the channel can be presented as

$$v_x \frac{\partial f_{\mathbf{p}}}{\partial x} + e\frac{\partial \varphi}{\partial x}\frac{\partial f_{\mathbf{p}}}{\partial p_x} = \int d^2\mathbf{q}\, w(q)(f_{\mathbf{p+q}} - f_{\mathbf{p}})\delta(\varepsilon_{\mathbf{p+q}} - \varepsilon_{\mathbf{p}}) \qquad (5.1)$$

Here $e = |e|$ is the electron charge, $\varepsilon_{\mathbf{p}} = p^2/2m$, m is the electron effective mass in GBL, $\mathbf{p} = (p_x, p_y)$ is the electron momentum in the GBL plane ($z = 0$), $w(q)$ is the probability of the electron scattering on disorder and acoustic phonons with the variation of the electron momentum by quantity $\mathbf{q} = (q_x, q_y)$, $v_x = p_x$, and axis x is directed in this plane (from the source contact to the drain contact, i.e., in the direction of the current). For simplicity, we disregard the effect of "Mexican hat" (see, for instance, Ref. [34]) and a deviation of the real energy spectrum in the GBL from the parabolic one (the latter can be marked in the source and drain sections with relatively high Fermi energies). The effective mass in GBLs $m = \gamma_1/2v_W$, where $\gamma_1 \simeq 0.35$ – 0.43 eV is the inter-layer hopping integral and $v_W \simeq 10^8$ cm/s is the characteristic velocity of electrons and holes in GLs [3, 34–36], so that $m \simeq (0.03 - 0.04)m_0$, where m_0 is the bare electron mass. One of the potential advantages of GBL-FETs is the possibility of ballistic transport even if the top-gate length L_t is not small. In such GBL-FETs, one can neglect the right-hand side term in Eq. (5.1).

As in Refs. [10, 15, 37, 38], we use the following equation for the electric potential $\varphi = \varphi(x) = \psi(x, z)|_{z=0}$ in the GBL plane:

$$\frac{(W_b + W_t)}{3}\frac{\partial^2\varphi}{\partial x^2} - \frac{\varphi - V_b}{W_b} - \frac{\varphi - V_t}{W_t} = \frac{4\pi e}{k}(\Sigma_- - \Sigma_+). \tag{5.2}$$

Here, Σ_- and Σ_+ are the electron and hole sheet densities in the channel, respectively, k is the dielectric constant of the layers between the GBL and the gates and W_b and W_t are the thicknesses of these layers. In the following, we put $W_b = W_t = W$ (except Section 5.5). Equation (5.2) is a consequence of the two-dimensional Poisson equation for the electric potential $\psi(x, z)$ in the GBL-FET gated section ($-L_t/2 \le x \le L_t/2$ and $-W_b \le z \le W_t$, where L_t is the length of the top gate) in the weak nonlocality approximation [37]. This equation provides the potential distributions, which can be obtained from the two-dimensional Poisson equation by expansion in powers of the parameter $\delta = (W_b^3 + W_t^3)/15(W_b + W_t)/L^2 = W^2/15L^2$, where L is the characteristic scale of the lateral inhomogeneities (in the x-direction) assuming that $\delta \ll 1$, i.e., L is not too small. The lowest approximation in such an expansion leads to the Shockley's gradual channel approximation, in which the first term on the left side of Eq. (5.2) is neglected [39, 40]. The factor 1/3 appeared due to features of the Green function of the Laplace operator in the case of the geometry under consideration.

The boundary conditions for Eqs. (5.1) and (5.2) are presented as

$$f_{\mathbf{p}}\Big|_{p_x \ge 0, x=-L_t/2} = f_{s,\mathbf{p}}, \quad f_{\mathbf{p}}\Big|_{p_x \le 0, x=L_t/2} = f_{d,\mathbf{p}}, \tag{5.3}$$

$$\varphi\big|_{x=-L_t/2} = 0,\quad \varphi\big|_{x=L_t/2} = V_d + (\varepsilon_{F,d} - \varepsilon_{F,s})/e = V_d^*, \tag{5.4}$$

where $f_{s,\mathbf{p}}$ and $f_{d,\mathbf{p}}$ are the electron distribution functions in the source and drain sections of the channel. The functions $f_{s,\mathbf{p}}$ and $f_{d,\mathbf{p}}$ are the Fermi distribution functions with the Fermi energies $\varepsilon_{F,s}$ and $\varepsilon_{F,d}$ which are determined by the back gate and drain voltages, V_b and V_d [17, 18] (see also the Appendix):

$$\varepsilon_{F,s} \simeq eV_b \frac{b}{(1+b)},\quad \varepsilon_{F,d} \simeq e(V_b - V_d)\frac{b}{(1+b)}, \tag{5.5}$$

where $b = a_B/8W$, $a_B = k\hbar^2/me^2$ is the Bohr radius, and $\hbar$ is the reduced Planck constant. In the following, we shall assume that $b \ll 1$, so that $\varepsilon_{F,s} \simeq beV_b$ and $\varepsilon_{F,d} \simeq be(V_b - V_d)$. In particular, if $a_B = 4$ nm (GBL on SiO_2) and $W = 10$ nm, one obtains $b \simeq 0.05$. Due to a smallness of b, we shall disregard a distinction between V_d^* and V_d because $V_d - V_d^* \simeq bV_d \ll V_d$ (as shown in the Appendix). Restricting ourselves by the consideration of GBL-FETs operation at not too high drain voltages, we also neglect the difference in the Fermi energies in the source and drain sections, i.e., put $\varepsilon_{F,d} \simeq \varepsilon_{F,s} = \varepsilon_F$.

The source-drain dc current density (current per unit length in the direction perpendicular its flow) can be calculated using the following formulas:

$$J = \frac{4e}{(2\pi\hbar)^2}\int d^2\mathbf{p} v_x f_{\mathbf{p}}$$

$$= \frac{e}{\pi^2\hbar^2}\int_{-\infty}^{\infty} dp_y \int_0^{\infty} dp_x v_x (f_{\mathbf{p}} - f_{-\mathbf{p}}). \tag{5.6}$$

In this case, Eq. (5.1) with boundary conditions (5.3) yields

$$f_{\mathbf{p}} - f_{-\mathbf{p}} \simeq \frac{\Theta(p_x^2/2m + e\varphi) - \Theta(p_x^2/2m + e\varphi - eV_d)}{1 + \exp[(p^2/2m + e\varphi - \varepsilon_F)/k_BT]}, \tag{5.7}$$

where T is the temperature, k_B is the Boltzmann constant, and $\Theta(\varepsilon)$ is the unity step function. Using Eqs. (5.6) and (5.7), we obtain

$$J = \frac{e}{\pi^2\hbar^2}\int_{-\infty}^{\infty} dp_y \int_{\Delta_m}^{\infty} d\xi \left\{ \frac{1}{1 + \exp[(p_y^2/2m + \xi - \varepsilon_F)/k_BT} \right.$$

$$\left. - \frac{1}{1 + \exp[(p_y^2/2m + \xi - \varepsilon_F + eV_d)/k_BT]} \right\}$$

$$= \frac{ek_BT}{\pi^2\hbar^2}\int_{-\infty}^{\infty} dp_y \left\{ \ln\left[\exp\left(\frac{\varepsilon_F - p_y^2/2m - \Delta_m}{k_BT}\right) + 1\right] \right.$$
$$\left. - \ln\left[\exp\left(\frac{\varepsilon_F - p_y^2/2m - \Delta_m - eV_d}{k_BT}\right) + 1\right] \right\}. \qquad (5.8)$$

Equation (5.8) can be presented in the following form:

$$J = J_0 \int_0^{\infty} dz \left\{ \ln\left[\exp(\delta_m - z^2) + 1\right] \right.$$
$$\left. - \ln\left[\exp(\delta_m - U_d - z^2) + 1\right] \right\}. \qquad (5.9)$$

Here (see, for instance, Ref. [23])

$$J_0 = \frac{2\sqrt{2me}(k_BT)^{3/2}}{\pi^2\hbar^2}, \qquad (5.10)$$

is the characteristic current density, and $\delta_m = (\varepsilon_F - \Delta_m)/k_BT$, and $U_d = eV_d/k_BT$ are the normalized voltage swing and drain voltage, respectively. At $m = 4 \times 10^{-29}$ g and $T = 300$ K, $J_0 \simeq 2.443$ A/cm.

Figure 5.2 shows the dependences of the source-drain current J normalized by the value J_0 as a function of the U_d calculated using Eq. (5.9) for different values of δ_m.

The GBL-FET transconductance g is defined as

$$g = \frac{\partial J}{\partial V_t}. \qquad (5.11)$$

Equations (5.9) and (5.11) yield

$$g = J_0 \int_0^{\infty} dz \left\{ \left[\exp(z^2 - \delta_m) + 1\right]^{-1} \right.$$
$$\left. - \left[\exp(z^2 - \delta_m + U_d) + 1\right]^{-1} \right\} \left(-\frac{\partial \delta_m}{\partial V_t} \right). \qquad (5.12)$$

The obtained formulas for the source-drain current and transconductance can be simplified in the following limiting cases.

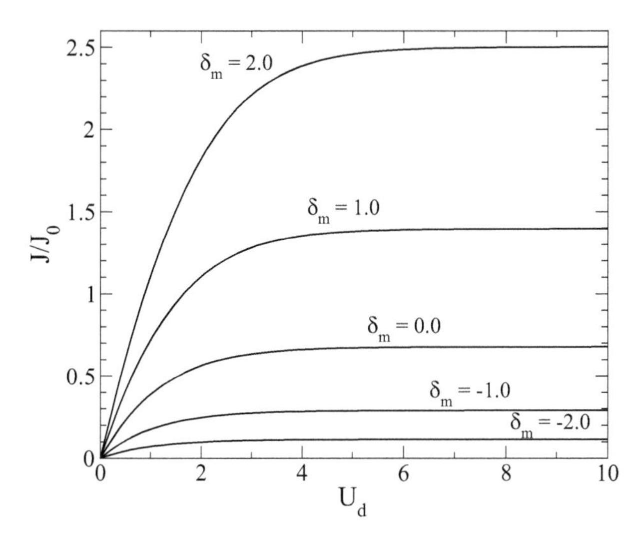

Figure 5.2 Normalized source-drain current J/J_0 versus normalized drain voltage U_d at different values of normalized top-gate voltage swing δ_m.

5.3.1 High Top-Gate Voltages

At high top-gate voltages, which correspond to the sub-threshold voltage range, the barrier height exceeds the Fermi energy ($\Delta_m \gg \varepsilon_F$), so that $\delta_m \gg 1$. In this case (the electron system in the gated section is nondegenerate), using Eqs. (5.9) and (5.12), we obtain

$$J = \frac{\sqrt{\pi}}{2} J_0 \exp\left(\frac{\varepsilon_F - \Delta_m}{k_B T}\right)\left[1 - \exp\left(-\frac{eV_d}{k_B T}\right)\right], \tag{5.13}$$

$$g = \frac{\sqrt{\pi}}{2}\frac{J_0}{k_B T} \exp\left(\frac{\varepsilon_F - \Delta_m}{k_B T}\right)\left[1 - \exp\left(-\frac{eV_d}{k_B T}\right)\right]\left(-\frac{\partial \Delta_m}{\partial V_t}\right). \tag{5.14}$$

5.3.2 Near Threshold Top-Gate Voltages

In this case, $\Delta_m \gtrsim \varepsilon_F$, i.e., $|\delta_m| \lesssim 1$, Eqs. (5.9) and (5.12) yield

$$J \simeq J_0 \frac{eV_d}{k_B T}\left[\zeta_1 + \zeta_2\left(\frac{\varepsilon_F - \Delta_m}{k_B T}\right)\right]\left[1 - \exp\left(-\frac{eV_d}{k_B T}\right)\right], \tag{5.15}$$

$$g \simeq \frac{J_0 e V_d}{(k_B T)^2}\zeta_2\left(-\frac{\partial \Delta_m}{\partial V_t}\right) \tag{5.16}$$

at low drain voltages $eV_\mathrm{d} \lesssim k_\mathrm{B}T$ ($U_\mathrm{d} \lesssim 1$), and

$$J \simeq J_0\left[\zeta_0 + \zeta_1\left(\frac{\varepsilon_\mathrm{F} - \Delta_\mathrm{m}}{k_\mathrm{B}T}\right)\right], \tag{5.17}$$

$$g \simeq \frac{J_0}{k_\mathrm{B}T}\zeta_1\left(-\frac{\partial \Delta_\mathrm{m}}{\partial V_\mathrm{t}}\right) \tag{5.18}$$

at high drain voltages $eV_\mathrm{d} \gg k_\mathrm{B}T$ ($U_\mathrm{d} \gg 1$). Here, $\zeta_0 = \int_0^\infty d\xi\ \ln[\exp(-\xi^2) + 1] \simeq 0.678$, $\zeta_1 = \int_0^\infty d\xi\ /[\exp(\xi^2) + 1] \simeq 0.536$, and $\zeta_2 = \int_0^\infty d\xi\ \exp(\xi^2)/[\exp(\xi^2) + 1]^2 \simeq 0.337$. In the limit $\varepsilon_\mathrm{F} = \Delta_\mathrm{m}$, Eqs. (5.13) and (5.14) provide the values J and g close to those obtained from Eqs. (5.17) and (5.18), which are rigorous in such a limit.

5.3.3 Low Top-Gate Voltages

At low top-gate voltages, $\Delta_\mathrm{m} < \varepsilon_\mathrm{F,}$ from Eqs. (5.9)

$$J \simeq \frac{2}{3}I_0[(\varepsilon_\mathrm{F} - \Delta_\mathrm{m})^{3/2}\Theta(\varepsilon_\mathrm{F} - \Delta_\mathrm{m}) - (\varepsilon_\mathrm{F} - \Delta_\mathrm{m} - eV_\mathrm{d})^{3/2}\Theta(\varepsilon_\mathrm{F} - \Delta_\mathrm{m} - eV_\mathrm{d})], \tag{5.19}$$

$$g \simeq I_0[(\varepsilon_\mathrm{F} - \Delta_\mathrm{m})^{1/2}\Theta(\varepsilon_\mathrm{F} - \Delta_\mathrm{m}) - (\varepsilon_\mathrm{F} - \Delta_\mathrm{m} - eV_\mathrm{d})^{1/2}\Theta(\varepsilon_\mathrm{F} - \Delta_\mathrm{m} - eV_\mathrm{d})]\left(-\frac{\partial \Delta_\mathrm{m}}{\partial V_\mathrm{t}}\right). \tag{5.20}$$

Here,

$$I_0 = \frac{2\sqrt{2me}}{\pi^2\hbar^2}, \tag{5.21}$$

with $J_0 = I_0\ (k_\mathrm{B}T)^{3/2}$.

Using Eqs. (5.19) and (5.20), one obtains

$$J \simeq I_0 eV_\mathrm{d}\sqrt{(\varepsilon_\mathrm{F} - \Delta_\mathrm{m})} \simeq I_0 eV_\mathrm{d}\sqrt{\varepsilon_\mathrm{F}}\left(1 - \frac{\Delta_\mathrm{m}}{2\varepsilon_\mathrm{F}}\right), \tag{5.22}$$

$$g \simeq \frac{1}{2}I_0\frac{eV_\mathrm{d}}{\sqrt{\varepsilon_\mathrm{F}}}\left(-\frac{\partial \Delta_\mathrm{m}}{\partial V_\mathrm{t}}\right) \tag{5.23}$$

at $eV_\mathrm{d} \ll \varepsilon_\mathrm{F} - \Delta_\mathrm{m}$, and

$$J = \frac{2}{3} I_0 (\varepsilon_F - \Delta_m)^{3/2} \simeq \frac{2}{3} I_0 \varepsilon_F^{3/2} \left(1 - \frac{3\Delta_m}{2\varepsilon_F} \right), \tag{5.24}$$

$$g \simeq I_0 \sqrt{(\varepsilon_F - \Delta_m)} \left(-\frac{\partial \Delta_m}{\partial V_t} \right) \tag{5.25}$$

at $eV_d \gg \varepsilon_F - \Delta_m$.

The dependences shown in Fig. 5.2 describe implicitly the dependences of J calculated using the universal Eq. (5.9) on the back-gate, top-gate, and drain voltages as well as on the geometrical parameters. Equations (5.13)–(5.25) provide these dependences in most interesting limits. However, to obtain the explicit formulas for J as well as for g, one needs to determine the dependences of the barrier height Δ_m on all voltages and geometrical parameters. Since the electron densities in the gated section in the limiting cases under consideration are different, the screening abilities of the electron system in this section and the potential distributions are also different. The latter leads to different Δ_m vs V_t relations.

5.4 Potential Distributions, Source-Drain Current, and Transconductance

To obtain the explicit dependences of the source-drain current and the transconductance on the gate voltages V_b and V_t as well as on the drain voltage V_d, one needs to find the relationship between the barrier height Δ_m and these voltages. This necessitates the calculations of the potential distribution in the channel. The latter can be found from Eq. (5.2) in an analytical form in the following limiting cases.

5.4.1 High Top-Gate Voltages: Sub-threshold Voltage Range ($\Delta_m >> \varepsilon_F$)

When the barrier height Δ_m exceeds the Fermi energy ε_F, the electron density is low in the gated section and, hence, one can disregard the contribution of the electron charge in this section. In such a limit, we arrive at the following equation for the potential:

$$\frac{d^2\varphi}{dx^2} - \frac{\varphi}{\Lambda_0^2} = \frac{F_0}{\Lambda_0^2}, \tag{5.26}$$

where $\Lambda_0 = \sqrt{2/3}W$ and $F_0 = -(V_b + V_t)/2$. Solving Eq. (5.26) considering boundary conditions (5.4), for the case of high top-gate voltages we obtain

$$\varphi = F_0\left[\frac{\cosh(x/\Lambda_0)}{\cosh(L_t/2\Lambda_0)} - 1\right] + V_d\frac{\sinh[(2x + L_t)/2\Lambda_0]}{\sinh(L_t/\Lambda_0)}. \tag{5.27}$$

Limiting our consideration by the GBL-FETs with not too short top gate ($L_t \gg W$), Eq. (5.27) can be presented as

$$\varphi \simeq -F_0\left[1 - 2\exp\left(-\frac{L_t}{2\Lambda_0}\right)\cosh\left(\frac{x}{\Lambda}\right)\right]$$
$$+ V_d\exp\left(-\frac{L_t}{2\Lambda_0}\right)\exp\left(\frac{x}{\Lambda_0}\right). \tag{5.28}$$

Equation (5.28) yields

$$\Delta_m \simeq eF_0\left(1 - \frac{1}{\eta_0}\right) - \frac{eV_d}{2\eta_0}$$
$$= -\frac{e(V_b + V_t)}{2}\left(1 - \frac{1}{\eta_0}\right) - \frac{eV_d}{2\eta_0}, \tag{5.29}$$

where $\eta_0 = \exp(L_t/2\Lambda_0)/2$. Simultaneously, for the position of the barrier top one obtains

$$x_m = -\frac{\Lambda_0}{2}\ln\left(1 + \frac{V_d}{F_0}\right)$$
$$= -\frac{\Lambda_0}{2}\ln\left(1 - \frac{2V_d}{V_b + V_t}\right). \tag{5.30}$$

The terms in the right-hand side of Eq. (5.29) containing parameter η_0 reflect the effect of the top-gate geometry (finiteness of its length). This effect is weakened with increasing top barrier length L_t. The effect of drain-induced barrier lowering in the case under consideration is described by the last term in the right-hand side of Eq. (5.29).

Equation (5.29) yields $(\partial\Delta_m/\partial V_t) = -(e/2)(1 - \eta_0^{-1})$. Invoking Eqs. (5.13) and (5.14), we obtain

$$J = \frac{\sqrt{\pi}}{2} J_0 \exp\left[\frac{e(V_t - V_{th})}{2k_B T}\left(1 - \frac{1}{\eta_0}\right)\right]$$
$$\times\left[1 - \exp\left(-\frac{eV_d}{k_B T}\right)\right]\exp\left(\frac{eV_d}{2\eta_0 k_B T}\right), \tag{5.31}$$

$$g \simeq \frac{\sqrt{\pi}}{4}\frac{eJ_0}{k_B T}\exp\left[\frac{e(V_t - V_{th})}{2k_B T}\left(1 - \frac{1}{\eta_0}\right)\right]$$
$$\times\left[1 - \exp\left(-\frac{eV_d}{k_B T}\right)\right]\exp\left(\frac{eV_d}{2\eta_0 k_B T}\right). \tag{5.32}$$

Here, $V_{th} = -[1 + 2b/(1 - \eta_0^{-1})]V_b \simeq -(1 + 2b)V_b$. The rightmost factors in the right-hand sides of Eqs. (5.31) and (5.32), associated with the effect of drain-induced barrier lowering, lead to an increase in g with increasing V_d not only at $eV_d \sim k_B T$ but at $eV_d \gg k_B T$: $g \propto \exp(eV_d/2\eta_0 k_B T)$. One can see that in the range of the top-gate voltages under consideration, the GBL-FET transconductance exponentially decreases with increasing $|V_t + V_b|$ and

$$g \simeq \frac{Je}{2k_B T} \ll g_0 = \frac{\sqrt{\pi}}{4}\frac{eJ_0}{k_B T}, \tag{5.33}$$

where at $T = 300$ K the characteristic value of the transconductance $g_0 \simeq 4330$ mS/mm.

5.4.2 Near Threshold Top-Gate Voltages ($\Delta_m \gtrsim \varepsilon_F$)

At $eV_d \lesssim k_B T \ll \varepsilon_F$, taking into account that the electron distribution is characterized by the equilibrium Fermi distribution function, the electron density in the gated section can be presented in the following form:

$$\Sigma \simeq \frac{2m}{\pi\hbar^2}(\varepsilon_F + e\varphi). \tag{5.34}$$

Considering this, we reduce Eq. (5.2) to

$$\frac{d^2\varphi}{dx^2} - \frac{\varphi}{\Lambda^2} = \frac{F}{\Lambda^2}. \tag{5.35}$$

Here,

$$\Lambda = \sqrt{\frac{a_B W}{12(1+2b)}} \simeq \sqrt{\frac{a_B W}{12}} = W\sqrt{\frac{2}{3}b},$$

$$F = \frac{[\varepsilon_F/e - b(V_b + V_t)]}{(1+2b)} \simeq -b(bV_b + V_t) \simeq -bV_t,$$

so that $\Lambda/\Lambda_0 \simeq \sqrt{b} < 1$. The solution of Eq. (5.35) with boundary condition (5.4) is given by

$$\varphi = F\left[\frac{\cosh(x/\Lambda)}{\cosh(L_t/2\Lambda)} - 1\right] + V_d \frac{\sinh[(2x+L_t)/2\Lambda]}{\sinh(L_t/2\Lambda)}$$

$$\simeq -F\left[1 - 2\exp\left(-\frac{L_t}{2\Lambda}\right)\cosh\left(\frac{x}{\Lambda}\right)\right]$$

$$+V_d \exp\left(-\frac{L_t}{2\Lambda}\right)\exp\left(\frac{x}{\Lambda}\right). \tag{5.36}$$

From Eq. (5.36) we obtain

$$\Delta_m \simeq eF\left(1 - \frac{1}{\eta}\right) - \frac{eV_d}{2\eta}$$

$$\simeq [\varepsilon_F - eb(V_b + V_t)]\left(1 - \frac{1}{\eta}\right) - \frac{eV_d}{2\eta}$$

$$\simeq -ebV_t\left(1 - \frac{1}{\eta}\right) - \frac{eV_d}{2\eta}, \tag{5.37}$$

where $\eta = \exp(L_t/2\Lambda)/2$, and the position of the barrier top is

$$x_m \simeq -\frac{\Lambda}{2}\frac{V_d}{F} \simeq -\frac{\Lambda}{2b}\frac{V_d}{V_b}. \tag{5.38}$$

Since $\Lambda < \Lambda_0$, one obtains $\eta \gg \eta_0$, and the terms in Eq. (5.37) containing parameter η can be disregarded. This implies that the effects of top-gate geometry and drain-induced barrier lowering are much weaker (negligible) in the case of the top-gate voltages in question in comparison with the case of high top-gate voltages.

Substituting Δ_m from Eq. (5.37) into Eq. (5.16), for low drain voltages we arrive at

$$g \simeq \frac{J_0 e^2 V_d \zeta_2}{(k_B T)^2}\left(1 - \frac{1}{\eta}\right)b. \tag{5.39}$$

At relatively high drain voltages ($eV_d > \varepsilon_F \gg k_BT$), the electron charge in the source portion of the gated section ($x \leq x_m$, where x_m is the coordinate of the barrier top) is primarily determined by the electrons injected from the source. The electron injection from the drain at high drain voltages is insignificant. Hence, the electron charge in the drain portion of the gated section can be disregarded. In this case, Eq. (5.2) can be presented as

$$\frac{d^2\varphi}{dx^2} - \frac{\varphi}{\Lambda^2} = \frac{F}{\Lambda^2} \tag{5.40}$$

at $-L_t/2 \leq x \leq x_0$, and

$$\frac{d^2\varphi}{dx^2} - \frac{\varphi}{\Lambda_0^2} = \frac{F_0}{\Lambda_0^2} \tag{5.41}$$

at $x_0 \leq x \leq L_t/2$.

At the point $x = x_0$ corresponding to the condition $e\varphi|_{x=x_0} + \varepsilon_F = 0$, the solutions of Eqs. (5.40) and (5.41) should be matched:

$$\varphi\Big|_{x=x_0-0} = \varphi\Big|_{x=x_0+0} = -\frac{\varepsilon_F}{e},$$

$$\frac{d\varphi}{dx}\bigg|_{x=x_0-0} = \frac{d\varphi}{dx}\bigg|_{x=x_0+0}. \tag{5.42}$$

Solving Eqs. (5.40) and (5.41) with conditions (5.4) and (5.42), we obtain the following formulas for the potential φ at $-L_t/2 \leq x \leq x_0$ and $x_0 \leq x \leq L_t/2$ as well as an equation for x_0:

$$\varphi = F\left[\frac{\cosh(x/\Lambda)}{\cosh(L_t/2\Lambda)} - 1\right] - \left[\frac{\varepsilon_F}{e} + F\frac{\cosh(x_0/\Lambda)}{\cosh(L_t/2\Lambda)} - F\right] \times \frac{\sinh[(x + L_t/2)/\Lambda)]}{\sinh[(x_0 + L_t/2)/\Lambda)]}, \tag{5.43}$$

$$\varphi = F_0\left[\frac{\cosh(x/\Lambda_0)}{\cosh(L_t/2\Lambda_0)} - 1\right] + V_d\frac{\sinh[(x + L_t/2)/\Lambda_0]}{\sinh(L_t/\Lambda_0)} - \left[\frac{\varepsilon_F}{e} + F_0\frac{\cosh(x_0/\Lambda_0)}{\cosh(L_t/2\Lambda_0)} - F_0 + V_d\frac{\sinh[(x_0 + L_t/2)/\Lambda_0]}{\sinh(L_t/\Lambda_0)}\right] \times \frac{\sinh[(x - L_t/2)/\Lambda_0]}{\sinh[(x_0 - L_t/2)/\Lambda_0]}, \tag{5.44}$$

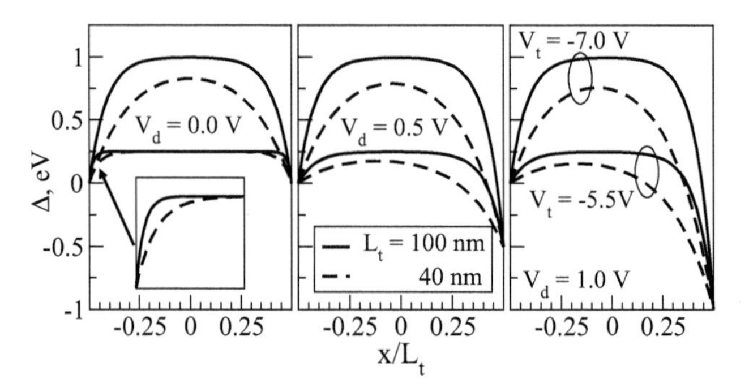

Figure 5.3 Barrier profile $\Delta = -e\varphi$ at different top-gate voltage V_t and drain voltage V_d for GBL-FETs with different top-gate length L_t. Upper and lower pairs of curves correspond to $V_t - V_{th} = -1.5$ V and $V_t - V_{th} \simeq 0$, respectively; $W = 10$ nm, $b = 0.05$, and $V_b = 5.0$ V.

$$\frac{1}{\Lambda}\left\{F\frac{\sinh(x_0/\Lambda)}{\cosh(L_t/2\Lambda)} - \left[\frac{\varepsilon_F}{e} + F\frac{\cosh(x_0/\Lambda)}{\cosh(L_t/2\Lambda)} - F\right]\right.$$

$$\times\frac{\cosh[(x_0 + L_t/2)/\Lambda)]}{\sinh[(x_0 + L_t/2)/\Lambda)]}$$

$$= \frac{1}{\Lambda_0}\left\{F_0\frac{\sinh(x_0/\Lambda_0)}{\cosh(L_t/2\Lambda_0)} + V_d\frac{\cosh[(x_0 + L_t/2)/\Lambda_0]}{\sinh(L_t/\Lambda_0)}\right.$$

$$-\left[\frac{\varepsilon_F}{e} + F_0\frac{\cosh(x_0/\Lambda_0)}{\cosh(L_t/2\Lambda_0)} - F_0 + V_d\frac{\sinh[(x_0 + L_t/2)/\Lambda_0]}{\sinh(L_t/\Lambda_0)}\right]$$

$$\left.\times\frac{\cosh[(x_0 - L_t/2)/\Lambda_0]}{\sinh[(x_0 - L_t/2)/\Lambda_0]}\right\}. \tag{5.45}$$

In the cases $V_b + V_t \simeq 0$ and $-(V_b + V_t) \gtrsim V_b \gg \varepsilon_F/e$, Eq. (5.45) yields $x_0 = -L_t/2 + \Lambda\ \ln[4bV_t/(\sqrt{b} + 2b)(V_b + V_t)]$ and $x_0 \simeq -L_t/2 + 2\Lambda_0\varepsilon_F/[-e(V_b + V_t)] \simeq -L_t/2 + 2b\Lambda_0 V_b/[-(V_b + V_t)]$, respectively. When $-(V_b + V_t) \to +0$, the matching point shifts toward the channel center. If $-(V_b + V_t)$ increases, the matching point tends to the source edge of the channel. In this case, the role of the electron charge in the vicinity of the source edge diminishes, and the potential distribution tends to that given by Eq. (5.27).

At the threshold, the matching point x_0 coincides with the position of the barrier maximum x_m. Considering that at $x_0 = x_m$, both the left-hand and right-hand sides of Eq. (5.45) are equal to zero, for the barrier top height near the threshold at relatively high drain voltages we obtain the following:

$$\Delta_{\rm m} \simeq eF_0\left(1-\frac{1}{\eta_0}\right)-\frac{eV_{\rm d}}{\eta_0}$$

$$=-\frac{e(V_{\rm b}+V_{\rm t})}{2}\left(1-\frac{1}{\eta_0}\right)-\frac{eV_{\rm d}}{\eta_0}, \qquad (5.46)$$

$$x_{\rm m} \simeq -\frac{L_{\rm t}}{2}+\Lambda\ln\left(\frac{2F}{F-F_0}\right) \simeq -\frac{L_{\rm t}}{2}+\Lambda\ln\frac{1}{b}. \qquad (5.47)$$

Both Eqs. (5.29) and (5.46) correspond to the situations when the electron density in a significant portion of the channel is fairly low. However there is a distinction in the dependence of $\Delta_{\rm m}$ on $V_{\rm d}$ (the pertinent coefficients differ by factor of two). This is because in the first case the barrier top is located near the channel center, whereas in the second case it is shifted to the vicinity of the source edge [compare Eqs. (5.30) and (5.47)].

Using Eqs. (5.19) and (5.50), we obtain $(\partial\, \Delta_{\rm m}/\partial\, V_{\rm t}) = -(e/2)(1 - \eta_0^{-1})$ and arrive at the following formula for the transconductance near the threshold, i.e., when $V_{\rm t} \simeq V_{\rm th}$

$$J \simeq J_0\left[\zeta_0+\zeta_1\frac{e(V_{\rm t}-V_{\rm th})}{2k_{\rm B}T}\left(1-\frac{1}{\eta_0}\right)\right], \qquad (5.48)$$

$$g \simeq \frac{J_0 e}{k_{\rm B}T}\frac{\zeta_1}{2}\left(1-\frac{1}{\eta_0}\right)=g_{\rm th}. \qquad (5.49)$$

Here, as above, $V_{\rm th} \simeq -(1 + 2b)V_{\rm b}$. In particular, Eqs. (5.48) and (5.49) at $V_{\rm t} = V_{\rm th}$, yield $J_{\rm th} \simeq J_0\zeta_0$. For a GBL-FET with $L_{\rm t}$ = 40 nm, W = 10 nm, at T = 300 K, one obtains $J_{\rm th} \simeq$ 1.656 A/cm and $g_{\rm th} \simeq$ 2167 mS/mm.

5.4.3 Low Top-Gate Voltages ($\Delta_{\rm m} < \varepsilon_{\rm F}$)

At relatively low top-gate voltages when $\Delta_{\rm m} < \varepsilon_{\rm F}$, the electron system is degenerate not only in the source and drain sections but in the gate section as well. In this top-gate voltage range, the spatial variation of the potential is characterized by $\Lambda \simeq W\, 2b/3$. As a result, for $\Delta_{\rm m}$ one obtains an equation similar to Eq. (5.37). Since $\Lambda < \Lambda_0 \ll L_{\rm t}$, the parameter determining the effect of the top-gate geometry and the effect of drain-induced barrier lowering is $\eta = \exp(L_t/2\Lambda)/2 \gg \eta_0$. As a consequence, one can neglect the effects in question in the top-gate voltage range under consideration. As a result, one can arrive at

$$J \simeq \frac{2}{3} I_0 e^{3/2} [b(V_t - V_{th})]^{3/2} - [b(V_t - V_{th} - V_d]^{3/2} \tag{5.50}$$

when $V_d \leq b(V_t - V_{th})$,

$$J \simeq \frac{2}{3} I_0 e^{3/2} [b(V_t - V_{th})]^{3/2}, \tag{5.51}$$

when $V_d > b(V_t - V_{th})$, and

$$\frac{\partial \Delta_m}{\partial V_t} = -be. \tag{5.52}$$

Considering Eq. (5.51), at low top-gate voltages we obtain

$$g \simeq \frac{beI_0}{2} \frac{eV_d}{\sqrt{\varepsilon_F}} \simeq \frac{\sqrt{b}e^{3/2} I_0}{2} \frac{V_d}{\sqrt{V_b}} \tag{5.53}$$

when $eV_d \gg \varepsilon_F - \Delta_m \simeq be(V_b + V_t)$, and

$$g \lesssim beI_0 \sqrt{\varepsilon_F} \simeq b^{3/2} e^{3/2} I_0 \sqrt{V_b} = g_{on} \tag{5.54}$$

when $eV_d \ll \varepsilon_F - \Delta_m \simeq be(V_b + V_t)$. As follows from Eqs. (5.33) and (5.34), the transconductance is proportional to a small parameter $b^{3/2}$. This is because the effect of the top-gate potential is weakened due to a strong screening by the degenerate electron system in the gated section. As a result, the transconductance at low top-gate voltages is smaller than that at the top-gate voltage corresponding to the threshold. Assuming that $b = 0.05$ and $V_b = 5$ ($\varepsilon_F \simeq 0.25$ eV), we obtain $g_{on} \simeq 1467$ mS/mm. Comparing Eqs. (5.49) and (5.54), we find $g_{on}/g_{th} \propto b^{3/2} \sqrt{eV_b/k_BT} \simeq 0.158$ and $g/g_{th} \lesssim g_{on}/g_{th} \simeq 0.68$.

Since the source-drain current at high top-gate voltages decreases exponentially when $-V_t$ increases, the transconductance decreases as well.

Figure 5.3 shows the barrier (conduction band) profile $\Delta = -e\varphi$ in the GBL-FETs calculated using Eqs. (5.27), (5.36), and (5.44) for different applied voltages and top-gate lengths. As demonstrated, the barrier height naturally decreases with increasing $V_t - V_{th}$. At the threshold ($V_t = V_{th}$), the barrier height is equal to the Fermi energy (at $b = 0.05$ and $V_b = 5$ V, $\varepsilon_F \simeq 0.25$ eV). One can see that shortening of the top-gate leads to a marked decrease in the barrier height (the short-gate effect). The source-drain current as a function of the drain voltage for different top-gate voltage swings is demonstrated in Fig. 5.4. The dependences corresponding to $V_t - V_{th} < 0$, $V_t - V_{th} = 0$, and $V_t - V_{th} > 0$ were calculated using formulas from subsections A, B, and C, respectively. Figure 5.5 shows that the transconductance as a function of the top-gate voltage swing exhibits a pronounced maximum at $V_t \simeq V_{th}$. This maximum is attributed to the following. At high top-gate voltages, the effect of screening is insignificant due to low electron density in the channel. As a result, the height of the barrier top is rather sensitive to the top-gate voltage variations. The source-drain current in this case is exponentially small, so that the transconductance is small. In contrast, at low top-gate voltages, the screening by the electrons in the channel is effective, leading to a much weaker control of the barrier height by the top voltage [pay attention to parameter $b \ll 1$ in Eqs. (5.50)–(5.54)]. Despite, a strong source-drain current provides a moderate values of the transconductance. However, in the near threshold voltage range, both the sensitivity of the barrier height and the source-drain current to the top-gate voltage are fairly large.

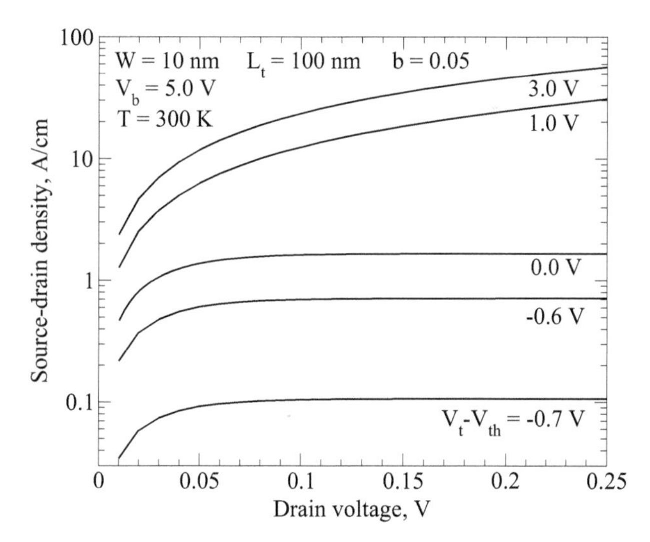

Figure 5.4 Source-drain current J versus drain voltage V_d at different values of the top-gate voltage swing $V_t - V_{th}$.

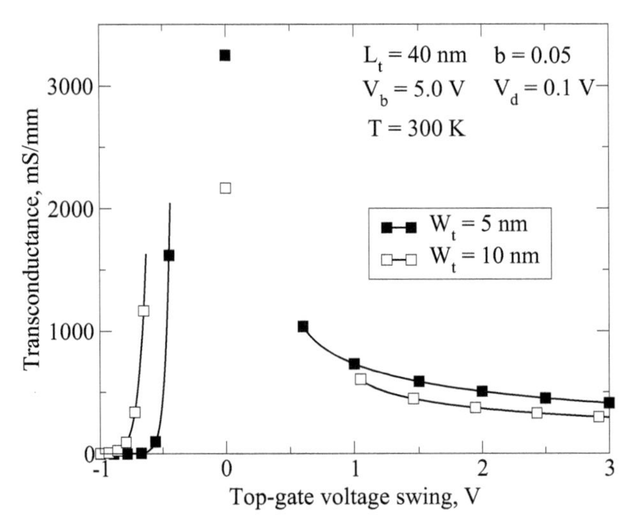

Figure 5.5 Transconductance g versus top-gate voltage swing $V_t - V_{th}$.

5.5 Discussion

5.5.1 Role of Geometrical Parameters

In the main part of the chapter, we assumed that the thicknesses of the gate layers are equal to each other: $W_b = W_t = W$. If $W_b \neq W_t$, the formulas obtained above should be slightly modified. In particular, near the threshold $(\partial\Delta_m/\partial V_t) = -(e/(1 + W_t/W_b)(1 - \eta_0^{-1})$, so that the transconductance instead of Eq. (5.49) is given by

$$g \simeq \frac{J_0 e}{k_B T} \frac{\zeta_1}{(1+W_t/W_b)\zeta_0}\left(1-\frac{1}{\eta_0}\right)$$

$$\simeq \frac{J_0 e}{k_B T} \frac{\zeta_1}{(1+W_t/W_b)\zeta_0}. \tag{5.55}$$

As follows from the comparison of Eqs. (5.49) and (5.55), changing W_t, in particular, from $W_t = W_b$ to $W_t = W_b/2$, leads to an increase in the transconductance at the top-gate voltages near the threshold of 50% (compare the g versus $V_t - V_{th}$ dependences in Fig. 5.5 for $W_b = W_t$ =10 nm and W_b =10 nm and W_t = 5 nm).

As demonstrated above, shortening of the top gate can result in deterioration of the GBL-FET characteristics. This effect is characterized by parameters η and η_0, which strongly depends on the top-gate length L_t. This effect gives rise to a decrease in the transconductance if L_t becomes smaller. However, in the range of large L_t the electron collisions can play a substantial role. This leads to the transconductance roll-off with increasing L_t. The effect of drain-induced barrier lowering is also characterized by parameters η and η_0. As follows from Section 5.4, at sub-threshold, the effect of drain-induced barrier lowering results in the appearance of the factor $\exp(eV_d/2\eta_0 k_B T)$ [see Eqs. (5.31) and (5.32)]. This factor can provide a marked increase in J and g with increasing drain voltage in GBL-FETs with relatively short top gates. For example, at $W = 40$ nm, $L_t = 40$ nm, $T = 300$ K, and $V_d = 0.25$ V, this factor is about 2.37.

The obtained characteristics comprise parameter b, which is proportional to the Bohr radius a_B. An increase in a_B when a large-k substrate, say, HfO_2 ($a_B = 20$ nm) is used instead of the SiO_2 substrate ($a_B = 4$ nm), and, consequently, an increase in b leads to higher sensitivity of the barrier height to the applied voltages and, hence, to higher transconductance.

5.5.2 Effect of Electron Scattering

As shown above, the potential distributions in the main part of the gated section are fairly flat. This implies that to determine the effect of electron scattering associated with disorder and acoustic phonons one can use Eq. (5.2) with the collisional term following the approach applied in Ref. [17]. The interaction of electrons with optical phonons, particularly with optical phonons in the substrate, can be effective in the gate-drain section at elevated drain voltages. However, even relatively strong scattering on optical phonons in this section should not markedly affect the dc characteristics under consideration, although it can be essential for the ac characteristics. Considering here the case when the elastic scattering mechanisms under consideration are strong, so that they lead to an effective isotropization of the electron distribution, one can find that the values of the source-drain current and the transconductance obtained in the previous section for the ballistic transport should be multiplied by a collision factor C. This factor is equals to $C_\infty = \sqrt{2\pi k_B T/m}/L_t v$, where $\nu = mw/2$ is the collision frequency (we put $w(q) = w$ = const. It characterizes the fraction of the electrons injected into the gated section and those reflected back due to the collisions. To obtain the GBL-FET characteristics with the top-gate lengths in a wide range (to follow the transition from the ballistic electron transport to the collision-dominated transport), we use for the collision factor the following interpolation formula:

$$C = \frac{1}{1 + L_t/L_{scat}}, \tag{5.56}$$

where $L_{scat} = \sqrt{2\pi k_B T/m}/\nu$ is the characteristic scattering length. Figures 5.6 and 5.7 show the dependences of the transconductance maximum (approximately at $V_t = V_{th}$) on the top-gate length calculated for GBL-FETs with different W_t. The scattering length is assumed to be ∞ (ballistic transport), 500 nm, and 75 nm. At $T = 300$ K, this corresponds to the collision frequencies $\nu = 0$, $\nu \simeq 1.14 \times 10^{12}$ s^{-1} (the electron mobilities $\mu \simeq 1.75 \times 10^5$ cm^2/V s) and $\nu \simeq 7.6 \times 10^{12}$ s^{-1} ($\mu \simeq 2.63 \times 10^4$ cm^2/V s), respectively). One can see that in the case of essential electron collisions, g versus L_t dependences exhibit pronounced maxima. This is attributed to an interplay of two effects: the short-gate effect (weakening of the barrier controllability by the

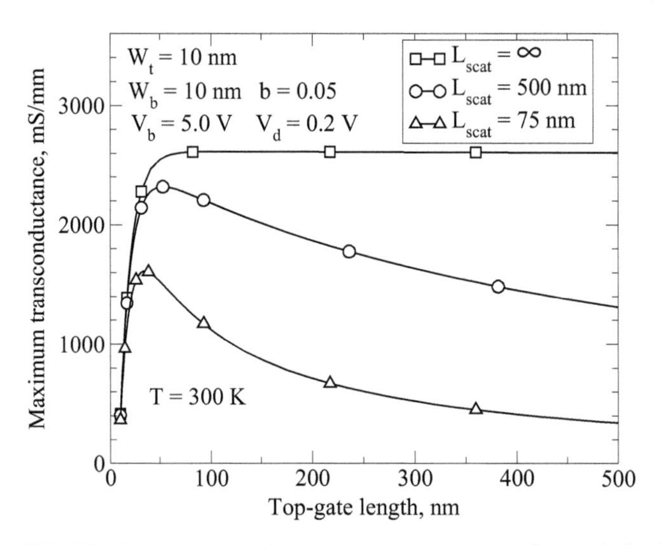

Figure 5.6 Maximum transconductance g versus top-gate length L_t for W_t = 10 nm.

top-gate voltage when L_t decreases) and the effect of collisions, which reinforces when L_t becomes larger (which decreases the current). As follows from Figs. 5.6 and 5.7, the electron collisions can lead to a dramatic decrease in the transconductance.

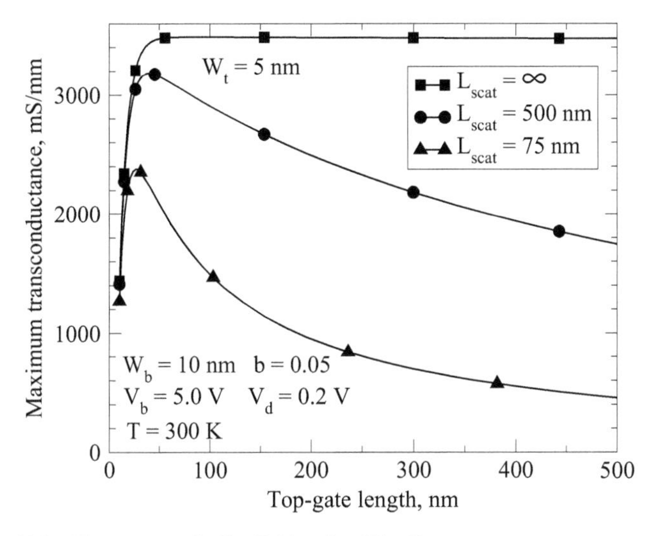

Figure 5.7 The same as in Fig. 5.6 but for W_t = 5 nm.

5.5.3 Charge Inversion in the Gated Section

At sufficiently high top-gate voltages $V_t < V_{in}$ when $\Delta_m > \varepsilon_F + E_g$, the top of the valence band in the gated section of the channel can be markedly populated by holes (inversion of the gated section charge), so that the term in the right-hand side of Eq. (5.2) becomes negative. The latter inequality corresponds to the following value of the inversion voltage [see Eqs. (5.A4) and (5.A5)]:

$$V_{in} = -V_b\left(1+2b+\frac{d_0}{W}\right) \simeq V_{th}\left(1+\frac{d_0}{W}\right). \tag{5.57}$$

The hole charge provides an effective screening of the transverse electric field in the gated section. This leads to weakening of the sensitivity of the barrier height and the source-drain current on the top gate voltage V_t [15] and, hence, to a decrease in the transconductance. This pattern is valid at the dc voltages or when the characteristic time of their variation is long in comparison with the characteristic times of the thermogeneration of holes and the tunneling between the channel side regions and the gated section [15, 17]. In the situation when the hole recharging of the gated section of the channel is a relatively slow process, the ac transconductance at the frequencies higher than the characteristic recharging frequency can substantially exceed the dc transconductance.

5.5.4 Interband Tunneling

At elevated top-gate voltages ($V_t < V_{in}$), the interband tunneling of electrons from the conduction band in the source to the valence band in the gated section as well as from the valence band in the gated section to the conduction band in the drain can be essential. At high drain voltages the latter tunneling processes can be particularly pronounced. This can result in elevated source-drain current despite rather high barrier. To limit the tunneling, the back-gate voltage, which mainly determines the energy gap, should be high enough. A decrease in the gate layer thicknesses also promotes the tunneling suppression. For example, if W = (5 – 10) nm, $V_b = - V_t = 5$ V, the energy gap in the gated section of the channel $E_g \simeq 0.17 - 0.34$ eV [see Eq. (5.A4)]. Despite some attempts to calculate the tunneling currents in G-FETs and GBL-FETs (see, for instance, Refs. [11, 19], the problem for GBL-FETs remains open. This is because the spatial nonuniformity of the energy gap in the channel and its nonlinear spatial dependence (particularly near the drain edge) associated with the features of the potential distribution under the applied voltages. Generalizing Eq. (5.A4), we obtain $E_g = ed[V_b - \varphi(x)]/2W$, hence the energy gap varies from $E_g = E_{g,s} \simeq edV_b/W$ at $x = -L_t/2$ to $E_g = E_{g,s} \simeq ed(V_b - V_d/2W$ at $x = L_t/2$. It reaches a maximum $E_g = d(eV_b + \Delta_m)/2W$ at $x = x_m$. One can see that $E_{g,d}$ can be markedly smaller than E_g, especially at not too small drain voltages. Due to this, the deliberation of the tunneling in GBL-FETs requires sufficiently rigorous device model (which could include the formulas for the potential distribution obtained above) and numerical approach.

5.6 Conclusions

We demonstrated that the developed device model of allows to derive the GBL-FET characteristics: the potential distributions along the channel and the dependences of the source-drain current and the transconductance on the applied voltages and the geometrical parameters as closed-form analytical expressions. The key element of the model, which provides an opportunity to solve the problem analytically, is the use of the Poisson equation in the weak nonlocality approximation. In particular, the model accounts for the effect of screening of the transverse electric field by the electron charge in the channel, the short-gate effect, and the effect of drain-induced barrier lowering. The parameters η_0 and η characterizing the strength of these effects in the cases of essentially depleted channel and strong screening were expressed via the geometrical parameters and the Bohr radius. As shown, the GBL-FET transconductance exhibits a pronounced maximum as a function of the top-gate voltage swing. The interplay of the short-gate effect and the electron collisions results in a nonmonotonic dependence of the transconductance on the top-gate length. The obtained analytical formulas for the potential barrier height, the source-drain current, and the transconductance can be used for GBL-FET optimization by proper choice of the thicknesses of gate layers, the top-gate length, and the bias voltages.

Acknowledgments

The authors are grateful to H. Watanabe for stimulating comments and to V. V. V'yurkov for providing Ref. [31]. The work was supported by the Japan Science and Technology Agency, CREST, Japan.

Appendix: Voltage Dependences of the Fermi Energy and the Energy Gap

Disregarding the effect of "Mexican-hat" and the nonparabolicity of the electron energy spectrum, the density of states can be considered independent of the energy (in the energy range under consideration). Taking this into account, the electron Fermi energies and the energy gaps in the source and drain sections of the GBL-FET channel are, respectively, given by Refs. [17, 18] ($W_b = W_t = W$),

$$\varepsilon_{F,s} = \frac{k_B T}{(1+b)} \ln\left[\exp\left(\frac{beV_b}{k_B T}\right) - 1\right], \tag{5.A1}$$

$$\varepsilon_{\rm F,d} = \frac{k_{\rm B}T}{(1+b)}\ln\left[\exp\left(\frac{be(V_{\rm b}-V_{\rm d})}{k_{\rm B}T}\right)-1\right], \tag{5.A2}$$

$$E_{\rm g,s} = \frac{edV_{\rm b}}{2W}, \quad E_{\rm g,s} = \frac{ed(V_{\rm b}-V_{\rm d})}{2W}. \tag{5.A3}$$

Here $a_{\rm B} = k\hbar^2/me^2$, $b = a_{\rm B}/8W$, and $d \lesssim d_0$, where $d_0 \simeq 0.34$ nm [32] is the spacing between the graphene layers in the GBL, while d stands for the effective spacing accounting for the screening of the transverse electric field by GBL (polarization effect). In portion of the gated section essentially occupied by electrons and its depleted portion, one obtains

$$E_{\rm g} = \frac{ed(V_{\rm b}-V_{\rm t})}{2W}, \; E_{\rm g} = \frac{ed_0(V_{\rm b}-V_{\rm t})}{2W}, \tag{5.A4}$$

respectively. Due to $a_{\rm B} \gg d$, from Eqs. (5.A1)–(5.A4) one obtains $\varepsilon_{\rm F,s} \geq \varepsilon_{\rm F,d} > E_{\rm g,s} \geq E_{\rm g,d}$. At $V_{\rm t} < 0$, $E_{\rm g}$ can significantly exceed $E_{\rm g,s}$ and $E_{\rm g,d}$. In the case of strong degeneracy of the electron system, Eqs. (5.A1) and (5.A2) yield

$$\varepsilon_{\rm F,s} \simeq eV_{\rm b}\frac{a_{\rm B}/8W}{(1+b)}, \quad \varepsilon_{\rm F,d} \simeq e(V_{\rm b}-V_{\rm d})\frac{b}{(1+b)}. \tag{5.A5}$$

The quantity $\varepsilon_{\rm F,d}$ is given by the same equations in which, however, $V_{\rm b}$ is substituted by $V_{\rm b} - V_{\rm d}$. As a result,

$$\varepsilon_{\rm F,s} - \varepsilon_{\rm F,d} \simeq \frac{beV_{\rm d}}{(1+b)} \simeq beV_{\rm d}, \tag{5.A6}$$

$$eV_{\rm d}^* = eV_{\rm d} + \varepsilon_{\rm F,d} - \varepsilon_{\rm F,s} \simeq \frac{eV_{\rm d}}{1+b}. \tag{5.A7}$$

Since parameter b in reality is small, so that $\varepsilon_{\rm F,s} - \varepsilon_{\rm F,d} \simeq beV_{\rm d} \ll eV_{\rm d}$ and $V_{\rm d}^* \simeq V_{\rm d}$ we put $\varepsilon_{\rm F,s} = \varepsilon_{\rm F,d} = \varepsilon_{\rm F}$ and substitute $V_{\rm d}^*$ by $V_{\rm d}$.

References

1. C. Berger, Z. Song, T. Li, X. Li, A. Y. Ogbazhi, R. Feng, Z. Dai, A. N. Marchenkov, E. H. Conrad, P. N. First, and W. A. de Heer, *J. Phys. Chem.*, **108**, 19912 (2004).
2. K. S. Novoselov, A. K. Geim, S. V. Morozov, D. Jiang, M. I. Katsnelson, I. V. Grigorieva, S. V. Dubonos, and A. A. Firsov, *Nature*, **438**, 197 (2005).

3. A. H. Castro Neto, F. Guinea, N. M. R. Peres, K. S. Novoselov, and A. K. Geim, *Rev. Mod. Phys.*, **81**, (2009).
4. J. Bai, X. Zhong, S. Jiang, Y. Huang, and X. Duan, *Nat. Nanotechnol.*, **5**, 190 (2010).
5. V. Ryzhii, M. Ryzhii, and T. Otsuji, *J. Appl. Phys.*, **101**, 083114 (2007).
6. F. Rana, *IEEE Trans. Nanotechnol.*, **7**, 91 (2008).
7. F. Xia, T. Murleer, Y.-M. Lin, A. Valdes-Garsia, and P. Avouris, *Nat. Nanotechnol.*, **4**, 839 (2009).
8. V. Ryzhii, M. Ryzhii, V. Mitin, and T. Otsuji, *J. Appl. Phys.*, **107**, 054512 (2010).
9. V. Ryzhii, A. A. Dubinov, T. Otsuji, V. Mitin, and M. S. Shur, *J. Appl. Phys.*, **107**, 054505 (2010).
10. V. Ryzhii, M. Ryzhii, and T. Otsuji, *Appl. Phys. Express*, **1**, 013001 (2008).
11. V. Ryzhii, M. Ryzhii, and T. Otsuji, *Phys. Status Solidi A*, **205**, 1527 (2008).
12. Y. Quyang, Y. Yoon, J. K. Fodor, and J. Guo, *Appl. Phys. Lett.*, **89** 203107 (2006).
13. G. Fiore and G. Iannaccone, *IEEE Electron Device Lett.*, **28**, 760 (2007).
14. G. Liang, N. Neophytou, D. E. Nikonov, and M. S. Lundstrom, *IEEE Trans. Electron Devices*, **54**, 677 (2007).
15. V. Ryzhii, M. Ryzhii, A. Satou, and T. Otsuji, *J. Appl. Phys.*, **103**, 094510 (2008).
16. M. Ryzhii, A. Satou, V. Ryzhii, and T. Otsuji, *J. Appl. Phys.*, **104**, 114505 (2008).
17. V. Ryzhii, M. Ryzhii, A. Satou, T. Otsuji, and N. Kirova, *J. Appl. Phys.*, **105**, 104510 (2009).
18. M. Ryzhii and V. Ryzhii, *Phys. Rev. B*, **79**, 245311 (2009).
19. M. Cheli, G. Fiori, and G. Iannaccone, *IEEE Trans. Electron Devices*, **56**, 2979 (2009).
20. S. O. Koswatta, S. Hasan, M. S. Lundstrom, M. P. Anantram, and D. E. Nikonov, *Appl. Phys. Lett.*, **89**, 023125 (2006).
21. M. Lenzi, P. Palestri, E. Gnani, S. Reggiani, A. Gnudi, D. Esseni, L. Selmi, and G. Baccarani, *IEEE Trans. Electron Devices*, **55**, 2087 (2008).
22. S. Fregonese, J. Gouet, C. Manex, and T. Zimmer, *IEEE Trans. Electron Devices*, **56**, 1184 (2009).
23. K. Natori, *J. Appl. Phys.*, **76**, 4881 (1994).

24. A. Rahman, J. Guo, S. Datta, and M. S. Lundstrom, *IEEE Trans. Electron Devices*, **50**, 1853 (2003).
25. F. G. Pikus and K. K. Likharev, *Appl. Phys. Lett.*, **71**, 3661 (1997).
26. V. A. Sverdlov, T. J. Walls, and K. K. Likharev, *IEEE Trans. Electron Devices*, **50**, 1926 (2003).
27. N. Sano, *Phys. Rev. Lett.*, **93**, 246803 (2004).
28. G. Mugnaini and G. Iannaccone, *IEEE Trans. Electron Devices*, **52**, 1802 (2005).
29. R. Kim, P. A. Neophytou, G. Klimeck, and M. S. Lundstrom, *J. Vac. Sci. Technol. B*, **26**, 1628 (2008).
30. R. Akis, N. Faralli, D. K. Ferry, S. M. Goodnick, K. A. Phatak, and M. Saratini, *IEEE Trans. Electron Devices*, **56**, 2935 (2009).
31. A. N. Khomyakov and V. V. V'yurkov, *Russian Microelectron.*, **38**, 393 (2009).
32. T. Ohta, A. Q. Bostwick, T. Seyller, K. Horn, and E. Rotenberg, *Science*, **333**, 951 (2006).
33. E. McCann, *Phys. Rev. B*, **74**, 161403 (2006).
34. E. V. Castro, K. S. Novoselov, S. V. Morozov,N. M. R. Peres, J. M. B. Lopes dos Santos, L. Nils-son, F. Guinea, A. K. Geim, and A. H. Castro Neto, *J. Phys.: Condens. Matter*, **22**, 175503 (2010).
35. E. A. Henriksen, Z. Jiang, L.-C. Tung, M. E. Schwartz, M. Takita, Y. J. Wang, P. Kim, and H. Stormer, *Phys. Rev. Lett.*, **100**, 087403 (2008).
36. M. Orlita and M. Potemski, *Semicond. Sci. Technol.*, **25**, 063001 (2010).
37. A. A. Sukhanov and Y. Y. Tkach, *Sov. Phys. Semicond.*, **18**, 797 (1984).
38. A. O. Govorov, V. M. Kowalev, and A. V. Chaplik, *JETP Lett.*, **70**, 488 (1999).
39. S. M. Sze, *Physics of Semiconductor Devices* (Wiley, New York, 1981).
40. M. Shur, *Physics of Semiconductor Devices* (Prentice-Hall, New Jersey, 1990).

Chapter 6

Hydrodynamic Model for Electron-Hole Plasma in Graphene*

D. Svintsov,[a] V. Vyurkov,[a] S. Yurchenko,[b] T. Otsuji,[c,d] and V. Ryzhii[b,d,e]

[a]*Institute of Physics and Technology, Russian Academy of Sciences, Moscow 117218, Russia*

[b]*Scientific Educational Center "Photonics and Infrared Technics," Bauman Moscow State Technical University, Moscow 105005, Russia*

[c]*Research Institute for Electrical Communication, Tohoku University, Sendai 980-8577, Japan*

[d]*Japan Science and Technology Agency, CREST, Tokyo 107-0075, Japan*

[e]*Computational Nanoelectronics Laboratory, University of Aizu, Aizu-Wakamatsu 965-8580, Japan*

vyurkov@ftian.ru

We propose a hydrodynamic model describing steady-state and dynamic electron and hole transport properties of graphene structures which accounts for the features of the electron and hole spectra. It is intended for electron-hole plasma in graphene

---

*Reprinted with permission from D. Svintsov, V. Vyurkov, S. Yurchenko, T. Otsuji, and V. Ryzhii (2012). Hydrodynamic model for electron-hole plasma in graphene, *J. Appl. Phys.*, **111**, 083715. Copyright © 2012 American Institute of Physics.

Graphene-Based Terahertz Electronics and Plasmonics: Detector and Emitter Concepts
Edited by Vladimir Mitin, Taiichi Otsuji, and Victor Ryzhii

ISBN 978-981-4800-75-4 (Hardcover), 978-0-429-32839-8 (eBook)
www.jennystanford.com

characterized by high rate of inter-carrier scattering compared to external scattering (on phonons and impurities), i.e., for intrinsic or optically pumped (bipolar plasma), and gated graphene (virtually monopolar plasma). We demonstrate that the effect of strong interaction of electrons and holes on their transport can be treated as a viscous friction between the electron and hole components. We apply the developed model for the calculations of the graphene dc conductivity; in particular, the effect of mutual drag of electrons and holes is described. The spectra and damping of collective excitations in graphene in the bipolar and monopolar limits are found. It is shown that at high gate voltages and, hence, at high electron and low hole densities (or vice-versa), the excitations are associated with the self-consistent electric field and the hydrodynamic pressure (plasma waves). In intrinsic and optically pumped graphene, the waves constitute quasineutral perturbations of the electron and hole densities (electron-hole sound waves) with the velocity being dependent only on the fundamental graphene constants.

6.1 Introduction

The hydrodynamic approach is quite reasonable for the description of dense electron-hole plasma in semiconductor systems in which the electron-electron, electron-hole, and hole-hole collisions dominate over the collisions of electrons and holes with disorder. In particular, such a situation can occur in intrinsic graphene at the room temperature, particularly, in the structures with low or moderate permittivity of the layers between which the graphene layer is clad, when the characteristic inter-carrier collision frequency can reach high values. As a result, this frequency can be much greater than that of collisions with impurities or phonons. Similar situation occurs in the gated graphene at sufficiently high gate voltages when large electron or hole densities can be induced as well as at strong optical pumping of graphene. The hydrodynamic models of the electron-hole systems in different structures and devices based on the standard semiconductors with parabolic and near-parabolic energy spectra of electrons and holes are widely used (see, for instance, Ref. [1]). However, in case of graphene, such models and the pertinent equations should be revised due to the linear energy spectra of electrons and holes. A hydrodynamic approach was recently used to describe the stationary transport processes in graphene [2]. However, the role of a strong electron-hole scattering, which, as shown in our work, is crucial, was not addressed.

In this chapter, we develop a strict hydrodynamic model for electron-hole plasma in graphene and demonstrate its workability in some applications. The chapter is organized as follows. In Section 6.2, we derive the hydrodynamic equations (continuity equations, Euler equations, and energy transfer equations) for graphene from the Boltzmann–Vlasov kinetic equations for massless electrons and holes assuming high carrier-carrier collision frequencies. Section 6.3 deals with the application of the derived equations for the calculations of graphene dc conductivity taking into account the effect of electron-hole drag. In Section 6.4, the collective excitations in graphene at different conditions are considered using the obtained hydrodynamic equations. In particular, we show that two types of weakly damping excitations can exist in the electron-hole plasma: (1) electron (or hole) plasma waves in gated graphene in the state with the Fermi level far from the Dirac point and (2) quasi-neutral electron-hole sound waves in the bipolar electron-hole plasma. The damping of both plasma waves (in virtually monopolar plasma) and electron-hole sound waves (in bipolar plasma) is determined by the scattering on disorder, while the electron-hole scattering almost does not affect the damping. In contrast, the plasma waves in bipolar electron-hole systems exhibit a strong damping due to electron-hole scattering processes. In Section 6.5, we draw the main conclusions. Some cumbersome formulas are singled out to the Appendix.

6.2 Derivation of Hydrodynamic Equations

For the massless electrons in holes in graphene the spectrum is linear $\varepsilon(p) = v_F p$. Hence, the kinetic equations governing the distribution functions $f_e = f_e(\mathbf{p})$ and $f_h = f_h(\mathbf{p})$ read, respectively,

$$\begin{aligned}\frac{\partial f_e}{\partial t} + v_F \frac{\mathbf{p}}{p}\frac{\partial f_e}{\partial \mathbf{r}} + e\frac{\partial \varphi}{\partial \mathbf{r}}\frac{\partial f}{\partial \mathbf{p}} \\ = St\{f_e, f_e\} + St\{f_e, f_h\} + St_i\{f_e\},\end{aligned} \tag{6.1}$$

$$\frac{\partial f_{\mathrm{h}}}{\partial t}+\upsilon_{\mathrm{F}}\frac{\mathbf{p}}{p}\frac{\partial f_{\mathrm{h}}}{\partial \mathbf{r}}-e\frac{\partial \varphi}{\partial \mathbf{r}}\frac{\partial f}{\partial \mathbf{p}}$$
$$=St\{f_{\mathrm{h}},f_{\mathrm{h}}\}+St\{f_{\mathrm{h}},f_{\mathrm{e}}\}+St_{\mathrm{i}}\{f_{\mathrm{h}}\}. \tag{6.2}$$

Here $\upsilon_{\mathrm{F}}\mathbf{p}/p = \partial\varepsilon(p)/\partial\mathbf{p}$ is the electron (hole) velocity, $\upsilon_{\mathrm{F}} \simeq 10^8$ cm/s is the characteristic velocity of electrons and holes in graphene (Fermi velocity), $e = |e|$ is the absolute value of the electron charge, $\mathbf{E} = -\partial\varphi/\partial\mathbf{r}$ is the electric field, $St_i\{f_{\mathrm{e}}\}$ and $St_i\{f_{\mathrm{h}}\}$ are the collision integrals of electrons and holes, respectively, with disorder (impurities and phonons); $St\{f_{\mathrm{e}}, f_{\mathrm{e}}\}$, $St\{f_{\mathrm{h}}, f_{\mathrm{h}}\}$, and $St\{f_{\mathrm{e}}, f_{\mathrm{h}}\}$ are the inter-carrier collision integrals. In Eqs. (6.1) and (6.2) we have neglected the recombination terms assuming that the recombination rate is much smaller than that of collisions and the frequency of the plasma waves under consideration.

As known [3], the Fermi distribution functions of electrons and holes

$$f_{\mathrm{e}}(\mathbf{p})=\left[1+\exp\left(\frac{\varepsilon(p)-\mathbf{p}\cdot\mathbf{V}_{\mathrm{e}}-\mu_{\mathrm{e}}}{T}\right)\right]^{-1}, \tag{6.3}$$

$$f_{\mathrm{h}}(\mathbf{p})=\left[1+\exp\left(\frac{\varepsilon(p)-\mathbf{p}\cdot\mathbf{V}_{\mathrm{h}}-\mu_{\mathrm{h}}}{T}\right)\right]^{-1}, \tag{6.4}$$

turn the electron-electron and hole-hole collision integrals to zero owing to the conservation of momentum and energy in the inter-carrier collisions. In Eqs. (6.3) and (6.4) $\mathbf{V}_{\mathrm{e}}$ and $\mathbf{V}_{\mathrm{h}}$ are the average (drift) velocities of electrons and holes, respectively, μ_{e} and μ_{h} are electron and hole chemical potentials, and the temperature T is measured in energy units.

Hereafter we consider rather small drift velocities and perform an expansion of Eqs. (6.3) and (6.4) over $\mathbf{V}_{\mathrm{e}}$ and $\mathbf{V}_{\mathrm{h}}$:

$$f_{\mathrm{e}}(\mathbf{p})=f_{\mathrm{e},0}-\frac{\partial f_{\mathrm{e},0}}{\partial\varepsilon}\mathbf{p}\cdot\mathbf{V}_{\mathrm{e}}, \tag{6.5}$$

$$f_{\mathrm{e}}(\mathbf{p})=f_{\mathrm{h},0}-\frac{\partial f_{\mathrm{h},0}}{\partial\varepsilon}\mathbf{p}\cdot\mathbf{V}_{\mathrm{h}}. \tag{6.6}$$

Here $f_{\mathrm{e},0}$ and $f_{\mathrm{h},0}$ stand for the functions, given by Eqs. (6.3) and (6.4), with $\mathbf{V}_{\mathrm{e}} = \mathbf{V}_{\mathrm{h}} = 0$. The distribution functions (6.5) and (6.6) still turn the electron-electron and hole-hole collision integrals to zero and, at the same moment, conserve the number of particles (the density is the same in moving and stationary frames).

If the electric field is sufficiently weak and the inhomogenity of the electron-hole plasma is small (as it will be assumed in the following), the electron-hole collision integrals can be expanded over $\mathbf{V}_e$ and $\mathbf{V}_h$ and presented as $St\{f_e, f_h\} = -St\{f_h, f_e\} \simeq (\mathbf{V}_h - \mathbf{V}_e) \cdot \mathbf{S}\{f_{e,0}, f_{h,0}\}$ (for details see the Appendix). In this form, the electron-hole collision terms describe the friction between the electron and hole plasma components. Similarly, the terms corresponding to the collisions of electrons and holes with disorder can be presented in the forms $St_i\{f_e\} = -\mathbf{V}_e \cdot \mathbf{S}_i\{f_{e,0}\}$ and $St_i\{f_h\} = -\mathbf{V}_h \cdot \mathbf{S}_i\{f_{h,0}\}$. Here $\mathbf{S}$ and $\mathbf{S}_i$ are the functionals of the distribution functions $f_{e,0}$ and $f_{h,0}$.

At small values of electron and hole average velocities $\mathbf{V}_e$ and $\mathbf{V}_h$, the friction terms can be considered as perturbations in comparison with the electron-electron and hole-hole collision terms. Thus the distribution functions given by Eqs. (6.3) and (6.4) are the approximate solutions of Eqs. (6.1) and (6.2). Then the quantities $\mathbf{V}_e$, $\mathbf{V}_h$, μ_e, and μ_h (or the electron and hole sheet densities, Σ_e and Σ_h) can be found considering the terms in the left-hand sides of Eqs. (6.1) and (6.2) and the friction terms as perturbations, using the standard procedure (akin to the Chapman–Enskog method [4] for the derivation of the hydrodynamic equations from the kinetic equations).

On integrating Eqs. (6.1) and (6.2) over $d\Gamma_{\mathbf{p}} = gd^2\mathbf{p}/(2\pi\hbar)^2$ (where $g = 4$ is the electron degeneracy factor in graphene), we obtain the continuity equations for electrons and holes

$$\frac{\partial \Sigma_e}{\partial t} + \frac{\partial \Sigma_e \mathbf{V}_e}{\partial \mathbf{r}} = 0, \quad \frac{\partial \Sigma_h}{\partial t} + \frac{\partial \Sigma_h \mathbf{V}_h}{\partial \mathbf{r}} = 0. \tag{6.7}$$

To derive the Euler equations, one should integrate Eqs. (6.1) and (6.2) times $\mathbf{p}$ over $d\Gamma_{\mathbf{p}}$. In the case of parabolic dispersion, one could multiply the Boltzmann equation either by velocity or momentum as they are proportional to each other. For linear dispersion the choice of momentum is crucial. Just the momentum is conserved in particle-particle collisions unlike to the velocity which can be changed. After integration one obtains the system of Euler equations for electrons and holes:

$$\frac{3}{2}\frac{\partial}{\partial t}\frac{\langle p_\mathrm{e}\rangle \mathbf{V}_\mathrm{e}}{\upsilon_\mathrm{F}}+\frac{\partial}{\partial \mathbf{r}}\frac{\upsilon_\mathrm{F}\langle p_\mathrm{e}\rangle}{2}-e\Sigma_\mathrm{e}\frac{\partial\varphi}{\partial \mathbf{r}} = -\beta_\mathrm{e}\mathbf{V}_\mathrm{e}-\beta_\mathrm{eh}(\mathbf{V}_\mathrm{e}-\mathbf{V}_\mathrm{h}), \tag{6.8}$$

$$\frac{3}{2}\frac{\partial}{\partial t}\frac{\langle p_\mathrm{h}\rangle \mathbf{V}_\mathrm{h}}{\upsilon_\mathrm{F}}+\frac{\partial}{\partial \mathbf{r}}\frac{\upsilon_\mathrm{F}\langle p_\mathrm{h}\rangle}{2}+e\Sigma_\mathrm{h}\frac{\partial\varphi}{\partial \mathbf{r}} = -\beta_\mathrm{h}\mathbf{V}_\mathrm{h}-\beta_\mathrm{eh}(\mathbf{V}_\mathrm{h}-\mathbf{V}_\mathrm{e}). \tag{6.9}$$

Here the angle brackets denote an integration over the equilibrium Fermi distribution functions, in particular,

$$\langle p_\mathrm{e}\rangle = \int_0^\infty \left[1+\exp\left(\frac{\upsilon_\mathrm{F}p-\mu_\mathrm{e}}{T}\right)\right]^{-1}\frac{2\pi g p^2 dp}{(2\pi\hbar)^2}$$

is the momentum modulus per unit area, the friction coefficients β_eh, β_e, and β_h are the functions of the non-perturbed (steady-state) values of the chemical potentials $\mu_{\mathrm{e},0}$ and $\mu_{\mathrm{h},0}$.

One can rewrite Euler equations in a classical form on introducing the fictitious carrier masses

$$M_\mathrm{e} = \frac{\langle p_\mathrm{e}\rangle}{\upsilon_\mathrm{F}\Sigma_\mathrm{e}},\quad M_\mathrm{h} = \frac{\langle p_\mathrm{h}\rangle}{\upsilon_\mathrm{F}\Sigma_\mathrm{h}},$$

which are estimated as 0.016 of the free electron mass at $\mu_\mathrm{e} = \mu_\mathrm{h} = 0$ and $T = 300$ K.

The detailed derivation of the friction coefficients can be found in the Appendix; here, we write down only the final expressions in several limits. The electron-hole friction coefficient β_eh can be represented as

$$\beta_\mathrm{eh} = A\frac{T^4e^4}{\hbar^5\upsilon_\mathrm{F}^6\kappa^2}\cdot I\left(\frac{\mu_\mathrm{e}}{T},\frac{\mu_\mathrm{h}}{T}\right). \tag{6.10}$$

where κ is the effective permittivity of environment (the substrate and gate dielectric), the dimensionless constant A of the order of unity and the function I $(\mu_\mathrm{e}/T,\ \mu_\mathrm{h}/T)$ can be obtained after proper linearization of the electron-hole collision integral (see the Appendix). For intrinsic graphene $I\,(0, 0) = 1$, while for monopolar plasma in gated graphene $I\,(\mu/T, \mu/T) \propto e^{-\mu/T}$ due to the exponentially small number of holes. In the current chapter, we shall adopt the following interpolation for $\beta_{\rm eh}$ which is valid in the limiting cases of monopolar plasma and intrinsic graphene:

$$\beta_{\rm eh} = \frac{\nu_{\rm eh}}{v_{\rm F}} \frac{\langle p_{\rm e}\rangle\langle p_{\rm h}\rangle}{\langle p_{\rm e}\rangle+\langle p_{\rm h}\rangle} = \nu_{\rm eh} \frac{\Sigma_{\rm e} M_{\rm e} \Sigma_{\rm h} M_{\rm h}}{\Sigma_{\rm e} M_{\rm e} + \Sigma_{\rm h} M_{\rm h}}. \tag{6.11}$$

Here we have introduced the electron-hole collision frequency $\nu_{\rm eh}$ to be estimated below in the Section 6.3.

When the acoustic phonon scattering dominates, the coefficients $\beta_{\rm e}$ and $\beta_{\rm h}$ can be presented as [5, 7]

$$\beta_{\rm e} = \frac{D^2 T\langle p_{\rm e}^2\rangle}{4\rho s^2 \hbar^3 v_{\rm F}^2}, \quad \beta_{\rm h} = \frac{D^2 T\langle p_{\rm h}^2\rangle}{4\rho s^2 \hbar^3 v_{\rm F}^2},$$

where D is the deformation potential constant, ρ is the sheet density of graphene, and s is the sound velocity. Due to the existence of several acoustic phonon branches and considerable discrepancy in experimental data of graphene constants [5–7], it is reasonable to use semi-phenomenological formulas:

$$\beta_{\rm e} = \lambda T\langle p_{\rm e}^2\rangle, \quad \beta_{\rm h} = \lambda T\langle p_{\rm h}^2\rangle, \tag{6.12}$$

and extract the numerical value of λ from experimental data on dc conductivity (see Section 6.3).

For scattering on charged impurities, the friction coefficients $\beta_{\rm e}$ and $\beta_{\rm h}$ are reasonably proportional to the carrier densities and the density of charged impurities Σ_i

$$\beta_{\rm e} \propto \Sigma_{\rm e}\Sigma_{\rm i}, \qquad \beta_{\rm h} \propto \Sigma_{\rm h}\Sigma_{\rm i}. \tag{6.13}$$

In general, the density of charged impurities depends on the chemical potentials and temperature.

At last, the system of hydrodynamic equations should be supplemented with the energy transfer equations. Introducing the energy density $\langle \varepsilon_{\rm e,h}\rangle = v_{\rm F}\langle p_{\rm e,h}\rangle$, the pertinent equations can be written down as

$$\frac{\partial\langle\varepsilon_{\rm e}\rangle}{\partial t} + \frac{3}{2}\frac{\partial\langle\varepsilon_{\rm e}\rangle\mathbf{V}_{\rm e}}{\partial\mathbf{r}} - e\Sigma_{\rm e}\mathbf{V}_{\rm e}\frac{\partial\varphi}{\partial\mathbf{r}} = Q_{\rm e}, \tag{6.14}$$

$$\frac{\partial\langle\varepsilon_{\rm h}\rangle}{\partial t} + \frac{3}{2}\frac{\partial\langle\varepsilon_{\rm h}\rangle\mathbf{V}_{\rm h}}{\partial\mathbf{r}} + e\Sigma_{\rm h}\mathbf{V}_{\rm h}\frac{\partial\varphi}{\partial\mathbf{r}} = Q_{\rm h}, \tag{6.15}$$

where the heat sink rates Q_e and Q_h are proportional to the difference between electron and phonon temperatures. However, in the present chapter we neglect the heating as we deal with low-field dc conductivity and high-frequency plasma waves.

6.3 Effect of Electron-Hole Drag and DC Conductivity of Gated Graphene

As one of the demonstrations of the hydrodynamic equations applications we calculate the dc conductivity of gated graphene. It is assumed that the electric potential and the charge density are related by the Poisson equation. In the gradual channel approximation [8]

$$C(V_G - \varphi) = e(\Sigma_e - \Sigma_h). \tag{6.16}$$

Here $C = \kappa/4\pi d$ is the specific capacitance per unit area, d and κ are the thickness and permittivity of the gate dielectric, respectively, and V_G is the gate voltage.

To calculate the dc conductivity we rewrite the Euler equations for the steady-state situation in terms of electrochemical potentials:

$$\Sigma_e \frac{\partial(\mu_e - e\varphi)}{\partial \mathbf{r}} = -\beta_e \mathbf{V}_e - \beta_{eh}(\mathbf{V}_e - \mathbf{V}_h), \tag{6.17}$$

$$\Sigma_h \frac{\partial(e\varphi - \mu_h)}{\partial \mathbf{r}} = -\beta_h \mathbf{V}_h - \beta_{eh}(\mathbf{V}_h - \mathbf{V}_e), \tag{6.18}$$

where the derivative of electric potential is associated with drift current, as well as the derivative of chemical potential is associated with diffusion current. In the following we restrict our calculations by the consideration of the drift current only and, therefore, omit the terms $\partial\mu_{e,h}/\partial\mathbf{r}$ in Eqs. (6.17) and (6.18). As a result, we arrive at the following expressions for the mean (drift) velocities of electrons and holes:

$$\mathbf{V}_e = -\left[\frac{(\Sigma_e - \Sigma_h)\beta_{eh} + \beta_h\Sigma_e}{\beta_{eh}(\beta_e + \beta_h) + \beta_e\beta_h}\right]\frac{\partial\varphi}{\partial\mathbf{r}}, \tag{6.19}$$

$$\mathbf{V}_h = \left[\frac{(\Sigma_h - \Sigma_e)\beta_{eh} + \beta_e\Sigma_h}{\beta_{eh}(\beta_e + \beta_h) + \beta_e\beta_h}\right]\frac{\partial\varphi}{\partial\mathbf{r}}. \tag{6.20}$$

Substituting the quantities $\mathbf{V}_e$ and $\mathbf{V}_h$ from Eqs. (6.19) and (6.21) into the general expression for current density $\mathbf{j} = e(\Sigma_h \mathbf{V}_h - \Sigma_e \mathbf{V}_e) = G(-\partial\varphi/\partial\mathbf{r})$, one obtains the following formula for the conductivity G:

$$G = \frac{e^2(\Sigma_e - \Sigma_h)^2}{\beta_e + \beta_h + \beta_e\beta_h / \beta_{eh}} + \frac{e^2(\Sigma_e^2\beta_h + \Sigma_h^2\beta_e)}{\beta_{eh}(\beta_e + \beta_h) + \beta_e\beta_h}. \tag{6.21}$$

The first term in the right-hand side of Eq. (6.21) is associated with the scattering of electrons and holes on impurities and phonons. This term turns into zero at the Dirac point. However, far from the Dirac point, i.e., in purely electron or hole plasma, this term dominates. Meanwhile, the second term is due to the contribution of the electron-hole friction. The latter results in a high resistivity of graphene at the Dirac point and its vicinity.

To calculate the graphene minimal conductivity, we reasonably assume that at the Dirac point scattering between electrons and holes prevails: $\beta_e \ll \beta_{eh}$, $\beta_h \ll \beta_{eh}$. Plugging Eq. (6.10) for β_{eh} into Eq. (6.21) one obtains an expression for intrinsic graphene conductivity

$$G_0 = \frac{e^2\Sigma_0^2}{\beta_{eh}} \propto \frac{\hbar v_F^2 k^2}{e^2}. \tag{6.22}$$

Strikingly, the intrinsic graphene conductivity (minimum conductivity) does not depend on temperature. This is in agreement with the experimental results [9] demonstrating a constant conductivity over a broad range of temperature from 0.3 to 300 K in which the carrier density varies by 6 orders of magnitude. Worth mentioning the Eq. (6.22) was previously obtained using the scaling theory, [10], and also via thorough description of electron-hole scattering [11].

Earlier, the minimum conductivity of graphene was calculated under the assumption of strong interaction among carriers [12–14]. The strength of interaction is governed by the "fine-structure constant" $\alpha = e^2/\kappa\hbar v_F$ which is equal to 2.2 for intrinsic (suspended) graphene ($\kappa = 1$). The strong-interaction theories bind the conductivity of graphene to the conductance quantum e^2/h, regardless of the permittivity of environment. As the parameter α is about unity, both theories of weak and strong interaction give rise to close values of conductivity. However, in prospective graphene structures the screening caused by dielectrics and nearby gates make the value of α less than unity, therefore, the weak-interaction theories become more adequate to the situation.

After substituting Eq. (6.11) for β_{eh} into Eq. (6.22) one can rewrite that equation in the conventional form appropriate for the estimation of the electron-hole collision frequency ν_{eh}:

$$G_0 = \frac{2\Sigma_0 e^2}{M_0 \nu_{\mathrm{eh}}}, \qquad (6.23)$$

where M_0 is the fictitious mass of electrons and holes in intrinsic graphene dependent on temperature. Comparing G_0 given by Eq. (6.23) with the experimental value $G_0 \simeq (6\ \mathrm{k}\Omega)^{-1}$, one can estimate the electron-hole collision frequency as $\nu_{\mathrm{eh}} = 3 \times 10^{13}\ \mathrm{s}^{-1}$. Such a high value justifies an employment of the hydrodynamic model.

In the monopolar limit ($\Sigma_{\mathrm{e}} \gg \Sigma_{\mathrm{h}}$), the expression for the conductivity could be also simplified:

$$G = \frac{e^2 \Sigma_{\mathrm{e}}^2}{\beta_{\mathrm{e}}}. \qquad (6.24)$$

Equation (6.24) allows to estimate the coefficient λ in the phonon collision term given by Eq. (6.12). It is widely assumed that in suspended graphene samples or twisted graphene stacks the conductivity and mobility are limited by phonon scattering only. For instance [16], the carrier mobility B, defined as

$$B = \frac{G}{e\left|\Sigma_{\mathrm{e}} - \Sigma_{\mathrm{h}}\right|},$$

reaches the value of 120,000 $\mathrm{cm}^2\ \mathrm{V}^{-1}\ \mathrm{s}^{-1}$ at $\Sigma_{\mathrm{e}} - \Sigma_{\mathrm{h}} = 2 \times 10^{11}\ \mathrm{cm}^{-2}$ and the temperature $T = 240$ K (when the chemical potential $\mu \simeq$ 0.028 eV). Hence, in the monopolar limit, one obtains

$$B \simeq \frac{e}{\beta_{\mathrm{e}} \Sigma_{\mathrm{e}}} = \frac{e \Sigma_{\mathrm{e}}}{\lambda T \left\langle p_{\mathrm{e}}^2 \right\rangle}. \qquad (6.25)$$

Equation (6.25) yields $\lambda \simeq 2.4 \times 10^{54}\ \mathrm{cm\ s^{-1} J^{-2}}$. Accordingly, the characteristic collision frequency $\nu_{\mathrm{e}} = 2\upsilon_{\mathrm{F}}\beta_{\mathrm{e}}/3\langle p_{\mathrm{e}}\rangle$ at room temperature varies from $8.6 \times 10^{11}\ \mathrm{s}^{-1}$ at the Dirac point to $3.5 \times 10^{12}\ \mathrm{s}^{-1}$ at $V_{\mathrm{G}} = 10$ V and $d = 10$ nm.

Figure 6.1 demonstrates the dependence of the low-field dc conductivity G of graphene on the gate voltage V_{G}, plotted using Eqs.

(6.11), (6.12), and (6.21). The voltage dependence of the graphene resistivity and the dependence of the carrier mobility on the chemical potential μ [given by Eq. (6.25)] are shown in the insets in Fig. 6.1. In the case when scattering on acoustic phonons dominates, the conductivity tends to saturation at high gate voltages and the phonon-limited mobility decreases. To describe the conductivity of realistic graphene structures, one should account for other scattering objects: charged impurities, interfacial phonons, bulk phonons, etc. For example, scattering on charged impurities results in the following trends of conductivity curves in monopolar case: $G(V_G) \propto V_G$ if the density of impurities is constant and $G(V_G) \propto \sqrt{V_G}$ if the distribution of impurity levels is uniform in an energy scale [10]. The latter fact manifests a decrease in defect scattering at low temperatures.

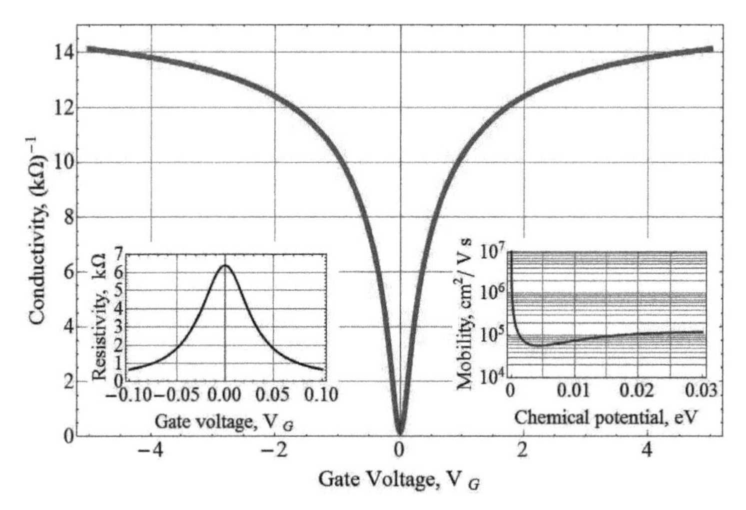

Figure 6.1 Conductivity of graphene G vs. gate voltage at $T = 300$ K, $d = 10$ nm, and $\kappa = 4$. Insets: left panel—resistivity G^{-1} near the Dirac point vs. gate voltage, right panel—mobility B vs. chemical potential.

The conductivity and mobility curves in Fig. 6.1 exhibit a good agreement with the experimental works, where phonon-limited mobility was investigated [16–18]. As for the theoretical approaches to the description of graphene conductivity, significant efforts have been focused on the calculation of the minimum conductivity and the conductivity far from the neutrality point [5, 7, 19]. The conductivity at an arbitrary value of carrier density was considered in Refs. [14, 15] without specifying the mechanism of scattering. Our model provides an opportunity to calculate the transport characteristics of graphene at any charge carrier density taking into account the definite scattering processes. Above all, we show that the effect of electron-hole drag significantly impacts the transport properties of graphene and leads to an abrupt drop of graphene resistivity outside of the Dirac point.

It is remarkable that the drag effect is crucial even at a small mismatch in the electron and hole densities [10]. The conductivity in the regime of drag almost does not depend on electron-hole scattering, although it still remains the strongest one in the system. Indeed, the phonon (impurity) scattering term in the expression for conductivity (6.21) dominates under the condition

$$\frac{|\Sigma_{\mathrm{h}} - \Sigma_{\mathrm{e}}|}{\Sigma_{\mathrm{e}}} > \sqrt{\frac{\beta_{\mathrm{h}}}{\beta_{\mathrm{eh}}}}\Bigg|_{\mu=0}. \tag{6.26}$$

For the preceding estimations, the ratio given by Eq. (6.26) is evaluated as 0.3 at room temperature. Providing Eq. (6.26) is satisfied, the difference between electron and hole velocities [Eq. (6.19)] becomes rather small:

$$\frac{|V_{\mathrm{h}} - V_{\mathrm{e}}|}{V_{\mathrm{h}} + V_{\mathrm{e}}} \simeq \frac{1}{2}\sqrt{\frac{\beta_{\mathrm{h}}}{\beta_{\mathrm{eh}}}}\Bigg|_{\mu=0}. \tag{6.27}$$

The difference between velocities still exists because of the opposite charges of electrons and holes and, therefore, the opposite directions of the electric forces. In principle, for the drag regime the two-fluid description of graphene system might be reduced to a single electron-hole fluid with renormalized charge of particles according to Eq. (6.26). However, it seems preferable to retain two-fluid model to cover all possible situations.

It is readily seen from above mentioned that a narrow resistivity peak in Fig. 6.1 could be explained by the drag effect. Moreover, stronger scattering leads to a wider resistivity peak. The latter is in concordance with the experimental data [17].

6.4 Plasma and Electron-Hole Sound Waves in Graphene

6.4.1 General Dispersion Relation for Collective Excitations

Below we demonstrate the application of the hydrodynamic equations for the calculation of the spectra of collective excitations in electron-hole system. We use the same gradual channel approximation as in the previous section [Eq. (6.16)]. To derive the spectra, one can apply the common technique of small-signal analysis assuming that

$$\mu_{\mathrm{e}} = \mu_{\mathrm{e},0} + \delta\mu_{\mathrm{e}} e^{i(kx-\omega t)}, \qquad \mu_{\mathrm{h}} = \mu_{\mathrm{h},0} + \delta\mu_{\mathrm{h}} e^{i(kx-\omega t)},$$
$$\delta V_{\mathrm{e}} = \delta V_{\mathrm{e},0} e^{i(kx-\omega t)}, \qquad \delta V_{\mathrm{h}} = \delta V_{\mathrm{h},0} e^{i(kx-\omega t)},$$
$$\varphi = \varphi_0 + \delta\varphi e^{i(kx-\omega t)},$$

where $\delta\mu_{\mathrm{e}}$, $\delta\mu_{\mathrm{h}}$, $\delta V_{\mathrm{e},0}$, $\delta V_{\mathrm{h},0}$, and $\delta\varphi$ are the amplitudes of alternating variations. The non-perturbed chemical potentials $\mu_{\mathrm{e},0}$ and $\mu_{\mathrm{h},0}$ coincide if the steady state is an equilibrium one and are determined by the doping and the gate potential. In particular, at $V_{\mathrm{G}} = 0$ and zero doping the Fermi level is located in the Dirac point. In intrinsic graphene under optical interband pumping the chemical potentials of electrons and holes can be rather different. For example, at $V_{\mathrm{G}} = 0$, $\mu_{\mathrm{e},0} = \mu_0 > 0$, and $\mu_{\mathrm{h}} = -\mu_0 < 0$, where μ_0 is determined by the intensity of pumping.

From the linearized versions of Eqs. (6.7–6.9) as well as Eq. (6.16) the following system of algebraic equations arises:

$$\left[-i\omega \frac{3}{2}\frac{\langle p_{\mathrm{e}} \rangle}{v_{\mathrm{F}}} + \frac{ik^2\Sigma_{\mathrm{e}}^2}{\omega}\left(\frac{v_F}{\langle p_{\mathrm{e}}^{-1} \rangle} + \frac{e^2}{C} \right) + \beta_{\mathrm{e}} + \beta_{\mathrm{eh}} \right] \delta V_{\mathrm{e}}$$
$$- \left[\frac{ik^2\Sigma_{\mathrm{e}}\Sigma_{\mathrm{h}}}{\omega}\frac{e^2}{C} + \beta_{\mathrm{eh}} \right] \delta V_{\mathrm{h}} = 0, \tag{6.28}$$

$$\left[-i\omega\frac{3}{2}\frac{\langle p_{\mathrm{h}}\rangle}{\upsilon_{\mathrm{F}}}+\frac{ik^2\Sigma_{\mathrm{h}}^2}{\omega}\left(\frac{\upsilon_F}{\langle p_{\mathrm{h}}^{-1}\rangle}+\frac{e^2}{C}\right)+\beta_{\mathrm{h}}+\beta_{\mathrm{eh}}\right]\delta V_{\mathrm{h}}$$

$$-\left[\frac{ik^2\Sigma_{\mathrm{e}}\Sigma_{\mathrm{h}}}{\omega}\frac{e^2}{C}+\beta_{\mathrm{eh}}\right]\delta V_{\mathrm{h}}=0. \tag{6.29}$$

The solvability condition for Eqs. (6.28) and (6.29) results in the general dispersion relation for the collective excitations:

$$\left[-i\omega+\frac{ik^2\upsilon_{\mathrm{e}}^2}{\omega}(1+r_{\mathrm{e}})+\nu_{\mathrm{e}}+\upsilon_{\mathrm{F}}\frac{2\beta_{\mathrm{eh}}}{3\langle p_{\mathrm{e}}\rangle}\right]$$

$$\left[-i\omega+\frac{ik^2\upsilon_{\mathrm{h}}^2}{\omega}(1+r_{\mathrm{h}})+\nu_{\mathrm{h}}+\upsilon_{\mathrm{F}}\frac{2\beta_{\mathrm{eh}}}{3\langle p_{\mathrm{h}}\rangle}\right]$$

$$=\left[\frac{ik^2\upsilon_{\mathrm{e}}^2\Sigma_{\mathrm{h}}}{\omega\Sigma_{\mathrm{e}}}r_{\mathrm{e}}+\frac{\beta_{\mathrm{eh}}\upsilon_{\mathrm{F}}}{\langle p_{\mathrm{e}}\rangle}\right]\left[\frac{ik^2\upsilon_{\mathrm{h}}^2\Sigma_{\mathrm{e}}}{\omega\Sigma_{\mathrm{h}}}r_{\mathrm{h}}+\frac{\beta_{\mathrm{eh}}\upsilon_{\mathrm{F}}}{\langle p_{\mathrm{h}}\rangle}\right] \tag{6.30}$$

Here for brevity we have introduced the dimensionless constants

$$r_{\mathrm{e}}=\frac{e^2\langle p_{\mathrm{e}}^{-1}\rangle}{C\upsilon_{\mathrm{F}}},\quad r_{\mathrm{h}}=\frac{e^2\langle p_{\mathrm{h}}^{-1}\rangle}{C\upsilon_{\mathrm{F}}},$$

the squared characteristic velocities

$$\upsilon_{\mathrm{e}}^2=\frac{2\Sigma_{\mathrm{e}}^2\upsilon_{\mathrm{F}}^2}{3\langle p_{\mathrm{e}}\rangle\langle p_{\mathrm{e}}^{-1}\rangle},\quad \upsilon_{\mathrm{h}}^2=\frac{2\Sigma_{\mathrm{h}}^2\upsilon_{\mathrm{F}}^2}{3\langle p_h\rangle\langle p_{\mathrm{h}}^{-1}\rangle},$$

and the frequencies

$$\nu_{\mathrm{e}}=\frac{2\upsilon_{\mathrm{F}}\beta_{\mathrm{e}}}{3\langle p_{\mathrm{e}}\rangle},\quad \nu_{\mathrm{h}}=\frac{2\upsilon_{\mathrm{F}}\beta_{\mathrm{h}}}{3\langle p_{\mathrm{h}}\rangle},$$

The quantities r_{e} and r_{h} determine the ratio of the electrostatic energy to the kinetic energy in the wave. In the particular gated structures we have considered above (κ = 4, d = 10 nm) the dimensionless constant r_{e} varies from 1.9 at V_{G} = 0 to 45 at V_{G} = 10 V. The electron-phonon collision frequencies ν_{e} and ν_{h}, as shown below, determine the damping of the waves.

6.4.2 Analytical Solutions for Symmetric Bipolar and Monopolar Systems

Exact solutions of the general dispersion equation (6.30) can be acquired for symmetric bipolar plasma (optically pumped or intrinsic graphene) and monopolar plasma.

In symmetric bipolar plasma all quantities characterizing the electron and hole systems coincide (e.g., $\upsilon_e = \upsilon_h = \upsilon$, $\Sigma_e = \Sigma_- = \Sigma$, ...), therefore, we shall omit the subscripts in this certain case. In symmetric systems Eq. (6.30) provides two solutions:

$$\omega_- = -i\frac{\nu}{2} + \sqrt{k^2\upsilon^2 - \left(\frac{\nu}{2}\right)^2}, \tag{6.31}$$

$$\omega_+ = -i\left(\frac{\nu}{2} + \frac{\nu_{eh}}{3}\right) + \sqrt{k^2\upsilon^2(1+2r)^2 - \left(\frac{\nu}{2} + \frac{\nu_{eh}}{3}\right)^2}. \tag{6.32}$$

The second solution ω_+ represents the waves with opposite motion of electrons and holes leading to the strong damping. These waves correspond to the perturbations of charge density (plasma waves).

The branch given by Eq. (6.31) is of the most interest. Electrons and holes in these waves move in the same direction, i.e., the electron-hole plasma remains quasi-neutral. Due to the co-directional motion of the electron and hole components, their mutual collisions do not affect the wave damping. In the case, it originates from the collisions of electrons and holes with impurities and phonons. We can conclude that those waves are associated solely with the pressure gradient. In the following, such waves will be referred to as electron-hole sound waves in analogy with electron-ion sound waves in classical plasma.

The velocity of electron-hole sound is

$$s_- = \upsilon = \upsilon_F\sqrt{\frac{2\Sigma^2}{3\langle p\rangle\langle p^{-1}\rangle}}. \tag{6.33}$$

It does not depend on structure parameters and does not exceed the Fermi velocity for any electron and hole densities. If the Fermi level crosses the Dirac point the analytical expression for velocity is

$$s_- = \upsilon_F\frac{\pi^2}{18\sqrt{\ln 2\zeta(3)}} \simeq 0.6\upsilon_F,$$

where $\zeta(x)$ stands for Riemann zeta function.

The dispersion laws for monopolar plasma can be derived providing the inequalities $\Sigma_e \gg \Sigma_h$, $\beta_e \gg \beta_h$, and $r_e \gg 1 \gg r_h$ are satisfied:

$$\omega_+ = -i\frac{\nu_e}{2} + \sqrt{k^2 \upsilon_e^2 (1+r_e) - \left(\frac{\nu_e}{2}\right)^2}, \tag{6.34}$$

$$\omega_- = -i\left(\frac{\nu_h}{2} + \frac{\nu_{eh}}{3}\right) + \sqrt{k^2 \upsilon_h^2 (1+2r_h)^2 - \left(\frac{\nu_h}{2} + \frac{\nu_{eh}}{3}\right)^2}. \tag{6.35}$$

A detailed analysis of these solutions shows that the branches ω_+ and ω_- correspond to the oscillations of majority (electrons) and minority (holes) carriers, respectively. The minority carrier oscillations are strongly damped by electron-hole friction and, therefore, they are omitted in the further consideration.

The wave velocity of the electronic plasma oscillations s_+ is

$$s_+ = \upsilon_e \left(1+r_e\right)^{1/2} \simeq \upsilon_F \sqrt{4\alpha k_F d} \propto V_G^{1/4} d^{1/4}, \tag{6.36}$$

where $\alpha = e^2/\kappa\hbar\upsilon_F$ is the coupling constant ("fine-structure constant") and $k_F = \mu_e/\hbar\upsilon_F$ is the Fermi momentum of electrons. This velocity markedly exceeds the Fermi one and is primarily determined by the Coulomb interaction of carriers with the gate.

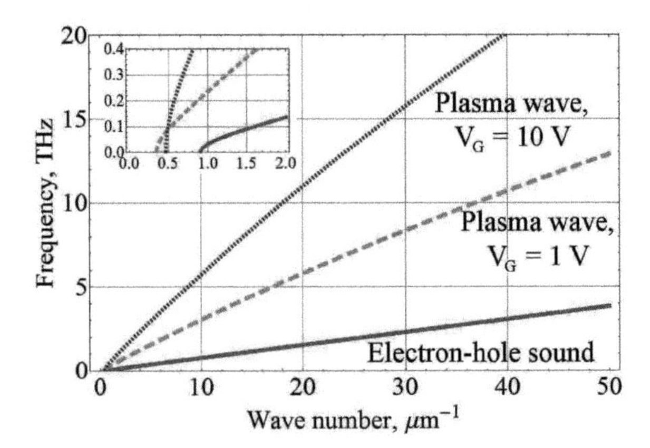

Figure 6.2 Dispersion of plasma and electron-hole sound waves at different gate voltages. Inset: non-linear dispersion of the waves at small frequencies resulting from collisions.

The dispersions for electron-hole sound in intrinsic graphene and plasma waves in gated graphene at high gate voltages are depicted in Fig. 6.2 in the THz range of frequencies. The curves are calculated under the assumption of equal chemical potentials $\mu_{e,0} = \mu_{h,0}$, i.e., when the non-perturbed state is an equilibrium one. The spectra exhibit non-linear dispersion at low frequencies originating from the damping (see the inset in Fig. 6.2).

Worth mentioning the dispersions are also non-linear at short wavelengths ($kd \sim 1$) when the gradual channel approximation becomes inapplicable. However, this limitation can be overcome if we use the rigorous solution of Poisson equation [21] instead of Eq. (6.16). For the considered gated structure the transition to the rigorous solution can be performed by the substitution:

$$C \to \frac{2\kappa k}{4\pi(1-e^{-2kd})}. \tag{6.37}$$

Equation (6.37) restrains the unlimited growth of the plasma wave velocities for large distances d between graphene sheet and metal gate.

6.4.3 Velocities and Damping Rates of the Waves

The wave velocities and damping rates can be calculated analytically in the linear domain of spectra, i.e., at ($v_{e,h} < \omega < s/d$) at an arbitrary value of electron and hole densities. Assuming $\omega = sk$, where s is the wave velocity, and plugging this into Eq. (6.30), one obtains two solutions s_- and s_+:

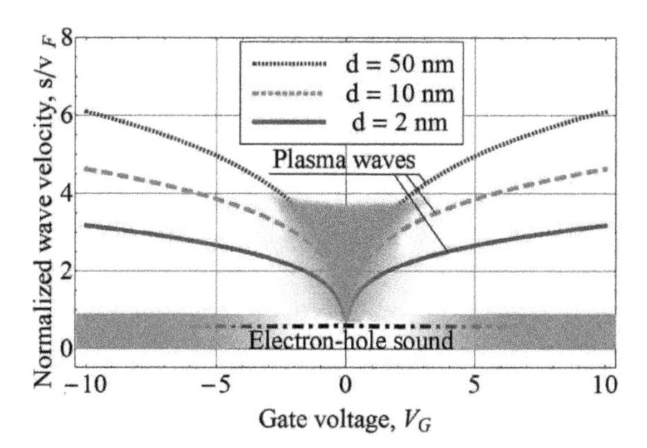

Figure 6.3 Velocities of plasma waves vs. gate voltage calculated for different gate layer thicknesses [Eq. (6.36)]. Dash-dotted line corresponds to the electron-hole sound velocity in the vicinity of the neutrality point. Regions of strong damping are filled.

$$s_{\pm}^{2}=\frac{1}{2}\Big[v_{e}^{2}(1+r_{e})+v_{h}^{2}(1+r_{h})\Big]$$

$$\pm\frac{1}{2}\sqrt{\Big[v_{e}^{2}(1+r_{e})-v_{h}^{2}(1+r_{h})\Big]^{2}+4v_{e}^{2}v_{h}^{2}r_{e}r_{h}}, \quad (6.38)$$

consistent with Eqs. (6.33) and (6.36).

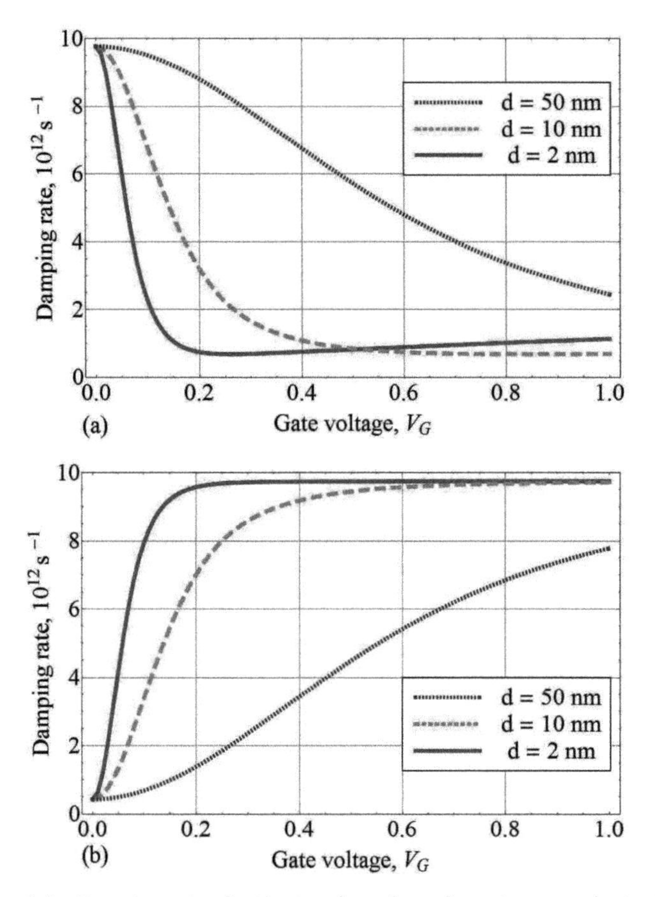

Figure 6.4 Damping rates for the two branches of spectrum: ω_+ (a, top) and ω_- (b, bottom).

In the vicinity of Dirac point the waves with the lower velocity s_- correspond to the electron-hole sound, while in the monopolar plasma they turn into the oscillations of minority carriers. The waves with the higher velocity s_+ behave as plasma waves at any electron and hole densities. The solutions given by Eq. (6.38) are plotted in Fig. 6.3 as the functions of gate voltage with the assumption of equal steady-state chemical potentials ($\mu_{e,0} = \mu_{h,0}$). One can readily see that the velocity s_- almost does not depend on the gate voltage applied, while the velocity s_+ exhibits unlimited growth at high gate voltages $s_+ \propto V_G^{1/4}$.

To calculate the damping rates of the waves we assume $\omega_\pm = s_\pm k + i\gamma_\pm$, where $\gamma_\pm$ characterizes the damping rate of the waves. The obtained damping rates $\gamma_\pm$ vs. the gate voltage are demonstrated in Fig. 6.4. It is clearly seen that in symmetric bipolar plasma the electron-hole sound branch ω_- exhibits weak damping. In the monopolar case ($\Sigma_e \gg \Sigma_h$) the plasma wave, corresponding to the oscillations of the majority carriers, is weakly damped. The minimum damping rate for the waves considered is of the order of 5×10^{11} s^{-1} and is determined by electron-phonon (or hole-phonon) collision frequency if the only scattering mechanism is acoustic phonon scattering. At high gate voltages the damping coefficient γ_+ grows linearly in accordance with the expression for characteristic collision frequency ν_e. The maximum damping rate of the waves is of the order of electron-hole collision frequency ν_{eh}. In accordance with these calculations, the regions of strong damping in Fig. 6.3 are marked with filling.

6.4.4 Comparison with Other Models

Plasma waves in gated graphene were discussed in Refs. [20, 21] using the kinetic approach under assumption of collisionless transport. As expected, formally setting the collision frequency zero in Eq. (6.34), we obtain the spectrum of plasma waves, derived previously [21]. Meanwhile, in the kinetic approach the existence of the electron-hole sound could not be predicted as it was assumed that the wave velocity should overcome Fermi velocity.

The dispersion curves of the plasma waves obtained via the hydrodynamic and kinetic theories are compared in Fig. 6.5 for the same parameters of the gated structure ($\kappa = 4$, d =10 nm), gate voltage $V_G = 10$ V, and three different collision frequencies. The only difference between the curves is in the non-linear part of spectra originating from collisions.

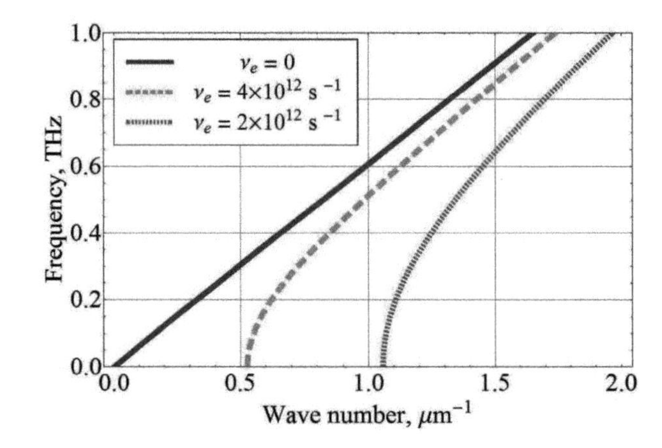

Figure 6.5 Comparison of dispersions for plasma waves, calculated using kinetic model [21] (disregarding collisions with impurities and phonons, solid line) and hydrodynamic model for $V_G = 10$ V, $d = 10$ nm, $\kappa = 4$, and three different collision frequencies.

Plasmons in graphene were also intensively studied within the analysis of the polarizability function $\prod(q, \omega)$ in random-phase approximation [22–24]. In particular, the square-root $\omega \propto \sqrt{k}$ and linear $\omega \propto k$ dispersions were obtained for the non-gated and gated graphene, consequently. To obtain the plasma wave dispersion for the non-gated graphene in hydrodynamic model we tend d to infinity in Eq. (6.37) and plug it into Eq. (6.34). Thus, plasma waves in non-gated graphene at high electron densities exhibit the following dispersion:

$$\omega_+ = \sqrt{\frac{\upsilon_F^2 k^2}{2}(1+4\alpha k_F) - \left(\frac{\nu_e}{2}\right)^2} - i\frac{\nu_e}{2}. \tag{6.39}$$

In the low frequency limit the above equation is simplified

$$\omega_+ \simeq \upsilon_F \sqrt{2\alpha k k_F}, \tag{6.40}$$

and the result quantitatively coincides with that predicted in Ref. [24]. What concerns the result obtained in Ref. [22] for plasmons in gated graphene, it coincides both with that obtained within kinetic approach [21] and, hence, with our result (6.34).

The transition from almost linear to square-root plasma wave dispersion at different gate dielectric thicknesses is shown in

Fig. 6.6 for the given value of electron density $\Sigma_e = 5 \times 10^{12}$ cm^{-2}. In the short-wave limit $kd \gg 1$ (providing the applicability of the hydrodynamic model) the dispersion of plasma waves is linear again with the velocity depending only on the fundamental constants of graphene. Worth mentioning that the electron-hole sound waves are insensitive to the gate and exhibit almost linear dispersion at any gated structure parameters.

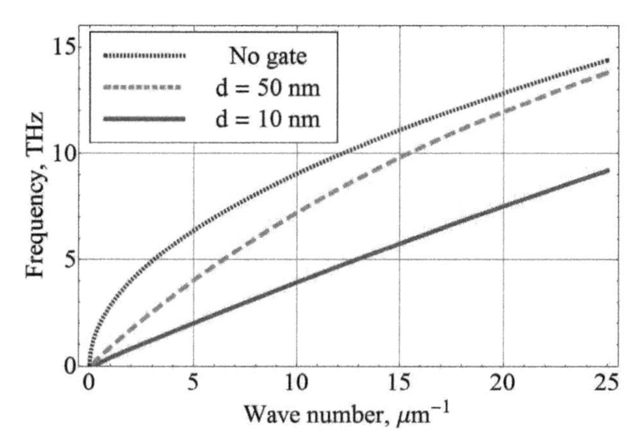

Figure 6.6 Dispersions of plasma waves at a given carrier density $\Sigma_e = 5 \times 10^{12}$ cm^{-2} and different thicknesses of gate dielectric.

6.5 Discussion of the Results

The hydrodynamic model of electron and hole transport in graphene which takes into account the linearity of carrier spectra has been developed. Strong interactions between electrons and holes result in mutual frictional forces proportional to the mismatch in electron and hole drift velocities. The interaction of carriers with acoustic phonons and impurities is also governed by a friction term in Euler equations. The estimated electron-hole collision rate is greater than that between carriers and phonons (impurities) at room temperature.

The model has been applied to the solution of two challenging problems: dc graphene conductivity and spectra and damping of the collective excitations in graphene structures.

The hydrodynamic equations derived provide an opportunity to calculate the conductivity of graphene sheet in a wide range of gate voltages including the Dirac point, where the influence of electron-hole collisions on the charge transport is dominant. However, when there is even a small difference (proportional to the ratio of external scattering rate to that of carrier-carrier) between electron and hole densities the majority carriers drag the minority ones. In this case, the conductivity is governed by scattering on phonons and impurities. The effect of drag is particularly pronounced in high-purity (suspended) graphene samples.

The spectra of collective excitations in electron-hole system in graphene have been calculated. The existence of two types of the excitations has been revealed: quasi-neutral electron-hole sound waves and plasma waves.

In the vicinity of the Dirac point or in the optically pumped intrinsic graphene the sound waves undergo weak damping of the order of 5×10^{11} s^{-1} in perfect structures. The damping of such waves is caused by weak scattering on acoustic phonons and is insensitive to strong electron-hole scattering. The latter fact is due to co-directional motion of electrons and holes in quasi-neutral sound waves, which almost eliminates electron-hole friction.

The quasi-neutral electron-hole sound waves under consideration are akin to those in bipolar electron-hole plasma in semimetals considered by Konstantinov and Perel a long time ago [25]. Later such waves were studied in semiconductors [26], two-band metals [27], and superconducting compounds [28]. Recently the acoustic plasma waves were studied in two dimensional semimetals [30] (CdHgTe/HgTe/CdHgTe quantum wells). It was shown that those waves can contribute to the thermodynamic properties of the materials, e.g., carrier relaxation rates in bulk semiconductors [29] and critical temperatures of superconducting compounds [28]. In graphene such waves could be feasibly excited by optical spots or modulated optical radiation. In both situations quasi-neutral non-uniform distributions of electrons and holes arise.

The propagation of plasma waves in symmetric electron-hole systems is strongly suppressed by electron-hole collisions. Strong damping of such waves can significantly decrease the rate of electron-hole recombination via plasmon emission, discussed in Ref.

[31]. In its turn, the recombination via the emission of electron-hole sound waves is prohibited as their velocity is smaller than v_F [Eq. (6.38)].

In sufficiently monopolar systems (electron or hole plasma, e.g., in gated graphene at high gate voltages) the damping of plasma waves, corresponding to the oscillations of majority carriers, is associated solely with the scattering on disorder (phonons and impurities). It can be rather weak in perfect structures, particularly at low temperatures. The dispersion of plasma waves in gated graphene structures is linear in a wide range of frequencies [32]. The dependence of the plasma wave velocity on the gated structure parameters is quantitatively similar to that obtained for the two-dimensional electron gas of massive electrons [33].

The hydrodynamic and kinetic models yield the same dispersions of plasma waves if one formally tends the collision frequency to zero. However, the hydrodynamic approach provides a regular way to describe the damping of the waves, associated with carrier scattering.

In conclusion, the hydrodynamic equations for bipolar graphene system were derived and applied for calculation of dc conductivity and spectra of collective excitations in graphene structures. The model opens the prospects to the simulation of graphene-based transistors, THz-range detectors and generators, and light emitting devices.

Acknowledgment

The research was supported via the grants 11-07-00464-a of the Russian Foundation for Basic Research, F793/8-05 of Computer Company NIX (science@nix.ru), and by the Japan Science and Technology Agency, CREST, Japan.

Appendix A: Dissipative Terms in the Euler Equations

In this appendix we derive explicit expression for the friction forces, associated with carrier-carrier, carrier-impurity, and carrier-phonon scattering.

A.1 Mutual Electron-Hole Friction

To derive the expression for electron-hole friction force one should linearize the collision integral

$$St\{f_e, f_h\} = 4\int W(q)\{f_e(\mathbf{p}-\mathbf{q})f_h(\mathbf{p}_1+\mathbf{q})[1-f_e(\mathbf{p})]$$
$$[1-f_h(\mathbf{p}_1)] - f_e(\mathbf{p})f_h(\mathbf{p}_1)[1-f_e(\mathbf{p}-\mathbf{q})]$$
$$[1-f_h(\mathbf{p}_1+\mathbf{q})]\}\frac{d^2\mathbf{p}_1 d^2\mathbf{q}}{(2\pi\hbar)^4}, \qquad (6.A1)$$

where the non-equilibrium distribution functions for electrons and holes are determined via Eqs. (6.3) and (6.4), the factor of 4 arises from 2 possible spins and valleys for particle $\mathbf{p}_1$. $W(q)$ is the Coulomb scattering probability which can be written down as

$$W(q) = \frac{2\pi}{\hbar v_F}\left[\frac{2\pi\hbar e^2}{\kappa(q+q_{TF})}\right]^2 \left|\langle u_{\mathbf{p}}|u_{\mathbf{p}'}\rangle\right|^2 \left|\langle u_{\mathbf{p}_1}|u_{\mathbf{p}_1'}\rangle\right|^2$$
$$\times\,\delta(p+p_1-|\mathbf{p}-\mathbf{q}|+|\mathbf{p}_1+\mathbf{q}|), \qquad (6.A2)$$

where $q_{TF} = 4\alpha T/v_F \ln(1+e^{\mu_e/T})(1+e^{-\mu_h/T})$ is the Thomas–Fermi momentum describing screening in graphene [24, 34], $\left|\langle u_{\mathbf{p}}|u_{\mathbf{p}'}\rangle\right|^2 = (1+\cos\theta_{pp'})/2$ is the matrix element of two Bloch functions in honeycomb lattice. In Eq. (6.A2) the Thomas–Fermi momentum could be replaced by the reciprocal gate-dielectric thickness for the qualitative description of gate screening.

In case of small velocities $\mathbf{p}\cdot\mathbf{V}_e \ll T$, $\mathbf{p}\cdot\mathbf{V}_h \ll T$, the electron-hole collision integral can be transformed using the common linearization technique for Fermi-systems [35]

$$St\{f_e, f_h\} = 4\int W(q)f_e(\mathbf{p})f_h(\mathbf{p}_1)[1-f_e(\mathbf{p}-\mathbf{q})]$$
$$[1-f_h(\mathbf{p}_1+\mathbf{q})]\times\frac{q\cdot(V_h-V_e)}{T}\frac{d^2\mathbf{q}d^2\mathbf{p}_1}{(2\pi\hbar)^4}, \qquad (6.A3)$$

where the energy and momentum conservation laws were used.

Equation (6.A3) timed by $\mathbf{q}/2$ and integrated over $d\Gamma_p$ is electron-hole friction force $\mathbf{f}_{eh}$ per unit area in the right hand side of the Euler equations (6.8) and (6.9). After averaging over angle between $\mathbf{q}$ and $\mathbf{V}_h - \mathbf{V}_e$ it is presented as

$$\mathbf{f}_{\rm eh} = \beta_{\rm eh}\,(\mathbf{V}_{\rm h} - \mathbf{V}_{\rm e}),$$

where

$$\beta_{\rm eh} = \int W(q) f_{\rm e}(\mathbf{p}) f_{\rm h}(\mathbf{p}_1)[1 - f_{\rm e}(\mathbf{p}-\mathbf{q})][1 - f_{\rm h}(\mathbf{p}_1+\mathbf{q})] \times \frac{4q^2}{T}\frac{d^2\mathbf{q}d^2\mathbf{p}_1 d^2\mathbf{p}}{(2\pi\hbar)^6}. \quad (6.A4)$$

In Refs. [11, 36], it was shown that almost collinear vectors $\mathbf{p}$, $\mathbf{p}_1$, and $\mathbf{q}$ give the leading contribution to the collision integral. This facts originates from the linearity of the carrier spectrum, i.e., particles moving with the same velocity and the same direction interact infinitely long. Following the technique, described in Ref. [11], the quantity $\beta_{\rm eh}$ in the collinear limit is presented as

$$\beta_{\rm eh} \propto \frac{T^4 e^4 \ln(1/\alpha)}{\hbar^5 \kappa^2 v_{\rm F}^6}\int_0^\infty dP \int_0^\infty dP_1 \int_{-P_1}^{P} F(P - z_{\rm e})F(P_1 + z_{\rm h})$$
$$\left[1 - F(P - Q - z_{\rm e})\right] \times [1 - F(P_1 + Q + z_{\rm h})]$$
$$\sqrt{PP_1(P-Q)(P_1+Q)}\frac{Q^2 dQ}{(|Q| - 4\alpha \ln F(z_{\rm e})F(-z_{\rm h}))^2}, \quad (6.A5)$$

where $F(x) = (1 + e^x)^{-1}$, $z_{\rm e} = \mu_{\rm e}/T$, $z_{\rm h} = \mu_{\rm h}/T$, and the term $\ln 1/\alpha$ originates from the corrections to the electronic sperctrum due to electron-electron interactions. In this form the friction coefficient $\beta_{\rm eh}$ can be simply computed numerically.

A.2 Friction Caused by Charged Impurities and Phonons

Calculation of friction force caused by charged impurities can be drawn analytically in the limit of monopolar plasma. The momentum relaxation rate is given by

$$\tau_{p,i}^{-1} = \int \Sigma_i W(q)(1 - \cos\theta_{pp'})\left|\langle u_{\mathbf{p}} | u_{\mathbf{p}'}\rangle\right|^2 \frac{d^2\mathbf{p}'}{(2\pi\hbar)^2}, \quad (6.A6)$$

where $\Sigma_{\rm i}$ is the sheet density of charged impurities. The Coulomb scattering probability $W(q)$ is given by Eq. (6.A2) with the following delta-function $\delta(pv_{\rm F} - p'v_{\rm F})$. Here we put into consideration only self-screening caused by carriers in graphene.

To calculate the friction force per unit area $\mathbf{f}_{\mathrm{ei}}$ one should integrate the momentum relaxation rate (6.A6) timed by $\mathbf{p}$ with the non-equilibrium part of distribution function $\delta f_{\mathrm{e}} = -\frac{\partial f_{\mathrm{e},0}}{\partial \varepsilon} \mathbf{p} \cdot \mathbf{V}_{\mathrm{e}}$:

$$\mathrm{f}_{ei} = \int \mathbf{p} \frac{\partial f}{\partial \varepsilon} \mathbf{p} \cdot \mathbf{V}_{\mathrm{e}} \tau_{p,i}^{-1} d\Gamma_{\mathbf{p}}. \tag{6.A7}$$

In monopolar case one can set $\partial f_{\mathrm{e},0}/\partial \varepsilon = -\delta(p v_{\mathrm{F}} - \mu_{\mathrm{e}})$ and perform trivial integration

$$\mathrm{f}_{ei} = -\frac{\pi e^4 \Sigma_{\mathrm{e}} \Sigma_{\mathrm{i}} V_{\mathrm{e}}}{\hbar v_{\mathrm{F}}^2 \kappa^2} \int_0^{2\pi} \frac{\sin^2 \theta_{pp'} d\theta_{pp'}}{(2\sin(\theta_{pp'}/2) + 4\alpha)^2},$$

where the dimensionless integral actually depends only on the permittivity κ of the environment.

We can use the same technique to derive the expression for the friction force, caused by phonon scattering. We start from the expression

$$\tau_{p,ph}^{-1} = \frac{D^2 T p}{4\rho_s s^2 \hbar^3 v_F},$$

derived in Ref. [5], and, using Eq. (6.A7), arrive at the following friction term

$$\mathbf{f}_{e\,ph} = -\frac{D^2 T \langle p_{\mathrm{e}}^2 \rangle \mathbf{V}_{\mathrm{e}}}{4\rho_s s^2 v_F^2 \hbar^3}. \tag{6.A8}$$

References

1. A. Jungel, *Quasi-Hydrodynamic Semiconductor Equations* (Birkauser Verlag, 2001).
2. R. Bistritzer and A. H. MacDonald, *Phys. Rev. B*, **80**, 085109 (2009).
3. V. F. Gantmacher and Y. B. Levinson, *Carrier Scattering in Metals and Semiconductors* (North-Holland, Amsterdam, 1987).
4. G. E. Uhlenbeck and G. W. Ford, *Lectures in Statistical Mechanics* (Providence, AMS, 1963).
5. F. Vasko and V. Ryzhii, *Phys. Rev. B*, **76**, 233404 (2007).
6. L. A. Falkovsky, *Phys. Lett. A*, **372**, 31 (2008).

7. E. H. Hwang and S. Das Sarma, *Phys. Rev. B*, **77**, 115449 (2008).
8. M. Shur, *Physics of Semiconductor Devices* (Pentice-Hall, Englewood Clifs, NJ, 1990).
9. K. S. Novoselov, et al., *Nature*, **438**, 197–200 (2005).
10. V. Vyurkov and V. Ryzhii, *JETP Lett.*, **88**, 322–325 (2008).
11. A. Kashuba, *Phys. Rev. B*, **78**, 085415 (2008).
12. M. I. Katsnelson, *Eur. Phys. J. B*, **51**, 157–160 (2006).
13. K. Ziegler, *Phys. Rev. B*, **75**, 233407 (2007).
14. L. A. Falkovsky and A. A. Varlamov, *Eur. Phys. J. B*, **56**, 281–284 (2007).
15. K. Ziegler, *Phys. Rev. Lett.*, **97**, 266802 (2006).
16. K. I. Bolotin, et al., *Phys. Rev. Lett.*, **101**, 096802 (2008).
17. K. I. Bolotin, et al., *Solid State Commun.*, **146**, 351 (2008).
18. R. S. Shishir, et al., *J. Comput. Electron.*, **8**(2), 43–50 (2009).
19. R. S. Shishir and D. K. Ferry, *J. Phys.: Condens. Matter*, **21**, 232204 (2009).
20. V. Ryzhii, *Jpn. J. Appl. Phys.*, **45**, L923 (2006).
21. V. Ryzhii, A. Satou, and T. Otsuji, *J. Appl. Phys.*, **101**, 024509 (2007).
22. A. Principi, R. Asgari, and M. Polini, *Solid State Commun.*, **151**, 21 (2011).
23. B. Wunsch, et al., *New J. Phys.*, **8**, 318 (2006).
24. S. D. Sarma and E. H. Hwang, *Phys. Rev. Lett.*, **102**, 206412 (2009).
25. O. V. Konstantinov and V. I. Perel, *Sov. Phys.-Solid State*, **9**, 3051 (1967).
26. A. S. Esperidiao, A. R. Vasconcellos, and R. Luzzi, *Solid State Commun.*, **73**(4), 275–279 (1990).
27. H. Gutfreunda and Y. Unnaa, *J. Phys. Chem. Sol.*, **34**, 9 (1973).
28. J. Ruvalds, *Adv. Phys.*, **30**(5), 677–695 (1981).
29. J. F. Lampin, F. X. Camescasse, A. Alexandrou, M. Bonitz, and V. Thierry-Mieg, *Phys. Rev. B*, **60**, R8453–R8456 (1999).
30. A. V. Chaplik, *JETP Lett.*, **91**, 4 (2010).
31. F. Rana, *Phys. Rev. B*, **84**, 045437 (2011).
32. A. V. Chaplik, *Sov. Phys. JETP*, **35**, 395 (1972).
33. M. Dyakonov and M. Shur, *Phys. Rev. Lett.*, **71**, 2465–2468 (1993).
34. A. H. C. Neto, F. Guinea, N. M. R. Peres, K. S. Novoselov, and A. K. Geim, *Rev. Mod. Phys.*, **81**, 109–162 (2009).

35. E. M. Lifshitz and L. P. Pitaevsky, *Physical Kinetics* (Pergamon Press, Oxford, 1981).
36. L. Fritz, J. Schmalian, M. Muller, and S. Sachdev, *Phys. Rev. B*, **78**, 085416 (2008).

Chapter 7

Interplay of Intra- and Interband Absorption in a Disordered Graphene*

F. T. Vasko,[a] V. V. Mitin,[a] V. Ryzhii,[b] and T. Otsuji[b]
[a]*Department of Electrical Engineering, University at Buffalo, Buffalo, New York 1460-1920, USA*
[b]*Research Institute of Electrical Communication, Tohoku University, Sendai 980-8577 and Japan Science and Technology Agency, CREST, Tokyo 107-0075, Japan*
ftvasko@yahoo.com

The absorption of heavily doped graphene in the terahertz (THz) and midinfrared (MIR) spectral regions is considered taking into account both the elastic scattering due to finite-range disorder and the variations of concentration due to long-range disorder. The interplay between intra- and interband transitions is analyzed for the high-frequency regime of response, near the Pauli blocking threshold. The gate voltage and temperature dependencies of the

*Reprinted with permission from F. T. Vasko, V. V. Mitin, V. Ryzhii, and T. Otsuji (2012). Interplay of intra- and interband absorption in a disordered graphene, *Phys. Rev. B*, **86**, 235424. Copyright © 2012 The American Physical Society.

Graphene-Based Terahertz Electronics and Plasmonics: Detector and Emitter Concepts
Edited by Vladimir Mitin, Taiichi Otsuji, and Victor Ryzhii

ISBN 978-981-4800-75-4 (Hardcover), 978-0-429-32839-8 (eBook)
www.jennystanford:com

absorption efficiency are calculated. It is demonstrated that for typical parameters, the smearing of the interband absorption edge is determined by a partly screened contribution to long-range disorder while the intraband absorption is determined by finite-range scattering. The latter yields the spectral dependencies which deviate from those following from the Drude formula. The obtained dependencies are in good agreement with recent experimental results. The comparison of the results of our calculations with the experimental data provides a possibility to extract the disorder characteristics.

7.1 Introduction

Both the gapless (massles) energy band structure and peculiarities of scattering mechanisms determine the response of graphene on the in-plane polarized radiation in the terahertz (THz) and midinfrared (MIR) spectral regions. Such a response is caused by both the direct interband transitions and the intraband transitions of free carriers accompanied by scattering processes (the Drude mechanism), see the reviews in Refs. [1–3] and references therein. The contributions of these two mechanisms in heavily doped graphene are described by the inter- and intraband dynamic conductivities, σ_{inter} and σ_{intra}, respectively, given by

$$\text{Re}\sigma_{\text{inter}} \approx \frac{e^2}{4\hbar}\theta(\hbar\omega - 2\varepsilon_{\text{F}}), \quad \sigma_{\text{intra}}(\omega) \approx \frac{\sigma_{\text{F}}}{1 + i\omega\tau_{\text{F}}}, \tag{7.1}$$

where ε_{F} and τ_{F} stand for the Fermi energy and the momentum relaxation time, $\sigma_{\text{F}} = (e^2/4\hbar)\varepsilon_{\text{F}}\tau_{\text{F}}/\hbar$ is the static conductivity, and $\theta(\varepsilon)$ is a unity step-like fuction. Here we have neglected a weak imaginary contribution $\text{Im}\sigma_{\text{inter}}$ [4] and the smearing of the interband absorption edge associated with the finiteness of the temperature T, and the collisional damping of the electron spectrum. When such a damping can be neglected at $T = 0$, the expression for $\text{Re}\sigma_{\text{inter}}$ describes an abrupt jump at photon energy $\hbar\omega = 2\varepsilon_{\text{F}}$ due to the Pauli blocking effect. The interband and intraband contributions are of the same order of magnitude at $\omega\tau_{\text{F}} \sim \sqrt{4\varepsilon_{\text{F}}\tau_{\text{F}}/\pi\hbar - 1}$. Therefore, the overlap of these contributions disappears if $\varepsilon_{\text{F}}\tau_{\text{F}}/\hbar \geq 20$ (see Fig. 7.1a). Thus, the description of interplay between both contributions is necessary in the THz (or MIR, depending on the doping level) spectral region for the typical samples with $\hbar/\varepsilon_F\tau_F \geq 0.2$. As shown below, the quasiclassical description of the intraband absorption and, therefore, the expression for $\sigma_{\text{intra}}(\omega)$ based on the Drude formula is not valid if $\hbar\omega$ exceeds temperature of carriers [5, 6]. This is because of the variation of the relaxation time as a function of the electron energy over energy intervals $\sim \hbar\omega$, where the intraband transitions take place (see Fig. 7.1b).

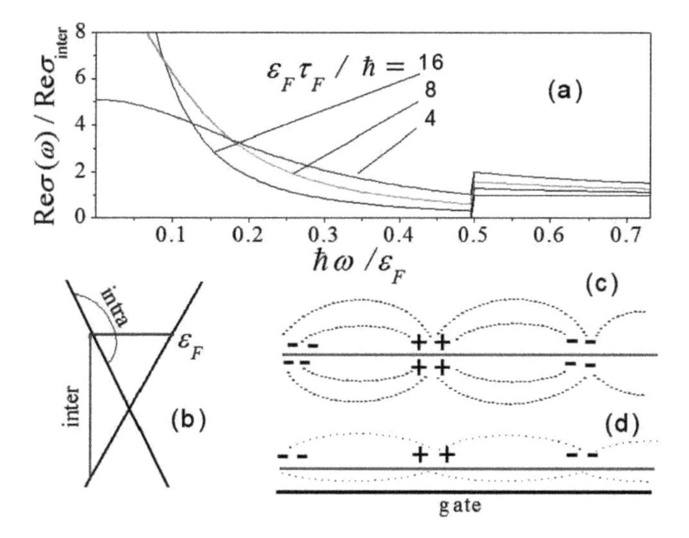

Figure 7.1 (a) Spectral dependencies of $\text{Re}\sigma(\omega)$ described by simplified Eq. (7.1) for different doping levels determined by $\varepsilon_F\tau_F/\hbar$. (b) Intra- and interband transitions between linear branches of gapless energy spectra. (c) Screening of nonuniform charge in ungated graphene with field distribution shown by dashed curves. (d) Gated graphene structure with partly screened charge and distribution of weak field shown by dotted curves.

The constant value for the interband conductivity $\text{Re}\sigma_{\text{inter}}$ associated with the frequency-independent absorption was observed for undoped suspended graphene in the visible range of the spectrum [7]. By analogy with electrical transport in field-effect transistors, MIR spectroscopy allows for the control of the Pauli blocking effect using the electrical gating [8]. In sufficiently heavily doped graphene, the Drude mechanism of absorption was observed near the interband threshold, up to the MIR spectral region [9–11].

Despite of a series of theoretical studies (see Refs. [12–14] and references therein), which have treated the problem of intra- and interband absorption at different levels of approximation, there are two drawbacks in the interpretation of the above-listed experimental data. First, the quantum description of the intraband absorption was not connected with the scattering parameters determined from the conductivity measurements and, second, an inhomogeneous smearing of the interband threshold due to long-range disorder was not considered. Thus, a reexamination of the intra- and interband absorption processes in a disorded graphene is timely now.

In this chapter, we present a general consideration which is based on the Kubo formula and takes into account both the scattering due to finite-range disorder and *a long-range partly screened variation of concentration*. The latter contribution appears to be essential because the screening of the potential spatial variations associated with the disorder with the in-plane scales longer than (or comparable to) the distance to gate is suppressed. This mechanism is schematically illustrated in Figs. 7.1c and 7.1d. As a result, the inhomogeneous mechanism of smearing of the interband absorption at $\hbar\omega \sim 2\varepsilon_F$ should be taken into account, together with the thermal smearing of the Fermi distribution and broadening of the spectral density due to the elastic scattering on the finite-range disorder. The dynamic conductivity in the high-frequency range (if $\omega\tau_F > 1$) is expressed via the Green's function, which is averaged over the finite-range disorder, and the carrier distribution averaged over the long-range partly screened disorder. The averaged characteristics of Green's function were obtained using the phenomenological description of the static conductivity similarly to Ref. [15]. Considering the interplay of the intra- and interband contributions, we analyze the spectral, temperature, and gate voltage dependencies of the absorption efficiency. The obtained results are compared with the recent experimental data [10, 11]. The fitting procedure demonstrates that the disorder characterization, including the evaluation of the parameters of high-energy scattering and long-range inhomogeneties, can be performed using the THz and MIR spectroscopy.

The chapter is organized as follows. In the next section (Section 7.2), we present the basic equations which describe the inter- and intraband mechanisms of absorption and consider partly screened long-range disorder in gated graphene. In Section 7.3 we analyze the spectra of relative absorption and their dependencies on the temperature disorder and as well as on the doping conditions. The last section includes the discussion of experimental data and the approximations used, and the conclusions.

7.2 Basic Equations

We start with consideration of the dynamic conductivity based on the Kubo formula which takes into account the elastic scattering and the partly screened long-range disorder. Since the absorption is determined by the real part of conductivity, $\mathrm{Re}\sigma_\omega$, we neglect the $\mathrm{Im}\sigma_\omega$ contribution [4].

7.2.1 Dynamic Conductivity

Absorption of the THz (MIR) radiation propagating along the normal to the graphene layer is described by the real part of dynamic conductivity [6], given by

$$\mathrm{Re}\sigma_\omega = \frac{2\pi e^2}{\omega L^2}\int dE(f_E - f_{E+\hbar\omega})$$
$$\times\left\langle \sum_{\alpha\alpha'} |(\alpha|\hat{v}_{\|}|\alpha')|^2 \delta(E-\varepsilon_\alpha)\delta(E-\varepsilon_{\alpha'}+\hbar\omega) \right\rangle. \tag{7.2}$$

Here L^2 is the normalization area, $f(\varepsilon_\alpha)$ is the equilibrium distribution over the states $|\alpha)$ with energies ε_α, and $|(\alpha|\hat{v}_{\|}|\alpha')|^2$ is the matrix element of the in-plane velocity operator $\hat{v}_{\|}$ between the states α and α' connected by the energy conservation law described by the δ function. The α states are defined by the eigenvalue problem $(\hat{h}+V_{\mathbf{x}})\psi_{\mathbf{x}}^{(\alpha)} = \varepsilon_\alpha \psi_{\mathbf{x}}^{(\alpha)}$ expressed via the single-particle Hamiltonian $\hat{h}$, a random potential $V_{\mathbf{x}}$, and the two-component wave function $\psi_{l\mathbf{x}}^{(\alpha)}$ with l = 1, 2. The statistical averaging of Eq. (7.2) over the random potential is denoted as $\langle\ldots\rangle$. To perform this averaging, we introduce the spectral density function

$$A_\varepsilon(l\mathbf{x},l'\mathbf{x}') = \sum_\alpha \delta(\varepsilon-\varepsilon_\alpha)\psi_{l\mathbf{x}}^{(\alpha)}\psi_{l'\mathbf{x}'}^{(\alpha)*}. \tag{7.3}$$

Thus, $\mathrm{Re}\sigma_\omega$ given by Eq. (7.1) is transformed into

$$\mathrm{Re}\sigma_\omega = \frac{2\pi(ev)^2}{\omega L^2}\int dE(f_E - f_{E+\hbar\omega})\int d\mathbf{x}_1\int d\mathbf{x}_2$$
$$\times\left\langle \mathrm{tr}\left[\hat{A}_{\mathrm{E}}(\mathbf{x}_1,\mathbf{x}_2)\hat{\sigma}_{\|}\hat{A}_{\mathrm{E}+\hbar\omega}(\mathbf{x}_2,\mathbf{x}_1)\hat{\sigma}_{\|}\right]\right\rangle, \tag{7.4}$$

where tr . . . means trace over the isospin variable, $\upsilon = 10^8$ cm/s is the characteristic velocity, and $\hat{\sigma}_{\|} = (\hat{\sigma}\cdot\mathbf{e})$ is written through the 2 × 2 Pauli matrix $\hat{\sigma}$ and the polarization ort **e**.

Further, the spectral density function is expressed through the exact retarded and advanced Green's functions $\hat{G}^R_{\mathrm{E}}(\mathbf{x}_1,\mathbf{x}_2)$ and $\hat{G}^A_{\mathrm{E}}(\mathbf{x}_1,\mathbf{x}_2)$ as follows $\hat{A}_{\mathrm{E}}(\mathbf{x}_1,\mathbf{x}_2) = i\left[\hat{G}^R_{\mathrm{E}}(\mathbf{x}_1,\mathbf{x}_2) - \hat{G}^A_{\mathrm{E}}(\mathbf{x}_1,\mathbf{x}_2)\right]/2\pi$. These Green's functions are governed by the standard equation

$$[E - i0 - v(\hat{\sigma}\cdot\mathbf{p}_1) - u_{\mathbf{x}_1} - v_{\mathbf{x}_1}]\hat{G}^R_{\mathrm{E}}(\mathbf{x}_1,\mathbf{x}_2)$$
$$= \hat{1}\,\delta(\mathbf{x}_1 - \mathbf{x}_2), \tag{7.5}$$

which is written through the potential energy with separated contributions from finite- and long-range disorder, u_{x1} and υ_{x1} labeled as *fr*- and *lr*- respectively. The averaging $\langle\ldots\rangle$ in Eq. (7.4) should also be separated as $\langle\ldots\rangle_{lr}$ and $\langle\ldots\rangle_{fr}$. It is convenient to introduce the variables $(\mathbf{x}_1 + \mathbf{x}_2)/2 = \mathbf{x}$ and $\mathbf{x}_1 - \mathbf{x}_2 = \Delta\mathbf{x}$ in Eqs. (7.4) and (7.5). Using these variables and neglecting a weak contribution of $\nabla\upsilon_{\mathbf{x}}$ we obtain $\upsilon_{\mathbf{x}+\Delta\mathbf{x}/2} \simeq \upsilon_{\mathbf{x}}$. As a result, after the replacement $E \longrightarrow E + \upsilon p$ one can separate the averaging over *fr*- and *lr*-disorder contributions and Eq. (7.4) takes the form

$$\mathrm{Re}\sigma_\omega = \frac{2\pi}{\omega}(ev)^2\int dE(\langle f_{E+vp}\rangle_{lr} - \langle f_{E+vp+\hbar\omega}\rangle_{lr})$$
$$\times\int d\Delta\mathbf{x}\left\langle \mathrm{tr}\left[\hat{A}_{\mathrm{E}}(\mathbf{x}_1,\mathbf{x}_2)\hat{\sigma}_{\|}\hat{A}_{\mathrm{E}+\hbar\omega}(\mathbf{x}_2,\mathbf{x}_1)\hat{\sigma}_{\|}\right]\right\rangle_{\mathrm{fr}}, \tag{7.6}$$

where we take into account that $\langle\ldots\rangle_{fr}$ depends only on $|\Delta\mathbf{x}|$ and use $\int d\mathbf{x} = L^2$.

We consider below the high-frequency spectral region, $\omega\tau_{\mathrm{F}} \gg 1$, when the correlation function $\langle\ldots\rangle_{fr}$ can be factorized through the averaged spectral functions $\hat{A}_{\Delta\mathbf{x}E} \equiv \left\langle\hat{A}_{\mathrm{E}}(\mathbf{x}_1,\mathbf{x}_2)\right\rangle_{\mathrm{fr}}$. The averaging over *fr* disorder gives the contribution

$$\int d\Delta\mathbf{x}\langle\mathrm{tr}[\cdots]\rangle_{\mathrm{sr}} = L^{-2}\sum_{\mathbf{p}}\mathrm{tr}\left(\hat{A}_{pE}\hat{\sigma}_{\|}\hat{A}_{pE+\hbar\omega}\hat{\sigma}_{\|}\right). \tag{7.7}$$

Here the spectral densities $\hat{A}_{pE}$ should be written through the averaged Green's functions, $\hat{G}^{A}_{pE} = \hat{G}^{R+}_{pE}$ and $\hat{G}^{R}_{pE}$, which are given by the matrix expression

$$\hat{G}^{R}_{\text{pE}} = \left[E - v(\hat{\sigma}\cdot\mathbf{p}_1) - \hat{\Sigma}^{R}_{\text{pE}}\right]^{-1},$$

$$\hat{\Sigma}^{R}_{\text{pE}} = \zeta^{R}_{\text{pE}} + \frac{(\hat{\sigma}\cdot\mathbf{p})}{p}\eta^{R}_{\text{pE}}, \tag{7.8}$$

where the self-energy function is determined in the Born approximation through the functions

$$\left|\begin{matrix}\zeta^{R}_{\text{pE}} \\ \eta^{R}_{\text{pE}}\end{matrix}\right| = \int \frac{d\mathbf{p}_1}{(2\pi\hbar)^2} W_{|\mathbf{p}-\mathbf{p}_1|}$$

$$\times \left|\begin{matrix} \dfrac{1}{E+i0-vp_1} + \dfrac{1}{E+i0+vp_1} \\ \dfrac{(\mathbf{p}\cdot\mathbf{p}_1)}{pp_1}\left(\dfrac{1}{E+i0-vp_1} - \dfrac{1}{E+i0+vp_1}\right)\end{matrix}\right|. \tag{7.9}$$

Using the Fourier transformation of $\left\langle u_{\mathbf{x}_1} u_{\mathbf{x}_2}\right\rangle_{fr} = \bar{u}^2 \exp[-(\mathbf{x}_1 - \mathbf{x}_2)^2 / 2l_{fr}]$, one obtains in (7.9) the Gaussian correlation function $W_{\Delta\text{P}}$ expressed through the strength of potential and correlation length $\bar{u}$ and l_{fr}. Because $\hat{G}^{R}_{\text{pE}}$ depends only on the matrix $(\hat{\sigma}\cdot\mathbf{p})$, the averaged spectral density of Eq. (7.7) takes the form

$$\hat{A}_{\mathbf{p}\text{E}} = \frac{A^{(+)}_{\text{pE}} + A^{(-)}_{\text{pE}}}{2} + \frac{(\hat{\sigma}\cdot\mathbf{p})}{2p}\left(A^{(+)}_{\text{pE}} - A^{(-)}_{\text{pE}}\right)$$

$$A^{(\pm)}_{\text{pE}} = \frac{i}{2\pi}\left(\frac{1}{\varepsilon_{\text{pE}} \mp v_{\text{pE}}p} - \frac{1}{\varepsilon_{\text{pE}} \pm v_{\text{pE}}p}\right), \tag{7.10}$$

where we introduced the renormalized dispersion law and velocity $\varepsilon_{\text{pE}} = E - \zeta^{R}_{\text{pE}}$ and $v_{\text{pE}} = v(1 + \eta^{R}_{\text{pE}} / vp)$, respectively.

For the averaging over *lr* disorder in Eq. (7.6) we introduce the population factor $\left\langle f_{\text{E+v}}\right\rangle_{\text{lr}} = \int_{-\infty}^{\infty} dE' \phi_{\text{E}-\text{E}'} f_{\text{E}'}$ with the kernel

$\phi_{E-E'}$ determined by the characteristics of disorder, see Section 7.2.2. Thus, based on the two assumptions employed (smoothness of *lr*-potential $\nabla \upsilon \to 0$ and high-frequency approach $\omega\tau_F \gg 1$) we obtain $\mathrm{Re}\sigma_\omega$ written through the triple integral

$$\mathrm{Re}\sigma_\omega = \frac{\pi(ev)^2}{\omega L^2}\int dE \int dE' \phi_{E-E'}(f_{E'} - f_{E'+\hbar\omega}) \times \sum_{\mathbf{p}} \left(A_{\mathrm{pE}}^{(+)} + A_{\mathrm{pE}}^{(-)}\right)\left(A_{\mathrm{pE}+\hbar\omega}^{(+)} + A_{\mathrm{pE}+\hbar\omega}^{(-)}\right), \tag{7.11}$$

where the contributions $\propto A_{\mathrm{pE}}^{(\pm)} A_{\mathrm{pE}}^{(\pm)}$ and $A_{\mathrm{pE}}^{(\pm)} A_{\mathrm{pE}}^{(\mp)}$ correspond to the intra- and interband absorption processes, respectively. Similar expressions for the homogeneous case, without *lr* contribution, were considered in Ref. [13].

7.2.2 Partly Screened Long-Range Disorder

Here we evaluate the square-averaged potential of *lr* disorder which is caused by a built-in random potential $w_{\mathbf{x}}$. We take into account the screening effect in the gated structure with the graphene sheet placed at $z = 0$ on the substrate of the width d and the static dielectric permittivity ϵ, see Fig. 7.1d. The electrostatic potential $V_{\mathbf{x}z}$ is defined by the Poisson equation with the induced charge at $z = 0$ determined by the concentration of electrons in the heavily doped graphene $n_{\mathbf{x}} = (4/L^4)\Sigma_{\mathbf{p}}\theta(\varepsilon_F - \varepsilon_{\mathbf{x}\mathbf{p}}) = \left[(\varepsilon_F - v_{\mathbf{x}})/\sqrt{\pi}v\hbar\right]^2$. The dispersion law $\varepsilon_{\mathbf{p}\mathbf{x}} = \upsilon_{\mathbf{p}} + \upsilon_{\mathbf{x}}$ is written here through the screened potential $\upsilon_{\mathbf{x}} = w_{\mathbf{x}} + V_{\mathbf{x}z=0}$. Performing the Fourier transformation over $\mathbf{x}$ plane one obtains the second-order equations for $V_{\mathbf{q}z}^{<}$ and $V_{\mathbf{q}z}^{>}$ which correspond to the substrate $0 > z > -d$ and to the upper half-space $z > 0$

$$\left(\frac{d^2}{dz^2} - q^2\right)\left|\begin{matrix} V_{\mathbf{q}z}^{>} \\ V_{\mathbf{q}z}^{<} \end{matrix}\right| = 0, \quad \begin{matrix} z > 0 \\ 0 > z > -d \end{matrix}. \tag{7.12}$$

The jump of electric field at $z = 0$ is defined by the charge distribution over graphene ($z \to 0$)

$$\left.\frac{dV_{\mathbf{q}z}^{>}}{dz}\right|_{z=0} - \epsilon \left.\frac{dV_{\mathbf{q}z}^{<}}{dz}\right|_{z=0} = -\psi_{\mathbf{q}},$$

$$\psi_{\mathbf{q}} = 4\pi e^2 \int_{L^2} d\mathbf{x} e^{-i\mathbf{q}\mathbf{x}} n_{\mathbf{x}}, \tag{7.13}$$

so that the potential remains homogeneous at $z = 0$: $V^{<}_{\mathbf{q}z=0} = V^{>}_{\mathbf{q}z=0} \equiv V_{\mathbf{q}z=0}$. The boundary condition at $z = -d$ is defined by the in-plane homogeneous back-gate voltage V_g as follows: $V^{<}_{\mathbf{q}z=-d} = V_g \delta_{\mathbf{q},0}$. The last boundary condition is the requirement that the potential $V^{>}_{\mathbf{q}z}$ vanishes at $z \longrightarrow \infty$.

The solution of this electrostatic problem can be written through $\exp(\pm qz)$ and $\Psi_{\mathbf{q}}$ which depends on $V_{\mathbf{q}z=0}$ through the concentration $n_{\mathbf{x}}$. Solving the system of the boundary conditions, we find that the screening potential at $z = 0$ should satisfy the requirement

$$V_{\mathbf{q}z=0} = -\psi_{\mathbf{q}} dK(qd),$$

$$k(y) = \frac{\sinh y}{y(\sinh y + \in \cosh y)}, \tag{7.14}$$

where the dimensionless function K describes the gate-induced quenching of $V_{\mathbf{q}z=0}$ due to the fixed potential at $z = -d$. According to Eq. (7.13), $\Psi_{\mathbf{q}}$ is the nonlinear function of the screened potential written through $n_{\mathbf{x}}$ and the electrostatic problem given by Eqs. (7.12) and (7.13) is transformed into the integral equation for $V_{\mathbf{q}z=0}$ determined by Eq. (7.14). Further, we consider the case of heavily doped graphene, when $\varepsilon_F \gg |\upsilon_{\mathbf{x}}|$ and the electron-hole puddles are absent, so that $n_{\mathbf{x}} \approx n_F(1 - 2\,\upsilon_{\mathbf{x}}/\varepsilon_F)$ is written through the averaged concentration n_F. As a result, the linearized dependency between $\Psi_{\mathbf{q}}$ and the screened potential $\upsilon_{\mathbf{q}}$ takes place

$$\psi_{\mathbf{q}} \approx 4\pi e^2 n_F \left(\delta_{\mathbf{q},0} - \frac{2v_{\mathbf{q}}}{\varepsilon_F} \right). \tag{7.15}$$

Therefore, Eqs. (7.14) and (7.15) give the linear relation between $V_{\mathbf{q}z=0} = \upsilon_{\mathbf{q}} - w_{\mathbf{q}}$ and the Fourier component of the screened potential $\upsilon_{\mathbf{q}}$. The solution for $\upsilon_{\mathbf{q}}$ takes form

$$v_{\mathbf{q}} = \frac{w_{\mathbf{q}}}{1 + 8(e/\hbar v)^2 \varepsilon_F dK(qd)}, \tag{7.16}$$

where screening is suppressed (i.e., $\upsilon_{\mathbf{q}} \longrightarrow w_{\mathbf{q}}$) if $d \longrightarrow 0$ and the complete screening ($|\upsilon_{\mathbf{q}}| \ll |w_{\mathbf{q}}|$) takes place at $d \longrightarrow \infty$, when Eq. (7.16) coincides with the Thomas–Fermi approximation.

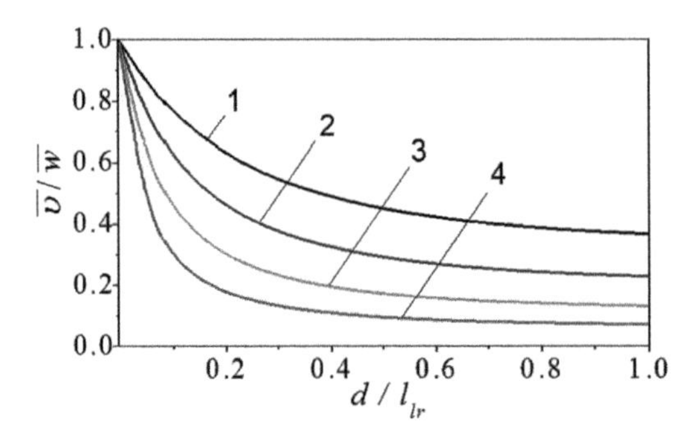

Figure 7.2 Dimensionless averaged potential $\overline{v}$ versus d/l_{lr} for different g_F = (1) 12.5, (2) 25, (3) 50, and (4) 100.

The averaging over *lr* disorder in the distribution $\langle f_{E+\upsilon p}\rangle_{lr}$ is performed after the Fourier transform of f_E with the use of the relation $\langle \exp[iv_x\tau/\hbar]\rangle_{lr} = \exp[-(\overline{v}\tau/\hbar)^2/2]$, where $\overline{v} = \sqrt{\langle v_x^2\rangle_{lr}}$ is the averaged *lr* potential. Under the inverse Fourier transformation (i.e., the integration over time) of the averaged distribution one obtains the result $\langle f_{E+\upsilon}\rangle_{lr} = \int dE' \phi_{E-E'} f_{E'}$ which is written through the Gaussian kernel

$$\phi_{\Delta E} = \exp[-(\Delta E/\overline{v})^2/2]/(\sqrt{2\pi}\overline{v}) \tag{7.17}$$

with the half-width $\overline{v}$. Using Eq. (7.16) one can write $\overline{v}$ through the averaged built-in potential $w_{\mathbf{q}}$. We consider here a simple case of the Gaussian distribution of $w_{\mathbf{x}}$ described by the correlation function

$$\langle w_{\mathbf{q}} w_{\mathbf{q}'}\rangle_{lr} = L^2\delta_{\mathbf{q}+\mathbf{q}',0}2\pi l_{lr}^2\overline{w}^2\exp[-(ql_{lr})^2/2], \tag{7.18}$$

which is written through the correlation length l_{lr} and the averaged built-in potential $\overline{w} = \sqrt{\langle w_{\mathbf{x}}^2\rangle_{lr}}$. Using Eqs. (7.16) and (7.18) one obtains

$$\frac{\overline{v}}{\overline{w}} = \left\{\int_0^\infty \frac{dx x e^{-x^2/2}}{[1+g_F(d/l_{lr})K(xd/l_{lr})]^2}\right\}^{1/2}, \tag{7.19}$$

and this ratio depends on d/l_{lr} and on the coupling constant, $g_F = (8e^2/\hbar\upsilon)l_{lr}/\lambda_F \propto \varepsilon_F\, l_{lr}$, written through the Fermi wavelength $\lambda_F = \upsilon\hbar/\varepsilon_F$. If $d/l_{lr} \ll 1$ one obtains $\overline{v}/\overline{w} \to 1$ and in the region $d/l_{lr} \geq 1$ one obtains $\overline{v}/\overline{w} \simeq \sqrt{(1+\epsilon)/g_F}$ which is independent on d/l_{lr}. In Fig. 7.2 we plot the function (7.19) for the case of SiO_2 substrate if the coupling constant $g_{\rm F} \simeq 10$–100 corresponds to the concentration range $\sim 4 \times 10^{11}$ - 4×10^{13} cm^{-2} and if $l_{\rm lr} \sim 50$ nm. With increasing of $n_{\rm F}$ or $l_{\rm lr}$ at $g_{\rm F} > 100$, the relation $\bar{v}/\bar{w} \simeq \sqrt{(1+\epsilon)/g_{\rm F}}$ appears to be valid for the interval of $d/l_{\rm lr} > 0.1$.

7.3 Relative Absorption

In this section we consider the absorption of radiation propagated along the normal to graphene sheet placed on semi-infinite substrate with the high-frequency dielectric permittivity κ. Here we analyze the relative absorption coefficient

$$\xi_\omega \simeq \frac{16\pi \, \mathrm{Re}\sigma_\omega}{\sqrt{\kappa}(1+\sqrt{\kappa})^2 c} \tag{7.20}$$

neglecting the second-order corrections with respect to the parameter $4\pi|\sigma_\omega|/c\sqrt{\kappa}$. Note that the reflection and transmission coefficients R_ω and T_ω (these values are connected by the energy conservation requrement $R_\omega + T_\omega + \xi_\omega = 1$), involve the first-order correction $\propto 4\pi|\sigma_\omega|/c\sqrt{\kappa}$ and the $\mathrm{Re}\sigma_\omega$ contributions should be involved for description of R_ω and T_ω.

7.3.1 Spectral Dependencies

First, we consider ξ_ω given by Eq. (7.20) after substituting the dynamic conductivity determined by Eqs. (7.9)–(7.11). Performing the multiple integrations in $\mathrm{Re}\sigma_\omega$, one obtains the spectral, gate voltage, and temperature dependencies of relative absorption. For calculations of the renormalized energy spectra and velocity defined by Eq. (7.9) we use here the typical parameters of *fr* disorder corresponding to the sample with the maximal sheet resistance ~3.5 kΩ and the correlation length $l_{fr} \sim 7.5$ nm, see Ref. [15] for details. Level of long-range disorder $\bar{v}$, corresponding to $l_{\rm lr} \gg \lambda_{\rm F}$, is determined by the ratio $d/l_{\rm lr}$, see Fig. 7.2; here we consider heavily doped graphene because a more complicated analysis is necessary for the case when electron-hole puddles are formed [16].

The spectral dependencies of ξ_ω normalized to the high-frequency interband contribution $\xi_{\rm inter}$ are plotted in Fig. 7.3 for the homogeneous graphene case $\bar{v} = 0$ at concentrations 5×10^{11} cm^{-2}

and 10^{12} cm^{-2} for temperatures T = 77 and 300 K. Similarly to the schematic spectra in Fig. 7.1a, one can separate contributions from intra- and interband transitions at $\hbar\omega \sim$ 100 meV and ~150 meV for lower- and higher-doped samples with the minimal value of $\xi_\omega/\xi_{\rm inter}$ ~0.15 and ~0.08, respectively. A visible deviation from the Drude spectral dependence ($\propto \omega^{-2}$) takes place due to quantum character of intraband transitions if $\hbar\omega \geq T$. Different signs of these deviations, which depend on $\hbar\omega$, T, and $n_{\rm F}$, appear due to complicated energy and momentum dependencies in Eqs. (7.9) and (10). The smearing of interband absorption around the Fermi energies ($2\varepsilon_{\rm F}$ are marked by arrows) is defined by both the scattering-induced broadening of the spectral density (7.10) and thermal effect (c.f. solid and dashed curves in Fig. 7.3).

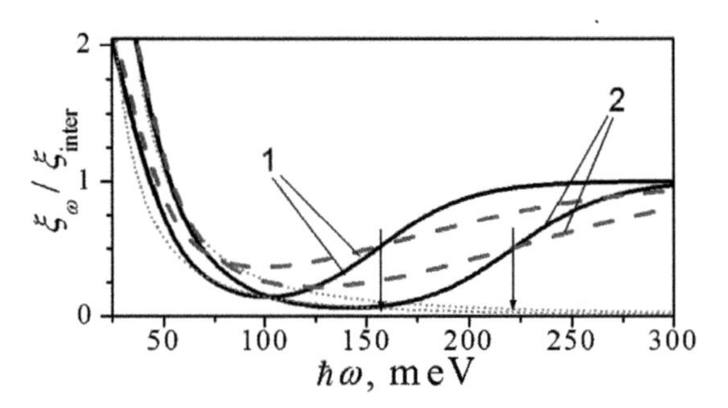

Figure 7.3 Spectral dependencies of relative absorption in homogeneous graphene (normalized to $\xi_{\rm inter} = \xi_{\omega\to\infty}$) at temperatures 77 and 300 K (solid and dashed curves, respectively) for concentrations (1) 5 × 10^{11} cm^{-2} and (2) 10^{12} cm^{-2}. Arrows correspond to the doubled Fermi energies and dotted curves are $\propto \omega^{-2}$ asymptotics for Drude absorption.

In addition, the inhomogeneous smearing of the interband absorption threshold due to partly screened long-range variations of concentration is essential, as it is shown in Fig. 7.4 for T = 77 K. Here we considered a strong *lr*-disorder case with $\bar{v} \simeq 30$ meV for $n_{\rm F}$ = 5 × 10^{11} cm^{-2} and $\bar{v} \simeq 20$ meV for $n_{\rm F}$ = 10^{12} cm^{-2} (we used the dependencies of Fig. 7.2 with $\bar{w} \simeq 50$ meV and $d/l_{\rm lr} \leq 0.1$). Similarly to the thermally induced effect, the partly screened disorder effect results both in smearing of the interband absorption threshold and in enhancement of the minimal absorption in the region between intra- and interband contributions.

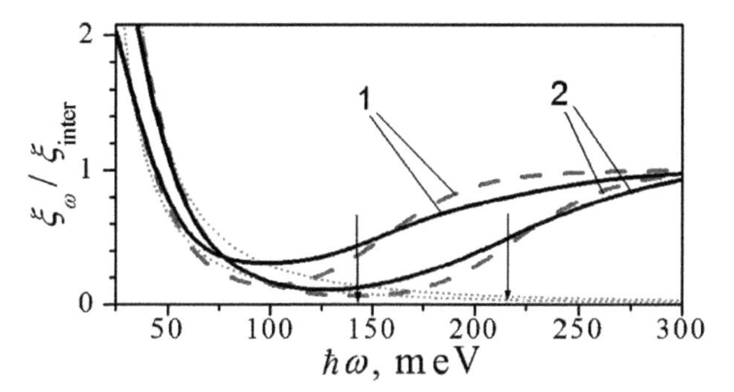

Figure 7.4 The same as in Fig. 7.3 at $T = 77$ K taking into account inhomogeneous smearing of threshold due to partly screened variations of concentrations around (1) 5×10^{11} cm^{-2} and (2) 10^{12} cm^{-2}. Dashed curves are plotted for the homogeneous case $\overline{v} \simeq 0$.

7.3.2 Comparison with Experiment

We turn now to comparison of our calculations with the recent experimental data [10, 11] where both the spectral dependencies of relative absorption in large-area samples and the static conductivity σ versus gate voltage V_g were measured. Using the phenomenological momentum relaxation rate v_p suggested in Ref. [15] and the experimental data for hole conductivity from Refs. [10, 11], one can fit the dependency $\sigma(V_g)$ as it is shown in Fig. 7.5. This approach gives us the *fr*-disorder scattering parameters for energies $\upsilon p \leq 200$ meV and we can extrapolate these data up to ~400 meV energies, which are necessary for description of the MIR response measured in Refs. [10, 11].

In Fig. 7.6 we plot the spectral dependencies of relative absorption measured in Ref. [12] for graphene with hole concentrations ~2.2, ~ 4.3, and ~5.8 × 10^{12} cm^{-2} at T = 100 K. Solid curves are plotted for the case of homogeneous sample with the scattering parameters determined from the transport data shown in Fig. 7.5. These dependencies are in agreement with experiment both in the intraband absorption region (at $\hbar\omega \leq 150$ meV) and above threshold of interband absorption, at $\hbar\omega > 2\varepsilon_F$. But the absorption in the intermediate region $\hbar\omega$ ~200–300 meV appears to be suppressed in comparison to the experimental data. The minimal absorption increases if we take into account the partly screened *lr*-disorder contribution with $\bar{\nu} \sim 80$, 75, and 60 meV for curves 1, 2, and 3, respectively. Such values of $\bar{\nu}$ are realized if $\bar{w}$ is comparable to ε_F and $d/l_{lr} \leq 0.1$, i.e., micrometer scale of inhomogeneities takes place, see Fig. 7.2. Thus, we have obtained a reasonable agreement with the experimental data. But the smearing of the interband threshold is stronger in comparison to experimental data because the Gaussian model does not describe a real lr distribution. A more accurate description is possible with the use of lr-disorder parameters taken from additional structure measurements.

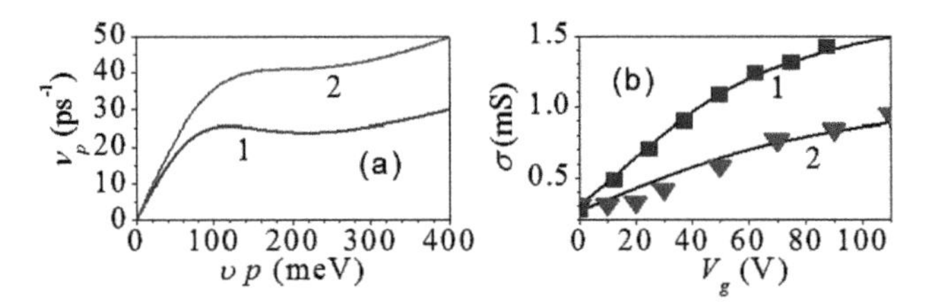

Figure 7.5 Fitting of experimental data from Refs. [10, 11] marked as 1 and 2: (a) momentum relaxation rate ν_p versus energy υp and (b) sheet conductivity versus gate voltage. Squares and triangles are experimental points from Refs. (1) [10] and (2) [11], respectively.

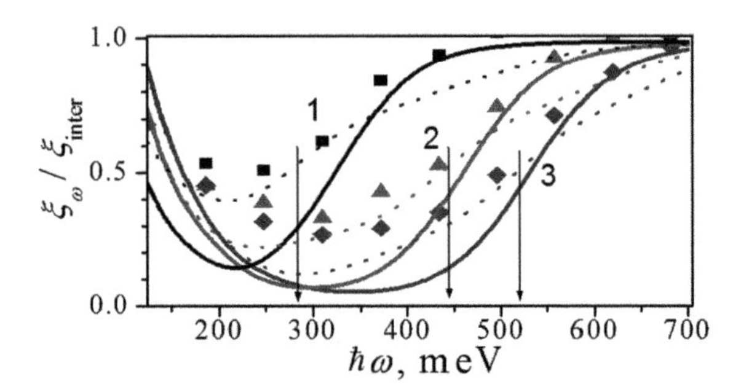

Figure 7.6 Fitting of relative absorption spectra (normalized to ξ_{inter}) to conditions of Ref. [10] at gate voltages (1) −30, (2) −60, and (3) −80 V. Solid and dotted curves correspond to homogeneous and inhomogeneous (with $\bar{w} \sim 150$ meV) samples. Squares, triangles, and diamonds are experimental points for the cases 1, 2, and 3, respectively. Arrows correspond to the doubled Fermi energies.

A similar fitting of ξ_ω/ξ_{inter} measured in Ref. [11] for hole concentrations ~4.3, ~6.5, and ~8.6, and 11×10^{12} cm^{-2} at $T = 300$

K is presented in Fig. 7.7. We plot the spectral dependencies both for the homogeneous sample (solid curves) and for the inhomogeneous sample taking into account the strong *lr* disorder (dotted curves) with $\bar{v} \sim 120$, 105, 90, and 75 meV for curves 1, 2, 3, and 4, respectively. Now, reasonable agreement with experimental data for interband absorption takes place (see the above discussion of Fig. 7.6 about the *lr*-disorder contribution) while the intraband absorption appears to be 2–3 times stronger in comparison to the disorder-induced contributions calculated. An additional contribution may appear due to relaxation via emission of optical phonons of energy $\hbar\omega_0$. Such a channel of intraband absorption is allowed if $\varepsilon_F > \hbar\omega_0$ and the rate of transitions is proportional to the density of states at $\varepsilon_F - \hbar\omega_0$. As a result, this process may be essential for the conditions of Ref. [11], where $\varepsilon_F \geq 250$ meV, while this mechanism should be negligible at the lower concentrations and temperatures used in Ref. [10]. Recent calculations of the phonon-induced intraband contribution [17] gives an additional absorption $\sim 0.2\xi_{inter}$ for the conditions of Ref. [11]. This contribution is of the order of the disagreement between calculations and experimental data presented in Fig. 7.7. More detailed experimental data, for the spectral range 200–400 meV at temperatures ≤ 300 K, are necessary in order to verify this contribution.

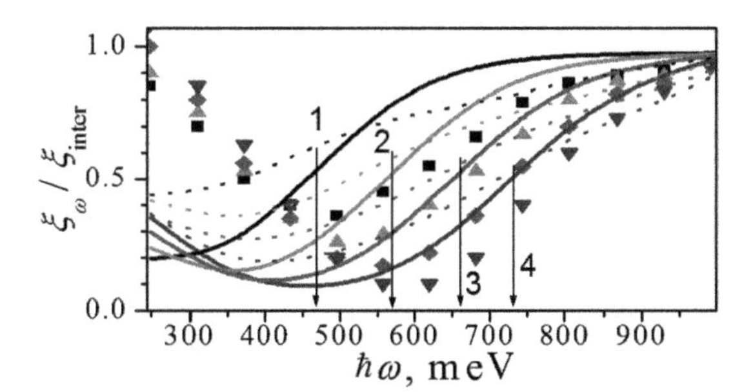

Figure 7.7 The same as in Fig. 7.6 for conditions of Ref. [11] at gate voltages (1) −30, (2) −60, (3) −90, and (4) −120 V. Squares, triangles, diamonds, and inverse triangles are experimental points for the cases 1, 2, 3, and 4, respectively.

Overall, the comparison performed here demonstrates reasonable agreement with the experimental data, which opens a way for the verification of scattering mechanisms and long-range disorder parameters. The Drude spectral dependencies (for the region ≤

100 meV) were fitted to the experimental data in Refs. [10, 11]; our calculations are in good agreement with these data because we used the scattering parameters taken from the transport measurements. A complete verification can be performed under complex analysis which should include comparisons of spectroscopic data with (i) steady-state transport measurements of conductivity and Hall effect versus V_g and T, (ii) a structural (e.g., scanning tunneling microscopy) characterization of long-range disorder, and (iii) effect of large-area inhomogeneties (ripples, wrinkles, holes, etc.) in centimeter-sized samples.

7.4 Conclusions

We have examined the spectral, concentration (gate voltage), and temperature dependencies of the relative absorption due to intra- and interband transitions in graphene taking into account both elastic scattering and partly screened long-range variations of concentration. The smearing of the interband absorption edge, deviation from the Drude spectral dependence due to quantum regime of intraband transitions, and interplay of these two contributions are demonstrated. The results are in agreement with the recent experimental data obtained for heavily doped graphene [10, 11]. This analysis allows to extract disorder parameters in the high-energy region (up to $\hbar\omega$ above the Fermi energy).

Let us discuss the assumptions used in the calculations performed. The main restriction of the results is due to consideration of the high-frequency spectral region, when $\mathrm{Re}\sigma_\omega$ can be written through the matrix product of the averaged spectral density function (7.10). A more complicate description, based on the Bethe–Salpeter equation, is necessary for the low-frequency region $\omega\tau_F \leq 1$, which is beyond the scope of our consideration. This case is realized if $\hbar\omega < 20$ meV, where the Drude dispertsion takes place, see Figs. 7.3 and 7.4. We also restrict ourselves by the phenomenological models for the finite-range disorder scattering [15] and for the partly screened long-range variations of concentration. It is enough for a description of spectral, gate voltage, and temperature dependencies of ξ_ω because they are expressed through simple correlation functions which are similar for any microscopic nature of disorder. In addition, a long-range random strain [18] may have affect on the interband transitions in large-area samples. The scattering processes caused by phonons and carrier-carrier collisions should be analyzed under calculations of the intra- and interband absorption in clean samples. A possible contribution of optical phonons at room temperature [11] is discussed in Section 7.3.2 (more detailed analysis is beyond of the scope of this chapter) while the interaction with acoustic phonons is negligible. The Coulomb interaction changes the interband absorption in the UV spectral region due to the excitonic effect [19]. The comparison with the experiment was performed here for the hole doping case and electron-hole asymmetry reported in Refs. [10, 11] is unclear. According to Ref. [20], the Coulomb-induced renormalization of the response and scattering processes give a weak contribution for typical disorder levels. This disagreement requires an additional investigation.

To conclude, we believe that our results open a way for characterization of scattering processes and long-range inhomogeneities using THz and MIR spectroscopy. More important, that effect of disorder on THz and MIR response is the main factor which determines characteristics of different devices (e.g., lasers [21] and photodetectors [22]) in this spectral region. We believe that our study will stimulate a further investigation of these device applications.

Acknowledgment

This research is supported by the NSF-PIRE-TeraNano grant (USA), by the JSPS Grant in Aid for Specially Promoting Research (No. 23000008, Japan), and by the JSPS Core-to-Core Program (No. 23004, Japan).

References

1. N. M. R. Peres, *Rev. Mod. Phys.*, **82**, 2673 (2010).
2. M. Orlita and M. Potemski, *Semicond. Sci. Technol.*, **25**, 063001 (2010); F. Bonaccorso, Z. Sun, T. Hasan, and A. C. Ferrari, *Nat. Photonics*, **4**, 611 (2010).
3. L. A. Falkovsky, *Phys. Usp.*, **51**, 887 (2008); K. F. Mak, L. Ju, F. Wangb, and T. F. Heinz, *Solid State Commun.*, **152**, 1341 (2012).

4. The imaginary contribution $\mathrm{Im}\sigma_\omega$ is expressed through $\mathrm{Re}\sigma_{\mathrm{inter}}$ with the use of the Kramers-Kronig relation. Using a cutoff frequency correspondent to the width of valence band ε_{m} one obtains a weak contribution $\mathrm{Im}\sigma_\omega \sim (\hbar\omega/2\pi\varepsilon_{\mathrm{m}})e^2/\hbar$.

5. K. Zeeger, *Semiconductor Physics* (Springer, New York, 1973).
6. G. D. Mahan, *Many-Particle Physics* (Plenum Press, New York, 1990); F. T. Vasko and O. E. Raichev, *Quantum Kinetic Theory and Applications* (Springer, New York, 2005).
7. R. R. Nair, P. Blake, A. N. Grigorenko, K. S. Novoselov, T. J. Booth, T. Stauber, N. M. R. Peres, and A. K. Geim, *Science*, **320**, 1308 (2008); K. F. Mak, M. Y. Sfeir, Y. Wu, C. H. Lui, J. A. Misewich, and T. F. Heinz, *Phys. Rev. Lett.*, **101**, 196405 (2008).
8. F. Wang, Y. Zhang, C. Tian, C. Girit, A. Zettl, M. Crommie, and Y. R. Shen, *Science*, **320**, 206 (2008); Z. Q. Li, E. A. Henriksen, Z. Jiang, Z. Hao, M. C. Martin, P. Kim, H. L. Stormer, and D. N. Basov, *Nat. Phys.*, **4**, 532 (2008).
9. J. M. Dawlaty, S. Shivaraman, J. Strait, P. George, Mvs. Chandrashekhar, F. Rana, M. G. Spencer, D. Veksler, and Y. Chen, *Appl. Phys. Lett.*, **93**, 131905 (2008).
10. J. Horng, C.-F. Chen, B. Geng, C. Girit, Y. Zhang, Z. Hao, H. A. Bechtel, M. Martin, A. Zettl, M. F. Crommie, Y. R. Shen, and F. Wang, *Phys. Rev. B*, **83**, 165113 (2011).
11. L. Ren, Q. Zhang, J. Yao, Z. Sun, R. Kaneko, Z. Yan, S. Nanot, Z. Jin, I. Kawayama, M. Tonouchi, J. M. Tour, and J. Kono, *Nano Lett.*, **12**, 3711 (2012).
12. T. Ando, Y. Zheng, and H. Suzuura, *J. Phys. Soc. Jpn.*, **71**, 1318 (2002); T. G. Pedersen, *Phys. Rev. B*, **67**, 113106 (2003); A. Gruneis, R. Saito, G. G. Samsonidze, T. Kimura, M. A. Pimenta, A. Jorio, A. G. Souza Filho, G. Dresselhaus, and M. S. Dresselhaus, *ibid.*, **67**, 165402 (2003); L. A. Falkovsky and A. A Varlamov, *Eur. Phys. J. B*, **56**, 281 (2007).
13. V. P. Gusynin, S. G. Sharapov, and J. P. Carbotte, *Phys. Rev. Lett.*, **96**, 256802 (2006); T. Stauber, N. M. R. Peres, and A. H. Castro Neto, *Phys. Rev. B*, **78**, 085418 (2008).
14. N. M. R. Peres, F. Guinea, and A. H. Castro Neto, *Phys. Rev. B*, **73**, 125411 (2006); S. Yuan, R. Roldan, H. D. Raedt, and M. I. Katsnelson, *Phys. Rev. B*, **84**, 195418 (2011).
15. P. N. Romanets and F. T. Vasko, *Phys. Rev. B*, **83**, 205427 (2011); F. T. Vasko and V. Ryzhii, *Phys. Rev. B*, **76**, 233404 (2007).
16. J. Martin, N. Akerman, G. Ulbricht, T. Lohmann, J. H. Smet, K. von Klitzing, and A. Yacoby, *Nat. Phys.*, **4**, 144 (2007); Y. Zhang, V. W. Brar, C. Girit, A. Zettl, and M. F. Crommie, *Nat. Phys.*, **6**, 722 (2009).

17. B. Scharf, V. Perebeinos, J. Fabian, and P. Avouris, arXiv:1211.3329.
18. V. M. Pereira, R. M. Ribeiro, N. M. R. Peres, and A. H. Castro Neto, *Europhys. Lett.*, **92**, 67001 (2010); V. K. Dugaev and M. I. Katsnelson, *Phys. Rev. B*, **86**, 115405 (2012).
19. L. Yang, J. Deslippe, C.-H. Park, M. L. Cohen, and S. G. Louie, *Phys. Rev. Lett.*, **103**, 186802 (2009); K. F. Mak, J. Shan, and T. F. Heinz, *ibid.*, **106**, 046401 (2011).
20. A. G. Grushin, B. Valenzuela, and M. A. H. Vozmediano, *Phys. Rev. B*, **80**, 155417 (2009); N. M. R. Peres, R. M. Ribeiro, and A. H. Castro Neto, *Phys. Rev. Lett.*, **105**, 055501 (2010); S. H. Abedinpour, G. Vignale, A. Principi, M. Polini, W.-K. Tse, and A. H. MacDonald, *Phys. Rev. B*, **84**, 045429 (2011).
21. V. Ryzhii, M. Ryzhii, and T. Otsuji, *J. Appl. Phys.*, **101**, 083114 (2007); B. Dora, E. V. Castro, and R. Moessner, *Phys. Rev. B*, **82**, 125441 (2010).
22. A. Urich, K. Unterrainer, T. Mueller, *Nano Lett.*, **11**, 2804 (2011); V. Ryzhii, M. Ryzhii, V. Mitin, and T. Otsuji, *J. Appl. Phys.*, **107**, 054512 (2010).

Chapter 8

Voltage-Controlled Surface Plasmon-Polaritons in Double Graphene Layer Structures*

D. Svintsov,[a,b] V. Vyurkov,[a,b] V. Ryzhii,[c,d] and T. Otsuji[c,d]
[a]*Institute of Physics and Technology, Russian Academy of Sciences, Moscow 117218, Russia*
[b]*Department of Physical and Quantum Electronics, Moscow Institute of Physics and Technology, Dolgoprudny 141700, Russia*
[c]*Research Institute for Electrical Communication, Tohoku University, Sendai 980-8577, Japan*
[d]*Japan Science and Technology Agency, CREST, Tokyo 107-0075, Japan*
svintcov.da@mipt.ru

The spectra and damping of surface plasmon-polaritons (SPPs) in double graphene layer structures are studied. It is proved that SPPs in those structures exhibit an outstanding voltage tunability of velocity and damping, inherent to gated graphene, and a pronounced low-

Graphene-Based Terahertz Electronics and Plasmonics: Detector and Emitter Concepts
Edited by Vladimir Mitin, Taiichi Otsuji, and Victor Ryzhii

ISBN 978-981-4800-75-4 (Hardcover), 978-0-429-32839-8 (eBook)
www.jennystanford.com

frequency coupling with photons inherent to non-gated structures. It is also shown that the spatial dispersion of conductivity significantly augments the free path and cutoff frequency of SPPs, which is of great importance for practical applications.

8.1 Introduction

Recently the fabrication of double graphene layer (GL) structures was reported [1]. These structures consist of two GLs separated by a dielectric. If a bias voltage V_b is applied between GLs, the layers serve as a gate for each other. Such a possibility of tuning electronic properties of graphene without any metal gates has already found its application for tunnel field-effect transistors [2] and optical modulators [3]. Moreover, the devices based on 2GL seem promising for resonant THz detectors [4] and near-field amplifiers [5].

Plasmonic properties of 2GL structures are of particular interest too. Firstly, they can operate as modulators of plasma waves (recently proposed for gated graphene in Ref. [6]). Secondly, plasma resonances can affect the characteristics of optical modulators, resulting in greater modulation depth near the plasmon peak [7]. Apart from the plasmons in GL-based structures [8–12], the surface plasmon-polaritons (SPPs) in GLs are of considerable practical interest too [13–15]. The spectra of plasmons in adjacent GLs were previously considered in Ref. [9].

In the present communication, we obtain the spectra of SPPs in double-GL structures. Those spectra exhibit a pronounced coupling with photons at low frequencies in contrast to gated GLs [8]. Moreover, just like in the gated structures, the velocity and damping of SPPs can be tuned by the voltage applied between layers. In particular, at low bias voltages the SPPs are strongly damped due to interband transitions of electrons. At high bias voltages those transitions are forbidden by the Pauli blocking and the propagation length of SPPs is limited only by the Drude absorption. This opens the prospects of using 2GL structures as efficient SPP modulators.

In contrast to many previous works on plasma waves in GL-based structures [9, 12, 16], in our calculations the spatial dispersion of conductivity is rigorously taken into account. It shows up in a major alteration of SPP spectra at the Fermi energies μ comparable with the temperature T or higher. For instance, the spectra $k(\omega)$ (k is the wave vector and ω is the frequency) are no longer scalable with the Fermi energy, and the edge of SPP absorption associated with the interband transitions is significantly shifted from $\omega = 2\mu/\hbar$, expected from the analysis of the Pauli blocking, to higher values.

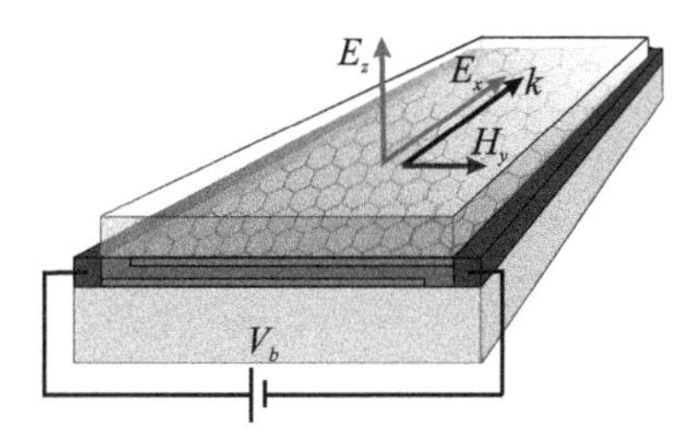

Figure 8.1 Schematic view of the 2GL structure and directions of vectors **E**, **H**, and **k** in the TM SPP.

8.2 Device Model

The double-GL structure under consideration is sketched in Fig. 8.1. The two GLs are separated by a thin insulating layer of width d and permittivity ε_m, the permittivities of the media under and over the GLs can be generally different, they are denoted by ε_b and ε_t, respectively. One side of each GL is connected to the contact while the other side is isolated. A bias voltage V_b can be applied between contacts, thus, one GL can serve as a gate for another. The x-axis coincides with the direction of the SPP propagation parallel to the metal contacts, the z-axis is orthogonal to the GL plane.

The applied bias voltage V_b between GLs (as shown in Fig. 8.1) induces an excess electron density in the top layer and hole density in the bottom one. If the residual doping of GLs is weak, the excess electron and hole densities in layers are equal. Meanwhile, the Fermi energies μ in GLs coincide in absolute values but have the opposite sign. The band diagram of the structure under the bias voltage is shown in Fig. 8.2. The potential difference between GLs is equal to $V_b - 2\mu/e$. The distinction of this value from V_b is attributed to the effect of quantum capacitance. The Gauss theorem binds the bias voltage V_b and the charge density $\pm e\Sigma$ in GLs as follows:

$$\frac{\varepsilon_m}{4\pi d}(V_b - 2\frac{\mu}{e}) = e\Sigma. \tag{8.1}$$

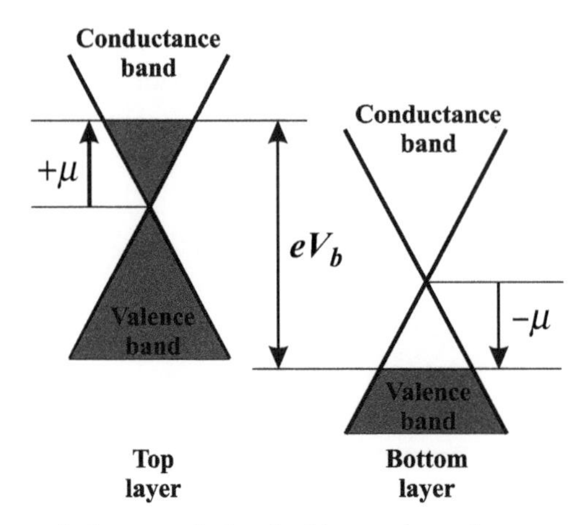

Figure 8.2 Band diagram of the double graphene layer structure at bias voltage V_b.

Provided the Fermi energy μ substantially exceeds the temperature T (measured in energy units), the charge density can be explicitly written,

$$e\Sigma = \frac{e\mu^2}{\pi\hbar^2 v_F^2}. \tag{8.2}$$

Equations (8.1) and (8.2) allow to obtain the value of Fermi energy at given bias voltage and, thus, derive the kinetic characteristics of GLs, in particular, the conductivity necessary for the dispersion of SPPs.

8.3 SPP Dispersion in Double Graphene Layer

Hereafter the spectrum (or dispersion law) will be regarded as the dependence of complex wave vector $\mathbf{k}$ on real frequency ω. Such statement of problem is consistent with the description of wave modulator.

To obtain the dispersion law of SPP one has to solve the complete set of Maxwell equations in free space and then match the solutions on conducting graphene layers. As the imaginary part of graphene conductivity is positive in a wide range of frequencies, the propagation of transverse electric (TE) SPP modes along its surface is prohibited [17]. Thus, we focus only on transverse magnetic (TM) modes with the ac electric and magnetic fields $\mathbf{E} = \{E_x, 0, E_z\}$, $\mathbf{H} = \{0, H_y, 0\}$. The field of SPP is localized near the graphene layers, and, therefore, is searched in the form,

$$\mathbf{E}_t = \mathbf{E}_{0t} e^{i(kx-\omega t)-\kappa_t z}, \tag{8.3}$$

$$\mathbf{E}_m = (\mathbf{E}_{0m\uparrow} e^{\kappa_m z} + \mathbf{E}_{0m\downarrow} e^{-\kappa_m z})\, e^{i(kx-\omega t)}, \tag{8.4}$$

$$\mathbf{E}_b = \mathbf{E}_{0b} e^{i(kx-\omega t)+\kappa_b z}, \tag{8.5}$$

where the indices t, m, and b refer to the top, middle, and bottom media, respectively, $\kappa_i = \sqrt{k^2 - \varepsilon_i \omega^2 / c^2}$, and c is the speed of light.

Matching the vectors **E** and **H** on the graphene layers (top and bottom layer conductivities are σ_t and σ_b, respectively), one arrives at the dispersion law,

$$\left[\frac{i\omega}{4\pi}\left(\frac{\varepsilon_t}{\kappa_t} + \frac{\varepsilon_m}{\kappa_m}\right) - \sigma_t\right]\left[\frac{i\omega}{4\pi}\left(\frac{\varepsilon_b}{\kappa_b} + \frac{\varepsilon_m}{\kappa_m}\right) - \sigma_b\right] e^{2\kappa_m d}$$
$$= \left[\frac{i\omega}{4\pi}\left(\frac{\varepsilon_b}{\kappa_b} - \frac{\varepsilon_m}{\kappa_m}\right) - \sigma_b\right]\left[\frac{i\omega}{4\pi}\left(\frac{\varepsilon_t}{\kappa_t} - \frac{\varepsilon_m}{\kappa_m}\right) - \sigma_t\right]. \tag{8.6}$$

As pointed out above, the Fermi energies in GLs are equal in modulus, hence, due to electron-hole symmetry the layer conductivities are equal $\sigma_t = \sigma_b = \sigma$. If, therewith, the top and bottom media are identical (i.e., $\varepsilon_t = \varepsilon_b = \varepsilon$), the general dispersion relation decouples into two independent modes,

$$\left(\frac{i\omega\varepsilon}{\kappa} - 4\pi\sigma\right) + \frac{i\omega\varepsilon_m}{\kappa_m}\tanh\left(\frac{\kappa_m d}{2}\right) = 0, \tag{8.7}$$

$$\left(\frac{i\omega\varepsilon}{\kappa} - 4\pi\sigma\right)\tanh\left(\frac{\kappa_m d}{2}\right) + \frac{i\omega\varepsilon_m}{\kappa_m} = 0. \tag{8.8}$$

Equation (8.7) corresponds to a symmetric (optical) mode when the current oscillates in phase in both layers, while Eq. (8.8) represents an antisymmetric (acoustic) mode. We are especially interested in symmetric oscillations, because the antisymmetric ones are strongly damped [9]. For small separation between layers ($\kappa_m d \ll 1$), the dispersion of symmetric mode depends only on the total conductivity of two layers [16],

$$\frac{\varepsilon_t}{\kappa_t} + \frac{\varepsilon_b}{\kappa_b} = -\frac{4\pi i}{\omega} \cdot 2\sigma. \tag{8.9}$$

To perform a numerical analysis of SPP spectra using Eq. (8.7), one needs to know the conductivity of GLs. The latter itself is a function of frequency and wave vector. A number of models of the GL conductivity has been developed so far [18], each being valid in a certain range of frequencies and carrier densities. For high-frequency SPPs (from THz to infrared range), we use the relation for the GL conductivity derived from the Kubo formula [19],

$$\sigma_{\mathbf{k}\omega} = \frac{ie^2}{\hbar\pi^2} \sum_{a=1,2} \int \frac{d^2\mathbf{p} v_x^2 \{f[\epsilon_a(\mathbf{p}_-)] - f[\epsilon_a(p_+)]\}}{[\epsilon_a(\mathbf{p}_+) - \epsilon_a(\mathbf{p}_-)][\hbar\omega - \epsilon_a(\mathbf{p}_+) + \epsilon_a(\mathbf{p}_-)]}$$

$$+ \frac{2ie^2\hbar\omega}{\hbar\pi^2} \int \frac{d^2\mathbf{p} v_{21} v_{12} \{f[\epsilon_1(\mathbf{p}_-)] - f[\epsilon_2(\mathbf{p}_+)]\}}{[\epsilon_2(\mathbf{p}_+) - \epsilon_1(\mathbf{p}_-)][(\hbar\omega)^2 - [\epsilon_2(\mathbf{p}_+) - \epsilon_1(\mathbf{p}_-)]^2]}. \tag{8.10}$$

Here the indices 1 and 2 refer to conductance and valence bands, in particular, $\epsilon_1(\mathbf{p}) = |\mathbf{p}|v_F$ and $\epsilon_2(\mathbf{p}) = -|\mathbf{p}|v_F$, $v_F \simeq 10^6$ m/s, $\mathbf{p}_\pm = \mathbf{p} \pm \hbar\mathbf{k}/2$, $f(\epsilon)$ is the electron distribution function (the equilibrium Fermi function $f(\epsilon) = [1 + e^{(\epsilon-\mu)/T}]^{-1}$ is assumed), $v_x = v_F \cos\theta_\mathbf{p}$ and $v_{12} = iv_F \sin\theta_\mathbf{p}$ are the matrix elements of velocity operator. The upper line in Eq. (8.10) corresponds to the intraband transitions, while the lower one accounts for the interband transitions. To allow for the momentum relaxation one should treat the frequency as $\omega \to \omega + i\tau^{-1}$. The definite expressions for τ^{-1} for different scattering processes can be found in Ref. [18]. Here, we restrict our consideration with perfect samples when electron-phonon scattering is the dominating mechanism. The corresponding relaxation time is given by the following formula [20]:

$$\tau^{-1} = \tau_0^{-1} \frac{\mu}{T}, \tag{8.11}$$

where the parameter $\tau_0^{-1} \simeq 3 \times 10^{11}$ s^{-1} at room temperature $T = 26$ meV.

It turns out that in the case under consideration, the spatial dispersion of the conductivity is significant. It can be neither omitted [16] nor treated as a perturbation [12]. As shown in the following, at fairly high Fermi energies the condition $kv_F \ll \omega$ is broken because the electron transitions in momentum space become substantially indirect. In the situation, it is the most convenient to calculate the conductivity (Eq. (8.10)) in elliptic coordinates: $2p_x = k \cosh u \cos v$, $2p_y = k \sinh u \sin v$. It should be noted that the wave vector is complex and the functions $|\mathbf{p} \pm \hbar\mathbf{k}/2|$ should be regarded as absolute values of complex vectors. Extracting the real part from Eq. (8.10) with the help of the Sokhotski theorem in the limit $\tau^{-1} \to 0$, one can arrive at the following expressions for the dissipation parts of the GL conductivity associated with the interband ($\mathrm{Re}\sigma_{\mathrm{inter}}$) and intraband ($\mathrm{Re}\sigma_{\mathrm{intra}}$) transitions, respectively,

$$\mathrm{Re}\,\sigma_{\mathrm{inter}} = \frac{e^2}{2\pi\hbar}\sqrt{1-q^2}\int_0^{\pi}\left\{f\left[-\frac{\hbar\omega}{2}(1+q\cos v)\right]\right.$$
$$\left.-f\left[\frac{\hbar\omega}{2}(1-q\cos v)\right]\right\}\frac{1-q^2\cos^2 v}{1-q^2\sin^2 v}\sin^2 v\,dv, \tag{8.12}$$

$$\mathrm{Re}\,\sigma_{\mathrm{intra}} = \frac{e^2}{2\pi\hbar}\left[q^2-1\right]^{-1/2}\int_0^{\infty}\left\{f\left[\frac{\hbar\omega}{2}(q\cosh u-1)\right]\right.$$
$$\left.-f\left[\frac{\hbar\omega}{2}(q\cosh u+1)\right]\right\}\frac{q^2\cosh^2 u-1}{q^2\sinh^2 u+1}\cosh^2 u\,du, \tag{8.13}$$

where the dimensionless parameter of spatial dispersion $q = |k|v_F/\omega$ is introduced. The reduction of interband absorption with rising q obvious from Eq. (8.12) results in lower damping of SPPs.

8.4 Results and Discussion

The spectra $\mathrm{Re}k(\omega)$ and $\mathrm{Im}k(\omega)$ of SPPs calculated numerically with Eqs. (8.7) and (8.10) are presented in Figs. 8.3 and 8.4. The structure parameters are as follows $\varepsilon_t = \varepsilon_m = \varepsilon_b = 5$ (boron nitride [21]), $d = 2$ nm. The following three characteristic ranges can be singled out in the SPP spectra.

Small wave vectors and frequencies $k \ll 8\alpha\mu/(\hbar c)$, where $\alpha = e^2/(\hbar c)$ is the fine structure constant. In this range of wave vectors, the dispersion is purely "photonic" (see the inset in Fig. 8.3), i.e., the dispersion law reads

$$k = \frac{\omega\sqrt{\varepsilon}}{c}. \tag{8.14}$$

This region of spectra is specific to SPPs in non-gated two-dimensional electron gas, while in the gated structures the velocity of plasma waves at low frequencies is much lower than speed of light [8, 22], and the coupling with photons is impossible.

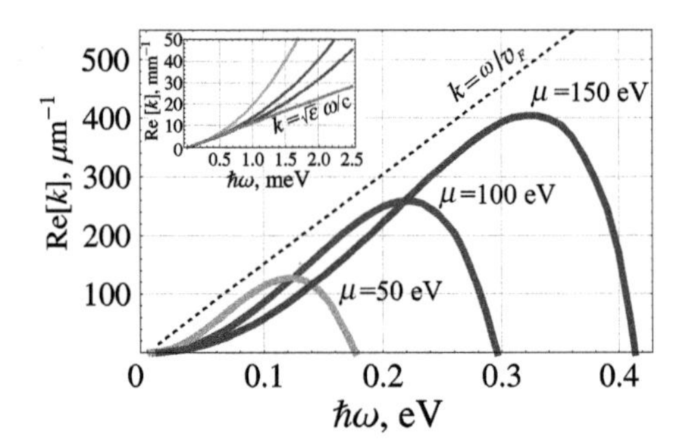

Figure 8.3 Dispersions Re$k(\omega)$ of SPPs in double GL at different Fermi energies. Inset: low-energy "photonic" behavior of spectra.

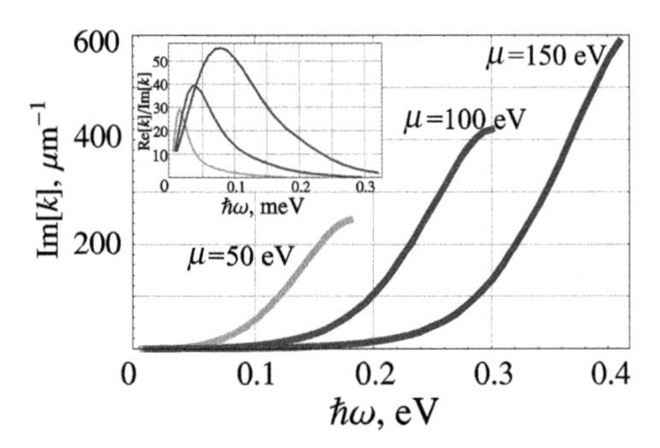

Figure 8.4 Damping Im$k(\omega)$ of SPPs in double GL at different Fermi energies. Inset: quality factor of SPPs defined as Rek/Imk at the same Fermi energies.

Intermediate wave vectors and frequencies $8\alpha\mu/(\hbar c) \ll k \ll \omega/v_{\mathrm{F}}$, $\omega \ll 2\mu/\hbar$. In this situation the coupling with photons is negligible and the spatial dispersion of conductivity could be omitted as well. For this reason, we can accept that $\kappa \approx k$ and simplify the expression for the SPP dispersion,

$$k = \frac{i\omega\varepsilon}{8\pi\sigma_{\mathbf{k}=0,\omega}}, \tag{8.15}$$

where the long-wavelength conductivity at $\mu \gg T$ is

$$\sigma_{\mathbf{k}=0,\omega} = \frac{e^2}{\pi\hbar^2}\frac{\mu\tau}{1-i\omega\tau} + \frac{ie^2}{2\pi\hbar}\ln\left|\frac{2\mu-\hbar\omega}{2\mu+\hbar\omega}\right| + \frac{e^2}{4\hbar}\left[f\left(-\frac{\hbar\omega}{2}\right) - f\left(\frac{\hbar\omega}{2}\right)\right]. \quad (8.16)$$

Equations (8.15) and (8.16) yield a well-known square-law dependence [22] Re $k \propto \omega^2$ until the interband transitions become significant ($\omega \simeq 2\mu/\hbar$) or until the spatial dispersion of conductivity starts playing role ($k \simeq \omega/v_F$).

High frequencies $\omega \sim 2\mu/\hbar$ or high wave vectors $k \sim \omega/v_F$. A simple analysis based on the Pauli exclusion principle states that the direct interband transitions are allowed at frequencies $\omega > 2\mu/\hbar$. At such frequencies, the light incident on GLs is effectively absorbed. This lies in the basis of graphene optical modulators [3]. One could expect that SPPs, just like photons, could also be strongly absorbed in the frequency range $\omega > 2\mu/\hbar$. However, this is not true. The situation is different for SPPs as their wave vectors can be large and the interband transitions are markedly indirect. From Eq. (8.12), it is readily seen that the interband absorption at finite $\mathbf{k}$ is not as strong as it is at $\mathbf{k} = 0$. In the threshold case $|\mathbf{k}| > \omega/v_F$, the interband transitions are forbidden at all due to the non-conservation of the energy and momentum. On the contrary, the intraband transitions at $|\mathbf{k}| > \omega/v_F$ are extremely strong because the quantity Reσ_{intra} exhibits a singularity Re$\sigma_{intra} \propto [(|k|v_F/\omega)^2 - 1]^{-1/2}$.

The imaginary parts of SPP spectra obtained with and without accounting for the spatial dispersion of conductivity are compared in Fig. 8.5, upper panel. The models disregarding the spatial dispersion were widely used before [9, 12, 16]; though, they can overestimate the damping of SPPs by an order of magnitude. The difference in damping rates obtained with these two approaches is especially pronounced at high Fermi energies and frequencies $\omega \sim 2\mu/\hbar$.

The real part of dependence $k(\omega)$ is also altered by indirect transitions, which is shown in Fig. 8.5, lower panel. The dispersions obtained without accounting for spatial dispersion cross the threshold line $k = \omega/v_F$ and penetrate in the region of strong intraband damping $k > \omega/v_F$; but in point of fact, the spatial dispersion of conductivity keeps the spectrum under the threshold.

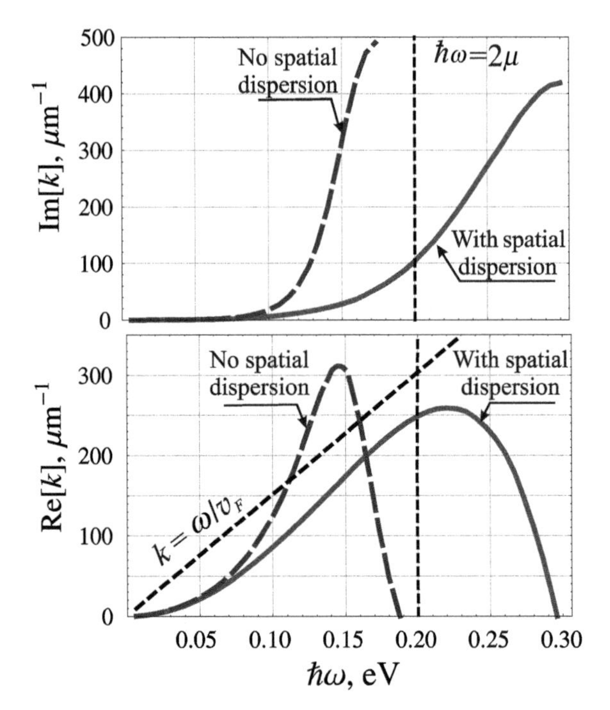

Figure 8.5 Comparison of SPP dispersions obtained with and without taking into account the spatial dispersion of conductivity. Upper panel: Im$k(\omega)$, lower panel: Re$k(\omega)$.

For practical applications, it is important to study the dependence of SPP propagation length l_p = (Imk)$^{-1}$ on the bias voltage V_b. A high sensitivity of propagation length to the bias voltage near the threshold is required for efficient plasmonic modulators.

The dependence $l_p(V_b)$ obtained from Eqs. (8.1) and (8.7) is plotted in Fig. 8.6. At high voltages (such that $2\mu > \hbar\omega$), the propagation length of SPPs is limited by the Drude absorption. If the only scattering mechanism is acoustic phonon scattering, the propagation length at $\mu \gg T$, $2\mu > \hbar\omega$ is voltage-independent and given by

$$l_p = 4\frac{e^2}{\hbar\omega\varepsilon}\frac{T}{\hbar\tau_0^{-1}}. \tag{8.17}$$

Worth noting the propagation length of SPP in double graphene layer is twice higher than in a single layer.

Providing the bias voltage decreases the interband transitions go in, resulting in strong damping. Near the threshold voltage the dependence $l_p(V_b)$ is nearly exponential with the slope $d\lg l_p/dV_b \simeq$ 20 dB/V for the interlayer thickness d = 2 nm. At almost zero bias, the SPP damping might be even greater than that predicted by our present consideration because the electron-hole scattering plays a significant role in the vicinity of the Dirac point [23].

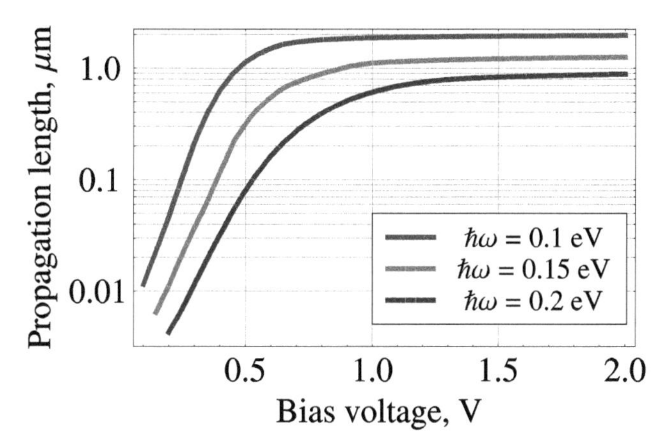

Figure 8.6 SPP propagation length vs. bias voltage V_b between layers at different frequencies.

8.5 Conclusions

The spectra of surface plasmon-polaritons in double graphene layer structures are obtained. Those excitations combine the voltage control inherent to gated structures and pronounced low-frequency coupling with photons intrinsic to non-gated graphene. The bias voltage applied between graphene layers shifts their Fermi energies and, hence, changes the rate of interband plasmon absorption. At low bias voltages, the SPPs are strongly damped, while at high voltages their propagation length is large since it is limited only by Drude absorption. This effect can be used in double-GL plasmonic modulators.

The substantial influence of the spatial dispersion of the GL conductivity (i.e., the indirect electron transitions) on the SPP spectra in double-GL structures is revealed. In particular, the interband cutoff frequency of SPPs is shifted to much higher values. The propagation length of SPPs turns out to be by an order of magnitude higher than that obtained from previous models neglecting the spatial dispersion.

Acknowledgement

The work of D. Svintsov was supported by the Grant 14.132.21.1687 of the Russian Ministry of Education and Science. The work of V. Vyurkov was supported by the Grant 11-07-00464 of the Russian Foundation for Basic Research. The work at RIEC was supported by the Japan Science and Technology Agency, CREST and by the Japan Society for Promotion of Science.

References

1. K. Kim, J. Choi, T. Kim, S. Cho, and H. Chung, *Nature*, **479**, 338–344 (2011).
2. L. Britnell, R. Gorbachev, R. Jalil, B. Belle, F. Schedin, A. Mishchenko, T. Georgiou, M. Katsnelson, L. Eaves, S. Morozov, N. Peres, J. Leist, A. Geim, K. Novoselov, and L. Ponomarenko, *Science*, **335**, 947 (2012).
3. M. Liu, X. Yin, and X. Zhang, *Nano Lett.*, **12**(3), 1482–1485 (2012).
4. V. Ryzhii, T. Otsuji, M. Ryzhii, and M. S. Shur, *J. Phys. D: Appl. Phys.*, **45**, 302001 (2012).
5. T. Stauber and G. Gomez-Santos, *Phys. Rev. B*, **85**, 075410 (2012).
6. D. R. Andersen, *J. Opt. Soc. Am. B*, **27**(4), 818–823 (2010).
7. V. Ryzhii, T. Otsuji, M. Ryzhii, V. G. Leiman, S. O. Yurchenko, V. Mitin, and M. S. Shur, *J. Appl. Phys.*, **112**, 104507 (2012).
8. V. Ryzhii, A. Satou, and T. Otsuji, *J. Appl. Phys.*, **101**, 024509 (2007).
9. E. H. Hwang and S. Das Sarma, *Phys. Rev. B*, **80**, 205405 (2009).
10. F. Koppens, D. Chang, and F. Javier Garcia de Abajo, *Nano Lett.*, **11**(8), 3370–3377 (2011).
11. L. Ju, B. Geng, J. Horng, C. Girit, M. Martin, Z. Hao, H. A. Bechtel, X. Liang, A. Zettl, Y. Ron Shen, and F. Wang, *Nat. Nanotechnol.*, **6**, 630–634 (2011).
12. M. Jablan, H. Buljan, and M. Soljacic, *Phys. Rev. B*, **80**, 245435 (2009).
13. Yu. Bludov, N. M. R. Peres, and M. I. Vasilevskiy, *Phys. Rev. B*, **85**, 245409 (2012).

14. A. Ferreira and N. M. R. Peres, *Phys. Rev. B*, **86**, 205401 (2012).
15. Yu. V. Bludov, M. I. Vasilevskiy, and N. M. R. Peres, *Europhys. Lett.*, **92**, 68001 (2010).
16. A. A. Dubinov, V. Ya. Aleshkin, V. Mitin, T. Otsuji, and V. Ryzhii, *J. Phys.: Condens. Matter*, **23**, 145302 (2011).
17. G. W. Hanson, *J. Appl. Phys.*, **103**, 064302 (2008).
18. S. Das Sarma, S. Adam, E. H. Hwang, and E. Rossi, *Rev. Mod. Phys.*, **83**, 407–470 (2011).
19. L. A. Falkovsky and A. A. Varlamov, *Eur. Phys. J. B*, **56**, 281–284 (2007).
20. F. Vasko and V. Ryzhii, *Phys. Rev. B*, **76**, 233404 (2007).
21. R. Geick, C. H. Perry, and G. Rupprecht, *Phys. Rev.*, **146**, 543–547 (1966).
22. S. Das Sarma and E. H. Hwang, *Phys. Rev. Lett.*, **102**, 206412 (2009).
23. D. Svintsov, V. Vyurkov, S. Yurchenko, V. Ryzhii, T. Otsuji, *J. Appl. Phys.*, **111**(8), 083715-083715-10 (2012).

Chapter 9

Hydrodynamic Electron Transport and Nonlinear Waves in Graphene*

D. Svintsov,[a] V. Vyurkov,[a] V. Ryzhii,[b] and T. Otsuji[b]
[a]*Institute of Physics and Technology, Russian Academy of Science, Moscow 117218 and Moscow Institute of Physics and Technology, Dolgoprudny 141 700, Russia*
[b]*Research Institute for Electrical Communication, Tohoku University, Sendai 980-8577, Japan*
v-ryzhii@riec.tohoku.ac.jp

We derive the system of hydrodynamic equations governing the collective motion of massless fermions in graphene. The obtained equations demonstrate the lack of Galilean and Lorentz invariance, and contain a variety of nonlinear terms due to quasirelativistic nature of carriers. Using those equations, we show the possibility of soliton formation in electron plasma of gated graphene. The quasirelativistic effects set an upper limit for soliton amplitude, which marks graphene out of conventional semiconductors. The

Graphene-Based Terahertz Electronics and Plasmonics: Detector and Emitter Concepts
Edited by Vladimir Mitin, Taiichi Otsuji, and Victor Ryzhii

ISBN 978-981-4800-75-4 (Hardcover), 978-0-429-32839-8 (eBook)
www.jennystanford.com

mentioned noninvariance of equations is revealed in spectra of plasma-waves in the presence of steady flow, which no longer obey the Doppler shift. The feasibility of plasma-wave excitation by direct current in graphene channels is also discussed.

9.1 Introduction

The models of carrier transport in graphene should account for strong carrier-carrier interaction [1, 2], which is governed by a large 'fine-structure constant' $e^2/(\hbar \upsilon_F) \sim 1$ and logarithmically divergent collision integral for collinear scattering [3, 4]. The relative strength of carrier-carrier scattering compared to other relaxation mechanisms was also proved in the transparency measurements of optically pumped graphene [5, 6]. The corresponding relaxation time is estimated to be less than 100 fs.

The most natural way to account for carrier-carrier interactions in transport models is to use local equilibrium (hydrodynamic) distribution functions as a first approximation to the solution of kinetic equation [7]. Several approaches for description of hydrodynamic transport in graphene were presented in Refs. [8–15]. Within hydrodynamic models, it is possible to explain the temperature-independent dc conductivity of graphene in charge neutrality point [8] and the strong Coulomb drag between electrons and holes [8, 15, 16]. The other predictions of hydrodynamic transport, such as preturbulent current flow due to low viscosity [11], existence of electron-hole sound [8], and current saturation at high electric fields due to heating of electrons [10], still expect their experimental verification.

In recent works on graphene hydrodynamics [9, 11 ,14], the equations were obtained under assumption of low drift velocity u of electron plasma ($u \ll \upsilon_F$). Several nonlinear terms were inevitably lost under such assumption. In several other works [12, 13], the hydrodynamics of massless quasiparticles in graphene was obtained from hydrodynamics of ultrarelativistic plasma by a simple replacement of the speed of light, c, by the Fermi velocity υ_F. Such a spurious analogy is misleading in this particular case as electrons in graphene are neither Galilean- [17] nor truly Lorentz-invariant system (in general, this refers to any electrons in solids).

The reason is that for velocities less and of the order of $v_F \simeq c/300$, the distortion of spacetime metrics is negligible. Hence, dealing with quasirelativistic particles in Galilean spacetime, one will obtain hydrodynamic equations that are neither Galilean nor Lorentz invariant.

In this chapter, we present an explicit derivation of hydrodynamic equations for massless electrons in graphene, following the general strategy put forward by Achiezer et al. [18]. In addition, we eliminate the restriction on the flow velocity requiring it to be much less than the Fermi velocity. This opens up an opportunity to study a wide variety of nonlinear phenomena, such as propagation of large-amplitude waves [19, 20], photovoltaic response [21], transport at high current flows [22], and acousto-electronic interactions [23]. The peculiarities of electron interactions with impurities and phonons along with electron-hole interactions have already been studied within a hydrodynamic approach in Refs. [8, 10, 13]. For this reason, in the present work we focus mainly on nondissipative nonlinear hydrodynamic transport.

In graphene it is impossible to introduce a constant electron effective mass m as a proportionality coefficient binding the momentum to the velocity. Consequently, the Euler equation can be no longer presented in canonical form $\partial_t \mathbf{u} + (\mathbf{u}\nabla)\mathbf{u} + (\nabla P)/\rho = 0$, where $\mathbf{u}$ is the drift velocity, ρ is the mass density of electrons, and P is the pressure. However, a fictitious (hydrodynamic) mass $\mathcal{M}$ depending on the particle density n and the drift velocity naturally arises in the Euler equation. One more unusual feature of Euler equation is an appearance of density- and velocity-dependent factor before the convection term $(\mathbf{u}\nabla)\mathbf{u}$. We show that in the degenerate electron system this term vanishes, giving way to a weaker fourth-order nonlinearity proportional to $u^2(\mathbf{u}\nabla)\mathbf{u}$.

Using the derived equations, we study the nonlinear effects in plasma-wave propagation in graphene. Hydrodynamics proved to be an extremely efficient tool for the study of electron plasma in two-dimensional (2D) electron systems [24–27]. Collective dynamics of electrons in two dimensions has a rich analogy with the hydrodynamics of liquids, including the phenomena of electron flow choking [27] and formation of shallow- and deep-water plasma waves in gated and nongated systems, respectively [26]. Despite huge efforts in the field of graphene plasmonics [28–32], the problem of nonlinear plasma waves stayed beyond the scope of recent works.

We show that the balance between nonlinearities and dispersion allows the formation of solitary plasma waves in gated graphene. We find that 'relativistic' terms in Euler equation set an upper limit for the soliton amplitude and broaden its profile. This much differs from the solitons in systems of massive 2D electrons [25, 33], which behave similarly to the solitary waves in water.

The features of electron hydrodynamics are also pronouncedly revealed in the spectra of collective (plasma) excitations in graphene in the presence of stationary electron flow with velocity u_0. It is natural to expect that velocities of forward and backward plasma waves are $u_0 \pm s_0$, where s_0 is the wave velocity in electron fluid at rest. We demonstrate, however, that in graphene the dependence of the wave velocities on the flow velocity is more complicated due to the lack of Galilean and Lorentz invariance. In particular case of strongly degenerate electron system, these velocities are $\frac{1}{2}u_0 \pm s_0$.

We also revealed the potentiality of plasma instability in gated graphene in the presence of steady current. We prove that this effect (predicted for high-mobility 2D electron systems based on the conventional semiconductors by Dyakonov and Shur [25]) persists for graphene with its unusual hydrodynamics. We find the ultimate increment of plasma waves and show that this instability could be realized in graphene channels of submicron length.

The work is organized as follows. In Section 9.2 we derive the set of hydrodynamic equations and discuss the terms arising due to the massless nature of Dirac fermions. In Section 9.3 we demonstrate several solutions of those equations revealing the features of electrons in graphene. In particular, we obtain the profiles of solitary waves in gated graphene, find the spectra of plasma waves in the presence of steady flow, and show the possibility of plasma-wave self-excitation under certain boundary conditions. The main results are discussed in Section 9.4. Some mathematical details concerning the derivation of equations are singled out into Appendix.

9.2 Derivation of Hydrodynamic Equations

We consider 2D plasma of massless electrons in graphene with a linear dispersion law $\epsilon_{\mathbf{p}} = pv_{\mathrm{F}}$. We assume the Fermi level to be above the Dirac point and neglect the contribution of holes.

The starting point for derivation of hydrodynamic equations lies in the construction of the distribution function of carriers which turns the collision integral to zero. Regardless of the energy spectrum $\epsilon_{\mathbf{p}}$ this function is

$$f(\mathbf{p}) = \left[1 + \exp\left(\frac{\epsilon_{\mathbf{p}} - \mathbf{pu} - \mu}{T}\right)\right]^{-1}, \tag{9.1}$$

where the quantities defined from the hydrodynamic equations are the chemical potential μ, the drift velocity $\mathbf{u}$, and the temperature T (measured in the energy units).

The set of hydrodynamic equations is obtained by integrating the kinetic equation timed by 1, p_i, and $\epsilon_{\mathbf{p}}$ over the phase space. The kinetic equation for massless electrons reads

$$\frac{\partial f}{\partial t} + \upsilon_{\mathrm{F}} \frac{\mathbf{p}}{p} \frac{\partial f}{\partial \mathbf{r}} + \mathbf{F} \frac{\partial f}{\partial \mathbf{p}} = St_{\mathrm{e-i}}\{f\} + St_{\mathrm{e-e}}\{f\}. \tag{9.2}$$

Here $St_{\mathrm{e-i}}\{f\}$ includes the electron-impurity and electron-phonon collision integrals, $St_{\mathrm{e-e}}\{f\}$ is the electron-electron collision integral, $\mathbf{F} = e\partial\varphi/\partial\mathbf{r}$ is the force acting on electron, $e = |e|$. The dissipative terms in hydrodynamic equations due to the electron-impurity and electron-phonon collisions were discussed in previous works [8, 10], and further will be omitted.

It is instructive that all statistical average values like the electron density $n = 4\Sigma_{\mathbf{p}} f_{\mathbf{p}}$, flux $\mathbf{j} = 4\Sigma_{\mathbf{p}} \mathbf{v}_{\mathbf{p}} f_{\mathbf{p}}$, and internal energy density $\varepsilon = 4\Sigma_{\mathbf{p}} \epsilon_{\mathbf{p}} f_{\mathbf{p}}$ can be calculated exactly with the distribution function (9.1) when the spectrum $\epsilon_{\mathbf{p}}$ is linear (see the derivation in Appendix); namely,

$$n = \frac{n_0}{\left[1 - u^2/\upsilon_{\mathrm{F}}^2\right]^{3/2}}, \tag{9.3}$$

$$\mathbf{j} = n\mathbf{u}, \tag{9.4}$$

$$\varepsilon = \frac{\varepsilon_0}{\left[1 - u^2/\upsilon_{\mathrm{F}}^2\right]^{5/2}}. \tag{9.5}$$

Here n_0 and ε_0 are the steady-state particle density and energy density, respectively, given by

$$n_0 = \frac{2T^2}{\pi\hbar^2 \upsilon_F^2} \int_0^\infty \frac{tdt}{1+e^{t-\mu/T}}, \tag{9.6}$$

$$\varepsilon_0 = \frac{2T^2}{\pi\hbar^2 \upsilon_F^2} \int_0^\infty \frac{t^2 dt}{1+e^{t-\mu/T}}. \tag{9.7}$$

For brevity, we introduce the pressure of electron plasma

$$P = \frac{\varepsilon}{2}, \tag{9.8}$$

and an analogue of the electron mass density

$$\rho = \frac{3\varepsilon}{2\upsilon_F^2}. \tag{9.9}$$

In these notations, the set of hydrodynamic equations takes on the form

$$\frac{\partial n}{\partial t} + \frac{\partial (n u_i)}{\partial x_i} = 0, \tag{9.10}$$

$$\frac{\partial (\rho u_i)}{\partial t} + \frac{\partial \Pi_{ij}}{\partial x_j} - en\frac{\partial \varphi}{\partial x_i} = 0. \tag{9.11}$$

Equations (9.10) and (9.11) represent the continuity and the Euler equation, respectively. The elements of stress tensor are (the velocity **u** is directed along x axis)

$$\Pi_{xx} = P\left[1 + 2(u/\upsilon_F)^2\right], \tag{9.12}$$

$$\Pi_{yy} = P\left[1 - 2(u/\upsilon_F)^2\right]. \tag{9.13}$$

The heat transfer equation (which will not be discussed here in detail) reads

$$\frac{\partial \varepsilon}{\partial t} + \upsilon_F^2 \frac{\partial (\rho u_i)}{\partial x_i} + enu_i \frac{\partial \varphi}{\partial x_i} = 0. \tag{9.14}$$

It is important that the electron sheet density n arises in the continuity equation, while the mass density ρ arises in the Euler equation. Those quantities are not directly proportional to each other as it is impossible to introduce a constant electron effective mass m. In other words, the fictitious particle mass $\mathcal{M} = \rho/n$ is density- and velocity-dependent and cannot be factored out of the differential operator. In the degenerate electron system ($\mu \gg T$) the expression for mass reads

$$\mathcal{M} = \frac{\mu / \upsilon_{\mathrm{F}}^2}{\sqrt{1-u^2/\upsilon_{\mathrm{F}}^2}}. \tag{9.15}$$

To recognize the peculiarities of the Euler equation obtained and analyze the emerging nonlinearities, it would be convenient to present it in the canonical form (hereafter we restrict ourselves to one-dimensional motion). Excluding the time derivatives of density ρ with the use of Eqs. (9.3) and (9.5) [see also Eqs. (9.A5) and (9.A6)], we arrive at the following equation:

$$\begin{aligned}\frac{\partial u}{\partial t}\left[1+\frac{\beta^2(5-6\xi)}{1-\beta^2}\right]+u\frac{\partial u}{\partial x}\left[(3-4\xi)-\frac{\beta^2(5-6\xi)}{1-\beta^2}\right]\\+\frac{2\xi\upsilon_{\mathrm{F}}^2}{3n}\frac{\partial n}{\partial x}(1-\beta^2)-\frac{n}{\rho}\frac{\partial(e\varphi)}{\partial x}=0.\end{aligned} \tag{9.16}$$

Here we have introduced the relativistic factor $\beta = u/\upsilon_{\mathrm{F}}$ and the dimensionless function ξ characterizing the thermodynamic state of the electron system,

$$\xi = \frac{n^2}{\varepsilon\langle\varepsilon^{-1}\rangle}, \tag{9.17}$$

where $\langle\varepsilon^{-1}\rangle \neq \varepsilon^{-1}$ is the density of inverse energy [see Eq. (9.A4)]. The function ξ varies from 1/2 at $\mu/T \to -\infty$ to 3/4 at $\mu/T \to +\infty$. At $\mu \gg T$ it is given by the following asymptotic relation:

$$\xi_{\mu\gg T} = \frac{3}{4}\left(1-2\frac{n_T}{n}\right). \tag{9.18}$$

Here n_{T} is the density of thermally activated electrons at $\mu = 0$.

In hydrodynamic equations for massive particles, two sources of nonlinearities exist: the current-density term in continuity equation, $\partial_{\mathrm{x}}(nu)$, and the nonlinear-convection term in Euler equation, $u\partial_{\mathrm{x}}u$. Much greater variety of nonlinearities is involved in the Euler equation for electrons in graphene (9.16). They can be classified as relativistic nonlinearities due to high drift velocities and nonlinearities due to density dependence of hydrodynamic mass

$\mathcal{M}$. To compare their 'strength,' we consider small perturbations of density, velocity, and electric potential: $n = n_0 + \delta n(x, t)$, $u = \delta u(x, t)$, $\varphi = \varphi_0 + \delta\varphi(x, t)$.

It is easy to see that the relativistic nonlinearities are, at least, the third-order terms. Dropping them, we can rewrite the Euler equation as

$$\frac{\partial u}{\partial t}+u\frac{\partial u}{\partial x}(3-4\xi)+\frac{1}{\mathcal{M}}\left[\frac{1}{n}\frac{\partial P}{\partial x}-\frac{\partial(e\varphi)}{\partial x}\right]=0. \tag{9.19}$$

The nonlinear convection term in Euler equation is weakened due to the factor $3 - 4\xi < 1$. In degenerate electron system this term can be estimated as

$$6\frac{n_T}{n_0}\delta u\frac{\partial\delta u}{\delta x}\approx 6\frac{n_T}{n_0}\frac{s_0^2}{n_0^2}\delta n\frac{\partial\delta n}{\delta x}. \tag{9.20}$$

Here s_0 is the velocity of collective excitations (plasma waves) in graphene. At low temperatures and elevated Fermi energies this term becomes infinitesimal and surrenders to the higher-order nonlinearity $u^3\delta_x u$.

In comparison with the convective term, the nonlinearity due to the density dependent mass ($\mathcal{M} \propto n^{1/2}$) is much stronger. The corresponding term could be evaluated as

$$-\frac{1}{2\mathcal{M}}\frac{\delta n}{n_0}\delta\left[\frac{1}{n_0}\frac{\partial P}{\partial x}-\frac{\partial(e\varphi)}{\partial x}\right]\approx-\frac{s_0^2}{2n_0^2}\delta n\frac{\partial\delta n}{\partial x}. \tag{9.21}$$

It is readily seen from the above considerations that nonlinear transport phenomena in graphene are rather governed by the density dependence of mass $\mathcal{M}$ than by nonlinear convection, which used to occur in common semiconductors.

9.3 Nonlinear Effects in Plasma-Wave Propagation

9.3.1 Formation of Solitons in Gated Graphene

In gated structures the electron density n is related to the local electric field $-\partial\varphi/\partial x$ in graphene via a weak nonlocality approximation (see derivation in Appendix B, and also Ref. [25])

$$\frac{\partial(e\varphi)}{\partial x}=-\frac{4\pi e^2 d_1 d_2}{\kappa(d_1+d_2)}\frac{\partial n}{\partial x}-\frac{4\pi e^2 d_1^2 d_2^2}{3\kappa(d_1+d_2)}\frac{\partial^3 n}{\partial x^3}. \tag{9.22}$$

Here d_1 and d_2 are distances from the graphene layer to the top and bottom gates, respectively, and κ is the gate dielectric permittivity. The first term on the right-hand side simply follows from Gauss's theorem for uniform electron density. The third derivative term is associated with a weak nonuniformity of charge in 2D layer; it accounts for a weak dispersion of plasma waves in gated structures. A subtle balance between dispersion and nonlinearity results in the formation of solitary waves.

We search for the solutions of hydrodynamic equations in the form $n = n_0 + \delta n(z)$, $u = \delta u(z)$, where $z = x - u_0 t$ is the running coordinate, and u_0 is the soliton velocity being slightly different from the plasma-wave velocity s_0 due to the dispersion. Within the hydrodynamic model, the expression for plasma-wave velocity can be represented as [8]

$$s_0^2=\upsilon_{\rm F}^2\frac{2\xi}{3}\left(1+\frac{4\pi e^2}{\kappa}\frac{d_1 d_2}{d_1+d_2}\left\langle\varepsilon^{-1}\right\rangle\right). \tag{9.23}$$

Integration of the continuity equation (9.10) provides the relation between u and n:

$$u=u_0\frac{n-n_0}{n}. \tag{9.24}$$

Eliminating the drift velocity u and the electric potential φ from the Euler equation (9.16) with the help of Eqs. (9.22) and (9.24), we arrive at the dynamic equation for a solitary wave. The latter is concisely represented in the limit $\mu \gg T$ using the dimensionless variables $\varsigma=z\sqrt{3/(d_1 d_2)}$, $v=\delta n/n_0$, $\beta=u/\upsilon_{\rm F}$, $\beta_0=u_0/\upsilon_{\rm F}$, $\tilde{s}_0=s_0/\upsilon_{\rm F}$ as follows:

$$F(v)\frac{\partial v}{\partial\varsigma}+\left(\tilde{s}_0^2-\frac{1}{2}\right)\frac{\partial^3 v}{\partial\varsigma^3}=0, \tag{9.25}$$

$$F(v)=\tilde{s}_0^2-\frac{1}{2}-\frac{\beta_0\beta^2(\beta+\beta_0)}{2(v+1)^{3/2}(1-\beta^2)}$$

$$+\frac{1-\beta^2}{2\sqrt{v+1}}-\frac{\beta_0^2}{(v+1)^{3/2}}. \tag{9.26}$$

Upon expanding $F(v)$ in series over v, one arrives at the well-known Korteweg–de Vries [34] (KdV) equation

$$\left(\tilde{s}_0^2-\beta_0^2\right)\frac{\partial v}{\partial \zeta}+\left(\tilde{s}_0^2-\frac{1}{2}\right)\frac{\partial^3 v}{\partial \zeta^3}+\left(\frac{3}{2}\beta_0^2-\frac{1}{4}\right)v\frac{\partial v}{\partial \zeta}=0. \tag{9.27}$$

The solutions of this equation correspond to the so-called bright solitons; their shape is given by [35]

$$\delta n(z)=\delta n_{\max}\cosh^{-2}\left[\frac{z}{2}\sqrt{\frac{3}{d_1 d_2}\frac{s_0^2}{2s_0^2-\upsilon_F^2}\frac{\delta n_{\max}}{n_0}}\right]. \tag{9.28}$$

The maximum soliton height $\delta n_{\max}$ is bound to its velocity u_0 via

$$\delta n_{\max}=\frac{n_0}{2}\frac{u_0^2-s_0^2}{s_0^2}. \tag{9.29}$$

The soliton width W is

$$W=2\sqrt{\frac{d_1 d_2}{3}\frac{2s_0^2-\upsilon_F^2}{s_0^2}\frac{\delta n_{\max}}{n_0}}. \tag{9.30}$$

Because $\delta n_{\max} \ll n_0$, and $s_0^2>\upsilon_F^2/2$, the soliton width can be much greater than the distance to the gates, which justifies the applicability of weak nonlocality approximation. Besides, the soliton width should markedly exceed the inelastic (electron-electron) free path to justify the validity of hydrodynamic approach. The free path is less than 100 nm, which follows from the experimental [5, 6] and theoretical [8] estimates of collision frequencies.

Apart from the numerical coefficients, the obtained parameters of solitons coincide with those in 2D plasma of massive electrons in Ref. [25]. To reveal the unique features of graphene electron hydrodynamics, one should go beyond the condition $\delta n \ll n_0$ and analyze the general expression for $F(v)$. The necessity of rigorous treatment arises when the velocity u approaches the Fermi velocity, i.e., already at $\delta n_{\max}/n_0 \approx \upsilon_F/s_0$.

The numerical solution of Eq. (9.25) shows that solitons exist when the maximum particle density $\delta n_{\max}$ lies below a certain critical density. The higher is the plasma-wave velocity s_0, the lower is the critical density. The relation between δn_{max} and soliton velocity $u_0 - s_0$ is plotted in Fig. 9.1, the termination of the curves corresponds to the critical density and velocity. It is seen from Eq. (9.28) that the soliton width shrinks as its amplitude increases. The numerical results of solving the rigorous KdV equation (9.25) plotted inside the insets in Fig. 9.1 indicate that the width of soliton decreases only slightly as its amplitude grows. Given the value of δn_{max}, the profile of real soliton is broader than that obtained from 'nonrelativistic' approximation (9.27).

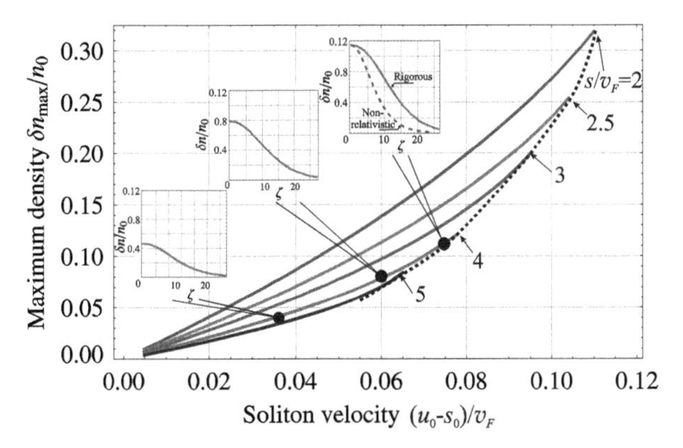

Figure 9.1 Dependence of soliton amplitude $\delta n_{max}/n_0$ on its velocity $(u_0 - s_0)/\upsilon_F$ at different velocities of plasma waves, s_0. The dashed black line indicates the boundary of soliton existence. The profiles of solitary waves obtained from numerical integration of Eq. (9.25) are plotted in the insets.

9.3.2 'Shallow-Water' Plasma Waves in the Presence of Steady Electron Flow

To obtain the spectra of plasma waves in the presence of steady flow with velocity u_0 we linearize the hydrodynamic equations assuming a harmonic time dependence of perturbations

$$n = n_0 + \delta n(x) e^{-i\omega t}, \tag{9.31}$$

$$u = u_0 + \delta u(x) e^{-i\omega t}. \tag{9.32}$$

This procedure leads to the 'equations of motion' for plasma oscillations,

$$-i\omega\delta n + n_0\frac{\partial \delta u}{\partial x} + u_0\frac{\partial \delta n}{\partial x} = 0, \tag{9.33}$$

$$-i\omega[1+\gamma]\delta u + u_0\frac{\partial \delta u}{\partial x}[3-4\xi_0-\gamma] + \frac{s_0^2}{n_0}\frac{\partial \delta n}{\partial x} = 0, \tag{9.34}$$

where we have introduced another relativistic factor γ:

$$\gamma = \frac{\beta_0^2}{1-\beta_0^2}(5-6\xi_0). \tag{9.35}$$

A weak dispersion of plasma waves was neglected here.

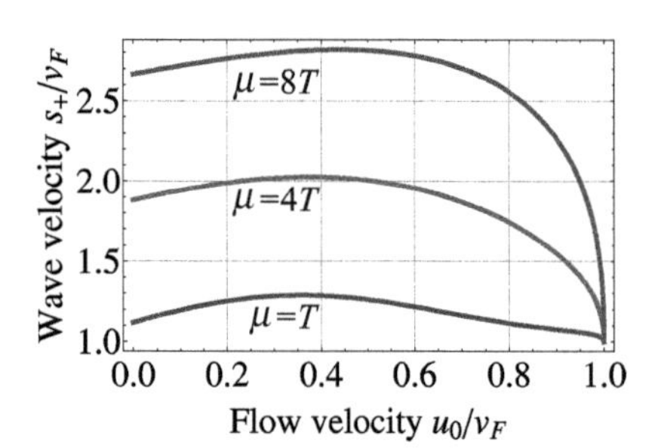

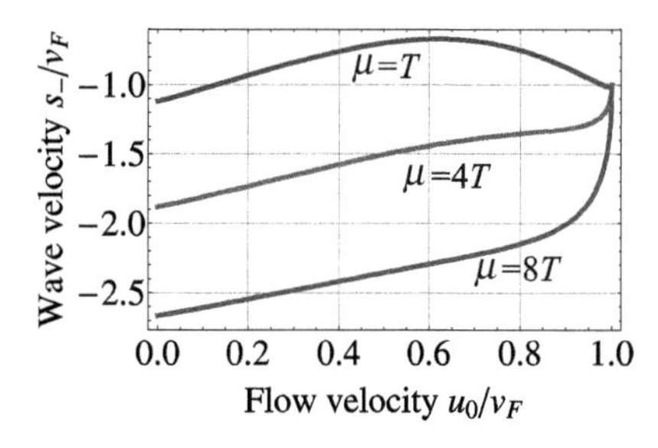

Figure 9.2 Dependencies of plasma-wave velocities $s_\pm$ in the presence of steady flow on the flow velocity u_0.

Assuming a harmonic dependence of all quantities on the coordinate, that is $\propto e^{ikx}$, we obtain the linear law of plasma-wave dispersion $\omega_\pm = s_\pm k$. The velocities of forward (s_+) and backward (s_-) waves are given by

$$s_\pm = \frac{2u_0(1-\xi) \pm \sqrt{s_0^2(1+\gamma) + u_0^2(2\xi-1+\gamma)^2}}{1+\gamma}. \tag{9.36}$$

This relation distinctly manifests the lack of Galilean invariance in the graphene hydrodynamic equations. On the contrary, in 2D plasma of massive electrons, the velocities $s_{m\pm}$ are given by

$$s_{m\pm} = u_0 \pm s_0. \tag{9.37}$$

In graphene, the presence of steady flow modifies the properties of electron system; therefore, the plasma-wave velocity $s_\pm$ depends on the flow velocity in a quite complicated manner described by Eq. (9.36). Particularly, when the flow velocity is small ($\beta_0 \ll 1$), we obtain

$$s_\pm = 2u_0(1-\xi) \pm s_0. \tag{9.38}$$

In a limit of strongly degenerate electrons ($\xi = 3/4$), the velocity of plasma waves is reduced to

$$s_\pm = \frac{1}{2}u_0 \pm s_0. \tag{9.39}$$

The factor 1 /2 formally originates due to the vanishing nonlinear convective term in the Euler equation. In case of large drift velocities u_0, the sign of convective term in Eq. (9.16) can switch from positive to negative. This, in its turn, leads to a decrease in wave velocity s_+ with rising flow velocity. An increase in fictitious mass $\mathcal{M}$ also contributes to this process.

The wave velocities $s_\pm$ obtained from Eq. (9.36) are plotted in Fig. 9.2 for different values of chemical potential, and for flow velocities ranging from zero up to υ_F. An unusual 'halved' drag by the flow at small velocities turns to a decrease in wave velocity at large u_0. Formally, at $u_0 = \upsilon_F$ the velocities of both branches converge to $s_\pm = \pm\upsilon_F$. However, this case is of purely academic interest as such fast flows are unattainable owing to velocity saturation [36].

9.3.3 Excitation of Electron Plasma-Waves by Direct Current

A special kind of plasma-wave instability (Dyakonov–Shur instability) occurs in high-mobility field-effect transistors under the condition of constant drain current [26]. An amplification of plasma-wave amplitude occurs after the reflection from the drain end. The corresponding increment ω''_m (for 2D plasma of massive electrons) was shown to be governed by the ratio of forward- and backward-wave velocities:

$$\omega_m'' = \frac{s_0^2 - u_0^2}{2Ls_0} \ln\left[\frac{s_0 + u_0}{s_0 - u_0}\right]. \tag{9.40}$$

This plasma-wave instability leads to radiation of electromagnetic waves due to oscillations of image charges in metal electrodes [37]. To find out whether such instability persists for the unusual electron dynamics in graphene, we solve Eqs. (9.33) and (9.34) with the boundary conditions

$$\delta n\big|_{x=0} = 0, \quad \left[n_0 \delta u + u_0 \delta n\right]\big|_{x=L} = 0. \tag{9.41}$$

The latter condition corresponds to a constant drain current. This can be realized either for transistors operating in the current-saturation mode [36], or with the help of an external circuit sustaining the constant current.

It is easy to show that the complex eigenfrequencies $\omega_n = \omega'_n + i\omega''_n$ of Eqs. (9.33) and (9.34) with boundary conditions (9.41) are

$$\omega_n' = \frac{\pi n}{2L} \frac{s_0^2 - u_0^2(3 - 4\xi - \gamma)}{\sqrt{s_0^2(1+\gamma) + u_0^2(2\xi - 1 + \gamma)^2}}, \tag{9.42}$$

$$\omega_{n=1}'' = \frac{\omega_{n=1}'}{\pi} \ln\left|\frac{s_+}{s_-}\right|$$

$$\approx \frac{2u_0(1-\xi)}{L} \frac{s_0^2 - u_0^2(3 - 4\xi - \gamma)}{s_0^2(1+\gamma) + u_0^2(1 - 2\xi - \gamma)^2}. \tag{9.43}$$

The imaginary part of the complex frequency is positive, which means an amplification of the waves. We also see that for the existence of instability it does not matter whether the spectrum of electrons is parabolic or linear. The instability persists if only the velocities of forward and backward plasma waves are different. At the same time, compared with massive particles, the wave increment (9.43) is smaller due to a smaller difference in wave velocities [$|s_+| - |s_-| \approx u_0$ for degenerate massless electrons instead of $2u_0$ for massive electrons]. As the flow velocity u_0 increases ($\beta_0 \gtrsim 1/2$), the wave increment begins to fall down because the velocity difference decreases.

The ultimate wave increment in such kind of instability is estimated as $\upsilon_F/(4L)$, which is attained at $u_0 \simeq \upsilon_F/2$ (see Fig. 9.3). For the self-excitation to arise, it should exceed $(2\tau_p)^{-1}$, where τ_p is the momentum relaxation time. Considering the high-quality samples, where scattering from acoustic phonons dominates, we can estimate $\tau_p^{-1} \approx 3 \times 10^{11}$ s^{-1} at room temperature [38]. The self-excitation turns out to be possible for channel lengths $L \lesssim 1.5$ μm. This optimistic anticipation could be hampered by the presence of impurity scattering, velocity saturation, and dependence of relaxation time on electron density. Nevertheless, even for shorter channels the hydrodynamic approach is valid and the self-excitation seems plausible.

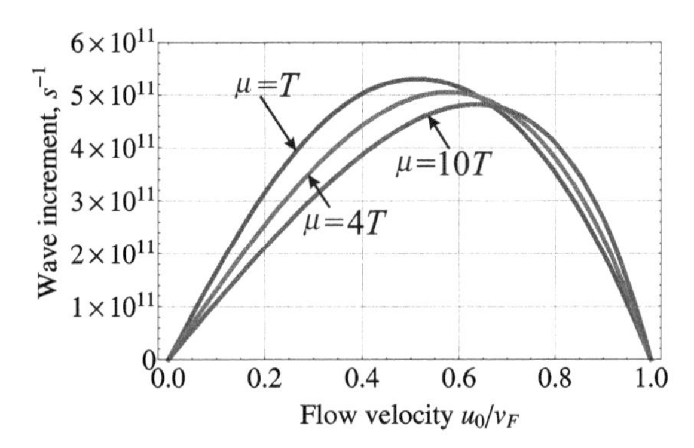

Figure 9.3 Dependence of wave increment ω_1'' on flow velocity u_0 at different values of Fermi energy.

9.4 Discussion of the Results

We have derived hydrodynamic equations describing the transport of massless electrons in graphene. A linear energy spectrum of carriers should be taken into account from the very beginning of derivation. It cannot be introduced as a small correction to the parabolic dispersion (like p^4 terms in Si, Ge, A_3B_5 [39]). On the other hand, the hydrodynamic equations for ultrarelativistic plasmas cannot be also directly applied to graphene as the Fermi velocity is much smaller compared to that of light.

The dependence of particle density n on drift velocity u [(Eq. (9.3)] may look confusing. However, it is the immediate consequence of the particular choice of distribution function (9.1). At the same time, one can choose the distribution function in the form

$$f(\mathbf{p})=\left[1+\exp\left(\frac{p\upsilon_{\mathrm{F}}-\mathbf{pu}}{T(1-u^2/\upsilon_{\mathrm{F}}^2)^{3/4}}-\frac{\mu}{T}\right)\right]^{-1}. \qquad (9.44)$$

This function turns collision integral to zero, reduces to the equilibrium Fermi function at $u \to 0$, and the corresponding particle density does not depend on u (but the internal energy still does). It is easy to show that Euler and continuity equations derived with this function and written in terms of n and u coincide with Eqs. (9.11) and (9.10). The equation of state also holds its view. The boundary conditions for hydrodynamic equations are imposed on measurable quantities n and u. Hence, the solutions of hydrodynamic equations do not depend on the choice between distribution functions (9.1) and (9.44).

In the obtained hydrodynamic equations for electrons in graphene, the effect of linear spectrum is clearly visible. First, the drift velocity u cannot overcome the Fermi velocity υ_{F}. Secondly, a varying fictitious hydrodynamic mass $\mathcal{M} \approx (\mu/\upsilon_{\mathrm{F}}^2)/\sqrt{1-u^2/\upsilon_{\mathrm{F}}^2}$ originates in the Euler equation. The obtained equations are neither Lorentz nor Galilean invaraint, which is directly revealed in the spectra of plasma waves in the presence of steady electron flow [Eq. (9.36)]. Our main conclusions concerning the spectra can be verified experimentally using the techniques of plasmon nano-imaging [30] in gated graphene under applied bias.

As we became aware recently, the spectra of plasma waves in the presence of steady flow and Dyakonov–Shur instability in graphene were analyzed in Ref. [9]. The form of Euler equation used was different from our Eq. (9.16) even in the limit $u \ll \upsilon_{\mathrm{F}}$; particularly, in Ref. [9] the gradient term $u\partial_x u$ did not vanish for degenerate electron systems. This led to different expression for plasma-wave velocities ($3u_0/4 \pm u_0$ instead of our $s_{\pm}$), and to higher estimate of the plasma-wave instability increment. One possible reason for the distinction of the Euler equations lies in dissimilar expressions for the electron plasma pressure P, which is substantially velocity dependent [see Eqs. (9.A2) and (9.A3): $P = \varepsilon/2 \propto \mu^3/(1-\beta^2)^{5/2}$].

The set of problems which could be solved via nonlinear hydrodynamic equations is not restricted within plasma waves. It would be also interesting to study the effects of velocity saturation associated with the upper limit of drift velocity u equal to υ_F. Those effects could be pronounced in graphene samples on substrates with high optical phonon energy. If it is the case, the velocity saturation caused by emission of optical phonons [36] seems as irrelevant.

For rigorous simulation of emerging graphene-based devices for THz generation and detection [21] one can as well employ the derived nonlinear equations. In the case, however, an Euler equation for holes and electron-hole friction terms should be supplied [8]. With large electron mobility and new hydrodynamic nonlinearities, graphene-based THz devices could outperform those based on conventional semiconductors.

9.5 Conclusions

The hydrodynamic equations governing the collective motion of massless electrons in graphene were derived. The validity of those equations is not restricted to small drift velocities. A variable fictitious mass depending on density and velocity arises in the hydrodynamic equations. It results in several nonlinear terms specific to graphene.

The possibility of soliton formation in electron plasma of the gated graphene was shown. The quasirelativistic terms in the dynamic equations set an upper limit of the soliton amplitude and stabilize its shape.

The obtained hydrodynamic equations demonstrate a lack of Galilean and true Lorentz invariance. This noninvariance is pronouncedly revealed in the spectra of plasma waves in the presence of steady flow with velocity u_0. The difference in velocities of forward and backward waves turns out to be u_0 instead of $2u_0$, expected for massive electrons in conventional semiconductors.

The possibility of plasma-wave self-excitation in high-mobility graphene samples under certain boundary conditions (Dyakonov–Shur instability) was demonstrated. The increment of such instability in graphene is less than that in common semiconductors due to smaller difference in velocities of forward and backward waves. However, the high mobility of electrons in graphene allows plasma-wave self-excitation for micron-length and shorter channels.

Acknowledgement

The work of D.S. and V.V. was supported by the Russian Ministry of Education and Science (Grant No. 14.132.21.1687), the Russian Foundation for Basic Research (Grant No. 11-07-00464), and Russian Academy of Science. The work at RIEC was supported by the Japan Science and Technology Agency, CREST and by the Japan Society for Promotion of Science. The authors are grateful to S.O. Yurchenko for helpful discussions, and to Prof. M. Polini for valuable comments.

Appendix A: Calculation of Average Values

The statistical average values can be exactly calculated with the distribution function (9.1) for linear energy spectrum. The particle density is given by

$$n = \frac{4}{(2\pi\hbar)^2}\int \frac{pdpd\theta}{1+e^{\frac{p(\upsilon_F - u\cos\theta)-\mu}{T}}}$$

$$= \frac{T^2}{(\pi\hbar\upsilon_F)^2}\left\{\int_0^\infty \frac{2\pi t dt}{1+e^{t-\mu/T}}\right\}\left\{\frac{1}{2\pi}\int_0^{2\pi}\frac{d\theta}{[1-\beta\cos\theta]^2}\right\}. \quad (9.A1)$$

The last term could be evaluated as $[1 - \beta^2]^{-3/2}$, while the remainder is nothing more but the particle density in the absence of flow n_0. Similarly, the energy density reads as follows:

$$\varepsilon = \frac{T^3}{(\pi\hbar\upsilon_F)^2}\left\{\int_0^\infty \frac{2\pi t^2 dt}{1+e^{t-\mu/T}}\right\}\left\{\frac{1}{2\pi}\int_0^{2\pi}\frac{d\theta}{[1-\beta\cos\theta]^3}\right\}$$

$$= \frac{\varepsilon_0}{[1-\beta^2]^{5/2}}. \quad (9.A2)$$

The stress tensor is

$$\Pi_{xx} = \frac{4}{(2\pi\hbar)^2}\int \frac{\upsilon_F p^2\cos^2\theta dpd\theta}{1+e^{\frac{p(\upsilon_F - u\cos\theta)-\mu}{T}}}$$

$$=\frac{T^3}{(\pi\hbar\upsilon_{\rm F})^2}\left\{\int_0^\infty \frac{2\pi t^2 dt}{1+e^{t-\mu/T}}\right\}\left\{\frac{1}{2\pi}\int_0^{2\pi}\frac{\cos^2\theta d\theta}{[1-\beta\cos\theta]^3}\right\}$$

$$=\frac{\varepsilon}{2}(1+2\beta^2). \qquad (9.A3)$$

The density of inverse energy $\langle\varepsilon^{-1}\rangle$ can be expressed in terms of elementary functions

$$\left\langle\varepsilon^{-1}\right\rangle=\frac{2T\ln[1+e^{\mu/T}]}{\pi\hbar^2\upsilon_{\rm F}^2\sqrt{1-u^2/\upsilon_{\rm F}^2}}. \qquad (9.A4)$$

The following relations for the derivatives of average values are required to represent the Euler equation in the canonical form:

$$dn=\frac{1}{1-\beta^2}(\left\langle\varepsilon^{-1}\right\rangle d\mu+3n\beta d\beta), \qquad (9.A5)$$

$$d\varepsilon=\frac{1}{1-\beta^2}(2nd\mu+5\varepsilon\beta d\beta). \qquad (9.A6)$$

Appendix B: Weak Nonlocality Approximation for Poisson Equation

One solves the Poisson equation for the gated 2D electron gas (2DEG). The 2DEG plane is $z = 0$, the grounded top and bottom gates are placed at $z = d_1$ and $z = -d_2$, respectively, and the top and bottom dielectric permittivities are κ_1 and κ_2. We assume that electron density in 2DEG varies only in the x direction and write the Poisson equation as

$$\frac{\partial^2\varphi}{\partial x^2}+\frac{\partial^2\varphi}{\partial z^2}=0, \qquad (9.B1)$$

with boundary conditions

$$\varphi|_{z=d_1}=\varphi|_{z=-d_2}=0, \qquad (9.B2)$$

$$\varphi|_{z=+0}=\varphi|_{z=-0}, \qquad (9.B3)$$

$$\kappa_1\partial\varphi/\partial z|_{z=+0}-\kappa_2\partial\varphi/\partial z|_{z=-0}=-4\pi\sigma. \qquad (9.B4)$$

Here $\sigma = -en$ is the 2D charge density, and n is the electron density. After the Fourier transform $\varphi_k = \int_{-\infty}^{+\infty} \varphi(x,z)e^{ikx}dx$ the Poisson equation becomes

$$-k^2\varphi_k + \frac{\partial^2 \varphi_k}{\partial z^2} = 0. \tag{9.B5}$$

Solving Eq. (9.B5) with boundary conditions (9.B2)–(9.B4), which apply to the Fourier components as well, we obtain

$$\varphi_k|_{z=0} = \frac{4\pi\sigma_k}{k[\kappa_1 \coth(kd_1) + \kappa_2 \coth(kd_2)]}. \tag{9.B6}$$

Assuming that the 2D electron density varies slowly ($kd_1 \ll 1$, $kd_2 \ll 1$), we expand Eq. (9.B6) in series over k and arrive at the final solution after inverse Fourier transform:

$$\varphi|_{z=0} = \frac{4\pi\sigma}{\kappa_2/d_2 + \kappa_1/d_1} + \frac{4\pi}{3}\frac{d_1\kappa_1 + d_2\kappa_2}{(\kappa_2/d_2 + \kappa_1/d_1)^2}\frac{\partial^2\sigma}{\partial x^2}. \tag{9.B7}$$

Equation (9.B7) is further simplified to Eq. (9.22) for equal permittivities of top- and bottom gate dielectrics $\kappa_1 = \kappa_2 = \kappa$.

References

1. V. N. Kotov, B. Uchoa, V. M. Pereira, F. Guinea, and A. H. Castro Neto, *Rev. Mod. Phys.*, **84**, 1067 (2012).
2. M. Schutt, P. M. Ostrovsky, I. V. Gornyi, and A. D. Mirlin, *Phys. Rev. B*, **83**, 155441 (2011).
3. A. B. Kashuba, *Phys. Rev. B*, **78**, 085415 (2008).
4. L. Fritz, J. Schmalian, M. Müller, and S. Sachdev, *Phys. Rev. B*, **78**, 085416 (2008).
5. J. M. Dawlaty, S. Shivaraman, M. Chandrashekhar, F. Rana, and M. G. Spencer, *Appl. Phys. Lett.*, **92**, 042116 (2008).
6. S. Boubanga-Tombet, S. Chan, T. Watanabe, A. Satou, V. Ryzhii, and T. Otsuji, *Phys. Rev. B*, **85**, 035443 (2012).
7. G. E. Uhlenbeck and G. W. Ford, *Lectures in Statistical Mechanics* (Providence, AMS, 1963).
8. D. Svintsov, V. Vyurkov, S. Yurchenko, V. Ryzhii, and T. Otsuji, *J. Appl. Phys.*, **111**(8), 083715 (2012).
9. A. Tomadin and M. Polini, Phys. Rev. B, **88**, 205426 (2013).

10. R. Bistritzer and A. H. MacDonald, *Phys. Rev. B,* **80**, 085109 (2009).
11. M. Müller, J. Schmalian, and L. Fritz, *Phys. Rev. Lett.,* **103**, 025301 (2009).
12. M. Mendoza, H. J. Herrmann, and S. Succi, *Sci. Rep.*, **3**, 1052 (2013).
13. M. Müller, L. Fritz, and S. Sachdev, *Phys. Rev. B,* **78**, 115406 (2008).
14. R. Roldan, J. N. Fuchs, and M. O. Goerbig, *Solid State Commun.*, (2013).
15. M. Schütt, P. M. Ostrovsky, M. Titov, I. V. Gornyi, B. N. Narozhny, and A. D. Mirlin, *Phys. Rev. Lett.,* **110**, 026601 (2013).
16. J. C. W. Song, D. A. Abanin, and L. S. Levitov, *Nano Lett.,* **13**(8), 3631 (2013).
17. S. H. Abedinpour, G. Vignale, A. Principi, M. Polini, W.-K. Tse, and A. H. MacDonald, *Phys. Rev. B,* **84**, 045429 (2011).
18. A. I. Akhiezer, V. F. Aleksin, and V. D. Khodusov, *Low Temp. Phys.*, **20**, 939 (1994).
19. E. Vostrikova, A. Ivanov, I. Semenikhin, and V. Ryzhii, *Phys. Rev. B,* **76**, 035401 (2007).
20. S. Rudin, G. Samsonidze, and F. Crowne, *J. Appl. Phys.,* **86**, 2083 (1999).
21. L. Vicarelli, M. S. Vitiello, D. Coquillat, A. Lombardo, A. C. Ferrari, W. Knap, M. Polini, V. Pellegrini, and A. Tredicucci, *Nat. Mater.*, **11**, 865 (2012).
22. A. Barreiro, M. Lazzeri, J. Moser, F. Mauri, and A. Bachtold, *Phys. Rev. Lett.,* **103**, 076601 (2009).
23. C. X. Zhao, W. Xu, and F. M. Peeters, *Appl. Phys. Lett.,* **102**, 222101 (2013).
24. A. V. Chaplik, Zh. Eksp. Teor. Fiz., **62**, 746 (1972) [*Sov. Phys. JETP,* **35**, 395 (1972)].
25. A. O. Govorov, V. M. Kovalev, and A. V. Chaplik, JETP Lett., **70**, 488 (1999).
26. M. I. Dyakonov and M. S. Shur, *Phys. Rev. Lett.,* **71**, 2465 (1993).
27. M. I. Dyakonov and M. S. Shur, *Phys. Rev. B,* **51**, 14341 (1995).
28. A. N. Grigorenko, M. Polini, and K. S. Novoselov, *Nat. Photonics,* **6**, 749 (2012).
29. L. Ju, B. Geng, J. Horng, C. Girit, M. Martin, Z. Hao, H. A. Bechtel, X. Liang, A. Zettl, Y. Ron Shen, and F. Wang, *Nat. Nanotechnol.*, **6**, 630 (2011).
30. Z. Fei, A. S. Rodin, G. O. Andreev, W. Bao, A. S. McLeod, M. Wagner, L. M. Zhang, Z. Zhao, M. Thiemens, G. Dominguez, M. M. Fogler, A. H. Castro Neto, C. N. Lau, F. Keilmann, and D. N. Basov, *Nature,* **487**, 82 (2012).

31. S. Das Sarma and E. H. Hwang, *Phys. Rev. Lett.,* **102**, 206412 (2009).
32. V. Ryzhii, A. Satou, and T. Otsuji, *J. Appl. Phys.,* **101**, 024509 (2007).
33. A. Nerses, and E. E. Kunhardt, *J. Math. Phys.*, **39**, 6392 (1998).
34. D. J. Korteweg and G. de Vries, *Philos. Mag.*, **5**, 422 (1895).
35. E. M. Lifshitz and L. P. Pitaevsky, *Physical Kinetics* (Pergamon Press, Oxford, 1981).
36. I. Meric, M. Y. Han, F. A. Young, B. Ozyilmaz, P. Kim, and K. L. Shepard, Nat. Nanotechnol., **3**, 654 (2008).
37. W. Knap, J. Lusakowskia, T. Parenty, S. Bollaert, A. Cappy, V. V. Popov, and M. S. Shur, *Appl. Phys. Lett.,* **84**, 2331 (2004).
38. F. T. Vasko and V. Ryzhii, *Phys. Rev. B,* **76**, 233404 (2007).
39. D. L. Woolard, H. Tian, R. J. Trew, M. A. Littlejohn, and K. W. Kim, *Phys. Rev. B,* **44**, 11119 (1991).

Chapter 10

Effect of Self-Consistent Electric Field on Characteristics of Graphene p-i-n Tunneling Transit-Time Diodes*

V. L. Semenenko,[a] V. G. Leiman,[a] A. V. Arsenin,[a] V. Mitin,[b] M. Ryzhii,[c,d] T. Otsuji,[d,e] and V. Ryzhii[d,e]

[a]*Department of General Physics, Moscow Institute of Physics and Technology, Dolgoprudny, Moscow Region 141700, Russia*
[b]*Department of Electrical Engineering, University at Buffalo, Buffalo, New York 1460-1920, USA*
[c]*Computational Nanoelectronics Laboratory, University of Aizu, Aizu-Wakamatsu 965-8580, Japan*
[d]*Japan Science and Technology Agency, CREST, Tokyo 107-0075, Japan*
[e]*Research Institute for Electrical Communication, Tohoku University, Sendai 980-8577, Japan*
v-ryzhii@riec.tohoku.ac.jp.

*Reprinted with permission from V. L. Semenenko, V. G. Leiman, A. V. Arsenin, V. Mitin, M. Ryzhii, T. Otsuji, and V. Ryzhii (2013). Effect of self-consistent electric field on characteristics of graphene p-i-n tunneling transit-time diodes, *J. Appl. Phys.*, 113, 024503. Copyright © 2013 American Institute of Physics.

Graphene-Based Terahertz Electronics and Plasmonics: Detector and Emitter Concepts
Edited by Vladimir Mitin, Taiichi Otsuji, and Victor Ryzhii

ISBN 978-981-4800-75-4 (Hardcover), 978-0-429-32839-8 (eBook)
www.jennystanford.com

We develop a device model for p-i-n tunneling transit-time diodes based on single- and multiple graphene layer structures operating at the reverse bias voltages. The model of the graphene tunneling transit-time diode (GTUNNETT) accounts for the features of the interband tunneling generation of electrons and holes and their ballistic transport in the device i-section, as well as the effect of the self-consistent electric field associated with the charges of propagating electrons and holes. Using the developed model, we calculate the dc current-voltage characteristics and the small-signal ac frequency-dependent admittance as functions of the GTUNNETT structural parameters, in particular, the number of graphene layers and the dielectric constant of the surrounding media. It is shown that the admittance real part can be negative in a certain frequency range. As revealed, if the i-section somewhat shorter than one micrometer, this range corresponds to the terahertz frequencies. Due to the effect of the self-consistent electric field, the behavior of the GTUNNETT admittance in the range of its negativity of its real part is rather sensitive to the relation between the number of graphene layers and dielectric constant. The obtained results demonstrate that GTUNNETTs with optimized structure can be used in efficient terahertz oscillators.

10.1 Introduction

Pioneering papers by Shur and Eastman [1, 2] have stimulated extensive studies (which continue already for the fourth decade) of ballistic electron and hole transport (BET and BHT, respectively), i.e., collision free transport in short semiconductor structures. The main incentive is the realization of fastest velocities of electrons/holes and, hence, achievement of the operation of diodes and transistors in terahertz range (THz) of frequencies and low power consumption. Even at the initial stage of the ballistic transport research, several concepts of ballistic THz sources have been put forward (see, for instance, an early review [3]). However, the realization THz generation in different semiconductor devices associated with BET/BHT, in particular, analogous to vacuum devices, meets the problems associated with electron scattering in real semiconductor structures. Creation of heterostructures with selective doping with a two-dimensional electron gas (2DEG) spatially separated from the donors has resulted in achievement of very long mean free path of electrons, at least at low temperatures. Recent discoveries of unique properties of graphene [4, 5], in particular, the demonstration of possibility of very long electron and hole mean free path in graphene layers (GLs) and what is even more interesting in multiple graphene layers (MGLs) [6, 7] add optimism in building graphene based THz devices using BET. The concept of graphene tunneling transit-time (GTUNNETT) p-i-n diode, which exhibits a negative dynamic conductivity in the THz range, was proposed and substantiated in Refs. [8, 9]. This concept based not only on BET or quasi-BET (as well as BHT or quasi-BHT) in GLs and MGLs but also on a strong interband tunneling under the electric field with a pronounced anisotropy [10, 11] due to the gapless energy spectrum, and constant absolute value of electrons and holes velocities [4]. Due to this, the electrons in the conduction band and the holes in the valence band generated owing to the interband tunneling in the electric field propogate primarily in the electric field direction with the velocity (in this direction) virtually equal to the characteristic velocity $v_W \simeq 10^8$cm/s. A large value of the directed velocity in GLs and MGLs promotes the device operation at elevated frequencies.

As shown (Refs. [8, 9]), for the self-excitation of THz oscillations in a circuit with GTUNNETT diode, this circuit should serve as a resonator. However, at elevated tunneling currents in GTUNNETTs considered previously [8, 9] the self-consistent charge associated with propagating electron and hole streams can affect the spatial distribution of the self-consistent electric field and the electric potential in the i-section. As a result, the self-consistent electric field near the p-i- and i-n-juctions can be substantially reinforced. This, in turn, influences the tunneling generation of electrons and holes and their transit conditions and, hence, the GTUNNETT dc and ac characteristics.

In this chapter, in contrast to the previous treatment [8, 9], we account for the self-consistent electric field associated with the variations of the electron and hole lateral charges in the i-section and their effect on the injection and the dc and ac characteristics. The effects of the space charge in planar TUNNETTs with the propagation of carriers perpendicular to the structure plane were considered by Gribnikov et al. [12]. The problems of calculation of the dc and ac characteristics of devices based on lateral structures accounting for the self-consistent electric field are substantially complicated by the features of the structure geometry (2D electron and hole channels and blade-like contact regions). In particular, as shown in the following, the related mathematical problems are reduced to a system of rather complex nonlinear integral-differential equations. Using the GTUNNETT device model, we derive these equations, solve them numerically, and find the characteristics.

10.2 Equations of the Model

The device under consideration is based on a GL or MGL structures with p- and n-side sections with the contacts and an undoped transit i-section. For definiteness, we assume that p- and n-sections in a GL or in an MGL are created by chemical doping [13], although in similar devices but with extra gates, these sections can be formed electrically [10, 11]. Under the reverse bias voltage V_0, the potential drops primarily across the i-section A schematic view of the structures in question and their band-diagram under sufficiently strong reverse bias ($V_0 > 0$) corresponding to the potential distribution in the i-section with a markedly nonuniform electric field are shown in Fig. 10.1.

The electrons and holes injected into the i-section of one GL are characterized by the electron and hole sheet concentrations, $\Sigma^- = \Sigma^-(t, x)$ and $\Sigma^+ = \Sigma^+(t, x)$, respectively, and the potential $\varphi = \varphi\ (t, x, y)$. Here, the axis x is directed along the GL (or MGL) plane, i.e., in the direction of the current, the axis y is directed perpendicular to this plane, and t is the time. In the general case when the i-section is based on MGL, $\Sigma^- = \Sigma^-(t, x)$ and $\Sigma^+ = \Sigma^+(t, x)$ are the electron and hole densities in each GLs. In the case of BET and BHT, on which we focus mainly, the electron and hole sheet densities in each GL obey the continuity equations

$$\frac{\partial \Sigma^{\mp}}{\partial t} \pm v_w \frac{\partial \Sigma^{\mp}}{dx} = g, \tag{10.1}$$

and the Poisson equation

$$\frac{\partial^2 \Phi}{\partial x^2} + \frac{\partial^2 \Phi}{\partial y^2} = \frac{4\pi\, eK}{æ}(\Sigma^- - \Sigma^+)\cdot \delta(y), \tag{10.2}$$

respectively. Here, $e = |e|$ is the electron charge, æ is the dielectric constant of the media surrounding GL (or MGL structure), K is the number of GLs in the GTUNNETT structure, and $\delta(y)$ is the delta function reflecting a narrowiness of GL and MGL structures even with rather large number of GLs in the y-direction. Equation (10.1) corresponds to the situation when the electrons and holes generated due to the interband tunneling in the i-section obtain the directed velocities $v_x = v_W$ and $v_x = -v_W$ and preserve them

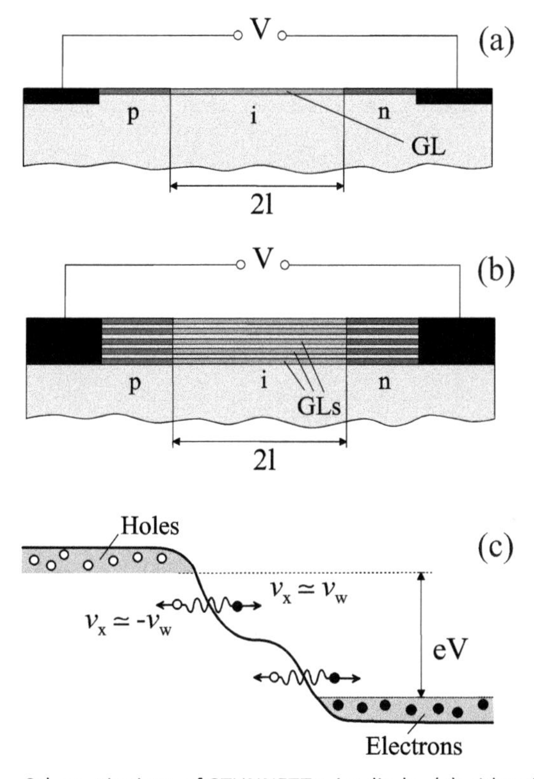

Figure 10.1 Schematic views of GTUNNETT p-i-n diodes (a) with a single GL, (b) with an MGL structure, and (c) their band diagram at reverse bias. Arrows show the propagation directions of electrons and holes generated due to interband tunneling (mainly in those regions, where the electric field is relatively strong).

during the propagation. The boundary conditions correspond to the assumption that the electrons and holes appear in the i-section only due to the interband tunneling (the injection of electrons from the p-section and holes from the n-section is negligible) and that the highly conducting side contacts to the p- and n-sections are of blade type (the thicknesses of GL and MGL and the contacts to them are much smaller than the spacing between the contacts)

$$\Sigma^-\Big|_{x=-l} = 0, \quad \Sigma^+\Big|_{x=+l} = 0, \tag{10.3}$$

$$\Phi\Big|_{x\le -l, y=0} = -V/2, \qquad \Phi\Big|_{x\ge +l, y=0} = V/2, \tag{10.4}$$

where $2l$ is the length of the i-section and $V = V_0 + \delta V_\omega \exp(-i\omega t)$ is the net voltage, which comprises the bias voltage V_0 and the signal component with the amplitude δV and the frequency ω. The interband tunneling generation rate of electrons and holes in each GL (per unit of its area) is given by [8–11]

$$g = g_0\left(\frac{2l|\partial\varphi/\partial x|}{V_0}\right)^{3/2}, \tag{10.5}$$

where $\varphi(x) = \Phi(x,y)|_{y=0}$, $\hbar$ is the Planck constant

$$g_0 = \frac{v_W}{2\pi^2}\left(\frac{eV_0}{2l\hbar v_W}\right)^{3/2},$$

so that the characteristic tunneling dc current (per unit length in the transverse z-direction) and the characteristic electron and hole sheet density can be presented as

$$J_{00} = 4el_{g0}, \qquad \Sigma_0 = 2l_{g0}/v_W.$$

From Eq. (10.2) with boundary condition (10.4), for the potential $\varphi(x)$ in the GL (or MGL structure) plane, we obtain

$$\varphi(x) = \frac{V}{\pi}\sin^{-1}\left(\frac{x}{l}\right) + \frac{Ke}{æ}\int_{-l}^{l}[\Sigma^+(x') - \Sigma^-(x')]G\left(\frac{x}{l},\frac{x'}{l}\right)dx', \tag{10.6}$$

where

$$G\left(\frac{x}{l},\frac{x'}{l}\right) = \frac{1}{2\pi}\left|\frac{\sin[(\cos^{-1}\xi + \cos^{-1}\xi')/2]}{\sin[(\cos^{-1}\xi - \cos^{-1}\xi')/2]}\right|. \tag{10.7}$$

Introducing the dimensionless quantities: $n^\mp = \Sigma^\mp/\Sigma_0$, $\xi = x/l$, $\xi' = x'/l$, and $\tau = v_W t/l$, Eqs. (10.1) and (10.6) can be reduced to the following system of nonlinear integro-differential equations:

$$\frac{\partial n^{\mp}}{\partial \tau} \pm \frac{\partial n^{\mp}}{\partial \xi} = \left[\frac{V/V_0}{\pi\sqrt{1-\xi^2}} + \gamma \int_{-1}^{1} [n^{+}(\xi',\tau) - n^{-}(\xi',\tau)] \right.$$

$$\left. \times \frac{\partial}{\partial \xi} G(\xi,\xi') d\xi' \right]^{3/2}. \qquad (10.8)$$

Here,

$$\gamma = \frac{2Kel\Sigma_0}{æV_0} = \frac{Ke^2}{4\pi^2 æ\hbar v_W}\sqrt{\frac{2leV_0}{\hbar v_W}}, \qquad (10.9)$$

$$\frac{\partial}{\partial \xi} G(\xi,\xi') = \frac{1}{\pi(\xi'-\xi)} \frac{\sqrt{1-\xi'^2}}{\sqrt{1-\xi^2}}. \qquad (10.10)$$

The term with the factor γ in the right-hand side of Eq. (10.8) is associated with the contributions of the electron and hole charges to the self-consistent electric field in the i-section.

The dc and ac components of the terminal dc and ac currents, J_0 and δJ_ω, are expressed via the dc and ac components $n_0^{\mp}$ and $\delta n_\omega^{\mp}$ as follows:

$$J_0 = ev_W K\Sigma_0 (n_0^{+} + n_0^{-}), \qquad (10.11)$$

$$\delta J_\omega = ev_W K\Sigma_0 \int_{-1}^{1} \rho(\xi)(\delta n_\omega^{+} + \delta n_\omega^{-})\, d\xi - i\omega C\delta V_\omega. \qquad (10.12)$$

Here, $\rho(\xi) = 1/\pi\sqrt{1-\xi^2}$ is the form factor and $C \sim æ/2\pi^2$ is the geometrical capacitance [14, 15]. The explicit coordinate dependence of the form factor is a consequence of the Shocley–Ramo theorem [16, 17] for the device geometry under consideration.

Hence the GTUNNETT small-signal admittance $Y_\omega = \delta J_\omega/\delta V_\omega$ is presented in the form

$$Y_\omega = ev_W K\Sigma_0 \int_{-1}^{1} \rho(\xi) \frac{d}{d\delta V_\omega}(\delta n_\omega^{+} + \delta n_\omega^{-}) d\xi - i\omega C. \qquad (10.13)$$

As follows from Eq. (10.8), the problem under consideration is characterized by the parameter γ.

If $æ = 1.0$, $K = 1$, and $2l = 0.7$ μm, in the voltage range $V_0 = 100$ – 200 mV, one obtains $\Sigma_0 \simeq (0.6 - 1.7) \times 10^{10}$ cm^{-2} and $\gamma \simeq 0.61 - 0.87$.

10.3 Spatial Potential Distributions and Current-Voltage Characteristics

If the charges created by the propagating electrons and holes in the i-section are insignificant, that corresponds to $\gamma \ll 1$. In this case, the potential distribution is given by

$$\varphi_0(x) \simeq \frac{V_0}{\pi} \sin^{-1}\left(\frac{x}{l}\right), \tag{10.14}$$

and Eqs. (10.8), neglecting the term with γ, can be solved analytically. Taking into account boundary conditions Eq. (10.3), from Eqs. (10.8), we obtain

$$n_0^+ + n_0^- = \frac{1}{\pi^{3/2}} \int_{-1}^{1} \frac{d\xi}{(1-\xi^2)^{3/4}} = const. \tag{10.15}$$

After that, using Eq. (10.11) and considering Eq. (10.15), one can find the following formula for the dc current-voltage characteristic:

$$\begin{aligned} J_0 &= \frac{KJ_{00}}{\pi^{3/2}} \int_0^1 \frac{d\xi}{(1-\xi^2)^{3/4}} \\ &= K \frac{\Gamma(1/4)\Gamma(1/2)}{\Gamma(3/4)} \frac{ev_W}{\pi^{7/2} 2\sqrt{2l}} \left(\frac{eV_0}{\hbar v_W}\right)^{3/2}, \end{aligned} \tag{10.16}$$

where $\Gamma(x)$ is the Gamma-function. A distinction between J_0 and J_{00} is due to the nonuniformity of the electric field in the i-section associated with the feature of the device geometry taken into account calculating J_0. At $K = 1$, $2l = 0.7$ μm, and $V_0 = 100 - 200$ mV, Eq. (10.16) yields $J_0 \simeq 0.18 - 0.51$ A/cm.

To take the effect of the self-consistent electric field on the tunneling, one needs to solve system of Eqs. (10.8). Due to a complexity of the nonlinear integro-differential equations in question, a numerical approach is indispensable. Equations (10.8) were solved numerically using successive approximation method, which is valid when $\gamma < \gamma_c = 5$. In the cases $\gamma > 5$, the method of the parameter evolution was implemented.

Using Eqs. (10.8) and (10.11) and setting $V = V_0$, we calculated numerically the GTUNNETT dc characteristics: spatial distributions of the dc electric potential and the dc components of the electron and hole densities, as well as the dc current-voltage characteristics. The pertinent results are shown in Figs. 10.2–10.5.

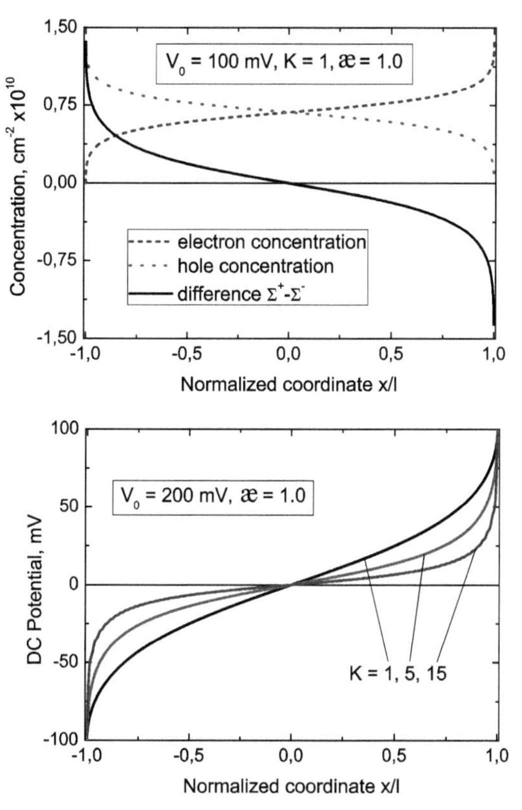

Figure 10.2 Spatial distributions of electron and hole concentrations (upper panel) and of electric potential (lower panel) in the GTUNNETT i-section.

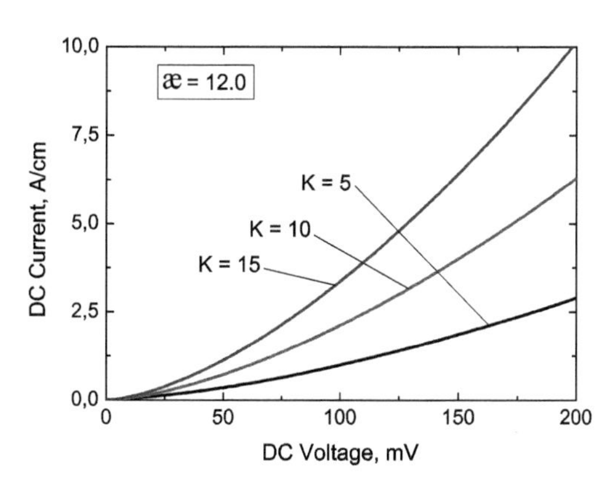

Figure 10.3 Current-voltage characteristics for GTUNNETTs with different number of GLs *K*.

Upper panel in Fig. 10.2 shows examples of the spatial distributions of the electron and hole sheet concentrations in the i-section. The spatial distributions of the dc electric potential calculated for GTUNNETTs with different numbers of GLs K at fixed voltage are shown in Fig. 10.2 (lower panel). One can see that an increase in K results in a marked concentration of the electric field near the doped sections. This is because at larger K, the net tunneling generation rate becomes stronger. This, in turn, results in higher charges of propagating electron and hole components, particularly, near the p-i- and i-n-junctions, respectively.

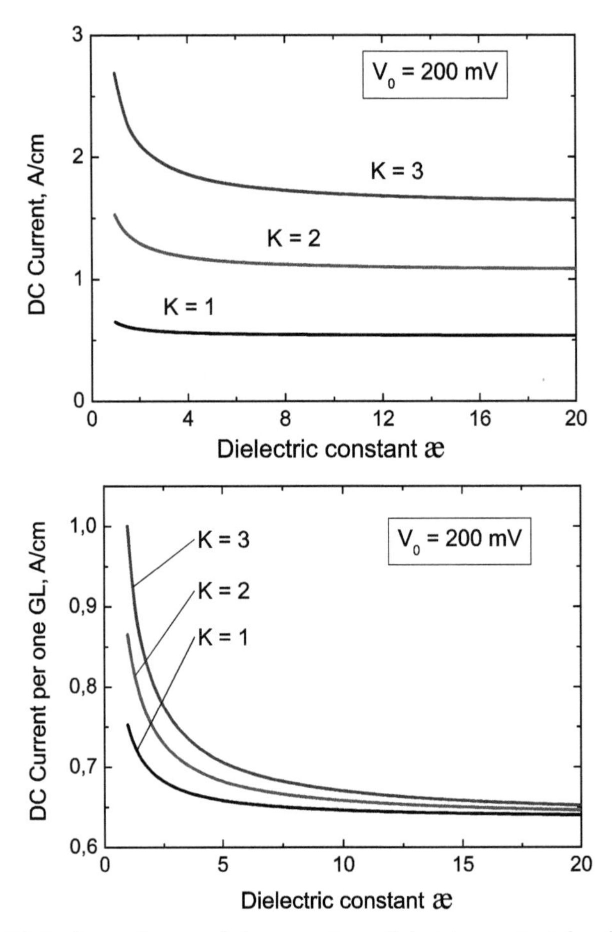

Figure 10.4 Dependences of dc current on dielectric constant for different numbers of GLs K: net current—upper panel and current per one GL—lower panel.

Figure 10.3 demonstrates a difference in the dc current-voltage characteristics in GTUNNETTs with different numbers of GLs *K*. As can be seen in Fig. 10.3, the dc current becomes larger with increasing *K* at more than *K*-fold rate. In paricular, Fig. 10.4 (upper panel) shows the three-fold increase in *K* leads to more than three-fold increase in the dc current particularly at relatively small dielectric constants. This is confirmed by plots in Fig. 10.4 (lower panel), from which it follows that the dc current in each GL is larger in the devices with larger number of GLs. Such a behavior of the dc current-voltage characteristics is attributed to the following. An increase in *K* leads to an increase in the dc current not only because of the increase in the number of current channels (this would provide just the *K*-fold rise in the dc current) but because of the reinforcement of the self-consistent electric field near the edges of the i-section (as mentioned above) and, hence, strengthening of the tunneling injection.

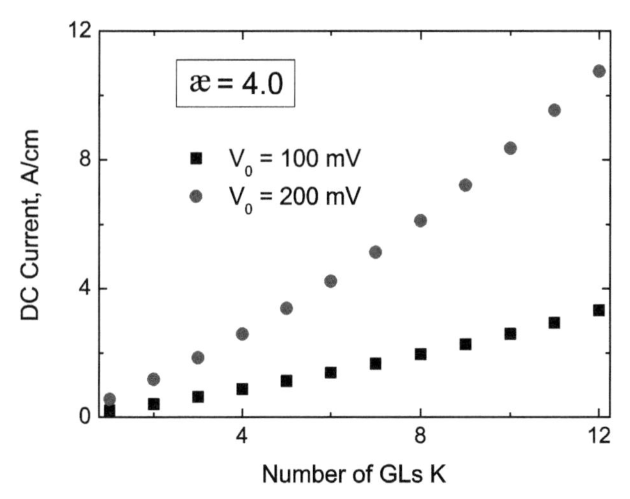

Figure 10.5 Dependences of dc current on number of GLs *K* at different voltages.

It is worth noting that when the dielectric constant is sufficiently large (about 15–20), the dc current in one GL is virtually the same in the devices with different number of GLs, because in this case, the charge effect is suppressed. In this case, the value of the dc current (per one GL) is approximately the same as that obtained in the above numerical estimate using analytical formula given by Eq. (10.16). In

GTUNNETTs with a moderate value of a, the charge effect leads to superlinear dependences of the net dc current vs number of GLs as seen in Fig. 10.5.

10.4 GTUNNETT Admittance

Disregarding the term in Eq. (10.8) associated with the electron and hole charges in the i-section, and using Eq. (10.12), one can obtain analytically the following formulas for the ac potential distribution in the i-section, and the imaginary and real parts of the GTUNNETT admittance [9]

$$\delta\varphi(x)_\omega = \frac{\delta V_\omega}{\pi}\sin^{-1}\left(\frac{x}{l}\right), \tag{10.17}$$

$$\mathrm{Im}Y_\omega = Y_0\sin(\omega\tau_t)\cdot\mathcal{J}_0(\omega\tau_t) - \omega C, \tag{10.18}$$

$$\mathrm{Re}Y_\omega = Y_0\cos(\omega\tau_t)\cdot\mathcal{J}_0(\omega\tau_t), \tag{10.19}$$

where $Y_0 = dJ_0/dV_0 = 3J_0/2V_0 \propto \sqrt{V_0}$ is the dc differential conductivity, C is the geometrical capacitance, $\tau_t = l/\upsilon_W$ is the characteristic transit time of electrons and holes across the i-section, and $\mathcal{J}_0(\omega\tau)$ is the Bessel function. As follows from Eqs. (10.18) and (10.19), the admittance imaginary and real parts oscillate as functions of the transit angle $\omega\tau_t$ with Re $Y_\omega < 0$ at $\omega\tau_t$ near the transit time resonances $\omega\tau_t = (2n - 1/2)\pi$, where $n = 1, 2, 3\ldots$ is the resonance index.

The numerical solution of Eqs. (10.8), accounting for the ac components of the applied voltage δV_ω and the self-consistent charges and electric field, shows that the oscillatory frequency dependences are preserved in this more realistic case although they are be quantitatively modified. In particular, these dependences vary with varying such parameters as the number of GLs K and dielectric constant æ. The results of numerical calculations using Eq. (10.8) are demonstrated in Figs. 10.6–10.9.

Figure 10.6 shows the dependence of the imaginary and real parts, Im Y_ω and Re Y_ω, of the admittance on the signal frequency calculated for GTUNNETTs with $2l = 0.7$ μm and $K = 1$ at $V_0 = 200$ mV at different values of dielectric constant æ. As follows from Fig. 10.6, the imaginary and real parts of the GTUNNETT admittance

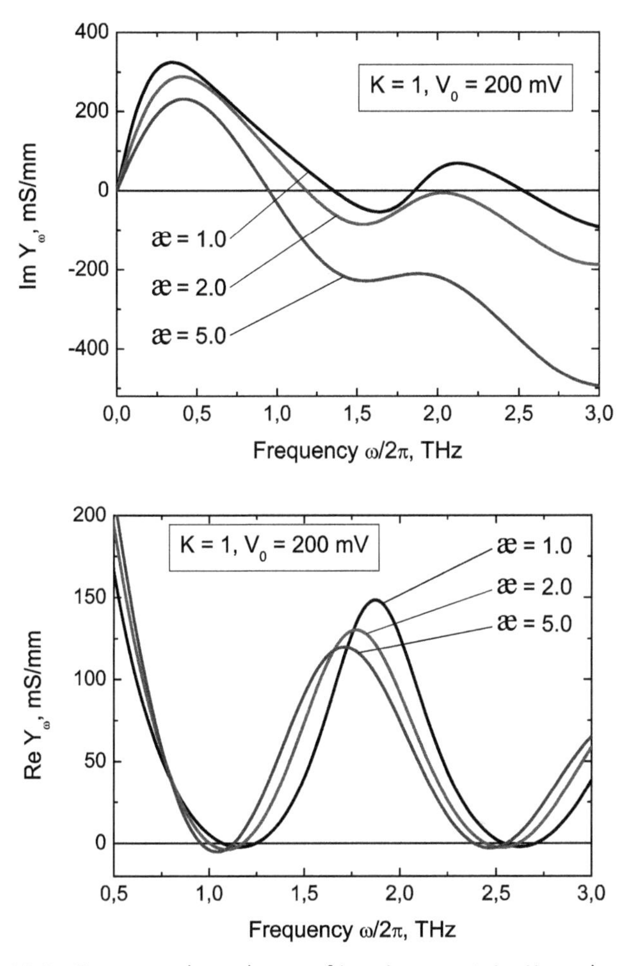

Figure 10.6 Frequency dependences of imaginary part, Im Y_ω, and real part, Re Y_ω, of GTUNNETT admittance (upper and lower panels, respectively) at different values of dielectric constant æ (K = 1 and $2l$ = 0.7 μm).

are oscillatory functions of the signal frequency [see also Eqs. (10.18) and (10.19)]. The oscillatory behavior is due to the transit-time resonances. At sufficiently high frequencies, the admittance imaginary part is negative because it is determined primarily by the geometrical capacitance, so that Im $Y_\omega \simeq -\omega C$. In the certain frequency ranges, the admittance real part is negative as shown in Fig. 10.6 (lower panel). These ranges correspond to the transit-time resonances. As seen from Fig. 10.7, the minima of the admittance real part (where the latter is negative) become deeper and the minima shift toward smaller frequencies with increasing dielectric constant æ. This is because an increase in æ leads to weakening of the effect of the electron and hole charges on the injection and propagation processes. The dashed line corresponds to the case when the role of these charges is diminished (very large æ) considered previously [9].

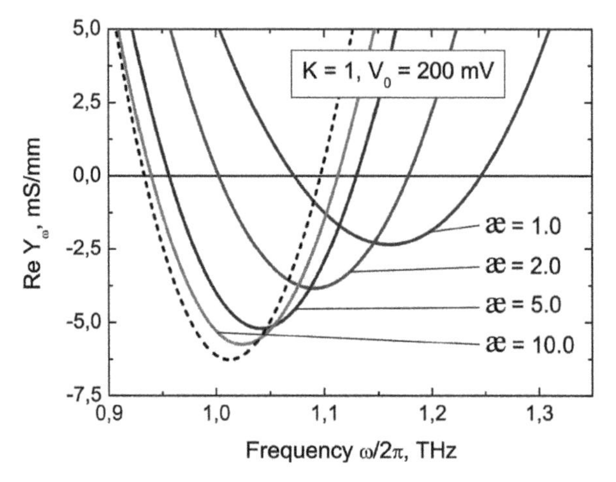

Figure 10.7 Frequency dependences of real part, Re Y_ω, of GTUNNETT admittance in the frequency range, where Re $Y_\omega < 0$ at different values of dielectric constant æ ($K = 1$ and $2l = 0.7$ µm) calculated accounting for electron and hole charges (solid lines) and neglecting them (dashed line) [9].

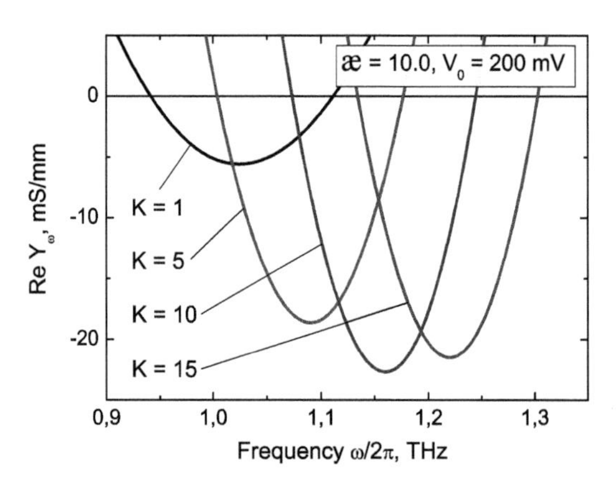

Figure 10.8 Frequency dependences of real part of admittance Re Y_ω for GTUNNETTs with different number of GLs K (æ = 10.0 and $2l = 0.7$ µm).

Figure 10.8 shows the admittance real part vs signal frequency calculated for GTUNNETTs with different number of GLs *K*. One can see that GTUNNETTs with larger number of GLs *K* exhibit much more deep minima (compare the curves which correspond to *K* = 1 and æ = 10.0 in Fig. 10.7 and that for *K* = 10 and æ = 10.0 in Fig. 10.8.). As follows from this comparison, the ten-fold increase in *K* leads to about fourfold increase in the minimum depth.

Figure 10.9 shows the dependence of the minimum value of the real part of the GTUNNETT admittance min $\{\mathrm{Re}Y_\omega\}$ on the number of GLs *K* calculated for different dielectric constants æ. The nonmonotonic character of these dependences is associated with the interplay of the factors determined by the influence of electron and hole charges: an increase in æ leads to a suppression of these charges role, while an increase in *K* results in proportional reinforcement of the effect of these charges accompanied by proportional increase in the current.

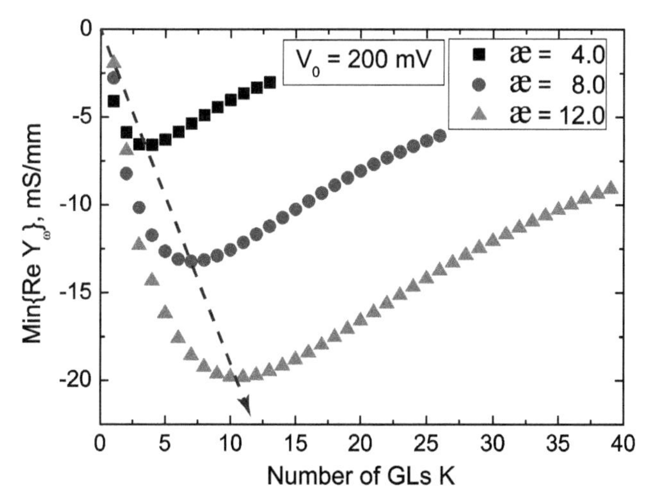

Figure 10.9 Minimum value of real part of GTUNNETT admittance as a function of number of GLs *K* for GTUNNETTs with different dielectric constant æ.

10.5 Discussion

Calculating the dc current, we disregarded the contributions of the thermogeneration of electron-hole pairs in the i-section and the injection of minority carriers from the n- and p-sections.

The thermogeneration dc current is given by

$$J_0^{\text{therm}} = 4Kelg_0^{\text{therm}}. \tag{10.20}$$

The thermogeneration rate g_0^{therm} at the temperature $T = 300$ K is estimated as [18] $g_0^{\text{therm}} \simeq 10^{20} - 10^{21}$ cm^{-2}s^{-1}. As a result, at $K = 1$ and $2l = 0.7$ μm from Eq. (10.20), one obtains $J_0^{\text{therm}} \simeq 2(10^{-3} - 10^{-2})$ A/cm.

The injection dc current can be estimated as

$$J_0^{\text{inj}} = \frac{2KeT^2}{\pi^2\hbar^2 v_w}\exp\left(-\frac{\mu}{T}\right), \tag{10.21}$$

where μ is the Fermi energy of electrons in the n-section and holes in the p-section. Setting $\mu = 50$ meV, from Eq. (10.21) for $K = 1$, we find $J_0^{\text{inj}} \simeq 0.077$ A/cm. Thus, $J_0^{\text{therm}}, J_0^{\text{inj}} \ll J_0$, at least in the voltage range considered above.

The typical values of the real part of the GTUNNETT small-signal admittance in the vicinity of the first transit-time resonance ($\omega\tau_t \sim 3/2$) is in the range Re $Y_\omega \simeq -5$ to -20 mS/mm depending on K and æ. This implies that the ac resistance, $R_\omega = H/Y_\omega$ of optimized GTUNNETTs with Re $Y_\omega = -20$ mS/mm ($K = 10$, æ = 10, and $V_0 = 200$ mV) and the width $H = (0.5 - 1)$ mm, is $R_\omega \simeq 50 - 100$ Ω. Assuming that in this case $J_0 \simeq 0.6 - 0.8$ A/mm at $V_0 = 200$ mV, the dc power, P_0, generated by the propagating electrons and holes in the contact n- and p-regions of the device can be estimated as $P_0 \simeq 150$ mW. When Re $Y_\omega = -5$ mS/mm ($K = 1$ and æ = 10), the same resistance is provided if $H = 2 - 4$ mm with the dc power $P_0 \simeq 20 - 40$ mW.

The depth of the minimum of the real part of the GTUNNETT admittance can be increased by applying higher bias voltage. This is because the admittance is proportional to the dc differential conductivity which, in turn, is approximately proportional to $\sqrt{V_0}$. However, an increase in V_0 results in the pertinent rise of P_0. If eV_0 exceeds the energy of optical phonons in GLs $\hbar\omega \simeq 200$ meV, their fast emission disrupts BET and BHT in the i-section. The emission of acoustic and, particularly, optical phonons leads to some isotropization of the angular distribution of the electron and hole velocities of electrons and holes. This results in some decrease in their mean directed velocities (they become somewhat smaller than $\pm v_W$) and, hence, in some increase in the electron and hole charges and the transit time. Such factors promote more steep dc current-voltage characteristics and modify the frequency dependence of the admittance. Although they should not change the GTUNNETT performance qualitatively.

The depth of the first resonance minimum of the real part of the GTUNNETT admittance with the parameters used in the above estimate is of the same order of magnitude as that of the small area (about 2 μm^2) THz resonant-tunneling diodes considered by Asada et al. [19]. It is worth noting that the frequency at which Im Y_ω turns zero can fall into the frequency range, where Re $Y_\omega < 0$. In such a case, GTUNNETTs can exhibit the self-excitation of THz oscillations even without the external resonant circuit. Moreover, the contact n- and p-sections (both ungated as in the case of their chemical doping or gated in the case of the electrical "doping") can serve as plasma resonant cavities [9, 15]. The combination of the transit-time and the plasma resonances can substantially liberalize the self-excitation conditions and enhance the THz emission efficiency. The matching of the transit-time and plasma resonant frequencies requires a proper choice of the length of the n- and p-sections (from several micrometers in the ungated structures to about one micrometer in the gated ones). The plasmons and related resonances in the gated GLs were predicted and analyzed several years ago [20, 21] (see also Ref. [22]). Recently, the plasmons in gated GLs were detected experimentally [23, 24]. It is interesting that as a gate for the GL plasmon cavity, another GL can be used [25] when the contact regions constitute double-GL structures [26, 27].

10.6 Conclusions

We have developed a self-consistent device model for p-i-n GTUNNETTs with different number of GLs, which enables the calculation of their realistic dc and ac characteristics. Our calculations have shown that

(i) The charges of electrons and holes propagating in the i-section and their effect on the spatial distribution of the self-consistent electric field increase the steepness of the dc current-voltage characteristics. This effect is stronger in GTUNNETTs with larger number of GLs and can be weakened by using the media surrounding GL or MGL structure (substrates and top dielectric layers) with elevated values of the dielectric constant;

(ii) The imaginary and real parts of the GTUNNETT admittance are oscillatory functions of the signal frequency due to the transit-time resonances. In the certain frequency ranges, which correspond to the THz frequencies, the admittance real part can be negative that enables the use of GTUNNETTs in the sources of THz radiation;

(iii) The charges of electrons and holes influence on the spatial distribution of the self-consistent electric field and the interband tunneling and can substantially affect the GTUNNETT admittance, particularly, its frequency dependence near the transit-time resonances. The role of this effect enhances with increasing number of GLs in the device structure and can be markedly weakened in the devices with relatively large dielectric constant;

(iv) The sensitivity of the GTUNNETT characteristics to the structural parameters (number of GLs, dielectric constant, and length of the i-section) and the bias voltage opens up wide opportunity for the optimization of GTUNNETTs for THz oscillators.

Acknowledgments

This work was partially supported by the Russian Foundation for Basic Research (Grants 11-07-12072, 11-07-00505, and 12-07-00710), by grants from the President of the Russian Federation (Russia), the Japan Science and Technology Agency, CREST, the Japan Society for Promotion of Science (Japan), and the TERANO-NSF Grant (USA).

References

1. M. S. Shur and L. F. Eastman, *IEEE Trans. Electron Devices*, **ED-26**, 1677 (1979).
2. M. S. Shur, *IEEE Trans. Electron Devices*, **ED-28**, 1120 (1981).
3. V. I. Ryzhii, N. A. Bannov, and V. A. Fedirko, *Sov. Phys. Semicond.*, **18**, 481 (1984).

4. A. K. Geim and K. S. Novoselov, *Nat. Mater.*, **6**, 183 (2007).
5. A. H. Castro Neto, F. Guinea, N. M. R. Peres, K. S. Novoselov, and A. K. Geim, *Rev. Mod. Phys.*, **81**, 109 (2009).
6. M. Sprinkle, D. Suegel, Y. Hu, Hicks, J. A. Tejeda, A. Taleb-Ibrahimi, P. Le Fevre, F. Bertran, S. Vizzini, H. Enriquez, S. Chiang, P. Soukiassian, C. Berger, W. A. de Heer, A. Lanzara, and E. H. Conrad, *Phys. Rev. Lett.*, **103**, 226803 (2009).
7. M. Orlita and M. Potemski, *Semicond. Sci. Technol.*, **25**, 063001 (2010).
8. V. Ryzhii, M. Ryzhii, V. Mitin, and M. S. Shur, *Appl. Phys. Express*, **2**, 034503 (2009).
9. V. Ryzhii, M. Ryzhii, M. S. Shur, and V. Mitin, *Physica E*, **42**, 719 (2010).
10. V. V. Cheianov and V. I. Fal'ko, *Phys. Rev. B*, **74**, 041403(R) (2006).
11. A. Ossipov, M. Titov, and C. V. J. Beenakker, *Phys. Rev. B*, **75**, 241401(R) (2007).
12. Z. S. Gribnikov, N. Z. Vagidov, V. V. Mitin, and G. I. Haddad, *J. Appl. Phys.*, **93**, 5435 (2003).
13. D. Farmer, Y.-M. Lin, A. Afzali-Ardakani, and P. Avouris, *Appl. Phys. Lett.*, **94**, 213106 (2009).
14. G. Khrenov and V. Ryzhii, *IEEE Trans. Electron Devices*, **42**, 166 (1995).
15. V. Ryzhii, A. Satou, I. Khmyrova, M. Ryzhii, T. Otsuji, V. Mitin, and M. S. Shur, *J. Phys.: Conf. Ser.*, **38**, 228 (2006).
16. W. Shocley, *J. Appl. Phys.*, **9**, 635 (1938).
17. S. Ramo, *Proc. IRE*, **27**, 584 (1939).
18. F. Rana, P. A. George, J. H. Strait, S. Shivaraman, M. Chandrashekar, and M. G. Spencer, *Phys. Rev. B*, **79**, 115477 (2009).
19. M. Asada, S. Suzuki, and N. Kishimoto, *Jpn. J. Appl. Phys., Part 1*, **47**, 4375 (2008).
20. V. Ryzhii, *Jpn. J. Appl. Phys., Part 2*, **45**, L923 (2006).
21. V. Ryzhii, A. Satou, and T. Otsuji, *J. Appl. Phys.*, **101**, 024509 (2007).
22. D. Svintsov, V. Vyurkov, S. O. Yurchenko, T. Otsuji, and V. Ryzhii, *J. Appl. Phys.*, **111**, 083715 (2012).
23. Z. Fei, A. S. Rodin, G. O. Andreev, W. Bao, et al., *Nature*, **487**, 82 (2012).
24. J. Chen, M. Badioli, P. Alonco-Gonzales, S. Thongrattanasari, et al., *Nature*, **487**, 77 (2012).
25. V. Ryzhii, T. Otsuji, M. Ryzhii, and M. S. Shur, *J. Phys. D: Appl. Phys.*, **45**, 302001 (2012).

26. M. Liu, X. Yin, and X. Zhang, *Nano Lett.*, **12**, 482 (2012).

27. L. Britnel, R. V. Gorbachev, R. Jalil, B. D. Belle, F. Shedin, A. Mishenko, T. Georgiou, M. I. Katsnelson, L. Eaves, S. V. Morozov, N. M. R. Peres, J. Leist, A. K. Geim, K. S. Novoselov, and L. A. Ponomarenko, *Science*, **335**, 947 (2012).

Chapter 11

Damping Mechanism of Terahertz Plasmons in Graphene on Heavily Doped Substrate*

A. Satou,[a,b] Y. Koseki,[a] V. Ryzhii,[a,b] V. Vyurkov,[c] and T. Otsuji[a,b]
[a]*Research Institute of Electrical Communication, Tohoku University, Sendai 980-8577, Japan*
[b]*CREST, Japan Science and Technology Agency, Tokyo 107-0075, Japan*
[c]*Institute of Physics and Technology, Russian Academy of Sciences, Moscow 117218, Russia*
a-satou@riec.tohoku.ac.jp

Coupling of plasmons in graphene at terahertz frequencies with surface plasmons in a heavily doped substrate is studied theoretically. We reveal that a huge scattering rate may completely damp out the plasmons, so that proper choices of material and geometrical parameters are essential to suppress the coupling effect and to

*Reprinted with permission from A. Satou, Y. Koseki, V. Ryzhii, V. Vyurkov, and T. Otsuji (2014). Damping mechanism of terahertz plasmons in graphene on heavily doped substrate, *J. Appl. Phys.*, **115**, 104501. Copyright © 2014 AIP Publishing LLC.

Graphene-Based Terahertz Electronics and Plasmonics: Detector and Emitter Concepts
Edited by Vladimir Mitin, Taiichi Otsuji, and Victor Ryzhii

ISBN 978-981-4800-75-4 (Hardcover), 978-0-429-32839-8 (eBook)
www.jennystanford.com

obtain the minimum damping rate in graphene. Even with the doping concentration 10^{19} – 10^{20} cm^{-3} and the thickness of the dielectric layer between graphene and the substrate 100 nm, which are typical values in real graphene samples with a heavily doped substrate, the increase in the damping rate is not negligible in comparison with the acoustic-phonon-limited damping rate. Dependence of the damping rate on wavenumber, thicknesses of graphene-to-substrate and gate-to-graphene separation, substrate doping concentration, and dielectric constants of surrounding materials are investigated. It is shown that the damping rate can be much reduced by the gate screening, which suppresses the field spread of the graphene plasmons into the substrate.

11.1 Introduction

Plasmons in two-dimensional electron gases (2DEGs) can be utilized for terahertz (THz) devices. THz sources and detectors based on compound semiconductor heterostructures have been extensively investigated both experimentally and theoretically [1–8]. The two-dimensionality, which gives rise to the wavenumber-dependent frequency dispersion, and the high electron concentration on the order of 10^{12} cm^{-2} allow us to have their frequency in the THz range with submicron channel length. Most recently, a very high detector responsivity of the so-called asymmetric double-grating-gate structure based on an InP-based high-electron-mobility transistor was demonstrated [9]. However, resonant detection as well as single-frequency coherent emission have not been accomplished so far at room temperature, mainly owning to the damping rate more than 10^{12} s^{-1} in compound semiconductors.

Plasmons in graphene have potential to surpass those in the heterostructures with 2DEGs based on the standard semiconductors, due to its exceptional electronic properties [10]. Massive experimental and theoretical works have been done very recently on graphene plasmons in the THz and infrared regions (see review papers Refs. [11, 12] and references therein). One of the most important advantages of plasmons in graphene over those in heterostructure 2DEGs is much weaker damping rate close to 10^{11} s^{-1} at room temperature in disorder-free graphene suffered only from acoustic-phonon scattering [13]. That is very promising for the realization of the resonant THz detection [14] and also of plasma instabilities, which can be utilized for the emission. In addition, interband population inversion in the THz range was predicted [15, 16], and it has been investigated for the utilization not only in THz lasers in the usual sense but also in THz active plasmonic devices [17, 18] and metamaterials [19].

Many experimental demonstrations of graphene-based devices have been performed on graphene samples with heavily doped substrates, in order to tune the carrier concentration in graphene by the substrate as a back gate. Typically, either peeling or CVD graphene transferred onto a heavily doped p^+ -Si substrate, with a SiO_2 dielectric layer in between, is used (some experiments on graphene plasmons have adapted undoped Si/SiO_2 substrates [20, 21]). Graphene-on-silicon, which is epitaxial graphene on doped Si substrates [22], is also used. For realization of THz plasmonic devices, properties of plasmons in such structures must be fully understood. Although the coupling of graphene plasmons to surface plasmons in perfectly conducting metallic substrates with/without dielectric layers in between have been theoretically studied [23, 24], the influence of the carrier scattering in a heavily doped semiconductor substrate (with finite complex conductivity) has not been taken into account so far. Since the scattering rate in the substrate increases as the doping concentration increases, it is anticipated that the coupling of graphene plasmons to surface plasmons in the heavily doped substrate causes undesired increase in the damping rate.

The purpose of this chapter is to study theoretically the coupling between graphene plasmons and substrate surface plasmons in a structure with a heavily doped substrate and with/without a metallic top gate. The chapter is organized as follows. In the Section 11.2, we derive a dispersion equation of the coupled modes of graphene plasmons and substrate surface plasmons. In Section 11.3, we study coupling effect in the ungated structure, especially the increase in the plasmon damping rate due to the coupling and its dependences on the doping concentration, the thickness of graphene-to-substrate separation, and the plasmon wavenumber. In Section 11.4, we show that the coupling in the gated structures can be less effective due to the gate screening. We also compare the effect in structures having different dielectric layers between the top gate, graphene layer, and substrate, and reveal the impact of values of their dielectric constants. In Section 11.5, we discuss and summarize the main results of this chapter.

11.2 Equations of the Model

We investigate plasmons in an ungated graphene structure with a heavily-doped p^+-Si substrate, where the graphene layer is exposed on the air, as well as a gated graphene structure with the substrate and a metallic top gate, which are schematically shown in Figs. 11.1a and b, respectively. The thickness of the substrate is assumed to be sufficiently larger than the skin depth of the substrate surface plasmons. The top gate can be considered as perfectly conducting metal, whereas the heavily doped Si substrate is characterized by its complex dielectric constant.

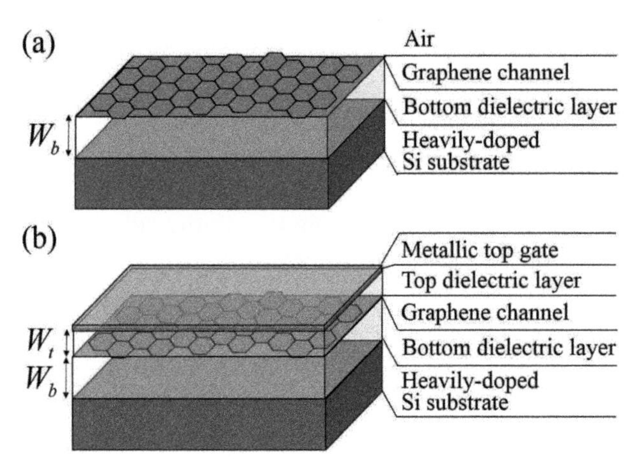

Figure 11.1 Schematic views of (a) an ungated graphene structure with a heavily doped Si substrate where the top surface is exposed on the air and (b) a gated graphene structure with a heavily doped Si substrate and a metallic top gate.

Here, we use the hydrodynamic equations to describe the electron motion in graphene [26], while using the simple Drude model for the hole motion in the substrate (due to virtual independence of the effective mass in the substrate on the electron density, in contrast to graphene). In addition, these are accompanied by the self-consistent

2D Poisson equation (the formulation used here almost follows that for compound semiconductor high-electron-mobility transistors, see Ref. [25]). Differences are the hydrodynamic equations accounting for the linear dispersion of graphene and material parameters of the substrate and dielectric layers. In general, the existence of both electrons and holes in graphene results in various modes such as electrically passive electron-hole sound waves in intrinsic graphene as well as in huge damping of electrically active modes due to the electron-hole friction, as discussed in Ref. [26]. Here, we focus on the case where the electron concentration is much higher than the hole concentration and therefore the damping associated with the friction can be negligibly small. Besides, for the generalization purpose, we formulate the plasmon dispersion equation for the gated structure; that for the ungated structure can be readily found by taking the limit $W_t \to \infty$ (see Fig. 11.1).

Then, assuming the solutions of the form $\exp(ikx - i\omega t)$, where $k = 2\pi/\lambda$ and ω are the plasmon wavenumber and frequency (λ denotes the wavelength), the 2D Poisson equation coupled with the linearized hydrodynamic equations can be expressed as follows:

$$\frac{\partial^2 \varphi_\omega}{\partial z^2} - k^2 \varphi_\omega = -\frac{8\pi e^2 \Sigma_e}{3m_e \epsilon} \frac{k^2}{\omega^2 + i\nu_e \omega - \frac{1}{2}(v_F k)^2} \varphi_\omega \delta(z), \qquad (11.1)$$

where φ_ω is the ac (signal) component of the potential, Σ_e, m_e, and ν_e are the steady-state electron concentration, the hydrodynamic "fictitious mass," and the collision frequency in graphene, respectively, and ϵ is the dielectric constant which is different in different layers. The electron concentration and fictitious mass are related to each other through the electron Fermi level, μ_e, and electron temperature, T_e:

$$\Sigma_e = \int_0^\infty \frac{2\varepsilon}{\pi \hbar^2 v_F^2} \left[1 + \exp\left(\frac{\varepsilon - \mu_e}{k_B T_e}\right)\right]^{-1} d\varepsilon, \qquad (11.2)$$

$$m_e = \frac{1}{v_F^2 \Sigma_e} \int_0^\infty \frac{2\varepsilon^2}{\pi \hbar^2 v_F^2} \left[1 + \exp\left(\frac{\varepsilon - \mu_e}{k_B T_e}\right)\right]^{-1} d\varepsilon. \qquad (11.3)$$

In the following we fix T_e and treat the fictitious mass as a function of Σ_e. The dielectric constant can be represented as

$$\epsilon = \begin{cases} \epsilon_t & 0 < z < W_t, \\ \epsilon_b & -W_b < z < 0, \\ \epsilon_s\,[1 - \Omega_s^2 / \omega(\omega + i\nu_s)], & z < -W_b, \end{cases} \tag{11.4}$$

where ϵ_t, ϵ_b, and ϵ_s are the *static* dielectric constants of the top and bottom dielectric layers and the substrate, respectively, $\Omega_s = \sqrt{4\pi e^2 N_s / m_h \epsilon_s}$ is the bulk plasma frequency in the substrate with N_s and m_h being the doping concentration and hole effective mass, and ν_s is the collision frequency in the substrate, which depends on the doping concentration. The dielectric constant in the substrate is a sum of the static dielectric constant of Si, ϵ_s = 11.7 and the contribution from the Drude conductivity. The dependence of the collision frequency, ν_s, on the doping concentration, N_s, is calculated from the experimental data for the hole mobility at room temperature in Ref. [27].

We use the following boundary conditions: vanishing potential at the gate and far below the substrate, $\varphi_\omega|_{z=W_t} = 0$ and $\varphi_\omega|_{z=-\infty} = 0$; continuity conditions of the potential at interfaces between different layers, $\varphi_\omega|_{z=+0} = \varphi_\omega|_{z=-0}$ and $\varphi_\omega|_{z=-W_b+0} = \varphi_\omega|_{z=-W_b-0}$; a continuity condition of the electric flux density at the interface between the bottom dielectric layer and the substrate in the z-direction, $\epsilon_b \partial\varphi_\omega/\partial z|_{z=-W_b+0} = \epsilon_s \partial\varphi_\omega/\partial z|_{z=-W_b-0}$; and a jump of the electric flux density at the graphene layer, which can be derived from Eq. (11.1). Equation (11.1) together with these boundary conditions yield the following dispersion equation

$$F_{gr}(\omega) F_{sub}(\omega) = A_c, \tag{11.5}$$

where

$$F_{gr}(\omega) = \omega^2 + i\nu_e\omega - \frac{1}{2}(v_F k)^2 - \Omega_{gr}^2, \tag{11.6}$$

$$F_{sub}(\omega) = \omega(\omega + i\nu_s) - \Omega_{sub}^2, \tag{11.7}$$

$$A_c = \frac{\epsilon_b^2 (H_b^2 - 1)}{(\epsilon_b H_b + \epsilon_t H_t)\,(\epsilon_s + \epsilon_b H_b)} \Omega_{gr}^2 \Omega_{sub}^2, \tag{11.8}$$

$$\Omega_{gr} = \sqrt{\frac{8\pi e^2 \Sigma_e k}{3 m_e \epsilon_{gr}(k)}}, \quad \epsilon_{gr}(k) = \epsilon_t H_t + \epsilon_b \frac{\epsilon_b + \epsilon_s H_b}{\epsilon_s + \epsilon_b H_b}, \tag{11.9}$$

$$\Omega_{\mathrm{sub}} = \sqrt{\frac{4\pi e^2 N_{\mathrm{s}}}{m_{\mathrm{h}} \epsilon_{\mathrm{sub}}(k)}}, \quad \epsilon_{\mathrm{sub}}(k) = \epsilon_{\mathrm{s}} + \epsilon_{\mathrm{b}} \frac{\epsilon_{\mathrm{b}} + \epsilon_{\mathrm{t}} H_{\mathrm{t}} H_{\mathrm{b}}}{\epsilon_{\mathrm{b}} H_{\mathrm{b}} + \epsilon_{\mathrm{t}} H_{\mathrm{t}}}, \quad (11.10)$$

and $H_{\mathrm{b,t}} = \coth kW_{\mathrm{b,t}}$. In Eq. (11.5), the term A_{c} on the right-hand side represents the coupling between graphene plasmons and substrate surface plasmons. If A_{c} were zero, the equations $F_{\mathrm{gr}}(\omega) = 0$ and $F_{\mathrm{sub}}(\omega) = 0$ would give independent dispersion relations for the former and latter, respectively. Qualitatively, Eq. (11.8) indicates that the coupling occurs unless $kW_{\mathrm{b}} \gg 1$ or $kW_{\mathrm{t}} \ll 1$, i.e., unless the separation of the graphene channel and the substrate is sufficiently large or the gate screening of graphene plasmons is effective. Note that the non-constant frequency dispersion of the substrate surface plasmon in Eq. (11.10) is due to the gate screening, which is similar to that in the structure with two parallel metal electrodes [28]. Equation (11.5) yields two modes which have dominant potential distributions near the graphene channel and inside the substrate, respectively. Hereafter, we focus on the oscillating mode primarily in the graphene channel; we call it "channel mode," whereas we call the other mode "substrate mode."

11.3 Ungated Plasmons

First, we study plasmons in the ungated structure. Here, the temperature, electron concentration, and collision frequency in graphene are fixed to $T_{\mathrm{e}} = 300$ K, $\Sigma_{\mathrm{e}} = 10^{12}$ cm^{-2}, and $\nu_{\mathrm{e}} = 3 \times 10^{11}$ s^{-1}. With these values of the temperature and concentration the fictitious mass is equal to 0.0427 m_0, where m_0 is the electron rest mass. The value of the collision frequency is typical to the acoustic-phonon scattering at room temperature [13]. As for the structural parameters, we set $\epsilon_{\mathrm{t}} = 1$ and $W_{\mathrm{t}} \to \infty$, and we assume an SiO_2 bottom dielectric layer with $\epsilon_{\mathrm{b}} = 4.5$. Then Eq. (11.5) is solved numerically.

Figures 11.2a and b show the dependences of the plasmon damping rate and frequency on the substrate doping concentration with the plasmon wavenumber $k = 14 \times 10^3$ cm^{-1} (i.e., the wavelength $\lambda = 4.5$ μm) and with different thicknesses of the bottom dielectric layer, W_{b}. The value of the plasmon wavelength is chosen so that it gives the frequency around 1 THz in the limit $N_{\mathrm{s}} \to 0$. They clearly demonstrate that there is a huge resonant increase in the damping rate at around $N_s = 3 \times 10^{17}$ cm^{-3} as well as a drop of the frequency. This is the manifestation of the resonant coupling of the graphene plasmon and the substrate surface plasmon. The resonance corresponds to the situation where the frequencies of graphene plasmons and substrate surface plasmons coincide, in other words, where the exponentially decaying tail of electric field of graphene plasmons resonantly excite the substrate surface plasmons.

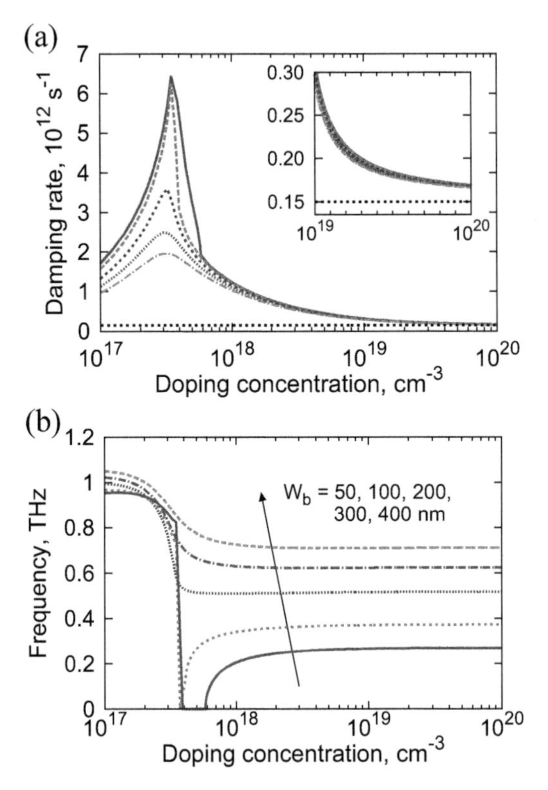

Figure 11.2 Dependences of (a) the plasmon damping rate and (b) frequency on the substrate doping concentration, N_s, with the plasmon wavenumber k = 14×10^3 cm^{-1} (the wavelength λ = 4.5 μm) and with different thicknesses of the bottom dielectric layer, W_b, in the ungated graphene structure. The inset in (a) shows the damping rate in the range $N_s = 10^{19} - 10^{20}$ cm^{-3} (in linear scale).

At the resonance, the damping rate becomes larger than 10^{12} s^{-1}, over 10 times larger than the contribution from the acoustic-phonon scattering in graphene, $\nu_e/2 = 1.5 \times 10^{11}$ s^{-1}. For structures with W_b =

50 and 100 nm, even the damping rate is so large that the frequency is dropped down to zero; this corresponds to an overdamped mode. It is seen in Figs. 11.2a and b that the coupling effect becomes weak as the thickness of the bottom dielectric layer increases. The coupling strength at the resonance is determined by the ratio of the electric fields at the graphene layer and at the interface between the bottom dielectric layer and substrate. In the case of the ungated structure with a relatively low doping concentration, it is roughly equal to $\exp(-kW_b)$. Since $\lambda = 4.5$ μm is much larger than the thicknesses of the bottom dielectric layer in the structures under consideration, i.e., $kW_b \ll 1$, the damping rate and frequency in Figs. 11.2a and b exhibit the rather slow dependences on the thickness.

Away from the resonance, we have several nontrivial features in the concentration dependence of the damping rate. On the lower side of the doping concentration, the damping rate increase does not vanish until $N_s = 10^{14} - 10^{15}$ cm^{-3}. This comes from the wider field spread of the channel mode into the substrate due to the ineffective screening by the low-concentration holes. On the higher side, one can also see a rather broad linewidth of the resonance with respect to the doping concentration, owning to the large, concentration-dependent damping rate of the substrate surface plasmons, and a contribution to the damping rate is not negligible even when the doping concentration is increased two-orders-of-magnitude higher. In fact, with $N_s = 10^{19}$ cm^{-3}, the damping rate is still twice larger than the contribution from the acoustic-phonon scattering. The inset in Fig. 11.2a indicates that the doping concentration must be at least larger than $N_s = 10^{20}$ cm^{-3} for the coupling effect to be smaller than the contribution from the acoustic-phonon scattering, although the latter is still non-negligible. It is also seen from the inset that, with very high doping concentration, the damping rate is almost insensitive to W_b. This originates from the screening by the substrate that strongly expands the field spread into the bottom dielectric layer.

As for the dependence of the frequency, it tends to a lower value in the limit $N_s \to \infty$ than that in the limit $N_s \to 0$, as seen in Fig. 11.2b, along with the larger dependence on the thickness W_b. This corresponds to the transition of the channel mode from an ungated plasmon mode to a gated plasmon mode, where the substrate effectively acts as a back gate.

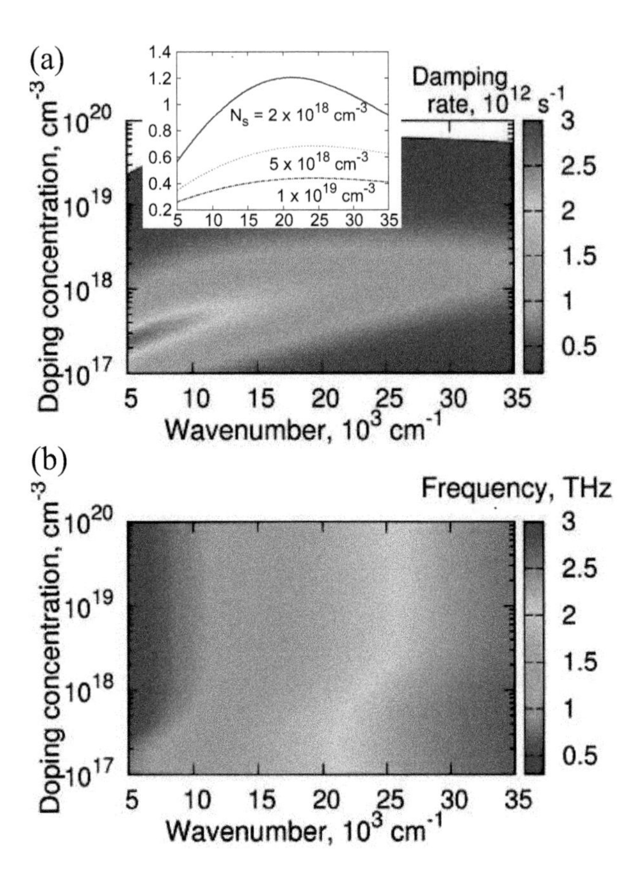

Figure 11.3 Dependences of (a) the plasmon damping rate and (b) frequency on the substrate doping concentration, N_s, and the plasmon wavenumber, k, with different the thickness of the bottom dielectric layer W_b = 300 nm in the ungated graphene structure. The inset of (a) shows the wavenumber dependence of the damping rate with certain doping concentrations. The region with the damping rate below 0.2×10^{12} s^{-1} is filled with white in (a).

To illustrate the coupling effect with various frequencies in the THz range, dependences of the plasmon damping rate and frequency on the substrate doping concentration and plasmon wavenumber with W_b = 300 nm are plotted in Figs. 11.3a and b. In Fig. 11.3a, the peak of the damping rate shifts to the higher doping concentration as the wavenumber increases, whereas its value decreases. The first feature can be understood from the matching condition of the wavenumber-dependent frequency of the ungated graphene plasmons and the doping-concentration-dependent frequency of the substrate surface plasmons, i.e., $\Omega_{gr} \propto k^{1/2}$, roughlly speaking, and $\Omega_{sub} \propto N_s^{1/2}$. The second feature originates from the exponential decay factor, $\exp(-kW_b)$, of the electric field of the channel mode at the interface between the bottom dielectric layer and the substrate; since the doping concentration is $\lesssim 10^{18}$ cm^{-3} at the resonance for any wavevector in Fig. 11.3, the exponential decay is valid. Also, with a fixed doping concentration, say $N_a > 10^{19}$ cm^{-3}, the damping rate has a maximum at a certain wavenumber, resulting from the first feature (see the inset in Fig. 11.3a).

11.4 Gated Plasmons

Next, we study plasmons in the gated structures. We consider the same electron concentration, ficticious mass, and collision frequency, $\Sigma_e = 10^{12}$ cm^{-2}, $m_e = 0.0427\ m_0$, and $\nu_e = 3 \times 10^{11}$ s^{-1}, as Section 11.3. As examples of materials for top/bottom dielectric layers, we examine Al_2O_3/SiO_2 and diamond-like carbon (DLC)/3C-SiC. These materials choices not only reflect the realistic combination of dielectric materials available today but also demonstrate two distinct situations for the coupling effect under consideration, where $\epsilon_t > \epsilon_b$ for the former and $\epsilon_t < \epsilon_b$ for the latter.

Figures 11.4a and b show the dependences of the plasmon damping rate and frequency on the substrate doping concentration with the wavenumber $k = 37 \times 10^3$ cm^{-1} (the plasmon wavelength $\lambda = 1.7$ μm), with thicknesses of the Al_2O_3 top dielectric layer W_t = 20 and 40 nm, and with different thicknesses of the SiO_2 bottom dielectric layer, W_b. As seen, the resonant peaks in the damping rate as well as the frequency drop due to the coupling effect appear, although the peak values are substantially smaller than those in the ungated structure (cf. Fig. 11.2). The peak value decreases rapidly as the thickness of the bottom dielectric layer increases; it almost vanishes when $W_b \geq 300$ nm. These reflect the fact that in the gated structure the electric field of the channel mode is confined dominantly in the top dielectric layer due to the gate screening effect. The field only weakly spreads into the bottom dielectric layer, where its characteristic length is roughtly proportional to W_t, rather than the wavelength λ as in the ungated structure. Thus, the coupling

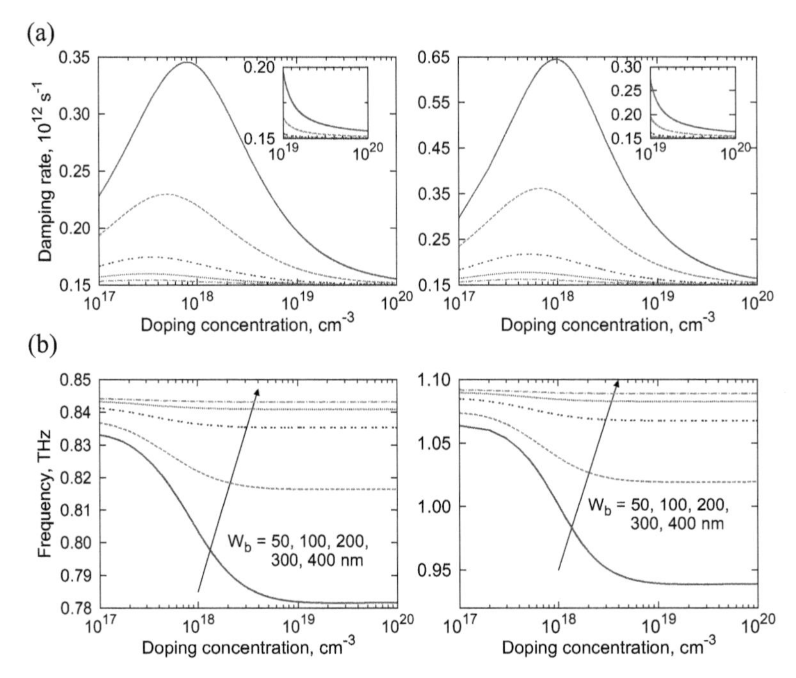

Figure 11.4 Dependences of (a) the plasmon damping rate and (b) frequency on the substrate doping concentration, N_s, with the plasmon wavelength λ = 1.7 μm (the wavenumber $k = 37 \times 10^3$ cm^{-1}), with thicknesses of the Al_2O_3 top dielectric layer W_t = 20 and 40 nm (left and right panels, respectively), and with different thicknesses of the SiO_2 bottom dielectric layer, W_b, in the gated graphene structure. The insets in (a) show the damping rate in the range $N_s = 10^{19} - 10^{20}$ cm^{-3} (in linear scale).

effect on the damping rate together with on the frequency vanishes quickly as W_b increases, even when the wavenumber is small and $kW_b \ll 1$. More quantitatively, the effect is negligible when the first factor of A_c given in Eq. (11.8) in the limit $kW_b \ll 1$ and $kW_t \ll 1$,

$$\frac{\epsilon_b^2 (H_b^2 - 1)}{(\epsilon_b H_b + \epsilon_t H_t)(\epsilon_s + \epsilon_b H_b)} \simeq \frac{1}{1 + (W_b/\epsilon_b)/(W_t/\epsilon_t)} \tag{11.11}$$

is small, i.e., when the factor $(W_b/\epsilon_b)/(W_t/\epsilon_t)$ is much larger than unity. A rather strong dependence of the damping rate on W_b can be also seen with high doping concentration, in the insets of Fig. 11.4a.

Figure 11.5 shows the dependence of the plasmon damping rate on the substrate doping concentration and plasmon wavenumber, with dielectric layer thicknesses W_t = 20 and W_b = 50 nm. As compared with the case of the ungated structure (Fig. 11.3a), the peak of the damping rate exhibits a different wavenumber dependence; it shows a broad maximum at a certain wavenumber (around 150 $\times 10^3$ cm^{-1} in Fig. 11.5) unlike the case of the ungated structure, where the resonant peak decreases monotonically as increasing the wavenumber. This can be explained by the screening effect of the substrate against that of the top gate. When the wavenumber is small and the doping concentration corresponding to the resonance is low, the field created by the channel mode is mainly screened by the gate and the field is weakly spread into the bottom direction. As the doping concentration increases (with increase in the wavevector which gives the resonance), the substrate begins to act as a back gate and the field spreads more into the bottom dielectric layer, so that the coupling effect becomes stronger. When the wavenumber becomes so large that $kW_b \ll 1$ does not hold, the field spread is no longer governed dominantly by the substrate or gate screening, i.e., the channel mode begins to be "ungated" by the substrate. Eventually, the coupling effect on the damping rate again becomes weak, with the decay of the field being proportional to $\exp(-kW_b)$.

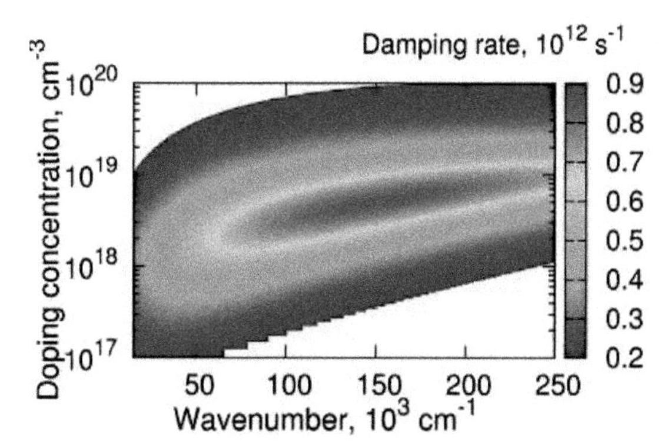

Figure 11.5 Dependence of the plasmon damping rate on the substrate doping concentration, N_s, and the plasmon wavenumber, k, with different the thicknesses of the Al_2O_3 top dielectric layer W_t = 20 nm and the SiO_2 bottom dielectric layer W_b = 50 nm in the gated graphene structure. The region with the damping rate below 0.2×10^{12} s^{-1} is filled with white.

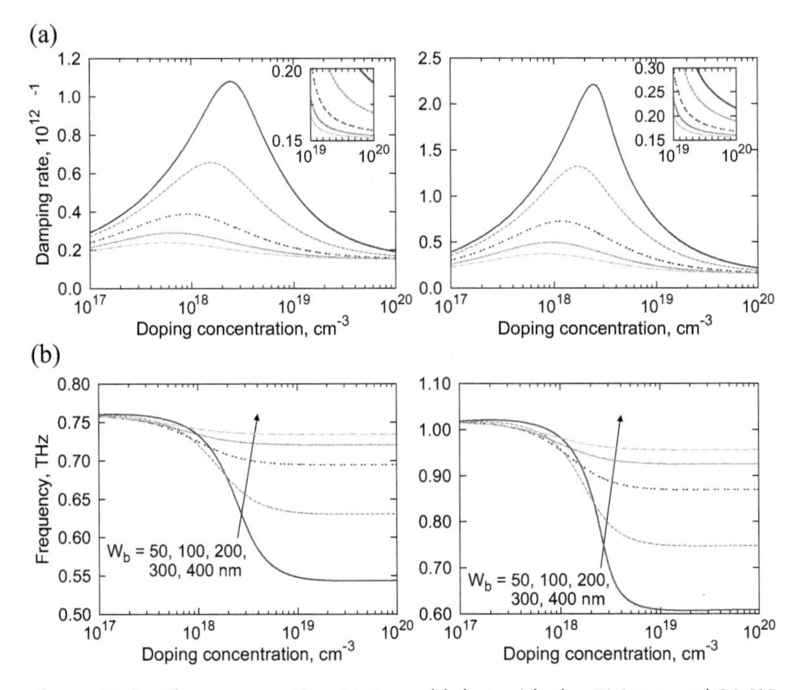

Figure 11.6 The same as Figs. 11.4a and b but with the DLC top and 3C-SiC bottom dielectric layer.

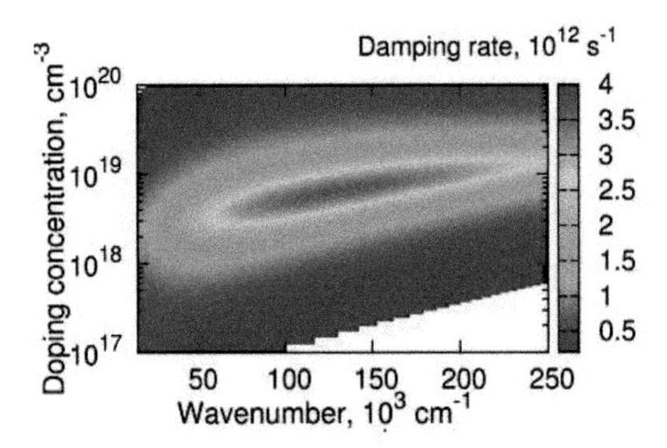

Figure 11.7 The same as Fig. 11.5 but with the DLC top and 3C-SiC bottom dielectric layer.

As illustrated in Eq. (11.11), the coupling effect in the gated strcture is characterized by the factor $(W_b/\epsilon_b)/(W_t/\epsilon_t)$ when the conditions $kW_b \ll 1$ and $kW_t \ll 1$ are met. This means that not only the thicknesses of the dielectric layers but also their dielectric constants are very important parameters to determine the coupling strength. For example, if we adapt the high-k material, e.g., HfO_2 in the top dielectric layer, it results in the more effective gate screening than in the gated structure with the Al_2O_3 top dielectric layer, so that the coupling effect can be suppressed even with the same layer thicknesses. The structure with the DLC top and 3C-SiC bottom dielectric layers (with $\epsilon_t = 3.1$ (Refs. [29, 30]) and $\epsilon_b = 9.7$) corresponds to the quite opposite situation, where the gate screening becomes weak and the substrate screening becomes more effective, so that stronger coupling effect is anticipated. Figures 11.6a, b, and 11.7 show the same dependences as in Figs. 11.4a, b, and 11.5, respectively, for the structure with the DLC top and 3C-SiC bottom dielectric layers. Comparing with those for the structure with the Al_2O_3 top and SiO_2 bottom dielectric layers, the damping rate as well as the frequency are more influenced by the coupling effect in the entire ranges of the doping concentration and wavevector. In particular, the increase in the damping rate with high doping concentration $N_s = 10^{19} - 10^{20}$ cm^{-3} and the thickness of the bottom layer $W_b = 50 - 100$ nm, which are typical values in real graphene samples, is much larger. However, this increase can be avoided by adapting thicker bottom layer, say, $W_b \gtrsim 200$ nm or by increasing the doping concentration.

11.5 Conclusions

In summary, we studied theoretically the coupling of plasmons in graphene at THz frequencies with surface plasmons in a heavily doped substrate. We demonstrated that in the ungated graphene structure there is a huge resonant increase in the damping rate of the "channel mode" at a certain doping concentration of the substrate ($\sim 10^{17}$ cm^{-2}) and the increase can be more than 10^{12} s^{-1}, due to the resonant coupling of the graphene plasmon and the substrate surface plasmon. The dependences of the damping rate on the doping concentration, the thickness of the bottom dielectric layer, and the plasmon wavenumber are associated with the field spread of the channel mode into the bottom dielectric layer and into the substrate. We revealed that even with very high doping concentration (10^{19} - 10^{20} cm^{-2}), away from the resonance, the coupling effect causes non-negligible increase in the damping rate compared with the acoustic-phonon-limited damping rate. In the gated graphene structure, the coupling effect can be much reduced compared with that in the ungated structure, reflecting the fact that the field is confined dominantly in the top dielectric layer due to the gate screening. However, with very high doping concentration, it was shown that the screening by the substrate effectively spreads the field into the bottom dielectric layer and the increase in the damping rate can be non-negligible. These results suggest that the structural parameters such as the thicknesses and dielectric constants of the top and bottom dielectric layers must be properly chosen for the THz plasmonic devices in order to maximize the active plasmonic effect onto real device implementations.

Acknowledgments

Authors thank M. Suemitsu and S. Sanbonsuge for providing information about the graphene-on-silicon structure and Y. Takakuwa, M. Yang, H. Hayashi, and T. Eto for providing information about the diamond-like-carbon dielectric layer. This work was supported by JSPS Grant-in-Aid for Young Scientists (B) (#23760300), by JSPS Grant-in-Aid for Specially Promoted Research (#23000008), and by JST-CREST.

References

1. M. Dyakonov and M. Shur, *Phys. Rev. Lett.*, **71**, 2465 (1993).
2. M. Dyakonov and M. Shur, *IEEE Trans. Electron Devices*, **43**, 380 (1996).
3. M. S. Shur and J.-Q. Lu, *IEEE Trans. Micro. Theory Tech.*, **48**, 750 (2000).
4. M. S. Shur and V. Ryzhii, *Int. J. High Speed Electron. Syst.*, **13**, 575 (2003).
5. F. Teppe, D. Veksler, A. P. Dmitriev, X. Xie, S. Rumyantsev, W. Knap, and M. S. Shur, *Appl. Phys. Lett.*, **87**, 022102 (2005).
6. T. Otsuji, Y. M. Meziani, T. Nishimura, T. Suemitsu, W. Knap, E. Sano, T. Asano, V. V. and Popov, *J. Phys.: Condens. Matter*, **20**, 384206 (2008).
7. V. Ryzhii, A. Satou, M. Ryzhii, T. Otsuji, and M. S. Shur, *J. Phys.: Condens. Matter*, **20**, 384207 (2008).

8. V. V. Popov, D. V. Fateev, T. Otsuji, Y. M. Meziani, D. Coquillat, and W. Knap, *Appl. Phys. Lett.*, **99**, 243504 (2011).
9. T. Watanabe, S. Boubanga Tombet, Y. Tanimoto, Y. Wang, H. Minamide, H. Ito, D. Fateev, V. Popov, D. Coquillat, W. Knap, Y. Meziani, and T. Otsuji, *Solid-State Electron.*, **78**, 109 (2012).
10. V. Ryzhii, *Jpn. J. Appl. Phys., Part 2*, **45**, L923 (2006); V. Ryzhii, A. Satou, and T. Otsuji, *J. Appl. Phys.*, **101**, 024509 (2007).
11. T. Otsuji, S. A. Boubanga Tombet, A. Satou, H. Fukidome, M. Suemitsu, E. Sano, V. Popov, M. Ryzhii, and V. Ryzhii, *J. Phys. D: Appl. Phys.*, **45**, 303001 (2012).
12. A. N. Grigorenko, M. Polini, and K. S. Novoselov, *Nat. Photonics*, **6**, 749 (2012).
13. E. Hwang and S. Das Sarma, *Phys. Rev. B*, **77**, 115449 (2008).
14. V. Ryzhii, T. Otsuji, M. Ryzhii, and M. S. Shur, *J. Phys. D: Appl. Phys.*, **45**, 302001 (2012).
15. V. Ryzhii, M. Ryzhii, and T. Otsuji, *J. Appl. Phys.*, **101**, 083114 (2007).
16. M. Ryzhii and V. Ryzhii, *Jpn. J. Appl. Phys., Part 2*, **46**, L151 (2007).
17. A. A. Dubinov, V. Ya. Aleshkin, V. Mitin, T. Otsuji, and V. Ryzhii, *J. Phys.: Condens. Matter*, **23**, 145302 (2011).
18. V. Popov, O. Polischuk, A. Davoyan, V. Ryzhii, T. Otsuji, and M. Shur, *Phys. Rev. B*, **86**, 195437 (2012).
19. Y. Takatsuka, K. Takahagi, E. Sano, V. Ryzhii, and T. Otsuji, *J. Appl. Phys.*, **112**, 033103 (2012).
20. L. Ju, B. Geng, J. Horng, C. Girit, M. Martin, Z. Hao, H. Bechtel, X. Liang, A. Zettl, Y. R. Shen, and F. Wang, *Nat. Nanotechnol.*, **6**, 630 (2011).
21. J. H. Strait, P. Nene, W.-M. Chan, C. Manolatou, S. Tiwari, F. Rana, J. W. Kevek, and P. L. McEuen, *Phys. Rev. B*, **87**, 241410 (2013).
22. M. Suemitsu, Y. Miyamoto, H. Handa, and A. Konno, *e-J. Surf. Sci. Nanotechnol.*, **7**, 311 (2009).
23. N. J. M. Horing, *Phys. Rev. B*, **80**, 193401 (2009).
24. J. Yan, K. Thygesen, and K. Jacobsen, *Phys. Rev. Lett.*, **106**, 146803 (2011).
25. A. Satou, V. Vyurkov, and I. Khmyrova, *Jpn. J. Appl. Phys., Part 1*, **43**, L566 (2004).
26. D. Svintsov, V. Vyurkov, S. Yurchenko, T. Otsuji, and V. Ryzhii, *J. Appl. Phys.*, **111**, 083715 (2012).
27. C. Bulucea, *Solid-State Electron.*, **36**, 489 (1993).

28. S. A. Maier, *Plasmonics: Fundamentals and Applications* (Springer Science, NY, 2007).
29. The dielectric constant of DLC varies in the range between 3.1 and 7.8, depending on its growth condition (Ref. [30]). Here, we choose the lowest value for demonstration of the case where $\epsilon_t \ll \epsilon_b$.
30. H. Hayashi, S. Takabayashi, M. Yang, R. Jesko, S. Ogawa, T. Otsuji, and Y. Takakuwa, "Tuning of the dielectric constant of diamond-like carbon films synthesized by photoemission-assisted plasma-enhanced CVD," in *2013 International Workshop on Dielectric Thin Films for Future Electron Devices Science and Technology (2013 IWDTF)*, Tokyo, Japan, November 7–9, 2013.

Chapter 12

Active Guiding of Dirac Plasmons in Graphene*

M. Yu. Morozov,[a] A. R. Davoyan,[b] I. M. Moiseenko,[c] A. Satou,[d] T. Otsuji,[d] and V. V. Popov[a,c,e]
[a]*Kotelnikov Institute of Radio Engineering and Electronics (Saratov Branch), Russian Academy of Sciences, Saratov 410019, Russia*
[b]*Department of Electrical and Systems Engineering, University of Pennsylvania, Philadelphia, Pennsylvania 19104, USA*
[c]*Physics Department, Saratov State University, Saratov 410012, Russia*
[d]*Research Institute of Electrical Communication, Tohoku University, 2-1-1 Katahira, Aoba-ku, Sendai 980-8577, Japan*
[e]*Saratov Scientific Center of the Russian Academy of Sciences, Saratov 410028, Russia*
mikkym@mail.ru

The Dirac plasmon propagation in active pristine graphene with the carrier population inversion created by the diffusion of the photoexcited carriers from a semiconductor substrate is studied

*Reprinted with permission from M. Y. Morozov, A. R. Davoyan, I. M. Moiseenko, A. Satou, T. Otsuji, and V. V. Popov (2015). Active guiding of Dirac plasmons in graphene, *Appl. Phys. Lett.*, **106**, 061105. Copyright © 2015 AIP Publishing LLC.

Graphene-Based Terahertz Electronics and Plasmonics: Detector and Emitter Concepts
Edited by Vladimir Mitin, Taiichi Otsuji, and Victor Ryzhii

ISBN 978-981-4800-75-4 (Hardcover), 978-0-429-32839-8 (eBook)
www.jennystanford.com

theoretically. It is shown that an order of magnitude smaller pump power can be used for the diffusion pumping as compared to direct optical pumping of graphene for obtaining the same plasmon gain in graphene. We find that the field of the amplified plasmons remains strongly confined in the vicinity of graphene similarly to the case of the attenuated plasmons. Remarkably, the diffusion pumping is characterized by low insertion losses due to small photoexcited carrier concentration in the carrier-supplying semiconductor substrate in the region of the plasmon field confined near graphene.

Recent advances in the areas of nanophotonics, plasmonics, and metamaterials open promising possibilities for high-speed, broadband, and power-saving electromagnetic signal processing. Further progress in these directions could be based on merging and unifying principles of electronics and integrated optics, and employing recently emerging electronic technologies. Terahertz (THz) plasmonics that incorporates the advantages of both high-speed electronic semiconductor technology and subwavelength nanophotonics may bring optics and electronics together for designing highly functional THz devices. Recently, ideas of graphene plasmonics have been actively discussed, and a number of important graphene-based THz integrated optoelectronic devices have been proposed theoretically and demonstrated experimentally (see, e.g., review papers, Refs. [1, 2], and references therein).

Excitation of highly confined plasmons in graphene allows for the guiding and manipulation of THz radiation in a deep subwavelength regime [3–6]. Such a strong THz field localization may be utilized for the design of high-speed plasmonic interconnects and electrically controllable graphene plasmonic devices on a single chip. However, the plasmon damping due to the Drude absorption by free carriers in graphene and semiconductor substrate might hinder their application in THz integrated nanocircuits [7].

Previously, the plasmons in graphene charged by the doping or electrostatic gating have been mostly considered while the thermoactivated Dirac plasmons in pristine (neutral) graphene [8, 9] have not received much of the experimental attention due to small carrier density in pristine graphene at the thermal equilibrium. On the other hand, the gapless linear energy spectrum and high carrier mobility make pristine graphene a promising medium for

THz amplification and lasing. Optical pumping of pristine graphene may lead to an effective carrier population inversion and THz gain in graphene caused by the dominance of the interband carrier transitions with THz photon emission over the intraband (Drude-like) carrier absorption [10]. Various designs of graphene based THz lasers and THz plasmonic amplifiers with the optical pumping were proposed recently (see review paper Ref. [11] and references therein). Under optical pumping, the carrier density in pristine graphene enhances considerably so that the plasmon frequency can reach THz range. Amplification of the Dirac plasmons in optically pumped pristine graphene was theoretically considered in Ref. [12] and experimentally observed in Ref. [13]. However, because graphene absorbs below 2.3% of the incident optical pump power [14], the efficiency of optically pumped graphene laser is questionable. A concept of the diffusion pumping that allows for much weaker pump power as compared with direct optical pumping of graphene was proposed recently [15].

In this chapter, we employ the diffusion pumping concept for suppressing plasmon losses and amplification of Dirac plasmons in graphene. We study the plasmon propagation in pristine graphene deposited on a widegap semiconductor substrate and demonstrate that graphene pumping by diffusion of the electron-hole pairs photoexcited in the semiconductor substrate by incoming light suppresses the plasmonic losses and can lead to the plasmonic gain in graphene at THz frequencies. This phenomenon may be used as a tool for controlling plasmon propagation and creating the "loss-free" and active graphene-based THz nanocircuits.

The proposed concept of diffusion-pumping amplification of the plasmons in graphene is schematically shown in Fig. 12.1. A monolayer of pristine graphene is deposited on the top surface ($y = d$ in Fig. 12.1) of wide-gap semiconductor (GaAs in our particular case) substrate of thickness d. An incident optical pump beam illuminates the bottom surface ($y = 0$ in Fig. 12.1) of the semiconductor substrate generating the electron-hole pairs inside semiconductor. The photoexcited carriers diffuse across the semiconductor slab and, approaching the top surface of the substrate slab, are captured into graphene creating the carrier population inversion in graphene. It is shown below that the plasmonic gain in the population inverted graphene can totally suppress the plasmon losses and can lead to strong amplification of the plasmons guided along the graphene layer at practically achievable input optical pump power.

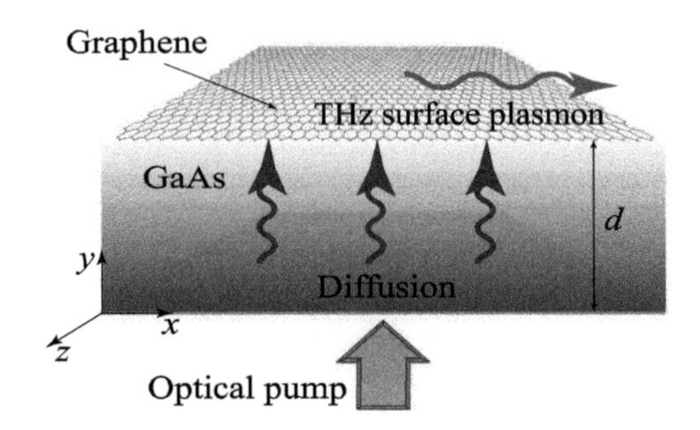

Figure 12.1 Diffusion pumping of THz plasmons in graphene. The electron-hole pairs in GaAs are generated by illuminating the bottom surface of GaAs substrate with optical beam. The photoexcited carriers diffuse towards the top surface of the substrate and, being captured into graphene, create the carrier population inversion that leads to the plasmon gain in graphene.

In order to get a deeper insight into the physical processes in such a system, we use a system of coupled equations describing (i) the plasmon propagation in graphene and (ii) diffusion-pumping of graphene.

We consider the plasmon propagation in pristine (neutral) graphene with the carrier population inversion. Because the plasmon wavelength is two or even three orders of magnitude shorter than the free-space electromagnetic radiation wavelength at the same frequency [3–5], we utilize the quasi-electrostatic approach and, representing the spatio-temporal dependence of the plasmon field as $\exp(-i\omega t + ik_x x + ik_y y)$, write the plasmon dispersion relation as [16]

$$k_x = i\frac{\varepsilon_0(\varepsilon_{\text{air}} + \varepsilon_{\text{s}})\omega}{\sigma(\omega)}, \tag{12.1}$$

where k_x and ω are the x-component of the plasmon wave-vector and angular frequency, respectively, ε_{air} and ε_{s} are the dielectric constants of the ambient medium (air) and semiconductor material of the substrate, respectively, ε_0 is the electric constant, and $\sigma(\omega)$ is the dynamic conductivity of the population inverted graphene. Equation (12.1) allows us to determine the plasmon wavelength, $2\pi/\text{Re}\ k_x$, and power decrement for the attenuated plasmons,

2 Im $k_x > 0$, (or power increment for the amplified plasmons, 2 Im $k_x < 0$). Equation (12.1) describes the plasmon propagation in graphene on the substrate of infinite thickness. This is a justified assumption because, as shown below, the plasmon wavelength is much shorter than the substrate thickness so that the plasmons "do not feel" the bottom surface of the substrate.

General equation for the plasmon dispersion, Eq. (12.1), is valid for either charged or pristine (neutral) graphene. For the population inverted pristine graphene, the dynamic conductivity is given by [12]

$$\sigma(\omega)=\frac{e^2}{4\hbar}\left\{\frac{8k_BT\tau}{\pi\hbar(1-i\omega\tau)}\ln\left[1+\exp\left(\frac{E_F}{k_BT}\right)\right]\right.$$
$$+\tanh\left(\frac{\hbar\omega-2E_F}{4k_BT}\right)$$
$$\left.-\frac{4\hbar\omega}{i\pi}\int_0^\infty\frac{G(\varepsilon,E_F)-G(\hbar\omega/2,E_F)}{(\hbar\omega)^2-4\varepsilon^2}d\varepsilon\right\}, \tag{12.2}$$

where e is the electron charge, $\hbar$ is the reduced Planck constant, k_B is the Boltzmann constant, τ and T are the mean free time and temperature of the carriers in graphene, respectively, E_F is the quasi-Fermi energy in graphene ($+E_F$ and $-E_F$ for electrons and holes, respectively), and $G(\varepsilon, \varepsilon') = \sinh(\varepsilon/k_BT)/[\cosh(\varepsilon/k_BT) + \cosh(\varepsilon'/k_BT)]$.

It is assumed in Eq. (12.2) that the diffusively populated carriers in graphene are in the thermal equilibrium within the carrier gas described by the Fermi-Dirac distribution function with the carrier temperature T. At the thermal equilibrium within the ensemble of carriers, the density of the electron-hole pairs in graphene n_{gr} can be related to the quasi-Fermi energy as [17]

$$n_{gr}=\frac{2}{\pi\hbar^2V_F^2}\int_0^\infty\frac{\varepsilon d\varepsilon}{1+\exp\left(\frac{\varepsilon-E_F}{k_BT}\right)}, \tag{12.3}$$

where $V_F \approx 10^8$ cm/s is the characteristic velocity of electrons and holes in graphene. Thus, upon knowing the carrier density n_{gr}, we can calculate the quasi-Fermi energy in graphene, E_F, by the numerical reversion of integral in Eq. (12.3).

We find the density of the electron-hole pairs in graphene, n_{gr}, created by their diffusion from the semiconductor substrate by solving the equation for the ambipolar diffusion of the photoexcited electron-hole pairs in the substrate [18]

$$D_{\text{a}}\frac{d^2 n_{\text{s}}}{dy^2}-\frac{n_{\text{s}}}{\tau_{\text{R}}}+\frac{\alpha}{\hbar\omega_{\text{p}}}P_{\text{p}}e^{-\alpha y}=0, \qquad (12.4)$$

where n_{s} is the bulk concentration of electron-hole pairs in the semiconductor, D_{a} is the ambipolar diffusion coefficient, τ_{R} is the electron-hole pair recombination time in semiconductor, P_{p} and ω_{p} are the input optical pump power and frequency, respectively, and α is the light absorption coefficient in semiconductor. We solve Eq. (12.4) with the boundary conditions $dn_{\text{s}}/dy = 0$ for $y = 0$ and $n_{\text{s}} = 0$ for $y = d$. The first condition implies the absence of carrier flux through the bottom (graphene-free) surface of the substrate, whereas the second one describes the infinitely fast capture of carriers reaching the top semiconductor-substrate surface into graphene. This assumption is reasonable as a first approximation due to a significant energy gap offset between semiconductor and graphene.

Populated carrier density in graphene n_{gr} is related to the diffusion carrier flux density from the semiconductor substrate, $J_{\text{in}} = -D_{\text{a}}(dn_{\text{s}}/dy)|_{y=d}$, by the rate equation $n_{\text{gr}}/\tau_{\text{gr}} - J_{\text{in}} = 0$, where τ_{gr} is the carrier spontaneous recombination time in graphene. (In principle, the populated carrier density in graphene also changes (decreases) due to the stimulated emission of plasmons. However, within the linear problem of calculating plasmon dispersion and plasmon gain in this chapter, this effect is neglected.)

Moving further, we calculate the plasmon propagation and amplification in a diffusion-pumped graphene. We solved Eq. (12.1) assuming the following characteristic parameters: $\varepsilon_{\text{s}} = 12.8$ (GaAs) and $\varepsilon_{\text{air}} = 1$, $T = 300$ K, the mean free time of carriers in graphene is $\tau = 1$ ps [19], and $\tau_{\text{gr}} = 5$ ps [20]. The diffusion equation Eq. (12.4) was solved for $\hbar\omega_{\text{p}} = 2.5 \times 10^{-19}$ J (which corresponds to the wavelength 808 nm of commercially available laser diodes), $D_{\text{a}} = 20$ cm^2/s, $\tau_{\text{R}} =$ 5 ns, and $\alpha = 10^4$ cm^{-1} [18].

First, we calculate the dispersion of THz plasmons for different pump power values (see Fig. 12.2a) neglecting the losses in the substrate. One can see that the THz plasmon wavelength is from two to three orders of magnitude shorter than the free-space THz radiation wavelength (at the same frequency). This justifies the use of the quasi-electrostatic approach for describing the dispersion of graphene plasmons. We also note that with the increase of the pump power, the plasmon wavelength also increases for a given frequency due to the enhancement of carrier concentration in graphene with increasing the pump power.

We plot in Fig. 12.2b the power gain per the plasmon wavelength, $\Gamma = \exp(-4\pi \, \mathrm{Im}\, k_x/\mathrm{Re}\, k_x)$, versus the pump power for different substrate thicknesses ($\Gamma > 1$ and $\Gamma < 1$ correspond to the amplification and damping of plasmons, respectively) at frequency 5 THz. It is seen from Fig. 12.2b that the same maximal plasmon gain (about 2.2) is reached for any substrate thickness but at different pump power values for different substrate thicknesses. That can be explained as the following. For each substrate thickness, the quasi-Fermi energy steadily grows with increasing the pump power. At small quasi-Fermi energy values, the Drude absorption in graphene surpasses the stimulated emission of the plasmons that corresponds to the plasmon absorption $\Gamma < 1$. For greater values of E_F, the stimulated emission of the plasmons starts prevailing over the Drude absorption, which results in the plasmon gain $\Gamma > 1$. For further increasing the quasi-Fermi energy level, the Drude losses in graphene (described by the real part of the first term in the curly brackets of the conductivity expression Eq. (12.2)) grow roughly proportional to the quasi-Fermi energy value (for $E_F \gg k_B T$) while the second term in the curly brackets of Eq. (12.2) responsible for the stimulated emission of plasmons in graphene saturates for high values of E_F. Therefore, sooner or later, the Drude absorption again prevails over the stimulated emission which results in diminishing the plasmon gain for large E_F. In the process of increasing the pump power, the quasi-Fermi energy passes through the same value corresponding to the maximal power released from graphene due to the carrier population inversion, which results in the same maximal plasmon gain for any substrate thickness.

One can notice the existence of optimal substrate thickness (see curve 2 in Fig. 12.2b) corresponding to the smallest optical pump power value yielding a maximum plasmon gain in graphene. The optimal value of the substrate thickness is determined by the interplay of carrier generation and recombination in the substrate.

For the substrate thickness much smaller than the optical-pump-power attenuation length in semiconductor, $1/\alpha$, which is about 1 μm for GaAs, only a small part of the optical pump power is absorbed in the semiconductor substrate generating small number of the electron-hole pairs which results in a weak population inversion (see Fig. 12.2c) and hence small plasmon gain in graphene. When the thickness of the substrate by far exceeds the ambipolar carrier diffusion length $L_a = \sqrt{D_a \tau_R}$ ($L_a \approx 3$ μm for GaAs), only a small number of the photoexcited carriers reach graphene due to their recombination in the substrate which also leads to a weak population inversion and hence small plasmon gain in graphene. Therefore, the optimal substrate thickness (2 μm in Fig. 12.2b) yielding the maximum population inversion and greatest plasmon gain in graphene is in between $1/\alpha$ and L_a.

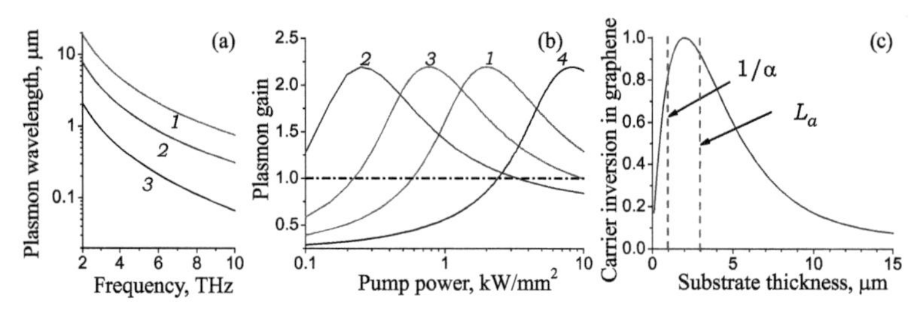

Figure 12.2 (a) Plasmon wavelength versus plasmon frequency for different pump power values 1000 W/mm^2 (curve 1), 500 W/mm^2 (curve 2), and 100 W/mm^2 (curve 3) for the substrate thickness d = 2 μm. (b) Power gain per the plasmon wavelength versus input optical pump power for different substrate thicknesses 0.1 μm (curve 1), 2 μm (curve 2), and 7 μm (curve 3) at frequency 5 THz. Curve 4 corresponds to direct optical pumping of suspended graphene (with no substrate). (c) Carrier density in graphene as a function of the substrate thickness. The optical pump power attenuation length, $1/\alpha$, and diffusion lengths, L_a, are shown by the vertical dashed straight lines.

The main finding of this work is that, with using the diffusion pumping mechanism, the maximal plasmon gain can be reached for rather small optical pump power. This value of the pump power is by an order of magnitude smaller than that for direct optical pumping of suspended (with no substrate) graphene (cf. curves 2 and 4 in Fig. 12.2b). (The generation rate of the photoexcited carriers in suspended graphene for direct optical pumping was estimated as $\kappa P_p/\hbar\omega_p$, where the light absorption coefficient in graphene was put at $\kappa = 2.3\%$. It is worth mentioning that the optical absorption in graphene on the semiconductor substrate illuminated from the graphene side is even smaller than that in suspended graphene due to partial optical reflection from the substrate surface [21]. Therefore, the direct optical pumping of graphene layer on the semiconductor substrate from the graphene side is even less efficient than the direct optical pumping of suspended graphene.)

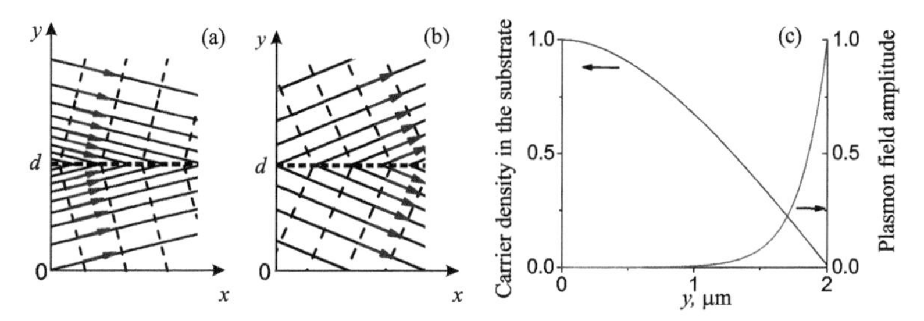

Figure 12.3 Amplitude-phase patterns of the plasmon field for (a) attenuated and (b) amplified plasmons. The equi-amplitude and equi-phase lines of the plasmon field are shown by solid and dashed straight lines, respectively. The equi-amplitude line density corresponds to the plasmon field amplitude at a given point of space. The energy-flux vectors of the plasmon field are shown by blue arrows. The plane of graphene is shown by the thick dashed line. (c) Normalized photoexcited electron-hole pair concentration (left ordinate axis) and the plasmon electric field amplitude at frequency 5 THz (right ordinate axis) versus the substrate thickness. The bottom surface of the substrate is at $y = 0$ and graphene is located on the top surface of the substrate at $y = 2$ μm.

The last important question to be answered is whether the plasmon amplification affects the plasmon confinement in the vicinity of the graphene layer. The amplitude-phase patterns of the plasmon field for attenuated and amplified plasmons are shown in Figs. 12.3a and b, respectively. The equi-amplitude and equi-phase lines of the plasmon field (shown by solid and dashed straight lines, respectively, in Figs. 12.3a and b) are perpendicular to each other in the media above and below graphene. The equi-amplitude line density corresponds to the plasmon field amplitude at a given point of space. The energy-flux vectors of the plasmon field (shown by arrows in Figs. 12.3a and b) are directed at the normal to the equi-phase lines. For attenuated plasmons (Fig. 12.3a), the energy-flux vectors have the normal-to-graphene components directed towards graphene in both media above and below graphene, which corresponds to the absorption of the plasmon energy in graphene. On the contrary, for the amplified plasmons (Fig. 12.3b), the energy-flux vectors have the normal-to-graphene components directed outwards graphene in both media above and below graphene, which corresponds to the plasmon gain due to the energy flux effused from the population inverted graphene. Surprisingly, in spite of the fact that the energy leaks from graphene in this case, the amplified plasmon field still decays away from the graphene layer. This happens because the plasmon field farther away from graphene is created by the energy flux effused from graphene at earlier moments of time when the plasmon energy in graphene was much smaller. It is known that the plasmon field of attenuated (lossless in the limiting case) plasmons rapidly decays away from graphene [3, 16]. Actually, the same decay length (of the order the plasmon wavelength shown in Fig. 12.2a) is also characteristic of the amplified plasmons, which means that the amplified plasmons are strongly confined in the vicinity of graphene (Fig. 12.3c). The photoexcited carrier distribution across the substrate thickness is also shown in Fig. 12.3c. One can see from Fig. 12.3c that the bulk free carrier density in semiconductor n_s is very small in the vicinity of the graphene layer due to fast capture of the carriers into graphene. At the same time, the plasmon field rapidly decays in the substrate away from graphene. Therefore, the Drude losses in the substrate have a marginal effect on the plasmon damping in graphene. (This validates the assumption of a lossless substrate while calculating the plasmon dispersion and damping (gain) in graphene.)

In conclusion, we have developed a concept of the plasmonic graphene amplifier employing the carrier diffusion pumping from a semiconductor substrate. It is shown that an order of magnitude smaller pump power can be used with the diffusion pumping as compared to direct optical pumping of graphene. We find that the field of the amplified plasmons remains strongly confined in the vicinity of graphene similarly to the case of the attenuated plasmons. Diffusion pumping mechanism combines an effective pumping of graphene with low insertion losses due to small photoexcited carrier concentration in the carrier-supplying semiconductor substrate in the region of the plasmon field confined near graphene. Amplified plasmons can be used as low-loss interconnects and active elements in the plasmonic graphene THz nanocircuits.

The work was supported by the Russian Foundation for Basic Research through Grant No. 13-02-12070 and by JSPS Grant-in-Aid for Specially Promoting Research (No. 23000008), Japan.

References

1. A. N. Grigorenko, M. Polini, and K. S. Novoselov, *Nat. Photonics*, **6**, 749 (2012).
2. X. Luo, T. Qiu, W. Lu, and Z. Ni, *Mater. Sci. Eng. R*, **74**, 351 (2013).
3. F. H. L. Koppens, D. E. Chang, and F. J. García de Abajo, *Nano Lett.*, **11**, 3370 (2011).
4. J. Chen, M. Badioli, P. Alonso-González, S. Thongrattanasiri, F. Huth, J. Osmond, M. Spasenović, A. Centeno, A. Pesquera, P. Godignon, A. Z. Elorza, N. Camara, F. J. García de Abajo, R. Hillenbrand, and F. H. Koppens, *Nature*, **487**, 77 (2012).
5. Z. Fei, A. S. Rodin, G. O. Andreev, W. Bao, A. S. McLeod, M. Wagner, L. M. Zhang, Z. Zhao, M. Thiemens, G. Dominguez, M. M. Fogler, A. H. Castro Neto, C. N. Lau, F. Keilmann, and D. N. Basov, *Nature*, **487**, 82 (2012).
6. J. T. Kim and S.-Y. Choi, *Opt. Express*, **19**(24), 24557 (2011).
7. A. Satou, Y. Koseki, V. Ryzhii, V. Vyurkov, and T. Otsuji, *J. Appl. Phys.*, **115**, 104501 (2014).
8. O. Vafek, *Phys. Rev. Lett.*, **97**, 266406 (2006).
9. S. Das Sarma and Q. Li, *Phys. Rev. B*, **87**, 235418 (2013).
10. V. Ryzhii, M. Ryzhii, and T. Otsuji, *J. Appl. Phys.*, **101**, 083114 (2007).
11. T. Otsuji, S. A. Boubanga Tombet, A. Satou, H. Fukidome, M. Suemitsu, E. Sano, V. Popov, M. Ryzhii, and V. Ryzhii, *MRS Bull.*, **37**, 1235 (2012).
12. A. A. Dubinov, V. Ya. Aleshkin, V. Mitin, T. Otsuji, and V. Ryzhii, *J. Phys.: Condens. Matter*, **23**, 145302 (2011).
13. T. Watanabe, T. Fukushima, Y. Yabe, S. A. Boubanga Tombet, A. Satou, A. A. Dubinov, V. Ya. Aleshkin, V. Mitin, V. Ryzhii, and T. Otsuji, *New J. Phys.*, **15**, 075003 (2013).
14. R. R. Nair, P. Blake, A. N. Grigorenko, K. S. Novoselov, T. J. Booth, T. Stauber, N. M. R. Peres, and A. K. Geim, *Science*, **320**, 1308 (2008).
15. A. R. Davoyan, M. Yu. Morozov, V. V. Popov, A. Satou, and T. Otsuji, *Appl. Phys. Lett.*, **103**, 251102 (2013).
16. M. Jablan, H. Buljan, and M. Soljacic, *Phys. Rev. B*, **80**, 245435 (2009).
17. V. Ya. Aleshkin, A. A. Dubinov, and V. Ryzhii, *JETP Lett.*, **89**, 63 (2009).

18. Y. Morozov, T. Leinonen, M. Morozov, S. Ranta, M. Saarinen, V. Popov, and M. Pessa, *New J. Phys.*, **10**, 063028 (2008).
19. T. Otsuji, S. A. Boubanga-Tombet, A. Satou, H. Fukidome, M. Suemitsu, E. Sano, V. Popov, M. Ryzhii, and V. Ryzhii, *J. Phys. D: Appl. Phys.*, **45**, 303001 (2012).
20. P. A. George, J. Strait, J. Dawlaty, S. Shivaraman, M. Chandrashekhar, F. Rana, and M. G. Spencer, *Nano Lett.*, **8**, 4248 (2008).
21. J. M. Dawlaty, S. Shivaraman, J. Strait, P. George, M. Chandrashekhar, F. Rana, M. G. Spencer, D. Veksler, and Y. Chen, *Appl. Phys. Lett.*, **93**, 131905 (2008).

Chapter 13

Vertical Electron Transport in van der Waals Heterostructures with Graphene Layers*

V. Ryzhii,[a,b] T. Otsuji,[a] M. Ryzhii,[c] V. Ya. Aleshkin,[d] A. A. Dubinov,[d] V. Mitin,[a,e] and M. S. Shur[f]

[a]*Research Institute for Electrical Communication, Tohoku University, Sendai 980-8577, Japan*
[b]*Center for Photonics and Infrared Engineering, Bauman Moscow State Technical University and Institute of Ultra High Frequency Semiconductor Electronics of RAS, Moscow 111005, Russia*
[c]*Department of Computer Science and Engineering, University of Aizu, Aizu-Wakamatsu 965-8580, Japan*
[d]*Institute for Physics of Microstructures of RAS and Lobachevsky State University of Nizhny Novgorod, Nizhny Novgorod 603950, Russia*
[e]*Department of Electrical Engineering, University at Buffalo, Buffalo, New York 1460-1920, USA*
[f]*Department of Electrical, Electronics, and Systems Engineering and Department of Physics, Applied Physics, and Astronomy, Rensselaer Polytechnic Institute, Troy, New York 12180, USA*
v-ryzhii@riec.tohoku.ac.jp

---

*Reprinted with permission from V. Ryzhii, T. Otsuji, M. Ryzhii, V. Ya. Aleshkin, A. A. Dubinov, V. Mitin, and M. S. Shur (2015). Vertical electron transport in van der Waals heterostructures with graphene layers, *J. Appl. Phys.*, **117**, 154504. Copyright © 2015 AIP Publishing LLC.

Graphene-Based Terahertz Electronics and Plasmonics: Detector and Emitter Concepts
Edited by Vladimir Mitin, Taiichi Otsuji, and Victor Ryzhii

ISBN 978-981-4800-75-4 (Hardcover), 978-0-429-32839-8 (eBook)
www.jennystanford.com

We propose and analyze an analytical model for the self-consistent description of the vertical electron transport in van der Waals graphene-layer (GL) heterostructures with the GLs separated by the barriers layers. The top and bottom GLs serve as the structure emitter and collector. The vertical electron transport in such structures is associated with the propagation of the electrons thermionically emitted from GLs above the inter-GL barriers. The model under consideration describes the processes of the electron thermionic emission from and the electron capture to GLs. It accounts for the nonuniformity of the self-consistent electric field governed by the Poisson equation which accounts for the variation of the electron population in GLs. The model takes also under consideration the cooling of electrons in the emitter layer due to the Peltier effect. We find the spatial distributions of the electric field and potential with the high-electric-field domain near the emitter GL in the GL heterostructures with different numbers of GLs. Using the obtained spatial distributions of the electric field, we calculate the current-voltage characteristics. We demonstrate that the Peltier cooling of the two-dimensional electron gas in the emitter GL can strongly affect the current-voltage characteristics resulting in their saturation. The obtained results can be important for the optimization of the hot-electron bolometric terahertz detectors and different devices based on GL heterostructures.

13.1 Introduction

Van der Waals heterostructures based on graphene-layers (GLs) separated by the barrier layers made of hBN, MoS_2,WS_2, and others have recently attracted a considerable interest (see, for example, the review paper [1]). These heterostructures can be used in novel electron and optoelectron devices. Several devices using these structures (among other GL-structures) have been proposed and realized [2–16]. In particular, we proposed and evaluated the hot-electron bolometric GL-based detectors of terahertz (THz) radiation consisting of multiple-GL structures with the inter-GL barrier layers [1]. In these devices (see Fig. 13.1), the extreme (top and bottom) GLs serve as the emitter and collector, respectively. In our previous work [17], considering the vertical electron transport in the multiple-GL heterostructures, we used a simplified model in which the spatial distribution of the electric field across the main portion of the GL detector structure was assumed virtually uniform. However, the features of the electron injection from the emitter GL and the electron capture into the inner GLs in the multiple-GL heterostructures, leading to a charging of the GLs adjacent to the emitter and deviation of the self-consistent electric-field distributions from uniform ones, can affect the characteristics of the vertical electron transport in such heterostructures. The cooling of the two-dimensional electron gas (2DEG) in the emitter GL due to the Peltier effect can add complexity to the characteristics. In this chapter, we develop a model for the vertical electron transport in van der Waals GL-based heterostructures accounting for the thermionic electron emission from and the electron capture into GLs, self-consistent electric-field distributions, and the electron cooling in the emitter GL. The developed model and obtained results might be useful for the optimization of the performance of the hot-electron bolometric THz detectors, particularly, their dark current characteristics and, hence, the detector detectivity, as well as the optimization of other devices based multiple-GL heterostructures.

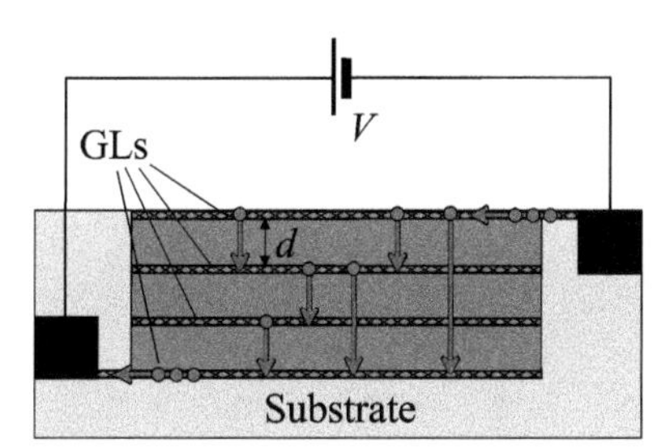

Figure 13.1 Schematic structure of the multiple-GL heterostructure under consideration. The horizontal arrows correspond to the electrons injected and absorbed by contacts, whereas the vertical arrows correspond to the electrons crossing and captured into GLs.

13.2 General Equations of the Model

We consider multiple-GL heterostructures with the n-doped GLs and the sufficiently thick inter-GL barrier layers (tunneling non-transparent), made, for example, of WS_2. Figure 13.1 shows the

GL-structure under consideration. The top and bottom GLs are connected to the pertinent electrodes. These GLs serve as the emitter and collector of electrons. The bias voltage V is applied between the latter. The inner GLs are disconnected from any contacts, so that their potentials are floating. The bottom electrode serves as the collector of the electrons passed through all GLs (not captures into the inner GLs) and as the heat sink. The heterostructure can be supplied by the back electrode (not shown in Fig. 13.1), can provide the collection of the electrons passed through all GL, and plays the role of the extra heat sink. The terminal current is due to the electrons thermionically emitted from GLs and propagating above the tops of the barriers.

Since the thickness of the inter-GL barriers d and the net thickness of the GL-heterostructure dN, where N is the number of the barriers in the heterostructure under consideration, is relatively large [the number of GLs is equal to $(N + 1)$], the geometrical inter-GL capacitance is much smaller than the quantum capacitance, so that the effect of quantum ca-pacitance [18] can be disregarded.

The Poison equation governing the electric potential distribution $\varphi = \varphi(z)$ in the direction perpendicular to the GL plane (the z-direction) and the equation governing the electron balance in each GL are presented in the following form:

$$\frac{d^2\varphi}{dz^2} = \frac{4\pi e}{\kappa}\sum_{n=0}^{N}(\Sigma_{\rm n} - \Sigma_{\rm D})\cdot\delta(z - nd), \tag{13.1}$$

$$\frac{jp_{\rm n}}{e} = \Theta_{\rm n}, \quad \Theta_{\rm n} = \Theta_{\rm D}\exp\left(\frac{\mu_{\rm n} - \Delta_{\rm C}}{T_{\rm n}}\right) \tag{13.2}$$

Here, $e = |e|$ is the electron charge, $\Sigma_{\rm n}$, $\Sigma_{\rm D}$, and $\mu_{\rm n}$, are the 2D densities of electrons and donors (assumed to be the same in all GLs) and the electron Fermi energy in the GL with the index $n = 1, 2, \ldots, N$, $\delta(z)$ is the Dirac delta function, κ and d are the barrier dielectric constant and thickness, j is the density of the electron current across GLs, $p_{\rm n}$ and $\Theta_{\rm n}$ are the capture parameter [19–21] (proportional to the probability of the electron capture into GLs), and the rate of the electron thermoemission from the n-th GL ($p_0 = 1, p_1, p_2, \ldots, p_{\rm N} \leq 1$), respectively, $\Delta_{\rm C}$ is the band offset between the GL and the barrier material, and $T_{\rm n}$ is the electron temperature (in the energy unit) in the n-th GL. In Eq. (13.2), we put

$$\Theta_\mathrm{D} = \frac{2T_\mathrm{l}^2}{\pi\tau\hbar^2 v_\mathrm{W}^2}\left(\frac{\Delta_\mathrm{C}}{T_\mathrm{l}} + 1\right) \simeq \frac{\Sigma_\mathrm{D}}{\tau_\mathrm{esc}}, \tag{13.3}$$

where T_l is the lattice temperature, τ is the characteristic time of the momentum relaxation of electrons with the energy exceeding Δ_C in GLs, $\hbar$ is the Plank constant, $v_\mathrm{W} \simeq 10^8$ cm/s is the characteristic electron velocity in GLs, and τ_esc is the characteristic escape time from GLs. We disregard small variations of the electron density and the temperature in the pre-exponential factors.

Equation (13.2) is valid if the density of electrons above the barriers substantially exceeds its value in equilibrium.

The boundary conditions for the potential governed by Eq. (13.1) can be set as

$$\varphi\big|_{z=0} = 0, \quad \varphi\big|_{z=Nd} = V. \tag{13.4}$$

As follows from Eq. (13.1), the electric field in the n-th barrier $(d\varphi/dx)_\mathrm{n} = E_\mathrm{n}$ (n = 1, 2, 3, . . ., N), i.e., at $(n - 1)d < z < nd$, is equal to:

$$E_{\mathrm{n}+1} = E_\mathrm{n} + \frac{4\pi e}{\kappa}(\Sigma_\mathrm{n} - \Sigma_\mathrm{D}). \tag{13.5}$$

According to Eq. (13.4), we obtain

$$d\sum_{n=1}^{n=N} E_\mathrm{n} = V. \tag{13.6}$$

Assuming that the 2DEG in all GLs, except those fully depleted, is degenerate due to sufficiently high donor density ($\mu_\mathrm{n} \gg k_\mathrm{B}T$), one can use the following equation which relates the Fermi energy μ_n and the electron density Σ_n:

$$\mu_\mathrm{n} = \hbar v_W \sqrt{\pi\Sigma_\mathrm{n}}. \tag{13.7}$$

It is commonly believed that the probability of the electron capture to quantum wells (QWs) in the customary heterostructures is determined by the emission of optical phonons [22–25]. The scattering of the electrons propagating over the barriers on the electrons located in GLs can also be rather effective mechanism of the capture of the former. Generally, the capture parameter p_n, which relates the current density j and the rate of the capture and is proportional to the probability of the capture of electrons [23–25], depends on the electric-field distribution across the GL-structure.

In the case of ballistic or near ballistic transport of electrons above the barriers, p_n depends mainly on E_n. In the GL-structures with the collision-dominated electron transport, one can assume that p_n is determined by the electric field around the n-th GL, in particular, by the electric fields E_n and E_{n+1}. For the definiteness, we use the following approximation for the electric-field dependence (see the Appendix):

$$p_n = E_C/E_n. \tag{13.8}$$

Here, $E_C = \langle W_C \rangle \upsilon_T/b$ is the characteristic capture field, which is determined by the barrier layer and GL structural parameters: the average (averaged over the electron Maxwellian distribution in the barrier layer) probability of the capture of electrons crossing a GL, $\langle W_C \rangle$, the electron mobility across the barrier, $b = b(E)$, and the electron thermal velocity and the electron effective mass in the barrier layer, $\upsilon_T = \sqrt{T_b / 2m_b}$ and m_b, respectively.

Using Eqs. (13.2) and (13.6), we obtain equations which relates Σ_0 and Σ_n and the current density j

$$\exp\left(\frac{\mu_D\sqrt{\Sigma_0 / \Sigma_D} - \Delta_C}{T_0}\right) = \frac{j}{j_D}, \tag{13.9}$$

for the emitter GL, and

$$\exp\left(\frac{\mu_D\sqrt{\Sigma_n / \Sigma_0} - \Delta_C}{T_n}\right) = \frac{jp_n}{j_D}, \tag{13.10}$$

for the inner GLs. Here, $\mu_D = \hbar \upsilon_W \sqrt{\pi \Sigma_D}$ and

$$j_D \simeq \frac{e\Sigma_D}{\tau_{esc}} \simeq const. \tag{13.11}$$

Generally, the electron temperature in the emitter GL T_0 can differ from the electron temperatures in the inner GLs T_n and the lattice temperature T_l, in particular, due to the Peltier effect (see below). Neglecting for simplicity the electron heating in the barrier layers (see below), we put the electron temperatures in all inner GLs to be equal to the lattice temperature: $T_n = T_l$.

Due to the smallness of the ratio E_C/E_D, where $E_D = 4\pi e \Sigma_D / \kappa$, monotonic electric-field distributions can be partitioned into a smooth quasi-neutral ($\Sigma_n = \Sigma_D$) portion with relatively small electric field in a wide range adjacent to the collector and a narrow high-field portion (high-electric-field domain) near the emitter GL. At $\Sigma_D = 1.8 \times 10^{12}$ cm^{-2} and $\kappa = 4$, one obtains $E_D = 270$ kV/cm and $\mu_D = 150$ meV. Due to relatively strong doping of GLs, the high-electric field is concentrated primarily in the emitter barrier layer (similar to that in the multiple-quantum well structures based on the standard semiconductors [26–30]). Considering this and using Eqs. (13.8) and (13.10), for the electric field in the quasi-neutral region E_{Bulk} (in the barrier layers with $n > M$, where M is the number of charged, i.e., fully and partially depleted GLs), we find

$$E_{\mathrm{Bulk}} = E_{\mathrm{C}} \frac{j}{j_{\mathrm{D}}} \exp\left(\frac{\Delta_{\mathrm{C}} - \mu_{\mathrm{D}}}{T_{\mathrm{l}}} \right). \tag{13.12}$$

Taking into account that $E_{\mathrm{Emitter}} = E_1 = E_{\mathrm{D}}(\Sigma_0/\Sigma_{\mathrm{D}} - 1)$, from Eq. (13.10), we obtain

$$E_{\mathrm{Emitter}} = E_{\mathrm{D}} \left[\left(\frac{\Delta_{\mathrm{C}} + T_0 \ln \frac{j}{j_D}}{\mu_D} \right)^2 - 1 \right]. \tag{13.13}$$

Thus, the spatial distribution of the electric field is given by

$$E_{\mathrm{n}} = E_{\mathrm{Emitter}} - (n-1)E_{\mathrm{D}}, \quad 1 \le n \le M, \tag{13.14}$$

$$E_{\mathrm{n}} = E_{\mathrm{Bulk}}, \quad M < n \le N. \tag{13.15}$$

Using Eq. (13.6), we arrive at the following equation:

$$ME_{\mathrm{Emitter}} - (M-1)E_{\mathrm{D}} + (N-M)E_{\mathrm{Bulk}} = \frac{V}{d}, \tag{13.16}$$

with M is determined by the following inequalities:

$$\frac{(E_{\mathrm{Emitter}} - E_{\mathrm{Bulk}})}{E_{\mathrm{D}}} < M < \frac{(E_{\mathrm{Emitter}} - E_{\mathrm{Bulk}})}{E_{\mathrm{D}}} + 1. \tag{13.17}$$

The first two terms in the left-hand side of Eq. (13.16) correspond to the potential drop at the depleted region, whereas the third term is equal to the potential drop across the bulk of the heterostructure (across all other barrier layers). Equation (13.16) with Eqs. (13.12) and (13.13) yields the following current voltage characteristics (with the electron temperature in the emitter GL T_0 as a parameter):

$$\left[\frac{\Delta_C + T_0\left(\frac{\mu_D - \Delta_C}{T_1} + \ln J\right)^2}{\mu_D}\right] + \frac{(N-M)}{M}\frac{E_C}{E_D}J$$

$$= \frac{V/V_D + 2M - 1}{M}, \tag{13.18}$$

where $J = (j/j_D)\exp[(\Delta_C - \mu_D)/T_1]$.

13.3 The Peltier Cooling

The electrons leaving the emitter GL extract some energy from 2DEG in this GL. This can lead to the 2DEG cooling (the Peltier effect) and affect the electron thermionic emission from the emitter GL, the spatial distributions of the electric field and electron population in GLs, and the current density.

The density of the energy flux from the emitter GL is equal to $Q = j(\Delta_C - 2\mu_0/3)/e$. We assume that the main mechanism transferring the energy from the emitter GL lattice to 2DEG in this GL is associated with the GL's optical phonons. In this case, the electron effective temperature T_0 in the emitter GL is governed by the following equation describing the balance between the energy taken away by the emitted electrons from the emitter GL and the energy which the 2DEG in this GL receives from the lattice:

$$\frac{\hbar\omega_0}{\tau_0}\left[\exp\left(-\frac{\hbar\omega_0}{T_0}\right) - \exp\left(-\frac{\hbar\omega_0}{T_1}\right)\right] \simeq -\frac{(\Delta_C - 2\mu_D/3)}{\tau_{esc}}\frac{j}{j_D}. \tag{13.19}$$

Here, $\hbar\omega_0$ and τ_0 are the optical phonon energy in GL and their characteristic spontaneous emission time, respectively. Equation (13.19) leads to

$$\frac{1}{T_0} = \frac{1}{T_1} - \frac{1}{\hbar\omega_0}\ln\left(1 - \frac{J}{J_S}\right), \tag{13.20}$$

with

$$J_S = \frac{\tau_{esc}}{\tau_0}\left(\frac{\hbar\omega_0}{\Delta_C - 2\mu_D/3}\right)\exp\left(\frac{\Delta_C - \mu_D - \hbar\omega_0}{T_1}\right). \tag{13.21}$$

Figure 13.2 shows the dependences of the normalized electron temperature in the emitter GL T_0/T_l on the normalized current J calculated using Eq. (13.20). It is assumed that $\tau_{esc} \sim \tau_0 \sim 10^{-12}$ s, Δ_C = 400 meV, $\hbar\omega_0$ = 200 meV, μ_D = 150 meV ($\Sigma_D = 1.8 \times 10^{12}$ cm^{-2}), and T_l = 25 meV ($T_l \simeq 300$ K). At the above parameters, $J_S \simeq 5$, $j_D \exp[(\mu_D - \Delta_C)/T_l] \simeq 14.5$ A/cm^2 and $j_S = J_S j_D \exp[(\mu_D - \Delta_C)/T_l] \simeq 72.6$ A/cm^2. As follows from Eq. (13.20) andseen from Fig. 13.2, the electron temperature in the emitter GL is lower than the lattice temperature due to the effect of thermoelectric (Peltier) cooling. When the current J approaches to J_S (the saturation current), T_0 drastically decreases.

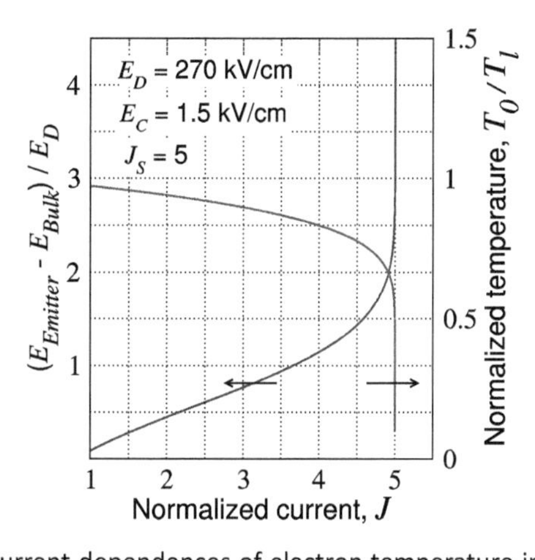

Figure 13.2 Current dependences of electron temperature in the emitter GL and difference between the emitter and bulk electric field.

13.4 Electric Fields versus Current and Current-Voltage Characteristics

Equation (13.12) yields the following expression for E_{Bulk} as a function of the normalized current J:

$$E_{\text{Bulk}} = E_C J. \tag{13.22}$$

Using Eqs. (13.13) and (13.20), we find the following $E_{\text{Emitter}} - J$ relation:

$$E_{\text{Emitter}} = E_{\text{D}} \left\{ \left[\frac{\Delta_{\text{C}} + \dfrac{(\mu_{\text{D}} - \Delta_{\text{C}} + T_{\text{l}} \ln J)}{1 - (T_{\text{l}} / \hbar\omega_0) \ln(1 - J/J_{\text{S}})}}{\mu_{\text{D}}} \right]^2 - 1 \right\}. \quad (13.23)$$

Apart from the T_0 - J relation, Fig. 13.2 shows also the dependence of $(E_{\text{Emitter}} - E_{\text{Bulk}})/E_{\text{D}}$ on the normalized current J correspond calculated using Eqs. (13.19), (13.22), and (13.23) (for E_{D} = 270 kV/cm and E_{C} = 1.5 kV/cm). As seen, the difference E_{Emitter} - E_{Bulk} steeply increases when J becomes close to the normalized saturation current J_{S}. Using the value of the difference (E_{Emitter} - E_{Bulk}) and inequalities (13.17), one can find the intervals of the normalized current corresponding to different numbers, M, of the charged inner GLs.

Considering the above, we arrive at the following current-voltage characteristic:

$$\left[\frac{\Delta_{\text{C}} + \dfrac{(\mu_{\text{D}} - \Delta_{\text{C}} + T_{\text{l}} \ln J)}{1 - (T_{\text{l}} / \hbar\omega_0) \ln(1 - J/J_{\text{S}})}}{\mu_{\text{D}}} \right]^2$$

$$+ \frac{(N-M)}{M} \frac{E_{\text{C}}}{E_{\text{D}}} J = \frac{V/V_{\text{D}} + 2M - 1}{M}. \quad (13.24)$$

Figure 13.3 shows the current-voltage characteristics of the heterostructures with the numbers of the inter-GL barriers equal to N = 6 and N = 11 (the numbers of GLs are five and ten, respectively) for different values of E_{C}. Other parameters are the same as for Fig. 13.2. The obtained characteristics clearly demonstrate the tendency to saturation with increasing voltage. At a relatively small E_{C}, the current-voltage characteristics of the heterostructures with different N are rather close to each other. This is because at such values of E_{C} and N, the potential drop across the quasi-neutral region is small in comparison with that across the depleted region, so that $E_{\text{Emitter}} \simeq V/d + (M - 1)E_{\text{D}}$ and, hence, the current is virtually independent on N. However, when E_{C} is not so small, the potential drop across the quasi-neutral region becomes essential and dependent on N. In this case, the current-voltage characteristics at different N can be markedly different (see curves for E_{C} = 7. 5 kV/cm in Fig. 13.3).

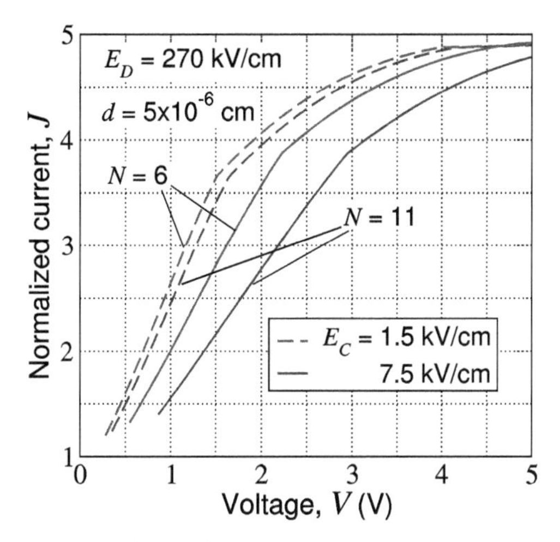

Figure 13.3 Current-voltage characteristics in GL structures with different numbers of inter-GL barrier layers N [with ($N - 1$) inner GLs] and different characteristic capture field E_C.

13.5 Potential Profiles

Considering the obtained current-voltage characteristics, one can find the voltage dependences of E_{Emitter} and E_{Bulk} corresponding to their current dependences shown in Fig. 13.2. Figure 13.4 shows these voltage dependences for the heterostructures with different E_C. One can see that in both cases $E_{\mathrm{Emitter}} \gg E_{\mathrm{Bulk}}$, the difference between this fields at E_C = 7. 5 kV/cm is markedly smaller.

Using Eqs. (13.14) and (13.15) and accounting for the calculated data for the $E_{\mathrm{Emitter}} - V$ and $E_{\mathrm{Bulk}} - V$ relations (see Fig. 13.4), one can find the spatial distributions of the electric potential (potential profiles) between the emitter and collector GLs. Figures 13.5 and 13.6 show examples of the potential profiles found for heterostructures with $N = 11$, $E_D = 270$ kV/cm, and $d = 5 \times 10^{-6}$ cm with E_C = 1.5 kV/cm and E_C = 7.5 kV/cm, respectively, at different voltages. One can see that the potential spatial distributions are substantially nonuniform: they steeply vary near the emitter GL (in the depleted, i.e., charged region) and are fairly smooth in the heterostructure bulk. It is also seen that an increase in the voltage leads to a change of the number of the charged GLs M (from $M = 1$ to $M = 3$ at the parameters used for Fig. 13.5 and in the chosen voltage range). The transition from the voltage corresponding a certain M to $M + 1$ results in the occurrence of jogs (actually rather weak) on the characteristics shown in Figs. 13.3 and 13.4. It is also seen that in a heterostructure with relatively large E_C and moderate voltages, the potential distribution is rather smooth except the immediate vicinity of the emitter GLs, where the electric field is only slighter higher (see the line for $V = 1$ V in Fig. 13.6). This can justify the simplified model [15] at not too small p_n.

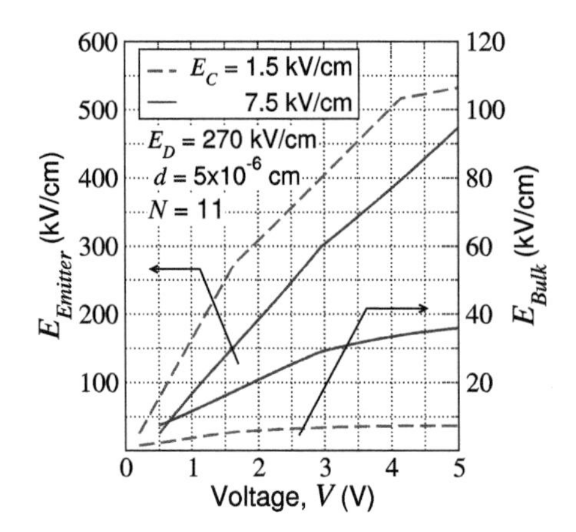

Figure 13.4 Electric fields in the emitter GL and in the heterostructure bulk versus voltage for GL structures with different values of the characteristic capture field E_C.

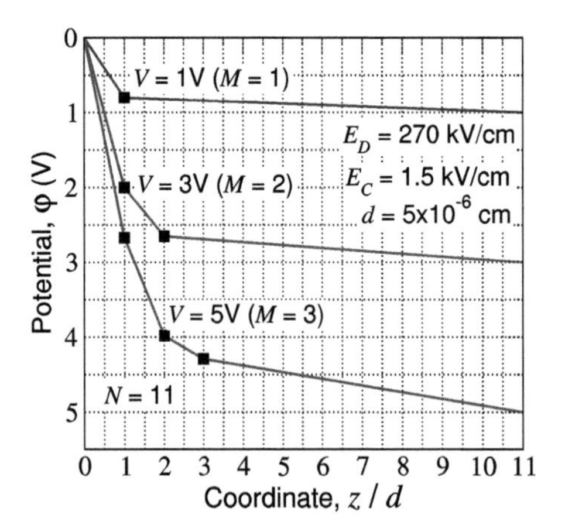

Figure 13.5 Potential profiles at different voltages. Markers correspond to positions of charged GLs.

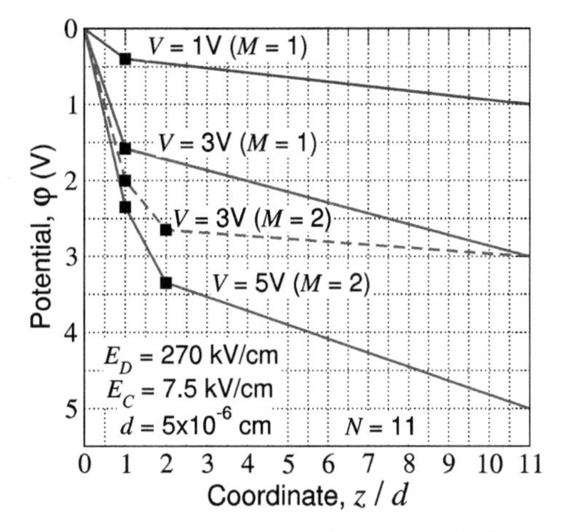

Figure 13.6 The same as in Fig. 13.5, but for $E_C = 7.5$ kV/cm and with dashed line corresponding to $E_C = 1.5$ kV and $V = 3$ V (from Fig. 13.5).

13.6 Discussions

13.6.1 Limiting Cases

In the voltages range not so close to the saturation, the Peltier cooling can be disregarded. In this range, the current-voltage characteristics can be expressed by simple equations. Taking into account that at moderate voltages only one GL is charged and the potential drops primarily across the barrier layer closest to the emitter GL (see Fig. 13.5), we obtain

$$E_{\text{Emitter}} \lesssim V/d, \tag{13.25}$$

$$J \simeq \exp\left[\frac{\mu_{\text{D}}}{T_{\text{l}}}\left(\sqrt{1+\frac{V}{V_{\text{D}}}}-1\right)\right] \simeq \exp\left(\frac{\mu_{\text{D}}}{2T_{\text{l}}}\frac{V}{V_{\text{D}}}\right), \tag{13.26}$$

i.e.,

$$j \simeq j_{\text{D}} \exp\left(\frac{\mu_{\text{D}}-\Delta_{\text{C}}}{T_{\text{l}}}\right) \exp\left[\frac{\mu_{\text{D}}}{2T_{\text{l}}}\left(\sqrt{1+\frac{V}{V_{\text{D}}}}-1\right)\right],$$

$$\simeq j_{\text{D}} \exp\left(\frac{\mu_{\text{D}}-\Delta_{\text{C}}}{T_{\text{l}}}\right) \exp\left(\frac{\mu_{\text{D}}}{2T_{\text{l}}}\frac{V}{V_{\text{D}}}\right),$$

$$= j_{\mathrm{D}} \exp\left(\frac{\mu_{\mathrm{D}} - \Delta_{\mathrm{C}}}{T_{\mathrm{l}}}\right) \frac{1}{p_{\mathrm{C}}^{\mathrm{eff}}}. \tag{13.27}$$

Here, $p_{\mathrm{C}}^{\mathrm{eff}} = \exp(\mu_{\mathrm{D}} V / 2 T_{\mathrm{l}} V_{\mathrm{D}})$.

In the heterostructures with not too small E_{C} and sufficiently large N, one has $E_{\mathrm{Emitter}} < (N - 1)E_{\mathrm{C}}$, i.e., the potential drops predominantly across the quasi-neutral region (bulk), and the following relation is true:

$$E_{\mathrm{Emitter}} > E_{\mathrm{Bulk}} \simeq \frac{V}{Nd}. \tag{13.28}$$

In this limiting cases, the current density can be presented as

$$J \simeq \frac{V}{NdE_{\mathrm{C}}} \gtrsim 1. \tag{13.29}$$

Hence,

$$j \simeq j_{\mathrm{D}} \exp\left(\frac{\mu_{\mathrm{D}} - \Delta_{\mathrm{C}}}{T_{\mathrm{l}}}\right) \frac{V}{NdE_{\mathrm{C}}},$$

$$= j_{\mathrm{D}} \exp\left(\frac{\mu_{\mathrm{D}} - \Delta_{\mathrm{C}}}{T_{\mathrm{l}}}\right) \frac{1}{p_{\mathrm{C}}^{\mathrm{eff}}} \tag{13.30}$$

where $p_{\mathrm{C}}^{\mathrm{eff}}$ is the effective capture parameter, which, in this case, is determined by the electric field in the main portion of the heterostructure, namely, by E_{Bulk}: $p_{\mathrm{C}}^{\mathrm{eff}} = E_{\mathrm{C}} / E_{\mathrm{Bulk}}$ [see Eq. (13.8)].

It is instructive that in the latter limit, the current density can be presented in the Ohmic form

$$j \simeq e n_{\mathrm{b}} b V / Nd, \tag{13.31}$$

with the electron density in the barrier layers $n_{\mathrm{b}} \propto \Sigma_{\mathbf{D}} / \langle W_{\mathrm{C}} \rangle)$ $\exp[(\mu_{\mathrm{D}} - \Delta_{\mathrm{C}}) / T_{\mathrm{l}})]$.

Equations (13.27) and (13.29) coincide in form with those obtained previously using a simplified model [17]. However, $p_{\mathrm{C}}^{\mathrm{eff}}$ in these equations is determined by the electric-field spatial distribution and the number of GLs (instead of $p_{\mathrm{C}} = const$ or p_{C}) determined by the average electric field in Ref. [17]. Moreover, $p_{\mathrm{C}}^{\mathrm{eff}}$ can explicitly depend on the capture probability W_{C} and characteristic capture field E_C (see the Appendix) as in the case of Eq. (13.30), or be independent of these parameters as in the case of Eq. (13.27).

13.6.2 Role of the p_n–E_n Relation

Above, we used the $p_n - E_n$ relation given by Eq. (13.8). The deviation from this relation does not lead to qualitative change in the obtained result. To demonstrate this, let us assume now that (see, for example, Refs. [27, 28])

$$p_n = \exp(-E_n / E_C^*), \tag{13.32}$$

where the characteristic field E_C* generally differs from E_C. In this case, Eq. (13.12) should be replaced by the following equation:

$$E_{bulk} = E_C^* \left(\frac{\Delta_C - \mu_D}{T_l} + \ln \frac{j}{j_D} \right). \tag{13.33}$$

The second term in the left-hand side of Eq. (13.18) (and the only this term) should also be modified accordingly. Such a replacement does not affect the saturation of the current voltage characteristics and the value J_s (and j_S) and again leads to Eqs. (13.27) and (13.30). In the latter equation, there should be now $p_C^{eff} \simeq \exp(-V / NdE_C^*) \lesssim 1$.

The current across the GL-heterostructures under consideration essentially depends on the effective capture parameter p_C^{eff} [see, Eqs. (13.27) and (13.28)], which, in turn, depends on the electric-field distribution and the number of GLs in the heterostructure. Hence, the voltage dependence of the dark current, i.e., the current in the absence of irradiation in the hot-electron bolometric THz detectors on the base of these heterostructures proposed in Ref. [17], is more complex than that follows from the simplified model of such photodetectors. In particular, the dark current-voltage characteristics calculated using the present model explicitly depend on the number of GLs (see Fig. 13.3) in contrast to more simple model [17]. The photocurrent caused by the THz radiation in the photodetectors in question and their responsivity also depend on p_C^{eff}. Qualitative reasonings show that since the photodetector detectivity is determined by the responsivity, dark current, and the photoelectric gain $g \propto 1 / p_C^{eff}$, the effects of the electric-field nonuniformity (as well as the Peltier cooling effect) can pronouncedly influence the detectivity as a function of the voltage and number of GLs. Therefore, such effects should be taken into account considering the photodetector quantitative optimization. One of the parameter which should be optimized is the number of GLs in the photodetector. The quantitative study of the impact of the nonuniformity effects and the effect of the Peltier cooling on the photodetector responsivity and detectivity require a separate detailed treatment.

13.6.3 Electron Heating in the Barrier Layers

In contrast to the cooling of the 2DEG in the emitter GL, the electron heating in the inner GLs, associated with the capture of relatively hot electrons heated in the barrier layers and cooled due the electron emission from these GLs, is relatively weak. This is because the energy brought to a GL by a captured electron and the energy swept away by an emitted electron are close to each other (these energies are close to Δ_C unless the electron temperature in the barrier layers $T_b \ll \Delta_C$). Due to this, $T_n - T_l \ll T_l - T_0$.

The effect of heating of the electrons propagating in the barrier layers is also should not be too marked. Indeed, assuming the relaxation of the electron energy obtained from the electric field (Joule heating power in the whole heterostructure is equal to jV) in the barrier layers is mainly due to the interaction of the electrons with the optical phonons in the barrier layers, for the estimate of the electron effective temperature, T_b, in these barriers, one can use the following equation:

$$\frac{\hbar\omega_0^{(b)}}{\tau_0^{(b)}}\left[\exp\left(-\frac{\hbar\omega_0^{(b)}}{k_B T_b}\right)-\exp\left(-\frac{\hbar\omega_0^{(b)}}{k_B T_l}\right)\right]=eb\left(\frac{V}{Nd}\right)^2. \quad (13.34)$$

Here, $\hbar\omega_0^{(b)}$ and $\tau_0^{(b)}$ are the optical phonon energy in the barrier layers and the characteristic time of their spontaneous emission. Equation (13.31) yields

$$\frac{1}{T_b}=\frac{1}{T_l}-\frac{1}{\hbar\omega_0^{(b)}}\ln\left[1+B_N\exp\left(\frac{\hbar\omega_0^{(b)}}{T_l}\right)\right], \quad (13.35)$$

where

$$B_{\mathrm{N}} = \frac{\tau_0^{(\mathrm{b})} eb}{\hbar\omega_0^{(\mathrm{b})}} \left(\frac{V}{Nd} \right)^2.$$

Assuming $\hbar\omega_0^{(b)}$ = 50 meV, $\tau_0^{(b)}$ 10^{-2}s, b = (10 – 15) $\mathrm{cm}^2/\mathrm{V} \cdot \mathrm{s}$, d = 10 – 50 nm, T_{l} = 300 K, N = 11, and V = 1 V, we obtain $B_{\mathrm{N}} \exp(\hbar\omega_0^{(b)} / T_l) \simeq 0.4885 - 0.7328$. As a result, from Eq. (13.35), we get the following estimate: $T_{\mathrm{b}}/T_{\mathrm{l}} \simeq 1.248 - 1.378$ (or $T_{\mathrm{b}} \simeq 374 -$ 413 K). Since the average capture probability and the characteristic capture electric field slowly vary with the temperature (see Fig. 13.8), the pertinent effect should not essentially change the obtained characteristics. In principle, the effect of electron heating in the barrier layers can be accounted for by the inclusion of the dependence of E_{C} on V (because T_{b} increases with increasing voltage). This might lead to somewhat more complex characteristics at elevated voltages.

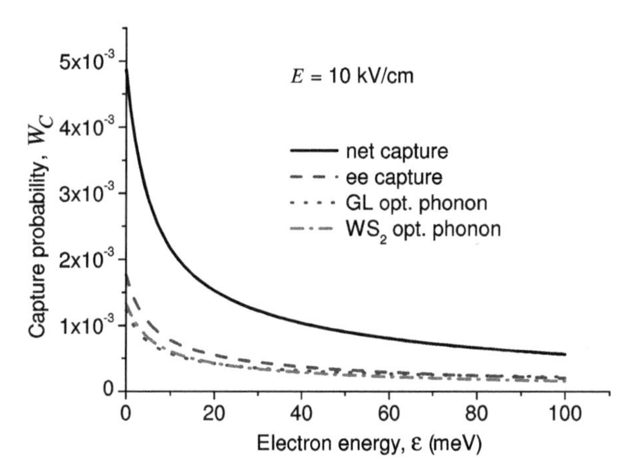

Figure 13.7 Energy dependence of the net capture probability as well as the contributions of different capture mechanisms.

13.6.4 Quantum Capacitance

Above, we disregarded the effect of quantum capacitance [16]. The ratio of the inter-GL capacitance C and the GL quantum capacitance C_{Q} is equal to

$$\frac{C}{C_Q} = \frac{\kappa\hbar^2 v_W^2}{8\mu_D d}. \tag{13.36}$$

Assuming that $\kappa = 4$, $\mu_D = 150$ meV, and $d = 10 - 50$ nm, we obtain $C/C_Q \simeq 0.09 - 0.018$, i.e., $C/C_Q \ll 1$ and, hence the quantum capacitance plays an insignificant role in comparison with the geometrical capacitance.

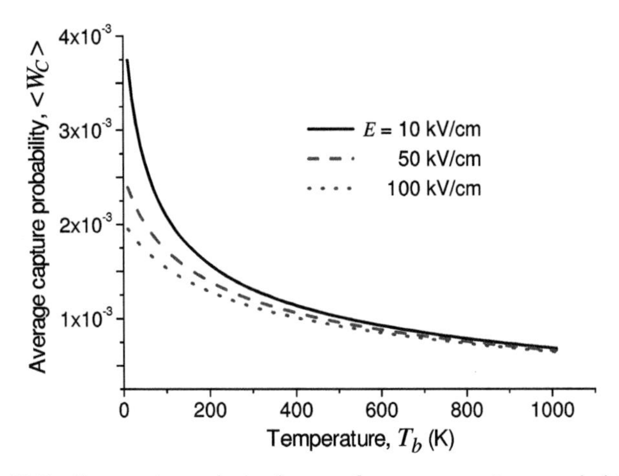

Figure 13.8 Temperature dependence of average capture probability for different barrier electric fields.

13.6.5 Contact Effects

The voltage drop across the contacts, V_C, to the GL heterostructures under consideration can be fairly small. Indeed, setting $j_D \exp[(\mu_D - \Delta_C)/T_l] = 14.5$ A/cm^2 (as above), considering the voltage range V = 1–2 V in Fig. 13.3 , for a device with the length (spacing between the side contacts) $L = 100$ μm^2, for the current $I = jL \simeq$, we obtain $I \simeq 0.3 - 0.6$ A/cm. According to the experimental data [31–34] for contacts to GLs, the contact resistance can be estimated as R_C = (100 - 500) Ω μm. As a result, we find $V_C \simeq (1 - 15) \times 10^{-3}$ V. These values are much smaller than the applied voltage (voltage drop between the emitter and collector GLs). Hence, the contact effects can be neglected. The necessity to apply voltages about few volts is due to a relatively high emitter-collector resistance, which is associated with current created by relatively small fraction of electrons overcoming the inter-GL barriers.

13.7 Conclusions

Using an analytical model for the self-consistent description of the vertical electron transport in the van der Waals GL heterostructures with the GLs separated by the barrier layers, we calculated current-voltage characteristics and potential distribution in those heterostructures. The thermionic current over the barriers is controlled by the electron thermionic emission from the GLs and by the capture of electrons into GLs. At small applied voltages, the electric field is practically homogeneous through the structure with equal potential drop on each barrier. As the voltage increases, the probability of electrons capture into GLs decreases and the potential distributions (see Figs. 13.5 and 13.6) turn to be substantially inhomogeneous with well pronounced two regions: with high electric field near the emitter and low electric field region in the heterostructure bulk (high-field domain and quasi-neutral low-field quasi-neutral domains, respectively). An increase in the applied voltage gives rise to an increase of the number of charged, fully or partially depleted GLs and, therefore, to an increase in the width of the high-electric field (charged) region. The electric field in the bulk is practically homogeneous. We demonstrated that the thermionic emission from the emitter GL can lead to a substantial cooling of the 2DEG in this GL (the Peltier effect). This effect results in the saturation of the current-voltage characteristics with increasing voltage. The high and low field domains were observed in conventional semiconductor heterostructures with quantum wells, but the current saturation in the van der Waals GL heterostructures under consideration is new phenomenon associated with the specifics of the emitter with 2DEG. The obtained results, which reveal the origin of the current-voltage characteristics nonlinearity in the van der Waals GL heterostructures, can be important for the optimization of the dark current characteristics of the THz hot-electron bolometers based on those heterostructures. An extended version of the developed model can also be applied for the consideration of the GL heterostructures response to incoming THz radiation and strict calculations of the responsivity and detectivity of such bolometric detectors.

Acknowledgments

This work was supported by the Japan Society for promotion of Science (Grant-in-Aid for Specially Promoting Research 23000008), Japan and by the Russian Scientific Fund Foundation (Project #14 29 00277). The works at UB and RPI were supported by the U.S. Air Force Award No. FA9550-10-1-391 and the U.S. Army Research Laboratory Cooperative Research Agreement, respectively.

Appendix: Capture Parameter

Let us introduce the probability $W_c(\varepsilon)$ of the capture of an electron crossing a GL. Considering the interaction of the electrons crossing the GL with the optical phonons in GL, optical phonons in the barrier layer material (WS_2 [35]), and with the electrons located in the GLs, one can obtain the dependence of $W(\varepsilon)$ on the electron energy ε. Figure 13.7 shows an example of the energy dependence of the capture probability $w_C(\varepsilon, E)$ for the electron density in GLs equal to 10^{12} cm^{-2} and the electric field of 10^4 V/cm.

At not too low voltages when the electron system in the barrier layers is sufficiently far from equilibrium, the rate of the capture C_{GL} into a GL of electrons incident from left and right is

$$C_{\rm GL} \propto \int_0^\infty dp_z \upsilon_z W_C(\varepsilon_z)$$

$$\times\left\{\exp\left[-\frac{(p_z - m_b u_b)^2}{2m_b T_b}\right] + \exp\left[-\frac{(p_z + m_b u_b)^2}{2m_b T_b}\right]\right\}$$

$$\simeq 2\int_0^\infty d\varepsilon W_C(\varepsilon, E)\exp\left(-\frac{\varepsilon}{T_b}\right). \tag{13.A1}$$

The current through the GL (assuming that the capture probability $W_C(\varepsilon)$ is small) is given

$$j \propto \int_0^\infty dp_z \upsilon_z$$

$$\times\left\{\exp\left[-\frac{(p_z - m_b u_b)^2}{2m_b T_b}\right] - \exp\left[-\frac{(p_z + m_b u_b)^2}{2m_b T_b}\right]\right\}$$

$$\simeq 2\int_0^{\infty} d\varepsilon \exp\left(-\frac{\varepsilon}{T_{\mathrm{b}}}\right)\left(\frac{\sqrt{2m_{\mathrm{b}}\varepsilon}u_{\mathrm{b}}}{T_{\mathrm{b}}}\right). \qquad (13.A2)$$

Here, T_{b}, m_{b}, and $b(E)$ are the effective electron temperature, the effective mass, and mobility in the material of the barrier layers, respectively, and $u_{\mathrm{b}} = b(E)E$ is the drift velocity across the barrier.

Thus introducing the capture parameter p_{C} as $C = p_{\mathrm{C}}j/e$, we obtain

$$p_{\mathrm{C}} = \frac{\int_0^{\infty} d\varepsilon W_C(\varepsilon)\exp\left(-\frac{\varepsilon}{T_{\mathrm{b}}}\right)}{\int_0^{\infty} d\varepsilon \exp\left(-\frac{\varepsilon}{T_{\mathrm{b}}}\right)\left(\frac{\sqrt{2m_{\mathrm{b}}\varepsilon}u_{\mathrm{b}}}{T_{\mathrm{b}}}\right)},$$

$$= \frac{2\upsilon_T}{\sqrt{\pi}u_b}\int_0^{\infty}\frac{d\varepsilon W_C(\varepsilon)}{T_{\mathrm{l}}}\exp\left(-\frac{\varepsilon}{T_{\mathrm{b}}}\right). \qquad (13.A3)$$

As it should follow from Eq. (13.A3):

$$p_{\mathrm{C}} \simeq \frac{\langle W_{\mathrm{C}}\rangle \upsilon_T}{b(E)E}, \qquad (13.A4)$$

where $\langle W_{\mathrm{C}}\rangle$ is the average capture probability, $\upsilon_T = \sqrt{T_{\mathrm{b}}/2m_{\mathrm{b}}}$ is the thermal velocity (in the barrier layer), and $b(E)$ is the electron mobility in the barrier layers. Figure 13.8 shows the average capture probability as function of the electron temperature in the barrier T_{b}. As seen, the average capture probability is weakly dependent on the electric field (affecting the electron wave functions in the barrier layers) in a wide range of the latter (at least up to 10^5 V/cm).

If the electric field E is smaller than that corresponding to the electron velocity saturation in the material of the barriers E_{S}, Eq. (13.A4) yields $p_{\mathrm{C}} = E_{\mathrm{C}}/E$, i.e., Eq. (13.8), where $E_{\mathrm{C}} \simeq \langle W_{\mathrm{C}}\rangle \upsilon_{\mathrm{T}}/b$. Setting the low-field mobility perpendicular to the layers equal to b = 10 – 15 cm^2/V · s, i.e., somewhat lower than the achieved values of lateral mobility of WS_2 [36–39], and $\upsilon_{\mathrm{T}} = 10^7$ cm/s, for T_{b} = 300–400 K, we obtain $W_{\mathrm{C}}(\varepsilon) \sim 1.5 \times 10^{-3}$ (see Fig. 13.8). This leads to the following estimate: $E_{\mathrm{C}} \simeq$ (1.0–1.5) kV/cm. Lower transverse mobility, higher doping of GLs, and the inclusion of other mechanisms (say, the capture associated with the plasmon emission) can result in some increase in $\langle W_{\mathrm{C}}\rangle$ and E_{C}. In the main text, we assume that E_{C} = 1.5 and 7.5 kV/cm. Note that Eq. (13.8) is valid at $E_{\mathrm{n}} \gtrsim E_{\mathrm{C}}$.

References

1. A. K. Geim and I. V. Grigorieva, *Nature*, **499**, 419 (2013).
2. M. Liu, X. Yin, and X. Zhang, *Nano Lett.*, **12**, 1482 (2012).
3. L. Britnell, R. V. Gorbachev, R. Jalil, B. D. Belle, F. Shedin, A. Mishenko, T. Georgiou, M. I. Katsnelson, L. Eaves, S. V. Morozov, N. M. R. Peres, J. Leist, A. K. Geim, K. S. Novoselov, and L. A. Ponomarenko, *Science*, **335**, 947 (2012).
4. L. Britnell, R. V. Gorbachev, R. Jalil, B. D. Belle, F. Schedin, M. I. Katsnelson, L. Eaves, S. V. Morozov, A. S. Mayorov, N. M. R. Peres, A. H. Castro Neto, J. Leist, A. K. Geim, L. A. Ponomarenko, and K. S. Novoselov, *Nano Lett.*, **12**, 1707 (2012).
5. T. Georgiou, R. Jalil, B. D. Bellee, L. Britnell, R. V. Gorbachev, S. V. Morozov, Y.-J. Kim, A. Cholinia, S. J. Haigh, O. Makarovsky, L. Eaves, L. A. Ponomarenko, A. K. Geim, K. S. Nonoselov, and A. Mishchenko, *Nat. Nanotechnol.*, **8**, 100 (2013).
6. L. Britnell, R. V. Gorbachev, A. K. Geim, L. A. Ponomarenko, A. Mishchenko, M. T. Greenaway, T. M. Fromhold, K. S. Novoselov, and L. Eaves, *Nat. Commun.*, **4**, 1794 (2013).
7. V. Ryzhii, T. Otsuji, M. Ryzhii, V. G. Leiman, S. O. Yurchenko, V. Mitin, and M. S. Shur, *J. Appl. Phys.*, **112**, 104507 (2012).
8. V. Ryzhii, T. Otsuji, M. Ryzhii, and M. S. Shur, *J. Phys. D: Appl. Phys.*, **45**, 302001 (2012).
9. V. Ryzhii, A. Satou, T. Otsuji, M. Ryzhii, V. Mitin, and M. S. Shur, *J. Phys. D: Appl. Phys.*, **46**, 315107 (2013).
10. V. Ryzhii, M. Ryzhii, V. Mitin, M. S. Shur, A. Satou, and T. Otsuji, *J. Appl. Phys.*, **113**, 174506 (2013).
11. V. Ryzhii, A. A. Dubinov, V. Ya. Aleshkin, M. Ryzhii, and T. Otsuji, *Appl. Phys. Lett.*, **103**, 163507 (2013).
12. V. Ryzhii, A. A. Dubinov, T. Otsuji, V. Ya. Aleshkin, M. Ryzhii, and M. S. Shur, *Opt. Express*, **21**, 31567 (2013).
13. V. Ryzhii, T. Otsuji, V. Ya. Aleshkin, A. A. Dubinov, M. Ryzhii, V. Mitin, and M. S. Shur, *Appl. Phys. Lett.*, **104**, 163505 (2014).
14. C. Oh Kim, S. Kim, D. H. Shin, S. S. Kang, J. M. Kim, C. W. Jang, S. S. Joo, J. S. Lee, J. H. Kim, S.-H. Choi, and E. Hwang, *Nat. Commun.*, **5**, 3249 (2014).
15. A. Mishchenko, J. S. Tu, Y. Cao, R. V. Gorbachev, J. R. Wallbank, M. T. Greenaway, V. E. Morozov, S. V. Morozov, M. J. Zhu, S. L. Wong, F. Withers, C. R. Woods, Y.-J. Kim, K. Watanabe, T. Taniguchi, E. E. Vdovin,

O. Makarovsky, T. M. Fromhold, V. I. Fal'ko, A. K. Geim, L. Eaves, and K. S. Novoselov, *Nat. Nanotechnol.*, **9**, 808 (2014).

16. C.-H. Liu, Y.-C. Chang, T. B. Norris, and Z. Zhong, *Nat. Nanotechnol.*, **9**, 273 (2014).
17. V. Ryzhii, A. Satou, T. Otsuji, M. Ryzhii, V. Mitin, and M. S. Shur, *J. Appl. Phys.*, **116**, 114504 (2014).
18. S. Luryi, *Appl. Phys. Lett.*, **52**, 501 (1988).
19. H. C. Liu, *Appl. Phys. Lett.*, **60**, 1507 (1992).
20. E. Rosencher, B. Vinter, F. Luc, L. Thibaudeau, P. Bois, and J. Nagle, *IEEE J. Quantum Electron.*, **30**, 2875 (1994).
21. V. Ryzhii, *J. Appl. Phys.*, **81**, 6442 (1997).
22. J. A. Brum and G. Bastard, *Phys. Rev. B*, **33**, 1420 (1986).
23. L. Thibaudeau and B. Vinter, *Appl. Phys. Lett.*, **65**, 2039 (1994).
24. D. Bradt, Y. M. Sirenko, and V. Mitin, *Semicond. Sci. Technol.*, **10**, 260 (1995).
25. M. Ryzhii and V. Ryzhii, Jpn. *J. Appl. Phys., Part 1*, **38**, 5922 (1999).
26. M. Ershov, V. Ryzhii, and C. Hamaguchi, *Appl. Phys. Lett.*, **67**, 3147 (1995).
27. L. Thibaudeau, P. Bois, and J. Y. Duboz, *J. Appl. Phys.*, **79**, 446 (1996).
28. M. Ershov, H. C. Liu, M. Buchanan, Z. R. Wasilewski, and V. Ryzhii, *Appl. Phys. Lett.*, **70**, 414 (1997).
29. V. Ryzhii and H. C. Liu, Contact and space-charge effects in quantum well infrared photodetectors, *Jpn. J. Appl. Phys., Part 1*, **38**, 5815 (1999).
30. V. Ryzhii, M. Ryzhii, and H. C. Liu, *J. Appl. Phys.*, **92**, 207 (2002).
31. E. Watanabe, A. Conwill, D. Tsuya, and Y. Koide, *Diamond Relat. Mater.*, **24**, 171 (2012).
32. J. S. Moon, M. Antcliffe, H. C. Seo, D. Curtis, S. Lin, A. Schmitz, I. Milosavljevic, A. A. Kiselev, R. S. Ross, D. K. Gaskill, P. M. Campbell, R. C. Fitch, K.-M. Lee, and P. Asbeck, *Appl. Phys. Lett.*, **100**, 203512 (2012).
33. A. Hsu, H. Wang, K. K. Kim, J. Kong, and T. Palacios, *IEEE Electron Device Lett.*, **32**, 1008 (2011).
34. L. Wang, I. Meric, P. Y. Huang, Q. Gao, Y. Gao, H. Tran, T. Taniguchi, K. Watanabe, L. M. Campos, D. A. Muller, J. Guo, P. Kim, J. Hone, K. L. Shepard, and C. R. Dean, *Science*, **342**, 614 (2013).
35. H. Shi, H. Pan, Y.-W. Zhang, and B. Yakobson, *Phys. Rev. B*, **87**, 155304 (2013).
36. D. Braga, I. Gutiérrez Lezama, H. Berger, and A. F. Morpurgo, *Nano Lett.*, **12**, 5218 (2012).

37. N. Huo, Sh. Yang, Zh. Wei, S.-S. Li, J.-B. Xia, and J. Li, *Sci. Rep.*, **4**, 5209 (2014).
38. S. Jo, N. Ubrig, H. Berger, A. B. Kuzmenko, and A. F. Morpurgo, *Nano Lett.*, **14**, 2019 (2014).
39. D. Ovchinnikov, A. Allain, Y.-S. Huang, D. Dumcenco, and A. Kis, *ACS Nano*, **8**, 8174 (2014).

Chapter 14

Giant Plasmon Instability in Dual-Grating-Gate Graphene Field-Effect Transistor*

Y. Koseki,[a] V. Ryzhii,[a] T. Otsuji,[a] V. V. Popov,[b,c,d] and A. Satou[a]

[a]*Research Institute of Electrical Communication, Tohoku University, Sendai 980-8577, Japan*
[b]*Kotelnikov Institute of Radio Engineering and Electronics (Saratov Branch), 410019 Saratov, Russia*
[c]*Saratov State University, Saratov 410012, Russia*
[d]*Saratov Scientific Center of the Russian Academy of Sciences, Saratov 410028, Russia*
a-satou@riec.tohoku.ac.jp

We study instability of plasmons in a dual-grating-gate graphene field-effect transistor induced by dc current injection using self-consistent simulations with the Boltzmann equation. With only the acoustic-phonon-limited electron scattering, it is demonstrated that

Graphene-Based Terahertz Electronics and Plasmonics: Detector and Emitter Concepts
Edited by Vladimir Mitin, Taiichi Otsuji, and Victor Ryzhii

ISBN 978-981-4800-75-4 (Hardcover), 978-0-429-32839-8 (eBook)
www.jennystanford.com

a total growth rate of the plasmon instability, with the terahertz/midinfrared range of the frequency, can exceed 4×10^{12} s^{-1} at room temperature, which is an order of magnitude larger than in two-dimensional electron gases based on usual semiconductors. By comparing the simulation results with existing theory, it is revealed that the giant total growth rate originates from simultaneous occurrence of the so-called Dyakonov–Shur and Ryzhii–Satou–Shur instabilities.

14.1 Introduction

Electronic, hydrodynamic, and electromagnetic properties of two-dimensional (2D) plasmons in channels of field-effect transistors (FETs) have been investigated extensively for their utilization to terahertz (THz) devices [1–6] (see also review papers [7–10] and references therein). Especially, plasmon instability is one of the most important properties to realize compact, room-temperature operating THz sources. Self-excitation of plasmons due to instability induces ac voltages in the gate electrodes and, in turn, leads to the emission of THz waves.

The so-called Dyakonov–Shur (DS) instability [2, 11–13] and Ryzhii–Satou–Shur (RSS) instability [14–16] in single-gate FETs were proposed theoretically as mechanisms of plasmon instability by dc current injection through the transistor channel. The DS instability originates from the Doppler shift effect at asymmetric boundaries in the channel, i.e., zero time variation of potential at the source contact and zero time variation of electron velocity near the drain contact, which are naturally realized by operating the FET in the saturation regime. In the same saturation regime, the RSS instability takes place due to the transit-time effect of fast-moving electrons in the high-field region in the drain side. Alternatively, the so-called dual-grating-gate structure (see Fig. 14.1a), in which two types of interdigitately placed gates form a very efficient grating coupler between THz waves and 2D plasmons [17, 18], has been proposed for direct THz emission without antenna integration [17, 19]. The RSS instability in this structure has been investigated analytically [16]. In addition, asymmetry of the gate placement expects to lead to partial realization of the asymmetric boundary conditions and, in turn, of the DS instability [9, 10].

However, in FETs or high-electron-mobility transistors (HEMTs) based on usual semiconductors (Si and compound semiconductors such as InGaAs and GaN), growth rates of the instabilities are of the order of 10^{11} s^{-1}, which are limited by electron saturation velocities ($\lesssim 2 \times 10^7$ cm/s in the GaAs channel). With such low growth rates the plasmons are easily damped out at room temperature by a large damping rate ($\gtrsim 10^{12}$ s^{-1}) associated with electron scattering.

Plasmons in graphene have then attracted much attention owning to its gapless energy spectrum and massless carriers [20–29]. The most straightforward yet striking advantage of graphene plasmons over those in usual semiconductors is its ultimately low scattering rate at room temperature, if external scattering sources in graphene such as impurities and defects and those induced by the substrate and gate insulator are excluded and the electron scattering is limited only by the acoustic-phonon scattering [30] in graphene. Then, the plasmon damping rate can be down to 10^{11} s^{-1} [31], together with the electron drift velocity up to $\lesssim 10^8$ cm/s. These lead to an expectation that the plasmon instabilities in graphene are very strong and can take place at room temperature. Although the technology of graphene fabrication with ulitimately high quality still needs to be progressed, a recent experimental report on graphene encapsulated into hexagonal boron nitride layers [32], which demonstrated the electron mobility at room temperature comparable to the acoustic-phonon-limited value, supports its feasibility.

In this chapter, we conduct simulations of the plasmon instabilities in the dual-grating-gate graphene FET with dc current injection, assuming the acoustic-phonon-limited scattering rate at room temperature, and demonstrate occurrence of giant instabilities with their total growth rate (which we define as the growth rate subtracting the damping rate) exceeding 4×10^{12} s^{-1} and with the plasmon frequency ranging in the THz/midinfrared range. We show the gate-length dependence of the plasmon frequency and the total growth rate extracted from the simulations, and we found distinct dependences of growth rates specific to the DS and RSS instabilities [2, 14], thus identifying the giant total growth rate as simultaneous occurrence, or more specifically, a linear superposition of those instabilities.

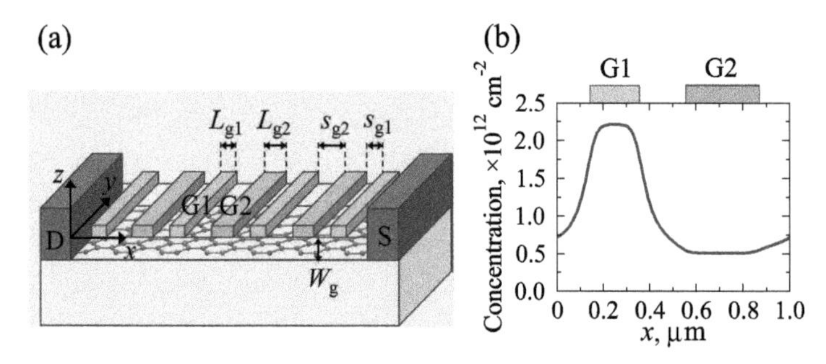

Figure 14.1 Schematic views of (a) a dual-grating-gate graphene FET and (b) a profile of steady-state electron concentration under consideration.

14.2 Mechanisms of Dyakonov–Shur and Ryzhii–Satou–Shur Instabilities

The mechanisms of the DS and RSS instabilities and their features are summarized as follows. The DS instability originates from the Doppler shift effect at asymmetric boundaries. In case of FETs in the saturation regime, the potential is fixed at the source side (the short boundary), while the current, i.e., the electron velocity is fixed at the drain side (the open boundary). Between these asymmetric boundaries, a resonant cavity for plasmons is formed. Then, the open boundary reflects traveling plasmons towards it with the current amplitude preserved, whereas the short boundary reflects them with amplification. The resonant frequency of the excited plasmons is determined by geometrical factors (gate length, gate dielectric thickness, and lengths of ungated regions in the channel) and gate voltage through the electron concentration. The latter enables the tuning of the frequency. On the other hand, the RSS instability takes place in the same saturation regime due to the transit-time effect of fast-moving electrons in the high-field region (i.e., the low-concentration region) in the drain side. The electric field created by those electrons modulates the concentration at the edge of the high-concentration region and results in the bunching of electrons in the low-concentration region. Depending on the length of the low-concentration region and the electron velocity, it interferes with plasmons in the high-concentration region either constructively or destructively. The constructive interference corresponds to the instability. The plasmon frequency is again determined by the geometrical factors in the high-concentration region and the gate voltage.

In case of dual-grating-gate structures, both instabilities can be realized by following configurations. First, the resonant plasmon cavity can be formed by modulating the concentration profile through gate voltages in such a way that high- and low-concentration regions are created. This also enables the occurrence of the RSS instability. Second, asymmetry of the gate placement leads to partial realization of the asymmetric boundary conditions and, in turn, of the DS instability.

14.3 Simulation Model

Figure 14.1a shows a dual-grating-gate graphene FET under consideration. In our simulation model, we assume that the number of periods of the grating gates is sufficiently large, i.e., the total channel length is longer than the THz/midinfrared wavelength, so that we can ignore effects of nonperiodicity, such as the presence of the source and drain contacts and we can consider one period as a unit cell. In addition, we assume that the channel width in the y direction is so long compared with the length of a period that the electron concentration in the channel, together with the geometry, can be treated as uniform in the y direction. We fix the thickness of the gate dielectric, W_g = 50 nm, the left spacings of the gates 1 and 2, s_{g1} = 200 nm, s_{g2} = 300 nm, whereas we vary the lengths of the gate 1 and 2, L_{g1} and L_{g2}, in the range between 100 and 400 nm in order to reveal characteristics of the instabilities obtained in this work by comparison with existing theory. The values of those parameters were chosen such a way that frequencies of self-excited plasmons fall into the THz/midinfrared range. The dielectric constant of media surrounding the graphene channel is set to ϵ = 4. Figure 14.1b is a profile of the steady-state electron concentration without current flow, and it was calculated self-consistently with a uniform electron doping $\Sigma_e = 5 \times 10^{11}$ cm^{-2}, together with certain gate voltages. The slight electron doping is introduced to avoid plasmon damping due to the electron-hole friction [27]; even at the point with the lowest electron concentration, the hole concentration is negligibly low,

$\Sigma_h = 6.8 \times 10^9$ cm^{-2}. On the other hand, the highest electron concentration is set not so high that the difference between maximum and minimum Fermi energies (166 and 68.5 meV, respectively) does not exceed optical-phonon energies in graphene and thus electrons injected quasiballistically from the region with high electron concentration do not experience optical-phonon emission, which would critically hinder the RSS instability.

We use the quasiclassical Boltzmann equation to describe the electron transport in the channel:

$$\frac{\partial f}{\partial t} + \upsilon_F \frac{p_x}{|p|}\frac{\partial f}{\partial x} - eE_x \frac{\partial f}{\partial p_x} = J_{LA}(f \mid p), \tag{14.1}$$

where $\upsilon = 10^8$ cm/s is the Fermi velocity in graphene, E_x is the self-consistent electric field in graphene, and $p = (p_x, p_y)$ is the momentum. The boundary condition for Eq. (14.1) is periodic, i.e., $f|_{x=0} = f|_{x=L}$. On the right-hand side of Eq. (14.1), we take into account collision integrals for the acoustic-phonon scattering J_{LA}, where

$$J_{LA}(f|p) = \frac{1}{(2\pi\hbar)^2}\int dp'\, W_{LA}(p'-p)[f(p') - f(p)], \tag{14.2}$$

and an explicit expression of the transition probability W_{LA} can be found in Ref. [30]. We use the so-called weighted essentially nonoscillatory finite-difference scheme [33, 34] to solve Eq. (14.1), which is demonstrated to be applicable for graphene transport simulation [35]. The time step of the simulation was set to $\Delta t = 0.05$ fs to avoid numerical instabilities.

Equation (14.1) is accompanied by the self-consistent 2D Poisson equation

$$\nabla \cdot [\epsilon \nabla \varphi(t,x,z)] = 4\pi e[\Sigma(t,x) - \Sigma_e]\delta(z), \tag{14.3}$$

where $\nabla = (\partial/\partial x, \partial/\partial z)$, φ is the electric potential, $\Sigma = \int dpf/\pi^2\hbar^2$ is the electron concentration in the channel, and Σ_e is the electron doping concentration. The electric field in the channel can be found by $E_x = -(\partial\varphi/\partial x)_{z=0}$. The boundary condition at $x = 0$ and $L = 0$ is periodic, while we set the natural boundary conditions, $\partial\varphi/\partial z = 0$, at $z = W_t$ and $z = -W_b$, where W_t and W_b are sufficiently large so that they do not affect the numerical results. Besides, the electron concentration in the channel expressed as a delta function in Eq. (14.3) is accounted for as a boundary condition at $z = 0$:

$$\left.\frac{\partial\varphi}{\partial z}\right|_{z=+0}-\left.\frac{\partial\varphi}{\partial z}\right|_{z=-0}=\frac{4\pi e}{\epsilon}(\Sigma-\Sigma_e). \tag{14.4}$$

We use a finite-element library called LIBMESH [36] to solve Eq. (14.3) with those boundary conditions. More details of the simulation model are given elsewhere [31].

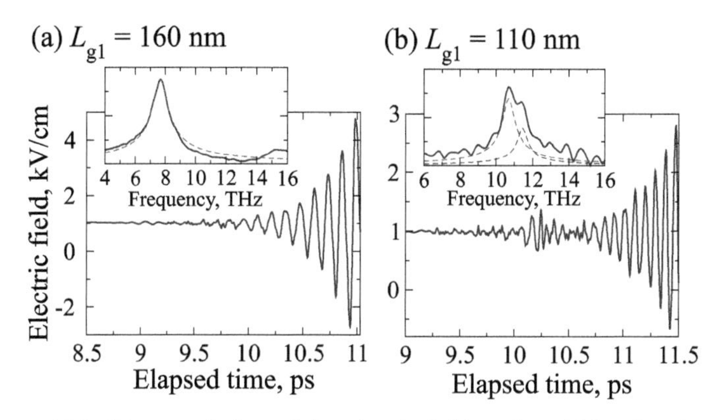

Figure 14.2 Time evolution of the electric field at the middle of a channel region under the gate 2 with (a) L_{g1} = 160 nm and (b) 110 nm. In both cases, the external electric field is set to E_{ext} = 0.8 kV/cm. The insets show corresponding absolute value of the Fourier transform.

A simulation starts by applying a uniform external dc electric field in the channel direction E_{ext} to inject positive source-drain dc current. To avoid unwanted plasmon excitation associated with an abrupt turn-on of the dc electric field, an artificially large damping factor is set in front of the collision integral in Eq. (14.1), and it is gradually decreased to unity within 6 ps, which is much longer than the inverse of plasmon frequency obtained in the simulation. Then, the electron concentration, electric field, etc., begin to oscillate and their amplitudes increase with time, as shown in Fig. 14.2a. This is identified as occurrence of plasmon instability.

In general, simulated oscillations contain several modes. Distinct fundamental modes can be obtained with certain sets of parameters, and the frequency and total growth rate can be easily extracted. However, with other sets they contain fundamental and second modes with very close frequencies and/or higher harmonics which have comparable amplitudes with the fundamental modes, resulting in the beating (see Fig. 14.2b) and/or the distortion of wave forms.

To extract the frequency and total growth rate of the fundamental mode, we perform the Fourier transform of the oscillation with respect to time, pick up the first peak and another close peak, if any, and perform a curve fitting to them with the following function:

$$F(\omega)=\sum_{i=1,2}\frac{E_{\mathrm{i}}}{4\pi}\frac{|\gamma_{\mathrm{i}}|}{\sqrt{\gamma_{\mathrm{i}}^{2}+(\omega-\omega_{\mathrm{i}})^{2}}}. \tag{14.5}$$

Equation (14.5) is equal to the absolute value of the Fourier transform of the summation of two exponentially growing harmonic functions $\Sigma_{i=1,2}\, E_i \exp(\gamma_i t)\cos(\omega_i t + \theta_i)$ around its peak(s). As seen in the insets of Figs. 14.2a and b, the fundamental and adjacent second modes can be well separated by this method.

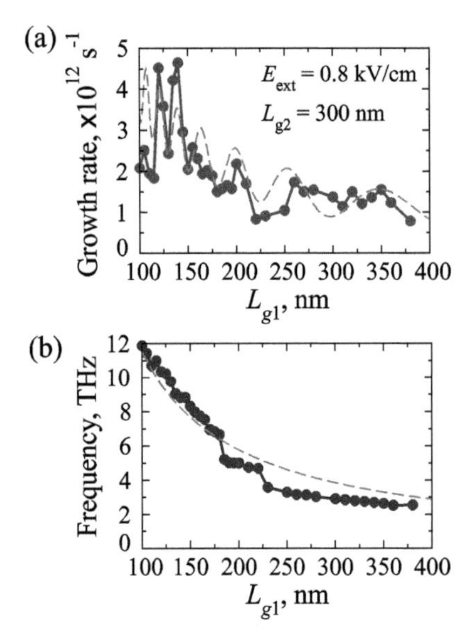

Figure 14.3 (a) Total growth rate and (b) frequency of the fundamental plasmon mode as functions of the length of the gate 1 (circles with solid lines) and theoretical fitting curves (dashed lines).

14.4 Results and Discussion

Figures 14.3a and b show the total growth rate and frequency of the fundamental mode as a function of the length of the gate 1 with E_{ext} = 0.8 kV/cm and with L_{g2} = 300 nm. They clearly demonstrates the total growth rate of the instability exceeding 4×10^{12} s^{-1} with the frequency in the THz/midinfrared range. This value is an order of magnitude larger than achievable in FETs or HEMTs based on usual semiconductors at room temperature. This giant total growth rate of the plasmon instability in the dual-grating-gate graphene FET is attributed to the lower plasmon damping and also to the larger drift velocity, both originating from the lower scattering rate limited only by acoustic phonons. This point shall be discussed in more detail shortly later.

The gate-length dependence of the total growth rate exhibits an oscillatory behavior; the oscillation period becomes shorter and the amplitude becomes larger as the gate length becomes shorter. It is specific to the RSS instability, and it corresponds to the constructive/destructive interference between plasmons in the high-concentration region (under gate 1) and the bunched electrons in the low-concentration region (under gate 2). This effect is illustrated in Fig. 14.4, where an oscillation of the electron velocity is built up at the left edge of the high-concentration region (near x = 150 nm), where the electrons are injected to the low-concentration region. Besides, there is a monotonically increasing portion of the total growth rate with decreasing L_{g1} in Fig. 14.3a. This can be attributed to the DS instability, in which the growth rate is inversely proportional to the traveling time of plasmons in the gated region and therefore to the gate length [2]. Those characteristics signify the simultaneous occurrence of the DS and RSS instabilities. In fact, the overall characteristics of the growth rate can be described qualitatively well by the formula according to Ref. [15], which is a linear superposition of the growth rates of those instabilities:

$$\gamma = \frac{\upsilon_{\mathrm{DS}}}{L_{\mathrm{g1}}} - \frac{\upsilon_{\mathrm{RSS}}}{L_{\mathrm{g1}}} J_0\left(\frac{L_\gamma}{L_{\mathrm{g1}}}\right), \tag{14.6}$$

where $\upsilon_{\mathrm{DS}} = 4 \times 10^7$ cm/s, $\upsilon_{\mathrm{RSS}} = 7.5 \times 10^7$ cm/s, and L_γ = 5800 nm are fitting parameters (see the fitting curve in Fig. 14.3a).

As seen in Fig. 14.3b, the frequency is almost inversely proportional to L_{g1}, and it obeys the well-known dispersion of gated 2D plasmons,

$$f = \frac{s_{\mathrm{g1}}}{L_{\mathrm{g1}}}, \tag{14.7}$$

where s_{g1} = 1.15 × 10^8 cm/s is the plasmon phase velocity which is extracted as a fitting parameter and which is quantitatively consistent with the value calculated analytically [20] for the region under gate 1.

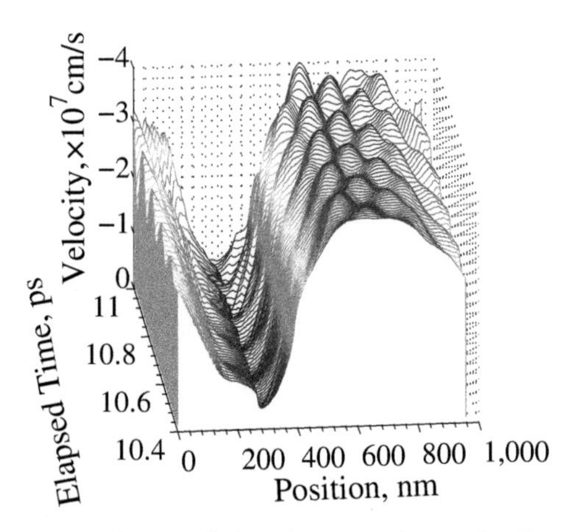

Figure 14.4 Time evolution of the electron velocity distribution after the onset of plasmon instability, in the case with L_{g1} = 160 nm and L_{g2} = 300 nm.

The giant instability found here at room temperature originates from the ultimately weak electron scattering rate in graphene, and there are two factors for this: (1) the weak damping of plasmons and (2) the large drift velocity in both high- and low-concentration regions that leads to large DS and RSS instabilities, respectively. First, the estimated damping rate was around 1.2 × 10^{11} s^{-1} in the low-concentration region and 2.5 × 10^{11} s^{-1} in the high-concentration region; note that the electron scattering rate, and thus the plasmon damping rate, for the acoustic-phonon scattering is proportional to the square root of the electron concentration [30, 31]. Such low values of the plasmon damping rate can be achieved only at nitrogen temperature or lower in other materials. It is worth mentioning that the low-concentration region with adjacent ungated regions acts as a better (passive) plasmon resonant cavity, while the instabilities take place primarily for plasmons in the high-concentration region, as is evident from the dependence of the frequency on L_{g1} and its consistency with Eq. (14.7), and the frequency is insensitive to the length of the gate 2, as shown in Fig. 14.5 (in contrast, the total growth rate can vary with it). This is similar to the situation discussed in

Ref. [37], where the ungated plasmon resonance can be effectively tuned by the gated region of the channel. In fact, it can be seen in Fig. 14.4 that a higher harmonic oscillation with more than one node is excited in the low-concentration region. This, together with the oscillation in the high-concentration region, form the fundamental mode of the whole period.

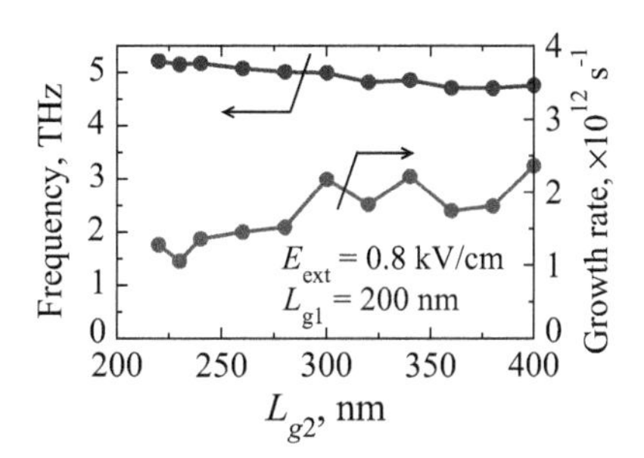

Figure 14.5 Total growth rate and frequency of the fundamental mode as functions of the length of gate 2.

In addition, values of drift velocities obtained in the simulation, $\simeq 4 \times 10^7$ cm/s in the low-concentration region and $\simeq 10^7$ cm/s in the high-concentration region, confirm the factor (2) for the giant growth rate. Especially the former is more than twice larger than the saturation velocity in Si and InGaAs channels. However, the velocities extracted by Eq. (14.6) as fitting parameters differ from these values by several factors. A reasonable explanation to this discrepancy is that the DS instability also takes place not only in the high-concentration region but also in the low-concentration region. Figure 14.4 exhibits waves propagating in opposite directions in the latter, suggesting the occurrence of the DS instability there. This should add a constant term in Eq. (14.6), and then the fitting velocities should become closer to the simulated values.

14.5 Conclusion

In conclusion, we have conducted simulations of plasmon instability driven by dc current injection in the dual-grating-gate graphene FET. We have obtained a giant total growth rate of the instability at room temperature exceeding 4×10^{12} s^{-1}. Through the dependences of the total growth rate and frequency on the gate lengths, we have revealed that the giant total growth rate originates from simultaneous occurrence of the DS and RSS instabilities. The obtained result strongly suggests that a graphene FET with the dual-grating-gate structure is very promising for the realization of a high-power, compact, room-temperature operating THz source.

Acknowledgments

This work was financially supported by Japan Society for the Promotion of Science (JSPS) Grant-in-Aid for Young Researcher (No. 26820122), by JSPS Grant-in-Aid for Specially Promoted Research (No. 23000008), and by JSPS and Russian Foundation for Basic Research under Japan-Russia Research Cooperative Program. The simulation was carried out using the computational resources provided by Research Institute for Information Technology Center, Nagoya University, through the HPCI System Research Project (No. hp140086), by the Information Technology Center, the University of Tokyo, and by Research Institute for Information Technology, Kyushu University.

References

1. S. J. Allen, D. C. Tsui, and R. A. Logan, *Phys. Rev. Lett.*, **38**, 980 (1977).
2. M. Dyakonov and M. Shur, *Phys. Rev. Lett.*, **71**, 2465 (1993).
3. M. Dyakonov and M. Shur, *IEEE Trans. Electron Devices*, **43**, 380 (1996).
4. D. Veksler, F. Teppe, A. P. Dmitriev, V. Y. Kachorovskii, W. Knap, and M. S. Shur, *Phys. Rev. B*, **73**, 125328 1 (2006).
5. G. C. Dyer, G. R. Aizin, S. J. Allen, A. D. Grine, D. Bethke, J. L. Reno, and E. A. Shaner, *Nat. Photonics*, **7**, 925 (2013).
6. I. V. Rozhansky, V. Y. Kachorovskii, and M. S. Shur, *Phys. Rev. Lett.*, **114**, 246601 (2015).
7. M. S. Shur and V. Ryzhii, *Int. J. High Speed Electron. Syst.*, **13**, 575 (2003).
8. W. Knap, M. Dyakonov, D. Coquillat, F. Teppe, N. Dyakonova, J. Lusakowski, K. Karpierz, M. Sakowicz, G. Valusis, D. Seliuta, I. Kasalynas, A. Fatimy, Y. M. Meziani, and T. Otsuji, *J. Infrared, Millimeter, Terahertz Waves*, **30**, 1319 (2009).
9. T. Otsuji and M. Shur, *IEEE Microwave Mag.*, **15**, 43 (2014).

10. T. Otsuji, *IEEE Trans.* Terahertz Sci. Technol., **5**, 1110 (2015).
11. A. P. Dmitriev, A. S. Furman, V. Y. Kachorovskii, G. G. Samsonidze, and G. G. Samsonidze, *Phys. Rev. B*, **55**, 10319 (1997).
12. M. Dyakonov and M. S. Shur, *Appl. Phys. Lett.*, **87**, 111501 (2005).
13. V. V. Popov, G. M. Tsymbalov, and M. S. Shur, *J. Phys.: Condens. Matter*, **20**, 384208 (2008).
14. V. Ryzhii, A. Satou, and M. S. Shur, *Phys.* Status Solidi A, **202**, 113 (2005).
15. V. Ryzhii, A. Satou, and M. S. Shur, *IEICE Trans. Electron.* **E89-C**, 1012 (2006).
16. V. Ryzhii, A. Satou, M. Ryzhii, T. Otsuji, and M. S. Shur, *J. Phys.: Condens. Matter*, **20**, 384207 (2008).
17. T. Otsuji, M. Hanabe, T. Nishimura, and E. Sano, *Opt. Express*, **14**, 4815 (2006).
18. V. V. Popov, D. V. Fateev, T. Otsuji, Y. M. Meziani, D. Coquillat, and W. Knap, *Appl. Phys. Lett.*, **99**, 243504 (2011).
19. T. Otsuji, Y. M. Meziani, T. Nishimura, T. Suemitsu, W. Knap, E. Sano, T. Asano, and V. V. Popov, *J. Phys.: Condens. Matter*, **20**, 384206 (2008).
20. V. Ryzhii, *Jpn. J. Appl. Phys.*, **45**, L923 (2006).
21. V. Ryzhii, A. Satou, and T. Otsuji, *J. Appl. Phys.*, **101**, 24509 (2007).
22. S. A. Mikhailov, *Phys. Rev. B*, **84**, 045432 (2011).
23. A. A. Dubinov, V. Y. Aleshkin, V. Mitin, T. Otsuji, and V. Ryzhii, *J. Phys.: Condens. Matter*, **23**, 145302 (2011).
24. L. Ju, B. Geng, J. Horng, C. Girit, M. Martin, Z. Hao, H. a. Bechtel, X. Liang, A. Zettl, Y. R. Shen, and F. Wang, *Nat. Nanotechnol.*, **6**, 630 (2011).
25. L. Vicarelli, M. S. Vitiello, D. Coquillat, A. Lombardo, A. C. Ferrari, W. Knap, M. Polini, V. Pellegrini, and A. Tredicucci, *Nat. Mater.*, **11**, 865 (2012).
26. B. Wang, X. Zhang, F. J. García-Vidal, X. Yuan, and J. Teng, *Phys. Rev. Lett.*, **109**, 073901 (2012).
27. D. Svintsov, V. Vyurkov, S. Yurchenko, T. Otsuji, and V. Ryzhii, *J. Appl. Phys.*, **111**, 083715 (2012).
28. A. N. Grigorenko, M. Polini, and K. S. Novoselov, *Nat. Photonics*, **6**, 749 (2012).
29. V. V. Popov, O. V. Polischuk, S. A. Nikitov, V. Ryzhii, T. Otsuji, and M. S. Shur, *J. Opt.*, **15**, 114009 (2013).

30. E. H. Hwang and S. Das Sarma, *Phys. Rev. B*, **77**, 115449 (2008).
31. A. Satou, V. Ryzhii, F. T. Vasko, V. V. Mitin, and T. Otsuji, *Proc. SPIE*, **8624**, 862412 (2013).
32. L. Wang, I. Meric, P. Y. Huang, Q. Gao, Y. Gao, H. Tran, T. Taniguchi, K. Watanabe, L. M. Campos, D. A. Muller, J. Guo, P. Kim, J. Hone, K. L. Shepard, and C. R. Dean, *Science*, **342**, 614 (2013).
33. J. A. Carrillo, I. M. Gamba, A. Majorana, and C.-w. Shu, *J. Comput. Phys.*, **184**, 498 (2003).
34. M. Galler and F. Schürrer, *J. Comput. Phys.*, **212**, 778 (2006).
35. P. Lichtenberger, O. Morandi, and F. Schürrer, *Phys. Rev. B*, **84**, 045406 (2011).
36. B. S. Kirk, J. W. Peterson, R. H. Stogner, and G. F. Carey, *Eng. Comput.*, **22**, 237 (2006).
37. V. V. Popov, A. N. Koudymov, M. Shur, and O. V. Polischuk, *J. Appl. Phys.*, **104**, 024508 (2008).

Part 2

Detectors Based on Lateral Transport in Graphene and Graphene Structures

Chapter 15

Plasma Mechanisms of Resonant Terahertz Detection in a Two-Dimensional Electron Channel with Split Gates*

V. Ryzhii,[a] A. Satou,[a] T. Otsuji,[b] and M. S. Shur[c]
[a]*Computer Solid State Physics Laboratory, University of Aizu, Aizu-Wakamatsu 965-8580, Japan*
[b]*Research Institute of Electrical Communication, Tohoku University, Sendai 980-8577, Japan*
[c]*Department of Electrical, Computer, and Systems Engineering, Rensselaer Polytechnic Institute, Troy, New York 12180, USA*
v-ryzhii@u-aizu.ac.jp

We analyze the operation of a resonant detector of terahertz (THz) radiation based on a two-dimensional electron gas (2DEG) channel with split gates. The side gates are used for the excitation of plasma oscillations by incoming THz radiation and control of the resonant plasma frequencies. The central gate provides the potential barrier

*Reprinted with permission from V. Ryzhii, A. Satou, T. Otsuji, and M. S. Shur (2008). Plasma mechanisms of resonant terahertz detection in a two-dimensional electron channel with split gates, *J. Appl. Phys.*, **103**, 014504. Copyright © 2008 American Institute of Physics.

Graphene-Based Terahertz Electronics and Plasmonics: Detector and Emitter Concepts
Edited by Vladimir Mitin, Taiichi Otsuji, and Victor Ryzhii

ISBN 978-981-4800-75-4 (Hardcover), 978-0-429-32839-8 (eBook)
www.jennystanford.com

separating the source and drain portions of the 2DEG channel. Two possible mechanisms of the detection are considered: (1) modulation of the ac potential drop across the barrier and (2) heating of the 2DEG due to the resonant plasma-assisted absorption of THz radiation followed by an increase in thermionic dc current through the barrier. Using the device model we calculate the frequency and temperature dependences of the detector responsivity associated with both dynamic and heating (bolometric) mechanisms. It is shown that the dynamic mechanisms dominates at elevated temperatures, whereas the heating mechanism provides larger contribution at low temperatures, $T \lesssim 35$–40 K.

15.1 Introduction

Plasma oscillations in heterostructures with a two-dimensional electron gas (2DEG) and some devices using the excitation of these oscillations have been experimentally studied over decades (see, for instance, Refs. [1–17]). There are also many theoretical papers on different aspects of these plasma waves and their potential applications. The gated and ungated 2DEG channels in heterostructures can serve as resonant cavities for terahertz (THz) electron plasma waves [18]. The resonant properties of such channels can be utilized in different THz devices, in particular, resonant detectors [3–5, 9, 10, 12–17]. The mechanism of the THz detection observed in transistor structures [3, 4, 6, 7, 9, 10, 15–17] might be attributed to the nonlinearity of the plasma oscillations as suggested previously [18]. The variation of the conductivity of 2DEG with periodic gate system can be possibly explained by the heating of 2DEG by absorbed THz radiation (heating or bolometric mechanism) [5, 13]. Recently, a concept of resonant detectors in which a plasma resonant cavity is integrated with a Schottky junction has been proposed and substantiated [19, 20]. As shown, when the frequency of incoming THz radiation is close to one of the plasma resonant frequencies, the ac potential drop across the barrier at the Schottky junction becomes resonantly large. This leads to rather large values of the rectified component of the current through the junction, which is used as output signal of the detector. Since the resonant excitation of the plasma oscillations by absorbed THz radiation

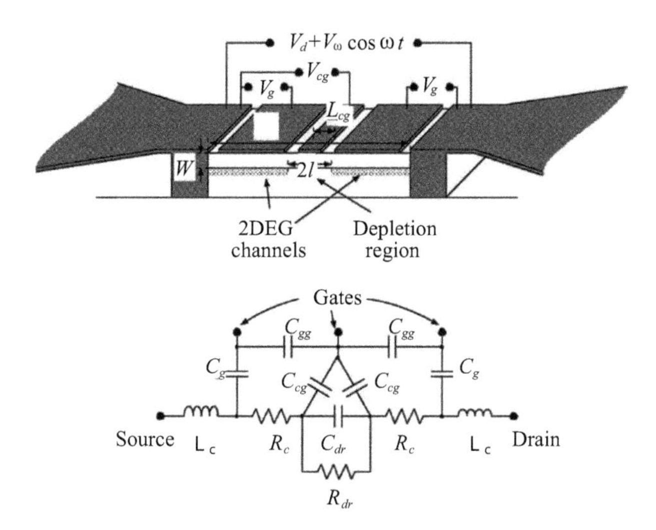

Figure 15.1 Schematic view of the THz detector under consideration and its simplified equivalent circuit.

leads to the heating of 2DEG, the heating (bolometric) mechanism can also contribute to the rectified current over the Schottky barrier. Thus, two mechanisms can be responsible for the operation of this detector: the dynamic mechanism considered previously [19] and the heating mechanism. Similar mechanisms can work in the detectors utilizing the excitation of plasma oscillations but having the barrier of another origin. In particular, the barrier formed by an additional gate which results in an essential depletion under it was used in recent experiments [14]. In this chapter, we develop a device model for a THz resonant detector based on a heterostructure with the 2DEG channel and the barrier region utilizing the plasma oscillations excitation. We consider a device with three gates (see Fig. 15.1): the 2DEG under two extreme gates forms the plasma resonant cavities, while the central gate, to which a sufficiently large negative bias voltage is applied, forms the barrier. The excitation of the plasma oscillations by the incident THz radiation is assumed to be associated with an antenna connected with the source and drain contacts to the 2DEG channel. In the following, we demonstrate that each of the mechanisms in question can dominate in different temperature ranges (the dynamic mechanism is predominant at elevated temperatures, whereas the heating mechanism provides a larger contribution at low temperatures, $T \lesssim 35$–40 K). The device structure under consideration differs from that studied in Ref. [14], in which the THz radiation input was realized by a periodic grating. However, the physical mechanisms of detection are the same.

15.2 Equations of the Model

We assume that the net drain-source voltage V includes both the dc bias voltage V_d and the ac component $V_\omega \cos \omega t$ induced by incoming THz radiation via an antenna connected with the source and drain contacts. Thus, $V = V_d + V_\omega \cos \omega t$, where ω is the THz radiation frequency. The bias voltage V_g is applied between the side gates and the pertinent contacts. This voltage affects the electron density in the quasineutral portions of the 2DEG channel, so that the dc electron density is given by

$$\Sigma_0 = \Sigma_d + \frac{æ V_g}{4\pi e W}, \tag{15.1}$$

where e is the electron charge, W is the gate layer thickness, and æ is the dielectric constant. It is also taken to be that the potential drop, V_{cg}, between the central gate and the source contact is negative and its absolute value is sufficiently large to deplete some portion of the 2DEG channel underneath this gate and create a potential barrier. This implies that $V_{cg} < -4\pi e W/æ\Sigma_d = V_{th}$, where V_{th} is the 2DEG threshold voltage.

The ac component of the electric potential in the channel $\delta\varphi$ satisfies the boundary conditions which follow from the fact that at the source and drain contacts its values coincide with the ac components of the incoming signal voltage:

$$-\delta\varphi\big|_{x=-L} = \delta\varphi\big|_{x=+L} = \frac{1}{2} V_\omega \cos \omega t. \tag{15.2}$$

Assuming that at the conductivity of the quasineutral sections is much larger than that of depletion region (including the thermionic current over the barrier and the displacement current associated with the capacitances shown in the equivalent circuit of Fig. 15.1), one may use the following additional conditions:

$$\frac{\partial}{\partial x}\delta\varphi\bigg|_{x=\pm l} \simeq 0. \tag{15.3}$$

Here $2L$ is the length of the 2DEG channel between the source and drain contacts, $2l > L_{cg}$ is the length of the depletion region under the central gate, and L_{cg} is the length of the latter (see Fig. 15.1). The quantity l is determined by the potential of the central gate V_{cg}. Definitely, the conductivity of the quasineutral sections markedly exceeds the real part of the depleted region admittance. The role of the displacement current across the depleted region is discussed in Section 15.8.

The electron transport in the channel is governed by the hydrodynamic equations coupled with the Poisson equation for the self-consistent electric potential [18]. At sufficiently small intensities of THz radiation, this system of equations can be linearized and reduced to the following equation [21, 22] valid in the quasineutral regions of the channel ($-L \leq x < -l$ and $l < x \leq L$):

$$\frac{\partial}{\partial t}\left(\frac{\partial}{\partial t}+\nu\right)\delta\varphi = s^2 \frac{\partial^2}{\partial x^2}\delta\varphi, \tag{15.4}$$

where ν is the frequency of electron collisions with impurities and phonons, $s=\sqrt{4\pi e^2\Sigma_0 W / m æ}$ is the plasma wave velocity, and m is the electron effective mass. The effect of electron pressure results in some renormalization of the plasma wave velocity [23–25]: $s=\sqrt{(4\pi e^2\Sigma_0 W / m æ)+s_0^2}$, where s_0 is the velocity of the electron "sound" (usually $s_0 \ll s$).

Assuming that the central gate length is comparable with the gate layer thickness, the shape of the barrier under the central gate can be considered as parabolic: $\Delta_b(x) = \Delta_b[1 - (x/l)^2] - e(V_d + \delta V^b)(l + x)/2l$, where Δ_b is the barrier height in equilibrium (it depends on potential, V_{cg}, of the central gate), and $V_d + \delta V^b$ is the lateral potential drop across the depletion (barrier) region associated with the dc bias voltage V_d and the ac voltage δV^b. The variation of the barrier height $\delta\Delta_b$ due to the lateral potential drop is given by

$$\delta\Delta_b = -\frac{e(V_d+\delta V^b)}{2}\left[1-\frac{e(V_d+dV^b)}{8\Delta_b}\right] \simeq -\frac{e(V_d+\delta V^b)}{2}. \tag{15.5}$$

Hence, at $e|V_d + \delta V^b| \ll \Delta_b$, the net source-drain current can be calculated using the following formula:

$$J = J_m \exp\left(-\frac{\Delta_b}{k_B T}\right)\left\{\exp\left[\frac{e(V_d+\delta V^b)}{2k_B T}\right]-1\right\}, \tag{15.6}$$

where J_m is the maximum current (for a 2DEG $J_m \propto T^{3/2}$), the ac potential drop, δV^b, across the depletion (barrier) region is given by

$$\delta V^{\mathrm{b}} = \delta\varphi\big|_{x=+l} - \delta\varphi\big|_{x=-l} = 2\delta\varphi\big|_{x=+l}, \tag{15.7}$$

and $T = T_0 + \delta T$ is the electron temperature, T_0 is the lattice temperature, and δT is the variation of the electron temperature due to THz irradiation.

In this case, at $e\delta V^b < k_B T < \Delta_b$, Eq. (15.6) yields the following equation for the dc source-drain current J_0:

$$\Delta J_0 = J_0 - J_{00} = J_{\mathrm{m}} \exp\left(-\frac{\Delta_{\mathrm{b}}}{k_{\mathrm{B}}T_0}\right)\exp\left(\frac{eV_{\mathrm{d}}}{2k_{\mathrm{B}}T_0}\right) \times \left[\frac{e^2\overline{(\delta V^{\mathrm{b}})^2} + 8(\Delta_{\mathrm{b}} - eV_{\mathrm{d}}/2)k_{\mathrm{B}}\overline{\delta T}}{8(k_{\mathrm{B}}T_0)^2}\right] \tag{15.8}$$

at $eV_d > k_B T_0$ and

$$\Delta J_0 = J_0 - J_{00} = J_{\mathrm{m}} \exp\left(-\frac{\Delta_{\mathrm{b}}}{k_{\mathrm{B}}T_0}\right)\left(\frac{eV_{\mathrm{d}}}{2k_{\mathrm{B}}T_0}\right) \times \left[\frac{e^2\overline{(\delta V^{\mathrm{b}})^2} + 8(\Delta_{\mathrm{b}} - eV_{\mathrm{d}}/2)k_{\mathrm{B}}\overline{\delta T}}{8(k_{\mathrm{B}}T_0)^2}\right] \tag{15.9}$$

at $eV_d \ll k_B T_0$. Here,

$$J_{00} = J_{\mathrm{m}} \exp\left(-\frac{\Delta_{\mathrm{b}}}{k_{\mathrm{B}}T_0}\right)\left[\exp\left(\frac{eV_d}{2k_{\mathrm{B}}T_0}\right) - 1\right] \tag{15.10}$$

is the dc source-drain current without THz irradiation (dark current).

15.3 Plasma Resonances

Solving Eq. (15.4) with boundary conditions (15.2) and (15.3) and taking into account Eq. (15.7), we obtain

$$\delta V^{\mathrm{b}} = \frac{V_\omega}{2}\left\{\frac{e^{i\omega t}}{\cos\left[q_\omega(L-l)\right]} + \frac{e^{-i\omega t}}{\cos\left[q_{-\omega}(L-l)\right]}\right\}. \tag{15.11}$$

Here

$$q_{\pm\omega} = \sqrt{\frac{m \text{æ} \omega(\omega \mp i\nu)}{4\pi e^2 \Sigma_0 W}}. \tag{15.12}$$

Equation (15.11) yields

$$\overline{(\delta V^{\mathrm{b}})^2} = \frac{V_\omega^2}{2}\left[\cos^2\left(\frac{\pi\omega}{2\Omega}\right) + \sinh^2\left(\frac{\pi\nu}{4\Omega}\right)\right]^{-1}, \tag{15.13}$$

where

$$\Omega = \sqrt{\frac{\pi^3 e^2 \Sigma_0 W}{\text{æ}\, m(L-l)^2}} \tag{15.14}$$

is the characteristic plasma frequency. Due to the dependence (given by Eq. (15.1)) of the dc electron density on the gate voltage V_g, the characteristic plasma frequency Ω can be tuned by this voltage. Because of the dependence of Ω on the length, $L - l$, of the quasineutral sections of the getated 2DEG channel (serving as the resonant plasma cavities) and, hence, on the potential of the central gate V_{cg}, Ω is somewhat varied with varying V_{cg}.

15.4 Resonant Electron Heating

The power absorbed by the channel can be calculated using the following equation:

$$\delta P_\omega = \frac{2e^2 \Sigma_0 \nu}{m(\omega^2 + \nu^2)} \int_l^L dx \left|\frac{\partial \delta\varphi}{\partial x}\right|^2. \tag{15.15}$$

Considering Eq. (15.15), the energy balance equation governing the averaged electron temperature can be presented as

$$\frac{k_{\mathrm{B}} \overline{\delta T}}{\tau_\varepsilon} = \frac{e^2 \nu}{m(\nu^2 + \omega^2)} \overline{\left(\frac{\delta V^{\mathrm{b}}}{L-l}\right)^2} K, \tag{15.16}$$

where τ_ε is the electron energy relaxation time. The factor K is associated with the nonuniformity of the ac electric field under the side gates. This factor is given by

$$K = \left(\frac{\pi}{8}\right)^2 \frac{\omega\sqrt{\omega^2 + \nu^2}}{\Omega^2}$$
$$\left[\left(\frac{2\Omega}{\pi\nu}\right)\sinh\left(\frac{\pi\nu}{2\Omega}\right) - \left(\frac{\Omega}{\pi\omega}\right)\sin\left(\frac{\pi\omega}{\Omega}\right)\right]. \tag{15.17}$$

At $\nu \ll \omega, \Omega$, Eq. (15.17) yields

$$K = \left(\frac{\pi}{8}\right)^2 \left(\frac{\omega}{\Omega}\right)^2 \left[1 - \left(\frac{\Omega}{\pi\omega}\right)\sin\left(\frac{\pi\omega}{\Omega}\right)\right], \tag{15.18}$$

so that one can use the following simplified formula:

$$k_B \overline{\delta T} \simeq \left(\frac{\pi}{8}\right)^2 \frac{e^2 \nu \tau_\varepsilon}{mL^2\Omega^2} \overline{(\delta V^b)^2}. \tag{15.19}$$

15.5 Rectified Current and Detector Responsivity

Using Eqs. (15.8) and (15.15), we obtain

$$\Delta J_0 = J_m \exp\left(\frac{V_d - 2\Delta_b}{2k_B T_0}\right) \cdot \frac{(1+H)e^2 \overline{(\delta V^b)^2}}{8(k_B T_0)^2}, \tag{15.20}$$

where the term

$$H = \frac{4(2\Delta_b - eV_d)\nu\tau_\varepsilon}{m(\nu^2 + \omega^2)(L-l)^2} K \tag{15.21}$$

is associated with the contribution of the electron heating to the rectified current.

Combining Eqs. (15.13) and (15.20), we arrive at

$$\Delta J_0 = J_m \cdot \frac{\exp\left(\frac{V_d - 2\Delta_b}{2k_B T_0}\right)(1+H)}{4\left[\cos^2\left(\frac{\pi\omega}{2\Omega}\right) + \sinh^2\left(\frac{\pi\nu}{4\Omega}\right)\right]} \left(\frac{eV_\omega}{2k_B T_0}\right)^2. \tag{15.22}$$

Since V_ω^2 is proportional to the incoming THz power, the detector responsivity as a function of the signal frequency and the structural parameters (except the antenna parameters) can be in the following form:

$$R \propto \frac{\exp\left(\frac{V_{\rm d} - 2\Delta_{\rm b}}{2k_{\rm B}T_0}\right)}{\left[\cos^2\left(\frac{\pi\omega}{2\Omega}\right) + \sinh^2\left(\frac{\pi\nu}{4\Omega}\right)\right]} \frac{(1+H)}{\sqrt{k_{\rm B}T_0}}. \tag{15.23}$$

As follows from Eq. (15.23), the responsivity as a function of the frequency of incoming THz radiation exhibits sharp peaks at the plasma resonant frequencies $\omega = \Omega(2n - 1)$, where $n = 1, 2, 3, \ldots$ is the resonance index, provided that the quality factor of resonances $\Omega/\nu \gg 1$.

15.6 Comparison of Dynamic and Heating Mechanisms

As seen from Eq. (15.22), the ratio of the heating and dynamic components of the rectified dc source-drain current is given by the quantity H (heating parameter). At $\nu \ll \omega$, Ω, considering Eqs. (15.18) and (15.19), the quantity H can be estimated as

$$H = \frac{\pi^2}{16} \frac{(2\Delta_{\rm b} - eV_{\rm b})\nu\tau_\varepsilon}{m\Omega^2(L-l)^2} \simeq \frac{\pi^2}{8} \frac{\Delta_{\rm b}\nu\tau_\varepsilon}{m\Omega^2(L-l)^2}. \tag{15.24}$$

Considering that $\Omega = \pi s/2(L - l)$, Eq. (15.24) can be presented in the following form:

$$H \simeq \frac{1}{4}\left(\frac{2\Delta_{\rm b} - eV_{\rm d}}{ms^2}\right)\nu\tau_\varepsilon \simeq \left(\frac{\Delta_{\rm b}}{2ms^2}\right)\nu\tau_\varepsilon. \tag{15.25}$$

The first factor in Eq. (15.25) is usually small, while the second one can be large, so that H can markedly exceed unity, particularly, at low temperatures. In line with the above assumptions, Eqs. (15.24) and (15.25) are valid at moderate drain voltages when $\Delta_{\rm b} - eV_{\rm d}/2$ is positive and, moreover, not too small in comparison with $\Delta_{\rm b}$.

It is natural to assume that in a wide range of the temperature (from liquid helium to room temperatures), the electron collision frequency (the inverse momentum relaxation time) is determined by the electron interaction with acoustic phonons and charged impurities, while the electron energy relaxation is due to the interaction with both acoustic and polar optical phonons. Hence, $\nu = \nu^{(ac)} + \nu^{(i)}$ and $\tau_\varepsilon = \tau_\varepsilon^{(ac)}\tau_\varepsilon^{(op)} / [\tau_\varepsilon^{(ac)} + \tau_\varepsilon^{(op)}]$, where $\nu^{(ac)}$ and $\nu^{(i)}$ are related to the electron scattering on acoustic phonons and impurities (charged), respectively, whereas $\tau_\varepsilon^{(ac)}$ and $\tau_\varepsilon^{(op)}$ are the electron energy relaxation times associated with the acoustic and optical phonons. As a result, the product $\nu\tau_\varepsilon$ can be presented as

$$\nu\tau_\varepsilon = \frac{\left[\nu^{(ac)} + \nu^{(i)}\right]\tau_\varepsilon^{(ac)}\tau_\varepsilon^{(op)}}{\left[\tau_\varepsilon^{(ac)} + \tau_\varepsilon^{(op)}\right]}. \tag{15.26}$$

We shall use the following formulas for the temperature dependences of the parameters in this equation:

$$\nu^{(ac)} = \frac{1}{\bar{\tau}_{PA}^{(ac)}}\left(\frac{k_B T_0}{\hbar\omega_0}\right)^{1/2} + \frac{1}{\bar{\tau}_{DA}^{(ac)}}\left(\frac{k_B T_0}{\hbar\omega_0}\right)^{3/2}, \tag{15.27}$$

$$\frac{1}{\tau_\varepsilon^{(ac)}} = \frac{2ms_a^2}{\hbar\omega_0}\left[\frac{1}{\bar{\tau}_{PA}^{(ac)}}\left(\frac{k_B T_0}{\hbar\omega_0}\right)^{-1/2} + \frac{1}{\bar{\tau}_{DA}^{(ac)}}\left(\frac{k_B T_0}{\hbar\omega_0}\right)^{1/2}\right], \tag{15.28}$$

$$\frac{1}{\tau_\varepsilon^{(op)}} = \frac{1}{\bar{\tau}^{(op)}}\frac{2}{3}\left(\frac{\hbar\omega_0}{k_B T_0}\right)^2 \exp\left(-\frac{\hbar\omega_0}{k_B T_0}\right), \tag{15.29}$$

where $\bar{\tau}_{PA}^{(ac)}$ = 8 ps, $\bar{\tau}_{DA}^{(ac)}$ = 4 ps, and $\bar{\tau}^{(op)}$ = 0.14 ps (Ref. [26]) are the characteristic scattering times for the acoustic phonon scattering (polar and deformation mechanisms, respectively) and for the optical phonon scattering, and $\hbar\omega_0$ is the optical phonon energy. The temperature dependence of the collision frequency of 2D electrons with charged impurities is assumed to be as follows:

$$\nu^{(i)} = \bar{\nu}^{(i)}\left(\frac{\hbar\omega_0}{k_B T_0}\right) \cdot S(|z_i|, k_B T_0 / \hbar\omega_0). \tag{15.30}$$

Here the function [27]

$$S(z_i, \xi) = \int_0^1 \exp(-\alpha |z_i| \sqrt{\xi} x)\frac{dx}{\sqrt{1-x^2}}$$

characterizes a decrease in the collision frequency with increasing thickness of the spacer, $|z_i|$, between the 2DEG channel and the charged donor sheet, and $\alpha = 4\sqrt{2m\hbar\omega_0} / \hbar \simeq 10^7$ cm^{-1}. The characteristic frequency $\overline{\nu}_i$ is proportional to the donor sheet density. We set $\overline{\nu}^{(i)} = (10^8 - 10^9)$ s^{-1}. At T_0 = 4.2 K, these values correspond to $\nu^{(i)} = 10^{10} - 10^{11}$ s^{-1} and the electron mobilities $\mu \simeq (10^5 - 10^6)$ cm^2/V s. When $|z_i|$= 0, Eq. (15.30) yields $\nu_i \propto T_0^{-1}$, while at sufficiently thick spacers, one obtains $\nu_i \propto T_0^{-3/2}$. In the case of relatively thick electron channels, in which the quantization is insignificant, $\nu_i \propto T_0^{-3/2}$.

Figure 15.2 shows the temperature dependence of the "heating" parameter H, which determines the relative contribution of the heating mechanism to the detector responsivity, calculated for different values of parameter $\overline{\nu}^{(i)}$, i.e., different doping levels. It is assumed that for a GaAs channel $\hbar\omega_0/k_B$ = 421 K [26]. We set s = 1×10^8 cm/s, $s_a = 5 \times 10^5$ cm/s, the barrier height Δ_b = 110 meV, and the bias voltage V_d = 20 mV, so that the effective barrier height $\Delta_b^{(eff)} = \Delta_b - eV_b/2$ = 100 meV. As seen from Fig. 15.2, the heating mechanism can provide significantly larger contribution to the detector responsivity at low temperatures ($T_0 \lesssim$ 35–40 K). However at elevated temperatures, this mechanism becomes relatively inefficient. This is attributed to weak electron heating due to strong energy relaxation on optical phonons at elevated temperature.

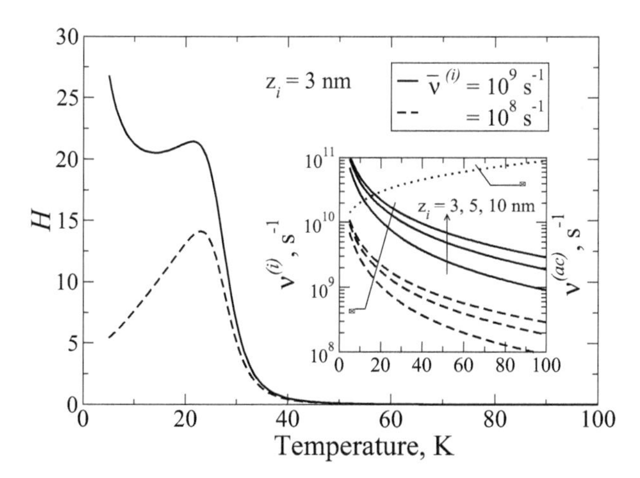

Figure 15.2 Temperature dependences of heating parameter H. The inset shows the temperature dependences of the electron collision frequencies associated with impurity and acoustic scattering mechanism.

15.7 Temperature Dependences of the Detector Responsivity and Detectivity

As mentioned in Section 15.5, the detector resonsivity exhibits sharp peaks at the plasma resonant frequencies $\omega = \Omega(2n - 1)$. The sharpness of these peaks depends on the electron collision frequency ν, which in turn depends on the temperature.

As follow from Eq. (15.23), the maximum values of the detector responsivity at the fundamental resonance ($n = 1$) are given by

$$\max R \propto \frac{\exp\left(\dfrac{V_{\rm d} - 2\Delta_{\rm b}}{2k_BT_0}\right)}{\sinh^2\left(\dfrac{\pi\nu}{4\Omega}\right)}\frac{(1+H)}{\sqrt{k_BT_0}}. \qquad (15.31)$$

As a result, for the value of the ratio of the detector responsivity at the resonance and the dark current one obtains

$$\frac{\max R}{J_{00}} \propto \frac{1}{\sinh^2\left(\dfrac{\pi\nu}{4\Omega}\right)}\frac{1+H}{(k_{\rm B}T)^2}. \qquad (15.32)$$

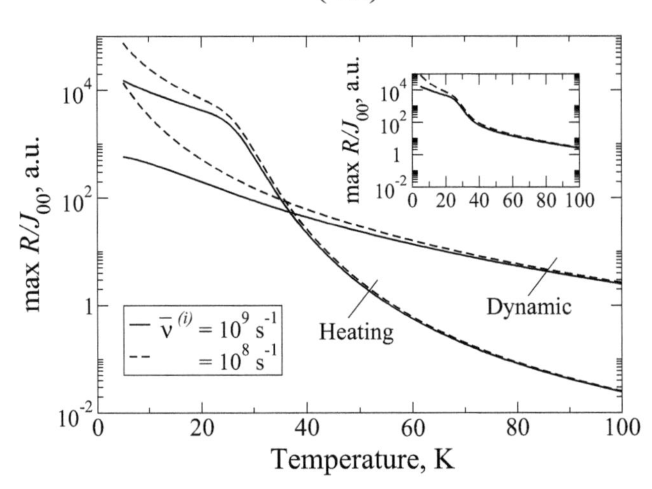

Figure 15.3 Temperature dependences of dynamic and heating contributions to max R/J_{00}, of the detector responsivity maximum (resonant) value and the dark current. The inset shows the net value max R/J_{00} as a function of temperature.

Figure 15.3 shows the temperature dependences of the maximum (resonant) values of the dynamic and heating contributions to the detector responsivity devided by the dark current value calculated using Eq. (15.32) for the barrier height $\Delta_b^{(eff)}$ = 100 meV, the fundamental plasma frequencies $\Omega/2\pi = 1$ THz, and different values of the parameter $\overline{\nu}^{(i)}$. The inset shows the temperature dependence of max R/J_{00}, which accounts for both mechanisms.

Taking into account that the detector detectivity $D^* \propto R/\sqrt{J_{00}}$, we arrive at the following formula:

$$\max D^* \propto \frac{\exp\left(\dfrac{V_d - 2\Delta_b}{4k_B T_0}\right)}{\left[\sinh^2\left(\dfrac{\pi\nu}{4\Omega}\right)\right]} \frac{(1+H)}{(k_B T_0)^{5/4}}. \tag{15.33}$$

Figure 15.4 shows the temperature dependences of the detector detectivity calculated for different values of the effective barrier height $\Delta_b^{(eff)} = \Delta_b - eV_d/2$.

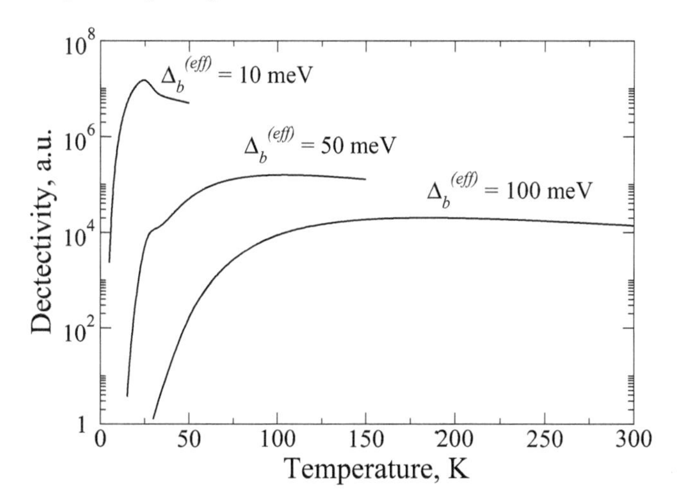

Figure 15.4 Temperature dependences of the detector responsivity D^* for different values of the effective barrier height $\Delta_b^{(eff)}$.

As follows from the above formulas, the detector responsivity and detectivity increase with decreasing effective barrier height. However, this quantity cannot be set too small (at a given temperature

T_0) in the frame of our model, which is valid if $\Delta_b^{(\text{eff})} \gg k_B T_0$, and when the 2DEG channel can be partitioned into the quasineutral and deleted sections.

15.8 Discussion

The displacement current across the depleted region can be, in principle, essential. To take this current into consideration, one needs to modify Eq. (15.3) [28]. Introducing the admittance, Y_ω, of the barrier region, we can use the following boundary conditions at the edge of the quasineutral sections of the 2DEG channel (for the asymmetrical plasma modes for which $\delta\varphi_\omega\big|_{x=-1} = -\delta\varphi_\omega\big|_{x=l}$):

$$-\sigma_\omega \frac{d\varphi_\omega}{dx}\bigg|_{x=\pm l} = 2Y_\omega\, \delta\varphi_\omega\big|_{x=\pm l}. \tag{15.34}$$

Here, $\sigma_\omega = -i[e^2\Sigma_0 / m(\omega + i\nu)]$ is the ac conductivity of the quasineutral sections of the channel accounting for the electron collisions and inertia. Since the real part of the conductivity of the depleted (barrier) region is definitely small in comparison with the conductivity of the quasineutral sections, one can take into account only the capacitive component of the current across the depleted region. In this case, $Y_\omega = -i\omega C$, where C is the pertinent capacitance, which is determined by the capacitances C_{cg} and C_{dr}: $C = C_{dr} + C_{gs}/2$ (see the equivalent circuit in Fig. 15.1). This equivalent circuit accounts for the inductance of the quasineutral sections $\mathcal{L}_c$, due to the inertia of the electron transport along the channel, the channel resistance due to the electron scattering, the resistance of the barrier region R_{dr}, and different capacitances. Equation (15.34) can be presented in the form:

$$\frac{\Omega^2(L-l)}{\omega(\omega+i\nu)} \frac{d\delta\varphi_\omega}{dx}\bigg|_{x=\pm l} = c\delta\varphi_\omega\big|_{x=\pm l}, \tag{15.35}$$

where

$$c = \frac{2\pi^2 CW}{æ(L-l)} \simeq \frac{2\pi^2 CW}{æ\, L}. \tag{15.36}$$

If the parameter $c \ll 1$, Eqs. (15.3) and (15.35) practically coincide (except the case $\omega \gg \Omega$). According to Ref. [28], $C_{\rm dr} = (æ/2\pi^2)\Lambda_{\rm dr}$ and $C_{\rm cg} = (æ/2\pi^2)\Lambda_{\rm cg}$, where $\Lambda_{\rm dr}$ and $\Lambda_{\rm dr}$ are logarithmic factors which are determined by the geometry of the planar conducting areas (the gates and quasineutral sections of the channel). These factors can be estimated as [28] $\Lambda_{\rm dr} \sim \Lambda_{\rm cg} \sim \ln(2L/l)$. Hence, $c \simeq (W/L)\Lambda$, where $\Lambda = \Lambda_{\rm dr} + \Lambda_{\rm cg}/2$. The factor Λ somewhat exceeds unity, although it is not too large. Thus, taking into account that $W \ll L$, one can conclude that for real device structures, $c \ll 1$. The condition of the smallness of c can be also presented in the form $C_{\rm g} \gg C$, where $C_{\rm g} = æ4\pi\, W \simeq L/W$ is the capacitance of the side gates (see Fig. 15.1). Relatively small capacitances $C_{\rm dr}$ and $C_{\rm cg}$ can, to some extent, affect the plasma oscillations. However, their role reduces mainly to a small modification in the resonant plasma frequencies [28].

If c would be large, the ac potential near the edges of the quasineutral regions is small as well as the ac potential drop across the depleted region. This implies that the boundary condition (15.34) could be $\delta\varphi_{\omega}\big|_{x=\pm l} \simeq 0$. In such a case, the fundamental plasma frequency is doubled and dynamic mechanism is weakened. However, the details of the ac potential distribution near the edges of the quasineutral regions are not crucial for the heating mechanism.

Actually, there is some delay in the electron transit across the barrier. The delay time can be estimated as $\tau_{\rm b} \simeq 2l/v_{\rm T}$, where $v_T \simeq \sqrt{k_{\rm B}T_0/m}$ is the thermal electron velocity. The delay in the electron transit and, therefore, the electron transit time effects can be disregarded if $\omega\tau_{\rm b} \sim \Omega\tau_{\rm b} < 1$ [29]. Assuming $\Omega/2\pi = 1$ THz and $v_{\rm T} = 10^7$ cm/s, from the last inequality we obtain $2l < 0.6$ μm.

15.9 Conclusions

We developed a device model for a resonant detector of THz radiation based on the gated 2DEG channel with an electrically induced barrier. The model accounts for the resonant excitation of the plasma oscillations in the gated 2DEG channel as well as the rectified dc current through the barrier. As shown, this current comprises two components: the dynamic component associated with the ac potential drop across the barrier and the heating (bolometric) component due to a change in the electron temperature, which stems from the resonant plasma-assisted absorption of THz radiation. Using our model, we calculated the frequency and temperature dependences of the detector responsivity. The detector responsivity exhibits sharp resonant peaks at the frequencies of incoming THz radiation corresponding to the plasma resonances in the gated 2DEG channel. The plasma resonances can tuned by the gate voltage V_{cg} and, to some extent, by the potential of the central gate V_{cg}. It is demonstrated that the dynamic mechanism dominates at elevated temperatures, whereas at low lattice temperatures ($T_0 \lesssim 35$–40 K for AlGaAs/GaAs-based detectors), the heating mechanism prevails. This can be attributed to a marked increase in the electron energy relaxation time τ_ε with decreasing lattice temperature.

At rather low temperatures when the ratio $\Delta_{\rm b}^{\rm (eff)} / k_{\rm B}T_0$ is large, the thermionic electron current over the barrier can be surpassed by the tunneling current. The rectified portion of this current can be associated with both dynamic and heating mechanisms (the thermo-assisted tunneling current in the case of the latter mechanism). This, however, requires a separate detailed study.

In the case of detection of THz radiation modulated at some frequency $\omega_{\rm m} \ll \omega$, the relative contributions of the dynamic and heating mechanisms to the responsivity, $R^{\rm m}$, characterizing the detector response at the frequency $\omega_{\rm m}$ can be different than those considered above. This is due to different inertia of these mechanisms. In particular, if $\omega_{\rm m} > \tau_\varepsilon^{-1}$(but $\omega_{\rm m} \ll \omega$), the variation of the electron effective temperature averaged over the THz oscillations $\overline{\delta T_{\rm m}} \simeq \overline{\delta T} / \omega_m \tau_\varepsilon$, i.e., $\overline{\delta T_{\rm m}} \ll \overline{\delta T}$. As a result, $R_{\rm m}/R \simeq (\omega_{\rm m}\tau_\varepsilon)^{-1}$. This implies that even at low temperatures, the heating mechanism can be inefficient if $\omega_{\rm m} > \tau_\varepsilon^{-1}$.

Both dynamic and heating mechanisms might be responsible for the THz detection in the plasmonic resonant detectors utilizing another barrier structure (for instance, with the lateral Schottky barrier [19, 20] or with electron transport through the gate barrier [30–32]) and another types of the plasma resonant cavity (with ungated quasineutral regions as in high-electron mobility transistors with relatively long ungated source-gate and gate-drain regions), as well as in those with other methods of the excitation of plasma oscillations (utilizing periodic gate structures [14]).

Acknowledgments

This work was supported by the Grant-in-Aid for Scientific Research (S) from the Japan Society for Promotion of Science, Japan. The work at RPI was partially supported by the Office of Naval Research, USA.

References

1. S. J. Allen, Jr., D. C. Tsui, and R. A. Logan, *Phys. Rev. Lett.*, **38**, 980 (1977).
2. D. C. Tsui, E. Gornik, and R. A. Logan, *Solid State Commun.*, **35**, 875 (1980).
3. W. Knap, Y. Deng, S. Rumyantsev, J.-Q. Lu, M. S. Shur, C. A. Saylor, and L. C. Brunel, *Appl. Phys. Lett.*, **80**, 3433 (2002).
4. W. Knap, Y. Deng, S. Rumyantsev, and M. S. Shur, *Appl. Phys. Lett.*, **81**, 4637 (2002).
5. X. G. Peralta, S. J. Allen, M. C. Wanke, N. E. Harff, J. A. Simmons, M. P. Lilly, J. L. Reno, P. J. Burke, and J. P. Eisenstein, *Appl. Phys. Lett.*, **81**, 1627 (2002).
6. W. Knap, J. Lusakowski, T. Parently, S. Bollaert, A. Cappy, V. V. Popov, and M. S. Shur, *Appl. Phys. Lett.*, **84**, 2331 (2004).
7. J. Lusakowski, W. Knap, N. Dyakonova, L. Varani, J. Mateos, T. Gonzales, Y. Roelens, S. Bullaert, A. Cappy, and K. Karpierz, *J. Appl. Phys.*, **97**, 064307 (2005).
8. T. Otsuji, M. Hanabe, and O. Ogawara, *Appl. Phys. Lett.*, **85**, 2119 (2004).
9. F. Teppe, D. Veksler, V. Yu. Kacharovskii, A. P. Dmitriev, X. Xie, X.-C. Zhang, S. Rumyantsev, W. Knap, and M. S. Shur, *Appl. Phys. Lett.*, **87**, 022102 (2005).
10. F. Teppe, W. Knap, D. Veksler, M. S. Shur, A. P. Dmitriev, V. Yu. Kacharovskii, and S. Rumyantsev, *Appl. Phys. Lett.*, **87**, 052105 (2005).
11. M. Hanabe, T. Otsuji, T. Ishibashi, T. Uno, and V. Ryzhii, *Jpn. J. Appl. Phys., Part 1*, **44**, 3842 (2005).
12. M. Lee, M. C. Wanke, and J. L. Reno, *Appl. Phys. Lett.*, **86**, 033501 (2005).
13. E. A. Shaner, M. Lee, M. C. Wanke, A. D. Grine, J. L. Reno, and S. J. Allen, *Appl. Phys. Lett.*, **87**, 193507 (2005).
14. E. A. Shaner, A. D. Grine, M. C. Wanke, M. Lee, J. L. Reno, and S. J. Allen, *IEEE Photonics Technol. Lett.*, **18**, 1925 (2006).
15. D. Veksler, F. Teppe, A. P. Dmitriev, V. Yu. Kacharovskii, W. Knap, and M. S. Shur, *Phys. Rev. B*, **73**, 125328 (2006).

16. A. El Fatimy, F. Teppe, N. Dyakonova, W. Knap, D. Seliuta, G. Valusis, A. Shcherepetov, Y. Roelens, S. Bollaert, A. Cappy, and S. Rumyantsev, *Appl. Phys. Lett.*, **89**, 131926 (2006).
17. J. Torres, P. Nouvel, A. Akwaoue-Ondo, L. Chusseau, F. Teppe, A. Shcherepetov, and S. Bollaert, *Appl. Phys. Lett.*, **89**, 201101 (2006).
18. M. Dyakonov and M. Shur, *IEEE Trans. Electron Devices*, **43**, 1640 (1996).
19. V. Ryzhii and M. S. Shur, *Jpn. J. Appl. Phys., Part 2*, **45**, L1118 (2006).
20. A. Satou, N. Vagidov, and V. Ryzhii, *IEICE Techn. Rep.*, ED2006-197, p. 77 (2006).
21. A. Satou, V. Ryzhii, I. Khmyrova, M. Ryzhii, and M. S. Shur, *J. Appl. Phys.*, **95**, 2084 (2004).
22. V. Ryzhii, A. Satou, W. Knap, and M. S. Shur, *J. Appl. Phys.*, **99** (2006) 084507.
23. S. Rudin and T. L. Reinecke, *Phys. Rev. B*, **54**, 2791 (1996).
24. F. J. Crowne, *J. Appl. Phys.*, **82**, 1242 (1997).
25. S. Rudin and G. Samsonidze, *Phys. Rev. B*, **58**, 16369 (1998).
26. V. F. Gantmakher and Y. B. Levinson, *Carrier Scattering in Metals and Semoconductors* (North-Holland, Amsterdam, 1987).
27. A. Shik, *Quantum Wells* (World Scientific, Singapore, 1997).
28. V. Ryzhii, A. Satou, I. Khmyrova, M. Ryzhii, T. Otsuji, V. Mitin, and M. S. Shur, *J. Phys.: Conf. Ser.*, **38**, 228 (2006).
29. V. Ryzhii and M. S. Shur, *Phys. Status Solidi A*, **202**, R113 (2005).
30. V. Ryzhii, I. Khmyrova, and M. Shur, *J. Appl. Phys.*, **88**, 2868 (2000).
31. I. Khmyrova and V. Ryzhii, *Jpn. J. Appl. Phys., Part 1*, **39**, 4727 (2000).
32. A. Satou, V. Ryzhii, I. Khmyrova, and M. S. Shur, *Semicond. Sci. Technol.*, **18**, 460 (2003).

Chapter 16

Terahertz and Infrared Photodetection Using p-i-n Multiple-Graphene-Layer Structures*

V. Ryzhii,[a,b] M. Ryzhii,[a,b] V. Mitin,[c] and T. Otsuji[b,d]
[a]*Computational Nanoelectronics Laboratory, University of Aizu, Aizu-Wakamatsu 965-8580, Japan*
[b]*Japan Science and Technology Agency, CREST, Tokyo 107-0075, Japan*
[c]*Department of Electrical Engineering, University at Buffalo, State University of New York, New York 14260, USA*
[d]*Research Institute for Electrical Communication, Tohoku University, Sendai 980-8577, apan*
v-ryzhii@u-aizu.ac.jp

We propose to utilize multiple-graphene-layer structures with lateral p-i-n junctions for terahertz and infrared (IR) photodetection and substantiate the operation of photodetectors based on these structures. Using the developed device model, we calculate the

*Reprinted with permission from V. Ryzhii, M. Ryzhii, V. Mitin, and T. Otsuji (2010). Terahertz and infrared photodetection using p-i-n multiple-graphene-layer structures, *J. Appl. Phys.*, **107**, 054512. Copyright © 2010 American Institute of Physics.

Graphene-Based Terahertz Electronics and Plasmonics: Detector and Emitter Concepts
Edited by Vladimir Mitin, Taiichi Otsuji, and Victor Ryzhii

ISBN 978-981-4800-75-4 (Hardcover), 978-0-429-32839-8 (eBook)
www.jennystanford.com

detector dc responsivity and detectivity as functions of the number of graphene layers and geometrical parameters and show that the dc responsivity and detectivity can be fairly large, particularly, at the lower end of the terahertz range at room temperatures. Due to relatively high quantum efficiency and low thermogeneration rate, the photodetectors under consideration can substantially surpass other terahertz and IR detectors. Calculations of the detector responsivity as a function of modulation frequency of THz and IR radiation demonstrate that the proposed photodetectors are very fast and can operate at the modulation frequency of several tens of gigahertz.

16.1 Introduction

Unique properties of graphene layers (GLs) [1–3] make them promising for different nanoelectronic device applications. The gapless energy spectrum of GLs, which is an obstacle for creating transistor-based digital circuits, opens up prospects to use GLs in terahertz and infrared (IR) devices. Novel optoelectronic terahertz and IR devices were proposed and evaluated, in particular, in Refs. [4–11]. Recent success in fabricating multiple-GL structures with long momentum relaxation time of electrons and holes [12] promises a significant enhancement of the performance of future graphene optoelectronic devices [13].

In this chapter, we study the operation of terahertz and IR photodetectors based on multiple-GL structures with reverse biased p-i-n junctions. We refer to the photodetectors in question as to GL-photodetectors (GLPDs). We focus on GLPDs (a) with p- and n-doped sections in GLs near the side contacts [14, 15], p^+ and n^+ contacts (for example, made of doped poly-Si) [16], and multiple GL-structures with the Ohmic side contacts and (b) with split gates which provide the formation of the electrically induced p- and n-sections [17–22]. These gates are separated from a multiple-GL structure by a insulating layer (gate layer) made of, for instance, SiO_2 or HfO_2. Its thickness W_g should be chosen (sufficiently small) to provide an effective formation of the electron and hole regions under the gates in sufficiently large number of GLs under the applied gate voltages. The device structures under considerations are shown in Figs. 16.1a and b. It is assumed that the highly conducting GL(s) between the SiC substrate and the top GLs is removed. Multiple GL-structures without this highly conducting GL can be fabricated using chemical/mechanical reactions and transferred substrate techniques (chemically etching the substrate and the highly conducting bottom GL (Ref. [23]) or mechanically peeling the upper GLs, then transferring the upper portion of the multiple-GL structure on a Si substrate). The operation of GLPDs is associated with the interband photogeneration of the electrons and holes in the i-sections of GLs by incoming THz or IR radiation. The photogenerated electrons and holes propagate under the electric field created by the bias voltage (which provides the p-i-n junction reverse bias) in the directions toward the n-section and p-section, respectively. This results in the dc or ac photocurrent induced in the circuit. The GLPD operation principle is actually similar to that in the standard p-i-n photodiodes based on bulk semiconductor materials.

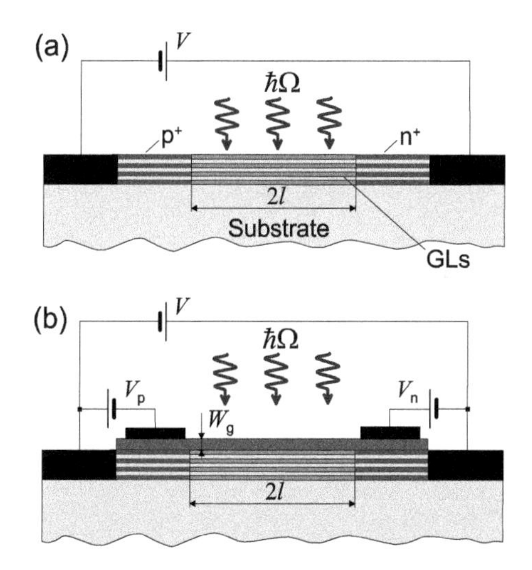

Figure 16.1 Device structures of GLPDs with (a) doped and (b) electrically induced p-i-n junctions.

Using the developed device model, we calculate the GLPD responsivity and detectivity as terahertz or IR photodetector (Sections 16.2 and 16.3), evaluate its dynamic response (Section 16.4), and compare GLPDs with some other terahertz and IR photodetectors (Section 16.5), in particular, with quantum-well

IR photodetectors (QWIPs) and quantum-dot IR photodetectors (QDIPs). In Section 16.6, we discuss possible role of the Pauli blocking in the spectral characteristics of GLPDs and draw main conclusions. The calculations related to the effect of screening of the vertical electric field on the formation of the electrically induced p- and n-sections in multiple-GL structures are singled out in the Appendix.

16.2 Responsivity and Detectivity

We assume that the intensity of the incident terahertz or IR radiation with the frequency Ω apart from the dc component I_0 includes the ac component: $I(t) = I_0 + \delta I_\omega \exp(-i\omega)t$, where δI_ω and ω are the amplitude of the latter component and its modulation frequency, respectively. In such a situation, the net dc current (per unit width of the device in the direction perpendicular to the current) can be presented in the following form:

$$J_0 = J_0^{\text{dark}} + J_0^{\text{photo}}, \tag{16.1}$$

with

$$J_0^{\text{dark}} = 4Kel(g_{\text{th}} + g_{\text{tunn}}) \tag{16.2}$$

and

$$J_0^{\text{photo}} = 4el\sum_{k=1}^{K}\frac{\beta_\Omega I_0^{(k)}}{\hbar\Omega}. \tag{16.3}$$

Here e is the electron charge, K is the number of GLs, $2l$ is the length of the GL i-section, g_{th} and $g_{\text{tunn}} = (eV/2lv_W^{2/3}\hbar)^{3/2}/8\pi^2$ are the rates of thermal and tunneling generation of the electron-hole pairs (per unit area) [22], V is the reverse bias voltage, $v_W \simeq 10^8$ cm/s is the characteristic velocity of electrons and holes in graphene (see, for instance, Ref. [3]), $\beta_\Omega = \beta[1-2f(\hbar\Omega/2)]$ is the absorption coefficient of radiation in a GL due to the interband transitions [24], where $\beta_\Omega = (\pi e^2/c\hbar) \simeq 0.023$, $f(\varepsilon)$ is the distribution function of electrons and holes in the i-section, $\hbar$ is the Planck constant, and c is the speed of light. The term g_{tunn} in Eq. (16.2) is associated with the interband tunneling in the electric field in the i-section. The quantity $I_0^{(k)} = I_0(1-\beta_\Omega)^{K-k}$ is the intensity of terahertz or IR radiation at the k-th GL ($1 \le k \le K$).

At a sufficiently strong reverse bias, electrons and holes are effectively swept out from the i-section to the contacts and heated. Due to this, one can assume that under the GLPD operation conditions, $f(\hbar\Omega/2) \ll 1$, so that the distinction between β_Ω and β can be disregarded (see below).

Using Eqs. (16.1) and (16.3), we arrive at the following formula for the GLPD dc responsivity:

$$R_0 = \frac{J_0^{\text{photo}}}{2\hbar I_0} = K^* \frac{2e\beta_\Omega}{\hbar\Omega}, \quad (16.4)$$

where $K^* = \Sigma_{k=1}^{K}(1-\beta_\Omega)^{K-k} = [1-(1-\beta_\Omega)^K]/\beta_\Omega$. Equation (16.4) can also be rewritten as

$$R_0 = [1-(1-\beta_\Omega)^K]\frac{2e}{\hbar\Omega} = \eta\frac{e}{\hbar\Omega} \propto \frac{1}{\hbar\Omega}, \quad (16.5)$$

where $\eta = 2[1-(1-\beta_\Omega)^K]$ is the GLPD quantum efficiency and the factor 2 appears because each photon absorbed due to the interband transition generates two carriers, an electron and a hole. As a result, η can exceed unity. However, $\eta < 2$ because there is no multiplication of the photoelectrons and photoholes propagating in the i-section, so that the photoelectric gain G in the photodetectors under consideration is equal (or, at least, close) to unity. If $K = 1$, Eq. (16.5) at $\Omega/2\pi = 1$ THz, yields, $R_0 \simeq 12$ A/W. For $K = 50$–100 ($K^* \simeq 30$–39), setting $K = 50$, in the frequency range $\Omega/2\pi = 1$–10 THz, from Eq. (16.4) we obtain $R_0 \simeq 35$–350 A/W. As follows from Eqs. (16.4) and (16.5), the GLPD responsivity is virtually independent of the temperature. This results from the temperature-independent absorption coefficient and $G \simeq 1$.

The dark-current limited detectivity, D^*, defined as $D^* = (J^{\text{photo}}/NP)\sqrt{A\cdot\Delta f}$, where N is the noise (in Amperes), P is the power received by the photodetector (in watts), A is the area of the photodetector (in cm^2) (see, for instance, Refs. [25, 26]), and Δf is the bandwidth, can be expressed via the responsivity R_0 and the dark current $J_0^{\text{dark}}H$, where H is the device width in the direction perpendicular to the current, and be presented as

$$D^* = R_0\sqrt{\frac{A}{4eJ_0^{\text{dark}}H}}. \quad (16.6)$$

Using Eqs. (16.2)–(16.6), we arrive at

$$D^* = \frac{K^*}{\sqrt{K}}\frac{\beta_\Omega}{\hbar\Omega}\frac{1}{\sqrt{2(g_{th}+g_{tunn})}}$$
$$= \frac{[1-(1-\beta_\Omega)^K]}{\sqrt{K}\hbar\Omega}\frac{1}{\sqrt{2(g_{th}+g_{tunn})}} \propto \frac{1}{\hbar\Omega}. \quad (16.7)$$

Assuming that $K = 1$ and $g_{th} = 10^{21}$ cm^{-2} s^{-1} [27], at $T = 300$ K for $\Omega/2\pi$ = 1–2 THz we obtain $D^* \simeq (4.1–8.2) \times 10^8$ cm Hz$^{1/2}$/W. Setting K = 50, we arrive at $D^* \simeq (1.7–3.4) \times 10^9$ cm Hz$^{1/2}$/W. Due to a significant decrease in the thermogeneration rate at lower temperatures, the detectivity markedly increases with decreasing temperature. Indeed, at $T = 77$ K, setting $g_{th} = 10^{13}$ cm^{-2} s^{-1} [27], we obtain $D^* \simeq (1.7–3.4) \times 10^{13}$ cm Hz$^{1/2}$/W.

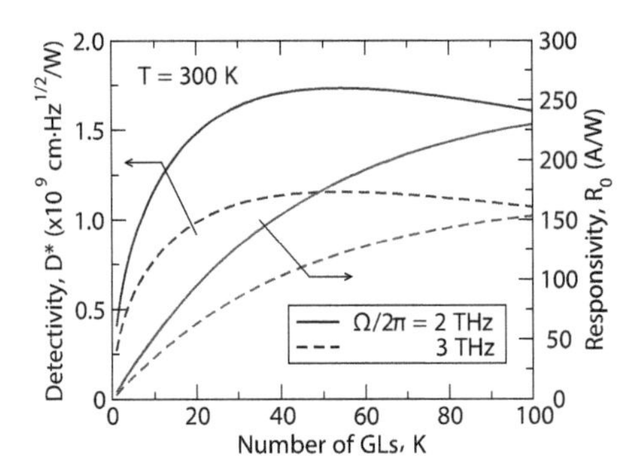

Figure 16.2 Responsivity R_0 and detectivity D^* vs. number of GLs K for $\Omega/2\pi$ = 2 and 3 THz.

Figure 16.2 shows the dependences of the dc responsivity and detectivity on the number of GLs K calculated for $\Omega/2\pi$ = 2 and 3 THz using Eqs. (16.5) and (16.7) for GLPDs with sufficiently large l in which the tunneling is weak (see below). Equation (16.7) shows that the GLPD detectivity decreases with increasing photon frequency. This is because the spectral dependence of the GLPD detectivity is determined by that of the responsivity. As follows from Eqs. (16.6) and (16.7), the detectivity drops at elevated bias voltages when the interband tunneling prevails over the thermogeneration of the electron and hole pairs. According to Eqs. (16.4) and (16.7), the GLPD responsivity is independent of the length of the i-section $2l$ and the bias voltage V (at least if the latter is not too small), whereas the detectivity increases with increasing l and decreases with V. This is because the component of the dark current associated with the thermogeneration rate increases linearly with increasing l. At the same time, the tunneling component is proportional to $V^{3/2}/\sqrt{l}$. As a result, from Eq. (16.7) one can arrive at

$$D^* \propto \frac{1}{\sqrt{1+b(V/l)^{3/2}}}, \tag{16.8}$$

where $b \propto 1/g_{\text{th}}$. Thus, at elevated bias voltages when the tunneling generation surpassed the thermal generation, $D^* \propto V^{-3/4}$. Figure 16.3 shows the voltage-dependence of the detectivity D^*_{tunn} with the interband tunneling normalized by the detectivity D^* without tunneling calculated for different values of the i-section length $2l$.

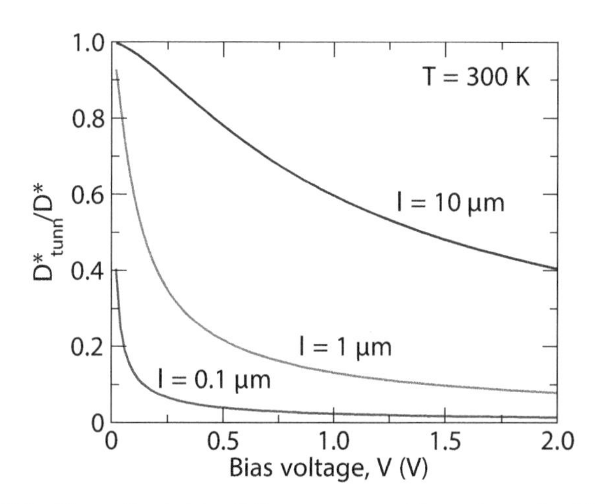

Figure 16.3 Normalized detectivity vs. reverse bias voltage for GLDPs with different length of the i-section $2l$.

Since the value of D^* is determined by both l and V, the latter values should be properly chosen. There is also a limitation associated with the necessity to satisfy the condition $2l \lesssim l_{\text{R}}$, where l_{R} is the recombination length. Otherwise, the recombination in the i-section can become essential resulting in degrading of the GLPD performance. The recombination length can be defined as $l_{\text{R}} = L_{\text{V}}$, where $L_V \simeq \langle v \rangle \tau_R$, τ_{R} is the recombination time, and $\langle v \rangle$ is the average drift velocity in the i-region. Assuming that for the electron and hole densities close to that in the intrinsic graphene at $T = 300$

K, one can put $\tau_R = 5 \times 10^{-10}$ s. Setting for sufficiently large bias $\langle v \rangle$ = 5×10^7 cm/s [28], one can find $L_V \simeq 250$ μm.

16.3 Features of GLPD with Electrically Induced Junction

The main distinctions of the detectors with doped and electrically induced p-i-n junction are that the Fermi energies of electrons and holes in the latter depend on the gate voltages $V_p < 0$ in the p-section and $V_n > 0$ in the n-section [see Fig. 16.1b] and that these energies and, hence, the heights of the barriers confining electrons and holes in the pertinent section are different in different GLs. This is due to the screening of the transverse electric field created by the gate voltages by the GL charges. This screening results in a marked decrease in the barrier heights of the GLs located in the MGL-structure depth (with large indices k) in comparison with the GLs near the top. In the following, we set $V_n = -V_p = V_g$. The barriers in question effectively prevent the electron and hole injection from the contacts under the reverse bias if the barrier heights are sufficiently large. However, the electron and hole injection into the GLs with large k leads to a significant increase in the dark current (in addition to the current of the electrons and hole generated in the i-section). This can substantially deteriorate the GLPD detectivity. As for the responsivity, it can still be calculated using Eq. (16.4) or Eq. (16.5). To preserve sufficiently high detectivity, the number of GLs should not be too large. Indeed, taking into account the contributions of the current injected from the p- and n-regions to the net dark current, the detectivity of a GLPD with the electrically induced p-i-n junction can be presented as

$$D^* = \frac{K^* \beta_\Omega}{\hbar\Omega\sqrt{2[\Sigma_{k=1}^{K} j_i^{(k)} / el + K(g_{\rm th} + g_{\rm tunn})]}}, \qquad (16.9)$$

where $j_{\rm i}^{(\rm k)} \propto \exp(-\varepsilon_{\rm F}^{(\rm k)} / k_{\rm B}T)$ is the electron current injected from the p-region and the hole current injected from the n-region, and $\varepsilon_{\rm F}^{(k)}$ is the Fermi energy of holes (electrons) in the p-region (n-region) of the k-th GL ($1 \leq k \leq K$).

The injected currents are small in comparison with the current of the electrons and holes thermogenerated in the i-region if

$$\frac{\upsilon_{\mathrm{W}}\Sigma_0}{\pi}\exp\left(-\frac{\varepsilon_{\mathrm{F}}^{(k)}}{k_{\mathrm{B}}T}\right) < 2lg_{\mathrm{th}}. \tag{16.10}$$

Here Σ_0 is the electron and hole density in the intrinsic graphene (at given temperature).

Considering Eqs. (16.A5) and (16.A6) from the Appendix, the latter imposes the following limitation on the number of GLs in GLDs with the electrically induced p-i-n junctions and the gate voltage V_{g}:

$$K < K^{\max} = \frac{1}{\gamma}\sqrt{\frac{\varepsilon_{\mathrm{F}}^{\mathrm{T}}}{k_{\mathrm{B}}T\,\ln(\upsilon_{\mathrm{W}}\Sigma_0/2\pi lg_{\mathrm{th}})}} \propto V_{\mathrm{g}}^{1/12}. \tag{16.11}$$

The quantity γ is defined in the Appendix. For $\gamma = 0.12$, assuming that the Fermi energy in the topmost GL $\varepsilon_{\mathrm{F}}^{(1)} = \varepsilon_{\mathrm{F}}^{\mathrm{T}} = 100$ meV, $T = 300$ K, $\Sigma_0 \simeq 8 \times 10^{10}$ cm^{-2}, $l = 10$ µm, and $g_{\mathrm{th}} \simeq 10^{21}$ cm^{-2} s^{-1}, we obtain $K^{\max} \simeq 34$, i.e., fairly large. A decrease in l results in smaller $K^{\max}$ due to a substantial increase in the tunneling current. For instance, if $l = 0.1$ µm [11], we obtain $K^{\max} \simeq 7$.

Thus, by applying gate voltages one can form the p- and n-sections with sufficiently high densities and large Fermi energies of holes and electrons in GL-structures with a large number of GLs.

16.4 Dynamical Response

To describe the dynamic response to the signals modulated with the frequency $\omega \ll \Omega$, one needs to find the ac components of the net density of photogenerated electron and holes, $\delta\Sigma_{\omega}^{-}$ and $\delta\Sigma_{\omega}^{+}$, respectively. These components are governed by the following equation:

$$-i\omega\delta\Sigma_{\omega}^{\mp} \pm \langle\upsilon\rangle\frac{d\delta\Sigma_{\omega}^{\mp}}{dx} = \sum_{k=1}^{K}\frac{\beta_{\Omega}\delta I_{\omega}^{k}}{\hbar\Omega}. \tag{16.12}$$

As in above, $\delta I_{\omega}^{(k)} = \delta I_{\omega}(1-\beta)^{K-k}$, so that

$$\sum_{k=1}^{K}\frac{\beta_{\Omega}\delta I_{\omega}^{(k)}}{\hbar\Omega} = K^{*}\frac{\beta_{\Omega}\delta I_{\omega}}{\hbar\Omega}.$$

In the case of ballistic transport of electrons and holes across the i-section, $\langle\upsilon\rangle \simeq \upsilon_{\mathrm{W}}$. If the electron and hole transport is substantially affected by quasielastic scattering, so that the momentum distributions of electrons and holes are virtually semi-isotropic, one can put $\langle v\rangle \simeq v_W/2$ (see also Ref. [28]). Solving Eq. (16.12) with the boundary conditions $\delta\Sigma_\omega^\mp\big|_{x=\pm l} = 0$, we obtain

$$\delta\Sigma_\omega^\mp = K * \frac{\beta_\Omega}{\hbar\Omega}\cdot\frac{\exp[i\omega(l \pm x)/\langle v\rangle]-1}{i\omega}\delta I_\omega. \qquad (16.13)$$

Using Eq. (16.13) and considering the Ramo–Shockley theorem [28, 29] applied to the case of specific contact geometry [30] the ac current induced in the side contacts (terminal current) can be presented as

$$\frac{\delta J_\omega^{\text{photo}}}{\delta I_\omega} = K * \frac{el\beta_\Omega}{\pi\hbar\Omega}\int_{-1}^{1}\frac{d\xi}{\sqrt{1-\xi^2}}\frac{[e^{i\omega\tau_t}\cos(\omega\tau_t\xi)-1]}{i\omega\tau_t}. \qquad (16.14)$$

Here we introduced the characteristic transit time $\overline{\tau}_t = l/\langle v\rangle$. The feature of the contact geometry (the bladelike contacts) was accounted for by using the form factor $g(\xi) = 1/\pi\sqrt{1-\xi^2}$ [31]. This is valid because the thickness of the side contacts and multiple-GL structure under consideration $Kd \ll l$ even at rather large numbers of GLs K, where d is the spacing between GLs. Similar approach was used previously to analyze the dynamic response of the lateral p-n junction photodiodes made of the standard semiconductors [32] and the graphene tunneling transit-time terahertz oscillator [22]. Integrating in Eq. (16.14), we arrive at the following:

$$\frac{\delta J_\omega^{\text{photo}}}{\delta I_\omega} = K * \frac{4el\beta_\Omega}{\hbar\Omega}\left[\frac{\sin(\omega\tau_t)\mathcal{J}_0(\omega\tau_t)}{\omega\tau_t} + i\frac{1-\cos(\omega\tau_t)\mathcal{J}_0(\omega\tau_t)}{\omega\tau_t}\right], \qquad (16.15)$$

where $\mathcal{J}_0(\xi)$ is the Bessel function. Equation (16.15) yields

$$\frac{\left|\delta J_\omega^{\text{photo}}\right|}{\delta I_\omega} = K * \frac{4el\beta_\Omega}{\hbar\Omega}\frac{\sqrt{1-2\cos(\omega\tau_t)\mathcal{J}_0(\omega\tau_t)+\mathcal{J}_0^2(\omega\tau_t)}}{\omega\tau_t}. \qquad (16.16)$$

Using Eq. (16.16), for the frequency dependent responsivity $R_\omega = \left|\delta J_\omega^{\text{photo}}\right|/2l\delta I_\omega$, we obtain

$$R_\omega = K * \frac{2e\beta}{\hbar\Omega}\frac{\sqrt{1-2\cos(\omega\tau_t)\mathcal{J}_0(\omega\tau_t)+\mathcal{J}_0^2(\omega\tau_t)}}{\omega\tau_t}$$

$$= R_0 \frac{\sqrt{1 - 2\cos(\omega\tau_t)\mathcal{J}_0(\omega\tau_t) + \mathcal{J}_0^2(\omega\tau_t)}}{\omega\tau_t}. \quad (16.17)$$

Figure 16.4 shows the responsivity R_ω normalized by its dc value versus the modulation frequency ω calculated using Eq. (16.17) for different values of the transit time $\tau_t \propto l$. At $\langle v \rangle = v_W / 2 \simeq 5 \times 10^7$ cm/s, the range τ_t = 10–30 ps corresponds to the length of the i-section $2l$ = 10–30 μm. As seen from Fig. 16.4, $R_\omega / R_0 = 1/\sqrt{2}$ when the modulation cutoff frequency $f_t = \omega_t/2\pi \simeq$ 8.3–24.39 GHz. Naturally, in GLPDs with shorter i-sections, f_t can be markedly larger (although, at the expense of a substantial increase in the tunneling current). Due to this, GLPDs can be used as ultrafast terahertz and IR photodetectors (Fig. 16.4).

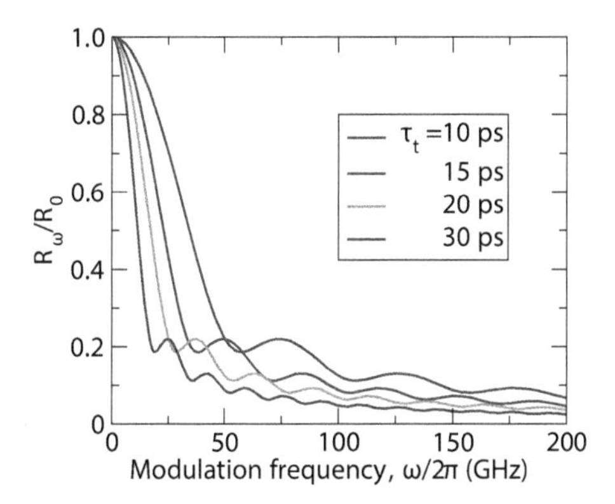

Figure 16.4 Dependence of responsivity (normalized) on modulation frequency $\omega/2\pi$ for different τ_t.

16.5 Comparison with QWIPs and QDIPs

Now we compare the responsivity and detectivity of GLPDs with K GLs calculated above and those of QWIPs (properly coupled with the incident terahertz or IR radiation) and QDIPs with the same number of QWs. The fraction of the absorbed photon flux in one QW $\beta_\Omega^{(QW)} \simeq \sigma_\Omega^{(QW)} \Sigma_0^{(QW)}$, where $\sigma^{(QW)}$ is the cross section of the photon absorption due to the intersubband transitions and Σ_d the donor sheet density. Setting the usual values $\sigma_\Omega^{(QW)} \simeq 2 \times 10^{-15}$ cm^2

and $\Sigma_0^{(QW)} \simeq 10^{12}$ cm^{-2}, one obtains $\beta_\Omega^{(QW)} \simeq 0.002$. This value is one order of magnitude smaller that β_Ω. Hence, one can neglect the attenuation of radiation in QWIPs with $K \lesssim 100$ (whereas in GLPDs it can be essential). In such a case, the responsivity of QWIPs is independent of the number of QWs K in the QWIP structure [33] and given by

$$R_0^{(QW)} = \frac{e\beta_\Omega^{(QW)}}{\hbar\Omega p_c}, \tag{16.18}$$

where p_c is the so-called capture probability which relates to the QWIP gain $G^{(QW)}$ as $G^{(QW)} = (1-p_c)/Kp_c \simeq 1/Kp_c$ [33]. Using Eqs. (16.4) and (16.18), we obtain

$$\frac{R_0}{R_0^{(QW)}} \simeq \frac{2\beta_\Omega p_c}{R_\Omega^{(QW)}} K^*. \tag{16.19}$$

Assuming that K = 50–100, ($K^* \simeq$ 30–39), $R_0^{(QW)}$ = 0.002, and p_c = 0.1, we obtain $R_0/R_0^{(QW)} \simeq 70-90$.

The ratio of the detectivities can be presented as

$$\frac{D^*}{D^{*(QW)}} \simeq \frac{2\beta_\Omega\sqrt{p_c}}{\beta_\Omega^{(QW)}} \frac{K^*}{K} \sqrt{\frac{g_{th}^{(QW)}}{g_{th}}} \tag{16.20}$$

As can be extracted from Ref. [26], $g_{th} \propto \exp(-\hbar\omega_0 / k_B T)$, where $\hbar\omega_0 \simeq 0.02$ eV is the energy of optical phonon in GLs, whereas $g_{th}^{(QW)} \propto \exp(-\varepsilon^{(QW)} / k_B T)$, where $\varepsilon^{(QW)}$ is the QW ionization energy ($\hbar\Omega \gtrsim \varepsilon^{(QW)}$), Eq. (16.17) yields

$$\frac{D^*}{D^{*(QW)}} \propto \frac{2\beta_\Omega\sqrt{p_c}}{\beta_\Omega^{(QW)}} \frac{K^*}{K} \exp\left[\frac{\hbar(\omega_0 - \Omega)}{2k_B T}\right]. \tag{16.21}$$

Different dependences of the responsivities and detectivities of GLDs and QWIPs on K are due different directions of the dark current and photocurrent: parallel to the GL plane in the former case and perpendicular to the QW plane in the latter case. The product of the factors in the right-hand side of Eq. (16.21) except the last one can be on the order of unity. This is because a large ratio $2\beta_\Omega / \beta_\Omega^{(QW)}$ can be compensated by relatively small capture parameter p_c. However, the exponential factor in Eq. (16.21) is large in the terahertz range: $\Omega \lesssim \omega_0$, i.e., at $\Omega/2\pi \lesssim 50$ THz. In particular, at T = 300 K and $\Omega/2\pi$ = 5–10 THz, the exponential factor in question is about 24–36.

The GLPD responsivity and detectivity can also markedly exceed those of QDIPs [for which formulas similar to Eqs. (16.21) can be used] despite lower capture probability and thermoexcitation rate in QDIPs in comparison with QWIPs (see, for instance, Refs. [34–36]). The ratios $D^*/D^{*(QW)}$ and $D^*/D^{*(QD)}$ dramatically increases with decreasing $\hbar\Omega$ and T. This is attributed to the fact that the spectral dependence of the GLPDs detectivity is similar to the spectral dependence of the responsivity ($D^* \propto \beta_\Omega/\hbar\Omega$), whereas the GLPDs intended for the photodetection in different spectral ranges exhibit the same dark current (due to the gapless energy spectrum). In contrast, in QWIPs, QDIPs, and some other photodetectors (see, for example, Refs. [9, 10]) the transition to lower photon frequency requires to utilize the structures with lower ionization energy and, hence, exponentially higher dark current. The latter leads to quite different spectral dependencies of the GLPD detectivity and the detectivity of QWIPs and QDIPs (which drops with creasing photon frequency).

GLPDs can surpass QWIPs and QDIPs in responsivity even at relatively high photon frequencies. For example, for a GLPD with K =25 at $\Omega/2\pi$ = 75 THz (wavelength λ about 4 μm), we arrive at $R_0 \simeq 3$ A/W. This value is three times larger than $R_0^{(QD)}$ obtained experimentally for a QDIP with 25 InAs QD layers [37]. As for the detectivity, comparing a GLPD with $K = 70$ operating at $T = 300$ K and $\Omega/2\pi \simeq 15$ THz ($\lambda \simeq 20$ μm) and a QDIP with 70 QD layers [38], we obtain $D^* \sim 2 \times 10^8$ cm Hz$^{1/2}$/W and $D^{*(QD)} \sim 10^7$ cm Hz$^{1/2}$/W, respectively.

GLPDs can surpass also photodetectors on narrow-gap and gapless bulk semiconductors like HgCdTe (BSPDs). Apart from advantages associated with potentially simpler fabrication, GLPDs might exhibit higher detectivity (compare the data above and those from Ref. [39]). This can be attributed to relatively low thermogeneration rate in GLPDs compared to BSPDs with very narrow or zeroth gap. The point is that thermogeneration rate in GLPDs at room temperatures is primarily due absorption of optical phonons [27] which have fairly large energy $\hbar\omega_0$ (the Auger processes in GLs are forbidden) [40], whereas this rate in BSPDs is essentially determined by the Auger processes which are strong [39].

16.6 Discussion and Conclusions

Analyzing Eqs. (164)–(16.7), we accepted that $f(\hbar\Omega) \ll 1$ and, therefore, disregarded the frequency dependence of the absorption coefficient β_Ω associated with the population of the low energy states by electrons and holes. For a rough estimate, the value of distribution function $f(0)$ can be presented as $f(0) \lesssim f_0(0)\Sigma/\Sigma_0$, where Σ is the electron and hole density in the i-section under the reverse bias. This density, in turn, can be estimated as $\Sigma = 2g_{th}\tau_t = g_{th}2l/\langle v \rangle$. Using the same parameters as above, for l =10 μm and T = 30 K, we obtain Σ/Σ_0 = 0.25, so that $f(0) \lesssim 0.125$ and $\beta_\Omega|_{\Omega\to 0} \gtrsim 0.75\beta$. It implies that the numerical data obtained above for the frequencies at the lower end of the terahertz range might be slightly overestimated. However, since the electron-hole system in the i-section can be pronouncedly heated by the electric field, the actual values of $f(\hbar\Omega/2)$ can be smaller than in the latter estimate and, hence, β_Ω can be rather close to β. The electron and hole heating in intrinsic GLs under the electric field was studied recently [28, 41]. As shown in Ref. [41] solving the kinetic equations, the electron and hole distribution functions in a GL in the range of low energies becomes $f(0)$ become small at fairly small electric fields. Indeed, at T =300 K, $f(0)$ varies from $f(0) \simeq 0.2$ to $f(0) \simeq 0.05$ when the electric field E increases from E = 3 V/cm to E =30 V/cm. For a GLPD with the length of the i-section $2l$ = 10–30 μm (as in the estimates in Sections 16.3 and 16.4), these values of the electric field correspond to the reverse bias voltage $V \simeq 3 \times 10^{-3}$–$9 \times 10^{-2}$ V. Hence, the electron and hole populations in the i-section should not lead to a marked deviation of the GL absorption β_Ω from β = const. In contrast with the cases considered in Refs. [28, 41]. The finiteness of the transit time of electrons and holes in the i-section of the GL-structures under consideration might affect the electron and hole heating resulting in minor modification of the absorption coefficient. Therefore, more careful calculation of β_Ω at relatively low Ω can, in principle, be useful.

In summary, we proposed and evaluated GLPDs based multiple-GL p-i-n structures. It was shown that GLPDs can exhibit high responsivity and detectivity in the terahertz and IR ranges at room temperatures. Due to relatively high quantum efficiency and low thermogeneration rate, the GLPD responsivity and detectivity can substantially exceed those of other photodetectors.

Acknowledgments

The authors are grateful to M. S. Shur, A. A. Dubinov, V. V. Popov, A. Satou, M. Suemitsu, and F. T. Vasko for fruitful discussions and comments. This work was supported by the Japan Science and Technology Agency, CREST, Japan.

Appendix: Effect of Vertical Screening in Multiple GL-Structures

As was pointed out above, the thickness of multiple GL-structures even with rather large number of GLs K is in reality small in comparison with the lateral sizes of the device, namely, the lengths of all the section and, hence, the gates. Owing to this, the distribution of the dc electric potential $\psi = 2\varphi_0/V_g$ normalized by $V_g/2$ (where $V_g = V_n = -V_p$) in the direction perpendicular to the GL plane (corresponding to the axis z) can be found from the one-dimensional Poisson equation:

$$\frac{d^2\psi}{dz^2} = \frac{8\pi e}{æV_g}\sum_{k=1}^{K}\Sigma^{(k)}\delta(z - kd + d). \qquad (16.A1)$$

Here $\Sigma_0^{(k)}$ is the electron (hole) density in the k-th GL in the n-section (p-section), d is the spacing between GLs, and $\delta(z)$ is the Dirac delta function. Considering that $\Sigma_0^{(k)} = (\varepsilon_F^k)^2/\pi\hbar^2 v_F^2 = (e^2V_g^2/4\pi\hbar^2 v_F^2)\psi^2\big|_{z=kd}$, where ε_F^k is the Fermi energy in the n-section (p-section) of the k-th GL, and replacing the summation in Eq. (16.A1) by integration (that is valid if K is not too small), we reduce Eq. (16.A1) to the following:

$$\frac{d^2\psi}{dz^2} = \frac{\psi^2}{L_s^2}, \qquad (16.A2)$$

with the characteristic screening length $L_s = \hbar v_F\sqrt{æd/2e^3V_g} \propto V_g^{-1/2}$. One can assume that $\psi\big|_{z=0} = 2 + W_g(d\psi/dz)\big|_{z=0}$ and $\psi\big|_{z=\infty} = 0$ [as well as $(d\psi/dz)\big|_{z=\infty} = 0$], where W_g is the thickness of the layer separating the multiple GL-structure and the gates. Solving Eq. (16.A2) with the latter boundary conditions, we arrive at

$$\psi = \frac{1}{(C + z/\sqrt{6}L_s)^2}, \tag{16.A3}$$

where C satisfies the following equation:

$$C^3 - C/2 = (W_g/\sqrt{6}L_s). \tag{16.A4}$$

Since in reality $W_g \gg \sqrt{6}L_s$, one obtains $C \simeq (W_g/\sqrt{6}L_s)^{1/3} \propto V_g^{1/6}$. Setting d = 0.35 nm, W_g = 10 nm, $æ$ = 4, and V_g = 2 V, one can obtain $L_s \simeq$ 0.31 nm, and $C \simeq$ 2.4.

Equation (16.A3) yields

$$\varepsilon_F^{(k)} \simeq \frac{eV_g}{2[C + (k-1)d/\sqrt{6}L_s]^2} = \varepsilon_F^T a^{(k)}. \tag{16.A5}$$

Here $\varepsilon_F^T = \varepsilon_F^{(1)} = eV_g/2C^2 \propto (V_g/W_g)^{2/3}$ is the Fermi energy of electrons in the topmost GL in the n-section (holes in the p-section) and

$$a^{(k)} = [1 + (k-1)\gamma]^{-2}, \tag{16.A6}$$

where $\gamma = d/\sqrt{6}L_sC \propto V_g^{1/3}$. At the above parameters (in particular, W_g = 10–50 nm and V_g = 2 V) $\gamma = d/\sqrt{6}L_sC \simeq 0.12-0.20$.

References

1. C. Berger, Z. Song, T. Li, X. Li, A. Y. Ogbazhi, R. Feng, Z. Dai, A. N. Marchenkov, E. H. Conrad, P. N. First, and W. A. de Heer, *J. Phys. Chem.*, **108**, 19912 (2004).
2. K. S. Novoselov, A. K. Geim, S. V. Morozov, D. Jiang, M. I. Katsnelson, I. V. Grigorieva, S. V. Dubonos, and A. A. Firsov, *Nature (London)*, **438**, 197 (2005).
3. A. H. Castro Neto, F. Guinea, N. M. R. Peres, K. S. Novoselov, and A. K. Geim, *Rev. Mod. Phys.*, **81**, 109 (2009).
4. F. T. Vasko and V. Ryzhii, *Phys. Rev. B*, **77**, 195433 (2008).
5. V. Ryzhii, M. Ryzhii, and T. Otsuji, *J. Appl. Phys.*, **101**, 083114 (2007).
6. F. Rana, *IEEE Trans. Nanotechnol.*, **7**, 91 (2008).
7. A. Satou, F. T. Vasko, and V. Ryzhii, *Phys. Rev. B*, **78**, 115431 (2008).

8. A. A. Dubinov, V. Ya. Aleshkin, M. Ryzhii, T. Otsuji, and V. Ryzhii, *Appl. Phys. Express,* **2**, 092301 (2009).
9. V. Ryzhii, V. Mitin, M. Ryzhii, N. Ryabova, and T. Otsuji, *Appl. Phys. Express,* **1**, 063002 (2008).
10. V. Ryzhii and M. Ryzhii, *Phys. Rev. B,* **79**, 245311 (2009).
11. F. Xia, T. Murller, Y.-M. Lin, A. Valdes-Garsia, and F. Avouris, *Nat. Nanotechnol.,* **4**, 839 (2009).
12. P. Neugebauer, M. Orlita, C. Faugeras, A.-L. Barra, and M. Potemski, *Phys. Rev. Lett.,* **103**, 136403 (2009).
13. V. Ryzhii, M. Ryzhii, A. Satou, T. Otsuji, A. A. Dubinov, and V. Ya. Aleshkin, *J. Appl. Phys.,* **106**, 084507 (2009).
14. Yu.-M. Lin, D. B. Farmer, G. S. Tulevski, S. Xu, R. G. Gordon, and P. Avouris, Device Research Conf., Tech. Dig., p. 27 (2008).
15. D. Wei, Y. Liu, Y. Wang, H. Zhang, L. Huang, and G. Yu, *Nano Lett.,* **9**, 1752 (2009).
16. J. Zhu and J. C. S. Woo, Ext. Abstracts of the 2009 Int. Conf. on Solid State Devices and Materials, Sendai, 2009 (unpublished), pp. G-9–G-2.
17. V. V. Cheianov and V. I. Fal'ko, *Phys. Rev. B,* **74**, 041403(R) (2006).
18. L. M. Zhang and M. M. Fogler, *Phys. Rev. Lett.,* **100**, 116804 (2008).
19. B. Huard, J. A. Sulpizio, N. Stander, K. Todd, B. Yang, and D. Goldhaber-Gordon, *Phys. Rev. Lett.,* **98**, 236803 (2007).
20. B. Ozyilmaz, P. Jarillo-Herrero, D. Efetov, D. Abanin, L. S. Levitov, and P. Kim, *Phys. Rev. Lett.,* **99**, 166804 (2007).
21. M. Ryzhii and V. Ryzhii, *Jpn. J. Appl. Phys., Part 2,* **46**, L151 (2007).
22. V. Ryzhii, M. Ryzhii, V. Mitin, and M. S. Shur, *Appl. Phys. Express,* **2**, 034503 (2009).
23. A. Bostwick, T. Ohta, T. Seyller, K. Horn, and E. Rotenberg, *Nat. Phys.,* **3**, 36 (2007).
24. L. A. Falkovsky and A. A. Varlamov, *Eur. Phys. J. B,* **56**, 281 (2007).
25. A. Rose, *Concepts in Photoconductivity and Allied Problems* (Wiley, New York, 1963).
26. *Long Wavelength Infrared Detectors,* edited by M. Razeghi (Gordon and Breach, Amsterdam, 1996).
27. F. Rana, P. A. George, J. H. Strait, S. Shivaraman, M. Chanrashekhar, and M. G. Spencer, *Phys. Rev. B,* **79**, 115447 (2009).
28. R. S. Shishir, D. K. Ferry, and S. M. Goodnick, *J. Phys.: Conf. Ser.,* **193**, 012118 (2009).

29. S. Ramo, *Proc. IRE,* **27**, 584 (1 3).
30. C. K. Jen, *Proc. IRE,* **29**, 345 (1941).
31. V. Ryzhii and G. Khrenov, *IEEE Trans. Electron Devices,* **42**, 166 (1995).
32. N. Tsutsui, V. Ryzhii, I. Khmyrova, P. O. Vaccaro, H. Taniyama, and T. Aida, *IEEE J. Quantum Electron.,* **37**, 830 (2001).
33. H. Schneider and H. C. Liu, *Quantum Well Infrared Photodetectors* (Springer, Berlin, 2007).
34. V. Ryzhii, *Semicond. Sci. Technol.,* **11**, 759(1996).
35. V. Ryzhii, I. Khmyrova, M. Ryzhii, and V. Mitin, *Semicond. Sci. Technol.,* **19**, 8 (2004).
36. A. Rogalski, J. Antoszewski, and L. Faraone, *J. Appl. Phys.,* **105**, 091101 (2009).
37. H. Lim, S. Tsao, W. Zhang, and M. Razeghi, *Appl. Phys. Lett.,* **90**, 131112 (2007).
38. S. Chakrabarti, A. D. Siff-Roberts, X. H. Su, P. Bhattacharya, G. Ariyawansa, and A. G. U. Perera, *J. Phys. D,* **38**, 2135 (2005).
39. P. Martyniuk, S. Krishna, and A. Rogalski, *J. Appl. Phys.,* **104**, 034314 (2008).
40. M. S. Foster and I. L. Aleiner, *Phys. Rev. B,* **79**, 085415 (2009).
41. O. G. Balev, F. T. Vasko, and V. Ryzhii, *Phys. Rev. B,* **79**, 165432 (2009).

Chapter 17

Negative and Positive Terahertz and Infrared Photoconductivity in Uncooled Graphene*

V. Ryzhii,[a,b,c,d] **D. S. Ponomarev,**[b,c] **M. Ryzhii,**[e] **V. Mitin,**[a,f] **M. S. Shur,**[g,h] **and T. Otsuji**[a]

[a]*Research Institute of Electrical Communication, Tohoku University, Sendai 980-8577, Japan*
[b]*Institute of Ultra High Frequency Semiconductor Electronics of RAS, Moscow 117105, Russia*
[c]*Center of Photonics and Two-Dimensional Materials, Moscow Institute of Physics and Technology, Dolgoprudny 141700, Russia*
[d]*Center for Photonics and Infrared Engineering, Bauman Moscow State Technical University, Moscow 111005, Russia*
[e]*Department of Computer Science and Engineering, University of Aizu, Aizu-Wakamatsu 965-8580, Japan*
[f]*Department of Electrical Engineering, University at Buffalo, Buffalo, NY 1460-192, USA*
[g]*Department of Electrical, Computer, and Systems Engineering and Department of Physics, Applied Physics, and Astronomy, Rensselaer Polytechnic Institute, Troy, NY 12180, USA*
[h]*Electronics of the Future, Inc., Vienna, VA22181, USA*
v-ryzhii@ riec.tohoku.ac.jp

*Reprinted with permission from V. Ryzhii, D. S. Ponomarev, M. Ryzhii, V. Mitin, M. S. Shur, and T. Otsuji (2019). Negative and positive terahertz and infrared photoconductivity in uncooled graphene, *Opt. Mater. Express*, **9**(2), 585–597. Copyright © 2019 Optical Society of America.

Graphene-Based Terahertz Electronics and Plasmonics: Detector and Emitter Concepts
Edited by Vladimir Mitin, Taiichi Otsuji, and Victor Ryzhii

ISBN 978-981-4800-75-4 (Hardcover), 978-0-429-32839-8 (eBook)
www.jennystanford.com

We develop the model for the terahertz (THz) and infrared (IR) photoconductivity of graphene layers (GLs) at room temperature. The model accounts for the linear GL energy spectrum and the features of the energy relaxation and generation-recombination mechanisms inherent at room temperature, namely, the optical phonon absorption and emission and the Auger interband processes. Using the developed model, we calculate the spectral dependences of the THz and IR photoconductivity of the GLs. We show that the GL photoconductivity can change sign depending on the photon frequency, the GL doping and the dominant mechanism of the carrier momentum relaxation. We also evaluate the responsivity of the THz and IR photodetectors using the GL photoconductivity. The obtained results along with the relevant experimental data might reveal the microscopic processes in GLs, and the developed model could be used for the optimization of the GL-based photodetectors.

17.1 Introduction

The gapless energy carrier spectrum and high carrier mobility in graphene layers (GLs) [1] enable unique optical and transport properties. The GL conductivity dependence on the carrier effective temperature has been analyzed and reported in a number of publications (for example, [2–9]). Due to the specifics of the carrier scattering in the GLs, the variation of the GL DC conductivity under the terahertz (THz) and infrared (IR) irradiation (THz and IR photoconductivity) can exhibit interesting behavior [6, 10]. The response of the carrier system in the GLs caused by the THz and IR radiation enables their utilization in the bolometric photodetectors [11–15]. Extensive studies of the carrier dynamics in the GLs in the response to ultrashort optical pulses (for example, [16–24]) stimulate the creation of novel ultrafast optoelectronic devices. The GL ac (dynamic) conductivity under the optical pumping can become negative in the THz range due to the population inversion [25–28] (see also [16, 19]).

The THz and IR photoconductivity in GLs is associated with the interplay of the variations of the carrier effective temperature and density (the quasi-Fermi energy). Depending on the dominating scattering mechanisms and the photon energies, the effective temperature can either exceed or be smaller than the lattice temperature. This can be accompanied by a splitting of the quasi-Fermi levels [25, 29]. The interplay of the intra- and interband absorption of the incident THz and IR radiation [30] adds complexity to the processes governing the GL conductivity.

The GL photoconductivity was experimentally studied in a number of works demonstrating the effect of negative photoconductivity [10, 31, 32]. However, it was also shown that the sign of the THz photoconductivity of the GLs changes with the radiation frequency depending on the environment [32].

Despite a variety of the works aimed to describe and explain the features of the THz and IR photoconductivity, there is still no transparent model of the phenomena. The previous studies mainly were focused on the deeply cooled GLs and related bolometric photodetectors. Due to this, the main energy relaxation and generation-recombination mechanisms were associated with the carrier interaction with acoustic phonons and ambient thermal radiation. However, in the uncooled GLs, the carrier intraband and interband transitions due to the recombination and emission of the optical phonons [33] (see also [10]) and the carrier-carrier interaction (Auger processes) [34, 35] can dominate. The inclusion of these mechanisms is necessary for the explanation of the nontrivial spectral characteristics of the GL photoconductivity. Namely, the observed inversion of the photoconductivity sign requires an adequate but transparent model. In this chapter, we develop such a model, which accounts not only for the variations of the carrier temperature but also for the possible splitting of the quasi-Fermi levels associated with the THz and IR radiation absorption. This approach allows us to consider the effect of the THz and IR photoconductivity at room temperatures in more details. We also evaluate the application of this effect for uncooled THz and IR photodetectors. The obtained results can be used for the optimization of these photodetectors and promote a deeper insight into the GL microscopic properties.

We focus on both the intrinsic (or compensated) and p-doped GLs at relatively elevated temperatures (such as room temperature). We assume that the carrier momentum relaxation is associated with impurity scattering, scattering on defects, and and on the acoustic phonons. The carrier scattering on the clustered impurities inside the GLs and the substrate inclusions in the decorated GLs is accounted for as well. The intraband and interband transitions of the carriers due to the emission and absorption of the optical phonons are considered as the crucial mechanism of the energy relaxation and the generation-recombination. Despite the specifics of the Auger generation-recombination in the GLs associated with the carrier dispersion law [34] and the carrier collinear scattering (see [35] and references therein), these generation-recombination mechanisms are also included into the model under consideration.

17.2 GL Conductivity Dependence on the Carrier Temperature

Due to relatively high carrier densities in GLs at the room temperature, their pair collisions lead to the establishment of the quasi-Fermi energy distributions, $f^{\mathrm{e}}(\varepsilon)$ and $f^{\mathrm{h}}(\varepsilon)$, with the same effective temperature T. Therefore, we use the following formula for the real part of the intraband conductivity of the GL- channel σ_Ω at the frequency Ω (see [2–4]):

$$\sigma_\Omega = -\frac{e^2 T_0 \tau_0}{\pi\hbar^2}\left(\frac{T}{T_0}\right)^{l+1}\int_0^\infty \frac{d\xi\xi^{l+1}}{(1+\Omega^2\tau_0^2(T/T_0)^{2l}\xi^{2l})}\frac{\mathrm{d}}{d\xi}[f^{\mathrm{e}}(\xi)+f^{\mathrm{h}}(\xi)]. \tag{17.1}$$

Here it is assumed that the momentum relaxation time as a function of the momentum p is equal to $\tau_{\mathrm{p}} = \tau_0(pv_W/T_0)^l = \tau_0(T/T_0)^l\xi^l$, where τ_0 is its characteristic value, $\xi = pv_W/T_0$ is the normalized carrier energy in the GL, $v_W \simeq 10^8$ cm/s is the characteristic velocity in GLs, and $\hbar$ is the Planck constant.

If the scattering on the weakly screened charged objects (weakly screened Coulomb, long-scale scattering) prevails, $\tau_{\mathrm{p}} \propto p$. This case corresponds to the superscript l in Eq. (17.1) equal to unity ($l = 1$). In the case of scattering on the neutral impurities and defects, the strongly screened charged impurities, and the acoustic phonons, $\tau \propto p^{-1}$ (i.e., $l = -1$). If several mechanisms contribute to the carrier scattering, τ_{p} can be described by a combination the linear and the inverse dependences. For a simplified description, in some papers $\tau_{\mathrm{p}} = const$ is used.

The scattering processes can depend on the presence of large-scale charged defects, clustered impurities [36, 37], and various particles in the GL or near the GL interface with the substrate. The latter can be important in the so-called decorated GLs [38, 39].

When the carrier momentum relaxation is determined by the short–range scattering, Eq. (17.1) yields for the pertinent conductivity σ_Ω is given by

$$\sigma_\Omega = \frac{\sigma_{00}}{(1+\Omega^2\tau^2)}\left[\frac{1}{\exp(-\mu^{\rm e}/T)+1}+\frac{1}{\exp(-\mu^{\rm h}/T)+1}\right]. \quad (17.2)$$

Here $\sigma_{00} = (e^2 T_0 \tau_0/\pi\hbar^2)$ is the characteristic intraband conductivity (it is equal to the low electric-field conductivity in the case of short-range scatterers). In the intrinsic (or compensated) GLs, the electron-hole symmetry leads to the equality of the pertinent quasi-Fermi energies $\mu^{\rm e} = \mu^{\rm h} = \mu$ (this does not generally imply that the positions of the quasi-Fermi levels $\varepsilon_{\rm F}^{\rm e} = \mu^{\rm e}/T$ and $\varepsilon_{\rm F}^{\rm h} = -\mu^{\rm h}/T$ coincide; such a coincidence only occurs if $\mu = 0$). In this case, Eq. (17.2) transforms to the following:

$$\sigma_\Omega = \frac{2\sigma_{00}}{(1+\Omega^2\tau^2)}\frac{1}{[\exp(-\mu/T)+1]}. \quad (17.3)$$

In the case of the short–range scattering, one can find for the characteristic time τ from Eq. (17.1) $\tau \simeq \tau_0\sqrt{3/\pi}$. If the long-scale momentum relaxation is the dominant (l = 1), we obtain from Eq. (17.1)

$$\sigma_\Omega = \frac{4\sigma_{00}}{(1+\Omega^2\tau^2)}\left(\frac{T}{T_0}\right)^2 \mathcal{F}_1\left(\frac{\mu}{T}\right), \quad (17.4)$$

where

$$\mathcal{F}_1(\eta) = \int_0^\infty \frac{d\xi\xi}{\exp(\xi-\eta)+1}$$

is the Fermi-Dirac integral. Equations (17.3) and (17.4) show that the variations of both the quasi-Fermi energy μ and the effective temperature T caused by the radiation absorption determine the variation of the conductivity, i.e., the effect of photoconductivity. As demonstrated in the following, the quasi-Fermi energy changes not only due to the carrier density change but due to the heating.

17.3 Generation-Recombination and Energy Balance Equations

The equation governing the interband balance of the carriers can be generally presented as

$$G_{\text{Auger}} + G_{\text{Opt}} + G_{\text{Ac}} + G_{\text{Rad}} = 0. \tag{17.5}$$

The terms in Eq. (17.5) correspond to the interband Auger generation-recombination processes and the processes associated with optical-phonon, acoustic-phonon, and radiative transitions (in particular, indirect transitions, for which some selection restrictions are lifted). In the situations under consideration (in particular in the temperature range in question), the characteristic times of the processes associated with acoustic phonons and radiative transitions (as was mentioned in Section 17.1) are much longer than those related to the optical-phonon and Auger mechanisms [6, 9, 33, 35]. Due to this, we disregard the terms G_{Ac} and G_{Rad} in Eq. (17.5).

Due to the fast processes of the optical phonon decay into acoustic phonons and their effective removal, confirmed by high values of the G-layer thermal conductivity [40–42], the optical phonon system in the GLs is assumed to be in equilibrium with the lattice having the temperature T_0.

For the rate, G_{Opt}, of the interband and intraband transitions assisted by the optical phonons emission and absorption one can use rather general equations accounting for the transitions matrix elements, the features of carrier energy spectrum in the GLs, and the conservation of the carrier momentum and energy [33]. As shown [29] (see also [11, 43]), at $\hbar\omega_0 \gg \mu, T, T_0$, where $\hbar\omega_0 \simeq 200$ meV is the energy of optical phonons in the GL, the general equations for G_{Opt} [33] can be simplified by separating of the exponential factors and relatively slow varying factors depending on μ and T_0. Similar procedure leads to a simplified formula for G_{Auger}. As the result, in the situation under consideration, we present Eq. (17.5) describing the carrier balance in the conduction and valence bands of the GL in the following form:

$$\frac{1}{\tau_{\text{Opt}}}\left\{\exp\left[\frac{2\mu}{T}+\hbar\omega_0\left(\frac{1}{T_0}-\frac{1}{T}\right)\right]-1\right\}+\frac{1}{\tau_{\text{Auger}}}\left[\exp\left(\frac{2\mu}{T}\right)-1\right]=\beta^{\text{inter}}\frac{I_{\Omega}}{\Sigma_0} \tag{17.6}$$

The equation governing the energy electron and hole balance is given by:

$$\frac{1}{\tau_{\mathrm{Opt}}}\left\{\exp\left[\frac{2\mu}{T}+\hbar\omega_0\left(\frac{1}{T_0}-\frac{1}{T}\right)\right]-1\right\}+\frac{1}{\tau_{\mathrm{Opt}}^{\mathrm{intra}}}\left\{\exp\left[\hbar\omega_0\left(\frac{1}{T_0}-\frac{1}{T}\right)\right]-1\right\}$$
$$=\left(\beta^{\mathrm{intra}}+\beta^{\mathrm{inter}}\right)\frac{\hbar\Omega I_{\Omega}}{\hbar\omega_0\Sigma_0}. \tag{17.7}$$

Here τ_{Opt} and $\tau_{\mathrm{Opt}}^{\mathrm{intra}}$ are the characteristic recombination and intraband relaxation times associated with the carrier interaction with the optical phonons, τ_{Auger} is the Auger recombination time, $\beta^{\mathrm{inter}} = (\pi\alpha/\sqrt{\kappa})\ \tanh(\hbar\Omega/4T)$ [2] (for a small μ) and $\beta^{\mathrm{intra}} = (4\pi\sigma_{\Omega}/\sqrt{\kappa}c)$ are the GL interband and intraband absorption coefficients (determined by the interband and intraband ac conductivities), respectively, Σ_0 is the characteristic carrier density, $\alpha = (e^2/c\hbar) \simeq 1/137$ is the fine structure constant, c is the speed of light in vacuum, and κ is the background dielectric constant. The factors $\Sigma_0/\tau_{\mathrm{Opt}}, \Sigma_0/\tau_{\mathrm{Opt}}^{\mathrm{inter}}$, and $\Sigma_0/\tau_{\mathrm{Auger}}$ are weak functions of μ and T (mentioned above, see, for example, [29]). These factor can be considered as phenomenological parameters of the model.

The exponential terms in the left-hand sides of Eqs. (17.6) and (17.7) are proportional to the rate of the generation-recombination and the rate of the energy transfer into the lattice, respectively. The terms in the right-hand sides of these equations correspond to the carrier photogeneration and to the power obtained by the electron-hole system due to the radiation absorption. The latter is proportional to $(\beta^{\mathrm{intra}} + \beta^{\mathrm{inter}})\hbar\Omega I_{\Omega}$.

Limiting our consideration by low radiation intensities I_{Ω}, Eqs. (17.6) and (17.7) are presented as

$$\frac{1}{\tau_{\mathrm{Opt}}}\frac{2\mu}{T}+\left(\frac{1}{\tau_{\mathrm{Opt}}}+\frac{1}{\tau_{\mathrm{Opt}}^{\mathrm{intra}}}\right)\frac{\hbar\omega_0(T-T_0)}{T_0^2}\simeq(\beta_0^{\mathrm{intra}}+\beta_0^{\mathrm{inter}})\frac{\Omega I_{\Omega}}{\omega_0\Sigma_0}. \tag{17.8}$$

$$\left(\frac{1}{\tau_{\mathrm{Opt}}}+\frac{1}{\tau_{\mathrm{Auger}}}\right)\frac{2\mu}{T}+\frac{1}{\tau_{\mathrm{Opt}}}\frac{\hbar\omega_0(T-T_0)}{T_0^2}=\beta_0^{\mathrm{inter}}\frac{I_{\Omega}}{\Sigma_0}, \tag{17.9}$$

where, due to a smallness of I_Ω, we put $\Sigma_G = \Sigma_0 = \pi T_0^2 / 3\hbar^2 v_W^2$ and $\beta^{inter} \simeq \beta_0^{inter} = (\pi\alpha / \sqrt{\kappa}) \tanh(\hbar\Omega/4T_0)$ and $\beta^{intra} \simeq \beta_0^{intra} = \pi\alpha D_0/\sqrt{\kappa}(1+3\Omega^2\tau_0^2/\pi^2)$. Here $D_0 = (4T_0\tau_0/\pi\hbar)$ is the Drude factor. At T_0 = 25 meV and $\tau_0 = (10^{-12} - 10^{-13})$ s, one obtains $D_0 \simeq 5.3 - 52.7$.

17.4 GL Photoconductivity

Using Eqs. (17.8) and (17.9), we express the normalized carrier quasi-Fermi energy μ/T and temperature variation $(T - T_0)/T_0$ via the radiation intensity I_Ω, its frequency Ω, and the characteristic times $\tau_{Opt}, \tau_{Opt}^{intra}$, and τ_{Auger} as

$$\frac{\mu}{T} \simeq -\left[(\beta_0^{inter} + \beta_0^{intra})\left(\frac{\Omega}{\omega_0}\right) - \beta_0^{inter}(1+a)\right]\frac{\tau_R I_\Omega}{2\Sigma_0}. \tag{17.10}$$

$$\frac{T - T_0}{T_0} \simeq \left[(\beta_0^{inter} + \beta_0^{intra})\left(\frac{\Omega}{\omega_0}\right)(1+b) - \beta_0^{inter}\right]\frac{T_0}{\hbar\omega_0}\frac{\tau_R I_\Omega}{\Sigma_0}. \tag{17.11}$$

Here

$$\frac{1}{\tau_R} = \frac{(1+a)}{\tau_{Auger}} + \frac{a}{\tau_{Opt}},$$

is the characteristic relaxation-recombination time, which accounts for the interband transitions associated with the optical phonons and the Auger generation recombination processes, as well as the intraband energy relaxation on the optical phonons, $a = \tau_{Opt}/\tau_{Opt}^{intra}$ and $b = \tau_{Opt}/\tau_{Auger}$. The parameters a and b can be estimated using the data found previously [29, 33] and [33, 35], respectively. Equations (17.10) and (17.11) can be rewritten as

$$\frac{T - T_0}{T_0} \simeq \frac{1}{(a+b+ab)}\frac{T_0}{\hbar\omega_0}\left[\tanh\left(\frac{\hbar\Omega}{4T_0}\right)\left(\frac{\Omega}{\omega_0}(1+b) - 1\right)\right.$$
$$\left. + \frac{D_0(1+b)}{(1+3\Omega^2\tau_0^2/\pi^2)}\left(\frac{\Omega}{\omega_0}\right)\right]\frac{I_\Omega}{\overline{I}_\Omega}, \tag{17.12}$$

$$\frac{\mu}{T} \simeq -\frac{1}{2(a+b+ab)}\left[\tanh\left(\frac{\hbar\Omega}{4T_0}\right)\left(\frac{\Omega}{\omega_0} - 1 - a\right) + \frac{D_0}{(1+3\Omega^2\tau_0^2/\pi^2)}\left(\frac{\Omega}{\omega_0}\right)\right]\frac{I_\Omega}{\overline{I}_\Omega}, \tag{17.13}$$

where

$$\overline{I}_{\Omega} = \frac{\sqrt{\kappa}\,\Sigma_0}{\pi\alpha\tau_{\text{Opt}}}.$$

The first term in the square brackets of Eq. (17.12) is negative when $\Omega < \omega_0/(1 + b)$. In this case, the interband transitions promote a decrease in the carrier temperature. The point is that every act of the photon absorption brings the energy $\hbar\Omega$ to the carrier system, while the interband emission of optical phonon decreases the carrier system energy by the value $\hbar\omega_0$, which is a smaller value. If $b \ll 1$, these processes lead to the carrier heating. The Auger recombination (in the gapless GLs) does not directly change the carrier energy, but decreases the cooling role of the optical photon emission.

In the limit of very strong Auger processes ($\tau_{\text{Auger}} \ll \tau_{\text{Opt}}$, i.e., $b \gg 1$), Eq. (17.12) yields

$$\frac{T - T_0}{T_0} \simeq \frac{1}{(1+a)}\frac{T_0}{\hbar\omega_0}\left(\frac{\Omega}{\omega_0}\right)\left[\tanh\left(\frac{\hbar\Omega}{4T_0}\right) + \frac{D_0}{(1+3\Omega^2\tau_0^2/\pi^2)}\right]\frac{I_\Omega}{\overline{I}_\Omega} \geq 0. \tag{17.14}$$

In the latter case, the radiative interband processes supply the energy, while the interband optical phonon emission does not work (only the intraband optical phonon processes contribute to the carrier energy relaxation). In this limit, we obtain from Eq. (17.13) $\mu = 0$. The latter implies that the electron and hole quasi-Fermi levels coincide ($\varepsilon_F^e = \varepsilon_F^h = 0$).

Figure 17.1 shows the carrier temperature variation $(T - T_0)/T_0$ and the quasi-Fermi energy μ/T normalized by $I_\Omega/\overline{I}_\Omega$ (i.e., the quantities $[(T - T_0)/T_0](\overline{I}\Omega/I\Omega)$ and $(\mu/T)(\overline{I}_\Omega/I_\Omega))$, versus the photon energy $\hbar\Omega$ in the GLs with dominating short-range scattering calculated using Eqs. (17.12) and (17.13) for different b (i.e., different τ_{Auger}) and different τ_0. We set $\hbar\omega_0 = 200$ meV, $T_0 = 25$ meV, $a \simeq \pi^2(T_0/\hbar\omega_0)^2(1 + 2.19T_0/\hbar\omega_0) \simeq 0.2$ [24].

As seen from Fig. 17.1 that the photon energies where μ and $(T - T_0)$ change their signs do not coincide.

Using Eqs. (17.3) and (17.13) when $|\mu| \ll 2T$, for the difference $\Delta\sigma_\Omega = \sigma_{\Omega\to 0} - \sigma_{00}$ between the dc conductivity under irradiation and the dc dark conductivity we obtain

$$\frac{\Delta\sigma_\Omega}{\sigma_{00}} \simeq \frac{\mu}{2T}, \tag{17.15}$$

so that

$$\frac{\Delta\sigma_\Omega}{\sigma_{00}} \simeq -\frac{1}{4(a+b+ab)}\left[\tanh\left(\frac{\hbar\Omega}{4T_0}\right)\left(\frac{\Omega}{\omega_0}-1-a\right)+\frac{D_0}{(1+3\Omega^2\tau_0^2/\pi^2)}\left(\frac{\Omega}{\omega_0}\right)\right]\frac{I_\Omega}{\overline{I}_\Omega}. \tag{17.16}$$

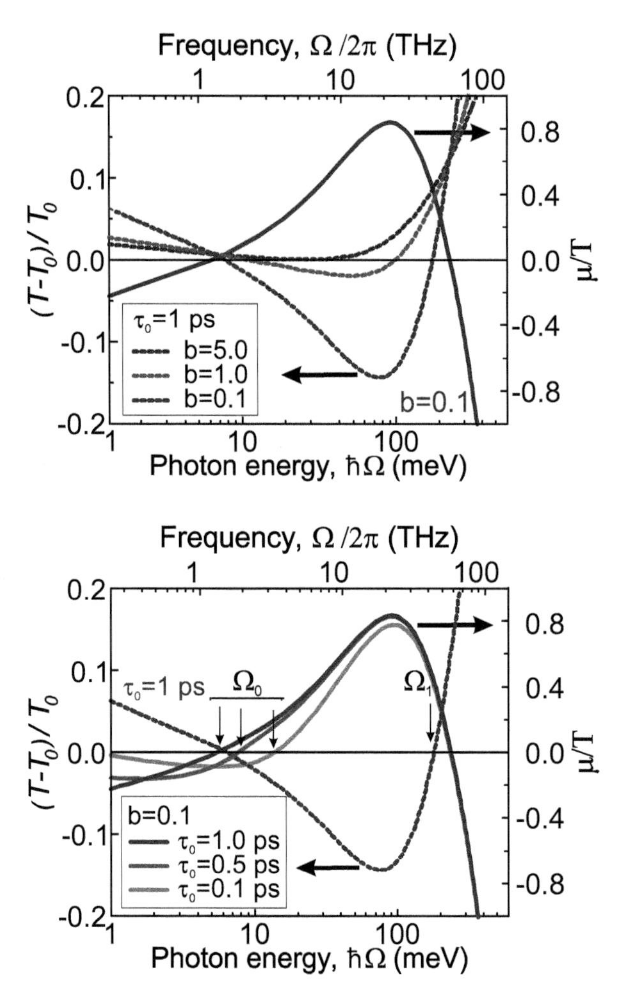

Figure 17.1 Upper panel: the normalized carrier temperature variation $(T - T_0)/T_0$ (dashed lines) for different values of the parameter b (upper panel) and τ_0 = 1 ps and the quasi-Fermi energy μ/T (solid line) for b = 0.1 and τ_0 = 1 ps. Lower panel: the normalized carrier temperature variation $(T - T_0)/T_0$ (dashed line) for b = 0.1 and τ_0 = 1 ps and the quasi-Fermi energy μ/T (solid lines) for b = 1 and different τ_0.

Figure 17.2 shows the GL photoconductivity $(\Delta\sigma_\Omega/\sigma_{00})$ normalized by a factor $(I_\Omega/\overline{I}_\Omega))$ calculated using Eq. (17.16) for different parameters. The GL photoconductivity as a function of the photon energy changes its sign twice at $\hbar\Omega = \hbar\Omega_0$ and $\hbar\Omega = \hbar\Omega_1$ which correspond to the points, where $\mu/T = \varepsilon^e = \varepsilon^h = 0$. One can see that the absolute value of the GL photoconductivity steeply decreases with increasing b, i.e., with the decreasing Auger recombination time τ_{Auger} (and, hence, decreasing τ_R). If τ_{Auger} tends to zero, the quantities μ/T and $\Delta\sigma_0$ also tend to zero, despite $(T - T_0) \neq 0$, as follows from Eqs. (17.13) and (17.16). All this is because at very intensive Auger processes the splitting of the quasi-Fermi levels vanishes, and the factor in the square brackets in the right-side of Eq. (17.2) is equal to unity. Hence, the conductivity becomes independent of the carrier temperature.

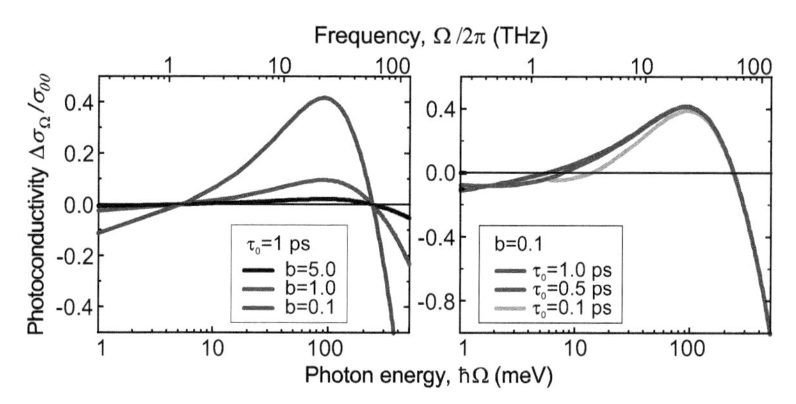

Figure 17.2 The normalized GL photoconductivity $\Delta\sigma_\Omega/\sigma_{00}$ as functions of the photon energy $\hbar\Omega$ (dominant short-range scattering) for different $b = \tau_{Opt}/\tau_{Auger}$ and τ_0 = 1 ps (left panel) and different τ_0 and b = 0.1 (right panel).

As seen from Fig. 17.2, the GL photoconductivity is negative $\Delta\sigma_\Omega < 0$ in the photon energy ranges $0 < \hbar\Omega < \hbar\Omega_0$ and $\hbar\Omega_1 > \hbar\Omega$. In the intermediate range $\hbar\Omega_0 < \hbar\Omega < \hbar\Omega_1$, the GL photoconductivity is positive.

Taking into account that in reality $D_0 > 4T_0(1 + a)/ \hbar\omega_0$, i.e., $\omega_0\tau_0 > \pi(1 + a)$, from Eq. (17.16) we obtain

$$\Omega_0 \simeq \frac{\pi}{\sqrt{3}\tau_0}\sqrt{\frac{D_0}{(1+a)}\frac{4T_0}{\omega_0} - 1} \simeq \sqrt{\frac{16\pi}{3(1+a)\omega_0\tau_0}\frac{T_0}{\hbar}} \propto \tau_0^{-1/2} \quad (17.17)$$

and

$$\Omega_1 \simeq (1+a)\omega_0\left[1-\frac{4\pi^2}{3(1+a)^2\omega_0\tau_0}\frac{T_0}{\hbar\omega_0}\right] \simeq (1+a)\omega_0. \quad (17.18)$$

For $\tau_0 = (10^{-12} - 10^{-13})$ s^{-1} and $a \simeq 0.2$, from Eqs. (17.17) and (17.18) one obtains $\Omega_0/2\pi \simeq (1.35 - 3.16)$ THz and $\Omega_1/2\pi \simeq (59.29 - 60.92)$ THz. Thus, the frequencies, at which the GL photoconductivity changes sign, correspond to the THz range $(\Omega_0/2\pi)$ and to the Mid-IR range $(\Omega_1/2\pi)$.

It should be noted that if the momentum relaxation time τ_p is independent of the momentum p (as was assumed in some GL conductivity models [10, 44, 45]), the conductivty variation $\Delta\sigma_0/\sigma_{00} \simeq \mu/4T$ [46] (compare with Eq. (17.15). Therefore, the results for the case of $\tau_p = const$ are qualitatively close to those obtained in this section.

17.5 Clustered Impurities and Decorated GLs: Long-Range Scattering

In this case, using Eq. (17.4), at relatively low intensities of the incident radiation I_Ω when the deviation of μ from zero is small, we obtain

$$\sigma_\Omega = \frac{\sigma_{00}}{(1+\Omega^2\tau_0^2)}\left(\frac{T}{T_0}\right)^2\left[\frac{\pi^2}{3}+\frac{4\mu}{T}\ln 2+\left(\frac{\mu}{T}\right)^2\right] \simeq \frac{\pi^2}{3}\frac{\sigma_{00}}{(1+\Omega^2\tau_0^2)}\left(\frac{T}{T_0}\right)^2. \quad (17.19)$$

Hence,

$$\frac{\Delta\sigma_\Omega}{\sigma_{00}} \simeq \frac{2\pi^2}{3}\frac{(T-T_0)}{T_0}. \quad (17.20)$$

The latter is quite different from Eq. (17.15), which is valid in the case of short-range momentum relaxation. Equation (17.19) with an equation similar to Eq. (17.11) results in (compare with Eq. (17.14))

$$\frac{\Delta\sigma_\Omega}{\sigma_{00}} \simeq \frac{2\pi^2}{3(a+b+ab)}\frac{T_0}{\hbar\omega_0}\left[\tanh\left(\frac{\hbar\Omega}{4T_0}\right)\left(\frac{\Omega}{\omega_0}(1+b-1)\right)+\frac{D_0(1+b)}{(1+\Omega^2\tau_0^2)}\left(\frac{\Omega}{\omega_0}\right)\right]\frac{I_\Omega}{\overline{I}_\Omega}. \quad (17.21)$$

Figure 17.3 shows the normalized responsivity of the photodetectors based on the GLs with the dominant long-range scattering calculated using Eq. (17.21).

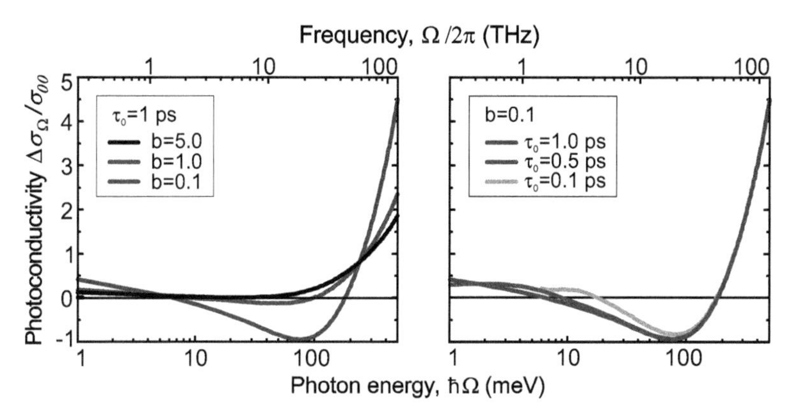

Figure 17.3 The same as in Fig. 17.2 but in the case of dominant long-range scattering: for $b = \tau_{Opt}/\tau_{Auger}$ and τ_0 = 1 ps (left panel) and for τ_0 and b = 0.1 (right panel).

As follows from the comparison of Eqs. (17.16) and (17.21) (see also Figs. 17.2 and 17.3), the photoconductivities involving the short- and long-range scattering mechanisms exhibit fairly different spectral dependences: the photoconductivity at long-range scattering is positive at low and high photon energy and negative in the range of the intermediate photon energies. This is because in the case of the dominant short-range scattering, the variation of the dc conductivity stimulated by the interband and intraband radiative transitions is associated primarily with the variation of μ (more exactly with the variation of μ/T), i.e., with the shifts of the electron and hole quasi-Fermi levels. The latter is in contrast with the case of the dominant long-scale scattering in which the variation of the dc conductivity is due to the variation of the carrier effective temperature. This can be seen from the comparison of Eqs. (17.3) and (17.19) describing a decrease in the conductivity with increasing $-\mu/T$ and an increase with increasing T, respectively.

17.6 Responsivity of the GL-Based Photodetectors

Using the obtained values of the photoconductivity, $\Delta\sigma_{\Omega}$, of the GLs one can find the responsivity $R_{\Omega} = |\Delta\sigma_{\Omega}|E/\hbar\Omega I_{\Omega}$ of the photodetectors based on undoped GL. Using Eqs. (17.16) and (17.21), for the cases of the dominant short- and long-scale scattering we arrive at

$$\frac{R_{\Omega}}{R} \simeq \frac{D_0}{(a+b+ab)}\left(\frac{T_0}{\hbar\Omega}\right)\left|\tanh\left(\frac{\hbar\Omega}{4T_0}\right)\left(\frac{\Omega}{\omega_0}-1-a\right)+\frac{D_0}{(1+3\Omega^2\tau_0^2/\pi^2)}\left(\frac{\Omega}{\omega_0}\right)\right|, \tag{17.22}$$

$$\frac{R_{\Omega}}{R} \simeq \frac{2\pi^2 D_0}{3(a+b+ab)}\left(\frac{T_0^2}{\hbar^2\omega_0\Omega}\right)\left|\tanh\left(\frac{\hbar\Omega}{4T_0}\right)\left(\frac{\Omega}{\omega_0}(1+b)-1\right)+\frac{D_0(1+b)}{(1+3\Omega^2\tau_0^2/\pi^2)}\left(\frac{\Omega}{\omega_0}\right)\right|, \tag{17.23}$$

respectively, where

$$R = \frac{3\alpha}{16\sqrt{\kappa}}\frac{e^2\hbar\tau_{\mathrm{Opt}}v_{\mathrm{W}}^2 E}{T_0^3 L} \tag{17.24}$$

Here E is the electric field along the GL plane and L is the spacing between the contacts. For example, setting $\tau_0 = 10^{-12}$ s, $\tau_{\mathrm{Opt}} = 10^{-12}$ s, $\kappa = 4$, $L = 10^{-4}$ cm, and $E = 100$ V/cm, from Eq. (17.24) we obtain $R \simeq 2.47 \times 10^{-4}$ A/W. One needs to note that the expression for R_{Ω} includes a large parameter D_0. In particular, for the responsivity of the photodetector using high quality GLs ($\tau_0 = 10^{-11}$ s) at $\hbar\Omega = 10$ meV (i.e., $\Omega/2\pi \simeq 2.5$ THz) Eq. (17.21) yields $R_{\Omega} \simeq 1.8$ A/W.

Figures 17.4 and 17.5 show the R_{Ω}/R versus $\hbar\Omega$ dependences calculated using Eqs. (17.12) and (17.22) for the GL-photodetectors with different τ_0 and b at $a = 0.2$. As the GL photoconductivity, the detector responsivity R_{Ω} also turns to zero at certain frequencies.

The responsivity of the GL-based photodetectors is proportional to the fundamental constant $\pi\alpha \simeq 0.023$, which determines the coupling of the carriers and the incident radiation. The inclusion of the plasmonic effects associated with the excitation of the plasmons in the optical couplers (metal gratings and so on) and in the GLs can lead to a substantial reinforcement of the carrier-radiation coupling (see, for example [47] and references therein). As a result, the GL-detector responsivity can be markedly enhanced. The responsivity spectral characteristic can exhibit pronounced plasmonic resonances superimposed on the spectral dependences obtained above.

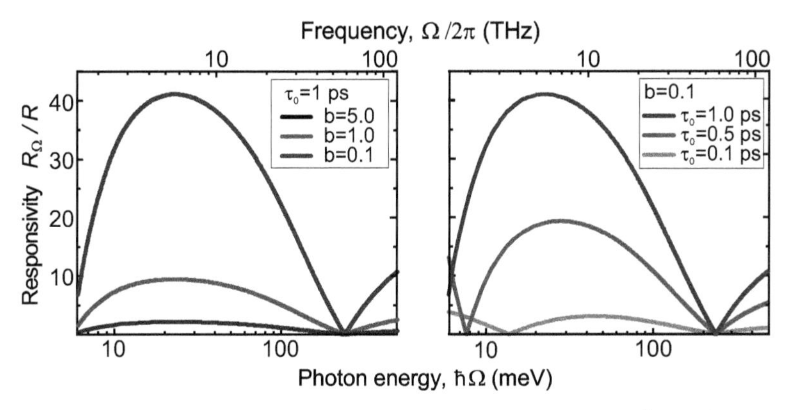

Figure 17.4 The spectral characteristics of the responsivity, R_{Ω}/R, of the GL-based photodetectors with dominant short-range scattering at different b and τ_0 = 1 ps (left panel) and different τ_0 and b = 0.1 (right panel).

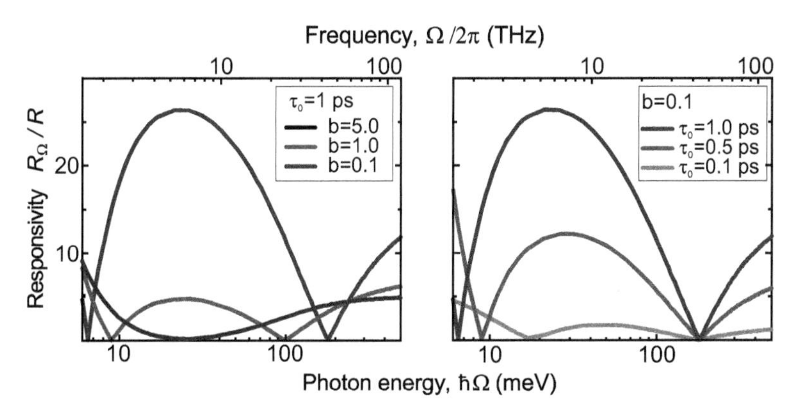

Figure 17.5 The same as in Fig. 17.4 but for the photodetectors with the dominant long-range scattering.

17.7 Heavily Doped GLs

Since the pristine GLs are usually p-type, we will consider the GLs with sufficiently high acceptor and hole densities Σ_A. In the limit of short Auger recombination time, $\mu^e = -\mu^h$ and the variations of the carrier temperature do not lead to the variations of the GL conductivity and the effect of the photoconductivity vanishes. In a more realistic situation one can to assume the long recombination time τ_{Auger}, therefore, the term with τ_{Auger} can be omitted. In this case, we can use Eqs. (17.6) and (17.7) but with 2μ replaced by (μ^{e} + μ^{h}). Apart from this, the quantities β_0^{inter} and β_0^{intra} should be replaced by

$$\beta_{\text{A}}^{\text{inter}} = \frac{\pi\alpha}{2\sqrt{\kappa}}\left[\tanh\left(\frac{\hbar\Omega/2 - \mu_0}{2T_0}\right) + \tanh\left(\frac{\hbar\Omega/2 + \mu_0}{2T_0}\right)\right]$$

and

$$\beta_{\text{A}}^{\text{intra}} = \frac{4\pi\alpha D_0}{\sqrt{\kappa}(1 + \Omega^2\tau_0^2 T_0^2/\mu_0^2)},$$

respectively, and instead of Σ_0 one needs to put $\Sigma = \Sigma_{\text{A}} \gg \Sigma_0$. As a result, we arrive at the following equations:

$$\exp\left[\hbar\omega_0\left(\frac{1}{T_0} - \frac{1}{T}\right)\right] = 1 + \left[\beta_{\text{A}}^{\text{intra}}\left(\frac{\Omega}{\omega_0}\right) + \beta_{\text{A}}^{\text{inter}}\left(\frac{\Omega}{\omega_0} - 1\right)\right]\frac{I_\Omega}{\Sigma_{\text{A}}}\tau_{\text{Opt}}^{\text{intra}}. \tag{17.25}$$

$$\exp\left[\frac{\mu^{e} + \mu^{h}}{T} + \hbar\omega_0\left(\frac{1}{T_0} - \frac{1}{T}\right)\right] = 1 + \beta_{\text{A}}^{\text{inter}}\frac{I_\Omega}{\Sigma_{\text{A}}}\tau_{\text{Opt}}. \tag{17.26}$$

From Eqs. (17.25) and (17.26), we obtain

$$\frac{T - T_0}{T_0} \simeq \frac{T_0}{\hbar\omega_0}\left[\beta_{\text{A}}^{\text{intra}}\left(\frac{\Omega}{\omega_0}\right) + \beta_{\text{A}}^{\text{inter}}\left(\frac{\Omega}{\omega_0} - 1\right)\right]\frac{I_\Omega}{\Sigma_{\text{A}}}\tau_{\text{Opt}}^{\text{intra}}, \tag{17.27}$$

$$\frac{\mu^{e} + \mu^{h}}{T} + \frac{\hbar\omega_0(T - T_0)}{T_0 T} = \beta_{\text{A}}^{\text{inter}}\frac{I_\Omega}{\Sigma_{\text{A}}}\tau_{\text{Opt}}. \tag{17.28}$$

The hole Fermi energies μ^{e} and μ^{h} are related to the effective temperature T as

$$U = \left(\frac{T}{T_0}\right)^2\left[\mathcal{F}_1\left(\frac{\mu^{h}}{T}\right) - \mathcal{F}_1\left(\frac{\mu^{e}}{T}\right)\right]. \tag{17.29}$$

Here $U \propto \Sigma_{\text{A}} = const.$

Taking into account that at $\mu \gg T$, $\mathcal{F}_1(x) \simeq (x^2/2 + \pi^2/6) + x\ln(1 + \exp(-|x|)$, at $T \sim T_0$ from Eq. (17.28) for $\mu^{h} = \mu$ we obtain

$$\frac{\mu}{T} \simeq \frac{\mu_0}{T_0}\left[1 - \left(1 + \frac{\pi^2}{3}\frac{T_0^2}{\mu_0^2}\right)\frac{(T - T_0)}{T_0}\right] \simeq \frac{\mu_0}{T_0}\left[1 - \frac{(T - T_0)}{T_0}\right]. \tag{17.30}$$

The screening length of the charged scatterers in the doped GLs with $\mu \gg T$, is much shorter than that in the intrinsic GLs. Therefore, it is natural to assume that in the former GLs the short-range scattering dominates. Considering Eq. (17.2) with the carrier temperature and the quasi-Fermi energies given by Eqs. (17.27), (17.28), and (17.30), we get

$$\frac{\Delta\sigma_\Omega}{\sigma_{00}} \simeq -\frac{\Sigma_0}{a\Sigma_A}\exp\left(-\frac{\mu_0}{T_0}\right)\left\{\left[\tanh\left(\frac{\hbar\Omega/2-\mu_0}{2T_0}\right)+\tanh\left(\frac{\hbar\Omega/2+\mu_0}{2T_0}\right)\right]\right.$$

$$\left.\times\left(\frac{\Omega}{\omega_0}-1-a\right)+\frac{D_0}{(1+\Omega^2\tau_0^2T_0^2/\mu_0^2)}\left(\frac{\Omega}{\omega_0}\right)\right\}\frac{I_\Omega}{\bar{I}_\Omega}, \qquad (17.31)$$

If $\mu_0 > \hbar\Omega_1/2 \simeq (1 + a)\,\hbar\omega_0/2$, Eq. (17.15) yields $\Delta\sigma_\Omega < 0$ at all the photon energies. This is in line with the experimental results for the p-GLs with $\Sigma_A = (2 - 11) \times 10^{12}$ cm^{-2}, i.e., with $\mu_0 = 192 - 367$ meV [10].

Equation (17.31) resembles Eq. (17.16). However, the latter includes a small factor $[\exp(-\mu_0/T_0)/\Sigma_A]$, different frequency-dependent factor of the interband absorption coefficient, and much shorter $\tau = \tau_0 T_0/\mu_0 \ll \sqrt{3}\tau_0/\pi$. Due to the former, the responsivity of the photodetectors based on doped GLs is smaller than that of the intrinsic GLs. The factor $[\exp(-\mu_0/T_0)/\Sigma_A]$ reflects the facts that the GL conductivity is a rather weak function of the carrier temperature at $\mu_0 \gg T_0$ and that heat received by the carrier system from the absorbed radiation is distributed among a larger number of the carriers ($\Sigma_A \gg \Sigma_0$).

17.8 Conclusions

We developed an analytical model describing the THz and IR photoconductivity in the GLs at room temperature. Using this model we showed that the energy relaxation and the generation-recombination associated with the optical phonons and the Auger processes are crucial for determining the behavior of GL photoconductivity. We found that in the GLs with dominating short-range scattering mechanism of the carrier momentum relaxation the photoconductivity of the intrinsic GLs changes sign at a certain photon frequency Ω_0 in the THz range. Depending on the GL parameters (the transition from the negative to positive photoconductivity) occurs at a certain photon frequency Ω_1 in the mid-IR range (the photoconductivity is negative at $\Omega < \Omega_0$ and $\Omega > \Omega_1$, while it is positive in the range $\Omega_0 < \Omega < \Omega_1$. In contrast, in the case of the long-range scattering dominances, the photoconductivity is positive at $\Omega < \Omega_0$ and $\Omega > \Omega_1$ and negative at $\Omega_0 < \Omega_1$. The change in the GL environment might result in the change from the short-range to long-range scattering. The evaluated responsivity of the GL-based THz and IR photodetectors indicates their prospects for different applications.

Funding

Japan Society for Promotion of Science (16H06361); Russian Science Foundation (14-2900277); Russian Foundation for Basic Research (16-37-60110, 18-07-01379); RIEC Nation-Wide Collaborative Research Project, Japan; Office of Naval Research (Project Monitor Dr. Paul Maki).

Acknowledgments

The authors are grateful to A. Arsenin, V. E. Karasik, V. G. Leiman, and P. P. Maltsev for useful discussions.

Disclosures

The authors declare that there are no conflicts of interest related to this article.

References

1. A. H. Castro Neto, F. Guinea, N. M. R. Peres, K. S. Novoselov, and A. K. Geim, The electronic properties of graphene, *Rev. Mod. Phys.*, **81**(1), 109–162 (2009).
2. L. A. Falkovsky and A. A. Varlamov, Space-time dispersion of graphene conductivity, *Eur. Phys. J. B*, **56**(4), 281–284 (2007).
3. V. T. Vasko and V. Ryzhii, Voltage and temperature dependence of conductivity in gated graphene, *Phys. Rev. B*, **76**(23), 233404 (2007).

4. E. H. Hwang, S. Adam, and S. D. Sarma, Carrier transport in two-dimensional graphene layers, *Phys. Rev. Lett.*, **98**(18), 186806 (2007).
5. V. Vyurkov and V. Ryzhii, Effect of Coulomb scattering on graphene conductivity, *JETP Lett.*, **88**(5), 370–373 (2008).
6. V. T. Vasko and V. Ryzhii, Photoconductivity of intrinsic graphene, *Phys. Rev. B*, **77**(19), 195433 (2008).
7. E. H. Hwang and S. das Sarma, Acoustic phonon scattering limited carrier mobility in two-dimensional extrinsic graphene, *Phys. Rev. B*, **77**(11), 115449 (2008).
8. E. H. Hwang and S. das Sarma, Screening induced temperature dependent transport in 2D graphene, *Phys. Rev. B*, **79**(16), 165404 (2009).
9. O. G. Balev, V. T. Vasko, and V. Ryzhii, Carrier heating in intrinsic graphene by a strong dc electric field, *Phys. Rev. B*, **79**(16), 165432 (2009).
10. J. N. Heyman, J. D. Stein, Z. S. Kaminski, A. R. Banman, A. M. Massari, and J. T. Robinson, Carrier heating and negative photoconductivity in graphene, *J. Appl. Phys.*, **117**(1), 015101 (2015).
11. V. Ryzhii, T. Otsuji, M. Ryzhii, N. Ryabova, S. O. Yurchenko, V. Mitin, and M. S. Shur, Graphene terahertzuncooled bolometers, *J. Phys. D: Appl. Phys.*, **46**, 065102 (2013).
12. Xu Du, D. E. Prober, H. Vora, and C. Mckitterick, Graphene-based bolometers, *2D Mater.*, **1**(1), 1–22 (2014).
13. Qi Han, T. Gao, R. Zhang, Yi Chen, J. Chen, G. Liu, Y. Zhang, Z. Liu, X. Wu, and D. Yu, Highly sensitive hot electron bolometer based on disordered graphene, *Sci Rep.*, **3**, 3533 (2013).
14. G. Skoblin, J. Sun, and A. Yurgens, Graphene bolometer with thermoelectric readout and capacitive coupling to an antenna, *Appl. Phys. Lett.*, **112**(6), 063501 (2018).
15. Y. Wang, W. Yin, Q. Han, X. Yang, H. Ye, Q. Lv, and D. Yin, Bolometric effect in a waveguide-integrated graphene photodetector, *Chin. Phys. B*, **25**(11), 118103 (2016).
16. T. Li, L. Luo, M. Hupalo, J. Zhang, M. C. Tringides, J. Schmalian, and J. Wang, Femtosecond population inversion and stimulated emission of dense Dirac fermions in graphene, *Phys. Rev. Lett.*, **108**(16), 167401 (2012).
17. S. Boubanga-Tombet, S. Chan, T. Watanabe, A. Satou, V. Ryzhii, and T. Otsuji, Ultrafast carrier dynamics and terahertz emission in optically pumped graphene at room temperature, *Phys. Rev. B*, **85**(3), 035443 (2012).

18. I. Gierz, J. C. Petersen, M. Mitrano, C. Cacho, I. E. Turcu, E. Springate, A. Stohr, A. Kohler, U. Starke, and A. Cavalleri, Snapshots of nonequilibrium Dirac carrier distributions in graphene, *Nat. Mater.*, **12**(12), 1119–1124 (2013).
19. T. Watanabe, T. Fukushima, Y. Yabe, S. A. Boubanga-Tombet, A. Satou, A. A. Dubinov, V. Ya. Aleshkin, V. Mitin, V. Ryzhii, and T. Otsuji, The gain enhancement effect of surface plasmonpolaritons on terahertz stimulated emission in optically pumped monolayer graphene, *New J. Phys.*, **15**(7), 07503 (2013).
20. K. J. Tielrooij, J. C. W. Song, S. A. Jensen, A. Centeno, A. Pesquera, A. Z. Elorza, M. Bonn, L. S. Levitov, and F. H. L. Koppens, Photoexcitation cascade and multiple hot-carrier generation in graphene, *Nat. Phys.*, **9**(4), 248–252 (2013).
21. E. Gruber, R. A. Wilhelm, R. Petuya, V. Smejkal, R. Kozubek, A. Hierzenberger, B. C. Bayer, I. Aldazabal, A. K. Kazansky, F. Libish, A. V. Krasheninnikov, M. Schleberger, S. Facsko, A. G. Borisov, A. Arnau, and F. Aumayr, Ultrafast electronic response of graphene to a strong and localized electric field, *Nat. Commun.*, **7**, 13948 (2016).
22. G. X. Ni, L. Wang, M. D. Goldflam, M. Wagner, Z. Fei, A. S. McLeod, M. K. Liu, F. Keilmann, B. Ozyilmaz, A. H. Castro Neto, J. Hone, M. M. Fogler, and D. N. Basov, Ultrafast optical switching of infrared plasmon polaritons in high-mobility graphene, *Nat. Photonics*, **10**, 244–247 (2016).
23. A. Mousavian, B. Lee, A. D. Stickel, and Y.-S. Lee, Ultrafast photocarrier dynamics in single-layer graphene driven by strong terahertz pulses, *J. Opt. Soc. Am. B*, **35**(6), 1255–1259 (2018).
24. M. Baudisch, A. Marini, J. D. Cox, T. Zhu, F. Silva, S. Teichmann, M. Massicotte, F. Koppens, L. S. Levitov, F. J. G. de Abajo, and J. Biegert, Ultrafast nonlinear optical response of Dirac fermions in graphene, *Nat. Commun.*, **9**, 1018 (2018).
25. V. Ryzhii, M. Ryzhii, and T. Otsuji, Negative dynamic conductivity of graphene with optical pumping, *J. Appl. Phys.*, **101**(8), 083114(2007).
26. V. Ryzhii, M. Ryzhii, A. Satou, T. Otsuji, A. A. Dubinov, and V. Y. Aleshkin, Feasibility of terahertz lasing in optically pumped epitaxial multiple graphene layer structures, *J. Appl. Phys.*, **106**(8), 084507 (2009).
27. D. Svintsov, V. Ryzhii, A. Satou, T. Otsuji, and V. Vyurkov, Carrier-carrier scattering and negative dynamic conductivity in pumped graphene, *Opt. Express*, **22**(17), 19873–19686 (2014).
28. D. Svintsov, V. Ryzhii, and T. Otsuji, Negative dynamic Drude conductivity in pumped graphene, *Appl. Phys. Express*, **7**, 115101 (2014).

29. V. Ryzhii, M. Ryzhii, V. Mitin, A. Satou, and T. Otsuji, Effect of heating and cooling of photogenerated electron-hole plasma in optically pumped graphene on population inversion, *Jpn. J. Appl. Phys.*, **50**(9), 094001 (2011).

30. F. T. Vasko, V. V. Mitin, V. Ryzhii, and T. Otsuji, Interplay of intra- and interband absorption in disordered graphene, *Phys. Rev. B*, **86**(23), 235424 (2012).

31. S. Zhuang, Y. Chen, Y. Xia, N. Tang, X. Xu, J. Hu, and Z. Chen, Coexistence of negative photoconductivity and hysteresis in semiconducting graphene, *AIP Adv.*, **6**(4), 045214 (2016).

32. C. J. Docherty, C. T. Lin, H. J. Joyce, R. J. Nicholas, L. M. Hertz, L. J. Li, and M. B. Johnston, Extreme sensitivity of graphene photoconductivity to environmental gases, *Nat. Commun.*, **3**, 1228 (2012).

33. F. Rana, P. A. George , J. H. Strait, S. Sharavaraman, M. Charasheyhar, and M. G. Spencer, Carrier recombination and generation rates for intravalley and intervalley phonon scattering in graphene, *Phys. Rev. B*, **79**(11), 115447 (2009).

34. M. S. Foster and I. L. Aleiner, Slow imbalance relaxation and thermoelectric transport in graphene, *Phys. Rev. B*, **79**(8), 085415 (2009).

35. G. Alymov, V. Vyurkov, V. Ryzhii, A. Satou, and D. Svintsov, Auger recombination in Dirac materials: a tangle of many-body effects, *Phys. Rev. B*, **97**(20), 205411 (2018).

36. K. M. McCreary, K. Pi, A. G. Swartz, Wei Han, W. Bao, C. N. Lau, F. Guinea, M. I. Katsnelson, and R. K. Kawakami, Effect of cluster formation on graphene mobility, *Phys. Rev. B*, **81**(11), 115453 (2010).

37. N. Sule, S. C. Hagness, and I. Knezevic, Clustered impurities and carrier transport in supported graphene, *Phys. Rev. B*, **89**(16), 165402 (2014).

38. T. Stauber, G. Gomez-Santos, and F. Javier Garcia de Abajo, Extraordinary absorption of decorated undoped graphene, *Phys. Rev. Lett.*, **112**(7), 077401 (2014),

39. V. Ryzhii, T. Otsuji, M. Ryzhii, V. Mitin, and M. S. Shur, Effect of indirect interband transitions on terahertz conductivity in "decorated" graphene bilayer heterostructures, *Lithuanian J. Phys.*, **55**(4), 243–248 (2015).

40. A. A. Balandin, S. Ghosh, W. Bao, I. Calizo, D. Teweldebrhan, F. Miao, and C. N. Lau, Superior thermal conductivity of single-layer graphene, *Nano Lett.*, **8**(3), 902–907 (2008).

41. S. Ghosh, I. Calizo, D. Teweldebrhan, E. P. Pokatilov, D. L Nika, A. A. Balandin, W. Bao, F. Miao, and C. N. Lau, Extremely high thermal conductivity of graphene: prospects for thermal management applications in nano-electronic circuits, *Appl. Phys. Lett.*, **92**(15), 151911 (2008).
42. A. A. Balandin, Thermal properties of graphene and nanostructured carbon materials, *Nat. Mater.*, **10**, 569–581 (2011).
43. V. Ryzhii, M. Ryzhii, V. Mitin, and T. Otsuji, Toward the creation of terahertz graphene injection laser, *J. Appl. Phys.*, **110**(9), 094503 (2011).
44. K. F. Mak, M. Y. Sfeir, Y. Wu, Ch. H. Lui, J. A. Misewich, and T. F. Heinz, Measurement of the optical conductivity of graphene, *Phys. Rev. Lett.*, **101**(19), 196405 (2008).
45. H. M. Dong, W. Xu, and F. M. Peters, Electrical generation of terahertz blackbody radiation from graphene, *Opt. Express*, **26**(19), 24621–24626 (2018).
46. V. Ryzhii, M. Ryzhii, D. S. Ponomarev, V. G. Leiman, V. Mitin, M. S. Shur, and T. Otsuji, Negative photoconductivity and hot-carrier bolometric detection of terahertz radiation in graphene-phosphorene hybrid structures, *J. Appl. Phys.*, **125**(15), 151608 (2019).
47. B. Zhao, J. M. Zhao, and Z. M. Zhang, Resonance enhanced absorption in a graphene monolayer using deep metal gratings, *J. Opt. Soc. Am. B*, **32**(6), 1176–1185 (2018).

Part 3

Population Inversion and Negative Conductivity in Graphene and Graphene Structures

Chapter 18

Negative Dynamic Conductivity of Graphene with Optical Pumping*

V. Ryzhii,[a] M. Ryzhii,[a] and T. Otsuji[b]
[a]*Computer Solid State Physics Laboratory, University of Aizu, Aizu-Wakamatsu 965-8580, Japan*
[b]*Research Institute of Electrical Communication, Tohoku University, Sendai 980-8577, Japan*
v-ryzhii@u-aizu.ac.jp

We study the dynamic ac conductivity of a nonequilibrium two-dimensional electron-hole system in optically pumped graphene. Considering the contribution of both interband and intraband transitions, we demonstrate that at sufficiently strong pumping the population inversion in graphene can lead to the negative net ac conductivity in the terahertz range of frequencies. This effect might be used in graphene-based coherent sources of terahertz radiation.

*Reprinted with permission from V. Ryzhii, M. Ryzhii, and T. Otsuji (2007). Negative dynamic conductivity of graphene with optical pumping, *J. Appl. Phys.*, **101**, 083114. Copyright © 2007 American Institute of Physics.

Graphene-Based Terahertz Electronics and Plasmonics: Detector and Emitter Concepts
Edited by Vladimir Mitin, Taiichi Otsuji, and Victor Ryzhii

ISBN 978-981-4800-75-4 (Hardcover), 978-0-429-32839-8 (eBook)
www.jennystanford.com

18.1 Introduction

A significant progress in the fabrication of *graphene*, demonstration of its exceptional electronic properties, and the possibility of creating electronic and optoelectronic devices on the basis of graphene heterostructures have attracted widespread attention [1] (see also, Refs. [2–16]). The gapless energy spectrum of electrons and holes in graphene can lead to specific features of its transport, plasma, and optical properties. One might expect that under optical pumping resulting in photogeneration of electron-hole pairs, the population inversion can be achieved. In such a case, lasing of rather long-wave (terahertz) radiation in graphene may be possible. In this work, using the developed model, we calculate the dynamic ac conductivity of a nonequilibrium two-dimensional electron-hole system in graphene under interband optical excitation. Both interband and intraband transitions are taken into account in the model. It is assumed that the energy of the photogenerated electrons and holes is close to the energy of several optical phonons, so that the photogeneration of electrons and holes followed by the emission of the cascade of optical phonons leads to an essential population of the bottom of the conduction band (by electrons) and the top of the valence band (by holes). We demonstrate that sufficiently strong optical excitation may result in the population inversion in graphene, so that the real part of the net ac conductivity (which is proportional to the absorption coefficient of radiation) can be negative. Due to the gapless energy spectrum, the ac conductivity is negative in the range of terahertz frequencies.

18.2 Equations of the Model

Under optical excitation, electrons and holes are photogenerated with the energy $\varepsilon_0 = \hbar\Omega/2$. Here $\hbar\Omega$ is the energy of the incident photons. Due to rather short characteristic time, τ_0, of the optical phonon emission, the photogenerated electrons and holes emit a cascade of optical phonons populating the states with the energies $\varepsilon_N = \hbar\Omega/2 - N\hbar\omega_0$, where $\hbar\omega_0$ is the optical phonons, and $N = 1, 2, 3, \ldots$ ($N \leq N_{max}$). Upon emitting the cascade of optical phonons, the photogenerated electrons and holes are thermalized. The band diagram (corresponding to linear gapless dispersion relations for electrons and holes), the processes under consideration, and the energy distributions of the photogenerated electrons and holes are shown schematically in Fig. 18.1. If the characteristic times of the electron-electron, electron-hole, and hole-hole interactions $\tau_{ee} = \tau_{hh}$ are small compared to the recombination time τ_R, these interactions result in the Fermi distributions of both electrons and holes. In the optically pumped ungated graphene in which the equilibrium electron and hole concentrations are equal to each other, the Fermi distributions of electrons and holes are characterized by the same Fermi energy ε_F and the effective temperature T, i.e., $f_e(\varepsilon) = f_h(\varepsilon) = f(\varepsilon)$, where

$$f(\varepsilon) \simeq \left[1 + \exp\left(\frac{\varepsilon - \varepsilon_F}{k_B T}\right)\right]^{-1}. \tag{18.1}$$

Here $\varepsilon = v_F p$ and ε_F are the electron (hole) kinetic energy and the Fermi energy, v_F is the characteristic velocity of the electron and hole energy spectra ($v_F \simeq 10^8$ cm/s), p is the absolute value of the electron (hole) momentum, and k_B is the Boltzmann constant. In the ungated graphene heterostructures (and gated heterostructures with the gate potential $V_g = 0$) in the absence of irradiation, $\varepsilon_F = 0$ and $T = T_0$, where T_0 is the lattice temperature. Assuming that $\Omega/2 - N_{max}\omega_0 \ll \omega_0$, where $N_{max} \simeq \Omega/2\omega_0$, i.e., assuming that the energy transferring by the optical radiation to the two-dimensional (2D) electron-hole system is small, we may disregard the difference between T and T_0.

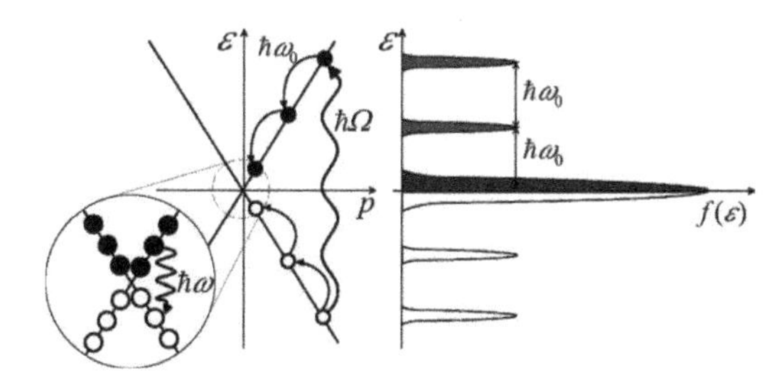

Figure 18.1 Schematic view of graphene band structure and energy distributions of photogenerated electrons and holes. Arrows show the transitions corresponding to optical excitation by photons with energy $\hbar\Omega$, cascade emission of optical phonons with energy $\hbar\omega_0$, and radiative recombination with emission of photons with energy $\hbar\omega$.

We shall calculate the real part of the ac conductivity Re σ_ω of a nonequilibrium electron-hole system in graphene. The quantity Re σ_ω is proportional to the absorption coefficient of photons with the frequency ω. It comprises the contributions of both interband and intraband transitions accompanied by the absorption and emission of photons:

$$\mathrm{Re}\sigma_\omega = \mathrm{Re}\sigma_\omega^{(\mathrm{inter})} + \mathrm{Re}\sigma_\omega^{(\mathrm{intra})}. \tag{18.2}$$

Here the contributions of the interband and intraband (the Drude-mechanism) transitions into the real part of the ac conductivity of the electron-hole system are given, respectively, by (see, for instance, Refs. [12, 16])

$$\begin{aligned}\mathrm{Re}\sigma_\omega^{(\mathrm{inter})} &= \frac{e^2}{4\hbar}\left[1 - f_\mathrm{e}\left(\frac{\hbar\omega}{2}\right) - f_\mathrm{h}\left(\frac{\hbar\omega}{2}\right)\right] \\ &= \frac{e^2}{4\hbar}\left[1 - 2f\left(\frac{\hbar\omega}{2}\right)\right] = \frac{e^2}{4\hbar}\tanh\left(\frac{\hbar\omega - 2\varepsilon_\mathrm{F}}{4k_\mathrm{B}T}\right)\end{aligned} \tag{18.3}$$

and

$$\begin{aligned}\mathrm{Re}\sigma_\omega^{(\mathrm{intra})} &= \frac{e^2 v_\mathrm{F}\tau}{\pi\hbar^2(1+\omega^2\tau^2)}\int_0^\infty dpp\left(-\frac{df_0}{dp}\right) \\ &= \frac{e^2 k_\mathrm{B}T\tau}{\pi\hbar^2(1+\omega^2\tau^2)}\Phi\left(\frac{\varepsilon_\mathrm{F}}{k_\mathrm{B}T}\right),\end{aligned} \tag{18.4}$$

where

$$\Phi(\eta_\mathrm{F}) = \int_0^\infty \frac{dxx\exp(x-\eta_\mathrm{F})}{[1+\exp(x-\eta_\mathrm{F})]^2}, \tag{18.5}$$

and τ is the momentum relaxation time of electrons and holes associated with their scattering on impurities and acoustic phonons.

In the case of generation of the electron-hole pairs by normally incident optical radiation, the normalized electron and hole Fermi energy $\eta_\mathrm{F} = \varepsilon_\mathrm{F}/k_\mathrm{B}T$ obeys the following balance equations:

$$R(\eta_F) - R(0) = \alpha_\Omega I_\Omega, \tag{18.6}$$

where $R(\eta_\mathrm{F})$ is the rate of electron-hole recombination, α_Ω and I_Ω are the efficiency of the photogeneration of the electron-hole pairs and the intensity of incident optical radiation with the photon energies $\hbar\Omega$. The term in the right-hand side of Eq. (18.6) is the rate of the electron-hole pair photogeneration

$$\alpha_{\Omega} = \frac{4\pi}{c} \frac{\mathrm{Re}\,\sigma_{\Omega}^{(\mathrm{inter})}}{\hbar\Omega}, \tag{18.7}$$

where c is the speed of light and σ_{Ω} is the conductivity of graphene associated with the interband transitions induced by the photons with the frequency Ω. Considering that for optical photons $\hbar\Omega \gg K_{\mathrm{B}}T$, ε_{F}, the real part of the conductivity can be set as [12]

$$\mathrm{Re}\,\sigma_{\Omega}^{(\mathrm{inter})} = \frac{e^2}{4\hbar}\tanh\left(\frac{\hbar\Omega - 2\varepsilon_{\mathrm{F}}}{4k_{\mathrm{B}}T}\right) \simeq \frac{e^2}{4\hbar}, \tag{18.8}$$

so that

$$æ_{\Omega} = \frac{\pi\alpha}{\hbar\Omega}, \tag{18.9}$$

where $\alpha = (e^2/c\hbar) \simeq 1/137$.

18.3 Weak Pumping

When optical pumping is not strong, one might assume that $\eta_{\mathrm{F}} < 1$. In this case, introducing the characteristic recombination time τ_{R}, determined by the dominant recombination mechanism, the recombination term in the left-hand side of Eq. (18.2) can be presented as

$$R(\eta_{\mathrm{F}}) - R(0) \simeq \frac{\Sigma_0}{\tau_{\mathrm{R}}}\eta_{\mathrm{F}}, \tag{18.10}$$

where

$$\Sigma_0 = \frac{1}{\pi\hbar^2}\int_0^{\infty} \frac{dpp}{[1+\exp(v_{\mathrm{F}}p/T)]}$$

$$= \frac{1}{\pi}\left(\frac{k_{\mathrm{B}}T}{\hbar v_{\mathrm{F}}}\right)^2 \int_0^{\infty} \frac{dxx}{(\exp x + 1)} = \frac{\pi}{12}\left(\frac{k_{\mathrm{B}}T}{\hbar v_{\mathrm{F}}}\right)^2, \tag{18.11}$$

is the electron (hole) concentration in the absence of the pumping (i.e., when $\varepsilon_{\mathrm{F}} = 0$). In this particular case, the electron (hole) concentration $\Sigma = \Sigma_0 + \Delta\Sigma$, where $\Delta\Sigma \simeq (\ln 2/\pi)(k_{\mathrm{B}}T/\hbar v_{\mathrm{F}})^2\eta_{\mathrm{F}}$, so that the term in the right-hand side of Eq. (18.10) $\Sigma_0\eta_{\mathrm{F}}/\tau_{\mathrm{R}} \simeq (12\ln 2/\pi^2)$ $\Delta\Sigma_0/\tau_{\mathrm{R}}$. Using Eqs. (18.5) and (18.9), we obtain

$$\eta_F = \frac{\tau_R æ_\Omega}{\Sigma_0} I_\Omega = 12\alpha \left(\frac{\hbar v_F}{k_B T}\right)^2 \frac{\tau_R I_\Omega}{\hbar\Omega}. \quad (18.12)$$

If $\varepsilon_F \neq 0$, Eq. (18.8) for $\hbar\omega/2$, $\varepsilon_F < k_B T$ can be presented in the following form:

$$\mathrm{Re}\,\sigma_\omega^{(\mathrm{inter})} \simeq \frac{e^2}{8\hbar}\left(\frac{\hbar\omega}{2k_B T} - \eta_F\right). \quad (18.13)$$

Taking into account that at $\eta_F < 1$, $\Phi(\eta_F) \simeq \ln 2 + \eta_F/2$, the contribution of the intraband transitions can be estimated as

$$\mathrm{Re}\,\sigma_\omega^{(\mathrm{intra})} \simeq \frac{(\ln 2 + \eta_F/2)e^2}{\pi\hbar} \frac{k_B T\tau}{\hbar\left(1+\omega^2\tau^2\right)}. \quad (18.14)$$

Considering Eqs. (18.2), (18.13), and (18.14), we arrive at the following equation for the net ac conductivity of the electron-hole system under consideration:

$$\mathrm{Re}\,\sigma_\omega \simeq \frac{e^2}{8\hbar}\left\{\frac{\hbar\omega}{2k_B T} + \frac{8\ln 2}{\pi}\left(\frac{k_B T\tau/\hbar}{1+\omega^2\tau^2}\right) - \left[1 - \frac{4}{\pi}\left(\frac{k_B T\tau/\hbar}{1+\omega^2\tau^2}\right)\right]\eta_F\right\}, \quad (18.15)$$

Using Eq. (18.12) in the case $k_B T\tau/\hbar < \omega^2\tau^2$, Eq. (18.15) can be presented as

$$\mathrm{Re}\,\sigma_\omega \simeq \frac{e^2}{8\hbar}\left[g_\omega - 12\alpha\left(\frac{\hbar v_F}{k_B T}\right)^2 \frac{\tau_R I_\Omega}{\hbar\Omega}\right], \quad (18.16)$$

where

$$g_\omega = \frac{\hbar\omega}{2k_B T} + \frac{8\ln 2}{\pi}\left(\frac{k_B T\tau/\hbar}{1+\omega^2\tau^2}\right). \quad (18.17)$$

Accounting for that function g_ω exhibits a minimum at $\omega = \bar{\omega}$, where

$$\bar{\omega} = \left(\frac{32\ln 2}{\pi}\right)^{1/3}\left(\frac{k_B T\tau}{\hbar}\right)^{2/3}\frac{1}{\tau} \simeq \left(\frac{k_B T\tau}{\hbar}\right)^{2/3}\frac{1.5}{\tau}, \quad (18.18)$$

the net ac conductivity near the threshold at which it changes sign can be presented as

$$\mathrm{Re}\,\sigma_\omega \simeq \frac{e^2 \bar{g}}{8\hbar}\left[1 + \frac{3}{2}\left(\frac{\omega - \bar{\omega}}{\bar{\omega}}\right)^2 - \frac{I_\Omega}{\bar{I}_\Omega}\right], \quad (18.19)$$

where

$$\overline{g} = \left(\frac{4\ln 2}{\pi}\right)^{1/3} \left(\frac{\hbar}{k_{\rm B}T\tau}\right)^{1/3} \tag{18.20}$$

and

$$\overline{I}_\Omega = \frac{(4\ln 2/\pi)^{1/3}}{12\alpha} \left(\frac{\hbar}{k_{\rm B}T\tau}\right)^{1/3} \left(\frac{k_{\rm B}T}{\hbar v_{\rm F}}\right)^2 \frac{\hbar\Omega}{\tau_{\rm R}}$$

$$\simeq 11 \left(\frac{\hbar}{k_{\rm B}T\tau}\right)^{1/3} \left(\frac{k_{\rm B}T}{\hbar v_{\rm F}}\right)^2 \frac{\hbar\Omega}{\tau_{\rm R}}, \tag{18.21}$$

is the threshold intensity of optical radiation.

As follows from Eq. (18.19), Re $\sigma_\omega < 0$ in the frequency range $\omega_{\rm min} < \omega < \omega_{\rm max}$, where

$$\omega_{\rm max} = \overline{\omega}\left[1 - \sqrt{\frac{2}{3}}\sqrt{\frac{I_\Omega}{\overline{I}_\Omega} - 1}\right],$$

$$\omega_{\rm max} = \overline{\omega}\left[1 + \sqrt{\frac{2}{3}}\sqrt{\frac{I_\Omega}{\overline{I}_\Omega} - 1}\right]. \tag{18.22}$$

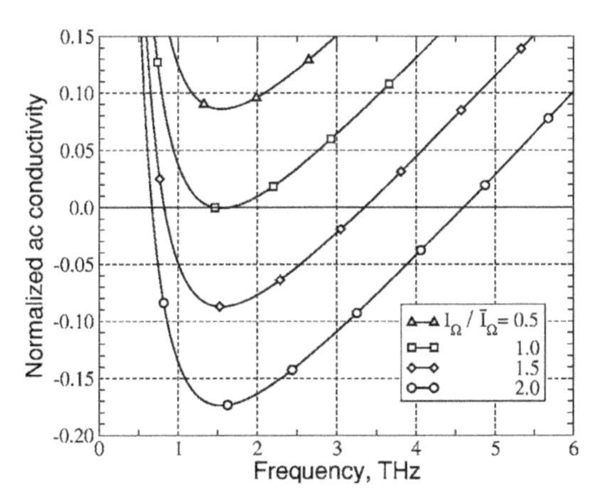

Figure 18.2 Frequency dependence of normalized ac conductivity at different intensities of optical radiation.

At $v_{\rm F} = 10^8$ cm/s, $T = 77$ K, $\tau = 1 \times 10^{-12}$ s, $\hbar\Omega \simeq 0.8$ eV ($\hbar\Omega = 4\hbar\omega_0$, where the optical phonon energy in graphene is assumed to be $\hbar\omega_0$ = 0.2 eV), and τ_R =10^{-7}–10^{-9} s, we obtain $\bar{\omega} \simeq 1.5$ THz, $\bar{g} \simeq 0.4$, and $\bar{I}_\Omega \simeq 0.07 \times (1\text{–}100)$ W/cm^2. If τ = 3 × 10^{-12} s and the same as above other parameters, $\bar{\omega} \simeq 1\text{THz}$, $\bar{g} \simeq 0.3$, and $\bar{I}_\Omega \simeq 0.05 \times (1-100)$ W/cm^2. Figure 18.2 shows the frequency dependence of the ac conductivity normalized by the characteristic conductivity $e^2/4\hbar$ at different intensities of optical radiation. It is assumed that T = 77 K and $\tau = 1 \times 10^{-12}$ s. One can see that the range of frequencies in which Re $\sigma_\omega \leq 0$ widens when the intensity of incident optical radiation increases beyond its threshold value. Figure 18.3 shows the variation of this range (confined by the frequencies ω_{max} and ω_{min}) with increasing optical intensity calculated for different values of the electron and hole momentum relaxation time τ (at T = 77 K). As follows from Fig. 18.3, the graphene-based structures with longer momentum relaxation time exhibit smaller threshold frequency $\bar{\omega}$ but slower expansion of the interval ($\omega_{max} - \omega_{min}$). This is due to the dependence of both $\bar{\omega}$ and $\bar{I}_\Omega$ on τ.

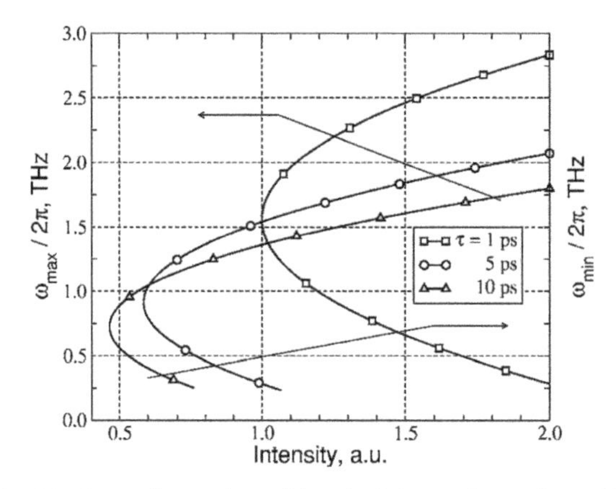

Figure 18.3 Maximum (upper branch) and minimum (lower branch) frequency as functions of optical radiation intensity for different values of momentum relaxation time.

Combining Eqs. (18.8) and (18.18), one can obtain

$$\eta_F = \left(\frac{4\ln 2}{\pi}\right)^{1/3} \left(\frac{\hbar}{k_B T \tau}\right)^{1/3} \frac{I_\Omega}{\bar{\bar{I}}_\Omega} = \gamma \frac{I_\Omega}{\bar{\bar{I}}_\Omega}, \qquad (18.23)$$

where γ = $(4 \ln 2/\pi)^{1/3}(\hbar/k_B T\tau)^{1/3}$. As follows from Eq. (18.23), the condition when Eqs. (18.10)–(18.22) are valid ($\eta_F < 1$) can be presented as $I_\Omega < \overline{I_\Omega}/\gamma$. At T = 77 K and $\tau = (1–3) \times 10^{-12}$ s, $\gamma^{1} \simeq$ 2.3 - 3.3. Hence, at elevated intensities of incident optical radiation when $I_\Omega > \overline{I_\Omega}/\gamma$, Eq. (18.10) should be replaced by the more general equation that is valid at $\eta_F > 1$.

18.4 Strong Pumping

Under strong illumination, the electron and hole Fermi energy ε_F can be relatively large. If $\Omega/2 - N_{max}\omega_0 \ll \omega_0$, the optical radiation absorbed by the electron-hole system should not, nevertheless, result in a significant heating of the latter. In this case, $\eta_F \gg 1$ and Eq. (18.10) is replaced by

$$R(\eta_F) - R(0) = \frac{\Sigma}{\tau_R} \simeq \frac{1}{2\pi\tau_R}\left(\frac{k_B T}{\hbar v_F}\right)^2 \eta_F^2 = \frac{1}{2\pi\tau_R}\left(\frac{\varepsilon_F}{\hbar v_F}\right)^2. \quad (18.24)$$

Equations (18.6), (18.9), and (18.24) lead to

$$\varepsilon_F = \hbar v_F \sqrt{\frac{2\pi^2 \alpha \tau_R I_\Omega}{\hbar\Omega}}. \quad (18.25)$$

Since at strong optical pumping one may expect that in the terahertz range of frequencies $\hbar\omega < \varepsilon_F$, the interband contribution to the ac conductivity in the case under consideration is

$$\mathrm{Re}\sigma_\omega^{(\mathrm{inter})} = \frac{e^2}{4\hbar}\tanh\left(\frac{\hbar\omega - 2\varepsilon_F}{4k_B T}\right) \simeq -\frac{e^2}{4\hbar}. \quad (18.26)$$

Simultaneously, taking into account that at $\eta_F \gg 1$ Eq. (18.5) yields $\Phi(\eta_F) \simeq \eta_F$, the contribution of the intraband transitions accompanied with the electron and hole scattering on impurities and acoustic phonons can be presented in the following form [compare with Eq. (18.14)]:

$$\mathrm{Re}\sigma_\omega^{(\mathrm{intra})} \simeq \frac{e^2}{\pi\hbar}\left(\frac{k_B T\tau/\hbar}{1+\omega^2\tau^2}\right)\eta_F = \frac{e^2}{\pi\hbar^2}\frac{\tau\varepsilon_F}{\left(1+\omega^2\tau^2\right)}. \quad (18.27)$$

As a result, we arrive at

$$\mathrm{Re}\sigma_\omega \simeq \frac{e^2}{4\hbar}\left[-1 + \frac{\tau v_F}{\left(1+\omega^2\tau^2\right)}\sqrt{\frac{32\alpha\tau_R I_\Omega}{\hbar\Omega}}\right]. \quad (18.28)$$

For $\omega, \bar{\omega} \gg \tau^{-1}$, Eq. (18.28) can be presented as

$$\mathrm{Re}\,\sigma_\omega \simeq \frac{e^2}{4\hbar}\left[-1+\xi\left(\frac{\bar{\omega}}{\omega}\right)^2\sqrt{\frac{I_\Omega}{\overline{I}_\Omega}}\right], \tag{18.29}$$

where $\xi = \sqrt{\pi/24\ln 2} \simeq 0.43$. Equation (18.29) implies that when the intensity becomes too large, the ac conductivity changes its sign from negative to positive. It occurs when $I_\Omega > \overline{I}_\Omega/\xi$. This is associated with the saturation of the contribution of the interband transitions [see Eq. (18.24)], while the contribution of the intraband transitions increases with increasing intensity (due to an increase in the electron and hole concentration). This effect is weaker for relatively high frequencies ($\omega \gtrsim \bar{\omega}$). Assuming, for instance, $\omega = \bar{\omega}$ we find that Re $\sigma_\omega < 0$ when $\overline{I}_\omega < I_{\text{omega}} < 5\overline{I}_\Omega$.

18.5 Conclusions

We developed a model of a nonequilibrium two-dimensional electron-hole system in optically pumped graphene. The model accounts for the feature of the energy spectrum of electrons and holes and the processes determining the ac conductivity. Using the developed model, we calculated the ac conductivity of graphene under optical excitation followed by emission of a cascade of optical phonons. We demonstrated that at sufficiently strong pumping the population inversion in graphene can lead to the negative net ac conductivity in the terahertz range of frequencies. The main effect determining the low end of the frequency range where the ac conductivity is negative is associated with the intraband transitions electrons and holes due to their scattering on impurities and acoustic phonons (free carrier intraband absorption). The effect of population inversion and negative ac conductivity might be used in graphene-based coherent sources of terahertz electro-magnetic radiation utilizing stimulated photon emission. A similar effect can occur in gated graphene structures, for example, in the structures with graphene separated from a highly conducting substrate playing the role of the gate. In these structures, varying the gate voltage one can control the equilibrium concentrations of electrons and holes and, hence, control the ac conductivity and output terahertz radiation. Since a two-dimensional electron-hole system can serve as a resonant plasma cavity [17, 18], the negative ac conductivity in the terahertz range can also lead to the self-excitation of plasma waves [12–15] followed by their conversion into electro-magnetic radiation. We considered the case of normally incident optical radiation photogenerating electrons and holes. The lateral (waveguide) optical input can also be used for the pumping of graphene-based structures.

Acknowledgment

This work was supported by the Grant-in-Aid for Scientific Research (S) from the JSPS, Japan.

References

1. K. S. Novoselov, A. K. Geim, S. V. Morozov, D. Jiang, M. I. Katsnelson, I. V. Grigorieva, S. V. Dubonos, and A. A. Firsov, *Nature*, **438**, 197 (2005).
2. C. Berger, Z. Song, T. Li, X. Li, A. Y. Ogbazhi, R. Feng, Z. Dai, A. N. Marchenkov, E. H. Conrad, P. N. First, and W. A. de Heer, *J. Phys. Chem.*, **108**, 19912 (2004).
3. B. Obradovic, R. Kotlyar, F. Heinz, P. Matagne, T. Rakshit, M. S. Giles, M. A. Stettler, and D. E. Nikonov, *Appl. Phys. Lett.*, **88**, 142102 (2006).
4. K. Nomura and A. H. MacDonald, cond-mat/0606589 (2006).
5. I. L. Aleiner and K. B. Efetov, cond-mat/0607200 (2006).
6. B. Trauzettel, Ya. M. Blanter, and A. F. Morpurgo, cond-mat/0606505 (2006).
7. V. P. Gusynin and S. G. Sharapov, *Phys. Rev. Lett.*, **95**, 146801 (2005).
8. Y. Zhang, Y.-W. Tan, H. Stormer, and P. Kim, *Nature*, **438**, 201 (2005).
9. D. Abanin, P. A. Lee, and L. S. Levitov, *Phys. Rev. Lett.*, **96**, 176803 (2006).
10. V. P. Gusynin, S. G. Sharapov, and J. P. Carbotte, *Phys. Rev. Lett.*, **96**, 256802 (2006).
11. V. P. Gusynin and S. G. Sharapov, *Phys. Rev. B*, **73**, 245411 (2006).
12. L. A. Falkovsky and A. A. Varlamov, cond-mat/0606800 (2006).
13. V. Ryzhii, *Jpn. J. Appl. Phys., Part 2*, **45**, L923 (2006).
14. O. Vafek, *Phys. Rev. Lett.*, **97**, 266406 (2006).
15. V. Ryzhii, A. Satou, and T. Otsuji, *J. Appl. Phys.*, **101**, 024509 (2007).
16. L. A. Falkovsky, *Phys. Rev. B*, **75**, 033409 (2007).

17. M. Dyakonov and M. Shur, *IEEE Trans. Electron. Devices,* **43**, 1640 (1996).
18. M. S. Shur and V. Ryzhii, *Int. J. High Speed Electron. Syst.,* **13**, 575 (2003).

Chapter 19

Population Inversion of Photoexcited Electrons and Holes in Graphene and Its Negative Terahertz Conductivity*

V. Ryzhii,[a] M. Ryzhii,[a] and T. Otsuji[b]
[a]*Computer Solid State Physics Laboratory, University of Aizu, Aizu-Wakamatsu 965-8580, Japan*
[b]*Research Institute of Electrical Communication, Tohoku University, Sendai 980-9577, Japan*
v-ryzhii@u-aizu.ac.jp

We demonstrate that sufficiently strong optical excitation may result in the population inversion in graphene, so that the real part of the ac conductivity can be negative in the terahertz range of frequencies. We study also how the heating of the electron-hole system influences the effect of negative ac conductivity. The effect of population inversion and negative ac conductivity might be used in graphene-based coherent sources of terahertz electromagnetic radiation.

*Reprinted with permission from V. Ryzhii, M. Ryzhii, and T. Otsuji (2008). Population inversion of photoexcited electrons and holes in graphene and its negative terahertz conductivity, *Phys. Status Solidi C*, **5**(1), 261–264. Copyright © 2008 WILEY-VCH Verlag GmbH & Co. KGaA.

Graphene-Based Terahertz Electronics and Plasmonics: Detector and Emitter Concepts
Edited by Vladimir Mitin, Taiichi Otsuji, and Victor Ryzhii

ISBN 978-981-4800-75-4 (Hardcover), 978-0-429-32839-8 (eBook)
www.jennystanford.com

19.1 Introduction

A significant progress in fabrication of *graphene*, demonstration of its exceptional electronic properties and possibility of creating novel electronic and optoelectronic devices on the basis of graphene heterostructures have attracted widespread attention [1].

The gapless energy spectrum of electrons and holes in graphene ($\varepsilon = \pm v_{\mathrm{F}}\,|\mathbf{p}|$, where $v_{\mathrm{F}} \simeq 10^8$ cm/s is the characteristic velocity and $\mathbf{p}$ is the momentum) can lead to specific features of its transport, plasma, and optical properties. Due to the gapless energy spectrum of graphene, the Fermi energy ε_{F} in equilibrium is equal to zero. Hence, the electron and hole distribution functions at the bottom of the conduction band and the top of the valence band $f_{\mathrm{e}}(0) = f_{\mathrm{h}}(0) = 1/2$. This implies that at even weak photoexcitation, the values of the distribution functions at low energies can be $f_{\mathrm{e}}(\varepsilon) = f_{\mathrm{h}}(\varepsilon) > 1/2$, that corresponds to the population inversion [2]. Such a population inversion might lead to the negative ac conductivity at relatively small frequencies associated with the interband transitions. However, the intraband processes determined the Drude conductivity provide the positive contribution to the ac conductivity. Nevertheless, one might expect that under sufficiently strong optical excitation resulting in photogeneration of electron-hole pairs, the population inversion can be achieved. In such a case, lasing of rather long-wave (terahertz) radiation in graphene may be possible. In this work, which generalizes our previous paper [2], using the developed model, we study the ac conductivity of a nonequilibrium two-dimensional electron-hole system in graphene under interband optical excitation. Both interband and intraband transitions are taken into account in the model. It is assumed that the energy of the photogenerated electrons and holes is close to the energy of several optical phonons, so that the photogeneration of electrons and holes followed by emission of the cascade of optical phonons leads to an essential population of the bottom of the conduction band (by electrons) and the top of the valence band (by holes). We show that the ac conductivity can be negative in certain (terahertz) range of frequencies. The heating of the electron-hole system by absorbed optical radiation is also considered. It is demonstrated that the role of this heating essentially depends on the ratio of the photon energy and the energy of optical phonons.

19.2 Qualitative Pattern

We consider a graphene-based heterostructure illuminated by light with the photon energy $\hbar\Omega$, where $\hbar$ is the reduced Planck constant and Ω is the photon frequency. The illumination results in the photogeneration in graphene of electrons and holes with kinetic energy $\varepsilon_0 = \hbar\Omega/2$. Due to a very short time of the optical phonon emission τ_0, the photogeneration of electrons and holes leads to the emission of a cascade of optical phonons, so that the photogenerated electrons and holes occupy the states with the energies close to $\varepsilon_N = \hbar(\Omega/2 - N\omega_0)$. Here $\hbar\omega_0$ is the optical phonon energy and N is the number of emitted optical phonons ($N \leq N_m$, where N_m is determined by the integer part of $\Omega/2\omega_0$). When $\varepsilon_{Nm} = \hbar(\Omega/2 - N_m\omega_0) \simeq 0$, the photogenerated electrons and holes after the cascade emission of optical phonons are concentrated near the bottom of the graphene conduction band and the top of the graphene valence band, respectively. In this case, the energy distribution of electrons and holes corresponds to that shown schematically in Fig. 19.1. In such a

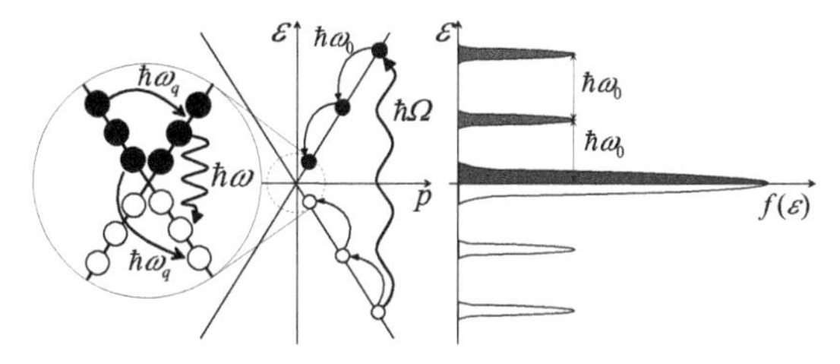

Figure 19.1 Schematic view of graphene band structure (left-side figure). The arrows show different interband and intraband transitions: transitions associated with absorption of photons with energy $\hbar\Omega$, emission of terahertz photons with energy $\hbar\omega$, interaction with optical and acoustic phonons (with energies $\hbar\omega_0$ and $\hbar\omega_q$). Right-side figure shows energy distributions of photogenerated electrons and holes.

case, the processes of the electron and hole photogeneration do hot heat the electron-hole system, and one can assume that the effective temperature of the latter $T \simeq T_0$, where T_0 is the lattice temperature. However, if ε_{Nm} is not small, the processes in question can result in a marked heating of the electron-hole system ($T > T_0$). Taking into account that in graphene $\hbar\omega_0 \simeq 0.2$ eV, the first case can occur when the graphene structure is illuminated by a semiconductor laser with the photon energy $\hbar\Omega = 0.8$ eV ($N_m = 2$). If the graphene structure is photoexcited by a CO_2 laser with the photon energy $\hbar\Omega = 0.1$ eV ($N_m = 0$), a significant part of the absorbed optical energy goes to the heating of the electron-hole system because the electron and hole energy relaxation on optical phonons is "switched off."

19.3 Frequency Dependence of the Conductivity

We shall analyze the real part of the ac conductivity Re σ_ω of a nonequilibrium electron-hole system in graphene. For this purpose, the general formulae for the ac conductivity of graphene obtained previously are used [2, 3]. The quantity Re σ_ω is proportional to the absorption coefficient of photons with the frequency ω. It comprises the contributions of both interband and intraband transitions accompanied by the absorption and emission of photons:

$$\mathrm{Re}\sigma_\omega = \mathrm{Re}\sigma_\omega^{\mathrm{inter}} + \mathrm{Re}\sigma_\omega^{\mathrm{intra}}. \tag{19.1}$$

Here the first and the second terms represents the contributions of the interband and intraband (the Drude-mechanism) transitions into the real part of the ac conductivity of the electron-hole system [2, 3, 5]. The interband contribution can be expressed via the normalized Fermi quasi-energy $\eta_F = \varepsilon_F/k_BT$, of the nonequilibrium electron (hole) system, where k_B is the Boltzmann constant. In equilibrium, i.e., without optical excitation, $\eta_F \propto \varepsilon_F = 0$. At relatively weak optical excitation when the concentration of photogenerated electrons and holes is small in comparison with their thermal concentration, one may assume that $\varepsilon_F < k_BT$, i.e., $\eta_F < 1$. In this case, for the frequencies $\omega < 2k_BT/\hbar$, the interband contribution to the real part of the ac conductivity can be presented in the following form:

$$\mathrm{Re}\sigma_\omega^{(\mathrm{inter})} \simeq \frac{e^2}{8\hbar}\left(\frac{\hbar\omega}{2k_BT} - \eta_F\right), \tag{19.2}$$

where e is the electron charge. In equilibrium, when $\eta_F = 0$, and Eq. (19.2) reduces to that obtained previously [3]. At $\eta_F < 1$, the contribution of the intraband transitions can be estimated as [4]

$$\mathrm{Re}\,\sigma_{\omega}^{(\mathrm{intra})} \simeq \frac{\left(\ln 2+\eta_{\mathrm{F}}/2\right)e^{2}}{\pi\hbar}\frac{T\tau}{\hbar\left(1+\omega^{2}\tau^{2}\right)}. \tag{19.3}$$

Here τ is the momentum relaxation time of electrons and holes. As shown previously [2], at weak optical excitation,

$$\eta_{\mathrm{F}} = \frac{\tau_{\mathrm{R}}æ_{\Omega}}{\Sigma_{0}}I_{\Omega} = 12\alpha\left(\frac{\hbar v_{\mathrm{F}}}{k_{\mathrm{B}}T}\right)^{2}\frac{\tau_{\mathrm{R}}I_{\Omega}}{\hbar\Omega}. \tag{19.4}$$

Here $æ_{\Omega} = \pi\alpha/\hbar\Omega$ is the interband absorption coefficient, $\Sigma_0 = (\pi/12)$ $(k_{\mathrm{B}}T/\hbar v_{\mathrm{F}})^2$ is the equilibrium density of electrons and holes, τ_{R} is the recombination time (the lifetime of nonequilibrium electrons and holes), and $\alpha = (e^2/\hbar c) \simeq 1/137$, where c is the speed of light. Using Eqs. (19.1)–(19.4) at $k_{\mathrm{B}}T\tau/\hbar \ll \omega^2\tau^2$, we arrive at the following equation for the ac conductivity of the electron-hole system under consideration:

$$\mathrm{Re}\,\sigma_{\omega} \simeq \frac{e^{2}}{8\hbar}\left[g_{\omega} - 12\alpha\left(\frac{\hbar v_{\mathrm{F}}}{k_{\mathrm{B}}T}\right)^{2}\frac{\tau_{\mathrm{R}}I_{\Omega}}{\hbar\Omega}\right]. \tag{19.5}$$

Here

$$g_{\omega} = \frac{\hbar\omega}{2k_{\mathrm{B}}T} + \frac{8\ln 2}{\pi}\left(\frac{k_{\mathrm{B}}T\tau/\hbar}{1+\omega^{2}\tau^{2}}\right). \tag{19.6}$$

One can see that g_{ω} as a function of ω exhibits a minimum at $\omega = \bar{\omega}$, where

$$\bar{\omega} = \left(\frac{32\ln 2}{\pi}\right)^{1/3}\left(\frac{k_{\mathrm{B}}T\tau}{\hbar}\right)^{2/3}\frac{1}{\tau}. \tag{19.7}$$

Considering this, the ac conductivity near the threshold, at which the ac conductivity changes its sign, can be presented as

$$\mathrm{Re}\,\sigma_{\omega} \simeq \frac{e^{2}\bar{g}}{8\hbar}\left[1+\frac{3}{2}\left(\frac{\omega-\bar{\omega}}{\bar{\omega}}\right)^{2} - \frac{I_{\Omega}}{\bar{I}_{\Omega}}\right], \tag{19.8}$$

where

$$\bar{g} = \left(\frac{4\ln 2}{\pi}\right)^{1/3}\left(\frac{\hbar}{k_{\mathrm{B}}T\tau}\right)^{1/3}, \tag{19.9}$$

$$\bar{I}_{\Omega} = \frac{(4\ln 2/\pi)^{1/3}}{12\alpha}\left(\frac{\hbar}{k_{\mathrm{B}}T\tau}\right)^{1/3}\left(\frac{k_{\mathrm{B}}T}{\hbar v_{\mathrm{F}}}\right)^{2}\frac{\hbar\Omega}{\tau_{\mathrm{R}}} \tag{19.10}$$

is the threshold intensity of optical radiation, and

$$\omega_{\min} = \bar{\omega}\left[1 - \sqrt{\frac{2}{3}}\sqrt{\frac{I_\Omega}{\overline{I_\Omega}} - 1}\right], \tag{19.11}$$

$$\omega_{\max} = \bar{\omega}\left[1 + \sqrt{\frac{2}{3}}\sqrt{\frac{I_\Omega}{\overline{I_\Omega}} - 1}\right], \tag{19.12}$$

As follows from Eq. (19.8) and seen in Fig. 19.2, Re $\sigma_\omega < 0$ in the frequency range $\omega_{\min} < \omega < \omega_{\max}$.

Let us set $v_F = 10^8$ cm/s, $T = 77$ K, $\tau = 1 \times 10^{-12}$ s, $\hbar\Omega \simeq 0.8$ eV ($\hbar\Omega = 4\hbar\omega_0$, where the optical phonon energy in graphene is assumed to be $\hbar\omega_0 = 0.2$ eV). The recombination time associated with the emission of acoustic phonons can be roughly estimated as $\tau_R \simeq \tau(v_F/s)$ where s is the speed of sound. Hence, $\tau_R \simeq 10^3\,\tau \simeq 10^{-9}$ s. As a result, we obtain $(\hbar/k_BT\tau)^{1/3} \simeq 0.45$, $\bar{\omega} \simeq 1.5$ THz, and $\overline{I_\Omega} \simeq 7\ \text{W/cm}^2$. If $\tau = 3 \times 10^{-12}$ s, assuming the same as above other parameters, $\bar{\omega} \simeq 1$ THz, $\bar{g} \simeq 0.3$ and $\overline{I_\Omega} \simeq 1.7\ \text{W/cm}^2$. Figure 19.2 shows the frequency dependence of the ac conductivity normalized by the characteristic conductivity $e^2/2\hbar$ at different intensities of optical radiation. It is assumed that $T = 77$ K and $\tau = 1 \times 10^{-12}$ s. One can see that the range of frequencies in which $\text{Re}\sigma_\omega \leq 0$ widens when the intensity of incident optical radiation increases beyond its threshold value.

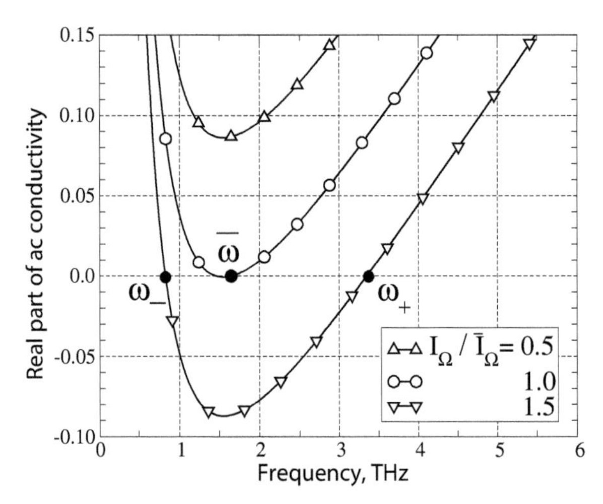

Figure 19.2 Normalized ac conductivity versus frequency for different intensities of optical excitation.

19.4 Heating of Electrons and Holes

As follows from Eqs. (19.7) and (19.10), $\bar{\omega}$ and $\overline{I_\Omega}$ depend on the effective temperature of the electron-hole system. This leads also to the temperature dependences of $\omega_{\rm min}$ and $\omega_{\rm max}$. Let assume that $\tau(T) \propto T^s$ and $\tau_{\rm R}(T) \propto T^r$, where s and q are some numbers. Therefore

$$\bar{\omega} \propto T^{(2-s)/3}, \quad \overline{I_\Omega} \propto T^{(5-s)/3-r} \tag{19.13}$$

Considering the impurity and acoustic phonon scattering mechanisms at $k_{\rm B}T \ll \hbar\omega_0$, one may set [4] $s = 1$.

The effective temperature of the electron-hole system is governed the energy balance equation

$$\frac{k_{\rm B}(T - T_0)\Sigma_0}{\tau_\varepsilon} = \hbar(\Omega/2 - N_{\rm m}\omega_0)æ_\Omega I_\Omega$$
$$= \frac{\pi\alpha}{2}\left(\frac{\Omega - 2N_{\rm m}\omega_0}{\Omega}\right)I_\Omega. \tag{19.14}$$

Here τ_ε is the energy relaxation time. If the electron and hole momentum and energy relaxation is associated with acoustic phonons, one can use the following estimate $\tau_\varepsilon \simeq \tau(v_{\rm F}/s) \gg \tau$. At moderate intensities of optical excitation, the effective temperature is close to T_0, so that $T = T_0 + \Delta T$, where

$$\Delta T = \left[\frac{\pi\alpha}{2k_{\rm B}}\left(\frac{\Omega - 2N_{\rm m}\omega_0}{\Omega}\right)\frac{\tau_\varepsilon I_\Omega}{\Sigma_0}\right]. \tag{19.15}$$

Considering Eqs. (19.13) and (19.15), one obtains the following formula, which replaces Eq. (19.8):

$${\rm Re}\,\sigma_\omega \simeq \frac{e^2\bar{g}}{8\hbar}\left[1 + \frac{3}{2}\left(\frac{\omega - \bar{\omega}^*}{\bar{\omega}^*}\right)^2\right.$$
$$\left. - \frac{I_\Omega}{\overline{I_\Omega}} + \left(\frac{\Omega - 2N_{\rm m}\omega_0}{\Omega}\right)\frac{I_\Omega^2}{\overline{I_\Omega I_\Omega^*}}\right]. \tag{19.16}$$

Here

$$\frac{\bar{\omega}^*}{\bar{\omega}} \simeq 1 + \left(\frac{2-s}{5-s-3r}\right)\left(\frac{\Omega - 2N_{\rm m}\omega_0}{\Omega}\right)\frac{I_\Omega}{\overline{I_\Omega^*}} \tag{19.17}$$

and

$$\overline{I_\Omega^*} = \frac{6}{(5-s-3r)\pi\alpha} \frac{k_B T \Sigma_0}{\tau_\varepsilon}. \quad (19.18)$$

The ratio of the characteristic intensities $\overline{I_\Omega}$ and $\overline{I_\Omega^*}$, as follows from Eqs. (19.10) and (19.18), is given by

$$\frac{\overline{I_\Omega}}{\overline{I_\Omega^*}} = \beta\left(\frac{\Omega}{\omega_0}\right), \quad (19.19)$$

where

$$\beta \sim \left(\frac{\hbar}{k_B T \tau}\right)^{1/3} \left(\frac{\tau_\varepsilon}{\tau_R}\right)\left(\frac{\hbar\omega_0}{k_B T}\right). \quad (19.20)$$

Thus, the heating of the electron-hole system due to optical excitation results in a shift of the frequency corresponding to the minimum of Re σ_ω and in a nonlinear dependence of the latter quantity on the intensity. The first and second factors in the right-hand side of Eq. (19.19) are of the order unity. In contrast, the second factor is rather large. Hence generally $\beta \gg 1$, i.e., $\overline{I_\Omega} \gg \overline{I_\Omega^*}$. This implies that if Ω is sufficiently close to $2N_m\omega_0$, the terms associated with the heating can be disregarded. However, when Ω markedly deviates from $2N_m\omega_0$, the role of the nonlinear term in Eq. (19.16) rises drastically. In such a case, the frequency range where Re $\sigma_\omega < 0$ shrinks and can disappear.

19.5 Discussion

The effect of negative ac conductivity can also occur in gated graphene structures [1, 5], for example, in the structures with graphene separated from a highly conducting substrate playing the role of the gate. In these structures, varying the gate voltage one can control the equilibrium concentrations of electrons and holes and, hence, control the ac conductivity not only in equilibrium but under optical excitation as well. However, an increase in the equilibrium density of electrons (or holes) might lead to a significant increase in the intraband contribution to the ac conductivity, which can prevail the contribution of the interband transitions. If the real part of ac conductivity is negative, it can be used for generating of terahertz radiation.

One option is associated with the utilization of the optically pumped graphene as an active region for a terahertz laser provided that the pertinent conditions for effective reflection of the generated terahertz photons are satisfied by proper design of the graphene heterostructure edges. Another option is associated with the self-excitation of plasma oscillations. This is because a two-dimensional electron-hole system, in particular in the graphene, can serve as a resonant plasma cavity [5, 6]. The self-excitation of plasma waves due to the negative ac conductivity in the terahertz range followed by their conversion into electro-magnetic radiation can also be used in sources of terahertz radiation.

Acknowledgements

The authors are grateful to Professor F. T. Vas'ko for stimulating comments. One of the authors (V. R.) thanks Professor A. A. Varlamov for numerous discussions. This work was supported by the Grant-in-Aid for Scientific Research (S) from the Japan Society for Promotion of Science, Japan.

References

1. K. S. Novoselov, A. K. Geim, S. V. Morozov, et al., *Nature*, **438**, 197 (2005).
2. V. Ryzhii, M. Ryzhii, and T. Otsuji, *J. Appl. Phys.*, **101**, 083114(2007).
3. L. A. Falkovski and A. A. Varlamov, *Eur. Phys. J.*, **56**, 281 (2007).
4. L. A. Falkovski, *Phys. Rev. B*, **75**, 033409 (2007).
5. V. Ryzhii, A. Satou, and T. Otsuji, *J. Appl. Phys.*, **101**, 024509 (2007).
6. M. S. Shur and V. Ryzhii, *Int. J. High Speed Electron. Syst.*, **13**, 575 (2003).

Chapter 20

Negative Terahertz Dynamic Conductivity in Electrically Induced Lateral p-i-n Junction in Graphene*

V. Ryzhii,[a] M. Ryzhii,[a] M. S. Shur,[b] and V. Mitin[c]
[a]*Computational Nanoelectronics Laboratory, University of Aizu, Aizu-Wakamatsu 965-8580, and Japan Science and Technology Agency, CREST, Tokyo 107-0075, Japan*
[b]*Department of Electrical, Electronics, and Systems Engineering, Rensselaer Polytechnic Institute, Troy 12180, USA*
[c]*Department of Electrical Engineering, University at Buffalo, Buffalo 14260, USA*
v-ryzhii@u-aizu.ac.jp

We analyze a graphene tunneling transit-time device based on a heterostructure with a lateral p-i-n junction electrically induced in the graphene layer and calculate its ac characteristics. Using the developed device model, it is shown that the ballistic transit

*Reprinted with permission from V. Ryzhii, M. Ryzhii, M. S. Shur, and V. Mitin (2010). Negative terahertz dynamic conductivity in electrically induced lateral p-i-n junction in graphene, *Physica E*, **42**, 719–721. Copyright © 2009 Elsevier B.V.

Graphene-Based Terahertz Electronics and Plasmonics: Detector and Emitter Concepts
Edited by Vladimir Mitin, Taiichi Otsuji, and Victor Ryzhii

ISBN 978-981-4800-75-4 (Hardcover), 978-0-429-32839-8 (eBook)
www.jennystanford.com

of electrons and holes generated due to interband tunneling in the i-section results in the negative ac conductance in the terahertz frequency range. The device can serve as an active element of terahertz oscillators.

20.1 Introduction

Theoretical predictions and experimental observations of unique properties of graphene-based structures, in particular, the possibility of ballistic electron and hole transport in fairly large samples at room temperatures (see, for instance, Refs. [1–3]) stimulate proposals of novel graphene electron and optoelectron devices. In this chapter, we consider a tunneling transit-time device based on a gated graphene with electrically induced lateral p-i-n junction and present the device dynamic characteristics, the frequency dependences of the real and imaginary parts of the ac conductance, calculated using the developed device model. In the graphene tunneling transit-time (G-TUNNETT) device under consideration, the depleted i-section of the graphene layer play the role of both the tunneling injector and the transit region at the same time.

Extending the device model used by us recently [4] (mainly by the inclusion into the consideration of the nonuniformity of the electric field in the i-section and the possibility of the self-excitation of plasma oscillations in the gated p- and n-sections), we show that the G-TUNNETT can exhibit negative ac conductance in the terahertz (THz) range of frequencies and, hence, serve as an active element in THz oscillators.

20.2 Device Model and Terminal AC Conductance

The G-TUNNETT structure under consideration is shown schematically in Fig. 20.1a. It is assumed that the gate-source voltage ($V_p < 0$) and the drain-gate voltage ($V_n > 0$) are applied. This results in creating the hole and electron two-dimensional systems (with the Fermi energy ε_F), i.e., the p- and n-regions in the channel sections under the pertinent gates. Under the reverse source-drain voltage

($V > 0$), a lateral p-i-n junction is formed, which corresponds to the band diagram shown in Fig. 20.1b. It is assumed that apart from a dc component $V_0 > 0$ which corresponds to a reverse bias, the net source-drain voltage V comprises also an ac component $\delta V \exp(-i\omega t)$: $V = V_0 + \delta V_\omega \exp(-i\omega t)$, where δV_ω and ω are the signal amplitude and frequency, respectively.

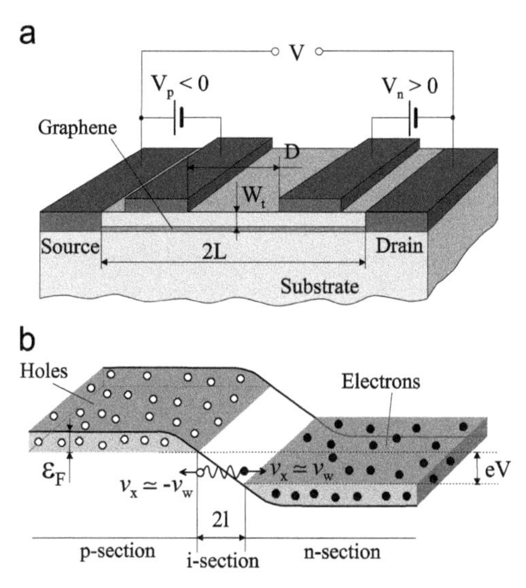

Figure 20.1 Structure of a G-TUNNETT device (a) and its band diagram under the reverse bias (b). Arrows with wavy tail indicate the directions of propagation of the generated electrons and holes generated due to interband tunneling.

The probability of interband tunneling in graphene is a fairly sharp function of the angle between the direction of the electron (hole) motion and the x-direction (from the source to the drain) [5]. Taking this into account, we disregard some spread in the x-component of the velocity of the injected electrons and assume that all the generated electrons and holes propagate in the x-direction with the velocity $v_x \simeq v_W$, where $v_W \simeq 10^8$ cm/s is the characteristic velocity of the electron and hole spectra in graphene. Considering the formula for the interband tunneling probability [5], we assume that the rate of the local tunneling generation of electron hole pairs $G/2l \propto \mathcal{E}^{3/2}$, where $2l$ is the length of the depleted i-section and $\mathcal{E}$ is the electric field in the i-section. In the structures with the spacing between the top gates $D \gg W_t, l_s$, where W_t is the thickness of the layers separating the graphene layer and the gates and l_s is the characteristic screening lengths, one can assume $2l \simeq D$ (as for the problem of screening in two-dimensional electron and hole systems, see, for instance, Refs. [6, 7]). One can expect that this is valid when the dc conductivity of the p- and n-sections is much higher than the dc conductivity of the intrinsic graphene. Actually, the length of the depleted high-electric field i-section $2l$ depends also on V_0. However, we disregard this effect. In the case of ballistic electron and hole transport in the i-section when the generated electrons and holes do not change the directions of their propagation, the continuity equations governing the ac components of the electrons and holes densities $\delta\Sigma_\omega^-$ and $\delta\Sigma_\omega^+$ can be presented as

$$-i\omega\delta\Sigma_\omega^\mp \pm v_W \frac{d\delta\Sigma_\omega^\mp}{dx} = \frac{3G_0}{4l}\frac{\delta V_\omega}{V_0}. \quad (20.1)$$

Here $G_0 = G_0(x)$ and δG_ω are the dc and ac components of the local tunneling rate, respectively. The boundary conditions are as follows: $\delta\Sigma_\omega^\mp\big|_{x=\mp l} = 0$.

We shall consider Eq. (20.1) in the following two limiting cases: (a) the nonuniformity of the electric field in the i-section is insignificant, so that the components of the latter can be estimated as $\mathcal{E}_0 \simeq V_0/2l$ and $\delta\mathcal{E}_\omega \simeq \delta V_\omega/2l$, and (b) the electric field exhibits strong maxima near the edges of the n- and p-sections. In the former case, the tunneling generation of electrons and holes takes place in the whole i-section, whereas in the latter case, the electrons and holes are primarily generated at the points $|x| \lesssim l$.

Solving Eq. (20.1) in the case "a" and using the Shockley–Ramo theorem for the induced current, the real and imaginary parts of the ac (dynamic) component of the source-drain (terminal) conductance $\sigma_\omega^{sd} = \delta J_\omega^{sd}/\delta V_\omega$ can be presented as

$$\mathrm{Re}\sigma_\omega^{sd} = \frac{3\sigma_0}{2}\frac{\sin(\omega t)}{\omega t}\mathcal{J}_0(\omega\tau), \quad (20.2)$$

$$\mathrm{Im}\sigma_\omega^{sd} = \frac{3\sigma_0}{2}\left[\frac{1-\cos(\omega\tau)\mathcal{J}_0(\omega\tau)}{\omega\tau} - c\omega\tau\right]. \quad (20.3)$$

Here $\sigma_0 = J_0/V_0$ is the dc conductance, J_0 is the dc current, $c = 2C/3\sigma_0\tau$, C is the lateral p-i-n junction capacitance, $\mathcal{J}_0(\xi)$ is the Bessel function, and $\tau = l/v_W$ is the characteristic transit time of electrons and holes across the i-section. In the case "b," instead of Eq. (20.3), we arrive at

$$\mathrm{Re}\sigma_{\omega}^{\mathrm{sd}} = \frac{3\sigma_0}{2}\cos(\omega\tau)\mathcal{J}_0(\omega\tau), \tag{20.4}$$

$$\mathrm{Im}\sigma_{\omega}^{\mathrm{sd}} = \frac{3\sigma_0}{2}[\sin(\omega\tau)\mathcal{J}_0(\omega\tau) - c\omega\tau]. \tag{20.5}$$

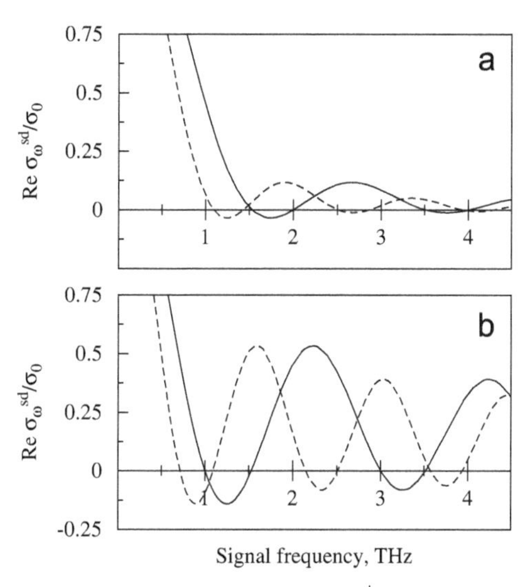

Figure 20.2 Real part of the ac conductance $\sigma_{\omega}^{\mathrm{sd}}$ versus the signal frequency $f = \omega/2\pi$ calculated for uniform (a) and strongly nonuniform (b) electric-field spatial distributions in the i-section for different lengths of i-section: solid lines correspond to $2l = 0.5$ μm ($\tau = 0.25$ ps) and dashed lines correspond to $2l = 0.7$ μm ($\tau = 0.35$ ps).

Figure 20.2 shows the real part of the ac conductance $\sigma_{\omega}^{\mathrm{sd}}$ (normalized by its ac value) as a function of the signal frequency $f = \omega/2\pi$ calculated for G-TUNNETTs with $2l = 0.5$ and 0.7 μm (this corresponds to $\tau = 0.25$ and 0.35 ps, respectively) and using Eqs. (20.2) and (20.4). As seen, the real part of the ac conductance exhibits a pronounced oscillatory behavior with the frequency ranges where it has different signs. The real part of the ac conductance can be negative in the THz range of frequencies in the G-TUNNETT structures with the i-section length only moderately smaller than 1 μm. As follows from Eq. (20.2) and Fig. 20.2 in the case "a," the relative value of the real part of the ac conductance, corresponding to the first minima, for l = 0.25 μm is $\mathrm{Re}\sigma^{sd}_{\omega=2\pi f_1}/\sigma_0 \simeq -0.034$ at the frequency $\omega/2\pi = f_1 \simeq$ 1.75 THz and $\mathrm{Re}\sigma^{sd}_{\omega=2\pi f_1}/\sigma_0 \simeq -0.141$ at the frequency $\omega/2\pi = f_1 \simeq$ 1.25 THz in the case "b," respectively. At V_0 = 0.2 V, one can obtain $\sigma_0 \simeq 3\times 10^{12}\,\mathrm{s}^{-1}$ and $\mathrm{Re}\sigma^{sd}_{\omega=2\pi f_1} \simeq -1\times 10^{11}\,\mathrm{s}^{-1}$. In the case "b," Eq. (20.4) (see also Fig. 20.2) for the same parameters and the same dc current yields $f_1 \simeq$ 1.25 THz, $\mathrm{Re}\sigma^{sd}_{\omega=2\pi f_1}/\sigma_0 \simeq -0.141$, and $\mathrm{Re}\sigma^{sd}_{\omega=2\pi f_1} \simeq -4\times 10^{11}\,\mathrm{s}^{-1}$. A significant difference in the absolute values of $\mathrm{Re}\sigma^{sd}_{\omega=2\pi f_1}/\sigma_0$ in the limiting cases under consideration is associated with the following. In the case "a," there is a marked spread Δt_{tr} in the transit times of electrons and holes generated at different points of the i-section: $\Delta t^{(a)}_{tr} \simeq \tau$. However, in the case "b," $\Delta t^{(b)}_{tr} \ll \tau$, so that the oscillations of Re σ_ω are more pronounced. Smaller value of f_1 in the case "b" is due a longer average transit time: $t^{(b)}_{tr} \simeq 2\tau$, while $t^{(a)}_{tr} \simeq \tau$.

20.3 Self-Excitation of Plasma Oscillations

The fundamental plasma frequency in the gated p- and n-sections, in which the spectrum of the plasma waves is a sound-like with the characteristic velocity s and the characteristic plasma frequency $\Omega = \pi s/2(L-l)$. At sufficiently high gate voltages when the electron (hole) Fermi energy in the n-section (p-section) $\varepsilon_F = (\hbar v_W/2\sqrt{2})\sqrt{æV_t/eW_t} \gg k_B T$, one obtains $s \simeq \sqrt{2e^2 W_t \varepsilon_F/æ\hbar^2}$ $> v_W$ [8], where $2L$ is the spacing between the source and drain contacts, T is the temperature, and k_B is the Boltzmann constant. Since the plasma frequency in graphene can be rather large (owing to a large s), the condition $\omega < \Omega$ assumed above can be easily fulfilled for the signal frequencies in the THz range (as well as the condition $\omega \simeq \Omega$) by a proper choice of the device structure length $2L$ and the gate voltage V_t. Indeed, setting, for example, $s = 8\times 10^8$ cm/s, and $L-l$ = 1 μm, one can obtain $\Omega/2\pi$ = 4 THz.

If the frequencies $f_1, f_2, \ldots$, which corresp6ond to the pertinent minimum in the Re σ_ω versus ω dependence, are far from the resonant plasma frequency $\Omega/2\pi$, the quasi-neutral p- and n-sections serve as the highly conducting contacts. However, when $f_1 \sim \Omega/2\pi$ (or $f_2 \sim \Omega/2\pi$), i.e., when the frequency of the plasma oscillations falls into the range where $\mathrm{Re}\sigma_\omega^{\mathrm{sd}} < 0$, the hole and electron systems in the gated p- and n-sections, respectively, can play the role of the plasma resonators [9]. In this situation, the self-excitation of the plasma oscillations (plasma instability) can be possible [10]. Indeed, taking into account that the Drude conductance of the gated p- and n-sections is equal to $\sigma_\omega^{\mathrm{D}} = \sigma_0^{\mathrm{D}}[i/(\omega + i\nu)\tau]$, where $\sigma_0^{\mathrm{D}} = (e^2 p_{\mathrm{F}}/\pi\hbar^2)[l/(L-l)]$ is the characteristic Drude dc conductance of the gated graphene section with the Fermi momentum p_{F} and ν is the frequency of hole and electron collisions with defects and acoustic phonons, one can arrive at the following dispersion equation for the plasma oscillations in the gated p- and n-sections with the frequency ω:

$$\sqrt{\frac{\omega}{\omega + i\nu}}\cot\left[\frac{\pi\sqrt{\omega(\omega+i\nu)}}{2\Omega}\right] = i\mathcal{R}\mathcal{F}(\omega\tau), \tag{20.6}$$

where $\mathcal{R} = \sigma_0/\sigma_0^{\mathrm{D}} = (6\hbar^2\sigma_0/e^2 p_{\mathrm{F}})[(L-l)/l]$, $\mathcal{F}(\omega\tau) = \sin(\omega\tau)\mathcal{J}_0(\omega\tau)$, and $\mathcal{F}(\omega\tau) = (\omega\tau)\cos(\omega\tau)\mathcal{J}_0(\omega\tau)$ in the cases "a" and "b," respectively. Using the expression for the dc tunneling current [4], $\mathcal{R}$ can be expressed via the dc source-drain voltage V_0 and the electron and hole density in the gated sections Σ_0 as $\mathcal{R} = (3/2)^{3/2}\pi^3\sqrt{(eV_0/v_{\mathrm{W}} l\Sigma_0\hbar)}[(L-l)/l]$ (since it is assumed that the dc voltage drops primarily across the depleted i-section, the $R \ll 1$). Assuming that $\Sigma_0 = 5 \times 10^{11}$ cm^{-2}, $l = 0.25$ μm, $L-l = 2.35$ μm, (for $s = 6 \times 10^8$ cm/s, this corresponds to $f = 1.25$ THz) and $V_0 = 0.2$–0.8 V, one obtains $R \simeq 0.16$–0.32. Here we have neglected the contribution of electrons and holes in the i-section to its capacitance as well as the geometrical capacitance of this section, which are small in comparison with the gated p- and n-sections capacitance. When $\Omega \gg \nu$ (i.e., $Q = 4\Omega/\pi\nu \gg 1$), Eq. (20.7) yields $\mathrm{Re}\,\omega \simeq (2n-1)\Omega$, where $n = 1, 2, 3, \ldots$ is the plasma mode index. Simultaneously, from Eq. (20.6) we obtain the following expressions for the damping/growth rate $\gamma = \mathrm{Im}\,\omega$ of plasma oscillations (with $n = 1$): $\gamma = -\nu/2 - \mathcal{R}\Omega\mathcal{F}(\Omega\tau)\pi$ for the oscillations with the asymmetrical distribution of the ac potential $\delta\varphi_\omega$ corresponding to the upper curve in Fig. 20.3 with the in-plane ac electric field in the i-section ($-l < x < l$) $\delta\mathcal{E}_\omega \neq 0$ (the phases of oscillation in the p- and n-sections are opposite), and $\gamma \simeq -\nu/2$ for the oscillations (damped) with the symmetrical distributions of the ac potential (see the lower curve in Fig. 20.3) and $\delta\mathcal{E}_\omega = 0$. Using this, we arrive at the following condition of the plasma oscillations self-excitation: $Q > 2[\mathcal{R}|\min\mathcal{F}(\Omega\tau)|]^{-1}$. As follows from the latter inequality, the self-excitation of plasma oscillations (plasma instability) can occur if their quality factor Q is sufficiently large. Setting $|\min\mathcal{F}(\Omega\tau)| = 0.28$, and $\mathcal{R} = 0.16$–0.32, we obtain that Q should be fairly large: $Q > 25$–50. The latter requires $\nu \lesssim (1–2) \times 10^{11}$ s^{-1}. As a result, the self-excitation of the plasma oscillations in the G-TUNNETTs under consideration is possible only in sufficiently perfect graphene layers.

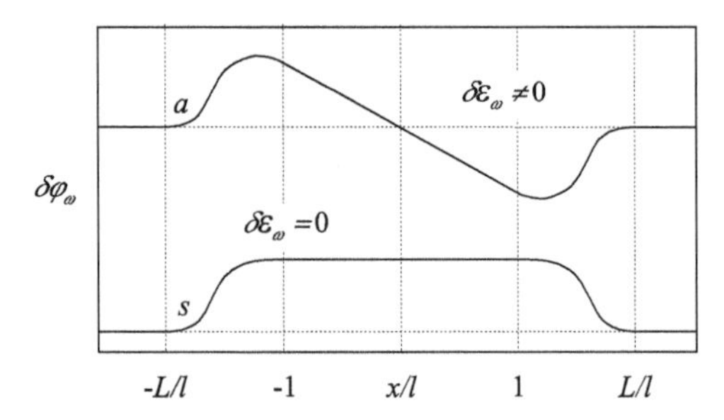

Figure 20.3 Qualitative view of spatial distributions of the ac potential for asymmetrical and symmetrical modes.

20.4 Conclusions

We demonstrated that the G-TUNNETT under consideration can exhibit the negative terminal dynamic conductivity in the THz range of frequencies due to the tunneling and transit-time effects in its i-section. Owing to this, G-TUNNETTs can be used in THz oscillators with a complementary resonant cavity. The generation of THz radiation is also possible due to the self-excitation of the plasma oscillations in the device p- and n-sections.

Acknowledgments

The authors are grateful to Professors T. Otsuji and M. Asada for stimulating discussions. The work was supported by the Japan Science and Technology Agency, CREST and the Japan Society for Promotion of Science, Japan, as well by the Airforce Office of Scientific Research, USA.

References

1. C. Berger, Z. Song, T. Li, X. Li, A. Y. Ogbazhi, R. Feng, Z. Dai, A. N. Marchenkov, E. H. Conrad, P. N. First, and W. A. de Heer, *J. Phys. Chem.*, **108**, 19912 (2004).
2. S. V. Morozov, K. S. Novoselov, M. I. Katsnelson, F. Schedin, D. C. Elias, J. A. Jaszczak, and A. K. Geim, *Phys. Rev. Lett.*, **100**, 016602 (2008).
3. X. Du, I. Skachko, A. Barker, and E. Y. Andrei, *Nat. Nanotechnol.*, **3**, 491 (2008).
4. V. Ryzhii, M. Ryzhii, V. Mitin, and M. S. Shur, *Appl. Phys. Express*, **2**, 034503 (2009).
5. V. V. Cheianov and V. I. Fal'ko, *Phys. Rev. B*, **74**, 041403 (R) (2006).
6. V. Ryzhii, M. Ryzhii, and T. Otsuji, *Phys. Status Solidi A*, **205**, 1527 (2008).
7. L. M. Zhang and M. M. Fogler, *Phys. Rev. Lett.*, **100**, 116804 (2008).
8. V. Ryzhii, A. Satou, and T. Otsuji, *J. Appl. Phys.*, **101**, 024509 (2007).
9. M. Dyakonov and M. Shur, *IEEE Trans. Electron. Devices*, **43**, 1640 (1996).
10. V. Ryzhii, A. Satou, and M. Shur, *Phys. Status Solidi A*, **202**, R113 (2005).

Chapter 21

Threshold of Terahertz Population Inversion and Negative Dynamic Conductivity in Graphene Under Pulse Photoexcitation*

A. Satou, V. Ryzhii, Y. Kurita, and T. Otsuji
Research Institute of Electrical Communication, Tohoku University, 2-1-1 Katahira, Aoba-ku, Sendai 980-8577, Japan
a-satou@riec.tohoku.ac.jp

We present a theoretical study of population inversion and negative dynamic conductivity in intrinsic graphene in the terahertz (THz) frequency range upon pulse photoexcitation at near-/mid-infrared wavelengths. The threshold pulse fluence required for population inversion and negative dynamic conductivity can be orders of magnitude lower when the pulse photon energy is lower, because of

*Reprinted with permission from A. Satou, V. Ryzhii, Y. Kurita, and T. Otsuji (2013). Threshold of terahertz population inversion and negative dynamic conductivity in graphene under pulse photoexcitation, *J. Appl. Phys.*, **113**, 143108. Copyright © 2013 AIP Publishing LLC.

Graphene-Based Terahertz Electronics and Plasmonics: Detector and Emitter Concepts
Edited by Vladimir Mitin, Taiichi Otsuji, and Victor Ryzhii

ISBN 978-981-4800-75-4 (Hardcover), 978-0-429-32839-8 (eBook)
www.jennystanford.com

the inverse proportionality of the photoexcited carrier concentration to the pulse photon energy and because of the weaker carrier heating. We also investigate the dependence of dynamic conductivity on momentum relaxation time. Negative dynamic conductivity takes place either in high- or low-quality graphene, where Drude absorption by carriers in the THz frequency is weak.

21.1 Introduction

Graphene has attracted much attention for a wide variety of devices, owning to its exceptional electronic and optical properties [1–3]. In particular, THz devices such as lasers [4–9] and photodetectors [10–12] which take advantage of the high carrier mobility and gapless dispersion of graphene have been investigated [13]. We have demonstrated that population inversion can occur in optically pumped graphene in the THz/far-infrared range of frequencies; hence lasing in this range is possible, utilizing the gapless linear energy spectrum and relatively high optical phonon (OP) energy in graphene [4–8]. Due to the energy spectrum $\varepsilon = \pm v_F \hbar k$, where $v_F \simeq 10^8$ cm/s is the Fermi velocity and k is the wavenumber, the Fermi energy ε_F at equilibrium in intrinsic graphene is equal to zero. Hence, the electron and hole distribution functions at the bottom of the conduction band and the top of the valence band have values of $f_e(0) = f_h(0) = 1/2$. This implies that upon photoexcitation, the values of the distribution functions at low energies can be greater than one half, $f_e(\varepsilon) = f_h(\varepsilon) > 1/2$, corresponding to the population inversion. This type of population inversion means that the total dynamic conductivity in graphene at THz/far-infrared frequencies is negative, and that lasing in graphene at these frequencies is possible.

Recently, we measured the carrier relaxation and recombination dynamics in optically pumped epitaxial graphene on silicon [14] and in exfoliated graphene [15] by THz time-domain spectroscopy based on an optical pump/THz & optical probe technique, and we observed the amplification of THz radiation by stimulated emission from graphene under pulse excitation. To our knowledge, those are the first observation of THz amplification using optically pumped graphene, and they demonstrate the possibility of realizing THz lasers based on graphene.

The carrier dynamics in optically pumped graphene strongly depend on the initial temperature of carriers and the intensity of the optical pumping. For sufficiently low carrier concentrations, that is, at low temperatures under weak pumping, photoexcited carriers accumulate effectively near the Dirac point via the cascade emission of OPs. It is predicted that population inversion can be achieved efficiently under these conditions [4, 5, 8]. In contrast, at room temperature or under stronger pumping, where the carrier concentration is high ($\sim 10^{12}$ cm^{-2}), carrier-carrier (CC) scattering plays a crucial role in the dynamics after pulse excitation, because of the fast quasi-equilibration of the carriers. Ultrafast optical pump-probe spectroscopy on graphene has indicated that the quasi-equilibration by CC scattering occurs on a time scale of 10 - 100 fs [16–19], which is much faster than a single OP emission. In this case, the pulse excitation makes carriers initially very hot initially, and the energy relaxation and recombination via OP emission follow. The short time scale for the quasi-equilibration is partly caused by the ineffective dielectric screening of the Coulomb potential by relatively low dielectric layers surrounding graphene, which is typically supported by a SiC or SiO_2 substrate and exposed to air.

We have previously shown that population inversion and negative dynamic conductivity in intrinsic graphene at room temperature under pulse excitation with a photon energy of 0.8 eV, which corresponds to a wavelength of 1.55 μm, can be achieved with a pulse energy fluence above a certain threshold [7, 20].

In the present chapter, we investigate the dependence on the pulse photon energy of the threshold pulse fluence for THz negative dynamic conductivity, by extending our previous model to take into account the effect of the Pauli blocking [7, 20]. We also examine the dependence of dynamic conductivity on momentum relaxation time, τ. We consider the two limiting cases where $\omega\tau \geq 1$, which correspond to high-quality graphene, and where $\omega\tau \ll 1$, which corresponds to low-quality graphene.

21.2 Formulation of the Model

Assuming a quasi-Fermi distribution of carriers due to quasi-equilibration through CC scattering, such that f_ε = {1 + exp[(ε –

$\varepsilon_F)/k_BT_c]\}^{-1}$, the time evolution of the distribution is represented by the quasi-Fermi level, $\varepsilon_F(t)$, and the carrier temperature, $T_c(t)$. In intrinsic graphene, both the electron distribution and the hole distribution can be expressed by a single carrier distribution; the symmetric dispersion of electrons and holes in the energy range under consideration, $\varepsilon < 1$ eV, means that the distributions remain identical even under the pulse excitation. The rate equations that governs the quasi-Fermi level and carrier temperature can be obtained from the quasi-classical Boltzmann equation [7, 20, 21]

$$\frac{dn}{dt}=\frac{2}{\pi}\sum_{i=\Gamma,\mathrm{K}}\int_0^{\infty}kdk\left[\frac{\left(1-f_{\hbar\omega_i-\varepsilon}\right)\left(1-f_{\varepsilon}\right)}{\tau_{\mathrm{i,inter}}^{(+)}}-\frac{f_{\varepsilon}f_{\hbar\omega_i-\varepsilon}}{\tau_{\mathrm{i,inter}}^{(-)}}\right], \tag{21.1}$$

$$\frac{d\varepsilon}{dt}=\frac{2}{\pi}\sum_{i=\Gamma,\mathrm{K}}\int_0^{\infty}kdk\left\{\hbar\omega_i\left[\frac{f_{\varepsilon}\left(1-f_{\varepsilon+\hbar\omega_i}\right)}{\tau_{\mathrm{i,intra}}^{(+)}}-\frac{f_{\varepsilon}\left(1-f_{\varepsilon-\hbar\omega_i}\right)}{\tau_{\mathrm{i,intra}}^{(-)}}\right]\right.$$
$$\left.+\varepsilon\left[\frac{\left(1-f_{\hbar\omega_i-\varepsilon}\right)\left(1-f_{\varepsilon}\right)}{\tau_{\mathrm{i,inter}}^{(+)}}-\frac{f_{\varepsilon}f_{\hbar\omega_i-\varepsilon}}{\tau_{\mathrm{i,inter}}^{(-)}}\right]\right\}, \tag{21.2}$$

where $n = n(\varepsilon_F, T_c)$ and $\varepsilon = \varepsilon(\varepsilon_F, T_c)$ are the concentration and energy density of either type of carriers, which are obtained by integrating the distribution function multiplied by proper factors over k. The index i includes the long-wavelength Γ-OP and short-wavelength K-OP, which result in intravalley and intervalley transitions, respectively; $\tau_{\mathrm{i,intra}}^{(\pm)}$ and $\tau_{\mathrm{i,inter}}^{(\pm)}$, which are the intraband and interband scattering rates for OPs, where (+) indicates absorption and (-) indicates emission. All the possible transitions via OP scattering are illustrated in Fig. 21.1a. The expression for $\tau_{\mathrm{i,intra}}^{(\pm)}$ and $\tau_{\mathrm{i,inter}}^{(\pm)}$ have been reported by Suzuura and Ando [22]. Figure 21.1b shows that these rates are on the order of subpicoseconds for high-energy carriers.

The first term (intraband transition) in Eq. (21.2) in the curly brackets contains the factor $\hbar\omega_i$, whereas the second term (interband transition) contains ε. These terms describe the energy that a carrier loses or acquires during their transitions. The energy conservation for the latter transition is implicitly satisfied because the rest of the OP energy, $\hbar\omega_i - \varepsilon$, is taken into account in the same equation for the other kind of carrier. Here, we assume the OPs are at equilibrium and neglect the nonequilibrium OP population, because of their emission via OP scattering of carriers. The effects of the nonequilibrium OP population have been discussed elsewhere [8].

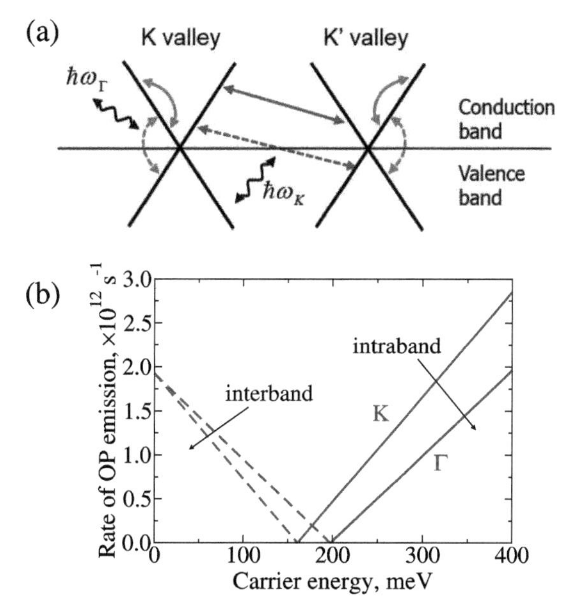

Figure 21.1 (a) Schematic view of the intra-/inter-band, intra-/inter-valley transitions via OP scattering and (b) their rates for OP emission. Note that the rates for OP absorption can be neglected compared with those for the emission because of the negligible OP occupation number at room temperature.

Equations (21.1) and (21.2) and the initial conditions form a nonlinear system of equations for $\varepsilon_F(t)$ and $T_c(t)$, which can be solved numerically. Equations (21.1) and (21.2) are accompanied by the initial carrier concentration and energy density from which the initial quasi-Fermi level and carrier temperature can be found

$$n\big|_{t=0} = n_0 + \Delta n,\ \varepsilon\big|_{t=0} = \varepsilon_0 + \Delta\varepsilon, \tag{21.3}$$

where n_0 and ε_0 are the intrinsic carrier concentration and energy density, and Δn and $\Delta\varepsilon$ are the contributions of photogenerated carriers. These are equal to the concentration and energy density generated by pumping

$$\Delta n \simeq \frac{\pi\alpha\Delta J}{\sqrt{\epsilon}\hbar\Omega}\left(1 - 2f_{\hbar\Omega/2}\big|_{t=0}\right),\ \Delta\varepsilon \simeq \frac{\hbar\Omega}{2}\Delta n, \tag{21.4}$$

where $\alpha \sim 1/137$, ϵ is the dielectric constant surrounding graphene, $\hbar\Omega$ is the pulse photon energy, and ΔJ is the pulse fluence.

In Eq. (21.4), the last factor $1 - 2f_{\hbar\Omega/2}$ roughly takes into account the Pauli blocking of photoabsorption, and its distribution function is approximated by the quasi-Fermi distribution after the photoabsorption by $\varepsilon_F|_{t=0}$ and $T_c|_{t=0}$, calculated self-consistently from Eqs. (21.3) and (21.4). This approximation underestimates the effect of Pauli blocking by the *nonequilibrium* photogenerated carriers when the pulse is so intense that the rate of the photogeneration is comparable to that of the quasi-equilibration by the CC scattering. However, it overestimates the effect of the time-dependent increase in the distribution function *during* pulse excitation, which partly compensates for the underestimation. In addition, we assume the pulse has sufficiently narrow such that OP scattering is not effective during the pulse absorption: $\Delta t \lesssim \tau^{(+)}_{\mathrm{i,inter}}(\hbar\Omega/2), \tau^{(+)}_{\mathrm{i,intra}}(\hbar\Omega/2)$.

To study not only the gain acquired from the population inversion but also the net gain, including the loss from Drude absorption of electromagnetic waves by carriers in graphene, we introduce the real part of dynamic conductivity [23] because it is related to the net gain

$$\mathrm{Re}\,\sigma_\omega \simeq \frac{e^2}{4\hbar}\left(1 - 2f_{\hbar\omega/2}\right) + \frac{2e^2 v_F^2}{\pi}\int_0^\infty dk k \frac{\tau}{1+(\omega\tau)^2}\left(-\frac{df_\varepsilon}{d\varepsilon}\right), \qquad (21.5)$$

where τ is the momentum relaxation time. The first term in Eq. (21.5) corresponds to the interband contribution, which depends on only frequency through the distribution function and can be negative when the rate of stimulated emission exceeds the rate of absorption during the population inversion. Because the distribution functions the quasi-Fermi distribution, the condition of the population inversion for photon energy $\hbar\omega$ is represented as $\varepsilon_F > \hbar\omega/2$. However, the second term corresponds to the intraband contribution (Drude conductivity) which is always positive in this system.

When the real part of dynamic conductivity becomes negative at a particular frequency, the electromagnetic wave at that frequency passing through optically pumped graphene is amplified and graphene acts as a gain medium. Equation (21.5) shows that the largest achievable negative value of the conductivity is $-e^2/4\hbar$, which corresponds to the amplification of $\left(2.3/\sqrt{\epsilon}\right)$ % of the incoming electromagnetic wave.

21.3 Results and Discussion

21.3.1 Time Evolution of Quasi-Fermi Level and Carrier Temperature

We use our model to investigate population inversion and time-dependent dynamic conductivity in intrinsic graphene at room temperature after pulse excitation. We varied the pulse photon energy, $\hbar\Omega$, between 0.1 and 0.8 eV, corresponding to infrared wavelengths of 12.4–1.55 μm. In addition, $\epsilon = 5.5$, which corresponds to the effective dielectric constant of the interface between the air and SiC. We use pulse fluence rather than pulse intensity as a measure of pulse strength because the former does not depend explicitly on the pulse photon energy, whereas the latter does, through the pulse width.

Figure 21.2 shows the time evolution of the quasi-Fermi level and carrier temperature after the pulse excitation for different pulse fluences and pulse photon energies. Population inversion at THz frequencies of up to ~10 THz, which corresponds to $\hbar\omega/2 \sim 20$ meV, is achieved at certain threshold pulse fluences, and it lasts for around 10 ps.

The mechanism of population inversion in the THz range is explained as follows. Shortly after pulse excitation, the carrier temperature rapidly increases, except for excitation with low-energy photons, and the quasi-Fermi level can even be negative because of the quasi-equilibration of carriers at very high temperatures induced by the photogenerated carriers. After several picoseconds, the high-energy carriers relax through intraband OP emission and accumulate in the low-energy region, which is shown by the increase in the quasi-Fermi level. Satou et al. reported that population inversion lasts for around 10 ps due to the imbalance between the time scales of intra- and inter-band OP emission [20].

Figures 21.2a–d show that the threshold fluence of population inversion decreases as the pulse photon energy decreases. This is also visible in Fig. 21.3, which illustrates the dependence of the threshold fluence for population inversion on the pulse photon energy for different THz photon frequencies. There are several factors which contribute to this decrease. First, a lower fluence is required for the lower photon energy pulse to have the same carrier concentration; more precisely, Eq. (21.4) shows that $\Delta J \propto \hbar\Omega$. Second, the less heating of the carriers reduces the carrier temperature, resulting in the carriers accumulating in the low-energy region (even the cooling can occur as shown in Fig. 21.2d; see below for more details). Third, recombination which arises from interband OP emission is suppressed by the lower carrier temperature. These factors result in the nonlinear lowering of the threshold as pulse photon energy decreases; Fig. 21.3 shows an almost linear dependence of the threshold in the high-energy photon region, caused by the first factor, and the sharp drop in the region of photon energy below 0.25 eV, caused by the second and third factors.

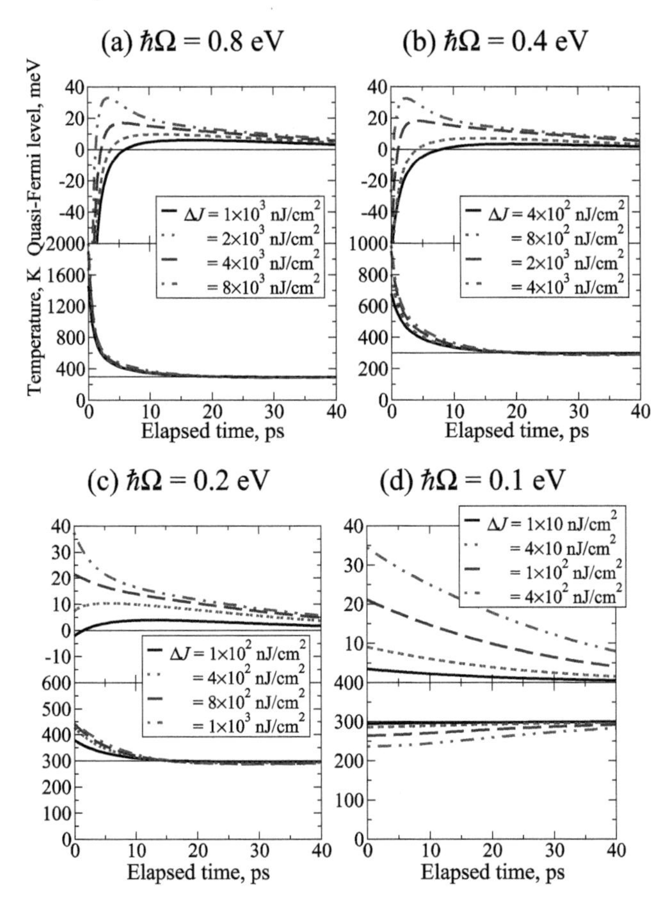

Figure 21.2 The time evolution of the quasi-Fermi level and carrier temperature for different pulse fluences and pulse photon energies. (a) $\hbar\Omega$ = 0.8 eV, (b) 0.4 eV, (c) 0.2 eV, (d) 0.1 eV. Room temperature (300 K) is indicated by the thin solid lines.

The effect of Pauli blocking depends strongly on the pulse photon energy. For $\hbar\Omega = 0.8$ eV, the changes in the quasi-Fermi level and the carrier temperature are negligible for the pulse fluences when Pauli blocking is taken into account. In contrast, for $\hbar\Omega = 0.1$ eV, the threshold fluence is 1.3- to 1.8-fold larger depending on the frequency, and the quasi-Fermi level that can be achieved is also smaller.

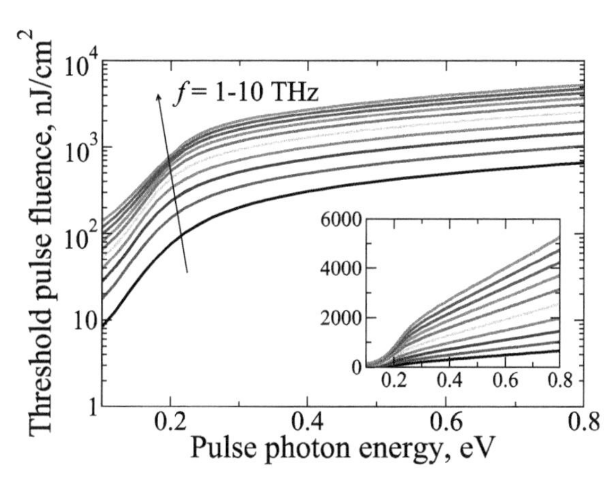

Figure 21.3 Dependence of threshold fluence for population inversion on pulse photon energy for different THz photon frequencies. Inset shows the same dependence but with a linear scale.

As has been discussed in Refs. [5, 8], the cooling for low pulse photon energies occurs because the initial photoexcitation can give a lower energy to the carriers than at thermal equilibrium. In fact, the energy of the photogenerated carriers, $\hbar\Omega/2$, is lower than the average energy of the carriers in intrinsic graphene at room temperature, $\varepsilon/n \approx 0.0567$ eV, when $\hbar\Omega < 0.13$ eV. Furthermore, a small amount of cooling after the heating caused by the photoexcitation can also occur (Figs. 21.2a–c). This arises from the discrete nature of the OP emission and absorption which means the carrier temperature cannot exactly match the lattice temperature. The lattice temperature must be reached via acoustic phonon scattering or radiative generation/recombination on a substantially longer time scale.

21.3.2 Time Evolution of Dynamic Conductivity for High- and Low-Quality Graphene

Dynamic conductivity in Eq. (21.5) depends primarily on the carrier distribution, i.e., on the carrier concentration, and also on the momentum relaxation time, τ. Because photoexcitation leads to population inversion and also to higher carrier concentration, the negative dynamic conductivity arises if the increase in the absolute value of interband conductivity overcomes the increase in Drude absorption in Eq. (21.5). To achieve a low threshold for negative dynamic conductivity in the THz range, it is necessary to suppress the Drude absorption. This can be done by using either high-quality graphene, where $\omega\tau \geq 1$, or low-quality graphene, where $\omega\tau \ll 1$. As can be seen in Eq. (21.5), in high-quality graphene, the high-frequency cut-off of the Drude conductivity occurs. In low-quality graphene, the Drude conductivity ceases over the entire frequency range.

Figure 21.4 shows the time evolution of normalized dynamic conductivity, Re $\sigma_{\omega}/(e^2/4\hbar)$, in the THz range with different pulse photon energies and momentum relaxation times. The pulse fluence for each pulse photon energy was chosen so that sufficiently large negative dynamic conductivity is obtained. It demonstrates that the negative dynamic conductivity can be achieved in the THz range with sufficiently high pulse fluence, and that the same level of negative dynamic conductivity can be achieved with a much lower pulse fluence for lower pulse photon energies. Furthermore, the magnitude, frequency range, and time duration of negative dynamic conductivity depend strongly on the momentum relaxation time.

For τ = 10 ps, negative dynamic conductivity occurs from 1 to greater than 10 THz, and it duration is 5–30 ps, depending on the frequency. The duration is longer for lower frequencies, as a result of the relaxation of the quasi-Fermi level shown in Fig. 21.2. For τ = 1 ps, negative dynamic conductivity occurs between 4 and 10 THz with a smaller magnitude and a much shorter duration. This indicates that the high-quality graphene with a longer momentum relaxation time is preferable. Figure 21.4 shows that the minimum frequency for negative dynamic conductivity is primarily determined by the momentum relaxation time, although it is slightly reduced for lower pulse photon energies.

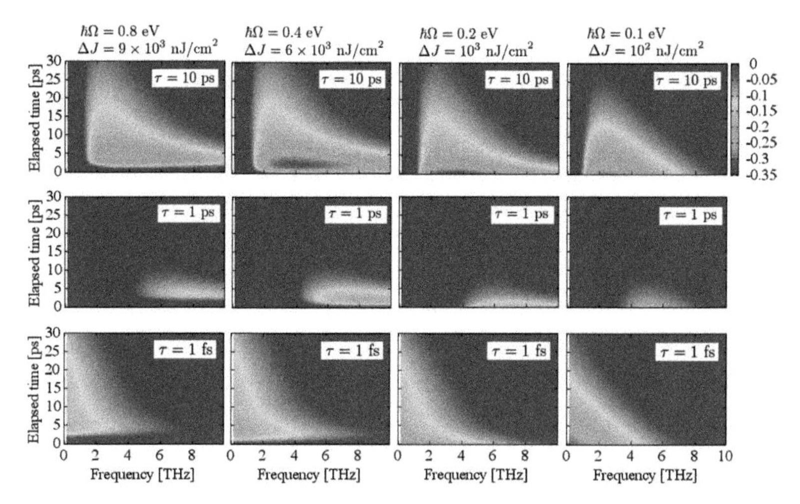

Figure 21.4 Time-evolution of dynamic conductivity in the THz range normalized to $e^2/4\hbar$ and cut off above zero. Columns from left to right correspond to $\hbar\Omega$ = 0.8, 0.4, 0.2, and 0.1 eV and $\Delta J = 9 \times 10^3$, 6×10^3, 1×10^3, and 1×10^2 nJ/cm^2, respectively. Rows from top to bottom correspond to τ = 10 ps, 1 ps, and 1 fs, respectively.

In contrast, for τ = 1 fs, negative dynamic conductivity occurs in the low frequency region, below 6 THz. However, that the meaning of the "time-dependent dynamic conductivity" is valid only if its variation is slower than the frequency under consideration, which is above several hundred GHz in this study. This means the negative dynamic conductivity can be achieved as low as several hundred GHz. The upper limit of the frequency range for this negative dynamic conductivity is lower than for the case where $\omega\tau \ll 1$. This is because the Drude conductivity is almost constant in the THz range, whereas it vanishes due to the high-frequency cut-off for $\omega\tau \gg 1$.

Our calculation show that even for $\hbar\Omega$ = 0.1 eV the threshold momentum relaxation time of negative dynamic conductivity is τ = 5 fs. Above this, dynamic conductivity does not occur in the THz range, regardless of how large the pulse fluence is. This threshold roughly corresponds to the value at which the Drude term in Eq. (21.5) at the zero-frequency limit exceeds the largest negative value of the interband contribution, $\tau \sim (\pi/8 \log 2)(\hbar/k_B T_l)$.

Figure 21.5 shows the maximum negative value of dynamic conductivity over its duration as a function of frequency and pulse fluence for different pulse photon energies and momentum relaxation times. It shows that the threshold pulse fluence and the pulse fluence required to obtain the same negative dynamic conductivity are decreased by several orders of magnitude for any frequency as the pulse photon energy is decreased. Comparing the cases where τ = 10 and 1 ps, it is seen that the decrease in the momentum relaxation time causes a large increase in the minimum frequency for the negative dynamic conductivity and an increase in the threshold pulse fluence for frequencies near the minimum. However, this is insignificant for frequencies sufficiently above the minimum because of the high-frequency cut-off of Drude conductivity. The minimum frequency for the negative dynamic conductivity is reduced for the lower pulse photon energy. The frequency can be as low as about 700 GHz for $\hbar\Omega$ = 0.1 eV and τ = 10 ps, whereas it is around 1.1 THz for $\hbar\Omega$ = 0.8 eV and τ = 10 ps, demonstrating that pumping with a lower photon energy is superior. For τ = 1 fs, the threshold pulse fluence for each photon energy is on the same order of magnitude as that for τ = 10 ps.

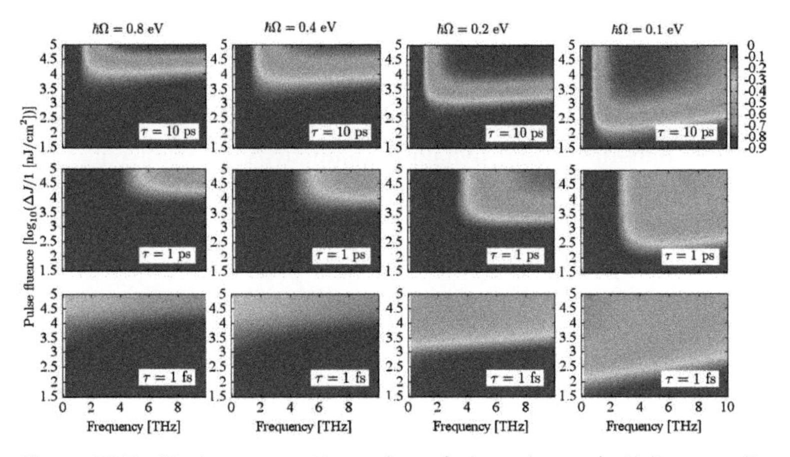

Figure 21.5 Maximum negative value of dynamic conductivity over its duration as a function of frequency and pulse fluence. Columns from left to right correspond to $\hbar\Omega$ = 0. 8, 0.4, 0.2, and 0.1 eV, respectively. Rows from top to bottom correspond to τ = 10 ps, 1 ps, and 1 fs, respectively.

In addition, the maximum is saturated below the fundamental limit, $-e^2/4\hbar$, for very high pulse fluences. For example, for τ = 10 ps and $\hbar\Omega$ = 0.1 eV, it is saturated with the value Re $\sigma_\omega/(e^2/4\hbar) \approx -0.4$ at f = 1 THz and ≈ -0.8 at f = 2 THz. The reason is that, for very high pulse fluences, the increase in Drude absorption caused by the increase in carrier concentration exceeds the interband contribution, even though a very high quasi-Fermi level is obtained, and the maximum negative value of dynamic conductivity, which corresponds to the saturation value, is reached at a later time after the relaxation of the carrier concentration. The Pauli blocking also contributes to the saturation at lower photon energies. The saturation values for $\tau = 1$ fs are smaller than those for $\tau = 10$ ps; in particular, it is about 3-fold smaller for $\hbar\Omega = 0.4$ and 0.8 eV. Again, this is due to the nonzero, almost constant value of Drude conductivity in the THz range with the limit $\omega\tau \ll 1$.

Above, we have considered a very short momentum relaxation time, ~1 fs, to demonstrate the possibility of negative dynamic conductivity in low-quality graphene. This might be achieved with high concentration of impurities, disorders, or grain boundaries, which correspond to graphene samples with carrier mobility roughly $10–10^2$ cm^2/Vs with electron concentration $\sim 10^{12}$ cm^{-2}. In such a case, modification of the graphene band structure caused by these imperfections is no longer negligible, and more sophisticated discussion is necessary. However, our discussion above should be a good starting point for negative dynamic conductivity in low-quality graphene.

21.3.3 Optical-Phonon Emission versus Other Recombination Mechanisms

In this chapter, we have considered interband OP emission alone as the dominant recombination mechanism. There are other recombination mechanisms allowed in graphene, which we will briefly compare with interband OP emission.

Radiative recombination in graphene takes place on a submicrosecond time scale, which is much slower than that of the interband OP emission, and can be neglected. Recombination via acoustic-phonon emission in disordered graphene also occurs on a submicrosecond time scale [24]. Recombination via interband plasmon emission in intrinsic graphene or optically pumped intrinsic graphene with the same amount of electrons and holes should be strongly suppressed by electron-hole friction [25], which results in a very short quasiparticle lifetime of plasmons, on the order of 0.1 ps.

The recombination via the interband tunneling between the boundaries of adjacent electron-hole puddles can be effective in graphene samples with puddle sizes of ~30 nm and an average carrier concentration fluctuation of $\sim 4 \times 10^{10}$ cm^{-2} [27]. However, it should be naturally suppressed in graphene samples with a larger puddle size and smaller carrier concentration fluctuation, such as in graphene on a very clean SiO_2 substrate [15] or graphene on hBN [28].

It has been suggested that Auger processes, which are forbidden by the vanishing phase space for the case of strictly linear band dispersion with no energy-level broadening, could be allowed in heavily doped, strongly pumped graphene with an electron concentration $>10^{13}$ cm^{-2} as a consequence of the appearance of the finite energy level broadening due to electron-electron interaction [29–31]. However, in intrinsic graphene with a moderate photocarrier concentration of $\sim 10^{12}$ cm^{-2} and a dielectric constant of $\epsilon = 5.5$, the screening is inefficient and the characteristic parameter of the electron-electron interaction, $\alpha^2_{ee} = (e^2/\epsilon\hbar v_F)^2 = 0.16$, is relatively small. In this case, the broadening should be small [32], at least in the energy range under consideration. Therefore, the characteristic time scale for Auger processes should be neglected compared with that for the interband OP emission. Moreover, Auger processes can be further suppressed by introducing a dielectric layer above graphene with a high dielectric constant. Other extrinsic mechanisms of energy level broadening, such as disorder, must be avoided to suppress Auger processes. Trigonal warping of the graphene band structure also allows the finite phase space for Auger processes [26], although the generation and recombination rates are expected to be negligible in the energy range under consideration.

This validates our assumption that only the interband OP emission is dominant compared with other recombination mechanisms. In conjunction with the discussion in preceding subsections, this suggests that the use of high-quality graphene with a long momentum relaxation time, low disorder, and few electron-hole puddles is required for achieving low-threshold negative dynamic conductivity. Even where the other recombination mechanisms cannot be neglected, the dependence of the threshold on the pulse photon energy should be unchanged. Quantitatively, the threshold pulse fluence should increase as the recombination rate increases.

21.4 Conclusions

We have investigated population inversion and negative dynamic conductivity in intrinsic graphene under pulse photoexcitation with pulse photon energy of 0.1–0.8 eV and their dependences on pulse fluence. We showed that the threshold pulse fluence can be lowered by two orders of magnitude with a lower pulse photon energy. This was attributed to the inverse proportionality of the photogenerated carrier concentration to the photon energy, as well as to the weaker carrier heating, which results in more carriers accumulating in the low-energy region and in the suppression of recombination via interband OP emission. We also showed that negative dynamic conductivity occurs within the long relaxation time limit, $\omega\tau \geq 1$, where Drude absorption in the THz range is reduced by the high-frequency cut-off, and within the very short relaxation time limit, $\omega\tau \ll 1$, where it ceases for the entire frequency range. We also demonstrated that pulse excitation with lower pulse photons has a lower threshold pulse fluence and a lower minimum frequency for negative dynamic conductivity, as low as 700 GHz with $\hbar\Omega = 0.1$eV and $\tau = 10$ ps.

Acknowledgments

This work was supported by JSPS Grant-in-Aid for Specially Promoted Research (#23000008) and by JSPS Grant-in-Aid for Young Scientists (B) (#23760300).

References

1. K. S. Novoselov, A. K. Geim, S. V. Morozov, D. Jiang, M. I. Katsnelson, I. V. Grigorieva, S. V. Dubonos, and A. A. Firsov, *Nature*, **438**, 197 (2005).
2. A. K. Geim and A. H. MacDonald, *Phys. Today*, **60**(8), 35 (2007).
3. F. Bonaccorso, Z. Sun, T. Hasan, and A. C. Ferrari, *Nat. Photonics*, **4**, 611 (2010).
4. V. Ryzhii, M. Ryzhii, and T. Otsuji, *J. Appl. Phys.*, **101**, 083114 (2007).
5. V. Ryzhii, M. Ryzhii, and T. Otsuji, *Phys. Status Solidi C*, **5**, 261 (2008).
6. V. Ryzhii, M. Ryzhii, A. Satou, T. Otsuji, A. A. Dubinov, and V. Ya. Aleshkin, *J. Appl. Phys.*, **106**, 084507 (2009).

7. A. Satou, T. Otsuji, and V. Ryzhii, *Jpn. J. Appl. Phys., Part 1*, **50**, 070116 (2011).
8. V. Ryzhii, M. Ryzhii, V. Mitin, A. Satou, and T. Otsuji, *Jpn. J. Appl. Phys., Part 1*, **50**, 094001 (2011).
9. V. Ryzhii, M. Ryzhii, V. Mitin, and T. Otsuji, *J. Appl. Phys.*, **110**, 094503 (2011).
10. V. Ryzhii, V. Mitin, M. Ryzhii, N. Ryabova, and T. Otsuji, *Appl. Phys. Express*, **1**, 063002 (2008).
11. V. Ryzhii, N. Ryabova, M. Ryzhii, N. V. Baryshnikov, V. E. Karasik, V. Mitin, and T. Otsuji, *Opto-Electron. Rev.*, **20**, 15 (2012).
12. L. Vicarelli, M. S. Vitiello, D. Coquillat, A. Lombardo, A. C. Ferrari, W. Knap, M. Polini, V. Pellegrini, and A. Tredicucci, *Nat. Mater.*, **11**, 865 (2012).
13. T. Otsuji, S. A. Boubanga Tombet, A. Satou, H. Fukidome, M. Suemitsu, E. Sano, V. Popov, M. Ryzhii, and V. Ryzhii, *J. Phys. D: Appl. Phys.*, **45**, 303001 (2012).
14. H. Karasawa, T. Komori, T. Watanabe, H. Fukidome, M. Suemitsu, V. Ryzhii, and T. Otsuji, *Int. J. Infrared Millimeter Waves*, **32**, 655 (2011).
15. S. Boubanga-Tombet, S. Chen, T. Watanabe, A. Satou, V. Ryzhii, and T. Otsuji, *Phys. Rev. B*, **85**, 035443 (2012).
16. D. Sun, Z.-K. Wu, C. Divin, X. Li, C. Berger, W. A. de Heer, P. N. First, and T. B. Norris, *Phys. Rev. Lett.*, **101**, 157402 (2008).
17. J. M. Dawlaty, S. Shivaraman, M. Chandrashekhar, F. Rana, and M. G. Spencer, *Appl. Phys. Lett.*, **92**, 042116 (2008).
18. M. Breusing, C. Ropers, and T. Elsaesser, *Phys. Rev. Lett.*, **102**, 086809 (2009).
19. K. Dani, J. Lee, R. Sharma, A. Mohite, C. Galande, P. Ajayan, A. Dattelbaum, H. Htoon, A. Taylor, and R. Prasankumar, *Phys. Rev. B*, **86**, 125403 (2012).
20. A. Satou, T. Otsuji, V. Ryzhii, and F. T. Vasko, in *Proceedings of the PIERS, Kuala Lumpur*, (2012), p. 486.
21. Although a more elaborated model based on the system of the Bloch equations which takes into account quantum many-particle effects has been developed in the literature [for instance, see E. Malic, T. Winzer, E. Bobkin, and A. Knorr, *Phys. Rev. B*, **84**, 205406 (2011)], the quasi-classical approach used here suffices for the qualitative analysis of the threshold behavior of the population inversion and negative dynamic conductivity.

22. H. Suzuura and T. Ando, *J. Phys. Soc. Jpn.*, **77**, 044703 (2008).
23. L. A. Falkovsky and A. A. Varlamov, *Eur. Phys. J. B*, **56**, 281 (2007).
24. F. Vasko and V. Mitin, *Phys. Rev. B*, **84**, 155445 (2011).
25. D. Svintsov, V. Vyurkov, S. Yurchenko, T. Otsuji, and V. Ryzhii, *J. Appl. Phys.*, **111**, 083715 (2012).
26. L. E. Golub, S. A. Tarasenko, M. V. Entin, and L. I. Magarill, *Phys. Rev. B*, **84**, 195408 (2011).
27. V. Ryzhii, M. Ryzhii, and T. Otsuji, *Appl. Phys. Lett.*, **99**, 173504 (2011).
28. J. Xue, J. Sanchez-Yamagishi, D. Bulmash, P. Jacquod, A. Deshpande, K. Watanabe, T. Taniguchi, P. Jarillo-Herrero, and B. J. LeRoy, *Nat. Mater.*, **10**, 282 (2011).
29. T. Winzer and E. Malić, *Phys. Rev. B*, **85**, 241404(R) (2012).
30. S. Tani, F. Blanchard, and K. Tanaka, *Phys. Rev. Lett.*, **109**, 166603 (2012).
31. D. Brida, A. Tomadin, C. Manzoni, Y. J. Kim, A. Lombardo, S. Milana, R. R. Nair, K. S. Novoselov, A. C. Ferrari, G. Cerullo, and M. Polini, e-print arXiv:1209.5729v1.
32. V. Kotov, B. Uchoa, V. Pereira, F. Guinea, and A. Castro Neto, *Rev. Mod. Phys.*, **84**, 1067 (2012).

Chapter 22

Double Injection in Graphene p-i-n Structures*

V. Ryzhii,[a] I. Semenikhin,[b] M. Ryzhii,[c] D. Svintsov,[b] V. Vyurkov,[b] A. Satou,[a] and T. Otsuji[a]

[a]*Research Institute for Electrical Communication, Tohoku University, Sendai 980-8577, Japan*

[b]*Institute of Physics and Technology, Russian Academy of Sciences, Moscow 117218, Russia*

[c]*Department of Computer Science and Engineering, University of Aizu, Aizu-Wakamatsu 965–8580, Japan*

v-ryzhii@u-aizu.ac.jp

We study the processes of the electron and hole injection (double injection) into the i-region of graphene-layer and multiple graphene-layer p-i-n structures at the forward bias voltages. The hydrodynamic equations governing the electron and hole transport in graphene coupled with the two-dimensional Poisson equation

*Reprinted with permission from V. Ryzhii, I. Semenikhin, M. Ryzhii, D. Svintsov, V. Vyurkov, A. Satou, and T. Otsuji (2013). Double injection in graphene p-i-n structures, *J. Appl. Phys.*, **113**, 244505. Copyright © 2013 AIP Publishing LLC.

Graphene-Based Terahertz Electronics and Plasmonics: Detector and Emitter Concepts
Edited by Vladimir Mitin, Taiichi Otsuji, and Victor Ryzhii

ISBN 978-981-4800-75-4 (Hardcover), 978-0-429-32839-8 (eBook)
www.jennystanford.com

are employed. Using analytical and numerical solutions of the equations of the model, we calculate the band edge profile, the spatial distributions of the quasi-Fermi energies, carrier density and velocity, and the current-voltage characteristics. In particular, we demonstrated that the electron and hole collisions can strongly affect these distributions. The obtained results can be used for the realization and optimization of graphene-based injection terahertz and infrared lasers.

22.1 Introduction

Due to unique properties, graphene-layer (GL) and multiple-graphene-layer (MGL) structures [1] with p-n and p-i-n junctions are considered as novel building blocks for a variety of electron, terahertz, and optoelectronic devices. Such junctions have been realized using both chemical doping [2–6] and "electrical" doping in the gated structures [7–10]. The formation of p-n and p-i-n junctions with the electrically induced p- and n-regions is possible even in MGL structures with rather large number of GLs [11]. Different devices based on p-i-n junctions in GL and MGL structures have been proposed recently. In particular, the GL and MGL structures with reversed biased p-i-n junctions can be used in infrared (IR) and terahertz (THz) detectors [12–19] and the tunneling transit-time THz oscillators [20, 21]. The GL and MGL p-i-n junctions under the forward bias can be the basis of IR and THz lasers exploiting the interband population inversion [22–25] (see also recent experimental results [26–28]), in which the injection of electrons and holes from the n- and p-regions is utilized [29, 30] instead of optical pumping. A simplified model of the GL and MGL injection lasers with the p-i-n junctions was considered recently [30]. The model in question assumes that the recombination of electrons and holes in the i-region and the leakage thermionic and tunneling currents at the p-i and i-n interfaces are relatively small. This situation can occur in the structures with sufficiently short i-regions or at relatively low temperatures when the recombination associated with the emission of optical phonons and thermionic leakage are weakened. In such a case, different components of the current can be considered as perturbations, and the spatial distributions of the electric potential and the carrier density along the i-region are virtually uniform. However, in the p-i-n structures with relatively long i-regions and at the elevated voltages the spatial distribution of the potential can be rather nonuniform, particularly, near the edges of the i-region, i.e., near the p-i and i-n interfaces. In this case, the electric field in the i-region can be sufficiently strong. Such an effect can markedly influence the density of the injected carriers, the conditions of the population inversion, and current-voltage characteristics.

In this chapter, we develop a model for the GL and MGL forward biased p-i-n structures which accounts for relatively strong recombination and consider its effect on the characteristics which can be important for realization of IR/THz injection lasers. The problem of the electron and hole injection (double injection) in GL and MGL p-i-n structures is complicated by the two-dimensional geometry of the device and the features of the carrier transport properties. This makes necessary to use two-dimensional Poisson equation for the self-consistent electric potential around the i-region and to invoke the hydrodynamic equations for carrier transport in GL and MGLs. Thus, the present model is a substantial generalization of the model applied in the previous paper [30]. To highlight the effects of strong recombination and the nonuniformity of the potential distributions, we, in contrast to Refs. [30, 31], disregard for simplicity the effects of the carrier heating or cooling and the effects associated with the nonequilibrium optical phonons. This can be justified by the fact that the injection of electrons and holes from the pertinent contacts does not bring a large energy to the electron-hole system in the i-region unless the applied bias voltage is large.

The chapter is organized as follows. In Section 22.2, we present the GL and MGL p-i-n structures under consideration and the device mathematical model. Section 22.3 deals with an analytical solutions of the equations of the model for a special case. Here the general equations are reduced to a simple equation for the quasi-Fermi energies of electrons and holes as functions of the applied bias voltage. In Section 22.4, the equations of the model are solved numerically for fairly general cases. The calculations in Sections 22.3 and 22.4 provide the spatial distributions of the band edges, the quasi-Fermi energies and densities of electrons and holes, and their average (hydrodynamic) velocities. It is shown that the obtained analytical distributions match well with those obtained numerically, except very narrow regions near the p-i and i-n interfaces. In Section 22.5, the obtained dependences are used for the calculation of the current-voltage characteristics of the GL and MGL p-i-n structures. Section 22.6 deals with the discussions of some limitations of the model and possible consequences of their lifting. In Section 22.7 we draw the main conclusions.

22.2 Model

22.2.1 The Structures Under Consideration

We consider the GL or MGL p-i-n structures with either chemically or electrically doped p- and n-region (see the sketches of the structures on the upper and lower panels in Fig. 22.1). We assume that the p- and n-regions (called in the following p- and n-contacts) in single- or

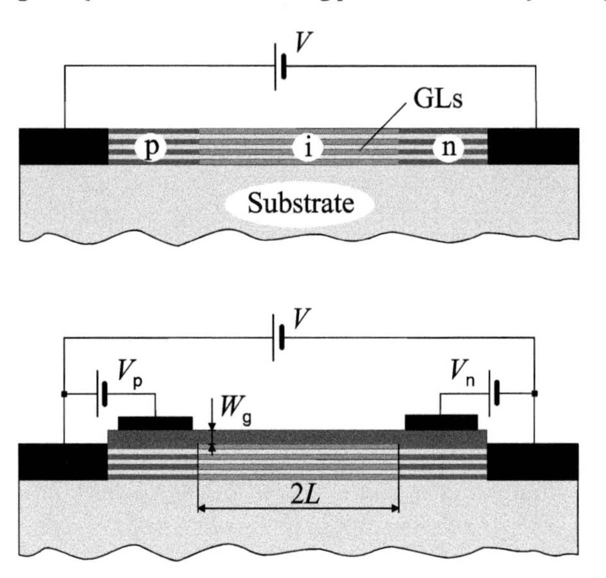

Figure 22.1 Schematic view of the cross-sections of MGL p-i-n structures with chemically doped n- and p-contact regions (upper panel) and with such regions electrically induced by the side gate-voltages $V_p = -V_g$ and $V_n = V_g > 0$ (lower panel).

multiple-GL structures are strongly doped, so that the Fermi energy in each GL in these structures ε_{Fc} and hence the built-in voltage V_{bi} are sufficiently large

$$\varepsilon_{\rm Fc} = eV_{\rm bi}/2 \gg k_{\rm B}T,$$

where T is the temperature and $k_{\rm B}$ is the Boltzmann constant. In the single-GL structures with the electrically induced p- and n-contact regions

$$\varepsilon_{\rm Fc} = eV_{\rm bi}/2 = \frac{\hbar \upsilon_{\rm W}}{2}\sqrt{\frac{\kappa V_{\rm g}}{eW_{\rm g}}},$$

where $\hbar$ and κ are the Planck constant and the dielectric constant, respectively, $\upsilon_{\rm W} \simeq 10^8$ cm/s is the characteristic velocity of electrons and holes in GLs, $W_{\rm g}$ is the gate layer thickness, and $V_{\rm g}$ is the gate voltage ($V_{\rm p} = -V_{\rm g}$ and $V_{\rm n} = V_{\rm g}$ (see Fig. 22.1)). Figure 22.2 shows the schematic views of the band profiles of a GL p-i-n structure at the equilibrium (see Fig. 22.2a) and at the forward bias (see Fig. 22.2b). The Fermi energies of the holes and electrons in the contact p- and n-regions are assumed to be $\varepsilon_{\rm Fc}$. It is shown that at $V = 0$, the densities of electron and hole (thermal) in the i-region are relatively small. However, at the forward bias these densities become

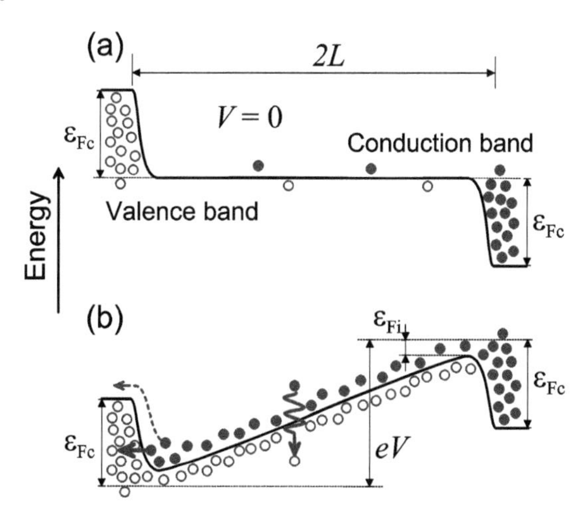

Figure 22.2 Qualitative view of band profiles of a GL p-i-n structure (a) at $V = 0$ and (b) at forward bias $V > 0$. Opaque and open circles correspond to electrons and holes, respectively. Wavy, straight, and dashed arrows correspond to the recombination in the i-region (assisted by optical phonon emission), tunneling at the contact, and thermionic leakage to the contact, respectively.

rather large due to the injection. The injection is limited by the hole charge near the p-i junction and by the electron charge near the i-n junction. Due to this, the Fermi energy of the injected electrons and holes, ε_{Fi}, at the pertinent barriers is generally smaller than ε_{Fc}. However, at the bias voltages comparable with the build-in voltage V_{bi}, these Fermi energy can be close to each other. The qualitative band profiles of Fig. 22.2 are in line with the band profiles found from the self-consistent numerical calculation shown in Section 22.4.

Due to high electron and hole densities in the i-region under the injection conditions, the electron and hole energy distribution are characterized by the Fermi functions with the quasi-Fermi (counted from the Dirac point) ε_{Fe} and ε_{Fh}, respectively, and the common effective temperature T, which is equal to the lattice temperature.

22.2.2 Equations of the Model

The model under consideration is based on the two-dimensional Poisson equation for the self-consistent electric potential and the set of hydrodynamic equations describing the electron and hole transport along the GL (or GLs) in the i-region. The Poisson equation for the two-dimensional electric potential $\psi = \psi(x, z)$ is presented as

$$\frac{\partial^2 \psi}{\partial x^2} + \frac{\partial^2 \psi}{\partial z^2} = \frac{4\pi e}{\kappa}(\Sigma_e - \Sigma_d - \Sigma_h + \Sigma_a) \cdot \delta(z), \tag{22.1}$$

where $e = |e|$ is the electron charge, Σ_e and Σ_h are the net electron and hole sheet densities in the system (generally consisting of K GLs), $\Sigma_d = \Sigma_d(x)$ and $\Sigma_a = \Sigma_a(x)$ are the densities of donors and acceptors in the i-region (located primarily near the contact regions), and $\delta(z)$ is the delta function reflecting the fact that the GL (MGL system) are located in a narrow layer near the plane $z = 0$. The axis x is directed in this plane.

The transport of electrons and holes is governed by the following system of hydrodynamic equations [32]:

$$\frac{d\Sigma_e u_e}{dx} = -R, \quad \frac{d\Sigma_h u_h}{dx} = -R, \tag{22.2}$$

$$\frac{1}{M}\frac{d(e\varphi - \varepsilon_{Fe})}{dx} = \nu u_e + \nu_{eh}(u_e - u_h), \tag{22.3}$$

$$-\frac{1}{M}\frac{d(e\varphi + \varepsilon_{Fh})}{dx} = \nu u_h + \nu_{eh}(u_h - u_e), \tag{22.4}$$

$$\Sigma_{\rm e} = \frac{2k_{\rm B}^2T^2K}{\pi\hbar^2 v_{\rm W}^2}\int_0^\infty \frac{dyy}{\left[\exp\left(y-\varepsilon_{\rm Fe}/k_{\rm B}T\right)+1\right]}, \tag{22.5}$$

$$\Sigma_{\rm h} = \frac{2k_{\rm B}^2T^2K}{\pi\hbar^2 v_{\rm W}^2}\int_0^\infty \frac{dyy}{\left[\exp\left(y-\varepsilon_{\rm Fh}/k_{\rm B}T\right)+1\right]}, \tag{22.6}$$

Here $\varphi = \psi|_{z=0}$ is the potential in the GL plane, $\varepsilon_{\rm Fe}$, $\varepsilon_{\rm Fh}$, $u_{\rm e}$, and $u_{\rm h}$ are the quasi-Fermi energies and hydrodynamic velocities of electrons and holes, respectively, ν is the collision frequency of electrons and holes with impurities and acoustic phonons, $\nu_{\rm eh}$ is their collision frequency with each other, and M is the fictitious mass, which at the Fermi energies of the same order of magnitude as the temperature can be considered as a constant. The recombination rate R in the case of dominating optical phonon mechanism can be presented in the following simplified form (see, for instance, Refs. [30, 31, 33]):

$$R = \frac{K\Sigma_{\rm T}}{\tau_{\rm R}}\left[\frac{(\mathcal{N}_0+1)}{N_0}\exp\left(\frac{\varepsilon_{\rm Fe}+\varepsilon_{\rm Fh}-\hbar\omega_0}{k_{\rm B}T}\right)-1\right]$$
$$\simeq \frac{K\Sigma_{\rm T}}{\tau_{\rm R}}\left[\exp\left(\frac{\varepsilon_{\rm Fe}+\varepsilon_{\rm Fh}}{k_{\rm B}T}\right)-1\right]. \tag{22.7}$$

Here $\Sigma_{\rm T} = \pi k_{\rm B}^2T^2/6\hbar^2 v_{\rm W}^2$ is the equilibrium density of electrons and holes in the i-region in one GL at the temperature T, $\tau_{\rm R} = \Sigma_{\rm T}/G_{\rm T}$ is the characteristic time of electron-hole recombination associated with the emission of an optical phonon, $G_{\rm T} \propto \mathcal{N}_0$ is the rate of thermogeneration of the electron-hole pairs due to the absorption of optical phonons in GL at equilibrium, $\hbar\omega_0$ is the optical phonon energy, and $\mathcal{N}_0 = [\exp(\hbar\omega_0/k_{\rm B}T) - 1]^{-1}$.

22.2.3 Boundary Conditions

The boundary conditions for the electron and hole velocities and Fermi energies are taken in the following form:

$$u_{\rm e}|_{x=-L} = -u_{\rm R}, \qquad u_{\rm h}|_{x=-L} = u_{\rm R}, \tag{22.8}$$

and

$$\varepsilon_{\rm Fe}|_{x=L} = \varepsilon_{\rm Fh}|_{x=-L} = \varepsilon_{\rm Fc}/2 = eV_{\rm bi}/2, \tag{22.9}$$

where $2L$ is the spacing between p- and n-region (length of the i-region), $u_{\rm R}$ is the recombination velocity of electrons and holes in narrow space-charge regions adjacent to the p- and n-region, respectively, due to the interband tunneling and due to the leakage over the barriers (edge recombination velocity) [7, 34] and V is the applied bias voltage (see Fig. 22.1). The quantity ε_{Fc} is the Fermi energy at the contact regions. The electric potential at the contact regions is determined by the applied voltage V

$$(e\psi + \varepsilon_{Fh})|_{x\le -L, z=0} = \frac{eV}{2}, \tag{22.10}$$

$$(e\psi - \varepsilon_{Fe})|_{x\ge L, z=0} = -\frac{eV}{2}, \tag{22.11}$$

where V is the applied bias voltage. Combining Eqs. (22.9)–(22.11), one obtains

$$\psi|_{x\le -L, z=0} = \varphi|_{x=-L} = \frac{e(V - V_{bi})}{2}, \tag{22.12}$$

$$\psi|_{x\ge L, z=0} = \varphi|_{x=L} = -\frac{e(V - V_{bi})}{2}. \tag{22.13}$$

The dependence $e\varphi = e\varphi(x)$ yields the coordinate dependence of the Dirac point or the band edge profile (with respect to its value in the i-region center, $x = 0$).

22.2.4 Dimensionless Equations and Boundary Conditions

To single out the characteristic parameters of the problem we introduce the following dimensionless variables: $\Psi = e\psi/k_BT$, $\Phi = e\varphi/T$, $\mu_e = \varepsilon_{Fe}/T$, $\mu_h = \varepsilon_{Fh}/k_BT$, $\mu_c = \varepsilon_{Fc}/k_BT$, $\sigma_e = \Sigma_e/\Sigma_T$, $\sigma_h = \Sigma_h/\Sigma_T$, $\sigma_d = \Sigma_d/\Sigma_T$, $\sigma_a = \Sigma_a/\Sigma_T$, $U_e = u_e\tau_R/L$, $U_h = u_h\tau_R/L$, $U_R = u_R\tau_R/L$, $\xi = x/L$, and $\zeta = z/L$. Using these variables, Eqs. (22.1) and (22.6) are presented as

$$\frac{\partial^2\Psi}{\partial^2\xi^2} + \frac{\partial^2\Psi}{\partial\zeta^2} = 4\pi Q(\sigma_e - \sigma_d - \sigma_h + \sigma_a)\cdot\delta(\zeta), \tag{22.14}$$

$$\frac{d\sigma_e U_e}{d\xi} = 1 - \exp(\mu_e + \mu_h), \tag{22.15}$$

$$\frac{d\sigma_h U_h}{d\xi} = 1 - \exp(\mu_e + \mu_h), \tag{22.16}$$

$$\frac{d(\Phi-\mu_e)}{d\xi}=q\left[U_e\left(\frac{\nu}{\nu_{eh}}\right)+U_e-U_h\right], \tag{22.17}$$

$$-\frac{d(\Phi+\mu_h)}{d\xi}=q\left[U_h\left(\frac{\nu}{\nu_{eh}}\right)+U_h-U_e\right], \tag{22.18}$$

$$\sigma_e=\frac{12}{\pi^2}\int_0^\infty\frac{dyy}{\left[\exp(y-\mu_e)+1\right]}, \tag{22.19}$$

$$\sigma_h=\frac{12}{\pi^2}\int_0^\infty\frac{dyy}{\left[\exp(y-\mu_h)+1\right]}. \tag{22.20}$$

Here Q and q are the "electrostatic" and "diffusion" parameters given, respectively, by the following formulas:

$$Q=K\frac{\pi e^2 L k_B T}{6\kappa\hbar^2 \upsilon_W^2},\ q=\frac{M\nu_{eh}L^2}{\tau_R k_B T}.$$

Assuming, L = 1 – 5 μm, K = 1 – 2, κ = 4, T = 300 K. ε_{Fc} = 100 meV, $\nu_{eh} = 10^{13}$ s^{-1}, and $\tau_R = 10^{-9}$ s, we obtain Q = 12.5–125, q = 0.04–1.0. The same values of q one obtains if $\nu_{eh} = 10^{12}$ s^{-1} and $\tau_R = 10^{-10}$ s. The parameters Q and q can also be presented as $4\pi Q \sim L/r_S^T$ and $q \sim (L^2/D\tau_R) \simeq (L/L_D)^2$, where $r_S^T = (\kappa\hbar^2\upsilon_W^2/4e^2T)$ and $L_D = \sqrt{(D\tau_R)}$ are the characteristic screening and the diffusion lengths in the i-region (in equilibrium when $\mu_i = 0$), respectively, and $D = \upsilon_W^2/2\nu_{eh}$ is the diffusion coefficient.

The boundary conditions for the set of the dimensionless Eqs. (22.14)–(22.20) are given by

$$\mu_e\,|_{\xi=1}=\mu_h\,|_{\xi=-1}=\mu_c=eV_{bi}/2k_BT, \tag{22.21}$$

$$U_e\,|_{\xi=-1}=-U_R,\quad U_h\,|_{\xi=1}=U_R, \tag{22.22}$$

$$\Psi\,|_{\xi\le-1,\zeta=0}=\Phi\,|_{\xi=-1}=eV/2k_BT-\mu_c, \tag{22.23}$$

$$\Psi\,|_{\xi\ge1,\zeta=0}=\Phi\,|_{\xi=1}=-eV/2k_BT+\mu_c. \tag{22.24}$$

In the following we focus our consideration mainly on the structures with very abrupt p-i and i-p junction neglecting terms σ_d and σ_a in Eq. (22.14). The effect of smearing of these junctions resulting in $\sigma_d \neq 0$ and $\sigma_a \neq 0$ will be briefly discussed in Section 22.6.

22.3 Injection Characteristics (Analytical Approach for a Special Case)

Consider the special case in which v/v_{eh}, $U_R \ll 1$. When $Q \gg 1$, Eq. (22.14) is actually a partial differential equation with a small parameter ($Q^{-1} \ll 1$) at highest derivatives. Usually the net solution of such equations can be combined from their solutions disregarding the left side (valid in a wide range of independent variables) and the solution near the edges (which are affected by the boundary conditions) in the regions of width $\eta \sim Q^{-1} \sim r_S^T/L \ll 1$, where the derivatives are large, provided a proper matching of these solutions [35]. Hence, since parameter $Q \gg 1$, for a wide portion of the i-region one can assume that $\sigma_e \simeq \sigma_h \simeq \sigma_i = const$ and, consequently, $\mu_e \simeq \mu_h = \mu_i$.

In this case, Eqs. (22.15) and (22.16) yield

$$U_e = -\frac{\left(e^{2\mu_i} - 1\right)}{\sigma_i}(\xi + 1), \tag{22.25}$$

$$U_h = -\frac{\left(e^{2\mu_i} - 1\right)}{\sigma_i}(\xi - 1). \tag{22.26}$$

It is instructive that the condition $\mu_e = \mu_h = \mu_i \simeq const$, assumed above, is fulfilled only if the collisions with impurities and acoustic phonons are disregarded. Therefore, in this section we assume that $v \ll v_{eh}$. Considering this, we find

$$\Phi = -2q\frac{\left(e^{2\mu_i} - 1\right)}{\sigma_i}\xi. \tag{22.27}$$

At the points $\xi = \pm(1 - \eta) \simeq \pm 1$, one can use the simplified matching conditions (see Fig. 22.2b) and obtain the following equation for μ_i:

$$\mu_i + 2q\frac{\left(e^{2\mu_i} - 1\right)}{\sigma_i} = \frac{eV}{2k_BT} \tag{22.28}$$

or

$$\mu_i + \frac{2q\pi^2\left(e^{2\mu_i} - 1\right)}{12\int_0^\infty \frac{dyy}{\left[\exp(y - \mu_i) + 1\right]}} = \frac{eV}{2k_BT}. \tag{22.29}$$

If $eV < 2k_BT$, one can expect that $2\mu_i < 1$, so that $\sigma_i \simeq 1$, and Eq. (22.29) yields

$$\mu_i \simeq \frac{eV}{2k_BT(1+4q)}. \tag{22.30}$$

At $eV > 2T$, $2\mu_i > 1$ (the electron and hole components are degenerate), from Eq. (22.29)

$$\mu_i + \frac{\pi^2 q}{3\mu_i^2} e^{2\mu_i} = \frac{eV}{2k_BT}. \tag{22.31}$$

If formally $q \to 0$ (very long diffusion length or near ballistic transport of electrons and holes in the i-region), one can see from Eqs. (22.29)–(22.31) that μ_i tends to $eV/2T$, i.e., the Fermi energies of electrons and holes tend to eV/2 (as in the previous paper [30]).

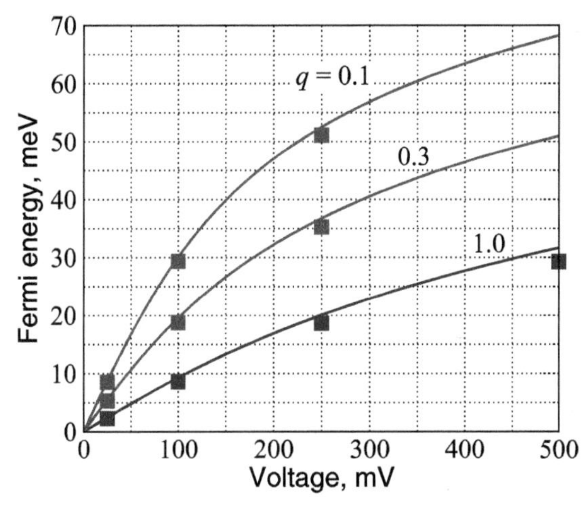

Figure 22.3 Fermi energy in the i-region ε_F as a function of the bias voltage V for different values of parameter q. Opaque squares correspond to electron (hole) Fermi energies at $x = 0$ calculated numerically in Section 22.4.

Figure 22.3 shows the voltage dependences of the Fermi energy, $\varepsilon_{Fe} = T\mu_i \simeq \varepsilon_{Fh}$, of electrons and holes in the main part of the i-region calculated using Eq. (22.29) for different values of parameter $q \propto (L/L_D)^2$. The dependences shown in Fig. 22.3 imply that the recombination leads to the natural decrease in the Fermi energies and the densities of electrons and holes in the i-region. These quantities increase with increasing applied voltage V, but such an increase is a slower function of V (logarithmic) than the linear one. The results of the calculations based on Eq. (22.29) (i.e., based on a simplified model valid for weak electron and hole collisions with impurities and acoustic phonons and the edge recombination) practically coincide with the results of numerical calculations involving a rigorous model (shown by the markers in Fig. 22.3). The results of these calculations are considered in Section 22.4.

22.4 Numerical Results

Finding of analytical solutions of Eq. (22.14) coupled with Eqs. (22.15)–(22.20) in the near-edge region and their matching with the smooth solutions in the main portion of the i-region is a difficult problem due to the nonlinearity of these systems of equations and its integro-differential nature. In principle, what could such solutions provide is an information on the width of the near-edge regions. The latter is not so important, because the main characteristics of the structures under consideration are determined by the electron-hole system in the structure bulk (in the main portion of the i-region). However, the net solution is particularly interesting to verify the results of the analytical model considered in Section 22.3. For this purpose, we solved Eqs. (22.14)–(22.20) numerically by iterations.

Due to large values of parameter Q, one could expect very sharp behavior of Ψ and Φ near the edges of the i-region, so that strongly nonuniform mesh was used. The potential Φ was found by reducing the Poisson Eq. (22.14) to the following:

$$\Phi(\xi) = -\frac{V/T - 2\mu_{\rm c}}{\pi}\sin^{-1}\xi + \frac{4\pi Q}{k}\int_{-1}^{1} d\xi' G(\xi,\xi')[\sigma_{\rm h}(\xi') - \sigma_{\rm e}(\xi')], \qquad (22.32)$$

where

$$G(\xi,\xi') = \frac{1}{4\pi}\ln\left\{\frac{1+\cos\left[\sin^{-1}(\xi)+\sin^{-1}(\xi')\right]}{1-\cos\left[\sin^{-1}(\xi)-\sin^{-1}(\xi')\right]}\right\}$$

is the Green function, which corresponds to the geometry and the boundary conditions under consideration.

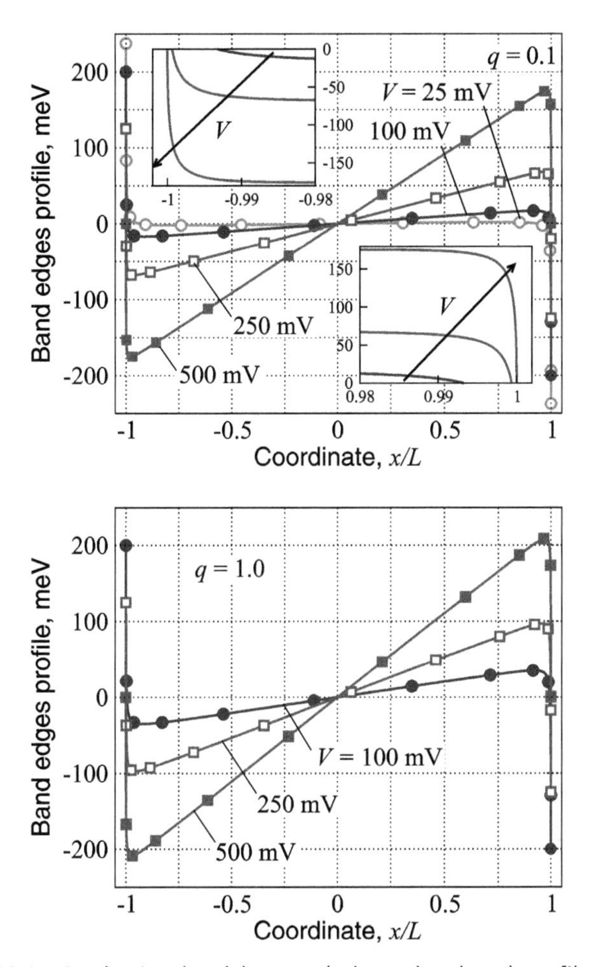

Figure 22.4 Conduction band bottom (valence band top) profiles in the i-region calculated for different bias voltages V at q = 0.1 (upper panel) and q = 1.0 (lower panel) at Q = 100. The extreme left and right markers (at x/L = $\mp$ 1) show the positions of the Dirac point at p-i and i-n junctions. Insets show detailed behavior in close vicinities near the i-region edges.

Figure 22.4 shows the profiles of the band edges, i.e., the conduction band bottom (valence band top) in the i-region ($-L \leq x \leq L$) obtained from numerical solutions of Eqs. (22.14)–(22.20) for different values of the bias voltage V and parameter q at Q = 100 (and v/v_{eh}, $U_R \ll 1$ as in Section 22.3). As follows from Fig. 22.4, an increase in the bias voltage V leads to the rise of the barrier for the electrons injected from the n-region and the holes from the p-region. These barriers are formed due to the electron and hole self-consistent surface charges localized in very narrow region (of the normalized width $\eta \sim Q^{-1} = 0.01$) in the i-region near its edges. The electric field near the edges changes the sign. In the main portion of the i-region, the electric field is almost constant. Its value increases with increasing parameter q, i.e., with reinforcement of the recombination. Figures 22.5 and 22.6 demonstrate the spatial distributions of the electron Fermi energy $\varepsilon_{Fe}(x)$ and the electron density $\Sigma_e(x)$ in the i-region for $q = 1.0$ at different bias voltages. The hole Fermi energy and the hole density are equal, respectively, to $\varepsilon_{Fh}(x) = \varepsilon_{Fe}(-x)$ and $\Sigma_h(x) = \Sigma_e(-x)$.

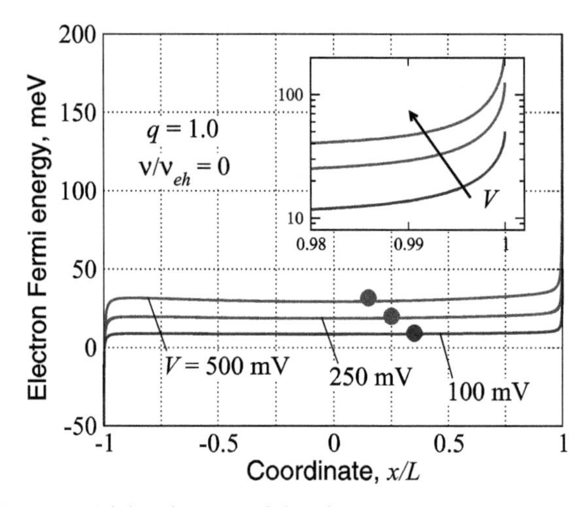

Figure 22.5 Spatial distributions of the electron Fermi energy ε_{Fe} at $q = 1$ and $Q = 100$ calculated for different values of bias voltage V at $U_R = 0$. Inset shows dependences near the n-contact. Opaque circles correspond to data obtained using analytical model.

One can see from Figs. 22.5 and 22.6 that the electron Fermi energy and the electron density steeply drop in fairly narrow region adjacent to the p-region. However, they remain virtually constant in the main portion of the i-region if the ratio of the collisions frequency of electrons and holes with impurities and acoustic phonons and the frequency of electron-hole collisions $\nu/\nu_{eh} \ll 1$. In the latter case, the values of these quantities across almost the whole i-region coincide with those obtained analytically above with a high accuracy (see

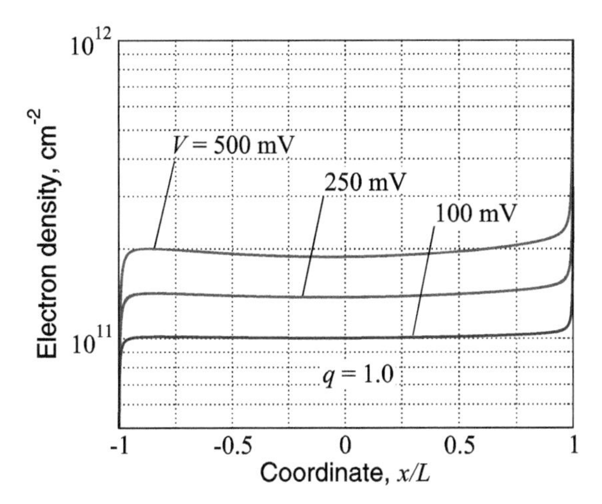

Figure 22.6 Spatial distributions of electron density Σ_e at different bias voltages V (v/v_{eh} = 0 and U_R = 0).

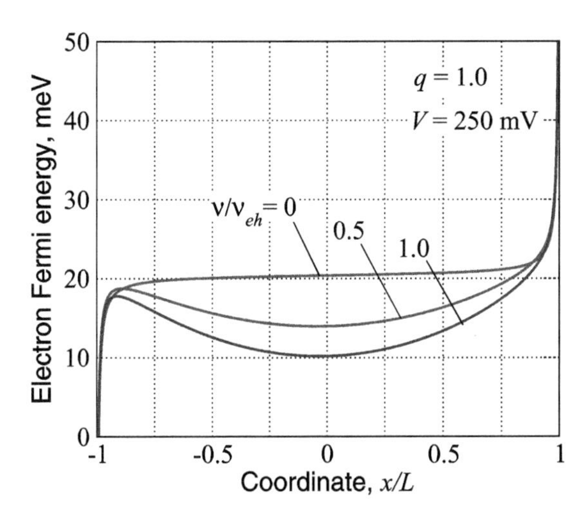

Figure 22.7 Spatial distributions of the electron Fermi energy ε_{Fe} at q = 1 and Q = 100 calculated for different values of ratio v/v_{eh} and V = 250 mV at U_R = 0.

the markers in Figs. 22.3 and 22.5). This justifies the simplified approach used in the analytical model for the case v/v_{eh}, $U_R \ll 1$ developed in Section 22.3. Nevertheless, at not too small values of parameter v/v_{eh}, the coordinate dependences of the electron (and hole) Fermi energy exhibit a pronounced sag (see Fig. 22.7). Similar sag was observed in the coordinate dependences of the electron and hole densities (not shown). As seen from Fig. 22.8, an increase in the ratio v/v_{eh} leads also to a pronounced modification of the coordinate dependences of the electron (hole) velocity: the absolute value of the electron velocity markedly decreases with increasing v/v_{eh}. However, the changes in the band edge profiles and hence the electric field in the main portion of the i-region are insignificant.

Above, we have considered the cases of the p-i-n structures with sufficiently long i-region, so that the normalized edge recombination velocity $U_R \ll 1$. In such cases, in accordance with the boundary conditions U_e (and U_h) tends to zero at the pertinent contact region (see Fig. 22.8).

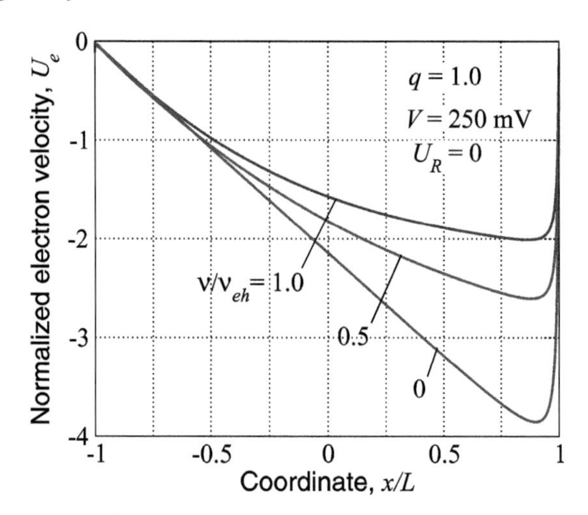

Figure 22.8 Coordinate dependences of normalized electron velocity U_e calculated for $q = 1.0$ and $V = 250$ mV at different values of v/v_{eh}.

The calculations of the spatial distributions of the band edge profiles, Fermi energies, and carrier concentrations showed that the variation of the edge recombination velocity U_R (at least in the range from zero to ten) does not lead to any pronounced distinction even in the near-edge regions. In all cases considered, as seen from Fig. 22.9, the absolute value of U_e steeply increases in a narrow region near the n-contact and the gradually drops (virtually linearly) across the main portion of the i-region. However, U_e (and, hence, the electron hydrodynamic velocity u_e), being virtually independent of U_R from the n-region to the near-edge region adjacent to the p-i interface, strongly depends on U_R in the latter region (see Fig. 22.9 (both left and right panels)). The hole velocity exhibits the same behavior with $U_\mathrm{h}(x) = -U_\mathrm{e}(-x)$.

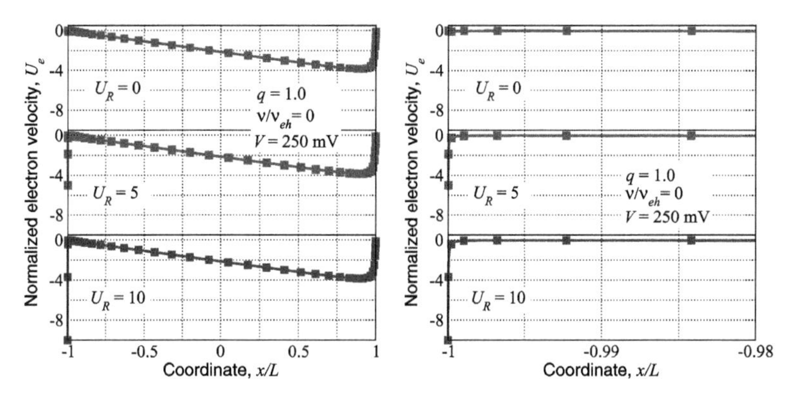

Figure 22.9 The normalized electron velocity U_e versus coordinate x for $q = 1.0$ and $V = 250$ mV at different values of normalized edge recombination velocity U_R (left panel). Right panel shows details of the same dependences in close vicinity of the p-i interface.

Due to rather short near-contact regions, the edge recombination does not affect substantially the integral characteristics of the p-i-n structures under consideration, in particular total number of the injected electrons and holes and their net recombination rate at least in the practical p-i-n structures with $Q \gg 1$ and not too large U_R (i.e., not too short i-regions).

22.5 Current-Voltage Characteristics

The current (associated with the recombination in the i-region) can be calculated as

$$J = e\int_{-\mathrm{L}}^{\mathrm{L}} dxR = \frac{eKL\Sigma_\mathrm{T}}{\tau_\mathrm{R}}\int_{-1}^{1} d\xi[e^{(\mu_\mathrm{e}+\mu_\mathrm{h})} - 1]. \qquad (22.33)$$

Using the obtained analytical formulas derived above, in which $\mu_\mathrm{e}(x) \simeq \mu_\mathrm{h}(x) = \mu_\mathrm{i} \simeq const$, we find

$$J = \frac{2eKL\Sigma_\mathrm{T}}{\tau_\mathrm{R}}(e^{2\mu_\mathrm{i}} - 1) \qquad (22.34)$$

or

$$J = K\, J_0 (e^{2\mu_i} - 1), \tag{22.35}$$

where

$$J_0 = \frac{\pi\, eLk_B^2 T^2}{3\hbar^2 v_w^2 \tau_R}. \tag{22.36}$$

Setting L = 1 – 5 μm and $\tau_R = 10^{-10} - 10^{-9}$ s at T = 300 K, from Eq. (22.34) one obtains $J_0 \simeq (3 - 150) \times 10^{-3}$ A/cm.

The current-voltage characteristics obtained for different parameters q (using Eqs. (22.29) and (22.35) at $v/v_{eh} \ll 1$) are shown in Fig. 22.10. As follows from Fig. 22.10, the injection current superlinearly increases with increasing bias voltage V. The steepness of such current-voltage characteristics decreases with increasing parameters q. This is because a larger q corresponds in part to larger v_{eh}, i.e., to a larger resistance. In the limit $q \to 0$, the current-voltage dependence approaches to $J \propto \exp(eV/2T)$. Thus, at the transition from near ballistic electron ($q \ll 1$) and hole transport to collision dominated transport ($q \sim 1$), the current-voltage characteristics transform from exponential to superlinear ones.

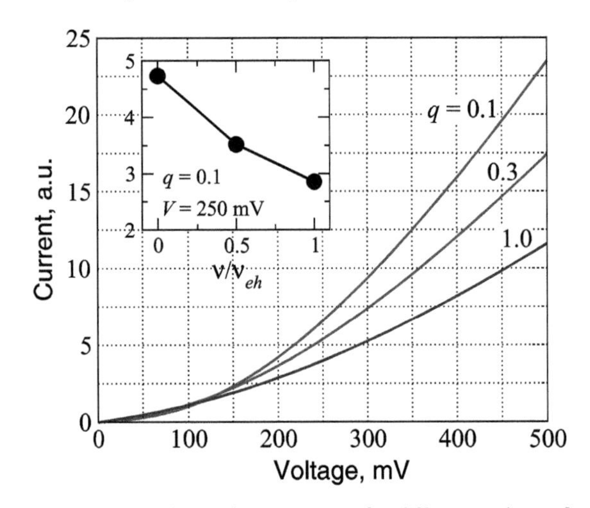

Figure 22.10 Current-voltage characteristics for different values of parameter q. Inset shows the current as a function of parameter v/v_{eh} calculated at V = 250 mV for q = 1.0.

However, at finite values of parameter v/v_{eh}, μ_i can be a pronounced function of the coordinate as seen from Fig. 22.7. This results in a dependence of the current on the parameter in question. Inset in Fig. 22.10 shows the dependence of the current on parameter v/v_{eh} at fixed bias voltage. A decrease in the current with increasing parameter v/v_{eh} is associated with a sag in the spatial dependence of the Fermi energy (see Fig. 22.7), which gives rise to weaker recombination in an essential portion of the i-region and, therefore, lower recombination (injection) current.

22.6 Discussion

The rigorous calculation of the edge recombination current, particularly, its tunneling component and parameter U_R requires special consideration. Here we restrict ourselves to rough estimates based of a phenomenological treatment. The characteristic velocity, u_R^{tunn}, of the edge recombination due to tunneling of electrons through the barrier at the p-i-interface (and holes through the barrier at the i-n-interface) can be estimated as $u_R^{tunn} = D^{tunn} v_W/\pi$. Here D^{tunn} is the effective barrier tunneling transparency, determined, first of all, by the shape of the barrier. The appearance of factor $1/\pi$ is associated with the spread in the directions of electron velocities in the GL plane. In the case of very sharp barrier, $D^{tunn} \simeq 1$. However, practically D^{tunn} is less than unity, although due to the gapless energy spectrum of carriers in GLs and MGLs one can not expect that it is very small. The edge recombination velocity associated with the thermionic current of electrons (holes) over the pertinent barrier is proportional to $D_{th} \propto \exp(-\mu_c)$, so that $u_R^{th} \propto (v_W/\pi)\exp(-\mu_c)$. The factor $\exp(-\mu_c)$ is very small when the contact n- and p-regions are sufficiently doped ($\mu_c \gg 1$), hence, $U_R^{th} \ll U_R^{tunn}$. In the case of the electrically induced p- and n-regions in MGL structures, the condition $\mu_c \gg 1$ can impose certain limitation on the number of GL K [11]. As a result, parameter U_R can be estimated as $U_R \simeq u_R^{tunn} \tau_R/L = D^{tunn} v_W \tau_R/\pi L$. At $\tau_R = 10^{-10}$ s, $L = 5$ μm, and $D^{tunn} = 0.5$ we find that $U_R \simeq 3$. As shown above, the edge recombination even at relatively large parameter U_R markedly influences the carrier densities and their Fermi energies only in immediate vicinity of the p-i and i-n interfaces (see Fig. 22.9). To find conditions when the surface (tunneling) recombination can be disregarded in comparison to the recombination assisted by optical phonons, we compare the contributions of this recombination to the net current. Taking into account Eq. (22.35) and considering that the edge recombination tunneling current $J^{\mathrm{tunn}} \simeq ev_{\mathrm{W}}D^{\mathrm{tunn}}\Sigma_{\mathrm{T}}$, we obtain

$$\frac{J}{J^{\mathrm{tunn}}} \gtrsim \frac{2\pi L}{D^{\mathrm{tunn}} v_{\mathrm{W}} \tau_{\mathrm{R}}}(e^{2\mu_{\mathrm{i}}} - 1) \simeq \frac{2}{U_{\mathrm{R}}^{\mathrm{tunn}}} e^{2\mu_{\mathrm{i}}}. \qquad (22.37)$$

Equation (22.37) yields the following condition when the edge recombination is provides relatively small contribution to the injection current:

$$L > \frac{D^{\mathrm{tunn}} v_{\mathrm{W}} \tau_{\mathrm{R}}}{2\pi} e^{-2\mu_{\mathrm{i}}}. \qquad (22.38)$$

Assuming that $\tau_{\mathrm{R}} = 10^{-10}$ s, $D^{\mathrm{tunn}} = 0.5$, and $\mu_{\mathrm{i}} = 1.0 - 1.4$ ($\varepsilon_{\mathrm{Fi}} = 25 - 35$ meV), from inequality (22.36) we obtain $2L > (1 - 2)$ μm. As follows from Eq. (22.38), the role of the edge recombination steeply drops when μ_{i}, i.e., the bias voltage V increase. This is because the recombination rate R, associated with optical phonons, rapidly rises with increasing μ_{i}. If $\tau_{\mathrm{R}} = (10^{-10} - 10^{-9})$ s and $\mu_{\mathrm{i}} = 2.5$ ($\varepsilon_{\mathrm{Fi}} = 62.5$ meV), one can arrive at the following condition: $2L > 0.1 - 1$ μm.

As demonstrated in Section 22.4, the spatial variations of the band edge profile, density of carriers and their Fermi energies are very sharp near the p-i and i-n interfaces. This is owing to large values of parameter Q and the assumption of abrupt doping profiles. In this case, the normalized width of the transition region is $\eta \sim Q^{-1} \sim r_S^T/L \ll 1$. In real GL and MGL p-i-n structures, the boundaries between p-, i-, and n-region are somewhat smeared (with the characteristic length $l \ll L$). This can be associated with the specifics of chemical doping or with the fringing effects. In the latter case, the length of the intermediate region is about the thickness of the gate layer $l \sim W_{\mathrm{g}}$ (see Fig. 22.1). In both cases, this can result in a smoothing of the dependences in near-contact regions in comparison with those shown in Figs. 22.4–22.9 and in a marked decrease in the electric field in the p-i and i-p junctions and in the edge tunneling current. Figure 22.11 demonstrates examples of the band edge profiles calculated for the abrupt p-i and i-n junctions and for that with smeared junction. In the latter case, it was assumed that the acceptor and donor densities varied exponentially with the characteristic length $l = 0.03L$: $\sigma_{\mathrm{d}} \propto \exp[-(\xi - 1)^2/\eta^2]$ and $\sigma_{\mathrm{d}} \propto \exp[-(\xi + 1)^2/\eta^2]$, with $\eta = l/L = 0.03$, $\Sigma_c = 5.6 \times 10^{12}$cm^{-2} ($\sigma_c = 62.8$, $\mu_c \simeq 10$). As seen from Fig. 22.11, an increase in smearing of the dopant distributions leads to a natural increase in the widths of the transition regions and a decrease (relatively small) in the electrical field in the i-region.

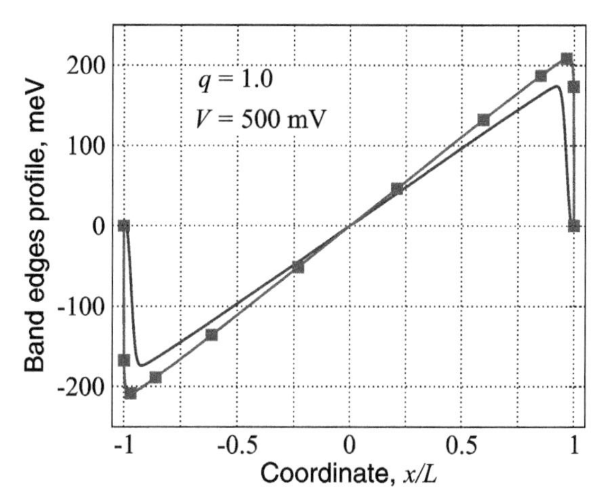

Figure 22.11 Comparison of band edge profiles in structures with abrupt (solid line with markers, the same as in Fig. 22.4) and with smeared p-i and n-i junctions (without markers) at V = 500 mV for q = 1.0.

At moderate number of GLs K in the p-i-n structures under consideration, the net thickness of the latter Kd, where $d \sim 0.35$ nm is the spacing between GLs, is smaller than all other sizes. In such a case, the localization of the space charges of electrons and holes in the i-region in the z-direction can be described by the delta function $\delta(z)$ as in Eq. (22.1). In the case of large K, the delta function should be replaced by a function with a finite localization width. This also should give rise to an extra smearing of the spatial distribu-tions in the x-direction. The pertinent limitation on the value K can be presented as follows: $Kd \ll r_s$, $l \ll L$, where $r_s = [r_S^T/\ln(1+e^{\mu_i})]$ is the screening length. At $r_s \lesssim l \ll L$, assuming that $\kappa = 4$ and $\mu_i \gtrsim 1$ one can obtain $r_s \simeq 10$ nm and arrive at the following condition $K \ll 30$. An increase in K results in an increase in parameters Q and, hence, in some modification of the spatial distributions in near contact region, although the latter weakly affects these distributions in the main portion of the i-region. As a result, the injection current increases proportionally to K (see Eq. (22.35)).

We disregarded possible heating (cooling) of the electron-hole system in the i-region. Under large bias voltages the electric field in this region can be fairly strong. In this case, the dependence of the drift electron and hole velocities on the electric field might be essential. However, relatively strong interaction of electrons and holes with optical phonons having fairly large energy prevents substantial heating [36] in the situations considered in Sections 22.4 and 22.5. Moreover, in the essentially bipolar system the recombination processes with the emission of large energy optical phonons provide a marked cooling of the system, so that its effective temperature can become even lower than the lattice temperature [30, 31].

The above consideration was focused on the double injection in GL and MGL p-i-n structures at the room temperatures. The obtained results are also applicable at somewhat lower temperatures if the recombination is primarily associated with optical phonons. However, at lower temperatures, other recombination mechanisms can become more crucial, for instance, the Auger recombination mediated by scattering mechanisms on impurities and acoustic phonons and due to trigonal warping of the GL band structure, which provide additional momentum and enables transitions to a broader range of final states [37–39], i.e., the indirect Auger recombination [40] (see also Refs. [41, 42]), acoustic phonon recombination, invoking the momentum transfer due to disorder [43], radiative recombination [44], tunneling recombination in the electron-hole puddles [45], and edge tunneling recombination (see Ref. [34] and Section 22.4). Relatively weak role of the optical phonon assisted processes at low temperatures can result in pronounced hot carrier effects (see, for instance, Refs. [46, 47). However, these effects are beyond the scope of our chapter and require special consideration.

22.7 Conclusions

We studied theoretically the double electron-hole injection in GL and MGL p-i-n structures at the forward bias voltages using the developed device model. The mathematical part of the model was based on a system of hydrodynamic equations governing the electron and hole transport in GLs coupled with the two-dimensional Poisson equation. The model is characterized primarily by the following two parameters: the electrostatic parameter Q and the diffusion parameter q. Large value of parameter Q in practical structures results in the formation of very narrow edge regions near p-i and i-n interfaces, in which all physical quantities vary sharply, and the formation of a wide region, which is stretched across the main portion of the i-region, where all spatial dependences are rather smooth. Using analytical and numerical solutions of the equations of the model, we calculated the band edge profiles, the spatial distributions of the quasi-Fermi energies, carrier density and velocity, and the current-voltage characteristics. It was demonstrated that the electron-hole collisions and the collisions of electrons and holes with impurities and acoustic phonons can strongly affect the characteristics of the p-i-n structures. In particular, such collisions result in a pronounced lowering of the injection efficiency (weaker dependences of the electron and hole Fermi energies and their densities on the bias voltage) and the modification of the current-voltage characteristics from an exponential to superlinear characteristics (at collision-dominated transport). It is also shown that for the case of relatively perfect GL and MGL structures, the developed analytical model provides sufficient accuracy in the calculations of spatial distributions in the main portion of the i-region and of the p-i-n structure overall characteristics. The effects associated with the edge recombination and smearing of the p-i and i-n junctions are evaluated.

The obtained results appear to be useful for the realization and optimization of GL- and MGL-based injection terahertz and infrared lasers.

Acknowledgments

The authors are grateful to M. S. Shur and V. Mitin for numerous fruitful discussions. The work was financially supported in part by the Japan Science and Technology Agency, CREST and by the Japan Society for Promotion of Science.

References

1. A. H. C. Neto, F. Guinea, N. M. R. Peres, K. S. Novoselov, and A. K. Geim, *Rev. Mod. Phys.*, **81**, 109 (2009).
2. D. Farmer, Y.-M. Lin, A. Afzali-Ardakani, and P. Avouris, *Appl. Phys. Lett.*, **94**, 213106 (2009).
3. E. C. Peters, E. J. H. Lee, M. Burghard, and K. Kern, *Appl. Phys. Lett.*, **97**, 193102 (2010).
4. K. Yan, Di Wu, H. Peng, L. Jin, Q. Fu, X. Bao, and Zh. Liu, *Nat. Commun.*, **3**, 1280 (2012).
5. H. C. Cheng, R. J. Shiue, C. C. Tsai, W. H. Wang, and Y. T. Chen, *ACS Nano*, **5**, 2051 (2011).
6. Yu Tianhua, C. Kim, C.-W. Liang, and Yu Bin, *Electron Device Lett.*, **32**, 1050 (2011).
7. V. V. Cheianov and V. I. Falko, *Phys. Rev. B*, **74**, 041403 (2006).
8. B. Huard, J. A. Silpizio, N. Stander, K. Todd, B. Yang, and D. Goldhaber-Gordon, *Phys. Rev. Lett.*, **98**, 236803 (2007).
9. J. R. Williams, L. DiCarlo, and C. M. Marcus, *Science*, **317**, 638 (2007).
10. H.-Y. Chiu, V. Perebeinos, Y.-M. Lin, and P. Avouris, *Nano Lett.*, **10**, 4634 (2010).
11. M. Ryzhii, V. Ryzhii, T. Otsuji, V. Mitin, and M. S. Shur, *Phys. Rev. B*, **82**, 075419 (2010).
12. F. Xia, T. Mueller, Y.-M. Lin, A. Valdes-Garcia, and P. Avoris, *Nat. Nanotechnol.*, **4**, 839 (2009).
13. V. Ryzhii, M. Ryzhii, V. Mitin, and T. Otsuji, *J. Appl. Phys.*, **107**, 054512 (2010).
14. T. Mueler, F. Xia, and P. Avouris, *Nat. Photonics*, **4**, 297 (2010).
15. M. Ryzhii, T. Otsuji, V. Mitin, and V. Ryzhii, *Jpn. J. Appl. Phys.*, **50**, 070117(2011).
16. N. M. Gabor, J. C. W. Song, Q. Ma, N. L. Nair, T. Taychatanapat, K. Watanabe, T. Taniguchi, L. S. Levitov, and P. Jarillo-Herrero, *Science*, **334**, 648 (2011).
17. V. Ryzhii, N. Ryabova, M. Ryzhii, N. V. Baryshnikov, V. E. Karasik, V. Mitin, and T. Otsuji, *Optoelectron. Rev.*, **20**, 15 (2012).
18. D. Sun, G. Aivazian, A. M. Jones, J. Ross, W. Yao, D. Cobden, and X. Xu, *Nat. Nanotechnol.*, **7**, 114 (2012).
19. M. Freitag, T. Low, and P. Avouris, *Nano Lett.*, **13**, 1644 (2013).

20. V. Ryzhii, M. Ryzhii, M. S. Shur, and V. Mitin, *Appl. Phys. Express*, **2**, 034503 (2009).
21. V. L. Semenenko, V. G. Leiman, A. V. Arsenin, V. Mitin, M. Ryzhii, T. Otsuji, and V. Ryzhii, *J. Appl. Phys.*, **113**, 024503 (2013).
22. V. Ryzhii, M. Ryzhii, and T. Otsuji, *J. Appl. Phys.*, **101**, 083114 (2007).
23. V. Ryzhii, M. Ryzhii, A. Satou, T. Otsuji, A. A. Dubinov, and V. Ya. Aleshkin, *J. Appl. Phys.*, **106**, 084507 (2009).
24. V. Ryzhii, A. A. Dubinov, T. Otsuji, V. Mitin, and M. S. Shur, *J. Appl. Phys.*, **107**, 054505 (2010).
25. A. A. Dubinov, V. Ya. Aleshkin, V. Mitin, T. Otsuji, and V. Ryzhii, *J. Phys.: Condens. Matter*, **23**, 145302 (2011).
26. S. Boubanga-Tombet, S. Chan, A. Satou, T. Otsuji, and V. Ryzhii, *Phys. Rev. B*, **85**, 035443 (2012).
27. T. Otsuji, S. Boubanga-Tombet, A. Satou, M. Ryzhii, and V. Ryzhii, J. Phys. *D*, **45**, 303001 (2012).
28. T. Li, L. Luo, M. Hupalo, J. Zhang, M. C. Tringides, J. Schmalian, and J. Wang, *Phys. Rev. Lett.*, **108**, 167401 (2012).
29. M. Ryzhii and V. Ryzhii, Jpn. *J. Appl. Phys.*, **46**, L151 (2007).
30. V. Ryzhii, M. Ryzhii, V. Mitin, and T. Otsuji, *J. Appl. Phys.*, **110**, 094503 (2011).
31. V. Ryzhii, M. Ryzhii, V. Mitin, A. Satou, and T. Otsuji, *Jpn. J. Appl. Phys.*, **50**, 094001 (2011).
32. D. Svintsov, V. Vyurkov, S. O. Yurchenko, T. Otsuji, and V. Ryzhii, *J. Appl. Phys.*, **111**, 083715 (2012).
33. F. Rana, P. A. George, J. H. Strait, J. Dawlaty, S. Shivaraman, M. Chandrashekhar, and M. G. Spencer, *Phys. Rev. B*, **79**, 115447 (2009).
34. V. Ryzhii, M. Ryzhii, and T. Otsuji, *Phys. Status Solid A*, **205**, 1527 (2008).
35. A. H. Nayfeh *Perturbation Methods* (Wiley-VCH, Weinheim, 2004).
36. R. S. Shishir and D. K. Ferry, *J. Phys.: Condens. Matter*, **21**, 344201 (2009).
37. M. S. Foster and I. Aleiner, *Phys. Rev. B*, **79**, 085415 (2009).
38. L. E. Golub, S. A. Tarasenko, M. V. Entin, and L. I. Magarill, *Phys. Rev. B*, **84**, 195408 (2011).
39. A. Satou, V. Ryzhii, Y. Kurita, and T. Otsuji, *J. Appl. Phys.*, **113**, 143108 (2013).
40. E. Kioupakis, P. Rinke, K. T. Delaney, and C. G. Van de Walle, *Appl. Phys. Lett.*, **98**, 161107 (2011).

41. P. A. George, J. Strait, J. Davlaty, S. Shivaraman, M. Chandrashekhar, F. Rana, and M. G. Spencer, *Nano Lett.*, **8**, 4248 (2008).
42. T. Winzer and E. Malic, *Phys. Rev. B*, **85**, 241404 (2012).
43. F. Vasko and V. Mitin, *Phys. Rev. B*, **84**, 155445 (2011).
44. V. Vasko and V. Ryzhii, *Phys. Rev. B*, **77**, 195433 (2008).
45. V. Ryzhii, M. Ryzhii, and T. Otsuji, *Appl. Phys. Lett.*, **99**, 173504 (2011).
46. O. G. Balev, F. T. Vasko, and V. Ryzhii, *Phys. Rev. B*, **79**, 165432 (2009).
47. O. G. Balev and F. T. Vasko, *J. Appl. Phys.*, **107**, 124312 (2010).

Chapter 23

Carrier-Carrier Scattering and Negative Dynamic Conductivity in Pumped Graphene*

D. Svintsov,[a,b,c] V. Ryzhii,[a,d] A. Satou,[a] T. Otsuji,[a] and V. Vyurkov[b]
[a]*Research Institute for Electrical Communication, Tohoku University, Sendai 980-8577, Japan*
[b]*Institute of Physics and Technology, Russian Academy of Sciences, Moscow 117218, Russia*
[c]*Department of General Physics, Moscow Institute of Physics and Technology, Dolgoprudny 141700, Russia*
[d]*Center for Photonics and Infrared Engineering, Bauman Moscow State Technical University, Moscow 105005, Russia*
svintcov.da@mipt.ru

We theoretically examine the effect of carrier-carrier scattering processes on the intraband radiation absorption and their contribution to the net dynamic conductivity in optically or electrically pumped

*Reprinted with permission from D. Svintsov, V. Ryzhii, A. Satou, T. Otsuji, and V. Vyurkov (2014). Carrier-carrier scattering and negative dynamic conductivity in pumped graphene, *Opt. Express*, **22**(17), 19873–19886. Copyright © 2014 Optical Society of America.

Graphene-Based Terahertz Electronics and Plasmonics: Detector and Emitter Concepts
Edited by Vladimir Mitin, Taiichi Otsuji, and Victor Ryzhii

ISBN 978-981-4800-75-4 (Hardcover), 978-0-429-32839-8 (eBook)
www.jennystanford.com

graphene. We demonstrate that the radiation absorption assisted by the carrier-carrier scattering is comparable with Drude absorption due to impurity scattering and is even stronger in sufficiently clean samples. Since the intraband absorption of radiation effectively competes with its interband amplification, this can substantially affect the conditions of the negative dynamic conductivity in the pumped graphene and, hence, the interband terahertz and infrared lasing. We find the threshold values of the frequency and quasi-Fermi energy of nonequilibrium carriers corresponding to the onset of negative dynamic conductivity. The obtained results show that the effect of carrier-carrier scattering shifts the threshold frequency of the radiation amplification in pumped graphene to higher values. In particular, the negative dynamic conductivity is attainable at the frequencies above 6 THz in graphene on SiO_2 substrates at room temperature. The threshold frequency can be decreased to markedly lower values in graphene structures with high-κ substrates due to screening of the carrier-carrier scattering, particularly at lower temperatures.

23.1 Introduction

Graphene, a two-dimensional carbon crystal, possesses no energy band gap and, hence, is promising for detection and generation of far-infrared and terahertz (THz) radiation [1–3]. Several concepts of graphene-based THz lasers with optical pumping [4] or electrical (injection) pumping [5, 6] have been proposed and analyzed. Recently, the possibility of THz-wave amplification by optically pumped graphene was shown experimentally [7].

Generally, one of the main reasons for low efficiency of THz semiconductor laser is the intraband (Drude) radiation absorption [8], which grows rapidly with decreasing the radiation frequency. This absorption process aggressively competes with the radiation amplification due to the stimulated interband electron transitions under the conditions of population inversion. This problem is crucial for graphene-based lasers as well.

The optical conductivity of clean undoped graphene is equal to the universal value $\sigma_q = e^2/4\hbar$, where e is the elementary charge and $\hbar$ is the Planck's constant. This value corresponds to the interband radiation absorption coefficient $\pi\alpha = 2.3\%$ [10], where $\alpha = e^2/\hbar c$ is the fine-structure constant (c is the speed of light in vacuum). The optical conductivity of pumped graphene can be negative due to prevailing stimulated electron transitions from the conduction to the valence band associated with the interband population inversion [11]. The interband dynamic conductivity appears to be negative at frequencies $\omega < 2\varepsilon_F/\hbar$, where ε_F is the quasi-Fermi energy of pumped carriers (see Fig. 23.1a and b). However, the coefficient of radiation amplification in pumped graphene is still limited by 2.3%. Hence, for the operation of graphene-based lasers the Drude absorption coefficient [12] should lie below 2.3%. In the previous models of negative dynamic conductivity in pumped graphene, it was usually assumed that the Drude absorption originates from electron and hole scattering on impurities, lattice defects, and phonons (Fig. 23.1d), which can, in principle, be removed almost completely in high-quality graphene samples and at low temperatures (see, for example, [3, 13]). However, there is an unavoidable mechanism leading to the intraband absorption, namely, the radiation absorption assisted by the carrier-carrier scattering (Fig. 23.1c).

The carrier-carrier scattering (for brevity denoted as c-c scattering) was shown to be a key factor in the relaxation kinetics of photoexcited electrons and holes in graphene [7, 9, 14, 15]. It can also be responsible for weakly temperature-dependent minimal dc conductivity of graphene [16–20]. The strength of c-c scattering in graphene is governed by the relatively large coupling constant, $\alpha_c = e^2/(\kappa_0 \hbar v_F) \sim 1$, where $v_F = 10^6$ m/s is the velocity of massless electrons and holes (Fermi velocity) and κ_0 is the background dielectric constant.

Among a variety of phenomena originating from the carrier-carrier scattering in graphene-based structures, there are modifications of quasiparticle spectra [20], relaxation [14, 15], thermalization, and recombination [9]. In this chapter, we focus on the scattering-assisted intraband radiation absorption. The latter appears to be a quite strong absorption mechanism, especially in the pumped graphene with population inversion. We derive the expression for the real part of the intraband dynamic conductivity $\mathrm{Re}\,\sigma_{cc}$ (which is proportional to the intraband contribution to the radiation absorption) arising due to carrier-carrier collisions in clean pumped graphene. The probability of corresponding intraband radiation absorption process is evaluated using the second-order perturbation theory and the Fermi golden rule. Comparing Reσ_{cc} with the interband conductivity Reσ_{inter}, we find the threshold values of the quasi-Fermi energies of pumped carriers, required to attain the net negative dynamic conductivity (negative absorption coefficient) at given frequency, and calculate the pertinent threshold frequencies. We also show that screening of the Coulomb potential by the carriers in graphene plays the significant role in the c-c scattering. Because of screening, the intraband conductivity due to the c-c collisions grows slowly (almost linearly) with increasing the quasi-Fermi energy of pumped carriers. We find that in the pumped graphene, the main contribution to the intraband conductivity arises from electron-hole (e-h) scattering, while the electron-electron (e-e) and hole-hole (h-h) collisions yield less than one tenth of its total value. The rate of c-c scattering and the corresponding radiation absorption can be reduced in graphene clad between materials with high dielectric constants.

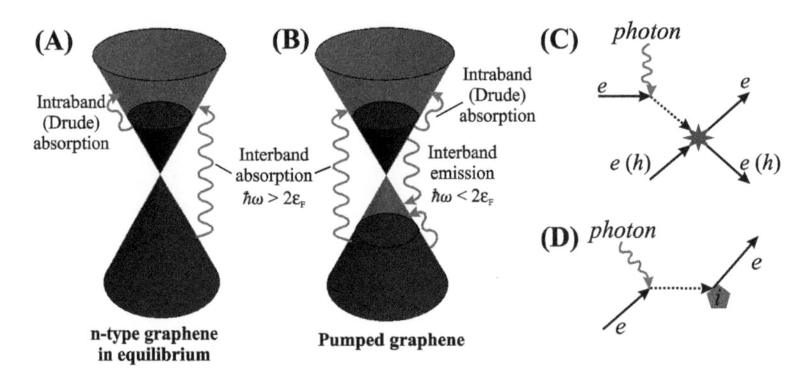

Figure 23.1 Schematic views of band diagrams of (A) n-type graphene in equilibrium and (B) pumped graphene. Wavy arrows indicate photon absorption and emission processes. Diagram of photon absorption by an electron (e) associated with (C) electron-electron (electron-hole) and (D) impurity (i) scattering.

In the present chapter, we shall restrict our consideration of dynamic conductivity within terahertz range (1–15 THz). At higher frequencies, the optical response is governed mainly by interband transitions [21] which lead to the absorption of radiation at $\hbar\omega > 2\varepsilon_F$, while the Drude absorption drops proportionally to ω^{-2}. At lower frequencies, one can readily use the static limit of Boltzmann/hydrodynamic theory to obtain the conductivity.

The chapter is organized as follows. In Section 23.2, we derive the general equations for the radiation power absorbed due to c-c scattering-assisted intraband transitions, and find the pertinent contribution to the dynamic conductivity. In Section 23.3, we obtain the dependences of the intraband, interband, and net dynamic conductivity on the radiation frequency, the quasi-Fermi energy, the background dielectric constant; and find the frequency threshold of the radiation amplification. Section 23.4 deals with discussion of the obtained results and their validity. In Section 23.5, we draw the main conclusions. Some mathematical details are singled out in Appendix.

23.2 Intraband Dynamic Conductivity General Equations

There are two mechanisms through which the c-c scattering affects the optical conductivity, and, hence, absorption of radiation. First, the finite lifetime of carriers leads simply to the smearing of the interband absorption/amplification edge [22]. Second, a free electron can absorb a photon and transfer the excess energy and momentum to the other carrier (Fig. 23.1c). The latter mechanism is quite similar to the conventional Drude absorption, where the excess momentum is transferred to an impurity or phonon (Fig. 23.1d). However, in the case of graphene, the radiation absorption due to c-c scattering exhibits a number of unusual features.

In semiconductors with parabolic bands, e-e scattering cannot directly affect the conductivity and absorption of radiation. The reason is that the total current carried by two electrons is not changed in the scattering process due to momentum conservation. In graphene, however, electron velocity and momentum are not directly proportional to each other (sometimes referred to as "momentum-velocity decoupling" [19]). Hence, the momentum conservation does not imply the current conservation, and e-e scattering can contribute to relaxation of current [23] and optical absorption. Such effects were studied in III–V semiconductors in the light of p^4-corrections to the parabolic bands [24]. In graphene, the carrier dispersion law is not parabolic "from the very beginning." Hence, those effects can play a more considerable role.

Electron-hole (e-h) scattering processes do not conserve the total current and can contribute to radiation absorption independent of the carrier spectrum. Apart from graphene with Fermi energy close to Dirac point, e-h scattering plays a minor role in conductivity due to vanishing number of holes. This is not the case of the pumped graphene, where e-h scattering is intensified due to the great number of carriers in both bands.

We consider a two-step quantum-mechanical process: the quasiparticle (electron or hole) absorbs light quantum passing to the virtual state, then it collides with other quasiparticle and transfers the excess energy (and momentum) to it. If the frequency of electromagnetic wave ω substantially exceeds the electron (or hole) collision frequency ν, one can use the Fermi golden rule to calculate the absorption probability in unit time. We will first consider the radiation absorption due to e-e collisions, and then generalize the result to account for e-h processes. The initial momenta of incoming quasiparticles are denoted by $\mathbf{p}_1$ and $\mathbf{p}_2$, while the final momenta are $\mathbf{p}_3$ and $\mathbf{p}_4$.

To work out the matrix elements appearing in the Fermi golden rule we apply the second order perturbation theory considering the two perturbations. The first one, $\hat{V}_F(t)$, is the interaction of electron with external electromagnetic field

$$\hat{\mathbf{V}}_F(t) = \frac{e}{2c}(\mathbf{A}_0, \hat{\mathbf{v}}_1 + \hat{\mathbf{v}}_2)(e^{i\omega t} + e^{-i\omega t}), \tag{23.1}$$

where $\hat{\mathbf{v}}_1$ and $\hat{\mathbf{v}}_2$ are the velocity operators of the two electrons, $\mathbf{A}_0$ is the amplitude of field vector potential, and ω is the frequency of the electromagnetic wave. The vector $\mathbf{A}_0$ and the corresponding electric field $\mathbf{E}_0 = i\omega\mathbf{A}_0/c$ lie in the plane of graphene layer.

The second perturbation, $\hat{V}_C$, is the Coulomb interaction between carriers. It has nonzero matrix elements connecting the two-particle states $|\mathbf{p}_1, \mathbf{p}_2\rangle$ and $|\mathbf{p}_3, \mathbf{p}_4\rangle = |\mathbf{p}_1 + \mathbf{q}, \mathbf{p}_2 - \mathbf{q}\rangle$, where $\mathbf{q}$ is the transferred momentum. We denote these matrix elements by $V_C(\mathbf{q}) \equiv \left\langle \mathbf{p}_3, \mathbf{p}_4 | \hat{V}_C | \mathbf{p}_1, \mathbf{p}_2 \right\rangle$,

$$V_C(\mathbf{q}) = \frac{2\pi e^2 \hbar}{q\kappa_0 \kappa(q)} \left\langle u_{\mathbf{p}_1}^{(e)} u_{\mathbf{p}_3}^{(e)} \right\rangle \left\langle u_{\mathbf{p}_2}^{(e)} u_{\mathbf{p}_4}^{(e)} \right\rangle, \tag{23.2}$$

where $\left\langle u_{\mathbf{p}_i}^{(e)} u_{\mathbf{p}_j}^{(e)} \right\rangle = \cos(\theta_{ij}/2)$ are overlap factors of the electron envelope functions in graphene, θ_{ij} is the angle between momenta $\mathbf{p}_i$ and $\mathbf{p}_j$, κ_0 is the background dielectric constant, and $\kappa(q)$ is the dielectric function of graphene itself. For our purposes, it can be taken in the static limit

$$\kappa(q) = 1 + q_{TF}/q, \tag{23.3}$$

q_{TF} is the Thomas-Fermi (screening) momentum [25].

A well-known relation for the second-order matrix element connecting initial $|i\rangle = |\mathbf{p}_1, \mathbf{p}_2\rangle$ and final $|f\rangle = |\mathbf{p}_3, \mathbf{p}_4\rangle$ states reads [26]

$$\langle f|\hat{V}|i\rangle = \sum_{m} \frac{\langle f|\hat{V}_{F\omega}|m\rangle\langle m|\hat{V}_C|i\rangle}{(\varepsilon_{p_1}+\varepsilon_{p_2})-(\varepsilon_{p_{1m}}+\varepsilon_{p_{2m}})} + \frac{\langle f|\hat{V}_C|m\rangle\langle m|\hat{V}_{F\omega}|i\rangle}{(\varepsilon_{p_1}+\varepsilon_{p_2}+\hbar\omega)-(\varepsilon_{p_{1m}}+\varepsilon_{p_{2m}})}, \tag{23.4}$$

where the index m counts the intermediate states, $\varepsilon_{\mathbf{p}} = pv_F$ is the electron dispersion law in graphene, and $\hat{V}_{F\omega} = (e/2c)(\mathbf{A}_0, \hat{\mathbf{v}}_1 + \hat{\mathbf{v}}_2)$ is the Fourier-component of electron-field interaction. The summation over intermediate momenta $\mathbf{p}_m$ is easily performed as photon momentum is negligible compared to electron momentum. Using also the energy conservation law

$$\varepsilon_{\mathbf{p}_1} + \varepsilon_{\mathbf{p}_2} + \hbar\omega = \varepsilon_{\mathbf{p}_3} + \varepsilon_{\mathbf{p}_4}, \tag{23.5}$$

we obtain a very simple relation for matrix element

$$\langle f|V|i\rangle = \frac{e}{2c}\frac{V_C(\mathbf{q})}{\hbar\omega}(\mathbf{A}_0, \mathbf{v}_{\mathbf{p}_1} + \mathbf{v}_{\mathbf{p}_2} - \mathbf{v}_{\mathbf{p}_3} - \mathbf{v}_{\mathbf{p}_4})\cos(\theta_{13}/2)\cos(\theta_{24}/2). \tag{23.6}$$

Equation (23.6) clearly demonstrates that for parabolic bands no absorption due to e-e scattering can occur as $\mathbf{v}_{\mathbf{p}_1} + \mathbf{v}_{\mathbf{p}_2} - \mathbf{v}_{\mathbf{p}_3} - \mathbf{v}_{\mathbf{p}_4}$ turns to zero as a result of momentum conservation.

The power, P, absorbed by an electron system is calculated using the Fermi golden rule, taking into account the occupation numbers of the initial and final states:

$$P = \frac{\hbar\omega}{2}\frac{2\pi g^2}{\hbar}\sum_{\mathbf{p}_1, \mathbf{p}_2, \mathbf{q}} |\langle f|V|i\rangle|^2 \delta(\varepsilon_{\mathbf{p}_1} + \varepsilon_{\mathbf{p}_2} + \hbar\omega - \varepsilon_{\mathbf{p}_3} - \varepsilon_{\mathbf{p}_4})$$
$$\times f_e(\mathbf{p}_1) f_e(\mathbf{p}_2)[1 - f_e(\mathbf{p}_3)][1 - f_e(\mathbf{p}_4)][1 - \exp(-\hbar\omega/k_B T)]. \tag{23.7}$$

Here $g = 4$ stands for the spin-valley degeneracy factor in graphene. The pre-factor 1/2 cuts off the equivalent scattering processes;

without it the indistinguishable collisions $\mathbf{p}_1 + \mathbf{p}_2$ and $\mathbf{p}_2 + \mathbf{p}_1$ are treated separately, which would be incorrect. The function $f_e(\mathbf{p})$ in Eq. (23.7) is the electron distribution function, which is assumed to be the quasi-equilibrium Fermi function:

$$f_e(\mathbf{p}) = \left[1 + \exp\left(\frac{\varepsilon_\mathbf{p} - \mu_e}{k_B T}\right)\right]^{-1}. \tag{23.8}$$

To treat the radiation absorption in pumped graphene, we introduce the different quasi-Fermi energies of electrons and holes, μ_e and μ_h. In symmetrically pumped systems $\mu_e = -\mu_h = \varepsilon_F > 0$. In what follows, if otherwise not stated, we will consider symmetrically pumped graphene. In such system, the occupation numbers of electrons and holes with energies $\varepsilon_p > 0$ are equal, hence, the subscripts e and h of the distribution functions can be omitted. In deriving Eq. (23.7) we have also neglected the exchange-type scattering, which occurs only between electrons with same spin and from the same valley. This assumption does not significantly affect the final numerical values.

Finally, to obtain the real part of the intraband dynamic conductivity due to e-e collisions $\mathrm{Re}\sigma_{ee}$, we express the vector-potential in Eq. (23.6) through electric field and equate (23.7) with $\mathrm{Re}\sigma_{ee}E_0^2/2$. As a result, this conductivity is expressed via the universal optical conductivity of clean graphene σ_q, the coupling constant $\alpha_c = e^2/(\kappa_0 \hbar v_F)$, and the dimensionless 'collision integral' $I_{ee,\omega}$:

$$\mathrm{Re}\sigma_{ee} = \sigma_q \frac{\alpha_c^2}{\pi^3}\left(\frac{k_B T}{\hbar\omega}\right)^3 \left[1 - \exp\left(-\frac{\hbar\omega}{k_B T}\right)\right] I_{ee,\omega}, \tag{23.9}$$

$$I_{ee,\omega} = \int \frac{d\mathbf{Q}\,d\mathbf{k}_1 d\mathbf{k}_2}{Q^2 \kappa^2(Q)} (\Delta\mathbf{n}_{ee})^2 \cos^2(\theta_{1\pm}/2)\cos^2(\theta_{2\pm}/2)$$
$$\times \delta[k_{1+} + k_{2-} + \hbar\omega/(k_B T) - k_{1-} - k_{2+}] F(k_{1+}) F(k_{2-})[1 - F(k_{1-})][1 - F(k_{2+})]. \tag{23.10}$$

Here, for the simplicity of further analysis, we have introduced the dimensionless momenta $\mathbf{k}_{1\pm} = (\mathbf{p}_1 \pm \mathbf{q}/2)v_F/(k_B T)$, $\mathbf{k}_{2\pm} = (\mathbf{p}_2 \pm \mathbf{q}/2)v_F/(k_B T)$, $Q = \mathbf{q}v_F/(k_B T)$, as well as the dimensionless change in the electron current $\Delta\mathbf{n}_{ee} = (\mathbf{k}_{1+}/k_{1+} + \mathbf{k}_{2-}/k_{2-}) - (\mathbf{k}_{1-}/k_{1-} + \mathbf{k}_{2+}/k_{2+})$; $\theta_{1\pm}$ and $\theta_{2\pm}$ are the angles between $\mathbf{k}_{1+}$ and $\mathbf{k}_{1-}$ and between $\mathbf{k}_{2+}$ and $\mathbf{k}_{2-}$, respectively, $F(x) = \{1 + \exp[x - \varepsilon_F/(k_BT)]\}^{-1}$ is the Fermi function of dimensionless argument.

At the frequencies exceeding the carrier collision frequency $\omega \gg \nu$, the contributions to the real part of dynamic conductivity from different scattering mechanisms are summed up. The intraband conductivity due to e-h collisions is given by expression similar to (9–10), with several differences. First, the change in current carried by electron and hole is given by $\Delta\mathbf{n_{eh}} = (\mathbf{k}_{1+}/k_{1+} - \mathbf{k}_{2-}/k_{2-}) - (\mathbf{k}_{1-}/k_{1-} - \mathbf{k}_{2+}/k_{2+})$ as the charge of hole is opposite to that of electron. Second, along with 'simple' electron-hole scattering, the annihilation-type interaction between electron and hole is also possible. In such process, electron and hole annihilate, emit a virtual photon, which produces again an electron-hole pair. The probability of such process is the same as of scattering process, but the overlap factors $\left\langle u_{\mathbf{p}_i}^{(e)} u_{\mathbf{p}_j}^{(e)} \right\rangle = \cos(\theta_{ij}/2)$ should be changed to $\left\langle u_{\mathbf{p}_i}^{(e)} u_{\mathbf{p}_j}^{(h)} \right\rangle = \sin(\theta_{ij})$[17]. As a result, the electron-hole scattering contribution to the optical conductivity becomes

$$\mathrm{Re}\sigma_{\mathrm{eh}} = \sigma_{\mathrm{q}} \frac{2\alpha_c^2}{\pi^3} \left(\frac{k_B T}{\hbar\omega} \right)^3 \left[1 - \exp\left(-\frac{\hbar\omega}{k_B T} \right) \right] I_{\mathrm{eh},\omega}, \qquad (23.11)$$

$$I_{\mathrm{eh},\omega} = \int \frac{d\mathbf{Q} d\mathbf{k}_1 d\mathbf{k}_2}{Q^2 \kappa^2} (\Delta\mathbf{n}_{\mathrm{eh}})^2 [\cos^2(\theta_{1\pm}/2)\cos^2(\theta_{2\pm}/2) + \sin^2(\theta_{1\pm}/2)\sin^2(\theta_{2\pm}/2)]$$
$$\times \delta[k_{1+} + k_{2-} + \hbar\omega/(k_B T) - k_{1-} - k_{2+}] F(k_{1+}) F(k_{2-}) [1 - F(k_{1-})][1 - F(k_{2+})]. \qquad (23.12)$$

Accordingly, the net intraband conductivity due to c-c collisions $\mathrm{Re}\sigma_{\mathrm{intra}}$ in symmetrically pumped graphene is

$$\mathrm{Re}\sigma_{\mathrm{cc}} = \sigma_{\mathrm{q}} \frac{2\alpha_c^2}{\pi^3} \left(\frac{k_B T}{\hbar\omega} \right)^3 \left[1 - \exp\left(-\frac{\hbar\omega}{k_B T} \right) \right] (I_{\mathrm{ee},\omega} + I_{\mathrm{eh},\omega}). \qquad (23.13)$$

Here we have noted that h-h scattering contribution equals e-e contribution. The coefficient of radiation absorption is readily obtained from Eq. (23.13) timing it by $4\pi/c$. The dimensionless 'collision integrals' $I_{\mathrm{ee},\omega}$ and $I_{\mathrm{eh},\omega}$ are given by Eqs. (23.10) and (23.12). They can be evaluated numerically passing to the elliptic coordinates [9]. In those coordinates, the delta-function can be analytically integrated, reducing the dimensionality of integral by unity. This procedure is described in Appendix.

23.3 Analysis of Intraband and Net Dynamic Conductivity

It is natural that the obtained expression (23.13) for the real part of conductivity is proportional to the universal ac conductivity of clean undoped graphene σ_q, and the square of the coupling constant α_c. The latter characterizes the strength of Coulomb interaction between carriers in graphene. The frequency behavior of Reσ_{cc} follows the well-known Drude-like dependence ω^{-2}. At frequencies $\hbar\omega \ll k_BT$, the pre-factor before the collision integral in Eq. (23.13) yields the ω^{-2}-dependence. At large frequencies $\hbar\omega \geq k_BT$, the pre-factor behaves as ω^{-3}, but the "collision integrals" $I_{ee,\omega}$ and $I_{eh,\omega}$ grow linearly due to the increasing number of final states.

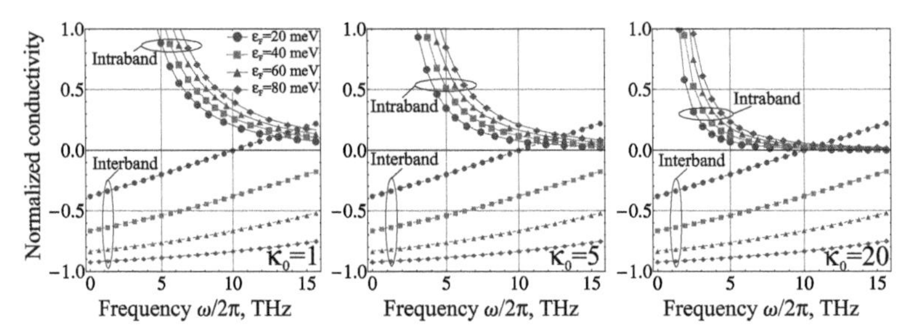

Figure 23.2 Real parts of the intraband (upper panels) and interband (lower panels) contributions, Reσ_{intra} and Reσ_{inter}, to dynamic conductivity normalized by σ_q at different quasi-Fermi energies ε_F in graphene structures with different background dielectric constants κ_0 (T = 300 K).

In what follows, we shall study the dynamic conductivity of sufficiently clean pumped graphene. In this situation, the positive intraband conductivity is due to carrier-carrier collisions only, thus Re$\sigma_{\mathbf{intra}}$ = Reσ_{cc}. Figure 23.2 shows the dependences of Reσ_{intra} given by Eq. (23.13) (normalized by σ_q) on frequency $\omega/2\pi$ for the symmetrically pumped graphene at different quasi-Fermi energies ε_F and different background dielectric constants κ_0 at room temperature T = 300 K. In Fig. 23.2, we also show the frequency-dependent real part of the dynamic conductivity associated with the interband transitions [11]:

$$\mathrm{Re}\sigma_{\text{inter}} = \sigma_q \tanh\left(\frac{\hbar\omega/2 - \varepsilon_F}{2k_BT}\right). \tag{23.14}$$

From Fig. 23.2 one can see that at frequencies above ~ 6.5 THz at κ_0 = 1 (and above ~ 2.5 THz at κ_0 = 20) the Drude conductivity due to c-c collisions lies below σ_q. Accordingly, the net conductivity $\mathrm{Re}(\sigma_{\mathrm{intra}} + \sigma_{\mathrm{inter}})$ can be negative in this frequency range at some level of pumping.

As the background dielectric constant κ_0 is increased, the intraband conductivity due to c-c collisions drops, which is illustrated in Fig. 23.2. *Ex facte*, one could expect that it scales as κ_0^{-2}, and increasing the dielectric constant one could reduce the radiation absorption due to c-c collisions almost to zero. Such considerations are actually irrelevant due to screening. The Thomas-Fermi screening momentum in pumped graphene is given by

$$q_{\mathrm{TF}} \approx 8\alpha_{\mathrm{c}} \frac{k_{\mathrm{B}}T}{v_{\mathrm{F}}} \ln\left[1+\exp\left(\frac{\varepsilon_{\mathrm{F}}}{k_{\mathrm{B}}T}\right)\right]. \tag{23.15}$$

Thus, at small momenta q the Coulomb scattering matrix element is independent of κ_0. At large momenta $q \approx \varepsilon_{\mathrm{F}}/v_{\mathrm{F}}$ the scattering matrix element $V_C^2(k_{\mathrm{F}}) \propto [\kappa_0^2 + 8e^2/(\hbar v_{\mathrm{F}})]^{-2}$, which slightly depends on κ_0 due to large value of bare coupling constant $e^2/\hbar v_{\mathrm{F}} \approx 2.2$.

In Fig. 23.3 we show the net dynamic conductivity including interband and intraband c-c contributions. At weak pumping ($\varepsilon_{\mathrm{F}} \simeq 20$ meV) the negative dynamic conductivity is attainable only in graphene clad between high-κ substrates. At elevated pumping ($\varepsilon_{\mathrm{F}} \simeq 40$ meV) the negative conductivity is possible at any value of κ_0. Further increase in the pumping level is not much efficient as the interband conductivity reaches its maximum value of $\sigma_{\mathbf{q}}$ at $\varepsilon_{\mathrm{F}} \gg \hbar\omega$, while the intraband conductivity continues to grow with increasing ε_{F}.

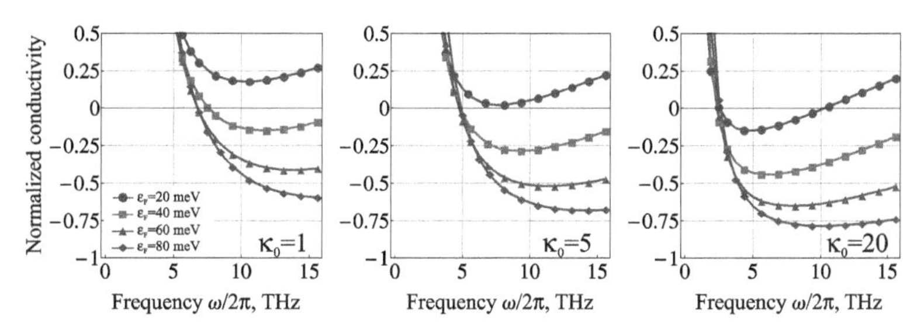

Figure 23.3 Real parts of net dynamic conductivity $\mathrm{Re}(\sigma_{\mathrm{intra}} + \sigma_{\mathrm{inter}})$ normalized by σ_q at different quasi-Fermi energies ε_F in graphene structures with different background dielectric constants κ_0 (T = 300 K).

Both the interband and intraband c-c conductivities are actually functions of two dimensionless combinations $\hbar\omega/k_BT$ and ε_F/k_BT. Once the temperature is reduced, $\text{Re}\sigma_{\text{intra}}$ drops as well due to reduced phase space for collisions near Fermi-surface. The dependence $\text{Re}\sigma_{\text{inter}}(\omega)$ becomes more abrupt in the vicinity of interband threshold $\hbar\omega = 2\varepsilon_F$. Thereby, cooling of graphene sample is advantageous for achieving the negative dynamic conductivity. This is confirmed by Fig. 23.4. In Figs. 23.4a and b, we show the real part of the net dynamic conductivity versus frequency $\omega/2\pi$ and quasi-Fermi energy ε_F (values of $\text{Re}\sigma$ are marked by colors) at the room temperature T = 300 K (A) and at T = 200 K (B). It is worth noting that the temperature T in the above equations is in fact the effective temperature of the electron-hole system. Therefore, the reduction of this temperature can be achieved not only by decreasing of the ambient temperature, but also by direct cooling of this system at a certain pumping conditions. In particular, the cooling of electron-hole system in the optically and electrically pumped graphene lasing structures due to the emission of optical phonons can be substantial [5, 27].

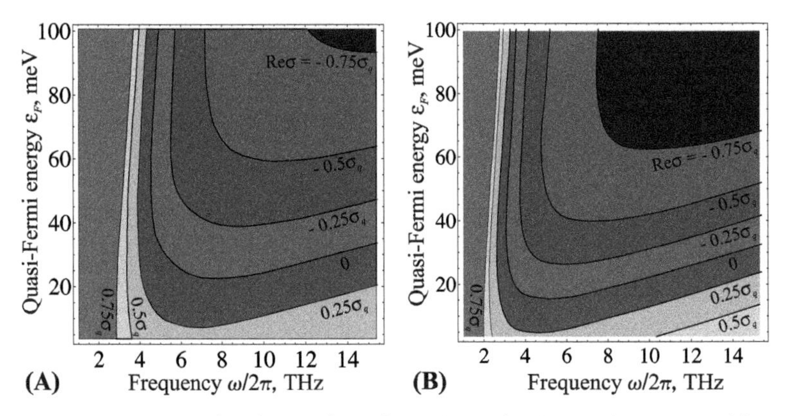

Figure 23.4 Map of real part of net dynamic conductivity $\text{Re}(\sigma_{\text{intra}} + \sigma_{\text{inter}})/\sigma_q$ vs frequency and quasi-Fermi energy for $\kappa_0 = 5$: (A) at T = 300 K and (B) at T = 200 K. The area $\text{Re}(\sigma_{\text{intra}} + \sigma_{\text{inter}})/\sigma_q < -0.75$ is filled in solid color.

In Fig. 23.5a we show the threshold lines of the negative net dynamic conductivity [which are solutions of the equation $\text{Re}\sigma(\omega, \varepsilon_F) = 0$] at different values of background dielectric constant. The normalized threshold frequency $\hbar\omega_0/k_BT$ slowly moves to lower values as κ_0 increases. This is illustrated in Fig. 23.5b. At room temperature and κ_0 = 1, the negative conductivity could be attained only at frequencies $\omega/2\pi \geq 6.5$ THz. For graphene on SiO_2 substrate [$\kappa_0 = (\kappa_{SiO_2} + 1)/2 = 2.5$], the room temperature threshold lies at $\omega_0/2\pi \approx 6$ THz. As discussed before, such slow decrease in threshold frequency is owing to the Thomas-Fermi screening of Coulomb interaction. Note that screening can be further enhanced by placing graphene close to the metal gate. In gated graphene, the Fourier component of Coulomb scattering potential $V_C(q)$ gains an additional factor $[1 - \exp(-2qd)]$, where d is the distance to the gate.

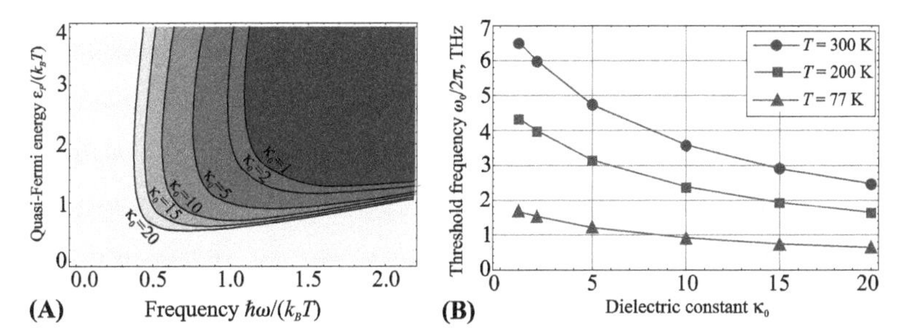

Figure 23.5 (A) Thresholds of negative dynamic conductivity at different values of background dielectric constant κ_0. Solid lines correspond to Re(σ_{intra} + σ_{inter}) = 0. (B) Threshold frequencies $\omega_0/2\pi$ vs background dielectric constant at different temperatures T ($\varepsilon_F = 3k_BT$).

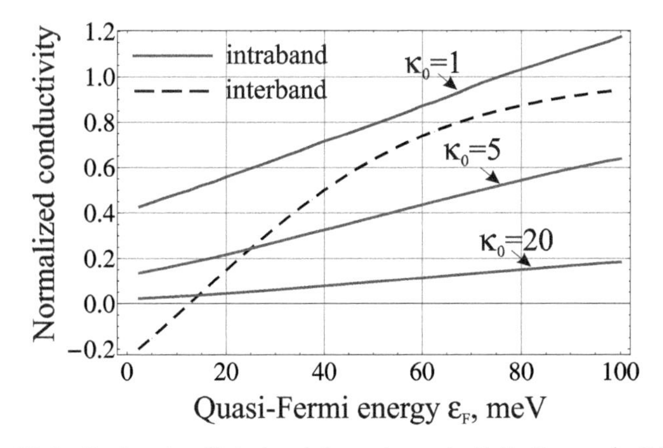

Figure 23.6 Real parts of intraband dynamic conductivity Reσ_{intra} (solid lines) and interband conductivity $-$Reσ_{inter} (dashed line) as functions of quasi-Fermi energy ε_F at fixed frequency $\omega/(2\pi)$ = 6 THz and different values of background dielectric constant κ_0.

The dependences of the inter- and intraband (carrier-carrier) conductivities on the quasi-Fermi energy at fixed frequency are shown in Fig. 23.6. The interband conductivity reaches its ultimate value of σ_q at $\varepsilon_F \gg \hbar\omega/2$. Intuitively, Re$\sigma_{intra}$ and the corresponding radiation absorption coefficient should be proportional to the number of particle-particle collisions in unit time. This number is proportional to the carrier density squared, at least for classical carrier systems. In reality, the dependence of Reσ_{intra} on the quasi-Fermi energy ε_F appears to be almost linear. This, first of all, results from the screening and also from the restricted phase space for collisions in Fermi-systems in a narrow layer near the Fermi-surface.

23.4 Discussion of the Results

The obtained relation for the intraband conductivity associated with the c-c scattering (Eq. (23.13)) corresponds to a Drude-like dependence Re$\sigma_{intra} \propto \omega^{-2}$. The latter tends to infinity at small frequencies. This tendency should be cut off at $\omega \sim \nu$, where ν is the electron collision frequency. At such small frequencies, the finite lifetime of quasiparticles leads to inapplicability of the Fermi golden rule. However, one can estimate the carrier-carrier collision frequency and the corresponding low-frequency conductivity by solving the kinetic equation for the massless electrons in graphene. We have generalized the variational approach to the solution of kinetic equation [28] developed in [16, 17] for intrinsic graphene, to the case of symmetrically pumped graphene. As a result, the following low-frequency dependence of intraband conductivity due to c-c scattering was obtained:

$$\mathrm{Re}\,\sigma_{intra} = \frac{2e^2}{\pi\hbar}\frac{k_B T}{\hbar}\ln\left[1+\exp\left(\frac{\varepsilon_F}{k_B T}\right)\right]\frac{\nu_{cc}}{\omega^2+\nu_{cc}^2}, \qquad (23.16)$$

where the carrier-carrier collision frequency ν_{cc} includes the contributions from the e-e and e-h scattering

$$\nu_{cc} = \frac{\alpha_c^2}{4\pi^2}\frac{k_B T}{\hbar}\frac{I_{ee,0}+I_{eh,0}}{\ln\left[1+\exp\left(\frac{\varepsilon_F}{k_B T}\right)\right]}, \qquad (23.17)$$

and the "zero" index at dimensionless collision integrals means that they are taken at $\omega = 0$.

At intermediate frequencies $\nu_{cc} \ll \omega \le k_B T/\hbar$, the results obtained from the kinetic equation (Eq. (23.16)) and the Fermi golden rule (Eq. (23.13)) coincide. At room temperature, zero Fermi energy, and background dielectric constant $\kappa_0 = 1$, the characteristic collision frequency is equal to $\nu_{cc} \approx 10^{13}$ s^{-1}. Once κ_0 or ε_F are increased, the collision frequency decreases due to screening of the Coulomb potential. Hence, for the 'target' frequencies $\omega/(2\pi) = 5 - 10$ THz one can readily use the golden rule result (Eq. (23.13)) to evaluate the carrier-carrier scattering contribution to the optical conductivity.

Until now we considered the intraband conductivity due to the c-c collisions only, as if other scattering processes were removed. In realty, the acoustic phonon contribution is unavoidable as well. The corresponding collision frequency can be presented as $(\tau_p^{-1})_{ph} = \nu_0 (p v_F / k_B T)$, where $\nu_0 \simeq 3 \times 10^{11}$ s^{-1} at room temperature [29]. Thus, at least for not very large values of κ_0, the c-c scattering contribution to intraband conductivity appears to be larger than the acoustic phonon contribution. The momentum relaxation time due to scattering by randomly distributed charged impurities with average density n_i is

$$\left(\tau_p^{-1}\right)i = 2\pi\alpha_c^2 \frac{p v_F}{\hbar} \frac{\hbar^2 n_i}{q_{TF}^2} \int_0^{\pi} \frac{\sin^2\theta d\theta}{\left[1 + (2p/q_{TF})\sin(\theta/2)\right]^2}. \quad (23.18)$$

The estimates for intrinsic graphene at room temperature, $\kappa_0 = 1$, impurity density $n_i = 10^{11}$ cm^{-2} and electron energy equal to the thermal energy ($p v_F = k_B T$) yield the electron-impurity collision frequency $(\tau_p^{-1})_i \approx 7 \times 10^{12} \text{s}^{-1}$, which is of the same order as the carrier-carrier collision frequency, and scales linearly with n_i. To obtain the real part of the net intraband dynamic conductivity, one should sum the collision frequencies due to various scattering mechanisms. The presented analysis shows that the carrier-carrier scattering contribution cannot be neglected in the case of intrinsic or symmetrically pumped graphene.

In our estimates for the Coulomb scattering probability, we have used the static (Thomas-Fermi) approximation for dielectric function of graphene. The use of the dynamic dielectric function can be mandatory to avoid the divergence in collision integrals for collinear scattering [9]. However, in Eqs. (23.10) and (23.12) such a divergence does not appear due to the presence of factor $\Delta\mathbf{n}^2$. The difference between $\mathrm{Re}\sigma_{\mathrm{intra}}$ obtained using static and dynamic screening is below 10%.

As known, the e-e and h-h scattering does not lead to the finite resistivity and the absorption of radiation by the carriers with the parabolic dispersion law (in the absence of umklapp processes [28]). In graphene, such an effect is possible due to the momentum-velocity decoupling. Formally, e-e collisions in pumped graphene are of the same "strength" as e-h collisions (see Eqs. (23.9) and (23.11)). However, the numerical comparison of Eqs. (23.9) and (23.11) shows that the e-e and h-h collisions play a minor role in the radiation absorption in the pumped graphene. The major contribution to $\mathrm{Re}\sigma_{\mathrm{cc}}$ comes from the electron-hole collisions, while $\mathrm{Re}\sigma_{\mathrm{ee}}$ actually constitutes less than 10% of $\mathrm{Re}\sigma_{\mathrm{cc}}$ (see Fig. 23.7a). This can be understood by analyzing the change in current in a single collision act. The corresponding change is given by the terms $(\Delta\mathbf{n}_{\mathrm{ee}})^2$ and $(\Delta\mathbf{n}_{\mathrm{eh}})^2$ in Eqs. (23.10) and (23.12). Expanding in powers of $\mathbf{q}$ and averaging over directions, one obtains $(\Delta\mathbf{n}_{\mathrm{ee}})^2 \propto q^2(k_1^{-1} - k_2^{-1})^2$ and $(\Delta\mathbf{n}_{\mathrm{eh}})^2 \propto q^2(k_1^{-1} + k_2^{-1})^2$. The latter implies that collisions of electrons with sufficiently different energies only play significant role in the e-e and h-h scattering-assisted absorption. For electron-hole collisions there is no such a restriction.

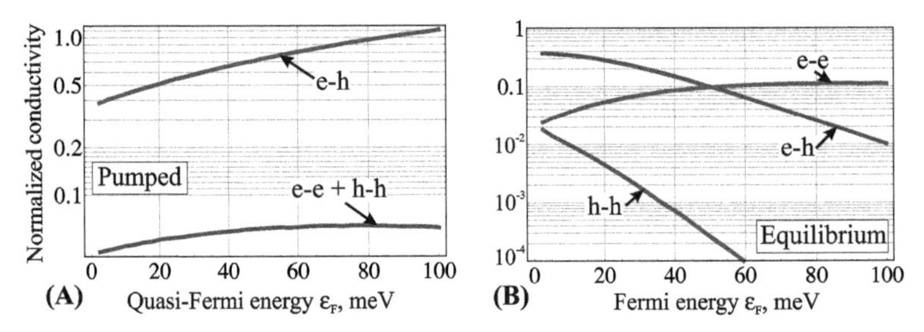

Figure 23.7 Separate contributions of e-h, e-e, and h-h scattering processes to real part of interband conductivity: (A) for pumped graphene and (B) for electron-doped graphene in thermodynamic equilibrium ($\kappa_0 = 1$, $\omega/2\pi = 6$ THz, and $T = 300$ K).

For the electron-hole system in equilibrium, the relative contributions of e-e and e-h scattering to the intraband conductivity are much different (see Fig. 23.7b). The h-h and e-h contributions drop quickly with increasing the Fermi energy due to vanishingly small number of holes. The remaining part of intraband conductivity due to the e-e collisions is of order of $0.1\sigma_q$ at $\omega/(2\pi) = 6$ THz and $\kappa_0 = 1$. The corresponding contribution to radiation absorption can be in principle detected for rather clean samples.

Recent experiments have shown that the THz radiation gain in the pumped graphene can greatly exceed 2.3% [30]. This phenomenon is attributed to the self-excitation of surface plasmon-polaritons in electron-hole system with population inversion [31]. The threshold of plasmon self-excitation in the pumped graphene is also governed by the ratio of the negative interband and positive intraband conductivities. The latter results mainly from the e-h collisions [18]. The presented calculations show that terahertz plasmons are rather amplified than damped in the pumped graphene, at least at high pumping levels and large background dielectric constants.

23.5 Conclusions

We have shown that in graphene there exists a strong mechanism of the intraband Drude-like radiation absorption, assisted by the carrier-carrier collisions. In such a process, a carrier absorbs a photon and transfers the excess energy and momentum to the other carrier, electron or hole. The radiation absorption assisted by the electron-electron and hole-hole collisions is possible only in semiconductors with non-parabolic bands, and graphene (with its linear energy spectrum of electrons and holes) is one of brightest representatives of such a family. We have evaluated the carrier-carrier scattering contribution to the intraband conductivity using the second-order perturbation theory and the Fermi golden rule. As demonstrated, the radiation absorption due to the carrier-carrier collisions can markedly surpass the Drude absorption due to the scattering on thermal or static disorder. This is especially pronounced in the pumped graphene with a large number of carriers in both bands.

We have studied the effect of carrier-carrier scattering on the conditions of the negative dynamic conductivity in pumped graphene by comparing the positive intraband dynamic conductivity due to the carrier-carrier collisions and the negative interband conductivity. We have shown that at room temperature, the net negative dynamic conductivity is attainable at frequencies above ~6.5 THz in suspended graphene. This frequency threshold can be shifted to lower values (~2.5 THz at room temperature) for graphene on high-κ substrates, as the presence of dielectric partially screens the Coulomb interaction. Cooling of the electron-hole system also reduces the negative conductivity threshold. Our results show that screening of Coulomb interaction by carriers in graphene itself (Thomas-Fermi screening) significantly affects the real part of intraband conductivity. Due to the screening, the c-c contribution to the intraband conductivity grows slowly (almost linearly) with increasing quasi-Fermi energy of nonequilibrium electrons and holes.

Appendix: Evaluation of Coulomb Integrals

We introduce the dimensionless energies of incident particles $\varepsilon_1 = k_{1+}$ and $\varepsilon_2 = k_{2-}$, and the transferred energy $\delta\varepsilon = k_{1+} - k_{1-} - w/2 = k_{2+} - k_{2-} - w/2$. Here, for brevity, $w = \hbar\omega/(k_B T)$. The kernel of the e-e Coulomb integral depends only on those energies and the transferred momentum Q:

$$C_{ee}(\varepsilon_1, \varepsilon_2, \delta\varepsilon, Q) = \int d\mathbf{k}_1 d\mathbf{k}_2 (\Delta\mathbf{n}_{ee})^2 \cos^2(\theta_{1\pm}/2)\cos^2(\theta_{2\pm}/2)\delta(\varepsilon_1 - k_{1+})\delta(\varepsilon_2 - k_{2-})$$
$$\times\delta[\delta\varepsilon - (k_{2+} - k_{2-} - w/2)]\delta[\delta\varepsilon - (k_{1+} - k_{1-} + w/2)]. \quad (23.19)$$

In these notations, $I_{ee,\omega}$ is rewritten as

$$I_{ee,\omega} = \int_{w/2}^{\infty} \frac{QdQ}{(Q+Q_{TF})^2} \int_{\delta\varepsilon_{min}}^{\delta\varepsilon_{max}} d\delta\varepsilon \int_{\varepsilon_{1min}}^{\infty} d\varepsilon_1 \int_{\varepsilon_{2min}}^{\infty} d\varepsilon_2 C_{ee}(\varepsilon_1, \varepsilon_2, \delta\varepsilon, Q) F(\varepsilon_1) F(\varepsilon_2)$$
$$\times[1 - F(\varepsilon_2 + \delta\varepsilon + w/2)][1 - F(\varepsilon_1 - \delta\varepsilon + w/2)]. \quad (23.20)$$

The limits of integration will be found further. In elliptic coordinates the Coulomb kernel is evaluated analytically. The change of variables is (i = 1, 2)

$$\mathbf{k}_i = \frac{Q}{2}\{\cosh u_i \cos v_i, \sinh u_i \sin v_i\}, \quad (23.21)$$

where $u_i > 0$, and $-\pi < v_i < \pi$. The product of four delta-functions in Eq. (23.19) can be rewritten as

$$\frac{4}{Q^4}\delta\left(\cos v_2-\frac{\delta\varepsilon+w/2}{Q}\right)\delta\left(\cos v_1-\frac{\delta\varepsilon-w/2}{Q}\right)\delta\left(\cosh u_1-\frac{w/2-\delta\varepsilon+2\varepsilon_1}{Q}\right)$$
$$\times\delta\left(\cosh u_2-\frac{w/2+\delta\varepsilon+2\varepsilon_2}{Q}\right). \quad (23.22)$$

Considering the arguments of delta-functions in Eq. (23.22), one finds the limits of integration over energies in Eq. (23.20) from general requirements $\cosh u_i > 1$, $|\cos v_i| < 1$, $i = 1, 2$:

$$\varepsilon_1>\frac{(Q+\delta\varepsilon-w/2)}{2};\varepsilon_2>\frac{(Q-\delta\varepsilon-w/2)}{2},Q>w/2;-Q+w/2<\delta\varepsilon<Q-w/2. \quad (23.23)$$

Integration over du_1, du_2, dv_1, and dv_2 removes the delta-functions, and the Coulomb kernel becomes

$$C(\varepsilon_1,\varepsilon_2,\delta\varepsilon,Q)=\frac{4(\Delta\mathbf{n})^2}{Q^4}\frac{\cos^2(\theta_{1\pm}/2)\cos^2(\theta_{2\pm}/2)}{|\sin v_1\sin v_2\sinh u_1\sinh u_2|}k_{1+}k_{1-}k_{2+}k_{2-}. \quad (23.24)$$

Here the sin- and sinh-terms in the denominator appear from the rule $\delta[f(x)] = \Sigma_i |f'(x_i)|^{-1}\delta(x - x_i)$, where the summation is performed over all zeros x_i of $f(x)$. In Eq. (23.24) $k_{1+}k_{1-}k_{2+}k_{2-}$ is the Jacobian determinant for elliptic coordinates.

The envelope function overlap factors are also concisely expressed in the elliptic coordinates:

$$\cos^2(\theta_{1\pm}/2)\cos^2(\theta_{2\pm}/2)k_{1+}k_{1-}k_{2+}k_{2-}$$
$$=\frac{1}{16}[(2\varepsilon_1-\delta\varepsilon+w/2)^2-Q^2][(2\varepsilon_2+\delta\varepsilon+w/2)^2-Q^2]. \quad (23.25)$$

As a result, the Coulomb kernel takes on the following form:

$$C_{ee}(\varepsilon_1,\varepsilon_2,\delta\varepsilon,Q)=\frac{(\Delta\mathbf{n})^2_{ee}}{4}\frac{\sinh u_1\sinh u_2}{\sin v_1\sin v_2}$$
$$=\frac{(\Delta\mathbf{n})^2_{ee}}{4}\sqrt{\frac{\left[(2\varepsilon_1-\delta\varepsilon+w/2)^2-Q^2\right][(2\varepsilon_2+\delta\varepsilon+w/2)^2-Q^2]}{[Q^2-(\delta\varepsilon+w/2)^2][Q^2-(\delta\varepsilon-w/2)^2]}}. \quad (23.26)$$

The expression for $(\Delta\mathbf{n})^2_{ee}$ in elliptic coordinates is quite lengthy, here we present in only in the limit $w \rightarrow 0$:

$$(\Delta\mathbf{n})^2_{ee} = \frac{2\left(Q^2 - \delta\varepsilon^2\right)}{\varepsilon_1\varepsilon_2\left(\varepsilon_1 - \delta\varepsilon\right)\left(\delta\varepsilon + \varepsilon_2\right)}$$

$$\times\left[\delta\varepsilon^2 + 4\delta\varepsilon(\varepsilon_2 - \varepsilon_1) + 2(\varepsilon_1 - \varepsilon_2)^2 - \frac{\delta\varepsilon^2\left(\delta\varepsilon - 2\varepsilon_1\right)\left(\delta\varepsilon + 2\varepsilon_2\right)}{Q^2}\right]. \quad (23.27)$$

Note that Eq. (23.26) contains the denominator divergent at $w \to 0$, which manifests the collinear scattering anomaly for massless particles [9, 16, 17]. This divergence is removed by the similar term in the expression for $(\Delta\mathbf{n})^2_{ee}$, which stands in the numerator.

Acknowledgments

The authors are grateful to M. Ryzhii for useful comments. The work at RIEC was supported by the Japan Society for Promotion of Science (Grant-in-Aid for Specially Promoting Research #23000008), Japan. The work at IPT RAS was supported by the Russian Foundation of Basic Research (grant #14-07-00937). The work by D.S was also supported by the grant of the Russian Foundation of Basic Research #14-07-31315 and by the JSPS Postdoctoral Fellowship For Foreign Researchers (Short-term), Japan.

References

1. F. Bonaccorso, Z. Sun, T. Hasan, and A. C. Ferrari, Graphene photonics and optoelectronics, *Nat. Photonics*, **4**, 611–622 (2010).
2. A. Tredicucci and M. S. Vitiello, Device concepts for graphene-based terahertz photonics, *IEEE J. Sel. Top. Quantum Electron.*, **20**, 8500109 (2014).
3. P. Weis, J. L. Garcia-Pomar, and M. Rahm, Towards loss compensated and lasing terahertz metamaterials based on optically pumped graphene, *Opt. Express*, **22**(7), 8473–8489 (2014).
4. A. Dubinov, V. Ya. Aleshkin, M. Ryzhii, T. Otsuji, and V. Ryzhii, Terahertz laser with optically pumped graphene layers and Fabri-Perot resonator, *Appl. Phys. Express*, **2**, 092301 (2009).
5. V. Ryzhii, M. Ryzhii, V. Mitin, and T. Otsuji, Toward the creation of terahertz graphene injection laser, *J. Appl. Phys.*, **110**, 094503 (2011).

6. V. Ryzhii, A. Dubinov, T. Otsuji, V. Ya. Aleshkin, M. Ryzhii, and M. Shur, Double-graphene-layer terahertz laser: concept, characteristics, and comparison, *Opt. Express*, **21**(25), 31567–31577 (2013).

7. S. Boubanga-Tombet, S. Chan, T. Watanabe, A. Satou, V. Ryzhii, and T. Otsuji, Ultrafast carrier dynamics and terahertz emission in optically pumped graphene at room temperature, *Phys. Rev. B*, **85**, 035443 (2012).

8. M. Martl, J. Darmo, C. Deutsch, M. Brandstetter, A. M. Andrews, P. Klang, G. Strasser, and K. Unterrainer, Gain and losses in THz quantum cascade laser with metal-metal waveguide, *Opt. Express*, **19**, 733 (2011).

9. D. Brida, A. Tomadin, C. Manzoni, Y. J. Kim, A. Lombardo, S. Milana, R. R. Nair, K. S. Novoselov, A. C. Ferrari, G. Cerullo, and M. Polini, Ultrafast collinear scattering and carrier multiplication in graphene, *Nat. Commun.*, **4**, article number: 1987 (2013).

10. K. F. Mak, M. Y. Sfeir, Y. Wu, C. H. Lui, J. A. Misewich, and T. F. Heinz, Measurement of the optical conductivity of graphene, *Phys. Rev. Lett.*, **101**, 196405 (2008).

11. V. Ryzhii, M. Ryzhii, and T. Otsuji, Negative dynamic conductivity of graphene with optical pumping, *J. Appl. Phys.*, **101**, 083114 (2007).

12. L. Ren, Q. Zhang, J. Yao, Z. Sun, R. Kaneko, Z. Yan, S. Nanot, Z. Jin, I. Kawayama, M. Tonouchi, J. M. Tour, and J. Kono, Terahertz and infrared spectroscopy of gated large-area graphene, *Nano Lett.*, **12**, 3711–3715 (2012).

13. A. Satou, V. Ryzhii, Y. Kurita, and T. Otsuji, Threshold of terahertz population inversion and negative dynamic conductivity in graphene under pulse photoexcitation, *J. Appl. Phys.*, **113**, 143108 (2013).

14. J. M. Dawlaty, S. Shivaraman, M. Chandrashekhar, F. Rana, and M. G. Spencer, Measurement of ultrafast carrier dynamics in epitaxial graphene, *Appl. Phys. Lett.*, **92**, 042116 (2008).

15. M. Mittendorff, T. Winzer, E. Malic, A. Knorr, C. Berger, W. A. de Heer, H. Schneider, M. Helm, and S. Winnerl, Anisotropy of excitation and relaxation of photogenerated charge carriers in graphene, *Nano Lett.*, **14**(3), 1504–1507 (2014).

16. A. B. Kashuba, Conductivity of defectless graphene, *Phys. Rev. B*, **78**, 085415 (2008).

17. L. Fritz, J. Schmalian, M. Müller, and S. Sachdev, Quantum critical transport in clean graphene, *Phys. Rev. B*, **78**, 085416 (2008).

18. D. Svintsov, V. Vyurkov, S. Yurchenko, V. Ryzhii, and T. Otsuji, Hydrodynamic model for electron-hole plasma in graphene, *J. Appl. Phys.*, **111**(8), 083715 (2012).
19. M. Schütt, P. M. Ostrovsky, I. V. Gornyi, and A. D. Mirlin, Coulomb interaction in graphene: relaxation rates and transport, *Phys. Rev. B*, **83**, 155441 (2011).
20. V. N. Kotov, B. Uchoa, V. M. Pereira, F. Guinea, and A. H. Castro Neto, Electron-electron interactions in graphene: current status and perspectives, *Rev. Mod. Phys.*, **84**, 1067 (2012).
21. L. A. Falkovsky and S. S. Pershoguba, Optical far-infrared properties of a graphene monolayer and multilayer, *Phys. Rev. B*, **76**, 153410 (2007).
22. F. T. Vasko, V. V. Mitin, V. Ryzhii, and T. Otsuji, Interplay of intra- and interband absorption in a disordered graphene, *Phys. Rev. B*, **86**, 235424 (2012).
23. D. Sun, C. Divin, M. Mihnev, T. Winzer, E. Malic, A. Knorr, J. E. Sipe, C. Berger, W. A. de Heer, and P. N. First, Current relaxation due to hot carrier scattering in graphene, *New J. Phys.*, **14**, 105012 (2012).
24. G. G. Zegrya and V. E. Perlin, Intraband absorption of light in quantum wells induced by electron-electron collisions, *Semiconductors*, **32**, 417–422 (1998).
25. E. H. Hwang and S. Das Sarma, Dielectric function, screening, and plasmons in two-dimensional graphene, *Phys. Rev. B*, **75**, 205418 (2007).
26. L. D. Landau and E. M. Lifshitz *Quantum Mechanics* (Pergamon Press, 1965).
27. V. Ryzhii, M. Ryzhii, V. Mitin, A. Satou, and T. Otsuji, Effect of heating and cooling of photogenerated electron-hole plasma in optically pumped graphene on population inversion, *Jpn. J. Appl. Phys.*, **50**, 094001 (2011).
28. J. M. Ziman, *Electrons and Phonons* (Oxford University Press, 1960).
29. F. T. Vasko and V. Ryzhii, Voltage and temperature dependencies of conductivity in gated graphene, *Phys. Rev. B*, **76**, 233404 (2007).
30. T. Watanabe, T. Fukushima, Y. Yabe, S. A. Boubanga Tombet, A. Satou, A. Dubinov, V. Ya. Aleshkin, V. Mitin, V. Ryzhii, and T. Otsuji, The gain enhancement effect of surface plasmon polaritons on terahertz stimulated emission in optically pumped monolayer graphene, *New J. Phys.*, **15**, 075003 (2013).
31. A. Dubinov, V. Aleshkin, V. Mitin, T. Otsuji, and V. Ryzhii, Terahertz surface plasmons in optically pumped graphene structures, *J. Phys.: Condens. Matter*, **23**, 145302 (2011).

Chapter 24

Negative Terahertz Conductivity in Disordered Graphene Bilayers with Population Inversion*

D. Svintsov,[a] T. Otsuji,[b] V. Mitin,[b,c] M. S. Shur,[d] and V. Ryzhii[b]
[a]*Laboratory of Nanooptics and Plasmonics, Moscow Institute of Physics and Technology, Dolgoprudny 141700, Russia*
[b]*Research Institute of Electrical Communication, Tohoku University, Sendai 980-8577, Japan*
[c]*Department of Electrical Engineering, University at Buffalo, SUNY, Buffalo, New York 1460–1920, USA*
[d]*Departments of Electrical, Electronics, and Systems Engineering and Physics, Applied Physics, and Astronomy, Rensselaer Polytechnic Institute, Troy, New York 12180, USA*
v-ryzhii@riec.tohoku.ac.jp

The gapless energy band spectra make the structures based on graphene and graphene bilayer with the population inversion to be promising media for the interband terahertz (THz) lasing. However,

*Reprinted with permission from D. Svintsov, T. Otsuji, V. Mitin, M. S. Shur, and V. Ryzhii (2015). Negative terahertz conductivity in disordered graphene bilayers with population inversion, *Appl. Phys. Lett.*, **106**, 113501. Copyright © 2015 AIP Publishing LLC.

Graphene-Based Terahertz Electronics and Plasmonics: Detector and Emitter Concepts
Edited by Vladimir Mitin, Taiichi Otsuji, and Victor Ryzhii

ISBN 978-981-4800-75-4 (Hardcover), 978-0-429-32839-8 (eBook)
www.jennystanford.com

a strong intraband absorption at THz frequencies still poses a challenge for efficient THz lasing. In this chapter, we show that in the pumped graphene bilayer, the indirect interband radiative transitions accompanied by scattering of carriers by disorder can provide a substantial negative contribution to the THz conductivity (together with the direct interband transitions). In the graphene bilayer on high-κ substrates with point charged defects, these transitions substantially compensate the losses due to the intraband (Drude) absorption. We also demonstrate that the indirect interband contribution to the THz conductivity in a graphene bilayer with the extended defects (such as the charged impurity clusters) can surpass by several times the fundamental limit associated with the direct interband transitions, and the Drude conductivity as well. These predictions can affect the strategy of the graphene-based THz laser implementation.

The absence of a band gap in the atomically thin carbon structures, such as graphene and graphene bilayer, enables their applications in different terahertz (THz) and infrared devices [1–4]. One of the most challenging and promising problems is the creation of the graphene-based THz lasers [5–8]. These lasers are expected to operate at room temperature, particularly, in the 6–10 THz range, where the operation of III–V quantum cascade lasers is hindered by the optical phonons [9]. Recent pump-probe spectroscopy experiments confirm the possibility of the coherent radiation amplification in the optically pumped graphene [10–16], enabled by a relatively long-living interband population inversion [16]. As opposed to optically pumped graphene lasers, graphene-based injection lasers are expected to operate in the continuous mode, with the inter-band population inversion maintained by the electron and hole injection from the n-and p-type contacts [6]. A single sheet of pumped graphene as the gain medium provides the maximum radiation amplification coefficient equal to $4\pi\sigma_{Q}/c = \pi\alpha = 2.3\%$, where $\sigma_{Q} = e^{2}/4\hbar$ is the universal optical conductivity of a single graphene layer, e is the electron charge, $\hbar$ is the Planck constant, and $\alpha \simeq 1/137$ is the fine-structure constant [16]. The THz gain in the graphene bilayer [17] or the non-Bernal stacked multiple-graphene layers [6] can be enhanced approximately proportional to the number of the layers. However, more crucial is the problem of competition between the interband radiation amplification and intraband (Drude) radiation absorption [5, 15, 18]. The latter scales with frequency ω approximately as $1/\omega^2\tau$, where τ is the momentum relaxation time. Hence, the onset of THz gain is typically believed to occur only in clean samples, where $\omega\tau \gg 1$ [19].

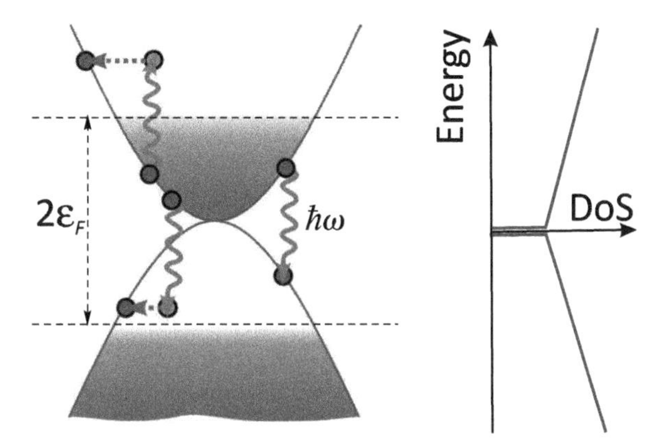

Figure 24.1 Energy band diagram of a graphene bilayer with population inversion (left) and schematic energy dependence of its density of states (right). Arrows indicate possible radiative transitions. Direct interband transitions involve only vertical photon emission (wavy arrow), while indirect emission and absorption processes are accompanied by carrier scattering (straight horizontal arrows).

In this letter, we show that the intraband radiation absorption in disordered graphene bilayer does not actually pose a problem for the THz lasing. On the contrary, the presence of a certain type of defects reinforces the negative contribution to the THz conductivity and, hence, improves the gain properties of the pumped graphene bilayer. The origin of such an effect is associated with the indirect interband radiative transitions (see Fig. 24.1), in which the electrons from the conduction band emit the photons being scattered by the defects, thus contributing to the radiation gain. The inclusion of such processes can be important in the indirect-gap materials and gapless semiconductors. However, in single-layer graphene with charged impurities, the indirect interband processes contribute to the net THz conductivity σ only moderately due to a low density of states (DoS) near the band edges [20]. The situation is remarkably different is the intrinsic graphene bilayer, where the electron-hole dispersion is gap-less and almost parabolic [21]. With constant DoS near the band edges, the indirect interband radiative processes significantly contribute to the net radiation gain (which is proportional to –Re σ) under the population inversion conditions. We analyze the spectral dependencies of the real part of the net dynamic THz conductivity Re σ in the pumped graphene bilayers for the scattering by the point charged impurities [22, 23] and by the impurity clusters [24, 25]. Our analysis shows that in the first case, the gain due to the indirect inter-band transitions substantially compensates the losses due to the Drude absorption. In the case of the cluster scattering, the gain due to the indirect interband transitions even exceeds the intraband losses. Depending on the cluster size and their density, the ratio $|\mathrm{Re}\,\sigma|/\sigma_Q$ can markedly exceed unity resulting in elevated net THz gain.

The real part of the net in-plane dynamic THz conductivity, Re σ, comprises the contributions of the direct (vertical) interband electron transitions, Re σ_d, and the contributions, Re σ_{ind}^{intra} and Re σ_{ind}^{inter}, of two types of the indirect electron transitions (inside both the bands and between the two bands), respectively

$$\mathrm{Re}\,\sigma = \mathrm{Re}\,\sigma_d + \mathrm{Re}\,\sigma_{ind}^{inter} + \mathrm{Re}\,\sigma_{ind}^{intra}. \quad (24.1)$$

The interband conductivity of graphene bilayers due to the direct transitions is given by [19, 21]

$$\mathrm{Re}\,\sigma_d = \sigma_Q \frac{\hbar\omega + 2\gamma_1}{\hbar\omega + \gamma_1}[f_\upsilon(-\hbar\omega/2) - f_c(\hbar\omega/2)], \quad (24.2)$$

where $f_\upsilon(\varepsilon)$ and $f_c(\varepsilon)$ are the carrier distribution functions in the valence and conduction bands, and $\gamma_1 \simeq 0.4$ eV is the hopping integral between carbon atoms in adjacent graphene planes [26]. Here and in the following, we assume that f_υ and f_c are the Fermi functions characterized by the quasi-Fermi energies $\mu_\upsilon = -\varepsilon_F$ and $\mu_c = \varepsilon_F$, respectively. This is justified by strong carrier-carrier scattering leading to a fast thermalization of excited carriers [10, 13]. In the THz frequency range of interest, $\hbar\omega \ll \gamma_1$, and the conductivity of bilayer is simply twice as large as that of a single layer. Under the pumping conditions, $\mu_c = -\mu_\upsilon = \varepsilon_F > 0$, and, according to Eq. (24.2), the conductivity is negative for photon energies $\hbar\omega$ below the double quasi-Fermi energy ε_F of pumped carriers [17].

To evaluate the real part of the THz conductivity due to the indirect transitions, we calculate the second-order transition amplitudes for the photon emission (absorption) accompanied by the electron scattering and apply the Fermi's golden rule. As a result, the general expressions for the indirect intra- and interband conductivities read as follows:

$$\mathrm{Re}\,\sigma_{\mathrm{ind}}^{\mathrm{intra}} = \frac{8\pi\sigma_{\mathrm{Q}}}{\hbar\omega^3}\sum_{\mathbf{k},\mathbf{k}',\lambda}[f_\lambda(\varepsilon_{\mathbf{k}}) - f_\lambda(\varepsilon_{\mathbf{k}'})]$$
$$\times\,\delta(\hbar\omega + \varepsilon_{\mathbf{k}} - \varepsilon_{\mathbf{k}'})\,|V_{\mathrm{S}}(\mathbf{k}-\mathbf{k}')|^2\, u_{\mathbf{k}\mathbf{k}'}^{\lambda\lambda}(\mathbf{v}_{\mathbf{k}'} - \mathbf{v}_{\mathbf{k}})^2, \quad (24.3)$$

$$\mathrm{Re}\,\sigma_{\mathrm{ind}}^{\mathrm{inter}} = \frac{8\pi\sigma_{\mathrm{Q}}}{\hbar\omega^3}\sum_{\mathbf{k},\mathbf{k}'}[f_\upsilon(\varepsilon_{\mathbf{k}}) - f_c(\varepsilon_{\mathbf{k}'})]$$
$$\times\,\delta(\hbar\omega - \varepsilon_{\mathbf{k}} - \varepsilon_{\mathbf{k}'})\,|V_{\mathrm{S}}(\mathbf{k}-\mathbf{k}')|^2\, u_{\mathbf{k}\mathbf{k}'}^{c\upsilon}(\mathbf{v}_{\mathbf{k}'} + \mathbf{v}_{\mathbf{k}})^2. \quad (24.4)$$

Here, $\lambda = \pm 1$ is the index corresponding to the conduction and valence bands, $\mathbf{v}_{\mathbf{k}} = 2\hbar\upsilon_0^2\mathbf{k}/\gamma_1$ is the electron velocity in the graphene bilayer, $\upsilon_0 \simeq 10^8$ cm/s is the velocity characterizing the energy spectra of graphene and graphene bilayer, $V_{\mathrm{S}}(\mathbf{q})$ is the $\mathbf{q}$-th Fourier component of the scattering potential, and $u_{\mathbf{k}\mathbf{k}'}^{\lambda\lambda'} = (1 + \lambda\lambda' \cos 2\theta_{\mathbf{k}\mathbf{k}'})/2$ is the overlap between the envelope wave functions in graphene bilayers. For a correct qualitative description of the low-frequency conductivity, we replace the frequency ω in the denominators of Eqs. (24.2) and (24.3) with $(\omega^2 + \tau^{-2})^{1/2}$, where τ is given by

$$\tau^{-1} = \frac{2\pi}{\hbar}\sum_{\mathbf{k}'}|V_{\mathrm{S}}(\mathbf{k}-\mathbf{k}')|^2\,(1-\cos\theta_{\mathbf{k}\mathbf{k}'})\cos^2\theta_{\mathbf{k}\mathbf{k}'}\delta(\varepsilon_{\mathbf{k}} - \varepsilon_{\mathbf{k}'}).$$

Equation (24.3) yields the well-known result for the dynamic conductivity in the high-frequency limit $\omega \gg \tau^{-1}$.

In the most practical situations, the Coulomb scattering by the substrate-induced charged impurities is the main factor determining the conductivity [27]. Considering impurities as random uncorrelated point scatterers with the average density n_{i}, we write the scattering matrix element as [28]

$$|V_{\mathrm{S}}(\mathbf{q})|^2 = n_i\left[\frac{2\pi e^2}{\kappa(q+q_{\mathrm{s}})}\right]^2, \quad (24.5)$$

where κ is the background dielectric constant and q_{s} is the Thomas-Fermi screening wave vector [30]. Using the Lindhard-type formula for the graphene bilayer polarizability [28] with the quasi-equilibrium Fermi functions for the conduction and valence bands, we obtain $q_s = 4\alpha_c\gamma_1(1 - \ln 2)/\hbar\, v_0$, where $\alpha_c = e^2/\hbar\kappa\, v_0$ is the coupling constant. Evaluating the integral in Eq. (24.4), we find the scattering-assisted interband conductivity

$$\mathrm{Re}\,\sigma_{\mathrm{ind}}^{\mathrm{inter}} \simeq 8\pi\alpha_c^2\sigma_Q \frac{v_0^2 n_i}{(\omega^2 + \tau^{-2})} \times \tanh\left(\frac{\hbar\omega - 2\varepsilon_F}{4T}\right)\Phi\left(\frac{\hbar v_0 q_s}{\sqrt{\hbar\omega\gamma_1}}\right), \tag{24.6}$$

where $\Phi(x) = \int_0^2 dy\,(1 - |1 - y|)(1 - y/2)/(y^{1/2} + x)^2$ with the following asymptotes: $\Phi(x) \approx 2\ln 2 - 1/2$, $x \ll 1$, $\Phi(x) \approx 1/(2x^2)$, $x \gg 1$. At intermediate frequencies, $1/\tau \ll \omega \ll \hbar(v_0 q_s)^2/\gamma_1$, the conductivity due to the indirect interband transitions scales as ω^{-1}. This is substantially different from the case of a single graphene layer, where it tends to a constant. A pronounced increase in the indirect interband contribution to the dynamic conductivity in a bilayer at low frequencies is attributed to the constant density of states in the vicinity of the band edges.

Figure 24.2 (upper panel) shows the spectral dependences of different contributions to the dynamic THz conductivity as well its net value for a moderate-quality exfoliated graphene bilayer [23] on the SiO_2 substrate with impurity density $n_i = 5 \times 10^{11}$ cm^{-2}, at temperature $T = 300$ K, and for the quasi-Fermi energy $\varepsilon_F = 60$ meV ($\tau = 5 \times 10^{-14}$ s). The bottom panel in Fig. 24.2 demonstrates the two-dimensional maps of the dynamic THz conductivity versus the frequency and the quasi-Fermi energy for the same structure. As seen, the contribution of the indirect interband transitions in such a sample is still weaker than those of both the indirect intra-band and direct interband transitions, although the former is essential for compensating, to some extent, the Drude absorption. This enables an increase in $|\mathrm{Re}\sigma|$ and some widening of the frequency range where $\mathrm{Re}\,\sigma < 0$. As follows from Fig. 24.2 (bottom panel), the dynamic THz conductivity is negative at $\varepsilon_F \gtrsim 30$ meV and $\omega/2\pi \gtrsim 10$ THz.

The relative contribution of the indirect interband transitions (compared to the "normal" intraband conductivity) increases with an increasing dielectric constant. Figure 24.3 shows the same characteristics as in Fig. 24.2, for a graphene bilayer being clad by the HfO_2 layers, i.e., immersed in a media with a fairly high dielectric

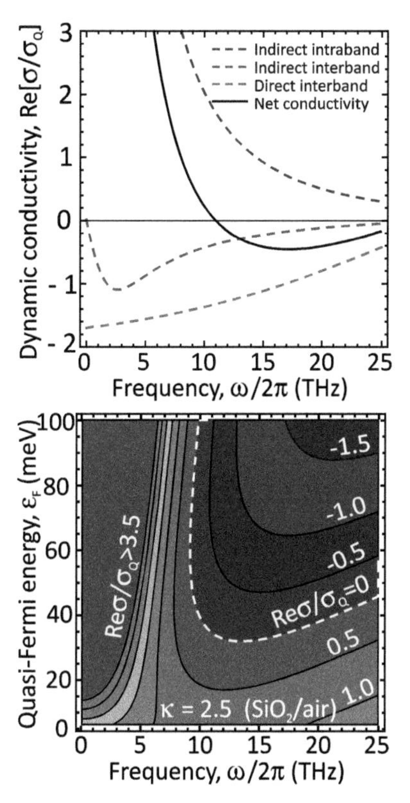

Figure 24.2 Contributions of different radiative transitions to the dynamic conductivity of a graphene bilayer on SiO_2 ($\kappa = 2.5$) and its net value (solid line) versus frequency for $n_i = 5 \times 10^{11}$ cm^{-2} and $\varepsilon_F = 60$ meV (upper panel) and the $\varepsilon_F - \omega$ map of the dynamic conductivity (bottom panel). Numbers above the lines indicate the values of Re $[\sigma/\sigma_Q]$. The region above the white dashed line corresponds to the negative dynamic conductivity.

constant ($\kappa = 20$, $n_i = 5 \times 10^{11}$ cm^{-2}, and $\tau = 5 \times 10^{-13}$ s). As seen from Fig. 24.3, the transition to the graphene bilayer structures with higher dielectric constant leads to a much stronger contribution of the indirect interband transitions and, hence, to a much larger value of |Re σ|. It is important to note that an increase is in the relative contribution, as the absolute values of both contributions go down at larger κ. The comparison of Figs. 24.2 and 24.3 shows that the ratio |Re $\sigma|/\sigma_Q$ for the minimum value of Re σ at $\kappa = 20$ is five-six time larger than that for $\kappa = 2.5$. Moreover, in the former case Re σ becomes negative starting from $\varepsilon_F = 20$ meV and $\omega/2\pi \gtrsim 5$ THz. At higher values of ε_F (i.e., at a stronger pumping), Re σ can be negative from the frequencies of about a few THz to a dozen THz.

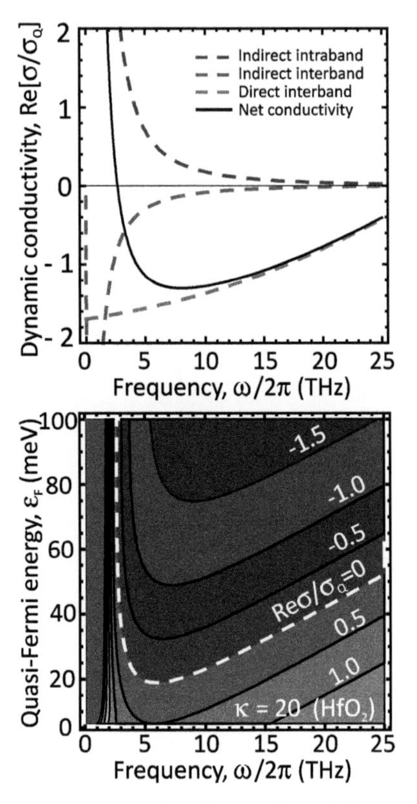

Figure 24.3 Same as in Fig. 24.2 but for graphene bilayer clad by HfO_2 layers ($\kappa = 20$).

The reinforcement of the negative dynamic THz conductivity effect with increasing background dielectric constant demonstrated above is interpreted as follows. An increase in κ results in the reduced the Thomas-Fermi wave vector q_s. This, in turn, leads to the switching from the strongly screened to almost bare Coulomb scattering. From the energy conservation laws (see Eqs. (24.3) and (24.4)), it follows that the indirect interband electron transitions are favored by a low momentum transfer q, while for the indirect transitions within one band q should be large, namely, $q > \omega/v_0$ as the electron velocity in graphene bilayer does not exceed v_0. Hence, for high values of κ, the scattering potential behaves approximately as $V_S(\mathbf{q}) \propto 1/q$, which supports the interband transitions with a low momentum transfer.

As shown above, the Coulomb scattering by the screened point defects cannot lead to the dominance of the indirect interband transitions over the indirect intraband ones. However, such a dominance can be realized in the case of the carrier scattering by extended scatterers. We consider the charged cluster of size l_c and charge Ze as a continuous two-dimensional distribution of charge density $\rho(\mathbf{r}) = Ze \exp(-r^2/l_c^2)/\pi l_c^2$ (we assume it to be Gaussian [25]). Solving the Poisson equation for this charge density and averaging over the random positions of the clusters with the average density n_c, we readily find the scattering matrix elements (instead of Eq. (24.5))

$$|V_S(\mathbf{q})|^2 = n_c \left[\frac{2\pi Ze^2}{\kappa(q+q_s)}\right]^2 \exp\left(-\frac{q^2 l_c^2}{2}\right). \tag{24.7}$$

In the limit of strong screening, $q \ll q_s$, our model of scattering coincides with the widely accepted model of the "Gaussian correlated disorder" [29–31] with the root-mean-square scattering potential $\sqrt{\overline{V^2}} \approx 2\pi\sqrt{n_c/\pi l_c^2}\, Ze^2/\kappa q_s$.

Figure 24.4 shows the spectral characteristics of the dynamic THz conductivity in the pumped graphene bilayer (ε_F = 60 meV) with clusters of charged impurities with different size l_c. If the scattering by the charged clusters is a dominating mechanism, the absolute value of the net dynamic THz conductivity, $|\mathrm{Re}\sigma|$, can markedly exceed the fundamental "direct interband limit" of $2\sigma_Q$. At reasonable values of $l_c \gtrsim 8$ nm and quasi-Fermi energy ε_F = 60 meV, one obtains $|\mathrm{Re}\ \sigma| > 4\sigma_Q$.

Above, we disregarded the processes of the radiative transitions accompanied by the electron and hole scattering by neutral point defects (with a short-range scattering potential). Such processes contribute to both Re $\sigma_{\rm ind}^{\rm intra}$ and Re $\sigma_{\rm ind}^{\rm inter}$ although the relative contribution of the latter is small ($\lesssim/\hbar\omega/4\varepsilon_F$). Hence, the obtained results are valid when the scattering by neutral point defects is weak compared with the scattering by charges objects (as well as by neutral but large-scale imperfections). In graphene bilayer with a large twist, the carrier spectrum is linear [32], and the optical conductivity is given by the expressions obtained for single layer

[20] with the renormalized band velocity v_0. The crossover between the parabolic and linear energy spectra at intermediate twisting angles requires a separate study.

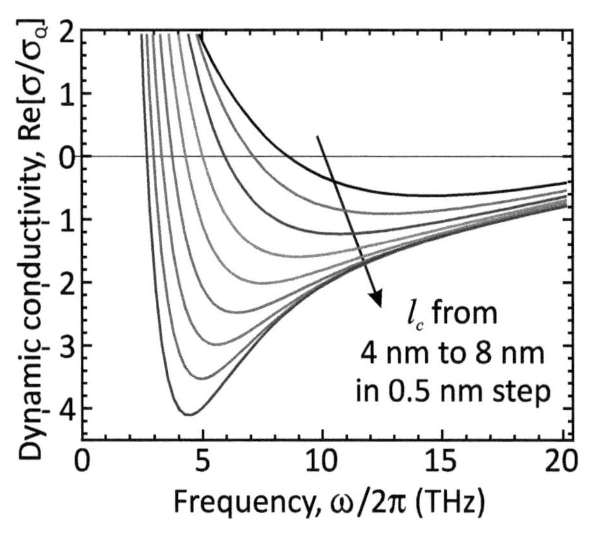

Figure 24.4 Net dynamic THz conductivity versus frequency calculated for graphene bilayer on SiO_2 with impurity clusters of different size l_c: cluster density and charge are $n_c = 10^{11}$ cm^{-2} and $Z = 7$, effective dielectric constant $\kappa = 2.5$, and quasi-Fermi energy $\varepsilon_F = 60$ meV.

In conclusion, we have demonstrated that the graphene bilayer with a long-range disorder (the nanometre-scale impurity clusters) can exhibit a strong negative THz conductivity with the span two times or more exceeding the fundamental limit $2\sigma_Q$. This effect is associated with the indirect interband transitions with the photon emission being accompanied by the disorder scattering. For the indirect interband photon emission to dominate over the "normal" Drude absorption, the long-wave length Fourier components of the scattering potential should prevail. Such kind of scattering potentials can be formed also by extended surface corrugations (wrinkles), multilayer patches, grain boundaries, quantum dots on the graphene bilayer surface [33, 34], and nano-hole arrays [35, 36].

The work was supported by the Japan Society for Promotion of Science (Grant-in-Aid for Specially Promoted Research No. 23000008) and by the Russian Scientific Foundation (Project No. 14-29-00277) and the grant of the Russian Foundation of Basic Research

No. 14-07-31315. The works at UP and RPI were supported by the U.S. Air Force Award No. FA9550-10-1-391 and by the U.S. Army Research Laboratory Cooperative Research Agreement, respectively.

References

1. A. H. Castro Neto, F. Guinea, N. M. R. Peres, K. S. Novoselov, and A. K. Geim, *Rev. Mod. Phys.*, **81**, 109 (2009).
2. A. Tredicucci and M. Vitiello, *IEEE J. Sel. Top. Quantum. Electron*, **20**, 130 (2014).
3. Q. Bao and K. P. Loh, *ACS Nano*, **6**, 3677 (2012).
4. F. Bonaccorso, Z. Sun, T. Hasan, and A. Ferrari, *Nat. Photonics*, **4**, 611 (2010).
5. V. Ryzhii, M. Ryzhii, and T. Otsuji, *J. Appl. Phys.*, **101**, 083114 (2007).
6. V. Ryzhii, M. Ryzhii, A. Satou, T. Otsuji, A. A. Dubinov, and V. Y. Aleshkin, *J. Appl. Phys.*, **106**, 084507 (2009).
7. V. Ryzhii, M. Ryzhii, V. Mitin, and T. Otsuji, *J. Appl. Phys.*, **110**, 094503 (2011).
8. V. Ryzhii, I. Semenikhin, M. Ryzhii, D. Svintsov, V. Vyurkov, A. Satou, and T. Otsuji, *J. Appl. Phys.*, **113**, 244505 (2013).
9. S. Kumar, *IEEE J. Sel. Top. Quantum. Electron*, **17**, 38 (2011).
10. S. Boubanga-Tombet, S. Chan, T. Watanabe, A. Satou, V. Ryzhii, and T. Otsuji, *Phys. Rev. B*, **85**, 035443 (2012).
11. T. Li, L. Luo, M. Hupalo, J. Zhang, M. C. Tringides, J. Schmalian, and J. Wang, *Phys. Rev. Lett.*, **108**, 167401 (2012).
12. S. Kar, D. R. Mohapatra, E. Freysz, and A. K. Sood, *Phys. Rev. B*, **90**, 165420 (2014).
13. I. Gierz, J. C. Petersen, M. Mitrano, C. Cacho, I. E. Turcu, E. Springate, A. Stohr, A. Kohler, U. Starke, and A. Cavalleri, *Nat. Mater.*, **12**, 1119–1124 (2013).
14. T. Winzer, E. Malic, and A. Knorr, *Phys. Rev. B*, **87**, 165413 (2013).
15. T. Watanabe, T. Fukushima, Y. Yabe, S. A. B. Tombet, A. Satou, A. A. Dubinov, V. Y. Aleshkin, V. Mitin, V. Ryzhii, and T. Otsuji, *New J. Phys.*, **15**, 075003 (2013).
16. I. Gierz, M. Mitrano, J. C. Petersen, C. Cacho, I. C. E. Turcu, E. Springate, A. Stöhr, A. Köhler, U. Starke, and A. Cavalleri, Population inversion in monolayer and bilayer graphene, e-print arXiv:1409.0211.

17. V. Y. Aleshkin, A. A. Dubinov, and V. Ryzhii, *JETP Lett.*, **89**, 63 (2009).
18. M. Martl, J. Darmo, C. Deutsch, M. Brandstetter, A. M. Andrews, P. Klang, G. Strasser, and K. Unterrainer, *Opt. Express*, **19**, 733 (2011).
19. A. Satou, V. Ryzhii, Y. Kurita, and T. Otsuji, *J. Appl. Phys.*, **113**, 143108 (2013).
20. D. Svintsov, V. Ryzhii, andT. Otsuji, *Appl. Phys. Express*, **7**, 115101 (2014).
21. E. McCann, D. S. Abergel, and V. I. Fal'ko, *Eur. Phys. J.: Spec. Top.*, **148**, 91 (2007).
22. E. H. Hwang, S. Adam, and S. D. Sarma, *Phys. Rev. Lett.*, **98**, 186806 (2007).
23. Y.-W. Tan, Y. Zhang, K. Bolotin, Y. Zhao, S. Adam, E. H. Hwang, S. Das Sarma, H. L. Stormer, and P. Kim, *Phys. Rev. Lett.*, **99**, 246803 (2007).
24. K. M. McCreary, K. Pi, A. G. Swartz, W. Han, W. Bao, C. N. Lau, F. Guinea, M. I. Katsnelson, and R. K. Kawakami, *Phys. Rev. B*, **81**, 115453 (2010).
25. N. Sule, S. C. Hagness, and I. Knezevic, *Phys. Rev. B*, **89**, 165402 (2014).
26. L. M. Zhang, Z. Q. Li, D. N. Basov, M. M. Fogler, Z. Hao, and M. C. Martin, *Phys. Rev. B*, **78**, 235408 (2008).
27. J.-H. Chen, C. Jang, S. Adam, M. Fuhrer, E. Williams, and M. Ishigami, *Nat. Phys.*, **4**, 377 (2008).
28. E. H. Hwang and S. Das Sarma, *Phys. Rev. Lett.*, **101**, 156802 (2008).
29. F. T. Vasko and V. Ryzhii, *Phys. Rev. B*, **76**, 233404 (2007).
30. S. Adam, P. W. Brouwer, and S. Das Sarma, *Phys. Rev. B*, **79**, 201404 (2009).
31. C. H. Lewenkopf, E. R. Mucciolo, and A. H. Castro Neto, *Phys. Rev. B*, **77**, 081410 (2008).
32. J. M. B. Lopes dos Santos, N. M. R. Peres, and A. H. Castro Neto, *Phys. Rev. Lett.*, **99**, 256802 (2007).
33. J. W. Klos, A. A. Shylau, I. V. Zozoulenko, H. Xu, and T. Heinzel, *Phys. Rev. B*, **80**, 245432 (2009).
34. G. Konstantatos, M. Badioli, L. Gaudreau, J. Osmond, M. Bernechea, F. P. G. de Arquer, F. Gatti, and F. H. L. Koppens, *Nat. Nanotechnol.*, **7**, 363 (2012).
35. J. Li, G. Xu, C. Rochford, R. Li, J. Wu, C. M. Edvards, C. L. Berne, Z. Chen, and V. A. Maroni, *Appl. Phys. Lett.*, **99**, 023111 (2011).
36. Yu. I. Latyshev, A. P. Orlov, V. A. Volkov, V. V. Enaldiev, I. V. Zagorodnev, O. F. Vyvenko, Yu. V. Petrov, and P. Monceau, *Sci. Rep.*, **4**, 7578 (2014).

Chapter 25

Negative Terahertz Conductivity in Remotely Doped Graphene Bilayer Heterostructures*

V. Ryzhii,[a,b] M. Ryzhii,[c] V. Mitin,[d] M. S. Shur,[e] and T. Otsuji[a]
[a]*Research Institute of Electrical Communication, Tohoku University, Sendai 980-8577, Japan*
[b]*Institute of Ultra High Frequency Semiconductor Electronics of RAS, and Center for Photonics and Infrared Engineering, Bauman Moscow State Technical University, Moscow 111005, Russia*
[c]*Department of Computer Science and Engineering, University of Aizu, Aizu-Wakamatsu 965-8580, Japan*
[d]*Department of Electrical Engineering, University at Buffalo, SUNY, Buffalo, New York 1460–1920, USA*
[e]*Departments of Electrical, Electronics, and Systems Engineering and Physics, Applied Physics, and Astronomy, Rensselaer Polytechnic Institute, Troy, New York 12180, USA*
v-ryzhii@u-aizu.ac.jp

*Reprinted with permission from V. Ryzhii, M. Ryzhii, V. Mitin, M. S. Shur, and T. Otsuji (2015). Negative terahertz conductivity in remotely doped graphene bilayer heterostructures, *J. Appl. Phys.*, **118**, 183105. Copyright © 2015 AIP Publishing LLC.

Graphene-Based Terahertz Electronics and Plasmonics: Detector and Emitter Concepts
Edited by Vladimir Mitin, Taiichi Otsuji, and Victor Ryzhii

ISBN 978-981-4800-75-4 (Hardcover), 978-0-429-32839-8 (eBook)
www.jennystanford.com

Injection or optical generation of electrons and holes in graphene bilayers (GBLs) can result in the interband population inversion enabling the terahertz (THz) radiation lasing. The intraband radiative processes compete with the interband transitions. We demonstrate that remote doping enhances the indirect interband generation of photons in the proposed GBL heterostructures. Therefore, such remote doping helps to surpass the intraband (Drude) absorption, and results in large absolute values of the negative dynamic THz conductivity in a wide range of frequencies at elevated (including room) temperatures. The remotely doped GBL heterostructure THz lasers are expected to achieve higher THz gain compared with previously proposed GBL-based THz lasers.

25.1 Introduction

The population inversion created by injection or optical pumping in graphene layers (GLs) and graphene bilayers (GBLs) [1] results in the negative dynamic conductivity in the terahertz (THz) range of frequencies [2, 3] and enables the graphene-based THz lasers [2–5]. Different research groups have demonstrated the THz gain in the pumped GLs [6–17] (see also Ref. [16]). GLs and GBLs can serve as the active regions of the THz lasers with the Fabry-Perot resonators, dielectric waveguides, and slot-lines including the plasmonic lasers [8, 18–23]. The GL and GBL heterostructure lasers can operate in a wide frequency range, including the 6–10 THz range, where using materials like A_3B_5 is hindered by the optical phonon effects [24–26]. The THz lasing in the GL and GBL structures is possible if the contribution of the interband radiative transitions to the real part of the dynamic conductivity Re σ in a certain range of frequencies surpasses the contribution of the intraband radiative processes associated with the Drude losses. The interband transitions include the direct (with the conservation of the electron momenta) and the indirect transitions (accompanied with the variation of the electron momentum due to scattering). The limitations imposed by the momentum and energy conservation laws allow only for the indirect intraband radiative transitions. In the GLs and GBLs with sufficiently long carrier momentum relaxation time τ, the direct interband radiative transitions dominate over the indirect intraband transitions and Re $\sigma < 0$. Recently [27, 28], we demonstrated that in GLs and GBLs with primarily long-range disorder scattering, the indirect inter-band transitions can prevail over the indirect intraband transitions, leading to fairly large absolute values of the negative conductivity $|\mathrm{Re}\sigma|$. These values could exceed the fundamental limits for the direct transitions, which are $\sigma_Q = e^2/4\hbar$ for GLs and $2\sigma_Q$ for GBLs [1] (here e is the electron charge and $\hbar$ is the Planck constant). As a result, in the GL and GBL structures with a long-range disorder, one can expect that the condition Re $\sigma < 0$ could be fulfilled in a wider range of the THz frequencies, including the low boundary of the THz range. When the momentum relaxation is associated with long-range scattering mechanism, moderate values of τ might be sufficient for achieving Re $\sigma < 0$, especially in GBLs where the density of states (DoS) $\rho(\varepsilon)$ as a function of energy ε near the band edges ($\rho(\varepsilon) \simeq const$) in the latter is considerably larger than in GLs (where $\rho(\varepsilon) \propto |\varepsilon|$).

This chapter deals with the GBL heterostructures with a remote impurity layer (RIL) located at some distance from the GBL plane, and incorporating donors and acceptors or donor and acceptor clusters. We consider the GBL-RIL heterostructure as an active region of an injection THz laser. We show that the electron and hole injection from the side n- and p-contact regions into the GBL-RIL heterostructures leads to the population inversion enabling the enhanced negative dynamic conductivity, which, in turn, can result in lasing in the THz range. Figure 25.1 shows the schematic view of an injection laser based on the GBL-RIL heterostructure at the reverse bias voltage V. The remote (selective) doping can markedly affect the scattering of electrons and holes (see, for example, Refs. [29, 30]), hence the indirect radiative transitions. The GL and GBL heterostructures using remote doping were already fabricated and used for device applications [31, 32]. The GBL-RIL heterostructures demonstrate the following distinct features. First, the net electron and hole densities determined by both the pumping and RIL and the electron and hole quasi-Fermi energies, $\varepsilon_{F,e}$ and $\varepsilon_{F,h}$, are generally not equal. Second, relatively large scale donor and acceptor density fluctuations result in smooth (long-range) spatial variations of the potential in the GBL created by the RIL, separated by a distance d from the GBL. This could result in the electron and hole scattering with a relatively small momentum change, and consequently, increase in the contribution of the indirect interband transitions in comparison with the intraband transitions. Thus, the main role of the RIL is not to induce extra carrier in the GBL (as in high-electron mobility transistors) but to provide the specific mechanism of the electron and hole scattering reinforcing the indirect interband radiative transitions. The intraband radiative transitions (leading to the Drude radiation absorption) are accompanied by a change in the electron or hole momentum. However, such transitions in the GBLs require larger variation of the momentum than the interband radiative transitions. As a result, an increase in the long-range scattering contributions results in an increase in the relative role of the indirect interband transitions and even, as shown in the following, their dominance. One needs to stress that all this takes place under the conditions of the interband population inversion and in the gapless materials. In the standard semiconductors with sufficiently wide energy gap and the lasers on the base of such semiconductors, the interband transitions (both direct and indirect) usually correspond to relatively large photon energies, at which the Drude absorption is much less important. In this regard, the situation under consideration (population inversion in gapless semiconductors) is, in some sense, unique. Although the domination of the long-range scattering and, therefore, a substantial increase in the relative contribution of the indirect interband transitions can occur in the GBL heterostructures with the clusterized charged impurities immediately in the GBL [28] (that is difficult to realize in a systematic engineering manner), the implementation of the RIL can intentionally provide a similar or stronger effect using the standard doping techniques.

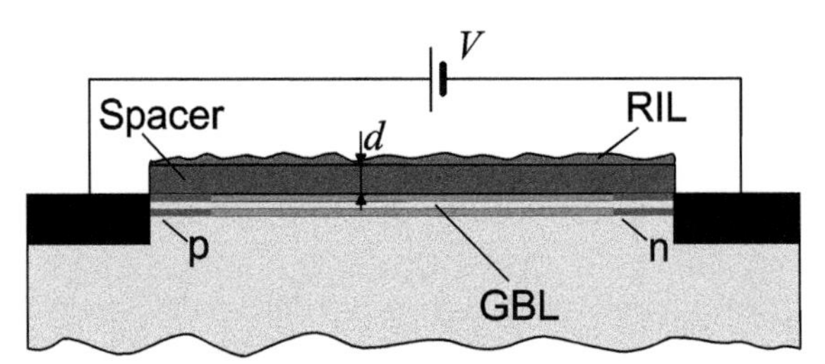

Figure 25.1 Schematic view of an injection laser based on the GBL-RIL heterostructure with the GBL doped with acceptors and the RIL doped with both donors and acceptors and with the chemically doped n- and p-injectors.

We calculate the spectral characteristics of the real part of the dynamic conductivity Re σ in the pumped GBL-RIL heterostructures as a function of the doping level and the spacer thickness d and compare the obtained characteristics with those of the undoped heterostructures. This analysis reveals the conditions for the selective doping enabling the negative dynamic conductivity in a wide frequency range, including the frequencies of a few THz. This might open new prospects of the GBL heterostructures for the efficient THz lasers.

25.2 Device Model

We study the GBL-RIL heterostructures pumped via the injection from the side p- and n-contacts as shown in Fig. 25.1. In the very clean pumped GBLs (or GLs) with a fairly high carrier mobility, the intraband (Drude absorption) at the radiation frequencies above 1 THz should not play a dominant role. Hence, adding the selective doping might not be needed. However, in the GBLs with a relatively low carrier mobility, caused by unavoidable (residual) charge impurities and imperfections, the RIL induced scattering could result in a substantial compensation of the Drude absorption, and therefore, enhancement of the negative dynamic conductivity. In the following, for definiteness, we consider the GBL-RIL heterostructures with the GBL doped with acceptors (with the density Σ_A) and the RIL formed by partially compensated donors (with the density $\Sigma_{D,R}$) and acceptors (with the density $\Sigma_{A,R}$). Hence, $\Sigma_{D,R} - \Sigma_{A,R} \simeq \Sigma_A$. We further assume that in these heterostructures, the RIL might comprise the clusterized acceptors and donors (with the correlated charge defects forming the charged clusters with the charge $Z_c e > e$ and the characteristic size l_c, where e is the electron charge). In the latter case, we assume that the cluster density is equal to $(\Sigma_{A,R} + \Sigma_{D,R})/Z_C$. Such GBL-RIL heterostructures are quasi-neutral even in the absence of the pumping. Pumping results in the formation of the two-dimensional (2D) electron and hole gases in the GBL with the equal densities $\Sigma_e = \Sigma_h = \Sigma$ and the quasi-Fermi energies $\varepsilon_{F,e} = \varepsilon_{F,h} = \varepsilon_F$. Figure 25.2 shows the GBL band diagram under pumping. In Fig. 25.2, the arrows indicate different radiative transitions: vertical interband transitions, indirect intraband transitions accompanied by the electron scattering on the impurities, phonons and holes (leading to the Drude absorption of radiation), and the interband indirect transitions.

The transverse electric field arising in the spacer due to the features of the doping can lead to the local opening of the energy gap [1, 33]. However, as shown below, for the doping levels under consideration, the energy gap is expected to be small, and the band opening and the energy spectrum nonparabolicity will be disregarded [33–35]. The energy spectrum is assumed to be $\varepsilon_{\mathbf{p}} = \varepsilon_{\mathbf{p}} = \pm p^2 v_W^2/\gamma_1 = \pm p^2/2m$ [1], where $\gamma_1 \simeq$ 0.35–0.43 eV is the inter-GL overlap integral in the GBLs, $v_W \simeq 10^8$ cm/s is the characteristic velocity, and $m = \gamma_1/2v_W^2$ is the effective mass of electrons and holes.

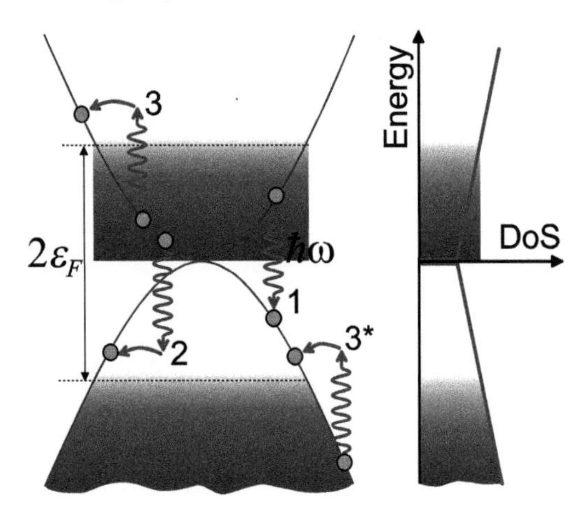

Figure 25.2 Energy band diagram of a pumped GBL (left panel) and energy dependency of its DoS (right panel). Arrows correspond to the direct and indirect interband transitions with the photon emission (wavy arrow "1" and wavy + smooth arrows "2") as well as to indirect intraband transitions with the photon absorption in the conduction and valence bands (wavy + smooth arrows "3" and "3*," respectively).

25.3 Main Equations

The real part of the net dynamic (THz) conductivity in the in-plane direction, Re σ, of GLs and GBLs, comprises the contributions from the direct (vertical), Re $\sigma_{\rm d}$, and two types of the indirect, $\mathrm{Re}\sigma_{\rm ind}^{\rm inter}$ and $\mathrm{Re}\sigma_{\rm ind}^{\rm intra}$, transitions, respectively,

$$\mathrm{Re}\sigma = \mathrm{Re}\left(\sigma_{\rm d} + \sigma_{\rm ind}^{\rm inter} + \sigma_{\rm ind}^{\rm intra}\right). \tag{25.1}$$

The first term in Eq. (25.1) corresponds to the transitions of type "1" in Fig. 25.2, the second term corresponds to type "2," and the third term corresponds to types "3" and "3*" transitions, respectively. The last term in the right-hand side of Eq. (25.1) corresponds to the processes responsible for the Drude absorption.

Due to the high frequencies of the inter-carrier scattering under sufficiently strong pumping, the energy distributions of the pumped carriers are characterized by the Fermi distribution functions $f_\upsilon(\varepsilon_{\mathbf{p}})$ and $f_c(\varepsilon_{\mathbf{p}})$, where $\varepsilon_{\mathbf{p}}$ is the dispersion relationship for the 2D carriers in GBLs, with the quasi-Fermi energy, $\varepsilon_F \simeq \pi\hbar^2\Sigma/2m$ and the effective temperature T.

The calculations of the GBL dynamic conductivity in the in-plain direction associated with the direct interband transitions accounting for the GBL energy spectra, following the approach developed in Refs. [36, 37] (see also Ref. [5]), yield

$$\mathrm{Re}\frac{\sigma_{\mathrm{d}}}{\sigma_Q} = \frac{(\hbar\omega + 2\gamma_1)}{(\hbar\omega + \gamma_1)}\tanh\left(\frac{\hbar\omega - 2\varepsilon_F}{4T}\right)$$
$$\simeq 2\tanh\left(\frac{\hbar\omega - 2\varepsilon_F}{4T}\right). \quad (25.2)$$

Applying the Fermi golden rule for the indirect intra- and interband electron radiative transitions and using the standard procedure [38], we obtain the following formulas for the respective components of the dynamic conductivity real parts (proportional to the absorption coefficients):

$$\mathrm{Re}\frac{\sigma_{\mathrm{ind}}^{\mathrm{intra}}}{\sigma_Q} = \frac{4\pi g}{\hbar\omega^3}\sum_{\mathbf{p},\mathbf{p}'}|V(\mathbf{p}-\mathbf{p}')|^2\, u_{\mathbf{pp}'}^{\lambda\lambda}(\mathbf{v}_{\mathbf{p}'} - \mathbf{v}_{\mathbf{p}})^2\,\delta(\hbar\omega + \varepsilon_{\mathbf{p}} - \varepsilon_{\mathbf{p}'})$$
$$\times\left\{\frac{1}{1+\exp[(\varepsilon_{\mathbf{p}} - \varepsilon_F)/T]} - \frac{1}{1+\exp[(\varepsilon_{\mathbf{p}'} - \varepsilon_F)]}\right\}, \quad (25.3)$$

$$\mathrm{Re}\frac{\sigma_{\mathrm{ind}}^{\mathrm{inter}}}{\sigma_Q} = \frac{2\pi g}{\hbar\omega^3}\sum_{\mathbf{p},\mathbf{p}'}|V(\mathbf{p}-\mathbf{p}')|^2\, u_{\mathbf{pp}'}^{\mathrm{cv}}(\mathbf{v}_{\mathbf{p}'} + \mathbf{v}_{\mathbf{p}})^2\,\delta(\hbar\omega - \varepsilon_{\mathbf{p}} - \varepsilon_{\mathbf{p}'})$$
$$\times\left\{\frac{1}{1+\exp[(\varepsilon_{\mathbf{p}} - \varepsilon_F)/T]} - \frac{1}{1+\exp[(\varepsilon_{\mathbf{p}'} + \varepsilon_F)/T]}\right\}. \quad (25.4)$$

Here, $g = 4$ is the spin-valley degeneracy factor, $\varepsilon_{\mathbf{p}}$, $\mathbf{v}_{\mathbf{p}} = 2\mathbf{p}\, v_W^2/\gamma_1 = \mathbf{p}/m$ is the velocity of electrons and holes with the momentum $\mathbf{p}$ in GBLs, $V(\mathbf{q})$ is the $\mathbf{q}$-the Fourier component of the scattering potential, $u_{\mathbf{pp}'}^{\lambda\lambda'} = (1 + \lambda\lambda' \cos 2\theta_{\mathbf{pp}'})/2$ is the overlap between the envelope wave functions in the GBLs, and $\lambda = \pm 1$ is the index indicating the conduction and valence bands. The extra factor of 2 in Eq. (25.3) accounts for the intraband electron and hole contributions.

Equation (25.3) yields the well-known result for the dynamic Drude conductivity in the high-frequency limit $\Gamma = \hbar/\tau < \hbar\omega < \varepsilon_F$, where Γ is the broadening of the carrier spectra.

25.4 Scattering Mechanisms

For the GBL-RIL heterostructures under consideration, we have the following formulas:

$$|V(\hbar\mathbf{q})|^2 = \left[\frac{2\pi e^2}{\kappa(q+q_{\mathrm{TF}})}\right]^2 [\Sigma_{\mathrm{A}} + \Sigma_{\mathrm{RCL}} \exp(-2qd)], \qquad (25.5)$$

for the discrete acceptors in the GBL and the discrete acceptors and donors in the RIL (see, for example, Ref. [30]) and

$$|V(\hbar\mathbf{q})|^2 = \left[\frac{2\pi e^2}{\kappa(q+q_{\mathrm{TF}})}\right]^2 [\Sigma_{\mathrm{A}} + Z_{\mathrm{c}}\Sigma_{\mathrm{RCL}} \exp(-2qd - q^2 l^2/2)], \qquad (25.6)$$

for the point charged defects (acceptors) in the GBL and the correlated clusterized acceptors and donors in the RIL, respectively. Here, $q = |\mathbf{p} - \mathbf{p}'|$, $\Sigma_{\mathrm{RCL}} = \Sigma_{\mathrm{A,R}} + \Sigma_{\mathrm{D,R}}$ is the net density of the charge impurities of both types located in the RIL (therefore, the cluster density in the "clusterized" RIL is equal to $\Sigma_{\mathrm{RCL}}/Z_{\mathrm{c}}$ with $\Sigma_{\mathrm{RCL}} > \Sigma_{\mathrm{A}}$), κ is the effective dielectric constant (which is the half-sum of the dielectric constants of the materials surrounding the GBL, i.e., of the spacer and the substrate), $q_{\mathrm{TF}} = (4e^2m/\hbar^2\kappa)$ is the Thomas-Fermi screening wavenumber [37, 39, 40], which is independent of the carrier density. Deriving Eq. (25.6), we have taken into account that the density of the charged clusters in the clusterized RCL is Z_{c}-times smaller than the donor and acceptor densities, but the cross-section of the carrier scattering on the clusters is proportional to $Z_{\mathrm{c}}^2 e^4$. The factor $\exp(-2qd)$ in Eqs. (25.5) and (25.6) is associated with the remote position of a portion of the charged scatterers (see, for example, Ref. [30]). Equation (25.6) corresponds to the Gaussian spatial distributions of the clusters that leads to the appearance of the factor $\exp(-q^2 l_c^2/2)$ (compare with Refs. [28, 41–43]).

One can see that the scattering matrix element explicitly depends on the background dielectric constant κ and via the dependence of q_{TF} on κ.

25.5 Net THz Conductivity

In the following, we consider the most practical case of the temperature, frequency range, and pumping in which $\hbar/\tau < \hbar\omega$, $T < \varepsilon_F$. For $T = 300$ K, the frequency range $\omega/2\pi = 1$–10 THz and $\varepsilon_F \gtrsim 50$ meV, these inequalities are satisfied when $\tau \gtrsim 0.05$–0.5 ps.

For GBL structures without the RIL ($\Sigma_{RIL} = 0$) and the totally screened charges in the GBL, in which $\mathrm{Re}\overline{\sigma_{ind}^{intra}} = \overline{\sigma_{ind}^{intra}}$ and $\mathrm{Re}\overline{\sigma_{ind}^{inter}} = \overline{\sigma_{ind}^{inter}}$, invoking Eqs. (25.3)–(25.6), one obtains

$$\mathrm{Re}\overline{\sigma_{ind}^{intra}} \simeq 2\frac{e^2\Sigma}{m\tau}\frac{1}{\omega^2} = 2\sigma_Q\left(\frac{\omega_D}{\omega}\right)^2, \tag{25.7}$$

where $\omega_D = \sqrt{4\hbar\Sigma/m\tau} = \sqrt{8\varepsilon_F/\pi\hbar\tau}$ is the Drude frequency. This quantity characterizes the relative strength of the Drude processes (absorption).

As the above takes place, Eqs. (25.3) and (25.4) lead to

$$\frac{\mathrm{Re}\overline{\sigma_{ind}^{inter}}}{\mathrm{Re}\overline{\sigma_{ind}^{intra}}} \simeq -\frac{\hbar\omega}{4\varepsilon_F} \tag{25.8}$$

To avoid cumbersome calculations and obtain transparent formulas suitable for a qualitative analysis, we use the mean value theorem for the integrals over $d\mathbf{q}$. As a result, we arrive at the following simplified formulas for $\mathrm{Re}\sigma_{ind}^{intra} + \mathrm{Re}\sigma_{ind}^{inter}$:

$$\begin{aligned}\mathrm{Re}\left(\sigma_{ind}^{intra} + \sigma_{ind}^{inter}\right) \\ \simeq \mathrm{Re}\overline{\sigma_{ind}^{intra}}\Bigg\{\int_{Q_{min}}^{Q_{max}} dqq\frac{[1+D\exp(-2qd-q^2l^2/2)]}{(Q_{max}^2 - Q_{min}^2)(q/q_{TF}+1)^2} \\ -\frac{\hbar\omega}{4\varepsilon_F}\int_{q_{min}}^{q_{max}} dqq\frac{[1+D\exp(-2qd-q^2l^2/2)]}{(q_{max}^2 - q_{min}^2)(q/q_{TF}+1)^2}\Bigg\}.\end{aligned} \tag{25.9}$$

Here, $D = Z_c(\Sigma_{RCL}/\Sigma_A) = Z_c(\Sigma_{D,R} + \Sigma_{A,R})/(\Sigma_{D,R} - \Sigma_{A,R}) \simeq Z_c(\Sigma_{D,R} + \Sigma_{A,R})/\Sigma_A \geq Z_c \geq 1$ is the doping parameter ($Z_c = 1$ and $l = 0$ for the RGL with the point charges (acceptors and donors) and $Z_c > 1$ and $l = l_c$ in the case of RIL with the clusterized acceptors and donors), $\hbar Q_{max} \simeq 2\sqrt{2m\varepsilon_F}(1 - \hbar\omega/4\varepsilon_F)$, $\hbar Q_{min} \simeq 2\sqrt{2m\varepsilon_F}(\hbar\omega/4\varepsilon_F)$, $\hbar q_{max} \simeq 2\sqrt{m\hbar\omega}, q_{min} \simeq 0$. The latter quantities follow from the conservation laws for the indirect intraband and interband radiative transitions. In particular, Eq. (25.9) gives rise to

$$\frac{\mathrm{Re}\left(\sigma_{ind}^{intra} + \sigma_{ind}^{inter}\right)}{\mathrm{Re}\sigma_{ind}^{intra}} \simeq 1 - \frac{1}{2}\frac{\int_{q_{min}}^{q_{max}} dqq \frac{[1 + D\exp(-2qd - q^2l^2/2)]}{(q/q_{TF} + 1)^2}}{\int_{Q_{min}}^{Q_{max}} dqq \frac{[1 + D\exp(-2qd - q^2l^2/2)]}{(q/q_{TF} + 1)^2}}. \quad (25.10)$$

For sufficiently pure GBLs and relatively strong RIL doping (with nearly complete compensation of the donor and acceptors), the doping parameter D can be fairly large. In this case, the contribution of the carrier scattering on the remote impurities to the net scattering rate can be dominant. In such a case, the absolute value of the net THz conductivity $|\mathrm{Re}\sigma|$ can markedly exceed the contribution solely from the direct interband transitions, i.e., $|\mathrm{Re}\sigma| > 2\sigma_Q$.

Considering Eqs. (25.1), (25.2), and (25.9), we arrive at the following formula for the net THz conductivity:

$$\frac{\mathrm{Re}\sigma}{2\sigma_Q} \simeq \tanh\left(\frac{\hbar\omega - 2\varepsilon_F}{4T}\right) + \left(\frac{\omega_D}{\omega}\right)^2 \left\{\int_{Q_{min}}^{Q_{max}} dqq \frac{[1 + D\exp(-2qd - q^2l^2/2)]}{(Q_{max}^2 - Q_{min}^2)(q/q_{TF} + 1)^2} - \frac{\hbar\omega}{4\varepsilon_F}\int_{q_{min}}^{q_{max}} dqq \frac{[1 + D\exp(-2qd - q^2l^2/2)]}{(q_{max}^2 - q_{min}^2)(q/q_{TF} + 1)^2}\right\}. \quad (25.11)$$

25.6 Results and Analysis

Using Eqs. (25.9) and (25.10), one can analyze the conditions when the interband radiative processes surpass the intraband ones, i.e., when $\mathrm{Re}\left(\sigma_{\mathrm{ind}}^{\mathrm{intra}}+\sigma_{\mathrm{ind}}^{\mathrm{inter}}\right)<0$.

For the practical GBL-RIL heterostructures with the HfO_2 spacer and the substrate material like Si and SiC, the effective dielectric constant (which is the half-sum of the dielectric constants of the spacer and the substrate) is in the range κ = 10–20. In the calculations, we assume set $m = 4 \times 10^{-29}$ g and ε_F = 60 meV ($\Sigma \sim 2.5 \times 10^{12}$ cm^{-2}). Equations (25.9)–(25.11) for d = 0 (no spacer) lead to approximately the same results as in Ref. [28] in both the cases l = 0 and $l \neq 0$ (clusterized charge impurities in the GBL). In particular, at d = 0 and l = 0, the indirect intraband radiative transitions can partially compensate the Drude absorption but not prevail over the latter. Because of this, we focus on the GBL-RIL heterostructures (i.e., the heterostructures with $d \neq 0$ and $D \neq 0$).

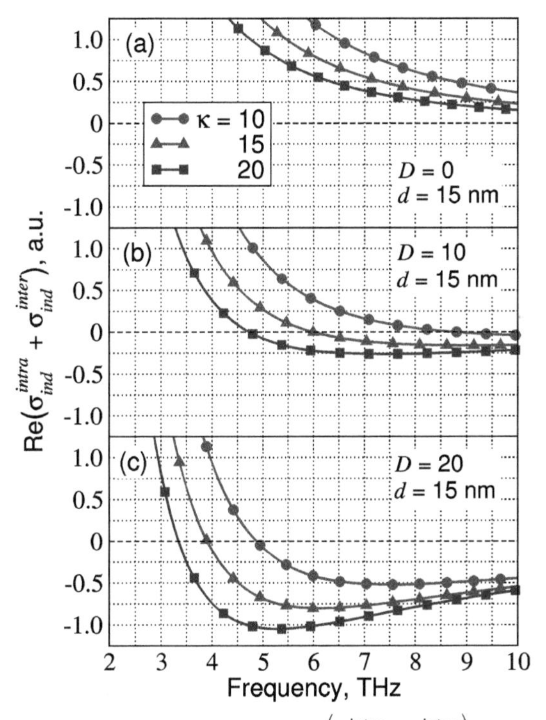

Figure 25.3 Frequency dependences of $\mathrm{Re}\left(\sigma_{\mathrm{ind}}^{\mathrm{intra}}+\sigma_{\mathrm{ind}}^{\mathrm{inter}}\right)$ calculated for the GBL-RIL heterostructures with spacer thickness d = 15 nm, different dielectric constants (κ = 10, 15, and 20), and different doping parameters D: (a) D = 0, (b) D = 10, and (c) D = 20.

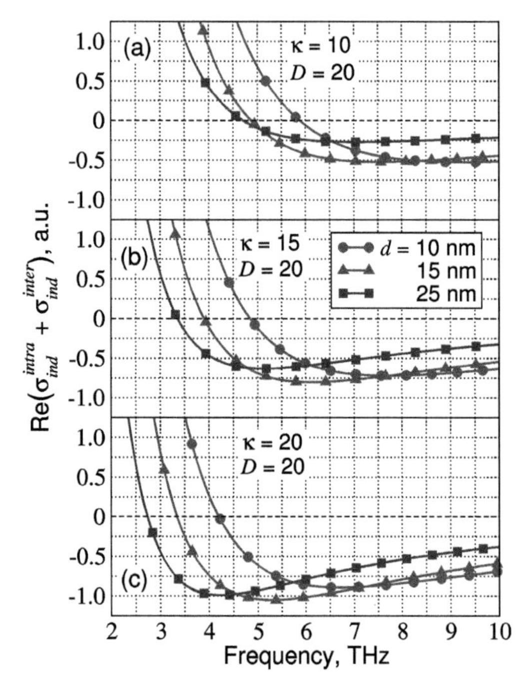

Figure 25.4 The same as in Fig. 25.3 but for doping parameter $D = 20$, different spacer thicknesses (d = 10, 15, and 25 nm), and different dielectric constants: (a) $\kappa = 10$, (b) $\kappa = 15$, and (c) $\kappa = 20$.

Figures 25.3 and 25.4 show the dependences of $\mathrm{Re}\left(\sigma_{\mathrm{ind}}^{\mathrm{intra}} + \sigma_{\mathrm{ind}}^{\mathrm{inter}}\right)$ on the frequency $f = \omega/2\pi$ calculated using Eq. (25.9) with $l = 0$ (uncorrelated impurities in the RIL) for different values of the doping parameter D, the spacer thickness d, and effective dielectric constant κ. As seen from Fig. 25.3a, in the absence of the RCL ($D = 0$), the indirect intraband transitions prevail over the indirect interband transitions. In such a case, the indirect transitions lead to the photon absorption in the entire frequency range, although the photon emission due to the interband transitions partially compensates the Drude absorption. This compensation is enhanced in the heterostructures with the doped RIL, i.e., for $D > 0$ (see Figs. 25.3b and c): an increase in the doping parameter D leads to the appearance of the frequency range where $\mathrm{Re}\left(\sigma_{\mathrm{ind}}^{\mathrm{intra}} + \sigma_{\mathrm{ind}}^{\mathrm{inter}}\right) < 0$. The latter range becomes wider (it starts from lower frequency) with increasing dielectric constant κ. These features of the transformation of the $\mathrm{Re}\left(\sigma_{\mathrm{ind}}^{\mathrm{intra}} + \sigma_{\mathrm{ind}}^{\mathrm{inter}}\right)$

frequency dependence are attributed to reinforcing of the long-range carrier scattering (with increasing D) and smoothing of the scattering potential (due to a stronger screening at higher κ). Both these effects result in an increase in the relative role of the indirect interband transitions. The scattering potential smoothing associated with an increase in the spacer thickness d also beneficial for the indirect interband transitions is seen in Fig. 25.4 as well.

Figure 25.5 shows the D-d plane where $\mathrm{Re}\left(\sigma_{\mathrm{ind}}^{\mathrm{intra}} + \sigma_{\mathrm{ind}}^{\mathrm{inter}}\right) > 0$ (below the D versus d lines) and $\mathrm{Re}\left(\sigma_{\mathrm{ind}}^{\mathrm{intra}} + \sigma_{\mathrm{ind}}^{\mathrm{inter}}\right) < 0$ (above these lines) are calculated considering Eq. (25.10) at different values of κ and frequency $f = \omega/2\pi$. As seen, the ranges of parameters D and d where the quantity $\mathrm{Re}\left(\sigma_{\mathrm{ind}}^{\mathrm{intra}} + \sigma_{\mathrm{ind}}^{\mathrm{inter}}\right)$ is negative are markedly wider in the case of a relatively large dielectric constant κ and the frequency f.

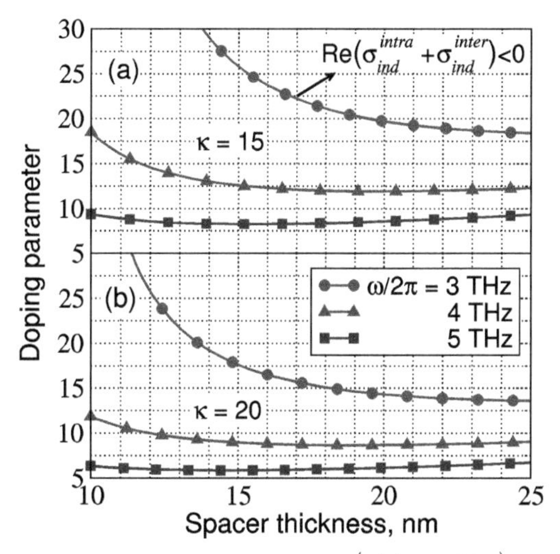

Figure 25.5 Areas in the *D–d* plane where $\mathrm{Re}\left(\sigma_{\mathrm{ind}}^{\mathrm{intra}} + \sigma_{\mathrm{ind}}^{\mathrm{inter}}\right) > 0$ (areas below lines corresponding to given frequencies) and $\mathrm{Re}\left(\sigma_{\mathrm{ind}}^{\mathrm{intra}} + \sigma_{\mathrm{ind}}^{\mathrm{inter}}\right) < 0$ (above these lines) at different values of dielectric constant κ: (a) $\kappa = 15$ and (b) $\kappa = 20$.

Figure 25.6 shows the threshold frequency $\omega_0/2\pi$ corresponding to $\mathrm{Re}\left(\sigma_{\mathrm{ind}}^{\mathrm{intra}} + \sigma_{\mathrm{ind}}^{\mathrm{inter}}\right) = 0$ as a function of the spacer thickness d calculated for $D = 20$ and different values of κ.

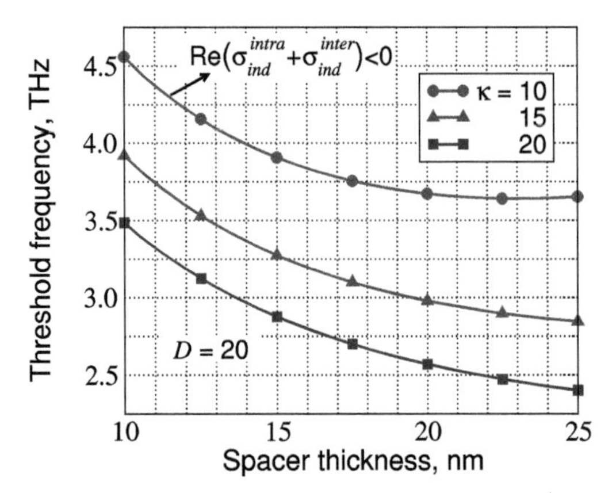

Figure 25.6 Threshold frequency $\omega_0/2\pi$ corresponding to $\mathrm{Re}\left(\sigma_{\mathrm{ind}}^{\mathrm{intra}} + \sigma_{\mathrm{ind}}^{\mathrm{inter}}\right) = 0$ as a function of the spacer thickness d calculated for $D = 20$ and different values of dielectric constant κ. As in Fig. 25.5, areas above the pertinent line are related to $\mathrm{Re}\left(\sigma_{\mathrm{ind}}^{\mathrm{intra}} + \sigma_{\mathrm{ind}}^{\mathrm{inter}}\right) < 0$.

Figure 25.7 shows the frequency dependences of the normalized net THz conductivity Re $\sigma/2\sigma_Q$ calculated using Eq. (25.11) for T = 300 K and different values of the doping parameter D, dielectric constant κ = 15, and Drude frequency $\Omega_D/2\pi$. The Drude frequencies are chosen to be $\omega_D/2\pi$ = 2.49 and 3.51 THz. At ε_F = 60 meV, this corresponds to the carrier momentum relaxation time τ = 1 ps and 0.5 ps, respectively. The latter values of τ corresponds to the carrier mobility in the GBL without the RIL, i.e., with $D = 0$, μ = 40000 cm^2/V s and 20 000 cm^2/V s, respectively. The large values of D imply rather strong doping of the RIL. For example, if $\Sigma_A = 1 \times 10^{11}$ cm^{-2} (that is consistent with the above value of τ), the parameter D is equal to 60 if $\Sigma_{D,R} = 3.05 \times 10^{12}$ cm^{-2} and $\Sigma_{D,R} = 2.95 \times 10^{12}$ cm^{-2} g. As seen in Fig. 25.7, the net THz conductivity is negative in a certain range of frequencies (above 1.5 THz) even for the case of $D = 0$, i.e., even in the GBL structures without the RIL. This is because, although the indirect intraband transitions (Drude transitions) prevail over the indirect interband transitions, the negative contribution of the direct interband transitions enable the negativity of the net conductivity. The latter is because the contribution of the direct transitions decreases with the increasing frequency much slowly (see the dashed line in Fig. 25.7) than the contribution due to the Drude processes. However, in the GBL-RIL heterostructures with

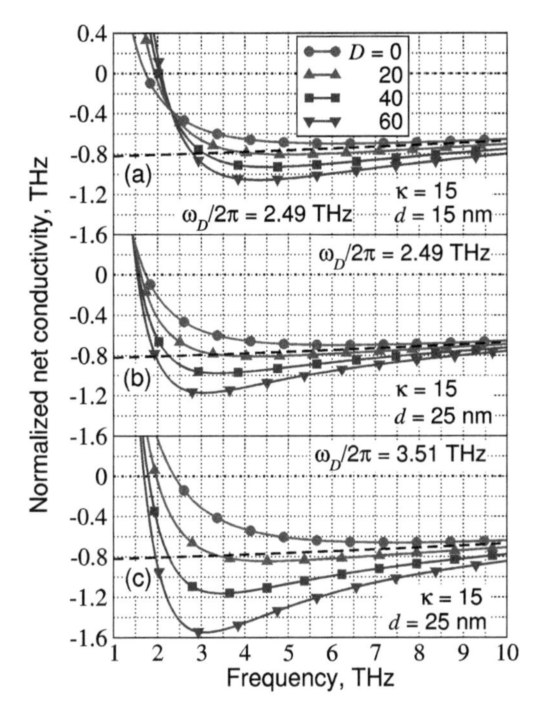

Figure 25.7 Frequency dependence of the normalized net THz conductivity Re$\sigma/2\sigma_Q$ of GBL-RIL heterostructures with $\kappa = 15$ and different values of doping parameter for: (a) d = 15 nm and $\omega_D/2\pi$ = 2.49 THz (τ = 1 ps), (b) d = 25 nm and $\omega_D/2\pi$ = 2.49 THz (τ = 1 ps), and (c) d = 25 and $\omega_D/2\pi$ = 3.51 THz (τ = 0.5 ps). Lines marked by circles correspond to normalize net THz conductivity without RIL (D = 0). Dashed lines show normalize THz conductivity due to solely direct interband transitions Re$\sigma_d/2\sigma_Q$.

sufficiently strong doping (large D), the absolute value of the net THz conductivity is pronounced larger than that in the similar heterostructures without the RIL. Moreover, the substantial contribution of the indirect interband transition can result in |Re σ| markedly exceeding its values |Re σ_d| associated with solely direct interband transitions (see the curves for D = 40 and 60, and the dashed lines in Fig. 25.7) and even exceeding the maximum value of the latter $2\sigma_Q$ (see the curve for D = 60). The comparison of the frequency dependences in Fig. 25.7 for D = 0 and D = 60 reveals that the RIL can provide a substantial increase in the absolute value of the THz conductivity (upto 4–6 times increase in the range of $\omega/2\pi$ = 3–5 THz, compare the curves for D = 60 and D = 0 in Fig. 25.7c), and hence, the corresponding enhancement of the laser THz gain (for the parameters under consideration). Such an increase can be even more large at somewhat lower frequencies.

The comparison of Figs. 25.7b and c also demonstrates that at larger Drude frequencies, i.e., at shorter carrier momentum relaxation time τ (and the same parameter D), the sag of the net THz conductivity frequency dependences becomes deeper. This is attributed to an increase in the intensity of the indirect processes (both intraband and interband) when τ decreases.

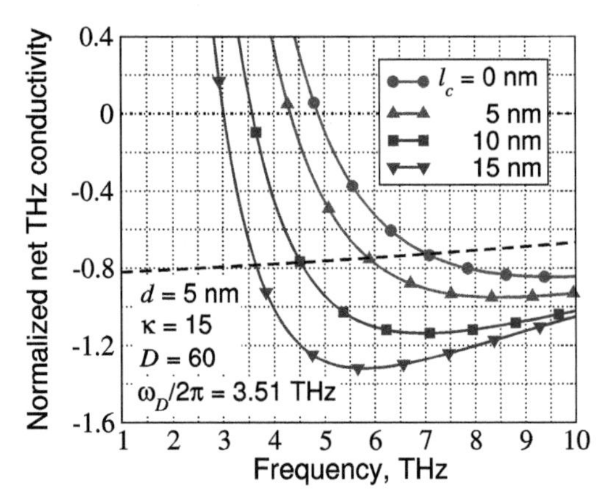

Figure 25.8 Frequency dependence of the normalized net THz conductivity Re $\sigma/2\sigma_Q$ of GBL-RIL heterostructures with clusterized impurities, $\kappa = 15$, $\omega/2\pi$ = 3.51 THz, d = 15 nm, D = 60, and different values of the cluster size l_c.

The plots in Fig. 25.8 were obtained using Eq. (25.11) with $l = l_c \neq 0$. Figure 25.8 shows the frequency dependences of the normalized net THz conductivity in the GBL-RIL heterostructures with clusterized (correlated) impurities in the RIL. As seen from Fig. 25.8, for relatively small spacer thickness (d = 5 nm), the clusterization ($l_c \neq 0$) leads to an additional enhancement of the indirect interband transitions, reinforcing the negative THz conductivity, particularly with an increasing cluster size. If d and l_c tend to be zero, the value of the net THz conductivity in the range of 8–10 THz (see the dashed line and the line marked by circles) approaches to that associated with the direct interband transitions. In this case, the indirect interband transitions merely compensate the indirect intraband transitions in line with previous predictions. However, at relatively large spacer thicknesses ($d \gtrsim 10$ nm), an increase in the absolute value of the net THz conductivity with the increasing size of the clusters is insignificant, since in this case, D is determined not only by the doping of the RIL but also by the cluster charge Z_c. This means that the chosen value $D = 60$ can correspond to different densities $\Sigma_{D,R}$ and $\Sigma_{A,R}$ depending on Z_c. The latter, in turn, depends on the degree of the compensation of donors by acceptors in the clusters.

25.7 Discussion

If the RIL doping does not compensate the acceptor system in the GBL, the electron and hole quasi-Fermi energies $\varepsilon_{F,e}$ and $\varepsilon_{F,h}$ are generally not equal to each other even in the pumping conditions. However, at sufficiently strong pumping when the density of the injected (or optically pumped) carriers is $\Sigma > \Delta\Sigma = \Sigma_A + \Sigma_{A,R} - \Sigma_{D,R}$, Eqs. (25.2), (25.9), and (25.10) are still approximately valid until $\hbar\omega/2 < \min\{\varepsilon_{F,e}, \varepsilon_{F,h}\}$.

The above model disregarded the opening of the band gap in the GBL under the transverse electric field arising due to the impurity charges in the RIL and GBL. Despite a fairly complex pattern of the opening gap in the GBLs by the electric field and doping [44, 45], the pertinent energy gap Δ_g can roughly be estimated in the manner as it was done in Ref. [34]. In the absence of the electron and hole injection from the side contacts, the carrier density in the GBL is rather small due to the compensation of the acceptor charges in the GBL and the donor and acceptor charges in the RIL (see Section 25.2). In this case, $\Delta_g \sim eE_\perp d_{GBL}$, where $E_\perp = 4\pi e\Sigma_A/\kappa$ is the transversal electric field created by the charged impurities and $d_{GBL} \simeq 0.36$ nm is the spacing between GLs in the GBL, so that $\Delta_g \sim 4\pi e^2\Sigma_A d_{GBL}/\kappa$. Under the pumping conditions, the 2D electron and hole gases partially screen the electric field in the GBL structure. As a result, one obtains $\Delta_g \sim eE_\perp d_{GBL}^{eff}$ with $d_{GBL}^{eff} < d_{GBL}$ [34]. Taking into account the screening of the transversal electric field $E_\perp$ by both electron and hole components, we obtain the following estimate: $\Delta_g \sim 4\pi e^2\Sigma_A d_{GBL}/\kappa[1 + (8d_{GBL}/a_B)]$, where $a_B = \kappa\hbar^2/me^2$ is the Bohr radius. Assuming $a_B = 10$–20 nm and $\Sigma_A = (5$–$10) \times 10^{11}$ cm^{-2}, we obtain $\Delta_g \sim 1.5$–5 meV. These values are in line with those estimated in Ref. [34] and extracted from the experimental data [44, 45]. This estimate validates our model.

In the case of the clusterized charged impurities, the long range variations of the potential associated with the clusters lead to the variation of the band gap. The latter variations reinforce the electron and hole scattering on the clusters, but this effect is not really essential.

In Eqs. (25.3) and (25.4), we have disregarded the indirect radiative processes associated with the electron-hole scattering in the GBL. The point is that the probability of such processes is proportional to $[2\pi e^2/\kappa(q + q_{TF})]^2\Sigma\eta$, where (in the degenerated two-dimensional electron and hole gases) the factors $\eta < 1$ is the fraction of electrons and holes effectively participated in the scattering processes $\eta = T/\varepsilon_F$ at $\hbar\omega < T$, so that $\Sigma\eta \simeq \Sigma_T = 2mT/\pi\hbar^2$. The values of q_{TF} and $\Sigma\eta$ are independent of the electron and hole densities and, hence, of the pumping conditions. This is because of the virtually constant density of state. At room temperatures, the value of Σ_T can be of the order of or less than Σ_A and much smaller than Σ_{RIL}. In the former case, one needs to replace the quantity D in Eqs. (25.9) and (25.10) by $D^* = D/(1 + 2\Sigma_T/\Sigma_A)$.

25.8 Conclusions

We proposed to use the GBL-RIL heterostructure with the population inversion due to the electron and hole injection as the active region of the interband THz lasers, and demonstrated that the incorporation of the RIL enables a substantial reinforcement of the effect of negative THz conductivity and the laser THz gain. This is associated with the domination of the indirect interband transitions in the GBL over the indirect intraband transitions (resulting in the Drude absorption) when the carrier scattering on a long-range potential prevails. As shown, the latter can be realized due to the remote doping and enhanced by the clusterization of impurities.

Using a simplified device model, we calculated the THz conductivity of the GBL-RIL heterostructures as a function of the frequency for different structural parameters (dopant densities, spacer thickness, impurity cluster size, and the injected carrier momentum relaxation time). We found that the absolute value of the THz conductivity in the GBL-RIL heterostructures, with a sufficiently highly doped RIL separated from the GBL by a spacer layer of properly chosen thickness, can markedly exceed the pertinent value in the GBL-heterostructures without the RIL. Thus, the remotely doped GBL heterostructures can be of interest for applications as the active media in the THz lasers.

Acknowledgments

The authors are grateful to D. Svintsov and V. Vyurkov for useful discussions and information. The work was supported by the Japan Society for Promotion of Science (Grant-in-Aid for Specially Promoted Research No. 23000008) and by the Russian Scientific Foundation (Project No. 14-29-00277). The works at UB and RPI were supported by the U.S. Air Force Award No. FA9550-10-1-391 and by the U.S. Army Research Laboratory Cooperative Research Agreement, respectively.

References

1. A. H. Castro Neto, F. Guinea, N. M. R. Peres, K. S. Novoselov, and A. K. Geim, *Rev. Mod. Phys.*, **81**, 109 (2009).
2. V. Ryzhii, M. Ryzhii, and T. Otsuji, *J. Appl. Phys.*, **101**, 083114 (2007).
3. M. Ryzhii and V. Ryzhii, *Jpn. J. Appl. Phys., Part 2*, **46**, L151 (2007).
4. V. Ryzhii, M. Ryzhii, V. Mitin, and T. Otsuji, *J. Appl. Phys.*, **110**, 094503 (2011).
5. V. Ya. Aleshkin, A. A. Dubinov, and V. Ryzhii, *JETP Lett.*, **89**, 63 (2009).
6. S. Boubanga-Tombet, S. Chan, A. Satou, T. Otsuji, and V. Ryzhii, *Phys. Rev. B*, **85**, 035443 (2012).
7. T. Otsuji, S. Boubanga-Tombet, A. Satou, H. Fukidome, M. Suemitsu, E. Sano, V. Popov, M. Ryzhii, and V. Ryzhii, *J. Phys. D*, **45**, 303001 (2012).
8. T. Li, L. Luo, M. Hupalo, J. Zhang, M. C. Tringides, J. Schmalian, and J. Wang, *Phys. Rev. Lett.*, **108**, 167401 (2012).
9. Y. Takatsuka, K. Takahagi, E. Sano, V. Ryzhii, and T. Otsuji, *J. Appl. Phys.*, **112**, 033103 (2012).
10. T. Watanabe, T. Fukushima, Y. Yabe, S. A. Boubanga Tombet, A. Satou, A. A. Dubinov, V. Ya. Aleshkin, V. Mitin, V. Ryzhii, and T. Otsuji, *New J. Phys.*, **15**, 075003 (2013).
11. T. Winzer, E. Maric, and A. Knorr, *Phys. Rev. B*, **87**,165413 (2013).

12. I. Gierz, J. C. Petersen, M. Mitrano, C. Cacho, I. C. Edmond Turcu, E. Springate, A. Stohr, A. Kohler, U. Starke, and A. Cavalleri, *Nat. Mater.*, **12**, 1119 (2013).
13. S. Kar, D. R. Mohapatra, E. Freysz, and A. K. Sood, *Phys. Rev. B*, **90**, 165420 (2014).
14. R. R. Hartmann, J. Kono, and M. E. Portnoi, *Nanotechnology*, **25**, 322001 (2014).
15. T. Otsuji, S. Boubanga-Tombet, A. Satou, M. Suemitsu, V. Ryzhii, and J. Infrared, *Millimeter, Terahertz Waves*, **33**, 825 (2012).
16. I. Gierz, M. Mitrano, J. C. Petersen, C. Cacho, I. C. E. Turcu, E. Springate, A. St0hr, A. K0hler, U. Starke, and A. Cavalleri, *J. Phys.: Condens. Matter*, **27**, 164204 (2015).
17. A. A. Dubinov, V. Ya. Aleshkin, M. Ryzhii, T. Otsuji, and V. Ryzhii, *Appl. Phys. Express*, **2**, 092301 (2009).
18. V. Ryzhii, M. Ryzhii, A. Satou, T. Otsuji, A. A. Dubinov, and V. Ya. Aleshkin, *J. Appl. Phys.*, **106**, 084507 (2009).
19. V. Ryzhii, A. A. Dubinov, T. Otsuji, V. Mitin, and M. S. Shur, *J. Appl. Phys.*, **107**, 054505 (2010).
20. F. Rana, *IEEE Trans. Nanotechnol.*, **7**, 91 (2008).
21. A. A. Dubinov, V. Ya. Aleshkin, V. Mitin, T. Otsuji, and V. Ryzhii, *J. Phys.: Condens. Matter*, **23**,145302 (2011).
22. V. V. Popov, O. V. Polischuk, A. R. Davoyan, V. Ryzhii, T. Otsuji, and M. S. Shur, *Phys. Rev. B*, **86**,195437 (2012).
23. A. Tredicucci and M. S. Vitiello, *IEEE J. Sel. Top. Quantum Electron.*, **20**, 8500109 (2014).
24. H. C. Liu, C. Y. Song, A. J. SpringThorpe, and J. C. Cao, *Appl. Phys. Lett.*, **84**, 4068 (2004).
25. B. S. Williams, *Nat. Photonics*, **1**, 517 (2007).
26. S. Kumar, *IEEE J. Sel. Top. Quantum Electron.*, **17**, 38 (2011).
27. D. Svintsov, V. Ryzhii, and T. Otsuji, *Appl. Phys. Express*, **7**, 115101 (2014).
28. D. Svintsov, T. Otsuji, V. Mitin, M. S. Shur, and V. Ryzhii, *Appl. Phys. Lett.*, **106**, 113501 (2015).
29. M. Shur, *Physics of Semiconductor Devices* (Prentice-Hall, New Jersey, 1990), ISBN: 0-13-666496-2.
30. A. Shik, *Quantum Wells: Physics and Electronics of Two-Dimensional Systems* (World Scientific, Singapore, 1997), ISBN: 981-02-3279-9.

31. T. Stauber, G. Gomez-Santos, and F. Javier Garcia de Abajo, *Phys. Rev. Lett.*, **112**, 077401 (2014).
32. G. Konstantatos, M. Badioli, L. Gaudreau, J. Osmond, M. Bernechea, F. P. Garcia de Arquer, F. Gatti, and F. H. L. Koppens, *Nat. Nanotechnol.*, **7**, 363–368 (2012).
33. E. McCann and V. Fal'ko, *Phys. Rev. Lett.*, **96**, 086805 (2006).
34. E. McCann, D. S. L. Abergel, and V. I. Fal'ko, *Eur. Phys. J.: Spec. Top.*, **148**, 91–103 (2007).
35. L. M. Zhang, Z. Q. Li, D. N. Basov, M. M. Fogler, Z. Hao, and M. C. Martin, *Phys. Rev. B*, **78**, 235408 (2008).
36. L. A. Falkovsky and S. S. Pershoguba, *Phys. Rev. B*, **76**, 153410 (2007).
37. S. Das Sarma, E. H. Hwang, and E. Rossi, *Phys. Rev. B*, **81**, 161407(R) (2010).
38. F. T. Vasko and A. V. Kuznetsov, *Electronic States and Optical Transitions in Semiconductor Heterostructures* (Springer, New York, 1998).
39. E. McCann and M. Koshino, *Rep. Prog. Phys.*, **76**, 056503 (2013).
40. H. Min, D. S. L. Abergel, E. H. Hwang, and S. Das Sarma, *Phys. Rev. B*, **84**, 041406(R) (2011).
41. F. T. Vasko and V. Ryzhii, *Phys. Rev. B*, **76**, 233404 (2007).
42. C. H. Lewenkopf, E. R. Mucciolo, and A. H. Castro Neto, *Phys. Rev. B*, **77**, 081410(R) (2008).
43. S. Adams, P. W. Brouwer, and S. Das Sarma, *Phys. Rev. B*, **79**, 201404 (2009).
44. A. J. Samuels and J. D. Carey, *ACS Nano*, **7**(3), 2790 (2013).
45. W. Zhang, Ch.-T. Lin, K.-K. Liu, T. Tite, Ch.-Y. Su, Ch.-H. Chang, Y.-H. Lee, Ch.-W. Chu, K.-H. Wei, J.-L. Kuo, and L.-J. Li, *ACS Nano*, **5**(9), 7517 (2011).

Part 4
Stimulated Emission and Lasing in Graphene and Graphene Structures

Chapter 26

Feasibility of Terahertz Lasing in Optically Pumped Epitaxial Multiple Graphene Layer Structures*

V. Ryzhii,[a,b] M. Ryzhii,[a,b] A. Satou,[a,b] T. Otsuji,[b,c] A. A. Dubinov,[d] and V. Ya. Aleshkin[d]

[a]*Computational Nanoelectronics Laboratory, University of Aizu, Aizu-Wakamatsu 965-8580, Japan*
[b]*Japan Science and Technology Agency, CREST, Tokyo 107-0075, Japan*
[c]*Research Institute of Electrical Communication, Tohoku University, Sendai 980-8577, Japan*
[d]*Institute for Physics of Microstructures, Russian Academy of Sciences, Nizhny Novgorod 603950, Russia*
v-ryzhii@u-aizu.ac.jp, a-satou@u-aizu.ac.jp

A multiple graphene layer (MGL) structure with a stack of GLs and a highly conducting bottom GL on SiC substrate pumped by optical radiation is considered as an active region of terahertz and far infrared

*Reprinted with permission from V. Ryzhii, M. Ryzhii, A. Satou, T. Otsuji, A. A. Dubinov, and V. Ya. Aleshkin (2009). Feasibility of terahertz lasing in optically pumped epitaxial multiple graphene layer structures, *J. Appl. Phys.*, **106**, 084507. Copyright © 2009 American Institute of Physics.

Graphene-Based Terahertz Electronics and Plasmonics: Detector and Emitter Concepts
Edited by Vladimir Mitin, Taiichi Otsuji, and Victor Ryzhii

ISBN 978-981-4800-75-4 (Hardcover), 978-0-429-32839-8 (eBook)
www.jennystanford.com

lasers with external metal mirrors. The dynamic conductivity of the MGL structure is calculated as a function of the signal frequency, the number of GLs, and the optical pumping intensity. The utilization of optically pumped MGL structures might provide the achievement of lasing with the frequencies of about 1 THz at room temperature due to a high efficiency of pumping.

26.1 Introduction

Since the first experimental demonstrations of the non-trivial properties of graphene (see, for instance, Refs. [1, 2] and numerous subsequent publications), due to the gapless energy spectrum of graphene, the latter can be used as an active region in the terahertz and far infrared (FIR) lasers with optical or injection pumping [3, 4]. The optical pumping with the photon energy $\hbar\Omega$ leads to the generation of electrons and holes with the energy $\varepsilon_0 = \hbar\Omega/2$. Since the interaction of electrons and holes with optical phonons is characterized by the fairly short time τ_0, the photogenerated electrons and holes quickly emit cascades of N optical phonons, where $N = [\varepsilon_0/\hbar\omega_0]$ and $[X]$ means the integer part of X. Due to relatively high energy of optical phonons in graphene ($\hbar\omega_0 \simeq 200$ meV), the number N can vary from zero (pumping by CO_2 lasers) to a few units (pumping by quantum cascade lasers or semiconductor injection diode lasers). Thus, the photogenerated electrons and holes populate the low energy regions of the conduction and valence bands of graphene (see also Refs. [5, 6]). As a result, the Fermi levels of electrons and holes are separated and shifted from the Dirac point to the conduction and valence bands, respectively. This corresponds to the population inversion for the interband transitions with absorption or emission of photons with relatively low energies $\hbar\omega < 2\varepsilon_F$, where ε_F is the quasi-Fermi energy of the electron and hole distributions, and leads to the negative contribution of the interband transitions to the real part of the dynamic conductivity Re σ_ω (which includes the contributions of both interband and intraband transitions). If Re $\sigma_\omega < 0$, the stimulated emission of photons with the relatively low energy $\hbar\omega$ (in the terahertz or FIR range) is possible. The intraband contribution is primarily associated with the Drude mechanism of the photon absorption. As shown previously, the realization of the condition of lasing is feasible [3, 6]. Some possible schemes of the graphene lasers (with metal waveguide structure or external metal mirrors) utilizing the above mechanism of optical pumping were considered recently (see, for instance, Ref. [7]).

In particular, the laser structure evaluated in Ref. [7] includes two graphene layers. Relatively weak absorption of optical radiation with the efficiency per each layer $\beta = \pi e^2/\hbar c \simeq 0.023$, where e is the electron charge, $\hbar$ is the reduced Planck constant, and c is the speed of light, necessitates the use of rather strong optical pumping. This drawback can be eliminated in the structures with multiple graphene layers (MGLs). Such epitaxial MGL structures including up to 100 very perfect graphene layers with a high electron mobility ($\mu \simeq 250\,000$ cm^2/s V) preserving up to the room temperatures and, hence, with a long momentum relaxation time were recently fabricated using the thermal decomposition from a 4*H*-SiC substrate [8, 9]. The incident optical radiation can be almost totally absorbed in these MGL structures providing enhanced pumping efficiency. A long momentum relaxation time (up to 20 ps) [10] implies that the intraband (Drude) absorption in the terahertz and FIR ranges can be weak. The optically pumped MGL structures in question with the momentum relaxation time about several picoseconds at room temperature might be ideal active media for interband terahertz lasers. The situation however is complicated by the presence of highly conducting bottom GL (and hence absorbing terahertz and FIR radiation due to the intraband processes) near the interface with SiC as a result of charge transfer from SiC.

In this chapter we analyze the operation of terahertz lasers utilizing optically pumped MGL structures by calculating their characteristics as function of the number of graphene layers K and optical pumping intensity I_Ω and demonstrate the feasibility of realization of such lasers operating at the frequencies from about 1 THz to several terahertz at room temperatures.

26.2 Model

We consider a laser structure with an MGL structure on a SiC substrate serving as its active region. Although the active media under consideration can be supplemented by different resonant cavities, for definiteness we address the MGL structure placed between the highly reflecting metal mirrors (made of Al, Au, or Ag) as shown in Fig. 26.1a. It is assumed that the MGL structure under consideration comprises K upper GLs (to which we refer to just as GLs) and a highly conducting bottom layer with a Fermi energy of electrons ε_F^B, which is rather large: $\varepsilon_F^B \simeq 400$ meV [8]. Apart from this, we briefly compare the MGL structure in question with that in which the bottom GL is absent. The latter structure can be fabricated by an additional peeling of upper GLs and depositing them on a Si substrate.

Due to the photogeneration of electrons and holes with the energy $\varepsilon_0 = \hbar\Omega/2$, followed by their cooling associated with the cascade emission of optical phonons, low energy states near the bottom of the conduction band and the top of the valence band can be essentially occupied. Taking into account that elevated electron and hole densities (at elevated temperatures and sufficiently strong optical pumping considered in the following), the electron and hole distributions in the range of energies $\varepsilon \ll \hbar\omega_0$ in the kth graphene layer ($1 \leq k \leq K$) can be described by the Fermi functions with the quasi-Fermi energies $\varepsilon_F^{(k)}$ (see Fig. 26.1b). The case of relatively weak pair collisions of the photogenerated electrons and holes (due to their low densities) in which the energy distributions deviate from the Fermi distributions was studied in Ref. [6]. Using the Falkovsky–Varlamov formula [11] for the dynamic conductivity of a MGL structure generalized for nonequilibrium electron-hole systems [3], one can obtain

$$\mathrm{Re}\,\sigma_\omega^B = \left(\frac{e^2}{4\hbar}\right)\left\{1 - \left[1 + \exp\left(\frac{\hbar\omega/2 - \varepsilon_F^B}{k_B T}\right)\right]^{-1} - \left[1 + \exp\left(\frac{\hbar\omega/2 + \varepsilon_F^B}{k_B T}\right)\right]^{-1}\right\} + \left(\frac{e^2}{4\hbar}\right)\frac{4k_B T\tau_B}{\pi\hbar(1+\omega^2\tau_B^2)}\ln\left[1 + \exp\left(\frac{\varepsilon_F^B}{k_B T}\right)\right] \tag{26.1}$$

for the bottom GL and

$$\mathrm{Re}\,\sigma_\omega^{(k)} = \left(\frac{e^2}{4\hbar}\right)\tanh\left(\frac{\hbar\omega - 2\varepsilon_F^{(k)}}{4k_B T}\right)$$

$$+\left(\frac{e^2}{4\hbar}\right)\frac{8k_{\rm B}T\tau}{\pi\hbar(1+\omega^2\tau^2)}\ln\left[1+\exp\left(\frac{\varepsilon_{\rm F}^{(k)}}{k_{\rm B}T}\right)\right] \tag{26.2}$$

for the GLs with $1 \leq k \leq K$. Here $\tau_{\rm B}$ and τ are the electron and hole momentum relaxation times in the bottom and other GLs, respectively, T is the electron and hole temperature, and $k_{\rm B}$ is the Boltzmann constant. The first and second terms in the right-hand sides of Eqs. (26.1) and (26.2) correspond to the interband and intraband transitions, respectively, which are schematically shown by arrows in Fig. 26.1b. For simplicity we shall disregard the variation in the electron and hole densities in the bottom GL under the optical pumping, so that the electron-hole system in this GL is assumed to be close to equilibrium with the Fermi energy $\varepsilon_{\rm F}^{\rm B}$ determined by the interaction with the SiC substrate.

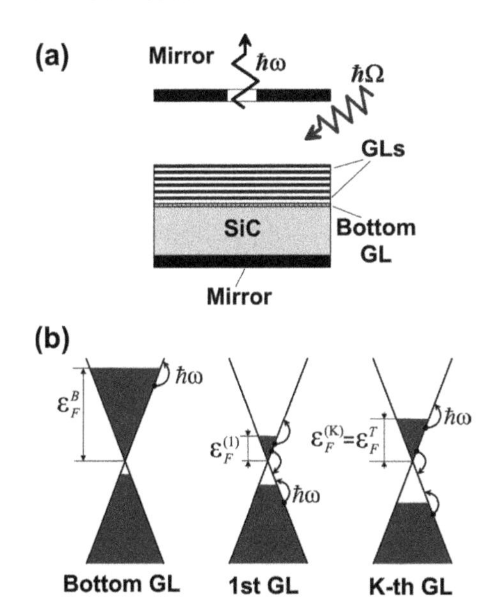

Figure 26.1 (a) Schematic view of a laser with a MGL structure. (b) Occupied (by electrons) and vacant states in different GLs under optical pumping. Arrows show transitions related to interband emission and intraband absorption of THz photons with energy $\hbar\omega$ (interband transitions related to optical pumping as well the processes of intraband relaxation of the photogenerated electrons and holes are not shown).

The quasi-Fermi energies in the GLs with $k \geq 1$ are mainly determined by the electron (hole) density in this layer Σ_k, i.e., $\varepsilon_F^{(k)} \propto \sqrt{\Sigma^{(k)}}$ and, therefore, by the rate of photogeneration $G_\Omega^{(k)}$ by the optical radiation (incident and reflected from the mirror) at the kth GL plane. Using Eq. (26.2) for $\hbar\omega = \hbar\Omega$, we obtain

$$G_\Omega^{(k)} = \frac{I_\Omega^{(k)}}{\hbar\Omega}\left(\frac{\pi e^2}{\hbar c}\right)\tanh\left(\frac{\hbar\Omega - 2\varepsilon_F^{(k)}}{4k_B T}\right). \tag{26.3}$$

Here, $I_\Omega^{(k)}$ is the intensity (power density) of the optical pumping radiation at the kth GL. At $\hbar\Omega > 2\varepsilon_F^{(k)}$ (for all GLs), Eq. (26.3) yields $G_\Omega^{(k)} \simeq \beta I_\Omega^{(k)}/\hbar\Omega$. Considering the attenuation of the optical pumping radiation due to its absorption in each GL, one can obtain

$$G_\Omega^{(k)} = \frac{I_\Omega}{\hbar\Omega}\beta[(1-\beta)^{K-k} + (1-\beta_B)^2(1-\beta)^{K+k-1}]. \tag{26.4}$$

Here I_Ω is the intensity of incident pumping radiation and $\beta_B = (4\pi/c)\,\sigma_\Omega^B$. The latter quantity accounts for the absorption of optical pumping radiation in the bottom layer.

A relationship between $\varepsilon_F^{(k)}$ and $G_\Omega^{(k)}$ is determined by the recombination mechanisms. We assume that $\varepsilon_F^{(k)} \propto [G_\Omega^{(k)}]^\gamma$, where γ is a phenomenological parameter. In this case,

$$\varepsilon_F^{(k)} = \varepsilon_F^B\left[(1-\beta)^{K-k}\frac{1+(1-\beta_B)^2(1-\beta)^{2k-1}}{1+(1-\beta_B)^2(1-\beta)^{2K-1}}\right]^\gamma, \tag{26.5}$$

where $\varepsilon_F^B = \varepsilon_F^{(K)}$ is the quasi-Fermi energy in the topmost GL.

26.3 MGL Structure Net Dynamic Conductivity

Taking into account that the thickness of the MGL structure is small in comparison with the wavelength of terahertz/FIR radiation, the generation and absorption of the latter is determined by the real part of the net dynamic conductivity

$$\mathrm{Re}\,\sigma_\omega = \mathrm{Re}\,\sigma_\omega^B + \mathrm{Re}\sum_{k=1}^{K}\sigma_\omega^{(k)}. \tag{26.6}$$

Taking into account that in the frequency range under consideration $\hbar\omega \ll \varepsilon_F^B$, so that one can neglect the first term in the right-hand side of Eq. (26.1) responsible for the interband absorption in the bottom GL, and using Eqs. (26.1), (26.2), and (26.6), we arrive at

$$\mathrm{Re}\,\sigma_\omega = \left(\frac{e^2}{4\hbar}\right)\left\{\frac{4k_\mathrm{B}\tau_\mathrm{B}}{\pi\hbar(1+\omega^2\tau_\mathrm{B}^2)}\ln\left[1+\exp\left(\frac{\varepsilon_\mathrm{F}^\mathrm{B}}{k_\mathrm{B}T}\right)\right]\right.$$
$$+\frac{8k_\mathrm{B}T\tau}{\pi\hbar(1+\omega^2\tau^2)}\sum_{\mathrm{k}=1}^{K}\ln\left[1+\exp\left(\frac{\varepsilon_\mathrm{F}^{(\mathrm{k})}}{k_\mathrm{B}T}\right)\right]$$
$$\left.+\sum_{\mathrm{k}=1}^{\mathrm{K}}\tanh\left(\frac{\hbar\omega-2\varepsilon_\mathrm{F}^{(\mathrm{k})}}{4k_\mathrm{B}T}\right)\right\}. \tag{26.7}$$

The first two terms in the right-hand side of Eq. (26.7) describe the intraband (Drude) absorption of terahertz radiation in all GLs, whereas the third term is associated with the interband transitions. When the latter is negative, i.e., when the inter-band emission prevails the interband absorption, the quantity Re σ_ω as a function of ω exhibits a minimum. At a strong optical pumping when the quantities $\varepsilon_\mathrm{F}^{(\mathrm{k})}$ are sufficiently large, Re $\sigma_\omega < 0$ in this minimum as well as in a certain range of frequencies $\omega_\mathrm{min} < \omega < \omega_\mathrm{max}$. Here ω_min and ω_max are the frequencies at which Re $\sigma_\omega = 0$; they are determined by τ_B, τ, and $\varepsilon_\mathrm{F}^\mathrm{T}$ (i.e., by the intensity of the incident optical pumping radiation).

Since in the MGL structure in question $\varepsilon_\mathrm{F}^\mathrm{B} \gg k_\mathrm{B}T$, considering such frequencies that $\omega^2\tau_\mathrm{B}^2$, $\omega^2\tau^2 \gg 1$, one can reduce Eq. (26.7) to the following:

$$\mathrm{Re}\,\sigma_\omega = \left(\frac{e^2}{4\hbar}\right)\left\{\frac{4\varepsilon_\mathrm{F}^\mathrm{B}}{\pi\hbar\omega^2\tau_\mathrm{B}}+\frac{8k_\mathrm{B}T}{\pi\hbar\omega^2\tau}\sum_{\mathrm{k}=1}^{\mathrm{K}}\ln\left[1+\exp\left(\frac{\varepsilon_\mathrm{F}^{(\mathrm{k})}}{k_\mathrm{B}T}\right)\right]\right.$$
$$\left.+\sum_{\mathrm{k}=1}^{\mathrm{K}}\tanh\left(\frac{\hbar\omega-2\varepsilon_\mathrm{F}^{(\mathrm{k})}}{4k_\mathrm{B}T}\right)\right\}. \tag{26.8}$$

Under sufficiently strong optical pumping when $\varepsilon_\mathrm{F}^{(\mathrm{k})} \gg \hbar\omega$, $k_\mathrm{B}T$, and, consequently, $\tanh[(\hbar\omega-2\,\varepsilon_\mathrm{F}^{(\mathrm{k})})/4k_\mathrm{B}T] \simeq -1$, setting $\Sigma_{\mathrm{k}=1}^{\mathrm{K}}\,\varepsilon_\mathrm{F}^{(\mathrm{k})} \simeq K*\varepsilon_\mathrm{F}^\mathrm{T}$, where $K^* < K$, from Eq. (26.8) one obtains

$$\mathrm{Re}\,\sigma_\omega\left(\frac{4\hbar}{e^2}\right) \simeq \frac{4}{\pi\hbar\omega^2}\left(\frac{\varepsilon_\mathrm{F}^\mathrm{B}}{\tau_\mathrm{B}}+\frac{2K*\varepsilon_\mathrm{F}^\mathrm{T}}{\tau}\right)-K. \tag{26.9}$$

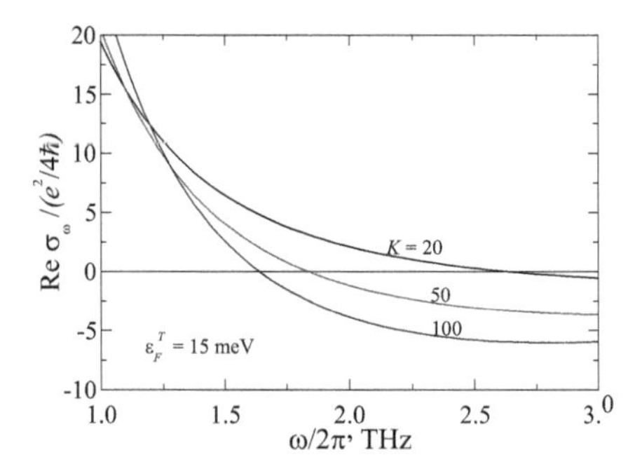

Figure 26.2 Frequency dependences of the real part of dynamic conductivity Re σ_ω normalized by quantity $e^2/4\hbar$ for MGL structures with different number of GLs K at modest pumping (ε_F^T = 15 meV).

26.4 Frequency Characteristics of Dynamic Conductivity

Figures 26.2 and 26.3 show the frequency dependences of Re σ_ω normalized by $e^2/4\hbar$ calculated for MGL structures with different K at different values of ε_F^T (i.e., different optical pumping intensities)

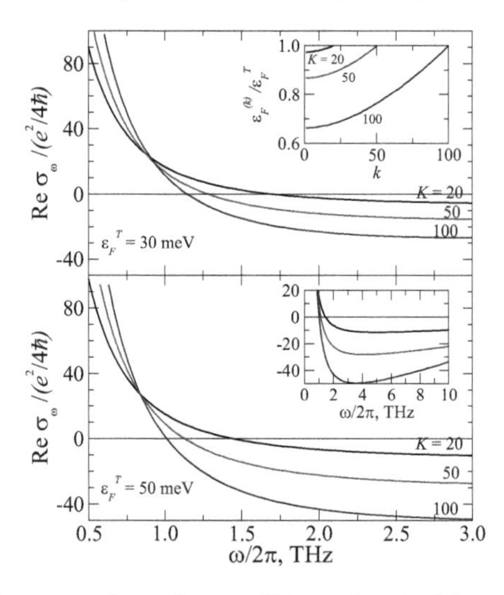

Figure 26.3 Frequency dependences of the real part of dynamic conductivity Re σ_ω normalized by quantity $e^2/4\hbar$ for MGL structures with different number of GLs K at ε_F^T = 30 and 50 meV. The inset on the upper panel shows how the GL population varies with the GL index k, whereas the inset on the lower panel demonstrates the dependences but in a wider range of frequencies.

using Eqs. (26.1), (26.2), (26.5), and (26.6) or Eqs. (26.5) and (26.7). We set ε_F^B = 400 meV [8], $\hbar\Omega$ = 920 meV, T = 300 K, τ_B = 1 ps, τ = 10 ps, and γ = 1/4. As seen from Figs. 26.2 and 26.3, Re σ_ω can be negative in the frequency range $\omega > \omega_{min}$ with ω_{min} decreasing with increasing quasi-Fermi energy ε_F^T in the topmost GL, i.e., with increasing optical pumping intensity (see below). In the MGL structures with K = 50–100 at ε_F^T = 30–50 meV, one has $\omega_{min}/2\pi \gtrsim 1$ THz (see Figs. 26.3 and 26.4). As follows from the inset on upper panel in Fig. 26.3, the quantities $\varepsilon_F^{(k)}$ are not too small in comparison with ε_F^T even in GLs with the indices $k \ll K$, i.e., in GLs near the MGL structure bottom. This implies that the pumping of such near bottom GLs is effective even in the MGL structures with $K \sim 100$. The quantity Re σ_ω as a function of ω exhibits a minimum (see the inset on lower panel in Fig. 26.3). The sign of Re σ_ω becomes positive at $\omega > \omega_{max}$, where ω_{max} is rather large (more than 10 THz). The MGL structures with a larger number K of GLs at stronger pumping exhibit smaller ω_{min} and deeper minima Re σ_ω. This is confirmed also by Fig. 26.5.

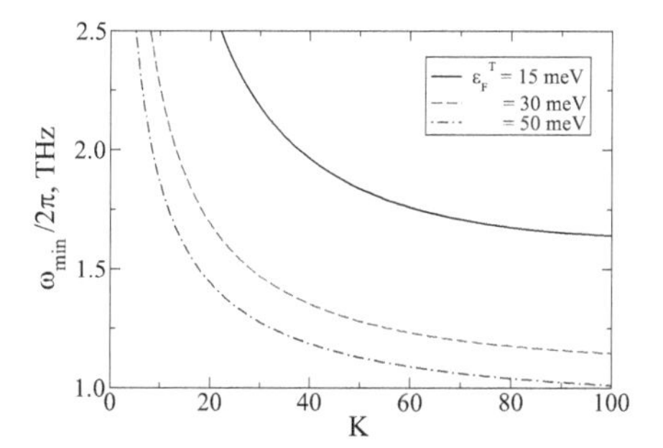

Figure 26.4 Dependences of ω_{min} on number of GLs.

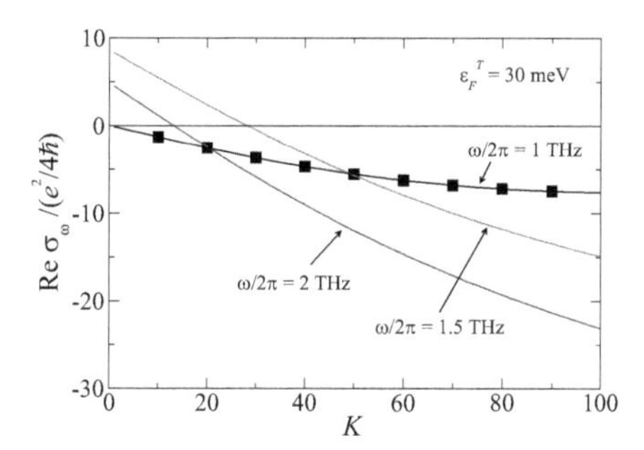

Figure 26.5 Real part of the normalized dynamic conductivity as a function of the number of GLs for different frequencies. The line with markers corresponds to a MGL structure without the bottom GL.

26.5 Role of the Bottom GL

The highly conducting bottom GL, whose existence is associated with the intrinsic features of the MGL structure growth, plays the negative role. This is mainly because it results in a marked absorption (due to the Drude mechanism) of the terahertz radiation emitted by other GLs. Since such an absorption increases with decreasing frequency, the achievement of the negative dynamic conductivity in the MGL structures with shorter momentum relaxation times τ_B and τ even at higher frequencies is significantly complicated by the Drude absorption. This is demonstrated by Fig. 26.6 (upper panel). In contrast, a decrease in the Fermi energy ε_F^B and the electron density in the bottom GL, of course, might significantly promote the achievement of negative dynamic conductivity in a wide frequency range, particularly at relatively low frequencies (compare the frequency dependences on upper and lower panels in Fig. 26.6). In this regard, the MGL structures with GLs exhibiting a long relaxation time τ (like that found in Ref. [10]) but with a lowered electron density in the bottom GL or without the bottom layer appears to be much more preferable. Such MGL structures can be fabricated using chemical/mechanical reactions and transferred substrate techniques (chemically etching the substrate and the highly conducting bottom GL (Ref. [12]) or mechanically peeling the upper GLs, then transferring the upper portion of the MGL structure on a Si or equivalent transparent substrate). The calculation of Re σ_ω for this MGL structure can be carried out by omitting the term Re σ_ω in Eq. (26.6). The pertinent results are shown in Fig. 26.5 (see the marked line) and Fig. 26.7. Here, as in Fig. 26.3, we assumed that $\hbar\omega$ = 920 meV, T = 300 K, τ_B = 1 ps, and τ = 10 ps. The obtained frequency dependences are qualitatively similar to those shown in Figs. 26.2 and 26.3. However the dependences for the MGL structures without the bottom layer exhibit a marked shift toward lower frequencies. In particular, as follows from Figs. 26.5 and 26.7, Re σ_ω can be negative even at $\omega/2\pi \lesssim 1$ THz (at chosen values of ε_F^T).

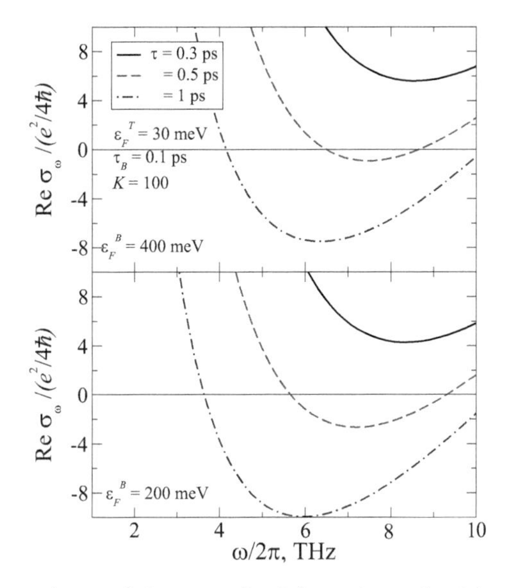

Figure 26.6 Real part of the normalized dynamic conductivity vs frequency calculated for MGL structures with $\tau_B = 0.1$ ps and different τ and ε_F^B.

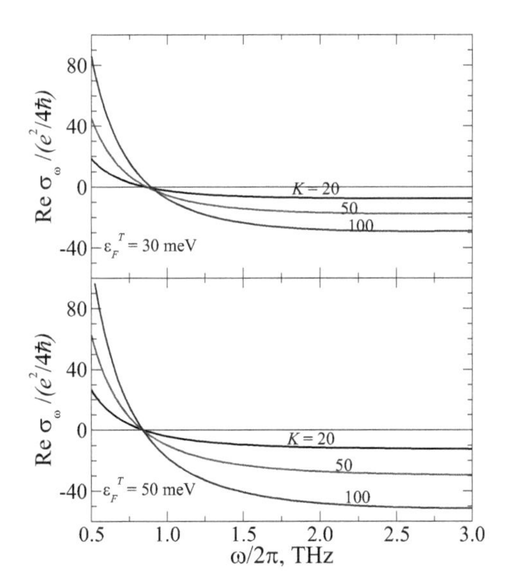

Figure 26.7 Real part of the normalized dynamic conductivity vs frequency calculated for MGL structures with different numbers of GLs and without bottom GL.

26.6 Condition of Lasing

To achieve lasing in the MGL structures under consideration, the following condition should be satisfied [7]:

$$\frac{8\pi}{c}|\mathrm{Re}\,\sigma_\omega|E_\mathrm{t}^2 > (1-r_1)E_1^2 + (1-r_2)E_2^2 + (a/R)^2E_1^2 + E_\mathrm{S}^2. \quad (26.10)$$

Here, E_t, E_1, and E_2 are maximum amplitudes of the terahertz electric field $E = E(z)$ at the MGL structure (placed at the distance t from the bottom mirror, where t is the thickness of the substrate), and near the pertinent mirror, respectively, $E_\mathrm{S}^2 = (\alpha_\mathrm{S} n_\mathrm{S}/2)\int_0^t E^2 dz$, α_S and n_S are the absorption coefficient of THz radiation in the substrate (SiC or Si) and real part of its refraction index, r_1 and r_2 are the reflection coefficients of THz radiation from the mirrors, and a/R is the ratio of the diameters of the output hole a and the mirror R. In deriving inequality (26.10), we neglected the finiteness of the MGL thickness (in comparison with t and the THz wavelength) and disregarded the diffraction losses. For simplicity, one can set $E_\mathrm{t} \sim E_1^2 \sim E_2^2$ and $E_\mathrm{S}^2 \sim (t\alpha_\mathrm{S} n_\mathrm{S}/2)E_2^2$, disregarding, in particular, partial reflection of THz radiation from the bottom GL. In this case, inequality (26.10) can be presented as

$$\frac{8\pi}{c}|\mathrm{Re}\,\sigma_\omega| > (1-r_1) + (1-r_2) + (a/R)^2 + t\alpha_\mathrm{S} n_\mathrm{S}/2 = L. \quad (26.11)$$

Assuming that $r_1 = r_2 = 0.99$, $(a/R) = 0.1$, $\alpha_\mathrm{S} \simeq 2$–$4\ \mathrm{cm}^{-1}$, $n_\mathrm{S} \simeq 3$ [13], and $t = 50\ \mu\mathrm{m}$, for L one obtains $L = 0.06$–0.09. However, as follows from Fig. 26.5, the quantity $(8\pi/c)|\mathrm{Re}\ \sigma_\omega|$ for a MGL structure with $K = 100$ at $\omega/2\pi = 1.5$ THz at $\varepsilon_\mathrm{F}^\mathrm{T} = 30$ meV is about $12\beta \simeq 0.275$. At $\omega/2\pi = 1.0$ THz but for the structure without the bottom GL (see the line with markers in Fig. 26.5), one obtains $(8\pi/c)|\mathrm{Re}\ \sigma_\omega| \simeq 0.345$. These values of $(8\pi/c)|\mathrm{Re}\ \sigma_\omega|$ well exceed the above value of L (in contrast with the structures with two GLs [7], for which the minimization of the losses is crucial). At the elevated frequencies, the ratio $(8\pi/c)|\mathrm{Re}\ \sigma_\omega|/L$ can be even much larger.

Considering that the electron (hole) density Σ^T in the topmost GL

$$\Sigma^\mathrm{T} = \frac{2}{\pi\hbar^2}\int_0^\infty \frac{dpp}{1+\exp[(v_\mathrm{F}p - \varepsilon_\mathrm{F}^\mathrm{T})/k_\mathrm{B}T]}$$

at $\varepsilon_\mathrm{F}^\mathrm{T} = 30$ meV at $T = 300$ K, we obtain $\Sigma^\mathrm{T} \simeq 2 \times 10^{11}\ \mathrm{cm}^{-2}$. Such a value (and higher) of the photogenerated electron and hole density is achievable experimentally (see, for instance, Ref. [14]). Assuming that $K = 100$, $\hbar\Omega = 920$ meV, and the recombination time $\tau_R \simeq 20$ ps at $T = 300$ K and $\Sigma^T \simeq 2 \times 10^{11}$ cm^{-2} [15], for the pertinent optical pumping power we obtain $I_\Omega \simeq 6.4 \times 10^4$ W/cm^2. One needs to point out that this value of I_Ω is much larger than the threshold of lasing at a certain frequency ($\omega_{min} < \omega < \omega_{max}$). At $T = 100$ K, the recombination time (due to optical phonon emission) is much longer [15], hence, the electron and hole densities in question can be achieved at much weaker optical pumping.

In the regime sufficiently beyond the threshold of lasing, when the stimulated radiative recombination becomes dominant, the pumping efficiency is determined just by the ratio of the energy of the emitted terahertz photons $\hbar\omega$ and the energy of optical photons $\hbar\Omega$, $\eta = \omega/\Omega$. However, the nonradiative recombination mechanisms can markedly decrease η. In the regime in question, the maximum output terahertz power can be estimated as max $P_\omega \simeq \pi R^2(\omega/\Omega)I_\Omega$. For example, at $\hbar\Omega = 920$ meV, $\hbar\omega/2\pi = 5.9$ meV ($\omega/2\pi \simeq 1.5$ THz), $2R = 0.1$ cm, and $I_\Omega = 3 \times 10^4$ W/cm^2, one obtains max $P_\omega \simeq 1.5$ W.

26.7 Conclusions

We studied real part of the dynamic conductivity of a MGL structure with a stack of GLs and a highly conducting bottom GL on SiC substrate pumped by optical radiation. It was shown that the negative dynamic conductivity in the MGL structures under consideration with sufficiently large number K of perfect upper GLs can be achieved even at room temperature provided the optical pumping is sufficiently strong. Due to large K, the absolute value of Re σ_ω in its minimum can significantly exceed the characteristic value of conductivity $e^2/4\hbar$. Thus, the MGL structures can serve as active media of terahertz lasers. This can markedly liberate the requirement for the quality of the terahertz laser resonant cavity. An increase in τ_B and τ promotes widening of the frequency range where Re $\sigma_\omega < 0$, particularly, at the low end of this range. This opens up the prospects of terahertz lasing with $\omega/2\pi \sim 1$ THz even at room temperature. The main obstacle appears to be the necessity of sufficiently long relaxation times τ_B and τ in GLs with rather high electron and hole densities: $\Sigma > 10^{11}$ cm^{-2}. Since the electron-hole recombination at the temperatures and densities under consideration might be attributed to the optical phonon emission [15], the optical pumping intensity required for terahertz lasing can be markedly lowered (by orders of magnitudes) with decreasing temperature. The conditions of lasing in the MGL structures without the bottom GL are particularly liberal.

Acknowledgments

One of the authors (V. R.) is grateful to M. Orlita and F. T. Vasko for useful discussions and information. This work was supported by the Japan Science and Technology Agency, CREST, Japan.

Appendix: Recombination

The rate of radiative recombination in the degenerate electron-hole system in the topmost GL due to spontaneous emission of photons at $\varepsilon_{\mathrm{F}}^{\mathrm{T}} \gg k_{\mathrm{B}}T$ can be calculated using the following formula [6, 16]:

$$R_{\mathrm{r}} \simeq \frac{2\upsilon_r}{\pi\hbar^3}\int_0^{\varepsilon_F^T/\upsilon_F} dpp^2 = \frac{2\upsilon_r(\varepsilon_{\mathrm{F}}^{\mathrm{T}})^3}{3\pi\hbar^3\upsilon_{\mathrm{F}}^3} \propto (\varepsilon_{\mathrm{F}}^{\mathrm{T}})^3. \tag{26.A1}$$

Here $\upsilon_{\mathrm{r}} = \sqrt{æ}(8e^2/3\hbar c)(\upsilon_{\mathrm{F}}/c)^2\upsilon_{\mathrm{F}}$ and æ is the dielectric constant $p_{\mathrm{F}} = \varepsilon_{\mathrm{F}}^{\mathrm{T}}/\upsilon_F$.

The rate of the electron-hole recombination associated with emission of optical phonons can be described the following equation [14]:

$$R_{\mathrm{ph}} \propto \int_0^{\hbar\omega_0/\upsilon_{\mathrm{F}}} \frac{dpp(\hbar\omega_0/\upsilon_{\mathrm{F}} - p)}{\left[1+\exp\left(\frac{\upsilon_{\mathrm{F}}p - \varepsilon_{\mathrm{F}}^{\mathrm{T}}}{k_{\mathrm{B}}T}\right)\right]\left[1+\exp\left(\frac{\hbar\omega_0 - \upsilon_{\mathrm{F}}p - \varepsilon_{\mathrm{F}}^{\mathrm{T}}}{k_{\mathrm{B}}T}\right)\right]}$$

$$\propto \exp\left(\frac{2\varepsilon_{\mathrm{F}}^{\mathrm{T}} - \hbar\omega_0}{k_{\mathrm{B}}T}\right). \tag{26.A2}$$

Equalizing $R_{\mathrm{r}}^{\mathrm{T}}$ and the rate of generation of electrons and holes by the optical pumping radiation G_{Ω}^{T} and the recombination rate, and considering Eqs. (26.A1) and (26.A2), one can find that

$\varepsilon_F^T \propto (G_\Omega^T)^{1/3}$ and $\varepsilon_F^T \propto \ln(G_\Omega^T)$, respectively. The radiative interband transitions stimulated by the thermal photons can also contribute to the recombination rate [6, 16] as well as the processes of electron-hole interaction in the presence of disorder. These processes provide different dependences of the recombination rate on the quasi-Fermi energy. Due to this, calculating the dependence of the latter in different GLs on the optical pumping intensity, we put, for definiteness, $\varepsilon_F^T \propto (G_\Omega^T)^\gamma$ with $\gamma = 1/4$. Such a dependence is less steep than that for the spontaneous radiative recombination and somewhat steeper than in the case of optical phonon recombination. The variation in parameter γ leads to some change in the $\varepsilon_F^{(k)} - k$ dependence, but it should not affect the main obtained results.

References

1. C. Berger, Z. Song, T. Li, X. Li, A. Y. Ogbazhi, R. Feng, Z. Dai, A. N. Marchenkov, E. H. Conrad, P. N. First, and W. A. de Heer, *J. Phys. Chem.*, **108**, 19912 (2004).
2. K. S. Novoselov, A. K. Geim, S. V. Morozov, D. Jiang, M. I. Katsnelson, I. V. Grigorieva, S. V. Dubonos, and A. A. Firsov, *Nature (London)*, **438**, 197 (2005).
3. V. Ryzhii, M. Ryzhii, and T. Otsuji, *J. Appl. Phys.*, **101**, 083114 (2007).
4. M. Ryzhii and V. Ryzhii, *Jpn. J. Appl. Phys., Part 2*, **46**, L151 (2007).
5. F. Rana, *IEEE Trans. Nanotechnol.*, **7**, 91 (2008).
6. A. Satou, F. T. Vasko, and V. Ryzhii, *Phys. Rev. B*, **78**, 115431 (2008).
7. A. A. Dubinov, V. Ya. Aleshkin, M. Ryzhii, T. Otsuji, and V. Ryzhii, *Appl. Phys. Express*, **2**, 092301 (2009).
8. F. Varchon, R. Feng, J. Hass, X. Li, B. Ngoc Nguyen, C. Naud, P. Mallet, J.-Y. Veuillen, C. Berger, E. H. Conrad, and L. Magaud, *Phys. Rev. Lett.*, **99**, 126805 (2007).
9. M. Orlita, C. Faugeras, P. Plochocka, P. Neugebauer, G. Martinez, D. K. Maude, A.-L. Barra, M. Sprinkle, C. Berger, W. A. de Heer, and M. Potemski, *Phys. Rev. Lett.*, **101**, 267601 (2008).
10. P. Neugebauer, M. Orlita, C. Faugeras, A.-L. Barra, and M. Potemski, *Phys. Rev. Lett.*, **103**, 136403 (2009).
11. L. A. Falkovsky and A. A. Varlamov, *Eur. Phys. J. B*, **56**, 281 (2007).

12. A. Bostwick, T. Ohta, T. Seyller, K. Horn, and E. Rotenberg, *Nat. Phys.*, **3**, 36 (2007).
13. J. H. Strait, P. A. George, J. M. Dawlaty, S. Shivaraman, M. Chanrashekhar, F. Rana, and M. G. Spencer, *Appl. Phys. Lett.*, **95**, 051912 (2009).
14. J. M. Dawlaty, S. Shivaraman, M. Chandrashekhar, F. Rana, and M. G. Spencer, *Appl. Phys. Lett.*, **92**, 042116 (2008).
15. F. Rana, P. A. George, J. H. Strait, S. Shivaraman, M. Chanrashekhar, and M. G. Spencer, *Phys. Rev. B*, **79**, 115447 (2009).
16. F. T. Vasko and V. Ryzhii, *Phys. Rev. B*, **77**, 195433 (2008).

Chapter 27

Terahertz Lasers Based on Optically Pumped Multiple Graphene Structures with Slot-Line and Dielectric Waveguides*

V. Ryzhii,[a,b] A. A. Dubinov,[a,c] T. Otsuji,[b,d] V. Mitin,[e] and M. S. Shur[f]

[a]*Computational Nanoelectronics Laboratory, University of Aizu, Aizu-Wakamatsu 965-8580, Japan*
[b]*Japan Science and Technology Agency, CREST, Tokyo 107-0075, Japan*
[c]*Institute for Physics of Microstructures, Russian Academy of Sciences, Nizhny Novgorod 603950, Russia*
[d]*Research Institute for Electrical Communication, Tohoku University, Sendai 980-8577, Japan*
[e]*Department of Electrical Engineering, University at Buffalo, State University of New York, New York 14260, USA*
[f]*Department of Electrical, Electronics, and Systems Engineering, Rensselaer Polytechnic Institute, Troy, New York 12180, USA*
v-ryzhii@u-aizu.ac.jp

---

*Reprinted with permission from V. Ryzhii, A. A. Dubinov, T. Otsuji, V. Mitin, and M. S. Shur (2010). Terahertz lasers based on optically pumped multiple graphene structures with slot-line and dielectric waveguides, *J. Appl. Phys.*, **107**, 054505. Copyright © 2010 American Institute of Physics.

Graphene-Based Terahertz Electronics and Plasmonics: Detector and Emitter Concepts
Edited by Vladimir Mitin, Taiichi Otsuji, and Victor Ryzhii

ISBN 978-981-4800-75-4 (Hardcover), 978-0-429-32839-8 (eBook)
www.jennystanford.com

Terahertz (THz) lasers on optically pumped multiple-graphene-layer (MGL) structures as their active region are proposed and evaluated. The developed device model accounts for the interband and intraband transitions in the degenerate electron-hole plasma generated by optical radiation in the MGL structure and the losses in the slot or dielectric waveguide. The THz laser gain and the conditions of THz lasing are found. It is shown that the lasers under consideration can operate at frequencies $\gtrsim 1$ THz at room temperatures.

27.1 Introduction

Graphene, graphene nanoribbons, and graphene bilayers (see, for instance, Ref. [1]) can be used in different terahertz (THz) devices [2–9]. Optical excitation of graphene can result in the interband population inversion and negative real part of the dynamic conductivity of a graphene layer σ_ω in the THz range of frequencies [7–9]. The negativity of the real part of the dynamic conductivity implies that the interband emission of photons with the energy $\hbar\omega$, where $\hbar$ is the reduced Planck constant, prevail over the intraband (Drude) absorption. If the THz photon losses in the resonant cavity are sufficiently small, the THz lasing can be realized in graphene-based devices with optical pumping [10, 11]. In Refs. [10, 11], the optically pumped THz lasers with a Fabry–Pérot resonator were considered. In this chapter, we propose and evaluate the THz lasers based on multiple-graphene-layer (MGL) structures with a metal slot-line waveguide (SLW) or a dielectric waveguide (DW) pumped by optical radiation. The specific features of characteristics of the MGL-based lasers under consideration are associated with the frequency dependences of the absorption in the waveguides and the gain-overlap factor, which is sensitive to the spatial distribution of the THz electric field. As shown in the following, the characteristics of the lasers with SLW and DW are fairly similar, so that we will mainly focus on the device with SLW. This is in part because the structures with SLW exhibit somewhat better confinement, and can be used not only for MGL-based lasers with optical pumping but also in MGL-based injection lasers with lateral p-i-n junctions (both electrically induced [12–14] and formed by pertinent doping of MGL structure).

27.2 Device Model

We consider lasers (a) with a MGL structure on a SiC or Si substrate with the side highly conducting metal strips and a highly conducting back electrode at the substrate bottom and (b) with a MGL structure on the top of a SiC or Si slab serving as a DW. The cross-sections (corresponding to the y–z plane) of the device structures under consideration are schematically shown in Fig. 27.1. The axis x corresponds to the direction of the electromagnetic wave propagation, whereas the y and z directions are in the MGL structure plane and perpendicular to it, respectively (see Fig. 27.1). The MGL plane corresponds to z = 0. The finiteness of the MGL structure thickness can be disregarded. It is assumed that the MGL structure under consideration comprises K upper GLs and a highly conducting bottom GL on a SiC substrate or K GLs (without the bottom GL) on a Si substrate. Epitaxial MGL structures with up to K = 100 GLs with very long momentum relaxation time of electrons and holes ($\tau \simeq 20$ ps) were recently fabricated using the thermal decomposition from 4H-SiC substrate [15]. MGL structures without the bottom GL can be fabricated using chemical/mechanical reactions and transferred substrate techniques, which include chemically etching the substrate and the highly conducting bottom GL (Ref. [16]) (or mechanically peeling the upper GLs) and transferring the upper portion of the MGL structure on a Si or equivalent transparent substrate. Since the electron density and the Fermi energy ε_F^B in the bottom GL is rather large ($\varepsilon_F^B \simeq 400$ meV) [17], the Drude absorption in this GL can be significant although it can be overcome by a strong emission from the upper GLs if their number is sufficiently large. The MGL structures without the bottom GL can exhibit significant advantages (see Ref. [11] and below).

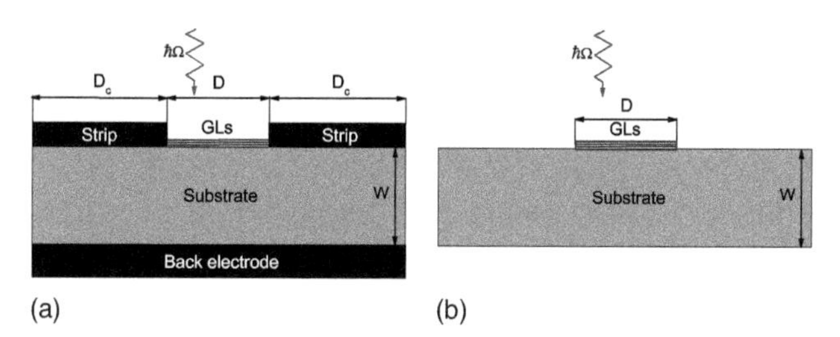

Figure 27.1 Schematic views of the device structures under consideration: (a) with a SLW and (b) with a DW.

It is assumed that the MGL structure is illuminated from the top by light with the energy of photons $\hbar\Omega$. The optical waveguide input of the pumping radiation is also possible. When $\hbar\Omega$ is close to $N\hbar\omega_0/2$, where $\hbar\omega_0 \simeq 0.2$ eV is the optical phonon energy and N is an integer, the photogeneration of electrons and holes and their cooling, associated with the cascade emission of optical phonons, result in an essentially occupation (population inversion) of low energy states near the bottom of the conduction band and the top of the valence band. At elevated electron and hole densities (i.e., at sufficiently strong optical pumping), the electron and hole distributions in the range of energies $\varepsilon \ll \hbar\omega_0$ in the kth GL ($1 \le k \le K$) can be described by the Fermi functions with the quasi-Fermi energies $\varepsilon_{\mathrm{F}}^{(\mathrm{k})}$.

The quasi-Fermi energies in the GLs with $k \geq 1$ are mainly determined by the electron (hole) density in this layer Σ_{k}, i.e., $\varepsilon_{\mathrm{F}}^{(\mathrm{k})} \propto \sqrt{\Sigma^{(\mathrm{k})}}$ and, therefore, by the rate of photogeneration $G_{\Omega}^{(\mathrm{k})}$ by the optical radiation at the kth GL plane. Considering the attenuation of the optical pumping radiation due to its absorption in each GL, one can obtain

$$G_{\Omega}^{(\mathrm{k})} = \frac{I_{\Omega}}{\hbar\Omega}\beta[(1-\beta)^{\mathrm{K-k}} + (1-\beta_{\mathrm{B}})^2(1-\beta)^{\mathrm{K+k-1}}]. \tag{27.1}$$

Here, I_{Ω} is the intensity of incident pumping radiation, $\beta = \pi e^2/\hbar c \simeq 0.023$, where e is the electron charge, c is the speed of light in vacuum, and $\beta_{\mathrm{B}} = (4\pi/c)\mathrm{Re}\ \sigma_{\Omega}^{\mathrm{B}}$. The latter quantity accounts for the absorption of optical pumping radiation in the bottom layer. A relationship between $\varepsilon_{\mathrm{F}}^{(\mathrm{k})}$ and $G_{\Omega}^{(\mathrm{k})}$ is determined by the recombination mechanisms [11]. As a result, $\varepsilon_{\mathrm{F}}^{(\mathrm{k})}$ can be expressed via the quasi-Fermi energy in the topmost GL $\varepsilon_{\mathrm{F}}^{\mathrm{T}} = \varepsilon_{\mathrm{F}}^{(\mathrm{K})}$, which, in turn, is a function of the intensity of incident pumping radiation I_{Ω}.

Since the thickness of the MGL structure is small in comparison with the wavelength of THz radiation, the generation and absorption are determined by the real part of the net dynamic conductivity, which is the sum of the real parts of the dynamic conductivity of the bottom GL Re $\sigma_{\omega}^{\mathrm{B}}$ and other GLs $\sigma_{\omega}^{(\mathrm{k})}$

$$\mathrm{Re}\,\sigma_{\omega} = \mathrm{Re}\,\sigma_{\omega}^{\mathrm{B}} + \mathrm{Re}\sum_{\mathrm{k=1}}^{\mathrm{K}}\sigma_{\omega}^{(\mathrm{k})}. \tag{27.2}$$

Considering the expressions for Re $\sigma_{\omega}^{\mathrm{B}}$ and Re $\sigma_{\omega}^{(\mathrm{k})}$ obtained previously [11], one can arrive at the following:

$$\mathrm{Re}\,\sigma_{\omega}=\left(\frac{e^2}{4\hbar}\right)\left\{\frac{4k_{\mathrm{B}}T\tau_{\mathrm{B}}}{\pi\hbar(1+\omega^2\tau_{\mathrm{B}}^2)}\ln\left[1+\exp\left(\frac{\varepsilon_{\mathrm{F}}^{\mathrm{B}}}{k_{\mathrm{B}}T}\right)\right]\right.$$

$$+\frac{8k_{\mathrm{B}}T\tau}{\pi\hbar(1+\omega^2\tau^2)}\sum_{\mathrm{k}=1}^{\mathrm{K}}\ln\left[1+\exp\left(\frac{\varepsilon_{\mathrm{F}}^{(\mathrm{k})}}{k_{\mathrm{B}}T}\right)\right]$$

$$\left.+\sum_{\mathrm{k}=1}^{\mathrm{K}}\tanh\left(\frac{\hbar\omega-2\varepsilon_{\mathrm{F}}^{(\mathrm{k})}}{4k_{\mathrm{B}}T}\right)\right\}. \tag{27.3}$$

Here, τ_{B} and τ are the electron and hole momentum relaxation times in the bottom and other GLs, respectively, T is the electron and hole temperature, and k_{B} is the Boltzmann constant. The first two terms in the right-hand side of Eq. (27.3) are associated with the intraband (Drude) absorption of THz radiation in all GLs, whereas the third term corresponds to the interband transitions. In the case of lasers without the bottom GL, one can use Eq. (27.3) formally setting $\tau_{\mathrm{B}} = \infty$.

The electromagnetic waves propagating along the SLW was considered using the following equation, which is a consequence of the Maxwell equations:

$$\frac{\partial^2 E_{\omega}(y,z)}{\partial y^2}+\frac{\partial^2 E_{\omega}(y,z)}{\partial z^2}+\left[\frac{\omega^2}{c^2}\eta(y,z)-q^2\right]E_{\omega}(y,z)=0. \tag{27.4}$$

Here, $E_{\omega}(y, z)$ the amplitude of the yth component of the THz electric field $E(t, x, y, z) = E_{\omega}(y, z)\exp[i(qx - \omega t)]$, $\eta(y, z)$ is the complex permittivity, q is the wavenumber of the propagating mode. The quantities $E_{\omega}(y, z)$, $dE_{\omega}(y, z)/dz$, and $\eta^{-1}(y, z)d[\eta(y, z)E_{\omega}(y, z)]/dy$ are continuous at the interfaces between the layers with different refractive indices. The boundary conditions for the guided mode correspond to the condition $E_{\omega}(y, z) \to 0$ at $y, z \to \pm\infty$. Equation (27.4) was solved numerically using the effective index and transfer-matrix methods (see, for instance, Refs. [18, 19]). The coefficient of absorption of the propagating mode α_{ω} and the coefficient of reflection from the interfaces between the laser structure edges and vacuum R were calculated using the following formulas, respectively: $\alpha_{\omega}= 2\,\mathrm{Im}\,q$ and $R = |(q - q_0)/(q + q_0)|^2$, where $q_0 = \omega/c$. The THz gain, which describes the attenuation or amplification of the propagating mode, under optical pumping can be calculated using the following formula:

$$g_\omega = \frac{4\pi \operatorname{Re} \sigma_\omega}{c\sqrt{\eta_S}} \Gamma_\omega - \alpha_\omega, \tag{27.5}$$

where

$$\Gamma_\omega = \frac{\int_{-D/2}^{D/2} |E_\omega(y,0)|^2 \, dy}{\int_{-\infty}^{\infty} \int_{-\infty}^{\infty} |E_\omega(y,z)|^2 \, dy dz} \tag{27.6}$$

is the gain-overlap factor (in the case when σ_ω is independent of coordinate y) and η_S is the permittivity of the substrate (SiC or Si).

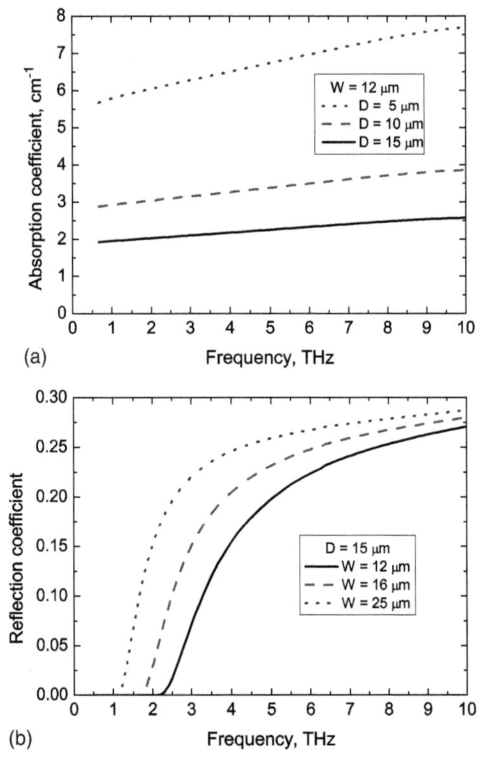

Figure 27.2 Coefficients of (a) absorption and (b) reflection vs frequency for SLWs with different geometrical parameters.

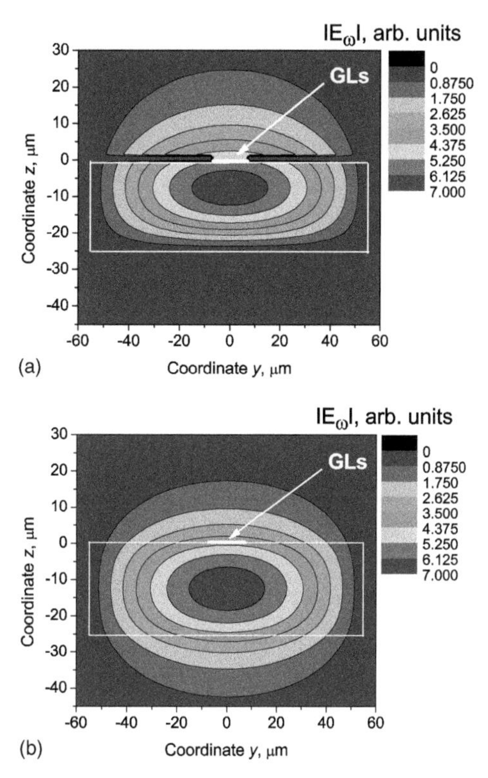

Figure 27.3 Spatial distributions of THz electric field (a) in SLW cross-section and (b) in DW cross-section.

27.3 Results and Discussion

Figure 27.2 shows the absorption coefficient α_{ω} of electromagnetic waves propagating along the SLW and the coefficient of reflection from the laser structure edges as function of frequency $\omega/2\pi$ calculated using Eq. (27.4) for different widths of the slot D and the strips D_c and the substrate thickness W. The strips and the back electrode are assumed to be made of Al. Examples of the spatial distributions of the THz electric field in the electromagnetic wave in laser structures with D =15 μm at $\omega/2\pi$ = 1.8 THz in question are show in Fig. 27.3. The electric field distributions with the frequency dependences of the dynamic conductivity of the MGL structures obtained by solving Eq. (27.4) were substituted to Eqs. (27.5) and (27.6) to find the gain-overlap factor and the THz gain. Figures

27.4 and 27.5 show the frequency dependences of the THz gain g_ω calculated for the laser structures with (Fig. 27.4) and without (Fig. 27.5) bottom GL and with different structural parameters at different pumping conditions (different values of the quasi-Fermi energy) and at $T = 300$ K. It was assumed that $\hbar\Omega = 920$ meV, $\tau = 10$ ps, and $\tau_B = 1$ ps. The chosen value of τ is twice as small as that found

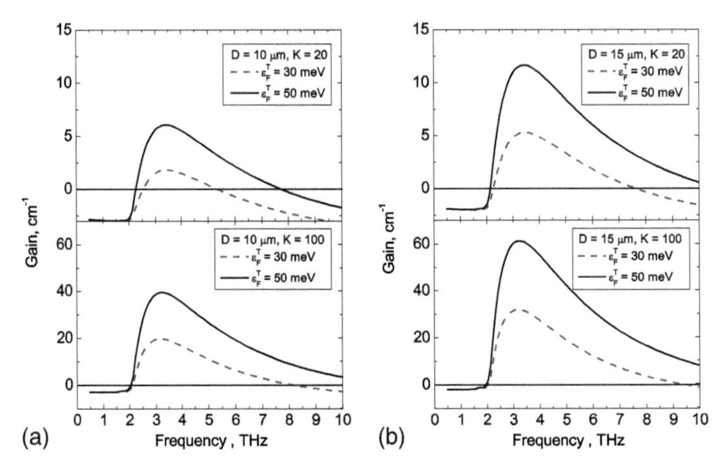

Figure 27.4 Frequency dependences of THz gain in laser structures with different width of the slot D and different number of GL K at different values of the quasi-Fermi energy ε_F^T : (a) $D = 10$ μm and (b) $D = 15$ μm.

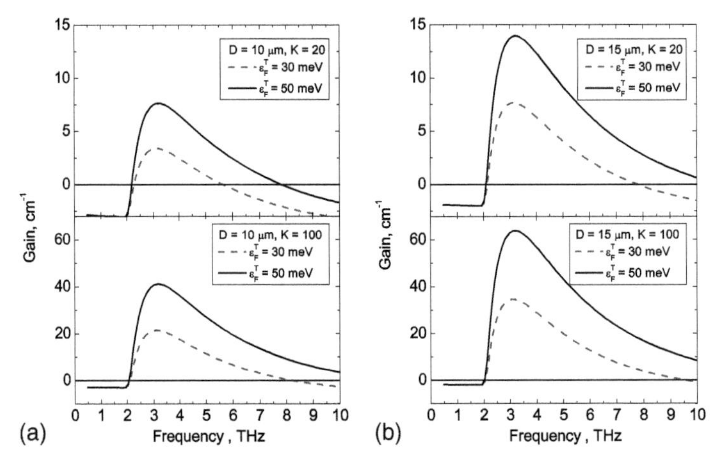

Figure 27.5 The same as in Fig. 27.4 but for laser structures without bottom GL.

experimentally [15] in MGL structures with different number of GLs (up to $K = 100$). To take into account a shortening of the momentum relaxation time with significant increasing electron density, this time in the bottom GL was chosen ten times shorter than in other GLs. As seen from comparison of the top panels of Figs. 27.4 and 27.5, the laser structures with and without the bottom GL (with $K = 20$) exhibit qualitatively similar frequency dependences of the THz gain g_ω (at chosen geometrical parameters) but with somewhat higher maxima in the latter structures. In the case of the structures with $K = 100$, these dependences (compare lower panels of Figs. 27.4 and 27.5) are virtually identical. The frequency $\omega_{\min}$ at which g_ω changes its sign is $\omega_{\min}/2\pi \simeq 2$ THz in the laser structures of both types.

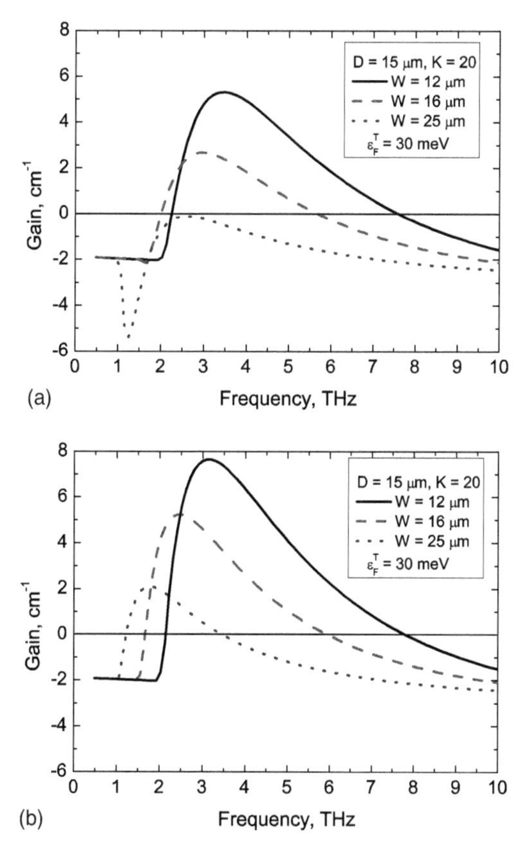

Figure 27.6 Comparison of THz gain vs frequency dependences in laser structures (a) with and (b) without bottom GL, and with different spacing between MGL structure and back electrode W.

The latter is different from the situation in the MGL lasers with the Fabry–Pérot resonator [11], in which eliminating the bottom GL can lead to a pronounced decrease in ω_{min}. This is attributed to the effect of SLW: the variation in the frequency results in a redistribution of the spatial distribution of the THz electric field and, hence, in a change in the gain-overlap factor. Due to this, the spacing between the MGL structure W and the back electrode affects the frequency dependences of the THz gain. Figure 27.6 shows these dependences calculated for different W. One can see that the height of the THz gain maxima decreases with increasing W. Simultaneously the frequency ω_{min} shifts toward lower values. This shift is more pronounced in the laser structures without the bottom GL (compare Figs. 27.6a and b). As seen from Fig. 27.6b, $\omega_{\text{min}}/2\pi$ can reach 1 THz. In the case of larger K and ε_F^T, as seen from Fig. 27.7, the THz gain in the range $\omega/2\pi \gtrsim 1$–2 THz can be rather large.

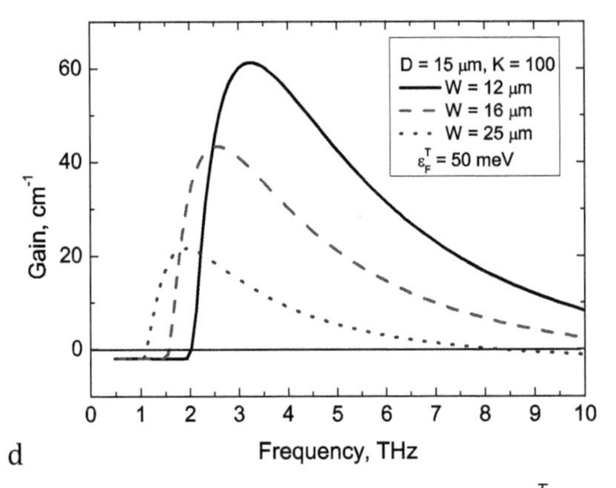

Figure 27.7 The same as in Fig. 27.6a but for $K = 100$ and $\varepsilon_F^T = 50$ meV.

A decrease in the momentum relaxation time τ (which results in an enhancement in the Drude absorption) leads to an increase in ω_{min} and to a decrease in $|\text{Re}\ \sigma_\omega|$ in the range of frequencies, where $\text{Re}\ \sigma_\omega < 0$. This is seen in Fig. 27.8.

The above results correspond to $T = 300$ K. Lowering of the temperature should lead to widening of the frequency range where $\text{Re}\ \sigma_\omega$ is negative and where g_ω can be positive. As a result, ω_{min} might decrease with decreasing temperature. The latter is confirmed by the calculated temperature dependences of ω_{min} shown in Fig. 27.9.

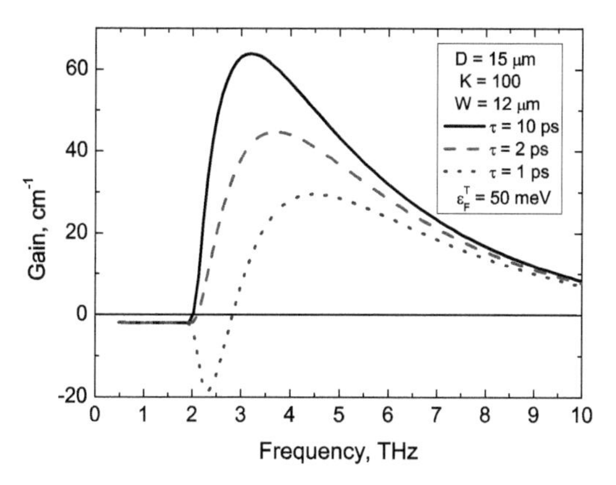

Figure 27.8 Frequency dependences of THz gain in laser structures (without the bottom GL) with different τ.

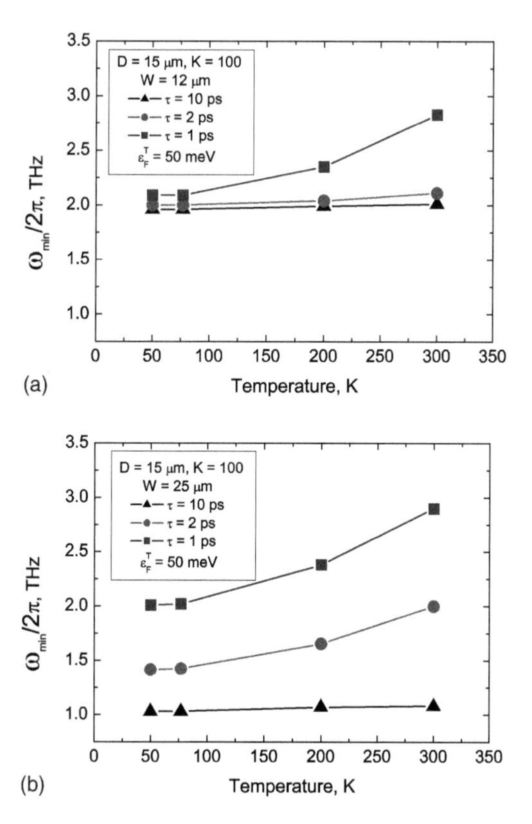

Figure 27.9 Temperature dependences of ω_{min} for laser structures with different τ: (a) $W = 12$ μm and (b) $W = 25$ μm.

Figure 27.10 shows the frequency dependences of the THz gain calculated for laser structures with SLW and DW for $T = 300$ K and $\tau = 10$ ps. One can see that the maximum of the THz gain in the laser with SLW is somewhat higher than that in the laser with DW. This can primarily be explained by the effect of the spatial distribution of the THz electric field on the gain-overlap factor (compare Figs. 27.3a and b). To maximize the THz gain at the desirable frequency, one needs carefully optimize the geometrical parameters.

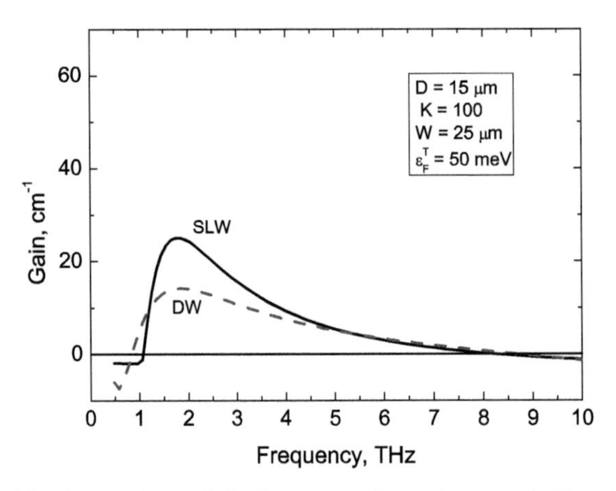

Figure 27.10 Comparison of the frequency dependences of THz gain in laser structures with SLW and DW.

Considering the propagation of the THz electromagnetic wave in the laser structure (with the length L in the x-direction) and its reflection from edges or from the external mirrors (with the reflective coefficients R_1 and R_2), the condition of lasing can be presented as

$$Lg_\omega > \ln \frac{1}{R_1 R_2}. \tag{27.7}$$

For a laser structure with SLW with $D = 15$ μm, $W = 12$ μm, and $K = 100$ at $\varepsilon_F^T = 30$–50 meV and $\omega/2\pi = 3.0$ THz, so that $R_1 = R_2 \simeq 0.05$ and $g_\omega \simeq 30$–60 cm^{-1} (see Figs. 27.2b and 27.4b, respectively), one obtains $L > 0.1$–0.2 cm. If $D = 15$ μm, $W = 25$ μm, and $K = 100$ at $\varepsilon_F^T = 50$ meV and $\omega/2\pi = 1.5$ THz ($R_1 = R_2 \simeq 0.05$ and $g_\omega \simeq 10$ cm^{-1} as follows from Figs. 27.2b and 27.7, respectively), one obtains $L > 0.6$ cm.

As shown above, at the quasi-Fermi in the topmost GL about $\varepsilon_F^T = 30$–50 meV, the achievement of the THz lasing in the devices under consideration at room temperatures is feasible. At T = 300 K the condition $\varepsilon_F^T \gtrsim 30$ meV corresponds to the electron and hole densities about 2×10^{11} cm^{-2}. Such densities can be obtained at reasonable optical powers (see Ref. [12] and the references therein).

27.4 Conclusions

We proposed THz lasers with optical pumping based on MGL structures with waveguides and calculated their characteristics. The feasibility of lasing in the devices under consideration at the low end of the THz frequency range at room temperatures was demonstrated.

Acknowledgments

The authors are grateful to V. V. Popov, M. Ryzhii, A. Satou, M. Suemitsu, and F. T. Vasko for fruitful discussions. This work was supported by the Japan Science and Technology Agency, CREST, Japan.

References

1. A. H. Castro Neto, F. Guinea, N. M. R. Peres, K. S. Novoselov, and A. K. Geim, *Rev. Mod. Phys.*, **81**, 109 (2009).
2. F. T. Vasko and V. Ryzhii, *Phys. Rev. B*, **77**, 195433 (2008).
3. V. Ryzhii, V. Mitin, M. Ryzhii, N. Ryabova, and T. Otsuji, *Appl. Phys. Express*, **1**, 063002 (2008).
4. F. Xia, T. Mueller, R. Golizadeh-Mojarad, M. Freitag, Y. Lin, J. Tsang, V. Perebeinos, and P. Avouris, *Nano Lett.*, **9**, 1039 (2009).
5. V. Ryzhii and M. Ryzhii, *Phys. Rev. B*, **79**, 245311 (2009).
6. Y. Kawano and K. Ishibashi, Proceedings of the 34th International Conference on Infrared, and Terahertz Waves, Busan, Korea, 21–25 September 2009, (unpublished), W4A04.0380.
7. V. Ryzhii, M. Ryzhii, and T. Otsuji, *J. Appl. Phys.*, **101**, 083114 (2007).
8. F. Rana, *IEEE Trans. Nanotechnol.*, **7**, 91 (2008).
9. A. Satou, F. T. Vasko, and V. Ryzhii, *Phys. Rev. B*, **78**, 115431 (2008).
10. A. A. Dubinov, V. Ya. Aleshkin, M. Ryzhii, T. Otsuji, and V. Ryzhii, *Appl. Phys. Express*, **2**, 092301 (2009).

11. V. Ryzhii, M. Ryzhii, A. Satou, T. Otsuji, A. A. Dubinov, and V. Ya. Aleshkin, *J. Appl. Phys.*, **106**, 084507 (2009).
12. V. V. Cheianov and V. I. Fal'ko, *Phys. Rev. B*, **74**, 041403(R) (2007).
13. M. Ryzhii and V. Ryzhii, *Jpn. J. Appl. Phys., Part 2*, **46**, L151 (2007).
14. M. Ryzhii and V. Ryzhii, *Physica E (Amsterdam)*, **40**, 317 (2007).
15. P. Neugebauer, M. Orlita, C. Faugeras, A.-L. Barra, and M. Potemski, *Phys. Rev. Lett.*, **103**, 136403 (2009).
16. A. Bostwick, T. Ohta, T. Seyller, K. Horn, and E. Rotenberg, *Nat. Phys.*, **3**, 36 (2007).
17. F. Varchon, R. Feng, J. Hass, X. Li, B. Ngoc Nguyen, C. Naud, P. Mallet, J.-Y. Veuillen, C. Berger, E. H. Conrad, and L. Magaud, *Phys. Rev. Lett.*, **99**, 126805 (2007).
18. K. J. Ebeling, *Integrated Optoelectronics: Waveguide Optics, Photonics, Semiconductors* (Springer-Verlag, Berlin, 1993).
19. M. Born and E. Wolf, *Principles of Optics: Electromagnetic Theory of Propagation, Interference and Diffraction of Light* (Pergamon, Oxford, 1964).

Chapter 28

Observation of Amplified Stimulated Terahertz Emission from Optically Pumped Heteroepitaxial Graphene-on-Silicon Materials*

H. Karasawa,[a] T. Komori,[a] T. Watanabe,[a] A. Satou,[a,c] H. Fukidome,[a,c] M. Suemitsu,[a,c] V. Ryzhii,[b,c] and T. Otsuji[a,c]

[a]*Research Institute of Electrical Communication, Tohoku University, 2-1-1 katahira, Aoba-ku, Sendai, Miyagi 980-8577, Japan*

[b]*Depertment of Computer Science and Engineering, University of Aizu, Aizu-Wakamatsu 965-8580, Japan*

[c]*JST-CREST, 3-5 Chiyoda-ku, Tokyo 102-0075, Japan*

otsuji@riec.tohoku.ac.jp

We experimentally observed the fast relaxation and relatively slow recombination dynamics of photogenerated electrons/holes in a heteroepitaxial graphene-on-Si material under pumping with a

*Reprinted with permission from H. Karasawa, T. Komori, T. Watanabe, A. Satou, H. Fukidome, M. Suemitsu, V. Ryzhii, and T. Otsuji (2011). Observation of amplified stimulated terahertz emission from optically pumped heteroepitaxial graphene-on-silicon materials, *J. Infrared Millimeter Terahertz Waves*, **32**, 655–665. Copyright © 2010 Springer Science+Business Media.

Graphene-Based Terahertz Electronics and Plasmonics: Detector and Emitter Concepts
Edited by Vladimir Mitin, Taiichi Otsuji, and Victor Ryzhii

ISBN 978-981-4800-75-4 (Hardcover), 978-0-429-32839-8 (eBook)
www.jennystanford.com

1550 nm, 80 fs pulsed fiber laser and probing with the corresponding terahertz beam generated by and synchronized with the pumping laser. The time-resolved electric-nearfield intensity originating from the coherent terahertz photon emission is electrooptically sampled in total-reflection geometry. The Fourier spectrum fairly agrees the product of the negative dynamic conductivity and the expected THz photon spectrum reflecting the pumping photon spectrum. This phenomenon is interpreted as an amplified stimulated terahertz emission.

28.1 Introduction

Graphene is a single-layer carbon-atomic honeycomb lattice crystal. Electrons and holes in graphene hold a linear dispersion relation with zero bandgap [1–3]. Since the discovery of the single-layer graphene by Geim et al. [4] in 2004, with their very primitive methodology of peeling the grapheite, graphene has made a great impact in academic and industry. In this situation, we have studied the dynamic ac conductivity of nonequilibrium two-dimensional electron-hole systems in optically pumped graphene and have expected negative-dynamic conductivity in far infrared and terahertz (THz) spectral ranges in optical pumping [5] at room temperature, due to the gapless energy spectrum. And considering the negative-dynamic conductivity, a very fast relaxation (≤1 ps) and relatively slow recombination (>>1 ps) [6], the population inversion can be achieved. We think this effect can be used the graphene-based coherent source of THz radiation [7, 8].

It is known that photoexcited carriers in graphene are first cooled and thermalized mainly by intraband relaxation processes on femtosecond to subpicosecond time scales, and then by interband recombination processes. Recently the ultrafast carrier relaxation via emissions from optical phonons has observed by Kamprath et al. [9], Wang et al. [10] and George et al. [11]. The measured optical phonon lifetimes found were ~7 ps [9], 2–2.5 ps [10], and ~1 ps [11], respectively. More precisely, Breusing et al. [12] revealed that photoelectons and photoholes in optically pumped exfoliated graphene and graphite loose a major fraction of their energy within 200~300 fs, mainly by emission of optical phonons.

In this work, we observe amplified stimulated THz emission in optically pumped and THz-probed epitaxial graphene heterostructures. The results reflect the recombination of photoelectrons and photoholes after the intraband ultrafast carrier relaxation via optical phonon emission, and provide evidence for the occurrence of negative dynamic conductivity in the terahertz spectral range.

28.2 Theoretical Model

When graphene is optically excited with photon energy $\hbar\Omega$, electrons and holes are photogenerated and first establish separate quasi-equilibrium distributions at around the level $\varepsilon_f \pm \hbar\Omega/2$ (ε_f: Fermi energy) within 20–30 fs after the excitation by the carrier-carrier (cc) scattering (see Fig. 28.1b). When the electronic temperature is not so high, the relaxation process is followed by cooling of these electrons and holes mainly by emission of a cascade of optical phonons ($\hbar\omega_0$)within 200~300 fs to occupy the states (see Fig. 28.1c)

$$\varepsilon_f \pm \varepsilon_N \approx \varepsilon_f \pm \hbar(\Omega/2 - N\omega_0) \tag{28.1}$$

where $\varepsilon_N < \hbar\omega_0$, $N = 1, 2, 3, \ldots$

Then, thermalization occurs via electron-hole recombination as well as intraband Fermization due to cc scattering and carrier-phonon (cp) scattering (as shown with energy $\hbar\omega_q$ in Fig. 28.1a) on a few picoseconds time scale (see Fig. 28.1d), while the interband cc scattering and cp scattering are slowed by the density of states effects and Pauli blocking [12].

The real part of the net ac conductivity, Re σ_ω, is proportional to the absorption of photons with frequency ω and comprises the contributions of both interband and intraband transitions.

$$\mathrm{Re}\,\sigma_\omega = \mathrm{Re}\,\sigma_\omega^{\mathrm{inter}} + \mathrm{Re}\,\sigma_\omega^{\mathrm{intra}} \tag{28.2}$$

Here the contribution of intraband transition corresponds to the Drude-like mechanism of THz photon absorption and interband transition corresponds to the generation/recombination rates [5, 6]. The real part of the conductivity corresponds to the imaginary part of the permittivity: Im ε. Im ε means the energy loss. Hence the negative-dynamic conductivity gives energy gain.

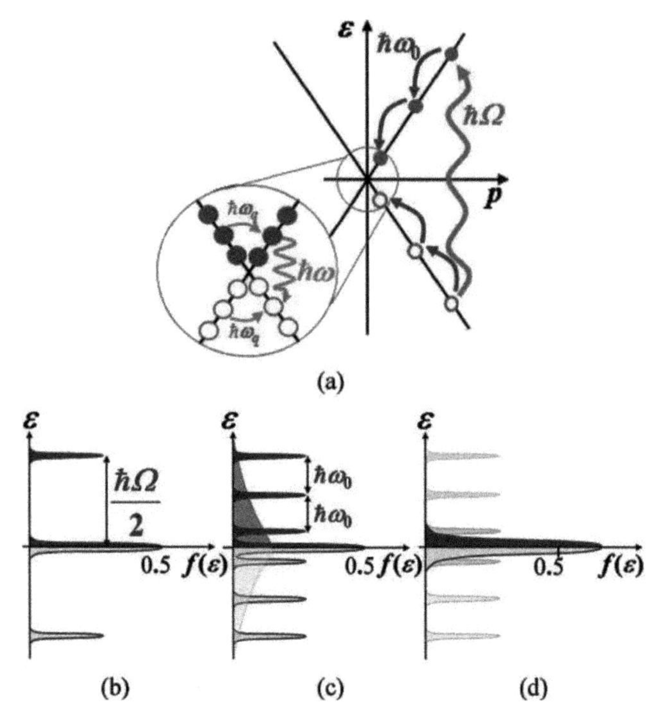

Figure 28.1 Schematic view of graphene band structure (a) and energy distributions of photogenerated electrons and holes when electronic temperature is comparable to the lattice temperature (b–d). Arrows denote transitions corresponding to optical excitation by photons with energy $\hbar\Omega$, cascade emission of optical phonons with energy $\hbar\omega_0$, and radiative recombination with emission of photons with energy $\hbar\omega$. (b) after ~20 fs from optical pumping, (c) after ~200 fs from optical pumping (shaded distribution is the case for high electronic temperature), (d) after a few ps from optical pumping.

Let us consider a situation under weak pumping with infrared photons:

$$T \approx T_0,\ \varepsilon_F < k_B T,\ \hbar\Omega >> k_B T,\ \tau_{rad} >> \tau,\ k_B T\tau/\hbar < \omega^2\tau^2,$$

where T is the electron temperature, T_0 is the lattice temperature, ε_F is the Fermi level, k_B is the Boltzmann constant, τ_{rad} is the electron-hole radiative recombination lifetime, τ is the intraband momentum relaxation time of electrons/holes, $\hbar$ is the reduced Plank constant, and ω is the angular frequency.

Re σ_ω^{inter} and Re σ_ω^{intra} become [9]

$$\mathrm{Re}\,\sigma_\omega^{\text{inter}} = \frac{e^2}{4\hbar}\left[1 - f_{\mathrm{e}}\left(\frac{\hbar\omega}{2}\right) - f_{\mathrm{h}}\left(\frac{\hbar\omega}{2}\right)\right]$$
$$\approx \frac{e^2}{8\hbar}\left(\frac{\hbar\omega}{2k_{\mathrm{B}}T} - \frac{\varepsilon_{\mathrm{F}}}{k_{\mathrm{B}}T}\right) = \frac{e^2}{8\hbar}\left(\frac{\hbar\omega}{2k_{\mathrm{B}}T} - \frac{\pi}{\hbar\Omega}\frac{e^2}{\hbar}\frac{\tau_{\mathrm{rad}}}{\Sigma_0}I_\Omega\right), \quad (28.3)$$

$$\mathrm{Re}\,\sigma_\omega^{\text{intra}} \approx \frac{(\ln 2 + \varepsilon_{\mathrm{F}}/2k_{\mathrm{B}}T)e^2}{\pi\hbar}\frac{k_{\mathrm{B}}T\tau}{\hbar(1+\omega^2\tau^2)}, \quad (28.4)$$

where e is the elementary charge, v_{F} is the Fermi velocity, Σ_0 is the carrier density in the dark, $f_{(\mathrm{e,h})}$ is the Fermi-Dirac distribution function (e: electrons, h: holes), and I_Ω is the pumping intensity. The contribution of the intraband transition: Re $\sigma_\omega^{\text{intra}}$ takes positive, decreasing monotonically and approaching 0 with increasing ω, while the interband term contributes to the linear decrease (increase) in the intraband term in the low (high) ω region. The break-even frequency depends on I_Ω and τ_{rad}. Therefore, $\mathrm{Re}\sigma_\omega$ is a minimum at a specific frequency: $\omega = \bar{\omega}$.

Around $\omega = \bar{\omega}$, $\mathrm{Re}\sigma_\omega$ becomes [5]

$$\mathrm{Re}\,\sigma_\omega \cong \frac{e^2\bar{g}}{8\hbar}\left[1 + \frac{3}{2}\left(\frac{\omega - \bar{\omega}}{\bar{\omega}}\right)^2 - \frac{I_\Omega}{\overline{I_\Omega}}\right] \quad (28.5)$$

where

$$\bar{g} = 2\left(\frac{4\ln 2}{\pi}\right)^{1/3}\left(\frac{\hbar}{k_{\mathrm{B}}T\tau}\right)^{1/3},$$
$$\bar{\omega} \approx \left(\frac{k_{\mathrm{B}}T\tau}{\hbar}\right)^{2/3}\frac{1.92}{\tau},$$
$$\overline{I_\Omega} \approx 11\left(\frac{\hbar}{k_{\mathrm{B}}T\tau}\right)^{1/3}\left(\frac{k_{\mathrm{B}}T}{\hbar v_{\mathrm{F}}}\right)^2\frac{\hbar\Omega}{\tau_{\mathrm{rad}}}, \quad (28.6)$$

and v_{F} is the Fermi velocity ($\sim 10^6$ m/s). As seen from Eq. 28.5, $\mathrm{Re}\sigma_\omega$ is a minimum at $\omega = \bar{\omega}$. When the pumping intensity exceeds the threshold $I_\Omega > \bar{I}_\Omega$, the dynamic conductivity becomes negative. When $T = 300$ K, $\tau = 10^{-12}$ s, and $\tau_{\mathrm{rad}} = 10^{-9} \sim 10^{-11}$ s, this threshold is

$\bar{I}_\Omega \approx 60 \sim 6000$ W/cm^2. Assuming a device size of 100 μm × 100 μm, we find that the pumping intensity required for negative dynamic conductivity is $\bar{I}_\Omega \approx 6 \sim 600$ mW. The value of 6 mW is a feasible pumping intensity (see Fig. 28.2).

When graphene is strongly pumped ($\varepsilon_F > k_B T$), Reσ_ω becomes [5]

$$\mathrm{Re}\,\sigma_\omega \cong \frac{e^2}{4\hbar}\left[-1+0.43\left(\frac{\bar{\omega}}{\omega}\right)^2\sqrt{\frac{I_\Omega}{\bar{I}_\Omega}}\right]. \tag{28.7}$$

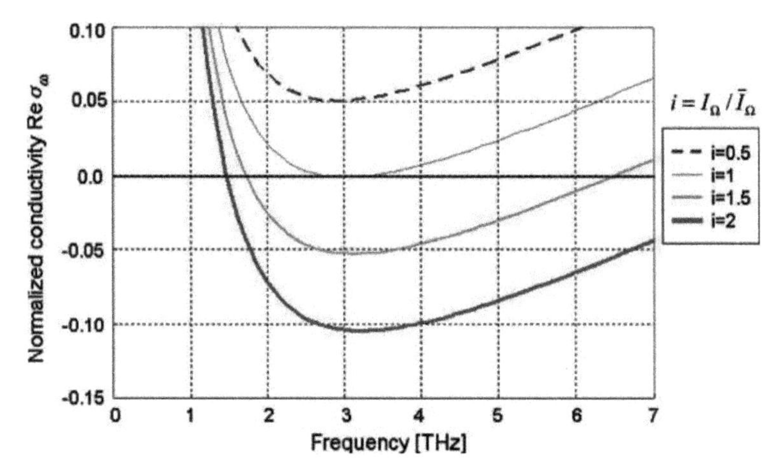

Figure 28.2 Calculated dynamic conductivity for various pumping intensities at 300 K. The vertical scale is normalized to the characteristic conductivity $e^2/4\hbar$.

Equation 28.7 shows that Re σ_ω changes from negative to positive at extremely high intensity; thus, at $\omega = \bar{\omega}$, for example, Re $\sigma_\omega < 0$ when $\bar{I}_\Omega < I_0 < 5\bar{I}_\Omega$ [5].

For the electron-hole recombination, radiative recombination via direct-transition to emit photons $\hbar\omega \approx 2\varepsilon_N$ and non-radiative recombination via Auger processes, plasmon emissions, and phonon emissions [11, 14–16] are considered. However, in the case of the radiative recombination, due to the relatively small values of $\hbar\omega \approx 2\varepsilon_N < 2\hbar\omega_0$, as well as the gapless energy spectrum, photon emissions over a wide THz frequency range are expected if the pumping photon energy is suitably chosen and the pumping intensity is sufficiently high. The incident photon spectra are expected to be reflected in the THz photoemission spectra as evidence that such a process occurs.

28.3 Graphene on Silicon Sample for the Experiment

In order to verify the proposed concept, we conduct an experimental study on the electromagnetic radiation emitted from an optically pumped graphene structure. The sample used in this experiment is heteroepitaxial graphene film grown on a 3C-SiC(110) thin film heteroepitaxially grown on a 300 μm thick Si(110) substrate via thermal graphitization of the SiC surface [17–19]. In the Raman spectrum of the graphene film, the principal bands of graphene, namely, the G (1595 cm^{-1}) and G′ (2730 cm^{-1}) bands, are observed, as shown in Fig. 28.3. Furthermore, transmission electron microscopy

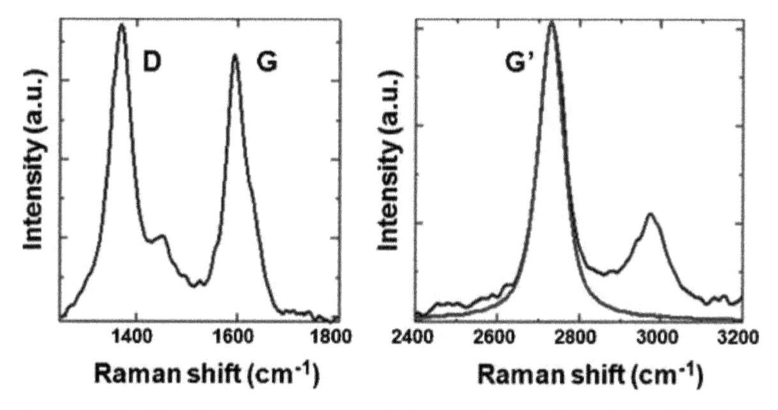

Figure 28.3 The Raman spectrum of the graphene, the principal bands of graphene, namely, D (1365 cm^{-1}), G (1595 cm^{-1}) and G′ (2730 cm^{-1}) bands. G band shows the structure of graphite and graphene. The peak at 1595 cm^{-1} corresponds to an optical phonon energy of 197.8 meV at the zone center. G′ band is expressed as a single component related to the two-dimensionality of the graphene film. The presence of D band indicates inclusion of defected graphite.

images indicate that the film is stratified. It is thus concluded that epitaxial graphene with a planar structure can be produced by this fabrication method. Furthermore, the epitaxial graphene layer is inferred to have a non-Bernal stacking arrangement because the G′ band in the Raman spectrum can be expressed as a single component related to the two-dimensionality of the graphene film [20, 21]. The non-Bernal stacked epitaxial graphene layers grown by our method can be treated as a set of isolated single graphene layers, as in the case of an epitaxial graphene layer on a C-terminated SiC bulk crystal [22]. The G-band peak at 1595 cm^{-1} corresponds to an optical phonon energy at the zone center of 197.8 meV.

28.4 Experimental

We measure the carrier relaxation and recombination dynamics in optically pumped epitaxial graphene-on-silicon (GOS) heterostructures using THz time-domain spectroscopy based on an optical pump/THz-and-optical-probe technique. The time-resolved field emission properties are measured by an electrooptic sampling method in total-reflection geometry [23]. To obtain the THz photon emissions from the above-mentioned carrier relaxation/recombination dynamics, the pumping photon energy (wavelength) is carefully selected to be around 800 meV (1550 nm). To perform intense pumping beyond the threshold, a femtosecond pulsed fiber laser with full width at half-maximum (FWHM) of 80 fs, pulse energy of 50 pJ/pulse, and frequency of 20 MHz was used as the pumping source. The setup is shown in Figs. 28.4a–c. The graphene sample is placed on the quartz base stage and a 80 μm thick (101)-oriented CdTe crystal is placed onto the sample; the CdTe crystal acts as a THz probe pulse emitter as well as an electrooptic sensor. The single femtosecond fiber laser beam is split into two beams: one for optical pumping and generating the THz probe beam, and the other for optical probing. The pumping laser, which is linearly polarized and mechanically chopped at ~1.2 KHz, is simultaneously focused at normal incidence onto the sample and the CdTe from below, while the probing laser, which is cross-polarized to the pumping beam, is focused from above. The incident pumping beam is defocused to about 50 μm in diameter on the sample to satisfy the above-mentioned pumping power requirements. The resulting photoexcited carrier density is $\sim 8 \times 10^{10}$ cm^{-2}, which is comparable to the background carrier density. Owing to nonlinear optical effects, the CdTe crystal can rectify the pumping laser pulse to emit THz envelope radiation. This THz emission irradiate the graphene sample, acting as THz probe signals to stimulate THz photon emission via electron-hole recombination in the GOS.

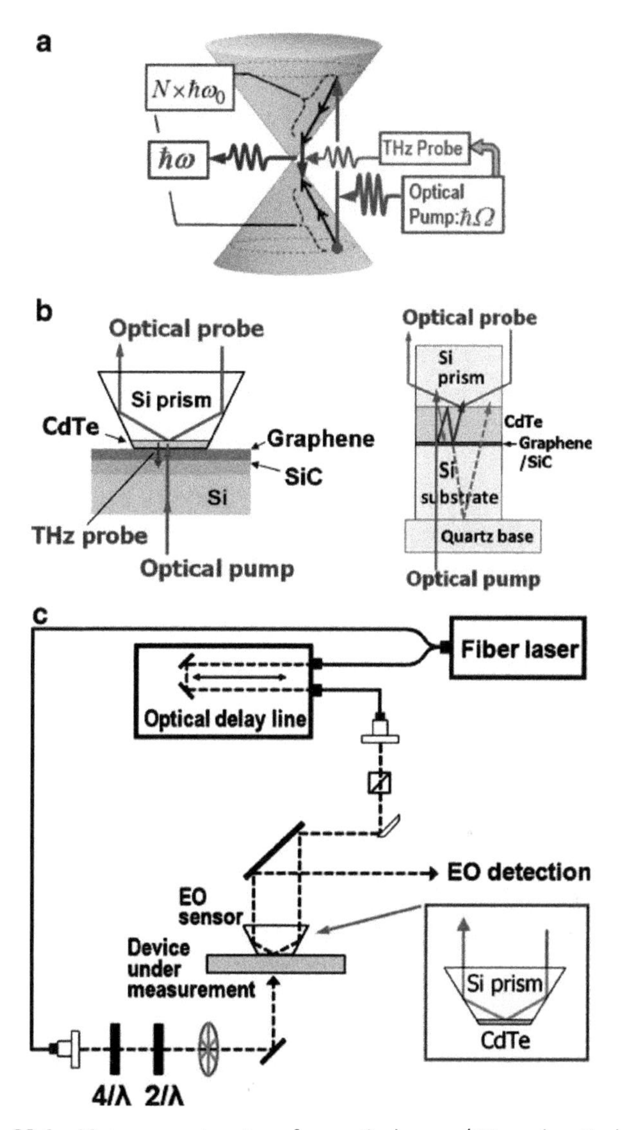

Figure 28.4 Measurement setup for optical-pump/THz-and-optical-probe spectroscopy. (a) Schematic image of stimulated THz emission from graphene by the optical-pump/THz-probe signals. (b) Pump/probe geometry. Left: CdTe crystal on top of the graphene sample generates THz probe pulse and allows electrooptic detection of THz electric field intensity. Right: optical pump beam (red line) generates the THz pulse making double reflection to work for the THz probe (blue line). The THz pulse also propagate through the GOS sample and reflect back (dashed blue line). (c) Electrooptic sampling setup.

The optical-pumping/THz-probing geometry and these beam propagations are schematically shown in Fig. 28.4b. The electric field intensity of the THz radiation is electroopticaly detected at the top surface of the CdTe crystal. Along with the propagation of the optical pump pulse through the CdTe crystal the THz pulse is generated and grows up. The THz pulse being emitted from the CdTe crystal is partially reflected at the top surface of the CdTe then subject back to the graphene, working as the THz probe pulse (shown with arrowed blue line in Fig. 28.4b). The graphene responds to this THz probe pulse radiation giving rise to stimulated THz emission. The emitted THz radiation reflects again at the heterointerface of the 80 nm thick SiC epilayer and the Si substrate, and then returns back to the top surface of the CdTe crystal. The propagation times of the THz probe pulse through the CdTe and graphene/SiC epilayer are 1000 fs and 0.75 fs, respectively (assuming the refractive index of CdTe and SiC in the THz frequencies to be 3.75 and 2.8, respectively). The total round-trip propagation time of the THz pulse through the CdTe crystal and graphene/SiC epilayer is 2001.5 fs (~2.0 ps). Therefore the original data of the temporal response that we observe in this experiment consist of the first forward propagating THz pulsation (no interaction with graphene) followed by the double reflected secondary THz pulsation (probing the graphene) just 2.0 ps after the first pulsation. The third THz pulsation comes ~6.6 ps after the second pulsation by reflecting at the bottom surface of the GOS as shown in Fig. 28.4b with dashed arrowed blue line, which is sufficiently separated in time from the second pulsation. We focus to observe the secondary THz pulsation that reflects the response of graphene.

The optical pump beam also partially reflects at the top surface of the CdTe crystal so that it may weakly pump the graphene. The time delay of the THz probing with respect to the first and second optical pumping is calculated to be 2.0 ps and 560 fs (assuming the infrared refractive index of the CdTe to be 2.69), respectively.

On the other hand, through the Si prism attached to the CdTe crystal, the optical probing beam is totally reflected back to the lock-in detection block, and its phase information reflecting the electric field intensity is lock-in amplified. By sweeping the timing of the optical probe using an optical delay line, the whole temporal profile of the field emission properties can be obtained. The system bandwidth is estimated to be around 6 THz, which is limited mainly by the Reststrahlen band of the CdTe sensor crystal.

Figure 28.5 shows the autocorrelations and spectral profiles of the pumping laser beam. The horizontal axis indicates the wavelength and frequency together with the estimated THz photon frequency to be emitted from the sample. The estimated THz photon spectrum $F_{\hbar\omega(=\Omega-2N\cdot\omega_0)}(\Omega - 2N \cdot \omega_0)$ is the replica of the pumping photon spectrum $F_{\hbar\Omega}(\Omega)$. The dotted line plots the dynamic conductivity at a pumping intensity twice as high as the threshold intensity calculated for 300 K using Eqs. 28.2–28.4 with a τ value of 2×10^{-12} ps. The shaded area shows the negative dynamic conductivity.

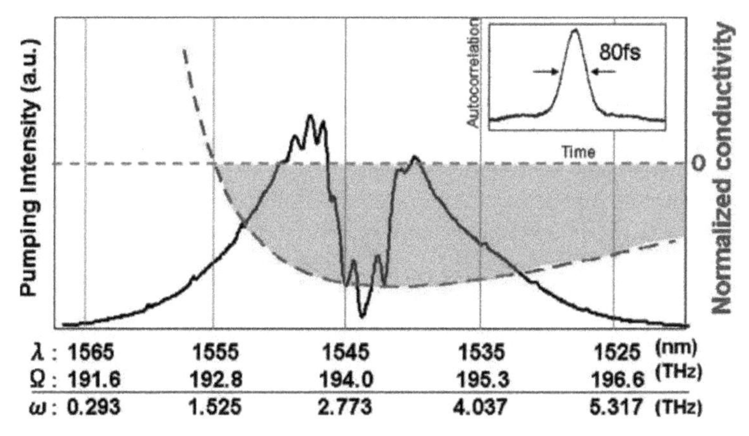

Figure 28.5 Temporal (autocorrelation) and spectral profiles of pumping laser pulse used in this experiment. Dotted line denotes the dynamic conductivity normalized to the characteristic conductivity $e^2/2\hbar$ at a pumping intensity of twice the threshold intensity at 300 K calculated using Eqs. 28.2–28.4. Shaded area denotes negative dynamic conductivity.

28.5 Results and Discussion

We observed THz emissions from the CdTe crystal only without the GOS sample and from the CdTe and the GOS sample. To confirm the effects of the THz probe, we replace the first CdTe crystal with another CdTe crystal having a high-reflectivity coating for IR on its bottom surface, in order to eliminate generation of the THz probe signal. In this case, no distinctive response is observed with or without graphene. Since the measurements are taken as an average, the observed response is undoubtedly a coherent process that cannot be obtained via spontaneous emission processes, providing clear evidence of stimulated emission.

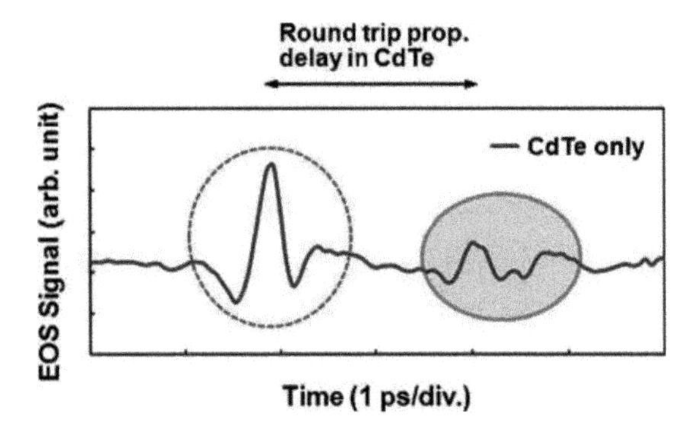

Figure 28.6 Typical measured temporal response of THz emission. The first pulsation (marked with a dashed circle line) is the direct propagating component, while the second pulsation (marked with a shaded solid circle line) is the double reflecting component interacting with graphene.

Typical raw data of observed temporal response is shown in Fig. 28.6. As is mentioned in IV, we extracted the temporal response of the secondary THz pulsation (designated with a shaded circle) from the measured raw data to identify the frequency response of the graphene. Figure 28.7a shows the measured temporal responses, in which repetitively measured results are overlaid, showing the measurement reproducibility. Figure 28.7b shows the Fourier spectra corresponding to the typical traces plotted with thick lines in Fig. 28.7a. Owing to the second-order nonlinear optical effects, the emission from "CdTe only" without the GOS sample exhibits a temporal response similar to optical rectification with a single peak at around 1 THz and an upper weak side lobe extending to around 5 THz as shown by the green lines in Fig. 28.7b. On the other hand, the temporal profile of the result from "CdTe and GOS" intensifies the pulsed response with higher frequency components so that its Fourier spectrum exhibits the growth of the main lobe around 2 to 4 THz and the lower side lobe around 1 THz. The main lobe fairly corresponds to the expected gain spectral region which is shown with the dashed line in Fig. 28.7b, while the side lobe fairly corresponds to the component of THz probe signal generated from the CdTe. It is inferred that the THz emissions from graphene are stimulated by the coherent THz probe radiation from the CdTe. Furthermore, the THz emissions are amplified via photoelectron/hole recombination in the range of the negative dynamic conductivity.

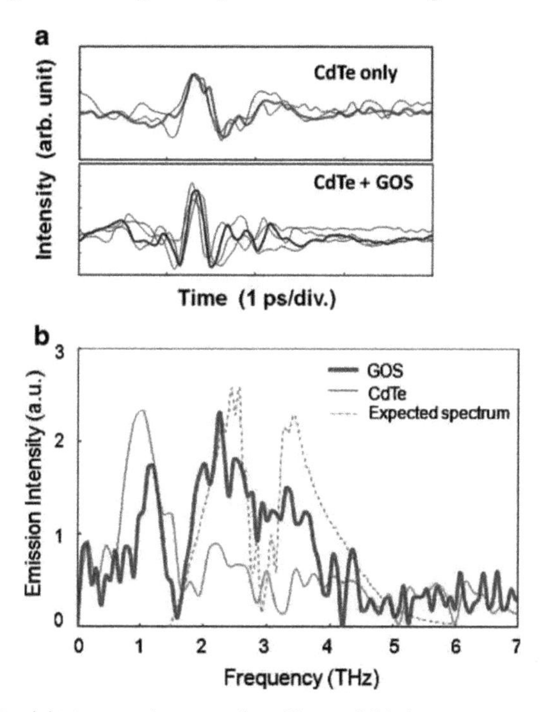

Figure 28.7 (a) Measured temporal profiles and (b) their Fourier spectra when the THz probe beam is generated by the optical rectification in the CdTe crystal. Dashed line in (b) is the photoemission spectrum predicted from the pumping laser spectrum and the negative dynamic conductivity.

The dashed line in Fig. 28.7b is the expected gain spectrum. The THz gain: g_ω is proportional to the negative dynamic conductivity $\mathrm{Re}\sigma_\omega(< 0)$ and is given by $g_\omega = 4\pi\mathrm{Re}\sigma_\omega/(c\sqrt{\varepsilon})\cdot\Gamma_\omega$, where Γ_ω is the gain overlap factor [24]. Therefore, the gain spectrum: $G(\omega)$ for the radiation of THz photon $\hbar\omega$ from graphene stimulated by the THz probe signal is calculated as the product of g_ω and the estimated THz photon spectrum $F_{\hbar(\Omega-2N\cdot\omega_0)}(\omega)$, the replica of the pumping photon spectrum: $G(\omega) = g(\omega) \cdot F_{\hbar(\Omega-2N\cdot\omega_0)}(\omega)$. The transfer function of the GOS response is obtained by the calculation of the "CdTe and GOS" spectrum divided by the "CdTe only" spectrum. If the calculated GOS transfer function corresponds to $G(\omega)$, the experimental results support the above mentioned assumption. The comparison between them is shown in Fig. 28.8. As is seen in Fig. 28.8, the experimental result fairly agrees with the calculated $G(\omega)$.

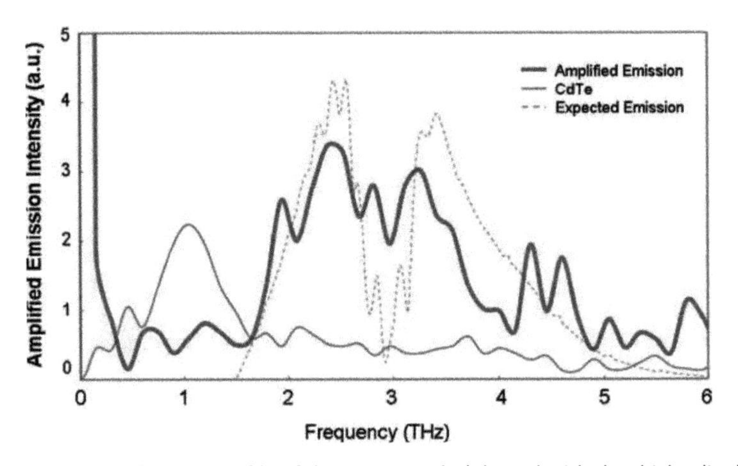

Figure 28.8 The gain profile of the GOS sample (plotted with the thicker line) which is calculated by the ratio of the emission spectrum of the "CdTe + GOS" to the spectrum of the "CdTe only". The expected spectrum (dashed line) and the emission spectrum from "CdTe only" (thinner line) are also plotted for comparison.

To see the Fourier spectrum shown in Figs. 28.7b and 28.8 in detail, its high-frequency part of the main lobe is weakened and lowered by ~0.5 THz with respect to the expected spectra. This result reflects the time-dependent energy relaxation of photoelectrons and photoholes in graphene as shown in Fig. 28.1b–d. As is mentioned in IV, the time duration from the optical pumping to the THz probing is 2.0 ps for the first pumping and 560 fs for the second weak pumping. Thus, it is considered that the photoelectrons/holes are stimulated when they start thermalization after losing their energy via the cascade of optical phonon emissions, but still stay in the nonequilibrium state with population inversion corresponding to Fig. 28.1b and d.

The THz probe timing with respect to the optical pump timing could be changed by using various CdTe crystals having different thickness. This will help further detail observation of the carrier relaxation/recombination dynamics. We should put a half-reflective coating at the surface (attached to the CdTe) of the Si prism like ITO to increase the S/N. These will be the future subject.

28.6 Conclusion

We measured the carrier relaxation and recombination dynamics in optically pumped epitaxial graphene-on-silicon (GOS) heterostructures using THz time-domain spectroscopy based on an optical pump/THz-and-optical-probe technique. We successfully observed coherent amplified stimulated THz emissions arising from the fast relaxation and relatively slow recombination dynamics of photogenerated electrons/holes in an epitaxial graphene heterostructures. The results provide evidence of the occurrence of negative dynamic conductivity, which can potentially be applied to a new type of THz laser.

Acknowledgments

The authors thank Yu Miyamoto and Hiroyuki Handa at RIEC, Tohoku University, Japan for their providing graphene-on-silicon samples, and also Maxim Ryzhii at University of Aizu for his valuable discussion. This work was financially supported in part by the JST-CREST program, Japan, and a Grant-in-Aid for Basic Research (S) from the Japan Society for the Promotion of Science.

References

1. A. K. Geim and K. S. Novoselov, *Nat. Mater.*, **6**, 183 (2007).
2. K. S. Novoselov, A. K. Geim, S. V. Morozov, D. Jiang, M. I. Katsnelson, I. V. Grigorieva, S. V. Dubonos, and A. Firsov, *Nature*, **438**, 197 (2005).
3. P. Kim, Y. Zhang, Y.-W. Tan, and H. L. Stormer, *Nature*, **438**, 201 (2005).
4. K. S. Novoselov, A. K. Geim, S. V. Morozov, D. Jiang, Y. Zhang, S. V. Dubonos, I. V. Grigorieva, and A. A. Frisov, *Science*, **306**, 666 (2004).

5. V. Ryzhii, M. Ryzhii, and T. Otsuji, *J. Appl. Phys.*, **101**, 083114 (2007).
6. L. A. Falkovsky and A. A. Varlamov, *Eur. Phys. J. B*, **56**, 281 (2007).
7. A. A. Dubinov, V. Y. Aleshkin, M. Ryzhii, T. Otsuji, and V. Ryzhii, *Appl. Phys. Express*, **2**, 092301 (2009).
8. V. Ryzhii, M. Ryzhii, A. Satou, T. Otsuji, A. A. Dubinov, and V. Y. Aleshkin, *J. Appl. Phys.*, **106**, 084507 (2009).
9. T. Kampfrath, L. Perfetti, F. Schapper, C. Frischkorn, and M. Wolf, *Phys. Rev. Lett.*, **95**, 187403 (2005).
10. H. Wang, J. H. Strait, P. A. George, S. Shivaraman, V. B. Shields, M. Chandrashekhar, J. Hwang, F. Rana, M. G. Spencer, C. S. Ruiz-Vargas, and J. Park, Arxiv 0909.4912 (2009).
11. P. A. George, J. Strait, J. Dawlaty, S. Shivaraman, M. Chandrashekhar, F. Rana, and M. G. Spencer, *Nano Lett.*, **8**, 4248 (2008).
12. M. Breusing, C. Ropers, and T. Elsaesser, *Phys. Rev. Lett.*, **102**, 086809 (2009).
13. By substituding Eqs. (9) and (12) and $\alpha = e^2/c\hbar$ (appearing just after Eq. (9)) into Eq. (13), all in [5], one can obtain the right-hand side of Eq. (28.3) in this chapter. Equation (14) in [5] corresponds to Eq. (28.4) in this chapter.
14. F. Rana, P. A. George, J. H. Strait, J. Dawlaty, S. Shivaraman, M. Chandrashekhar, and M. G. Spencer, *Phys. Rev. B*, **79**, 115447 (2009).
15. N. Bonini, M. Lazzeri, N. Marzari, and F. Mauri, *Phys. Rev. Lett.*, **99**, 176802 (2007).
16. F. Rana, *Phys. Rev. B*, **76**, 155431 (2007).
17. M. Suemitsu, Y. Miyamoto, H. Handa, and A. Konno, *e-J. Surf. Sci. Nanotechnol.*, **7**, 311 (2009).
18. Y. Miyamoto, H. Handa, E. Saito, A. Konno, Y. Narita, M. Suemitsu, H. Fukidome, T. Ito, K. Yasui, H. Nakazawa, and T. Endoh, *e-J. Surf. Sci. Nanotechnol.*, **7**, 107 (2009).
19. H. Fukidome, Y. Miyamoto, H. Handa, E. Saito, and M. Suemitsu, *Jpn. J. Appl. Phys.*, **49**, 01A1103 (2010).
20. L. G. Cançadoa, K. Takaia, T. Enokia, M. Endob, Y. A. Kimb, H. Mizusakib, N. L. Spezialic, A. Jorioc, and M. A. Pimentac, *Carbon*, **46**, 272 (2008).
21. C. Faugeras, A. Nerrière, M. Potemski, A. Mahmood, E. Dujardin, C. Berger, and W. A. de Heer, *Appl. Phys. Lett.*, **92**, 011914 (2008).

22. J. Hass, F. Varchon, J. E. Millán-Otoya, M. Sprinkle, N. Sharma, W. A. de Heer, C. Berger, P. N. First, L. Magaud, and E. H. Conrad, *Phys. Rev. Lett.*, **100**, 125504 (2008).

23. L. Min and R. J. D. Miller, *Appl. Phys. Lett.*, **56**, 524 (1999).

24. V. Ryzhii, A. A. Dubinov, T. Otsuji, V. Mitin, and M. S. Shur, *J. Appl. Phys.*, **107**, 054505 (2010).

Chapter 29

Toward the Creation of Terahertz Graphene Injection Laser*

V. Ryzhii,[a] M. Ryzhii,[a] V. Mitin,[b] and T. Otsuji[c]
[a]*Computational Nanoelectronics Laboratory, University of Aizu, Aizu-Wakamatsu 965-8580, Japan*
[b]*Department of Electrical Engineering, University at Buffalo, Buffalo, New York 14260-1920, USA*
[c]*Research Institute for Electrical Communication, Tohoku University, Sendai 980-8577, Japan*
v-ryzhii@u-aizu.ac.jp

We study the effect of population inversion associated with the electron and hole injection in graphene p-i-n structures at the room and slightly lower temperatures. It is assumed that the recombination and energy relaxation of electrons and holes are associated primarily with the interband and intraband processes assisted by optical phonons. The dependences of the electron-hole and optical phonon

*Reprinted with permission from V. Ryzhii, M. Ryzhii, V. Mitin, and T. Otsuji (2011). Toward the creation of terahertz graphene injection laser, *J. Appl. Phys.*, **110**, 094503. Copyright © 2011 American Institute of Physics.

Graphene-Based Terahertz Electronics and Plasmonics: Detector and Emitter Concepts
Edited by Vladimir Mitin, Taiichi Otsuji, and Victor Ryzhii

ISBN 978-981-4800-75-4 (Hardcover), 978-0-429-32839-8 (eBook)
www.jennystanford.com

effective temperatures on the applied voltage, the current-voltage characteristics, and the frequency-dependent dynamic conductivity are calculated. In particular, we demonstrate that at low and moderate voltages, the injection can lead to a pronounced cooling of the electron-hole plasma in the device i-section to the temperatures below the lattice temperature. However at higher voltages, the voltage dependences can be ambiguous exhibiting the *S*-shape. It is shown that the frequency-dependent dynamic conductivity can be negative in the terahertz (THz) range of frequencies at certain values of the applied voltage. The electron-hole plasma cooling substantially reinforces the effect of negative dynamic conductivity and promotes the realization of terahertz lasing. On the other hand, the heating of optical phonon system can also be crucial affecting the realization of negative dynamic conductivity and terahertz lasing at the room temperatures.

29.1 Introduction

The gapless energy spectrum of electrons and holes in graphene layers (GLs), graphene bilayers (GBLs), and non-Bernal stacked multiple graphene layers (MGLs) [1–3] opens up prospects of creating terahertz (THz) lasers based on these graphene structures. In such structures, GLs and MGLs with optical [4–10] and injection [11] pumping can exhibit the interband population inversion and negative dynamic conductivity in the THz range of frequencies and, hence, can serve as active media in THz lasers. The most direct way to create the interband population in GLs and MGLs is to use optical pumping [4] with the photon energy $\hbar\Omega_0$ corresponding to middle- and near-infrared (IR) ranges. In this case, the electrons and holes, photogenerated with the kinetic energy $\hbar Q_0/2$, transfer their energy to optical phonons and concentrate in the states near the Dirac point [4, 12, 13]. The amplification of THz radiation from optically pumped GL structures observed recently [14, 15] is attributed to the interband stimulated emission. However, the optical pumping with relatively high photon energies exhibits drawbacks. First of all, the optical pumping, which requires complex setups, might be inconvenient method in different applications of the prospective graphene THz lasers. Second, the excessive energy being received by the photogenerated electro-hole plasma from pumping source can lead to its marked heating because of the redistribution of the initial electron and hole energy $\hbar\Omega_0/2$ among all carries due to rather effective inter-carrier collisions. The latter results in a decrease of the ratio of the quasi-Fermi energies μ_e and μ_h to the electron-hole effective temperature T that, in turn, complicate achieving of sufficiently large values of the dynamic conductivity. As demonstrated recently [16, 17], the negative conductivity at the THz frequencies is very sensitive to the ratio of the photon energy $\hbar\Omega$ and the optical phonon energy $\hbar\omega_0$, as well as to the relative efficiency of the inter-carrier scattering and the carrier scattering on optical phonons. The decay of nonequilibrium optical phonons also plays an important role.

The abovementioned complications can be eliminated in the case of pumping resulting in the generation in GLs electrons and holes with relatively low initial energies. This in part can be realized in the case of optical pumping with $\hbar\Omega_0/2 < \hbar\omega_0$ [16]. Taking into account that in GLs $\hbar\omega_0 \simeq 0.2$ eV, in the case of CO_2 laser as a pumping source, $\Omega_0/\omega_0 \simeq 0.5$. As shown [16], in such a case, the electron-hole plasma can even be cooled, so that $T < T_0$, where T_0 is the lattice (thermostat) temperature. Another weakly heating or even cooling pumping method which can provide low effective temperature T (including $T < T_0$) is the injection pumping of electrons from n-section and holes from p-section in GL and MGL structures with p-i-n junctions.

In this chapter, we study the injection phenomena in GL and MGL p-i-n structures and calculate their characteristics important for THz lasers. The idea to use p-n junctions in GLs was put forward and briefly discussed by us previously [11]. Here, we consider more optimal designs of the structures (with a sufficiently long i-section) and account for realistic mechanisms of recombination at elevated temperatures (at the room temperature and slightly below).

This chapter is organized as follows. In Section 29.2, we describe the device structures under consideration and principles of their operation. The pertinent equations of the model governing the balance of electrons, holes, and optical phonons (rate equations) are presented in Section 29.3. These equations are reduced to an equation governing the electron-hole effective temperature. The solution of this equation in Sections 29.4 and 29.5 (both analytically in limiting cases and numerically) allows us to find the effective temperature of optical phonons and the current as functions of the applied voltage, the structural parameters, and the lattice temperature. In Section 29.6, the obtained characteristics of the injected electron-hole plasma are used to calculate the dependence of the dynamic conductivity of the latter as a function of the signal THz frequency and other quantities. Section 29.7 deals with the model limitations and discussion. In Section 29.8, we draw the main conclusions.

29.2 Device Model

We consider devices which comprise a GL or a MGL structure with several non-Bernal stacked GLs. It is assumed that the sections of GLs adjacent to the side contacts are doped (p- and n-sections). The device structure under consideration is shown in Fig. 29.1a. The dc voltage V is applied between the side contacts to provide the forward bias of the p-i-n junction. Due to doping of the side sections with the acceptor and donor sheet concentration Σ_i, the electron and hole Fermi energies counted from the Dirac point are $\mu_e = \mu_i$ and $\mu_h = -\mu_i$, where $\mu_i \simeq \hbar v_F \sqrt{\pi \Sigma_i}$, $\hbar$ is the reduced Planck constant and $v_F \simeq 10^8$ cm/s is the characteristic velocity of the carrier spectrum in GLs. Instead of doping of the side sections, the p- and n-sections can be created using highly conducting gates over these sections to which the gate voltages $V_p = -V_g < 0$ and $V_n = V_g > 0$ are applied [11]. In this case, the chemically doped p- and n-sections are replaced by the electrically induced sections (see, for instance [18, 19]), as shown in Fig. 292b. In single-GL structures, $\mu_i \propto \sqrt{V_g / W_g}$, where W_g is the spacing between the GL and the gate. In the case of MGL structures, the situation becomes more complex due to the screening of the transverse electric field in GLs (Ref. [19]) although the effective electric doping can be achieved in MGL structures with about dozen GLs. For definiteness, in the following, we shall consider primarily the devices with chemically doped p-i-n junctions. As shown below, under certain conditions, the frequency-dependent dynamic conductivity of the GL or MGL structures can be negative in a certain range of the signal frequencies. In this case, the self-excitation of THz modes propagating in the substrate serving as a dielectric waveguide (in the direction perpendicular to the injection current) and lasing is possible. The metal gates in the devices with electrically induced p-i and i-n junctions can also serve as the slot-line waveguides for

THz waves. Apart from this, the self-excitation of surface plasmons (plasmon-polaritons) is possible as well (see, for instance, Refs. [4, 6, 8–10]).

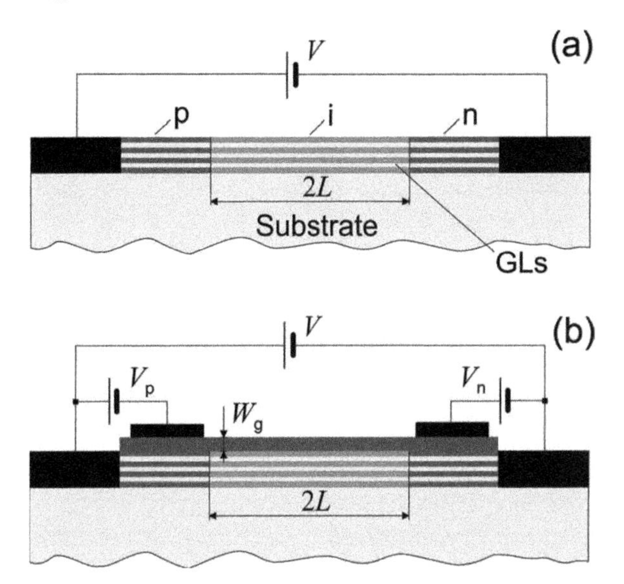

Figure 29.1 Schematic view of the cross-sections of MGL laser structures (a) with chemically doped n- and p-sections and (b) with such section electrically induced by the side gate-voltages $V_p = -V_g$ and $V_n = V_g > 0$.

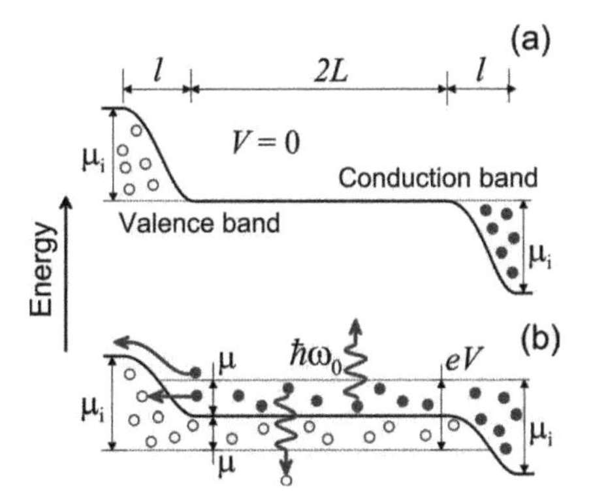

Figure 29.2 Band profiles of a GL in p-i-n junctions (a) at $V = 0$ and (b) at forward bias $V > 0$. Opaque and open circles correspond to electrons and holes, respectively. Wavy arrows show some interband (recombination) and intraband transitions assisted by optical phonons. Smooth arrows indicate tunneling and thermionic leakage processes.

When the p-i-n junction under consideration is forward biased by the applied voltage V, the electrons and holes are injected to the i-section from the pertinent doped side sections. The injected electrons and holes reaching the opposite doped section can recombine at it due to the interband tunneling or escape the i-section due to the thermionic processes. The band profiles in the structures under consideration at $V = 0$ and at the forward bias $V > 0$ are shown in Fig. 29.2. Since the probability of such tunneling is a very sharp function of angle of incidence, the leakage flux due to the tunneling electrons (holes) is much smaller that the flux of injected electrons (holes). The currents associated with the tunneling and thermionic leakage of electrons at the p-i-junction and holes at the i-n-junction depend on the electric field at the pertinent barriers and the applied voltage V. Because the relative role of the leakage currents diminishes with increasing length of the i-section $2L$, width of the p-i- and i-n-junctions l, and the barrier height at these junctions μ_i, we shall neglect it. The pertinent conditions will be discussed in the following. Thus, it is assumed that the main fractions of the injected electrons and holes recombine inside the i-section. The recombination of electrons and holes in GLs at not too low temperatures is mainly determined by the emission of optical photons [20]. Considering the sub-threshold characteristics (i.e., the states below the threshold of lasing) and focusing on the relatively high-temperature operation, we shall account for this recombination mechanism and disregard others [21–24] including the mechanism [25] associated with the tunneling between the electron-hole puddles (if any) [26–29]. Due to high net electron and hole densities in MGL structures with sufficient number of GLs, the latter mechanism can be effectively suppressed [25]. We also assume that the net recombination rate in the whole i-section is much smaller than the fluxes of injected electrons and holes.

29.3 Equations of the Model

Due to rather effective inter-carrier scattering, the electron and hole distribution functions (at least at not too high energies) can be very close to the Fermi distribution functions with quasi-Fermi energies μ_e and μ_h and the electron-hole effective temperature T. The latter quantities are generally different from those in equilibrium (without pumping) at which $\mu_e = \mu_h = 0$ and $T = T_0$. At the pumping of an intrinsic GL structure, $\mu_e = -\mu_h = \mu$, where generally $\mu > 0$. Under these conditions, the quasi-Fermi energy μ in the i-section (neglecting the leakage and recombination in the lowest approximation), is given by (see Fig. 29.2b)

$$\mu = eV/2, \tag{29.1}$$

where e is the electron charge.

The terminal current between the side contacts (per unit length in the lateral direction perpendicular to the current), which coincides with the recombination current, is given by

$$J = 2eLR_0^{\text{inter}} \tag{29.2}$$

The rate of the optical phonon-assisted interband transitions (recombination rate) R_0^{inter} and the rate of the intraband energy relaxation associated with optical phonons R_0^{intra} can be calculated using the following simplified formulas [9, 16] (see, also Ref. [20]):

$$\begin{aligned} R_0^{\text{inter}} &= \frac{\Sigma_0}{\tau_0^{\text{inter}}}\left[(\mathcal{N}_0+1)\exp\left(\frac{2\mu-\hbar\omega_0}{T}\right)-\mathcal{N}_0\right] \\ &= \frac{\Sigma_0}{\tau_0^{\text{inter}}}\left[(\mathcal{N}_0+1)\exp\left(\frac{eV-\hbar\omega_0}{T}\right)-\mathcal{N}_0\right], \end{aligned} \tag{29.3}$$

$$R_0^{\text{intra}} = \frac{\Sigma_0}{\tau_0^{\text{intra}}}\left[(\mathcal{N}_0+1)\exp\left(-\frac{\hbar\omega_0}{T}\right)-\mathcal{N}_0\right]. \tag{29.4}$$

Here, τ_0^{inter} and τ_0^{intra} are the pertinent characteristic times (relatively slow dependent on μ and T), Σ_0 is the equilibrium electron and hole density, and $\mathcal{N}_0$ is the number of optical phonons. Here and in all equations in the following, T and T_0 are in the energy units. When the optical phonon system is close to equilibrium, one can put $\mathcal{N}_0 = [\exp(\hbar\omega_0/T_0) - 1]^{-1} = \mathcal{N}_0^{\text{eq}}$. For numerical estimates, we set $R_0^{\text{inter}} = \Sigma_0/\tau_0^{\text{inter}} \simeq 10^{23}$ cm^{-2}s^{-1} [20]. Equations (29.2) and (29.3) yield the following general formula for the structure current-voltage characteristic:

$$J = \frac{2eL\Sigma_0}{\tau_0^{\text{inter}}}\left[(\mathcal{N}_0+1)\exp\left(\frac{eV-\hbar\omega_0}{T}\right)-\mathcal{N}_0\right] \tag{29.5}$$

Naturally, at $V = 0$, $T = T_0$, so that $\mathcal{N}_0 = [\exp(\hbar\omega_0/T_0)-1]^{-1}$ and $J = 0$. At $V > 0$, due to contributions of the recombination and injection to the energy balance of the electro-hole plasma in the i-section, the electron-hole effective temperature T can deviate from the lattice temperature T_0. The number of optical phonons $\mathcal{N}_0$ can also be different from its equilibrium value $\mathcal{N}_0^{\mathrm{eq}}$. Since $\hbar\omega_0$ is large, in a wide range of temperatures (including the room temperatures), $\hbar\omega_0 \gg T_0$.

The electron-hole plasma gives up the energy $\hbar\omega_0$ in each act of the optical phonon emission (interband and intraband) and receives the same energy absorbing an optical phonon. Hence, the net rate of the energy transfer from and to the electron hole-plasma due to the interaction with optical phonons is equal to $2L\hbar\omega_0(R_0^{\mathrm{inter}} + R_0^{\mathrm{intra}})$. Considering Eqs. (29.3) and (29.4) and taking into account that the Joule power associated with the injection current is equal to $Q = JV = 2eLR_0^{\mathrm{inter}}V$, an equation governing the energy balance in the electron-hole plasma in the i-section can be presented as

$$\frac{eV}{\tau_0^{\mathrm{inter}}}\left[(\mathcal{N}_0+1)\exp\left(\frac{eV-\hbar\omega_0}{T}\right)-\mathcal{N}_0\right]$$
$$=\frac{\hbar\omega_0}{\tau_0^{\mathrm{inter}}}\left[(\mathcal{N}_0+1)\exp\left(\frac{eV-\hbar\omega_0}{T}\right)-\mathcal{N}_0\right]$$
$$+\frac{\hbar\omega_0}{\tau_0^{\mathrm{intra}}}\left[(\mathrm{N}_0+1)\exp\left(-\frac{\hbar\omega_0}{T}\right)-\mathrm{N}_0\right]. \qquad (29.6)$$

Here, the left-hand side corresponds to the power received by the electron-hole plasma in the i-section from the pumping source, whereas the right-hand side corresponds to the power transferred to or received from the optical phonon system.

The number of optical phonons is governed by an equation which describes the balance between their generation in the interband and intraband transitions and decay due to the anharmonic contributions to the interatomic potential, leading to the phonon-phonon scattering and in the decay of optical phonons into acoustic phonons. This equation can be presented in the form

$$\frac{(\mathcal{N}_0 - \mathcal{N}_0^{\rm eq})}{\tau_0^{\rm decay}} = \frac{1}{\tau_0^{\rm inter}}\left[(\mathcal{N}_0 + 1)\exp\left(\frac{eV - \hbar\omega_0}{T}\right) - \mathcal{N}_0\right]$$
$$+ \frac{1}{\tau_0^{\rm intra}}\left[(\mathcal{N}_0 + 1)\exp\left(-\frac{\hbar\omega_0}{T}\right) - \mathcal{N}_0\right], \qquad (29.7)$$

where $\tau_0^{\rm decay}$ is the optical phonon decay time. This time can be markedly longer than $\tau_0^{\rm inter}$ and $\tau_0^{\rm intra}$, particularly in suspended GLs, so that parameter $\eta_0^{\rm decay} = \tau_0^{\rm decay}/\tau_0^{\rm inter}$ can exceed or substantially exceed unity. As shown [30–34], $\tau_0^{\rm decay}$ in GLs is in the range of 1–10 ps. As calculated recently [17], the characteristic times $\tau_0^{\rm inter}$ and $\tau_0^{\rm intra}$ can be longer than 1 ps. If so, the situation when $\eta_0^{\rm decay} < 1$ appears also to be feasible. The optical phonon decay time might be fairly short depending on the type of the substrate.

Instead of Eq. (29.7), one can use the following equation which explicitly reflexes the fact that the energy received by the electron-hole plasma from the external voltage source goes eventually to the optical phonon system

$$\eta_0^{\rm decay}\frac{eV}{\hbar\omega_0}\left[(\mathrm{N}_0 + 1)\exp\left(\frac{eV - \hbar\omega_0}{T}\right) - \mathrm{N}_0\right] = \mathrm{N}_0 - \mathrm{N}_0^{\rm eq}. \qquad (29.8)$$

Using Eqs. (29.6) and (29.7) or Eqs. (29.6) and (29.8), one can find T and $\mathcal{N}_0$ as functions of V and then calculate the current-voltage characteristic invoking Eq. (29.5), as well as the dynamic characteristics.

Equation (29.8) yields

$$\mathcal{N}_0 = \frac{\mathcal{N}_0^{\rm eq} + \eta_0^{\rm decay}\dfrac{eV}{\hbar\omega_0}\exp\left(\dfrac{eV - \hbar\omega_0}{T}\right)}{1 + \eta_0^{\rm decay}\dfrac{eV}{\hbar\omega_0}\left[1 - \exp\left(\dfrac{eV - \hbar\omega_0}{T}\right)\right]}. \qquad (29.9)$$

Substituting $\mathcal{N}_0$ given by Eqs. (29.9) to (29.6), we arrive at the following equation for *T*:

$$\frac{1 + \eta_0^{\rm dexay}\dfrac{eV}{\hbar\omega_0}\left[1 - \exp\left(\dfrac{eV - \hbar\omega_0}{T}\right)\right]}{\mathcal{N}_0^{\rm eq} + \eta_0^{\rm decay}\dfrac{eV}{\hbar\omega_0}\exp\left(\dfrac{eV - \hbar\omega_0}{T}\right)}$$

$$\times\left[\eta_0\frac{(\hbar\omega_0 - eV)}{\hbar\omega_0}\exp\left(\frac{eV}{T}\right)+1\right]$$

$$+\eta_0\frac{(\hbar\omega_0 - eV)}{\hbar\omega_0}\left[\exp\left(\frac{eV-\hbar\omega_0}{T}\right)-1\right]\exp\left(\frac{\hbar\omega_0}{T}\right)$$

$$-\exp\left(\frac{\hbar\omega_0}{T}\right)+1=0. \tag{29.10}$$

The ratio $\eta_0 = \tau_0^{\text{intra}}/\tau_0^{\text{inter}}$ is actually a function of μ and T. The $\eta_0 - \mu$ and $\eta_0 - T$ dependences are associated with the linearity of the density of states in GLs as a function of energy. To a good approximation, these dependences can be described [16] by function $\eta_0 = \hbar^2\omega_0^2/(6\mu^2+\pi^2T^2)$ with $\eta_0 \propto (\hbar\omega_0/T)^2$ at $\mu \ll T$ and $\eta_0 \propto (\hbar\omega_0/\mu)^2$ at $\hbar\omega_0 > \mu \gg T$. Thus, considering Eq. (29.1), η_0 in Eq. (29.10) is given by

$$\eta_0 = \frac{2\hbar^2\omega_0^2}{(3e^2V^2+2\pi^2T^2)}. \tag{29.11}$$

Introducing the effective temperature of the optical phonon system Θ such that $\mathcal{N}_0 = [\exp(\hbar\omega_0/\Theta) - 1]^{-1}$, i.e.,

$$\Theta = \frac{\hbar\omega_0}{\ln(1+\mathcal{N}_0^{-1})}, \tag{29.12}$$

and substituting $\mathcal{N}_0$ from Eq. (29.9) to Eq. (29.11), one can relate Θ and T. Then, calculating the T–V dependences using Eq. (29.10), one can find the pertinent $\Theta - V$ dependences.

29.4 Effective Temperatures and Current-Voltage Characteristics (Analytical Analysis)

29.4.1 Low Voltages

In particular, at $V = 0$, Eqs. (29.8) and (29.10) naturally yield $\mathcal{N}_0 = \mathcal{N}_0^{\text{eq}}$ and $T = T_0$. At sufficiently low voltages when $(\eta_0^{\text{decay}}eV/\hbar\omega_0) \ll 1$, the solutions of Eqs. (29.9) and (29.10) can be found analytically. In this case, $\mathcal{N}_0 \simeq \mathcal{N}_0^{\text{eq}}$. Considering this, at low voltages, Eq. (29.10) yields

$$T \simeq T_0 \left[1 - \frac{eV}{\hbar\omega_0} \frac{\eta_0^{\mathrm{eq}}}{(1+\eta_0^{\mathrm{eq}})} \right] < T_0, \tag{29.13}$$

where η_0^{eq} is the value of η_0 at $V \ll \hbar\omega_0/e$, i.e., $\eta_0^{\mathrm{eq}} \simeq 5$ [16]. As follows from Eq. (29.13), an increase in the applied voltage V leads to a decrease in the effective temperature of the electron-hole plasma (its cooling).

Using Eqs. (29.5) and (29.13) at low voltages, we also obtain

$$\frac{J}{J_0} \simeq \frac{eV}{T_0} \frac{\eta_0^{\mathrm{eq}}}{(1+\eta_0^{\mathrm{eq}})} \exp\left(-\frac{\hbar\omega_0}{T_0} \right). \tag{29.14}$$

Here, the current J is normalized by its characteristic value $J_0 = 2eL\Sigma_0/\tau_0^{\mathrm{inter}} = 2eL\overline{R}^{\mathrm{inter}}$.

29.4.2 Special Cases

In the special case $V = \hbar\omega_0/e$ (i.e., $V \simeq 0.2$ V), from Eqs. (29.9), (29.10), and (29.12), we obtain

$$T = \frac{\hbar\omega_0}{\ln\left(\dfrac{1+\eta_0^{\mathrm{decay}} + \mathcal{N}_0^{\mathrm{eq}}}{\eta_0^{\mathrm{decay}} + \mathcal{N}_0^{\mathrm{eq}}} \right)}, \tag{29.15}$$

$$\mathcal{N}_0 = \mathcal{N}_0^{\mathrm{eq}} + \eta_0^{\mathrm{decay}}. \tag{29.16}$$

One can see that in this case

$$\mathcal{N}_0 = \frac{1}{\exp(\hbar\omega_0/T) - 1}, \tag{29.17}$$

i.e., $\Theta = T$ and

$$J = J_0. \tag{29.18}$$

At $T_0 = 300$ K and $2L = 20$ µm, one obtains $J_0 \simeq 32$ A/cm. As follows from Eqs. (29.15)–(29.17) at $V = \hbar\omega_0/e$, T and Θ tend to T_0 if η_0^{decay} tends to zero and $T = \Theta$, and they both increase proportionally to η_0^{decay} as η_0^{decay} tends to infinity. Thus, at $V = \hbar\omega_0/e$ and $\eta_0^{\mathrm{decay}} \sim 1$, the effective temperatures are fairly high: $T = \Theta \sim \hbar\omega_0/\ln 2$ ($T = \Theta \sim 3300$ K). It is worth noting that J at $V = \hbar\omega_0/e$ is independent of parameter η_0^{decay}.

In interesting (but nonrealistic) limiting case $\eta_0^{\rm decay} = 0$, from Eq. (29.8), we immediately obtain $\mathcal{N}_0 = \mathcal{N}_0^{\rm eq}$. In such a case, $T = T_0$ both at $V = 0$ and $V = \hbar\omega_0/e$. If $V = \hbar\omega_0/2e$, one obtains

$$T \simeq \frac{T_0}{2[1-(T_0/\hbar\omega_0)\ln(2+\eta_0)]}, \tag{29.19}$$

$$J \simeq J_0(\eta_0+1)\exp\left(-\frac{\hbar\omega_0}{T_0}\right) << J_0. \tag{29.20}$$

At $T_0 = 300$ K, Eqs. (29.19) and (29.20) yield $T \simeq 0.60$, $T_0 = 180$ K and $J \simeq 31$ mA/cm at $T_0 = 300$ K and $T \simeq 0.569$, $T_0 = 114$ K and 0.64 mA/cm at $T_0 = 200$ K.

29.4.3 Long Optical Phonon Decay Time

In the case of relatively long optical decay time, when $\eta_0^{\rm decay} >> 1$ at $V \lesssim \hbar\omega_0/e$, neglecting terms of the order of $\mathcal{N}_0^{\rm eq}/\eta_0^{\rm decay} \simeq \exp(-\hbar\omega_0/T_0)/\eta_0^{\rm decay}$, from Eq. (29.10), we obtain

$$T \simeq \frac{eV}{\ln\left\{\dfrac{1+\eta_0^{\rm decay}(eV/\hbar\omega_0)}{\eta_0^{\rm decay}(eV/\hbar\omega_0)+\eta_0[(eV/\hbar\omega_0)-1]}\right\}}. \tag{29.21}$$

At $V = \hbar\omega_0/e$, Eq. (29.21) yields the same value of T as Eq. (29.15) provided $\eta_0^{\rm decay} \gg \mathcal{N}_0^{\rm eq}$.

29.5 Effective Temperatures and Current-Voltage Characteristics (Numerical Results)

To obtain T–V and Θ–V dependences in wide ranges of parameter $\eta_0^{\rm decay}$ and the applied voltage V, Eqs. (29.9)–(29.12) were solved numerically. Figures 29.3 and 29.4 show the voltage dependences of effective temperatures T and Θ calculated for different values $\eta_0^{\rm decay}$ and $T_0 = 300$ and 200 K. One can see from Fig. 29.3 that the electron-hole effective temperature markedly decreases with increasing voltage, so that $T < T_0$ in a certain voltage range [see also Eq. (29.13)] and then starts to rise. In the range of relatively high voltages ($V \lesssim \hbar\omega_0/e$), the T–V dependence is steeply rising [in line with analytical

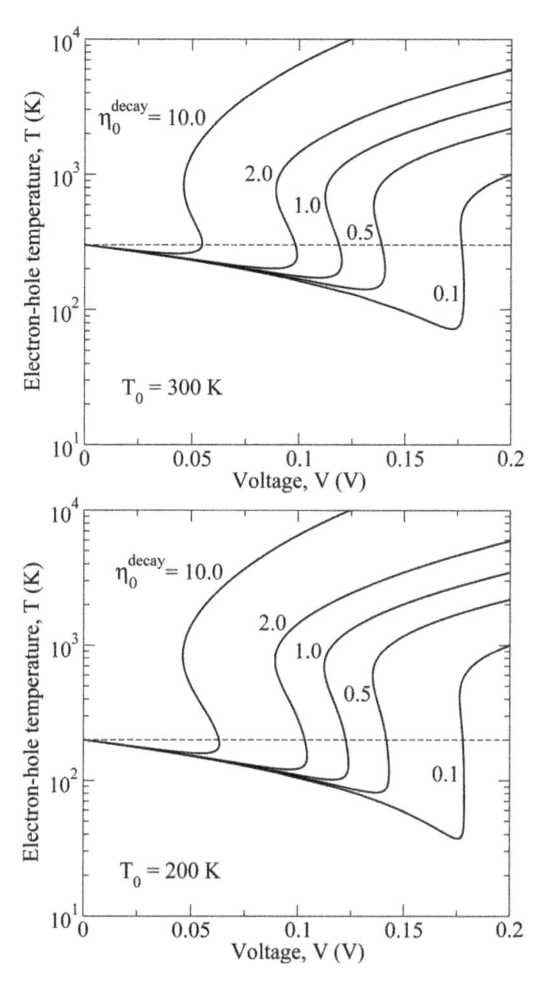

Figure 29.3 Electron-hole effective temperature T as function of applied voltage V for different values of η_0^{decay} and T_0. Dashed lines correspond to T = 300 K (upper panel) and T = 200 K (lower panel).

formula given by Eq. (29.21)] with $T > T_0$ or even $T \gg T_0$. However, it is intriguing that in a rather narrow voltage range where V is about some value V_d, the T–V dependences are ambiguous, so that these dependences as a whole are of the S-shape. The appearance of the S-shape characteristics can be attributed to a decrease in parameter η_0 with increasing T [see Eq. (29.11)]. This corresponds to a decrease in τ_0^{intra} and, hence, to an essential intensification of the intraband transitions, particularly, those associated with the reabsorption of nonequilibrium optical phonons when T increases. This is because at high electron-hole effective temperatures, the intraband transitions assisted by optical phonons take place between relatively high energy states with their elevated density. When V exceeds some "disruption" voltage V_d, the net power acquired by the electron-hole plasma can be compensated by the intraband energy relaxation on optical phonons only at sufficiently high T. As a result, in this case, the electron-hole temperature jumps to the values corresponding to higher branch of the T–V dependence. Thus, the "observable" T–V dependences and their consequences can as usual exhibit hysteresis instead of the S-behavior. One needs to point out that if the above temperature-dependent parameter η_0 is replaced in calculations by a constant, the calculated T–V dependences become unambiguous, although they exhibit a steep increase in the range $V \sim V_d$.

The behavior of T as a function of V markedly depends on parameters η_0^{decay} and η_0. The width of the voltage range where $T < T_0$ increases when parameter η_0^{decay} becomes smaller with increasing voltage. Simultaneously, the depth of the T–V sag dependent with $T < T_0$ increases with decreasing η_0^{decay} as well as with decreasing T_0. At small η_0^{decay}, the electron-hole cooling can be rather strong, particularly when T_0 = 200 K. This is natural because faster decay of optical phonons prevents their accumulation (heating) and promotes the electron-hole plasma cooling when the Joule power is smaller than the power transferred from electrons and holes to optical phonons. It worth noting that the voltage range where the T–V dependence is ambiguous widens with increasing η_0^{decay}.

As seen from Fig. 29.4, the optical phonon effective temperature also exhibits a S-shape voltage dependence. However, contrary to the electron-hole effective temperature, $\Theta \geq T_0$ at all the voltages is under consideration. The values of Θ at relatively high voltages steeply increase with increasing parameter η_0^{decay}.

Comparing the T–V dependences calculated for different lattice temperatures, one can find that at moderate and large values of η_0^{decay}, these dependences are virtually independent of T_0. This is because in such a case, the number of optical phonons $\mathcal{N}_0 \gg \mathcal{N}_0^{\text{eq}}$ and, hence, $\Theta \gg T_0$ even at not too high voltages, so that the role of equilibrium optical phonons is weak.

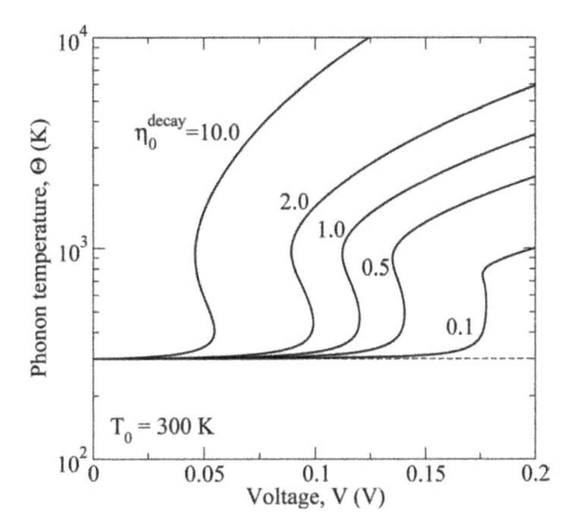

Figure 29.4 Optical phonon effective temperature Θ as function of applied voltage V for different values of η_0^{decay}. Dashed line corresponds to Θ = 300 K.

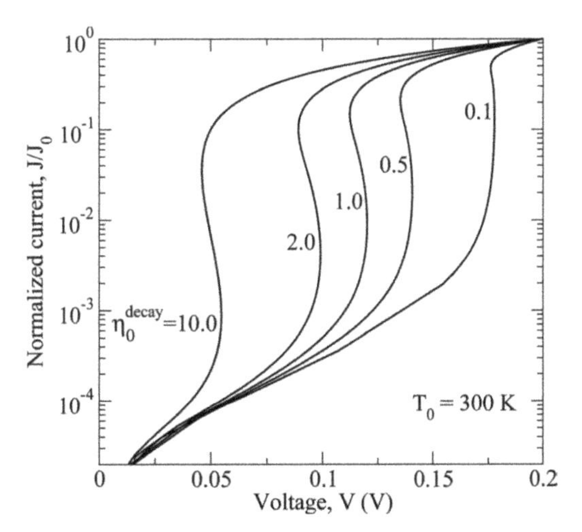

Figure 29.5 Normalized current-voltage characteristics for different values of η_0^{decay}.

Invoking Eq. (29.5), the T–V dependences obtained above can be used to find the current-voltage characteristics. Figure 29.5 shows the J–V characteristics calculated using Eq. (29.5) and the T–V and Θ–V dependences obtained numerically. As a consequence of the S-shape T–V and 0–V dependences, the J–V characteristics (as well as the voltage dependences of the dynamic conductivity considered in the following) are also of the S-shape. According to Figs. 29.3 and 29.5, the T–V and J–V characteristics in the range of low and moderate voltages are independent of parameter η_0^{decay}. This is in line with the results of the previous analytical analysis [see Eqs. (29.13) and (29.14)]. However, at relatively high voltages, distinctions in the J–V characteristics for different η_0^{decay} is significant although all of them tend to J_0 when V approaches to $\hbar\omega_0/e$.

29.6 Dynamic Conductivity

Knowing the T–V dependences, one can calculate the dynamic conductivity, σ_ω, of a GL under the injection pumping as a function of the signal frequency ω and the applied voltage V. To achieve lasing at the frequency ω, the real part of the complex dynamic conductivity at this frequency should be negative: $\text{Re}\sigma_\omega < 0$. As shown previously (see, for instance, Refs. [4, 35]), the interband contribution of the nonequilibrium electron-hole plasma with the quasi-Fermi energy μ and the effective temperature T is proportional to $\tanh[(\hbar\omega - 2\mu)/4T]$. The intraband contribution to $\text{Re}\sigma_\omega$, which corresponds to the Drude absorption, depends on ω, μ, and T as well. It also depends on the time of electron and hole momentum relaxation on impurities and phonons s. The latter is a function of the energy of electrons and holes ε. The main reason for the $\tau - \varepsilon$ dependence is a linear increase in the density of state in GLs with increasing ε. In this case, $\tau^{-1} = \nu_0(\varepsilon/T_0)$, where ν_0 is the collision frequency of electrons and holes in equilibrium at $T = T_0$. Considering this and taking into account Eq. (29.1), we can arrive at the following formula [16] approximately valid in the frequency range $\omega \gg \nu_0$:

$$\frac{\text{Re}\sigma_\omega}{\sigma_0} = \tanh\left(\frac{\hbar\omega - eV}{4T}\right) + C\frac{(e^2V^2 + 2\pi^2T^2/3)}{\hbar^2\omega^2}. \quad (29.22)$$

Here, $\sigma_0 = e^2/4\hbar$ and $C = 2\hbar\nu_0/\pi T_0$. In high quality MGLs (with $\nu_0^{-1} \simeq 20$ ps at $T_0 = 50$ K (Refs. [2, 3]), assuming that $\nu_0 \propto T_0$ (Ref. [36]) for $T_0 = 300$ K, one can set $\nu_0 \simeq 3 \times 10^{11}$ s^{-1} and hence, $C \simeq 0.005$. For substantially less perfect GLs with $\nu_0 \simeq 15 \times 10^{11}$ s^{-1}, one obtains $C \simeq 0.025$. These data are used for the calculations of $\text{Re}\sigma_\omega$.

Figure 29.6 demonstrates the frequency dependences of $\text{Re}\sigma_\omega/\sigma_0$ calculated using Eq. (29.22) for different values of parameter η_0^{decay}

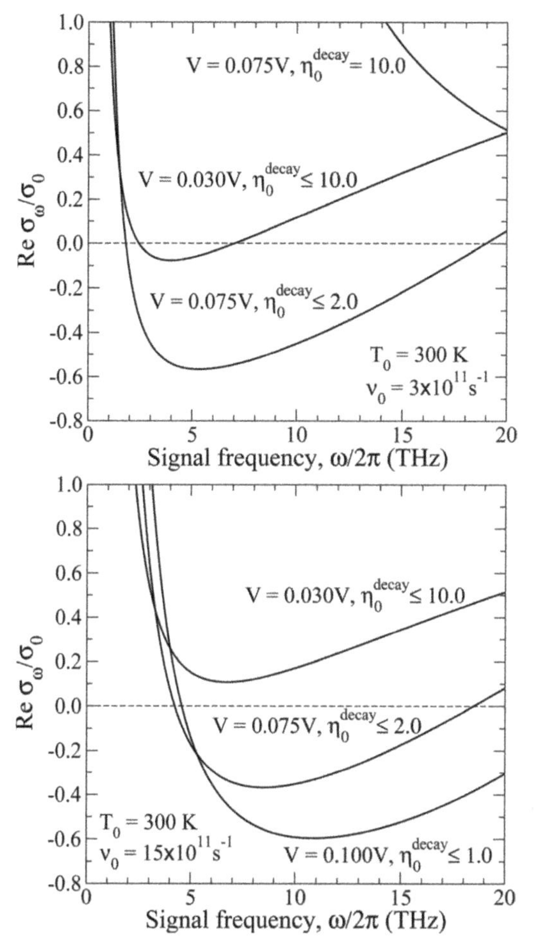

Figure 29.6 Dependence of normalized dynamic conductivity $\mathrm{Re}\sigma_\omega/\sigma_0$ on signal frequency $\omega/2\pi$ for different values of η_0^{decay} and $\nu_0 \simeq 3 \times 10^{11}$ s^{-1} (upper panel) and $\nu_0 \simeq 15 \times 10^{11}$ s^{-1} (lower panel).

at different voltages V for $\nu_0 \simeq 3 \times 10^{11}$ s^{-1} and $\nu_0 \simeq 15 \times 10^{11}$ s^{-1}. As seen from Fig. 29.6 at the injection conditions under consideration, the characteristic conductivity σ_0 is much smaller than the dc conductivity in the i-section $\sigma_\omega|_{\omega=0} = \sigma_{00}$. It is also seen that even at relatively large values of parameter η_0^{decay}, the dynamic conductivity can be negative in the THz range of frequencies provided the applied voltage is not so strong to cause the electron-hole plasma and optical phonon system overheating and the ambiguity of the voltage characteristics. This is confirmed by Fig. 29.7. Figure 29.7 shows $\mathrm{Re}\sigma_\omega/\sigma_0$ as a function of the applied voltage. As demonstrated, the range of the signal frequencies where $\mathrm{Re}\sigma_\omega < 0$ markedly shrinks and the quatity $|\mathrm{Re}\sigma_\omega|$ decreases when either η_0^{decay} or ν_0 increase. In particular, at large values of η_0^{decay}, the achievement of the negative dynamic conductivity and THz lasing can be complicated (at the temperatures $T_0 \sim 300$ K when the optical phonon recombination mechanism dominates). This is because when η_0^{decay} increases, the quantity V_d becomes small. As a result, the optical phonon system is overheated starting from relatively low voltages that leads to an "early" overheating of the electron-hole plasma (see Figs. 29.3 and 29.4). If the value of $\mathrm{Re}\sigma_\omega$ is insufficient to overcome the losses of the THz modes propagating along the GL structure, the structures with MGL can be used. In this case, the net dynamic conductivity of the MGL structure is given by $\mathrm{Re}\sigma_\omega \times K$, where K is the number of GLs [8–10].

29.7 Limitations of the Model and Discussion

High density of the electron-hole plasma in the i-section under the injection conditions promotes the quasi-neutrality of this section. The recombination does not significantly affect the uniform distribution of the electron and hole densities in the i-section (assumed above) if the recombination current J is much smaller than the maximum injection current J_m which can be provided by the p- and n-sections. This imposes the condition $J \ll 2J_\mathrm{m}$. The latter inequality is equivalent to the condition that the recombination length is longer than the i-section length $2L$. Considering a strong degeneracy of the electron and hole components in the p- and n-sections, respectively, the quantity J_m can be estimated as

$$J_m \simeq \frac{ev_F}{\pi^2\hbar^2}[\mu_i^2 - (\mu_i - eV)^2]. \tag{29.23}$$

At $T_0 = 300$ K, $2L = 20$ μm, $\mu_\mathrm{i} = 0.3$ eV, and $V = 0.1$ V, one obtains $J_0 \simeq 32$ A/cm and $2J_\mathrm{m} \simeq 65$ A/cm. As seen from Fig. 29.5, $J \ll J_0$ at least in the most interesting voltage range, where $\mathrm{Re}\sigma_\omega < 0$ (see Fig. 29.7), the condition $J \ll J_\mathrm{m}$ is satisfied.

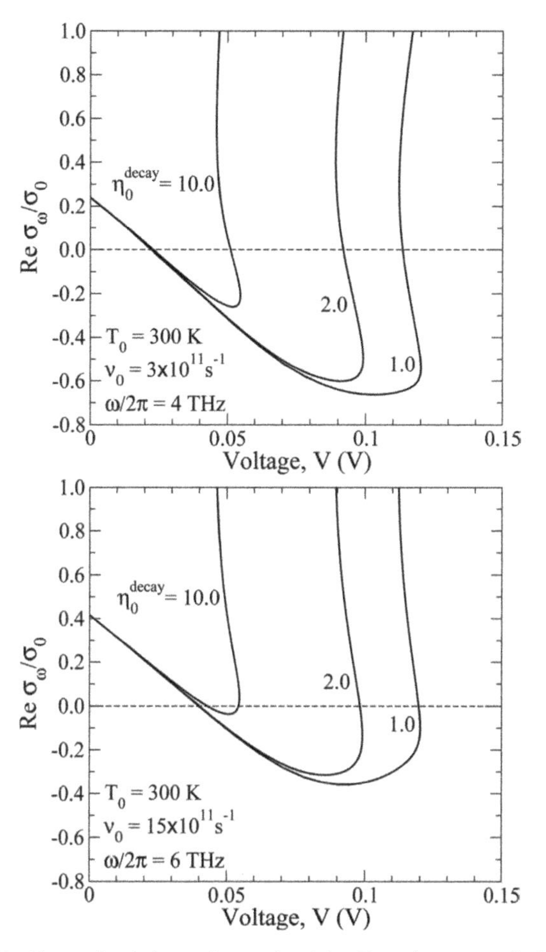

Figure 29.7 Normalized dynamic conductivity $\mathrm{Re}\sigma_\omega/\sigma_0$ vs applied voltage V for different values of η_0^{decay} and $\nu_0 \simeq 3 \times 10^{11}$ s^{-1}, $\omega/2\pi$ = 4 THz (upper panel) and $\nu_0 \simeq 15 \times 10^{11}$ s^{-1}, $\omega/2\pi$ = 6 THz (lower panel).

In the above consideration, we disregarded the leakage current from the i-section to the p- and n-sections. This current includes the tunneling and thermionic components. Both these components depend on the height, μ_{i}, of the barriers between the p- and i-sections and i- and n-sections and the applied voltage V (see Fig. 29.2). As shown previously [37], the tunneling current decreases with increasing width of the p-i or i-n junction l (see Fig. 29.2a), because it is sensitive to the electric field at the junction [18]. The width in question depends on the geometrical parameters of the structure, in particular on the spatial distributions of donor and acceptors near the junction, and the thickness of the gate layer W_g, as well as the shape of the gates (in the structures with the electrical doping). So, one can assume that this width can be sufficiently large to provide smooth potential distributions at the junctions. The effective height of the barriers at the p-i- and i-n-junctions, which determines the thermionic electron and hole current over these barriers, is equal to $\Delta = \mu_I - 2\mu = \mu_i - eV$. It can be small at elevated values of the quasi-Fermi energy μ which are necessary to achieve the negative dynamic conductivity (see, for instance, Refs. [4, 6, 9, 10]), i.e., in the most interesting case. This, in turn, implies that the electric field at the junctions and, hence, the tunneling current is decreased. In this case, the thermionic leakage current dominates over the tunneling leakage current and the latter is disregarded in the following estimates.

Taking into account, the height of the barrier Δ, for the contribution of thermionic current through the junctions to the net terminal current, one can obtain

$$J_{\text{th}} = \frac{4ev_{\text{F}}}{\pi^2}\left(\frac{T^2}{\hbar^2 v_{\text{F}}^2}\right)^2\left[\exp\left(\frac{eV-\mu_{\text{i}}}{T}\right)-1\right]$$
$$= 2ev_{\text{th}}\Sigma_0\left(\frac{T}{T_0}\right)^2\left[\exp\left(\frac{eV-\mu_{\text{i}}}{T}\right)-1\right]. \tag{29.24}$$

Here, $v_{\text{th}} = (12/\pi^3)v_{\text{F}}$. Comparing J and J_{th} given by Eqs. (29.5) and (29.24), respectively, one can conclude that the leakage (thermionic) current is small in comparison with the current associated with the optical phonon recombination if $(v_{\text{th}}\tau_0/L)\exp(-\mu_{\text{i}}/T) \ll \exp(-\hbar\omega_0/T)$ or

$$\mu_i - \hbar\omega_0 > T\ln\left[\left(\frac{v_{th}\tau_0}{L}\right)\left(\frac{T}{T_0}\right)^2\right]. \tag{29.25}$$

Assuming that $\tau_0 \sim 10^{-12}$ s, factor $(v_{\text{th}}\tau_0/L)$ in Eq. (29.21) is small when the length of the i-section $2L \gtrsim 1$ μm, i.e., at fairly practical values of $2L$. Therefore, neglect of the thermionic leakage in our calculations in the above sections is justified when $\mu_{\text{i}} > \hbar\omega_0 \simeq 0.2$ eV (more precisely, when $\mu_{\text{i}} - \hbar\omega_0 > T$). At $T \gtrsim T_0$, the latter inequality means that the donor and acceptor density in the pertinent sections should be

$$\Sigma_{\rm i} > \frac{1}{\pi}\left(\frac{\omega_0}{v_{\rm F}}\right)^2 \simeq 3.3\times 10^{12}\ {\rm cm}^{-2}. \qquad (29.26)$$

In the case of the devices with the electrically induced p- and n-section, the analogous condition sounds as

$$V_{\rm g} > \frac{4eW_{\rm g}}{æ}\left(\frac{\omega_0}{v_{\rm F}}\right)^2. \qquad (29.27)$$

Setting æ = 4 and $W_{\rm g}$ = 10 nm, the latter condition corresponds to $V_{\rm g} > 1.5$ V.

At a strong heating of the electron-hole plasma, say, at $V = \hbar\omega_0/e$, using Eq. (29.15), condition (29.25) is replaced by

$$\mu_{\rm i} > \hbar\omega_0 \frac{\ln(v_{\rm th}\tau_0/L)}{\ln\left(\dfrac{1+\eta_0^{\rm decay}+\mathcal{N}_0^{\rm eq}}{\eta_0^{\rm decay}+\mathcal{N}_0^{\rm eq}}\right)}. \qquad (29.28)$$

The latter inequality can impose somewhat stricter limitation on the values of $\mu_{\rm i}$ and $\Sigma_{\rm i}$ than those given by Eqs. (29.25) and (29.26) if $\eta_0^{\rm decay} > 1$.

It is notable that an effective confinement of the injected electrons and holes in the i-section by the barriers at p-i- and i-n-junctions in the GL structures under consideration can be realized by relatively low doping levels in p- and n-sections in comparison with structures with two-dimensional (2D) electron plasma in quantum wells on the base of the standard semiconductors. Indeed, the barrier height at $V = 0$ in the doped section of GL is equal to $\Delta = \mu_{\rm i} - eV$. In standard 2D systems with the electron effective mass m^* and the energy gap $\Delta_{\rm g}$, the barrier height is equal to $\Delta^* = \Delta_{\rm g} + \mu_{\rm i}^* - eV^*$, where $\mu_{\rm i}^* \simeq \pi \hbar^2\Sigma_{\rm i}^*/m^*$. To achieve the same value of the quasi-Fermi energy μ in the standard 2D system as in GLs, one needs to apply the voltage $V^* = V + \Delta_{\rm g}/e$. To provide, for example, the values $\mu_{\rm i} = \mu_{\rm i}^*$ = 03 eV, one needs $\Sigma_{\rm i} \simeq 7.33 \times 10^{12}$ cm^{-2} and $\Sigma_{\rm i}^* \simeq 3.82 \times 10^{13}$ cm^{-2}, respectively. Thus, in the standard 2D electron systems, the doping should be five times higher (for $m = 4 \times 10^{-29}$ g). This is due to lower density of states in massless GLs near the Dirac point compared to that in the standard 2D structures with $m \neq 0$. Since the effective mass of holes M^* in the standard semiconductors is markedly larger than m^*, the realization of the barrier height at the p-i-junction sufficient for the effective confinement of electrons at elevated temperatures requires fairly heavy doping. Thus, in contrast to the standard 2D structures, the thermionic leakage current in the GL or MGL p-i-n structures under consideration can be sufficiently small without the employment of wide-gap p- and n-sections. In passing, it should be mentioned that an extra confinement of the injected electrons and holes can be achieved if the p- and n-sections constitute arrays of graphene nanowires (doped or with electrically induced high electron and hole densities), so that the p-i-n structures considered above are replaced by the P-i-N structures.

One needs to stress that the assumptions (used in the above model) that the electron-hole plasma in the active region is virtually uniform as well as that the recombination current exceeds the leakage current are rather common in simplified models (the so-called rate-equation models) of in standard injection laser structures with the double injection (see, for instance, Refs. [38, 39]).

As follows from the above calculations, the effective electron-hole and optical phonon temperature can be very high at $V \sim \hbar\omega_0/e$ and be accompanied by effects associated with the S-shape characteristics. In this case, an expression for the rate of optical phonon decay $(\mathcal{N}_0 - \mathcal{N}_0^{\text{eq}})/\tau_0^{\text{decay}}$ used in Eq. (29.7) might be oversimplified due to a strong anharmonic lattice vibration. Possibly, the effects related to a strong anharmonism can be taken into account by a proper choice (renormalization) of parameters τ_0^{decay} and η_0^{decay}. In the above treatment, for simplicity, only one type of optical phonons with $\hbar\omega_0 \simeq 0.2$ eV was taken in to account. However, due to closeness of the optical phonon frequencies of different type in GLs and MGLs, the pertinent generalization of the model, adding computational complexity, should not lead to a marked change in the obtained results. Apart from this, at large effective temperatures, the radiative recombination and cooling (due to the radiative transfer of the energy outside the structure) can become essential [40–42], resulting in a limitation of these temperatures and affecting the S-shape dependences. This means that considering the range or relatively high applied voltages and, hence, strong injection, our purely "optical phonon" model should be generalized. However, this concerns not particularly interesting situations in which the dynamic conductivity is not negative.

As demonstrated, the main potential obstacles in the realization of negative dynamic conductivity and THz lasing in the injection GL and MGL structures at the room (or slightly lower) temperatures might be the intraband photon (Drude) absorption and the optical phonon heating. These effects are characterized by parameters $C \propto \nu_0$ and $\eta_0^{\mathrm{decay}} \propto \tau_0^{\mathrm{decay}}$ respectively. As for parameter C, it can be sufficiently small in perfect MGL structures like those studied in Ref. [2], so the problem of intraband absorption can be overcome. However, if real values of parameter η_0^{decay} cannot be decreased to an appropriate level ($\eta_0^{\mathrm{decay}} \lesssim 1$), the achievement of room temperature THz lasing in the structures under consideration might meet problems. In the case of such a scenario, the utilization of lower temperatures, at which the recombination and energy relaxation is associated with different mechanisms, can become indispensable.

29.8 Conclusions

In conclusion, we have studied theoretically the effect of population inversion associated with the electron and hole injection in GL and MGL p-i-n structures at the room and slightly lower temperatures when the interaction with optical phonons is the main mechanism of the recombination and energy relaxation. In the framework of the developed model, the electron-hole and optical phonon effective temperatures and the current-voltage characteristics have been calculated as functions of the applied voltage and the structure parameters. It has been demonstrated that the injection can lead to cooling of the injected electron-hole plasma in the device i-section to the temperatures lower than the lattice temperature at low and moderate voltages, whereas the voltage dependences can be ambiguous exhibiting the S-shape behavior at elevated voltages. The variations of the electron-hole effective temperature with increasing applied voltage are accompanied with an increase in the optical phonon effective temperature. Using the obtained voltage dependences, we have calculated the dynamic conductivity and estimated the ranges parameters and signal THz frequencies where this conductivity is negative. The electron-hole cooling might substantially promote the realization of THz lasing at elevated ambient temperatures. In summary, we believe that the obtained results instill confidence in the future of graphene-based injection THz lasers although their realization might require a thorough optimization.

Acknowledgments

The authors are grateful to A. Satou for numerous useful discussions. This work was supported by the Japan Science and Technology Agency, CREST and by the Japan Society for Promotion of Science, Japan.

References

1. A. H. C. Neto, F. Guinea, N. M. R. Peres, K. S. Novoselov, and A. K. Geim, *Rev. Mod. Phys.*, **81**, 109 (2009).
2. M. Sprinkle, D. Suegel, Y. Hu, J. Hicks, A. Tejeda, A. Taleb-Ibrahimi, P. Le Fevre, F. Bertran, S. Vizzini, H. Enriquez, S. Chiang, P. Soukiassian, C. Berger, W. A. de Heer, A. Lanzara, and E. H. Conrad, *Phys. Rev. Lett.*, **103**, 226803 (2009).
3. M. Orlita and M. Potemski, *Semicond. Sci. Technol.*, **25**, 063001 (2010).
4. V. Ryzhii, M. Ryzhii, and T. Otsuji, *J. Appl. Phys.*, **101**, 083114 (2007).
5. F. Rana, *IEEE Trans. Nanotechnol.*, **7**, 91 (2008).
6. A. Dubinov, V. Ya. Aleskin, M. Ryzhii, and V. Ryzhii, *Appl. Phys. Express*, **2**, 092301 (2009).
7. B. Dora, E. V. Castro, and R. Moessner, *Phys. Rev. B*, **82**, 125441 (2010).
8. A. A. Dubinov, V. Ya. Aleshkin, V. Mitin, T. Otsuji, and V. Ryzhii, *J. Phys.: Condens. Matter*, **23**, 145302 (2011).
9. V. Ryzhii, M. Ryzhii, A. Satou, T. Otsuji, A. A. Dubinov, and V. Ya. Aleshkin, *J. Appl. Phys.*, **106**, 084507 (2009).
10. V. Ryzhii, A. A. Dubinov, T. Otsuji, V. Mitin, and M. S. Shur, *J. Appl. Phys.*, **107**, 054505 (2010).
11. M. Ryzhii and V. Ryzhii, *Jpn. J. Appl. Phys.*, **46**, L151 (2007).
12. A. Satou, F. T. Vasko, and V. Ryzhii, *Phys. Rev. B*, **78**, 115431 (2008).
13. A. Satou, T. Otsuji, and V. Ryzhii, *Jpn. J. Appl. Phys.*, **50**, 070116 (2011).
14. T. Otsuji, S. A. Boubanga-Tombet, S. Chan, A. Satou, and V. Ryzhii, *Proc. SPIE*, **8023**, 802304 (2011).

15. S. Boubanga-Tombet, S. Chan, A. Satou, T. Otsuji, and V. Ryzhii, e-print arXiv:1011.2618.
16. V. Ryzhii, M. Ryzhii, V. Mitin, A. Satou, and T. Otsuji, *Jpn. J. Appl. Phys.*, **50**(9), 094001 (2011).
17. R. Kim, V. Perebeinos, and P. Avouris, *Phys. Rev. B*, **84**, 075449 (2011).
18. V. V. Cheianov and V. I. Fal'ko, *Phys. Rev. B*, **74**, 041403(R) (2006).
19. M. Ryzhii, V. Ryzhii, T. Otsuji, V. Mitin, and M. S. Shur, *Phys. Rev. B*, **82**, 075419 (2010).
20. F. Rana, P. A. George, J. H. Strait, S. Shivaraman, M. Chanrashekhar, and M. G. Spencer, *Phys. Rev. B*, **79**, 115447 (2009).
21. A. Satou, F. T. Vasko, and V. Ryzhii, *Phys. Rev. B*, **78**, 115431 (2008).
22. M. S. Foster and I. L. Aleiner, *Phys. Rev. B*, **79**, 085415 (2009).
23. D. M. Basko, S. Piscanec, and A. C. Ferrari, *Phys. Rev. B*, **80**, 165413 (2009).
24. F. T. Vasko and V. V. Mitin, e-print arXiv:1107.2708.
25. V. Ryzhii, M. Ryzhii, and T. Otsuji, e-print arXiv:1108.2077.
26. J. Martin, N. Akerman, G. Ulbricht, T. Lohmann, J. H. Smet, K. von Klitzing, and A. Yacoby, *Nat. Phys.*, **4**, 144 (2008).
27. Y. Zhang, V. W. Brar, C. Girit, A. Zett, and M. F. Cromme, *Nat. Phys.*, **5**, 722 (2009).
28. J. M. Poumirol, W. Escoffer, A. Kumar, M. Goiran, R. Raquet, and J. M. Broto, *New. J. Phys.*, **12**, 083006 (2010).
29. P. Parovi-Azar, N. Nafari, and M. R. R. Tabat, *Phys. Rev. B*, **83**, 165434 (2011).
30. H. Wang, J. H. Strait, P. A. George, S. Shivaraman, V. D. Shields, M. Chandrashekhar, J. Hwang, F. Rana, M. G. Spencer, C. S. Ruiz-Vargas, and J. Park, *Appl. Phys. Lett.*, **96**, 081917 (2010).
31. C. Auer, F. Schurer, and C. Ertler, *Phys. Rev. B*, **74**, 165409 (2006).
32. G. Pennigton, S. J. Kilpatrick, and A. E. Wickenden, *Appl. Phys. Lett.*, **93**, 093110 (2008).
33. M. Steiner, M. Freitag, V. Perebeinos, J. C. Tsang, J. P. Small, M. Kinoshita, D. Yuan, J. Liu, and P. Avouris, *Nat. Nanotechnol.*, **4**, 320 (2009).
34. P. A. George, J. Strait, J. Dawlaty, S. Shivaraman, M. Chandrashekhar, F. Rana, and M. G. Spencer, *Nano Lett.*, **8**, 4248 (2008).
35. L. A. Falkovsky and A. A. Varlamov, *Eur. Phys. J. B*, **56**, 281 (2007).
36. L. A. Falkovsky, *Phys. Rev. B*, **75**, 03349 (2007).

37. V. Ryzhii, M. Ryzhii, and T. Otsuji, *Phys. Status Solidi A*, **205**, 1527 (2008).
38. G. P. Agrawal and N. K. Dutta, *Semiconductor Lasers* (Van Nostrand Reinhold, New York, 1993).
39. L. A. Coldren and S. W. Corzine, *Diode Lasers and Photonic Integrated Circuits* (Wiley, New York, 1995).
40. F. T. Vasko and V. Ryzhii, *Phys. Rev. B*, **77**, 195433 (2008).
41. P. N. Romanets, F. T. Vasko, and M. V. Strikha, *Phys. Rev. B*, **79**, 033406 (2009).
42. O. G. Balev, F. T. Vasko, and V. Ryzhii, *Phys. Rev. B*, **79**, 165432 (2009).

Chapter 30

Ultrafast Carrier Dynamics and Terahertz Emission in Optically Pumped Graphene at Room Temperature*

S. Boubanga-Tombet,[a] S. Chan,[b] T. Watanabe,[a] A. Satou,[a,c] V. Ryzhii,[c,d] and T. Otsuji[a,c]
[a]*Research Institute of Electrical Communication, Tohoku University, 2-1-1 Katahira, Aoba-ku, Sendai 980-8577, Japan*
[b]*Nano-Japan Program, University of Pennsylvania, 3733 Spruce Street, Philadelphia, Pennsylvania, USA*
[c]*JST-CREST, Chiyoda-ku, Tokyo 1020075, Japan*
[d]*Computational Nanoelectronics Laboratory, University of Aizu, Tsuruga, Ikki-machi, Aizu-Wakamatsu 965-8580, Japan*
stephanealbon@hotmail.com

We report, within a picosecond time scale, fast relaxation and relatively slow recombination dynamics of photogenerated electrons and holes in an exfoliated graphene under infrared pulse excitation.

---

*Reprinted with permission from S. Boubanga-Tombet, S. Chan, T. Watanabe, A. Satou, V. Ryzhii, and T. Otsuji (2012). Ultrafast carrier dynamics and terahertz emission in optically pumped graphene at room temperature, *Phys. Rev. B*, **85**, 035443. Copyright © 2012 American Physical Society.

Graphene-Based Terahertz Electronics and Plasmonics: Detector and Emitter Concepts
Edited by Vladimir Mitin, Taiichi Otsuji, and Victor Ryzhii

ISBN 978-981-4800-75-4 (Hardcover), 978-0-429-32839-8 (eBook)
www.jennystanford.com

We conduct time-domain spectroscopic studies using an optical pump and terahertz probe with an optical probe technique and show that graphene sheet amplifies an incoming terahertz field. The graphene emission spectral dependency on laser pumping intensity shows a threshold-like behavior, testifying to the occurrence of the negative conductivity and the population inversion. The phase behavior of the measured terahertz electric field also shows clear Lorentzian-like normal dispersion around the gain peak, testifying to the amplification that can be attributed to stimulated emission of photocarriers in the inverted states. The emission spectra clearly narrow at a longer terahertz probe delay time, giving evidence that the quasi-Fermi energy moves closer to the equilibrium at this longer terahertz probe delay time.

30.1 Introduction

Graphene is a one-atom-thick planar sheet of carbon atoms that are densely packed in a honeycomb crystal lattice [1]. This material has many peculiar properties and potential applications, including the prediction and observation of half-integer quantum Hall effect [2], finite conductivity at zero-charge carrier concentration [3], perfect quantum tunneling effect [4], and ultrahigh carrier mobility [5], owing to massless and gapless energy spectra. The gapless and linear energy spectra of electrons and holes lead to nontrivial features such as negative dynamic conductivity in the terahertz (THz) spectral range [6], which may lead to the development of a new type of terahertz lasers [7, 8]. This has attracted intense interest due to the ongoing search for viable terahertz detectors and emitters. To realize such terahertz graphene-based devices, understanding the nonequilibrium carrier relaxation and recombination dynamics is critical. Recently, Karasawa et al. reported on terahertz emission in graphene that was heteroepitaxially grown on 3C-SiC/Si [9]. Time-resolved measurements of fast nonequilibrium carrier relaxation dynamics have also been carried out for multilayers and monolayers of graphene grown on SiC [10–13] and exfoliated from highly oriented pyrolytic graphite (HOPG) [14, 15]. Several methods for observing the relaxation processes have been reported. Dawlaty et al. [10] and Sun et al. [11] used an optical-pump and optical-probe technique, and George et al. [12] used an optical-pump and terahertz-probe technique to evaluate the dynamics starting with the main contribution of carrier-carrier (cc) scattering in the first 150 fs, followed by carrier-phonon (cp) scattering on the picosecond time scale. Ultrafast scattering of photoexcited carriers by optical phonons has been theoretically predicted by Ando [16], Suzuura [17], and Rana [18]. Kamprath et al. [14] observed strongly coupled optical phonons in the ultrafast carrier dynamics for a duration of 500 fs by optical-pump and terahertz-probe spectroscopy. Wang et al. [13] also observed ultrafast carrier relaxation by hot-optical phonon-carrier scattering for a duration of ~500 fs by using an optical-pump and optical-probe technique. The measured optical phonon lifetimes found in these studies were ~7ps [14], 2–2.5 ps [13], and ~1 ps [12], respectively, some of which agreed fairly well with theoretical calculations by Bonini et al. [19]. Recent studies by Breusing et al. [15, 20] more precisely revealed ultrafast carrier dynamics with a time resolution of 10 fs for exfoliated graphene and graphite.

In this chapter, we report on the fast relaxation and relatively slow recombination dynamics in optically pumped graphene. The recombination process is stimulated with terahertz probe pulse irradiation. The terahertz probe timing is set at ~2 ps and ~3.5 ps after the optical pumping. The observed results clearly show amplification of the terahertz probe pulse after graphene excitation. This amplification is supported by the measured amplitude of the terahertz electric field on graphene, which is larger than that measured on the area without graphene. Another indication of this amplification is given by the phase of the measured field that shows a clear Lorentzian-like normal dispersion around the gain peak. The graphene emission spectrum also shows a threshold behavior versus the optical pumping intensity, providing proof of the negative dynamic conductivity in the terahertz spectral range. The emission spectra also clearly narrows when the terahertz probe timing is set at ~3.5 ps, providing proof that the quasi-Fermi energy moves closer to the equilibrium state than it did at ~2 ps.

30.2 Theory and Background

When graphene is pumped with the infrared photon having an energy $\hbar\Omega$, electrons or holes are photogenerated via interband transitions. It has been shown that the photogenereted carriers first establish separate distributions around the level $\pm\hbar\Omega/2$ within 20–30 fs after excitation (see Fig. 30.1a) [11, 12, 14, 15]. At room temperature and/or strong pumping, cc scattering has a strong influence on the carrier relaxation dynamics. As discussed in Refs. [12, 14, 15], the distributions at around $\pm\hbar\Omega/2$ together with the intrinsic distributions are rapidly quasiequilibrated within 100–200 fs (Fig. 30.1b). Then optical phonons are emitted on the high-energy tail of the electron and hole distributions on a time scale of a few picoseconds (Fig. 30.1c). This intraband relaxation process is relatively fast and accumulates the nonequilibrium carriers around

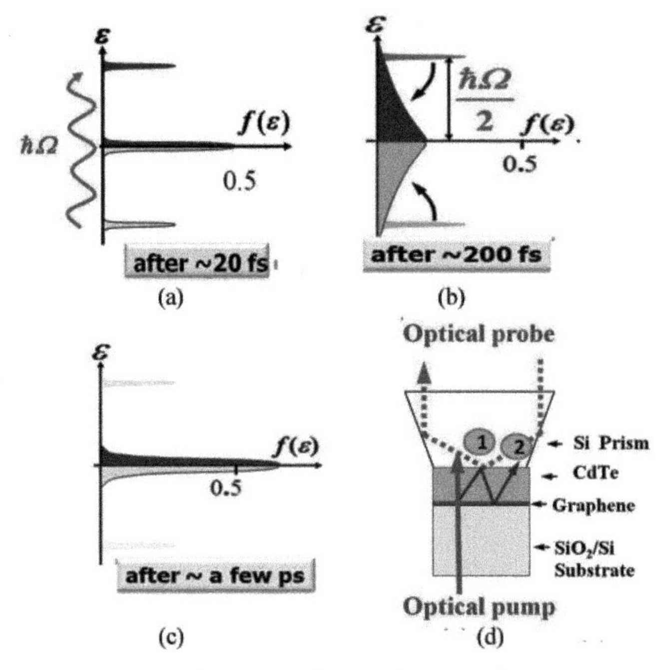

Figure 30.1 Carrier relaxation and recombination dynamics in optically pumped graphene from few tens of fermoseconds to picoseconds after pumping [panels (a), (b), and (c)]. The CdTe crystal on top of the graphene sample, the optical pump and probe, and the terahertz probe are represented (d).

the Dirac point. Then, further equilibration occurs via interband electron-hole recombination as well as intraband Fermization due to cc scattering and cp scattering on a time scale of a few picoseconds. The recombination process is mainly done by interband optical phonon scattering. It has also been shown that even in optically pumped graphene samples with disorder caused by strong fluctuation of the surface charges, resulting in the formation of the electron and hole puddles, the recombination can be associated with the interband tunneling between electron-hole puddles [21]. The fast intraband relaxation (ps or less) and the relatively slow interband recombination process ($\gg$1 ps) due to the linearity of the density of states and Pauli blocking lead to the population inversion [6]. In the case of the radiative recombination, due to the gapless symmetrical band structure, photon emission over a wide terahertz frequency range is expected if the pumping photon energy is suitably chosen and the pumping intensity is sufficiently high [22].

30.3 Sample and Characterization

In order to verify the above mentioned concept, we conduct time-domain spectroscopy experiments based on an optical pump/terahertz-and-optical-probe technique. The time-resolved field emission properties are measured by an electro-optic sampling (EOS) method in total-reflection geometry [23]. The sample used is exfoliated graphene on SiO_2/Si substrate. The sample structure as presented in Fig. 30.1d is made of (i) a 560-μm-thick highly doped Si (100) substrate having a resistivity of 0.005 Ωcm, used as the back gate; (ii) a thermally oxidized 300-nm-thick SiO_2 layer, and (iii) some islets of exfoliated monolayer, bilayer, and few- layers-thick graphene. The flake size of the monolayer graphene under measurement is about 7000 μm^2. In order to characterize the quality of our sample, we conduct back-gate current-voltage transfer characterization, as well as atomic and Kelvin force microscopy to the surface of the SiO_2/Si substrate and that of graphene on SiO_2/Si substrate. Figure 30.2 shows Atomic Force Microscopy (AFM) topographic images and the corresponding histograms. We report variances σ of 0.146 nm with SiO_2/Si substrate and 0.142 nm with graphene on SiO_2/Si, showing the good flatness of this graphene sample on the substrate. The inset of Fig. 30.2 shows transfer characteristics with a Dirac voltage close to 0 V, showing that the level of this intrinsic doping is very low in this sample. The Kelvin Force Microscopy (KFM) measurement (Fig. 30.3) shows a very smooth surface potential distribution on 10 × 10 μm area with a variance of 4.05 meV. We therefore carefully selected this good-quality sample, which has less substrate effect and a Dirac voltage that stays at the Dirac point, for the experiments.

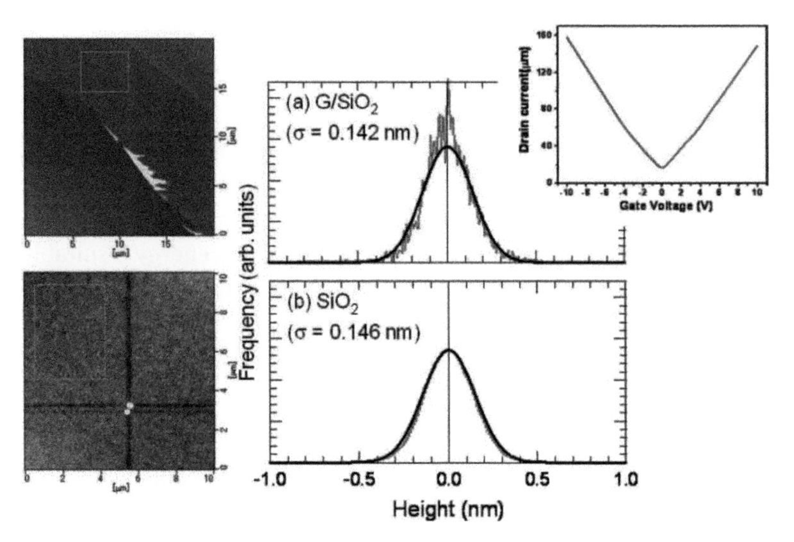

Figure 30.2 AFM topographic images and the corresponding histograms of graphene on (a) SiO_2/Si substrate and (b) SiO_2/Si substrate. The data shown in panels (a) and (b) are described by Gaussian distributions (black solid lines) with variances σ of 0.142 and 0.146 nm, respectively. Inset shows transfer characteristics of monolayer graphene.

30.4 Experiments

To obtain the terahertz photon emission from the abovementioned carrier relaxation and recombination dynamics, the pumping photon energy (wavelength) is selected to be around 800 meV (1550 nm), much higher than the optical phonon energy (~198 meV). A 2-mm-long, 0.5-mm-wide CdTe crystal with a Si prism is used for reflective electro-optic probing and placed onto the sample. The CdTe crystal acts as an electro-optic sensor as well as a terahertz probe pulse generator. A femtosecond-pulsed fiber laser with a full width at half-

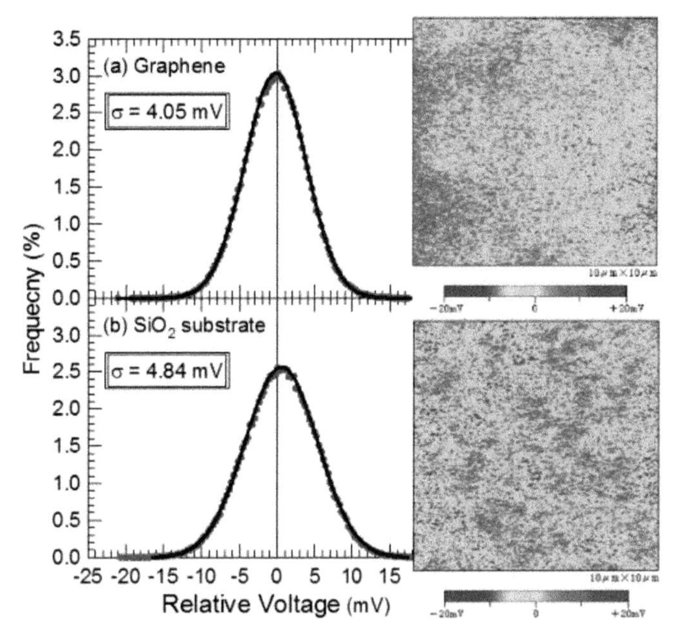

Figure 30.3 KFM topographic images and the corresponding histograms of graphene on (a) SiO_2/Si substrate and (b) SiO_2/Si substrate. The data shown in panels (a) and (b) are described by Gaussian distributions (black solid lines) with variances of 4.05 meV and 4.84 mV, respectively.

maximum (FWHM) of 80 fs, repetition rate of 20 MHz, and average power of about 4 mW was used as the optical pump and probe source. More precise description of the experimental setup can be found in Ref. [24]. The laser is split into two paths used for pump and probe. The pumping laser beam, being linearly polarized, is mechanically chopped at ~1.2 kHz (for lock-in detection) and focused with a beam diameter of about 40 μm onto the sample and the CdTe from the back side, while the probing beam is cross polarized with respect to the pump beam and focused from the top side (see Fig. 30.1d). Owing to the second-order nonlinear optical effects, the CdTe crystal can rectify the pump laser pulse to emit terahertz envelope radiation. The emitted terahertz pulse magnitude grows along the Cherenkov angle, preserving the phase-matching condition (between the infrared and the terahertz radiations are preserved) [25]. Its forward-propagating pulse is first electro-optically detected by the optical probe beam. The terahertz pulse is partially reflected at the top surface of the CdTe, then subjected back to the graphene, serving as a terahertz probe pulse (arrowed blue line in Fig. 30.1d) to stimulate terahertz photon emission via electron-hole recombination in the graphene. The terahertz probe pulse including stimulated emission reflects in most part at the SiO_2/Si-graphene interface and travels back to the Si prism, which is detected as a terahertz photon echo signal, as shown in Fig. 30.1d. Therefore, the original data of the experimental temporal response consists of the first forward-propagating terahertz pulsation (no interaction with graphene) followed by a photon echo signal (probing the graphene). The delay between these two pulsations is given by the total round-trip propagation time of the terahertz probe pulse through the CdTe. The system bandwidth is estimated to be around 6 THz, mainly limited by the reststrahlen band of the CdTe sensor crystal.

30.5 Results and Discussion

The experiments were done with two CdTe sensor crystals (A and B), having orientations (100) and (101) and thicknesses of 120 and 80 μm, respectively. Figure 30.4 shows temporal responses measured on monolayer graphene with the thinner (black line) and the thicker (red line) crystals for the pumping pulse intensity of 3×10^7 W/cm^2 (almost one order of magnitude below the level of Pauli blocking). These curves were plotted with the same origin for comparison. One can notice that, as predicted, each temporal profile is composed of two peaks from optical rectification (OR) in CdTe and the terahertz photon echo signal. The measured time delays between these two pulsations with crystal A (thinner) and crystal B (thicker) of around 2 and 3.5 ps respectively are in a good agreement with the round-trip propagation time of a terahertz pulse through the CdTe crystal. The refractive index of CdTe is obtained from Ref. [26]. The OR pulse is found to be broader in crystal B than in crystal A, owing to better phase-matching conditions in the thinner crystal. Indeed, the coherent length in CdTe is estimated to be around 100 μm at 0.8 eV [27, 28]. The inset of Fig. 30.4 presents the echo signal peak intensity measured with crystal B on graphene as well as the reference curve (grey line) measured on the area without graphene. One can notice that the peak obtained on graphene is more intense than that one obtained on the substrate without graphene. This suggests that graphene amplifies the terahertz echo (terahertz probe) signal. The possible origins of the observed amplification could be (i) the terahertz pulse that stimulates the emission from graphene by electron-hole radiative recombination whose energy falls in the range of the negative dynamic conductivity or (ii) the increase in reflectivity under pumping due to the increase in photocarrier-dependent Drude conductivity.

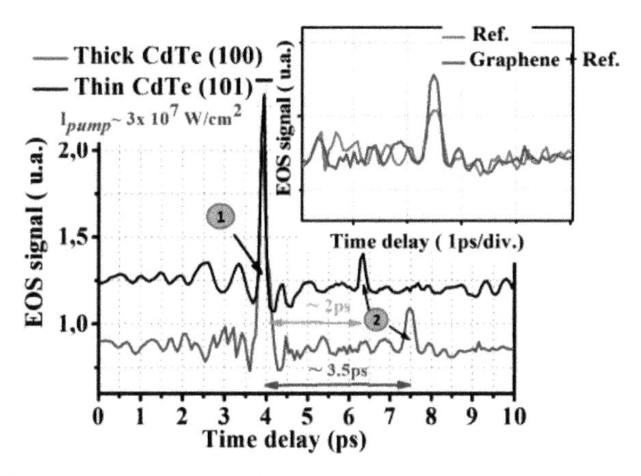

Figure 30.4 Measured temporal responses on monolayer graphene with thick (red line) and thin (black line) CdTe crystals for the pumping pulse intensity of about 3×10^7 W/cm^2. Inset: Temporal responses of echo photon signal measured on graphene (red line) and the area without graphene (grey line).

The graphene transfer function $H(\omega)$ is defined as $H(\omega) = G(\omega)/S(\omega)$, where $G(\omega)$ and $S(\omega)$ are the Fourier transforms of the photon echo signals measured on graphene and substrate (without graphene) respectively. Figure 30.5a shows the transfer function of monolayer graphene for different values of pumping pulse intensity. One can see from these results that decreasing I_{pump} drastically reduces the gain spectra, and below 10^7 W/cm^2 the emission completely disappears and only attenuation can be seen. The phase data (Fig. 30.5b) show clear Lorentzian-like normal dispersion around the gain peak when the pumping pulse intensity is beyond the threshold, demonstrating amplification that can be attributed to stimulated emission of photocarriers in the inverted states. Comparing the phase data for the pumping intensities beyond the

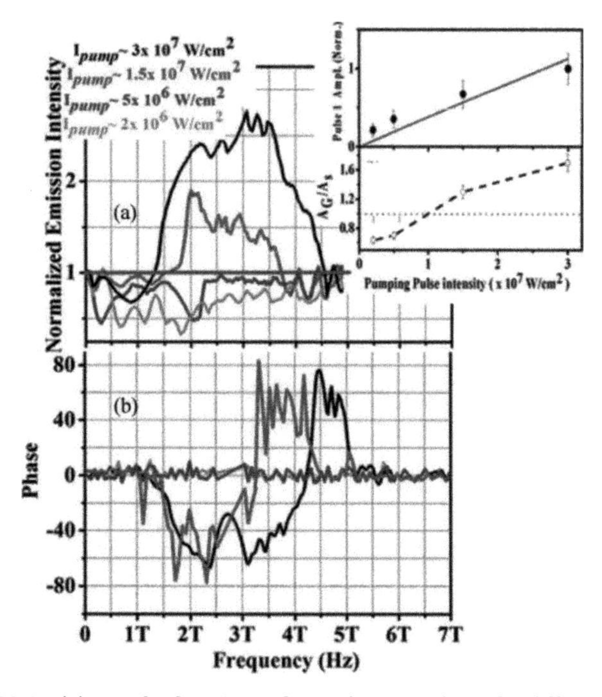

Figure 30.5 (a) Transfer functions of monolayer graphene for different values of pumping pulse intensities. Inset: normalized EOS signal amplitude of the first peak (upper panel) and A_G/A_S ratio (lower panel) for different values of the pumping pulse intensities. (b) Variation of the phase of the measured terahertz electric field for different values of pumping pulse intensities.

threshold, with an increase in the pumping intensity, the gain peak frequency (given by the zero point of the phase data) shows a blue shift and the upper cutoff frequency increases, whereas the lower cutoff frequency barely changes. These tendencies well reflect the dependence of the gain spectral profile on pumping intensity [6, 22]. We also present in the inset of Fig. 30.5a (lower panel) the corresponding ratio A_G/A_S, where A_G and A_S are the amplitudes of echo signal peaks measured on graphene and substrate respectively. The graphene transfer function and A_G/A_S show a clear threshold-like behavior, demonstrating population inversion and negative dynamic conductivity in optically pumped graphene. The threshold intensity is found to be around 10^7 W/cm^2. This is a good starting point for the realization of room-temperature terahertz lasers based on graphene. The inset of Fig. 30.5a (upper panel) presents the normalized EOS signal amplitude of the first peak (see Fig. 30.4) for different values of the pumping pulse intensities. This EOS signal is proportional to the terahertz electric field generated by OR in CdTe crystal. The terahertz emitted intensity is quadratically dependent of the infrared intensity ($I_{\mathrm{THz}} \propto I_{\mathrm{IR}}^2$). Since $I_{\mathrm{THz}} \propto E_{\mathrm{THz}}^2$, the linear dependency of the EOS signal amplitude with the pumping pulse intensity is a clear evidence of its OR source.

The obtained A_G/A_S value of the gain is larger than the theoretical limit of ~1.01 for a single process of stimulated emission estimated from Ref. [6], taking into account the absorbance of monolayer graphene and the permittivities of CdTe and SiO_2. Nevertheless, in the experimental setup, a multipath reflection effect is considered as a factor that increases the gain. The forward-propagating terahertz probe pulse is amplified, its large fraction is transmitted, and ~70% is reflected back at the SiO_2/Si interface. This reflected part will also interact with graphene and be amplified. Since the round-trip delay in the 300-nm-thick SiO_2 layer is ~4 fs, almost no distortion is seen in the temporal response. This multipath reflection effect may double the gain at most but cannot fully explain the obtained A_G/A_S.

We also considered an effect of photocarrier-dependent Drude conductivity on reflectivity. In case of our graphene/SiO_2/Si sample whose Si substrate has a low resistivity of 0.005 Ω cm, the increase in reflectivity under pumping due to the increase in photocarrier-dependent Drude conductivity is as small as 1–2% and quite monotonic over the entire terahertz spectral range of interest. A few percentages of such increase in the reflectivity should be de-embedded from the obtained photon-echo signal but barely affect the gain spectral profile.

Furthermore, to confirm the effects of the terahertz probe that stimulates the emission in graphene, the CdTe crystal was replaced by a CdTe crystal having a high-reflectivity coating for IR on its bottom surface in order to eliminate generation of the terahertz probe signal. In this case, no distinctive response was observed. Since the measurements are taken as an average, the observed response is undoubtedly a coherent process that cannot be obtained via spontaneous emission processes, which also supports the occurrence of the stimulated emission.

Figure 30.6 shows the Fourier transform of the photon echo signal measured on the area without graphene (reference) with crystal

A (black line) and crystal B (red line). The normalized dynamic conductivity for the pumping pulse intensity three times higher than the threshold pumping pulse intensity at 300 K is also presented (see Ref. [6]). The photon echo pulse interacts with graphene during the recombination process and induces emission of terahertz photon in graphene within the negative dynamic conductivity area (blue shaded area). The expected graphene emission spectral bandwidth is limited at lower frequencies by the Drude mechanism of terahertz absorption, and the higher frequency limit is given by the system bandwidth, which is estimated to be around 6 THz (see the horizontal solid lines in Fig. 30.6). The black and red shaded areas show the expected graphene emission bandwidth, from ~1.5 to ~5 THz using the thicker crystal and from ~1.5 to ~6 THz using the thinner crystal.

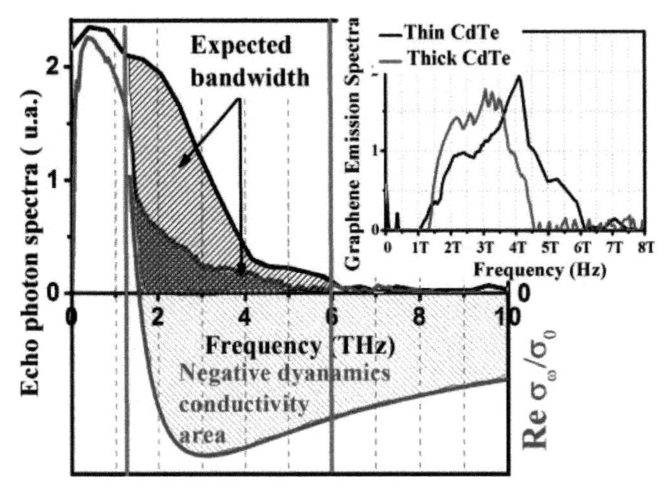

Figure 30.6 Terahertz emission spectra of the photon echo signal measured on the area without graphene (reference) with thick CdTe (red line) and thin CdTe (black line) crystals and the normalized dynamic conductivity (blue line). Inset: transfer functions of monolayer graphene obtained with thick CdTe (red line) and thin CdTe (black line) crystals.

The inset of Fig. 30.6 presents the transfer functions of monolayer graphene obtained with crystal A (black line) and crystal B (red line). The obtained spectra are in good agreement with the above mentioned expectations. It is possible to compare these spectra within the smallest bandwidth (from ~1.5 to ~5 THz). The graphene emission spectrum obtained with the thinner CdTe crystal is broader than that one obtained with the thicker crystal. In the thinner crystal, the graphene excitation with the terahertz probe is done earlier (2 ps) after the optical pump than in the thicker crystal (3.5 ps). This spectral narrowing at longer terahertz probe delay time is a clear trace of the progress of equilibration process from 2 to 3.5 ps. Indeed, at 3.5 ps, the quasi-Fermi energy is closer to the equilibrium compare to that at 2 ps. It is worth noting that this broadening may not be observed in the case of amplification due to an increase in reflectivity.

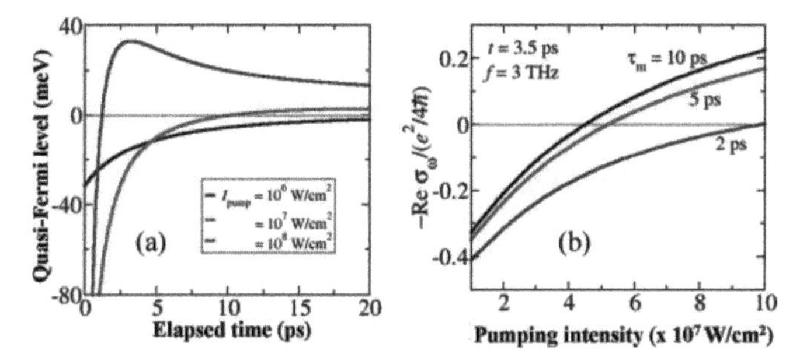

Figure 30.7 Time dependence of the quasi-Fermi energy of impulsively pumped graphene for different pulse intensities (a). Numerically simulated dynamic conductivity as a function of the pumping intensity at a fixed time of 3.5 ps and a fixed frequency of 3 THz (b).

Such a temporal dependency of the gain profile is theoretically testified [29]. Figure 30.7a shows numerically calculated temporal evolutions of the quasi-Fermi energy of impulsively pumped graphene in different pumping intensities. In this simulation, the system assumes that cc scattering is dominated at the earliest stage of the photocarrier energy relaxation so that the carrier distribution is always the quasi-Fermi distribution. Figure 30.7b shows the dynamic conductivity calculated as a function of the pumping intensity at a fixed time (3.5 ps) and a fixed frequency (3 THz). Compared with the experimentally observed threshold behavior, the measured results qualitatively agree with the simulated ones. The threshold intensity is on the same order. It is thought that the quantitative difference comes from the fact that we assume the complete quasiequilibration by the cc scattering, rasing the threshold pumping intensity to the highest level, and this leads to a rather pessimistic theoretical result.

From the above reported results and discussions, we conclude that the terahertz emission from graphene is stimulated by the coherent terahertz probe radiation and that the terahertz emission is amplified via electron-hole recombination in the range of the negative dynamic conductivity. Nevertheless, the authors want to open the field for further discussion if other viable explanations can be found.

30.6 Conclusion

We have successfully observed amplification of an incoming terahertz pulse during the relaxation-recombination process in graphene. We therefore studied (i) the pump-power-dependent measurements and showed the existence of a critical power at which the amplification regime starts. This behavior is generally believed to be evidence of the existence of a negative terahertz conductivity, which is a prerequisite for terahertz amplification. We investigated (ii) the amplitude and the phase of the measured terahertz field. The amplitude measured on graphene is larger than that measured on the area without graphene, while the phase shows clear inverted behavior with normal dispersion around the gain peak from that for the loss cases, demonstrating that the amplification can be attributed to stimulated emission of photo-carriers in the inverted states. We also confirmed (iii) the terahertz probe time-dependent emission spectra. The gain spectra show nonmonotonic frequency dependence with a clear narrowing when the terahertz probe timing is set at longer delay time (3.5 ps).

A possible interpretation is that we observed coherent amplified stimulated terahertz emissions and its threshold behavior against the pumping intensity arising from the fast relaxation and relatively slow recombination dynamics of photogenerated electrons or holes in exfoliated graphene, supporting the occurrence of negative dynamic conductivity, which can be applied to a new type of terahertz lasers.

Acknowledgments

The authors thank Prof. Massahiro Asada from the Tokyo Institute of Technology for his valuable discussion on the phase property of gain spectra. The authors are also grateful to S. Takabayashi, T. Fukushima, M. Kano, Y. Kurita, and K. Kobayashi from the Research Institute of Electrical Communication, Tohoku University, for their help with AFM and KFM measurements. This work was financially supported in part by the JST-CREST program, Japan, the Grant-in-Aid for Scientific Research (S), and specially promoted research programs from the Japan Society for the Promotion of Science.

References

1. K. S. Novoselov, A. K. Geim, S. V. Morozov, D. Jiang, Y. Zhang, S. V. Dubonos, I. V. Grigorieva, and A. A. Firsov, *Science*, **306**, 666 (2004).
2. Y. Zhang, Y. W. Tan, H. L. Stormer, and P. Kim, *Nature (London)*, **438**, 201 (2005).
3. K. S. Novoselov, A. K. Geim, S. V. Morozov, D. J. M. I. Katsnelson, I. V. Grigorieva, S. V. Dubonos, and A. A. Firsov, *Nature (London)*, **438**, 197 (2005).
4. M. I. Katsnelson, K. S. Novoselov, and A. K. Geim, *Nat. Phys.*, **2**, 620 (2006).
5. A. K. Geim and K. S. Novoselov, *Nat. Mater.*, **6**, 183 (2007).
6. V. Ryzhii, M. Ryzhii, and T. Otsuji, *J. Appl. Phys.*, **101**, 083114 (2007).
7. A. A. Dubinov, M. R. V. Y. Aleshkin, T. Otsuji, and V. Ryzhii, *Appl. Phys. Express*, **2**, 092301 (2009).
8. V. Ryzhii, M. Ryzhii, and T. Otsuji, *J. Appl. Phys.*, **106**, 084507 (2009).
9. H. Karasawa, T. Komori, T. Watanabe, A. Satou, H. Fukidome, M. Suemitsu, V. Ryzhii, and T. Otsuji, *J. Infrared Millimeter Terahertz Waves*, **32**, 655 (2011).
10. J. M. Dawlaty, S. Shivaraman, M. Chandrashekhar, F. Rana, and M. G. Spencer, *Appl. Phys. Lett.*, **92**, 042116 (2008).
11. D. Sun, Z. K. Wu, C. Divin, X. Li, C. Berger, W. A. deHeer, P. N. First, and T. B. Norris, *Phys. Rev. Lett.*, **101**, 157402 (2008).
12. P. A. George, J. Strait, J. Dawlaty, S. Shivaraman, M. Chandrashekhar, and F. R. M. G. Spencer, *Nano Lett.*, **8**, 4248 (2008).
13. H. Wang, J. H. Strait, P. A. George, S. Shivaraman, V. B. Shields, M. Chandrashekhar, J. Hwang, F. Rana, M. G. Spencer, C. S. Ruiz-Vargas, and J. Park, *Appl. Phys. Lett.*, **96**, 081917 (2010).
14. T. Kampfrath, L. Perfetti, F. Schapper, C. Frischkorn, and M. Wolf, *Phys. Rev. Lett.*, **95**, 187403 (2005).

15. M. Breusing, C. Ropers, and T. Elsaesser, *Phys. Rev. Lett.*, **102**, 086809 (2009).
16. T. Ando, *J. Phys. Soc. Jpn.*, **75**, 124701 (2006).
17. H. Suzuura and T. Ando, *J. Phys. Soc. Jpn.*, **77**, 044703 (2008).
18. F. Rana, P. A. George, J. H. Strait, J. Dawlaty, S. Shivaraman, M. Chandrashekhar, and M. G. Spencer, *Phys. Rev. B*, **79**, 115447 (2009).
19. N. Bonini, M. Lazzeri, N. Marzari, and F. Mauri, *Phys. Rev. Lett.*, **99**, 176802 (2007).
20. M. Breusing, S. Kuehn, T. Winzer, E. Malic, F. Milde, N. Severin, J. P. Rabe, C. Ropers, A. Knorr, and T. Elsaesser, *Phys. Rev. B*, **83**, 153410 (2011).
21. V. Ryzhi, M. Ryzhi, and T. Otsuji, *Appl. Phys. Lett.*, **99**, 173504 (2011).
22. V. Ryzhii, M. Ryzhii, V. Mitin, A. Satou, and T. Otsuji, *J. Appl. Phys.*, **50**, 094001 (2011).
23. L. Min and R. J. D. Miller, *Appl. Phys. Lett.*, **56**, 524 (1990).
24. S. Boubanga-Tombet, F. Teppe, J. Torres, A. Moutaouakil, D. Coquillat, N. Dyakonova, C. Consejo, P. Arcade, P. Nouvel, H. Marinchio, T. Laurent, C. Palermo, A. Penarier, T. Otsuji, L. Varani, and W. Knap, *Appl. Phys. Lett.*, **97**, 262108 (2010).
25. J. Hebling, G. Almasi, and I. Z. Kozma, *Opt. Express*, **10**, 1161 (2002).
26. M. Schall, H. Helm, and S. R. Keiding, *Int. J. Infrared Millimeter Waves*, **20**, 595 (1999).
27. B. Pradarutti, G. Matthaus, S. Riehemann, G. Notni, S. Nolte, and A. Tunnermann, *Opt. Commun.*, **281**, 5031 (2008).
28. M. Nagai, K. Tanaka, H. Ohtake, T. Bessho, T. Sugiura, T. Hirosumi, and M. Yoshida, *Appl. Phys. Lett.*, **85**, 3974 (2004).
29. A. Satou, T. Otsuji, and V. Ryzhii, *Jpn. J. Appl. Phys.*, **50**, 070116 (2011).

Chapter 31

Spectroscopic Study on Ultrafast Carrier Dynamics and Terahertz Amplified Stimulated Emission in Optically Pumped Graphene*

T. Otsuji,[a,c] S. Boubanga-Tombet,[a] A. Satou,[a,c] M. Suemitsu,[a] and V. Ryzhii[b,c]

[a]*Research Institute of Electrical Communication, Tohoku University, 2-1-1 Katahira, Aoba-ku, Sendai 980-8577, Japan*

[b]*Computational Nanoelectronics Laboratory, University of Aizu, Tsuruga, Ikki-machi, Aizu-Wakamatsu 965-8580, Japan*

[c]*JST-CREST, Chiyoda-ku, Tokyo 1020075, Japan*

otsuji@riec.tohoku.ac.jp

This chapter reviews recent advances in spectroscopic study on ultrafast carrier dynamics and terahertz (THz) stimulated emission in optically pumped graphene. The gapless and linear energy spectra

*Reprinted with permission from T. Otsuji, S. Boubanga-Tombet, A. Satou, M. Suemitsu, and V. Ryzhii (2012). Spectroscopic study on ultrafast carrier dynamics and terahertz amplified stimulated emission in optically pumped graphene, *J. Infrared Millimeter Terahertz Waves*, **33**, 825–838. Copyright © 2012 Springer Science+Business Media.

Graphene-Based Terahertz Electronics and Plasmonics: Detector and Emitter Concepts
Edited by Vladimir Mitin, Taiichi Otsuji, and Victor Ryzhii

ISBN 978-981-4800-75-4 (Hardcover), 978-0-429-32839-8 (eBook)
www.jennystanford.com

of electrons and holes in graphene can lead to nontrivial features such as negative dynamic conductivity in the THz spectral range, which may lead to the development of new types of THz lasers. First, the nonequilibrium carrier relaxation/recombination dynamics is formulated to show how photoexcited carriers equilibrate their energy and temperature via carrier-carrier and carrier-phonon scatterings and in what photon energies and in what time duration the dynamic conductivity can take negative values as functions of temperature, pumping photon energy/intensity, and carrier relaxation rates. Second, we conduct time-domain spectroscopic studies using an optical pump and a terahertz probe with an optical probe technique at room temperature and show that graphene sheets amplify an incoming terahertz field. Two different types of samples are prepared for the measurement; one is an exfoliated monolayer graphene on SiO_2/Si substrate and the other is a heteroepitaxially grown non-Bernal stacked multilayer graphene on a 3C-SiC/Si epi-wafer.

31.1 Introduction

Graphene is a one-atom-thick planar sheet of carbon atoms that are densely packed in a honeycomb crystal lattice [1–4]. Since the discovery of the monolayer graphene by Novoselov et al. [1] in 2004 graphene has made a great impact in academic and industry. This material has many peculiar properties and potential applications, including the half-integer quantum Hall effect [2], finite conductivity at zero-charge carrier concentration [3], perfect quantum tunneling effect [4], and ultrahigh carrier mobility [5] owing to massless and gapless energy spectra. The gapless and linear energy spectra of electrons and holes lead to nontrivial features such as negative dynamic conductivity in the terahertz spectral range [6–8], which may lead to the development of new types of terahertz lasers [8–11]. This has attracted intense interest due to the ongoing search for viable terahertz light sources.

To realize such terahertz graphene-based devices, understanding the nonequilibrium carrier relaxation and recombination dynamics is crucial. Intraband and interband carrier scatterings in graphene via optical phonon modes have been extensively studied [12–15]. Recently the effect of carrier-carrier scattering on carrier equilibration dynamics as well as electron-hole generation and recombination dynamics due to interband optical phonon scattering in optically excited graphene have been theoretically studied [15–17]. These theoretical studies gave us an general idea of the carrier relaxation and recombination dynamics; photoexcited carriers in graphene are firstly quasi-equilibrated via carrier-carrier scattering on ultrafast time scales of 20 to 200 fs and then cooled and thermalized mainly by intraband relaxation processes via optical phonon emission on subpicosecond to picosecond time scales, and then by interband recombination processes via interband optical phonon emission. Recently such ultrafast carrier relaxation via emissions of optical phonons has observed by several groups [18–24] and revealed the energy relaxation dynamics of photoexcited carriers. In particular, Breusing et al. [23] revealed that photoelectrons and photo-holes in optically pumped exfoliated graphene and graphite loose a major fraction of their energy within 200~300 fs, mainly by emission of optical phonons. At an earliest stage we theoretically discovered the possibility of negative dynamic conductivity in a wide terahertz frequency range in optically and/or electrically pumped graphene [6, 7]. Recently, we succeeded in observation of stimulated terahertz emission from graphene under femtosecond infrared laser pumping by using a terahertz photon echo method [25, 26].

In this chapter recent advances in spectroscopic study on ultrafast carrier dynamics and terahertz stimulated emission in optically pumped graphene are reviewed. First, the nonequilibrium carrier relaxation/recombination dynamics are formulated to show how photoexcited carriers equilibrate their energy and temperature via carrier-carrier and carrier-phonon scatterings and in what photon energies and in what time duration the dynamic conductivity can take negative values as functions of temperature, pumping photon energy/intensity, and carrier relaxation rates. Second, we conduct time-domain spectroscopic studies using an optical pump and a terahertz probe with an optical probe technique at room temperature and show that graphene sheets amply an incoming terahertz field. Two different types of samples are prepared for the measurement; one is an exfoliated monolayer graphene on SiO_2/Si substrate and the other is a heteroepitaxially grown non-Bernal stacked multilayer graphene on a 3C-SiC/Si epi-wafer. The graphene sample is first pumped by a femtosecond infrared pulsed laser. Then a terahertz probe pulse being synchronized with the pump pulse is impinged to the sample after a few picoseconds from the pumping. Its transmitted and reflected signal is electrooptically detected as a terahertz photon echo signal. The measured temporal response is Fourier transformed to characterize the gain spectral profiles of the graphene samples as functions of the pumping intensity and the probe delay timing.

31.2 Carrier Relaxation and Recombination Dynamics in Optically Pumped Graphene

Figure 31.1 presents the carrier relaxation/recombination processes and the nonequilibrium energy distributions of photoelectrons/ photoholes in optically pumped graphene at specific time scales from ~10 fs to picoseconds after pumping. When the photogenerated electrons and holes are heated in case of room temperature environment and/or strong pumping, collective excitations due to the carrier-carrier (CC) scattering, e.g., intraband plasmons should have a dominant play to perform an ultrafast carrier quasi-equilibration along the energy as shown in Fig. 31.1. Then optical phonons (OPs) are emitted by carriers on the high-energy tail of the electron and hole distributions. This energy relaxation process accumulates the nonequilibrium carriers around the Dirac points as shown in Fig. 31.1. Due to a fast intraband relaxation (ps or less) and relatively slow interband recombination (≫1 ps) of photoelectrons/ holes, one can obtain the population inversion under a sufficiently high pumping intensity [6, 7, 16]. Due to the gapless symmetric band structure of graphene, photon emissions over a wide terahertz frequency range are expected if the pumping infrared photon energy is properly chosen.

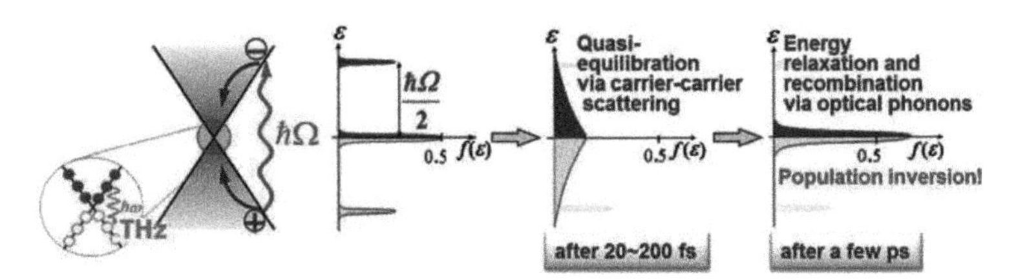

Figure 31.1 Carrier relaxation dynamics in optically pumped graphene.

We consider an intrinsic graphene under optical pulse excitation in the case where the CC scattering is dominant and carriers always take a quasi-equilibrium distribution [16]. We take into account both the intra and interband OPs. The carrier distribution (equivalent electron and hole distributions) is governed by the following equations for the total energy and concentration of carriers:

$$\left.\begin{aligned}
\frac{d\Sigma}{dt} &= \frac{1}{\pi^2}\sum_{\mathrm{i}=\Gamma,\mathrm{K}}\int d\mathbf{k}\Big[(1-f_{\hbar\omega_{\mathrm{i}}-\mathrm{v_w}\hbar\mathrm{k}})(1-f_{\mathrm{v_w}\hbar\mathrm{k}})/\\
&\qquad \tau^{(+)}_{\mathrm{i0,inter}} - f_{\mathrm{v_w}\hbar\mathrm{k}}f_{\hbar\omega_{\mathrm{i}}-\mathrm{v_w}\hbar\mathrm{k}}/\tau^{(-)}_{\mathrm{i0,inter}}\Big],\\
\frac{dE}{dt} &= \frac{1}{\pi^2}\sum_{\mathrm{i}=\Gamma,\mathrm{K}}\int d\mathbf{k}v_{\mathrm{w}}\hbar k\Big[(1-f_{\hbar\omega_{\mathrm{i}}-\mathrm{v_w}\hbar\mathrm{k}})(1-f_{\mathrm{v_w}\hbar\mathrm{k}})/\\
&\qquad \tau^{(+)}_{\mathrm{i0,inter}} - f_{\mathrm{v_w}\hbar\mathrm{k}}f_{\hbar\omega_{i}-\mathrm{v_w}\hbar\mathrm{k}}/\tau^{(-)}_{\mathrm{i0,inter}}\Big]\\
&\quad + \frac{1}{\pi^2}\sum_{\mathrm{i}=\Gamma,\mathrm{K}}\int d\mathbf{k}\hbar\omega_{\mathrm{i}}\Big[f_{\mathrm{v_w}\hbar\mathrm{k}}(1-f_{\mathrm{v_w}\hbar\mathrm{k}+\hbar\omega_{\mathrm{i}}})/\\
&\qquad \tau^{(+)}_{\mathrm{i0,intra}} - f_{\mathrm{v_w}\hbar\mathrm{k}}(1-f_{\mathrm{v_w}\hbar\mathrm{k}-\hbar\omega_{\mathrm{i}}})/\tau^{(-)}_{\mathrm{i0,intra}}\Big],
\end{aligned}\right\} \quad (31.1)$$

where Σ and E are the carrier concentration and energy density, f_ε is the quasi-Fermi distribution at energy ε, $\tau^{(\pm)}_{\mathrm{i0,inter}}$ and $\tau^{(\pm)}_{\mathrm{i0,intra}}$ are the inverses of the scattering rates for inter and intraband OPs (i = Γ for OPs near the zone center Γ point with ω_Γ = 198 meV, i = K for OPs near the zone boundary with ω_K = 161 meV, + for absorption, and – for emission). The values for ω_Γ and ω_K are typical theoretical values in the literatures [12–17]. Time evolution of quasi-Fermi energy ε_F and the carrier temperature T_c are determined by these equations. Figures 31.2a and b show the typical calculated results for time-dependent ε_F and T_c respectively after femtosecond-pulsed laser pumping with photon energy 0.8 eV [16]. It is clearly seen that (i) instantly after the pumping (t~0 ps) ε_F instantly falls down in negative due to CC scattering, and then (ii) ε_F is rapidly elevated by carrier cooling due to emission of OPs in carriers at high-energy tails, and in particular, (iii) ε_F becomes positive, i.e. the population becomes inverted, when the pumping intensity exceeds a certain threshold level. This result proves the occurrence of the population inversion. After that, the recombination process follows more slowly (~10 ps).

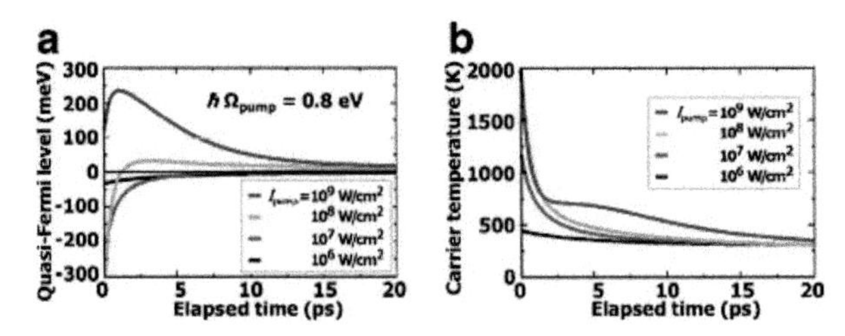

Figure 31.2 Simulated time evolution of (a) the quasi-Fermi level and (b) carrier temperature after impulsive infrared pumping for different pumping intensities.

It is noted that the population inversion is a prerequisite for but does not mean the gain, because we have the Drude absorption by carriers in graphene. A quantity that determines the gain at frequency ω is the real part of the net dynamic conductivity Re σ_ω; negative values of Re σ_ω implies the gain. The real part of the net *ac* conductivity Re σ_ω is proportional to the absorption of photons with frequency ω and comprises the contributions of both interband and intraband contributions [6]:

$$\begin{aligned}\operatorname{Re}\sigma_\omega &= \operatorname{Re}\sigma_\omega^{\text{inter}} + \operatorname{Re}\sigma_\omega^{\text{intra}} \\ &\approx \frac{e^2}{4\hbar}(1-2f_{\hbar\omega}) + \frac{(\ln 2 + \varepsilon_F/2k_BT)e^2}{\pi\hbar}\frac{k_BT\tau}{\hbar(1+\omega^2\tau^2)},\end{aligned} \quad (31.2)$$

where e is the elementary charge, $\hbar$ is the reduced Planck constant, k_B is the Boltzmann constant, and τ is the momentum relaxation time of carrier. The intraband contribution (the second term in Eq. 31.2) corresponds to the Drude absorption and depends on $\omega\tau$. Typical simulated results for Re σ_ω are shown in Fig. 31.3a and b as functions

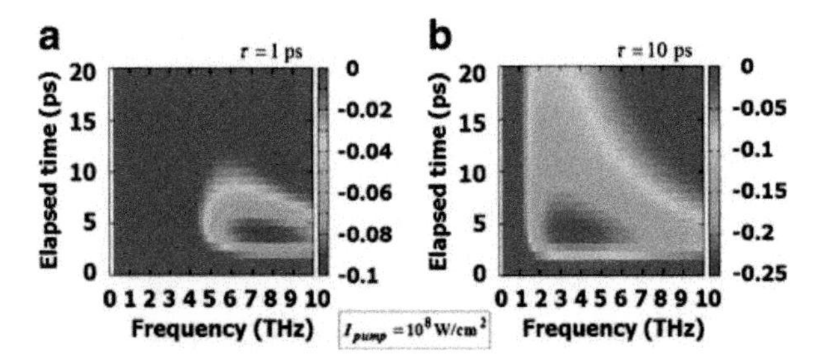

Figure 31.3 Simulated time-dependent terahertz dynamic conductivity when the pumping photon energy is 800 meV and its intensity is 10^8 W/cm^2. (a) Momentum relaxation time τ = 1 ps, (b) τ = 10 ps.

of time and frequency at a fixed pump intensity 1×10^8 W/cm^2 with different momentum relaxation times τ = 1, 10 ps. It is clearly seen that the gain spectral bandwidth widen with longer τ value; when τ = 10 ps, a broad terahertz gain bandwidth from ~1.5 to ~10 THz is expected in picoseconds time scale after pumping [27].

31.3 Experiments

31.3.1 Experimental Setup

In order to verify the above mentioned scenario to obtaining the terahertz gain we conduct time-domain spectroscopic experiments based on an optical pump/terahertz-and-optical-probe technique [25, 26]. Figure 31.4 shows the experimental setup and the pump/probe geometry. The time-resolved field emission properties are measured by an electrooptic sampling (EOS) method in total-reflection geometry. The pumping photon energy (wavelength) is selected to be around 800 meV (1550 nm), much higher than the optical phonon energy (~198 meV). A 2-mm-long, 0.5-mm-wide CdTe crystal with a Si prism is used for reflective electrooptic probing and placed directly onto the sample. The CdTe crystal acts as an electrooptic sensor as well as a terahertz probe pulse generator. A femtosecond-pulsed fiber laser with a full width at half-maximum (FWHM) of 80 fs, repetition rate of 20 MHz was used as the optical pump and probe source. The maximal available average power and fluence are about 4 mW and 0.2 nJ/pulse, respectively. The laser is split into two paths used for pump and probe. The pumping laser beam, being linearly polarized, is mechanically chopped at ~1.2 kHz (for lock-in detection) and focused with a beam diameter of about 40 μm onto the sample and the CdTe from the back side, while the probing beam is cross-polarized with respect to the pump beam and focused from the top side. The CdTe, a nonlinear optical crystal, can rectify the pump laser pulse to emit terahertz envelope radiation. The magnitude of the emitted terahertz envelope pulse grows along the Cherenkov angle, preserving the phase-matching condition between the infrared and the terahertz radiations [28]. Its forward-propagating pulse is first electrooptically detected by the optical probe beam. The terahertz pulse is partially reflected at the top surface of the CdTe, then subjects back to the graphene, serving as a terahertz probe pulse (arrowed blue line in Fig. 31.4) to stimulate terahertz photon emission. The terahertz probe pulse including stimulated emission reflects in most part at the interface between the SiO_2 (or SiC) layer and the Si substrate and travels back to the Si prism, which is detected as a terahertz photon echo signal, as shown in Fig. 31.4. Therefore, the original data of the experimental temporal response consists of the first forward-propagating terahertz pulsation (no interaction with graphene) followed by a photon echo signal (probing the graphene). The delay between these two pulsations is given by the total round-trip propagation time of the terahertz probe pulse through the CdTe. The system bandwidth is estimated to be around 6 THz, mainly limited by the Reststrahlen band of the CdTe sensor crystal.

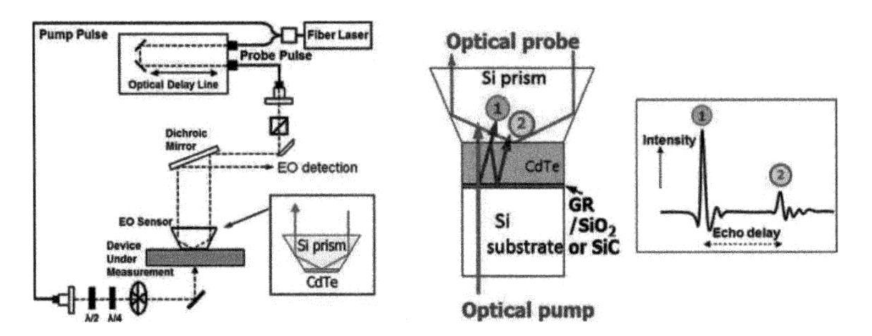

Figure 31.4 Experimental setup (left) and the pump/probe geometry.

31.3.2 Samples and Characterizations

31.3.2.1 Exfoliated graphene on SiO_2/Si

The first sample used is an exfoliated monolayer graphene on SiO_2/Si substrate. The sample consists of (i) a 560-μm-thick highly doped Si (100) substrate having a resistivity of 0.005 Ω-cm, used as the back gate, (ii) a thermally oxidized 300-nm-thick SiO_2 layer on top surface of the Si substrate, and (iii) some islets of exfoliated monolayer, bilayer, and few-layers-thick graphene that are transferred onto the SiO_2 layer. The flake size of the mono-layer graphene under measurement is about 7000 μm^2. The defect-free quality of the monolayer graphene flake was confirmed by Raman spectroscopy as shown in Fig. 31.5. Almost no peak is seen at D band whereas G peak stays at 1576 cm^{-1} with a sharp linewidth 9.8 cm^{-1}, resulting in a high G-peak-to-D-peak intensity ratio, I_G/I_D, greater than 35. As is experimentally examined in Ref. [18], the graphene crystal quality characterized by I_G/I_D strongly correlates to the carrier momentum relaxation time τ. The obtained I_G/I_D value (> 35) suggests that the assumption on the τ value of 10 ps in Section 31.2 is feasible even at room temperature. To see the G′ (2D) band it stays at 2670 cm^{-1} with mono but asymmetric peak and with a line width 18.6 cm^{-1}. Compared to the results for free-standing suspended monolayer graphene in Ref. [29], no degradation is seen in its crystal quality. Also very less doping effect is confirmed from the G′ peak position with no blue-shifting from the ideal position.

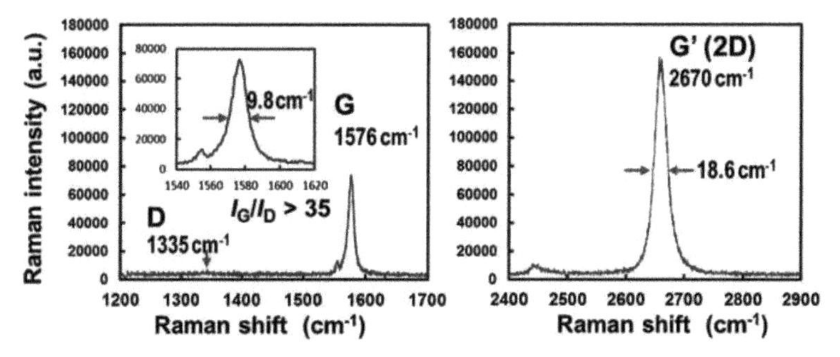

Figure 31.5 Raman spectrum of the monolayer exfoliated graphene on SiO_2/Si substrate, showing G and G′ (2D) peaks and excellent G-peak-to-D-peak ratio greater than 35.

In order to characterize the electrical quality of our sample, we conduct back-gate current-voltage transfer characterization, as well as atomic and Kelvin force microscopy to the surface of the SiO_2/Si substrate and that of graphene on SiO_2/Si substrate. Atomic Force Microscopy (AFM) proved a flat surface morphology of the monolayer graphene flake with a variances σ of 0.142 nm. Figure 31.6 shows the measured transfer characteristics with a Dirac voltage close to 0 V, showing that the level of intrinsic doping is very low in this sample. The Kelvin Force Microscopy (KFM) measurement shows a very smooth surface potential distribution on 10×10 μm^2 area with a variance of 4.05 meV as shown in Fig. 31.7. We therefore carefully selected this high-quality sample, which has less substrate effect [30], for the experiments.

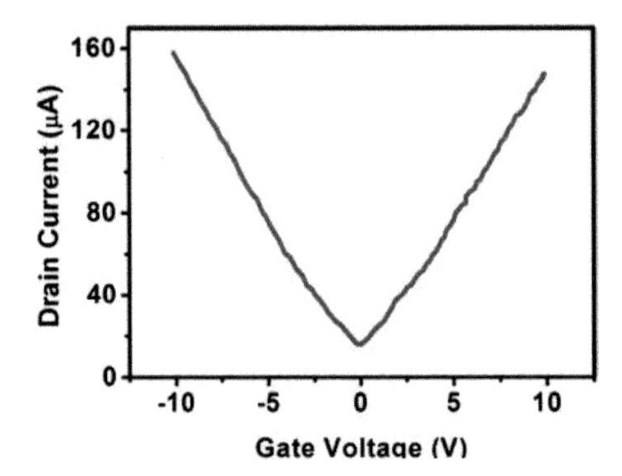

Figure 31.6 Measured ambipolar characteristics of the monolayer graphene flake sample on SiO_2/Si substrate.

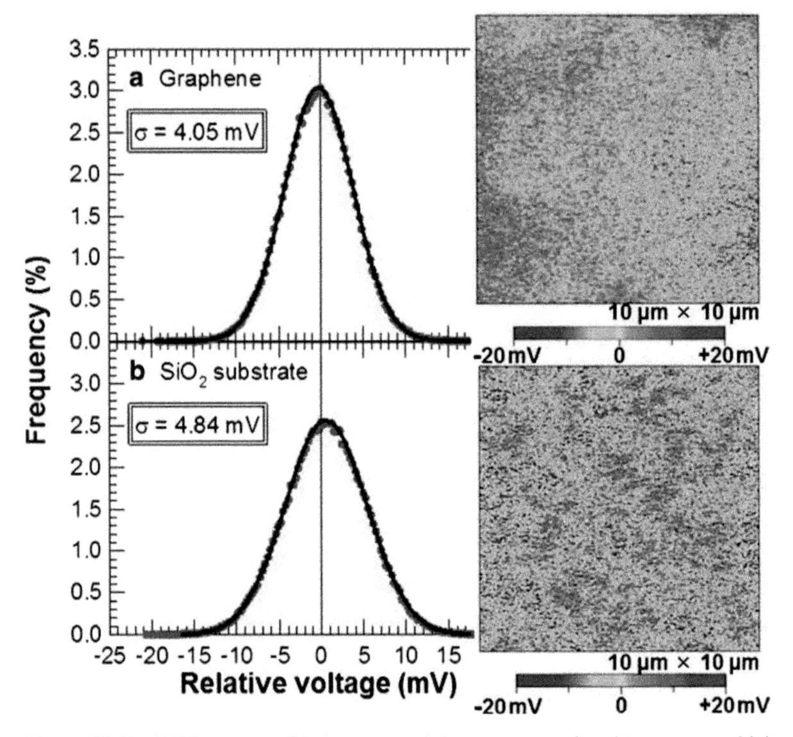

Figure 31.7 KFM topographic images and the corresponding histograms of (a) monolayer exfoliated graphene on SiO_2/Si substrate and (b) SiO_2/Si substrate. The data shown in panels (a) and (b) are described by Gaussian distributions (black solid lines) with variances of 4.05 meV and 4.84 mV respectively. (After Ref. [26].)

31.3.2.2 Heteroepitaxial graphene on 3C-SiC/Si

The second sample used in this experiment is a heteroepitaxial graphene film grown on a 3C-SiC(110) thin film heteroepitaxially grown on a 300-μm thick Si(110) substrate via thermal graphitization of the SiC surface [31, 32]. In the Raman spectrum of the graphene film, the principal bands of graphene, the G (1595 cm^{-1}) and G′ (2730 cm^{-1}) bands, are observed, as shown in Fig. 31.8. Furthermore, transmission electron microscopy images indicate that the film is stratified. It is thus concluded that epitaxial graphene with a planar structure can be produced by this fabrication method. Furthermore, the epitaxial graphene layer is inferred to have a non-Bernal stacking arrangement because the G′ band in the Raman spectrum can be expressed as a single component related to the two-dimensionality of the graphene film [32]. The non-Bernal stacked epitaxial graphene layers grown by our method can be treated as a set of isolated single graphene layers, as in the case of an epitaxial graphene layer on a C-terminated SiC bulk crystal [33]. The G-band peak at 1595 cm^{-1} corresponds to an optical phonon energy at the zone center of 197.8 meV.

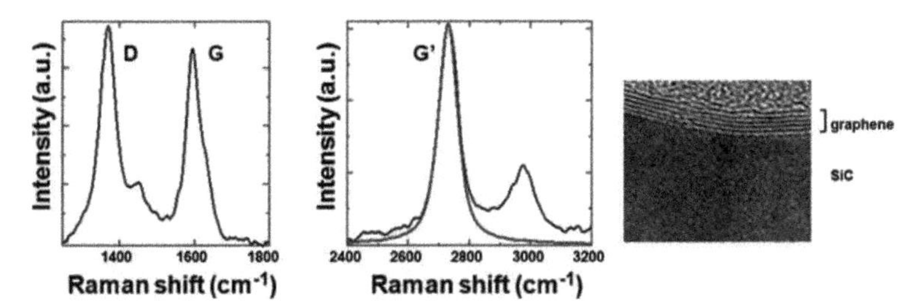

Figure 31.8 Raman spectra of heteroepitaxial graphene on 3C-SiC/Si(110) with the D (1365 cm^{-1}), G (1595 cm^{-1}) and G′ (2730 cm^{-1}) peaks. (After Ref. [25].) The right panel is a transferred electron microscopic image of the cross section of the part of multilayer graphene on 3C-SiC.

31.3.3 Results and Discussions

31.3.3.1 Exfoliated graphene on SiO_2/Si

The experiments were done with two CdTe sensor crystals (A and B), having orientations (100) and (101) and thicknesses of 120 and 80 μm, respectively. Figure 31.9 shows temporal responses

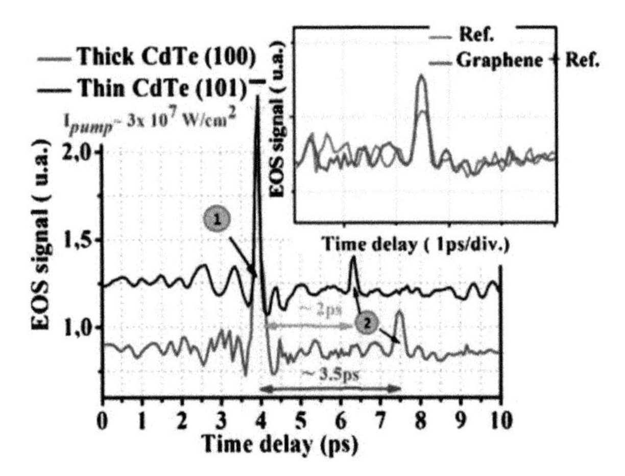

Figure 31.9 Measured temporal responses for monolayer exfoliated graphene on SiO_2/Si with thick (redline) and thin (black line) CdTe crystals for the pumping pulse intensity of about 3×10^7 W/cm^2. Inset: Temporal responses of photon echo signals measured on graphene (red line) and the area without graphene (grey line) for the measurement with thick CdTe crystal. (After Ref. [26].)

measured on monolayer graphene with the thinner (black line) and the thicker (red line) CdTe crystals for the pumping pulse intensity of 3×10^7 W/cm (almost one order of magnitude below the level of Pauli blocking). These curves are plotted with the same origin for comparison. One can notice that, as predicted, each temporal profile is composed of two peaks from optical rectification (OR) in CdTe and the terahertz photon echo signal. The measured time delays between these two pulsations with crystal A (thinner) and crystal B (thicker) of around 2 and 3.5 ps respectively are in good agreements with the round-trip propagation times of the terahertz pulse through the CdTe crystals. The refractive index of CdTe is obtained from Ref. [34]. The OR pulse is found to be narrower in crystal A than in crystal B, owing to better phase-matching conditions in the thinner crystal. Indeed, the coherent length in CdTe is estimated to be around 100 μm at a photon energy 0.8 eV [35]. The inset of Fig. 31.9 presents the echo signal peak intensity measured with crystal B on graphene as well as the reference curve (grey line) measured on the area without graphene. It is clearly seen that the peak obtained on graphene is more intense than that one obtained on the substrate without graphene. This suggests that graphene amplifies the terahertz echo (terahertz probe) signal. The possible origins of the observed amplification could be (i) the terahertz pulse that stimulates the emission from graphene by electron-hole radiative recombination whose energy falls in the range of the negative dynamic conductivity or (ii) the increase in reflectivity under pumping due to the increase in photocarrier-dependent Drude conductivity.

The graphene transfer function H(ω) is defined as H(ω) = G(ω)/S(ω), where G(ω) and S(ω) are the Fourier transforms of the photon echo signals measured on graphene and substrate (without graphene) respectively. Figure 31.10a shows the transfer function of monolayer graphene for different values of pumping pulse intensity I_{pump}. One can see from these results that decreasing I_{pump} drastically reduces the gain spectra, and below 10^7 W/cm the gain disappears and only attenuation can be seen. The phase data (Fig. 31.10b) show clear Lorentzian-like normal dispersion around the gain peak when the pumping pulse intensity is beyond the threshold, demonstrating amplification originated from stimulated emission of photocarriers in the inverted states. Comparing the phase data for the pumping intensities beyond the threshold, with an increase in the pumping intensity, the gain peak frequency (given by the zero point of the phase data) shows a blue shift and the upper cutoff frequency increases, whereas the lower cutoff frequency barely changes. These tendencies well reflect the dependence of the gain spectral profile on pumping intensity as shown in Fig. 31.3 [6, 36]. We also present in the inset of Fig. 31.10a (lower panel) the corresponding ratio AG/AS, where AG and AS are the amplitudes of echo signal peaks measured on graphene and substrate (without graphene) respectively. The graphene transfer function and AG/AS show a clear threshold-like behavior, demonstrating population inversion and negative dynamic conductivity in optically pumped graphene. The threshold intensity is found to be around 10^7 W/cm^2. The inset of Fig. 31.10a (upper panel) presents the normalized EOS signal amplitude of the first peak (see Fig. 31.9) for different values of the pumping pulse intensities. This EOS signal is proportional to the terahertz electric field generated by OR in CdTe crystal. The terahertz emitted intensity is quadratically dependent of the infrared intensity $(I_{\text{THz}} \propto I_{\text{IR}}^2)$. Since $I_{\text{THz}} \propto I_{\text{THz}}^2$, the linear dependency of the EOS signal amplitude with the pumping pulse intensity is a clear evidence of its OR source [26].

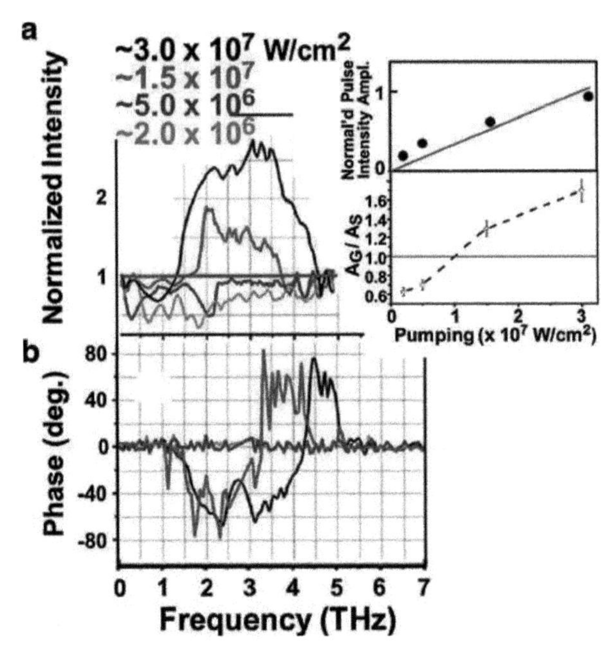

Figure 31.10 (a) Transfer functions of monolayer exfoliated graphene on SiO_2/Si for different values of pumping pulse intensities. Inset: normalized EOS signal amplitude of the first peak (upper panel) and AG/AS ratio (lower panel) for different values of the pumping pulse intensities. (b) Variation of the phase of the measured terahertz electric field for different values of pumping pulse intensities. (After Ref. [26].)

The obtained AG/AS value of the gain is larger than the theoretical limit of ~1.01 for a single process of stimulated emission estimated from Ref. [6], taking into account the absorbance of monolayer graphene and the permittivity's of CdTe and SiO_2. In the experimental setup, a multipath reflection effect is considered as a factor that increases the gain. The forward-propagating terahertz probe pulse is amplified, its large fraction is transmitted, and ~70% is reflected back at the SiO_2/Si interface. This reflected part will also interact with graphene and be amplified. Since the round-trip delay in the 300-nm-thick SiO_2 layer is ~4 fs, almost no distortion is seen in the temporal response. This multipath reflection effect may double the gain at most but cannot fully explain the obtained AG/AS.

We also considered an effect of photocarrier-dependent Drude conductivity on reflectivity. In case of our graphene/SiO_2/Si sample whose Si substrate has a low resistivity of 0.005 Ω-cm and whose carrier momentum relaxation time is long and close to 10 ps, the increase in reflectivity under pumping due to the increase in photocarrier-dependent Drude conductivity is as small as 1–2% and quite monotonic over the entire terahertz spectral range of interest. A few percentages of such increase in the reflectivity should be de-embedded from the obtained photon-echo signal but barely affect the gain spectral profile. We need further study to give a perfect quantitative interpretation for the observed results.

Figure 31.11 shows the Fourier transform of the photon echo signal measured on the area without graphene (reference) with crystal A (black line) and crystal B (red line). The normalized dynamic conductivity for the pumping pulse intensity three times higher than the threshold pumping pulse intensity at 300 K is also presented (see Ref. [6]). The photon echo pulse interacts with graphene during the recombination process and induces emission of terahertz photon in graphene within the negative dynamic conductivity area (blue shaded area). The expected graphene emission spectral bandwidth is limited at lower frequencies by the Drude mechanism of terahertz absorption, and the higher frequency limit is given by the system bandwidth, which is estimated to be around 6 THz (see the horizontal solid lines in Fig. 31.11). The black and red shaded areas show the expected graphene emission bandwidth, from ~1.5 to ~5 THz using the thicker crystal and from ~1.5 to ~6 THz using the thinner crystal.

The inset of Fig. 31.11 presents the transfer functions of monolayer graphene obtained with crystal A (black line) and crystal B (red line). The obtained spectra are in good agreement with the above mentioned expectations. It is possible to compare these spectra within the smallest bandwidth (from ~1.5 to ~5 THz). The graphene emission spectrum obtained with the thinner CdTe crystal is broader than that one obtained with the thicker crystal. In the thinner crystal, the graphene excitation with the terahertz probe is done earlier (2 ps) after the optical pump than in the thicker crystal (3.5 ps). This spectral narrowing at longer terahertz probe delay time is a clear trace of the progress of equilibration process from 2 to 3.5 ps as shown in Figs. 31.2 and 31.3. Indeed, at 3.5 ps, the quasi-Fermi energy is closer to the equilibrium compared to that at 2 ps. It is worth noting that this broadening may not be observed in the case of amplification due to an increase in reflectivity.

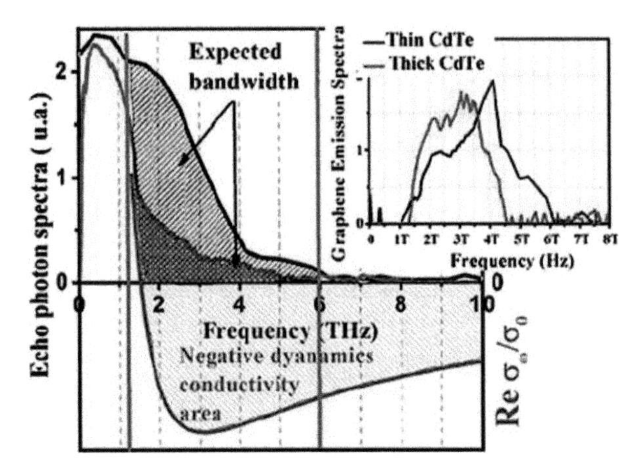

Figure 31.11 Terahertz emission spectra of the photon echo signal measured on the area without graphene (reference) with thick CdTe (red line) and thin CdTe (black line) crystals and the normalized dynamic conductivity (blue line). Inset: transfer functions of monolayer graphene obtained with thick CdTe (red line) and thin CdTe (black line) crystals. (After Ref. [26].)

Such a temporal dependency of the gain profile is theoretically testified [16]. As is shown in Fig. 31.2, temporal evolutions of the quasi-Fermi energy of impulsively pumped graphene rapidly recovers its level and exceeds the equilibrium level when pumping intensity exceeds a threshold and resumes to the equilibrium state. Figure 31.12 shows the dynamic conductivity calculated as a function of the pumping intensity at a fixed time (3.5 ps) and a fixed frequency (3 THz). Compared with the experimentally observed threshold behavior, the measured results qualitatively agree with the simulated ones. The threshold intensity is on the same order. It is thought that the quantitative difference comes from the fact that we assume the complete quasi-equilibration by the CC scattering, raising the threshold pumping intensity to the highest level, and this leads to excessively high threshold intensity.

Furthermore, to confirm the effects of the terahertz probe that stimulates the emission in graphene, the CdTe crystal was replaced by a CdTe crystal having a high-reflectivity coating for Infrared on its bottom surface in order to eliminate generation of the terahertz probe signal. In this case, no distinctive response was observed. Since the measurements are taken as an average, the observed response is undoubtedly a coherent process that cannot be obtained via spontaneous emission processes, which also supports the occurrence of the stimulated emission.

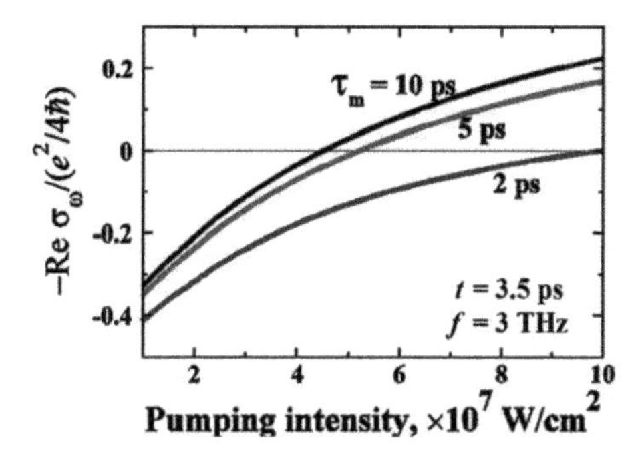

Figure 31.12 Numerically simulated dynamic conductivity as a function of the pumping intensity at a fixed time of 3.5 ps and a fixed frequency of 3 THz. (After Ref. [26].)

31.3.3.2 Heteroepitaxial graphene on 3C-SiC/Si

As mentioned above we focus to observe the secondary terahertz photon echo pulse that reflects the response of graphene. Typical raw data of observed temporal response is shown in Fig. 31.13a. As is mentioned in IV, we extracted the temporal response of the secondary terahertz pulsation from the measured raw data to identify the frequency response of the graphene. Figure 31.13a shows the measured temporal responses, in which repetitively measured results are overlaid, showing the measurement reproducibility. Figure 31.13b shows the Fourier spectra corresponding to the typical traces plotted with thick lines in Fig. 31.13a. Owing to the second-order nonlinear optical effects, the emission from "CdTe only" without the GOS sample exhibits a temporal response similar to optical rectification with a single peak at around 1 THz and an upper weak side lobe extending to around 5 THz as shown by the green lines in Fig. 31.13b. On the other hand, the temporal profile of the result from "CdTe and GOS" intensifies the pulsed response with higher frequency components so that its Fourier spectrum exhibits the growth of the main lobe around 2 to 4 THz and the lower side lobe around 1 THz. The main lobe fairly corresponds to the expected gain spectral region which is shown with the dashed line in Fig. 31.13b, while the side lobe fairly corresponds to the component of terahertz probe signal generated from the CdTe. It is inferred that the terahertz emissions from graphene are stimulated by the coherent terahertz probe radiation from the CdTe. Furthermore, the terahertz emissions are amplified via photoelectron/hole recombination in the range of the negative dynamic conductivity. Compared to the results from the exfoliated monolayer garaphene on SiO_2/Si, the measured responses show poor S/N and reproducibility, which is presumably attributed to poor crystal quality of the heteroepitaxial graphene sample. Further study is expected for more accurate measurements.

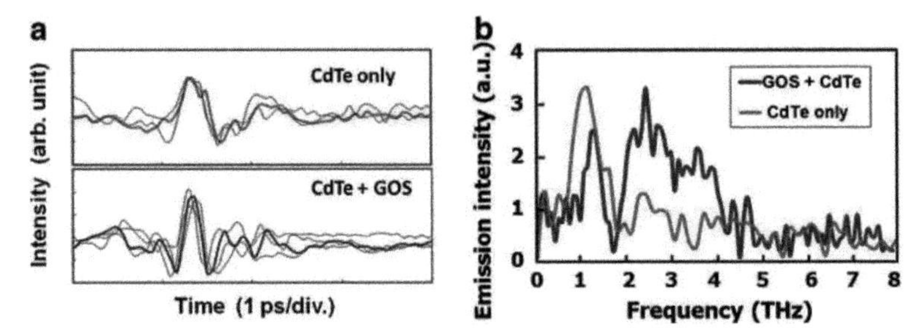

Figure 31.13 (a) Measured temporal profiles for heteroepitaxial graphene on 3C-SiC/Si (GOS) and (b) their Fourier spectra when the terahertz probe beam is generated by the optical rectification in the CdTe crystal. Dashed line in (b) is the photoemission spectrum predicted from the pumping laser spectrum and the negative dynamic conductivity. (After Ref. [25].)

31.4 Conclusion

Recent advances in spectroscopic study on ultrafast carrier dynamics and terahertz stimulated emission in optically pumped graphene were reviewed. When we consider the ultrafast carrier relaxation and relatively slow recombination dynamics in optically pumped graphene, one dramatic feature of negative dynamic conductivity in the terahertz range is derived. We have successfully observed amplification of an incoming terahertz pulse during the relaxation-recombination process in graphene. We studied the pump-power-dependent measurements and showed the existence of a critical pumping intensity at which the amplification regime starts. We investigated the amplitude and the phase of the measured terahertz field. The amplitude measured on graphene is larger than that measured on the area without graphene, while the phase shows clear inverted behavior with normal dispersion around the gain peak from that for the loss cases, demonstrating that the amplification can be attributed to stimulated emission of photocarriers in the inverted states. We also confirmed the dependence of the emission spectra on terahertz-probe timing. The gain spectra show non-monotonic frequency dependence with a clear narrowing when the terahertz probe timing is set at longer delay time (3.5 ps). A possible interpretation is that we observed coherent amplified stimulated terahertz emissions and its threshold behavior against the pumping intensity arising from the fast relaxation and relatively slow recombination dynamics of photogenerated electrons or holes in exfoliated graphene on SiO_2/Si as well as in heteroepitaxial graphene on 3C-SiC/Si. The obtained results support the occurrence of negative dynamic conductivity, which can be applied to new types of terahertz lasers.

Acknowledgements

The authors thank H. Karasawa, T. Watanabe, S. Chan, and T. Fukushima at Tohoku University, Japan, for their contribution on the experimental works. They also thank M. Ryzhii at University of Aizu, Japan, and V. Mitin at University at Buffalo, SUNY, USA, for their theoretical support, and J. Kono at Rice University, USA, for his valuable discussion. This work is financially supported by JST-CREST, Japan, JSPS Grant-in-Aid for Specially Promoting Research, JSPS Core-to-Core Programs, Japan, and the NSF-PIRE Teranano Nano-Japan Program, USA.

References

1. K. S. Novoselov, A. K. Geim, S. V. Morozov, D. Jiang, Y. Zhang, S. V. Dubonos, I. V. Grigorieva, and A. A. Firsov, *Science*, **306**, 666 (2004).
2. Y. Zhang, Y. W. Tan, H. L. Stormer, and P. Kim, *Nature*, **438**, 201 (2005).
3. K. S. Novoselov, A. K. Geim, S.V. Morozov, D. J. M. I. Katsnelson, I. V. Grigorieva, S. V. Dubonos, and A. A. Firsov, *Nature*, **438**, 197 (2005).
4. M. I. Katsnelson, K. S. Novoselov, and A. K. Geim, *Nat. Phys.*, **2**, 620 (2006).
5. A. K. Geim and K. S. Novoselov, *Nat. Mater.*, **6**, 183 (2007).

6. V. Ryzhii, M. Ryzhii, and T. Otsuji, *J. Appl. Phys.*, **101**, 083114 (2007).
7. M. Ryzhii and V. Ryzhii, *Jpn. J. Appl. Phys.*, **46**, L151–L153 (2007).
8. V. Ryzhii, M. Ryzhii, V. Mitin, and T. Otsuji, *J. Appl. Phys.*, **110**, 094503 (2011).
9. A. A. Dubinov, V. Y. Aleshkin, M. Ryzhii, T. Otsuji, and V. Ryzhii, *Appl. Phys. Express*, **2**, 092301 (2009).
10. V. Ryzhii, M. Ryzhii, A. Satou, T. Otsuji, A. A. Dubinov and V. Y. Aleshkin, *J. Appl. Phys.*, **106**, 084507 (2009).
11. V. Ryzhii, A. Dubinov, T. Otsuji, V. Mitin, and M. S. Shur, *J. Appl. Phys.*, **107**, 054505 (2010).
12. T. Ando, *J. Phys. Soc. Jpn.*, **75**, 124701 (2006).
13. H. Suzuura and T. Ando, *J. Phys. Soc. Jpn.*, **77**, 044703 (2008).
14. F. Rana, P. A. George, J. H. Strait, J. Dawlaty, S. Shivaraman, M. Chandrashekhar, and M. G. Spencer, *Phys. Rev. B*, **79**, 115447 (2009).
15. A. Satou, F. T. Vasko, and V. Ryzhii, *Phys. Rev. B*, **78**, 115431 (2008).
16. A. Satou, T. Otsuji, and V. Ryzhii, *Jpn. J. Appl. Phys.*, **50**, 070116 (2011).
17. R. Kim, V. Perebeinos, and P. Avouris, *Phys. Rev. B*, **84**, 075449 (2011).
18. J. M. Dawlaty, S. Shivaraman, M. Chandrashekhar, F. Rana, and M. G. Spencer, *Appl. Phys. Lett.*, **92**, 042116 (2008).
19. D. Sun, Z.-K. Wu, C. Divin, X. Li, C. Berger, W. A. de Heer, P. N. First, and T. B. Norris, *Phys. Rev. Lett.*, **101**, 157402 (2008).
20. P. A. George, J. Strait, J. Dawlaty, S. Shivaraman, M. Chandrashekhar, F. Rana, and M. G. Spencer, *Nano Lett.*, **8**, 4248 (2008).
21. H. Wang, J. H. Strait, P. A. George, S. Shivaraman, V. B. Shields, M. Chandrashekhar, J. Hwang, F. Rana, M. G. Spencer, C. S. Ruiz-Vargas, and J. Park, *Appl. Phys. Lett.*, **96**, 081917 (2010).
22. T. Kampfrath, L. Perfetti, F. Schapper, C. Frischkorn, and M. Wolf, *Phys. Rev. Lett.*, **95**, 187403 (2005).
23. M. Breusing, C. Ropers, and T. Elsaesser, *Phys. Rev. Lett.*, **102**, 086809 (2009).
24. S. Winner, M. Orlita, P. Plochocka, P. Kossacki, M. Potemski, T. Winzer, E. Malic, A. Knorr, M. Sprinkle, C. Berger, W. A. de Heer, H. Schneider, and M. Helm, *Phys. Rev. Lett.*, **107**, 237401 (2011).
25. H. Karasawa, T. Komori, T. Watanabe, A. Satou, H. Fukidome, M. Suemitsu, V. Ryzhii, and T. Otsuji, *J. Infrared Millimeter Terahertz Waves*, **32**, 655–665 (2011).

26. S. Boubanga-Tombet, S. Chan, T. Watanabe, A. Satou, V. Ryzhii, and T. Otsuji, *Phys. Rev. B*, **85**, 035443 (2012).

27. A. Satou, S. A. Boubanga Tombet, T. Otsuji, and V. Ryzhii, Study of threshold behavior of stimulated terahertz emission from optically pumped graphene, *OTST: Int. Conf. on Optical Terahertz Science and Technology, TuA3*, Santa Barbara, CA, USA, March 13–17, 2011.

28. J. Hebling, G. Alm'asi, and I. Z. Kozma, *Opt. Express*, **10**, 1161 (2002).

29. S. Berciaud, S. Ryu, L. E. Brus, and T. F. Heinz, *Nano Lett.*, **9**, 346–352 (2009).

30. J. Martin, N. Akerman, G. Ulbricht, T. Lohmann, J. H. Smet, K. Von Klitzing, and A. Yacoby, *Nat. Phys.*, **4**, 144–148, (2008).

31. M. Suemitsu and H. Fukidome, *J. Phys. D: Appl. Phys.*, **43**, 374012 (2010).

32. H. Fukidome, R. Takahashi, S. Abe, K. Imaizumi, H. Handa, H.-C. Kang, H. Karasawa, T. Suemitsu, T. Otsuji, Y. Enta, A. Yoshigoe, Y. Teraoka, M. Kotsugi, T. Ohkouchi, T. Kinoshita, and M. Suemitsu, *J. Mater. Chem.*, **21**, 17242–17248 (2011).

33. J. Hass, F. Varchon, J. E. Millán-Otoya, M. Sprinkle, N. Sharma, W. A. de Heer, C. Berger, P. N. First, L. Magaud, and E. H. Conrad, *Phys. Rev. Lett.*, **100**, 125504 (2008).

34. M. Schall, H. Helm, and S. R. Keiding, *Int. J. Infrared Millimeter Waves*, **20**, 595 (1999).

35. M. Nagai, K. Tanaka, H. Ohtake, T. Bessho, T. Sugiura, T. Hirosumi, and M. Yoshida, *Appl. Phys. Lett.*, **85**, 3974 (2004).

36. V. Ryzhii, M. Ryzhii, V. Mitin, A. Satou, and T. Otsuji, *Jpn. J. Appl. Phys.*, **50**, 094001 (2011).

Chapter 32

Gain Enhancement in Graphene Terahertz Amplifiers with Resonant Structures*

Y. Takatsuka,[a] K. Takahagi,[a] E. Sano,[a,b] V. Ryzhii,[b,c] and T. Otsuji[b,d]
[a]*Research Center for Integrated Quantum Electronics, Hokkaido University, Sapporo 060-8628, Japan*
[b]*Japan Science and Technology Agency, Core Research for Evolutional Science and Technology, Chiyoda, Tokyo 102-0075, Japan*
[c]*Computational Nanoelectronics Laboratory, University of Aizu, Aizu-Wakamatsu, Fukushima 965-8580, Japan*
[d]*Research Institute of Electrical Communication, Tohoku University, Sendai 980-8577, Japan*
esano@rciqe.hokudai.ac.jp

Terahertz (THz) devices have been investigated over the last decade to utilize THz waves for non-destructive sensing and high-speed

*Reprinted with permission from Y. Takatsuka, K. Takahagi, E. Sano, V. Ryzhii, and T. Otsuji (2012). Gain enhancement in graphene terahertz amplifiers with resonant structures, *J. Appl. Phys.*, **112**, 033103. Copyright © 2012 American Institute of Physics.

Graphene-Based Terahertz Electronics and Plasmonics: Detector and Emitter Concepts
Edited by Vladimir Mitin, Taiichi Otsuji, and Victor Ryzhii

ISBN 978-981-4800-75-4 (Hardcover), 978-0-429-32839-8 (eBook)
www.jennystanford.com

wireless communications. Graphene with gapless and linear energy spectra is expected to exhibit population inversion and has negative dynamic conductivity in the THz range when it is illuminated by infrared light. We analyze a THz amplifier utilizing this negative dynamic conductivity combined with electric field enhancements due to surface plasmon polaritons induced on a metal mesh and with a resonant structure. We evaluate its characteristics through finite-difference time-domain electromagnetic simulations. The amplifier is expected to remarkably enhance THz emissions compared with amplifiers without the resonant structure.

32.1 Introduction

The terahertz (THz) region, between the radio-wave and light-wave regions in electromagnetic waves, is expected to provide applications for high-speed wireless communications, non-destructive sensing, and other uses. However, functional devices like lasers operating with sufficient output power in this region at room temperature (RT) have not yet been achieved. The maximum operating frequency of electronic devices in this region is limited by the electron velocity in semiconductors. However, optical devices are affected by thermal noise at RT because the band-gap energy contributing to laser oscillation is small. Sufficient output power for THz waves has not been produced using conventional devices for these reasons. Therefore, we expect that THz devices will be developed by using novel operating principles.

Graphene, a monolayer of carbon atoms in a honeycomb lattice, has attracted considerable attention due to its electronic and optical properties [1]. Graphene has gapless and linear energy spectra. Ryzhii et al. predicted that optically pumped graphene would exhibit population inversion and have negative dynamic conductivity in the THz region [2, 3]. This negative dynamic conductivity led to THz lasers [4]. Very recently, Boubanga-Tombet et al. observed enhanced THz emissions from exfoliated graphene on an SiO_2/Si substrate by using electro-optic sampling and revealed threshold behavior against pumping intensity, which suggested the occurrence of negative dynamic conductivity [5].

Metallic sheets perforated with a periodic array of holes, referred to as metal mesh structures, have extremely high transmittance at wavelengths close to the period of the array. This phenomenon is known as extraordinary optical transmission (EOT). The holes generally limit the transmittance of electromagnetic waves with wavelengths longer than the hole diameter. EOT for metal mesh structures is thought to be caused by the field enhancement effect due to surface plasmon polaritons (SPPs), which are induced when electromagnetic waves are illuminated and that emit transmitted waves through the periodic hole array [6]. EOT with metal mesh structures has been reported in wide electromagnetic wave regions including the THz region [7–10].

Based on the electromagnetic properties of multiple layers of graphene (MLG) and metal mesh structures, we came up with an idea for THz amplifiers with remarkably enhanced gain at a particular frequency. We also predicted the useful characteristics of these amplifiers in our previous work [11]. We propose a THz amplifier with a resonant structure to increase gain in this chapter, and report these characteristics through finite-difference time-domain (FDTD) electromagnetic simulations.

32.2 FDTD Model

Figure 32.1a is a schematic of our THz amplifier with its resonant structure. Metal mesh structures with holes that are filled with MLGs demonstrate enhanced gain at a particular frequency. The metal substrate (high reflectance material like Au, Ag, and Al) acts as a mirror in the THz region. Since the THz wave is reflected at the mirror and again transmitted through the THz amplifier, gain will be increased more than with the original amplifier.

The FDTD simulation model we used is outlined in Fig. 32.1b. Only one unit cell needs to be simulated by imposing periodic boundary conditions. Perfect matching layers (PMLs) were placed on the top surface to avoid unwanted reflections from the surfaces. An electric wall was placed at the bottom surface corresponding to the mirror. The side area corresponded to the 90×90 μm period of the hole array, and the square holes were 45×45 μm in size. Distance L between the metal mesh structure and mirror ranged in size from

25 to 150 μm. Because the MLG only had negative conductivity in directions along the graphene layer (the x and y directions defined in Fig. 32.1b), the conductivity in these directions ($\sigma_x = \sigma_y = \sigma$) was set to a negative value, and σ_z was 0. MLG conductivity σ was limited to −10 S/m because negative conductivity with too large absolute value led to numerical divergence in the calculations. Moreover, to reduce the calculation time and memory size, thickness t of metal mesh structures with holes that were filled with MLG were fixed at 10 μm, and the spatial step size was 1 μm. It is a good approximation to regard almost all metals as perfect conductors in the THz region, and we treated the metal mesh structure as a perfect conductor. For simplicity, the permittivity of the region between the metal mesh and mirror was set to one.

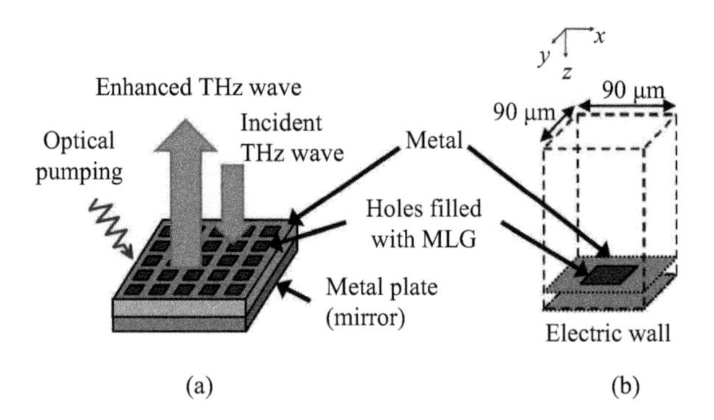

Figure 32.1 (a) Schematic of THz amplifier and (b) FDTD simulation model.

32.3 Results and Discussion

The emittance spectra of THz amplifiers with mirrors have been given in Fig. 32.2. The simulation models were L = 83 μm and L = 52 μm. The emittance exceeds one over almost all the frequencies. This is why there is no lossy medium in the simulation model and incident waves are reflected perfectly by the metal mesh and mirror. While one peak was observed at around 3 THz for L = 83 μm, two peaks were observed for L = 52 μm. These peaks of emittance were higher than the peak emittance for the basic amplifier without a mirror (~1.22) [11]. These results revealed that the gain of our THz amplifiers was increased with the introduction of mirrors. The insets in Fig. 32.2 show the temporal responses of the amplifiers to the input pulses. The long-continued oscillations were electromagnetic waves radiated from the SPPs and they contributed to the components of emittance peaks.

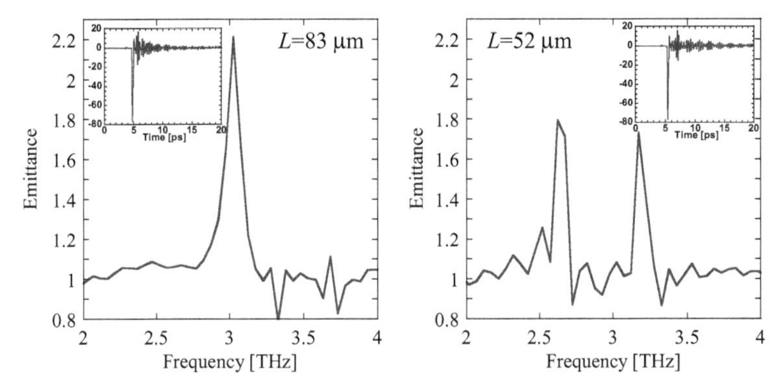

Figure 32.2 Emittance spectra of THz amplifier with mirror when *L*s are 83 μm and 52 μm. Insets: Temporal responses.

It seems peculiar that the shape of the emittance spectra changes with variation in *L* as can be seen in Fig. 32.2. Figure 32.3 summarizes the dependence of emittance characteristics on *L*. Distance *L* was changed from 25 to 150 μm in 1–5 μm steps. The closed circles denote the frequencies that emittance peaked and the open diamonds denote the values of emittance at these frequencies. Within the range of *L* that we simulated, the symbols were divided into three groups that were colored red, blue, and green. Two peaks were observed in the *L* regions, where the emittance values were decreased. In addition, the peak frequencies were limited in a particular range from 2.5 to 3.3 THz. We investigated the physical reason for these observations.

Figure 32.4 plots the simulated spectra for a mirror-less THz amplifier. The red and blue lines correspond to transmittance and reflectance. The green line, called emittance, equals the sum of transmittance and reflectance. The introduction of the mirror causes multiple reflections between the metal mesh and mirror. When the amplification of the reflected waves in the MLG is ignored, the amplitude of the wave reflected from the amplifier with the mirror should equal the emittance of the mirror-less amplifier. Drastically amplified waves were not expected to be emitted in the frequency region where emittance was smaller than one. The frequency where emittance was larger than one ranges from 2.47 to 3.33 THz in Fig. 32.4. This frequency region matched well with the particular range where emittance peaks appeared in Fig. 32.3.

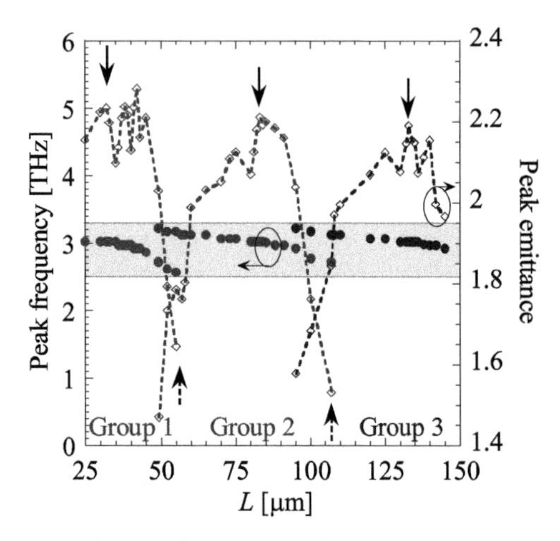

Figure 32.3 Dependence of emittance characteristics on L. Closed circles: Peak frequency. Open diamonds: Peak emittance. Solid arrows: Phase matching conditions for two waves incident to MLG. Dotted arrows: Depressing conditions.

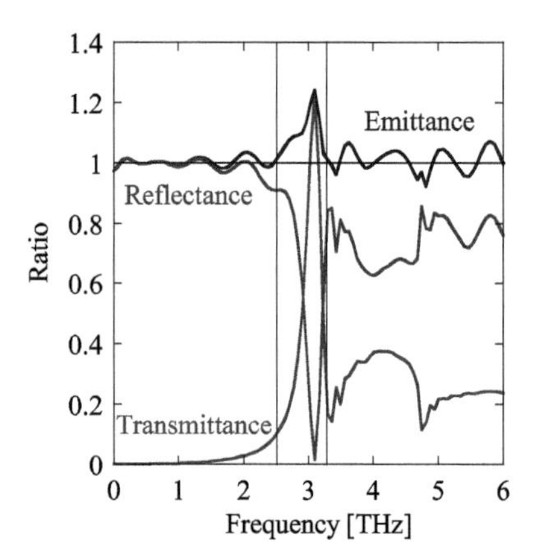

Figure 32.4 Transmittance (red line) and reflectance (blue line) for mirror-less THz amplifier. Green line is emittance corresponding to sum of these. Vertical lines have been drawn at 2.47 and 3.33 THz.

Next, let us discuss the dependence of emittance peaks on L based on the interference between incident and reflected THz waves. Figure 32.5 shows the schematic aspects of a THz wave incident on the amplifier, transmitted through MLG, and reflected from the mirror. The incident wave excites SPP on the metal mesh and the SPP radiates the amplified wave downward. The radiated wave corresponds to the transmitted wave. This process results in a time delay between the incident and transmitted waves [12]. We obtained the phase difference between the incident and transmitted waves by simulating the mirror-less amplifier. Figure 32.6 plots the phase difference as a function of frequency, $D_{\text{SPP}}(f)$. Using $D_{\text{SPP}}(f)$ at the peak frequency for distance L gives the phase difference as a function of L, $D_{\text{SPP}}(L)$. The phase difference between the two incident THz waves to MLG indicated by a and b in Fig. 32.5 is given by

$$\text{Phase delay}\,[^\circ] = 360\left[\frac{2L}{\lambda} - \text{int}\,(2L/\lambda)\right] + 180 + D_{\text{SPP}}(L), \quad (32.1)$$

where λ is the wavelength. The second term on the right-hand side of Eq. (32.1) describes the reflection phase at the mirror. Figure 32.7 plots the calculated phase difference between the two THz waves as function L along with $D_{\text{SPP}}(L)$. The phase difference equalled zero when L = 32, 82, and 132 μm. The emittance of the amplifier with the mirror was drastically increased under these phase matching conditions. The maximum peak emittance in the simulation results in Fig. 32.3 for the three groups appeared at L = 42, 83, and 134 μm. These values were in good agreement with the Ls under phase matching conditions. The bold arrows in Fig. 32.3 designate the Ls under phase matching conditions.

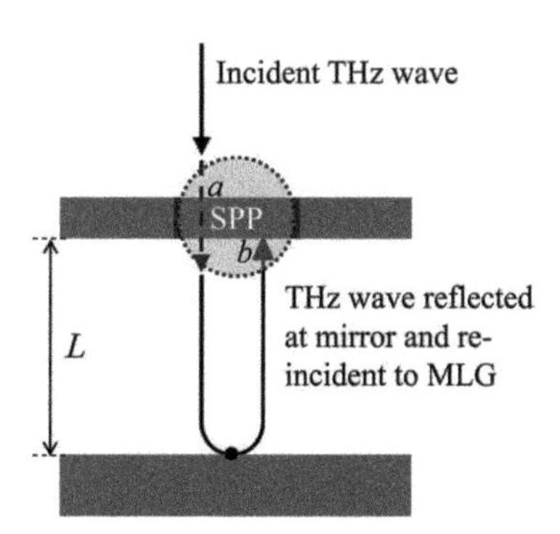

Figure 32.5 Schematic of THz waves in resonant structure.

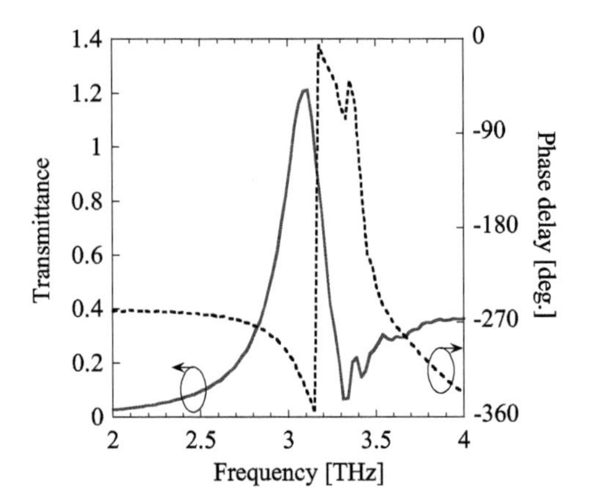

Figure 32.6 Transmittance and phase characteristics for mirror-less THz amplifier.

When the phase difference equaled 180°, the two incident waves were depressed. As we can see from Fig. 32.2, one peak at around 3 THz for L = 83 μm was depressed and split into two peaks for L = 52 μm. The distances Ls resulting in the phase difference of −180° at 3 THz were 57 and 107 μm. These two Ls are designated as open circles in Fig. 32.7. The dotted arrows in Fig. 32.3 designate the Ls under depressing conditions. The interference between two waves incident to MLG describes well the dependence of peak emittance on L.

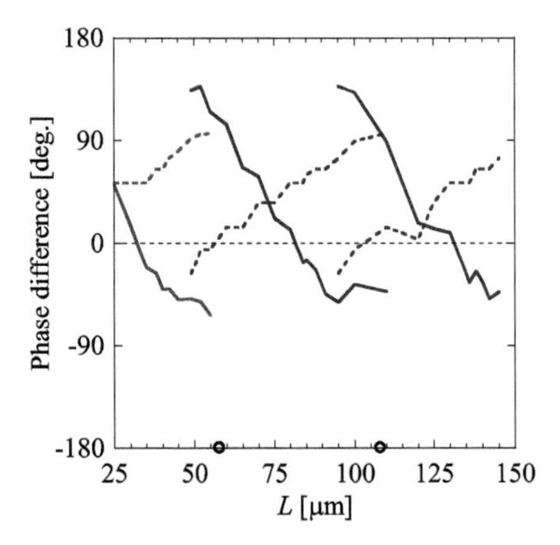

Figure 32.7 Calculated phase difference between two incident THz waves (solid lines) along with $D_{SPP}(L)$ (dotted lines). Open circles designate L resulting in phase difference of −180° at 3 THz.

We then investigated the dependence of peak emittance on conductivity σ and MLG thickness t for a THz amplifier with a resonant structure (L = 133 μm). Figure 32.8 plots the simulated results. The dependence of peak emittance on σ and t could be approximated by $\exp(A\sigma t)$ with an A of –7849.4 (dotted lines in Fig. 32.8). The optically pumped MLG (100 layers) was predicted to have conductivity σ of –48340 S/m, and its thickness t was 34 nm when the interlayer distance of the MLG was assumed to be 0.34 nm [3]. Since the peak emittance could not directly be obtained by using these parameters in the FDTD simulation, it was estimated by using the approximating equation. The peak emittance was calculated to be 3.66×10^5. This value was much larger than the peak transmittance of about five for the mirror-less THz amplifier [11]. A gain increase of about 7×10^4 times should be achieved by introducing a resonant structure.

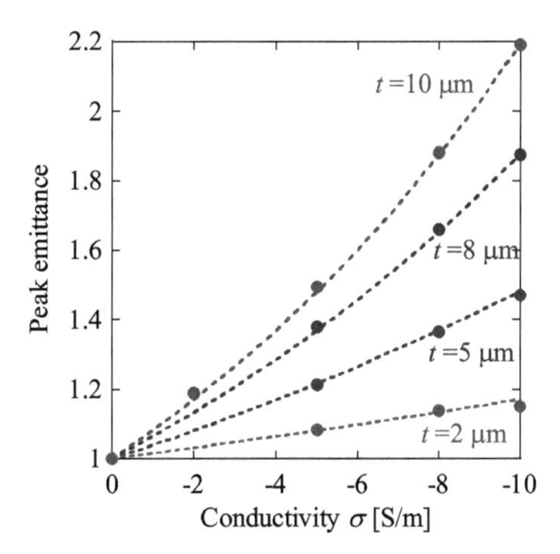

Figure 32.8 Dependence of peak emittance on σ and t. Closed circles: FDTD simulation. Dotted lines: Approximated with $\exp(A\sigma t)$.

32.4 Conclusions

We proposed MLG-based THz amplifiers with a resonant structure, and investigated our amplifiers through FDTD simulations. These amplifiers predicted remarkably enhanced emittance compared with amplifiers without a resonant structure. Consequently, our THz amplifier should be a very promising candidate for THz laser sources.

In addition, we investigated what effect the distance between the metal mesh and mirror had on the emittance characteristics based on the inference of the two waves incident to the MLG, and found the resonant structure design had outstanding features.

Acknowledgments

The present study was done with support from the JST CREST program and a Grant-in-Aid for Scientific Research from the Ministry of Education, Culture, Sports, Science and Technology, Japan.

References

1. K. S. Novoselov, A. K. Geim, S. V. Morozov, D. Jiang, Y. Zhang, S. V. Dubonos, I. V. Grigorieva, and A. A. Firsov, *Science*, **306**, 666 (2004).
2. V. Ryzhii and M. Ryzhii, *J. Appl. Phys.*, **101**, 083114 (2007).
3. V. Ryzhii, M. Ryzhii, A. Satou, T. Otsuji, A. A. Dubinov, and V. Ya. Aleshkin, *J. Appl. Phys.*, **106**, 084507 (2009).
4. V. Ryzhii, A. A. Dubinov, T. Otsuji, V. Mitin, and M. S. Shur, *J. Appl. Phys.*, **107**, 054505 (2010).
5. S. Boubanga-Tombet, S. Chen, T. Watanabe, A. Satou, V. Ryzhii, and T. Otsuji, *Phys. Rev. B*, **85**, 035443 (2012).
6. J. B. Pendry, L. Martín-Moreno, and F. J. Garcia-Vidal, *Science*, **305**, 847 (2004).
7. C. Winnewisser, F. Lewen, and H. Helm, *Appl. Phys. A*, **66**, 593 (1998).
8. D. W. Porterfield, J. L. Hesler, R. Densing, E. R. Mueller, T. W. Crowe, and R. M. Weikle II, *Appl. Opt.*, **33**, 6046 (1994).
9. C.-C. Chen, *IEEE Trans. Microwave Theory Tech.*, **21**, 1 (1973).
10. F. Miyamaru and M. Hangyo, *Appl. Phys. Lett.*, **84**, 2742 (2004).
11. Y. Takatsuka, E. Sano, V. Ryzhii, and T. Otsuji, *Jpn. J. Appl. Phys., Part 1*, **50**, 070118 (2011).
12. T. Tanaka, M. Akazawa, E. Sano, M. Tanaka, F. Miyamaru, and M. Hangyo, *Jpn. J. Appl. Phys., Part 1*, **45**, 4058 (2006).

Chapter 33

Plasmonic Terahertz Lasing in an Array of Graphene Nanocavities*

**V. V. Popov,[a,b] O. V. Polischuk,[a,b] A. R. Davoyan,[a]
V. Ryzhii,[c] T. Otsuji,[c] and M. S. Shur[d]**
[a]*Kotelnikov Institute of Radio Engineering and Electronics (Saratov Branch), Russian Academy of Sciences, Saratov 410019, Russia*
[b]*Saratov State University, Saratov 410012, Russia*
[c]*Research Institute for Electrical Communication, Tohoku University, Sendai 980-8577, Japan*
[d]*Department of Electrical, Computer, and Systems Engineering, Rensselaer Polytechnic Institute, Troy, New York 12180, USA*
popov_slava@yahoo.co.uk

We propose a novel concept of terahertz lasing based on stimulated generation of plasmons in a planar array of graphene resonant micro/nanocavities strongly coupled to terahertz radiation. Due to the strong plasmon confinement and superradiant nature of terahertz emission by the array of plasmonic nanocavities, the amplification of terahertz waves is enhanced by many orders of

---

*Reprinted with permission from V. V. Popov, O. V. Polischuk, A. R. Davoyan, V. Ryzhii, T. Otsuji, and M. S. Shur (2012). Plasmonic terahertz lasing in an array of graphene nanocavities, *Phys. Rev. B*, **86**, 195437. Copyright © 2012 American Physical Society.

Graphene-Based Terahertz Electronics and Plasmonics: Detector and Emitter Concepts
Edited by Vladimir Mitin, Taiichi Otsuji, and Victor Ryzhii

ISBN 978-981-4800-75-4 (Hardcover), 978-0-429-32839-8 (eBook)
www.jennystanford.com

magnitude at the plasmon resonance frequencies. We show that the lasing regime is ensured by the balance between the plasmon gain and plasmon radiative damping.

Fundamental limits reached by the available sources of electromagnetic radiation based on classical electronic oscillations radiating at radio and microwave frequencies, and on electron transitions between quantized energy levels corresponding to infrared and optical frequencies, give rise to the so-called terahertz (THz) gap [1, 2].

Extraordinary electronic properties of novel materials might help to go beyond the limits of the THz gap. In particular, graphene, a two-dimensional monolayer of graphite, has received a great deal of interest recently due to its unique electronic properties stemming from a linear (Dirac-type) gapless carrier energy spectrum $\varepsilon = \pm V_F|p|$ (see the inset in Fig. 33.1), where ε and p are the electron (hole) energy and momentum, respectively, $V_F \approx 10^8$ cm/s is the Fermi velocity, which is a constant for graphene, and upper and lower signs refer to the conduction and valence bands, respectively [3, 4]. Graphene is especially promising for THz photonics [5] due to its zero (or small in doped graphene) band gap.

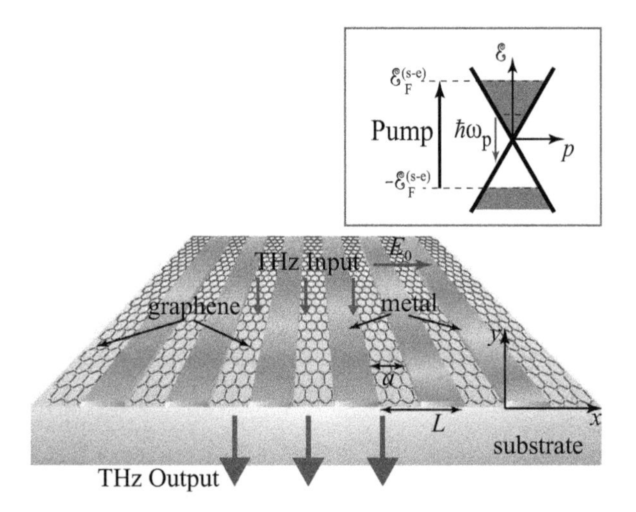

Figure 33.1 Schematic view of the array of graphene micro/nanocavities. The incoming electromagnetic wave is incident from the top at normal direction to the structure plane with the polarization of the electric field across the metal grating contacts. The energy band structure of pumped graphene is shown schematically in the inset.

Interband population inversion in graphene can be achieved by its optical pumping [6] or carrier injection [7]. At sufficiently strong excitation, the interband stimulated emission of photons can prevail over the intraband (Drude) absorption. In this case, the real part of the dynamic conductivity of graphene can be negative in the THz range. This effect can be used for THz photon lasing [8]. Stimulated emission of near-infrared [9] and THz [10] photons from population inverted graphene was recently observed.

Graphene exhibits strong plasmonic response due to both high density and small "relativistic" effective mass $m_{\mathrm{F}} = \varepsilon_{\mathrm{F}}/V_{\mathrm{F}}^2$ of free carriers, where ε_{F} is the Fermi energy. Dispersion of the plasma waves (plasmons) in graphene was studied for intrinsic (undoped) [11, 12] and doped (or gated) [13–17] graphene. Plasmon resonances in graphene can be controlled in graphene nanoribbon arrays [18–20] and tuned in the entire THz range, depending on the direction of the plasmon propagation in the array plane and/or by varying the nanoribbon width. Plasmons in patterned graphene strongly couple to electromagnetic waves, which makes graphene nanostructures very promising for the development of tunable graphene-based THz plasmonic metamaterials [19, 21–24].

Compared with the stimulated emission of electromagnetic modes (photons), the stimulated emission of plasmons by the interband transitions in population inverted graphene exhibits a much higher gain due to a small group velocity of the plasmons in graphene and strong confinement of the plasmon field in the vicinity of the graphene layer [25, 26]. Plasmon emission due to recombination of the electron-hole pairs in graphene was demonstrated experimentally recently [27]. However, a large plasmon gain in graphene leads to strong dephasing of the plasmon mode, hence preventing THz lasing. Also, strong coupling between the plasmons in graphene and electromagnetic radiation can hinder THz lasing from nonequilibrium plasmons. Therefore, to the best of our knowledge, neither plasmonic amplification of THz radiation nor THz plasmonic lasing in graphene has been reported so far.

In this chapter, we consider the amplification of a THz wave by the stimulated generation of resonant plasmons in a planar periodic array of graphene plasmonic micro/nanocavities strongly coupled to THz radiation. We show that, due to the strong confinement of the plasmon modes in the graphene micro/nanocavities and superradiant nature of electromagnetic emission from the array of the plasmonic micro/nanocavities, the amplification of THz waves is enhanced by several orders of magnitude at the plasmon resonance frequencies. It is shown that the plasmonic THz lasing becomes possible due to restoring the plasmon coherence in the graphene nanocavities strongly coupled to THz radiation at the balance between the plasmon gain and plasmon radiative damping.

Let us suppose that graphene micro/nanocavities are confined between the contacts of the metal grating located on a plane surface of a dielectric substrate, which can be high resistivity Si or SiC (see Fig. 33.1). We assume that the graphene is pumped either by optical illumination or by injection of electrons and holes from opposite metal contacts in each graphene nanocavity. In this case, the electron and hole densities in graphene can substantially exceed their equilibrium values and the electron and hole systems can be characterized by the quasi-Fermi energies $\pm\varepsilon_F$, respectively (see the inset in Fig. 33.1) and the effective temperature T. If the characteristic time of the emission of the optical phonon by an electron or a hole is much shorter than the time of the pair collisions, the nonequilibrium electrons and holes emit a cascade of optical phonons and occupy low energy states in the conduction and valence bands, respectively. In this case, the contribution of nonequilibrium carriers to the heating of the electron-hole system is small and their effective temperature T is close to the lattice temperature T_0 [28]. (If the effective temperature exceeds T_0, somewhat stronger pumping might be needed to ensure the population inversion in graphene [28].) For $T = T_0$, one can describe the response of pumped graphene by its complex-valued sheet conductivity in the local approximation [26] (see also Refs. [17, 29], as well as the recent review paper [30])

$$\sigma_{\mathrm{Gr}}(\omega)=\left(\frac{e^2}{4\hbar}\right)\left\{\frac{8k_BT\tau}{\pi\hbar(1-i\omega\tau)}\ln\left[1+\exp\left(\frac{\varepsilon_F}{k_BT}\right)\right]\right.$$

$$+\tanh\left(\frac{\hbar\omega-2\varepsilon_F}{4k_BT}\right)-\frac{4\hbar\omega}{i\pi}$$

$$\left.\times\int_0^{\infty}\frac{G(\varepsilon,\varepsilon_F)-G(\hbar\omega/2,\varepsilon_F)}{(\hbar\omega)^2-4\varepsilon^2}d\varepsilon\right\}. \tag{33.1}$$

Here ω is the frequency of the incoming electromagnetic wave, e is the electron charge, $\hbar$ is the reduced Planck constant, k_B is the Boltzmann constant, and

$$G(\varepsilon,\varepsilon')=\frac{\sinh(\varepsilon/k_BT)}{\cosh(\varepsilon/k_BT)+\cosh(\varepsilon'/k_BT)}.$$

The first term in the curly braces in Eq. (33.1) describes a Drude-model response for the intraband processes involving the phenomenological electron and hole scattering time τ, which can be estimated from the measured dc carrier mobility: $\tau=\mu\varepsilon_F/eV_F^2$ [31]. The temperature-independent carrier mobility μ > 250000 cm^2/V s observed recently in multilayer epitaxial graphene on 4H-SiC substrate [32, 33] corresponds to $\tau \approx 10^{-12}$ s for ε_F = 40 meV at room temperature. Carrier scattering times longer than 1 ps were observed recently by the Raman spectroscopy of optically pumped graphene [34, 35] in a quasiequilibrium regime (after the carriers are equilibrated due to fast carrier-carrier scattering). The remaining terms in Eq. (33.1) arise from the interband transitions. For sufficiently strong degeneracy of the electron and hole systems, the quasi-Fermi energy ε_F depends on the electron (hole) density, $N_{n(p)}(N_n = N_p)$, in graphene: $\varepsilon_F \sim \hbar V_F\sqrt{\pi N_{n(p)}}$ [4]. Hence, the quasi-Fermi energy is determined by the photogeneration rate [36] or by the carrier injection rate [7] under optical or carrier injection pumping, respectively. Of course, the simple estimates of the phenomenological parameters given above are to be considered only as rough approximations. Their exact values are to be measured [34, 35] or calculated by using a microscopic ab initio approach [37–39].

We assume that the external THz electromagnetic wave is incident upon the planar array of graphene micro/nanocavities at normal direction to its plane with the polarization of the electric field across the metal grating contacts as shown in Fig. 33.1. Then we solve the problem of the amplification of the THz wave by the array of graphene micro/nanocavities in a semianalytical self-consistent electromagnetic approach similar to that described in Ref. [40] (for more details of our theoretical approach see Ref. [41]).

Figure 33.2 shows the contour map of the calculated absorbance as a function of the quasi-Fermi energy (which corresponds to the pumping strength) and the THz wave frequency for an array of the graphene microcavities with period L = 4 μm and the length of each microcavity a = 2 μm. The absorbance is defined as the ratio between the absorbed or emitted (which corresponds to negative absorbance) THz power per unit area of the array and the energy flux density in the incoming THz wave. In the amplification regime, the negative value of the absorbance yields the amplification coefficient. The value of $\mathrm{Re}[\sigma_{\mathrm{Gr}}(\omega)]$ is negative above the solid black line in Fig. 33.2, corresponding to $\mathrm{Re}[\sigma_{\mathrm{Gr}}(\omega)] = 0$ (i.e., to transparent graphene). Above this boundary line, negative absorption (i.e., amplification) takes place at all frequencies and pumping strengths. The plasmon absorption resonances below the $\mathrm{Re}[\sigma_{\mathrm{Gr}}(\omega)] = 0$ line give way to the amplification resonances above this line. Plasmon resonances appear at frequencies $\omega = \omega_{\mathrm{p}}(q)$ determined by the selection rule for the plasmon wave vector $q_{\mathrm{n}} = (2n - 1)\pi/a_{\mathrm{eff}}$, where a_{eff} is the effective length of the graphene micro/nanocavity. The effective length of the graphene micro/nanocavity can be, in general, different from its geometric length a due to the effect of the metal contacts (see the related discussion for a conventional two-dimensional electron system in Ref. [42]). The frequency of the plasmon resonance is determined mainly by the imaginary part of the graphene conductivity in Eq. (33.1), while the real part of the conductivity is responsible for the energy loss (for $\mathrm{Re}[\sigma_{\mathrm{Gr}}(\omega)] > 0$) or energy gain (for $\mathrm{Re}[\sigma_{\mathrm{Gr}}(\omega)] < 0$).

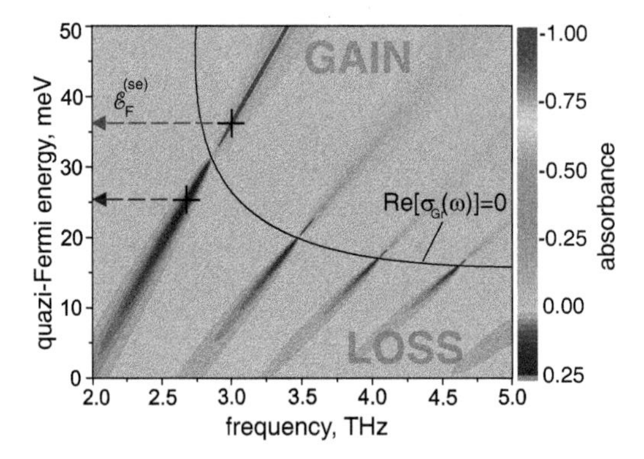

Figure 33.2 Contour map of the absorbance as a function of the quasi-Fermi energy and the frequency of incoming THz wave for the array of graphene microcavities with period L = 4 μm and the length of a graphene microcavity a = 2 μm. The electron scattering time in graphene is $\tau = 10^{-12}$ s. Blue and red arrows mark the quasi-Fermi energies for the maximal absorption and for the plasmonic lasing regime, respectively, at the fundamental plasmon resonance.

As seen from Fig. 33.2, the absorbance (including the negative absorbance in the amplification regime) at the plasmon resonance does not vary monotonously with increasing ε_F, but rather exhibits absorption and amplification maxima at some values of the quasi-Fermi energy. The reason is that the absorbed or amplified electromagnetic power depends not only on the plasmon loss due to its energy dissipation, $\gamma_{dis}(\varepsilon_F,\omega) < 0$, or plasmon gain, $g(\varepsilon_F,\omega) > 0$, respectively, but also on the coupling between the plasmons and electromagnetic wave. Coupling might be different for different structures but, basically, it is controlled by the radiative damping, $\gamma_{rad}(\varepsilon_F,\omega)$, due to plasmon radiative decay into electromagnetic waves. General phenomenological consideration of THz absorption in a planar periodic plasmonic structure [43–45] shows that the maximal absorption at the plasmon resonance takes place when $\gamma_{dis} = \gamma_{rad}$ irrespective of details of the plasmonic structure.

For zero pumping strength, the dissipative damping of the plasmons in graphene, γ_{dis}, comes from the energy loss due to the electron and hole scattering with the decay rate γ_{sc} and the energy loss due to generation of the electron-hole pairs with the decay rate $\gamma_{e\text{-}h}$. Hence $\gamma_{dis} = \gamma_{sc} + \gamma_{e\text{-}h}$. These two different energy loss mechanisms are accounted for by the first and second terms in the curly braces in Eq. (33.1), respectively. In the structure under consideration, the dissipative broadening of the plasmon resonance resulting from the carrier scattering in graphene for $\tau = 10^{-12}$ s is $2\gamma_{sc} = 1/2\pi\tau \approx 0.16$ THz. The plasmon loss resulting from the generation of the electron-hole pairs by the THz wave depends on frequency, see Eq. (33.1), and leads to a plasmon resonance broadening of about 0.014 THz at frequency 1.96 THz of the fundamental plasmon resonance in Fig. 33.2 for zero pumping strength, $\varepsilon_F = 0$. The radiative broadening of the plasmon resonance depends on both the carrier concentration in graphene and on the antenna properties of the planar periodic plasmonic structure [43–45]. In the "cold" structure (without pumping), the radiative broadening of the first plasmon resonance in Fig. 33.2 is about 0.01 THz.

For $2\varepsilon_F > \hbar\omega$, the energy loss due to the generation of the electron-hole pairs described by the second term in the curly braces in Eq. (33.1) becomes negative (which corresponds to the energy gain), while the net loss can be still positive due to the electron (hole) scattering contribution, so that the plasmon net dissipative damping becomes smaller. Because the electron scattering loss in graphene for $\varepsilon_F = 0$ is typically greater than the plasmon radiative loss in the array of the graphene micro/nanocavities, the plasmon net dissipative damping becomes equal to their radiative damping at some pumping strength $\varepsilon_F > 0$ that results in the maximal absorption at the plasmon resonance. The quasi-Fermi energy value corresponding to the maximal absorption at the first plasmon resonance is marked by the blue arrow in Fig. 33.2. The maximal theoretical value of the absorbance at the plasmon resonance is [44, 45] $A_{\rm res}^{\rm max} = 0.5(1-\sqrt{R_0})$, where R_0 is the reflectivity of a bare substrate, which yields $A_{\rm res}^{\rm max} \approx 0.23$ for a silicon substrate (see Fig. 33.3a).

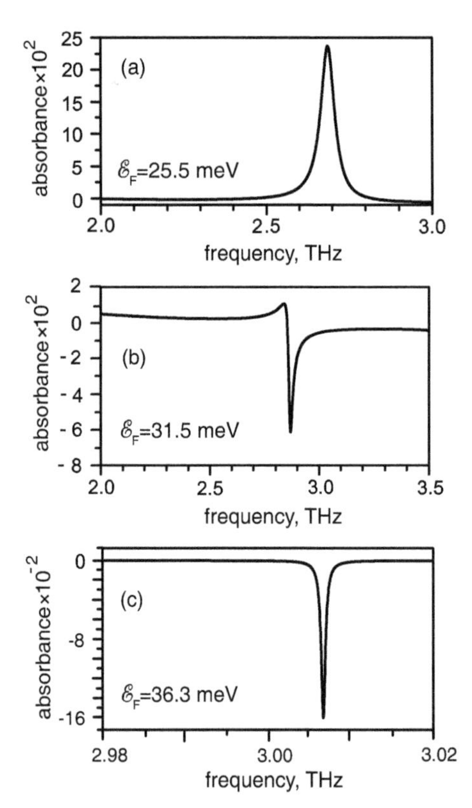

Figure 33.3 The fundamental plasmon resonance (a) in the maximal absorption regime, (b) near the graphene transparency regime $\mathrm{Re}[\sigma_{\rm Gr}(\omega)] = 0$, and (c) near the self-excitation regime.

With further increase of ε_F, the energy gain can balance the energy loss caused by the electron scattering in graphene resulting in zero net energy loss, $\mathrm{Re}[\sigma_{Gr}(\omega)] = 0$, and corresponding graphene transparency. In this case, the plasmon resonance line exhibits a nonsymmetric Fano-like shape [46, 47] shown in Fig. 33.3b, since the real part of graphene conductivity changes sign across the plasmon resonance. The linewidth of the Fano-like plasmon resonance is given mainly by its radiative broadening (since the dissipative damping is close to zero in this case). The radiative linewidth of the Fano-like resonance at $\mathrm{Re}[\sigma_{Gr}(\omega)] = 0$ is greater than that for zero pumping ($\varepsilon_F = 0$) because the carrier density in graphene is higher for $\varepsilon_F > 0$.

Above the graphene transparency line $\mathrm{Re}[\sigma_{Gr}(\omega)] = 0$, the THz wave amplification at the plasmon resonance frequency (see Fig. 33.3c) is several orders of magnitude stronger than away from the resonances (the latter corresponding to the photon amplification in population inverted graphene [9, 10]). Note that at a certain value of the quasi-Femi energy, $\varepsilon_F = \varepsilon_F^{(se)}$, the amplification coefficient at the plasmon resonance tends toward infinity with corresponding amplification linewidth shrinking down to zero. This corresponds to plasmonic lasing in the graphene micro/nanocavities in the self-excitation regime. The behavior of the amplification coefficient around the self-excitation regime is shown in Fig. 33.4a. The lasing occurs when the plasmon gain balances the electron scattering loss and the radiative loss, $g(\varepsilon_F^{(se)}, \omega) = -\ \gamma_{rad}(\varepsilon_F^{(se)}, \omega)\ +\ \gamma_{sc}]$; see Fig. 33.4b. It means that the plasmon oscillations are highly coherent in this case, with virtually no dephasing at all. When $g(\varepsilon_F,\omega) < -[\gamma_{rad}(\varepsilon_F^{(se)}, \omega)\ +\ \gamma_{sc}]$, a fast radiative decay hinders the plasmon stimulated generation in graphene, whereas when $g(\varepsilon_F,\omega) > -[\gamma_{rad}(\varepsilon_F^{(se)}, \omega)\ +\ \gamma_{sc}]$ a low radiative decay rate slows down the release of the plasmon energy into THz radiation. The quasi-Fermi energy corresponding to plasmonic lasing in the first plasmon resonance is marked by the red arrow in Fig. 33.2. Weaker plasmon gain is needed to meet the self-excitation condition at the higher-order plasmon resonances because of smaller radiative damping (due to a smaller oscillator strength) of the higher-order plasmon modes. Hence, plasmonic lasing takes place at quasi-Fermi energy values closer to the graphene-transparency line $\mathrm{Re}[\sigma_{Gr}(\omega)] = 0$ for the higher-order plasmon resonances (see Fig. 33.2). Therefore, dynamic and frequency ranges of plasmonic lasing decrease for higher order resonances. Plasmon radiative damping not only determines the plasmonic lasing condition, but also plays a constructive role in the amplification process. By conservation of energy, the electromagnetic power emitted from the array of graphene micro/nanocavities is proportional to the radiative decay rate of stored energy, $2|\gamma_{\text{rad}}|$. Hence, a much higher amplification coefficient can be reached at the first plasmon resonance as compared to that for the higher-order plasmon resonances (in the same frequency range around the self-excitation frequency).

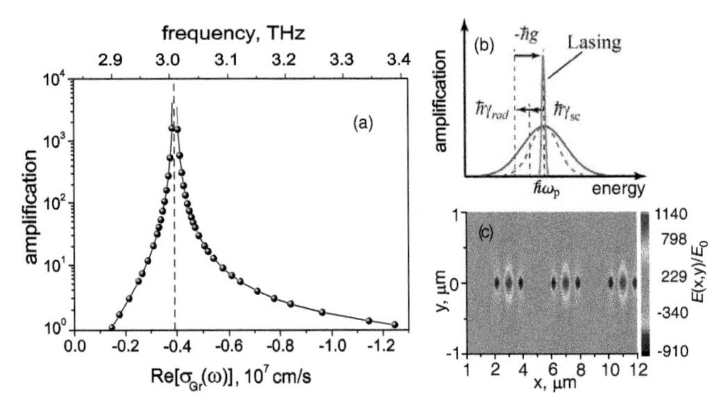

Figure 33.4 (a) The variation of the power amplification coefficient along the first-plasmon-resonance lobe (see Fig. 33.2) near the self-excitation regime. The frequency of the plasmon lasing is marked by the vertical dashed line. (b) Schematic illustration of the energy rate balance in the plasmon lasing regime. (c) The snapshot of the distribution of the normalized induced in-plane electric field at the moment of time corresponding to the maximal swing of plasma oscillations in the graphene microcavities at the fundamental plasmon amplification resonance shown in Fig. 33.3c.

Of course, the divergence of the amplification coefficient at $\varepsilon_{\text{F}} = \varepsilon_{\text{F}}^{(\text{se})}$ is unphysical. It is a consequence of the linear electromagnetic approach used in this work. It is natural to assume that the linear approach is quantitatively valid when the amplitude of amplified plasmon oscillations in graphene is much smaller than the unperturbed density of the photogenerated or injected electron-hole pairs. Estimating the unperturbed density of the nonequilibrium carriers as $(N_{\text{n}} + N_{\text{p}}) \sim 2\varepsilon_{\text{F}}^2/(\pi\hbar^2 V_{\text{F}}^2) \approx 4 \times 10^{11}$ cm^{-2} for $\varepsilon_{\text{F}} = 36$ meV, we can claim that our approach is quantitatively valid up to power amplification coefficients of 10^3 or even higher. (For example, the fluctuations corresponding to blackbody radiation at 300 K would be amplified above the mW/cm^2 level at the plasmon resonance frequencies.) This value of the amplification coefficient is five orders of magnitude higher than that away from the plasmon resonance. We expect that the main conclusions of this chapter concerning general properties and conditions of the amplification process and plasmon lasing remain valid qualitatively even for greater amplification coefficients.

Although we presented the above results of numerical calculations obtained for an array of graphene microcavities, similar results were obtained also for an array of graphene nanocavities. The plasmon resonance frequency is roughly proportional to the inverse value of the square root of the nanocavity length, so that, for example, the lasing at the fundamental plasmon resonance takes place at a frequency of about 6.7 THz at ε_F = 18 meV for the graphene nanocavity length 200 nm. It is worth mentioning that the main results of this chapter remain valid also for a shorter relaxation time. Only the corresponding values of the quasi-Fermi energy and the frequency of the plasmonic lasing increase somewhat in this case. For example, assuming the carrier relaxation time 0.1 ps, we obtain the plasmonic lasing at frequency 9.8 THz of the fundamental plasmon resonance for a quasi-Fermi energy value of about 55 meV in an array of the graphene nanocavities of length 200 nm each.

Enhanced THz emission from the graphene nanocavities is caused by the fact that plasmons in different nanocavities oscillate in phase (even without the incoming electromagnetic wave) because the metal contacts act as synchronizing elements between adjacent graphene nanocavities (applying a mechanical analogy, one may think of rigid crossbars connecting oscillating springs arranged in a chain). Therefore, the plasma oscillations in the array of graphene nanocavities constitute a single collective plasmon mode distributed over the entire area of the array, which leads to the enhanced superradiant electromagnetic emission from the array. The plasmon-mode locking regime among different graphene nanocavities is illustrated in Fig. 33.4c. Extraordinary properties of a collective mode in an array of synchronized dipole oscillators are well known in quantum optics: the power of electromagnetic emission from such an array grows as the square of the number of the oscillators in the array [48]. Superradiant plasmon resonances in concentric ring/disc gold nanocavities in the visible and near-infrared spectral ranges were experimentally demonstrated recently [49].

Giant THz amplification enhancement at the plasmon resonance is also ensured by strong plasmon confinement in the graphene micro/nanocavities; see Fig. 33.4c. A large plasmon gain in graphene would lead to the strong dephasing of a plasma wave over quite long propagation distance (which corresponds to the nonresonant stimulated generation of plasmons [25, 26]). Therefore, strong plasmon-mode confinement in a single-mode plasmonic cavity is required to ensure the resonant stimulated generation of plasmons. Plasmon confinement to a single-mode micro/nanocavity also enhances the rate of spontaneous electromagnetic emission by the plasmon mode due to the Purcell effect [50]. It is expected that the confinement of plasmons in a two-dimensional array of graphene micro/nanocavities could enhance the amplification even more.

In conclusion, we predict a giant amplification and lasing of THz radiation due to the stimulated generation of plasmons in an array of graphene resonant micro/nanocavities strongly coupled to THz radiation. The amplification of the THz wave at the plasmon resonance frequencies is several orders of magnitude stronger than away from the resonances. Giant THz wave amplification is due to the strong plasmon confinement and superradiant nature of THz emission by the array of plasmonic micro/nanocavities. The THz lasing at the plasmon resonance is achieved when the net plasmon gain in graphene approaches the negative of the radiative damping of the resonant plasmons, which ensures high plasmon coherence and self-excitation of plasmons in the graphene nanocavities. The amplification resonance line is mainly of Lorentzian shape except when the net energy gain in graphene is close to zero. In that case, a Fano-like resonant lineshape is formed due to a strong variation of the graphene conductivity across the resonance linewidth. These results might be of broad physical interest as revealing new general features of the strong interaction of an array of nonequilibrium oscillators with electromagnetic radiation. In terms of practical applications, these results can pave a way to creation of plasmonic graphene amplifiers and generators for THz frequencies.

The work was supported by the Russian Foundation for Basic Research (Grants No. 11-02-92101 and No. 12-02-93105) and by the Russian Academy of Sciences Program "Technological Fundamentals of Nanostructures and Nanomaterials." The work at RPI was supported by the US NSF under the auspices of I/UCRC

"CONNECTION ONE," NSF I-Corp, and by the NSF EAGER program. This work was financially supported in part by NPRP Grant No. NPRP 09-1211-2-475 from the Qatar National Research Fund, by the JSPS Grant-in-Aid for Specially Promoting Research (No. 23000008), Japan, by the JSPS-RFBR Japan-Russian Collaborative Research Program, and by JST-CREST, Japan.

References

1. X.-C. Zhang and J. Xu, *Introduction to THz Wave Photonics* (Springer-Verlag, New York, 2009).
2. M. Tonouchi, *Nat. Photonics*, **1**, 97 (2007).
3. K. S. Novoselov, A. K. Geim, S. V. Morozov, D. Jiang, Y. Zhang, S. V. Dubonos, I. V. Grigorieva, and A. A. Firsov, *Science*, **306**, 666 (2004).
4. A. H. Castro Neto, F. Guinea, N. M. R. Peres, K. S. Novoselov, and A. K. Geim, *Rev. Mod. Phys.*, **81**, 109 (2009).
5. B. Sensale-Rodriguez, R. Yan, M. M. Kelly, T. Fang, K. Tahy, W. S. Hwang, D. Jena, L. Liu, and H. G. Xing, *Nat. Commun.*, **3**, 780 (2012).
6. A. Satou, F. T. Vasko, and V. Ryzhii, *Phys. Rev. B*, **78**, 115431 (2008).
7. V. Ryzhii, M. Ryzhii, V. Mitin, and T. Otsuji, *J. Appl. Phys.*, **110**, 094503 (2011).
8. V. Ryzhii, M. Ryzhii, A. Satou, N. Ryabova, T. Otsuji, V. Mitin, F. T. Vasko, A. A. Dubinov, V. Y. Aleshkin, and M.S. Shur, *Future Trends in Microelectronics*, edited by S. Luryi, J. Xu, and A. Zaslavsky (John Wiley & Sons, New York, 2010).
9. T. Li, L. Luo, M. Hupalo, J. Zhang, M. C. Tringides, J. Schmalian, and J. Wang, *Phys. Rev. Lett.*, **108**, 167401 (2012).
10. S. Boubanga-Tombet, S. Chan, T. Watanabe, A. Satou, V. Ryzhii, and T. Otsuji, *Phys. Rev. B*, **85**, 035443 (2012).
11. O. Vafek, *Phys. Rev. Lett.*, **97**, 266406 (2006).
12. L. A. Falkovsky and A. A. Varlamov, *Eur. Phys. J. B*, **56**, 281 (2007).
13. V. Ryzhii, *Jpn. J. Appl. Phys.*, **45**, L923 (2006).
14. B. Wunsch, T. Stauber, F. Sols, and F. Guinea, *New J. Phys.*, **8**, 318 (2006).
15. E. H. Hwang and S. Das Sarma, *Phys. Rev. B*, **75**, 205418 (2007).
16. V. Ryzhii, A. Satou, and T. Otsuji, *J. Appl. Phys.*, **101**, 024509 (2007).
17. G. W. Hanson, *J. Appl. Phys.*, **103**, 064302 (2008).

18. V. V. Popov, T. Y. Bagaeva, T. Otsuji, and V. Ryzhii, *Phys. Rev. B*, **81**, 073404 (2010).

19. L. Ju, B. Geng, J. Horng, C. Girit, M. Martin, Z. Hao, H. A. Bechtel, X. Liang, A. Zettl, Y. R. Shen, and F. Wang, *Nat. Nanotechnol.*, **6**, 630 (2011).

20. A. Yu. Nikitin, F. Guinea, F. J. Garcia-Vidal, and L. Martin-Moreno, *Phys. Rev. B*, **84**, 161407 (2011).

21. S. Thongrattanasiri, F. H. L. Koppens, and F. J. Garcia de Abajo, *Phys. Rev. Lett.*, **108**, 047401 (2012).

22. A. Y. Nikitin, F. Guinea, F. J. Garcia-Vidal, and L. Martin-Moreno, *Phys. Rev. B*, **85**, 081405 (2012).

23. F. H. Koppens, D. E. Chang, and F. J. Garcia de Abajo, *Nano Lett.*, **11**, 3370 (2011).

24. A. Vakil and N. Engheta, *Science*, **332**, 1291 (2011).

25. F. Rana, *IEEE Trans. Nanotechnol.*, **7**, 91 (2008).

26. A. A. Dubinov, V. Y. Aleshkin, V. Mitin, T. Otsuji, and V. Ryzhii, *J. Phys.: Condens. Matter*, **23**, 145302 (2011).

27. A. Bostwick, T. Ohta, T. Seyller, K. Horn, and E. Rotenberg, *Nat. Phys.*, **3**, 36 (2007).

28. V. Ryzhii, M. Ryzhii, V. Mitin, A. Satou, and T. Otsuji, *Jpn. J. Appl. Phys.*, **50**, 094001 (2011).

29. L. A. Falkovsky and S. S. Pershoguba, *Phys. Rev. B*, **76**, 153410 (2007).

30. N. M. R. Peres, *Rev. Mod. Phys.*, **82**, 2673 (2010).

31. Y.-W. Tan, Y. Zhang, K. Bolotin, Y. Zhao, S. Adam, E. H. Hwang, S. Das Sarma, H. L. Stormer, and P. Kim, *Phys. Rev. Lett.*, **99**, 246803 (2007).

32. M. Orlita, C. Faugeras, P. Plochocka, P. Neugebauer, G. Martinez, D. K. Maude, A.-L. Barra, M. Sprinkle, C. Berger, W. A. de Heer, and M. Potemski, *Phys. Rev. Lett.*, **101**, 267601 (2008).

33. M. Sprinkle, D. Siegel, Y. Hu, J. Hicks, A. Tejeda, A. Taleb-Ibrahimi, P. Le Fevre, F. Bertran, S. Vizzini, H. Enriquez, S. Chiang, P. Soukiassian, C. Berger, W. A. de Heer, A. Lanzara, and E. H. Conrad, *Phys. Rev. Lett.*, **103**, 226803 (2009).

34. J. M. Dawlaty, S. Shivaraman, M. Chandrashekhar, F. Rana, and M. G. Spencer, *Appl. Phys. Lett.*, **92**, 042116 (2008).

35. T. Otsuji, S. A. Boubanga Tombet, A. Satou, H. Fukidome, M. Suemitsu, E. Sano, V. Popov, M. Ryzhii, and V. Ryzhii, *J. Phys. D: Appl. Phys.*, **45**, 303001 (2012).

36. V. Ryzhii, M. Ryzhii, A. Satou, T. Otsuji, A. A. Dubinov, and V. Y. Aleshkin, *J. Appl. Phys.*, **106**, 084507 (2009).
37. F. Rana, *Phys. Rev. B*, **76**, 155431 (2007).
38. R. Kim, V. Perebeinos, and P. Avouris, *Phys. Rev. B*, **84**, 075449 (2011).
39. T. Winzer and E. Malic, *Phys. Rev. B*, **85**, 241404(R) (2012).
40. D. V. Fateev, V. V. Popov, and M. S. Shur, *Fiz. Tekh. Poluprovodn.*, **44**, 1455 (2010) [*Semiconductors*, **44**, 1406 (2010)].
41. See Supplemental Material at http://link.aps.org/supplemental/10.1103/PhysRevB.86.195437 for more details of our theoretical approach.
42. V. V. Popov, A. N. Koudymov, M. Shur, and O. V. Polischuk, *J. Appl. Phys.*, **104**, 024508 (2008).
43. V. V. Popov, G. M. Tsymbalov, D. V. Fateev, and M. S. Shur, *Appl. Phys. Lett.*, **89**, 123504 (2006).
44. V. V. Popov, O. V. Polischuk, T. V. Teperik, X. G. Peralta, S. J. Allen, N. J. M. Horing, and M. C. Wanke, *J. Appl. Phys.*, **94**, 3556 (2003).
45. V. V. Popov, D. V. Fateev, O. V. Polischuk, and M. S. Shur, *Opt. Express*, **18**, 16771 (2010).
46. B. Luk'yanchuk, N. I. Zheludev, S. A. Maier, N. J. Halas, P. Nordlander, H. Giessen, and T. C. Chong, *Nat. Mater.*, **9**, 707 (2010).
47. A. E. Miroshnichenko, S. Flach, and Y. S. Kivshar, *Rev. Mod. Phys.*, **82**, 2257 (2010).
48. M. G. Benedict, A. M. Ermolaev, V. A. Malyshev, I. V. Sokolov, and E. D. Trifinov, *Superradiance: Multiatomic Coherent Emission* (IOP, Bristol, 1996).
49. Y. Sonnefraud, N. Verellen, H. Sobhani, G. A. E. Vandenbosch, V. V. Moshchalkov, P. Van Dorpe, P. Nordlander, and S. A. Maier, *ACS Nano*, **4**, 1664 (2010).
50. E. M. Purcell, *Phys. Rev.*, **69**, 681 (1946).

Chapter 34

The Gain Enhancement Effect of Surface Plasmon Polaritons on Terahertz Stimulated Emission in Optically Pumped Monolayer Graphene*

T. Watanabe,[a] T. Fukushima,[a] Y. Yabe,[a] S. A. Boubanga Tombet,[a] A. Satou,[a] A. A. Dubinov,[b] V. Ya. Aleshkin,[b] V. Mitin,[c] V. Ryzhii,[a] and T. Otsuji[a]

[a]*Research Institute of Electrical Communication, Tohoku University, Sendai 9808577, Japan*

[b]*Institute for Physics of Microstructures, Russian Academy of Sciences, Nizhny Novgorod 603950, Russia*

[c]*Department of Electrical Engineering, University at Buffalo, State University of New York, NY 14260, USA*

watanabe@riec.tohoku.ac.jp

*Reprinted with permission from T. Watanabe, T. Fukushima, Y. Yabe, S. A. Boubanga Tombet, A. Satou, A. A. Dubinov, V. Ya. Aleshkin, V. Mitin, V. Ryzhii, and T. Otsuji (2013). The gain enhancement effect of surface plasmon polaritons on terahertz stimulated emission in optically pumped monolayer graphene, *New J. Phys.*, **15**, 075003. Copyright © 2013 IOP Publishing Ltd and Deutsche Physikalische Gesellschaft.

Graphene-Based Terahertz Electronics and Plasmonics: Detector and Emitter Concepts
Edited by Vladimir Mitin, Taiichi Otsuji, and Victor Ryzhii

ISBN 978-981-4800-75-4 (Hardcover), 978-0-429-32839-8 (eBook)
www.jennystanford.com

Nonlinear carrier relaxation/recombination dynamics and the resultant stimulated terahertz (THz) photon emission with excitation of surface plasmon polaritons (SPPs) in photoexcited monolayer graphene has been experimentally studied using an optical pump/THz probe and an optical probe measurement. We observed the spatial distribution of the THz probe pulse intensities under linear polarization of optical pump and THz probe pulses. It was clearly observed that an intense THz probe pulse was detected only at the area where the incoming THz probe pulse takes a transverse magnetic (TM) mode capable of exciting the SPPs. The observed gain factor is in fair agreement with the theoretical calculations. Experimental results support the occurrence of the gain enhancement by the excitation of SPPs on THz stimulated emission in optically pumped monolayer graphene.

34.1 Introduction

Graphene has attracted increasing attention for terahertz (THz) and optoelectronic device applications due to its exceptional electronic and optical properties [1–5]. The conduction and valence bands of graphene have a symmetrical conical shape around the Brillouin zone edges, which are called K and K' points, and contact each other at 'Dirac points' at the K and K' points. Electrons and holes in graphene have a linear energy dispersion relation with zero bandgap, resulting in peculiar features such as massless relativistic fermions with back-scattering-free ultrafast transport [1–5] as well as the negative dynamic conductivity at THz frequencies under optical or electrical pumping [6–11]. Interband population inversion in graphene can be achieved by its optical pumping [6, 8] or carrier injection [7, 8] along with the ultrafast nonequilibrium carrier energy relaxation processes. At sufficiently strong excitation, the interband stimulated emission of photons can prevail over the intraband (Drude) absorption. In this case, the real part of the dynamic conductivity of graphene, $\mathrm{Re}[\sigma(\omega)]$, becomes negative at some frequencies ω. Owing to the gapless energy spectrum of graphene, $\mathrm{Re}[\sigma(\omega)]$ can be negative in THz range [6–11]. This effect can be exploited in graphene-based THz lasers with optical or injection pumping [8, 12–14].

The carrier dynamics in optically pumped graphene strongly depends on the initial temperature of carriers and the intensity of optical pumping. For sufficiently low carrier concentrations, that is, at low temperatures under weak pumping, photoexcited carriers accumulate effectively near the Dirac point via the cascade emission of optical phonons (OPs). Under these conditions population inversion can be achieved efficiently [6, 7, 9]. In contrast, at room temperature or under stronger pumping, where the carrier concentration is high (10^{12} cm^{-2}), carrier-carrier (CC) scattering plays a crucial role in the dynamics after pulse excitation, because of the fast quasi-equilibration of the carriers [15–17]. Ultrafast optical pump-probe spectroscopy on graphene has indicated that the quasi-equilibration by CC scattering occurs on a time scale of 10–100 fs [15–17], which is much faster than a single OP emission. In this case, the pulse excitation makes carriers very hot initially, and the energy relaxation and recombination via OP emission follow [9]. Under these conditions for intrinsic graphene at room temperature, population inversion can still be achievable when the pumping intensity exceeds a certain threshold level [9, 11].

Li et al. [18] observed stimulated emission of near-infrared photons in femtosecond (fs) infrared (IR) laser-pumped graphene at very short time durations just tens of femtoseconds after the pumping, which was before the photoexcited carriers were quasi-equilibrated by the CC scattering. The authors, on the other hand, observed stimulated emission of THz photons [19, 20] from fs-IR laser-pumped monolayer graphene at the time duration a few picoseconds after the pumping when photoexcited carriers were quasi-equilibrated and their populations were inverted at THz photon energies beyond the Dirac point. We utilized a time-resolved near-field reflective electro-optic sampling with a fs-IR laser pulse for optical pumping and a synchronously generated THz pulse for probing the THz dynamics of the sample in a THz photon-echo manner [19]. The gain spectral profiles showed qualitative agreement with theory in terms of threshold behavior against the pumping intensity, normal dispersion around the gain peak frequency and gain spectral narrowing particularly at the higher frequency band edge [6, 11, 13, 19]. However, the obtained gain factor exceeds the theoretical limit given by the quantum conductance [21] by more than one order of magnitude. We consider any artifact caused in experimental setup (gain multiplication due to multiple reflection, reflective index change due to increase in free carriers, etc.), [19] but cannot explain such a phenomenon.

One possibility is the amplified stimulated plasmon emission by the excitation of surface plasmon polaritons (SPPs), which was theoretically revealed by the authors [22]. Two-dimensional plasmons in graphene exhibit unique optoelectronic properties and mediate extraordinary light-matter interactions [5] which is expected to be exploited for advanced THz-active devices [5, 22–25]. In this chapter, we experimentally study the nonlinear carrier relaxation/recombination dynamics and resultant stimulated THz photon emission with the excitation of SPPs in photoexcited monolayer graphene. We observe the spatial distribution of the THz probe pulse intensities and its dependence on the polarizations of optical pump and THz probe pulses. Intense THz probe pulse is detected only at the area where the incoming THz probe pulse takes a TM mode capable of exciting the SPPs. The observed gain factor is in fair agreement with the theoretical calculations. Experimental results support the occurrence of the gain enhancement effect by the excitation of SPPs on THz stimulated emission in optically pumped monolayer graphene.

34.2 Stimulated Terahertz (THz) Photon and Plasmon Emission in Optically Pumped Graphene

In high-frequency (from THz to infrared) range, the graphene conductivity is derived from the Kubo formula in the following form [21]:

$$\sigma_{\mathbf{k}\omega} = \frac{ie^2}{\hbar\pi^2}\sum_{a=1,2}\int\frac{d^2\mathbf{p}\upsilon_x^2\{f[\varepsilon_a(\mathbf{p}_-)] - f[\varepsilon_a(\mathbf{p}_+)]\}}{[\varepsilon_a(\mathbf{p}_+) - \varepsilon_a(\mathbf{p}_-)][\hbar\omega - \varepsilon_a(\mathbf{p}_+) + \varepsilon_a(\mathbf{p}_-)]}$$

$$+\frac{2ie^2\omega}{\pi^2}\int\frac{d^2\mathbf{p}\upsilon_{21}\upsilon_{12}\{f[\varepsilon_1(\mathbf{p}_-)] - f[\varepsilon_2(\mathbf{p}_+)]\}}{[\varepsilon_2(\mathbf{p}_+) - \varepsilon_1(\mathbf{p}_-)][(\hbar\omega)^2 - [\varepsilon_2(\mathbf{p}_+) - \varepsilon_1(\mathbf{p}_-)]^2]}, \quad (34.1)$$

where the indices 1 and 2 refer to conduction and valence bands, respectively, $\varepsilon_1(\mathbf{p}) = |\mathbf{p}|v_F$ and $\varepsilon_2(\mathbf{p}) = -|\mathbf{p}|v_F$, $\upsilon_F \simeq 10^6\ \text{ms}^{-1}$, $\mathbf{p}_\pm = \mathbf{p} \pm \hbar\mathbf{k}/2$, $f(\varepsilon)$ is the electron distribution function (the equilibrium Fermi function $f(\varepsilon) = 1/[1 + e^{(\varepsilon-\varepsilon_F)/k_BT}]$ is assumed), $\upsilon_x = \upsilon_F \cos\theta_p$ and $\upsilon_{12} = i\upsilon_F \sin\theta_p$ is the matrix elements of the velocity operator. The first term in Eq. (34.1) corresponds to the intraband transitions, whereas the second term corresponds to the interband transitions. To allow for the electron (hole) momentum relaxation, one should treat the frequency in the intraband part of the conductivity expression as $\omega \to \omega + i\tau_m^{-1}$. Then, the real part of Eq. (34.1) can be expressed in the following simple form [21]:

$$\mathrm{Re}\,\sigma_\omega = \mathrm{Re}\,\sigma_\omega^{\mathrm{intra}} + \mathrm{Re}\,\sigma_\omega^{\mathrm{inter}}$$

$$\approx \frac{(\ln 2 + \varepsilon_F / 2k_BT)e^2}{\pi\hbar} \frac{k_BT\tau_m}{\hbar(1+\omega^2\tau_m^2)} + \frac{e^2}{4\hbar}(1-2f(\hbar\omega)), \quad (34.2)$$

where e is the elementary charge, $\hbar$ is the reduced Planck's constant, k_B is the Boltzmann constant, T is the temperature and τ_m is the momentum relaxation time of carriers. The intraband contribution, $\mathrm{Re}\,\sigma_\omega^{\mathrm{intra}}$, corresponds to the Drude absorption and is always positive. The THz frequency domain is situated in the intraband Drude conductivity dominated region.

The interband conductivity related to the second term of the right-hand side in Eq. (34.2) could take negative values by optical pumping [6, 8] or carrier injection [7, 8], giving rise to the population inversion. At sufficiently strong excitation, the interband stimulated emission of photons can prevail over the intraband (Drude) absorption. In this case, the real part of the dynamic conductivity of graphene, $\mathrm{Re}[\sigma(\omega)]$, becomes negative in THz range [6–11].

Figure 34.1 shows numerically simulated time evolution of $\mathrm{Re}[\sigma(\omega)]$ for intrinsic graphene at room temperature when graphene is impulsively pumped at the time of zero with photon energy around 0.8 eV [11, 20]. In this simulation quasi-Fermi distribution of carriers due to quasi-equilibration through CC scattering is assumed and Auger-type carrier recombination/multiplication processes [26–28] caused by the crystallographic imperfections or many-body effects under intense photoexcitation are ignored. It is clearly seen that a higher quality of graphene with longer momentum relaxation time and thus less crystallographic defects obtains larger values of negative conductivity in a wider frequency range.

It is worth noting that the negative THz conductivity of monolayer graphene is limited to the quantum conductivity ($e^2/4\hbar$) as seen in Eq. (34.2) [21]. This is because the absorption of THz photons that can contribute to the stimulated emission is only made via an interband transition process whose absorbance is limited by $\pi e^2/\hbar c \approx 2.3\%$ [29]. To overcome this limitation on quantum efficiency, a carrier recycling process like that exploited in quantum cascade lasers (QCLs) can be introduced. In this regard, waveguide structures with in-plane THz photon propagation along the graphene sheet [14] are preferable for comprising the laser cavities in order to maximize the gain overlapping and hence to overcome the quantum mechanical limit in comparison with vertical photon emitting cavity structures [12, 13]. However, even for 20 multiple-layered graphene that can increase the absorbance by almost 20 times, the absorption coefficient of THz photons along the inverted graphene is still relatively low, on the order of 1 cm^{-1} [14].

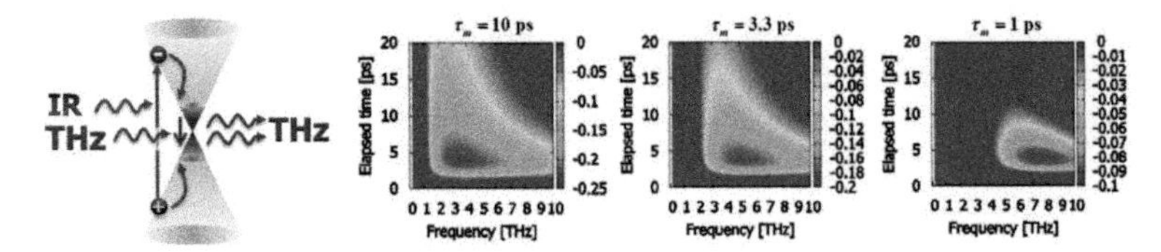

Figure 34.1 Schematic image of the stimulated THz emission in IR-pumped graphene (left) and numerically simulated temporal evolution of the real part of the dynamic conductivity Re[$\sigma(\omega)$] in graphene photoexcited with 0.8 eV, 80 fs pump fluence 8 μJcm^{-2} (average intensity 1×10^8 cm^{-2}) at the time of 0 ps having different momentum relaxation times τ_m of 10, 3.3 and 1 ps (right). Any positive values of Re[$\sigma(\omega)$] are clipped to the zero level (shown in red) to focus on the negative valued region.

As compared with the stimulated emission of the electromagnetic modes (i.e. photons), the stimulated emission of plasmons by the interband transitions in population inverted graphene can be a much stronger emission process. The plasmon gain under population inversion in intrinsic graphene has been theoretically studied in [22]. Nonequilibrium plasmons in graphene can be coupled to the TM modes of electromagnetic waves resulting in the formation and propagation of SPPs [22]. It is shown in [22] that the plasmon gain in pumped graphene can be very high due to small group velocity of the plasmons in graphene and strong confinement of the plasmon field in the vicinity the graphene layer. The propagation constant ρ of the graphene SPP along the z coordinate is derived from Maxwell's equations [22]

$$\sqrt{n^2-\rho^2}+n^2\sqrt{1-\rho^2}+\frac{4\pi}{c}\sigma_\omega\sqrt{1-\rho^2}\,\sqrt{n^2-\rho^2}=0, \quad (34.3)$$

where n is the refractive index, c is the speed of light in vacuum and σ_ω is the conductivity of graphene at frequency ω.

The absorption coefficient α is obtained as the imaginary part of the wave vector along the z coordinate: $\alpha = \mathrm{Im}(q_z) = 2\,\mathrm{Im}(\rho \cdot \omega/c)$. Figure 34.2 plots simulated α for monolayer graphene on a SiO_2/Si substrate ($\mathrm{Im}(n) \sim 3 \times 10^{-4}$) at 300 K. To drive graphene in the population inversion with a negative dynamic conductivity, quasi-Fermi energies are parameterized at ε_F = 10, 20, 30, 40, 50 and 60 meV and a carrier momentum relaxation time τ_m = 3.3 ps is assumed. The results demonstrate giant THz gain (negative values of absorption) of the order of 10^4 cm^{-1}. Since the absorption coefficients and the resultant gain coefficient (under the negative absorption conditions) directly reflect on the dynamic conductivity σ_ω as shown in Eq. (34.3), the gain spectra show similar dependence on momentum relaxation times and thus on the qualities of graphene to that for σ_ω as shown in Fig. 34.1.

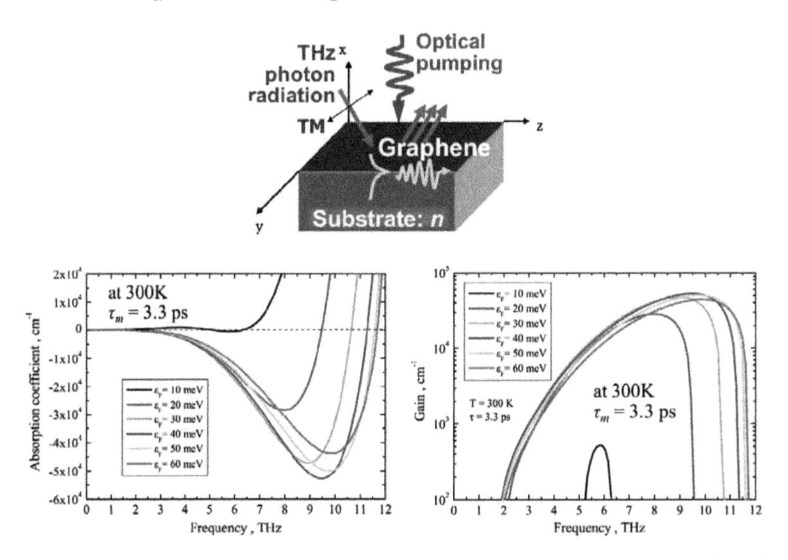

Figure 34.2 Frequency dependences of SPP absorption (left) and gain (right) for monolayer population-inverted graphene on SiO_2/Si substrate at 300 K for different quasi-Fermi energies ε_F = 10, 20, 30, 40, 50 and 60 meV. Carrier momentum relaxation time in graphene is τ_m = 3.3 ps. The results demonstrate giant THz gain (negative values of absorption) of the order of 10^4 cm^{-1}.

It is noted that the transverse electric (TE) modes of THz waves can be coupled with graphene plasmons, but basically the propagation along the z coordinate is not allowed so that they do not contribute to any gain enhancement. One possibility is the waveguide TE modes due to the CdTe thin layer (whose refractive index is lower than that of Si even in the THz range), which could propagate along the z coordinate. However, their group velocities are rather higher than those for usual TM SPPs and the overlapping factor for the graphene layer is lower than those for the TM SPPs because the field maximum is located near the middle of the CdTe layer. Therefore, the gain-enhancement effect of these TE modes must be sufficiently lower than those in the TE mode SPPs.

34.3 Experimental Observation of THz Stimulated Plasmon Emission in Optically Pumped Graphene

34.3.1 Experimental Setup and Sample Preparation

We conducted optical pump, THz probe and optical probe measurement at room temperature for intrinsic monolayer graphene on a SiO_2/Si substrate. The experimental setup is shown in Fig. 34.3, which is identical to the one described in [19] and is based on a time-resolved near-field reflective electro-optic sampling with an fs-IR laser pulse for optical pumping and a synchronously generated THz pulse for probing the THz dynamics of the sample in a THz photon-echo manner. A 140 μm thick CdTe crystal acting as a THz probe pulse emitter as well as an electro-optic sensor was placed on an exfoliated monolayer-graphene/SiO_2/Si sample. A femtosecond-pulsed fiber laser with a full-width at half-maximum of 80 fs, repetition rate of 20 MHz and average power of ~4 mW was used as the optical pump and probe source. The laser is split into two paths used for pump and probe. The pumping laser beam, being linearly polarized, is mechanically chopped at ~1.2 kHz (for lock-in detection) and focused with a beam diameter of about 120 μm onto the sample and the CdTe from the back side, while the probing beam is cross polarized with respect to the pump beam and focused from the top side. The CdTe can rectify the optical pump pulse to emit the envelope THz probe pulse. The emitted primary THz beam grows along the Cherenkov angle to be detected at the CdTe top surface as the primary pulse (marked with '①' in Fig. 34.3b), and then reflects being subject to the graphene sample. When the substrate of the sample is conductive, the THz probe pulse transmitting through graphene again reflects back to the CdTe top surface, which is electro-optically detected as a THz photon echo signal (marked with '②' in Fig. 34.3b). Therefore, the original temporal response consists of the first forward propagating THz pulsation (no interaction with graphene) followed by a photon echo signal (probing the graphene). The delay between these two pulsations is given by the total roundtrip propagation time of the THz probe pulse through the CdTe. The system bandwidth is estimated to be around 6 THz, mainly limited by the Reststrahlen band of the CdTe sensor crystal [19]. The roll off at higher frequency of the photon-echo spectrum starts at ~2 THz, continuing monotonically till ~4 THz and lasts at ~6 THz [19].

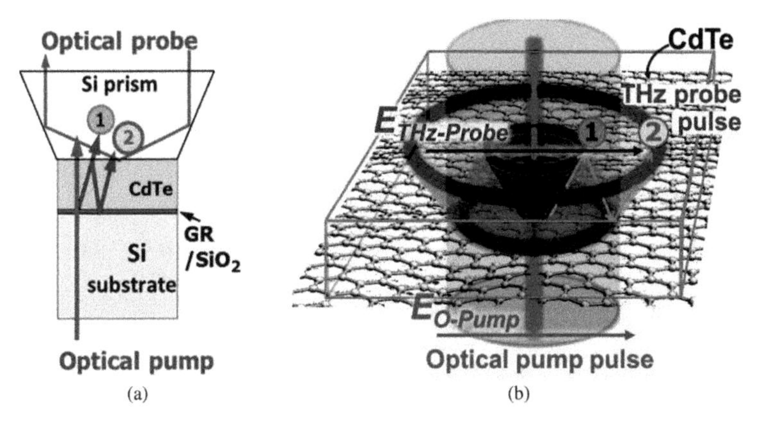

Figure 34.3 Experimental setup of the time-resolved optical pump, THz probe and optical probe measurement based on a near-field reflective electrooptic sampling. (a) Cross-sectional image of the pump/probe geometry, and (b) bird's-eye view showing the trajectories of the optical pump and THz probe beams. The polarization of the optical pump and the THz probe pulse are depicted with a red and a dark-blue arrow, respectively.

The sample prepared for this experiment is a monolayer graphene, which was exfoliated from graphite and transferred onto a SiO_2/Si substrate, and is identical to the one used in [19, 20]. The graphene sample was characterized by the surface morphology using the atomic-force microscopy, the crystallographic properties using the Raman spectroscopy, the doping effects using the Hall-effect measurement and the microscopic surface potential distribution (fluctuation) using the Kelvin-force microscopy as shown in [19, 20]. The flake size of the graphene was ~7000 μm^2. The momentum relaxation time was characterized to be 3.5 ps from the G-band to-D-band peak intensity ratio of the Raman spectra as also shown in [19, 20]. Table 34.1 summarizes these properties.

Table 34.1 Graphene sample properties

Synthesis:	Exfoliation from HOPG
Substrate:	300 nm thick SiO_2/560 μm thick Si (100) (resistivity: 0.005 Ωcm)
Flake size:	~ > 7000 μm^2
Surface height variation as variance:	0.142 nm (in 20 μm × 20 μm area)
Surface potential variation as variance:	4.02 meV (in 10 μm × 20 μm area)
Raman G-band peak to D-band peak intensity ratio:	~35
Estimated carrier momentum relaxation time:	3.3 ps at 300 K
Dirac voltage:	~0 V

34.3.2 Temporal Profile and Fourier Spectrum of the Population-Inverted Graphene to the THz Pulse Irradiation

Figure 34.4a shows temporal responses measured on the monolayer graphene with the CdTe crystal for different pumping pulse intensities up to the maximum intensity I_Ω = 3 × 10^7 W cm^{-2} (equivalently with pump fluence of 2.4 μJ cm^{-2}, almost one order of magnitude below the level of Pauli blocking). It is worth rephrasing that each temporal profile is composed of two peaks from optical rectification in CdTe and the THz photon echo signal. The measured time delay between these two pulsations of 3.5 ps is in good agreement with the roundtrip propagation time of the THz probe pulse through the CdTe crystals [19]. It is clearly seen that the peak obtained on graphene is more intense than that obtained on the substrate without graphene, and that the obtained gain factor exceeds the theoretical limit given by the quantum conductance [21] by more than one order of magnitude, as was measured in [19]. We calculated the voltage gain spectra obtained by Fourier transforming the temporal responses of the secondary pulse, normalized to the frequency response without graphene as shown in Fig. 34.4b. The measured waveforms and their corresponding gain spectra are well reproduced and showed similar pumping intensity dependence with the results shown in [19]. The roll off of the gain spectra at lower cutoff starts at ~2 THz and lasts at ~1.6 THz which is insensitive to the pumping intensity. On the other hand, the roll off at upper cutoff starts at ~3.5 THz at the maximum pumping intensity and lasts at ~4.1 THz, which seems to be limited by the measurement system bandwidth. The upper cutoff frequency decreases with decreasing the pumping intensity. Such a gain spectral dependence on pumping intensity qualitatively agrees with the theoretical calculations, but quantitatively exceeds by more than an order of the aforementioned quantum conductivity limit.

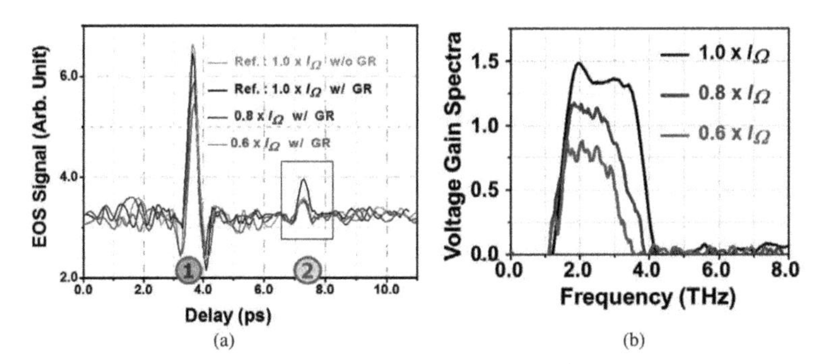

Figure 34.4 (a) Measured temporal responses of the THz photon-echo probe pulse (designated with '②') for different pumping intensities I_Ω (3×10^7 W cm^{-2}), $0.8 \times I_\Omega$ and $0.6 \times I_\Omega$. (b) Corresponding voltage gain spectra of graphene obtained by Fourier transforming the temporal responses of the secondary pulse measured at graphene flake, normalized to that at the position without graphene.

34.3.3 Spatial Field Distribution of the THz Probe Pulse Intensities

We observe the spatial distribution of the THz probe pulse under the linearly polarized optical pump and THz probe pulse conditions. To measure the in-plane spatial distributions of the THz probe pulse radiation, the optical probe pulse position (at the top surface of the CdTe crystal) was changed step-by-step by moving the incident point of the optical pump pulse. The pumping intensity I_{Ω} was fixed at the maximum level 3×10^7 W cm^{-2}. We measured ten times at every point with and without graphene, respectively, and took averages to obtain the electric field intensities and resultant Fourier transform spectra. Observed field distributions for the primary and secondary pulse intensities are shown in Fig. 34.5. The primary pulse field is situated along the circumference with diameter ~50 μm concentric to the center of optical pumping position. On the other hand, the secondary pulse (THz photon echo) field is concentrated only at the restricted spot area on and out of the concentric circumference with diameter ~150 μm, where the incoming THz probe pulse takes a TM mode capable of exciting the SPPs in graphene. The distance between the primary pulse position to the secondary pulse position is ~100 μm or longer. The observed field distribution reproduces the reasonable trajectory of the THz echo pulse propagation in the TM modes inside the CdTe crystal as shown in Fig. 34.2 when we assume the Cherenkov angle of 30°, which was determined by the fraction of the refractive indices between infrared and THz frequencies.

How to couple the incoming/outgoing THz pulse photons to the surface plasmons in graphene is the point of discussion, because the defectless and flat surface of graphene itself has no structural feature that can excite the SPPs. One possibility of the excitation of SPPs by the incoming THz probe pulse is the spatial charge-density modulation at the area of photoexcitation by optical pumping. The pump beam having a Gaussian profile with diameter ~120 μm may define the continuum SPP modes in a certain THz frequency range as seen in various SPPs waveguide structures [30, 31]. After a short propagation of the order of ~10 μm, the SPPs approach the edge boundary of the illuminated and dark area so that they could mediate the THz electromagnetic emission [32, 33]. The plasmon group velocity in graphene (exceeding the Fermi velocity) and propagation distance gives a propagation time of the order of 100 fs. According to the calculated gain spectra shown in Fig. 34.2b, the gain enhancement factor could reach or exceed ~10 at the gain peak frequency 4 THz, which is dominated in the optically probed secondary pulse signals. The obtained gain enhancement factor is ≳50, which is still somewhat higher than the calculated results whose causes should be clarified in a further study.

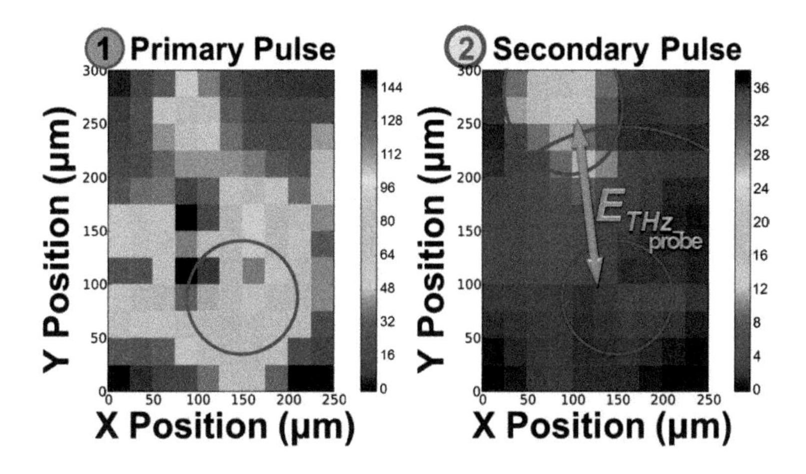

Figure 34.5 Spatial field distribution of the THz probe pulse intensities. The primary pulse shows nonpolar distribution, whereas the secondary pulse shows a strong localization to the area in which the THz probe pulse is impinged to the graphene surface in the TM modes.

34.4 Conclusion

Nonlinear carrier relaxation/recombination dynamics and the resultant stimulated THz photon emission with the excitation of SPPs in photoexcited monolayer graphene was experimentally studied using an optical-pump/THz-probe and optical probe measurement. The spatial distribution of the THz probe pulse intensities under linear polarization of the optical pump and THz probe pulses was measured. It was clearly observed that an intense THz probe pulse was detected only at the area where the incoming THz probe pulse takes a TM mode capable of exciting the SPPs. The observed gain factor is in fair agreement with theoretical calculations. Experimental results support the occurrence of the gain enhancement effect by the excitation of SPPs on THz stimulated emission in optically pumped monolayer graphene.

Acknowledgments

We thank V. V. Popov at Kotelnikov Institute of Radio Engineering and Electronics, Russia, M. S. Shur at Rensselaer Polytechnic Institute, USA, M. Ryzhii at CNEL, University of Aizu and E. Sano at RCIQE, Hokkaido University, Japan, for their contributions. This work was financially supported in part by JST-CREST, JSPS-GA-SPR (no. 23000008), JSPS Core-to-Core, Japan and NSF-PIRE-TeraNano, USA.

References

1. K. S. Novoselov, A. K. Geim, S. V. Morozov, D. Jiang, Y. Zhang, S. V. Dubonos, I. V. Grigorieva and A. A. Frisov, *Science*, **306**, 666 (2004).
2. K. Geim and K. S. Novoselov, *Nat. Mater.*, **6**, 183 (2007).
3. A. H. Castro Neto, F. Guinea, N. M. R. Peres, K. S. Novoselov, and A. K. Geim, *Rev. Mod. Phys.*, **81**, 109–162 (2009).
4. F. Bonaccorso, Z. Sun, T. Hasan, and A. C. Ferrari, *Nat. Photonics*, **4**, 611–622 (2010).
5. A. N. Grigorenko, M. Polini, and K. S. Novoselov, *Nat. Photonics*, **6**, 749–758 (2012).
6. V. Ryzhii, M. Ryzhii, and T. Otsuji, *J. Appl. Phys.*, **101**, 083114 (2007).
7. M. Ryzhii and V. Ryzhii, *Jpn. J. Appl. Phys.*, **46**, L151 (2007).
8. V. Ryzhii, M. Ryzhii, V. Mitin, and T. Otsuji, *J. Appl. Phys.*, **110**, 094503 (2011).
9. A. Satou, T. Otsuji, and V. Ryzhii, *Jpn. J. Appl. Phys.*, **50**, 070116 (2011).
10. V. Ryzhii, M. Ryzhii, V. Mitin, A. Satou, and T. Otsuji, *Jpn. J. Appl. Phys.*, **50**, 094001 (2011).
11. A. Satou, V. Ryzhii, Y. Kurita, and T. Otsuji, *J. Appl. Phys.*, **113**, 143108 (2013).
12. A. A. Dubinov, V. Y. Aleshkin, M. Ryzhii, T. Otsuji, and V. Ryzhii, *Appl. Phys. Express*, **2**, 092301 (2009).
13. V. Ryzhii, M. Ryzhii, A. Satou, T. Otsuji, A. A. Dubinov, and V. Y. Aleshkin, *J. Appl. Phys.*, **106**, 084507 (2009).

14. V. Ryzhii, A. Dubinov, T. Otsuji, V. Mitin, and M. S. Shur, *J. Appl. Phys.*, **107**, 054505 (2010).
15. J. M. Dawlaty, S. Shivaraman, M. Chandrashekhar, F. Rana, and M. G. Spencer, *Appl. Phys. Lett.*, **92**, 042116 (2008).
16. P. A. George, J. Strait, J. Dawlaty, S. Shivaraman, M. Chandrashekhar, and F. R. M. G. Spencer, *Nano Lett.*, **8**, 4248 (2008).
17. M. Breusing, C. Ropers, and T. Elsaesser, *Phys. Rev. Lett.*, **102**, 086809 (2009).
18. T. Li, L. Luo, M. Hupalo, J. Zhang, M. C. Tringides, J. Schmalian, and J. Wang, *Phys. Rev. Lett.*, **108**, 167401 (2012).
19. S. Boubanga-Tombet, S. Chan, T. Watanabe, A. Satou, V. Ryzhii, and T. Otsuji, *Phys. Rev. B*, **85**, 035443 (2012).
20. T. Otsuji, S. A. Boubanga Tombet, A. Satou, H. Fukidome, M. Suemitsu, E. Sano, P. Popov, M. Ryzhii, and V. Ryzhii, *J. Phys. D: Appl. Phys.*, **45**, 303001 (2012).
21. L. Falkovsky and A. Varlamov, *Eur. Phys. J. B*, **56**, 281 (2007).
22. A. A. Dubinov, Y. V. Aleshkin, V. Mitin, T. Otsuji, and V. Ryzhii, *J. Phys.: Condens. Matter*, **23**, 145302 (2011).
23. V. Ryzhii, A. Satou, and T. Otsuji, *J. Appl. Phys.*, **101**, 024509 (2007).
24. F. Rana, *IEEE Trans. Nanotechnol.*, **7**, 91–99 (2008).
25. A. R. Wright, J. C. Cao, and C. Zhang, *Phys. Rev. Lett.*, **103**, 207401 (2009).
26. J. H. Strait, H. Wang, S. Shivaraman, V. Shields, M. Spencer, and F. Rana, *Nano Lett.*, **11**, 4902–4906 (2011).
27. T. Winzer, A. Knorr, and E. Malic, *Nano Lett.*, **10**, 4839–4843 (2010).
28. T. Winzer and E. Malic, *Phys. Rev. B*, **85**, 241404 (2012).
29. R. R. Nair, P. Blake, A. N. Grigorenko, K. S. Novoselov, T. J. Booth, T. Stauber, N. M. R. Peres, and A. K. Geim, *Science*, **320**, 1308 (2008).
30. X. Y. He, Q. J. Wang, and S. F. Yu, *Plasmonics*, **7**, 571–577 (2012).
31. X. Y. He, *Opt. Express*, **17**, 15359–15371 (2009).
32. B. J. Lawrie, R. F. Haglund Jr., and R. Mu, *Opt. Express*, **17**, 2565 (2009).
33. S. Wedge and W. Barnes, *Opt. Express*, **12**, 3672 (2004).

Chapter 35

Graphene Surface Emitting Terahertz Laser: Diffusion Pumping Concept*

**A. R. Davoyan,[a,b] M. Yu. Morozov,[a] V. V. Popov,[a]
A. Satou,[c] and T. Otsuji[c]**
[a]*Kotelnikov Institute of Radio Engineering and Electronics (Saratov Branch), Russian Academy of Sciences, Saratov 410019, Russia*
[b]*Department of Electrical and Systems Engineering, University of Pennsylvania, Philadelphia, Pennsylvania 19104, USA*
[c]*Research Institute of Electrical Communication (RIEC), Tohoku University, Sendai, Miyagi 980-8577, Japan*
davoyan@seas.upenn.edu

We suggest a concept of a tunable graphene-based terahertz (THz) surface emitting laser with diffusion pumping. We employ significant difference in the electronic energy gap of graphene and a typical wide-gap semiconductor, and demonstrate that carriers generated in the semiconductor can be efficiently captured by graphene resulting in population inversion and corresponding THz

*Reprinted with permission from A. R. Davoyan, M. Yu. Morozov, V. V. Popov, A. Satou, and T. Otsuji (2013). Graphene surface emitting terahertz laser: diffusion pumping concept, *Appl. Phys. Lett.*, **103**, 251102. Copyright © 2013 AIP Publishing LLC.

Graphene-Based Terahertz Electronics and Plasmonics: Detector and Emitter Concepts
Edited by Vladimir Mitin, Taiichi Otsuji, and Victor Ryzhii

ISBN 978-981-4800-75-4 (Hardcover), 978-0-429-32839-8 (eBook)
www.jennystanford.com

lasing from graphene. We develop design principles for such a laser and estimate its performance. We predict up to 50 W/cm^2 terahertz power output for 100 kW/cm^2 pump power at frequency around 10 THz at room temperature.

Recent advances in science and engineering at terahertz (THz) frequencies have suggested a wide range of highly prospective applications ranging from harmless security systems and noninvasive medical treatment to time-domain spectroscopy [1]. At the same time, lack of an efficient, tunable, and compact source of a coherent radiation at frequencies of 1–15 THz enforces significant constrains in this research and development direction. Commonly available sources of coherent electromagnetic radiation based either on classical electron oscillations or on electron transitions in quantum systems experience fundamental limitations at THz frequencies, which are bound by short microwave range (≤0.1 THz) on one side and far-infrared (≥15 THz) on the other [2, 3].

Smart band-gap engineering, including quantum-cascade laser structures [4] and semiconductor multi-well lattice structures [5] as well as nonlinear frequency conversion mechanism [6] and plasma wave instabilities in semiconductor heterostructures [7], have been developed recently as possible routes for THz generation. Nevertheless, these methods lack in their functionality to this or that extent. For example, quantum-cascade lasers have a complicated structure design and require cryogenic cooling [4, 8], whereas THz generation with a nonlinear frequency mixing is sensitive to the phase matching conditions [6, 9].

Recent discovery of graphene exhibiting zero electronic band-gap [10] suggests unique opportunities for THz science [11–14]. In particular, several breakthroughs associated with unusual electronic transport [15] and electromagnetic response [11, 14] in graphene have been demonstrated. It has been speculated lately that due to zero electronic band-gap graphene can serve as a natural material for THz generation and lasing [14, 16]. A strong population inversion in pumped graphene was predicted theoretically [17–20], and the stimulated THz emission from graphene was observed experimentally [16, 21]. What is more important, the population inversion in graphene can be observed at room temperature for frequencies as low as 1 THz [17].

These unique properties of graphene have been employed for the design of THz lasing systems [22–26]. In particular, the stimulated emission of plasmons was discussed theoretically [18, 22], and several potential graphene-based laser designs with direct optical pumping of graphene were proposed [23, 24]. However, limited optical absorption rate of a single layer graphene (about 2.3% for visible light [15]) fundamentally limits the efficiency of graphene lasers with direct optical pumping [23].

In this Letter, we make use of an infinitesimally small electronic band-gap of graphene and suggest a concept for an efficient indirect pumping of graphene and develop principles for a tunable graphene-based THz surface emitting laser. We consider a diffusion-injection mechanism for graphene pumping and utilize this mechanism for an efficient THz laser design. We derive a phenomenological model, fully describing our THz laser, and applicable to other types of surface emitting graphene-based structures. Finally, we discuss the limitations of our approach and possibilities for further enhancement of the device performance.

Figure 35.1a shows schematically the design of the proposed graphene-based THz laser. We assume that a mono-layer graphene is deposited on a surface of a wide-gap semiconductor and is coated with a dielectric layer. Pumping of graphene might lead to a population inversion with a consequent THz emission, which, in turn, can be enhanced and stimulated by placing the structure into a resonator, for example, a Fabry-Perot cavity as shown in Fig. 35.1a. Here, we consider an indirect mechanism for graphene pumping. In particular, we consider the electron-hole pair generation in the semiconductor slab, which can be achieved by various techniques, for instance, by the optical pumping or carrier injection. The excited carriers reaching graphene due to the ambipolar diffusion are captured by graphene possessing lower energy states, see a schematic illustration of this process in Fig. 35.1c. Significant difference between the energy gap of a wide-gap semiconductor (that is, about $\simeq$1 eV) and that of graphene ($\ll$0.1 eV) prevents carrier escape from graphene back to the semiconductor. (Note that a potential barrier at the surface of the semiconductor formed due to the Fermi-level pinning may modify the carrier transport through the semiconductor-graphene interface. However, this effect is small for an undoped semiconductor and hence can be neglected [27].)

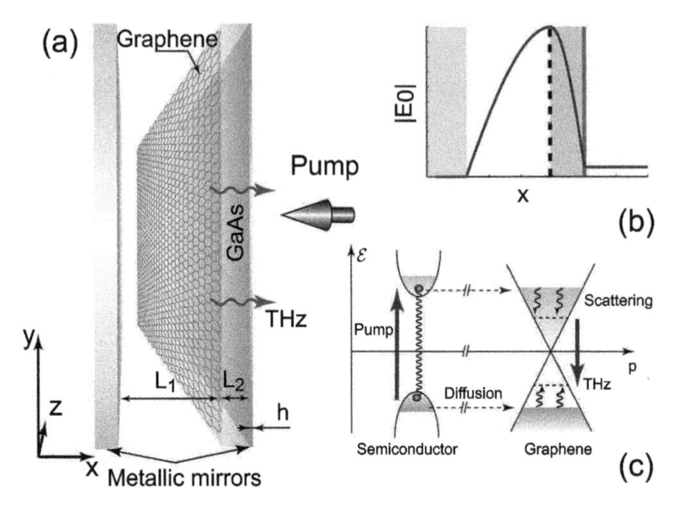

Figure 35.1 Schematic of the proposed graphene-based laser and its operation principles. (a) Sketch of the studied graphene-based laser structure, (b) typical eigen-mode profile in the resonant cavity, and (c) illustration of the suggested diffusion pumping mechanism with energy-band diagrams of a wide-gap semiconductor and graphene.

Ambipolar carrier diffusion in the semiconductor is described by a conventional diffusion equation [28], $D_{\mathrm{a}}\dfrac{d^2n_{\mathrm{s}}}{dx^2}-\dfrac{n_{\mathrm{s}}}{\tau_{\mathrm{R}}}+\dfrac{2\alpha e^{-2\alpha x}P_{\mathrm{in}}}{\hbar\omega_{\mathrm{P}}}=0$, where n_{s} is the concentration of electron-hole pairs in semiconductor, D_{a} is the ambipolar diffusion coefficient, τ_{R} is the electron-hole pair recombination time in semiconductor, P_{in} is the input optical power of pump light with frequency ω_{P}, and α is the light absorption coefficient in semiconductor (note that the electron-hole generation is due to the interband transitions). (Here, we consider the optical pumping of the semiconductor; however, in general, one might employ another more efficient method, such as the carrier injection mechanism.) Wherefrom, a diffusive current flowing into graphene can be expressed as $J_{\mathrm{in}}\big|_{x=x_{\mathrm{g}}}=D_{\mathrm{a}}\dfrac{\partial n_{\mathrm{s}}}{\partial x}\Big|_{x=x_{\mathrm{g}}}$, where the coordinate x_{g} corresponds to the position of graphene. Phenomenological equation for the nonequilibrium carrier concentration in graphene can be written as [28]

$$\frac{dn}{dt}=J_{\mathrm{in}}-\frac{n-n_0}{\tau}+\frac{1}{\hbar\omega}\mathrm{Re}(\sigma_{\mathrm{inter}})|E(x=x_{\mathrm{g}})|^2=0, \qquad (35.1)$$

where n_0 is the equilibrium two-dimensional carrier concentration in intrinsic graphene, τ is the effective nonradiative carrier recombination time in graphene. The last term in this equation describes the radiative generation/recombination of carriers in graphene due to emission/absorption of THz photons with frequency ω; σ_{inter} is the graphene interband conductivity [17]; and $E(x = x_g)$ is the THz electric field in graphene.

In order to enhance the THz emission from the pumped graphene, we assume that the structure under consideration is placed into a Fabry-Perot cavity, see Fig. 35.1a, which is comprised of two metallic mirrors, one of which, being thinner than the THz skin depth ($\simeq$50 nm), is semitransparent. (Note that we consider here metallic mirrors in order to reduce the THz resonator size as compared to conventional Bragg mirrors widely used in surface emitting lasers in optical range. Nevertheless, the concept and theoretical predictions discussed here can be easily extended to the latter case as well.) To achieve the strongest interaction between THz field and graphene we choose the thicknesses dielectric coating, L_1, and semiconductor cladding, L_2, as $L_1\sqrt{\varepsilon_1} \simeq L_2\sqrt{\varepsilon_2}$, where $\varepsilon_{1,2}$ are the dielectric permittivities of the coating and cladding at no pumping, respectively. Typical mode profile of THz electric field in the resonator is shown in Fig. 35.1b.

To account for the dynamics of the THz field in the Fabry-Perot resonator, we consider the energy balance equation $(E\dot{D} + H\dot{B}) + jE + \frac{\partial EH}{\partial x} = 0$ [29], where $j = \sigma\delta(x - x_g)E$ is the excited THz current in graphene, σ is the dynamic (frequency-dependent) two-dimensional conductivity of graphene. Representing the fields in the form $E(x, t) = \frac{1}{2}\psi(t)\,(E_0(x)e^{-i\omega t} + c.c.)$, where $\psi(t)$ is a dimensionless amplitude slowly varying in time, $E_0(x)$ is the resonator's eigen-mode profile, *c.c.* stands for the complex conjugation, and applying standard techniques accounting for a slowly varying envelope [29], we derive the following equation for the slowly varying amplitude:

$$\frac{d\psi^2(t)}{dt} = -\frac{2}{W_0}\psi^2(t)\Big[\text{Re}(\sigma)|E_0(x_g)|^2 + \omega\varepsilon_0\langle\text{Im}(\varepsilon(x))|E_0(x)|^2\rangle + S_0\Big], \qquad (35.2)$$

where $W_0 = 1/2\langle(\varepsilon_0\text{Re}(\varepsilon(x))|E_0(x)|^2 + \mu_0|H_0(x)|^2)\rangle$, $S_0 = 1/2(E_0(x) H_0^*(x) + c.c.)$ is the power outflow, E_0 and H_0 are electric and magnetic eigen-mode profiles, $\text{Im}(\varepsilon(x))$ is the imaginary part of the permittivity describing the absorption in the resonator, ε_0 is vacuum permittivity, and $\langle\cdot\rangle = \int_{-\infty}^{0} \cdot\, dx$.

Clearly, when the right hand side of Eq. (35.2) becomes positive, the slowly varying amplitude ψ increases, corresponding to the self-excitation of the system and THz lasing with the overall power output $\psi^2 S_0$. Obviously, such regime is possible only when $-\text{Re}(\sigma)|E_0(x_g)|^2 \geq [\varepsilon_0\omega\langle\text{Im}(\varepsilon(x))|E_0(x)|^2\rangle + S_0]$, i.e., when a strong population inversion is achieved in graphene. We note that Eq. (35.2) is universal for any configuration of vertical cavity structures with active graphene layers. It was shown in Ref. [17], that the conductivity of the population inverted graphene in a quasi-stationary limit (i.e., after ultra-fast nonequilibrium carrier relaxation has taken place) can be modeled by Fermi-Dirac distribution function $f(\xi) = \left[\exp\dfrac{\xi - \mu}{k_B T} + 1\right]^{-1}$ with an effective quasi-Fermi energy μ and effective carrier temperature T (in this case, the carrier concentration in graphene n is related to the quasi-Fermi level μ and temperature T as $n = \dfrac{2k_B^2}{\pi v^2 \hbar^2}\int_0^{\infty} \xi f(\xi)\, d\xi$). In our further analysis, we assume that the effective temperature T is fixed and does not depend on the quasi-Fermi energy [17]. In this case, the real part of graphene conductivity is given by the following relation [17]:

$$\text{Re}\sigma(\omega) = \text{Re}\sigma_{\text{intra}} + \text{Re}\sigma_{\text{inter}}$$

$$= \frac{e^2}{4\hbar}\left(\frac{8k_B T\tau_m \ln\left[1 + e^{\frac{\mu}{k_B T}}\right]}{\pi\hbar(1 + \omega^2\tau_m^2)} + \tanh\left(\frac{\hbar\omega - 2\mu}{4k_B T}\right)\right), \tag{35.3}$$

where the first term in the right-hand side describes the Drude-type intraband contribution, whereas the second term is associated with the interband transitions responsible for the carrier population inversion in graphene, τ_m is the characteristic carrier momentum relaxation time in graphene, k_B is the Boltzmann constant, and $\hbar$ is the reduced Planck constant. (Note that, in more general nonequilibrium case, the carrier distribution and corresponding dynamic conductivity of graphene should be described by a more accurate kinetic theory [17, 18].)

Figures 35.2a and b show the quasi-Fermi energy μ corresponding to the graphene transparency, i.e., Reσ = 0, depending on frequency at different values of temperature T (Fig. 35.2a) and the carrier momentum relaxation time τ_m (Fig. 35.2b). For $\tau_m \geq 1$ ps the population inversion takes place for frequencies 5–10 THz at relatively small quasi-Fermi energies $\mu < 30$ meV at room temperature, $T = 300$ K. In our further analysis, we fix the carrier momentum relaxation time in graphene at $\tau_m = 5$ ps for all temperatures and values of the quasi-Fermi energy.

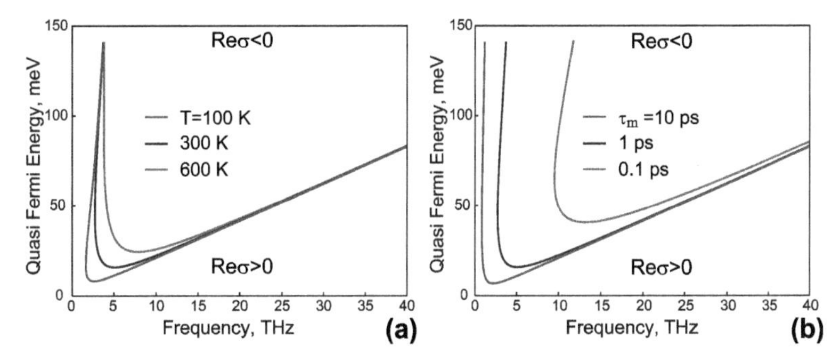

Figure 35.2 Plot of quasi-Fermi energy corresponding to the transparency of graphene, i.e., (σ) = 0, as a function of frequency: (a) for different temperatures at $\tau_m = 1$ ps and (b) for different momentum relaxation times at $T = 300$ K.

Equations (35.1) and (35.2) comprise a self-consistent system of balance equations fully describing THz stimulated emission in the proposed system. We solve these equations assuming that the semiconductor layer is gallium arsenide (GaAs) with the energy gap of $\simeq$1.42 eV. (The possibility of graphene deposition on top of GaAs was discussed in Ref. [30]. Note, however, that our concept is not limited to GaAs and one might choose other more suitable semiconductor.) The ambipolar diffusion coefficient and the nonradiative carrier recombination time for GaAs are $D_a \simeq$ 100 cm^2/s, $\tau_R \simeq 1$ ns, respectively [31], and $\alpha \simeq 1$ μm^{-1} at pump operating wavelength $\lambda_p = 850$ nm [32]. We also assume that the permittivity of the dielectric coating is $\varepsilon_1 = 2$ and metallic mirrors are made of silver described by a corresponding Drude-Lorentz model. Furthermore, we model the permittivity of GaAs based on the Drude model, i.e., $\varepsilon_{GaAs} = \varepsilon_2 - \langle n_s \rangle e^2/(\omega m^* \varepsilon_0(\omega + i\gamma))$, where $\varepsilon_2 = 12.8$ is GaAs permittivity at no pump, $m^* = 0.067$ m is the effective electron mass, e is the electron charge, ε_0 is the vacuum permittivity, $\gamma \simeq 2\pi \times 10^{12}$ s^{-1} is the collision frequency for the carriers in GaAs, which is responsible for THz wave damping, and $\langle n \rangle$ is the average concentration of photoexcited electron-hole pairs in the GaAs slab that depends on the pump power (see the diffusion equation). For the purpose of the concept demonstration, we neglect with the losses in metal and consider THz dissipation in the GaAs layer only. Due to the elevation of electron-hole concentration $\langle n_s \rangle$ with the increase of the pump power, the real part of the GaAs permittivity becomes smaller while its imaginary part grows leading to addition attenuation of THz wave in the Fabry-Perot resonator. The dependence of the real part of GaAs permittivity on the pump power implies that, even for the fixed geometrical dimensions of the resonator, the operation frequency changes with the pump power, i.e., $\omega = \omega(P_{pump})$.

First, we consider a stationary regime of the laser operation assuming $\frac{dn}{dt} = \frac{d\psi^2}{dt} = 0$. Generation of THz waves develops when the pump power exceeds the threshold level, which according to Eqs. (35.1) and (35.2) can be found from: $J_{in} \simeq (n(\mu_{th}) - n_0)/\tau$ and $\mathrm{Re}\sigma(\mu_{th})|E_0(x = x_g)|^2 + \omega\varepsilon_0\langle \mathrm{Im}(\varepsilon)|E_0|^2 \rangle + S_0 = 0$, where μ_{th} is the threshold quasi-Fermi energy and $\omega = \omega(P_{pump})$. The threshold power strongly depends on temperature T and the nonradiative recombination time of the electron-hole pairs in graphene τ.

In Fig. 35.3a, we plot the threshold optical pump power as function of the nonradiative recombination time τ for two different electron temperatures. The threshold power level dramatically increases with the decrease of the nonradiative recombination time. Increase in temperature also leads to the increase of the threshold power due to the enhancement of the intraband absorption in graphene, see Eq. (35.3). As was shown in Ref. [33], the nonradiative recombination time can exceed 10 ps at room temperature. This value of τ corresponds to the pump power threshold of about 10^4 W/cm^2 for $T = 300$ K.

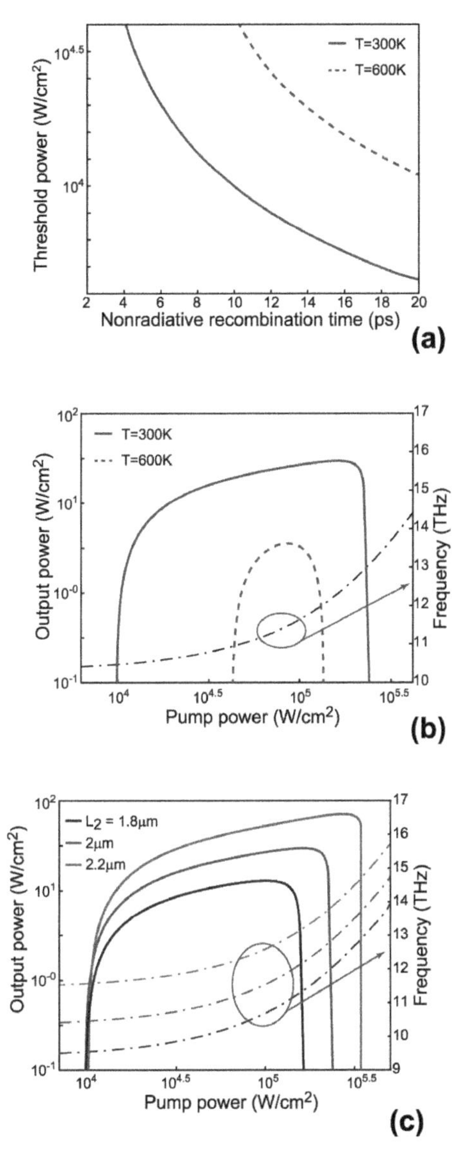

Figure 35.3 (a) Threshold optical pump power level versus the nonradiative recombination time in graphene for L_2 = 2 μm, (b) THz power output versus the optical pump power for two different temperatures, τ = 10 ps, and L_2 = 2 μm, and (c) the same as in panel (b) but for different thicknesses of GaAs layer for T = 300 K. Dashed-dotted thin lines in panels (b) and (c) denote the operating frequency shown in the right axis. The thickness of the semitransparent metallic mirror is h = 15 nm.

In Fig. 35.3b, we demonstrate the dependence of the output THz power P_{THz} upon the optical pump power P_{pump} for two different temperatures. The output power dramatically increases for values of pump power slightly exceeding the threshold value. With further increase of the pump power, the output THz power reaches its maximum and then the lasing quenches. Such a behavior is caused by the increase of the THz attenuation in GaAs due to photocarrier generation so that THz gain in graphene cannot overcome the THz wave losses in the Fabry-Perot resonator anymore. At the maximum of the power output, we can estimate the power conversion efficiency as $P_{THz}/P_{pump} \simeq 2 \times 10^{-4}$ for $T = 300$ K and $\tau = 10$ ps, which is similar to the conversion efficiency of THz sources using the difference-frequency generation schemes at room temperature [6, 34]. With increase of the temperature, the stronger intraband absorption of THz radiation in graphene leads to the decrease of lasing efficiency and narrows the dynamic range of lasing. As we have noted above, with the growth of the pump power, the operation frequency of our laser increases due to decreasing the real part of GaAs permittivity with the increase of the photocarrier density in GaAs, see Fig. 35.3b.

In Fig. 35.3c, we plot the THz output power versus the pump power for different thicknesses of GaAs layer at $T = 300$ K (assuming that $L_1\sqrt{\varepsilon_1} = L_2\sqrt{\varepsilon_2}$). The decrease of the optical length of the Fabry-Perot resonator leads to the increase of the lasing frequency, see Fig. 35.3c. We notice that the threshold power is practically the same, $P_{pump} \simeq 10^4$ W/cm^2, for the shown resonator dimensions. At the same time, the output power is smaller for a longer resonator, i.e., for lower operating frequencies, which we attribute to smaller gain in graphene at lower THz frequencies. The latter also results in narrowing the range the values of the pump power in which the lasing is possible.

Next, we solve the dynamic system of balance equations (35.1) and (35.2) searching for the time variation of the amplitude ψ and the quasi-Fermi energy μ in a self-excitation regime, see Fig. 35.4. After the optical pump power is switched on at zero moment of time, the quasi-Fermi energy grows in time and saturates after about 30 ps, Fig. 35.4a. This metastable quasi-Fermi energy level can be found from the condition $J_{in} \simeq (n(\mu) - n_0)/\tau$, which implies that the value of the metastable quasi-Fermi energy increases for greater pump power, Fig. 35.4a. (However, with increasing the population inversion, one should take into account the growth of probability of the non-equilibrium carrier escape back to the GaAs from graphene, as well as the increase in the effective carrier temperature.) With further time evolution of the system, the quasi-Fermi energy in graphene steeply decreases (Fig. 35.4a), which is followed by the generation of THz photons, see Fig. 35.4b. The duration of the metastable regime depends on the interplay between gain in graphene and the losses in the Fabry-Perot resonator. For a greater ratio between the gain and the losses, the duration of the metastable regime is shorter (see also Fig. 35.3b). Finally, the THz output power and the quasi-Fermi energy in graphene stabilize at the values corresponding to the stationary regime of the laser operation, see Fig. 35.3.

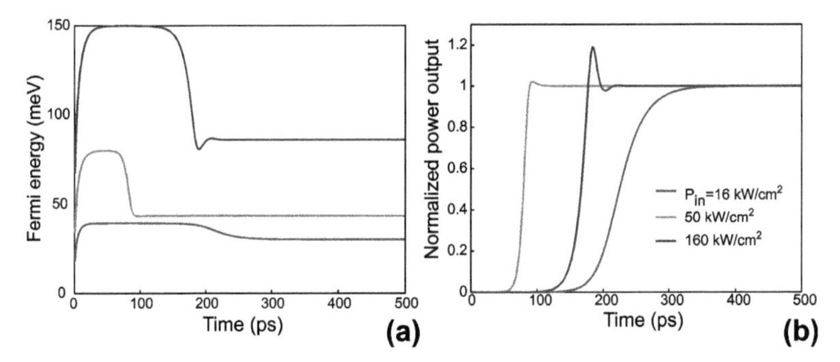

Figure 35.4 Transient dynamics of the laser parameters. (a) Quasi-Fermi energy variation in time and (b) output THz power normalized to the steady state output power level. Here, $T = 300$ K, $h = 15$ nm, and $L_2 = 2$ μm.

We note that our analysis is performed with certain approximations. In particular, we assume that the carrier temperature T, momentum relaxation time τ_m, and nonradiative recombination time τ in graphene, as well as the ambipolar diffusion coefficient and the recombination time in GaAs, are independent of the electron-hole concentration in respective media. Second, we consider that graphene bandgap exhibits Dirac-type energy spectrum even when deposited on GaAs surface. We neglect the losses in the metallic mirrors, which might play a significant role. Nevertheless, we believe that the main conclusions of this chapter remain valid qualitatively for more realistic parameters of the device. The performance of the proposed surface emitting graphene THz laser with the diffusion pumping can be improved further by using multilayer graphene structures [24], or by resonant THz plasmon excitation, for example, with the help of a perforated metal film [26] (thus also reducing losses in metal).

In conclusion, we have developed a concept of a tunable surface emitting graphene laser with an ambipolar carrier diffusion pumping from an adjacent wide-gap semiconductor. We have developed a phenomenological theoretical model describing surface emitting structures with graphene. Based on this model, we have estimated the threshold pump power levels as well as the efficiency of the laser operation. We predict that the efficiency of such systems can reach 0.1% at frequency 10 THz with proper material and structure engineering.

We are grateful to V. Ryzhii for valuable comments and discussions. The work was supported by the Russian Foundation for Basic Research (Grant Nos. 12-02-31888, 12-02-93105, and 13-02-12070), by the Russian Academy of Sciences Program "Fundamentals of Nanostructure Technologies and Nanomaterials," and by JSPS-RFBR Japan-Russian Collaborative Research Program.

References

1. X.-C. Zhang and J. Xu, *Introduction to THz Wave Photonics* (Springer-Verlag, New York, 2009).
2. M. Tonouchi, *Nat. Photonics*, **1**, 97 (2007).
3. G. P. Gallerano and S. Biedron, Overview of terahertz radiation sources, in *Proceedings of the Free Electron Laser Conference* (2004), pp. 216–221.
4. J. Faist, F. Capasso, D. L. Sivco, C. Sirtori, A. L. Hutchinson, and A. Y. Cho, *Science*, **264**, 553 (1994).
5. R. Kohler, A. Tredicucci, F. Beltram, H. E. Beere, E. H. Linfield, A. G. Davies, D. A. Ritchie, R. C. Iotti, and F. Rossi, *Nature*, **417**, 156 (2012).
6. M. Scheller, J. M. Yarborough, J. V. Moloney, M. Fallahi, M. Koch, and S. W. Koch, *Opt. Express*, **18**, 27112 (2010).
7. M. Dyakonov and M. Shur, *Phys. Rev. Lett.*, **71**, 2465 (1993).
8. S. Kumar, C. Wang, I. Chan, Q. Hu, and J. L. Reno, *Nat. Phys.*, **7**, 166 (2011).
9. M. A. Belkin, F. Capasso, and A. Belyanin, *Nat. Photonics*, **1**, 288 (2007).

10. K. S. Novoselov, A. K. Geim, S. V. Morozov, D. Jiang, Y. Zhang, S. V. Dubonos, I. V. Grigorieva, and A. A. Firsov, *Science*, **306**, 666 (2004).
11. F. H. Koppens, D. E. Chang, and F. J. Garcia de Abajo, *Nano Lett.*, **11**, 3370 (2011).
12. A. Vakil and N. Engheta, *Science*, **332**, 1291 (2011).
13. A. R. Davoyan, V. V. Popov, and S. A. Nikitov, *Phys. Rev. Lett.*, **108**, 127401 (2012).
14. T. Otsuji, S. A. B. Tombet, A. Satou, H. Fukidome, M. Suemitsu, E. Sano, V. Popov, M. Ryzhii, and V. Ryzhii, *J. Phys. D: Appl. Phys.*, **45**, 303001 (2012).
15. A. K. Geim and K. S. Novoselov, *Nat. Mater.*, **6**, 183 (2007).
16. T. Li, L. Luo, M. Hupalo, J. Zhang, M. C. Tringides, J. Schmalian, and J. Wang, *Phys. Rev. Lett.*, **108**, 167401 (2012).
17. V. Ryzhii, M. Ryzhii, and T. Otsuji, *J. Appl. Phys.*, **101**, 083114 (2007).
18. F. Rana, P. A. George, J. H. Strait, J. Dawlaty, S. Shivaraman, Mvs Chandrashekhar, and M. G. Spencer, *Phys. Rev. B*, **79**, 115447 (2009).
19. V. Ryzhii, M. Ryzhii, V. Mitin, A. Satou, and T. Otsuji, *Jpn. J. Appl. Phys., Part 1*, **50**, 094001 (2011).
20. A. Satou, V. Ryzhii, Y. Kurita, and T. Otsuji, *J. Appl. Phys.*, **113**, 143108 (2013).
21. S. Boubanga-Tombet, S. Chan, T. Watanabe, A. Satou, V. Ryzhii, and T. Otsuji, *Phys. Rev. B*, **85**, 035443 (2012).
22. F. Rana, *IEEE Trans. Nanotechnol.*, **7**, 91 (2008).
23. A. A. Dubinov, V. Ya. Aleshkin, M. Ryzhii, T. Otsuji, and V. Ryzhii, *Appl. Phys. Express*, **2**, 092301 (2009).
24. V. Ryzhii, A. A. Dubinov, T. Otsuji, V. Mitin, and M. S. Shur, *J. Appl. Phys.*, **107**, 054505 (2010).
25. V. Ryzhii, M. Ryzhii, V. Mitin, and T. Otsuji, *J. Appl. Phys.*, **110**, 094503 (2011).
26. V. V. Popov, O. V. Polischuk, A. R. Davoyan, V. Ryzhii, T. Otsuji, and M. S. Shur, *Phys. Rev. B*, **86**, 195437 (2012).
27. W. Mönch, *Semiconductor Surfaces and Interfaces*, 3rd ed. (Springer-Verlag, Berlin, 2001).
28. S. M. Sze, *Physics of Semiconductor Devices* (Wiley, New York, 1981).
29. L. D. Landau, L. P. Pitaevskii, and E. M. Lifshitz, *Electrodynamics of Continuous Media*, 2nd ed., Course of Theoretical Physics Vol. 8 (Butterworth-Heinemann, 1984).

30. M. Friedemann, K. Pierz, R. Stosch, and F. J. Ahlers, *Appl. Phys. Lett.*, **95**, 102103 (2009).
31. D. Streb, G. Klem, W. Fix, P. Kiesel, and G. H. Dhler, *Appl. Phys. Lett.*, **71**, 1501 (1997).
32. Y. Morozov, T. Leinonen, M. Morozov, S. Ranta, M. Saarinen, V. Popov, and M. Pessa, *New J. Phys.*, **10**, 063028 (2008).
33. P. A. George, J. Strait, J. Dawlaty, S. Shivaraman, M. Chandrashekhar, F. Rana, and M. G. Spencer, *Nano Lett.*, **8**, 4248 (2008).
34. K. Vijayraghavan, Y. Jiang, M. Jang, A. Jiang, K. Choutagunta, A. Vizbaras, F. Demmerle, G. Boehm, M. C. Amann, and M. A. Belkin, *Nat. Commun.*, **4**, 2021 (2013).

Chapter 36

Enhanced Terahertz Emission from Monolayer Graphene with Metal Mesh Structure*

T. Itatsu,[a] E. Sano,[a] Y. Yabe,[b] V. Ryzhii,[b] and T. Otsuji[b]
[a]*Research Center for Integrated Quantum Electronics, Hokkaido University, North13, West8, Sapporo-shi, 060-8628, Japan*
[b]*Research Institute of Electrical Communication, Tohoku University, 2-1-1 Katahira, Aoba-ku, Sendai-shi, Miyagi, 980-8577, Japan*
esano@rciqe.hokudai.ac.jp

The development of compact, low-cost sources and detectors operating at room temperature is needed to fully use terahertz (THz) electromagnetic waves in a variety of applications. Photo-excited graphene is expected to exhibit population inversion and enable the fabrication of a novel THz laser. We previously proposed a THz amplifier as a basic element in a THz laser by combining the population inversion in photo-excited graphene with the electric

*Reprinted with permission from T. Itatsu, E. Sano, Y. Yabe, V. Ryzhii, and T. Otsuji (2016). Enhanced terahertz emission from monolayer graphene with metal mesh structure, *Mater. Today: Proc.*, **3S**, S221–S226. Copyright © 2016 Elsevier Ltd.

Graphene-Based Terahertz Electronics and Plasmonics: Detector and Emitter Concepts
Edited by Vladimir Mitin, Taiichi Otsuji, and Victor Ryzhii

ISBN 978-981-4800-75-4 (Hardcover), 978-0-429-32839-8 (eBook)
www.jennystanford.com

field enhancement due to the spoof surface plasmon polaritons induced on the surface of a metal mesh structure. For this work, we fabricated prototype THz amplifiers composed of chemical-vapor-deposition-grown graphene and a metal mesh structure and observed extremely amplified THz emission by using an electro-optic sampling method.

36.1 Introduction

The terahertz (THz) electromagnetic (EM) wave region of 0.1–100 THz located between radio and light is expected to open up various applications including ultrahigh-speed wireless communications, material characterization, nondestructive evaluation, and imaging. However, the lack of commercially available microelectronic devices that can generate and detect THz waves at room temperature (RT) significantly hampers the development of the THz region. This situation is often called the "THz gap." The reason for opening the THz gap is that the maximum operating frequency of electronic devices is limited by the electron velocity in semiconductors and that optical devices are affected by thermal noise at RT because the bandgap energy contributing to laser oscillation is small. Therefore, we expect that THz devices will be developed using a novel operating principle and/or materials.

Graphene, a two-dimensional honeycomb structure of carbon atoms, has attracted a large amount of attention due to its unique nature and potential applications [1, 2]. In particular, the linear energy-band structure of graphene leads to the development of THz device applications [3]. Excitation of surface plasmon polaritons in population-inverted graphene can dramatically enhance the THz gain [4], which has recently been experimentally observed [5]. Based on the EM properties of photo-excited graphene and metal mesh structures described in the next section, we came up with an idea for THz amplifiers with remarkably enhanced gain at a particular frequency by combining these structures [6]. The purpose of this study was to fabricate prototype THz amplifiers composed of chemical-vapor-deposition (CVD)-grown graphene and a metal mesh structure and to characterize them by using an electro-optic sampling (EOS) method.

36.2 Principle

Theoretical investigations predicted that the electron-hole pairs generated by the illumination of infrared light (ex. 1.55 μm) on to graphene would exhibit population inversion and negative dynamic conductivity (or gain) in a wide range, 1–10 THz, with a peak at around 3 THz despite strong carrier-carrier scattering [7–9]. Recently, Boubanga-Tombet et al. observed enhanced THz emissions from exfoliated graphene on a SiO_2/Si substrate by using EOS and revealed threshold behavior against pumping intensity, which suggested the occurrence of negative dynamic conductivity [10].

A structure having a periodic array of holes in a metal sheet, referred to as metal mesh structure, has been known to exhibit extraordinary optical transmission (EOT) at wavelengths slightly longer than the period of the array. Surface plasmon polaritons (SPPs) are induced when EM waves are illuminated on to the metal mesh structure. These SPPs are often called "spoof" SPPs to differentiate them from SPPs on a flat metal [11]. The EM energy stored as SPPs on both sides of the metal mesh is reemitted into free space on the output surface. The SPP-mediated resonant mode couples to the non-resonant radiation mode through evanescent waves in the hole, which produces Fano-resonance transmission spectra [12–14]. The EOT with metal mesh structure has been reported in a wide range of the EM wave regions including that of THz [15, 16].

36.3 Experimental

36.3.1 Basic Structure of THz Amplifier

Figure 36.1a shows the previously proposed THz amplifier [6], and Fig. 36.1b shows a fabricated prototype THz amplifier. A semi-insulating GaAs substrate for supporting the graphene and metal mesh was used to generate THz waves by optical rectification and to avoid energy loss of both generated THz waves and induced SPPs. The CVD-grown monolayer graphene (Meijo Nano Carbon Co., Ltd.) was transferred onto the surface of the GaAs substrate with a thickness of 600 μm. A square metal mesh with a period of 60 μm and an edge length of holes of 30 μm was formed by depositing and lifting off 50-nm-thick Pt/Pd on the graphene. The period and edge length of holes were determined through finite-difference time-domain (FDTD) EM simulations so that the transmittance for the metal mesh peaked at 1.5 THz in consideration of the measurement system used in this work and THz gain spectrum shown by Ryzhii et al. [7]. We chose Pt/Pd because of better adhesion to graphene than other metals such as Au. This sample is referred to as GaAs + Metal + Graphene. For comparison, we fabricated two samples: monolayer graphene on a GaAs substrate (referred to as GaAs + Graphene) and a metal mesh with the same geometry on a GaAs substrate (referred to as GaAs + Metal).

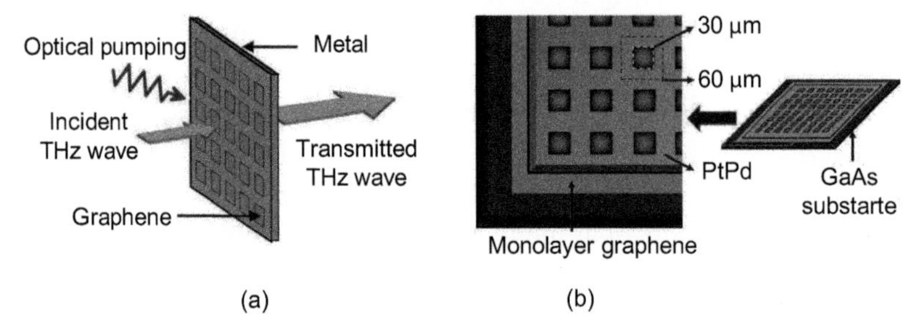

Figure 36.1 (a) Proposed THz amplifier; (b) prototype THz amplifier.

36.3.2 Measurement Method

The reflective EOS system shown in Fig. 36.2 was used to measure the THz time-domain responses of the fabricated samples [10]. A fiber laser with a pulse width (FWHM) of 80 fs and a wavelength of 1.55 μm was used for the pump and probe scheme. When the pump pulse enters the GaAs substrate, a THz pulse is generated by the optical rectification and propagates in almost the same direction as the pump pulse [17]. While the pump pulse enters the CdTe crystal, it is partly reflected at the sample/CdTe interface and again at the bottom of the GaAs substrate (orange lines in Fig. 36.2). A THz pulse with a Cherenkov angle of about 30 degrees is also generated in the CdTe crystal by optical rectification. The generated THz pulse is reflected at both surfaces of the CdTe crystal, as the red lines shows in Fig. 36.2. The blue lines in Fig. 36.2 indicate that the THz pulse generated by the pump pulse partly reflects at the sample/CdTe interface and the bottom of the GaAs substrate. The probe pulse detects a change in the reflection coefficient at the CdTe/Si prism interface, from which the change in the electric field caused by the electro-optic effect at the surface of CdTe crystal is obtained.

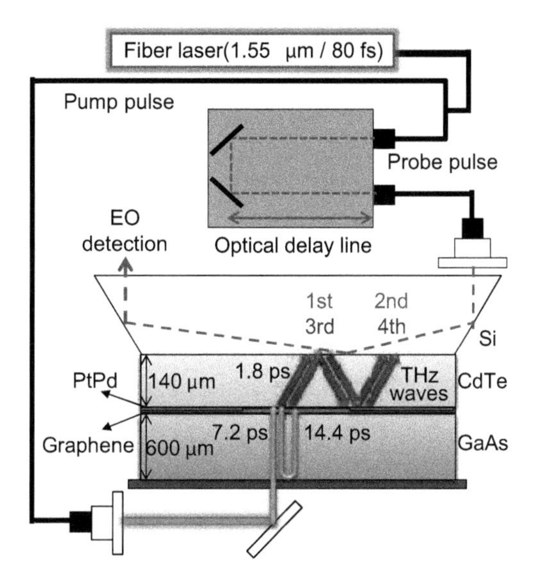

Figure 36.2 Schematic of measurement system.

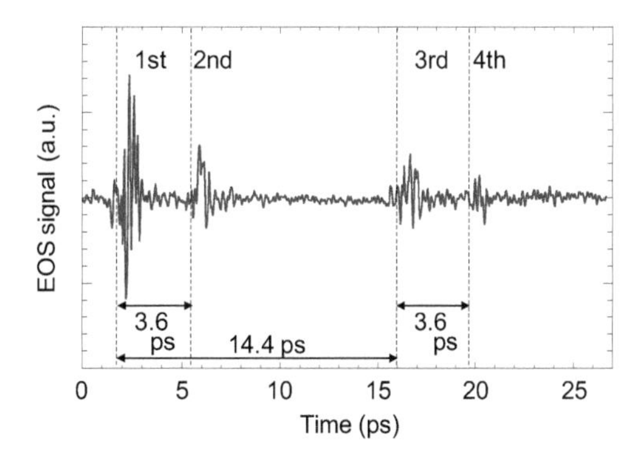

Figure 36.3 Field intensity time response of GaAs + Metal.

Figure 36.3 shows an example of the temporal response of the electric field for GaAs + Metal. We can see four distinctive waves. The following four waves were identified by calculating the propagation times of both pump optical and generated THz pulses using the thicknesses and indices of the GaAs substrate, those of the CdTe crystal, and Cherenkov angle (see Fig. 36.2). The 1st wave was generated in both the GaAs substrate and CdTe crystal by the pump pulse, the 2nd wave was the reflected wave of the 1st wave reflected at both surfaces of the CdTe crystal, the 3rd wave was generated by the pump pulse partly reflected at the sample/CdTe interface and the bottom of the GaAs substrate, and the 4th wave was the reflected wave of the 3rd wave.

36.3.3 EM Simulation Model

We used a commercially available FDTD EM simulator to investigate the transmission and reflection characteristics of the metal mesh structure without a gain medium (photo-excited graphene). The FDTD simulation model we used is shown in Fig. 36.4. By applying periodic boundary conditions, only one unit cell needed to be simulated. Perfectly matched layers were placed on the top and bottom surfaces to avoid unwanted reflections from them. The side distance corresponded to the 60 × 60 μm period of the hole array, and the square holes were 30 × 30 μm. Since almost all metals can be regarded as perfect conductors in the THz region, we treated the metal mesh structure as a perfect conductor. Both excitation and observation points were located far enough from the metal mesh to observe far-field distributions.

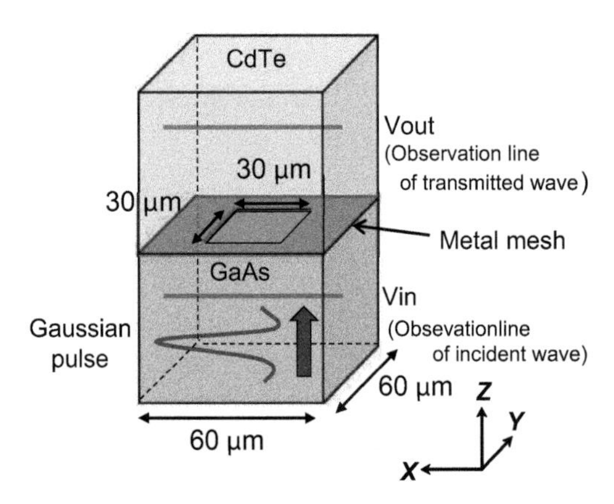

Figure 36.4 FDTD simulation model.

36.4 Results and Discussion

Figure 36.5 shows the temporal responses of GaAs + Graphene, GaAs + Metal, and GaAs + Metal + Graphene. In each sample, the field intensity of the 1st wave was largest and the 2nd wave was smaller than the 1st wave because it was a reflected wave. Since the propagation path of the 3rd and 4th waves was the same as that of 1st and 2nd waves, the field intensity of the 3rd wave was larger than that of the 4th wave in GaAs + Metal. However, the 4th wave was significantly larger than the 3rd wave in GaAs + Metal + Graphene. The 5th wave (the reflected wave of the 4th wave) was also observed. To investigate the THz responses in detail, we performed fast Fourier transform (FFT) in the range indicated by the arrows in Fig. 36.5.

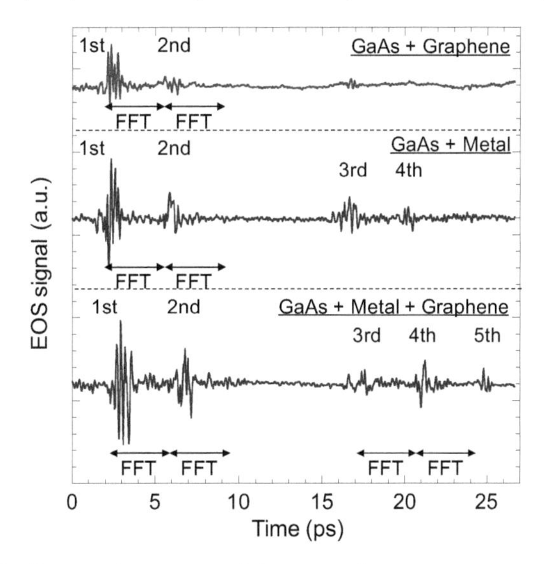

Figure 36.5 Temporal responses of three samples.

36.4.1 GaAs + Graphene

Figure 36.6 shows the frequency characteristics of the 1st and 2nd waves in GaAs + Graphene, along with the spectrum of the THz waves generated in the GaAs substrate and CdTe crystal (inset). By comparing these spectra, the 1st wave almost coincided with the spectrum of the THz waves generated in the GaAs substrate and

CdTe crystal. The 2nd wave (reflected wave) was greatly attenuated, and the 4th wave could be hardly observed. Therefore, the 1st wave (3rd wave) seemed to be absorbed when incident on the graphene. In contrast to the exfoliated high-quality graphene [10], the CVD-grown graphene we used could not exhibit population inversion due to possible grain boundaries and defects.

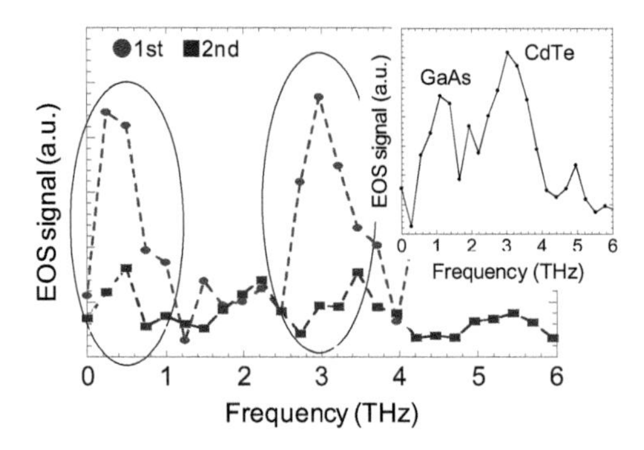

Figure 36.6 Frequency characteristics of GaAs + Graphene (inset) THz wave spectra generated in GaAs and CdTe.

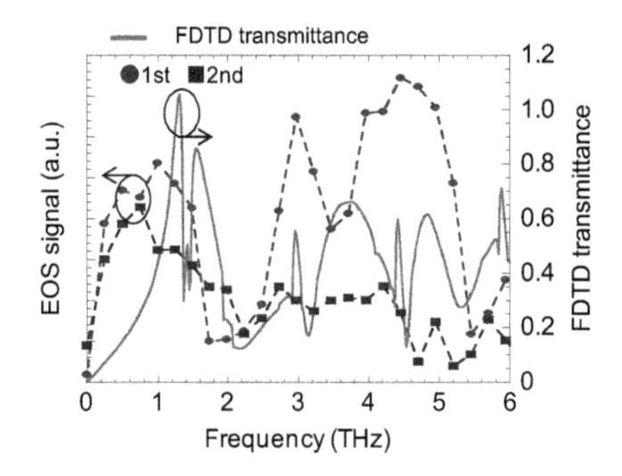

Figure 36.7 Frequency characteristics of GaAs + Metal along with simulated transmittance of metal mesh.

36.4.2 GaAs + Metal

Figure 36.7 shows the measured frequency characteristics of GaAs + Metal along with the transmittance of the metal mesh calculated using the FDTD simulator. The EOT band of the metal mesh was observed at 1.3 and 1.6 THz. The 1st wave was composed of the superposition of (a) the wave generated in the GaAs substrate and transmitted through the metal mesh and (b) the wave generated in the CdTe crystal. Although the peak measured at 1 THz slightly shifted to a lower frequency from the simulated EOT peak at 1.3 THz, the measured spectrum below 2 THz qualitatively matched the simulated spectrum. Since the 2nd wave was the reflected wave, the components of 1–1.3 and of 3–4 THz decreased and the frequency components of 2.1–2.5 THz remained almost unchanged. These results were qualitatively consistent with FDTD simulations (see also Fig. 36.9). The large components at 4–5 THz might be due to insufficient accuracy limited by the measurement system bandwidth.

36.4.3 GaAs + Metal + Graphene

The frequency characteristics of GaAs + Metal + Graphene are shown in Fig. 36.8. The 1st wave was composed of the same superposition with GaAs + Metal, where the 1.5-THz component decreased corresponding to the dip in the simulated transmittance of the metal mesh. It is obvious from the temporal responses shown in Fig. 36.6 that the 4th wave was amplified. The voltage ratios of the 2nd/1st and the 4th/3rd waves along with the simulated reflectance of the metal mesh are shown in Fig. 36.9. The measured voltage ratio of the 2nd/1st waves from 2 to 4 THz was in good agreement with the simulation. However, the component at 1.5 THz was amplified. This suggested that the photo-excitation produced population inversion in graphene although the time interval between the 1st and 2nd waves was slightly shorter than the time required for population inversion to build up (~4 ps) [18]. Since the interval from the first excitation to the 3rd and 4th waves were long enough for the formation of population inversion, the voltage ratio of the 4th/3rd waves at 1.7 THz was larger than that of the 2nd/1st waves. The population inversion was not observed in GaAs + Graphene, while it seemed to form by introducing the metal mesh structure. We speculate that the reason for these results might be the coupling between photo-generated electrons (holes) and the resonant field oscillation due to induced SPPs. A thicker metal (reduced resistance) and other types of metal with a work function close to that of graphene will be able to enhance the THz response.

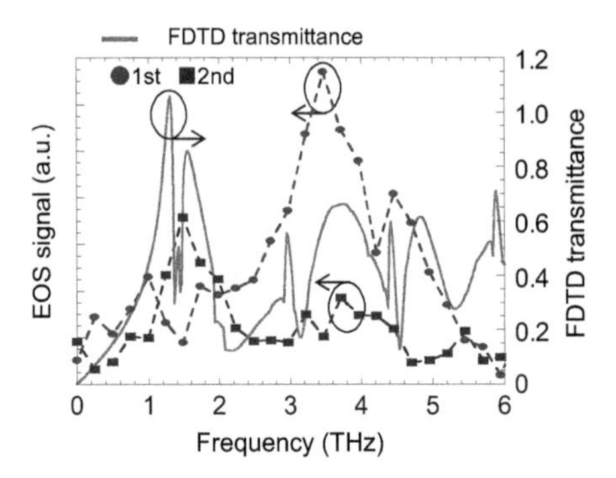

Figure 36.8 Frequency characteristics of GaAs + Metal + Graphene along with simulated transmittance of metal mesh.

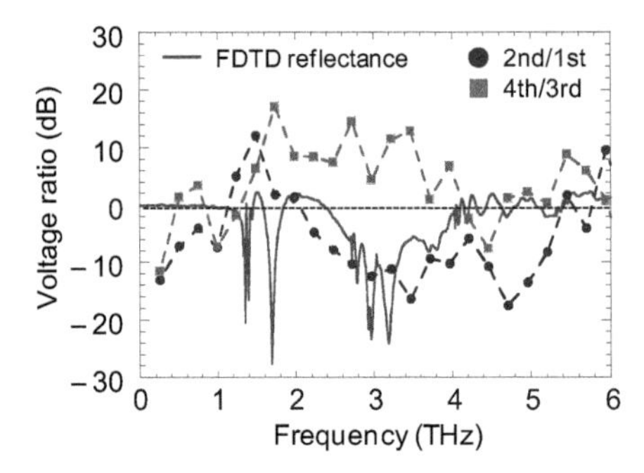

Figure 36.9 Measured voltage ratios along with simulated reflectance of metal mesh.

36.5 Conclusion

We fabricated prototype THz amplifiers composed of photo-excited graphene and a metal mesh structure and evaluated them by using the reflective EOS. An extremely amplified emission of THz radiation was observed for the amplifiers, although no amplification was exhibited using only graphene. We speculate that the amplification might be due to the coupling between photo-generated electrons (holes) and the resonant field oscillation due to induced SPPs. Further work is needed to verify this speculation.

References

1. K. S. Novoselov, A. K. Geim, S. V.Morozov, D. Jiang, Y.Zhang, S. V.Dubonos, I. V. Grigorieva, and A. A. Firsov, Electric field effect in atomically thin carbon films, *Science*, **306**, 666–669 (2004).
2. C. Berger, Z. Song, T. Li, X. Li, A. Y. Ogbazghi, R. Feng, Z. Dai, A. N. Marchenkov, E. H. Conrad, P. N. First, and W. A. de Heer, Ultrathin epitaxial graphite: 2D electron gas properties and a route toward graphene-based nanoelectronics, *J. Phys. Chem. B*, **108**, 19912–19916 (2004).
3. T. Otsuji, S. A. Boubanga Tombet , A. Satou, H. Fukidome, M. Suemitsu, E. Sano, V. Popov, M. Ryzhii, and V. Ryzhii, Graphene materials and devices in terahertz science and technology, *MRS Bull.*, **37**, pp. 1235–1243 (2012).
4. A. A. Dubinov, Y.V. Aleshkin, V. Mitin, T. Otsuji, and V. Ryzhii, Terahertz surface plasmons in optically pumped graphene structures, *J. Phys.: Condens. Matter*, **23**, 145302-1-8 (2011).
5. T. Watanabe, T. Fukushima, Y. Yabe, S. A. Boubanga Tombet, A. Satou, A. A. Dubinov, V. Ya. Aleshkin, V. Mitin, V. Ryzhii, and T. Otsuji, The gain enhancement effect of surface plasmon polaritons on terahertz stimulated emission in optically pumped monolayer graphene, *New J. Phys.*, **15**, 075003-1-11 (2013).
6. Y. Takatsuka, E. Sano, V. Ryzhii, and T. Otsuji, Terahertz amplifier based on multiple graphene layer with field-enhancement effect, *Jpn. J. Appl. Phys.*, **50**, 070118 (2011).
7. V. Ryzhii, M. Ryzhii, and T. Otsuji, Negative dynamic conductivity of graphene with optical pumping, *J. Appl. Phys.*, **101**, 083114 (2007).
8. A. Satou, T. Otsuji, and V. Ryzhii, Theoretical study of population inversion in graphene under pulse excitation, *Jpn. J. Appl. Phys.*, **50**, 070116 (2011).
9. A. Satou, V. Ryzhii, Y. Kurita, and T. Otsuji, Threshold of terahertz population inversion and negative dynamic conductivity in graphene under pulse photoexcitation, *J. Appl. Phys.*, **113**, 143108 (2013).
10. S. Boubanga-Tombet, S. Chan, T. Watanabe, A. Satou, V. Ryzhii, and T. Otsuji, Ultrafast carrier dynamics and terahertz emission in optically pumped graphene at room temperature, *Phys. Rev. B*, **85**, 035443 (2012).
11. J. B. Pendry, L. Martin-Moreno and F. J. Garcia-Vidal, Mimicking surface plasmons with structured surfaces, *Science*, **305**, 847–848 (2004).

12. C. Genet, M. P. van Exter and J. P. Woerdman, Fano-type interpretation of red shifts and red tails in hole array transmission spectra, *Opt. Commun.*, **225**, 331–336 (2003).
13. M. Sarrazin, J. P. Vigneron and J. M. Vigoureux, Role of Wood anomalies in optical properties of thin metallic films with a bidimensional array of subwavelength holes, *Phys. Rev. B*, **67**, 085415 (2003).
14. T. Tanaka, M. Akazawa, E. Sano, M. Tanaka, F. Miyamaru, and M. Hangyo, Transmission characteristics through two-dimensional periodic hole arrays perforated in perfect conductors, *Jpn. J. Appl. Phys.*, **45**, 4058–4063 (2006).
15. C. Winnewisser, F. Lewen and H. Helm, Transmission characteristics of dichroic filters measured by THz time-domain spectroscopy, *Appl. Phys. A*, **66**, 593-598 (1998).
16. D. W. Porterfield, J. L. Hesler, R. Densing, E. R. Mueller, T. W. Crowe, and R. M. Weikle II, Resonant metal-mesh bandpass filters for the far infrared, *Appl. Opt.*, **33**, 6046–6052 (1994).
17. M. Nagai, K. Tanaka, H. Ohtake, T. Bessho, T. Sugiura, T. Hirosumi, and M. Yoshida, Generation and detection of terahertz radiation by electro-optical process in GaAs using 1.56 μm fiber laser pulses, *Appl. Phys. Lett.*, **85**, 3974–3976 (2004).
18. E. Sano, Monte Carlo simulation of ultrafast electron relaxation in graphene, *Appl. Phys. Express*, **4**, 085101 (2011).

Chapter 37

Terahertz Light-Emitting Graphene-Channel Transistor Toward Single-Mode Lasing*

D. Yadav,[a] G. Tamamushi,[a] T. Watanabe,[a] J. Mitsushio,[a] Y. Tobah,[b] K. Sugawara,[a] A. A. Dubinov,[c] A. Satou,[a] M. Ryzhii,[d] V. Ryzhii,[a,e] and T. Otsuji[a]

[a]*Research Institute of Electrical Communication, Tohoku University, Sendai 980-8577, Japan*

[b]*Department of Electrical and Computer Engineering, University of Texas at Austin, Austin 78712, TX, USA*

[c]*Institute for Physics of Microstructures, Russian Academy of Sciences, Lobachevsky State University of Nizhny Novgorod, Nizhny Novgorod 603950, Russia*

[d]*Department of Computer Science and Engineering, University of Aizu, Aizu-Wakamatsu 965-8580, Japan*

[e]*Center for Photonics and Infrared Engineering, Bauman Moscow State Technical University, Moscow 105005, Russia*

otsuji@riec.tohoku.ac.jp

*Reprinted with permission from D. Yadav, G. Tamamushi, T. Watanabe, J. Mitsushio, Y. Tobah, K. Sugawara, A. A. Dubinov, A. Satou, M. Ryzhii, V. Ryzhii, and T. Otsuji (2018). Terahertz light-emitting graphene-channel transistor toward single-mode lasing, *Nanophotonics*, **7**(4), 741–752. Copyright © 2018 Taiichi Otsuji et al., published by De Gruyter.

Graphene-Based Terahertz Electronics and Plasmonics: Detector and Emitter Concepts
Edited by Vladimir Mitin, Taiichi Otsuji, and Victor Ryzhii

ISBN 978-981-4800-75-4 (Hardcover), 978-0-429-32839-8 (eBook)
www.jennystanford.com

A distributed feedback dual-gate graphene-channel field-effect transistor (DFB-DG-GFET) was fabricated as a current-injection terahertz (THz) light-emitting laser transistor. We observed a broadband emission in a 1–7.6-THz range with a maximum radiation power of ~10 μW as well as a single-mode emission at 5.2 THz with a radiation power of ~0.1 μW both at 100 K when the carrier injection stays between the lower cutoff and upper cutoff threshold levels. The device also exhibited peculiar nonlinear threshold-like behavior with respect to the current-injection level. The LED-like broadband emission is interpreted as an amplified spontaneous THz emission being transcended to a single-mode lasing. Design constraints on waveguide structures for better THz photon field confinement with higher gain overlapping as well as DFB cavity structures with higher Q factors are also addressed towards intense, single-mode continuous wave THz lasing at room temperature.

37.1 Introduction

Compact, high-power, and room-temperature terahertz (THz) sources are highly sought after to enable the practical realization of vast applications of THz waves [1–5], such as sensing and non-destructive imaging for safety and security [6–10], medical diagnosis [11–13], and ultra-broadband wireless communications [14–16]. Several types of THz light sources and/or oscillators (such as THz quantum cascade lasers [17-22], p-Ge lasers [23, 24], and resonant tunneling diode (RTD) oscillators [25]) have been developed so far. However, laser-type devices, on the one hand, suffer from phonon de-coherency, preventing room-temperature operation [26]. Electron devices like RTDs, on the other hand, suffer from their electron transit times and parasitic capacitance-resistance time constants. Their maximum available output power decreases with increasing frequency by a factor of three orders and is as small as 10 μW order at 1 THz at room temperature [25]. To break through these substantial limitations, graphene has attracted attention owing to its gapless and linear energy spectrum and massless Dirac Fermions, giving rise to superior carrier transport properties [27–31]. The latest studies have reported on the potential of relatively simpler multi-layer graphene devices [32, 33] to more complex graphene/h-BN heterostructures [34, 35] for highly sensitive detection as well as emission of THz radiation [35].

Optical and/or injection pumping of graphene can enable negative-dynamic conductivity in the THz spectral range, which may lead to new types of THz lasers [36–39]. The THz gain in optically pumped graphene has been experimentally confirmed [40, 41]. However, optical pumping suffers from carrier heating, preventing from obtaining carrier population inversion and eventual gain [38, 39]. On the other hand, in the graphene structures with p-i-n junctions, the injected electrons and holes have relatively low energies compared with those in optical pumping, so that the effect of carrier cooling can be rather pronounced, providing a significant advantage of the injection pumping in realization of graphene THz lasers [28, 42–44]. Recent extensive studies on Auger processes reveal the difficulty of carrier population inversion to be obtained [45–49], which could be dominated in carrier heating regime with breaking the linear dispersive conical band profiles and/or band broadening due to many body effects of the inter-carrier Coulomb scattering [50, 51]. In this regard, current-injection pumping is the best way to substantially suppress the carrier heating and the Auger processes towards the lasing operation.

Current-injection graphene THz laser, as theoretically proposed by Ryzhii et al. [42–44], can be constructed with a dual-gate graphene-channel field-effect transistor (DG-GFET) structure, in which electrons and holes are laterally injected between the dual-gated regions by applying complementary gate biases and the populations of electrons and holes are inverted by applying a weak positive drain-source dc bias. Due to the continuum of gapless energy spectra of conduction electrons and valence holes, there are no quantized upper/lower levels that define the photon energy of the spontaneous emission under population inversion. Thus, graphene under carrier population inversion will emit rather broadband photons according to its broad gain spectral profile [43]. Therefore, to make single-mode lasing, one needs to implement a pertinent high-Q laser cavity structure in which the gain medium of the graphene under carrier injection pumping is accommodated.

In this work, a prototype current-injection-type graphene laser device was fabricated in a distributed feedback (DFB) DG-GFET structure and challenged for the first observation of THz radiation oscillations. We experimentally observed amplified spontaneous broadband light emission from 1 to 7.6 THz at 100 K by carrier injection in one sample of the fabricated DFB-DG-GFETs, demonstrating the birth of a new type of THz light-emitting transistors (LETs). In another yet similar device, we also observed a single-mode emission at 5.2 THz corresponding to the fundamental mode of the DFB cavity at 100 K. Both devices exhibited peculiar nonlinear threshold-like behavior with respect to the current-injection level. The result is still at a preliminary level, but the linewidth for the single-mode emission fairly agrees with the calculation based on the DFB-Fabry-Perrot hybrid-mode modeling, suggesting that it could be single-mode lasing. The broadband emission in the former device is interpreted as an amplified spontaneous THz emission being transcended to a single-mode lasing.

37.2 Device Design and Fabrication

37.2.1 Design and Fabrication Details

The cross-sectional schematic of the device is shown in Fig. 37.1A. First, epitaxial graphene was synthesized by the thermal decomposition of a SiC substrate; this method provides both easiness of fabricating the device and high crystallinity of the film [53]. C-face is specifically chosen to obtain non-Bernal stacking of multiple undoped graphene layers excluding the heavily doped buffer layer that is substantially made for Si-face decomposition. Lattice structure and crystal quality were first characterized by Raman spectroscopy at 300 K. The Raman shift spectrum obtained is shown in Fig. 37.1B and C. No peak at the D band and sharp mono-peaks at the G and G′ bands with G′/G peak intensity ratio being 2.11 confirmed a high-quality, few-layer non-Bernal stacked graphene. Then the surface morphology of the synthesized epitaxial graphene was investigated by low-energy electron microscope (LEEM) at 300 K. Very large grains (>50 μm) consisting mainly of two-layer graphene were observed. Figure 37.1D shows that graphene is homogeneous in a rather large area; the lighter central area consists of two layers, whereas the darker area outside its edges contains three to six graphene layers [52]. This bilayer nature of the central grain area

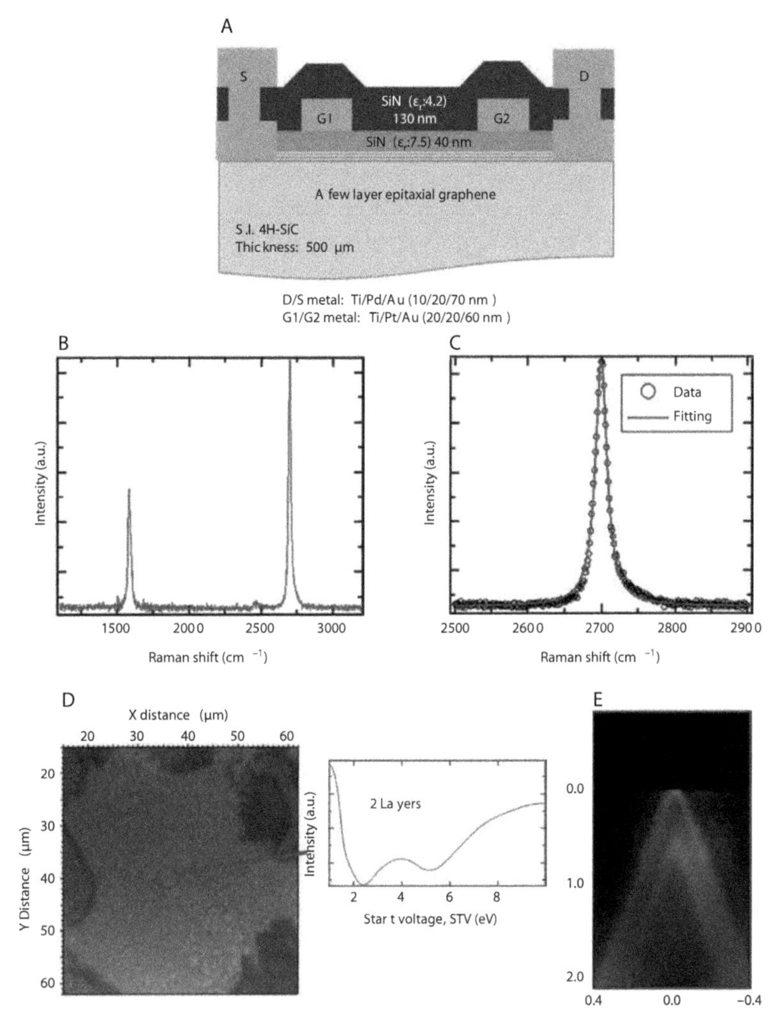

Figure 37.1 The Raman spectrum (B, C), the LEEM image (D), and the ARPES image (E) confirms D-peak free, high crystal quality, double-layered large (>50 μm) grain, and non-Bernal stacked linear dispersion with not unintentional doping, respectively. (A) Device cross-section, (B) Raman spectroscopy results of the epitaxial graphene on 4H-SiC substarte, (C) Lorentzian curve fitting of G′ peak confirming the presence of only one peak at 2700 cm^{-1} with FWHM = 28.52 cm^{-1}, (D) LEEM image and electron reflectance spectrum of epitaxial grapheme © *Jpn. Soc. Appl. Phys.* [52], and (E) ARPES image © *Jpn. Soc. Appl. Phys.* [52].

was also confirmed by the presence of the two dips in electron reflectance spectra between 0 and 8 eV, as shown in the inset in Fig. 37.1D. Figure 37.1E shows an angle resolved photoelectron spectroscopy (ARPES) image, manifesting that the main band holds a linear dispersion and is divided into two groups and Fermi level lies at 0 eV, confirming the high-quality, non-Bernal stacked, double-layer graphene with no unintentional doping [52, 54]. By using this epitaxial graphene, GFET was fabricated by a standard photolithography process involving a gate stack of SiN dielectric layer, providing an excellent intrinsic field-effect mobility exceeding 100,000 cm^2/Vs at 300 K at the maximal transconductance [55].

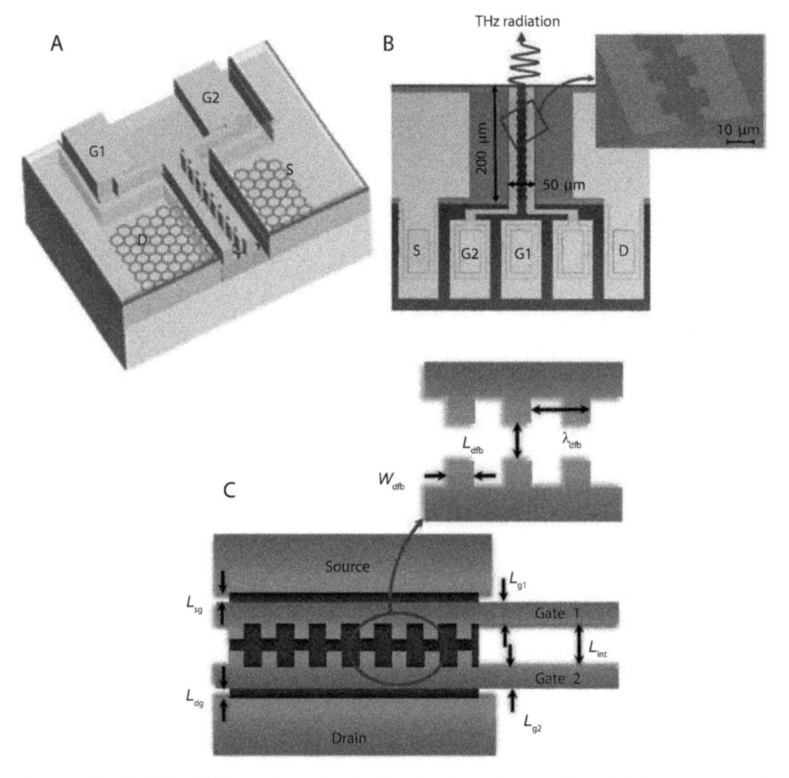

Figure 37.2 The DFB modulation index is given by L_{int}/L_{dfb}. (A) Schematic of DFB DG-GFET, (B) device SEM and photo images, (C) design parameters for the DFB structure: $L_{sg,dg} = 2$ μm, $L_{g1,2} = 15$ μm, $L_{int} = 20$ μm, $L_{dfb} = 18$ μm, $W_{dfb} = 6$ μm, $\lambda_{dfb} = 12$ μm, $f_{dfb} = 4.96$ THz, and $N_{dfb} = 16$.

Throughout this gate stack process, the epitaxial graphene layer was encapsulated between the SiC substrate and the SiN gate dielectric layer, preventing the exposure of the graphene surface to the air and thus from the chemical reactions of unintentional doping even under varying the temperatures.

A pair of toothbrush-shaped gate electrodes was patterned to form a DFB cavity in which the active gain area and corresponding gain coefficient are spatially modulated [56] (see Fig. 37.2A and B). The number of DFB periods N_{DFB} was designed to be N_{DFB} = 16, resulting in active graphene channel width of ~200 μm, as shown in Fig. 37.2B. Designed grating period Λ, the effective refractive index n_{eff}, the Bragg wavelength λ_{dfb}, and the principal mode f_P are 12 μm, 2.52, 60.5 μm, and 4.96 THz, respectively, as shown in Fig. 37.2C. In this work, we carried experiments on two devices that are similar in design but slightly differ in the substrate thickness and DFB design parameters. The DFB modulation indices for device 1 (showing broadband emission) and device 2 (showing single mode emission) are L_{int}/L_{dfb} = 20/15 and 20/18, respectively. Also, the carrier momentum relaxation time of these two devices, which were extracted from the slope of the ambipolar current-voltage curves, has different values: device 1 having longer values (ranging from 1.5 to 4 ps at temperatures 300~100 K) than those for device 2 (ranging from 0.8 to 2 ps at temperatures 300~100 K).

37.2.2 Principles of Operation

The operation principle called current injection THz lasing is as follows: when the dual-gate electrodes are applied with complementary biases ($V_g = -V_{g1} = +V_{g2}$), electrons or holes are injected underneath each gate electrode. The level of V_g determines the carrier injection level and corresponding Fermi level ε_F. With no applied source to drain bias V_d, the electrons and holes diffuse into the ungated *i* region and are recombined together to annihilate. Application of V_d, on the other hand, shifts the quasi-Fermi level of electrons and holes injected into the ungated *i* region to form a population inversion, and the recombination of those electrons and holes having an energy difference of several meV gives rise to spontaneous THz photon emission [42].

The real part of the dynamic conductivity Re $\sigma(\omega)$ in graphene under current-injection pumping is given by the sum of the intraband Drude-like component and the interband transition-related component as follows [42, 43]:

$$\mathrm{Re}\,\sigma(\omega) = \mathrm{Re}\,\sigma^{\mathrm{intra}}(\omega) + \mathrm{Re}\,\sigma^{\mathrm{inter}}(\omega), \tag{37.1}$$

$$\mathrm{Re}\,\sigma^{\mathrm{intra}}(\omega) \approx \frac{(\ln 2 + \varepsilon_{\mathrm{F}}/2k_{\mathrm{B}}T)e^2}{\pi\hbar^2} \frac{k_{\mathrm{B}}T\tau}{(1+\omega^2\tau^2)}, \tag{37.2}$$

$$\mathrm{Re}\,\sigma^{\mathrm{inter}}(\omega) = \frac{e^2}{4\hbar}(f(-\hbar\omega/2) - f(\hbar\omega/2))$$

$$\approx \frac{e^2}{2\hbar} \exp\left(\frac{eV_{\mathrm{d}} - 2e\sqrt{V_{\mathrm{F}}V_{\mathrm{g}}}}{2k_{\mathrm{B}}T}\right) \sinh\left(\frac{\hbar\omega - eV_{\mathrm{d}}}{2k_{\mathrm{B}}T}\right), \tag{37.3}$$

$$\varepsilon_{\mathrm{F}} = \hbar v_{\mathrm{F}} \sqrt{\frac{\kappa V_{\mathrm{g}}}{4ed}} \equiv e\sqrt{V_{\mathrm{F}}V_{\mathrm{g}}}, \tag{37.4}$$

where e is the elementary charge, $f(\varepsilon)$ is the Fermi-Dirac distribution function for electrons, $\hbar$ is the reduced Planck's constant, k_{B} is the Boltzmann constant, T is the carrier temperature, τ is the momentum relaxation time, v_{F} is the Fermi velocity, κ is the permittivity of the gate dielectric layer, and d is the gate-dielectric layer thickness. Re $\sigma^{\mathrm{intra}}(\omega)$ always takes positive values and has the dependence of $\frac{\tau}{1+\omega^2\tau^2}$ monotonically decreasing with increasing ω. The roll off frequency is given by $\omega_{\text{roll-off}} \sim 1/\tau$. Re $\sigma^{\mathrm{inter}}(\omega)$ causes its transition at around $\hbar\omega \sim 2\varepsilon_{\mathrm{F}}$ and takes the upper plateau approaching $\frac{e^2}{4\hbar}$ at high frequencies $\hbar\omega \gg 2\varepsilon_{\mathrm{F}}$ and the lower plateau approaching 0 at low frequencies $\hbar\omega \ll 2\varepsilon_{\mathrm{F}}$ when the carrier populations are equilibrated. When the carrier populations are inverted by the carrier-injection pumping with a certain value of V_{d}, on the contrary, the minimal plateau of Re $\sigma^{\mathrm{inter}}(\omega)$ may shift to the negative level (as low as $\frac{-e^2}{4\hbar}$). As a consequence, if $\hbar\omega_{\text{roll-off}} \ll 2\varepsilon_{\mathrm{F}}$, there exists a certain frequency range in which Re $\sigma(\omega)$ takes negative values. How widely and deeply the conductivity goes in negative at a given V_{d} depend directly on momentum relaxation time τ and carrier temperature T [42, 43]. A longer τ lowers the roll-off frequency and the conductivity values at higher frequencies ($\omega\tau \gg 1$). A lower T helps increase the population inversion to the higher levels and sharpens the transition on Re $\sigma^{\text{inter}}(\omega)$ at around $\hbar\omega \sim 2\varepsilon_F$, enlarging the upper cuttoff frequency of the negative conductivity (gain) spectrum. To obtain a rather wide gain spectrum in the THz range, τ should be as long as possible (at least picoseconds) and T should be as low as possible.

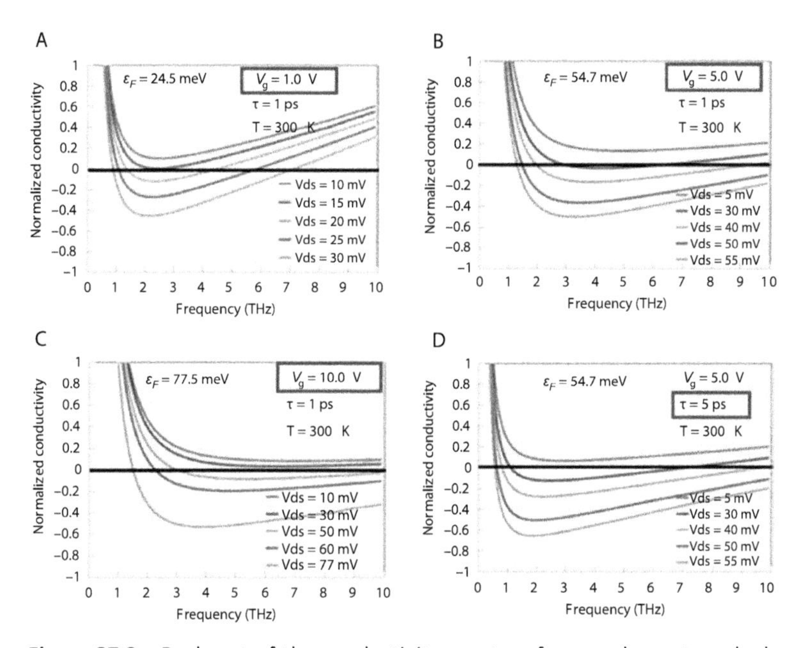

Figure 37.3 Real part of the conductivity spectra of a complementary dual-gate-biased DG-GFET at 300 K for different drain-source bias voltages. The vertical axis is normalized to the fundamental conductivity ($e^2/4h$) corresponding to 2.3% absorbance of monolayer graphene. The value of τ = 1 ps for (A) V_g (= $-V_{g1}$ = $+V_{g2}$) = 1.0 V, (B) V_g = 5.0 V, (C) V_g = 10.0 V, and (D) τ = 5 ps for V_g = 5.0 V.

Figure 37.3 plots the simulated real part of the dynamic conductivities of the DG-GFET at a constant carrier temperature of 300 K for different complementary gate biases ($V_g = -V_{g1} = +V_{g2}$) and drain-source biases V_d conditions according to Eqs. (37.1)–(37.4). The values of κ, d, τ, and T are assumed to be 4.7, 50 nm, 1 ps, and 300 K, respectively. When V_g = 1.0 V, a weakly biased condition giving rise to doping with a Fermi level ε_F of 24.5 meV, as shown in Fig. 37.3A, rather weak V_d beyond the threshold level of ~20 mV causes carrier population inversion and real net gain (negative dynamic conductivity) around 2 THz. Increasing the V_d value up to a certain level ($V_d \sim 30$ mV in Fig. 37.3A), which shifts the quasi-Fermi level equal to the magnitude of the original Fermi level at zero-V_d, can increase both the gain and its bandwidth. The lower and upper cutoff frequencies are situated between 0.7 and 7 THz. Further increase in V_d increases the channel electric field intensity resulting in carrier heating and weakening the gain. With increasing V_g, the carrier injection increases so that larger and wider gain bandwidths are obtained, as shown in Fig. 37.3B and C. As abovementioned, a longer τ, meaning a higher carrier mobility, gives a higher THz gain with its peak at lower frequency, as shown in Fig. 37.3D in comparison with Fig. 37.3B. Hence, to realize a current injection THz lasing, realization of GFET with high carrier mobility is indispensable. In real operating conditions, as is described in detail in Ref. [42], carrier temperature T may change depending on the current-injection levels, modifying the gain spectral profiles. This effect will be discussed to interpret the experimental results in the next section.

Another important criterion of the device operation is choosing the lasing frequency. The conductivity profile shows negative gain for a wide range of THz frequencies. To make single-mode lasing, one needs to implement a pertinent high-Q laser cavity structure in which the gain medium of the graphene under carrier injection pumping is accommodated. Such a laser cavity can be realized by using the gate electrodes as a DFB type waveguide structure. As shown in Fig. 37.2B, the DFB structure periodically modulates the distance between the two gate electrodes along the length of the device. This results in periodic modulation of the gain coefficient given by the conductivity and the gain overlapping factor (the ratio of the areal integration of the THz photon electric field overlapped with the gain area to that of the entire area) [55]. At the Bragg wavelength ($\lambda_B = 2n_{eff}\,\lambda_{dfb}$), determined by the modulation period λ_{dfb} and the effective refractive index n_{eff} of the gain coefficient, the spontaneously emitted THz photons intensify each other, in each period, along the dual-gate direction. At that resonance frequency, single-mode oscillation can be obtained if the current-injection gain at only the maximal mode exceeds the laser oscillation threshold. Figure 37.4A shows the quality factors for in-plane waveguided modes of the DFB-DG-GFET structure for different quality graphene samples having τ of 0.1, 1.0, 3.0, 5.0, and 7.0 ps at a weak drain bias of 30 mV. For longer τ of 5~7 ps, the spectra exhibit high Q factors up to 150 over a broader frequency range, suggesting LED-like broadband spontaneous emission. The modal gain for this case is ~ $-30\ cm^{-1}$ (losses are greater than gain, limiting the possible operation to spontaneous emission). Varying the DFB design parameters, quality of graphene layer, and thickness of substrate can largely affect the confinement of THz photon electric fields on to the active gain area, which can hugely alter the gain profile. The other important parameter is the drain bias V_d that determines the level of carrier population inversion. Figure 37.4B shows the numerically calculated quality factors for the case of device 2, with different V_d levels of 30, 50, 70, 100, and 150 mV for a relatively shorter fixed τ value of 3.5 ps. When V_d is relatively high at around 70~100 mV, the spectra exhibit single peak at the designed fundamental DFB mode at ~5 THz, suggesting the single-mode lasing operation. The modal gain in this case is ~5 cm^{-1}. Further increase in V_d to 150 mV totally vanishes the peak profile, reflecting the carrier heating and making the net THz gain disappear. These results suggest that graphene quality and carrier temperature reflecting τ values, drain bias reflecting the level of population inversion, and DFB modulation index reflecting the Q factor determine the overall spectral profile of the THz photon emission.

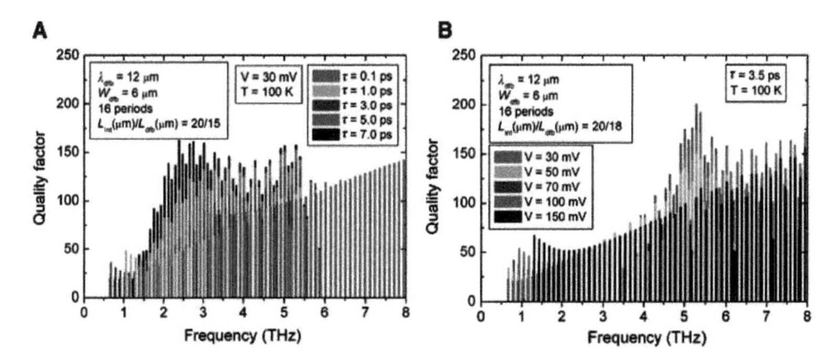

Figure 37.4 Simulated quality factors for in-plane waveguided modes (A) for different quality graphene samples having τ of 0.1, 1.0, 3.0, 5.0, and 7.0 ps at a fixed weak drain bias of 30 mV and (B) for different drain bias conditions at 30, 50, 70, 100, and 150 mV at a fixed τ of 3.5 ps. Numerically calculated Q factors for the case when τ is ~3.5 ps at relatively high drain bias conditions exhibit slightly better THz photon field confinement.

37.3 Results and Discussions

THz emission from the sample was measured at temperatures ranging from 300 K down to 100 K using a Fourier-transform far-infrared (FTIR) spectrometer with a 4.2 K-cooled Si bolometer (see Supplement 37.S1 for details). The background blackbody radiation was first observed under the zero-bias condition, which was subtracted from the one observed under biased conditions. The drain-current-to-gate-voltages (V_{g1}, V_{g2}) characteristics were first measured for the fabricated samples. They exhibited ambipolar characteristics as expected, as shown in Fig. 37.5A. The relationship

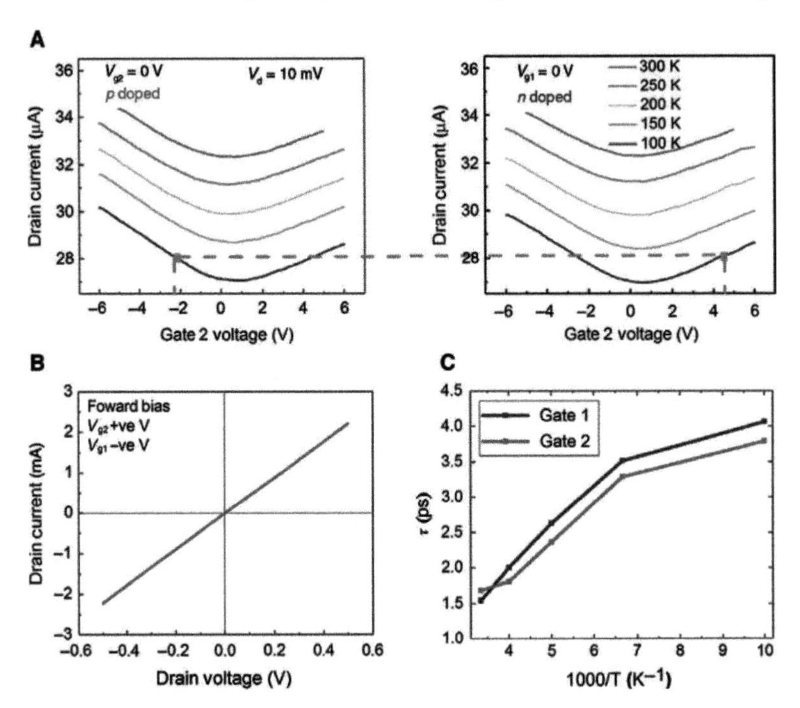

Figure 37.5 The slope of the ambipolar curves in (A), corresponding to the transconductance in proportion to the carrier momentum relaxation time, gets steeper with decreasing the temperature. (A) Measured ambipolar current-voltage characteristics of device 1 for G1 and G2 voltages with V_d = 10 mV. The square dots are typical points for symmetric electron/hole injections. (B) Measured uniform drain current-drain voltage characteristics at 300 K. (C) Temperature dependence of carrier momentum relaxation time (τ) for device 1.

between drain voltage (V_d) and drain current (I_d) when a positive voltage was applied to one gate and the negative voltage was applied to the other gate was found to be linear, as shown in Fig. 37.5B. Unlike ordinary semiconductor p-i-n diodes, immediate current flow is observed, irrespective of a positive or negative voltage change due to the gapless and symmetrical conical band dispersion for the conduction and valence bands. With complimentary biased gate 1 ($V_{g1} < 0$) and gate 2 ($V_{g2} > 0$), with a positive drain-source bias V_d (as designated by the square dots in Fig. 37.5A), the carrier population can be inverted at the intermediate channel region [42]. Figure 37.5C shows the temperature dependence of τ in these devices. With decreasing temperature, thermionic current reduces, i.e. the level of drain current decreases but the transconductance (g_m), the slope of the ambipolar curves in Fig. 37.5A, increases, reflecting the increase of τ. Hence, at low temperatures, τ increases so that one can expect negative conductivities to be obtained more easily (see Fig. 37.3D), enabling enhanced emission at low temperatures.

37.3.1 Broadband THz Light Emission

The emission results from device 1 are summarized in Fig. 37.6. A rather strong emission at 100 K, stronger than that at higher temperatures from 300 to 150 K (Fig. 37.6A), was observed in 1–7.6-THz range when V_d is forward-biased to a certain level under symmetric electron-hole injection conditions leading to carrier population inversion (see inset of Fig. 37.5B: V_{g1} = –2.28 V and V_{g2} = 4.56 V at 100 K). The emission power from the device integrated over 1–7.6 THz is of the order of ~80 μW at most. The shape of emission spectra reflects superimposed effects of the graphene-original conductivity profile, which gives larger gain at lower frequencies [43] and the DFB cavity effects that have the fundamental mode peak at ~4.96 THz. Regarding the former, as the device is kept at low temperatures, scattering processes in graphene are suppressed, which means longer momentum relaxation time of the carriers present in the channel (see Fig. 37.5D), thereby enabling the achievement of higher negative conductivity in the wide THz range and resultant broadband spontaneous THz emission (see Fig. 37.3D). To obtain single-mode lasing, improvement on the Q factor of the DFB cavity with larger numbers of the DFB periods and/

or deeper DFB modulation indices would be a possible solution (see Fig. 37.S2 in Supplemental Material). Further precise identification of the threshold temperature to the spontaneous THz emission is a next subject.

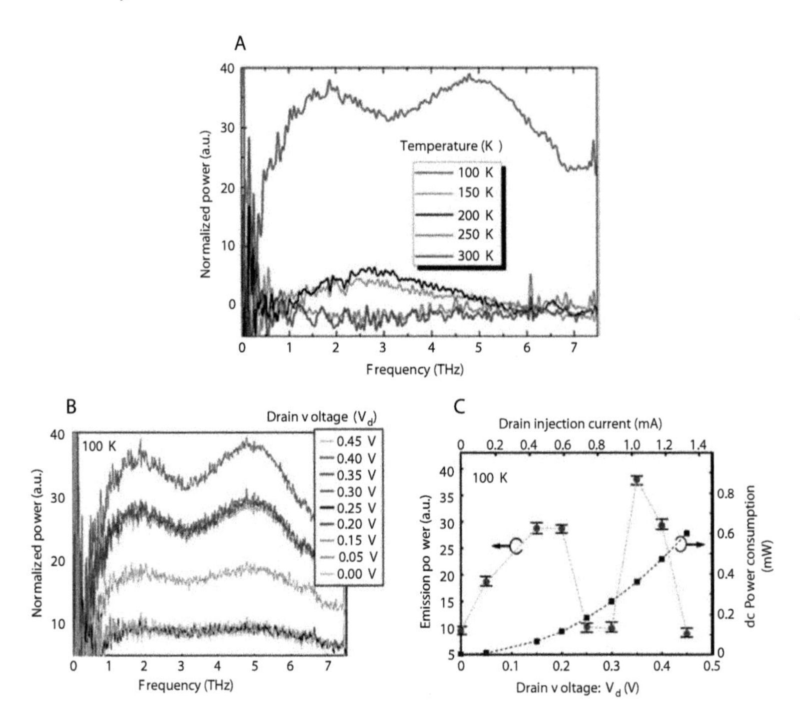

Figure 37.6 The threshold temperature for amplified spontaneous broadband THz emission stays between 100 and 150 K. The device also exhibits double-threshold-like behavior with respect to V_d and injection current. (A) Temperature dependence of the observed emission spectra in a fabricated DFB-DG-GFET (device 1) under carrier population inversion (V_{g2} = 4.56 V, V_{g1} = −2.28 V). (B) Observed emission spectra of the device under population inversion for different values of drain voltage at 100 K. (C) Measured THz emission power (left axis) and dc power consumption (right axis) versus drain voltage V_d and drain injection current.

With increasing V_d, the emission power increases non-monotonically while preserving its spectral shape, as shown in Fig. 37.6B. Apart from temperature-dependent spontaneous THz emission, the device also exhibited double-threshold-like behavior with respect to V_d and drain injection current, as shown in Fig. 37.6C.

The first threshold current to the spontaneous THz emission stays below 0.1 mA, whereas the second threshold current does around 0.9 mA. Such a double-threshold-like behavior may be due to the carrier overcooling effect [39, 42]. As theoretically predicted in Refs. [39] and [42], when carriers are excited, their energy is relaxed mainly via emission of optical phonons (including interband emission). The optical phonon decay rate, which depends on the overall thermal conductivity of the device, modifies the energy relaxation dynamics and carrier cooling decay. The existence of the scattering factors with a finite value of τ gives rise to the imaginary part of the conductivity, Im σ, which is known as the origin of the kinetic inductance of carriers. Such an inductive inertia of carriers may cause an overshoot on their energy transfer to the lattice phonons, resulting in carrier overcooling even below the lattice temperature in a limited time scale in a rather weak carrier-injection pumping regime [42]. Continuous injection pumping at a low V_d level may cause carrier overcooling steadily, which in turn enhances the carrier population inversion and contributes to promote the THz gain. When the carrier injection level rises further with increasing V_d and dc power consumption (see Fig. 37.6C), carrier heating is dominated, which prevails over the cooling effect, resulting in net carrier heating and reduction of the THz gain. Further increase in V_d increases the level of population inversion, recovering the net THz gain. As a consequence, the carrier temperature changes non-monotonically along with the increase in V_d [42]. This is thought to be a possible interpretation for the cause of such non-monotonic double-threshold-like behavior.

Also, the conductivity profile highly depends on the applied biases, as shown in Fig. 37.3. Increasing the gate voltage increases the number of injected charge carriers, which thereby increase the gain bandwidth and the lower cutoff for gain [43]. The drain bias is also very important as it determines the shifting of quasi-Fermi level; higher voltage causes more injection current and hence more gains. However, it also increases the band slope and causes carrier heating, which sacrifices the injection beyond the Fermi level [42, 44]. Hence, there is both a lower and upper limit on gain with respect to V_d, as also observed in our experimental results.

Another factor causing the broadband nature of emission from device 1 is the poor DFB cavity effects (substrate-thickness-dependent

THz photon field distribution [57] could not meet the maximal available gain-overlapping condition), which do not work properly as single-mode lasing but give a tendency of being transcended from spontaneous broadband THz emission to stimulated emission with a central peak at the DFB fundamental mode of ~5 THz. The simulated quality factor for the DFB cavity for excellent quality GFET at low value of V_d shows gain over a wide frequency range (as seen in Fig. 37.4A) supporting the possibility of observing broadband emission spectra from these devices. If the photon electric fields could be concentrated effectively on the graphene gain area, by optimizing the substrate thickness as well as additional thick high-K dielectric layer on top of the gate stack, one can expect single-mode lasing from these structures.

37.3.2 Towards Single-Mode Lasing

We repeated the experiment on another sample device 2 (having a bit smaller DFB modulation index and shorter momentum relaxation time than those in device 1, as mentioned in 2.1) and observed no emission at 300 K, but lowering the temperature down to 100 K showed current-injection-dependent (V_{g2} = 7.3 V, V_{g1} = –4.8 V) emission, as shown in Fig. 37.7A. This single-mode-like emission also exhibited a non-monotonic threshold behavior, i.e. the emission at 5.2 THz grows with increasing V_d from 0 to 0.1 V but weakens with further increasing V_d until 0.2 V. The emission grows again after increasing V_d from 0.2 to 0.5 V, giving a single-mode emission with the highest power of ~0.1 μW. Spectral narrowing with increasing the carrier injection around the threshold was also observed, as shown in Fig. 37.7B. The emission spectra at V_d = 0.5 V can fit to the Lorentzian curve with a Q factor of 170 (a linewidth of 30.6 GHz), which fairly agrees with the simulated values ~210, as shown in Fig. 37.4B. Compared to the broadband emission obtained from device 1, such a sharp single-mode emission is thought to be caused by the shorter momentum relaxation time of graphene carriers in device 2, which could allow the net THz gain only at around the DFB fundamental mode frequency ~5 THz, as is simulated in Fig. 37.4B. From the good correspondence between the experimental result and the analytical results for the gain spectra, it could be inferred that the obtained radiation is due to laser oscillation.

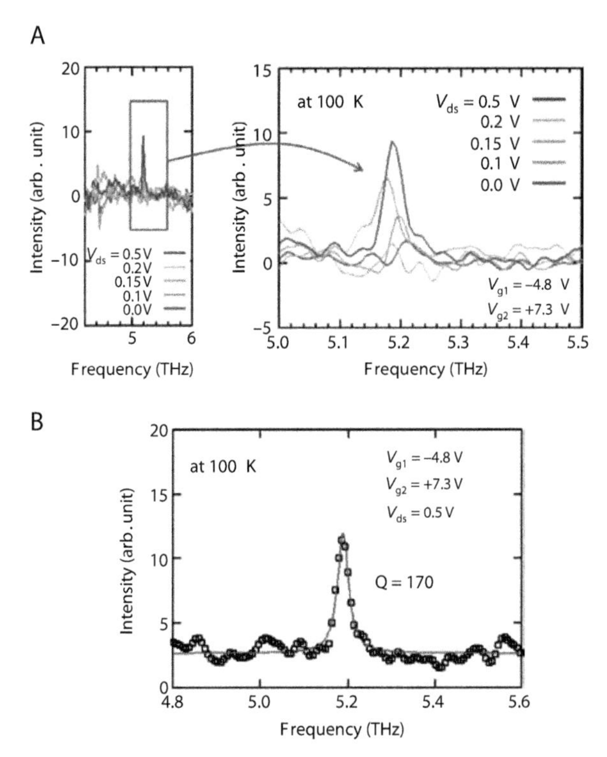

Figure 37.7 The single-mode like emission peak at ~5.2 THz with Q factor of 170 fairly agrees with the simulated results (at ~4.96 THz with Q ~210). (A) Observed spectra for device 2 at 100 K and (B) the fitting curve for the peak at ~5.2 THz showing Q factor to be 170.

Although device 2 shows single-mode emission, the peak power of emission is 3.5 times lower than that for device 1, which implies that high-quality GFET when paired with carefully designed DFB cavity could result in efficient THz lasing at higher THz frequencies. For example, numerical simulations suggest that an increase of N_{DFB} from 16 to 24 and an increase of the DFB cavity modulation index $L_{\mathrm{int}}/L_{\mathrm{dfb}}$ from 20/15 to 20/8 give rise to an increase of the quality factor of the DFB cavity by two orders of magnitude, respectively (see Supplement 37.S2 in detail). Also, pertinent waveguide structures with a thick dielectric layer on top of the GFET and/or ridge-waveguiding with thick dual-gate metal pillars as well as surface plasmon-polaritonic waveguiding may improve the THz photon field spatial confinement and the gain-overlapping factor.

Further investigation is needed to confirm if the observed emission is truly single-mode lasing. Also, as the drain bias dependence and device structure dependence of radiation spectrum are yet to be fully understood, we need to elucidate them by future experiments along with improving the device processing and integration to achieve high-power THz emission at room temperature.

37.4 Conclusions

A forward-biased graphene structure with a lateral p-i-n junction was implemented as a DFB-DG-GFET in which the current injection mechanism was realized using a dual-gate structure and the laser action mechanism was realized using a DFB cavity structure. Using our original process technology with epitaxial graphene and SiN gate dielectric, we achieved easy device integration and very high carrier mobilities exceeding 100,000 cm^2/Vs at room temperature. A rather intense carrier-injection level-dependent amplified spontaneous emission in the 1–7.6-THz range peaking at ~5 THz was observed at 100 K with integrated emission output power of ~80 μW. The results demonstrate the operation as a broadband THz light-emitting transistor. We also observed indications of single-mode emission at 5.2 THz at 100 K in another device. Although the results obtained are still preliminary level and a concrete dependence of emission power on the V_d is yet to be established, our results highlight that single-mode continuous wave THz lasing with an output power of the order of ~10 μW could be feasible by the carrier injection pumping in the DG-GFET with a carefully designed laser cavity structure.

Acknowledgments

The authors acknowledge T. Suemitsu, H. Fukidome, and M. Suemitsu for their contributions on device processes and graphene synthesis and characterization. The epitaxial graphene was synthesized by Prof. H. Fukidome, K. Tashima, and Prof. M. Suemitsu, for which detailed characterizations of the structural and electronic properties of the epitaxial graphene were performed as the contracted cooperation studies between Prof. H. Fukidome, Prof. M. Suemitsu, Photon Factory (2015S2-005, and 2015G536), and

SPring-8 (2015A1278 and 2015B1232) with devoted contributions from and insightful discussions with staffs at Photon Factory and SPring-8 (Prof. H. Kumigashira, Prof. K. Horiba, and Prof. M. Kotsugi). They also thank V. Ya. Aleshkin, S. Boubanga-Tombet, V. Mitin, and M. S. Shur for valuable discussions. The device process was carried out at the Laboratory for Nanoelectronics and Spintronics at RIEC in Tohoku University. The part of the works primarily contributed by D. Y., A. S., T. W., M. R., V. R., and T. O. was financially supported by JSPS KAKENHI (#23000008, #16H06361, and #16K14243), Japan. The part of the works primarily contributed by V. R. was supported by the Russian Scientific Foundation (#14-29-00277), and the part of the works by A. A. D. was supported by the Russian Foundation of the Basic Research (#18-52-50024).

References

1. S. S. Dhillon, M. S. Vitiello, E. H. Linfield, et al., The 2017 terahertz science and technology roadmap, *J. Phys. D: Appl. Phys.*, **50**, 43001 (2017).
2. M. Hangyo, Development and future prospects of terahertz technology, *Jpn. J. Appl. Phys.*, **54**, 120101 (2015).
3. H. A. Hafez, X. Chai, A. Ibrahim, et al., Intense terahertz radiation and their applications, *J. Opt.*, **18**, 93004 (2016).
4. M. Tonouchi, Cutting-edge terahertz technology, *Nat. Photonics*, **1**, 97–105 (2007).
5. N. Horiuchi, Terahertz technology: endless applications, *Nat. Photonics*, **4**, 140 (2010).
6. B. B. Hu and M. C. Nuss, Imaging with terahertz waves, *Opt. Lett.*, **20**, 1716–1718 (1995).
7. A. Redo-Sanchez, B. Heshmat, A. Aghasi, et al., Terahertz time-gated spectral imaging for content extraction through layered structures, *Nat. Commun.*, **7**, 12665 (2016).
8. C. Jansen, S. Wietzke, O. Peters, et al., Terahertz imaging: applications and perspectives, *Appl. Opt.*, **49**, E48–57 (2010).
9. K. Serita, S. Mizuno, H. Murakami, et al., Scanning laser terahertz near-field imaging system, *Opt. Express*, **20**, 12959–12965 (2012).
10. D. M. Mittleman, M. Gupta, R. Neelamani, R. G. Baraniuk, J. V. Rudd, and M. Koch, Recent advances in terahertz imaging, *Appl. Phys. B: Lasers Opt.*, **68**, 1085–1094 (1999).

11. A. J. Fitzgerald, E. Berry, N. N. Zinovev, G. C. Walker, M. A. Smith, and J. M. Chamberlain, An introduction to medical imaging with coherent terahertz frequency radiation, *Phys. Med. Biol.*, **47**, R67 (2002).
12. B. M. Fischer, M. Walther, and P. Uhd Jepsen, Far-infrared vibrational modes of DNA components studied by terahertz time-domain spectroscopy, *Phys. Med. Biol.*, **47**, 3807 (2002).
13. C. Yu, S. Fan, Y. Sun, and E. Pickwell-Macpherson, The potential of terahertz imaging for cancer diagnosis: a review of investigations to date, *Quant. Imaging Med. Surg.*, **2**, 33–45 (2012).
14. T. Nagatsuma, G. Ducournau, and C. C. Renaud, Advances in terahertz communications accelerated by photonics, *Nat. Photonics*, **10**, 371–379 (2016).
15. Y. Zhang, S. Qiao, S. Liang, et al., Gbps terahertz external modulator based on a composite metamaterial with a double-channel heterostructure, *Nano Lett.*, **15**, 3501–3506 (2015).
16. S. Koening, D. Lopez-Diaz, J. Antes, et al., Wireless sub-THz communication system with high data rate, *Nat. Photonics*, **7**, 977–981 (2013).
17. R. F. Kazarinov, and R. A. Suris, Possibility of the amplification of electromagnetic waves in a semiconductor with a superlattice, *BibSonomy. Sov. Phys. Semicond.*, **5**, 707 (1971).
18. J. Faist, F. Capasso, D. L. Sivco, C. Sirtori, A. L. Hutchinson, and A. Y. Cho, Quantum cascade laser, *Science*, **264**, 553–556 (1994).
19. R. Kbhler, A. Tredicucci, F. Beltram, et al., Terahertz semiconductor-heterostructure laser, *Nature*, **417**, 156–159 (2002).
20. B. S. Williams, Terahertz quantum-cascade lasers, *Nat. Photonics*, **1**, 517–525 (2007).
21. S. Fathololoumi, E. Dupont, C. W. I. Chan, et al. Terahertz quantum cascade lasers operating up to ~200 K with optimized oscillator strength and improved injection tunneling, *Opt. Express*, **20**, 3866–3876 (2012).
22. M. S. Vitiello, G. Scalari, B. Williams, and P. De Natale, Quantum cascade lasers: 20 years of challenges, *Opt. Express*, **23**, 5167–5182 (2015).
23. E. Briindermann and H. P. Rbser, First operation of a far-infrared p-germanium laser in a standard close-cycle machine at 15 K, *Infrared Phys. Technol.*, **38**, 201–203 (1997).
24. O. A. Klimenko, Y. A. Mityagin, S. A. Savinov, et al. Terahertz wide range tunable cyclotron resonance p-Ge laser, *J. Phys. Conf. Ser.*, **193**, 12064 (2009).

25. S. Suzuki, M. Asada, A. Teranishi, H. Sugiyama, and H. Yokoyama, Fundamental oscillation of resonant tunneling diodes above 1 THz at room temperature, *Appl. Phys. Lett.*, **97**, 242102 (2010).
26. Y. Chassagneux, Q. J. Wang, S. P. Khanna, et al., Limiting factors to the temperature performance of THz quantum cascade lasers based on the resonant-phonon depopulation scheme, *IEEE Trans. Terahertz Sci. Technol.*, **2**, 83–92 (2012).
27. R. R. Hartmann, J. Kono, and M. E. Portnoi, Terahertz science and technology of carbon nanomaterials, *Nanotechnology*, **25**, 322001 (2014).
28. A. Tredicucci and M. S. Vitiello, Device concepts for graphene-based terahertz photonics, *IEEE J. Sel. Top. Quantum Electron.*, **20**, 8500109 (2014).
29. A. K. Geim and K. S. Novoselov, The rise of graphene, *Nat. Mater.*, **6**, 183–191 (2007).
30. A. N. Grigorenko, M. Polini, and K. S. Novoselov, Graphene plasmonics, *Nat. Photonics*, **6**, 749–758 (2012).
31. K. S. Novoselov, A. K. Geim, S. V. Morozov, et al., Electric field effect in atomically thin carbon films, *Science*, **306**, 666–669 (2004).
32. K. M. Daniels, M. M. Jadidi, A. B. Sushkov, et al., Narrow plasmon resonances enabled by quasi-freestanding bilayer epitaxial graphene, *2D Mater.*, **4**, 25034 (2017).
33. I. H. Baek, J. M. Hamm, K. J. Ahn, et al., Boosting the terahertz nonlinearity of graphene by orientation disorder, *2D Mater.*, **4**, 25035 (2017).
34. V. Ryzhii, T. Otsuji, M. Ryzhii, et al., Graphene vertical cascade interband terahertz and infrared photodetectors, *2D Mater.*, **2**, 25002 (2015).
35. D. Yadav, S. B. Tombet, T. Watanabe, S. Arnold, V. Ryzhii, and T. Otsuji, Terahertz wave generation and detection in double-graphene layered van der Waals heterostructures, *2D Mater.*, **3**, 45009 (2016).
36. T. Otsuji, S. B. Tombet, A. Satou, M. Ryzhii, and V. Ryzhii, Terahertz-wave generation using graphene: toward new types of terahertz lasers, *IEEE J. Sel. Top. Quantum Electron.*, **19**, 8400209 (2013).
37. A. Satou, T. Otsuji, and V. Ryzhii, Theoretical study of population inversion in graphene under pulse excitation, *Jpn. J. Appl. Phys.*, **50**, 70116 (2011).
38. A. Satou, V. Ryzhii, Y. Kurita, and T. Otsuji, Threshold of terahertz population inversion and negative dynamic conductivity in graphene under pulse photoexcitation, *J. Appl. Phys.*, **113**, 143108 (2012).

39. V. Ryzhii, M. Ryzhii, V. Mitin, A. Satou, and T. Otsuji, Effect of heating and cooling of photogenerated electron-hole plasma in optically pumped graphene on population inversion, *Jpn. J. Appl. Phys.*, **50**, 094001 (2011).

40. S. Boubanga-Tombet, S. Chan, T. Watanabe, A. Satou, V. Ryzhii, and T. Otsuji, Ultrafast carrier dynamics and terahertz emission in optically pumped graphene at room temperature, *Phys. Rev. B*, **85**, 35443 (2012).

41. T. Li, L. Luo, M. Hupalo, et al., Femtosecond population inversion and stimulated emission of dense dirac fermions in graphene, *Phys. Rev. Lett.*, **108**, 167401 (2012).

42. V. Ryzhii, M. Ryzhii, V. Mitin, and T. Otsuji, Toward the creation of terahertz graphene injection laser, *J. Appl. Phys.*, **110**, 094503 (2011).

43. M. Ryzhii and V. Ryzhii, Injection and population inversion in electrically induced p-n junction in graphene with split gates, *Jpn. J. Appl. Phys.*, **46**, L151–L153 (2007).

44. V. Ryzhii, I. Semenikhin, M. Ryzhii, et al., Double injection in graphene p-i-n structures, *J. Appl. Phys.*, **113**, 244505 (2013).

45. J. H. Strait, H. Wang, S. Shivaraman, V. Shields, M. Spencer, and F. Rana, Very slow cooling dynamics of photoexcited carriers in graphene observed by optical-pump terahertz-probe spectroscopy, *Nano Lett.*, **11**, 4902–4906 (2011).

46. T. Winzer, A. Knorr, and E. Malic, Carrier multiplication in grapheme, *Nano Lett.*, **10**, 4839–4843 (2010).

47. R. Kim, V. Perebeinos, and P. Avouris, Relaxation of optically excited carriers in graphene, *Phys. Rev. B.*, **84**, 075449 (2011).

48. T. Winzer and E. Malic, Impact of Auger processes on carrier dynamics in graphene, *Phys. Rev. B.*, **85**, 241404(R) (2012).

49. D. Brida, A. Tomadin, C. Manzoni, et al., Ultrafast collinear scattering and carrier multiplication in graphene, *Nat. Commun.*, **4**, 1987 (2013).

50. D. C. Elias, R. V. Gorbachev, A. S. Mayorov, et al., Dirac cones reshaped by interaction effects in suspended graphene, *Nat. Phys.*, **7**, 701–704 (2011).

51. A. Bostowick, T. Ohta, T. Seyller, K. Horn, and E. Rotenberg, Quasiparticle dynamics in graphene, *Nat. Phys.*, **3**, 36–40 (2007).

52. K. Tajima, R. Suto, H. Fukidome, et al., Fabrication of ultrahigh-quality graphene on SiC (000-1) substrate and evaluation of Bernal-stacked

domain, The 76th JSAP (Japan Society of Applied Physics) Fall Meeting Abstracts, 14a-2T-6, p. 15-015 (2015), (in Japanese).

53. H. Fukidome, Y. Kawai, F. Fromm, et al., Precise control of epitaxy of graphene by microfabricating SiC substrate, *Appl. Phys. Lett.*, **101**, 41605 (2012).

54. T. Someya, H. Fukidome, H. Watanabe, et al., Suppression of supercollision carrier cooling in high mobility graphene on SiC (0001), *Phys. Rev. B*, **95**, 165303 (2017).

55. A. Satou, G. Tamamushi, K. Sugawara, J. Mitsushio, V. Ryzhii, and T. Otsuji, A fitting model for asymmetric I-V characteristics of graphene FETs for extraction of intrinsic mobilities, *IEEE Trans. Electron. Dev.*, **63**, 3300–3306 (2016).

56. B. S. Williams, S. Kumar, Q. Hu, and J. L. Reno, Distributed-feedback terahertz quantum-cascade lasers with laterally corrugated metal waveguides, *Opt. Lett.*, **30**, 2909–2911 (2005).

57. V. Ryzhii, A. A. Dubinov, T. Otsuji, V. Mitin, and M. S. Shur, Terahertz lasers based on optically pumped multiple graphene structures with slot-line and dielectric waveguides, *J. Appl. Phys.*, **107**, 54505 (2010).

Supplementary Material

This document provides supplementary information to "Terahertz light-emitting graphene-channel transistor toward single mode lasing."

37.S1 Experimental Methods

The emission experiments were carried out using Fourier-transformed far-infrared (FTIR) spectrometer with a Martin-Puplett-type interferometer, JASCO FARIS-1, at a series of temperatures starting from 300 K down to 100 K. The sample was installed onto a dedicated sample housing which was placed in the cryostat and the biases were controlled using voltage/current source-measure units, Keithley 2400, while the radiation intensity was measured using a 4.2 K-cooled silicon bolometer. Figure 37.S1 (a) shows the experiment setup along with the sample holder, a zoomed view of the device is shown in the inset of 37.S1 (b). The DFB-responsible

THz photons are to be emitted from the edge of the graphene active region. Thus, the sample was placed so that its edge was facing to the FTIR interferometer via the parabolic mirror. The THz electric field of the laser or LED emission is polarized along with the in-plane drain-source axis. The sample was, therefore, rotated so as to align its THz photon field being transparent to the metal mesh filter placed at the entrance of the Martin-Puplett interferometer. To carry out the experiments, we first measured the background spectrum, i.e. the spectrum with a sample unbiased. Then we measured bias dependent spectra with current flowing through the sample and normalized it with the background spectra to eliminate the undesired effects arising due to static blackbody radiation and any spectral artifacts emitted from the elements inside the spectrometer.

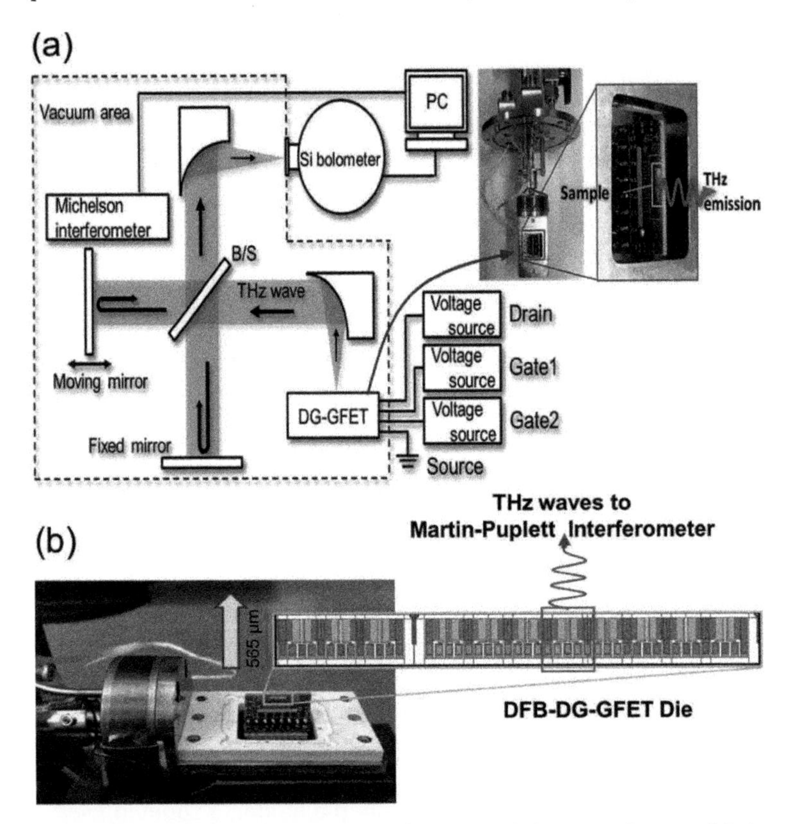

Figure 37.S1 (a) Measurement setup for THz emission experiments, (b) the sample die is installed in a cryostat with a dedicated mount holder.

37.S2 DFB Cavity Design Parameters

Figure 37.S2 (a) shows various DFB cavity design parameters. The improvement of quality factor hugely depends on these parameters.

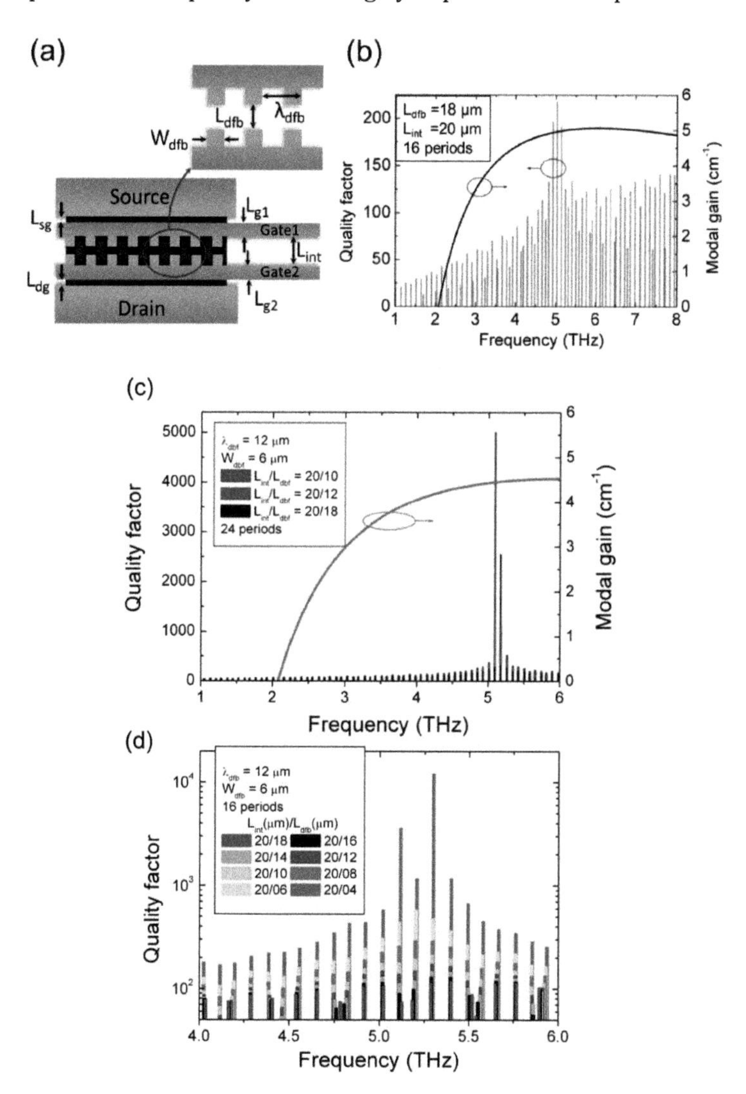

Figure 37.S2 (a) Design parameters for the DFB structure; numerical simulation plots for these DFB cavities having different design parameters (b) N_{DFB} = 16, (c) N_{DFB}= 24, and (d) comparison of quality factors for various DFB cavity modulation index.

For example, when we increase the number of periods of DFB (N_{DFB}) from 16 to 24 the Q value increases by 2 orders (see Fig. 37.S2 (b) and 37.S2 (c)). Figure 37.S2 (d) gives numerically simulated Q values when the value of L_{dfb}, that is, the degree of concavity and convexity of DFB is changed. Here also, we can observe that an increase of the DFB cavity modulation index L_{int}/L_{dfb} from 20/18 to 20/4 give rise to an increase of the quality factor of the DFB cavity by two orders of magnitude. From these plots, it can be summarized that Q value is dramatically improved by increasing the number of periods of DFB and by increasing the degree of modulation of the gain coefficient i.e. by increasing the degree of unevenness of DFB structure.

Part 5

Detectors and Emitters Based on Photon/Plasmon Assisted Tunneling between Graphene Layers

Chapter 38

Injection Terahertz Laser Using the Resonant Inter-Layer Radiative Transitions in Double-Graphene-Layer Structure*

V. Ryzhii,[a,b] A. A. Dubinov,[c] V. Ya. Aleshkin,[c] M. Ryzhii,[b,d] and T. Otsuji[a,b]

[a]*Research Institute for Electrical Communication, Tohoku University, Sendai 980-8577, Japan*

[b]*Japan Science and Technology Agency, CREST, Tokyo 107-0075, Japan*

[c]*Institute for Physics of Microstructures of Russian Academy of Sciences, and Lobachevsky State University of Nizhny Novgorod, Nizhny Novgorod 603950, Russia*

[d]*Department of Computer Science and Engineering, University of Aizu, Aizu-Wakamatsu 965-8580, Japan*

v-ryzhii@riec.tohoku.ac.jp

Graphene-Based Terahertz Electronics and Plasmonics: Detector and Emitter Concepts
Edited by Vladimir Mitin, Taiichi Otsuji, and Victor Ryzhii

ISBN 978-981-4800-75-4 (Hardcover), 978-0-429-32839-8 (eBook)
www.jennystanford.com

We propose and substantiate the concept of terahertz (THz) laser enabled by the resonant electron radiative transitions between graphene layers (GLs) in double-GL structures. We estimate the THz gain for TM-mode exhibiting very low Drude absorption in GLs and show that the gain can exceed the losses in metal-metal waveguides at the low end of the THz range. The spectrum of the emitted photons can be tuned by the applied voltage. A weak temperature dependence of the THz gain promotes an effective operation at room temperature.

The gapless energy spectrum of graphene layers (GLs) [1] enables the creation of different terahertz (THz) devices utilizing the interband transition. In particular, the interband population inversion and the pertinent negativity of the dynamic conductivity in GLs [2, 3] due to the optical or injection pumping can be used in GL-based THz lasers [4–9]. First experimental results on the THz emission from optically excited GLs [10] (see also review paper [11] and references therein) instill confidence in the realization of such lasers. One of the obstacles, limiting the achievement of the negative dynamic conductivity in the range of a few THz, is the reabsorption of the photons with the in-plane polarization emitted at the interband transitions due to the intraband transitions (the Drude absorption). Similar situation takes place in the quantum cascade lasers (QCLs) based on multiple quantum well (MQW) structures [12]. However, in the case of the photon polarization perpendicular to the QW plane the intraband (intrasubband) absorption can be much weaker than that following from the semi-classical Drude formula [13].

In this chapter, we propose a device structure based on a double-GL structure shown in Fig. 38.1 with the injection of electrons to one n-doped GL and to another p-doped GL, which can be used for lasing of THz photons with the electric field perpendicular to the GL plane due to the tunneling inter-GL radiative processes. The structure band diagram under the applied bias voltage and tunneling transitions assisted with the emission of photons with the energy $\hbar\omega \sim \Delta$, where Δ is the energy distance (gap) between the Dirac points in GLs, are demonstrated in Fig. 38.2. These transitions take place from the conduction band of the GL with 2DEG to the empty conduction band of GL with 2DHG. The transition from the filled valence band the former GL to the empty portion of the valence band of the latter GL also contributes to the emission of photons with $\hbar\omega \sim \Delta$. The structure comprises two GLs with the side contact at one of the GL edges. This double-GL structure plays the role of the laser active region. The opposite edge of GL is isolated from another contact. A narrow tunneling-transparent barrier separates GLs (its thickness d is about few nanometers). The applied bias voltage V provides the formation of the two-dimensional electron and hole gases (2DEG and 2DHG) in the upper and lower GLs, respectively, owing to the injection from the side contacts, so that the inter-GL population inversion occurs. The GL structures in question were fabricated recently [14–17]. The effective control of the GL population was, in particular, used for modulation of optical radiation [14, 15]. These structures can exhibit marked inter-GL tunneling current, which in the case of the Dirac point alignment, exhibit the negative differential inter-GL conductivity [18] (see also Refs. [19–21], where the latter effect was theoretically considered). The inter-GL barrier layers can be made of hBN, WS_2, or similar materials. The double-GL structures can also be used in different devices (THz detectors and photomixers) utilizing the plasmonic effects [22–24].

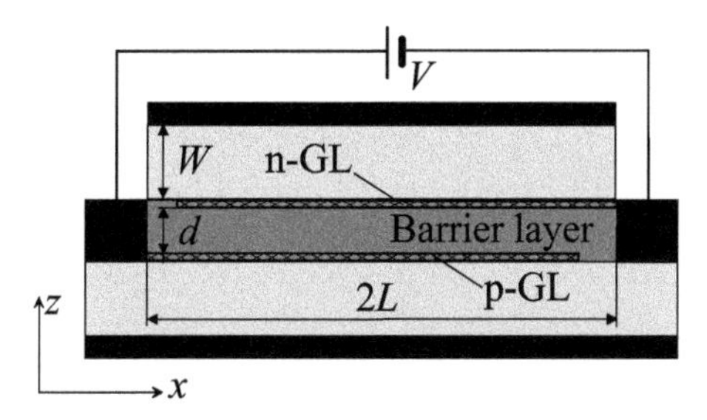

Figure 38.1 Schematic view of DG-laser with double-GL structure with MM wave-guide Dark regions corresponds to the highly conducting side contacts and strips (along the y-direction).

The THz laser metal-metal (MM) waveguide system consist of two parallel metal strips (see Fig. 38.1) as in some QCLs [12]. The net spacing between the strips ($2W + d$) is about 10 μm. The spatial distributions of the amplitudes of THz electric-field components $|E_z|$ and $|E_x|$ in the TM-mode propagating in the y-direction and the real part of refractive index in the MM wave-guide are shown in Fig. 38.3. It is assumed that the device structure is supplied with the proper mirrors reflecting the radiation propagating in the y-direction.

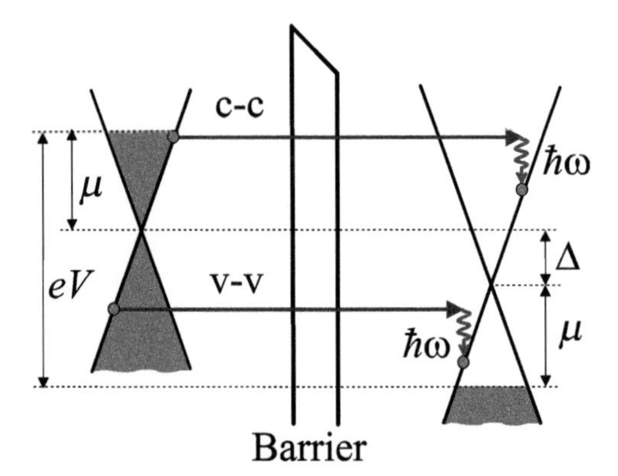

Figure 38.2 DG-laser band diagram. Arrows indicate the resonant-tunneling (with the conservation of electron momentum) inter-GL transitions between the conduction band ($c - c$) and the valence band ($v - v$) states assisted by the emission of photons (at weak depolarization shift).

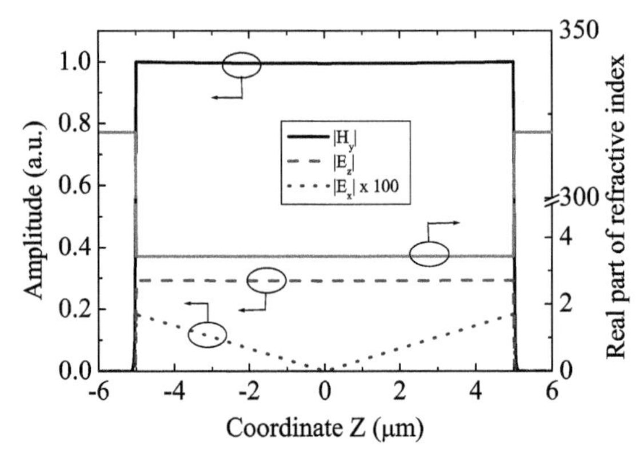

Figure 38.3 Spatial distributions of the amplitudes of THz electric- and magnetic-field components in TM mode and the real part of the refraction index in the MM wave-guide under consideration.

To enhance the laser output power, a more complex active region can used which includes several double-GL structures (with thin inter-GL layers and relatively thick layers separating the double-GL structures) or a multiple-GL structure with the tunneling-transparent inter-GL barrier layers. A wave guide system using the specific plasmon-polaritons [25] (see also Ref. [7]), associated the high conductivity of GLs and propagating along the structure (in the y-direction) can also be implemented.

Apart from weaker intraband reabsorption of the generated photons, the double-GL structures (and more complex ones) might exhibit the advantages in comparison with the p-i-n lasers with simultaneous injection of both electrons and holes into the same GLs [8, 26, 27] because of a more effective injection due to the absence of the electron-hole "friction" (Ref. [28]) and the weakening of nonradiative recombination mechanisms, in particular, that associated with the emission of optical phonons.

The electron and hole density in the pertinent GLs is given by

$$\Sigma = \Sigma_{\mathrm{i}} + \frac{\kappa\Delta}{4\pi e^2 d}, \tag{38.1}$$

where Σ_{i} is the density of donors in the upper GL (n-GL) and acceptors in the lower GL (p-GL), κ is the dielectric constant, and $e = |e|$ is the electron charge. If the length of the electron and hole diffusion exceeds the length of GLs, the local potential difference between GLs is close to V [8, 27]. In this situation, the quantity Δ (which determines the electric field $E = \Delta/ed$ in the inter-GL barrier) obeys the following equation (see Fig. 38.2):

$$\Delta + 2\mu = eV, \tag{38.2}$$

where μ is the value of the Fermi energy of 2DEG and 2DHG in the pertinent GL. In the case of the 2DEG and 2DHG strong degeneration ($\mu \gg T$, where T is the temperature in the energy units)

$$\mu \simeq \hbar v_{\mathrm{W}} \sqrt{\pi\Sigma}, \tag{38.3}$$

respectively, where $v_{\mathrm{W}} \simeq 10^8$ cm/s is the characteristic velocity of electrons and holes in GLs. Considering Eqs. (38.1)–(38.3), in the above limiting cases, we arrive at

$$\Delta/e = V + V_0 - \sqrt{2VV_0 + V_0^2 + V_t^2}, \tag{38.4}$$

where $V_0 = \hbar^2 v_{\mathrm{W}}^2 \kappa/2e^3 d$ and $V_{\mathrm{t}} = 2\hbar v_{\mathrm{W}} \sqrt{\pi\Sigma_{\mathrm{i}}}/e$. At $V \leq V_{\mathrm{t}}$, $\Delta \leq 0$, while at $V > V_{\mathrm{t}}$, $\Delta > 0$. The band diagram at the bias voltage $V > V_{\mathrm{t}}$ is shown in Fig. 38.2. In this case, the tunneling inter-GL transitions are possible only due to the scattering processes accompanying the tunneling [19–21]. However, the inter-GL transitions assisted by the emission of photons with the polarization corresponding to the photon electric field perpendicular to GLs (along the axis z) conserve the electron momentum and, hence, do not require any scattering (resonant-tunneling photon-assisted transitions). Assuming $\kappa = 4$, $d = 4$ nm, and $\Sigma_i = 10^{12}$ cm^{-2}, one obtains $V_0 \simeq 136$ mV and $V_t \simeq 221$ mV. The quantities $\Delta = 5 - 10$ meV correspond to $V \simeq 229 - 237$ mV and $\mu \simeq 112.0 - 113.5$ meV.

The real part of the transverse ac conductivity, Re $\sigma_{zz}(\omega)$, of the double-GL structure under consideration can be estimated using the following formula:

$$\mathrm{Re}\,\sigma_{zz}(\omega) = -\frac{2e^2}{\hbar}\frac{|z_{u,l}|^2\,\Sigma_i(1+\Delta/\Delta_i)\gamma\hbar\omega}{[\hbar^2(\omega-\omega_{max})^2+\gamma^2]}. \tag{38.5}$$

Here

$$\hbar\omega_{max} = \Delta - \frac{8\pi e^2|z_{u,l}|^2\,\Sigma_i(1+\Delta/\Delta_i)}{\kappa d} \tag{38.6}$$

is the inter-GL resonant transition energy, where the second term in the right-hand side constitutes the depolarization shift (see, for instance, Ref. [29]), which for media with the population inversion is negative, $z_{u,l} = \int \varphi_u^*(z) z \varphi_l(z)$, where $\varphi_u(z)$ and $\varphi_l(z)$ are the z-dependent factors of the wave functions in the upper and lower GLs, γ is the relaxation broadening, and $\Delta_i = 4\pi e^2 d\Sigma_i/\kappa$.

Figure 38.4 shows the energy gap between the Dirac points Δ as a function of the applied voltage V calculated using Eq. (38.4) at $\kappa = 4$, $d = 4$ nm, and different values of the dopant densities Σ_i.

At the resonance $\hbar\omega = \hbar\omega_{max}$, Eq. (38.5) yields

$$\mathrm{Re}\,\sigma_{zz}(\omega) = -\frac{2e^2|z_{u,l}|^2\Sigma_i(1+\Delta/\Delta_i)}{\hbar}\left(\frac{\hbar\omega_{max}}{\gamma}\right). \tag{38.7}$$

Introducing the gain-overlap factor for the THz mode in the wave guide under consideration $\Gamma = [\int_{-L}^{L} dx|E_z(x,0)|^2/\int_{-L}^{L}\int_{-W}^{W} dxdz|E_z(x,0)|^2]$, where $E_z(x, z)$ is the spatial distribution of the THz electric field in the TM mode propagating along the MM waveguide (in the y direction) and $2L$ is the length of GLs (distance between the side contacts), the maximum THz gain $g = 4\pi\mathrm{Re}\sigma\Gamma/c\sqrt{\kappa}$, where c is the speed of light in vacuum, can be estimated as

$$g = \frac{8\pi e^2|z_{u,l}|^2\Sigma_i(1+\Delta/\Delta_i)}{\hbar c\sqrt{\kappa}}\left(\frac{\hbar\omega_{max}}{\gamma}\right)\Gamma. \tag{38.8}$$

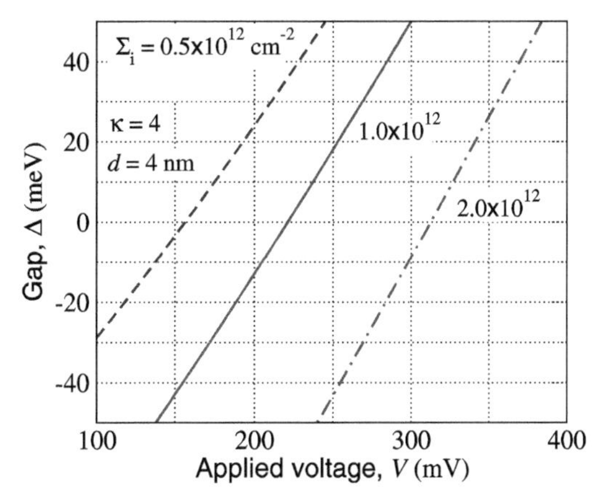

Figure 38.4 Energy gap Δ versus applied voltage V at different dopant densities Σ_i.

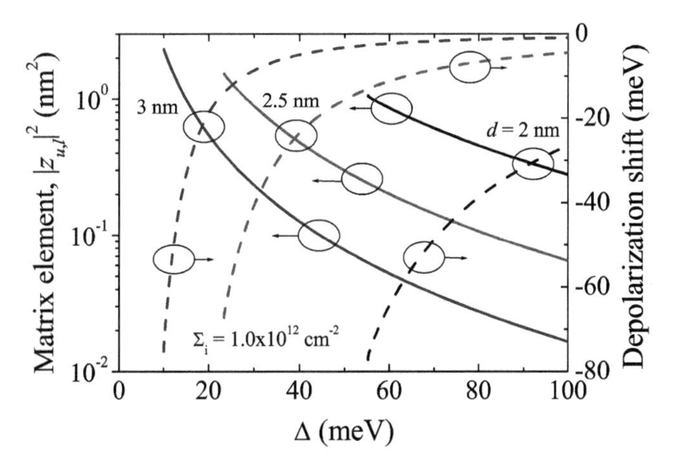

Figure 38.5 Inter-GL matrix element $|z_{u,l}|^2$ (solid lines) and depolarization shift (dashed lines) as functions of gap Δ.

The value of matrix element $|z_{u,l}|^2$ strongly depends on d and Δ. It is evaluated using a simple model, in which each GL is considered as a delta-layer separated by a barrier layer made of WS_2. Following Ref. [19], we assume that the conduction band offset between a GL and WS_2 barrier and the effective electron mass in WS_2 are equal to 0.4 eV and 0.27 of the free electron mass, respectively [30]. Figure 38.5 shows the dependence of the matrix element $|z_{u,l}|^2$ on the energy difference between the Dirac points Δ and the value of the depolarization shift (dependent of $|z_{u,l}|^2$) for different thicknesses of the WS_2 inter-GL barrier. As seen from Fig. 38.5, $|z_{u,l}|^2$ dramatically decreases with increasing Δ. According to Eqs. (38.5)–(38.8), this substantially affects the transverse ac conductivity Re $\sigma_{zz}(\omega)$, the resonant value of the photon energy $\hbar\omega_{max}$, and, hence the THz gain g.

Figures 38.6 and 38.7 show the resonant THz gain versus photon energy $\hbar\omega_{max}$ for different dopant densities Σ_i and the inter-GL barrier layer thicknesses. In Fig. 38.5, it is assumed that $\kappa = 4$, $d =$ 2 nm and $\gamma/\hbar = 1.6 \times 10^{12}$ s^{-1}. The gain-overlap factor for the MM waveguide with $W = 5$ μm is set to be $\Gamma = 10^3$ cm^{-1}. As seen from Fig. 38.6, an increase in the dopant density naturally increases the THz gain. However, this increase can be limited by an increase in broadening parameter γ. One can also see that the THz gain exceeds (or even well exceeds) the value 10 cm^{-1} in the range $\hbar\omega_{max} \gtrsim 10$ meV ($\omega_{max}/2\pi \gtrsim 2.5$ THz). Hence, g can exceed the TM-mode losses α in the copper MM wave-guides even at room temperature [31]. It is worth noting that the Drude absorption for the mode under consideration can be very weak because of relatively small THz electric field component E_x (as seen from Fig. 38.3, the ratio $|E_z|/|E_x|$ > 100), particularly at $z = 0$, where the double-GL structure is placed. The THz gain of the TM mode in the laser under consideration is at least comparable the maximum THz gain (without the Drude losses) in the injection lasers utilizing the TE mode and the intra-GL radiative transitions (see, for instance, Ref. [6]). Hence the former device can exhibit advantages due to the effective suppression of the Drude absorption. As follows from Eq. (38.8), the THz gain weakly depends on the temperature. Some temperature dependence appears due to the temperature dependence of parameter γ (or at low doping levels when $\mu \lesssim T$). Weak temperature dependence of the resonant tunneling in double-GL structures was observed experimentally [18] (see also Refs. [19, 20]). This can provide the superiority of the double-GL lasers under consideration over the GL-based lasers using the intra-GL transitions discussed previously [4–6] and QCL lasers [12, 31] at elevated temperatures in the low end of the THz range as well as in the range where the operation of QCLs is hampered by the optical phonon absorption.

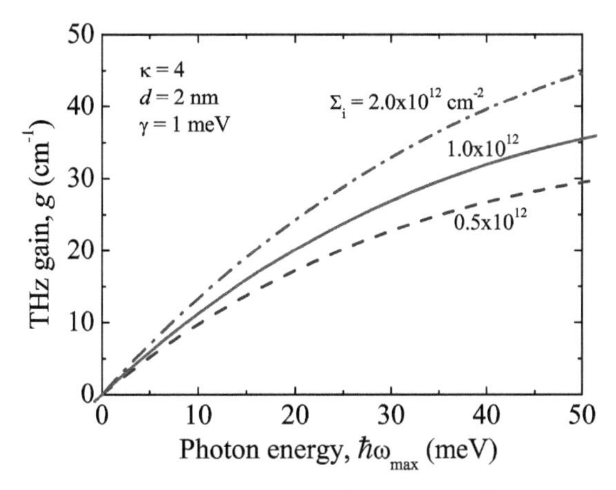

Figure 38.6 THz gain g versus resonant photon energy $\hbar\omega_{max}$ for double-GL structures with d = 2 nm and different values of Σ_i.

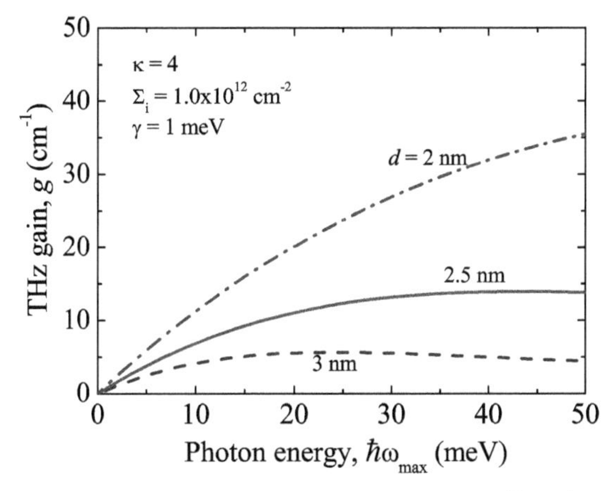

Figure 38.7 The same as in Fig. 38.6 but for $\Sigma_i = 1.0 \times 10^{12}$ cm^{-2} and different values of d.

The THz gain in the double-GL structures with thicker inter-GL layer is markedly smaller as seen from Fig. 38.7. In this case, to overcome the waveguide losses the device structure with an active region consisting of a system of several parallel double-GL structures can be used. Each the double-GL system can be separated from others by relatively thick, non-transparent barrier layer (with the thickness $D \gg d$. In such a case, the quantity g for the THz gain given by Eq. (38.8) should be multiplied by the number of the double-GL systems in the device structure. The multiple-GL structures in question can provide a much higher net THz gain without a marked increase in the Drude absorption enabling room temperature THz lasing in the range $\hbar\omega_{max} \simeq 5 - 10$ meV ($\omega_{max}/2\pi \simeq 1.25 - 2.5$ THz), where they might compete with the THz oscillators based on A_3B_5 resonant-tunneling structures [32].

In conclusion, we proposed THz lasers based on double-GL structures using the inter-GL radiative transitions and estimated their THz gain at different photon energies and applied voltages. It was demonstrated that the laser can exhibit advantages over other THz lasers due to the essential suppression of the Drude absorption in GL, weak temperature dependence of the THz gain even at room temperature, and voltage tuning of the spectrum of the emitted photons.

The authors are grateful to A. Satou and D. Svintsov for useful discussions. This work was supported by the Japan Society for Promotion of Science (Grant-in-Aid for Specially Promoting Research, No. 23000008) and the Japan Science and Technology Agency (CREST Project)), Japan, as well as by the Russian Foundation of Basic Research and the Dynasty Foundation, Russia.

References

1. A. H. Castro Neto, F. Guinea, N. M. R. Peres, K. S. Novoselov, and A. K. Geim, *Rev. Mod. Phys.*, **81**, 109 (2009).
2. V. Ryzhii, M. Ryzhii, and T. Otsuji, *J. Appl. Phys.*, **101**, 083114 (2007).
3. F. Rana, *IEEE Trans. Nanotechnol.*, **7**, 91 (2008).
4. A. A. Dubinov, V. Ya. Aleshkin, M. Ryzhii, T. Otsuji, and V. Ryzhii, *Appl. Phys. Express*, **2**, 092301 (2009).
5. V. Ryzhii, M. Ryzhii, A. Satou, T. Otsuji, A. A. Dubinov, and V. Ya. Aleshkin, *J. Appl. Phys.*, 106, 084507 (2009).
6. V. Ryzhii, A. A. Dubinov, T. Otsuji, V. Mitin, and M. S. Shur, *J. Appl. Phys.*, **107**, 054505 (2010).
7. A. A. Dubinov, V. Ya. Aleshkin, V. Mitin, T. Otsuji, and V. Ryzhii, *J. Phys.: Condens. Matter*, **23**, 145302 (2011).
8. V. Ryzhii, M. Ryzhii, V. Mitin, and T. Otsuji, *J. Appl. Phys.*, **110**, 094503 (2011).

9. V. V. Popov, O. V. Polischuk, A. R. Davoyan, V. Ryzhii, T. Otsuji, and M. S. Shur, *Phys. Rev. B*, **86**, 195437 (2012).

10. S. Boubanga-Tombet, S. Chan, T. Watanabe, A. Satou, V. Ryzhii, and T. Otsuji, *Phys. Rev. B*, 85, 035443 (2012).

11. T. Otsuji, S. A. Boubanga Tombet, A. Satou, M. Ryzhii, and V. Ryzhii, *IEEE J. Sel. Top. Quantum Electron.*, **19**, 8400209 (2013).

12. B. S. Williams, *Nat. Photonics*, **1**, 517 (2007).

13. F. Carosella, C. Ndebeka-Bandou, R. Ferreira, E. Dupont, K. Unterrainer, G. Strasser, A. WAcker, and G. Bastard, *Phys. Rev. B*, **85**, 085310 (2012).

14. M. Liu, X. Yin, E. Ulin-Avila, B. Geng, T. Zentgraf, L. Ju, F. Wang, and X. Zhang, *Nature*, **474**, 64 (2011).

15. M. Liu, X. Yin, and X. Zhang, *Nano Lett.*, **12**, 1482 (2012).

16. L. Britnell, R. V. Gorbachev, R. Jalil, B. D. Belle, F. Shedin, A. Mishenko, T. Georgiou, M. I. Katsnelson, L. Eaves, S. V. Morozov, N. M. R. Peres, J. Leist, A. K. Geim, K. S. Novoselov, and L. A. Ponomarenko, *Science*, **335**, 947 (2012).

17. T. Georgiou, R. Jalil, B. D. Bellee, L. Britnell, R. V. Gorbachev, S. V. Morozov, Y.-J. Kim, A. Cholinia, S. J. Haigh, O. Makarovsky, L. Eaves, L. A. Ponimarenko, A. K. Geim, K. S. Nonoselov, and A. Mishchenko, *Nat. Nanotechnol.*, **8**, 100 (2013).

18. L. Britnell, R. V. Gorbachev, A. K. Geim, L. A. Ponimarenko, A. Mishchenko, M. T. Greenaway, T. M. Fromhold, K. S. Nonoselov, and L. Eaves, *Nat. Commun.*, **4**, 1794 (2013).

19. R. M. Feenstra, D. Jena, and G. Gu, *J. Appl. Phys.*, 111, 043711 (2012).

20. P. Zhao, R. M. Feenstra, G. Gu, and D. Jena, *IEEE Trans. Electron Devices*, **60**, 951 (2013).

21. F. T. Vasko, *Phys. Rev. B*, **87**, 075424 (2013).

22. V. Ryzhii, T. Otsuji, M. Ryzhii, and M. S. Shur, *J. Phys. D: Appl. Phys.*, **45**, 302001 (2012).

23. V. Ryzhii, A. Satou, T. Otsuji, M. Ryzhii, V. Mitin, and M. S. Shur, *J. Phys. D: Appl. Phys.*, **46**, 315107 (2013).

24. V. Ryzhii, M. Ryzhii, V. Mitin, M. S. Shur, A. Satou, and T. Otsuji, *J. Appl. Phys.*, **113**, 174506 (2013).

25. D. Svintsov, V. Vyurkov, V. Ryzhii, and T. Otsuji, *J. Appl. Phys.*, **113**, 053701 (2013).

26. M. Ryzhii and V. Ryzhii, *Jpn. J. Appl. Phys., Part 2*, **46**, L151 (2007).

27. V. Ryzhii, I. Semenikhin, M. Ryzhii, D. Svintsov, V. Vyurkov, A. Satou, and T. Otsuji, *J. Appl. Phys.*, **113**, 244505 (2013).
28. D. Svintsov, V. Vyurkov, S. Yurchenko, T. Otsuji, and V. Ryzhii, *J. Appl. Phys.*, **111**, 083715 (2012).
29. F. T. Vasko and A. V. Kuznetsov, *Electronic States and Optical Transitions in Semiconductor Heterostructures* (Springer, New York, 1999).
30. H. Shi, H. Pan, Y.-W. Zhang, and B. Yakobson, *Phys. Rev. B*, **87**, 155304 (2013).
31. M. A. Belkin, J. A. Fan, S. Hormoz, F. Capasso, S. P. Khanna, M. Lachab, A. G. Davies, and E. Linfield, *Opt. Express*, **16**, 3242 (2008).
32. H. Kanaya, H. Shibayama, R. Sogabe, S. Suzuki, and M. Asada, *Appl. Phys. Express*, **5**, 124101 (2012).

Chapter 39

Double-Graphene-Layer Terahertz Laser: Concept, Characteristics, and Comparison*

V. Ryzhii,[a,b] A. A. Dubinov,[a,c] T. Otsuji,[a] V. Ya. Aleshkin,[c] M. Ryzhii,[d] and M. Shur[e]

[a]*Research Institute for Electrical Communication, Tohoku University, Sendai 980-8577, Japan*

[b]*Center for Photonics and Infrared Engineering, Bauman Moscow State Technical University, Moscow 105005, Russia*

[c]*Institute for Physics of Microstructures of Russian Academy of Sciences and Lobachevsky State University of Nizhny Novgorod, Nizhny Novgorod 603950, Russia*

[d]*Department of Computer Science and Engineering, University of Aizu, Aizu-Wakamatsu 965-8580, Japan*

[e]*Department of Electrical, Electronics, and System Engineering, Rensselaer Polytechnic Institute, Troy, NY 12180, USA*

v-ryzhii@riec.tohoku.ac.jp

Graphene-Based Terahertz Electronics and Plasmonics: Detector and Emitter Concepts
Edited by Vladimir Mitin, Taiichi Otsuji, and Victor Ryzhii

ISBN 978-981-4800-75-4 (Hardcover), 978-0-429-32839-8 (eBook)
www.jennystanford.com

We propose and analyze the concept of injection terahertz (THz) lasers based on double-graphene-layer (double-GL) structures utilizing the resonant radiative transitions between GLs. We calculate main characteristics of such double-GL lasers and compare them with the characteristics of the GL lasers with intra-GL interband transitions. We demonstrate that the double-GL THz lasers under consideration can operate in a wide range of THz frequencies and might exhibit advantages associated with the reduced Drude absorption, weaker temperature dependence, voltage tuning of the spectrum, and favorable injection conditions.

39.1 Introduction

Double-graphene-layer (double-GL) structures with narrow inter-GL barrier Boron Nitride (hBN), Tungsten Dioxide (WS_2), and other barrier layers [1–4] have recently been explored for applications in high speed electro-optical modulators [1, 5], transparent electronics [3], terahertz (THz) detectors [6, 7] and photomixers [8]. With the optical or injection pumping, various structures based a single-GL or multiple-GLs (twisted, non-Bernal stacked GLs or GLs separated by relatively thick barrier layers) can exhibit the interband population inversion and lasing of the TE-mode with the in-plane direction of the photon electric field [9–13]. Due to the gapless energy spectrum of GLs, such a lasing can occur in a wide THz range. The experimental studies of the THz emission from the pumped GLs [14, 15] have validated the concept of GL-based THz lasers (see also the review papers [16, 17]).

Recently [18], in contrast to GL-based lasers using the intra-GL interband radiative transitions [9–13], the use of the resonant radiative inter-GL transitions in double-GL structures with a sufficiently thin barrier between GLs was proposed to realize lasing. Potential advantages unclude much smaller Drude losses of the TM-mode, a weaker temperature sensitivity, the possibility of voltage tuning of the spectrum, and an increased injection efficiency. This can be particularly important at the low end of the THz frequency range (a few THz or less).

In this chapter, we develop a device model for the THz laser exploiting the inter-GL radiative transitions in the double-GL structure with a tunneling barrier layer. Such room temperature lasers might compete with the resonant-tunneling diodes [19] and quantum cascade lasers [20–22] at the low end of the THz gap. They can also find applications in the frequency range between 6 and 10 THz, which includes the optical phonon frequencies in A_3B_5 materials, in which the operation of quantum cascade lasers [20] is suppressed.

39.2 Device Structures and Principles of Operation

We primarily consider the device structure shown in Fig. 39.1a. It consists of two GLs chemically doped by donors (n-GL) and acceptors (p-GL), respectively, and separated by a narrow tunneling barrier layer of thickness d (about a few nanometers). The length of each GL is approximately equal to $2L$ (to the spacing between the side contact). The clad layers between the GLs and the pertinent metal plates are assumed of the same thickness equal to W. One of the edges of each GLs is connected to the side contact while the other one being isolated (independently-contacted GLs). This double-GL structure is similar to those fabricated recently [1–4]. The doping leads to the formation of the two-dimensional electron (2DEG) and the two-dimensional hole (2DHG) gases in the n-GL (the top GL) and the p-GL (the bottom GL), respectively. The bias voltage V between the side contacts induces the extra charges of opposite polarities at GLs and the electric field between GLs. This results in the band structure shown in Fig. 39.2a, where μ and Δ are the Fermi energy of 2DEG and 2DHG and the energy separation between the GL Dirac points, respectively [18]. As seen from Fig. 39.2a, the applied bias voltage causes the inter-GL population inversion between the conduction bands and between the valence band (c-c and v-v—transitions accompanied by the photon emission). Thus, the double-GL structure can serve as the laser active region with the lateral injection pumping. The nonradiative inter-GL tunneling and resonant-tunneling were observed in recent experiments [2–4] (see also theoretical papers [23, 24]). The electron and hole lateral injection from the pertinent contacts refills GLs loosing electrons and holes due to nonradiative and radiative inter-GL processes is realized by the electron and hole lateral injection from the pertinent contacts. The device structure is sandwiched by the clad layers (with thickness $W \gg d$), which, together with the metal layers, constitutc a metal-metal (MM) waveguide for the TM-modes propagating in the direction perpendicular to the x-z plane (see Fig. 39.1). Such waveguides are successfully used in particular in quantum cascade lasers [21, 22]. Considering this device, we will refer to it as for device "a." DC voltages $\pm V_g$ (gate voltages) applied between the GL edges and the metal strips of the MM waveguide, can induce sufficient densities of electrons and holes (the electrical induction of the extra electrons and holes by the gate voltages is sometimes dubbed as the "electrical" doping) even without chemical doping. We will keep in mind this option as well.

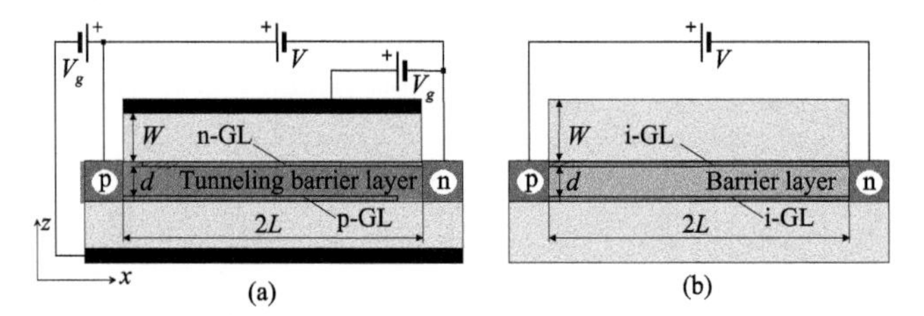

Figure 39.1 Schematic views of double-GL laser cross-sections (a) with the side monopolar injection to each independently contacted GL, tunneling barrier layer, and MM waveguide (device "a") and (b) with the injection of electrons and holes from p- and n-contacts to each GL and dielectric waveguide (device "b").

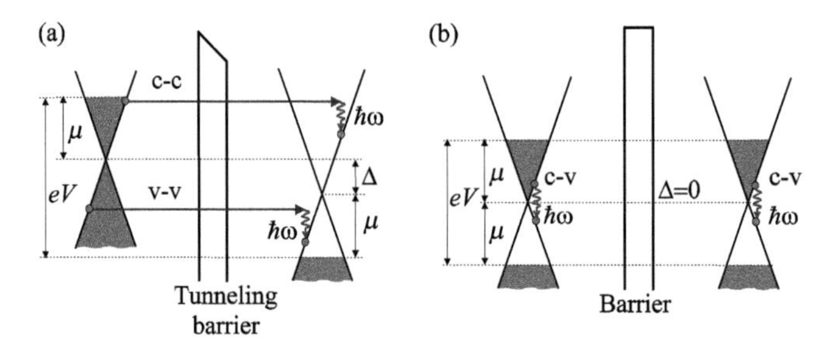

Figure 39.2 Band diagrams of laser structures with (a) inter-GL and (b) intra-GL radiative transitions. Wavy arrows indicate the inter-GL and intra-GL radiative $c-c$, $v-v$, and $c-v$ transitions in devices "a" and "b," respectively.

For comparison, we also consider the double GL-structure laser with the lateral injection of electrons and holes into each GL (and propagating toward each other along GL) and employing the interband transitions under the conditions of population inversion (see, for example, [9–12]). The device structure (device "b"), which, for the adequate comparison, is assumed to consist of two undoped GLs with the side p- and n-contacts forming a p-i-n-structure. One of the option to create a laser with such an active region is to use a dielectric waveguide. This device (device "b") and its band diagram are shown in Fig. 39.1b and Fig. 39.2b, respectively.

39.3 Inter-GL and Intra-GL Dynamic Conductivities

The electron and hole density in the pertinent GLs is given by

$$\Sigma = \Sigma_i + \frac{\kappa\Delta}{4\pi e^2 d}, \tag{39.1}$$

where Σ_i is the density of donors in the upper GL (n-GL) and acceptors in the lower GL (p-GL), κ is the dielectric constant of the barrier layer, and $e = |e|$ is the electron charge. In the case of "electrical" doping, Σ_i is the density of electrons and holes induced by the gate voltages $\pm V_g$ applied between the GLs and the metal strips of the MM waveguide: $\Sigma_i = æ V_g/4\pi e W$, where æ is the dielectric constant of the clad layers. In the case of relatively short GLs (such that the diffusion length of electrons and holes $L_d \gg L$) and strong degeneration of 2DEG and 2DHG ($\mu \gg k_B T$, k_B and T are the Boltzmann constant and the temperature, respectively), $\Delta = eV - 2\mu$ and $\mu \simeq \hbar v_W \sqrt{\pi\Sigma}$. Here $v_W \simeq 10^8$ cm/s is the characteristic velocity of electrons and holes in GLs. This yields

$$\Delta/e = V + V_0 - \sqrt{2VV_0 + V_0^2 + V_t^2}, \quad \mu = \frac{e}{2}\left(\sqrt{2VV_0 + V_0^2 + V_t^2} - V_0\right), \tag{39.2}$$

where $V_0 = \hbar^2 v_W^2 \kappa/2e^3 d$ and $V_t = 2\hbar v_W \sqrt{\pi\Sigma_i}/e$. At $V \leq V_t$, Eq. (39.2) yields $\Delta \leq 0$, while at $V > V_t$, $\Delta > 0$. Figure 39.2a shows the band diagram at the bias voltage $V > V_t$.

The inter-GL transitions assisted by the emission of photons with the polarization corresponding to the photon electric field perpendicular to GLs (along the axis z) conserve the electron momentum and, hence, do not involve scattering (resonant-tunneling photon-assisted transitions). Assuming $\kappa = 4$, $d = 4$ nm, and $\Sigma_i = 10^{12}$ cm^{-2}, one obtains $V_0 \simeq 136$ mV and $V_t \simeq 221$ mV. The quantities $\Delta = 5 - 10$ meV correspond to $V \simeq 229 - 237$ mV and $\mu \simeq 112.0 - 113.5$ meV.

The real part of $\sigma^{a}_{yy}(\omega)$ component of the tensor of the double-GL structure dynamic conductivity is given by [25] (see also [11, 26–28])

$$\mathrm{Re}\sigma^{a}_{yy}(\omega) \simeq 2\left(\frac{e^2}{4\hbar}\right)\left[2\exp\left(-\frac{\mu}{k_BT}\right)\sinh\left(\frac{\hbar\omega}{2k_BT}\right)+\frac{4\gamma\mu}{\pi(\hbar^2\omega^2+\gamma^2)}\right], \tag{39.3}$$

while the real part of $\sigma^{a}_{zz}(\omega)$ component can be presented as [18]

$$\mathrm{Re}\sigma^{a}_{zz}(\omega) = -\left(\frac{e^2}{\hbar}\right)\frac{2|z_{u,l}|^2\,\Sigma_i(1+\Delta/\Delta_i)\gamma\hbar\omega}{[\hbar^2(\omega-\omega_{max})^2+\gamma^2]}, \tag{39.4}$$

Here $z_{u,l} = \int \varphi^*_u(z)z\varphi_l(z)$, where $\varphi_u(z)$ and $\varphi_l(z)$ are the z-dependent factors of the wave functions in the upper and lower GLs, respectively, is the matrix element for the inter-GL transitions, $\gamma \simeq \hbar\nu$ is the relaxation broadening, ν is the collision frequency of electrons and holes, and $\Delta_i = 4\pi e^2 d\Sigma_i/\kappa$. The right-hand side of Eq. (39.3) is positive. It comprises the intra-GL interband and intraband (Drude) terms. The former is rather small at $\mu \gg k_BT$. The quantity $\hbar\omega_{max}$ in Eq. (39.4) corresponds to the maximum probability of the inter-GL radiative transitions:

$$\hbar\omega_{max} = \Delta - \frac{8\pi e^2|z_{u,l}|^2\,\Sigma_i(1+\Delta/\Delta_i)}{\kappa d} = \Delta + \Delta_{dep}, \tag{39.5}$$

The second term in the right-hand side of Eq. (39.5) describes the depolarization shift Δ_{dep} [29].

For the GL-structure with the intraband population inversion caused by the injection of both electrons and holes in each GL and with the suppressed the inter-GL transitions (device "b"), the conductivity tensor components are given by [9]

$$\mathrm{Re}\sigma^{b}_{xx}(\omega) = 2\left(\frac{e^2}{4\hbar}\right)\left[\tanh\left(\frac{\hbar\omega-2\mu}{4k_BT}\right)+\frac{4\gamma\mu}{\pi(\hbar^2\omega^2+\gamma^2)}\right],\quad \mathrm{Re}\sigma^{b}_{zz} = 0. \tag{39.6}$$

The sign of $\mathrm{Re}\sigma_{xx}^{b}(\omega)$ depends on the trade-off of the interband (negative) and intraband (positive) contributions. In a certain range of frequencies and Fermi energy, $\mathrm{Re}\sigma_{xx}^{b}(\omega) < 0$. This is to be exploited in the THz GL-based lasers.

39.4 Terahertz Gain and Gain-Overlap Factors

The THz gain for the TM-mode in device "a" and the TE-mode in device "b" is given by the following equation:

$$g^{a}(\omega) = -\frac{4\pi}{c\sqrt{\kappa}}[\mathrm{Re}\sigma_{yy}^{a}(\omega)\Gamma_{yy}^{a}(\omega) + \mathrm{Re}\sigma_{zz}^{a}(\omega)\Gamma_{zz}^{a}(\omega)] - \alpha^{a}(\omega), \quad (39.7)$$

$$g^{b}(\omega) = -\frac{4\pi}{c\sqrt{\kappa}}\mathrm{Re}\sigma_{xx}^{b}(\omega)\Gamma_{xx}^{b}(\omega) - \alpha^{b}(\omega), \quad (39.8)$$

where

$$\Gamma_{jj}^{a,b}(\omega) = \frac{\int_{-L}^{L} dx \left|E_{j}^{a,b}(x,0,\omega)\right|^{2}}{\int_{-L}^{L}\int_{-W}^{W} dxdz \left|E_{j}^{a,b}(x,z,\omega)\right|^{2}} \quad (j = x \cdot y, z), \quad (39.9)$$

are the gain-overlap factors, $E_{x}^{b}(x, z, \omega)$, $E_{y}^{a}(x, z)$ and $E_{z}^{a}(x, z, \omega)$ are the components of the electric field in the pertinent modes, and c is the speed of light in vacuum, and $\alpha^{a.b}(\omega) = 2\mathrm{Im}\, k_{y}(\omega)$ is the absorption coefficient of the propagating mode due to the losses in the pertinent waveguide. Considering Eqs. (39.3) and (39.4), Eq. (39.7) yields

$$g^{a}(\omega) = \frac{2\pi e^{2}}{\hbar c\sqrt{\kappa}}\left\{-\left[2\exp\left(-\frac{\mu}{k_{B}T}\right)\sinh\left(\frac{\hbar\omega}{2k_{B}T}\right) + \frac{4\gamma\mu}{\pi(\hbar^{2}\omega^{2}+\gamma^{2})}\right]\Gamma_{yy}^{a}(\omega)\right.$$
$$\left. + \frac{4|z_{u,l}|^{2}\,\Sigma_{i}(1+\Delta/\Delta_{i})\gamma\hbar\omega}{[\hbar^{2}(\omega-\omega_{max})^{2}+\gamma^{2}]}\Gamma_{zz}^{a}(\omega)\right\} - \alpha^{a}(\omega), \quad (39.10)$$

The solution of the Maxwell equations with the pertinent complex permittivity of the waveguide and metal strips yields the spatial distributions of the electric and magnetic fields, E_{x}, E_{y}, and E_{z} in the propagating modes and consequently, the gain-overlap factor. These equations were solved numerically using the effective index and transfer matrix methods (see, for instance [30, 31]). It was assumed that the cladding waveguide layers (above and below the double-GL structure and the inter-GL layers were made of hBN and WS_2, respectively. The hBN complex permittivity was extracted from [32] (æ ≃ $\varepsilon(x, 0, 0) \simeq 4$). It was also assumed that the metal strips in the MM waveguide are made of Au.

Figures 39.3a and b show the examples of the spatial distributions of the photon electric field components $|E_z^a(x, z, \omega)|$ and $|E_y^a(x, z, \omega)|$ in the TM mode of a MM waveguide with the sizes $L = W = 5$ μm in device "a." Figure 39.3c demonstrates the spatial distribution $|E_x^b(x, z, \omega)|$ in the dielectric waveguide. The obtained dependences correspond to $\omega/2\pi = 8$ THz. Figure 39.3d shows the spatial distributions of the electric field components $|E_z^a(x, 0, \omega)|$, $|E_y^a(x, 0, \omega)|$, and $|E_x^b(x, 0, \omega)|$ at the double-GL structure plane.

As follows from Fig. 39.3c, the TM-mode component E_y is small in comparison with the E_z component of this mode (compare the solid and dashed lines in Fig. 39.3d).

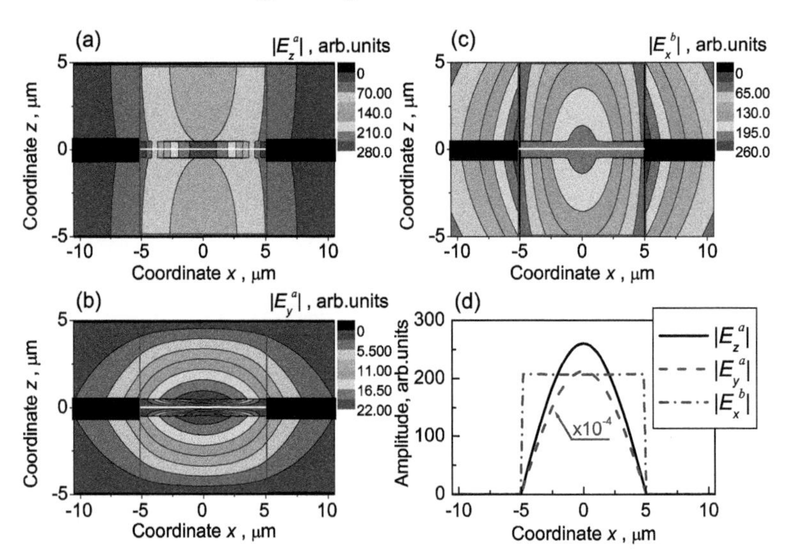

Figure 39.3 Spatial distributions of the photon electric field components: (a) $|E_z^a(x, z, \omega)|$, (b) $|E_y^a(x, z, \omega)|$ in the TM mode in MM waveguide (in device "a"), (c) $E_x^b(x, z, \omega)|$ in the TE mode (in device "b"), and (d) amplitudes of electric field components at GL plane ($z = 0$) for $\omega/2\pi = 8$ THz). White horizontal strips correspond to GLs with the barrier layer in between.

Figure 39.4a demonstrates the dependences of the inter-GL matrix element $|z_{u,l}|^2$ calculated using the Schrodinger equation for the wave functions accounting for the delta-function-like dependence in the z-direction of potentials of GLs separated by the barrier of width d. Following [32], we assume that the conduction band offset between GLs ans WS_2 barrier layer and the effective electron mass in the latter are equal to ΔE_g = 0.4 eV and m = 0.27m_0 (m_0 is the mass of free electron) [33]. Figure 39.4b shows the dependence of the depolarization shift Δ_{dep} on the energy separation between the Dirac points Δ calculated using the obtained values $|z_{u,l}|^2$ and Eq. (39.5). Due to the inter-GL population inversion, $\Delta_{dep} < 0$, and its value varies in a wide energy range with varying Δ.

As follows from Fig. 39.4a, sufficiently large values of $|z_{u,l}|^2$ can be achieved only in the double-GL structures with rather thin barrier layers (d about a few nm).

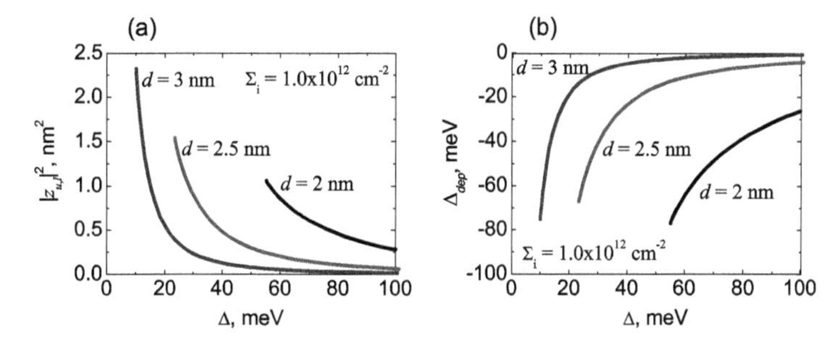

Figure 39.4 Dependences of (a) the inter-GL matrix element $|z_{u,l}|^2$ and (b) depolarization shift Δ_{dep} on the energy separation between the Dirac points Δ calculated for different values of spacing between GLs d.

Figure 39.5 shows the frequency dependences of the gain-overlap factors and the waveguide absorption coefficients calculated using Eq. (39.9) for different waveguide lateral sizes. As seen from Fig. 39.5, $\Gamma^a_{yy} \ll \Gamma^a_{zz}$. This is a consequence of a weak E_y component (see Fig. 39.3d). Hence, we can disregard the first term in right-hand side of Eq. (39.10) and reduce this equation to

$$g^a(\omega) = \left(\frac{8\pi e^2}{\hbar c\sqrt{\kappa}}\right)\frac{|z_{u,l}|^2\,\Sigma_i(1+\Delta/\Delta_i)\gamma\hbar\omega}{[\hbar^2(\omega-\omega_{max})^2+\gamma^2]}\Gamma^a_{zz}(\omega) - \alpha^a(\omega), \qquad (39.11)$$

This, in particular, implies that the Drude absorption of the TM mode in device "a" is insignificant.

As for device "b," taking into account Eq. (39.6), the THz gain can be calculated using the following equation:

$$g^{\mathrm{b}}(\omega)=\left(\frac{2\pi e^2}{\hbar c\sqrt{\kappa}}\right)\left[\tanh\left(\frac{2\mu-\hbar\omega}{4k_{\mathrm{B}}T}\right)-\frac{4\gamma\mu}{\pi(\hbar^2\omega^2+\gamma^2)}\right]\Gamma^{\mathrm{b}}_{\mathrm{xx}}(\omega)-\alpha^{\mathrm{b}}(\omega). \tag{39.12}$$

The obtained data for the MM waveguide absorption coefficients are in line with those found previously [22]. A significant increase in the absorption coefficient of the TM mode at relatively low frequencies seen in Fig. 39.5 is attributed to relatively small spacing between the metal strips in the MM waveguides under consideration compared to the radiation wavelength $\lambda = 2\pi c/\omega$.

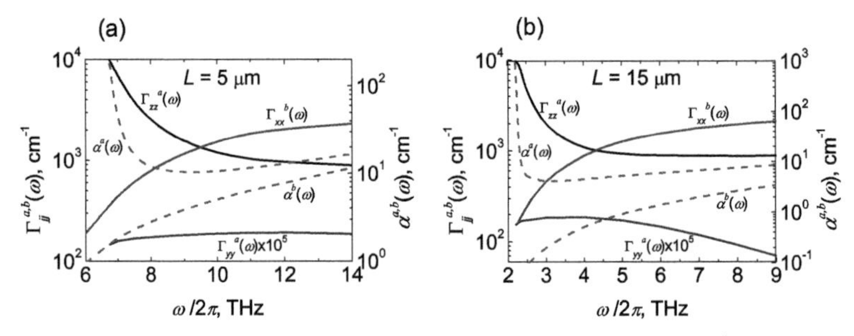

Figure 39.5 Frequency dependences of gain-overlap factors $\Gamma^{\mathrm{a,b}}_{\mathrm{jj}}(\omega)$ (solid lines) and waveguide absorption coefficients $\alpha^{\mathrm{a,b}}(\omega)$ (dashed lines): (a) $L = 5$ μm and $W = 5$ μm and (b) $L = 15$ μm and $W = 5$ μm.

39.5 Frequency and Voltage Dependences of the THz Gain

Figure 39.6 shows the frequency dependences of the THz gain $g^{\mathrm{a}}(\omega)$ a for double-GL structure (device "a") with $\Sigma_{\mathrm{i}} = 1 \times 10^{12}$ cm^{-2}, different values of the energy separation between the Dirac points Δ (i.e., different voltages V), and different waveguide geometrical parameters. The frequency dependences $g^{\mathrm{a}}(\omega)$ shown in Fig. 39.6 correspond to the relaxation broadening $\gamma = 1$ meV ($\nu = 1.6 \times 10^{12}$ s^{-1}) and doping level $\Sigma_{\mathrm{i}} = 1 \times 10^{12}$ cm^{-2}, i.e., $\mu_{\mathrm{i}} = \hbar v_{\mathrm{W}}\sqrt{\pi\Sigma_{\mathrm{i}}} \simeq 110$ meV. The actual value of the Fermi energy $\mu > \mu_{\mathrm{i}}$.

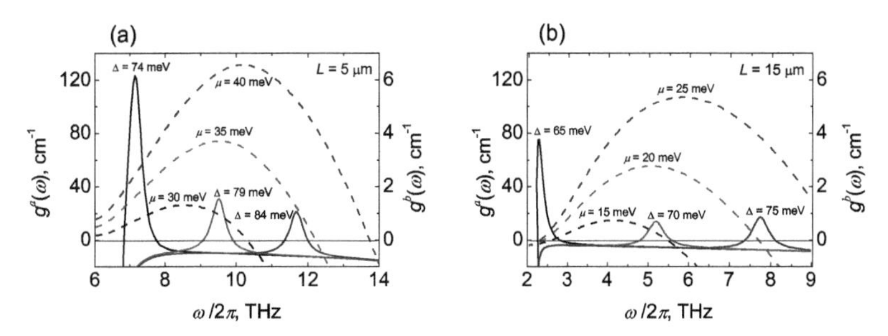

Figure 39.6 THz gain $g^a(\omega)$ versus frequency dependences (solid lines) for different values of energy separation between the Dirac points Δ and THz gain $g^b(\omega)$ (dashed lines) for different Fermi energies in GLs: (a) L = 5 μm and W = 5 μm and (b) L = 15 μm and W = 5 μm.

As follows from Fig. 39.6 (see solid lines), the THz gain as a function of the frequency exhibits a resonant behavior with the peak at $\omega = \omega_{\max}$. In this peak, $g^a(\omega)$ is positive and fairly large. The peak position is determined by Δ, in line with Eq. (39.5). The THz gain $g^a(\omega)$ outside the resonance is negative, practically coinciding with the absolute value of the MM waveguide absorption coefficient. Since, as follows from Eq. (39.2), Δ is a function of the bias voltage V, the position of the THz gain resonant peak, $\omega_{\max}$, of the THz gain is voltage tunable. The quantity Δ and, consequently, $\omega_{\max}$ depend on Σ_i. In device "a" with the "electrical" doping $\Sigma_i \propto V_g$, and, hence, in such a device, the position of the resonance can be also tuned by the gate voltage, V_g. Figure 39.7 demonstrates how the frequency $\omega_{\max}$ at which the THz gain $g^a(\omega)$ achieves a maximum value with varying bias voltage V and gate voltage V_g.

The height of the resonances $\max g^a(\omega)$ varies with increasing Δ. This can be attributed to the trade-off in a decrease in $|z_{u,l}|^2$ (see Fig. 39.4a) and in an increase of factor $(1 + \Delta/\Delta_i)$ in Eq. (39.11) with increasing Δ. The width of the resonant peaks is determined by parameter γ, i.e., by the collision frequencies of electrons and holes ν. At small ν the peaks in question can be very high, but large ν they are smeared.

It is worth noting that in the case of relatively large L, device "a" can exhibit rather high THz gain in a few THz range corresponding to the left peak in Fig. 39.6b. This is due to negligible Drude absorption in such a device structure.

For comparison, in Fig. 39.6 we also show the THz gain versus frequency calculated for device "b" with the same broadening parameter γ but different Fermi energies μ. In this device, the THz gain can be positive in a wide frequency range provided sufficiently large μ, i.e., sufficiently strong pumping [9–13]. However, as seen from Fig. 39.6, the THz gain markedly decreases (and becomes negative) when the frequency approaches the low end of the THz range. This is because of an increasing role of the Drude absorption with decreasing frequency of the TE mode.

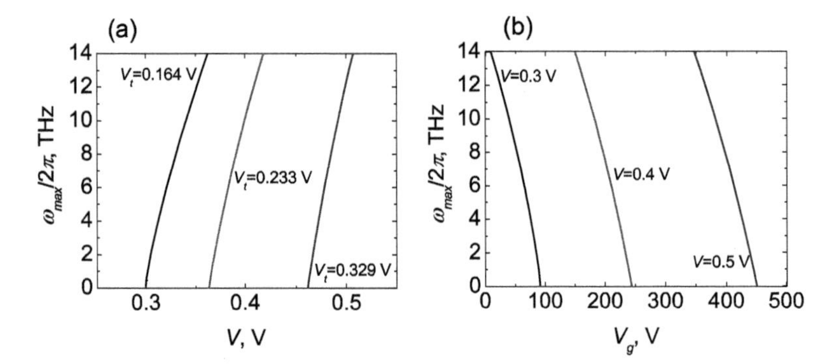

Figure 39.7 Tuning of frequency ω_{max} corresponding to peak of the THz gain $g^a(\omega)$ by (a) bias voltage V and (b) gate voltage V_g.

39.6 Discussion

The normal operation of the laser under consideration assumes sufficient densities of electrons in GLs. This is achieved by chemical or "electron" doping of GLs and the injection of electrons to the upper GL and holes to the lover GL. Reaching sufficient population in the entire GLs requires satisfying the condition $L \leq \mathscr{L}_i, \mathscr{L}_{eh}$ [34]. Here $\mathscr{L}_i \propto (D_i\tau_R^{inter})^{1/2}$ and $\mathscr{L}_{eh} \propto (D_{eh}\tau_R^{inter})^{1/2}$ are the diffusion lengths, $D_i \propto \nu_i^{-1}$ and $D_{eh} \propto \nu_{eh}^{-1}$ are the diffusion coefficients efficients, where ν_i and ν_{eh} are the electron and hole collision frequencies with impurities and with each other, respectively ($\nu_i + \nu_{eh} = \nu$), and τ_R^{inter} is the recombination time associated with the inter-GL processes. As shown [35], and references therein), the diffusion coefficient D_i in GLs can be very large. Considering that the electron-hole scattering and recombination in the double-GL structures might be weak due to the spatial separation of GLs with electrons and holes, one can expect that the condition $L \leq \mathscr{L}_{\rm i}, \mathscr{L}_{\rm eh}$ is satisfied at rather large values of L (tens on μm). Indeed, setting $D_{\rm eh} \sim D_{\rm i}$ = 40,000 cm^2/s [35] and $\tau_{\rm R}^{\rm inter} \sim 10^{-10}$ s, one obtains $L \leq \mathscr{L}_{\rm i} \sim \mathscr{L}_{\rm eh} \sim 20$ μm. The implementation of the "electrical" doping in the double-GL structures is beneficial because in such structures $\nu_{\rm i}$ can be markedly reduced.

In deriving Eq. (39.11), neglecting the processes of reabsorption of the emitted photons at the reverse transitions, we disregarded the deviation of factor $\{1 - \exp[(2\mu + \Delta)/k_{\rm B}T]\}$ from unity. This factor is really very close to unity at room and lower temperatures. Under these circumstances, the temperature dependence of the THz gain in device "a" is determined solely by the broadening factor γ. The value of the latter assumed above is realistic at room temperatures providing sufficient quality of GLs. At lower temperatures, this factor can be much smaller. This can leads to substantially higher and narrower peaks of the THz gain. The temperature dependence of the THz gain in device "b," is stronger because, apart from a marked change in the Drude absorption with the temperature variation of γ, the temperature variation also affects the population inversion [the first term in the right-hand side of Eq. (39.12)].

39.7 Conclusion

We have developed a device model for the proposed injection THz laser with an active region consisting of a double-GL structure with a barrier layer placed within a MM waveguide. The operation of this laser is associated with the inter-GL (intraband) tunneling photon-assisted transitions. Applying the model to the device structure with WS_2 inter-GL layer and hBN clad layers of the MM waveguide, we have derived general formulas for the dynamic conductivity of the double-GL structure and for the THz gain, estimated the values of the matrix element of the inter-GL radiative transitions for different thicknesses of the barrier layer, found the spatial distributions of the photon electric field in the TM mode propagating along the MM waveguide, and calculated the frequency dependences of the gain-overlap factor, the waveguide absorption coefficient, and the THz gain. For comparison, we analyzed the characteristics of a THz laser based on two GLs with simultaneous injection in each GL using the intra-GL (interband) radiative transitions (proposed and studied previously [9–13, 36, 37]). The results show that:

(i) The realization of THz lasing in the devices with the double-GL structure and WS_2 and hBN layers exploiting the inter-GL radiative transitions is feasible because of the possibility to achieve sufficiently high values of the THz gain;

(ii) Due to resonant-tunneling nature of the radiative transitions in the THz laser under consideration, the spectrum of the THz laser radiation can exhibit sharp maxima in the range from a few THz to a dozen THz and can be tuned by the applied voltages;

(iii) The double-GL inter-GL THz laser is rather insensitive to the temperature;

(iv) The proposed structures should exhibit advantages over the lasers utilizing the intra-GL transitions due to weaker temperature sensitivity, practically absent Drude absorption (crucial at the low end of the THz range), and possibly, higher injection efficiency;

(v) Both inter-GL and intra-GL THz lasers can be useful to cover the frequency range of several THz, in which the operation of A_3B_5 quantum cascade lasers is hampered by optical phonons.

Acknowledgments

This work was supported by the Japan Society for promotion of Science (Grant-in-Aid for Specially Promoting Research #23000008), Japan. The work by A. D. was also supported by the Russian Foundation of Basic Research and the Dynasty Foundation, Russia. The work at RPI was supported by the US Army Cooperative Research Agreement (Program Manager Dr. Meredith Reed). V. R. is grateful to M. J. Martin for sending recent Ref. [35].

References

1. M. Liu, X. Yin, and X. Zhang, Double-layer graphene optical modulator, *Nano Lett.*, **12**, 1482–1485 (2012).

2. L. Britnell, R. V. Gorbachev, R. Jalil, B.D . Belle, F. Shedin, A. Mishenko, T. Georgiou, M. I. Katsnelson, L. Eaves, S. V. Morozov, N. M. R. Peres, J.

Leist, A. K. Geim, K. S. Novoselov, and L. A. Ponomarenko, Field-effect tunneling transistor based on vertical graphene heterostructures, *Science*, **335**, 947–950 (2012).

3. T. Georgiou, R. Jalil, B. D. Bellee, L. Britnell, R. V. Gorbachev, S. V. Morozov, Y.-J. Kim, A. Cholinia, S. J. Haigh, O. Makarovsky, L. Eaves, L. A. Ponomarenko, A. K. Geim, K. S. Nonoselov, and A. Mishchenko, Vertical field-effect transistor based on graphene-WS2 heterostructures for flexible and transparent electronics, *Nat. Nanotechnol.*, **7**, 100–103 (2013).
4. L. Britnell, R. V. Gorbachev, A. K. Geim, L. A. Ponomarenko, A. Mishchenko, M. T. Greenaway, T. M. Fromhold, K. S. Novoselov, and L. Eaves, Resonant tunneling and negative differential conductance in graphene transistors, *Nat. Commun.*, **4**, 1794–1799 (2013).
5. V. Ryzhii, T. Otsuji, M. Ryzhii, V. G. Leiman, S. O. Yurchenko, V. Mitin, and M. S. Shur, Effect of plasma resonances on dynamic characteristics of double-graphene layer optical modulators, *J. Appl. Phys.*, **112**, 104507 (2012).
6. V. Ryzhii, T. Otsuji, M. Ryzhii, and M. S. Shur, Double graphene-layer plasma resonances terahertz detector, *J. Phys. D: Appl. Phys.*, **45**, 302001 (2012).
7. V. Ryzhii, A. Satou, T. Otsuji, M. Ryzhii, V. Mitin, and M. S. Shur, Dynamic effects in double-graphene-layer structures with inter-layer resonant-tunneling negative differential conductivity, *J. Phys. D: Appl. Phys.*, **46**, 315107 (2013).
8. V. Ryzhii, M. Ryzhii, V. Mitin, M. S. Shur, A. Satou, and T. Otsuji, Terahertz photomixing using plasma resonances in double-graphene-layerstructures, *J. Appl. Phys.*, **113**, 174506 (2013).
9. V. Ryzhii, M. Ryzhii, and T. Otsuji, Negative dynamic conductivity ofgraphene with optical pumping, *J. Appl. Phys.*, **101**, 083114 (2007).
10. A. A. Dubinov, V. Ya. Aleshkin. M. Ryzhii, T. Otsuji, and V. Ryzhii, Terahertz laser with optically pumped graphene layers and Fabry-Perot resonator, *Appl. Phys. Express*, **2**, 092301 (2009).
11. V. Ryzhii, M. Ryzhii, A. Satou, T. Otsuji, A. A. Dubinov, and V. Ya. Aleshkin, Feasibility of terahertz lasing in optically pumped expitaxial multiple graphene layer structures, *J. Appl. Phys.*, **106**, 084507 (2009).
12. V. Ryzhii, A. A. Dubinov, T. Otsuji, V. Mitin, and M. S. Shur, Terahertz lasers based on optically pumped multiple graphene structures with slot-line and dielectric waveguides, *J. Appl. Phys.*, **107**, 054505 (2010).

13. V. Ryzhii, M. Ryzhii, V. Mitin, and T. Otsuji, Toward the creation ofterahertz graphene injection laser, *J. Appl. Phys.*, **110**, 094503 (2011).

14. S. Boubanga-Tombet, S. Chan, T. Watanabe, A. Satou, V. Ryzhii, and T. Otsuji, Ultrafast carrier dynamics and terahertz emission in optically pumped graphene at room temperature, *Phys. Rev. B*, **85**, 035443 (2012).

15. T. Watanabe, T. Fukushima, Y. Yabe, S. A. Boubanga-Tombet, A. Satou, A. A. Dubinov, V. Ya. Aleshkin, V. Mitin, V. Ryzhii, and T. Otsuji, The gain enhancement effect of surface plasmon-polaritons on terahertz stimulated emission in optically pumped monolayer graphene, *New J. Phys.*, **15**, 075003 (2013).

16. T. Otsuji, S. A. Boubanga Tombet, A. Satou, M. Ryzhii, and V. Ryzhii, Terahertz-wave generation using graphene – toward new types of terahertz lasers, *IEEE J. Sel. Top. Quantum Electron.*, **19**, 8400209 (2013).

17. A. Tredicucci and M. S. Vitiello, Device concepts for graphene-based terahertz photonics, *IEEE J. Sel. Top. Quantum Electron.*, **20**, 8500109 (2014).

18. V. Ryzhii, A. A. Dubinov, V. Ya. Aleshkin, M. Ryzhii, and T. Otsuji, Injection terahertz laser using the resonant inter-layer radiative transitions in double-graphene-layer structure, *Appl. Phys. Lett.*, **103**, 163507 (2013).

19. H. Kanaya, H. Shibayama, R. Sogabe, S. Suzuki, and M. Asada, Fundamental oscillation up to 1.31 THz in resonant tunneling diodes with thin well and barriers, *Appl. Phys. Express*, **5**, 124101 (2012).

20. B. S. Williams, Terahertz quantum-cascade lasers, *Nat. Photonics*, **1**, 517 (2007).

21. K. Unterrainer, R. Colombelli, C. Gmachl, F. Capasso, H. Y. Hwang, D. L. Sivco, and A. Y. Cho, Quantum cascade lasers with double metal-semiconductor waveguide resonators, *Appl. Phys. Lett.*, **80**, 3060 (2002).

22. M. A. Belkin, J. A. Fan, S. Hormoz, F. Capasso, S. P. Khanna, M. Lachab, A. G. Davies, and E. Linfield, Terahertz quantum cascade lasers with copper metal-metal waveguides operating up to 178 K, *Opt. Express*, **16**, 3242–3248 (2008).

23. R. M. Feenstra, D. Jena, and G. Gu, Single-particle tunneling in doped graphene-insulator-graphene junctions, *J. Appl. Phys.*, **111**, 043711 (2012).

24. F. T. Vasko, Resonant and nondissipative tunneling in independently contacted graphene structures, *Phys. Rev. B*, **87**, 075424 (2013).
25. L. A. Falkovsky and A. A. Varlamov, Space-time dispersion of graphene conductivity, *Eur. Phys. J. B*, **56**, 281–284 (2007).
26. L. A. Falkovsky and S. S. Pershoguba, Optical far-infrared properties of a graphene monolayer and multilayer, *Phys. Rev. B*, **76**, 153410 (2007).
27. F. Carosella, C. Ndebeka-Bandou, R. Ferreira, E. Dupont, K. Unterrainer, G. Strasser, A. Wacker, and G. Bastard, Free carrier absorption in quantum cascade structures, *Phys. Rev. B*, **85**, 085310 (2012).
28. F. T. Vasko, V. V. Mitin, V. Ryzhii, and T. Otsuji, Interplay of intra- and interband absorption in a disordered graphene, *Phys. Rev. B*, **86**, 235424 (2012).
29. F. T. Vasko and A. V. Kuznetsov, *Electronic States and Optical Transitions in Semiconductor Heterostructures* (Springer, 1999).
30. K. J. Ebeling, *Integrated Optoelectronics: Waveguide Optics, Photonics, Semiconductors* (Springer, 1993).
31. M. Born and E. Wolf, *Principles of Optics: Electromagnetic Theory of Propagation, Interference and Diffraction of Light* (Pergamon, 1964).
32. D. M. Hoffman, G. L. Doll, and P. C. Eklund, Optical properties of pyrolytic boron nitride in the energy range 0.05–10 eV, *Phys. Rev. B*, **30**, 6051–6056 (1984).
33. H. Shi, H. Pan, Y.-W. Zhang, and B. Yakobson, Quasiparticle band structures and optical properties of strained monolayer MoS_2 and WS_2, *Phys. Rev. B*, **87**, 155304 (2013).
34. V. Ryzhii, I. Semenikhin, M. Ryzhii, D. Svintsov, V. Vyurkov, A. Satou, and T. Otsuji, Double injection in graphene p-i-n structures, *J. Appl. Phys.*, **113**, 244505 (2013).
35. R. Rengel and M. J. Martin, Diffusion coefficient, correlation function, and power spectral density of velocity fluctuations in monolayer graphene, *J. Appl. Phys.*, **114**, 143702 (2013).
36. M. Ryzhii and V. Ryzhii, Injection and population inversion in electrically induced p-n junction in graphene with split gates, *Jpn. J. Appl. Phys.*, **46**, L151–L153 (2007).
37. M. Ryzhii and V. Ryzhii, Population inversion in optically and electrically pumped graphene, *Physica E*, **40**, 317–320 (2007).

Chapter 40

Surface-Plasmons Lasing in Double-Graphene-Layer Structures*

A. A. Dubinov,[a,b] V. Ya. Aleshkin,[b] V. Ryzhii,[a,c] M. S. Shur,[d] and T. Otsuji[a]

[a]*Research Institute for Electrical Communication, Tohoku University, Sendai 980-8577, Japan*

[b]*Institute for Physics of Microstructures of Russian Academy of Sciences, and Lobachevsky State University of Nizhny Novgorod, Nizhny Novgorod 603950, Russia*

[c]*Center for Photonics and Infrared Engineering, Bauman Moscow State Technical University, Moscow 105005, Russia*

[d]*Department of Electrical, Electronics, and System Engineering, Rensselaer Polytechnic Institute, Troy, New York 12180, USA*

sanya@ipm.sci-nnov.ru

We consider the concept of injection terahertz lasers based on double-graphene-layer (double-GL) structures with metal surface-plasmon waveguide and study the conditions of their operation. The

---

*Reprinted with permission from A. A. Dubinov, V. Ya. Aleshkin, V. Ryzhii, M. S. Shur, and T. Otsuji (2014). Surface-plasmons lasing in double-graphene-layer structures, *J. Appl. Phys.*, **115**, 044511. Copyright © 2014 AIP Publishing LLC.

Graphene-Based Terahertz Electronics and Plasmonics: Detector and Emitter Concepts
Edited by Vladimir Mitin, Taiichi Otsuji, and Victor Ryzhii

ISBN 978-981-4800-75-4 (Hardcover), 978-0-429-32839-8 (eBook)
www.jennystanford.com

laser under consideration exploits the resonant radiative transitions between GLs. This enables the double-GL laser room temperature operation and the possibility of voltage tuning of the emission spectrum. We compare the characteristics of the double-GL lasers with the metal surface-plasmon waveguides with those of such laser with the metal-metal waveguides.

40.1 Introduction

There is a strong demand in compact and effective sources of terahertz (THz) radiation [1–7]. As demonstrated recently [8–10], resonant-tunneling diodes (RTDs) can cover the THz gap. However, the power of the generated THz radiation is still rather low. Quantum cascade lasers (QCLs) can be operated in the range of a few THz (Refs. [11–14])—but at lower temperatures. Another limitation of the QCL operation in the range of frequencies 5–12 THz is because GaAs and InP semiconductors exhibit a strong phonon absorption in the THz range [15]. Therefore, the search of new concepts of THz lasers based on non-traditional semiconductor heterostructures is indispensable. Various single-graphene layer (GL) or multiple-GLs (twisted, non-Bernal stacked GLs or GLs separated by relatively thick barrier layers) structures have been considered as candidates for the THz lasers. The gapless energy spectrum of GLs enables the interband radiative transitions involving photons with the frequencies from a few to dozens THz. Many publications discussed the ideas of GL-based THz lasers [16] utilizing the inter-band population inversion due to the optical or injection pumping, and different versions of such lasers (see, for example, Refs. [17–21]). Recent experimental results [22, 23] (see also Refs. [7, 24, 25]) demonstrated the feasibility of GL-based THz lasers realization. The lasing of the TE-mode with the in-plane direction of the photon electric field [17, 19, 20] is possible due to the domination of the interband photon emission over the intraband Drude absorption (under the conditions of the interband population inversion). The latter becomes relatively strong at the low end of the THz range. This can be an obstacle in the GL-lasers operation in the range of a few THz.

Applying bias between the two GLs in the double-GL structures, one can achieve the inter-GL population inversion. The double-GL structures with narrow inter-GL barrier Boron Nitride (hBN),

Tungsten Sulfide (WS_2), and other barrier layers have recently been fabricated and explored for applications [26–35]. Recently [36, 37], we proposed to use the resonant radiative inter-GL transitions in double-GL structures with a sufficiently thin barrier between GLs and with the side contact enabling the lateral injection of electrons and holes. One can expect the THz lasing of the TM-mode in such structures provided they are supplied with proper waveguides. We considered the double-GL structures with the side injecting contacts and the metal-metal waveguide [36, 37]. The latter is rather effective in QCLs [38, 39]. Potential advantages of the inter-GL structures include the TM-mode generation with much smaller Drude losses than in the TE mode (because of a very small lateral component of the photon electric field in the TM-mode). Other advantages include a weaker temperature sensitivity and the possibility of the spectrum tuning by the applied bias. The lateral injection in the devices in question allows replacing metal-metal waveguides with other waveguiding systems, including one metal plane with a dielectric slab (surface-plasmon waveguide) [40]. This can provide easy fabrication and better THz radiation output characteristics.

In this chapter, we develop a simplified device model of the THz laser exploiting the inter-GL radiative transitions in the double-GL structure with a tunneling barrier layer. We primarily consider the device structure shown in Fig. 40.1. It consists of two independently contacted GLs doped by donors (n-GL) and acceptors (p-GL), respectively, and separated by a narrow tunneling barrier layer of thickness d (about a few nanometers). GLs are also separated by narrow layers of thickness f (about a few nanometers of the same material) from other layers of the structure. The doping leads to the formation of the two-dimensional electron (2DEG) and the two-dimensional hole (2DHG) gases in the n-GL (the top GL) and the p-GL (the bottom GL), respectively. The bias voltage V between the side contacts induces the extra charges of opposite polarities in the GLs and the electric field between GLs. This causes the inter-GL population inversion between the conduction bands and between the valence bands. Consequently, there are conduction and valence intraband transitions accompanied by the photon emission [36, 37]. Thus, the double-GL structure can serve as the laser active region with the lateral injection pumping. The device structure is sandwiched by the cladding insulating layers (with thickness $W \gg w \gg f \gg d$) on a metal substrate, which constitutes a surface plasmon waveguide for the TM-modes propagating in the direction perpendicular to the x-z plane. We also compare the inter-GL THz lasers with lasers using the surface-plasmon waveguides considered here and with the metal-metal waveguides [38, 39].

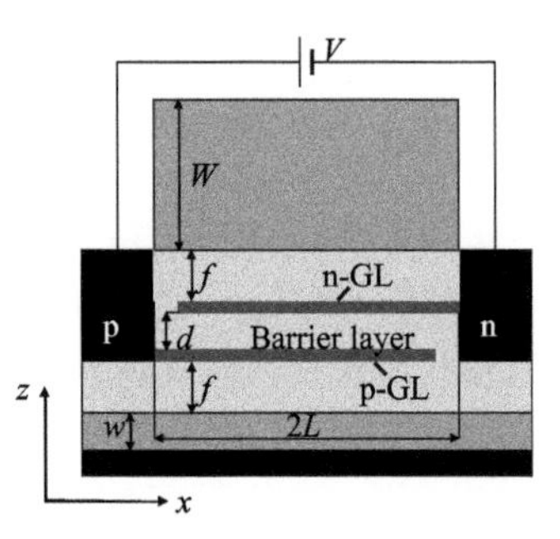

Figure 40.1 Schematic view of double-GL laser cross-sections with the side monopolar injection to each independently contacted GL, tunneling barrier layer, and surface plasmon waveguide. The laser active region is the double-GL structure and the surface plasmons propagate in the y-direction.

40.2 Dynamic Conductivity Tensor

As in the previous papers [26–35], we disregarded possible effect of the inter-GL layer on the energy spectrum of GLs. The real part of the component $\sigma_{yy}(\omega)$ of the dynamic conductivity tensor of the double-GL structure under consideration at the signal frequency ω comprises the intraband (Drude) and intra-GL interband terms (see, for example, Refs. [41, 42])

$$\mathrm{Re}\,\sigma_{yy}(\omega) \simeq 4\left(\frac{e^2}{4\hbar}\right) \times\left[\frac{2\gamma\mu}{\pi(\hbar^2\omega^2+\gamma^2)}+\exp\left(-\frac{\mu}{k_B T}\right)\sinh\left(\frac{\hbar\omega}{2k_B T}\right)\right], \quad (40.1)$$

where $e = |e|$, $\hbar$, and k_B are the electron charge and the reduced Planck and Boltzmann constants, respectively, μ is the Fermi energy, $\gamma = \hbar\nu$ is the relaxation broadening, ν is the collision frequency of electrons and holes. Both the intraband and interband intra-GL contributions to Re $\sigma_{yy}(\omega)$ are positive because the quasi-equilibrium energy distributions in each GL (characterized by the Fermi energy μ and the temperature T) do not exhibit intra-GL population inversion. For simplicity, here and in the following we assume a sufficiently strong degeneration of 2DEG and 2DHG.

The inter-GL transitions assisted by the emission of photons with the polarization corresponding to the photon electric field perpendicular to the GLs (along the z-axis) conserve the electron momentum and, hence, do not involve scattering (resonant-tunneling photon-assisted transitions). Therefore, the component Re $\sigma_{zz}(\omega)$ can be presented as [36, 37]

$$\mathrm{Re}\,\sigma_{zz}(\omega) = -\left(\frac{e^2}{\hbar}\right)\frac{2|z_{u,l}|^2\,\Sigma_i(1+\Delta/\Delta_i)\gamma\hbar\omega}{[\hbar^2(\omega-\omega_{max})^2+\gamma^2]}. \quad (40.2)$$

Here, $\Delta_i = 4\pi e^2 d\Sigma_i/\kappa$, Δ is the energy separation between the GL Dirac points, Σ_i is the density of donors in the upper GL (n-GL) and acceptors in the lower GL (p-GL), κ is the dielectric constant of the barrier layer, and $z_{u,l} = \int_{-\infty}^{\infty} \varphi_u^*(z) z \varphi_l(z) dz$ is the matrix element for the inter-GL transitions, where $\varphi_u(z)$ and $\varphi_l(z)$ are the z-dependent envelope wave functions in the upper and lower GLs, respectively. The quantities Δ and $\hbar\omega_{max}$ are given by

$$\Delta = e(V + V_0 - \sqrt{V_0^2 + V_t^2 + 2VV_0}\,) \quad (40.3)$$

$$\hbar\omega_{max} = \Delta + \Delta_{dep}. \quad (40.4)$$

Here,

$$\Delta_{dep} = -\frac{8\pi e^2\,|z_{u,l}|^2\,\Sigma_i(1+\Delta/\Delta_i)}{\kappa d} \quad (40.5)$$

is the depolarization shift [43, 44] $V_0 = \hbar^2 v_W^2 \kappa/2e^2 d$, $V_t = 2\hbar v_W \sqrt{\pi\Sigma_i}/e$, where $v_W \simeq 10^8$ cm/s. It is worth noting that due to the inter-GL population inversion, $\Delta_{dep} < 0$. If $\kappa = 4$, $d = 2$ nm, and $\Sigma_i = 3 \times 10^{11}$ cm^{-2}, the characteristic voltages $V_0 = 299$ mV and $V_t = 127$ mV. Assuming $V = 265 - 322$ mV and $\mu = 107 - 124$ meV, one can obtain $\Delta = 50 - 75$ meV.

In the simplest approach, one can consider each GL as a delta-layer separated by a barrier layer. We assume that this layer is made of WS_2. Following Ref. [28], we choose the conduction band offset between a GL and WS_2 and the effective electron mass in WS_2 to be equal to $\Delta_c = 0.4$ eV and $m = 0.27$ of the free electron mass, respectively [45]. To find $|z_{u,l}|^2$, we solved the Schrodinger equation for $\varphi_{u,l}(z) = \Psi_{u,l} \Big/ \sqrt{\int_{-\infty}^{\infty} |\Psi_{u,l}|^2 dz}$

$$-\frac{\hbar^2}{2m}\frac{d^2\Psi_{u,l}}{dz^2} + \xi\delta(z-d/2)\Psi_{u,l} + \tilde{\xi}\delta(z+d/2)\Psi_{u,l}$$
$$= \varepsilon_{u,l}\Psi_{u,l}, \tag{40.6}$$

where $\xi = -\hbar\sqrt{2\Delta_c/m}$ and $\tilde{\xi}$ is the parameter, which can be found from the equation $\Delta = \varepsilon_u - \varepsilon_l$. To solve the Schrodinger equation in our model, we search for the envelope wave function in the form

$$\Psi_{u,l}(z) = \begin{cases} \exp[\theta(\varepsilon_{u,l})(z+d/2)], & z < -d/2 \\ B(\varepsilon_{u,l})\exp[\theta(\varepsilon_{u,l})z] \\ +C(\varepsilon_{u,l})\exp[-\theta(\varepsilon_{u,l})z], & -d/2 \le z < d/2 \\ D(\varepsilon_{u,l})\exp[-\theta(\varepsilon_{u,l})(z-d/2)], & z \ge d/2 \end{cases}$$

where $\theta(\varepsilon_{u,l}) = \sqrt{2m|\varepsilon_{u,l}|}/\hbar$, ε_u and ε_l are the size quantization energies in a system of two tunnel-coupled delta layers, and

$$B(\varepsilon_{u,l}) = \frac{\xi + \eta(\varepsilon_{u,l})}{\eta(\varepsilon_{u,l})\exp[-\theta(\varepsilon_{u,l})d/2]},$$

$$C(\varepsilon_{u,l}) = \exp[-\theta(\varepsilon_{u,l})d/2] - B(\varepsilon_{u,l})\exp[-\theta(\varepsilon_{u,l})d],$$

$$D(\varepsilon_{u,l}) = B(\varepsilon_{u,l})\exp[\theta(\varepsilon_{u,l})d/2] + C(\varepsilon_{u,l})\exp[-\theta(\varepsilon_{u,l})d/2],$$

$$\eta(\varepsilon_{u,l}) = \hbar^2\theta(\varepsilon_{u,l})/m.$$

Then, ε_u and ε_l can be found from the following equation:

$$\xi\tilde{\xi}\exp[-2\theta(\varepsilon_{u,l})d] = [\xi + \eta(\varepsilon_{u,l})][\tilde{\xi} + \eta(\varepsilon_{u,l})]. \tag{40.7}$$

As seen from Eq. (40.3), Δ is a function of the bias voltage V. Therefore, the position of the resonant peak $\omega = \omega_{max}$ of the real part of the dynamic conductivity component Re $\sigma_{zz}(\omega)$ component is voltage tunable [see Eq. (40.4)]. The quantity Δ and, consequently, ω_{max} depend on d and Σ_i. Figure 40.2 demonstrates how the frequency ω_{max} varies with varying applied voltage V. The frequency ω_{max}

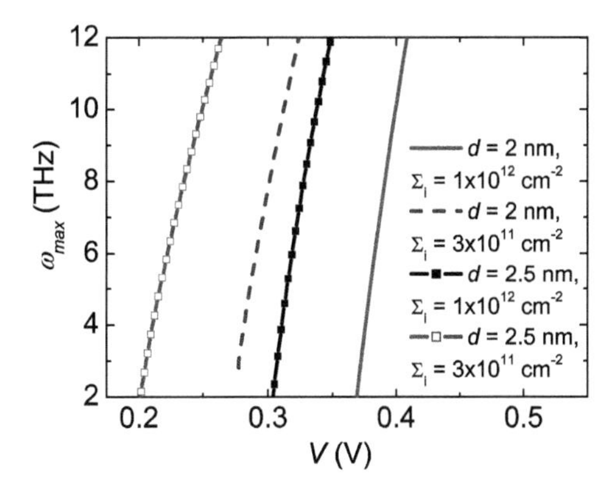

Figure 40.2 Tuning of frequency ω_{max} by bias voltage V.

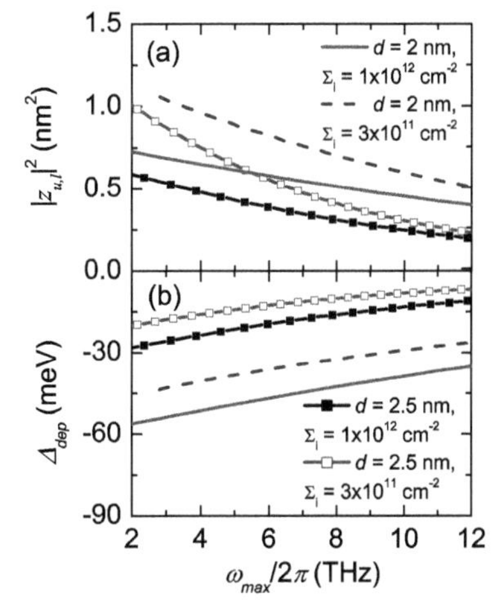

Figure 40.3 Dependences of (a) the square of the inter-GL matrix element $|z_{u,l}|^2$ and (b) depolarization shift Δ_{dep} on the ω_{max} calculated for different values of spacing between GLs d and the density of donors and acceptors in GLs Σ_i.

increases with increasing voltage V and can be relatively large. The maximum frequency is determined by the breakdown voltage, which is, by our estimate, about 1 V for nanometer thicknesses of materials such as hBN and WS_2. Figure 40.3 shows the dependences of the square of the inter-GL matrix element $|z_{u,l}|^2$ and the depolarization shift Δ_{dep} on ω_{max} calculated using Eqs. (40.5)–(40.7). As follows from Fig. 40.3a, sufficiently large values of $|z_{u,l}|^2$ can be achieved only in the double-GL structures with rather thin barrier layers (d is of the order of a few nm). Due to the inter-GL population inversion, $\Delta_{dep} < 0$ (see Fig. 40.3b), and its value varies in a wide frequency range with varying ω_{max}. The modulus of the depolarization shift Δ_{dep} increases significantly with decreasing barrier layer thickness d and with increasing the density of donors and acceptors in GLs, Σ_i.

40.3 Electric Field Distributions, Gain-Overlap Factor, and Modal Gain

The THz modal gain for the TM-mode in the device under consideration is given by the following expression:

$$\mathcal{G}(\omega) = -\frac{4\pi}{cn_{eff}(\omega)}[\sigma_{zz}(\omega)\Gamma_z(\omega) + \sigma_{yy}(\omega)\Gamma_y(\omega)], \qquad (40.8)$$

where

$$\Gamma_{z,y}(\omega) = \frac{\int_{-L}^{L} \left|E_{z,y}(x,0,\omega)\right|^2 dx}{\int_{-\infty}^{\infty}\int_{-\infty}^{\infty} \left|E_{z,y}(x,z,\omega)\right|^2 dxdz}$$

is the gain-overlap factors, $n_{eff}(\omega)$ is the effective refractive index of the mode, $E_z(x, z, \omega)$ and $E_y(x, z, \omega)$ are the components of the electric field in TM-mode, and c is the light speed in vacuum. The solution of the Maxwell equations with the pertinent complex permittivity of the surface plasmon waveguide and metal strips yields the spatial distributions of the electric fields, in the propagating mode and, consequently, the gain-overlap factors. These equations were solved numerically using the effective index and transfer matrix methods (see, for instance, Refs. [46, 47]). It was assumed that the cladding waveguide layers (above and below the double-GL structure) and the inter-GL layer were made of hBN and WS , respectively. The hBN complex permittivity was taken from Ref. [48]. It was also assumed that the metal strips and the substrate in the surface plasmon waveguide are made of Au. Figures 40.4a and b show the examples of the spatial distributions of the photon electric field components $E_z(x, z, \omega)$ and $E_y(x, z, \omega)$ in the TM mode of a surface plasmon waveguide with the sizes $L = W = 10$ μm, $w = 100$ nm, $f = 10$ nm in device from Fig. 40.1. The obtained dependences correspond to $\omega/2\pi = 8$ THz. Figure 40.5 shows the spatial distributions of the electric field components $E_z(x, 0, \omega)$ and $E_y(x, 0, \omega)$, at the double-GL structure plane ($z = 0$). As follows from Fig. 40.5, the TM-mode component $E_y(x, 0, \omega)$ is small in comparison with the $E_z(x, 0, \omega)$-component of this mode (compare the solid and dashed lines in Fig. 40.5).

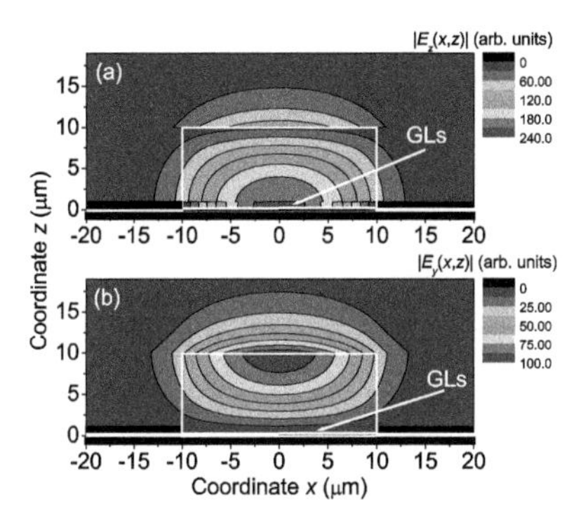

Figure 40.4 Spatial distributions of the photon electric field components: (a) $|E_z(x, z, \omega)|$, (b) $|E_y(x, z, \omega)|$ in the TM mode in surface plasmon waveguide for $\omega/2\pi = 8$ THz.

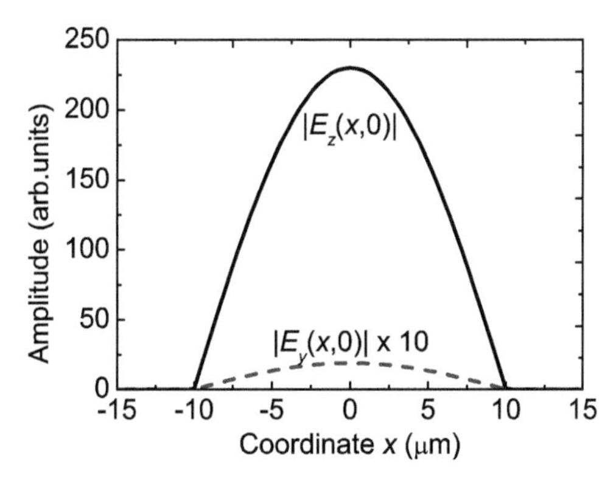

Figure 40.5 Spatial distributions of the amplitudes of electric field components at the GL plane ($z = 0$) for $\omega/2\pi = 8$ THz.

Figure 40.6 shows the frequency dependences of the gain-overlap factors. As seen from Fig. 40.6, $\Gamma_y(\omega) \ll \Gamma_z(\omega)$. This is the consequence of a weak $E_y(x, 0, \omega)$-component of the electric field of this mode (see Fig. 40.5). This, in particular, implies that the Drude absorption of the TM mode in our device is insignificant. Hence, we can disregard the second term in right-hand side of Eq. (40.8) and reduce this equation to

$$\mathcal{G}(\omega) \simeq \left[\frac{8\pi e^2}{\hbar c n_{\text{eff}}(\omega)}\right] \frac{|z_{\text{u,l}}|^2 \Sigma_{\text{i}}(1+\Delta/\Delta_{\text{i}})\gamma\hbar\omega}{[\hbar^2(\omega-\omega_{\max})^2+\gamma^2]} \Gamma_z(\omega). \quad (40.9)$$

Figure 40.7a shows the frequency dependences of the THz modal gain $\mathcal{G}(\omega)$ for a double-GL structure with $\Sigma_i = 3 \times 10^{11}$ cm^{-2}, $d = 2$ nm, different voltages V, and different values of the relaxation broadening constant γ. As follows from Fig. 40.7a (compare the solid and dashed lines), the modal gain as a function of the frequency exhibits a resonant behavior with the peak at $\omega = \omega_{\max}$. In this peak, $\mathcal{G}(\omega)$ is fairly large. The peak position is determined by the bias voltage V, as follows from Eqs. (40.2) and (40.5), and also depends on Σ_i and d. The height of the resonances max $\mathcal{G}(\omega)$ increases with increasing $\omega_{\max}$ and V. The width of the resonant peaks is determined by parameter γ, i.e., by the collision frequencies of electrons and holes ν. At small ν, the peaks can be very high, but they are smeared at large ν. Figure 40.7b shows the modal gain dependence on $\omega = \omega_{\max}$ for two values of the relaxation broadening γ. For the waveguide, Eq. (40.9) yields the threshold length $L_{\text{th}}(\omega)$ corresponding to the lasing threshold:

$$L_{\text{th}}(\omega) = \frac{\ln[1/R(\omega)]}{\mathcal{G}(\omega)-\alpha(\omega)}, \quad (40.10)$$

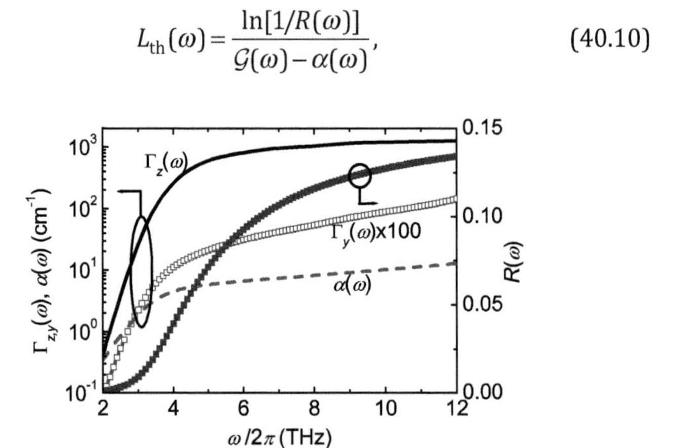

Figure 40.6 Frequency dependences of gain-overlap factors $\Gamma_{z,y}(\omega)$, waveguide absorption coefficient $\alpha(\omega)$, and the reflection coefficient $R(\omega)$.

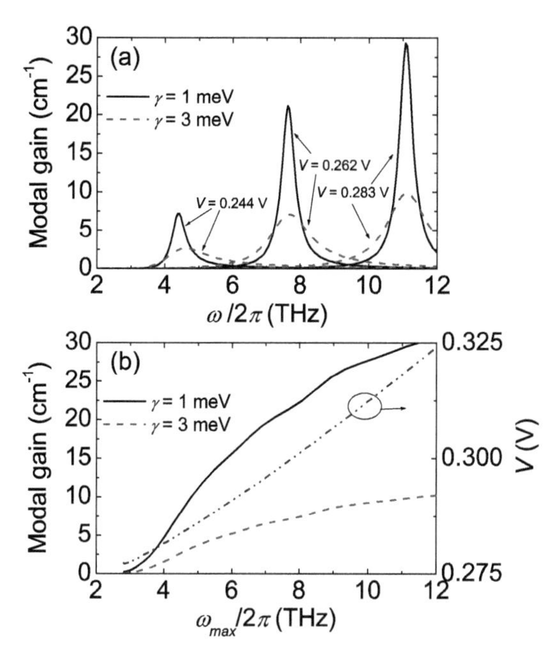

Figure 40.7 Modal gain $\mathcal{G}(\omega)$ as a function of (a) frequency ω for different values of bias voltages V and different relaxation broadening γ and (b) of frequency ω_{max} and bias voltage V.

where $\alpha(\omega)$ is the surface plasmon waveguide absorption coefficient and $R(\omega)$ is the reflection coefficient on structure edges. For the structure mirrors in the form of cleavage, $R(\omega)$ can be estimated as $R(\omega) = \left[\dfrac{\eta_{\text{eff}}(\omega)-1}{\eta_{\text{eff}}(\omega)+1}\right]^2$. Figure 40.6 shows the frequency dependence of the absorption coefficient $\alpha(\omega)$ and the reflection coefficient $R(\omega)$ in the structure under consideration. As follows from Fig. 40.6, $\alpha(\omega)$ and $R(\omega)$ increase considerably with the increasing surface plasmon frequency. Figure 40.8 shows the frequency dependence of the threshold length $L_{\text{th}}(\omega)$ at $\omega = \omega_{\text{th}}$ calculated for the double-GL structure with different parameters Σ_{i} and γ. As seen from Fig. 40.8, $L_{\text{th}}(\omega_{\text{max}})$ is about 1–5 mm in the frequency range of 5–12 THz. This length is typical for THz QCLs [11, 15]. However, the operation at lower frequencies can require the use of rather long device structures.

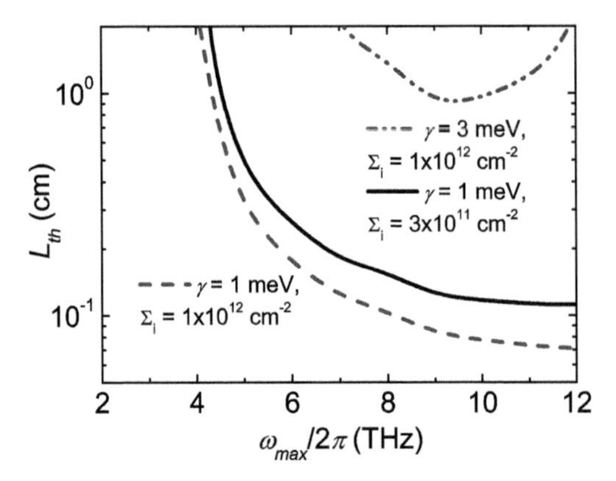

Figure 40.8 Waveguide threshold length $L_{th}(\omega_{max})$ for different parameters Σ_i and γ.

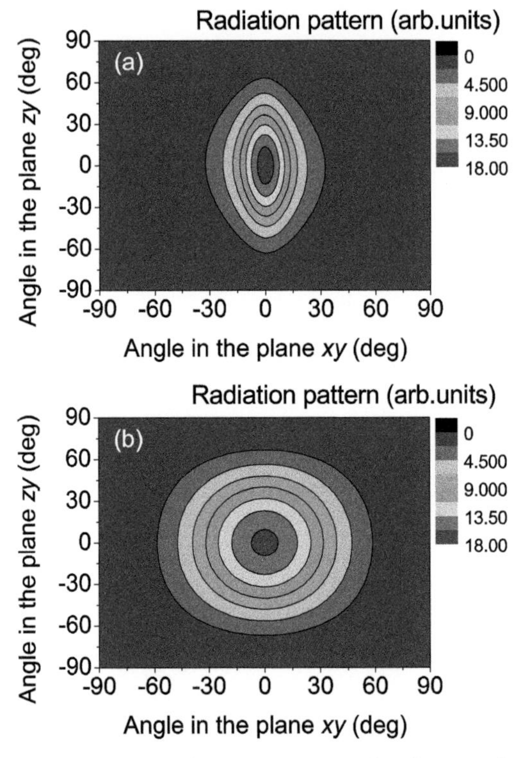

Figure 40.9 Far-field THz radiation patterns of a laser with the surface plasmon waveguide (a) and a laser with the metal-metal waveguide (b) for frequency $\omega/2\pi = 6$ THz.

The double-GL lasers with the surface plasmon waveguide considered above and the double-GL laser with metal-metal waveguide [36, 37] (with the same lateral sizes) are characterized by similar gain-overlap factors and the THz modal gains. However, the former exhibits better angular distributions of the output THz radiation. Figure 40.9 shows the angular dependence of far-field patterns of the output THz radiation generated by a laser with the surface plasmon waveguide and a laser with the metal-metal waveguide for frequency 6 THz (see Figs. 40.9a and b, respectively). These dependencies were calculated using the pertinent equation from Ref. [49]. As seen, the surface plasmon waveguide exhibits an advantage over the metal-metal waveguide, because the THz radiation beam divergence is much smaller for the surface plasmon waveguide than that for the metal-metal waveguide. Another advantage of the surface plasmon waveguide is a greater ratio of useful loss (the loss due to the output radiation) to the total loss as compared with metal-metal waveguides.

40.4 Conclusions

We propose and substantiated the concept of injection THz lasers based on the double-GL structures supplied by the surface plasmon waveguides and exploiting the inter-GL resonant radiative transitions. These lasers should be able to operate at room temperature. Achieving the THz lasing requires sharp inter-GL tunneling resonances, i.e., sufficiently small values of parameter γ. The collision frequency of electrons and holes ν should be relatively low and the inhomogeneous smearing of these resonances should be small in comparison with the collision broadening. The quantity $\gamma = 1$ meV (corresponding to fairly high values of the modal gain) corresponds to $\nu \simeq 1.6 \times 10^{12}$ s^{-1}, which is fairly realistic at room temperature. The double-GL lasers with the surface plasmon waveguides considered above and those using the metal-metal waveguides [36] have advantages associated with an effective voltage tuning and a low Drude absorption and optical phonon absorption. This should enable the operation at the low end of the THz range and in the range from 5 to 12 THz.

Acknowledgments

This work was supported by the Japan Society for Promotion of Science (Grant-in-Aid for Specially Promoting Research, No. 23000008), the Japan Science Technology Agency (CREST Project), Japan. The work by A. A. D. was also supported by the Dynasty Foundation, Russia. The work at RPI was supported by the US Army Cooperative Research Agreement (Program Manager Dr. Meredith Reed).

References

1. T. D. Dorney, R. G. Baraniuk, and D. M. Mittleman, *J. Opt. Soc. Am. A*, **18**, 1562 (2001).
2. P. Y. Han and X. C. Zhang, *Meas. Sci. Technol.*, **12**, 1747 (2001).
3. H. Takahashi and M. Hosoda, *Appl. Phys. Lett.*, **77**, 1085 (2000).
4. A. J. Fitzgerald, N. N. Berry, E. Zinovev, G. C. Walker, M. A. Smith, and J. M. Chamberlain, *Phys. Med. Biol.*, **47**, R67 (2002).
5. D. Arnone, C. Ciesla, and M. Pepper, *Phys. World*, **13**, 35 (2000).
6. M. Tonouchi, *Nat. Photonics*, **1**, 97 (2007).
7. A. Tredicucci and M. S. Vitiello, *IEEE J. Sel. Top. Quantum Electron.*, **20**, 8500109 (2014).
8. S. Suzuki, M. Asada, A. Teranishi, H. Sugiyama, and H. Yokoyama, *Appl. Phys. Lett.*, **97**, 242102 (2010).
9. M. Feiginov, C. Sydle, O. Cojocari, and P. Meisner, *Appl. Phys. Lett.*, **99**, 233506 (2011).
10. H. Kanaya, H. Shibayama, R. Sogabe, S. Suzuki, and M. Asada, *Appl. Phys. Express*, **5**, 124101 (2012).
11. F. Capasso, C. Gmachl, D. L. Sivco, and A. Y. Cho, *Phys. Today*, **55**(5), 34 (2002).
12. M. S. Vitiello and A. Tredicucci, *IEEE Trans. THz Sci. Technol.*, **1**, 76 (2011).
13. S. Kumar, *IEEE J. Sel. Top. Quantum Electron.*, **17**, 38 (2011).
14. S. Fathololoumi, E. Dupont, C. W. I. Chan, Z. R. Wasilewski, S. R. Laframboise, D. Ban, A. Matyas, C. Jirauschek, Q. Hu, and H. C. Liu, *Opt. Express*, **20**(4), 3866 (2012).

15. B. S. Williams, *Nat. Photonics*, **1**, 517 (2007).
16. V. Ryzhii, M. Ryzhii, and T. Otsuji, *J. Appl. Phys.*, **101**, 083114 (2007).
17. A. A. Dubinov, V. Ya. Aleshkin, M. Ryzhii, T. Otsuji, and V. Ryzhii, *Appl. Phys. Express*, **2**, 092301 (2009).
18. F. Rana, *IEEE Trans. Nanotechnol.*, **7**, 91 (2008).
19. V. Ryzhii, M. Ryzhii, A. Satou, T. Otsuji, A. A. Dubinov, and V. Ya. Aleshkin, *J. Appl. Phys.*, **106**, 084507 (2009).
20. V. Ryzhii, A. A. Dubinov, T. Otsuji, V. Mitin, and M. S. Shur, *J. Appl. Phys.*, **107**, 054505 (2010).
21. V. Ryzhii, M. Ryzhii, V. Mitin, and T. Otsuji, *J. Appl. Phys.*, **110**, 094503 (2011).
22. S. Boubanga-Tombet, S. Chan, T. Watanabe, A. Satou, V. Ryzhii, and T. Otsuji, *Phys. Rev. B*, **85**, 035443 (2012).
23. T. Li, L. Luo, M. Hupalo, J. Zhang, M. C. Tringides, J. Schmalian, and J. Wang, *Phys. Rev. Lett.*, **108**, 167401 (2012).
24. T. Otsuji, S. A. Boubanga Tombet, A. Satou, H. Fukidome, M. Suemitsu, E. Sano, V. Popov, M. Ryzhii, and V. Ryzhii, *J. Phys. D*, **45**, 303001 (2012).
25. T. Otsuji, S. A. Boubanga Tombet, A. Satou, M. Ryzhii, and V. Ryzhii, *IEEE J. Sel. Top. Quantum Electron.*, **19**, 8400209 (2013).
26. M. Liu, X. Yin, and X. Zhang, *Nano Lett.*, **12**, 1482 (2012).
27. L. Britnell, R. V. Gorbachev, R. Jalil, B. D. Belle, F. Shedin, A. Mishenko, T. Georgiou, M. I. Katsnelson, L. Eaves, S. V. Morozov, N. M. R. Peres, J. Leist, A. K. Geim, K. S. Novoselov, and L. A. Ponomarenko, *Science*, **335**, 947 (2012).
28. T. Georgiou, R. Jalil, B. D. Bellee, L. Britnell, R. V. Gorbachev, S. V. Morozov, Y.-J. Kim, A. Cholinia, S. J. Haigh, O. Makarovsky, L. Eaves, L. A. Ponomarenko, A. K. Geim, K. S. Nonoselov, and A. Mishchenko, *Nat. Nanotechnol.*, **7**, 100 (2013).
29. L. Britnell, R. V. Gorbachev, A. K. Geim, L. A. Ponomarenko, A. Mishchenko, M. T. Greenaway, T. M. Fromhold, K. S. Novoselov, and L. Eaves, *Nat. Commun.*, **4**, 1794 (2013).
30. V. Ryzhii, T. Otsuji, M. Ryzhii, V. G. Leiman, S. O. Yurchenko, V. Mitin, and M. S. Shur, *J. Appl. Phys.*, **112**, 104507 (2012).
31. R. M. Feenstra, D. Jena, and G. Gu, *J. Appl. Phys.*, **111**, 043711 (2012).
32. V. Ryzhii, T. Otsuji, M. Ryzhii, and M. S. Shur, *J. Phys. D: Appl. Phys.*, **45**, 302001 (2012).
33. F. T. Vasko, *Phys. Rev. B*, **87**, 075424 (2013).

34. V. Ryzhii, A. Satou, T. Otsuji, M. Ryzhii, V. Mitin, and M. S. Shur, *J. Phys. D: Appl. Phys.*, **46**, 315107 (2013).
35. V. Ryzhii, M. Ryzhii, V. Mitin, M. S. Shur, A. Satou, and T. Otsuji, *J. Appl. Phys.*, **113**, 174506 (2013).
36. V. Ryzhii, A. A. Dubinov, V. Ya. Aleshkin, M. Ryzhii, and T. Otsuji, *Appl. Phys. Lett.*, **103**, 163507 (2013).
37. V. Ryzhii, A. A. Dubinov, T. Otsuji, V. Ya. Aleshkin, M. Ryzhii, and M. Shur. *Opt. Express*, **21**, 31567 (2013).
38. K. Unterrainer, R. Colombelli, C. Gmachl, F. Capasso, H. Y. Hwang, D. L. Sivco, and A. Y. Cho, *Appl. Phys. Lett.*, **80**, 3060 (2002).
39. M. A. Belkin, J. A. Fan, S. Hormoz, F. Capasso, S. P. Khanna, M. Lachab, A. G. Davies, and E. Linfield, *Opt. Express*, **16**, 3242 (2008).
40. A. A. Tager, R. Gaska, I. A. Avrutsky, M. Fay, H. Chik, A. Spring-Thorpe, S. Eicher, J. M. Xu, and M. Shur, *IEEE J. Sel. Top. Quantum Electron.*, **5**, 664 (1999).
41. L. A. Falkovsky and A. A. Varlamov, Eur. Phys. J. B **56**, 281 (2007).
42. L. A. Falkovsky and S. S. Pershoguba, *Phys. Rev. B*, **76**, 153410 (2007).
43. A. Shik, *Quantum Wells: Physics and Electronics of Two-Dimensional Systems* (World Scientific, Singapore, 1998).
44. F. T. Vasko and A. V. Kuznetsov, *Electronic States and Optical Transitions in Semiconductor Heterostructures* (Springer, New York, 1999).
45. H. Shi, H. Pan, Y.-W. Zhang, and B. Yakobson, *Phys. Rev. B*, **87**, 155304 (2013).
46. K. J. Ebeling, *Integrated Optoelectronics: Waveguide Optics, Photonics, Semiconductors* (Springer, Berlin, 1993).[28]
47. M. Born and E. Wolf, *Principles of Optics: Electromagnetic Theory of Propagation, Interference and Diffraction of Light* (Pergamon, Oxford, 1964).
48. D. M. Hoffman, G. L. Doll, and P. C. Eklund, *Phys. Rev. B*, **30**, 6051 (1984).
49. H. C. Casey and M. B. Panish, *Heterostructure Lasers, Part A* (Academic, New York, 1978).

Chapter 41

Double Injection, Resonant-Tunneling Recombination, and Current-Voltage Characteristics in Double-Graphene-Layer Structures*

M. Ryzhii,[a] V. Ryzhii,[b,c] T. Otsuji,[b] P. P. Maltsev,[c] V. G. Leiman,[d] N. Ryabova,[a,e] and V. Mitin[f]

[a]*Department of Computer Science and Engineering, University of Aizu, Aizu-Wakamatsu 965-8580, Japan*
[b]*Research Institute for Electrical Communication, Tohoku University, Sendai 980-8577, Japan*
[c]*Institute of Ultra High Frequency Semiconductor Electronics, Russian Academy of Sciences, Moscow 111005, Russia*
[d]*Moscow Institute of Physics and Technology, Dolgoprudny, Moscow Region 141700, Russia*
[e]*Center for Photonics and Infrared Engineering, Bauman Moscow State Technical University, Moscow 105005, Russia*
[f]*Department of Electrical Engineering, University at Buffalo, Buffalo, New York 1460-1920, USA*
m-ryzhii@u-aizu.ac.jp

*Reprinted with permission from M. Ryzhii, V. Ryzhii, T. Otsuji, P. P. Maltsev, V. G. Leiman, N. Ryabova, and V. Mitin (2014). Double injection, resonant-tunneling recombination, and current-voltage characteristics in double-graphene-layer structures, *J. Appl. Phys.*, **115**, 024506. Copyright © 2014 AIP Publishing LLC.

Graphene-Based Terahertz Electronics and Plasmonics: Detector and Emitter Concepts
Edited by Vladimir Mitin, Taiichi Otsuji, and Victor Ryzhii

ISBN 978-981-4800-75-4 (Hardcover), 978-0-429-32839-8 (eBook)
www.jennystanford.com

We evaluate the effect of the recombination associated with interlayer transitions in ungated and gated double-graphene-layer (GL) structures on the injection of electrons and holes. Using the proposed model, we derive analytical expressions for the spatial distributions of the electron and hole Fermi energies and the energy gap between the Dirac points in GLs as well as their dependences on the bias and gate voltages. The current-voltage characteristics are calculated as well. The model is based on hydrodynamic equations for the electron and hole transports in GLs under the self-consistent electric field. It is shown that in undoped double-GL structures with weak scattering of electrons and holes on disorder, the Fermi energies and the energy gap are virtually constant across the main portions of GLs, although their values strongly depend on the voltages and recombination parameters. In contrast, the electron and hole scattering on disorder lead to substantial nonuniformities. The resonant inter-GL tunneling enables N-shaped current-voltage characteristics provided that GLs are sufficiently short. The width of the current maxima is much larger than the broadening of the tunneling resonance. In the double-GL structures with relatively long GLs, the N-shaped characteristics transform into the Z-shaped characteristics. The obtained results are in line with the experimental observations [1] and might be useful for design and optimization of different devices based on double-GL structures, including field-effect transistors and terahertz lasers.

41.1 Introduction

Double-graphene-layer (double-GL) structures, which consist of two GLs separated by a thin layer and independently connected with the side contacts, were fabricated and studied recently [1–4]. Such structures have drawn a considerable attention. This, in part, is due to potential applications of the double-GL structures in optically transparent transistor circuits [1–3], high speed modulators of optical and terahertz (THz) radiation [3–6], THz detectors and frequency multipliers [7–9], THz photomixers [10], and THz lasers [11]. Schematic views of the double-GL structures (both ungated and gated) and their band diagram under the bias voltage are shown in Fig. 41.1. Interesting features of devices based on independently contacted quantum wells, formed in the standard heterostructures, were discussed some time ago [12]. Under the operation conditions of different double-GL-based devices, the inter-GL transitions (tunneling or thermionic) tend to depopulate GLs. This can lead to a disruption of the device operation. The refilling of GLs is associated with the injection of electrons to one GL and holes another from the pertinent contacts. Thus, the injection of electrons and holes in the double-GL structures requires a careful consideration. In this chapter, we develop a model for transport phenomena in the double-GL structures as shown in Figs. 41.1a and b employing the hydrodynamic equations coupled with the Poisson equation for the self-consistent electric potential [13, 14]. Similar problem, but related to single- and multiple-GL (MGL) structures with injection of both electrons and holes into the same GL from the opposite n- to p-contacts, was considered recently [13, 14]. However, the double-GL structures with independently contacted GLs are characterized by important features associated with spatial separation of interacting two-dimensional electron gas (2DEG) and two-dimensional hole gas (2DHG). In particular, the existence of the energy gap between the Dirac points in GLs and the resonant tunneling between GLs should be addressed.

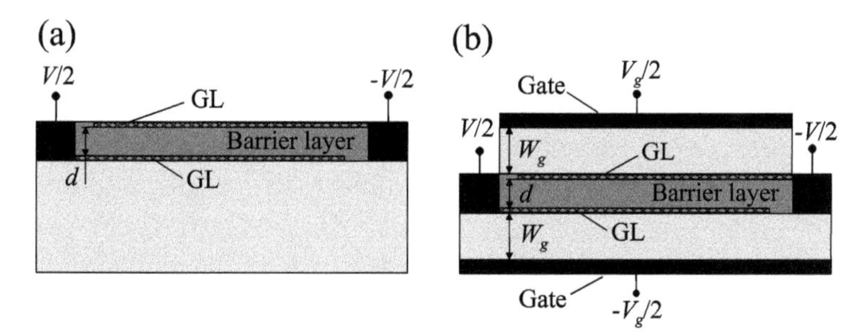

Figure 41.1 Schematic view of (a) the cross-sections of a double-GL structure and (b) the cross-section of its version with the top and bottom metal gates.

The chapter is organized as follows. In Section 41.2, we discuss the device model and write down the main equations, which govern the transport and recombination of the injected electrons in the main parts GLs, where the latter are quasi-neutral. The role of the near-contact regions is accounted by the boundary condition set at the edges of the quasi-neutral regions [14]. Section 41.3 deals with the calculations of the spatial distributions and voltage dependences of the Fermi energy, the energy gap, and electric potential assuming that the electron-hole scattering dominates over the scattering on disorder. In Section 41.4, we found how the disorder scattering affects the spatial distributions and voltage dependences. In Section 41.5, the current-voltage characteristics are calculated and discussed using the results of Section 41.4. In Section 41.6, we draw the main conclusions.

41.2 Model and the Pertinent Equations

We assume that GLs in the double-GL structures under consideration are ungated and undoped (see Fig. 41.1a), or gated and "electrically" doped. In the latter case, GLs are filled with 2DEG and 2DHG (see Fig. 41.1b) electrically induced by the gate voltages $\pm V_g/2$. The application of the bias voltage V between the opposite edges of the ungated GLs (as shown in Fig. 41.1a) results in the formation of 2DEG and 2DHG in the pertinent GLs. In the gated double-GL structure shown in Fig. 41.1b, the 2DEG and 2DHG densities are determined by both the bias and gate voltages, V and V_g. The double-GL structures under consideration are particularly interesting for devices utilizing the resonant tunneling between GLs and resonant tunneling assisted by the photon emission [11].

As shown previously by numerical solutions of the two-dimensional Poisson equation with realistic boundary conditions and structural parameters [14], in sufficiently long GL-structures (their length $2L$ significantly exceeds the characteristic screening length $r_s = (\kappa\hbar^2 v_W^2/4e^2T))$, i.e., the parameter $Q = 2L/r_s = (\pi e^2 LT/6\kappa\hbar^2 v_W^2) \gg 1$, in the main part of GLs, the electron-hole plasma is quasi-neutral. Here, T is the temperature (in the energy units), $v_W \simeq 10^8$ cm/s is the characteristic velocity of electrons and holes in GLs, κ is the dielectric constant, and $\hbar$ is the reduced Planck constants. Indeed, if $\kappa = 4$ and $2L = 10$ µm at the temperatures $T = 300$ K, parameter $Q \simeq 600$. Thus, the electron density in one GL, Σ_e, is with high accuracy equal to the hole density Σ_h in another GL: $\Sigma_e = \Sigma_h = \Sigma$. Figure 41.2 schematically shows the Dirac points spatial positions (separated by the energy gap Δ) and potential profiles in both GLs under the applied bias voltage V. This figure demonstrates qualitatively that the potential profiles being rather sharp near the contact regions are smooth in the main parts of GLs.

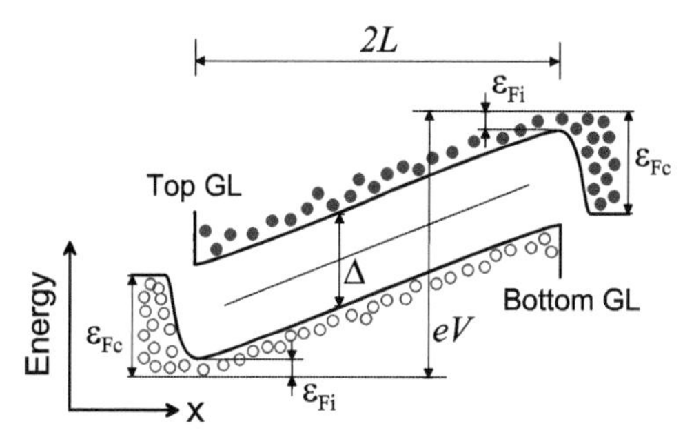

Figure 41.2 Schematic view of spatial variations of the Dirac points and potential profiles in top and bottom GLs, respectively, in a double-GL structure at $\Delta > 0$. Filled and open circles correspond to electrons in the conduction band of top GL and to holes in the valence band of bottom GL.

At sufficiently high bias and gate voltages, the 2DEG and 2DHG densities Σ are large, so that the electron and hole energy distributions are well characterized by the Fermi functions with the quasi-Fermi energies (counted from the Dirac points) $\varepsilon_{Fe} = \varepsilon_{Fh} = \varepsilon_F$ and the common effective temperature T. The latter is assumed to be equal to the lattice temperature.

The electron and hole densities in the pertinent GLs Σ and the energy gap between the Dirac points are related to each other as is

$$\Sigma = \Sigma_g + \frac{\kappa\Delta}{4\pi e^2 d}, \tag{41.1}$$

where $\Sigma_g = \kappa V_g/8\pi e W_g$ is the density of 2DEG and 2DHG induced by the gate voltages $\pm V_g/2$, d is the spacing between GLs, and W_g is the spacing between the GLs and gates (see Figs. 41.1a and b). Generally, the energy gap Δ, i.e., the energy separation between the Dirac points in GLs, (which determines the built-in electric field $E = \Delta/ed$ in the inter-GL barrier), as well as ε_F and the electric potential at the GL plane φ are functions of coordinate x (the x-axis is directed in the GL plane from the p- to n-contacts).

In the case of the 2DEG and 2DHG strong degeneration ($\varepsilon_F \gg T$)

$$\varepsilon_F \simeq \hbar v_W \sqrt{\pi\Sigma}, \tag{41.2}$$

$$\Delta \simeq \frac{4e^2 d}{\kappa \hbar^2 \upsilon_W^2}(\varepsilon_F^2 - \varepsilon_{Fg}^2), \tag{41.3}$$

where $\varepsilon_{Fg} \simeq \hbar \upsilon_W \sqrt{\pi \Sigma_g} = \hbar \upsilon_W \sqrt{\kappa V_g / 8eW_g}$.

In the absence of the inter-GL recombination, 2DEG and 2DHG are in equilibrium. In this case, Fermi energy in each GL and the energy gap are given by

$$\varepsilon_F^{eq} = \frac{e}{2}\left[\sqrt{(2VV_0 + V_0^2 + V_{bi}^2)} - V_0\right], \tag{41.4}$$

$$\Delta^{eq} = e\left[V + V_0 - \sqrt{(2VV_0 + V_0^2 + V_{bi}^2)}\right], \tag{41.5}$$

where $V_0 = \hbar^2 \upsilon_W^2 \kappa / 2e^3 d$ and $V_{bi} = 2\varepsilon_{F_g}/e$. At $V < V_{bi}$, $\Delta^{eq} < 0$, but at $V \geq V_{bi}$, $\Delta^{eq} \geq 0$. In particular, when $\Sigma_g = 0$, i.e., $V_{bi} = 0$, $\Delta_{eq} \geq 0$ at all V. Figure 41.3 shows the energy diagrams of a double-GL structure at $\Delta = 0$ and $\Delta > 0$.

High densities of 2DEG and 2DHG assume the validity of a hydrodynamic approach for the description of the electron and hole transports. We use the hydrodynamic model presented in Ref. [14].

The transport of electrons and holes is governed by the following system of hydrodynamic equations [14]

$$\frac{d\Sigma\, u_e}{dx} = -R, \quad \frac{d\Sigma\, u_h}{dx} = -R, \tag{41.6}$$

$$\frac{1}{M}\frac{d(e\varphi_e - \varepsilon_F)}{dx} = \nu u_e + \nu_{eh}(u_e - u_h), \tag{41.7}$$

$$-\frac{1}{M}\frac{d(e\varphi_h + \varepsilon_F)}{dx} = \nu u_h + \nu_{eh}(u_h - u_e), \tag{41.8}$$

$$\Sigma = \frac{12\Sigma_T}{\pi^2}\int_0^\infty \frac{dy\,y}{[\exp(y - \varepsilon_F/T) + 1]}. \tag{41.9}$$

Here, u_e and u_h are the value of the electron and hole hydrodynamic velocities and φ_e and φ_h are the potentials of the top GL (filled with electrons) and the bottom GL (filled with holes), respectively, ν is the collision frequency of electrons and holes with impurities and acoustic phonons (with the short-range disorder), ν_{eh} is their collision frequency with each other, M is the fictitious mass, which at the Fermi energies of the same order of magnitude as the temperature can be considered as a constant, and $\Sigma_{\rm T} = \pi T^2/6\hbar^2 v_{\rm W}^2$.

The recombination of the major carriers injected to the degenerate 2DEG and 2DHG in the double-GL structures under consideration is primarily associated with the inter-GL transitions. The rate of these processes depends on the Fermi energy (density of 2DEG and 2DHG) and the energy gap Δ. When $\Delta = 0$, the resonant-tunneling transitions with the conservation of momentum can dominate [1, 7, 15, 16]. Such transitions correspond to the arrow in Fig. 41.3a. We present the rate of the resonant-tunneling recombination in the following form:

$$R_{\rm rt} \simeq \frac{\Sigma_T}{\tau_{\rm rt}} \exp\left(-\frac{\Delta^2}{\gamma^2}\right), \tag{41.10}$$

where $\tau_{\rm rt}$ is the characteristic resonant-tunneling recombination time and γ is the resonance broadening parameter.

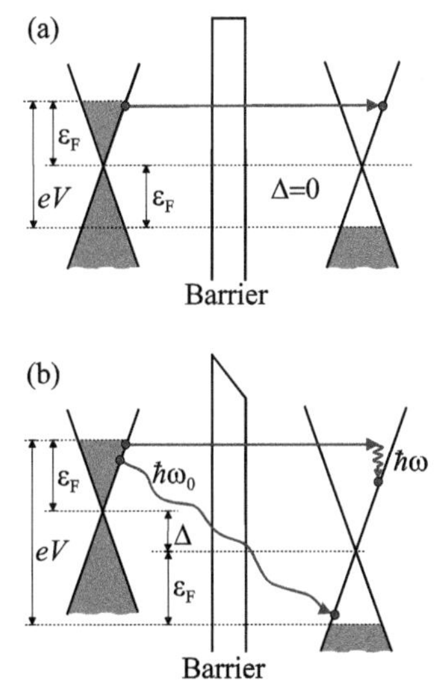

Figure 41.3 Energy band diagrams of double-GL MGL structure (a) at $\Delta = \Delta^{\rm eq} = 0$, $eV < \hbar\omega_0$ and (b) at $\Delta = \Delta^{\rm eq} > 0$, $eV > \hbar\omega_0$. Arrow in (a) indicates inter-GL resonant-tunneling transition (with conservation of electron energy and momentum), arrow in (b) corresponds to transitions assisted by photon and optical phonon emission (with energies $\hbar\omega$ and $\hbar\omega_0$, respectively).

The inter-GL transitions due to the scattering on impurities or acoustic phonons (non-resonant processes) and due to the processes mediated by photons (see Fig. 41.3b) in which the momentum or energy is not conserved, can also contribute to the recombination. Due to a strong coupling of clectrons and holes with optical phonons, the processes assisted by optical phonon emission can also greatly contribute to the rate of the inter-GL transitions (as it takes place in GL structures with bipolar injection into GLs [10, 14, 17–19]). For concreteness, we assume that the resonant-tunneling processes and the processes accompanied by optical phonon emission are the main mechanisms of the inter-GL recombination.

In the case $V < \hbar\omega_0/e \simeq 200$ mV, $2\varepsilon_F + \Delta < \hbar\omega_0$, and the transitions between the tails of the energy distributions are possible. At $eV > \hbar\omega_0 \simeq 200$ meV, the inter-GL transitions assisted by the optical phonon emission are not limited by the Pauli exclusion principle (see Fig. 41.3b). In this case, the inter-GL recombination can be rather strong restricting penetration of the injected electrons and hole sufficiently far from GL edges. Considering this, the rate of the nonresonant inter-GL recombination assisted by optical phonon emission can be presented as

$$R_{nr} \simeq \frac{\Sigma_T}{\tau_{nr}} \exp\left(\frac{2\varepsilon_F + \Delta}{T}\right), \tag{41.11}$$

if $\varepsilon_F + \Delta < \hbar\omega_0$, and

$$R_{nr} \simeq \frac{\Sigma_T}{\tau_{nr}} \exp\left(\frac{\hbar\omega_0}{T}\right), \tag{41.12}$$

if $\varepsilon_F + \Delta > \hbar\omega_0$. Here, $\tau_r \simeq (\Sigma_T/G_T)\exp(2d)$ is the characteristic recombination time associated with and with the inter-GL transitions assisted by the optical phonon emission, respectively, G_T is the rate thermogeneration of the electron-hole pairs due to the absorption of optical phonons in equilibrium (it is about $G_T = 10^{21}$ cm^{-2}s^{-1} at room temperature [16]), and æ is the tunneling decay factor characterizing the overlap of wavefunctions in the top and bottom GLs, The right-hand side of Eq. (41.11), which provides a somewhat simplified dependence of the recombination rate on ε_F and Δ, which differs from that used in Refs. [14, 15] (see also Refs. [16, 17]) for the recombination rate associated with the intra-GL transitions mediated by optical phonons by factors $\exp(-2æ\, d)$and $\exp(\Delta/T)$. The latter is due to the energy gap associated with the GL spatial separation and the potential difference between GLs. Invoking Ref. [17], we find $\Sigma_T/G_T \simeq (10^{-9} - 10^{-10})$ s. Hence, accounting for that $\exp(2æ\, d) \gg 1$, one obtains $\tau_{nr} \gg (10^{-9} - 10^{-10})$ s. According to the experimental results [1], the rate of the inter-GL transitions at the resonant tunneling $R_{rt} \simeq 10^{22}$ cm^{-2} s^{-1}. This yields $\tau_{rt} \simeq (1 - 3) \times 10^{-11}$ s^{-1}.

Thus, the net inter-GL recombination rate is assumed to be as

$$R = R_{nr} + R_{nr}. \tag{41.13}$$

Considering the same boundary conditions as in Ref. [14] set at the points near the contacts (outside narrow space-charge regions), i.e., at $x = \pm L^* = (L - r_s) \simeq \pm L$, but generalized by accounting for the energy gap (see Fig. 41.2), we have

$$(e\varphi_h + \varepsilon_F)|_{x=-L} = \frac{eV}{2}, \quad (e\varphi_e - \varepsilon_F)|_{x=L} = -\frac{eV}{2}. \tag{41.14}$$

The boundary conditions for the electron and hole velocities can be taken as (no electron and hole current at the disconnected GL edges)

$$u_e|_{x=-L} = u_h|_{x=L} = 0. \tag{41.15}$$

As in Ref. [14], we introduce the following dimensionless variables: $\Phi_{e,h} = e\varphi_{e,h}/k_BT$, $\mu = \varepsilon_F/T$, $\mu_g = \varepsilon_{F_g}/T$, $\upsilon = eV/T$, $\upsilon_0 = eV_0/T$, $\delta = \Delta/T$, $U_e = u_e\tau_{nr}/L$, $U_h = u_h\tau_{nr}/L$, and $\xi = x/L$. In these notations, the dimensionless generation-recombination term $r(\mu) = R\tau_{nr}/\Sigma_T$ and the dimensionless density $\sigma(\mu) = \Sigma/\Sigma_T$ are

$$\frac{d[\sigma(\mu)U_e]}{d\xi} = -r(\mu), \quad \frac{d[\sigma(\mu)U_h]}{dx} = -r(\mu), \tag{41.16}$$

$$\frac{d(\Phi_e - \mu)}{d\xi} = \frac{\beta_{eh}(\mu)}{\sigma(\mu)}\left[U_e\left(\frac{\nu}{\nu_{eh}}\right) + U_e - U_h\right], \tag{41.17}$$

$$-\frac{d(\Phi_h + \mu)}{d\xi} = \frac{\beta_{eh}(\mu)}{\sigma(\mu)}\left[U_h\left(\frac{\nu}{\nu_{eh}}\right) + U_h - U_e\right], \tag{41.18}$$

$$\sigma(\mu) = \frac{12}{\pi^2}\int_0^\infty \frac{dyy}{[\exp(y - \mu) + 1]}, \tag{41.19}$$

$$r(\mu) = \exp[2\mu + 2(\mu^2 - \mu_g^2)/\upsilon_0] + \eta \exp[-4b^2(\mu^2 - \mu_g^2)^2], \qquad (41.20)$$

where $\beta_{eh}(\mu) = [M(\mu)\nu_{eh}(\mu)\sigma(\mu))L^2/T\tau_r) = \overline{\beta}_{eh}\, I(\mu)$, where the function $I(\mu)$ is numerically calculated in Appendix, $\upsilon_0 = (\hbar^2\upsilon_W^2\kappa/2e^2dT)$, $b = (T/\upsilon_0\gamma)$, and $\eta = \tau_{nr}/\tau_{rt}$. The parameter β_{eh} varies in a wide range depending on the scattering and recombination parameters and the GL length.

The quantity $q_{eh} = \beta_{eh}/\sigma$ can be presented as $(L/\mathcal{L}_{eh})^2$, where $\mathcal{L}_{eh} = D_{eh}\tau_r$ is the diffusion length and $D_{eh} = \upsilon_W^2/2\nu_{eh}$ is the bipolar diffusion coefficient.

41.3 Spatial and Voltage Dependences of Fermi Energy, Energy Gap, and Potential

In the structures under consideration in which the density of 2DEG and 2DHG markedly exceeds the density residual impurities, one can assume that $\nu \ll \nu_{eh}$ (even despite the spatial separation of 2DEG and 2DHG). Taking this into account, we disregard in Eqs. (41.10) and (41.11), the terms proportional to a small parameter ν/ν_{eh}. In such a case, Eqs. (41.16)–(41.18), with boundary conditions which follow from Eqs. (41.14) and (41.15), yield $\mu = \mu_i = \varepsilon_{Fi}/T = const$, $\delta = \delta_i = const$, $\Phi_e = -\delta_i/2 + \Phi_i$, and $\Phi_h = \delta_i/2 + \Phi_i$, where $\Phi_i \propto \xi$,

$$U_e = -\frac{r(\mu_i)}{\sigma(\mu_i)}(\xi + 1), \quad U_h = -\frac{r(\mu_i)}{\sigma(\mu_i)}(\xi - 1), \qquad (41.21)$$

$$\Phi_i = -\frac{2\beta_{eh}(\mu_i)r(\mu_i)}{\sigma^2(\mu_i)}\xi, \qquad (41.22)$$

$$\delta_i = 2(\mu_i^2 - \mu_g^2)/\upsilon_0. \qquad (41.23)$$

As follows from the boundary conditions, μ_i and Δ_i are related to each other also as:

$$\mu_i + \frac{2\beta_{eh}(\mu_i)r(\mu_i)}{\sigma^2(\mu_i)} = \frac{\upsilon - \delta_i}{2}. \qquad (41.24)$$

Considering Eqs. (41.22) and (41.23), we arrive at the following equation for μ_i:

$$\mu_i + \frac{(\mu_i^2 - \mu_g^2)}{\upsilon_0} + \frac{2\beta_{eh}(\mu_i)r(\mu_i)}{\sigma^2(\mu_i)} = \frac{\upsilon}{2}. \qquad (41.25)$$

The quantity δ_i in Eq. (41.23), which is proportional to $v_0^{-1} \propto d$, explicitly describes the effect of spatial separation of GLs resulting in the appearance of the energy gap. According to the above formulas, the Fermi energy and the energy gap are independent of the coordinate (in the main parts of GLs except near-contact regions), while potential is a linear function of the coordinate [see Eq. (41.22)]. This corresponds to Fig. 41.2.

If the inter-GL recombination is insignificant, formally setting $r(\mu) = 0$, from Eq. (41.24), we obtain the equilibrium value of the Fermi energy in each GL, and hence, using Eq. (41.23), the equilibrium value of the energy gap coinciding with Eqs. (41.4) and (41.5).

Equation (41.25) yields the explicit relationship between the normalized Fermi energy μ_i in the main portions of GLs (except narrow regions near the contacts) on parameters q and v_0, the bias voltage V, and the gate voltage V_g [via the dependence $\mu_g(V_g)$]. At relatively weak electron-hole scattering (and the scattering on disorder) and recombination, when $\beta \ll 1$, and 2DEG and 2DHG are close to equilibrium, Eqs. (41.23)–(41.25) yield μ_i and δ_i close to $\mu_i^{eq} = \varepsilon_F^{eq}/T$ and $\delta_i^{eq} = \Delta^{eq}/T$ [see Eqs. (41.4) and (41.5)].

Figure 41.4 shows the dependences of the normalized energy gap δ_i and the normalized Fermi energy, μ_i in the main parts of GLs calculated using Eqs. (41.23)–(41.25) for the parameter $\overline{\beta}_{eh}$ in the range of $\overline{\beta}_{eh}$ = 0.001 – 0.01 ($\overline{\beta}_{eh}\,\eta$ = 0.1 – 1.0). It was assumed that $\kappa = 4$, $d = 2$ nm, $\eta = 100$, $T = 300$ K, and $\mu_g = 2$ (i.e., $V_{bi} = 100$ mV), so that $v_0 \simeq 10.5$, $\gamma = 1$ meV, $b \simeq 2.375$. We used the functions $\beta_{eh}(\mu)$ (for $\alpha = e^2/\kappa\hbar v_W = 0.505$) and $\sigma(\mu)$ calculated in Appendix.

Using the data obtained from Ref. [1] ($\tau_{rt} \sim (1 - 3) \times 10^{-11}$ s and $L^2 = 0.6\ \mu m^2$), one can find $\overline{\beta}_{eh} \sim 0.0017 - 0.0051$. The quantity η = 100 corresponds to $\tau_{nr} \sim (1 - 3) \times 10^{-9}$ s (see Section 41.2). As seen from Fig. 41.4, at relatively small values of parameter $\overline{\beta}_{eh}$, both δ_i and μ_i increase monotonically with increasing the bias voltage (see curves 1 and 2). At elevated values of parameter $\overline{\beta}_{eh}$, for example, at relatively long GLs (see curves 3 and 4), the voltage dependences of δ_i and μ_i are of the S-shape, i.e., multi-valued in a certain voltage range corresponding to the tunneling resonance δ_i = 0). The monotonic and multi-valued behavior correspond, in particular, to relatively short and long GLs, respectively ($\overline{\beta}_{eh} \propto L^2$).

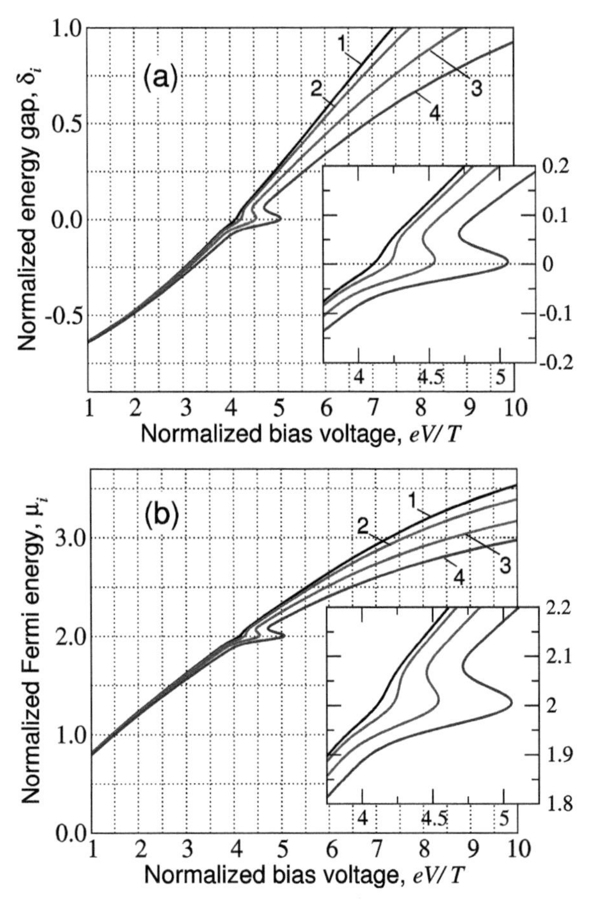

Figure 41.4 Voltage dependences of (a) normalized energy gap between the Dirac points and (b) normalized Fermi energy for different values of parameter β_{eh} : 1— β_{eh} = 0.001, 2—0.002, 3—0.005, and 4—0.01. Insets show zoom of the same characteristics near the tunneling resonance.

41.4 Role of Scattering on Disorder

Although the scattering of electrons and holes on residual impurities and acoustic phonons is comparably weak, it can lead to a small but qualitative modification of the injection characteristics. Considering the terms with ν as perturbations, one can search for solutions of Eqs. (41.15)–(41.18) in the form: $\mu = \mu_i + \delta\mu$, $\delta - \delta_i = 4q_0\mu_i\delta\mu$, and so on. In particular, we arrive at the following equation for the variation of the Fermi energy:

$$\frac{d\delta\mu}{d\xi} = \left(\frac{\nu}{\nu_{\rm eh}}\right)\frac{\beta_{\rm eh}(\mu_{\rm i})r(\mu_{\rm i})}{(1+2\mu_{\rm i}/\upsilon_0)\sigma^2(\mu_{\rm i})}\xi. \tag{41.26}$$

Taking into account that $\delta\mu|_{\xi=\pm1} = 0$, from Eq. (41.26) we obtain

$$\delta\mu = \frac{\beta_{\rm d}(\mu_{\rm i})r(\mu_{\rm i})}{2(1+2\mu_{\rm i}/\upsilon_0)\sigma^2(\mu_{\rm i})}(\xi^2-1). \tag{41.27}$$

Here, $\beta_{\rm d}(\mu) = (\nu/\nu_{\rm eh})\beta_{\rm eh}(\mu)$. Simultaneously, one can find

$$\delta - \delta_{\rm i} = \frac{2\beta_{\rm d}(\mu_{\rm i})r(\mu_{\rm i})}{\upsilon_0(1+2\mu_{\rm i}/\upsilon_0)\sigma^2(\mu_{\rm i})}(\xi^2-1). \tag{41.28}$$

As follows from Eqs. (41.27) and (41.28), the scattering of electrons and holes on disorder results in the spatial distributions of the Fermi energy μ (i.e., $\varepsilon_{\rm F}$) and the energy gap Δ with minima at the center of the structure ($\xi = 0$). The span of these quantities variations are as follows:

$$\delta\mu|_{\xi=0} = -\frac{\beta_{\rm d}(\mu_{\rm i})r(\mu_{\rm i})}{2(1+2\mu_{\rm i}/\upsilon_0)\sigma^2(\mu_{\rm i})},$$

$$= -\left(\frac{L}{\mathcal{L}}\right)^2\frac{r(\mu_{\rm i})}{2(1+2\mu_{\rm i}/\upsilon_0)\sigma(\mu_{\rm i})}, \tag{41.29}$$

$$(\delta - \delta_{\rm i})|_{\xi=0} = -\left(\frac{L}{\mathcal{L}}\right)\frac{2r(\mu_{\rm i})}{\upsilon_0(1+2\mu_{\rm i}/\upsilon_0)\sigma(\mu_{\rm i})}, \tag{41.30}$$

where we have introduced the diffusion lengths $\mathcal{L} = (\nu_{\rm eh}/\nu)\,\mathcal{L}_{\rm eh} = (\beta_{\rm eh}/\beta_{\rm d})\mathcal{L}_{\rm eh} \gg \mathcal{L}_{\rm eh}$ associated with the scattering on disorder.

41.5 Current-Voltage Characteristics

The inter-GL current is calculated using the expression for the rate of the pertinent transitions as a function of the local values of the Fermi energy integrated over the GL length

$$J = \frac{\overline{J}}{2}\int_{-1}^{1} d\xi r(\mu). \tag{41.31}$$

Here,

$$\overline{J} = (2e\Sigma_{\rm T}LH/\tau_{\rm nr}) = (\pi eLHT^2/3\hbar^2\upsilon_{\rm W}^2\tau_{\rm nr}), \tag{41.32}$$

where H is the width of GLs (in the direction perpendicular to the current), and $r(\mu)$ is given by Eq. (41.20). If $v \ll v_{eh}$, so that $\mu \simeq \mu_i \simeq$ *const* in the main parts of GLs, Eq. (41.31) can be rewritten as

$$J \simeq \bar{J} r(\mu_i) = J_i. \tag{41.33}$$

At the tunneling resonance at which $\delta_i = 0$ and $\mu_i = \mu_g$, the current is equal to (if $2\mu_g < \hbar\omega_0/T$, i.e., $\mu_g < 4$)

$$J_i^{max} \simeq \bar{J}(e^{2\mu_g} + \eta) = \bar{J}(e^{eV_{bi}/T} + \eta), \tag{41.34}$$

if $eV_{bi} < \hbar\omega_0$, and

$$J_i^{max} \simeq \bar{J}(e^{\hbar\omega_0/T} + \eta), \tag{41.35}$$

if $eV_{bi} > \hbar\omega_0$. When $\mu_g = 2$, i.e., $V_{bi} = 100$ mV, $\bar{\beta}_{eh} = 0.001$, and $\eta = 100$, Eq. (41.34) yields $J_I^{max}/\bar{J} \simeq 150$. At higher values of $\mu_g \propto V_{bi} \propto \sqrt{V_g}$, the voltage dependence of the current peak height tends to saturation or becomes relatively slow if a relatively weak (non-exponential) dependence on the non-resonant inter-GL transition rate is accounted for.

Figure 41.5 shows the current-voltage characteristics calculated using Eq. (41.30) with Eqs. (41.20), (41.23), and (41.25) for the same parameters as in Fig. 41.4. As seen from Fig. 41.5, the current-voltage characteristics exhibit pronounced maxima corresponding to the tunneling resonances. Indeed, the bias voltage at which the current reaches the maximum coincides with that corresponding to $\delta_i = 0$ (see Fig. 41.4a). It should be noted that the width, ΔV, of the maxima markedly exceeds the width, $\gamma/e = 1$ mV, of the resonant maximum of the inter-GL transition rate $r(\mu)$ by the order of magnitude. However, the most remarkable feature of the obtained current-voltage characteristics is the transformation of their shape from the N-type (curves 1 and 2) to the Z-type (curves 3 and 4) when the parameter $\bar{\beta}_{eh}$ increases. As seen from Fig. 41.5, the current-voltage characteristics become multi-valued (when the voltage dependences of the Fermi energy and the energy gap become of the S-shape) in certain voltage ranges if the parameter $\bar{\beta}_{eh}$ is sufficiently large (i.e., if the length of GL, L, is relatively large). The effect of broadening of the resonant maxima in the current-voltage characteristics is due to relatively slow variations of the energy gap with varying bias voltage. Indeed, the width of the current-voltage resonant peak can be estimated as $\Delta V/(\gamma/d\delta_i/dv) = \Delta V/(\gamma/d\Delta_i/dV)$. Since $d\delta_i/dv = d\Delta_i/$

$dV \ll 1$ (see Fig. 41.4a), $\Delta V \gg \gamma$. Even at $\overline{\beta}_{eh} \ll 1$, when 2DEG and 2DHG in the pertinent GLs are close to equilibrium, $d\delta_i/dv \simeq 2\mu_g/(v_0 + 2\mu_g) = V_{bi}/(V_0 + V_{bi}) \simeq 0.275$ for the parameters used in the above calculations. An increase in β results in further decreasing of $d\delta_i/dv$ and, hence, increasing the width of the current-voltage resonant peak. The occurrence of the voltage range, where they are multi-valued, i.e., the transformation of the N-shaped current-voltage characteristics to the Z-shaped ones can be attributed to the potential drop across GLs. Similar effect in quantum-well resonant-tunneling transistors and other tunneling devices was considered previously [21–25] (see also Ref. [26]). For the values $\tau_r = (1 - 3) \times 10^{-9}$ s^{-1} and $\tau_{rt} = (1 - 3) \times 10^{-11}$ s^{-1} used above and $2LH = 0.6$ μm^2 (as in Ref. [1]), the tunneling resonant peak current J_i^{max} ($J_I^{max}/\overline{J} \simeq 150$) is in the range of 50–150 nA, i.e., of the same order of magnitude as in the experiment [1].

An increase in the gate voltage V_g leads to an increase in the Fermi energy ε_F (i.e., μ) and, hence, to an increase in V_{bi} (i.e., μ_g). As a result, the position of the current peak shifts to higher bias voltages.

The shift of the current maxima associated with an increase in ε_F and, consequently, V_{bi} occurs when the levels of the chemical doping of GLs are elevated (the donor density in the top GL and the acceptor density in the bottom GL increase). In this case, the resonant condition is achieved at higher bias voltage V. Just such doped structures were studied in Ref. [1], in which the current peaks correspond to higher values of the bias voltages (V about several tenth of Volt) than in Fig. 41.5 ($V \sim 0.1$ V) for the chemically undoped GLs. However, an increase in the dopant densities leads to an increase in the collision frequency ν. This can result in a marked effect of the disorder scattering to the spatial distributions of the Fermi energy, the energy gap [see Eqs. (41.27) and (41.28)], and the current. Due to this, the function $r(\mu)$ in Eq. (41.31) becomes coordinate-dependent. This leads to lowering of the current peak and its additional broadening (inhomogeneous). Indeed, using Eqs. (41.27) and (41.31), at $\mu_i = \mu_g$ (when $\delta_i = 0$) and integrating over $d\xi$, we obtain

$$J^{max} \simeq J_i^{max}\left[1 + \frac{\beta_d^2(\mu_g)r^2(\mu_g)}{15(1+2\mu_g/v_0)^2\sigma^4(\mu_g)}r''(\mu_g)\right], \qquad (41.36)$$

where $r''(\mu_g) = d^2r(\mu_i)/d\mu_g^2$. Since $r''(\mu_g) \simeq -12b^2 d\mu_g^2$ is negative, from Eq. (41.33), we obtain $J^{max} < J_i^{max}$ with

$$J_i^{max} - J^{max} \simeq \frac{12b^2\beta_d^2(\mu_g)(e^{2\mu_g}+\eta)^2\mu_g^2}{15(1+2\mu_g/v_0)^2\sigma^4(\mu_g)}.$$

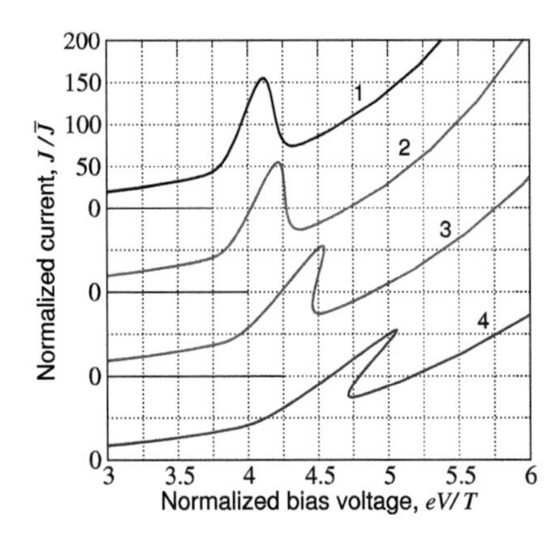

Figure 41.5 Current-voltage characteristics for different values of parameter $\bar{\beta}_{eh}$: 1—$\bar{\beta}_{eh}$ = 0.001, 2—0.002, 3—0.005, and 4—0.01; transition from N-shaped to Z-shaped characteristics.

Taking into account that at $\mu > 1\beta_d \simeq \bar{\beta}_d\mu^2$ (where $\bar{\beta}_d$ is independent of μ_g and $\sigma(\mu) \propto \mu^2$, Eq. (41.35) can be presented as

$$J_i^{max} - J^{max} \sim \bar{\beta}_d^2 b^2 \eta^2. \tag{41.37}$$

The latter quantity is proportional to a small factor $\bar{\beta}_d^2$ and two large factors b^2 and η^2. Hence, even relatively weak scattering on disorder can markedly affect the shape of the maxima in the current-voltage characteristics.

The current-voltage characteristics of the N-, S-, and Z-type can lead to the instability of uniform (or virtually uniform) spatial distributions of the electron and hole densities and the electric potential (in particular, to the formation of density domains or current filamentation). The stability of the uniform states in the main part of the double-GL structures under consideration when their current-voltage characteristics are of the N-shape was considered recently [9]. The problem of the stability in the case of the Z-shaped characteristics predicted above is going to be considered elsewhere.

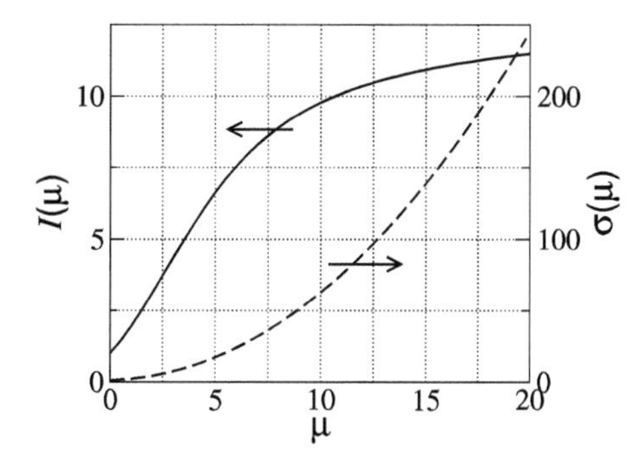

Figure 41.6 Electron-hole collision function $I(\mu)$ and normalized density $\sigma(\mu)$.

41.6 Conclusions

In conclusion:

(1) We have developed the device model for the double-GL structures with independently contacted GLs, which describes the processes of the electron and hole injections from the opposite contacts accompanied with the inter-GL electron-hole recombination.

(2) Using the model, we have derived analytical expressions for the spatial distributions of the electron and hole Fermi energies and the energy gap between the Dirac points in GLs as functions of the bias and gate voltages and the structural parameters. It has been shown that these quantities can be virtually coordinate-independent in the main parts of GLs (except narrow near contact regions) if the mutual scattering of electrons and holes dominate. An increase in relative strength of electron and hole scatterings on disorder can lead to substantial sag of the spatial dependences in question with the minima in the center of the GL structures. The shape of the Fermi energy versus voltage characteristics varies from the monotonic to the S-type with increasing length of GLs.

(3) We have calculated the current-voltage characteristics and revealed that they exhibit maxima associated with the resonant-tunneling inter-GL transitions (the N-type characteristics) in the GL-structures with relatively short GLs. In the structures with long GLs, the N-type characteristics

can transform to the Z-type characteristics. In the latter case, the effect of hysteresis can exist. This should be taken into account choosing the range of operation voltages of the lasers based on the double-GL structures [11, 27].

(4) The obtained results are in line with recent experimental observations [1]. They can be useful for the development and optimization of double-GL-based field-effect transistors terahertz detectors and terahertz lasers.

Acknowledgments

The authors are grateful to A. Satou and D. Svintsov for useful discussions. This work was supported by the Japan Society for Promotion of Science (Grant-in-Aid for Specially Promoting Research, No. 23000008), Japan. The work at MIPT was supported by the Russian Foundation for Basic Research (Grant Nos. 11-07-12072, 11-07-00505, and 12-07-00710). The work at UB was supported by the NSF-TERANO Program.

Appendix: Functions $I(\mu)$ and $\sigma(\mu)$

The function $\sigma(\mu)$ is given by Eq. (41.19). At $\mu \to 0$, $\sigma(\mu) \to 1$, while at large μ, one obtains $\sigma(\mu) \simeq 6\mu^2/\pi^2$. The function ($\beta_{eh} \propto$ $I(\mu) = i(\mu)/i(0)$, where according to Ref. [13] (see, also, Ref. [20]) and considering the Thomas-Fermi screening of the electron-hole interaction

$$i(\mu) = \int_0^\infty\int_0^\infty dx dx' F(x-\mu)F(x'-\mu) \times \int_{-x'}^{x} dQ[1-F(x-Q-\mu)][1-F(x'+Q-\mu)] \times \frac{\sqrt{xx'(x-Q)(x'+Q)}Q^2}{[|Q|-4\alpha\ln F^2(\mu)]^2}, \quad (41.A1)$$

where $F(x) = (1 + e^x)^{-1}$ and $\alpha = e^2/\kappa\hbar v_W$ (at $\kappa = 4$, $\alpha \simeq 0.505$). Figure 41.6 shows $\sigma(\mu)$ and $I(\mu)$ calculated numerically using Eqs. (41.16) and (41.A1), respectively.

References

1. L. Britnell, R. V. Gorbachev, A. K. Geim, L. A. Ponomarenko, A. Mishchenko, M. T. Greenaway, T. M. Fromhold, K. S. Novoselov, and L. Eaves, *Nat. Commun.*, **4**, 1794–1799 (2013).
2. L. Britnell, R. V. Gorbachev, R. Jalil, B. D. Belle, F. Shedin, A. Mishenko, T. Georgiou, M. I. Katsnelson, L. Eaves, S. V. Morozov, N. M. R. Peres, J. Leist, A. K. Geim, K. S. Novoselov, and L. A. Ponomarenko, *Science*, **335**, 947 (2012).
3. T. Georgiou, R. Jalil, B. D. Bellee, L. Britnell, R. V. Gorbachev, S. V. Morozov, Y.-J. Kim, A. Cholinia, S. J. Haigh, O. Makarovsky, L. Eaves, L. A. Ponomarenko, A. K. Geim, K. S. Novoselov, and A. Mishchenko, *Nat. Nanotechnol.*, **7**, 100 (2013).
4. M. Liu, X. Yin, and X. Zhang, *Nano Lett.*, **12**, 1482–1485 (2012).
5. V. Ryzhii, T. Otsuji, M. Ryzhii, V. G. Leiman, S. O. Yurchenko, V. Mitin, and M. S. Shur, *J. Appl. Phys.*, **112**, 104507 (2012).
6. V. Ryzhii, T. Otsiji, M. Ryzhii, V. G. Leiman, S. O. Yurchenko, V. Mitin, and M. S. Shur, *J. Appl. Phys.*, **112**, 104507 (2012).
7. P. Zhao, R. M. Feenstra, G. Gu, and D. Jena, *IEEE Trans. Electron Devices*, **60**, 951 (2013).
8. V. Ryzhii, T. Otsuji, M. Ryzhii, and M. S. Shur, *J. Phys. D: Appl. Phys.*, **45**, 302001 (2012).
9. V. Ryzhii, A. Satou, T. Otsuji, M. Ryzhii, V. Mitin, and M. S. Shur, *J. Phys. D: Appl. Phys.*, **46**, 315107 (2013).
10. V. Ryzhii, M. Ryzhii, V. Mitin, M. S. Shur, A. Satou, and T. Otsuji, *J. Appl. Phys.*, **113**, 174506 (2013).
11. V. Ryzhii, A. A. Dubinov, V. Y. Aleshkin, M. Ryzhii, and T. Otsuji, *Appl. Phys. Lett.*, **103**, 163507 (2013).
12. Z. S. Gribnikov, A. N. Korshak, and V. V. Mitin, *J. Appl. Phys.*, **83**, 1481 (1998).
13. D. Svintsov, V. Vyurkov, S. Yurchenko, T. Otsuji, and V. Ryzhii, *J. Appl. Phys.*, **111**, 083715 (2012).
14. V. Ryzhii, I. Semenikhin, M. Ryzhii, D. Svintsov, V. Vyurkov, A. Satou, and T. Otsuji, *J. Appl. Phys.*, **113**, 244505 (2013).
15. R. M. Feenstra, D. Jena, and G. Gu, *J. Appl. Phys.*, **111**, 043711 (2012).
16. F. T. Vasko, *Phys. Rev. B*, **87**, 075424 (2013).
17. F. Rana, P. A. George, J. H. Strait, S. Shivaraman, M. Chandrashekhar, and M. G. Spever, *Phys. Rev. B*, **79**, 115447 (2009).

18. V. Ryzhii, M. Ryzhii, V. Mitin, and T. Otsuji, *J. Appl. Phys.*, **110**, 094503 (2011).
19. V. Ryzhii, M. Ryzhii, A. Satou, T. Otsuji, A. A. Dubinov, and V. Y. Aleshkin, *J. Appl. Phys.*, **106**, 084507 (2009).
20. A. Kashuba, *Phys. Rev. B*, **78**, 085415 (2008).
21. V. Ryzhii, O. Kosatykh, B. Tolstikhin, and I. Khmyrova, *Sov. Microelectron.*, **18**, 84 (1989).
22. V. Ryzhii and I. Khmyrova, *Sov. Phys. Semicond.*, **25**, 387 (1991).
23. B. A. Glavin, V. A. Kochelap, and V. V. Mitin, *Phys. Rev. B*, **56**, 13346 (1997).
24. M. Meixner, P. Rodin, E. Scholl, and A. Wacker, *Eur. Phys. J. B*, **13**, 157 (2000).
25. O. V. Pupysheva, A. V. Dmitriev, A. A. Farajuan, H. Mizuseki, and Y. Kawazoe, *J. Appl. Phys.*, **100**, 033718 (2006).
26. A. Wacker and E. Scholl, *J. Appl. Phys.*, **78**, 7352 (1995).
27. V. Ryzhii, A. A. Dubinov, T. Otsuji, V. Y. Aleshkin, M. Ryzhii, and M. S. Shur, *Opt. Express*, **21**, 31567 (2013).

Chapter 42

Voltage-Tunable Terahertz and Infrared Photodetectors Based on Double-Graphene-Layer Structures*

V. Ryzhii,[a,b] T. Otsuji,[a] V. Ya. Aleshkin,[c] A. A. Dubinov,[c] M. Ryzhii,[d] V. Mitin,[e] and M. S. Shur[f]

[a]*Research Institute for Electrical Communication, Tohoku University, Sendai 980-8577, Japan*
[b]*Institute of Ultra High Frequency Semiconductor Electronics, Russian Academy of Sciences, Moscow 111005, Russia*
[c]*Institute for Physics of Microstructures of Russian Academy of Sciences, and Lobachevsky State University of Nizhny Novgorod, Nizhny Novgorod 603950, Russia*
[d]*Department of Computer Science and Engineering, University of Aizu, Aizu-Wakamatsu 965-8580, Japan*
[e]*Department of Electrical Engineering, University at Buffalo, Buffalo, New York 1460-1920, USA*
[f]*Department of Electrical, Electronics, and System Engineering, Rensselaer Polytechnic Institute, Troy, New York 12180, USA*
v-ryzhii@riec.tohoku.ac.jp

*Reprinted with permission from V. Ryzhii, T. Otsuji, V. Ya. Aleshkin, A. A. Dubinov, M. Ryzhii, V. Mitin, and M. S. Shur (2014). Voltage-tunable terahertz and infrared photodetectors based on double-graphene-layer structures, *Appl. Phys. Lett.*, **104**, 163505. Copyright © 2014 AIP Publishing LLC.

Graphene-Based Terahertz Electronics and Plasmonics: Detector and Emitter Concepts
Edited by Vladimir Mitin, Taiichi Otsuji, and Victor Ryzhii

ISBN 978-981-4800-75-4 (Hardcover), 978-0-429-32839-8 (eBook)
www.jennystanford.com

We propose and theoretically substantiate the concept of terahertz and infrared photodetectors using the resonant radiative transitions between graphene layers (GLs) in double-GL structures. The calculated absorption spectrum and the spectral characteristics of the photodetector responsivity exhibit sharp resonant maxima at the photon energies in a wide range. The resonant maxima can be tuned by the applied voltage. We compare the photodetector responsivity with that of the GL p-i-n photodiodes and quantum-well infrared photodetectors. Weak temperature dependences of the photocurrent and dark current enable the effective operation of the proposed photodetector at room temperature.

Fabrication and exploration of double-graphene-layer (double-GL) structures with narrow inter-GL barrier Boron Nitride (bBN), Tungsten Disulfide (WS_2), and other barrier layers [1–4] have recently attracted much attention due to their potential applications in high speed modulators of terahertz (THz) and infrared (IR) radiation [1, 5], transistors [3], THz photomixers [6], and lasers [7, 8]. The THz lasing in the latter devices is due to the inter-GL population inversion and the inter-GL resonant radiative transitions. The double-GL-based lasers might exhibit advantages over the THz lasers exploiting the intra-GL interband population inversion and vertical radiative transitions [9–14] (see also Refs. [15–18]). The nonlinearity of the inter-GL tunneling current in the double-GL structures [3, 4, 19, 20] can also be used for the detection of THz radiation [21, 22]. Various graphene-based structures are attractive for the interband THz and IR photodetectors [18, 23–33]. In this chapter, we propose and evaluate THz/IR photodetectors based on the double-GL structures using the inter-GL resonant optical absorption.

The double-GL photodetectors (DGL-PDs) can exhibit some advantages over the GL-based photodetectors using the intra-GL interband transitions. These advantages are: (1) resonant voltage-tunable spectrum and (2) lower dark current (due to its non-resonant tunneling nature). The latter property should result in large detectivity, particularly, at elevated (room) temperatures. The DGL-PDs under consideration can also find applications in the frequency range between 6 and 10 THz (corresponding to the photon energies $\hbar\omega \simeq 25 - 40$ meV and to the wavelength $\lambda = 30 - 50$ μm), which includes the optical phonon frequencies in A_3B_5 materials and not accessible for such photodetectors as quantum-well and quantum-dot infrared photodetectors (QWIPs and QDIPs) [34, 35] due to high absorption by optical phonons.

We consider the DGL-PD structure shown in Fig. 42.1. The structure in Fig. 42.1 consists of two independently contacted GLs in which the electrons and holes are induced by the voltages $\pm V_g/2$ applied to the highly conducting gates ("electrical" doping). The top electrode is a grated structure with the grating providing coupling to the incident THz or IR radiation. The bottom gate can serve not only as the gate controlling the carrier density but also as the reflector of the incident radiation. The bias voltage V is applied between the pertinent contacts. These structures are similar to those recently fabricated [1–4]. The grating is necessary because the inter-GL radiative transitions are associated with the component of the radiation electric field perpendicular to the GL plane as in n-type QWIPs (other methods of the radiation coupling can be used as well). The DGL-PDs can also use the chemically doped GLs with or without gates.

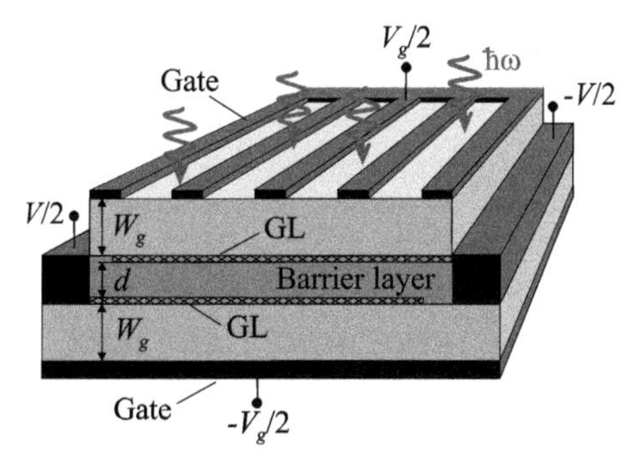

Figure 42.1 Schematic view of DGL-PD structure with "electrical doping" and top gate serving as the grating coupler.

Under operation conditions, the top GL (n-type GL) is filled the two-dimensional electron gas, while the bottom GL (p-type GL) is filled with the two-dimensional hole gas. The energy gap between the Dirac points in GLs is equal to

$$\frac{\Delta}{e} = V + V_0 - \sqrt{2VV_0 + V_0^2 + V_t^2}, \tag{42.1}$$

where $V_0 = \hbar^2 v_W^2 \kappa/2e^3 d$, $V_t = 2\hbar v_W \sqrt{\pi\Sigma_i}/e$, $\Sigma_i \propto V_g/W_g$, is the electron and hole densities induced by the gate voltages (or the densities of donors and acceptors in DGL-PDs with the chemical doping), d and W_g are the thicknesses of the barrier and gate layers, respectively (see Fig. 42.1), κ is the dielectric constant, $e = |e|$ is the electron charge, $\hbar$ is the Planck constant, and $v_W \simeq 10^8$ cm/s is the characteristic velocity of electrons and holes in GLs. Assuming $\kappa = 4$, $d = 4$ nm, and $\Sigma_i = 10^{12}$ cm^{-2}, one obtains $V_0 \simeq 136$ mV and $V_t \simeq 221$ mV. Figure 42.2 shows the band diagrams at $V = 0$ and $0 < V < V_t$. If $V = 0$, the Fermi levels in GLs are flat (see Fig. 42.2a) and $\Delta = -(\mu_e + \mu_h) = -2\mu$, where $\mu_e = \mu_h = \mu$ are the electron and hole Fermi energies at $V = 0$, respectively. At $V \leq V_t$, Eq. (42.1) yields $\Delta \leq 0$.

The operation of DGL-PDs is associated with the absorption of the incident IR radiation accompanied by the electron tunneling transitions between GLs and causing the electric terminal current. The inter-GL radiative transitions with the absorption of photons (shown schematically in Fig. 42.2) with the energy $\hbar\omega$ conserve the electron lateral momentum and, hence, do not involve scattering (resonant-tunneling photon-assisted transitions), if

$$\hbar\omega \simeq -\Delta + \hbar\omega_{dep} = \hbar\omega_{max}. \quad (42.2)$$

Here,

$$\hbar\omega_{dep} = \frac{8\pi e^2 |z_{u,l}|^2}{\kappa d}\left(\Sigma_i + \frac{\kappa\Delta}{4\pi e^2 d}\right) \quad (42.3)$$

is the depolarization shift (see, for example, Refs. [36, 37]), $z_{u,l} = \int \varphi_u^*(z) z \varphi_l(z) dz$ is the matrix element of the inter-GL transitions, where $\varphi_u(z)$ and $\varphi_l(z)$ are the z-dependent factors of the wave functions in the upper and lower GLs, respectively (the axis z is directed perpendicular to the GL plane).

The real part of the double-GL structure dynamic conductivity in the direction perpendicular to the GL plane $\sigma_{zz}(\omega)$ can be presented as (compare with Refs. [7, 8])

$$\mathrm{Re}\sigma_{zz}(\omega) = \left(\frac{e^2}{\hbar}\right)\frac{2|z_{u,l}|^2 \gamma\hbar\omega}{[\hbar^2(\omega - \omega_{max})^2 + \gamma^2]}\left(\Sigma_i + \frac{\kappa\Delta}{4\pi e^2 d}\right). \quad (42.4)$$

Here, $\gamma \simeq \hbar\nu$ is the relaxation broadening and ν is the collision frequency of electrons and holes. Equation (42.4) accounts for the transitions both between the conduction and valence band states. The quantity $\hbar\omega_{max}$ in Eq. (42.4) corresponds to the maximum probability of the inter-GL radiative transitions. In contrast to the double-GL-based devices considered in Refs. [7, 8], the doping level of GL Σ_i and the bias voltage V are chosen in such a way that $-\Delta + \hbar\omega_{dep} > 0$.

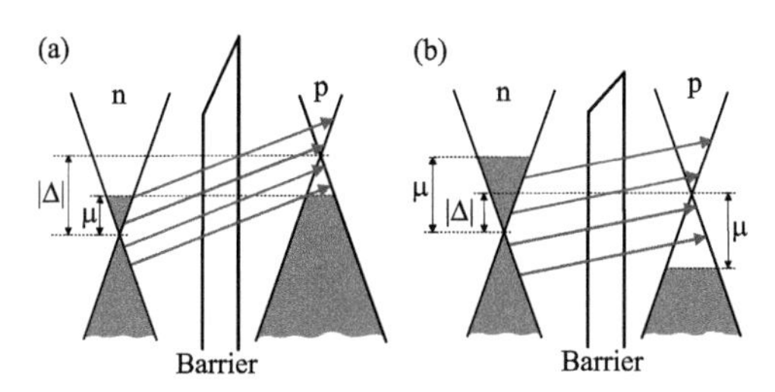

Figure 42.2 DGL-PD band diagrams at (a) $V = 0$ and (b) $0 < V < V_t$. Arrows schematically indicate the photon-assisted inter-GL resonant-tunneling transitions between the initial (in n-type GL) and final states (in p-type GL) in the conduction bands and between such states in the valence bands in GLs.

According to Eq. (42.4), the probability of the inter-GL transition with the absorption of an incident photon with the energy $\hbar\omega$ is given by

$$\beta_\omega = \left(\frac{\pi e^2}{c\hbar}\right)\frac{8e|z_{u,l}|^2\,\gamma\hbar\omega}{[\hbar^2(\omega-\omega_{max})^2+\gamma^2]}\left(\Sigma_i + \frac{\kappa\Delta}{4\pi\, e^2 d}\right)\theta. \tag{42.5}$$

Here, $\theta < 1$ is the input efficiency (determined by the properties of the grating, reflection, and so on). This yields the following spectral dependence of the DGL-PD responsivity:

$$R_\omega = \left(\frac{\pi e^2}{c\hbar}\right)\frac{8|z_{u,l}|^2\,\gamma}{[\hbar^2(\omega-\omega_{max})^2+\gamma^2]}\left(\Sigma_i + \frac{\kappa\Delta}{4\pi\, e^2 d}\right)\theta. \tag{42.6}$$

Using Eqs. (42.3) and (42.6), the maximum of the DGL-PD responsivity, which is achieved at $\hbar\omega = \hbar\omega_{max}$, can be presented as

$$R_{\omega_{max}} = \left(\frac{e\kappa d}{c\hbar}\right)\left(\frac{\hbar\omega_{dep}}{\gamma}\right)\theta. \tag{42.7}$$

Figure 42.3 shows the DGL-PD responsivity R_ω versus the photon energy $\hbar\omega$ calculated for different voltages V using Eq. (42.6) with Δ and $\hbar\omega_{max}$ given by Eqs. (42.1) and (42.2), respectively. The parameters used in the calculations were $\kappa = 4$, $\gamma = 1$ meV (i.e., $\nu = 1.6 \times 10^{12}\ \mathrm{s}^{-1}$), $\Sigma_i = 10^{12}\ \mathrm{cm}^{-2}$, $d = 2$ and 4 nm, ($V_0 = 136 - 272$ mV and $V_t = 221$ mV), and $\theta = 0.5$. The quantity $|z_{u,l}|^2$ was calculated for different inter-GL barrier layer (WS_2) thicknesses d as in Refs. [7, 8, 38]. As seen from Fig. 42.3, the DGL-PD responsivity exhibits fairly sharp peaks associated with the inter-GL resonant-tunneling transitions accompanied by the absorption of the incident photons. The peak values of the responsivity are rather high in a wide spectral range.

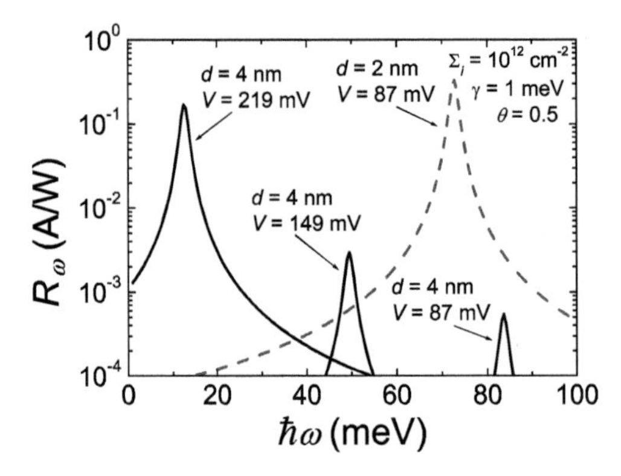

Figure 42.3 Dependences of DGL-PD responsivity R_ω on photon energy $\hbar\omega$ for the inter-GL barrier layer thickness d = 4 nm (solid lines) and d = 2 nm (dashed line) at different applied voltages V.

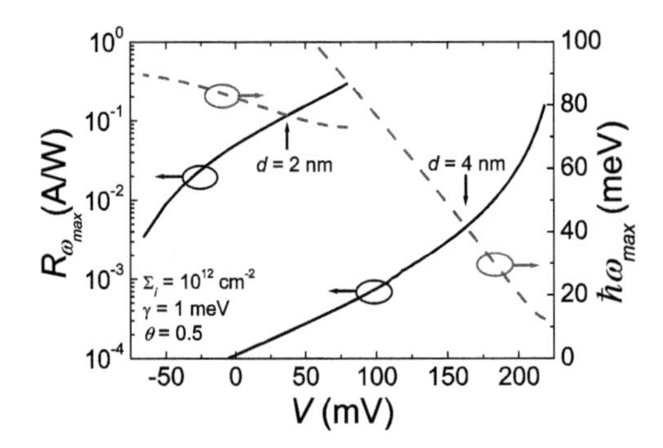

Figure 42.4 Dependences of the responsivity maximum $R_{\omega_{\max}}$ (solid lines) and corresponding photon energy $\hbar\omega_{\max}$ (dashed lines) on bias voltages V for different thicknesses d.

Figure 42.4 shows the dependences of $\hbar\omega_{\max}$ and $R_{\omega_{\max}}$ on the applied voltage V calculated for different thicknesses, d, of the inter-GL barrier layer. A marked shift of the responsivity maxima with varying bias voltage V (see Figs. 42.3 and 42.4) enables the DGL-PD spectrum voltage tuning. However, the height of the responsivity maxima is very sensitive to the bias voltage V (compare peaks for d = 4 nm in Fig. 42.3). This is also seen in Fig. 42.4. The maximum value of the DGL-PD responsivity markedly depends on the electrical doping determined by the gate voltage V_g (see Fig. 42.5).

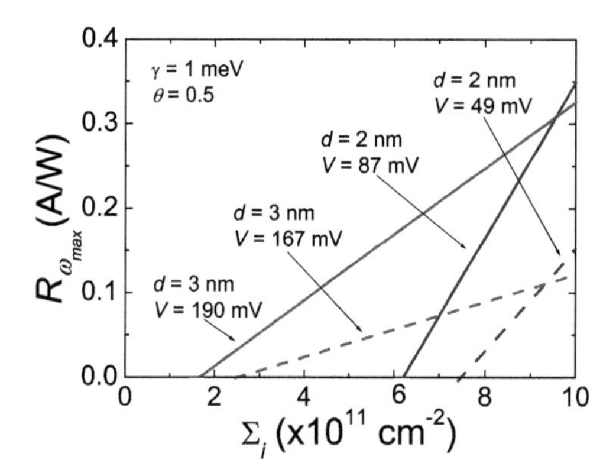

Figure 42.5 Dependence of the responsivity maximum $R_{\omega_{\max}}$ on the electron and hole density $\Sigma_i \propto V_g/W_g$ at different bias voltages V and thicknesses d.

For comparison, the responsivities of a single-GL p-i-n-PD and QWIPs are as follows [27, 34, 35]:

$$R_{\omega}^{\text{pin}} \simeq \left(\frac{\pi e^2}{c\hbar}\right)\frac{eg^{\text{pin}}}{\hbar\omega}, \quad R_{\omega}^{\text{qwip}} \simeq \left(\frac{e}{\hbar\omega}\right)\sigma_{\text{i}}\Sigma_{\text{i}}g^{\text{qwip}}\theta, \tag{42.8}$$

where g^{pin} and g^{qwip} are the GL p-i-n-PD and QWIP photoelectric gains, respectively, and σ_{i} is the QW photoionization cross-section by an incident photon (with the energy close to the gap between the edge of the QW and the subband bottom). In the DGL-PDs with relatively short i-region with the length of this region shorter than the electron and hole bipolar diffusion length l_{D}, one obtains $g^{\text{pin}} \lesssim 2$. But in the DGL-PDs with a long i-region, $g^{\text{pin}} \sim (l_{\text{D}}/l)^2 \ll 1$. The QWIP photoelectric gain g^{qwip} is approximately equal to the probability of the capture of the electron crossing the QW ($g^{\text{qwip}} \geq 1$).

Thus, at $\omega = \omega_{\max}$, from Eqs. (42.7) and (42.8), we obtain

$$\frac{R_{\omega_{\max}}}{R_{\omega_{\max}}^{\text{pin}}} \simeq \frac{\hbar^2\omega_{\max}\omega_{\text{dep}}\theta}{\varepsilon_d\gamma g^{\text{pin}}}, \quad \frac{R_{\omega_{\max}}}{R_{\omega_{\max}}^{\text{qwip}}} \simeq \frac{\hbar^2\omega_{\max}\omega_{\text{dep}}}{\varepsilon_{\text{i}}\gamma g^{\text{qwip}}}. \tag{42.9}$$

Here, $\varepsilon_{\text{d}} = 2\pi e^2/\kappa d$ and $\varepsilon_{\text{i}} = c\hbar\sigma_{\text{i}}\Sigma_{\text{i}}/\kappa d$.

Using the above formulas and assuming $\Sigma_i = 10^{12}$ cm^{-2}, $\sigma_i \sim 2 \times 10^{-15}$ cm^{-2} [39], d = 2 nm, and $\hbar\omega$ = 72 meV at which R_ω = 0.3 A/W (see Fig. 42.3), one obtains $R_{\omega_{\max}}/R^{\rm pin}_{\omega_{\max}} \sim \theta/g^{\rm pin}$ and $R_{\omega_{\max}}/R^{\rm qwip}_{\omega_{\max}} \sim 20/g^{\rm qwip}$. This shows that the peak responsivity of GBL-PDs can exceed that of GL p-i-n-PDs and QWIPs with a moderate photoelectric gain. Figure 42.6 shows the ratio of the responsivities of DGL-PDs and QWIPs (with $g^{\rm qwip}$ = 1) versus the photon energy at which the responsivity achieves the maxima calculated for different doping levels and the inter-GL barrier layer thicknesses.

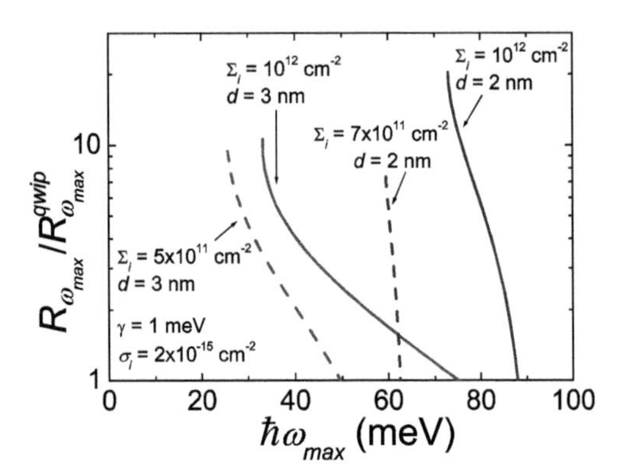

Figure 42.6 Ratio of DGL-PD and QWIP responsivities $R_{\omega_{\max}}/R^{\rm qwip}_{\omega_{\max}}$ versus $\hbar\omega_{\max}$ for different Σ_i and d.

Since the dark current in DGL-PDs is associated with the inter-GL tunneling [2, 3], at elevated temperatures it can be much smaller than the dark current in GL p-i-n- PDs [27]. This and a weak temperature dependence of the responsivity are beneficial for achieving high values of the detectivity at elevated temperatures, including the room temperature. However, at the voltage corresponding to Δ = 0, the inter-GL transitions become of the resonant-tunneling origin [3, 19, 20], and the dark current in DGL-PDs can exhibit a pronounced peak.

The grating gate period and the gate layer thickness should be optimized to maximize the radiation coupling and limit the spatial periodicity of the carrier (electron) density in the top GL. This periodicity might add more complexity to the DGL-PD spectral characteristics (see, for example, Ref. [40]). The pertinent effects require a separate treatment.

In conclusion, we evaluated the responsivity of the proposed DGL-PDs and demonstrated that the DGL-PD responsivity as a function of the photon energy can exhibit the voltage tunable resonant maxima in a wide spectral range. This provides the possibility of implementing of effective DGL-PDs competitive with other THz and IR photodetectors, especially at elevated (room) temperatures.

This work was supported by the Japan Society for Promotion of Science (Grant-in-Aid for Specially Promoting Research, No. 23000008), Japan, and by the Dynasty Foundation, Russia. The works at UB and RPI were supported by the NSF-TERANO Program and the U.S. Army Cooperative Research Agreement (Project Manager Dr. Meredith Reed), respectively.

References

1. M. Liu, X. Yin, and X. Zhang, *Nano Lett.*, **12**, 1482 (2012).
2. L. Britnell, R. V. Gorbachev, R. Jalil, B. D. Belle, F. Shedin, A. Mishenko, T. Georgiou, M. I. Katsnelson, L. Eaves, S. V. Morozov, N. M. R. Peres, J. Leist, A. K. Geim, K. S. Novoselov, and L. A. Ponomarenko, *Science*, **335**, 947 (2012).
3. T. Georgiou, R. Jalil, B. D. Bellee, L. Britnell, R. V. Gorbachev, S. V. Morozov, Y.-J. Kim, A. Cholinia, S. J. Haigh, O. Makarovsky, L. Eaves, L. A. Ponomarenko, A. K. Geim, K. S. Novoselov, and A. Mishchenko, *Nat. Nanotechnol.*, **8**, 100 (2013).
4. L. Britnell, R. V. Gorbachev, A. K. Geim, L. A. Ponomarenko, A. Mishchenko, M. T. Greenaway, T. M. Fromhold, K. S. Novoselov, and L. Eaves, *Nat. Commun.*, **4**, 1794 (2013).
5. V. Ryzhii, T. Otsuji, M. Ryzhii, V. G. Leiman, S. O. Yurchenko, V. Mitin, and M. S. Shur, *J. Appl. Phys.*, **112**, 104507 (2012).
6. V. Ryzhii, M. Ryzhii, V. Mitin, M. S. Shur, A. Satou, and T. Otsuji, *J. Appl. Phys.*, **113**, 174506 (2013).
7. V. Ryzhii, A. A. Dubinov, V. Ya. Aleshkin, M. Ryzhii, and T. Otsuji, *Appl. Phys. Lett.*, **103**, 163507 (2013).
8. V. Ryzhii, A. A. Dubinov, T. Otsuji, V. Ya. Aleshkin, M. Ryzhii, and M. S. Shur, *Opt. Express*, **21**, 31567 (2013).
9. V. Ryzhii, M. Ryzhii, and T. Otsuji, *J. Appl. Phys.*, **101**, 083114 (2007).
10. F. Rana, *IEEE Trans. Nanotechnol.*, **7**, 91 (2008).

11. A. A. Dubinov, V. Ya. Aleshkin, M. Ryzhii, T. Otsuji, and V. Ryzhii, *Appl. Phys. Express*, **2**, 092301 (2009).

12. V. Ryzhii, M. Ryzhii, A. Satou, T. Otsuji, A. A. Dubinov, and V. Ya. Aleshkin, *J. Appl. Phys.*, **106**, 084507 (2009).

13. V. Ryzhii, A. A. Dubinov, T. Otsuji, V. Mitin, and M. S. Shur, *J. Appl. Phys.*, **107**, 054505 (2010).

14. V. Ryzhii, M. Ryzhii, V. Mitin, and T. Otsuji, *J. Appl. Phys.*, **110**, 094503 (2011).

15. S. Boubanga-Tombet, S. Chan, T. Watanabe, A. Satou, V. Ryzhii, and T. Otsuji, *Phys. Rev. B*, **85**, 035443 (2012).

16. T. Watanabe, T. Fukushima, Y. Yabe, S. A. Boubanga-Tombet, A. Satou, A. A. Dubinov, V. Ya. Aleshkin, V. Mitin, V. Ryzhii, and T. Otsuji, *New. J. Phys.*, **15**, 075003 (2013).

17. T. Otsuji, S. A. Boubanga Tombet, A. Satou, M. Ryzhii, and V. Ryzhii, *IEEE J. Sel. Top. Quantum Electron.*, **19**, 8400209 (2013).

18. A. Tredicucci and M. S. Vitiello, *IEEE J. Sel. Top. Quantum Electron.*, **20**, 8500109 (2014).

19. R. M. Feenstra, D. Jena, and G. Gu, *J. Appl. Phys.*, **111**, 043711 (2012).

20. F. T. Vasko, *Phys. Rev. B*, **87**, 075424 (2013).

21. V. Ryzhii, T. Otsuji, M. Ryzhii, and M. S. Shur, *J. Phys. D: Appl. Phys.*, **45**, 302001 (2012).

22. V. Ryzhii, A. Satou, T. Otsuji, M. Ryzhii, V. Mitin, and M. S. Shur, *J. Phys. D: Appl. Phys.*, **46**, 315107 (2013).

23. F. Vasko and V. Ryzhii, *Phys. Rev. B*, **77**, 195433 (2008).

24. V. Ryzhii and M. Ryzhii, *Phys. Rev. B*, **79**, 245311 (2009).

25. V. Ryzhii, M. Ryzhii, N. Ryabova, V. Mitin, and T. Otsuji, Jpn. *J. Appl. Phys., Part 1*, **48**, 04C144 (2009).

26. F. Xia, T. Murller, Y.-M. Lin, A. Valdes-Garsia, and F. Avouris, *Nat. Nanotechnol.*, **4**, 839 (2009).

27. V. Ryzhi, M. Ryzhii, V. Mitin, and T. Otsuji, *J. Appl. Phys.*, **107**, 054512 (2010).

28. T. Mueller, F. Xia, and Ph. Avouris, *Nat. Photonics*, **4**, 297 (2010).

29. M. C. Lemme, F. H. L. Koppens, A. L. Falk, M. S. Rudner, H. Park, L. S. Levitov, and C. M. Marcus, *Nano Lett.*, **11**, 4134 (2011).

30. V. Ryzhii, N. Ryabova, M. Ryzhii, N. V. Baryshnikov, V. E. Karasik, V. Mitin, and T. Otsuji, *Opto-Electron. Rev.*, **20**, 15 (2012).

31. M. Freitag, T. Low, and Ph. Avouris, *Nano Lett.*, **4**, 1644 (2013).
32. S. Song, Q. Chen, L. Jina, and F. Suna, *Nanoscale*, **5**, 9615 (2013).
33. K. S. Novoselov, V. I. Falko, L. Colombo, P. R. Gellert, M. G. Schwab, and K. Kim, *Nature*, **490**, 192 (2012).
34. H. C. Liu, H. Luo, D. Ban, M. Wchter, C. Y. Song, Z. R. Wasilewski, M. Buchanan, G. C. Aers, A. J. SpringThorpe, J. C. Cao, S. L. Feng, B. S. Williams, and Q. Hu, *Proc. SPIE*, **6029**, 602901 (2005).
35. V. Ryzhii, I. Khmyrova, M. Ryzhii, and V. Mitin, *Semicond. Sci. Technol.*, **19**, 8 (2004).
36. A. Shik, *Quantum Wells: Physics and Electronics of Two-Dimensional Systems* (World Scientific, London, 1998).
37. F. T. Vasko and A. V. Kuznetsov, *Electronic States and Optical Transitions in Semiconductor Heterostructures* (Springer, New York, 1999).
38. A. A. Dubinov, V. Ya. Aleshkin, V. Ryzhii, M. S. Shur, and T. Otsuji, *J. Appl. Phys.*, **115**, 044511 (2014).
39. F. Luc, E. Rosencher, and Ph. Bois, *Appl. Phys. Lett.*, **62**, 2542 (1993).
40. V. V. Popov, O. V. Polischuk, S. A. Nikitov, V. Ryzhii, T. Otsuji, and M. S. Shur, *J. Opt.*, **15**, 114009 (2013).

Chapter 43

Plasmons in Tunnel-Coupled Graphene Layers: Backward Waves with Quantum Cascade Gain*

D. Svintsov,[a] Zh. Devizorova,[b,c] T. Otsuji,[d] and V. Ryzhii[d]
[a]*Laboratory of 2d Materials' Optoelectronics, Moscow Institute of Physics and Technology, Dolgoprudny 141700, Russia*
[b]*Department of Physical and Quantum Electronics, Moscow Institute of Physics and Technology, Dolgoprudny 141700, Russia*
[c]*Kotelnikov Institute of Radio Engineering and Electronics, Russian Academy of Science, Moscow 125009, Russia*
[d]*Research Institute of Electrical Communication, Tohoku University, Sendai 980-8577, Japan*
svintcov.da@mipt.ru

We theoretically demonstrate that graphene-insulator-graphene tunnel structures can serve as plasmonic gain media due to the possibility of stimulated electron tunneling accompanied by emission of plasmons under application of interlayer voltage. The probability

*Reprinted with permission from D. Svintsov, Zh. Devizorova, T. Otsuji, and V. Ryzhii (2016). Plasmons in tunnel-coupled graphene layers: backward waves with quantum cascade gain, *Phys. Rev. B*, **94**, 115301. Copyright © 2016 American Physical Society.

Graphene-Based Terahertz Electronics and Plasmonics: Detector and Emitter Concepts
Edited by Vladimir Mitin, Taiichi Otsuji, and Victor Ryzhii

ISBN 978-981-4800-75-4 (Hardcover), 978-0-429-32839-8 (eBook)
www.jennystanford.com

of plasmon-assisted tunneling is resonantly large at certain values of frequency and interlayer voltage corresponding to the transitions between chiral electron states with collinear momenta, which is a feature unique to the linear bands of graphene. The plasmon dispersion develops an anticrossing with the resonances in tunnel conductivity and demonstrates negative group velocity in several frequency ranges.

43.1 Introduction

The ultrarelativistic nature of electrons in graphene gives rise to the uncommon properties of their collective excitations–surface plasmons (SPs) [1–3]. The deep subwavelength confinement [2], the unconventional density dependence of frequency [3, 4], and the absence of Landau damping [4] are their most well-known features. Among more sophisticated predictions there stand the existence of transverse electric plasmon modes [5] and quasineutral electron-hole sound at the charge neutrality [6, 7]. It was not until the discovery of van der Waals heterostructures that the low-loss SPs supported by graphene could be observed [8]. The reported propagation length to wavelength ratio reaching 25 looks to be the fundamental limit of SP quality factor at room temperature governed by the electron-phonon interaction [9], which hinders further experimental studies of plasmonic effects. The perception of this fact has motivated the search for the methods to provide the gain of SPs in graphene [10–14].

Here, we theoretically demonstrate that the resonant tunneling structures composed of parallel graphene layers can act as plasmonic gain media by themselves. The very presence of negative differential resistance (NDR) in the static current-voltage characteristics of these structures [15] gives rise to the self-oscillation in electrical circuits [16] and might potentially lead to to the self-excitation of plasmons [17–19]. However, the *static* NDR of graphene tunnel diodes is insufficient to replenish the plasmon losses [20], which calls for the stability of electron plasma.

In this chapter, we show that the dynamic and nonlocal effects in the tunnel conductivity radically change the picture of plasmon propagation. The calculated dynamic tunnel conductivity of a double graphene layer biased by voltage V has a negative real part at frequencies $\omega < eV/\hbar$, even if the static NDR is absent in the structure. This can be viewed as a consequence of the "interlayer population inversion." Surprisingly, the dynamic nonlocal conductivity possesses sharp resonances at certain frequencies and wave vectors q due to the prolonged tunneling interaction between chiral states with collinear momenta in neighboring layers. The singularities in the tunnel conductivity emerge at a series of lines on the ω–q plane, whose pattern is especially rich in twisted layers. At finite bias V, the dispersion of acoustic SPs does not develop a low-frequency gap, as opposed to SPs in coupled layers of massive electrons in equilibrium [21]. Instead, the SP spectrum develops an anti-crossing with the tunnel resonances and demonstrates the parts with negative group velocity. At the same time, the dispersion is quite close to the tunnel resonances, and the tunnel gain can exceed the SP loss due to both inter- and intraband absorption.

The chapter is organized as follows. In Section 43.2 we derive the dispersion relation for plasmons in tunnel-coupled graphene layers. Section 43.3 is devoted to the calculation of high-frequency nonlocal conductivity, which possesses both tunneling and in-plane components. Both components of conductivity affect the spectrum and damping (or gain) of acoustic plasmons, which is analyzed in Section 43.4. Section 43.5 is devoted to the effect of interlayer twist on high-frequency conductivity and SP dispersion. Possible experimental manifestations of the predicted effects are discussed in Section 43.6. Some cumbersome calculations are relegated to the Appendices.

43.2 Dispersion Relation for Plasmons in Tunnel-Coupled Layers

We consider the propagation of plasmons in the graphene-insulator-graphene structure shown in Fig. 43.1a. The application of interlayer voltage V results in the electrical doping of layers; the corresponding filling of the bands is shown in Fig. 43.1b. In the absence of built-in voltage, the density of induced electrons in the top layer equals the density of holes in the bottom one. This results in equal in-plane conductivities of the layers $\sigma_{||}$.

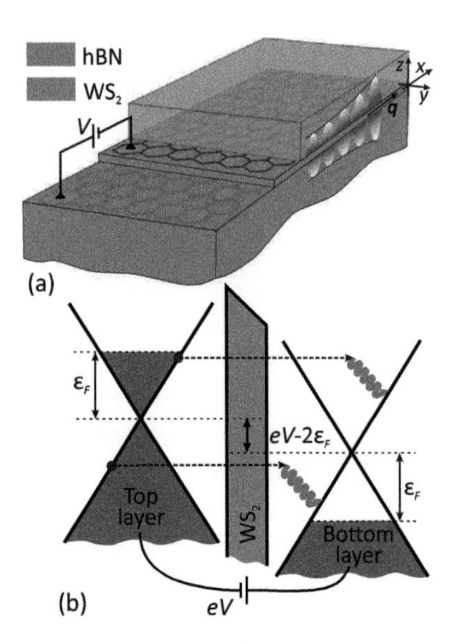

Figure 43.1 (a) Schematic view of the double graphene layer encapsulated in hexagonal boron nitride (hBN) overlaid by the image of acoustic SP amplified by the tunneling. (b) Band diagram of the structure biased by voltage *V*. Inelastic interlayer electron tunneling transitions accompanied plasmon emission (wavy arrows) are shown schematically.

The method of obtaining the SP dispersion relies on the solution of Poisson's equation and the carrier transport equations. The interlayer tunnel current, $\delta J_\perp$, appears as a source term in the continuity equation for the charge densities $\delta Q_{\rm t,b}$ on the top and bottom layers:

$$-i\omega\delta Q_{\rm t,b} = -i\mathbf{q}\delta\mathbf{j}_{\rm t,b} \mp \delta J_\perp. \tag{43.1}$$

Here $\delta\mathbf{j}_{\rm t,b}$ are the in-plane current densities which are related to the electric potentials $\delta\varphi_{\rm t,b}$ via $\delta\mathbf{j}_{\rm t,b} = -i\mathbf{q}\sigma_{||}(\mathbf{q}, \omega)\delta\varphi_{\rm t,b}$. The interlayer current density is

$$\delta J_\perp = G_\perp(\mathbf{q}, \omega)(\delta\varphi_{\rm t} - \delta\varphi_{\rm b}),$$

where $G_\perp(\mathbf{q}, \omega)$ is tunnel conductivity. By matching the solutions of the Poisson's equation (κ is the background dielectric constant)

$$-q^2\delta\varphi(z) + \frac{\partial^2\delta\varphi(z)}{\partial z^2} = -\frac{4\pi}{\kappa}[\delta Q_{\rm t}\delta(z-d/2) + \delta Q_{\rm b}\delta(z+d/2)] \tag{43.2}$$

on graphene planes located at $z = \pm d/2$, we arrive at the plasmon dispersion equation which can be written as

$$\epsilon(\mathbf{q},\omega) \equiv \left[1 + \frac{2\pi i q}{\omega\kappa}\sigma_{||}(1+e^{-qd})\right] \times \left[1 + \frac{2\pi i q}{\omega\kappa}\left(\sigma_{||} + \frac{2G_{\perp}}{q^2}\right)(1-e^{-qd})\right] = 0. \quad (43.3)$$

The zeros of the first and second terms in the dielectric function $\epsilon(\mathbf{q}, \omega)$ yield the dispersions of the optical and acoustic SPs, respectively. It is intuitive that the presence of tunneling does not affect the optical mode, while it can modify the spectrum of the acoustic branch considerably. The reason is that the average interlayer field in the optical mode is zero, and it cannot stimulate the interlayer carrier transfer, contrary to the strong field in the acoustic mode. We note here that the ratio of transverse and in-plane electric fields in the acoustic mode is approximately $2(qd)^{-1} \gg 1$, which also speaks in favor of the strong tunneling effects.

In addition to solving Eq. (43.3) for the mode dispersions $\omega(q)$, we shall study the loss function [22], $-\mathrm{Im}\,\epsilon^{-1}(\mathbf{q}, \omega)$ (also referred to as plasmon spectral weight [23]). Within the RPA treatment, this quantity is proportional to the dynamical structure factor measured with inelastic scattering experiments. To focus on acoustic plasmons, we define the acoustic contribution to the loss function:

$$-\mathrm{Im}\,\epsilon_{\mathrm{ac}}^{-1}(\mathbf{q},\omega) = -\mathrm{Im}\left\{1 + \frac{2\pi i q}{\omega\kappa}\left(\sigma_{||} + \frac{2G_{\perp}}{q^2}\right)(1-e^{-qd})\right\}^{-1}. \quad (43.4)$$

The peaks in $-\mathrm{Im}\,\epsilon_{\mathrm{ac}}^{-1}(\mathbf{q},\omega)$ correspond to the acoustic SPs, while the sign of the loss function determines whether the plasmons are damped or amplified.

43.3 High-Frequency Nonlocal Tunnel Current

The only missing ingredient required for the analysis of surface plasmon modes is the expression for the high-frequency nonlocal tunnel conductivity $G_{\perp}(\mathbf{q}, \omega)$. The theoretical studies of the latter have been limited to the dc [24, 25] or local ($q = 0$) ac cases [26].

Here, we consider the linear response of voltage-biased graphene layers to the propagating acoustic plasmon whose distribution of electric potential $\delta\varphi(z)e^{iqx-i\omega t}$ is highly nonuniform (see Appendix A for explicit expressions). The electrons in tunnel-coupled graphene layers are described with the tight-binding Hamiltonian

$$\widehat{H}_0 = \begin{pmatrix} \widehat{H}_{G+} & \widehat{T} \\ \widehat{T}^* & \widehat{H}_{G-} \end{pmatrix}, \tag{43.5}$$

where the blocks $\widehat{H}_{G\pm} = \upsilon_0 \sigma \hat{\mathbf{p}} \pm \hat{I}\Delta/2$ stand for isolated graphene layers, $\upsilon_0 = 10^6$ m/s is the Fermi velocity, Δ is the voltage-induced energy spacing between the Dirac points, $\hat{\mathbf{p}}$ is the in-plane momentum operator, $\hat{I}$ is the identity matrix, and $\widehat{T} = \Omega\hat{I}$ is the tunneling matrix. Such model of tunnel coupling applies to the AA-aligned graphene bilayer [27, 28] and has proved to be useful for the description of dc tunneling in van der Waals heterostructures [24, 29]. The effects of the small interlayer twist will be addressed at the end of the chapter.

The eigenstates of $\widehat{H}_0$ can be labeled by the in-plane momentum $\mathbf{p}$, the index $s = \{c, \upsilon\}$ for the conduction and valence bands, respectively, and the number $l = \pm 1$ governing the z localization of the wave function. At strong bias, $\Delta \gg \Omega$, the state with $l = +1(-1)$ is localized primarily on the top (bottom) layer, while at zero bias the states $l = -1$ and $l = +1$ are the symmetric and antisymmetric states of the coupled quantum wells, respectively. The states' energies are $\varepsilon_{\mathbf{p}}^{ls} = s\upsilon_0 p + l\tilde{\Delta}/2$, where $\tilde{\Delta} = \sqrt{4\Omega^2 + \Delta^2}$ is the level spacing governed by the application of voltage and the tunneling repulsion [30].

The electron interaction with ac plasmon field is described by the Hamiltonian $\delta\widehat{V}$ whose matrix elements are the overlap integrals between the wave functions of $\widehat{H}_0$ eigenstates $|\mathbf{p}sl\rangle$ and the potential of the acoustic plasmon. To explicitly estimate these matrix elements, we switch from a tight-binding to a continuum description of electron states, and model each graphene layer as a one-dimensional delta well (see Appendix B for details). The strength of the well is chosen to provide the correct value of the work function from graphene to the surrounding dielectric U_b (0.4 eV for graphene embedded in WS_2). Proceeding this way, we are able to present the matrix elements of electron-plasmon interaction as

$$\langle \mathbf{p}sl | \delta\widehat{V} | \mathbf{p}'s'l' \rangle = e\delta\varphi_0 u_{\mathbf{pp}'}^{ss'} S_{ll'} \delta_{\mathbf{p},\mathbf{p}'-\mathbf{q}}, \tag{43.6}$$

where $\delta\varphi_0$ is the plasmon potential amplitude at the graphene layer, $u^{ss'}_{\mathbf{pp}'}$ is the overlap between chiral wave functions of bands s and s', and $S_{ll'}$ are the dimensionless overlap integrals. Due to the antisymmetry of the plasmon mode distribution, $S_{++} = -S_{--}$ and $S_{\pm} = S_{\mp}$.

The evaluation of interlayer conductivity is based on the solution of the quantum Liouville equation for the electron density matrix $\hat{\rho}$. Being interested in the linear response to the plasmon field, we solve it in the form

$$i\hbar\frac{\partial\delta\hat{\rho}}{\partial t} = [\hat{H}_0, \delta\hat{\rho}] + [\delta\hat{V}, \hat{\rho}^{(0)}], \tag{43.7}$$

where $\hat{\rho}^{(0)}$ is the density matrix in the absence of the ac field (but in the presence of strong dc tunneling), and $\delta\hat{\rho}$ is the sought-for linear correction. The solution of Eq. (43.7) in the basis of $\hat{H}_0$ eigenstates is immediate, though it relies on a particular choice of $\hat{\rho}^{(0)}$. When the frequency of interlayer tunneling $\Omega/\hbar$ exceeds the energy relaxation frequency in a layer ν_ε, the electron is "collectivized" by the two layers, and the states $|\mathbf{p}sl\rangle$ have a well-defined occupancy [31]. For this reason, we choose $\hat{\rho}^{(0)}$ to be diagonal in this basis, and the diagonal elements are the Fermi functions $f^{sl}_{\mathbf{p}}$ with quasi-Fermi levels shifted by eV for different l [32]. The subsequent calculation is based on the statistical averaging of current operator, which is the time derivative of the charge operator $\hat{Q}_{\mathrm{t}}$:

$$\frac{\partial\delta Q_{\mathrm{t}}}{\partial t} = -\frac{i}{\hbar}\mathrm{Tr}([\hat{Q}_{\mathrm{t}}, \hat{H}_0]\delta\hat{\rho}). \tag{43.8}$$

Equation (43.8) with the density matrix obtained from (43.7) lets us evaluate both in-plane and tunnel conductivities at once, in accordance with the two terms on the right-hand side of the continuity equation (43.1). A lengthy but straightforward calculation leads to

$$\sigma_{||}(\mathbf{q},\omega) = -ig\frac{e^2}{\hbar}S_{++}\cos\theta_M \\ \times\sum_{ss'\mathbf{p}}\frac{\left|\mathbf{v}^{ss'}_{\mathbf{pp}'}\right|^2}{\varepsilon^s_{\mathbf{p}_-} - \varepsilon^s_{\mathbf{p}_+}}\frac{f^s_{\mathbf{p}_+} - f^{s'}_{\mathbf{p}_-}}{\varepsilon^s_{\mathbf{p}_+} - \varepsilon^s_{\mathbf{p}_-} - (\hbar\omega + i\delta)}, \tag{43.9}$$

$$G_{\perp}(\mathbf{q},\omega)=-ig\frac{e^2}{2\hbar}S_{\pm}\sin\theta_M\sum_{\substack{l\neq l'\\ ss'\mathbf{p}}}\left|u_{\mathbf{pp}'}^{ss'}\right|^2$$

$$\times\frac{\varepsilon_{\mathbf{p}-}^{s'l'}-\varepsilon_{\mathbf{p}+}^{sl}}{\varepsilon_{\mathbf{p}_+}^{sl}-\varepsilon_{\mathbf{p}_-}^{s'l'}-(\hbar\omega+i\delta)}(f_{\mathbf{p}_+}^{sl}-f_{\mathbf{p}_-}^{s'l'}). \qquad (43.10)$$

Above, $g = 4$ is the spin-valley degeneracy factor, θ_M is the "mixing angle" characterizing the strength of coupling, $\sin\theta_M = 2\Omega/\tilde{\Delta}$; $\mathbf{p}_{\pm}\equiv\mathbf{p}\pm\hbar\mathbf{q}/2$, and $\mathbf{v}_{\mathbf{pp}'}^{ss'}$ is the matrix element of the velocity operator between chiral states $|\mathbf{p}s\rangle$ and $|\mathbf{p}'s'\rangle$.

The expressions for conductivities (43.9) and (43.10) are the main results of this section. Though several limiting cases for these equations have been studied previously, the treatment of nonlocal effects with nonequilibrium population has been missing up to now. We first note that in the limit of large bias, $\Delta\gg\Omega$, Eq. (43.9) naturally yields the conductivity of a single graphene layer [33]. In the same limit, the factors $S_{\pm}$ and $\sin\theta_M$ are each proportional to the small tunneling exponent $e^{-k_b d}$, where k_b^{-1} is the decay length of the electron wave function. In the opposite limit $\Omega\lesssim\Delta$, the electron states of individual layers are highly mixed (the layer is not a good quantum number), and the distinction between in-plane and tunnel conductivities loses its meaning. In the ultimate case of zero bias, the antisymmetric plasmon field cannot induce the transitions between states of the same symmetry, which is reflected in the fact that $\cos\theta_M S_{++} = 0$. In this limit, $\sigma_{||} = 0$, and the spectrum of acoustic plasmons is governed fully by the tunneling [21].

In the local limit, the tunnel conductivity becomes

$$G_{\perp}(0,\omega)=2\frac{e^2}{2\hbar}S_{\pm}\sin\theta_M\hbar\omega[n_--n_+]$$

$$\times\left[\pi\delta(\tilde{\Delta}-\hbar\omega)-i\frac{2\tilde{\Delta}}{\tilde{\Delta}^2-(\hbar\omega)^2}\right], \qquad (43.11)$$

where n_+ and n_- are the carrier densities in the states with $l = +1$ and $l = -1$. This result has been first obtained by Kazarinov and Suris in the theory of inelastic tunneling in superlattices [31] and later rederived in the case of tunneling in double-layer [26] and AA-stacked bilayer graphene [28]. In equilibrium, the low-energy symmetric state has larger population, $n_- > n_+$, the imaginary part of tunnel conductivity is negative at $\hbar\omega < 2\Omega$, while the real part is positive. The situation at nonequilibrium and at finite wave vector is radically different.

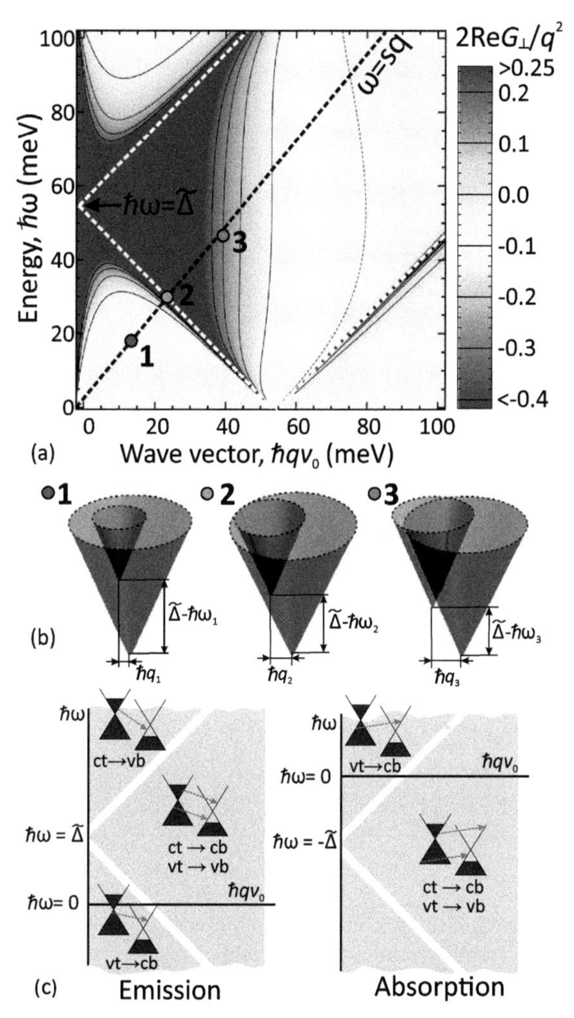

Figure 43.2 (a) Color map of the tunnel conductivity, $2\text{Re}G_{\perp}/q^2$ (in units of $e^2/4\hbar$), calculated at temperature T = 77 K and interlayer voltage V = 0.2 V. Dielectric layer is 3 nm WS_2 (effective mass m^* = $0.28m_0$, conduction band offset to graphene U_b = 0.4 eV [35]). Red dashed line corresponds to the zero conductivity; black dashed line shows the dispersion of acoustic SP in the absence of tunneling. (b) Band diagrams illustrating available electron states for plasmon-assisted tunneling at different frequencies and wave vectors. (c) Map of the frequency and wave vector ranges, for which various inelastic interlayer electron transitions are possible.

The first peculiarity of Eq. (43.10) in the presence of voltage bias is the negative value of the real part of tunnel conductivity at frequencies $\omega < eV/\hbar$. This negativity implies that the interlayer transitions accompanied by the emission of the quantum $(\omega, \mathbf{q})$ are more probable than the inverse absorptive transitions. The band filling providing the negative tunnel conductivity (Fig. 43.1b) can be viewed as an interlayer population inversion similar to that in quantum cascade lasers [34]. The frequency and wave vector dependence of $2\mathrm{Re}G_{\perp}/q^2$ is shown in Fig. 43.2a, where the "cold" colors stand for the negative and "warm" colors for the positive conductivity. An analysis of the energy-momentum conservation reveals distinct regions on the frequency-wave vector plane, where different types of radiative and absorptive tunnel transitions are relevant; see Fig. 43.2c. Among those, the most pronounced is the interlayer intraband emission allowed within the quadrant $qv_0 \geq |\tilde{\Delta}/\hbar - \omega|$. The interband transitions are generally weaker due to the small overlap of chiral wave functions of different bands [29].

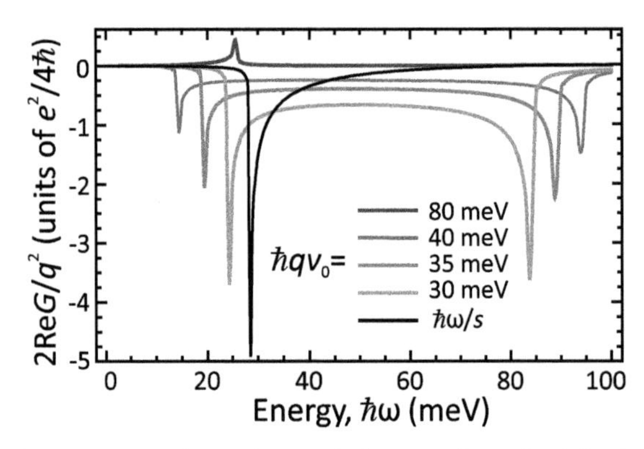

Figure 43.3 Frequency dependence of the tunnel conductivity $\mathrm{Re}2G_{\perp}/q^2$ at certain wave vectors, i.e., cuts of the plot of Fig. 43.2 along $\hbar q v_0$ = 80, 40, 35, 30 meV, as well as along the unperturbed dispersion of acoustic SP $q = \omega/s$, where $s \approx 1.1v_0$. Weak singularity corresponding to the absorptive transition between like bands is seen in the red curve; at other wave vectors the transitions corresponding to tunneling emission are more pronounced.

Another distinct feature of the tunnel conductivity is its large absolute value near the series of lines $qv_0 = |\omega \pm \tilde{\Delta}/\hbar|$. The effect is demonstrated also in Fig. 43.3, where we show the frequency dependence of tunnel conductivity at certain wave vectors. The origin of these resonances can be explained by analyzing the possible electron states involvedin plasmon-assisted tunneling at different frequencies and wave vectors, Fig. 43.2b. To be precise, we focus on the interlayer intraband tunneling. Above the resonance, at $qv_0 > \omega - \tilde{\Delta}/\hbar$, the electrons capable of tunneling occupy a hyperbolic cut of the mass shell in graphene (case B3 in Fig. 43.2). With decreasing the frequency and wave vector, the hyperbola degenerates into a line (case B2), and the tunneling occurs between states with collinear momenta and equal velocities, whose interaction lasts for an infinitely long time in the absence of scattering. Alternatively, the singularities in the tunnel conductivity can be traced back to the van Hove singularities in the joint density of states [24]. At even lower frequencies (case B1), the intraband transitions are impossible, but the weaker interband tunneling sets in. Without the carrier scattering, the collinear tunneling singularities are square root, and the real part of the intraband tunnel conductivity is given by

$$\mathrm{Re}G_{\perp}^{\mathrm{intra}} = -\frac{e^2}{\hbar}\frac{q^2}{2\pi}\omega\left\{\frac{\mathcal{I}\left(\frac{\hbar q v_0}{2kT},\frac{eV-\hbar\omega}{2kT}\right)}{\sqrt{q^2v_0^2-(\tilde{\Delta}/\hbar-\omega)^2}} - \frac{\mathcal{I}\left(\frac{\hbar q v_0}{2kT},\frac{\hbar\omega+eV}{2kT}\right)}{\sqrt{q^2v_0^2-(\tilde{\Delta}/\hbar+\omega)^2}}\right\}, \tag{43.12}$$

where we have introduced an auxiliary integral $\mathcal{I}(\alpha,\beta) = \int_1^\infty dt\sqrt{t^2-1}\,[f_F(\alpha t-\beta) - f_F(\alpha t+\beta)]$ and the dimensionless "Fermi function" $f_F(\zeta) = [1+e^{\zeta}]^{-1}$. The singularities in Eq. (43.12) are similar to the those of in-plane conductivity at the onset of the Landau damping [3]:

$$\mathrm{Re}_{\parallel}^{\mathrm{intra}} \propto \left[q^2v_0^2-\omega^2\right]^{-1/2}. \tag{43.13}$$

The actual value of the resonant conductivity in clean samples is limited by electron-acoustic phonon scattering [9]. We account for it by replacing the delta-peaked spectral functions of individual electrons in Eqs. (43.9) and (43.10) with Lorentz functions whose width is proportional to the imaginary part of electron self-energy [24]. With the scattering rate $\tau_{\mathrm{tr}}^{-1} \simeq (2\text{–}8)\times 10^{-11}\ \mathrm{s}^{-1}$ at

T = 77–300 K [36, 37] and electron density $n = 5 \times 10^{11}$ cm^{-2}, the tunnel resonances are pronounced up to the room temperature.

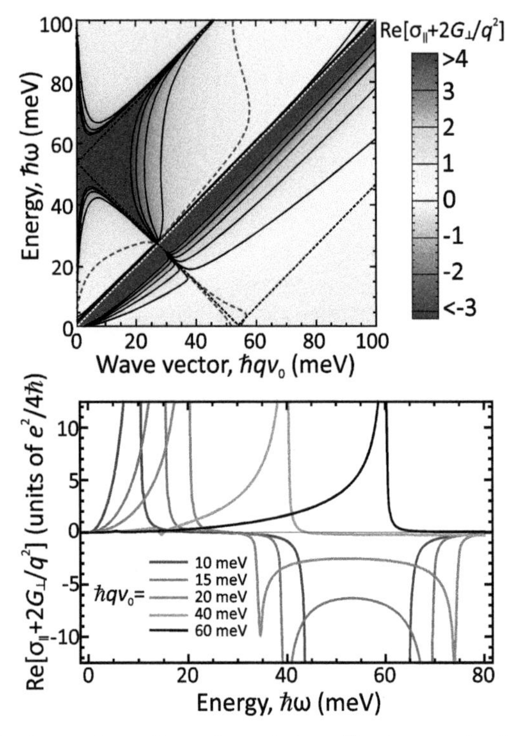

Figure 43.4 Space-time dispersion of the effective conductivity Re[$\sigma_{||}$ + $2G_{\perp}/q^2$] (normalized by $e^2/4\hbar$) governing the damping (or gain) of acoustic SPs. The structure parameters are the same as in Fig. 43.2. The contour of zero conductivity is highlighted with red dashed line. The bottom panel shows the cuts of the upper plot along certain wave vectors.

From a practical point of view, it is important that the real part of net "effective conductivity," $\sigma_{||}$ + $2G_{\perp}/q^2$, which enters the SP dispersion is negative in a wide range of frequencies and wave vectors (Fig. 43.4). This implies the possibility of amplified propagation plasmons instead of their damping. The sign of the net effective conductivity is governed by the competition of three processes: (1) emission of surface plasmons upon interlayer tunneling, (2) interband plasmon absorption, and (3) intraband (free-carrier) plasmon absorption. The latter two processes are intensified due to the finiteness of the SP wave vector, and use of the well-known local limit for the in-plane conductivity would lead to a considerable underestimate of the absorption (see Ref. [9] and Appendix C for details). Our numerical calculations show that the net negative conductivity is still possible as the collinear tunneling resonance can be tuned into the frequency range corresponding to the transparency of graphene, i.e., the frequencies where the Drude absorption is low while the interband absorption is suppressed by the Pauli blocking.

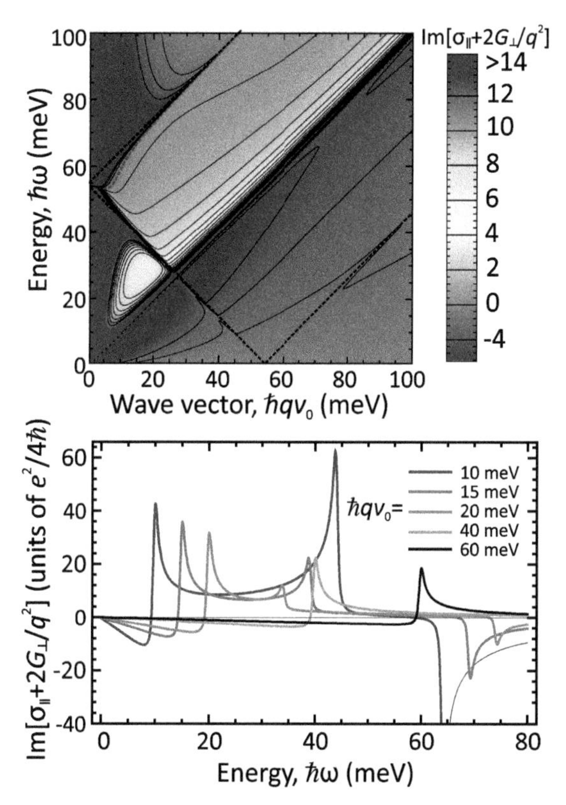

Figure 43.5 Space-time dispersion of the effective conductivity $\mathrm{Im}[\sigma_{||} + 2G_{\perp}/q^2]$ (normalized by $e^2/4\hbar$) governing the spectrum renormalization of acoustic SPs. The structure parameters are the same as in Fig. 43.2. The contour of zero conductivity is highlighted with red dashed line. The bottom panel shows the cuts of the upper plot along certain wave vectors.

The square-root singularities in $\mathrm{Re}\, G_{\perp}^{\mathrm{intra}}$ above the threshold of interlayer intraband transitions are mirrored in the singularities in $\mathrm{Im}\, G_{\perp}^{\mathrm{intra}}$ below the threshold. This situation is illustrated in Fig. 43.5, where we show the color map of the imaginary part of full "effective" conductivity, $\mathrm{Im}[2G_{\perp}/q^2 + \sigma_{||}]$, and its cuts at certain wave vectors. The presence of singularity in the imaginary part of tunnel conductivity at $\omega = \tilde{\Delta}/\hbar - qv_0$ suggests the strong renormalization of the acoustic SP dispersion, which will be shown in the next section. We note here that the mentioned singularities are absent in the case of tunneling between the layers of massive 2d electrons. In the massive case, the electron states with collinear momenta do not necessarily have the same velocity, and their interaction does not necessarily last for a very long time. Loosely speaking, for massive particles one should average the tunnel conductivity (43.12) over the absolute values of particle velocity v_0, which would wash the singularities out.

43.4 Plasmons in the Presence of Tunneling

The long-wavelength part of the acoustic SP mode in coupled graphene layers at equilibrium does not qualitatively differ from that for coupled massive 2d electrons. The SP dispersion develops a gap, such that

$$(\hbar\omega_{\mathrm{ac}})^2(q \to 0) = (2\Omega)^2 + 2\Omega E_{\mathrm{C}}, \tag{43.14}$$

where $E_{\mathrm{C}} = 8\pi g d e^2(n_- - n_+)/(\hbar\kappa)$. It is possible to show that the formation of such gap is generic for coupled layers, once the in-plane dispersion of electrons is not renormalized by tunneling [38]. Indeed, the real part of tunnel conductivity is positive for the photon energy equal to the tunnel splitting of levels, 2Ω. Using the Kramers-Kronig relations, one obtains that the imaginary part of conductivity is negative at $\hbar\omega < 2\Omega$, which prohibits the existence of TM plasmons. Apart from 2Ω, an extra contribution to the plasmon gap in Eq. (43.14) comes from the electrostatic energy of electrons, E_{C}.

At finite bias, both the tunneling and in-plane electron motion govern the plasmon spectrum. The imaginary part of in-plane conductivity $\mathrm{Im}\sigma_{||}$ is positive and singular above the domain of Landau damping, i.e., at $\omega \to qv_0$. Such behavior originating from the linearity of carrier dispersion leads to the existence of acoustic mode even in the limit $d \to 0$ [4, 39], contrary to the case of parallel 2d layers of massive carriers in the absence of tunneling [40]. Moreover, the imaginary part of tunnel conductivity at nonequilibrium is also positive below the threshold of interlayer transitions. Hence, already at a very small bias, the condition $\mathrm{Im}[\sigma_{||} + 2G_{\perp}/q^2] > 0$ can be fulfilled, and long-wavelength linear dispersion of SPs is restored. At higher frequencies, the tunneling resonance crosses the unperturbed SP dispersion $\omega = sq$. At this point, the effect of tunneling is the strongest one.

This is illustrated in the plot of the acoustic mode contribution to the loss function, $\mathrm{Im}\,\epsilon_{\mathrm{ac}}^{-1}(\mathbf{q}, \omega)$, Fig. 43.6a. The plasmon peak in the loss function develops an anticrossing with the tunnel resonance, and the bending of the SP dispersion is such that its group velocity increases and changes its sign at some point, passing through infinity. The "locking" of the long-wavelength SP dispersion in the domain $(\omega \geq qv_0) \cup (\omega \leq \tilde{\Delta}/\hbar - qv_0)$ comes, formally, from the interplay of two singular conductivities, in-plane and out-of-plane, which are positive and singularly large at the threshold of the respective transitions (see Fig. 43.5).

The interaction of unperturbed SP mode with tunnel resonance leading to abnormally large and negative group velocities is radically different from the interaction of plasmon modes with polar phonons [8], cyclotron resonances [41], etc. In the two latter cases, the interaction leads to the decrease in group velocity. The origin of the upward bending of SP dispersion is, actually, the inverted population between two layers [42]. Similar group velocity enhancement and divergence occurs for light interacting with a gas of inverted two-level systems [43], and can be analyzed with a simple Lorentz oscillator model [44].

The effects of tunneling on SP dispersion are more pronounced at low levels' spacing $\tilde{\Delta}$ and high carrier density n. By fixing $\tilde{\Delta}$ and increasing the carrier density (which can be achieved by extra gates), one can observe a large enhancement of the SP velocity below the resonance, as shown in Fig. 43.6b. At some critical density the long-wavelength branch of SP dispersion can disappear entirely. At large level spacing, the effects of plasmon spectrum renormalization are relevant just in a narrow vicinity of the threshold of intraband tunneling, and further broadening of the tunnel resonance by the carrier scattering can wash out the renormalization effects. However, our calculations show that the predicted effects of tunneling on plasmons can survive up to the room temperature and relatively high interlayer voltage. The plasmon dispersion at T = 300 K and V = 0.3 V shown by the orange line in Fig. 43.6b still demonstrates the backward-wave behavior as well as the possibility of plasmon gain.

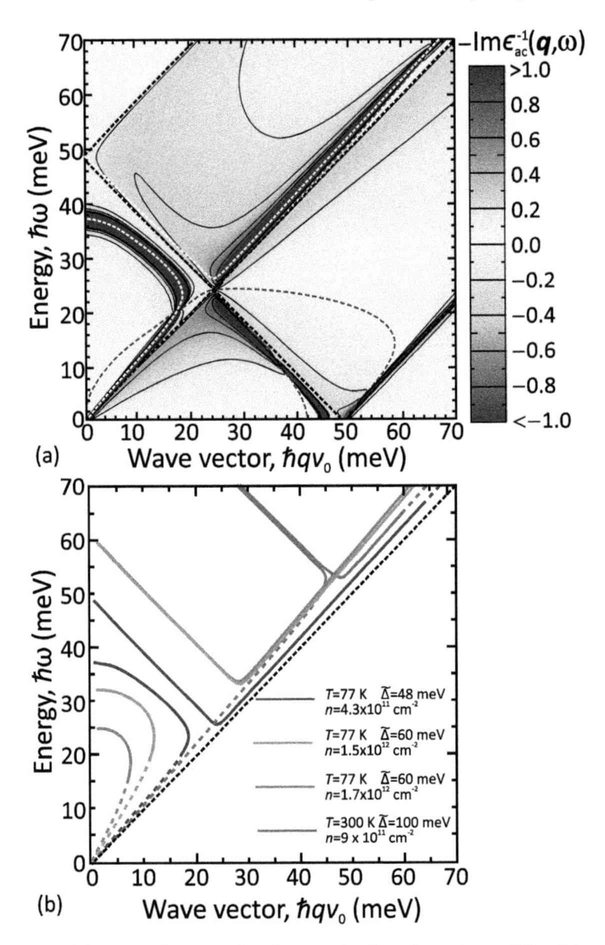

Figure 43.6 (a) Acoustic contribution to the loss function of double graphene layer calculated for 2.5 nm WS_2 barrier, T = 77 K and V = 0.2 V. The plasmon spectrum develops an anticrossing with the collinear tunneling resonance. (b) Plasmon spectra calculated for different temperatures, electron densities, and level spacing of tunnel coupled layers $\tilde{\Delta}$. Dashed parts of the spectra correspond to the damped plasmons and solid parts to the amplified plasmons. Black dashed line is $\hbar\omega = qv_0$.

The net gain of surface plasmons (i.e., negative real part of the net effective conductivity $\sigma_{||} + 2G_{\perp}/q^2$) is possible both below and above the tunnel resonance at $\omega = \tilde{\Delta}/\hbar - qv_0$. Below the resonance, the gain is due to the emission of SPs upon tunneling from the valence band of top layer to the conduction band of the bottom one (see Fig. 43.2c). In this domain, the renormalization of SP dispersion leads to the reduction of the wave vector at given frequency, and hence, to the reduction of in-plane loss. Above the resonance, the gain is due to the emission of SPs upon interlayer intraband transitions; the linear dispersion of SPs is almost unaffected by tunneling in this case.

43.5 Effects of Interlayer Twist

The effects of rotational twist of graphene layers in the tunneling heterostructures manifests itself in the relative shift of their Dirac points by the vectors $\Delta\mathbf{q}_{\mathrm{i}}$ in the reciprocal space ($i = 1 \ldots 6$; see the inset in Fig. 43.5). With the neglect of small off-diagonal elements of the $\mathcal{T}$ matrix, the tunnel conductivity of twisted layers $G_{\perp}^{\mathrm{T}}(\mathbf{q}, \omega)$ is related to the tunnel conductivity of the aligned layers $G_{\perp}$ via

$$G_{\perp}^{\mathrm{T}}(\mathbf{q},\omega) = \frac{1}{6}\sum_{i=1}^{6} G_{\perp}(\mathbf{q}+\Delta\mathbf{q}_{\mathrm{i}},\omega). \tag{43.15}$$

In the presence of twist, the locus of collinear scattering singularities on the ω–q plane breaks down into six hyperbolas (or less, for particular angles between $\mathbf{q}$ and $\Delta\mathbf{q}$). The acoustic plasmon dispersion develops an anticrossing with each of the hyperbolas, demonstrating several frequency ranges with negative group velocity and gain. An example of the loss function for the wave vector parallel to the twist vector $\Delta\mathbf{q}$ in one pair of valleys is shown in Fig. 43.7 for $\hbar|\Delta\mathbf{q}|v_0 = 18$ meV (twist angle $\theta_{\mathrm{T}} = 0.57°$). In this example, there exist four curves corresponding to the singular plasmon gain and four for the singular absorption. Remarkably, the plasmon gain in twisted layers for certain directions of propagation can be greater than that in aligned layers, because the tunnel resonances can come closer to the unperturbed SP dispersion. Generally, the spectrum of plasmons in twisted layers becomes anisotropic with sixfold rotational symmetry.

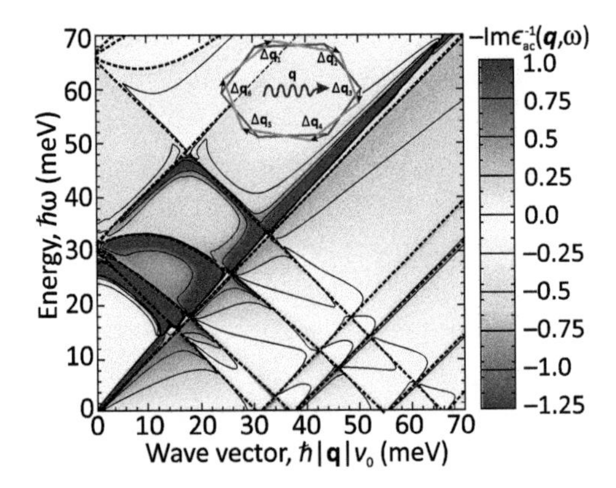

Figure 43.7 Acoustic contribution to the loss function in twisted layers (twist angle θ_T = 0.57°); the wave vector is parallel to the misalignment vector in one pair of valleys. Inset shows the positions of *K* points in the reciprocal space for twisted graphene layers.

43.6 Discussion and Conclusions

The renormalization of SP dispersion due to the tunneling can be observed with the recently introduced tool of scattering-type near-field microscopy [45]. A more practical and readily observable effect of tunneling is the enhancement of SP propagation length due to the partial compensation of in-plane absorption by the tunneling emission. Moreover, our calculations show that under proper choice of the barrier layer (2 . . . 3 nm WS_2), the full compensation of plasmon losses and the net plasmon gain are possible in the far-infrared range $\hbar\omega \approx 20 \ldots 60$ meV. The predicted effect of the net gain opens up the possibility of creating graphene-based sources of coherent plasmons—spacers—and nanoscale sources of photons based, e.g., on the evanescent coupling of photonic waveguides and graphene-based active media for plasmons.

The observation of net plasmon gain in coupled graphene layers poses strong constraints on the tunnel transparency of the barrier material. At the same time, the spontaneous emission of SPs upon tunneling [46, 47] is observable for a wide class of dielectrics. The tunneling SP emission with their subsequent conversion into free-space electromagnetic modes upon scattering might explain the terahertz electroluminescence from graphene-hBN-graphene diodes observed recently [48]. The presence of luminescence in

[48] correlates with the magnitude of NDR in the static $I(V)$ curve supporting the tunneling origin of the emission.

The tunneling assisted by the spontaneous emission of plasmons can also manifest itself as an extra peak in the bias-dependent differential conductivity dG/dV [49]. Such peaks have not yet been identified with tunneling spectroscopy of graphene-based structures [50]. The extension of the presented calculation of tunnel current to the case of spontaneous emission of surface plasmons would reveal the favorable conditions for plasmon-assisted resonant tunneling and guide further experimental work.

In conclusion, we have theoretically demonstrated a number of unique properties of surface plasmons in tunnel-coupled voltage-biased graphene layers, including the amplified propagation due to the resonant tunneling under interlayer population inversion, and a strong renormalization of dispersion law. The pronounced effect of tunneling on both spectrum and damping of plasmons results from singularities in the tunnel conductivity which are, in turn, inherited from the linear bands of graphene.

Acknowledgments

The work of D. S. was supported by Grant No. 16-19-10557 of the Russian Science Foundation. The work at RIEC was supported by Japan Society for Promotion of Science KAKENHI Grants No. 23000008 and No. 16H06361. The authors thank V. Vyurkov, S. Fillipov, A. Dubinov, A. Arsenin, and D. Fedyanin for helpful discussions.

Appendix A: Plasmon Modes Supported by the Double Layer

In this section, we review the general properties of plasmons in double-graphene-layer structures [22, 51] in the absence of tunneling. The acoustic SP dispersion in double layers is equivalent to the plasmon dispersion in graphene with a perfectly conducting gate [4, 39]. The low-frequency part of the SP spectrum can be obtained with the neglect of interband conductivity, while the intraband conductivity can be calculated in the Boltzmann limit ($\hbar q v_0 \ll \varepsilon_F$, $\hbar\omega \ll \varepsilon_F$):

$$\sigma_{\parallel}^{\text{intra}} \approx ig\frac{e^2}{\hbar}\frac{\tilde{\varepsilon}_{\text{F}}}{2\pi\hbar}\frac{\omega}{q^2 v_0^2}\left[\frac{\omega}{\sqrt{\omega^2 - q^2 v_0^2}} - 1\right], \tag{43.A1}$$

where $\tilde{\varepsilon}_{\text{F}} = T\ln\left(1+e^{\varepsilon_{\text{F}}/T}\right)$. The dispersion relations of plasmons (43.3) in the absence of tunneling ($G_{\perp} = 0$) can be solved analytically with conductivity given by Eq. (43.A1). This yields the linear dispersion of the acoustic mode

$$\omega_{\text{ac}} = v_0 \frac{1+4\alpha_{\text{c}} q_{\text{F}} d}{\sqrt{1+8\alpha_{\text{c}} q_{\text{F}} d}} q, \tag{43.A2}$$

and the square-root dispersion of the optical mode

$$\omega_{\text{opt}} \approx v_0 \sqrt{4\alpha_{\text{c}} q q_{\text{F}}}. \tag{43.A3}$$

Here, we have introduced the Fermi wave vector $q_{\text{F}} = \tilde{\varepsilon}_{\text{F}}/\hbar v_0$, the coupling constant $\alpha_{\text{C}} = e^2/\hbar\kappa v_0$, and neglected the spatial dispersion of conductivity for the optical mode. This is possible as the phase velocity of acoustic SPs greatly exceeds the Fermi velocity. On the other hand, the velocity of the acoustic mode always exceeds the Fermi velocity, but can be arbitrarily close to it. For this reason, the spatial dispersion of conductivity cannot be neglected for acoustic modes even in the formal long-wavelength limit $q \ll q_{\text{F}}$, $q \ll T/\hbar v_0$ as the inequality $q v_0 \ll \omega$ cannot be generally fulfilled.

The distinct dispersions of optical and acoustic plasmon modes are seen from the loss functions shown in Fig. 43.8. The acoustic part of the loss function is given by Eq. (43.4), while the optical part is given by

$$\operatorname{Im} \epsilon_{\text{opt}}^{1}(\mathbf{q},\omega) = \operatorname{Im}\left\{1+\frac{2\pi i q \sigma_{\parallel}}{\omega\kappa}\left(1+e^{-qd}\right)\right\}^{-1}. \tag{43.A4}$$

To account for the plasmon damping one has to go beyond Eq. (43.A1) and calculate the real parts of the in-plane conductivity (see Appendix C for analytical expressions). As seen from Fig. 43.8, the plasmon peak in the spectral function becomes broadened at $\hbar\omega \approx 2\varepsilon_{\text{F}} - \hbar q v_0$, which corresponds to the onset of interband transitions. Below the threshold of interband transitions, the plasmon damping is governed by the free-carrier absorption which can be calculated from the Boltzmann equation with electron-phonon scattering taken into account.

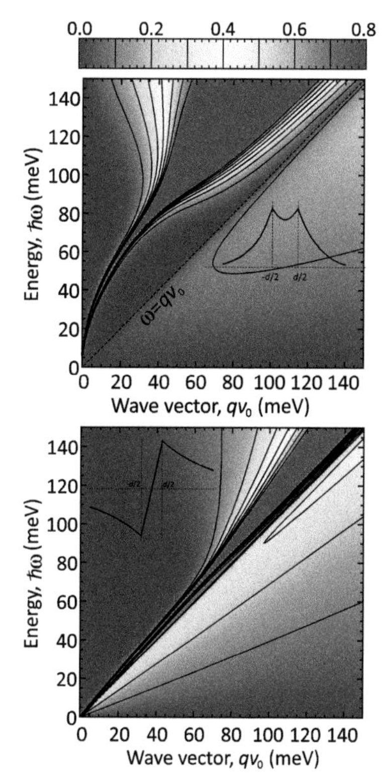

Figure 43.8 Optical (top) and acoustic (bottom) contributions to the loss function for double graphene layer in the absence of tunneling. Fermi energy ε_F = 100 meV, temperature T = 300 K, insulator thickness d = 3 nm, dielectric constant κ = 5.

The spatial distribution of the plasmon potential $\delta\varphi(z)$ in acoustic mode is governed solely by the wave vector q and can be obtained from the Poisson's equation (43.2). It is convenient to present it as

$$\delta\varphi(z) = \delta\varphi_0 s(z), \tag{43.A5}$$

where φ_0 is the electric potential on the top layer, and $s(z)$ is the dimensionless "shape function" having the following form:

$$s(z) = \begin{cases} e^{-q(z+d/2)}, & z < -d/2, \\ -\dfrac{\sinh(qz)}{\sinh(qd/2)}, & |z| < d/2, \\ -e^{-q(z-d/2)}, & z > d/2. \end{cases} \tag{43.A6}$$

The spatial dependence of the shape functions for acoustic and optical modes is shown in the inset of Fig. 43.8.

Appendix B: Estimate of the Tight-Binding Parameters

To estimate the tight-binding parameters Ω and $S_{ll'}$, we switch to the continuum description of electron states in the z direction. We model each graphene layer with a delta well [52]

$$U_{\mathrm{t,b}}(z)=2\sqrt{\frac{\hbar^2 U_{\mathrm{b}}}{2m^*}}\delta(z-z_{\mathrm{t,b}}), \tag{43.B1}$$

where the potential strength is chosen to provide a correct value of electron work function U_{b} from graphene to the surrounding dielectric, and m^* is the effective electron mass in the dielectric. The effective Schrödinger equation in the presence of voltage bias Δ/e between graphene layers takes on the following form:

$$-\frac{\hbar^2}{2m^*}\frac{\partial^2\psi(z)}{\partial z^2}+[U_{\mathrm{t}}(z)+U_{\mathrm{b}}(z)+U_{\mathrm{F}}(z)]\psi(z)=E\psi(z), \tag{43.B2}$$

where U_{F} is the potential energy created by the applied field

$$U_{\mathrm{F}}(z)=\frac{\Delta}{2}\begin{cases}1,\, z<-d/2,\\ 2z/d,\, |z|<d/2,\\ -1,\, z>d/2,\end{cases} \tag{43.B3}$$

The solutions of the effective Schrödinger equation represent decaying exponents at $|z| > d/2$, and a linear combination of Airy functions in the middle region $|z| < d/2$:

$$\psi_{\mathrm{M}}(z)=C\,\mathrm{Ai}(-z/a+\varepsilon)+D\,\mathrm{Bi}(-z/a+\varepsilon), \tag{43.B4}$$

where $\varepsilon = 2m^*|E|a^2/\hbar^2$ is the dimensionless energy and $a = (\hbar^2 d/2m^*\Delta)^{1/3}$ is the effective length in the electric field. A straightforward matching of the wave functions at the graphene layers yields the dispersion equation

$$\det\begin{pmatrix} e^{-k_1 d/2} & -\mathrm{Ai}(d/2a+\varepsilon) & -\mathrm{Bi}(d/2a+\varepsilon) & 0 \\ (2k_{\mathrm{b}}-k_1)e^{-k_1 d/2} & -\frac{1}{a}\mathrm{Ai}'(d/2a+\varepsilon) & -\frac{1}{a}\mathrm{Bi}'(d/2a+\varepsilon) & 0 \\ 0 & -\mathrm{Ai}(-d/2a+\varepsilon) & -\mathrm{Bi}(-d/2a+\varepsilon) & e^{-k_2 d/2} \\ 0 & -\frac{1}{a}\mathrm{Ai}'(-d/2a+\varepsilon) & -\frac{1}{a}\mathrm{Bi}'(-d/2a+\varepsilon) & (2k_{\mathrm{b}}-k_2)e^{-k_2 d/2} \end{pmatrix}=0, \tag{43.B5}$$

where $k_b = \sqrt{2m^* U_b/\hbar^2}$ is the decay constant of the bound state wave function in a single delta well, $k_1 = \sqrt{2m^*(E+\Delta/2)/\hbar^2}$, $k_2 = \sqrt{2m^*(E-\Delta/2)/\hbar^2}$. Equation (43.B5) yields two energy levels E_l ($l = \pm 1$) which can be found only numerically (see Fig. 43.9a). The respective wave functions are shown in Fig. 43.9b; at strong bias they are *almost* the wave functions localized on the different layers (see the discussion below). Despite the complexity of Eq. (43.B5), the dependence of E_l on the energy separation between layers Δ can be accurately modeled by

$$E_l(\Delta) = -U_b + \frac{l}{2}\sqrt{(E_{+1,\Delta=0} - E_{-1,\Delta=0})^2 + \Delta^2}. \tag{43.B6}$$

The same functional dependence of energy levels on Δ is naturally obtained by diagonalizing the block Hamiltonian (43.5),

$$E_l(\Delta) = -U_b + l\sqrt{\Omega^2 + \frac{\Delta^2}{4}}. \tag{43.B7}$$

This allows us to estimate the tunnel coupling Ω as half the energy splitting of states in the double-graphene-layer well in the absence of applied bias:

$$\Omega = \frac{1}{2}[E_{+1,\Delta=0} - E_{-1,\Delta=0}]. \tag{43.B8}$$

The wave functions corresponding to a relatively strong bias Δ = 200 meV are shown in Fig. 43.9. It is simple to relate the true eigenfunctions $\Psi_+(z)$ and $\Psi_-(z)$ to the functions located on the top and bottom layers $\Psi_t(z)$ and $\Psi_b(z)$:

$$\Psi_b = \cos\alpha\,\Psi_- + \sin\alpha\,\Psi_+, \tag{43.B9}$$

$$\Psi_t = -\sin\alpha\,\Psi_- + \cos\alpha\,\Psi_+, \tag{43.B10}$$

where

$$\cos\alpha = \frac{2\Omega}{\sqrt{(2\Omega)^2 + (\Delta - \tilde{\Delta})^2}}. \tag{43.B11}$$

Knowing the wave functions of the coupled layers, we can estimate the matrix elements of electron-plasmon interaction and present them in the following form:

$$\langle \mathbf{p}sl|\delta\hat{V}|\mathbf{p}'s'l'\rangle = \delta_{\mathbf{p},\mathbf{p}'-\mathbf{q}} u^{ss'}_{\mathbf{pp}'} e\delta\varphi_0 S_{ll'}. \tag{43.B12}$$

Here we have introduced the shorthand notations for the overlap factors of dimensionless plasmon potential and eigenfunctions of coupled layers,

$$S_{++} = \int_{-\infty}^{\infty} \psi_{+1}^{*}(z) s(z) \psi_{+1}(z), \qquad (43.B13)$$

$$S_{\pm} = \int_{-\infty}^{\infty} \psi_{+1}^{*}(z) s(z) \psi_{-1}(z), \qquad (43.B14)$$

and, obviously, $S_{--} = -S_{++}$.

The dependence of the overlap factors S_{++} and $S_{\pm}$ on the interlayer potential drop Δ is shown in Fig. 43.10. We note that these overlap factors weakly depend on the plasmon wave vector q as far as it ismuch smaller than electron wave function decay constant k_b. In this approximation, one can set $s(z) \approx 2z/d$ for $|z| < d/2$, $s(z) \approx 1$ at $|z| > d/2$.

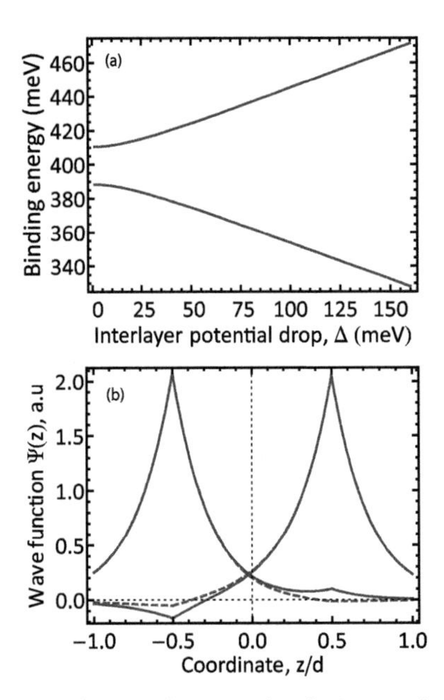

Figure 43.9 (a) Dependence of energy levels in coupled graphene layers on the interlayer potential drop Δ. (b) Wave functions of the coupled layers calculated for $\Delta = 200$ meV. The barrier is 2.5 nm WS_2. Solid lines in (b) show the wave functions corresponding to $l = +1$ (red) and $l = -1$ (blue), while the dashed lines show the wave functions of the top and bottom layers obtained as a linear combination (43.B9) of the eigenfunctions.

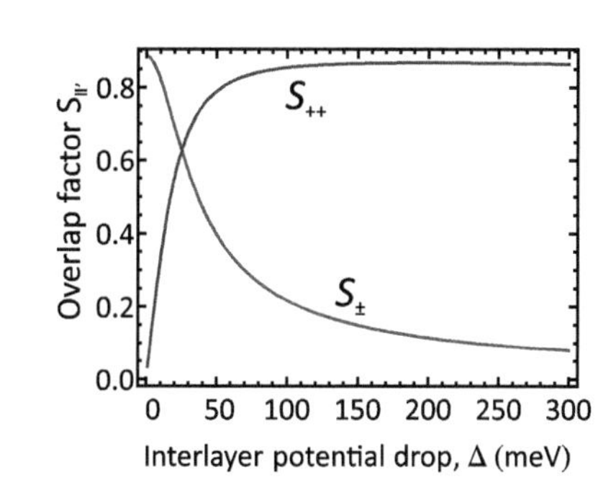

Figure 43.10 Dependence of the overlap factors S_{++} and $S_{\pm}$ on the interlayer voltage drop Δ calculated for the 2.5 nm WS_2 dielectric layer.

Appendix C: Analytical Results for the Conductivity: In-Plane Conductivity

Despite the complex structure of Eqs. (43.9) and (43.10), several analytical approximations can be made in the frequency range of interest $\hbar\omega < 2\varepsilon_F$, where the plasmons are weakly damped—at least, for the real part of conductivity that determines absorption or gain. For brevity, from here to the end of this chapter we work with "God-given units" $\hbar = \upsilon_0 \equiv 1$. We start with the evaluation of in-plane interband conductivity [the term with $s = \upsilon$ and $s' = c$ in Eq. (43.9)].

The interband velocity matrix element reads $\langle c\mathbf{p}_-|\, \hat{\upsilon}_x \,|\upsilon\mathbf{p}_+\rangle = i \sin [(\theta_{p+} + \theta_{p-})/2]$, where θ_{p-} and θ_{p+} are the angles between the momenta of initial and final states with the x axis. The subsequent calculations are conveniently performed in the elliptic coordinates

$$\mathbf{p} = \frac{q}{2}\{\cosh u \cos \upsilon, \sinh u \sin \upsilon\}. \tag{43.C1}$$

In these coordinates $|\mathbf{p}_\pm| = (q/2)[\cosh u \pm \cos \upsilon]$, $|\langle c\mathbf{p}_-|\, \hat{\upsilon}_x \,|\upsilon\mathbf{p}_+\rangle|^2 dp_x dp_y = (q^2/4) \cosh^2 u \sin^2 \upsilon du d\upsilon$. This leads us to (we henceforth omit the factor $S_{++} \cos \theta_M$)

$$\mathrm{Re}\sigma_{\parallel}^{\upsilon\to c} = \frac{e^2}{2\pi}\frac{\omega}{\sqrt{\omega^2 - q^2}}\int_0^{\pi} d\upsilon \sin^2 \upsilon \times \left\{ f^{\upsilon}\left[-\frac{\omega}{2} + \frac{q}{2}\cos \upsilon\right] - f^{c}\left[\frac{\omega}{2} + \frac{q}{2}\cos \upsilon\right]\right\}. \tag{43.C2}$$

To proceed further, we note that in the domain of interest $\omega > q$ one always has $q \cos \upsilon < \omega$. Due to this fact, the difference of distribution functions is a smooth function of υ, while the prefactor $\sin^2 \upsilon$ varies strongly. This allows us to integrate $\sin^2 \upsilon$ exactly, and replace the difference of distribution functions with its angular average. This leads us to

$$\mathrm{Re}\sigma_{\parallel}^{\upsilon\to c} \approx \frac{e^2}{4}\frac{T\omega}{q}\chi(q,\omega)\ln\frac{\cosh\frac{\varepsilon_{\mathrm{F}}}{T}+\cosh\frac{\omega+q}{2T}}{\cosh\frac{\varepsilon_{\mathrm{F}}}{T}+\cosh\frac{\omega-q}{2T}}, \tag{43.C3}$$

where we have introduced a resonant factor

$$\chi(q,\omega) = \frac{\theta(\omega)}{\sqrt{\omega^2 - q^2}}. \tag{43.C4}$$

Clearly, the neglect of spatial dispersion in the case of acoustic SPs with velocity slightly exceeding the Fermi velocity results in an underestimation of the real part of the interband conductivity and, hence, of the plasmon damping.

We now pass to the in-plane conductivity associated with the intraband transitions. Here, we can restrict ourselves to the classical description of the electron motion justified at frequencies $\omega \ll \varepsilon_{\mathrm{F}}$, $q \ll q_{\mathrm{F}}$; otherwise, strong interband SP damping takes place. Clearly, one could work out the terms with $s = s'$ and $l = l'$ in Eq. (43.9); however, the accurate inclusion of carrier scattering in such equations is challenging. Instead, we use the kinetic equation to evaluate $\sigma_{\parallel}^{c\to c}$; this formalism allows an inclusion of carrier scattering in a consistent manner. One should, however, keep in mind that in the nonlocal case $q \neq 0$ a simple τ_{p} approximation is not particle conserving. A particle-conserving account of collisions is achieved with the Bhatnagar–Gross–Krook collision integral [53] in the right-hand side of the kinetic equation,

$$-i\omega\delta f(\mathbf{p}) + i\mathbf{qv}\delta f(\mathbf{p}) + ie\mathbf{qv}\delta\varphi\frac{\partial f_0}{\partial\varepsilon}$$
$$= -\nu\left[\delta f(\mathbf{p}) + \frac{d\varepsilon_{\mathrm{F}}}{dn}\frac{\partial f_0}{\partial\varepsilon}\delta n\right]. \tag{43.C5}$$

Here $\delta f(\mathbf{p})$ is the sought-for field-dependent correction to the equilibrium electron distribution function f_0, δn_{q} is the respective correction to the electron density, $\mathbf{v} = \mathbf{p}/p$ is the quasiparticle velocity, and ν is the electron collision frequency which is assumed to be energy-independent. Solving Eq. (43.C5) and recalling the continuity equation $\omega\delta n = q\delta\mathbf{j}$ we obtain the in-plane intraband conductivity:

$$\sigma_{\parallel}^{\text{intra}} = \frac{ige^2\tilde{\varepsilon}_F}{(2\pi)^2 q} \frac{J_2\left(\frac{\omega+i\nu}{q}\right)}{1-\frac{i\nu}{2\pi\omega}J_1\left(\frac{\omega+i\nu}{q}\right)}, \tag{43.C6}$$

where

$$J_n(x) = \int_0^{2\pi} \frac{\cos^n\theta d\theta}{x-\cos\theta}. \tag{43.C7}$$

Similarly to the real part of the interband absorption, the intraband absorption is generally larger in the nonlocal case $q \neq 0$ compared to the local case. This difference is illustrated in Fig. 43.11, where the local ($q = 0$) and nonlocal expressions at the acoustic plasmon dispersion ($q = \omega/s$) are compared. This result is in agreement with the recent measurements of plasmon propagation length in graphene on hBN: the local Drude formula underestimated the plasmon damping, and the account of nonlocality was crucial to explain the experimental data [9].

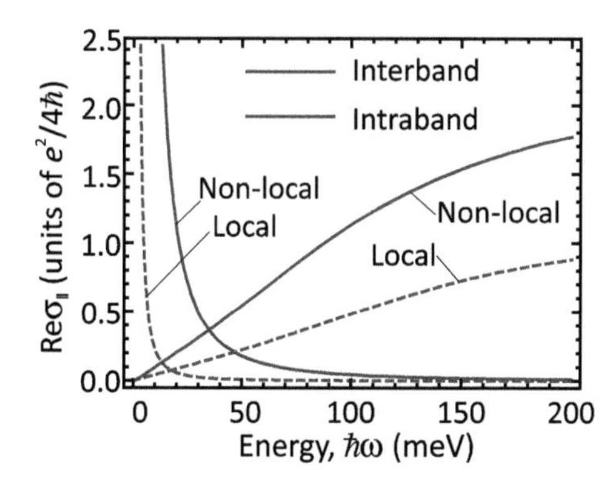

Figure 43.11 Comparison of the real parts of the interband (red) and intraband (blue) conductivities of a single graphene layer evaluated in the local limit (dashed) and at finite wave vector corresponding to the acoustic SP dispersion $q = \omega/s$ (solid). The parameters used in the calculation are $\varepsilon_F = 100$ meV, $T = 300$ K, $s = 1.2\upsilon_0$. Acoustic phonons are considered as the main carrier relaxation mechanism.

Appendix D: Tunnel Conductivity

Approximations similar to those used in deriving Eq. (43.C3) can be made to evaluate the interlayer interband conductivity. The only difference is that electrons in different layers have different chemical potentials. We present these results without derivation:

$$\frac{2G_{\perp}^{\upsilon\to c}}{q^2} = -e^2\frac{T\omega}{2q}\left\{\chi(q,\tilde{\Delta}-\omega)\ln\frac{\cosh\frac{q+eV-\omega}{4T}}{\cosh\frac{q-eV+\omega}{4T}} - \chi(q,\tilde{\Delta}+\omega)\ln\frac{\cosh\frac{q+eV+\omega}{4T}}{\cosh\frac{q-eV-\omega}{4T}}\right\}, \quad (43.D1)$$

$$\frac{2G_{\perp}^{c\to\upsilon}}{q^2} = -e^2\frac{T\omega}{2q}\left\{\chi(q,\omega-\tilde{\Delta})\ln\frac{\cosh\frac{q+eV-\omega}{4T}}{\cosh\frac{q-eV+\omega}{4T}} - \chi(q,-\tilde{\Delta}-\omega)\ln\frac{\cosh\frac{q+eV+\omega}{4T}}{\cosh\frac{q-eV-\omega}{4T}}\right\}. \quad (43.D2)$$

An analytical estimate of the tunnel conductivity associated with intraband transitions is possible only in the limit $\varepsilon_F \gg T$, which is not always the case in our calculations. Again, passing to the elliptic coordinates in Eq. (43.10) and keeping the terms with $s = s'$ one readily finds the result of Eq. (43.12). The value of the auxiliary integral at zero temperature is

$$\mathcal{I}(\alpha,\beta) = \frac{\theta(\beta)}{2}\left[\frac{\beta}{\alpha}\sqrt{\frac{\beta^2}{\alpha^2}-1} - \ln\left(\frac{\beta}{\alpha}+\sqrt{\frac{\beta^2}{\alpha^2}-1}\right)\right] - \{\beta\to-\beta\}. \quad (43.D3)$$

Finally, to obtain physically reasonable results one has to estimate the actual value of the resonant conductivity in the presence of carrier scattering. The result for in-plane intraband conductivity is given essentially by Eq. (43.C6). For all other terms, the classical approximation used in (43.C6) is invalid. To account for the scattering in those terms, we replace the delta-peaked spectral functions of individual particles in the expressions for conductivity with Lorentz-type functions using the following rule:

$$\sum_{\mathbf{p}} \frac{1}{\omega + i\delta - \left(\varepsilon_{\mathbf{p}}^{\mathrm{sl}} - \varepsilon_{\mathbf{p}'}^{\mathrm{s'l'}}\right)}$$

$$\Rightarrow \frac{1}{(2\pi)^2} \int d\varepsilon d\varepsilon' \sum_{\mathbf{p}} \frac{\mathcal{A}_{\mathrm{sl}}(\mathbf{p},\varepsilon)\mathcal{A}_{\mathrm{s'l'}}(\mathbf{p}',\varepsilon)}{\omega + i\delta - (\varepsilon - \varepsilon')}. \tag{43.D4}$$

The spectral function is given by

$$\mathcal{A}_{\mathrm{sl}}(\mathbf{p},\varepsilon) = \frac{2\gamma}{(\varepsilon - \varepsilon_{\mathbf{p}}^{\mathrm{sl}})^2 + \gamma^2}. \tag{43.D5}$$

In our calculations, the broadening factor γ equals the imaginary part of the self-energy due to the electron-phonon collisions evaluated at the Fermi surface [36]:

$$\gamma = \mathrm{Im}\,\Sigma(p_{\mathrm{F}}, \varepsilon_{\mathrm{F}}) = \frac{\varepsilon}{T} \frac{D^2 T^2}{4\rho s^2 v_0^2}\bigg|_{\varepsilon=\varepsilon_{\mathrm{F}}}. \tag{43.D6}$$

The approximation (43.D4) corresponds to the neglect of vertex corrections in the current-current correlator represented by the bubble diagram. The effect of the vertices is typically to reduce the collision frequency; hence, the approximation (43.D4) underestimates the resonant conductivity. All the more, careful calculations of the tunnel conductivity show that interference of the carrier scattering events in different layers leads to a further decrease of the effective scattering rate [31]. We leave the determination of the effective scattering rate γ to a future work, and use its upper estimate in the present chapter.

References

1. A. Grigorenko, M. Polini, and K. Novoselov, *Nat. Photonics,* **6**, 749 (2012).
2. F. H. L. Koppens, D. E. Chang, and F. J. Garcia de Abajo, *Nano Lett.,* **11**, 3370 (2011).
3. E. H. Hwang and S. Das Sarma, *Phys. Rev. B,* **75**, 205418 (2007).
4. V. Ryzhii, A. Satou, and T. Otsuji, *J. Appl. Phys.,* **101**, 024509 (2007).

5. S. A. Mikhailov and K. Ziegler, *Phys. Rev. Lett.*, **99**, 016803 (2007).
6. D. Svintsov, V. Vyurkov, S. Yurchenko, T. Otsuji, and V. Ryzhii, *J. Appl. Phys.*, **111**, 083715 (2012).
7. S. Gangadharaiah, A. M. Farid, and E. G. Mishchenko, *Phys. Rev. Lett.*, **100**, 166802 (2008).
8. A. Woessner, M. B. Lundeberg, Y. Gao, A. Principi, P. Alonso-Gonzalez, M. Carrega, K. Watanabe, T. Taniguchi, G. Vignale, M. Polini, J. Hone, R. Hillenbrand, and F. H. L. Koppens, *Nat. Mater.*, **14**, 421 (2015).
9. A. Principi, M. Carrega, M. B. Lundeberg, A. Woessner, F. H. L. Koppens, G. Vignale, and M. Polini, *Phys. Rev. B*, **90**, 165408 (2014).
10. A. Tomadin and M. Polini, *Phys. Rev. B*, **88**, 205426 (2013).
11. D. Svintsov, V. Vyurkov, V. Ryzhii, and T. Otsuji, *Phys. Rev. B*, **88**, 245444 (2013).
12. A. A. Dubinov, V. Y. Aleshkin, V. Mitin, T. Otsuji, and V. Ryzhii, *J. Phys.: Condens. Matter*, **23**, 145302 (2011).
13. F. Rana, *IEEE Trans. Nanotechnol.*, **7**, 91 (2008).
14. M. Sabbaghi, H.-W. Lee, T. Stauber, and K. S. Kim, *Phys. Rev. B*, **92**, 195429 (2015).
15. L. Britnell, R. Gorbachev, A. Geim, L. Ponomarenko, A. Mishchenko, M. Greenaway, T. Fromhold, K. Novoselov, and L. Eaves, *Nat. Commun.*, **4**, 1794 (2013).
16. A. Mishchenko, J. Tu, Y. Cao, R. Gorbachev, J. Wallbank, M. Greenaway, V. Morozov, S. Morozov, M. Zhu, S. Wong, F. Withers, C. R. Woods, Y.-J. Kim, K. Watanabe, T. Taniguchi, E. E. Vdovin, O. Makarovsky, T. Fromhold, V. Fal'ko, A. Geim, L. Eaves, and K. Novoselov, *Nat. Nanotechnol.*, **9**, 808 (2014).
17. M. Feiginov and V. Volkov, *JETP Lett.*, **68**, 662 (1998).
18. V. Ryzhii and M. Shur, *Jpn. J. Appl. Phys.*, **40**, 546 (2001).
19. B. Sensale-Rodriguez, *Appl. Phys. Lett.*, **103**, 123109 (2013).
20. V. Ryzhii, A. Satou, T. Otsuji, M. Ryzhii, V. Mitin, and M. S. Shur, *J. Phys. D: Appl. Phys.*, **46**, 315107 (2013).
21. S. Das Sarma and E. H. Hwang, *Phys. Rev. Lett.*, **81**, 4216 (1998).
22. E. H. Hwang and S. Das Sarma, *Phys. Rev. B*, **80**, 205405 (2009).
23. E. H. Hwang and S. Das Sarma, *Phys. Rev. B*, **64**,165409 (2001).
24. L. Brey, *Phys. Rev. Appl.*, **2**, 014003 (2014).
25. F. T. Vasko, *Phys. Rev. B*, **87**, 075424 (2013).

26. V. Ryzhii, A. A. Dubinov, V. Y. Aleshkin, M. Ryzhii, and T. Otsuji, *Appl. Phys. Lett.*, **103**, 163507 (2013).

27. R. Bistritzer and A. H. MacDonald, *Proc. Natl. Acad. Sci. U.S.A.*, **108**, 12233 (2011).

28. C. J. Tabert and E. J. Nicol, *Phys. Rev. B*, **86**, 075439 (2012).

29. M. Greenaway, E. Vdovin, A. Mishchenko, O. Makarovsky, A. Patane, J. Wallbank, Y. Cao, A. Kretinin, M. Zhu, S. Morozov, V. I. Fal'ko, K. Novoselov, A. Geim, T. Fromhold, and L. Eaves, *Nat. Phys.*, **11**, 1057 (2015).

30. F. T. Vasko and A. V. Kuznetsov, *Electronic States and Optical Transitions in Semiconductor Heterostructures* (Springer Science & Business Media, 2012).

31. R. Kazarinov and R. Suris, *Sov. Phys. Semicond.*, **6**, 148 (1972).

32. In the other limiting case, $\Omega \ll \nu_g$, the electron states of individual layers have a well-defined occupancy. The effects of plasmon spectrum modification in this case, however, persist. For the parameters used in the calculation, $\hbar\Omega \approx 10$ meV, which exceeds both electron-phonon and electron-electron scattering rate at nitrogen temperature.

33. L. A. Falkovsky and A. A. Varlamov, *Eur. Phys. J. B*, **56**, 281 (2007).

34. J. Faist, F. Capasso, D. L. Sivco, C. Sirtori, A. L. Hutchinson, and A. Y. Cho, *Science*, **264**, 553 (1994).

35. H. Shi, H. Pan, Y.-W. Zhang, and B. I. Yakobson, *Phys. Rev. B*, **87**, 155304 (2013).

36. F. T. Vasko and V. Ryzhii, *Phys. Rev. B*, **76**, 233404 (2007).

37. K. I. Bolotin, K. J. Sikes, J. Hone, H. L. S tormer, and P. Kim, *Phys. Rev. Lett.*, **101**, 096802 (2008).

38. The result can be applied to AA-stacked graphene bilayer, but not to the AB-stacked one. The carrier dispersion for AB stacking does not represent two parallel Dirac cones, and the expression for dynamic conductivity has a different structure [54].

39. A. Principi, R. Asgari, and M. Polini, *Solid State Commun.*, **151**, 1627 (2011).

40. A. Chaplik, *Sov. Phys. JETP*, **35**, 395 (1972).

41. R. Roldan, M. O. Goerbig, and J.-N. Fuchs, *Phys. Rev. B*, **83**, 205406 (2011).

42. L. J. Wang, A. Kuzmich, and A. Dogariu, *Nature (London)*, **406**, 277 (2000).

43. R. Y. Chiao, *Phys. Rev. A*, **48**, R34 (1993).

44. When relaxation processes are properly taken into account, the two parts of the split dispersion Re$\omega(q)$ should merge through the resonance. The divergence in the group velocity, defined as δReω/dq, does not occur in this case. For simplicity of calculation, we have ignored the real part of conductivity when plotting Fig. 43.6, and sought for the real solutions of the dispersion equation.
45. Z. Fei, E. G. Iwinski, G. X. Ni, L. M. Zhang, W. Bao, A. S. Rodin, Y. Lee, M. Wagner, M. K. Liu, S. Dai, M. D. Goldflam, M. Thiemens, F. Keilmann, C. N. Lau, A. H. Castro-Neto, M. M. Fogler, and D. N. Basov, *Nano Lett.,* **15**, 4973 (2015).
46. J. Lambe and S. L. McCarthy, *Phys. Rev. Lett.,* **37**, 923 (1976).
47. M. Parzefall, P. Bharadwaj, A. Jain, T. Taniguchi, K. Watanabe, and L. Novotny, *Nat. Nanotechnol.*, **10**, 1058 (2015).
48. D. Yadav, S. Tombet, T. Watanabe, V. Ryzhii, and T. Otsuji, in *73rd Annual Device Research Conference (DRC)* (IEEE, Piscataway, 2015), pp. 271–272.
49. C. Zhang, M. L. F. Lerch, A. D. Martin, P. E. Simmonds, and L. Eaves, *Phys. Rev. Lett.*, **72**, 3397 (1994).
50. E. E. Vdovin, A. Mishchenko, M. T. Greenaway, M. J. Zhu, D.Ghazaryan, A. Misra, Y. Cao, S. V. Morozov, O. Makarovsky, T. M. Fromhold, A. Patanè, G. J. Slotman, M. I. Katsnelson, A. K. Geim, K. S. Novoselov, and L. Eaves, *Phys. Rev. Lett.,* **116**, 186603 (2016).
51. D. Svintsov, V. Vyurkov, V. Ryzhii, and T. Otsuji, *J. Appl. Phys.,* **113**, 053701 (2013).
52. A. A. Dubinov, V. Y. Aleshkin, V. Ryzhii, M. S. Shur, and T. Otsuji, *J. Appl. Phys.,* **115**, 044511 (2014).
53. P. L. Bhatnagar, E. P. Gross, and M. Krook, *Phys. Rev.*, **94**, 511 (1954).
54. E. J. Nicol and J. P. Carbotte, *Phys. Rev. B,* **77**, 155409 (2008).

Chapter 44

Terahertz Wave Generation and Detection in Double-Graphene Layered van der Waals Heterostructures*

D. Yadav,[a] S. Boubanga Tombet,[a] T. Watanabe,[a] S. Arnold,[a] V. Ryzhii,[a,b] and T. Otsuji[a]
[a]*Research Institute of Electrical Communication, Tohoku University, Sendai 980-8577, Japan*
[b]*Institute of Ultra-High-Frequency Semiconductor Electronics, RAS, Moscow 111005, Russia*
otsuji@riec.tohoku.ac.jp

We report on the first experimental observation of terahertz emission and detection in a double graphene layered (GL) heterostructure which comprises a thin hexagonal-boron nitride tunnel-barrier layer sandwiched between two separately contacted GLs. Inter-GL

*Reprinted with permission from D. Yadav, S. Boubanga Tombet, T. Watanabe, S. Arnold, V. Ryzhii, and T. Otsuji (2016). Terahertz wave generation and detection in double-graphene layered van der Waals heterostructures, *2D Mater.*, **3**, 045009. Copyright © 2016 IOP Publishing Ltd.

Graphene-Based Terahertz Electronics and Plasmonics: Detector and Emitter Concepts
Edited by Vladimir Mitin, Taiichi Otsuji, and Victor Ryzhii

ISBN 978-981-4800-75-4 (Hardcover), 978-0-429-32839-8 (eBook)
www.jennystanford.com

population inversion is induced by electrically biasing the structure. Resonant tunneling and negative differential resistance is expected when the two graphene band structures are perfectly aligned. However, in the case of small misalignments we demonstrate that the photon-absorption/emission-assisted non-resonant- and resonant-tunneling causes all excess charges in the n-type GL to recombine with the holes in the p-type GL giving rise to an increased measured dc current. This work highlights a novel strategy for the realization of efficient voltage-tunable terahertz emitters and detectors.

44.1 Introduction

Graphene, because of its well-known unique characteristics, is a captivating material for many applications ranging from flexible displays [1, 2] photodetectors (PD) [3], ultrafast lasers [4], plasmonics oscillators [5] as well as optical modulators [6]. In particular, applications of graphene field effect transistors (GFET) operating in the terahertz (THz) range are appealing as it is one of the least explored frequency region and holds potential to revolutionize the fields of security, medical imaging, chemical sensing and high-speed wireless communication [7]. Recent research on vertically stacked heterostructures of graphene with hexagonal-boron nitride (h-BN) and other two-dimensional materials has opened a new realm for intriguing device physics making them ideal candidates for future high-frequency technology [8–12].

Our theoretical studies on double-graphene-layered (DGL) heterostructures deploying inter-graphene layer (GL) intraband transitions show that such structures can be exploited for efficient and tunable THz/IR lasers and PDs [13–15] with many advantages over devices based on intra-GL interband radiative transitions [16–18].

Here, we report on the fabrication and first experimental observation of THz emission and detection in the DGL heterostructures based on a thin h-BN tunnel-barrier sandwiched between two monolayers of graphene. The measured relative intensity of the emission spectra shows a clear gate bias dependence with an estimated emission power around 7 pW μm^{-2}. In the detection regime our device yielded measured responsivity of 1.55 AW^{-1} at 1 THz and 300 K. These results reasonably agree with some important aspects of the theoretical model reported in the recent work by Ryzhii et al. [19–22] allowing us to attribute the observed THz emission (detection) to the photon-assisted resonant (non-resonant) radiative inter-GL transitions. These results have shed light on one of the path that might lead to the realization of new, compact, tunable, coherent, and room-temperature operating THz sources and detectors.

44.2 Device Description and Principles of Operation

The fabrication of the proposed DGL heterostructure involves micromechanical cleavage and transfer of tiny flakes of graphene and h-BN on a Si/SiO_2 substrate. The fabrication process includes first transfer of a relatively thick h-BN layer (~50 nm) on top of an oxidized Si wafer used as the back gate. This first layer of h-BN acts as an atomically flat substrate on which the active part of the device is mounted and ensures high mobility in the GLs. The h-BN flakes used in this work were exfoliated from ultra-pure h-BN single bulk crystal. On this high-quality, atomically flat h-BN buffer layer, an exfoliated monolayer of graphene was transferred carefully using the standard dry transfer process and patterned using electron-beam lithography and oxygen plasma etching to define the contact bars. Next, a relatively thin layer of h-BN (around 2–7 atomic layers) was transferred. This thin layer served as the tunnel barrier. The second GL was then transferred on top of the thin h-BN layer. Lastly, Ti/Au contacts were fabricated on both GLs by using e-beam lithography, evaporation, and lift-off processes. The schematic of the device is shown in Fig. 44.1a while Figs. 44.1b and c depict the SEM images of the fabricated device. One can see the position of source and drain electrodes (light gray), the active area of the device (dark gray) and the SiO_2 substrate (black) in Fig. 44.1b. The line-like structure in between 'source' and 'drain' contacts, is the active area of the device containing overlapped graphene monolayers. This SEM image shows a top view of the device consisting of Si/SiO_2/h-BN/graphene/h-BN/graphene. Being a top view the stacked structure cannot be seen in this image. A zoomed image in 44.1c shows the active area of the device to be around 1.4–1.5 μm wide.

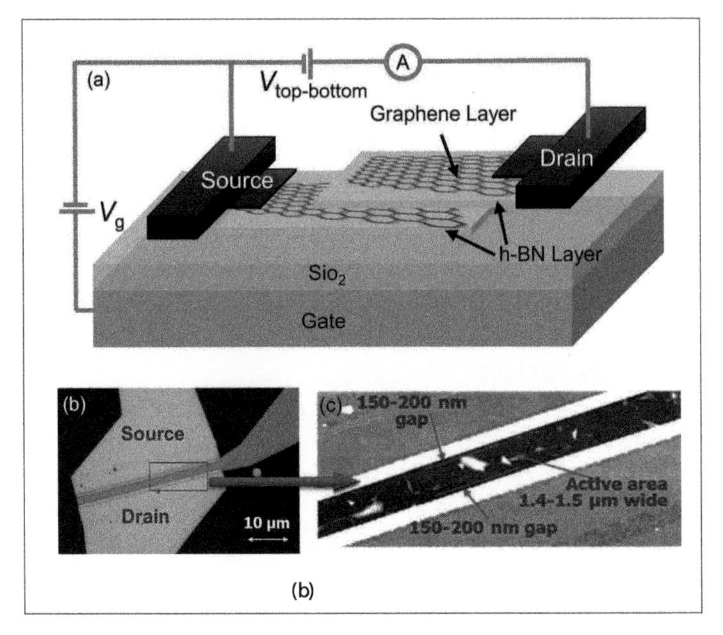

Figure 44.1 (a) The schematic of the fabricated DGL device. (b) SEM image of the device showing the position of source and drain electrodes (light gray), the active area of the device (dark gray) and the SiO_2 substrate (black) when viewed from top. (c) A zoomed SEM image showing active area of the device to be 1.4–1.5 μm wide.

Relative crystallographic orientation between the two GLs is vital to get resonant transitions from the device. Appearance of negative differential conductance (NDC) in the current-voltage (I–V) characteristics is one indication of how well the two GLs are aligned [23, 24]. In order to observe the NDC at low enough energies (or gate voltages) the two layers have to be aligned so that their crystallographic axes are parallel. Since most of the flakes are exfoliated along the crystallographic directions a straight-line edge of the 2nd GL flake was aligned to meet that of the 1st GL flake during the transfer process [25]. For aligning the edge of a flake in our devices, the substrate stage was rotated to match the line of the edge. With microzoom-scope eye and assistance from the computer, the best relative alignment possible is around ~1° [23]. There are two possibilities of the straight-line edges; one is 'zig-zag' and the other is 'armchair,' but they cannot be distinguished well via optical micro-zoom-scope observation [26]. When both the top and bottom GLs are aligned with the zig-zag edges or armchair edges, the misalignment could be very small (~1° or less), but the cases of one side with the zig-zag and the other with the armchair give rise to large misalignment (~30°). In this work, two devices were fabricated independently by using the aforementioned procedure. Both the devices have slight miss-alignment (>4°) between the crystallographic axes of the two GLs, while differing in terms of thickness of h-BN tunnel barrier layer, ~3 nm in device 1 and ~2 nm in the case of device 2. The device 1 is used for emission studies and device 2 for the THz detection studies. Some of the other fabricated samples have small miss-alignment ~1° (see supplementary information for details). However, a recent report by Kim et al. [27] showcases another technique to have highly accurate layer alignment by carefully picking the monolayers by a hemispherical epoxy handle, which should be introduced in future study.

The physical mechanism behind the device operation is the photon assisted tunneling transitions. The bias voltage (V) applied between the GL's contacts induces the electron gas in one GL and the hole gas in the other GL, forming a DGL capacitor, displacing the chemical potentials between the GLs. The electron and hole densities as well as the chemical potentials in the GLs are also modulated by the gate voltage (V_g).

Figures 44.2a and b schematize the device operation mechanism and show the inter-GL valence band to valence band (V–V) and conduction band to conduction band (C–C) resonant-tunneling.

When the Dirac points in the two GLs are aligned by applying V and V_g pertinently, the system is in resonance and the energy and momentum conservations are satisfied during the tunneling. This pronouncedly increases and maximizes the probability of quantum mechanical resonant tunneling (QM-RT) between the two GLs. Further application of bias V leads to bands misalignment and the reduction of tunneling current, producing the NDC shape in the I–V characteristics. Nevertheless, when the system is slightly off resonance with a bands-offset energy Δ in THz and IR range, TM polarized photons with energy $\hbar\omega \sim \Delta$ can mediate the inter-GL resonant tunneling between the p-type and the n-type GLs (see Figs. 44.2a and b). Therefore, in this case the device operation principle is associated with the absorption/emission of the TM polarized photon with energy $\hbar\omega \sim \Delta$ accompanied by the electron tunneling transitions between the GLs and causing the electric terminal current. The inter-GL radiative transitions with the absorption (Fig. 44.2a) or emission (Fig. 44.2b) of photons conserve the electron lateral momentum and are referred to as photon assisted quantum mechanical resonant tunneling (PA-RT).

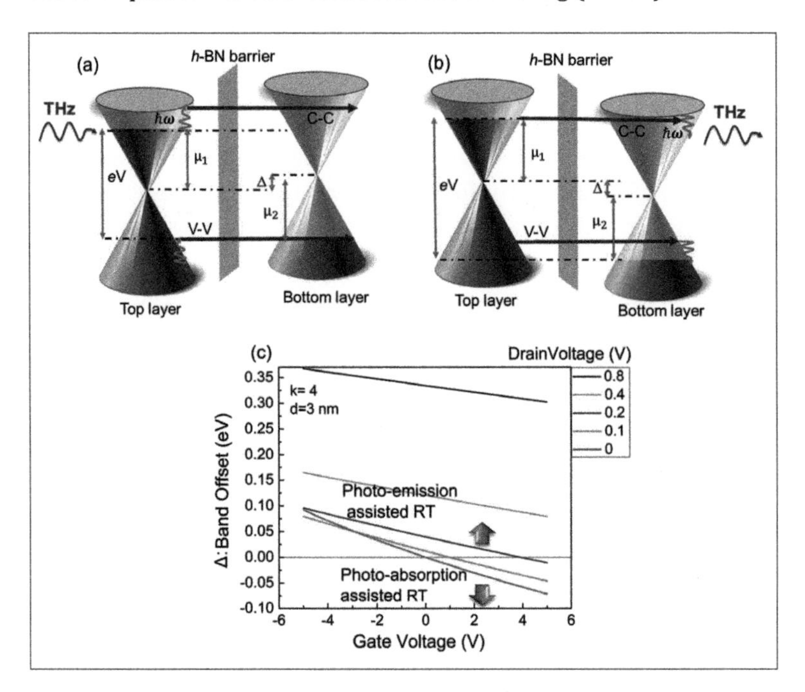

Figure 44.2 Band diagrams of the DGL emitter/detector structures with (a) photo-absorption-assisted inter-GL, and (b) photo-emission-assisted inter-GL radiative transitions. Wavy arrows indicate the inter-GL radiative V–V and C–C transitions both in devices (a) and (b). The inter-GL transitions work for the TM-mode THz photon radiations in our structures. (c) The calculation result for the bias voltage dependent band-offset energy Δ of the DGL device having a 3 nm thick h-BN tunnel barrier layer. $\kappa = 4$ is the permittivity of the h-BN layer.

The energies of the photons emitted or absorbed by the structure in the inter-GL resonant-tunneling (RT) transitions is determined by the voltage dependent band-offset energy $\Delta(V, V_g)$ between the Dirac points of the GLs and the depolarization shift. We calculated $\Delta(V, V_g)$ using the model developed in [21] in order to find the most appropriate condition for DGL to perform photo-absorption-assisted RT and/or photo-emission-assisted RT. The results, as shown in

Fig. 44.2c, indicate that the device operation strongly depends on the relative sign of Δ. For the band offset conditions when the energy of the final state of the tunneling inter-GL transition is lower than the initial state ($\Delta > 0$), the tunneling is associated with emission of the photons (see also Fig. 44.2b) whereas when $\Delta < 0$, the tunneling is associated with absorption of the photons (see also Fig. 44.2a). The resonant tunneling causes all excess charges in the n-type GL to recombine with the holes in the p-type GL. Our recent study reveals the possibility of plasmon-assisted inter-layer resonant/non-resonant tunneling as another mechanism of producing net gain in the THz range [15].

44.3 Experimental Methods

The detection experiments were carried out with our dedicated setup. It involves a uni-traveling-carrier photodiode (UTC-PD) module as a THz source. The UTC-PD works as a THz photomixer providing tunable output in the THz range (300 GHz–1 THz) in response to two continuous-wave (CW) laser beam inputs whose difference in frequencies are tuned to be equal to the output frequency. In case of 1 THz with 1 μW output power, the wavelengths of the two CW laser beams were set to be 1540.000 nm and 1547.946 nm, respectively, and their intensities were optically amplified to a level of 35 mW. Then the beams were coupled and fed to the UTC-PD. The THz photons were guided onto the active area of the DGL device through a set of parabolic and indium tin oxide mirrors. The gate and drain terminals were connected to a semiconductor parameter analyzer (Keysight: Model B1500A) while the source terminal was grounded. The dc current-voltage (I–V) characteristics of the DGL device were recorded for different V_g values. The dc photocurrent-voltage characteristics were obtained by subtracting the measured I–V curves without the THz irradiation from the ones with the THz irradiation. The experiment was performed at room temperature in oblique incidence geometry as shown in Fig. 44.3a to yield the THz electric field perpendicular to the GLs (TM polarized field) and thus the tunneling transitions associated with the absorption of the THz photons [28].

The emission experiments were carried out using Fourier-transformed far-infrared spectroscopy at 100 K and 300 K. The sample was placed in the vacuum chamber and the biases were controlled using voltage/current source-meters (Keithley 2400) while the radiation intensity was measured using a 4.2 K-cooled silicon bolometer. To carry out the experiments, we first measured the background spectrum, i.e. the spectrum with unbiased sample. Then we measured bias dependent spectra with current flowing through the sample and normalized it to the background spectra to eliminate the undesired effects arising due to static blackbody emission and any spectral artifacts emitted from the elements inside the spectrometer. Since the emitted TM polarized photons exit the sample from the side edges (see Fig. 44.3b), a set of metal prisms was used to collect and direct the emitted radiation towards the detector as shown in Fig. 44.3c.

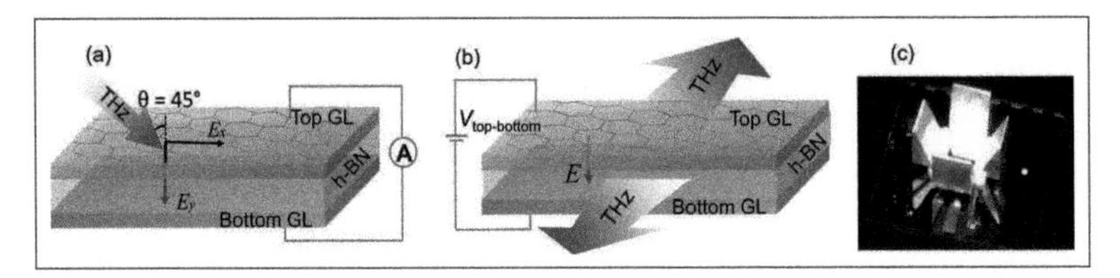

Figure 44.3 Experimental arrangement for (a) THz detection showing THz photons incident at an angle (45°) on the DGL device to maximize the induced electric field (E_y) perpendicular to the GLs and for (b) emission showing THz photons exiting the sample from the side edges. (c) The device chip mount base with metal prisms attached close to the device edges directing the emitted radiation towards the detector.

44.4 Results and Discussions

Figure 44.4 depicts the measured *I*–*V* curves for device 1 exhibiting no traces of the NDC at room temperature (Fig. 44.4a) but weak NDC traces at 100 K (Fig. 44.4b). Under the conditions that the weak NDCs are obtained, the bias voltage conditions allow the inter-GL population inversion and photon-assisted RT radiative transitions between the conduction bands and the valence bands of the two GLs. To verify such an expected emission of THz radiation, we measured emission spectra of the device by using the FT-IR setup as described in section 3 for different gate bias conditions at a fixed drain voltage (V_d = 0 V). Figure 44.5 depicts emission spectra of device 1 at 300 K (a) and at 100 K (b). Figure 44.5a shows no signs of increased THz emission beyond the background blackbody radiation as expected given the absence of NDCs in I-V curves at 300 K (see Fig. 44.4a). On the contrary, when the sample is cooled down to 100 K, one can see broadband THz emission from the DGL structure under specific gate bias conditions as depicted in Fig. 44.5b. The measured spectra show increasing emission intensity with increasing negative gate bias voltages while no emission is observed for positive gate voltages. This reasonably agrees with the calculations shown in Fig. 44.2c based on the theoretical prediction by Ryzhii et al. [21].

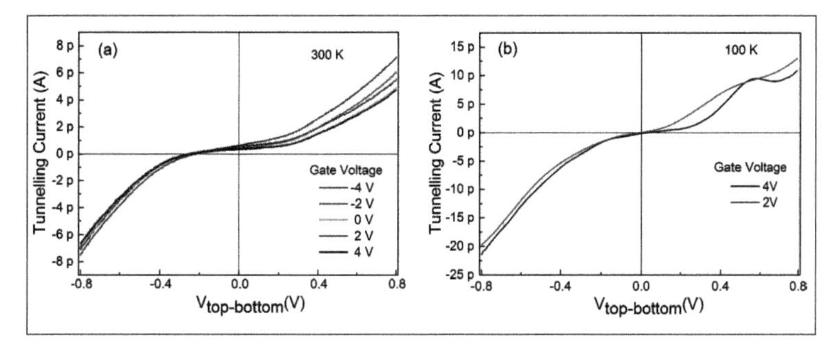

Figure 44.4 Measured *I–V* curves of device 1 (a) at 300 K and (b) at 100 K exhibiting weak NDCs.

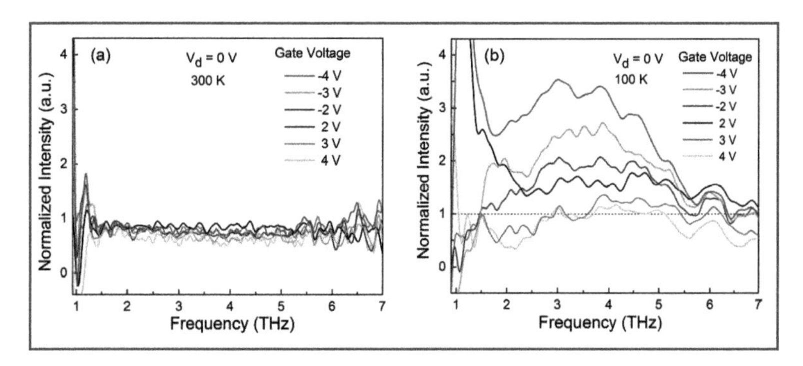

Figure 44.5 Measured emission spectra at different gate bias voltages at drain bias of 0 V (a) at 300 K, and (b) at 100 K.

The device exhibits relatively broad emission ranging from 2 to 6 THz, peaking around 3.1 THz, likely due to a weak and rather broad NDC in this sample at 100 K as shown in Fig. 44.4b. The linewidth of the emission spectra reflects the resonant tunneling quality factor Q which correlates with the carrier momentum relaxation time, rotational misalignment and non-uniform tunneling due to inhomogeneities present in GLs [23, 29]. The peak position is determined by Δ which is a function of V and V_g. This means that the emission frequencies from the DGL device is expected to be voltage tunable when the sample shows pronounced gate tunable NDC peaks in the I–V. The integrated emission intensity of device 1 in the 2–6 THz frequency range was estimated to be 7 pW μm^{-2} at V_g = −4 V. This estimation is based on the calculation of the black body radiated power at 100 K which serves as reference and multiplied by the peak values of intensity of normalized emission curve at a particular gate voltage. Recent paper has reported thermal emission from GFET of the order of 2.1 nW in 1–3 THz range by using a double-patch antenna along with a silicon lens attachment [30]. Our DGL source hold potentialities to emit THz with much higher power given a pronounced and stronger NDC (Fig. 44.S1 (b) in supplementary information) and coupling device structure. The observed results might include in part the effect of plasmon-assisted resonant tunneling, which could be performed under inter-GL misaligned conditions [15].

For biasing conditions when $\Delta < 0$, the RT is preceded by an absorption of photons. When the NDC cannot be obtained but only thermionic positive conductance is obtained as seen in I–V curves of device 2 at 300 K (Fig. 44.6a), the tunneling process takes place in a non-resonant manner so that the photon-assisted tunneling may take place in a broadband nature with the incident photon energy. The detection experimental results are shown in Fig. 44.6b, depicting the incoming radiation induced tunneling photocurrent versus DGL bias voltages V for different V_g levels under 1 THz photon irradiation at 300 K. The photo-current increases with increasing V both positively and negatively. The increase of the DGL photocurrent with the positive gate biases is more likely due to the non-resonant tunneling associated with absorption of THz photons. This nonlinear monotonic increase comes from the quadratic rectification whose responsivity corresponds to the nonlinearity of the I–V curve in Fig. 44.6a. The prior works of Vasko [31] and Ryzhii et al. [14, 32] give theoretical support for the photo-absorption-assisted resonant/non-resonant tunneling and the resultant tunneling current in the DGL. They also provide thorough discussion on effects of applied gate bias on sensitivity of these devices. The several peaks superposed onto the monotonic nonlinear increase of photocurrent as seen in Fig. 44.6b are not essentially arising from the device itself and not resonant peaks neither but most likely fluctuations caused by fluctuation in power of the THz emitter used for detection experiments and environmental noise. Since the photo-current is of the order of pico-amperes, the sources of noise mentioned above affects very much the experiments and results.

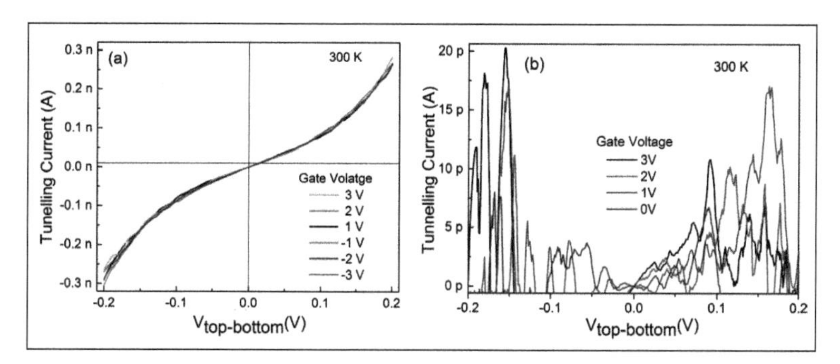

Figure 44.6 Measured *I–V* curves of device 2 at 300 K (a), tunneling current induced by the incoming THz photon irradiation at different gate biases (b).

The absorption of incident radiation is generally limited to the Fermi surface charge carriers in standard semiconductors. For instance, in the case of lateral GFET [35], only the electrons in top surface of graphene have freedom to receive the incident energy and contribute to the photo-generated current, thus limiting the sensitivity of such devices. However, in our case, current flows vertically down from one GL to the other counterpart. Not only the electrons present at Fermi surface but all the excess charge carriers in top GL can absorb the incoming radiation and thermionically tunnel to the other GL leading to very high photocurrents. We quantify the photocurrent generated by the 1 μW incoming THz irradiation in terms of photoresponsivity. Intrinsic responsivity for a detector can be defined as:

$$R_{\mathrm{j}} = \frac{S_{\mathrm{t}}\Delta j}{S_{\mathrm{d}}P_{\mathrm{in}}(\sin\theta)^2}, \tag{44.1}$$

where S_{t} is THz beam spot size (S_{t} = 2.83 mm^2), S_{d} is DGL active area (S_{d} = 54.75 μm^2), Δj is maximum induced photocurrent (Δj = 15 pA, see Fig. 44.6b), θ is THz incident angle ($\theta = 45°$, see Fig. 44.3a), and P_{in} is THz input power (P_{in} = 1 μW). It is assumed in the above definition that all the incoming THz power is coupled effectively to the device. However, due to absence of any antenna structure in this case, only a small fraction of incident radiation is channeled onto the device. Hence, this value of responsivity shall be considered as a lower limit.

Using equation (44.1), we obtained $R_j \sim 1.55$ A W^{-1} for non-resonant detection at room temperature. In the case of resonant tunneling, the current flowing through the device is about 10^3 times larger than the non-resonant case (see I–V characteristics with NDCs in Fig. 44.S1b in supplementary information, compare to Fig. 44.S1a). Since the incoming THz field is expected to induce resonant tunneling in a well aligned sample, three orders larger Δj can be expect in the resonant case. Responsivity is directly proportional to photocurrent Δj, its considerably larger value in the resonant case will likely result in drastic enhancement of responsivity. Indeed, a theoretical work by Rodriguez [34] also predict five orders higher responsivity (3×10^5 VW^{-1} in samples showing NDCs) in resonant detection case compared to the non-resonant (almost zero for samples without NDCs).

Our detectors hold great potentialities compared to other graphene based THz photodetectors. Indeed, graphene FET based THz detectors operating: on photo-thermoelectric effects; over-damped plasma waves excitation; and on both thermoelectric/plasmonic effects were recently reported with responsivity ~5 nAW^{-1} [35], 4.7 VW^{-1} [30] and 10 VW^{-1} [36]; ~1.3 mAW^{-1} [37]; 0.25 VW^{-1} [38] respectively. Our detector shows higher responsivity even in the non-resonant detection regime and is promised to about three orders of magnitude higher sensitivity in the resonant mode. The next step of this work to observe the resonant THz detection in the DGL device with a small crystallographic misalignment between the GLs (Fig. 44.S1a in supplementary information) will be a subject of future publication. Introduction of an antenna structure such as a grating gate structure proposed by Ryzhii et al. [14] or patch antenna and Si lens attached to the device would greatly improve the responsivity as experimentally reported by Tong et al. [30].

44.5 Conclusion

We fabricated a DGL van der Waals heterostructure device and experimentally observed its response to the THz radiation via non-resonant tunneling between the GLs. Since the device holds an undesired crystallographic rotational misalignment between the GLs, no NDCs was observed at room temperature. Correspondingly, non-resonant photoresponse was observed at room temperature in response to 1 THz photon irradiation, demonstrating a high responsivity of 1.55 AW^{-1}. In one of our devices we observed weak NDCs at 100 K, giving a broadband emission (2–6 THz) with an intensity 7 pW μm^{-2}. The emission was observed only when the band-offset levels between the GLs were set in negative so that the photo-emission-assisted resonant tunneling can take place. These results are the first demonstration proving experimentally that the gated DGL devices based on the photon-assisted resonant radiative inter-GL transitions can be exploited to create highly efficient THz sources and THz photo detectors.

Acknowledgments

The authors are grateful to Akira Satou at RIEC, Tohoku University, Japan for the valuable discussions.

Funding

This work was financially supported in part by JSPS KAKENHI (#23000008, and #16H06361), Japan. VR acknowledges also the support by the Russian Foundation for Basic Research (Grant No.16-29-03033).

References

1. Bonaccorso, F., Sun, Z., Hasan, T., and Ferrari, A. C. (2010). Graphene photonics and optoelectronics, *Nat. Photonics*, **4**, 611–622.
2. Bae, S., et al. (2010). Roll-to-roll production of 30-inch graphene films for transparent electrodes, *Nat. Nanotechnol.*, **4**, 574–578.
3. Xia, F., Mueller, T., Lin, Y. M., Valdes-Garcia, A., and Avouris, P. (2009). Ultrafast graphene photodetector, *Nat. Nanotechnol.*, **4**, 839–843.

4. Sun, Z., Hasan, T., Torrisi, F., Popa, D., Privitera, G., Wang, F., Bonaccorso, F., Basko, D. M., and Ferrari, A. C. (2010). Graphene mode-locked ultrafast laser, *ACS Nano*, **4**, 803–810.
5. Echtermeyer, T. J., Britnell, L., Jasnos, P. K., Lombardo, A., Gorbachev, R. V., Grigorenko, A. N., Geim, A. K., Ferrari, A. C., and Novoselov, K. S. (2011). Strong plasmonic enhancement of photo voltage in graphene, *Nat. Commun.*, **2**, 458.
6. Liu, M., Yin, X., Avila, E. U., Geng, B., Zentgraf, T., Ju, L., Wang, F., and Zhang, X. (2011). A graphene-based broadband optical modulator, *Nature*, **474**, 64–67.
7. Otsuji, T., Boubanga, Tombet, S. A., Satou, A., Fukidome, H., Suemitsu, M., Sano, E., Popov, V., Ryzhii, M., and Ryzhii, V. (2012). Graphene materials and devices in terahertz science and technology, *MRS Bull.*, **37**, 12.
8. Grigorenko, A. N., Polini, M., and Novoselov, K. S. (2010). Graphene plasmonics, *Nat. Photonics*, **6**, 749.
9. Dean, C. R., et al. (2010). Boron nitride substrates for high-quality graphene electronics, *Nat. Nanotechnol.*, **5**, 722–726.
10. Geim, A. K. and Grigorieva, I. V. (2013). Van der Waals heterostructures, *Nature*, **499**, 419–425.
11. Britnell, L., Gorbachev, R. V., Geim, A. K., Ponomarenko, L. A., Mishchenko, A., Greenaway, M. T., Fromhold, T. M., Novoselov, K. S., and Eaves, L. (2013). Resonant tunneling and negative differential conductance in graphene transistors, *Nat. Commun.*, **4**, 1794.
12. Gannett, W., Regan, W., Watanabe, K., Taniguchi, T., Crommie, M. F., and Zettl, A. (2011). Boron nitride substrates for high mobility chemical vapor deposited graphene, *Appl. Phys. Lett.*, **98**, 242105.
13. Ryzhii, V., Otsuji, T., Ryzhii, M., Aleshkin, V. Y., Dubinov, A. A., Svintsov, D., Mitin, V., and Shur, M. S. (2015). Graphene vertical cascade interband terahertz and infrared photodetectors, *2D Mater.*, **2**, 025002.
14. Ryzhii, V., Otsuji, T., Aleshkin, V. Y., Dubinov, A. A., Ryzhii, M., Mitin, V., and Shur, M. S. (2014). Voltage-tunable terahertz and infrared photodetectors based on double-graphene-layer structures, *Appl. Phys. Lett.*, **104**, 163505.
15. Svintsov, D., Devizorova, Zh., Otsuji, T., and Ryzhii, V. (2016). Plasmons in tunnel-coupled graphene layers: backward waves with quantum cascade gain, *Phys. Rev. B*, **94**, 115301.
16. Ryzhii, V., Ryzhii, M., Mitin, V., and Otsuji, T. (2011). Toward the creation of terahertz graphene injection laser, *J. Appl. Phys.*, **110**, 094503.

17. Ryzhii, V., Ryabova, N., Ryzhii, M., Baryshnikov, N. V., Karasik, V. E., Mitin, V., and Otsuji, T. (2012). Terahertz and infrared photodetectors based on multiple graphene layer and nanoribbon structures *Opto-Electron. Rev.*, **20**, 15–25.
18. Ryzhii, V., Dubinov, A. A., Otsuji, T., Mitin, V., and Shur, M. S. (2010). Terahertz lasers based on optically pumped multiple graphene structures with slot-line and dielectric waveguides, *J. Appl. Phys.*, **107**, 054505.
19. Dubinov, A. A., Aleshkin, V. Y., Ryzhii, V., Shur, M. S., and Otsuji, T. (2014). Surface-plasmons lasing in double-graphene-layer structures, *J. Appl. Phys.*, **115**, 044511.
20. Ryzhii, V., Dubinov, A. A., Aleshkin, V. Y., Ryzhii, M., and Otsuji, T. (2013). Injection terahertz laser using the resonant inter-layer radiative transitions in double-graphene-layer structure, *Appl. Phys. Lett.*, **103**, 163507.
21. Ryzhii, V., Dubinov, A. A., Otsuji, T., Aleshkin, V. Y., Ryzhii, M., and Shur, M. S. (2013). Double-graphene-layer terahertz laser: concept, characteristics, and comparison, *Opt. Express*, **21**, 31567.
22. Ryzhii, V., Satou, A., Otsuji, T., Ryzhii, M., Mitin, V., and Shur, M. S. (2013). Dynamic effects in double graphene-layer structures with inter-layer resonant-tunneling negative conductivity, *J. Phys. D: Appl. Phys.*, **46**, 315107.
23. Mishchenko, A., et al. (2014). Twist-controlled resonant tunneling in graphene/boron nitride/graphene heterostructures, *Nat. Nanotechnol.*, **9**, 808–813.
24. Brey, L. (2014). Coherent tunneling and negative differential conductivity in a graphene/h-BN/graphene heterostructure, *Phy. Rev. B*, **2**, 014003.
25. Neubeck, S., You, Y. M., Ni, Z. H., Blake, P., Shen, Z. X., Geim, A. K., and Novoselov, K. S. (2010). Direct determination of the crystallographic orientation of graphene edges by atomic resolution imaging, *Appl. Phys. Lett.*, **97**, 053110.
26. Tang, S., et al. (2013). Precisely aligned graphene grown on hexagonal boron nitride by catalyst free chemical vapor deposition, *Sci. Rep.*, **3**, 2666.
27. Kim, K., et al. (2016). van der Waals heterostructures with high accuracy rotational alignment, *Nano Lett.*, **16**, 1989–1995.
28. Popov, V. V., Bagaeva, T. Y., Otsuji, T., and Ryzhii, V. (2010). Oblique terahertz plasmons in graphene nanoribbon arrays, *Phy. Rev. B*, **81**, 073404.

29. Fallahazad, B., et al. (2015). Gate-tunable resonant tunneling in double bilayer graphene heterostructures, *Nano Lett.*, **15**, 428–433.

30. Tong, J., Muthee, M., Chen, S. Y., Yngveson, S. K., and Yan, J. (2015). Antenna enhanced graphene THz emitter and detector, *Nano Lett.*, **15**, 5295.

31. Vasko, F. T. (2013). Resonant and nondissipative tunneling in independently contacted graphene structures, *Phys. Rev. B*, **87**, 075424.

32. Ryzhii, V., Otsuji, T., Ryzhii, M., and Shur, M. S. (2012). Double graphene-layer plasma resonances terahertz detector, *J. Phys. D: Appl. Phys.*, **45**, 302001.

33. Vicarelli, L., Vitiello, M. S., Coquillat, D., Lombardo, A., Ferrari, A. C., Kanp, W., Polini, M., Pellegrini, V., and Tredicucci, A. (2012). Graphene field-effect transistors as room temperature detectors, *Nat. Mater.*, **11**, 3417.

34. Rodriguez, B. S. (2013). Graphene-insulator-graphene active plasmonic terahertz devices, *Appl. Phys. Lett.*, **103**, 123109.

35. Mittendorff, M., Winner, S., Kamann, J., Eroms, J., Weiss, D., Schneider, H., and Helm, M. (2013). Ultrafast graphene-based broadband THz detector, *Appl. Phys. Lett.*, **103**, 021113.

36. Cai, X., et al. (2014). Sensitive room-temperature terahertz detection via the photothermoelectric effect in graphene, *Nat. Nanotechnol.*, **9**, 814.

37. Spirito, D., Coquillat, D., De Bonis, S. L., Lombardo, A., Bruna, M., Ferrari, A. C., Pellegrini, V., Tredicucci, A., Knap, W., and Vitiello, M. S. (2014). High performance bilayer-graphene terahertz detector, *Appl. Phys. Lett.*, **104**, 061111.

38. Bianco, F., Perenzoni, D., Convertino, D., De Bonis, S. L., Spirito, D., Perenzoni, M., Coletti, C., Vitiello, M. S., and Tredicucci, A. (2015). Terahertz detection by epitaxial-graphene field-effect-transistors on silicon carbide, *Appl. Phys. Lett.*, **107**, 131104.

Supplementary Information

Graphene Layers' Rotation Angle Sensitive Current-Voltage Characteristics

Negative differential conductance (NDC) strongly depends on the relative crystallographic orientation between the two graphene layers (GLs) [S1]. The sample with *I–V* characteristics shown in Fig. 44.1a was fabricated with the rotation angle misalignment of ~30° while this angle is about 1° for the sample with *I–V* shown in Fig. 44.1b. In case of aligned top and bottom GLs (small values of θ), the Dirac cones are rather overlapped Fig. 44.1c which at appropriate bias voltages facilitates all the charge carriers from one layer to resonantly tunnel and recombine with the charges carriers of other layer leading to a giant gain in the tunneling current (Fig. 44.1d schematically depicts this situation). Figure 44.1a shows no NDC at room temperature however a very clear NDC peak can be seen in the sample with precisely aligned GLs at room temperature. Resonant tunneling can lead to $\pm 10^3$ times larger photocurrent than non-resonant process.

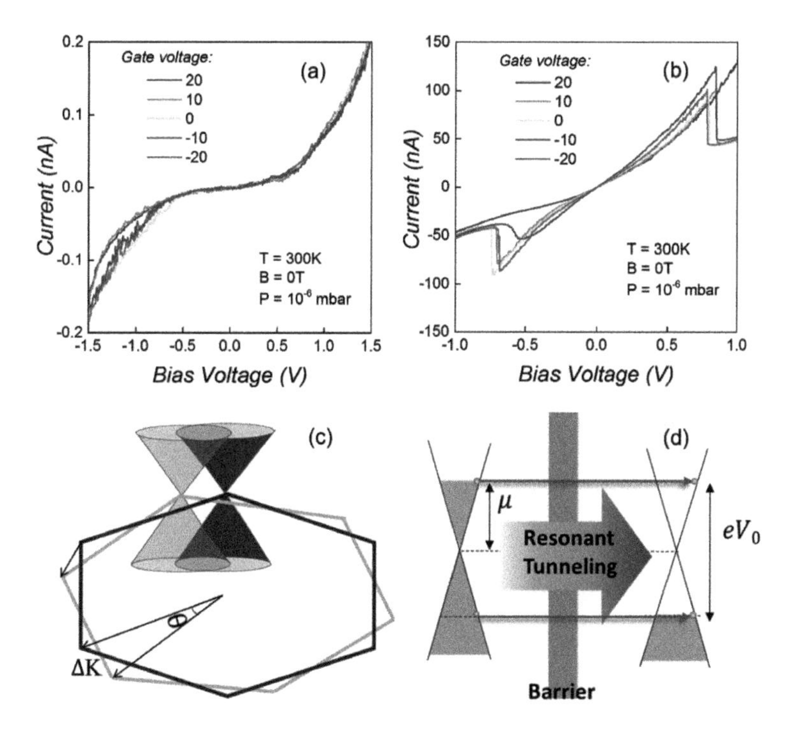

Figure 44.S1 Measured *I–V* curves of the device having ~30° misalignment between two GLs (a) and another device with ~1° misalignment exhibiting a clear NDC at 300 K (b). (c) shows twisted graphene Dirac cones and (d) shows that resonant tunneling occurs when two GLs gets aligned.

Reference

S1. Mishchenko, A., Tu, J. S., Cao, Y., Gorbachev, R. V., Wallbank, J. R., Greenaway, M. T., Morozov, V. E., Morozov, S. V., Zhu, M. J., Wong, S. L., Withers, F., Woods, C. R., Kim, Y.-J., Watanabe, K., Taniguchi, T., Vdovin, E. E., Makarovsky, O., Fromhold, Fal'ko, V. I., Geim, A. K., Eaves, L., and Novoselov, K. S. (2014). Twist-controlled resonant tunneling in graphene/boron nitride/graphene heterostructures, *Nat. Nanotechnol.*, **9**, 808–813.

Chapter 45

Ultra-Compact Injection Terahertz Laser Using the Resonant Inter-Layer Radiative Transitions in Multi-Graphene-Layer Structure*

A. A. Dubinov,[a,b,c] A. Bylinkin,[c] V. Ya. Aleshkin,[a,b]
V. Ryzhii,[d,e,f] T. Otsuji,[d] and D. Svintsov[c]

[a]*Institute for Physics of Microstructures of Russian Academy of Sciences, Nizhny Novgorod 603950, Russia*
[b]*Lobachevsky State University of Nizhny Novgorod, Nizhny Novgorod 603950, Russia*
[c]*Laboratory of 2D Materials' Optoelectronics, Moscow Institute of Physics and Technology, Moscow, Russia*
[d]*Research Institute for Electrical Communication, Tohoku University, Sendai 980-8577, Japan*
[e]*Institute of Ultra High Frequency Semiconductor Electronics of Russian Academy of Sciences, Moscow 117105, Russia*
[f]*Center for Photonics and Infrared Engineering, Bauman Moscow State Technical University, Moscow 105005, Russia*
sanya@ipmras.ru

*Reprinted with permission from A. A. Dubinov, A. Bylinkin, V. Ya. Aleshkin, V. Ryzhii, T. Otsuji, and D. Svintsov (2016). Ultra-compact injection terahertz laser using the resonant inter-layer radiative transitions in multi-graphene-layer structure, *Opt. Express*, **24**(26), 29603–29612. Copyright © 2016 Optical Society of America.

Graphene-Based Terahertz Electronics and Plasmonics: Detector and Emitter Concepts
Edited by Vladimir Mitin, Taiichi Otsuji, and Victor Ryzhii

ISBN 978-981-4800-75-4 (Hardcover), 978-0-429-32839-8 (eBook)
www.jennystanford.com

The optimization of laser resonators represents a crucial issue for the design of terahertz semiconductor lasers with high gain and low absorption loss. In this chapter, we put forward and optimize the surface plasmonic metal waveguide geometry for the recently proposed terahertz injection laser based on resonant radiative transitions between tunnel-coupled graphene layers. We find an optimal number of active graphene layer pairs corresponding to the maximum net modal gain. The maximum gain increases with frequency and can be as large as ~500 cm^{-1} at 8 THz, while the threshold length of laser resonator can be as small as ~50 μm. Our findings substantiate the possibility of ultra-compact voltage-tunable graphene-based lasers operating at room temperature.

45.1 Introduction

The compact tunable sources of terahertz radiation are highly demanded in security, medical, and telecommunication applications [1–3]. Quantum cascade lasers (QCLs) are considered among most promising candidates to bridge the terahertz gap [4–7], but their operation is currently limited to the cryogenic temperatures only. The reason for low-temperature QCL operation is the strong direct leakage current via closely located subbands in the 1–2 THz range, and phonon-assisted depopulation of the upper subbands in the 4–6 THz range [8]. Also, in the frequency region of 5–12 THz, the operation of QCLs is greatly hindered by the optical phonon absorption (restrallen band) in GaAs and InP [5]. The lead salt diode lasers can be considered as alternative to QCLs as they cover the very broad frequency region (6.5–26 THz) [9, 10], but their figures of merit, in particular, output power, are limited due to growth technology problems.

Therefore, the problem of creating the room-temperature terahertz semiconductor laser remains unresolved. In the past ten years, a considerable effort was focused on the creation of graphene-based terahertz lasers with optical [11–14] and electrical [15] pumping. The gapless electron-hole spectrum in graphene allows one to experimentally achieve optical gain under population inversion in the whole 1–5 THz range [16] which can be markedly enhanced by plasmon excitation [17, 18]. High optical phonon energy and suppression of the Auger process by the dynamic screening create favorable conditions for relatively long non-radiative recombination lifetimes [19, 20] in graphene. This has enabled the recent demonstration of THz lasing in electrically pumped graphene at 5 THz at liquid nitrogen temperature [21]. However, the prospects of the room-temperature operation of such lasers remain unclear due to the enhancement of recombination processes and self-absorption [22].

Both problems of Drude absorption and interband non-radiative recombination are avoided in the recently proposed lasers based on interlayer electron transitions in tunnel-coupled graphene layers (GLs) [23–25]. The population inversion between the layers in such structures is achieved by an application of the bias voltage, while the emission of THz photons is due to the photon-assisted interlayer tunneling, similar to that in QCLs [26] (see Fig. 45.1b). The resonant-tunneling photon emission which is the key effect enabling the proposed lasing scheme was recently observed in the measurements of THz electroluminescence from graphene-insulator-graphene structures [27]. The presence of luminescence in these structures correlated with the negative differential conductivity in the static current-voltage curves, thereby supporting the resonant-tunneling origin of the emission. Moreover, in agreement with the theory of inelastic resonant photon-assisted tunneling, the structures operated either as THz emitters or detectors, depending on the voltage bias and carrier densities in GLs. A related yet different experimentally demonstrated method to generate light from graphene-based tunneling structures is based on the elastic electron tunneling followed by the interband photon emission [28].

In this chapter, we study whether it is possible to make a step from experimentally observed spontaneous THz emission in tunnel-coupled GLs toward a full-scale THz lasing. We introduce a scheme of THz laser based on resonant photon-assisted transitions in multiple graphene layer structure embedded in a surface plasmonic waveguide enhancing the light-matter interactions [29–32]. We find the ultimate modal gain and threshold length for lasing, and show that a seemingly simple solution to increase the number of active graphene layer pairs does not always lead to the gain enhancement. The reason is that THz gain due to photon-assisted tunneling associated with the transverse component of electric field always competes with interband and Drude absorption in GLs aided by the lateral electric fields. A similar problem of the waveguide optimization aimed at the absorption reduction is well-known in the design of QCLs [33]. Actually, there exists a limited physical space inside the laser resonator, where the transverse electric field significantly exceeds the lateral field, so that the tunneling gain dominates over the absorption loss. In this chapter, we find the corresponding upper limit of the modal gain in the THz lasers based on the multiple tunnel-coupled GLs. To this end, we combine the microscopic model of photon-aided resonant tunneling in graphene double layers with electromagnetic simulations of the TM-mode surface plasmon waveguides. Depending on the lasing frequency, the ultimate gain can reach from ~50 cm^{-1} (at 4 THz) to ~500 cm^{-1} (at 10 THz) and is achieved for the waveguide structures with dozens of active graphene double layers. The corresponding threshold length can be as low as 40–300 μm, depending on the frequency of operation. Our findings support the possibility of creating ultra-compact graphene-based THz lasers.

45.2 Device Model

We consider the device structure shown in Fig. 45.1a. It consists of N pairs of independently-contacted GLs separated by a narrow tunneling barrier layer of thickness d (2.5 nanometers of WS_2). To increase the density of tunneling carriers, we assume the layers to be doped by donors (n-GL, marked with red in Fig. 45.1a) and acceptors (p-GL, marked with blue), respectively. Each pair of the active GLs is separated by a narrow insulating layer of thickness f (90 nanometers of hexagonal boron nitride, hBN). The doping with density of 10^{12} cm^{-2} leads to the formation of the two-dimensional electron (2DEG) and the two-dimensional hole (2DHG) gases in the n-GL (the top GL of each pair) and the p-GL (the bottom GL of each pair), respectively. In the structure under consideration, the height of the contacts H is related to the number of double graphene layers N and the distance f between them via $H = N \times (f + d)$.

The bias voltage V between the side contacts induces the extra charges of opposite sign in the layers and the electric field between them, which causes the inter-GL population inversion (Fig. 45.1b).

Under the population-inverted conditions, the stimulated electron tunneling accompanied by the photon emission is more probable than the inverse absorptive process. Thus, such a GL structure can serve as the laser gain medium with the lateral injection pumping. The active graphene structure is sandwiched by the hBN insulator cladding layers with total thickness W = 10 μm equal to half wavelength of THz photon of frequency $\omega/2\pi = 6$ THz in the insulator. The whole structure is placed atop a metal substrate constituting a surface plasmon waveguide for the TM-modes propagating in the y-direction in Fig. 45.1a. The electrical contacts are separated by an insulating gap with thickness w = 90 nm from the waveguiding surface.

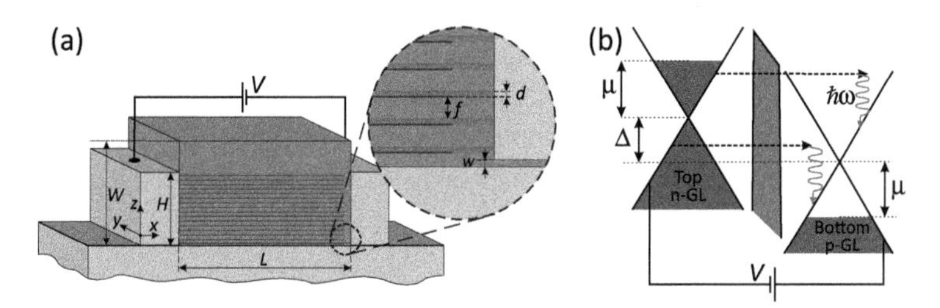

Figure 45.1 (a) Schematic view of the terahertz laser with a stack of tunnel-coupled graphene layers (N pairs) with side injection embedded in a surface plasmon waveguide (b) Band diagram illustrating the process of stimulated photon-assisted resonant tunneling under application of interlayer voltage V.

The net modal gain $g(\omega)$ for the TM-mode in the device under consideration is given by the following expression:

$$g(\omega) = -\frac{4\pi n_{\text{eff}}(\omega, N)}{c\kappa}[\text{Re}\sigma_{zz}(\omega)\Gamma_{\text{z}}(\omega) + \text{Re}\sigma_{yy}(\omega)\Gamma_{\text{y}}(\omega)] \quad (45.1)$$

where

$$\Gamma_{\text{z,y}}(\omega) = \frac{\sum_{j=0}^{N-1}\int_0^L \left|E_{\text{z,y}}(x, w + (f+d)j, \omega)\right|^2 dx}{\int_{-\infty}^{+\infty}\int_{-\infty}^{+\infty} \left|E_{\text{z}}(x, z, \omega)\right|^2 dxdz} \quad (45.2)$$

are the gain-overlap factors, $n_{\text{eff}}(\omega, N)$ is the effective mode index, $E_{\text{z,y}}$ are the components of the electric field in the TM-mode, c is the light speed in vacuum, L is the waveguide width, κ is the dielectric constant of the barrier layer, σ_{zz} and σ_{xx} are the transverse and lateral dynamic conductivities of tunnel-coupled graphene double layer structure.

The inter-GL transitions assisted by the emission of TM-photons with the dominant z-component of electric field conserve the in-plane electron momentum and, hence, do not involve scattering (resonant-tunneling photon-assisted transitions). The resonant character of tunneling transitions results in the Lorentzian shape of the real part of tunnel conductivity [26]

$$\mathrm{Re}\,\sigma_{zz}(\omega) \cong -2\frac{e^2}{\hbar}\left|z_{\mathrm{u,l}}\right|^2 \Sigma \frac{\omega\gamma}{(\omega-\omega_{\max})^2+\gamma^2}, \tag{45.3}$$

where Σ is the density of electrons (holes) in the pertinent GL, $ez_{\mathrm{u,l}}$ is the dipole matrix element between wave functions of carriers localized on the upper and lower layers, and γ is the relaxation broadening of the tunnel resonance which is found to be ~3 meV for the given carrier density (see Appendix for the detailed calculation). As seen from the band diagram in Fig. 45.1b, the maximum conductivity corresponds to the transitions with emission of photon with energy $\hbar\omega_{\max}$ equal to the spacing between Dirac points in the neighboring GLs Δ. This energy should corrected by the value of depolarization shift [34]

$$\Delta_{\mathrm{dep}} = -8\pi\frac{e^2\left|z_{\mathrm{u,l}}\right|^2 \Sigma}{\kappa d}, \tag{45.4}$$

where κ and d are the permittivity and thickness of the tunneling barrier layer. The resonant conductivity at $\omega = \omega_{\max}$ reads as follows

$$\mathrm{Re}\,\sigma_{zz}(\omega_{\max}) \cong -2\frac{e^2}{\hbar}\left|z_{\mathrm{u,l}}\right|^2 \Sigma\frac{\omega_{\max}}{\gamma}. \tag{45.5}$$

The dipole matrix element $z_{\mathrm{u,l}}$ is evaluated as

$$z_{\mathrm{u,l}} = \int_{-\infty}^{\infty} \phi_{\mathrm{u}}^{*}(z) z \phi_{\mathrm{l}}(z) dz, \tag{45.6}$$

where $\phi_{\mathrm{u}}(z)$ and $\phi_{\mathrm{l}}(z)$ are the z-dependent envelope wave functions of electrons located at the upper and lower GLs, respectively. These wave functions are found by solving the effective Schrödinger equation where each graphene layer is modeled as a one-dimensional delta-well [25]. The strength of the delta-well is chosen to provide the correct work function from graphene to the surrounding dielectric (0.4 eV for graphene embedded in WS_2 [35]). The effective mass of electron in WS_2 is taken to be 0.27 of the free electron mass [36].

The in-plane conductivity of graphene layers is given by [37]

$$\mathrm{Re}\,\sigma_{yy}(\omega) = \frac{e^2}{\hbar}\left\{\frac{2\gamma\mu}{\pi(\hbar^2\omega^2+\gamma^2)} + \frac{1}{4}[f(-\hbar\omega/2) - f(\hbar\omega/2)]\right\}, \quad (45.7)$$

where the first term in curly brackets is responsible for Drude absorption and the second one—for the interband absorption, μ is the Fermi energy of carriers (electrons in top GL and holes in bottom one), and $f(\varepsilon) = [1 + e^{(\varepsilon-\mu)/kT}]^{-1}$ is the Fermi function.

In the case of relatively short GLs (for the diffusion length of electrons and holes $L_d \gg L$), the carrier Fermi Fermi energy μ is constant throughout the layer and related to the applied voltage V and spacing between the Dirac points Δ as $\Delta = eV - 2\mu$. At strong doping and/or low temperatures such that $\mu \gg k_B T$, this Fermi energy is a square-root function of density, $\mu = \hbar v_W \sqrt{\pi\Sigma}$, where $v_W \approx 10^8$ cm/s is the characteristic velocity of electrons and holes in GLs. This yields the following relation between Δ and V:

$$\Delta = e\left(V + V_0 - \sqrt{V_0^2 + V_t^2 + 2VV_0}\right), \quad (45.8)$$

where $eV_0 = [\hbar^2 v_W^2 \kappa/2e^2 d]^{1/2}$, and $V_t = 2\hbar v_W \sqrt{\pi\Sigma_i}/e$ is the threshold voltage. At $V \leq V_t$, Eq. (45.8) yields $\Delta \leq 0$, while at $V > V_t$, $\Delta > 0$. In latest case, the photon emission upon interlayer intraband tunneling dominates over the absorption.

The solution of the Maxwell equations with the pertinent complex permittivity of the surface plasmon waveguide and metal strips yields the spatial distributions of the electric fields, $E_{z,y}(x, z, \omega)$ in the propagating mode and, consequently, the gain-overlap factors. These equations were solved numerically using 2D mode analysis in COMSOL. It is assumed that the metal strips and substrate in the surface plasmon waveguide are made of Au [38]. The h-BN complex permittivity was extracted from work [39]. In the calculations we have neglected the influence of GLs on the field distributions, because the electric field component along the GLs and, hence, the surface currents are small in the TM mode, as we shall discuss below.

45.3 Results and Discussion

Figures 45.2a–d show the examples of the spatial distributions of the photon electric field components $|E_z(x, z, \omega)|$ and $|E_y(x, z, \omega)|$ in the TM mode of a surface plasmon waveguide in device of Fig. 45.1. The obtained dependences correspond to $\omega/2\pi = 5$ THz. In the waveguide under consideration, the vertical component of electric field induced by the interlayer tunneling is dominantly localized near the bottom metal surface. The lateral component of electric is, on the contrary, localized at the air/dielectric interface. This creates favorable conditions for the tunneling gain if the graphene double-layers are localized close to the bottom metal surface.

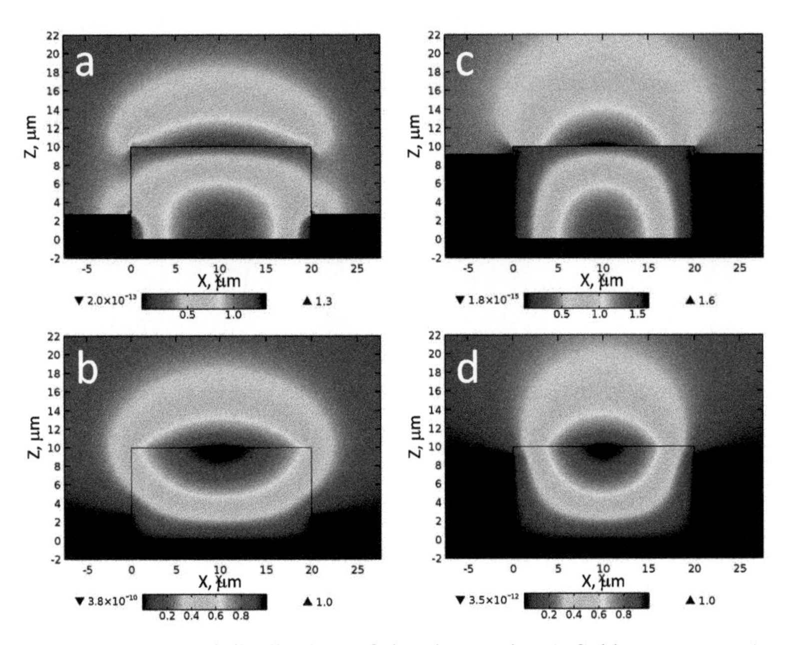

Figure 45.2 Spatial distributions of the photon electric field components in TM mode of the surface plasmonic waveguide under consideration for two different heights of the metal contacts, corresponding to $N = 30$ (a, b) and $N = 100$ (c, d). Panels (a, c) show the z-component of electric field, panels (b, d) show the y-component. The wave frequency is $\omega/2\pi = 5$ THz.

Increasing the number of active graphene layers and, hence, the height of metal contacts, one increases the gain due to interlayer tunneling transitions. On the other hand, the topmost layers are located closer to the maximum of in-plane field $|E_y(x, z, \omega)|$, as seen in Fig. 45.2, and increase in the layer number can significantly increase ohmic losses (due to both Drude and interband absorption). Thus, there should exist an optimal number of active layers for which the net modal gain reaches its maximum.

Figure 45.3a shows the frequency (for $\omega = \omega_{max}$) dependences of the modal gain calculated with Eq. (45.1) for several fixed numbers of active double layer structures. From Fig. 45.3a one can see that the modal gain is higher for structures with a small number of GL pairs in the lowfrequency region. For low frequencies Drude absorption increases faster than the gain with an increase in the number of GL pairs. The situation is reversed for the high frequency region. The gain threshold, i.e. the frequency for which the gain exceeds the loss in the GLs, shifts to the higher frequencies with increasing N.

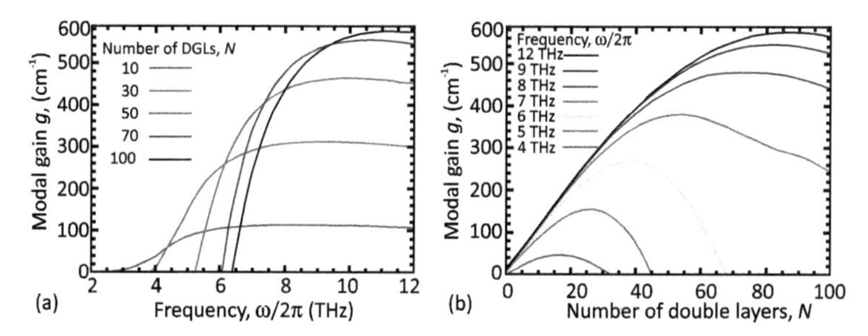

Figure 45.3 (a) Frequency dependence of modal gain at fixed numbers of double graphene layers N. (b) Dependence of modal gain on the number of double layers at fixed frequencies. The increase in N above the threshold value leads to increase in Drude absorption and reduction in modal gain.

The presence of the optimal number of active double layers N_{opt} is seen from Fig. 45.3b, where plot the dependence of modal gain on N at several fixed frequencies corresponding to the interlayer tunneling resonance. The modal gain maximum increases and shifts toward higher values of N with increasing frequency. For example, for $\omega/2\pi = 5$ THz $N_{opt} \sim 27$, and for $\omega/2\pi = 8$ THz $N_{opt} \sim 72$. Furthermore, with increasing frequency the range of the number of GL pairs at which the gain is possible becomes broader.

Figure 45.4 shows the frequency dependences of N_{opt} and the corresponding maximum modal gain. For structure considered in our work, the maximum modal gain becomes positive at $\omega/2\pi$

= 3 THz (N_{opt} ~ 18) and increases up to 580 cm^{-1} with increasing the frequency. The increase in maximum modal gain at moderate frequencies ($\omega \leq 10$ THz) is due to the reduction in Drude absorption and increase in the resonant tunnel conductivity [Eq. (45.5)] due to the factor ω_{max} in the numerator. At higher frequencies, the maximum attainable gain saturates because large resonant frequencies require the application of large interlayer voltage which, in turn, reduces the dipole matrix elements $ez_{u,l}$. This dependence of maximum tunnel conductivity is illustrated in the inset of Fig. 45.4, where we plot the dimensionless 'gain factor' $4\pi\sigma_{zz}(\omega_{max})/c\sqrt{\kappa}$.

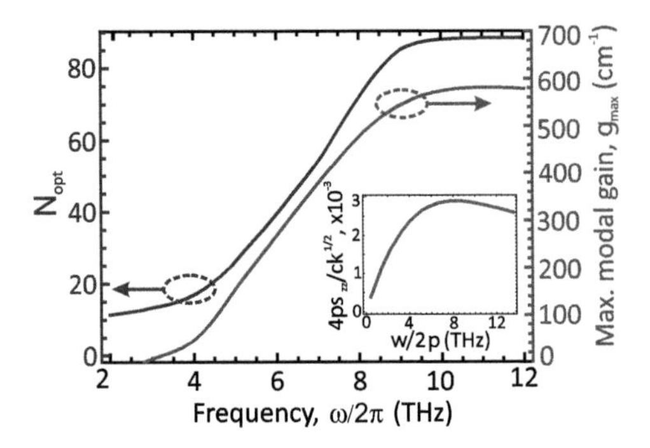

Figure 45.4 Frequency dependence of the optimal number of double graphene layers N_{opt} and the corresponding maximum modal gain. Inset: frequency dependence of dimensionless 'gain factor' $4\pi\sigma_{zz}(\omega_{max})/c\sqrt{\kappa}$.

Using Eq. (45.1), one can determine the threshold length $L_{th}(\omega)$ of the structure for which the lasing condition is fulfilled:

$$L_{th}(\omega) = \frac{\ln[1/R(\omega)]}{g(\omega) - \alpha(\omega)} \tag{45.9}$$

where $\alpha(\omega)$ is the surface plasmon waveguide absorption coefficient due to losses in metal, and $R(\omega)$ is the reflection coefficient of the structure edges. If the resonator mirrors represent the cleaved surfaces of h-BN, $R(\omega)$ can be estimated as follows:

$$R(\omega) = \left(\frac{n_{eff}(\omega) - 1}{n_{eff}(\omega) + 1}\right)^2. \tag{45.10}$$

Figure 45.5 shows the frequency dependence of the absorption coefficient $\alpha(\omega)$ and the reflection coefficient $R(\omega)$ in the structure under consideration for three different values of *N*. From Fig. 45.5 one can see that reflection coefficients increase with increasing the frequency for our waveguide structure. At the same time, a fairly weak dependence of reflection coefficients on *N* is observed. The absorption coefficients have maxima at 5.5 THz, and the frequency corresponding to the maximum absorption depends strongly on the waveguide geometry. Below 5.5 THz, the electromagnetic wave is weakly bound to the waveguide, which leads to the low Ohmic losses in metal. The absorption maximum corresponds to the wave with strong electric field at the side contacts. At higher frequencies, the electric field of the wave is localized primarily at the bottom metal surface, aside from the waveguide sidewalls, which again leads to the reduction in absorption.

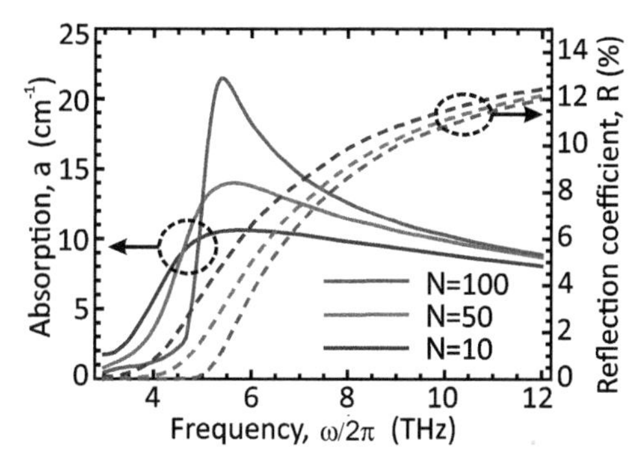

Figure 45.5 Frequency dependence of waveguide absorption and reflection coefficients different values of *N*.

Figure 45.6 shows the frequency dependence of laser threshold length $L_{th}(\omega)$ for different values of *N* in the given structure. The trends observed in Fig. 45.6 are similar to those in Fig. 45.3. For each frequency, there exists an optimal value of *N*. In the frequency range of interest (5–12 THz) where the operation of quantum cascade lasers is strongly hindered by the semiconductor restrallen bands, $L_{th}(\omega)$ can be reduced down to values of 40–300 μm by a proper selection of *N*. Such laser threshold lengths are considerably less than the characteristic lengths of quantum cascade lasers (over 1000 μm [40]). Furthermore, this means that one can use a sufficiently small area graphene layers for laser structures (20 × 40 μm) which typically demonstrate better electronic properties.

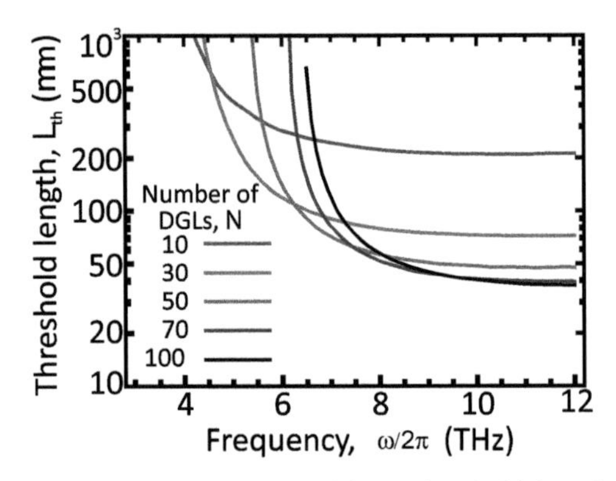

Figure 45.6 Frequency dependence of laser threshold length for different values number of double graphene layers (DGLs) *N*.

45.4 Conclusion

We have proposed and substantiated the concept of injection THz laser based on the multiple graphene-layer structure exploiting the interlayer resonant radiative transitions and embedded in plasmonic waveguide. An increase in the number of active pairs of tunnel-coupled graphene layers can lead to a dramatic increase in the modal gain (up to ~500 cm^{-1}) as compared with quantum cascade lasers. We have found, however, that there exists an optimal number of active double graphene layers for each frequency. Further increase of layer number would lead to an increase in interband and Drude absorption in the structure. The graphene-based lasers with the surface plasmon waveguides considered above have advantages associated with an efficient voltage tuning, low Drude absorption and optical phonon absorption. This would enable their operation in the frequency range from 5 to 12 THz.

Appendix: Calculation of the Tunnel Resonance Broadening

The broadening factor γ entering the expressions for tunnel conductivity is different from the collisional broadening of electron energy in a single layer and from the quantity $\hbar\tau_{tr}^{-1}$, where τ_{tr} is the transport relaxation time. The origin of this difference is the interference of carrier scattering events in the tunnel-coupled layers: if the action of impurity on carriers in both layers were the same, there would be no scattering and, hence, no interlayer current [26, 41]. The account for this interference leads to the following expression for γ

$$\gamma = 2\pi \sum_{\mathbf{q}} \left\langle \left| V_{\mathrm{t}}(\mathbf{q}) - V_{\mathrm{b}}(\mathbf{q}) \right|^2 \right\rangle \delta[v_{\mathrm{W}}(p - |\mathbf{p}-\mathbf{q}|)](1 - \cos\theta_{\mathbf{p},\mathbf{p}-\mathbf{q}})/2, \tag{45.11}$$

where $V_{\mathrm{t}}(\mathbf{q})$ and $V_{\mathrm{b}}(\mathbf{q})$ are the Fourier components of impurity potential acting on the electrons in bottom and top layers, the factor $(1 - \cos\theta_{\mathbf{p},\mathbf{p}-\mathbf{q}})/2$ comes from the overlap of chiral wave functions in graphene, and the brackets denote averaging over random impurity positions. Performing this averaging, we find

$$\gamma = 2\pi(\Sigma_{\mathrm{i,t}} + \Sigma_{\mathrm{i,b}}) \sum_{\mathbf{q}} V_0^2(\mathbf{q})(1 - e^{-qd})^2 \delta[v_{\mathrm{W}}(p - |\mathbf{p}-\mathbf{q}|)](1 - \cos\theta_{\mathbf{p},\mathbf{p}-\mathbf{q}})/2, \tag{45.12}$$

where $\Sigma_{\mathrm{i,t}}$ and $\Sigma_{\mathrm{i,b}}$ are the impurity densities in top and bottom layers, $V_0(\mathbf{q}) = 2\pi e^2/\kappa(q + q_{\mathrm{TF}})$ is the Fourier transform of the Coulomb potential created by a single impurity, and q_{TF} is the Thomas-Fermi screening wave vector. The small factor $1 - e^{-qd} \ll 1$ is due to the discussed interference of scattering events. The evaluation of Eq. (45.12) with the following parameters $\Sigma_{\mathrm{i,t}} = \Sigma_{\mathrm{i,b}} = 10^{12}$ cm^{-2}, $\kappa = 5$, $d = 2.5$ nm, carrier energy $pv_{\mathrm{W}} = 110$ meV (corresponding to the Fermi energy at given density) leads us to $\gamma \approx 3$ meV.

Funding

The theoretical work at MIPT was supported by Grant No. 16-19-10557 of the Russian Science Foundation. Experimental support by TO acknowledges JSPS KAKENHI No. 16H06361.

References

1. P. F.-X. Neumaier, K. Schmalz, J. Borngraber, R. Wylde, and H.-W. Hubers, Terahertz gas-phase spectroscopy: chemometrics for security and medical applications, *Analyst*, **140**, 213–222 (2015).
2. T. Hochrein, Markets, availability, notice, and technical performance of terahertz systems: historic development, present, and trends, *J. Infrared. Millimeter Terahertz Waves*, **36**, 235–254 (2015).

3. T. Nagatsuma, S. Horiguchi, Y. Minamikata, Y. Yoshimizu, S. Hisatake, S. Kuwano, N. Yoshimoto, J. Terada, and H. Takahashi, Terahertz wireless communications based on photonics technologies, *Opt. Express*, **21**(20), 23736– 23747 (2013).
4. S. Williams, Terahertz quantum-cascade lasers, *Nat. Photonics*, **1**, 517–525 (2007).
5. R. Colombelli, F. Capasso, C. Gmachl, A. L. Hutchinson, D. L. Sivco, A. Tredicucci, M. C. Wanke, A. M. Sergent, and A. Y. Cho, Far-infrared surface-plasmon quantum-cascade lasers at 21.5 μm and 24 μm wavelengths, *Appl. Phys. Lett.*, **78**(18), 2620–2622 (2001).
6. S. Vitiello and A. Tredicucci, Tunable emission in THz quantum cascade lasers, *IEEE Trans. THz Sci. Technol.*, **1**, 76–84 (2011).
7. M. A. Belkin and F. Capasso, New frontiers in quantum cascade lasers: high performance room temperature terahertz sources, *Phys. Scr.*, **90**(11), 118002 (2015).
8. Y. Chassagneux, Q. J. Wang, S. P. Khanna, E. Strupiechonski, J.-R. Coudevylle, E. H. Linfield, A. G. Davies, F. Capasso, M. A. Belkin, and R. Colombelli, Limiting factors to the temperature performance of THz quantum cascade lasers based on the resonant-phonon depopulation scheme, *IEEE Trans. THz Sci. Technol.*, **2**(1), 83–92 (2012).
9. L. N. Kurbatov, A. D. Britov, S. M. Karavaev, S. D. Sivachenko, S. N. Maksimovskii, I. I. Ovchinnikov, M. M. Rzaev, and P. M. Starik, Far-IR heterojunction lasers tunable to 46.2 μm, *JETP Lett.*, **37**(9), 499–502 (1983).
10. K.V. Maremyanin, A. V. Ikonnikov, A. V. Antonov, V. V. Rumyantsev, S. V. Morozov, L. S. Bovkun, K. R. Umbetalieva, E. G. Chizhevskiy, I. I. Zasavitskiy, and V. I. Gavrilenko, Long-wavelength injection lasers based on $Pb_{1-x}Sn_xSe$ alloys and their use in solid-state spectroscopy, *Semiconductors*, **49**(12), 1623–1626 (2015).
11. V. Ryzhii, M. Ryzhii, and T. Otsuji, Negative dynamic conductivity of graphene with optical pumping, *J. Appl. Phys.*, **101**, 083114 (2007).
12. V. Ya. Aleshkin, A. A. Dubinov, and V. Ryzhii, Terahertz laser based on optically pumped graphene: model and feasibility of realization, *JETP Lett.*, **89**(2), 63–67 (2009).
13. P. Weis, J. L. Garcia-Pomar, and M. Rahm, Towards loss compensated and lasing terahertz metamaterials based on optically pumped graphene, *Opt. Express*, **22**(7), 8473–8489 (2014).
14. V. Ryzhii, A. A. Dubinov, T. Otsuji, V. Mitin, and M. S. Shur, Terahertz lasers based on optically pumped multiple graphene structures with slot-line and dielectric waveguides, *J. Appl. Phys.*, **107**, 054505 (2010).

15. V. Ryzhii, I. Semenikhin, M. Ryzhii, D. Svintsov, V. Vyurkov, A. Satou, and T. Otsuji, Double injection in graphene p-i-n structures, *J. Appl. Phys.*, **113**, 244505 (2013).
16. S. Boubanga-Tombet, S. Chan, T. Watanabe, A. Satou, V. Ryzhii, and T. Otsuji, Ultrafast carrier dynamics and terahertz emission in optically pumped graphene at room temperature, *Phys. Rev. B*, **85**, 035443 (2012).
17. T. Watanabe, T. Fukushima, Y. Yabe, S. A. Boubanga Tombet, A. Satou, A. A. Dubinov, V. Ya. Aleshkin, V. Mitin, V. Ryzhii, and T. Otsuji, The gain enhancement effect of surface plasmon polaritons on terahertz stimulated emission in optically pumped monolayer graphene, *New J. Phys.*, **15**, 075003 (2013).
18. T. Otsuji, V. Popov, and V. Ryzhii, Active graphene plasmonics for terahertz device applications, *J. Phys. D: Appl. Phys.*, **47**, 094006 (2014).
19. F. Rana, P. A. George, J. H. Strait, J. Dawlaty, S. Shivaraman, M. Chandrashekhar, and M. G. Spencer, Carrier recombination and generation rates for intravalley and intervalley phonon scattering in graphene, *Phys. Rev. B*, **79**, 115447 (2009).
20. A. Tomadin, D. Brida, G. Cerullo, A. C. Ferrari, and M. Polini, Nonequilibrium dynamics of photoexcited electrons in graphene: collinear scattering, Auger processes, and the impact of screening, *Phys. Rev. B*, **88**, 035430 (2013).
21. G. Tamamushi, T. Watanabe, A. A. Dubinov, J. Mitsushio, H. Wako, A. Satou, T. Suemitsu, H. Fukidome, M. Suemitsu, M. Ryzhii, V. Ryzhii, and T. Otsuji, Single-mode terahertz emission from current-injection graphenechannel transistor under population inversion, in *74th Annual Device Research Conference* (IEEE, 2016), pp. 1–2.
22. D. Svintsov, V. Ryzhii, A. Satou, T. Otsuji, and V. Vyurkov, Carrier-carrier scattering and negative dynamic conductivity in pumped graphene, *Opt. Express*, **22**(17), 19873–19886 (2014).
23. V. Ryzhii, A. A. Dubinov, V. Ya. Aleshkin, M. Ryzhii, and T. Otsuji, Injection terahertz laser using the resonant inter-layer radiative transitions in double-graphene-layer structure, *Appl. Phys. Lett.*, **103**, 163507 (2013).
24. V. Ryzhii, A. A. Dubinov, T. Otsuji, V. Ya. Aleshkin, M. Ryzhii, and M. Shur, Double-graphene-layer terahertz laser: concept, characteristics, and comparison, *Opt. Express*, **21**(25), 31567–31577 (2013).
25. A. A. Dubinov, V. Ya. Aleshkin, V. Ryzhii, M. S. Shur, and T. Otsuji, Surface-plasmons lasing in double-graphenelayer structures, *J. Appl. Phys.*, **115**, 044511 (2014).

26. R. Kazarinov and R. Suris, Theory of electrical and electromagnetic properties of semiconductors with superlattice, *Sov. Phys. Semicond.*, **6**, 148 (1972).

27. D. Yadav, S. Boubanga Tombet, T. Watanabe, S. Arnold, V. Ryzhii, and T. Otsuji, Terahertz wave generation and detection in double-graphene layered van der Waals heterostructures, *2D Mater.*, **3**(4), 045009 (2016).

28. F. Withers, O. Del Pozo-Zamudio, S. Schwarz, S. Dufferwiel, P. M. Walker, T. Godde, A. P. Rooney, A. Gholinia, C. R. Woods, P. Blake, S. J. Haigh, K. Watanabe, T. Taniguchi, I. L. Aleiner, A. K. Geim, V. I. Fal'ko, A. I. Tartakovskii, and K. S. Novoselov, WSe_2 light-emitting tunneling transistors with enhanced brightness at room temperature, *Nano Lett.*, **15**, 8223–8228 (2015).

29. Y.Fan, N.-H. Shen, T. Koschny, and C. M. Soukoulis, Tunable terahertz meta-surface with graphene cut-wires, *ACS Photonics*, **2**, 151–156 (2015).

30. F. H. L. Koppens, D.E. Chang, and F. J. Garcia de Abajo, Graphene plasmonics: a platform for strong light–matter interactions, *Nano Lett.*, **11**, 3370–3377 (2011).

31. T. Low and P. Avouris, Graphene plasmonics for terahertz to mid-infrared applications, *ACS Nano*, **8**, 1086–1101 (2014).

32. Y. Fan, Z. Liu, F. Zhang, Q. Zhao, Z. Wei, Q. Fu, J. Li, C. Gu, and H. Li, Tunable mid-infrared coherent perfect absorption in a graphene meta-surface, *Sci. Rep.*, **5**, 13956 (2015).

33. A. A. Bogdanov and R. A. Suris, Mode structure of a quantum cascade laser, *Phys. Rev. B*, **83**, 125316 (2011).

34. F. T. Vasko and A. V. Kuznetsov, *Electronic States and Optical Transitions in Semiconductor Heterostructures* (Springer, New York, 1999).

35. T. Georgiou, R. Jalil, B. D. Bellee, L. Britnell, R. V. Gorbachev, S. V. Morozov, Y.-J. Kim, A. Cholinia, S. J. Haigh, O. Makarovsky, L. Eaves, L. A. Ponomarenko, A. K. Geim, K. S. Nonoselov, and A. Mishchenko, Vertical field-effect transistor based on graphene-WS_2 heterostructures for flexible and transparent electronics, *Nat. Nanotechnol.*, **7**, 100–103 (2013).

36. H. Shi, H. Pan, Y.-W. Zhang, and B. Yakobson, Quasiparticle band structures and optical properties of strained monolayer MoS_2 and WS_2, *Phys. Rev. B*, **87**, 155304 (2013).

37. L. A. Falkovsky and A. A. Varlamov, Space-time dispersion of graphene conductivity, *Eur. Phys. J. B*, **56**, 281 (2007).

38. L. Ward, *The Optical Constants of Bulk Materials and Films* (IOP Publishing, Bristol, 1994).

39. D. M. Hoffman, G. L. Doll, and P. C. Eklund, Optical properties of pyrolytic boron nitride in the energy range 0.05–10 eV, *Phys. Rev. B*, **30**, 6051 (1984).

40. M. S. Vitiello, G. Scalari, B. Williams, and P. De Natale, Quantum cascade lasers: 20 years of challenges, *Opt. Express*, **23**(4), 5167–5182 (2015).

41. L. Zheng and A. H. MacDonald, Tunneling conductance between parallel two-dimensional electron systems, *Phys. Rev. B*, **47**, 10619 (1993).

Part 6

Graphene Based van der Waals Heterostructures for THz Detectors and Emitters

Chapter 46

Graphene Vertical Hot-Electron Terahertz Detectors*

V. Ryzhii,[a,b] A. Satou,[a] T. Otsuji,[a] M. Ryzhii,[c] V. Mitin,[d] and M. S. Shur[e]

[a]*Research Institute for Electrical Communication, Tohoku University, Sendai 980-8577, Japan*
[b]*Center for Photonics and Infrared Engineering, Bauman Moscow State Technical University and Institute of Ultra High Frequency Semiconductor Electronics, Russian Academy of Sciences, Moscow 111005, Russia*
[c]*Department of Computer Science and Engineering, University of Aizu, Aizu-Wakamatsu 965-8580, Japan*
[d]*Department of Electrical Engineering, University at Buffalo, Buffalo, New York 1460-1920, USA*
[e]*Departments of Electrical, Electronics, and Systems Engineering and Physics, Applied Physics, and Astronomy, Rensselaer Polytechnic Institute, Troy, New York 12180, USA*
v-ryzhii@riec.tohoku.ac.jp

Graphene-Based Terahertz Electronics and Plasmonics: Detector and Emitter Concepts
Edited by Vladimir Mitin, Taiichi Otsuji, and Victor Ryzhii

ISBN 978-981-4800-75-4 (Hardcover), 978-0-429-32839-8 (eBook)
www.jennystanford.com

We propose and analyze the concept of the vertical hot-electron terahertz (THz) graphene-layer detectors (GLDs) based on the double-GL and multiple-GL structures with the barrier layers made of materials with a moderate conduction band off-set (such as tungsten disulfide and related materials). The operation of these detectors is enabled by the thermionic emissions from the GLs enhanced by the electrons heated by incoming THz radiation. Hence, these detectors are the hot-electron bolometric detectors. The electron heating is primarily associated with the intraband absorption (the Drude absorption). In the frame of the developed model, we calculate the responsivity and detectivity as functions of the photon energy, GL doping, and the applied voltage for the GLDs with different number of GLs. The detectors based on the cascade multiple-GL structures can exhibit a substantial photoelectric gain resulting in the elevated responsivity and detectivity. The advantages of the THz detectors under consideration are associated with their high sensitivity to the normal incident radiation and efficient operation at room temperature at the low end of the THz frequency range. Such GLDs with a metal grating, supporting the excitation of plasma oscillations in the GL-structures by the incident THz radiation, can exhibit a strong resonant response at the frequencies of several THz (in the range, where the operation of the conventional detectors based on A_3B_5 materials, in particular, THz quantum-well detectors, is hindered due to a strong optical phonon radiation absorption in such materials). We also evaluate the characteristics of GLDs in the mid- and far-infrared ranges where the electron heating is due to the interband absorption in GLs.

46.1 Introduction

The gapless energy spectrum of graphene [1] enables using single- or multiple graphene-layer (GL) structures for different terahertz (THz) and infrared (IR) photodetectors based on involving the interband transitions [1–7] (see also Refs. [8–18]), where different THz and IR photodetectors based on GLs were explored. The interband photodetectors use either the GLs serving as photoconductors or the lateral p-i-n junctions. In the latter case, the electrons and holes are generated in the depleted i-region and move to the opposite GL contacts driven by the electric field in the depletion region [3]. The multiple-GL structures with the lateral p-i-n junctions can consist of either several non-Bernal stacked twisted GLs as in Ref. [3] or GLs separated by the barrier layers such as thin layers of boron nitride (hBN), tungsten disulfide (WS_2), or similar materials. Such heterostructures have recently attracted a considerable interest and enabled several novel devices being proposed and realized [19–31]. The GL-photodetectors, especially those based on the multiple-GL structures, can combine a high responsivity with a relatively low dark current at elevated temperatures (up to room temperatures). This is because the dark current in the photodetectors in question is mainly determined by the absorption of the optical phonons. Since the optical phonon energy $\hbar\omega_0$ in GLs is rather large (about 0.2 eV), the number of optical phonons is small even at the room temperature. This results in a low thermal generation rate. The mechanisms of the thermal generation associated with the absorption of the acoustic phonons and the Auger processes are forbidden due to the features of the GL energy spectrum. However, the interband tunneling in strong lateral electric fields in the i-region can lead to an enhanced generation of the electron-hole pairs and an elevated dark current limiting the photodetector detectivity [4]. Effective THz detection can be achieved in the lateral diodes with the absorbing GL source and drain sections separated by an array of graphene nanoribbons (GNRs), which form the potential barriers for hot electrons injected from the source to the drain [22]. As shown in this chapter, an effective THz detection can be achieved in the photodetectors based on double-GL and cascade multiple-GL structures with the vertical transport of hot electrons over the barrier layers. We propose and evaluate such THz detectors operating in the regime of the thermionic emission of hot electrons from GLs and their vertical transport over the barrier layers. The advantages of the THz detectors under consideration include high responsivity and detectivity in a wide spectral range at room temperature and a relatively high-speed operation.

The chapter is organized as follows. In Section 46.2, we discuss the device structures under consideration and the GL detector (GLD) operation principle. Section 46.3 deals with general formulas for the dark current and photocurrent associated with the thermionic emission of electrons from GL and controlled by their capture into GLs. In Section 46.4, we calculate the variations of the electron temperature in GLs caused by the intraband (Drude) absorption of the incident THz radiation. In Sections 46.5 and 46.6, using the formulas obtained in Sections 46.3 and 46.4, we derive the expressions for the GLD responsivity and dark-current-limited detectivity, respectively. In Section 46.7, we discuss how the electron capture in the GLs affects the GLD responsivity and detectivity. In Section 46.8, we consider the possibility to use the plasmonic resonances and get an enhanced response at elevated frequencies. Section 46.9 deals with the analysis of the limitations of our model. In Section 46.10, we evaluate the GLD operation in the IR spectral range and compare GLDs with some other photodetectors. In Section 46.11, we summarize the main results of the chapter.

46.2 Device Structures and Principle of Operation

We consider two types of the GLDs: (a) based on the n-doped double-GL structure and (b) n-doped multiple-GL structure with the GLs separated by the barrier layers made of WS_2 or similar material with a relatively small conduction band off-set. As an example, Fig. 46.1 shows a GLD using a four-GL structure. The double-GLDs consist of only the top and bottom GLs serving as the emitter and collector, respectively (no inner GLs). In the multiple-GLDs, the inner GLs clad by the emitter and collector GLs are disconnected from the contacts. In the double-GLDs (with a single barrier), the bias voltage V applied between the top and bottom GLs induces the negative electron charge in the emitter GL the equal positive charge in the collector GL. If the equilibrium electron concentration is low and the bias voltage is sufficiently strong, the hole gas will be formed in the collector GL. In GLDs with multiple-GL structures, the inner GLs remain quasi-neutral, so that the electron gas in each GL is formed primarily due to the n-type doping, whereas the top and bottom GLs can be charged due to the bias voltage. Figure 46.2 shows the GLPD band diagrams under the bias. It is assumed that the GLDs under consideration are irradiated by the normally incident THz photons with the energy $\hbar\Omega$. The operation of GLDs is associated with the electron heating due to the intraband absorption (Drude absorption) and the interband absorption (see,

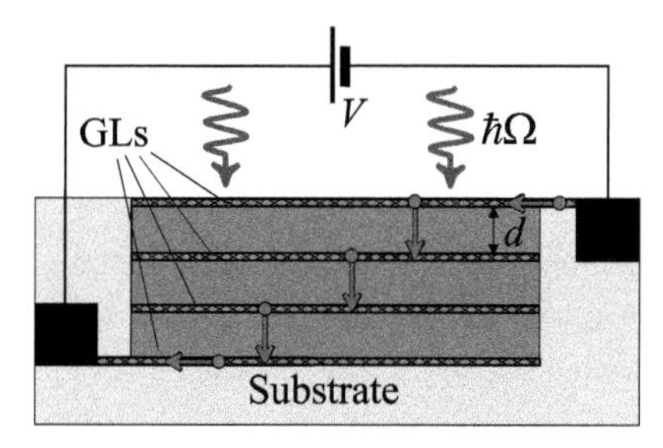

Figure 46.1 Schematic structure of vertical GLDs based on multiple-GL structure (with minimum of two GLs). The arrows show the current flow (for the case when all electrons crossing a GL are captured in it, i.e., for capture probability $p_c = 1$).

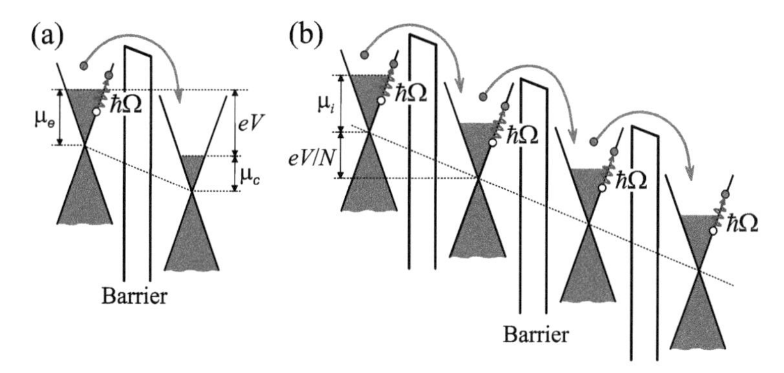

Figure 46.2 Band diagrams of (a) double-GLD and (b) multiple-GLD (with the cascade electron inter-GL transitions) under applied bias. The wavy arrows indicate the intraband (Drude) absorption, while smooth arrows correspond to thermionic emission processes resulting in the electron inter-GL transitions between neighboring GLs and providing the dark current and photocurrent. The inter-GL transitions between the distant GLs, which are possible at finite values of the capture probability, are not shown.

for example, Ref. [32]) of the incident radiation resulting in an increase of the thermionic current over the barrier layers. Thus, the proposed GLDs are the barrier hot-electron bolometers. In GLDs with the double-GL structures, the electrons entering from the emitter GL and exiting to the collector GL support the lateral current flowing via the contacts, so that the carrier densities in the GLs are maintained. In the multiple-GL structures, the electron density in each GL between the emitter and collector GLs is maintained due to the balance between the electrons leaving and entering GLs via the adjacent barriers due to the thermal emission and the capture processes. If the probability of the capture of an electron crossing a GL is smaller than unity, the GLD operation can exhibit the effect of photoelectric gain. The origin of this gain is of the same nature as in the vertical quantum-well infrared photodetectors (QWIPs) [33–36].

The donor density Σ_i and the bias voltage V determine the electron Fermi energies μ_e and μ_c in the top (emitter) and bottom (collector) GLs, respectively ($\mu_e > \mu_h$, because the bias voltage increases the electron density in the emitting GL and decreases it in the collecting GL). Considering the geometrical and quantum capacitances [37] and taking into account the energy gap between the Dirac points in GLs (see Fig. 46.2a), in the double-GL structure at relatively low bias voltages one can obtain

$$\mu_e \simeq \mu_i\left[1+\frac{eV}{2(eV_i+\mu_i)}\right], \quad \mu_c \simeq \mu_i\left[1-\frac{eV}{2(eV_i+\mu_i)}\right], \tag{46.1}$$

$$\mu_i = \hbar v_W\sqrt{\pi\Sigma_i}. \tag{46.2}$$

Here, $V_i = 4\pi\Sigma_i ed/\kappa$, e is the electron charge, $\hbar$ is the Planck constant, $v_W \simeq 10^8$ cm/s is the characteristic velocity of electrons and holes in GLs, and κ and d are the dielectric constant and the thickness of the barrier, respectively. At $\Sigma_i = (1.0–1.8) \times 10^{12}$ cm^{-2}, $\kappa = 4$ and $d =$ 10–50 nm, one obtains $\mu_i \simeq$ 100–150 meV and $V_i \simeq$ 452–3630 mV. Relatively, large values of V_i imply that for the realistic moderate values of V, considered in the following, the correction of the Fermi energies in the emitter and collector GLs is small in comparison with μ_i. In the multiple-GL structures (with a large number of GLs and the inter-GL barriers $N \gg 1$), all the GLs except the top and bottom one's are quasi-neutral. Although the electrically induced variation of the Fermi energies in the emitter and collector GLs can be essential (for the mechanism of the photoelectric gain), we will assume that in all GLs, including the top and bottom one's, the Fermi energies are close to each other and approximately equal to the value determined by the donor density

$$\mu \simeq \mu_i. \tag{46.3}$$

46.3 Vertical Electron Dark Current and Photocurrent

We restrict our consideration to the double- and multiple-GL structures with relatively thick inter-GL barriers, so that the tunneling current between the GLs can be neglected (the pertinent calculations can be done using the approach developed in Refs. [38, 39]). We assume that the main contribution to the vertical current is due to the thermoemission of electrons resulting in the inter-GL transitions (producing the dark current). The impinging THz irradiation heats the electron gas in GLs. This leads to an increase in the thermoemission rate intensifying of the inter-GL transitions, and, hence, the vertical current. The direct electron photoemission is insignificant when the energy of photons $\hbar\Omega$ is smaller than the GL-barrier conduction band off-set Δ_C (the height of the barrier with respect of the Dirac point). For the GL structures with the WS_2 barriers [40], it implies $\hbar\Omega < \Delta_C \simeq 0.4$ eV. Hence, this inequality is well satisfied for the THz radiation.

The rate of the thermionic emission from a GL (per unit of its area) is given by

$$\Theta = \frac{\Sigma_i}{\tau_{esc}} \exp\left(\frac{\mu_i - \Delta_C}{k_B T}\right), \tag{46.4}$$

where T is the effective electron temperature which (under the irradiation) is higher than the lattice temperature T_l, k_B is the Boltzmann constant, and τ_{esc} is the characteristic time of escape from the GLs of the electrons with the energy $\varepsilon > \Delta_C$, $\tau_{esc} \sim \tau$, where τ is the momentum relaxation time. Using Eq. (46.4) and assuming for simplicity that $eV/N > k_B T$ (V/N is the voltage drop across the barrier), and taking into account the electrons photoexcited from the emitter and the photoexcited from and captured to the internal GLs (in multiple-GLDs), we find the thermionic current density, j

$$j = \frac{e\Theta}{p_c} = \frac{e\Sigma_i}{p_c \tau_{esc}} \exp\left(\frac{\mu_i - \Delta_C}{k_B T}\right)$$

$$\simeq \frac{e\mu_i^2}{\pi\hbar^2 v_W^2 p_c \tau_{esc}} \exp\left(\frac{\mu_i - \Delta_C}{k_B T}\right). \tag{46.5}$$

Here, p_c is the probability of the capture of an electron crossing a GL.

In the GL structures with at least one internal GL (and in the multiple-GL structures), the effects of the balance of thermogeneration from and capture to each GL are taken into account by introducing the capture probability p_c, as in the standard models of QWIPs [33–36]. In such an approach, the rate of the electron capture into each GL is equal to $p_c j/e$. Equating the capture rate $p_c j/e$ and the thermogeneration rate Θ, one obtains $j = e\Theta/p_c$ [Eq. (46.5)]. The quantity p_c^{-1} can be relatively large if the capture probability is small. This quantity essentially determines the dark current and photocurrent gain $g \propto 1/p_c$.

Equation (46.5) yields the following formula for the current density j_0 without irradiation (i.e., for the dark current) when the dark electron temperature T is equal to the lattice temperature T_0:

$$j_0 \simeq \frac{e\mu_i^2}{\pi\hbar^2 v_W^2 p_c \tau_{esc}} \exp\left(\frac{\mu_i - \Delta_C}{k_B T_0}\right). \tag{46.6}$$

In the double-GLDs, all the electrons generated by the emitter GL are captured by the collector GL, so that in such a case $p_c = 1$.

Considering the variation of the electron temperature $T - T_0$, the photocurrent density $j - j_0$ can be presented as

$$j - j_0 = j_0 \left(\frac{\Delta_C - \mu_i}{k_B T_0}\right) \frac{(T - T_0)}{T_0}. \tag{46.7}$$

46.4 Electron Heating by Incoming THz Radiation

As previously [22, 32], we assume that the electron energy relaxation is associated with the processes of the emission and absorption of optical phonons. In this case, for the rate, $\hbar\omega_0 R$, of the energy transfer from the electron system to the optical phonon system is determined by (see, for example, Refs. [22, 32])

$$R = \frac{\Sigma_i}{\tau_0}\left[(\mathcal{N}_0 + 1)\exp\left(-\frac{\hbar\omega_0}{k_B T}\right) - \mathcal{N}_0\right]. \tag{46.8}$$

Here, $\hbar\omega_0$ and $\mathcal{N}_0$ are the energy and the number of optical phonons, respectively, τ_0 is the characteristics time of the optical phonon spontaneous emission for the electron energy $\varepsilon > \hbar\omega_0$.

If the characteristic time of the optical phonons $\tau_0^{\text{decay}} \ll \tau_0$, $\mathcal{N}_0$ is close to its equilibrium value: $\mathcal{N}_0 = [\exp(\hbar\omega_0/k_B T_0) - 1]^{-1} \simeq \exp(-\hbar\omega_0/k_B T_0)$. In the case of $\tau_0^{\text{decay}} > \tau_0$, the effective energy relaxation time τ_0 should be replaced by $\tau_0(1 + \xi_0)$ (where $\xi_0 = \tau_0^{\text{decay}}/\tau_0$) [14].

When the effective electron temperature in GLs deviates from its equilibrium value (due to the absorption of THz radiation), the energy relaxation rate can be presented as [see Eq. (46.8)]

$$R \simeq \frac{\Sigma}{\tau_0}\left(\frac{\hbar\omega_0}{k_B T_0}\right)\exp\left(-\frac{\hbar\omega_0}{k_B T_0}\right)\frac{(T - T_0)}{T_0}. \tag{46.9}$$

The rate of the energy transfer from the electron system to the optical phonon system $\hbar\omega_0 R$ is equal to the rate, $\hbar\Omega G$, of the energy transferred from the THz radiation to the electron system

$$\hbar\omega_0 R = \hbar\Omega G. \tag{46.10}$$

Considering the intraband, i.e., the so-called free electron absorption (the Drude absorption) and the interband absorption, the net absorption rate can approximately be presented as

$$G \simeq \beta I\left[\frac{D}{(1+\Omega^2\tau^2)} + \frac{\sinh(\hbar\Omega/2k_B T)}{\cosh(\hbar\Omega/2k_B T) + \cosh(\mu_i/k_B T)}\right]. \tag{46.11}$$

Here, $\beta = \pi e^2/c_0\hbar \simeq 0.023$, c_0 is the speed of light in vacuum, I is the THz photon flux entering into the device (or the incident photon flux in the case of the anti-reflection coating), and

$$D = \frac{4k_B T\tau}{\pi\hbar}\ln\left[\exp\left(\frac{\mu_i}{k_B T}\right) + 1\right] \simeq \frac{4\mu_i\tau}{\pi\hbar} \tag{46.12}$$

is the Drude weight, the factor determining the contribution of the Drude absorption (it is proportional for the real part of the intraband conductivity of GLs). For the realistic values of τ, the factor D can markedly exceed unity. Indeed, assuming μ_i = 100–150 meV and $\tau = 10^{-13}$ s, one obtains $D \simeq 20$–30. Strictly speaking, Eq. (46.11) is valid at not too strong absorption.

Since the Fermi energy in the GLD under consideration should be sufficiently large, the processes of the interband absorption of THz photons (their energy $\hbar\Omega \ll \mu_i$), corresponding to the second term in Eq. (46.11), are effectively suppressed due to the Pauli blocking. This implies that the electron heating by THz radiation is primarily associated with the intraband absorption (with the Drude or the so-called free-electron absorption). In Eq. (46.11) and in the following equations, we disregard the attenuation in the multiple-GLDs of the THz photon flux associated with the absorption in GLs, which are closer to the irradiated surface (emitter). This should be valid at not too large values of *N*.

Taking into account, the energy balance in each GL governed by Eq. (46.10) and using Eq. (46.11) (omitting the term describing the interband absorption), we arrive to the following expression for the variation of the effective electron energy caused by the THz of IR radiation of moderate intensity:

$$\frac{(T-T_0)}{T_0}=\frac{\beta D\tau_0(1+\xi_0)I}{\Sigma_{\rm i}(1+\Omega^2\tau^2)}\left(\frac{k_{\rm B}T_0}{\hbar\omega_0}\frac{\Omega}{\omega_0}\right)\exp\left(\frac{\hbar\omega_0}{k_{\rm B}T_0}\right). \qquad (46.13)$$

The above equation corresponds to the electron energy relaxation time (determined by the optical phonons), which is equal to [22]

$$\tau_0^{\varepsilon}=\tau_0(1+\xi_0)\left(\frac{k_{\rm B}T_0}{\hbar\omega_0}\right)^2\exp\left(\frac{\hbar\omega_0}{k_{\rm B}T_0}\right)\gg\tau_0. \qquad (46.14)$$

46.5 Responsivity

Using Eqs. (46.6) and (46.8), for the GLD responsivity $\mathcal{R}=(j-j_0)/\hbar\Omega I$, we obtain

$$\mathcal{R}=\frac{e\mu_{\rm i}^2}{\pi\hbar^2 v_{\rm W}^2 p_{\rm c}\tau_{\rm esc}\hbar\Omega I}\left(\frac{\Delta_{\rm c}-\mu_{\rm i}}{k_{\rm B}T_0}\right)$$
$$\times\exp\left(\frac{\mu_{\rm i}-\Delta_{\rm C}}{k_{\rm B}T_0}\right)\frac{(T-T_0)}{T_0}. \qquad (46.15)$$

Using Eqs. (46.13) and (46.15), we arrive at the following expressions for the responsivity:

$$\mathcal{R}=\frac{\overline{\mathcal{R}}}{(1+\Omega^2\tau^2)}\left(\frac{\mu_{\rm i}}{\hbar\omega_0}\right)\left(\frac{\Delta_{\rm c}-\mu_{\rm i}}{\hbar\omega_0}\right)\exp\left(\frac{\mu_{\rm i}+\hbar\omega_0-\Delta_{\rm c}}{k_{\rm B}T_0}\right). \qquad (46.16)$$

Here,

$$\overline{\mathcal{R}}=\frac{4e\beta(1+\xi_0)}{\pi p_{\rm c}\hbar}\left(\frac{\tau_0\tau}{\tau_{\rm esc}}\right). \qquad (46.17)$$

As seen from Eq. (46.16), the GLD responsivity is proportional to an exponential factor. To achieve reasonable GLD characteristics, the Fermi energy $\mu_{\rm i}$ should not be too small in comparison with the barrier height $\Delta_{\rm C}$. One can also see that $\mathcal{R} \propto \overline{\mathcal{R}} \propto 1/p_{\rm c}$. As stated above, in the GLDs with the multiple-GL structures, the factor $1/p_{\rm c}$ can be fairly large.

Equation (46.16) describes the GLD responsivity as a function of the THz radiation frequency Ω, the temperature T_0, and the GL doping (via the dependence of $\mu_{\rm i}$ on $\Sigma_{\rm i}$),

Assuming $\hbar\omega_0 = 200$ meV, $\tau_0^{\rm decay} + \tau_0 = 0.7$ ps, $\tau_{\rm esc}/\tau \sim 1.2$, and $p_{\rm c} = 1$ for $T = 300$ K, from Eq. (46.17), we obtain from Eq. (46.17) $\overline{\mathcal{R}} \simeq 27$ A/W.

Figure 46.3 shows the GLD responsivity versus the photon frequency $f = \Omega/2\pi$ calculated for different donor densities $\Sigma_{\rm i}$ using Eqs. (46.16) and (46.17) for $\Delta_{\rm C} = 400$ meV and the same other parameters as in the above estimate. This corresponds to the GLDs based on the double-GL structure or to the GLDs based on the multiple-GLDs with a strong electron capture in the internal GLs. The responsivity of the latter can be much higher than that shown in Fig. 46.3 if $p_{\rm c} \ll 1$ (see below).

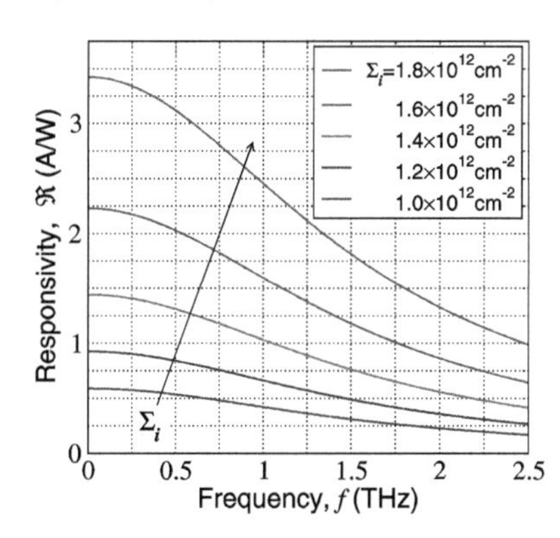

Figure 46.3 Spectral dependences of responsivity of GLDs with different donor densities.

46.6 Dark Current Limited Detectivity

Considering that the shot noise current (at the value of the signal current equal to the dark current) is given by $J_{\rm noise} = \sqrt{4egJ_{\rm dark}\Delta f}$, where Δf is the bandwidth and $g = 1/Np_c$ is the dark current and photoelectric gain, the dark current limited detectivity (see, for example, Ref. [36]) can be presented in the following form:

$$D^* = \frac{\mathcal{R}}{\sqrt{4egj_0}}. \tag{46.18}$$

Accounting for Eq. (46.16), we arrive at

$$D^* = \frac{\overline{D}^*}{(1+\Omega^2\tau^2)}\left(\frac{\Delta_C - \mu_i}{\hbar\omega_0}\right) \times \exp\left(\frac{\mu_i - \Delta_C}{2k_B T_0}\right)\exp\left(\frac{\hbar\omega_0}{k_B T_0}\right)\sqrt{\frac{N}{p_c}}, \tag{46.19}$$

where

$$\overline{D}^* = 2\sqrt{\pi}\beta\left(\frac{k_B T_0}{\hbar\omega_0}\right)\left[\frac{(1+\xi_0)\tau_0\tau v_W}{\hbar\omega_0\sqrt{\tau_{esc}}}\right]. \tag{46.20}$$

For $\tau_0^{decay} + \tau_0 = 0.7$ ps, $\tau \sim 0.1$ ps, $\tau_{esc} \sim 0.12$ ps, and $T = 300$ K, $\Delta_C = 400$ meV, $\Sigma_i = 1.8 \times 10^{12}$ cm^{-2} ($\mu_i = 150$ meV), $N/p_c = 1-25$, and $f = \Omega/2\pi \ll 1.6$ THz from Eqs. (46.19) and (46.20), we obtain $\overline{D}^* \simeq 1.3 \times 10^7$ cmHz$^{1/2}$/W and $D^* \simeq (0.35 - 1.75) \times 10^9$ cmHz$^{1/2}$/W. Figure 46.4 shows the spectral characteristics of GLDs with $\Sigma_i = 1.0 \times 10^{12} - 1.8 \times 10^{12}$ cm^{-2} ($\mu_i \simeq 100$–150 meV) calculated using Eqs. (46.19) and (46.20) for the same other parameters as from the latter estimate and Fig. 46.3.

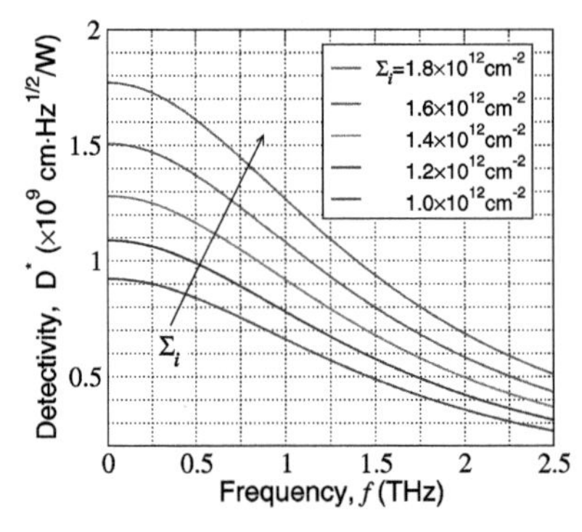

Figure 46.4 Spectral dependences of detectivity of GLDs with different donor densities and $N/p_c = 25$.

From Eqs. (46.16), (46.17), (46.19), and (46.20), one can see that the GLD responsivity is independent on N (in the framework of the present model), whereas the GLD detectivity is proportional to $\sqrt{N}$ (as in QWIPs [36]).

46.7 Role of the Electron Capture

As follows from Eqs. (46.16), (46.17), (46.19), and (46.20), both the responsivity and detectivity of the multiple GLDs increase with decreasing capture probability p_c, i.e., with increasing photoelectric gain. The latter quantity is determined by several factors, in particular, by the degree of the electron heating in the inter-GL barriers and, hence, by the potential drop across these barriers and their thickness. The detailed calculations of p_c require additional quantum-mechanical calculations of the electron transitions from the continuum states above the barriers to the bound states in GLs coupled with the ensemble Monte Carlo modeling of the electron propagation across the GL-structure similar to that made previously for multiple-QW structures based on the standard semiconductor heterostructures (see, for example, Refs. [35, 41, 42]). This is, however, beyond the scope of this work, so that here we consider p_S as a phenomenological parameter. Figures 46.5 and 46.6 show the GLD responsivity and detectivity as functions of the capture parameter. One can see that a decrease in the capture parameter p_c

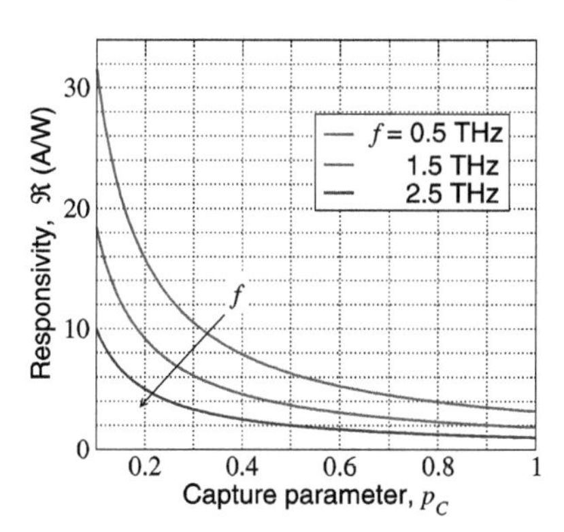

Figure 46.5 Responsivity of GLD as a function of the capture parameter p_c for different radiation frequencies ($\Sigma_i = 1.8 \times 10^{12}$ cm^{-2}).

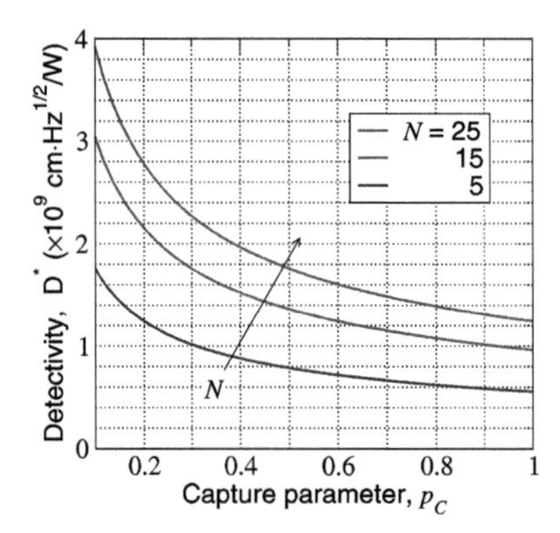

Figure 46.6 Detectivity as a function of the capture parameter p_c for GLDs with different number of the inter-GL barriers N ($\Sigma_i = 1.8 \times 10^{12}$ cm^{-2} and f = 1 THz).

leads to a substantial rise of $\mathcal{R}$. At low p_c, the GLD responsivity can be fairly high. The detectivity D^* of GLDs with the multiple-GL structure also rises with decreasing p_c as well as with increasing N. Since the capture probability p_c in the multiple-GL structures should markedly decrease with increasing electric field in the barrier layers $E = V/Nd$ (as in multiple-QWIPs [35]), the GLD responsivity and detectivity can be rising functions of the bias voltage if the proper heat removal is provided.

46.8 Effect of Plasmonic Resonances

Since the absorption of the incident THz radiation is associated with the Drude mechanism, the absorption efficiency and, hence, the GLD responsivity and detectivity can be relatively small in the frequency range $\Omega/2\pi > \tau^{-1}$. If $\tau \simeq 0.1$ ps, this corresponds to $\Omega/2\pi > 1.6$ THz. However, the operation of GLDs can be extended to much higher frequencies if the GLD structure is supplied by a metal grating over the top GL (not shown in Fig. 46.1). In this case, the incident THz radiation can result in an efficient excitation of plasma oscillations in the electron-hole system in the double-GL GLDs and in the system of electrons in all GLs (in the multiple-GL structures). Simplifying the equations from Ref. [43] for the GLDs with a metal grating, the rate of the THz radiation absorption G_n at the frequency near the n-th plasmon resonance ($\Omega \simeq \Omega_n$) can be presented as [43]

$$G_{\rm n}\frac{\beta IDA_{\rm n}}{1+(\Omega-\Omega_{\rm n})^2\tau^2 a_{\rm n}^2}. \tag{46.21}$$

Here, $A_{\rm n} = 1/(1 + \beta D/2\sqrt{\kappa})^2 \simeq 1$ and $a_{\rm n} = 4/(1 + \beta D/2\sqrt{\kappa})^2 \simeq 4$ are determined by the ratio of the collisional damping (which is actually close to $1/2\tau$) and the parameter of the radiative damping [44, 45]. Equation (46.21) does not contain any geometrical parameters such as the grating period, length of grating strips, and the spacing between the grating and the top GL. These parameters only determine the dependence of the resonant plasma frequencies $\Omega_{\rm n}$ on the device geometry. This is valid as long as those dimensions are much shorter than the THz radiation wavelength and the net length of the grating is of the same order of magnitude as the wavelength.

The quantities $\Omega_{\rm n}$ depend on the net electron density in all GLs $(N + 1)\Sigma_{\rm i}$, the spacing between the top GL and the metal grating W, and the period of the grating. The latter determines the "quantized" wavenumber $q_{\rm n}$ of the excited plasma modes (standing plasma waves). One can put $q_{\rm n} = (\pi/2L)n$, $2L$ is the length of the GL-structure in the lateral direction, and $n = 1, 2, 3, \ldots$ is the plasma mode index. For simplicity, one can use the following equation for the frequency of the plasma modes (corresponding to $q_{\rm n}W \gtrsim 1$):

$$\Omega_{\rm n} \sim \sqrt{\frac{e^2\mu_{\rm i}(N+1)}{\kappa\hbar^2}q_{\rm n}} \tag{46.22}$$

or

$$\Omega_{\rm n} \sim \sqrt{\frac{\pi e^2\mu_{\rm i}(N+1)}{2L\kappa\hbar^2}n}. \tag{46.23}$$

The square-root dependence of $\Omega_{\rm n}$ on N appears because the net electron density, which determines the contribution to the self-consistent electric field in the plasma waves by all the GLs is proportional to $(N + 1)$, whereas the electron fictitious mass m_f in GLs is proportional to $\mu_{\rm i} \propto \sqrt{\Sigma_{\rm i}}$ (see, for example, Ref. [39]). Setting $\mu_{\rm i} = 150$ meV, $2L/n = 0.5$–1.0 μm (i.e., $2L = 10$ μm and $n = 10$), and $N = 5$, from Eq. (46.23), we obtain $f_{10} = \Omega_{10}/2\pi \simeq 7.4$–$10.4$ THz. If $A_{\rm n} \sim 1$ and $a_{\rm n} \sim 1$, the GLD responsivity at the resonance is of the same order of magnitude as at the low edge of the THZ range $\Omega \ll \tau^{-1}$ (see Figs. 46.3 and 46.5). Thus, the resonant excitation of plasma oscillations results in a strong absorption of the incident THz radiation and, hence, in elevated values of the GLD responsivity (and detectivity) at relatively high frequencies (several THz). Such GLDs can cover the frequency range $f \simeq$ 6–10 THz ($\hbar\Omega \simeq 25 - 40$ meV), which is not accessible by A_3B_5-based detectors, in particular, THz quantum-well detectors (QWDs) [46–48].

46.9 Limitations of the Model

The model used in the above calculations has some simplifications. These simplifications are: (i) The capture probability is the same for all GLs in the GL-structures; (ii) The thermoassisted tunneling is insignificant; (iii) The Joule heating of the structure is negligible.

Since the capture probability p_c depends on the heating of electrons in the barriers, it can be determined only by the average electric field in the GL-structure but partially by the electric field in the adjacent barriers. In this case, the probability of the electron capture to the particular GL can depend on its index. Such kind of non-locality of the electric-field dependence can lead to more nontrivial spatial distributions (as in QWIPs [41, 42]). However, in the GL-structures with the barrier thickness much smaller than the characteristic energy relaxation length, the pertinent effect should be weak. This justifies the assumption that p_c is a constant (which generally depends on the average electric field).

At sufficiently high bias voltages (much higher than those assumed above), the electron escape from GLs can be associated with the thermoassisted tunneling from the bound states in GLs to the continuum states above the barriers. This tunneling can also be used in double- and multiple-GLDs with the structures similar to those considered above. Since the effective activation energy for this mechanism can be markedly smaller that $(\Delta_c - \mu)$, GLDs with the thermoassisted tunneling can comprise the barriers with larger conduction band offsets than between GLs and WS_2, for example, with the *hBN* barriers. However, this problem requires a separate consideration.

Above we considered the case of not too low bias voltages ($eV/N > k_BT$). The Joule power j_0V can result in an overheating of the GL structure if V is relatively strong. Such an overheating can be avoided either by decreasing μ_i (decreasing the GL doping level) or by lowering the bias voltage V. In the range of bias voltages $eV/N < k_BT$, the GLD responsivity and detectivity given by Eqs. (46.16) and (46.19) should be multiplied by the factors $\zeta = \{1 - \exp[-(eV/Nk_BT)]\} \simeq eV/Nk_BT$ and $\sqrt{\zeta} = \sqrt{1-\exp[-(eV/Nk_BT)]} \simeq \sqrt{eV/Nk_BT}$, respectively. The transfer to the range of relatively low bias voltages leads to a decrease in the Joule power as V^2, but at the expense of a decrease in the responsivity and detectivity ($\mathcal{R} \propto V/N$ and $D^* \propto \sqrt{V/N}$).

The Joule heating can lead to overheating of GLDs if the Joule power exceeds the maximum heat energy which can be removed from the GLD unit area, W_{max}, without a substantial heating. This results in the following limitation:

$$W^{\max} > j_0 VA = \frac{e\Sigma_i}{p_c \tau_{esc}} \exp\left(\frac{\mu_i - \Delta_C}{k_B T_0}\right) V, \tag{46.24}$$

where A is the device area. Assuming a typical voltage drop cross the GL-structure to be on the order of 50–500 mV and the thermal resistance of the package to be on the order of 10 K/W, we obtain that W^{max} and the current leading to the ten degrees overheating j_0^{max} are equal to 1 W and 2–20 A, respectively. For a typical 300 $\times$ 300 μm^2 device, this corresponds to a fairly reasonable current density of $j_0^{max} \sim 2 \times (10^3 - 10^4)$ A/cm^2. Setting $\Sigma_i = 2 \times 10^{12}$ cm^{-2}, $\tau_{esc} = 0.1$ ps, and $p_c = 0.5$, we obtain $j_0 \sim 3 \times 10^2$ A/cm^2 (i.e., $j_0 < j_0^{max}$). Much higher current densities could be achieved with improved heat sinks (see, for example, Ref. [49]) and/or in the pulsed regime of operation.

46.10 Discussion

Tunneling can be based on the materials with larger conduction band offsets than between GLs and WS_2.

In principle, GLDs can also effectively operate in the mid- and near-IR ranges. At sufficiently high photon energies, the intraband absorption is negligible, whereas the interband radiative processes, corresponding to the second term in the right-hand side of Eq. (46.11), can efficiently contribute to the heating of the electron gas in GLs if $\hbar\Omega \gtrsim 2\mu_i$. In such a case for the photon energies $2\mu_i < \hbar\Omega < 2\Delta_C$, the GLD responsivity is given by

$$\mathcal{R}_{\rm IR} \simeq \widetilde{\mathcal{R}}\left(\frac{\Delta_{\rm c} - \mu_{\rm i}}{\hbar\omega_0}\right)\exp\left(\frac{\mu_{\rm i} + \hbar\omega_0 - \Delta_{\rm C}}{k_{\rm B}T_0}\right), \tag{46.25}$$

$$\widetilde{\mathcal{R}}_{\rm IR} = \frac{\pi\beta e(1+\xi_0)}{p_{\rm c}\hbar\omega_0}\left(\frac{\tau_0}{\tau_{\rm esc}}\right)\left(\frac{k_{\rm B}T_0}{\hbar\omega_0}\right). \tag{46.26}$$

At $\Sigma_{\rm i}$ = (1.0–1.8) $\times$ 10^{12} cm^{-2} ($\mu_{\rm i} \simeq$ 100–150 meV), Eqs. (46.25) and (46.26) yield the values of the responsivity $\mathcal{R}_{\rm IR}$ about 20–30 times smaller than $\mathcal{R}$ in the range $\Omega \ll 1/\tau$ (see Figs. 46.3 and 46.5). In particular, at $\Sigma_{\rm i}$ = 1.8 $\times$ 10^{12} cm^{-2}, assuming $p_{\rm c}$ = 0.2–1.0, we obtain rather high values $\mathcal{R}_{\rm IR} \simeq 0.11 - 0.55$ A/W. The GLD detectivity in the mid- and near-IR range $D^*_{\rm IR}$, being much lower than D^* in the THz range, can be still relatively high (for room temperature). Note that $\widetilde{\mathcal{R}}_{\rm IR}$ and $D^*_{\rm IR}$ are independent of the photon energy in its wide range (from 200–300 meV to 800 meV).

Comparing the GLDs based on the vertical double-GL structure under consideration with the GLDs with a lateral structure and the barrier region consisting of an array of graphene nanoribbons using the electron heating in n-GL contact region [22], one can see that both types of THz detectors at the room temperature exhibit close spectral characteristics. However, the GLDs with the vertical multiple-GL structure can have much higher responsivity and, especially, detectivity if $p_{\rm c} < 1$ and $N \gg 1$.

In principle, room-temperature THz detectors utilizing the thermionic emission of electrons heated by the absorbed THz radiation from QWs can be made of A_3B_5 or Si-Ge heterostructures. Such detectors on the base of vertical multiple-QW structures were proposed and realized a long time ago (see Refs. [50, 51], respectively, as well as a recent chapter [52]). The THz detectors based on lateral structures with the barrier regions formed by the metal gates were also realized [53, 54] (see also Ref. [55]). However, the responsivity and detectivity of GLDs under consideration can be markedly higher than that using the A_3B_5 multiple-QW structures. Comparing the Drude factor D for GL-structures [see Eq. (46.12)] and the same factor D_{QW} for QW-structures with GaAs QWs, one can find the ratio of these factors at the equal electron density $\Sigma_{\rm i}$ and momentum relaxation time τ is given by

$$\frac{D}{D_{\rm QW}} \simeq \frac{mv_{\rm W}^2}{\mu_{\rm i}} \simeq \frac{m}{m_{\rm f}}, \tag{46.27}$$

where m and m_f are the effective and fictitious electron masses in QWs and GLs, respectively. For GaAs QWs and GLs with $\mu_i \simeq 150$ meV, these masses are approximately equal to each other. This implies that the THz power absorbed in QWs and GLs are close. However, the electron energy relaxation time in GLs is longer than that in GaAs-QWs and other standard semiconductor QWs. This is mainly due to relatively large optical phonon energy in GLs. Indeed, using Eq. (46.14) and assuming that $\tau_0^{decay} + \tau_0 = (0.7 - 1.4)$ ps at the room temperature, we obtain $\tau_0^{\varepsilon} \simeq (32.5 - 65)$ ps, while for GaAs ($\hbar\omega_0 \simeq$ 36 meV and $\tau_0 \simeq 0.14$ ps), InAs ($\hbar\omega_0 \simeq 30$ meV and $\tau_0 \simeq 0.2$ ps), and InSb ($\hbar\omega_0 \simeq 25$ meV and $\tau_0 \simeq 0.7$ ps) QWs one obtains $\tau_0^{\varepsilon} \simeq 0.56$, 0.93, and 3.93 ps, respectively. Longer electron energy relaxation time corresponds to more effective heating of the electron gas and, hence, higher responsivity. Another factor promoting higher responsivity (and detectivity) of GLDs is the possibility to achieve higher photoelectric gain due to smaller values of the expected capture parameter p_c.

The THz QWPs using the direct intersubband photoexcitation from QWs require the heterostructures with rather small band off-sets ($\Delta_C \sim \hbar\Omega$). They exhibit a modest responsivity (about few tens of mA/W or less [46–48]) with $D^* \simeq 5 \times 10^7$ cm Hz$^{1/2}$/W at $T_0 = 10$ K [46]. Hence, in the few-THz range, GLDs surpass QWPs. GLDs with the grating using the plasmonic effects although should exhibit advantages over QWPs in the range 6–10 THz (see above). Additional advantages of GLDs might be associated with better heat removal conditions [49, 56, 57] than in the case of different A_3B_5 devices.

Due to a substantial progress in fabrication and experimental studies of the multiple-GL structures with the inter-GL barrier layers made of transition metal dichalcogenides [19] (see also Refs. [58–62]), the realization of the proposed GLDs appears to be feasible. In particular, similar GL-structures with five periods and 20 nm thick barriers [58] and with ten periods [59] were demonstrated.

46.11 Conclusions

We proposed THz GLDs based on the double-GL and multiple-GL structures with the barrier layers made of WS_2 exploiting the enhanced thermionic electron emission from GLs due to the intraband (Drude) absorption, developed the device model, and calculate the GLD responsivity and detectivity at the room temperature. We demonstrated that GLDs, especially, those based on the multiple-GL structures can exhibit fairly high responsivity and detectivity surpassing hot-electron detectors based on the standard heterostructures. The main advantages of GLDs are associated with relatively long electron energy relaxation time and the pronounced effect of photoelectric gain at a low capture probability of the electron capture into GLs. As shown, GLDs using the resonant electron heating associated with the plasmonic effects and GLDs exploiting the electron heating due to the interband absorption can also operate in the far-, mid-, and near-IR ranges of the radiation spectrum.

Acknowledgments

This work was supported by the Japan Society for Promotion of Science (Grant-in-Aid for Specially Promoting Research No. 23000008), Japan. V. R. and M. R. acknowledge the support of the Russian Scientific Foundation (Project No. 14-29-00277). The work at the University at Buffalo was supported by the NSF TERANO grant and the US Air Force Office of Scientific Research. The work at RPI was supported by the US Army Cooperative Research Agreement.

References

1. A. H. Castro Neto, F. Guinea, N. M. R. Peres, K. S. Novoselov, and A. K. Geim, *Rev. Mod. Phys.*, **81**, 109 (2009).
2. F. T. Vasko and V. Ryzhii, *Phys. Rev. B*, **77**, 195433 (2008).
3. J. Park, Y. H. Ahn, and C. Ruiz-Vargas, *Nano Lett.*, **9**, 1742–1746 (2009).
4. V. Ryzhii, M. Ryzhii, V. Mitin, and T. Otsuji, *J. Appl. Phys.*, **107**, 054512 (2010).
5. T. Mueller, F. N. A. Xia, and P. Avouris, *Nat. Photonics*, **4**, 297–301 (2010).
6. M. Furchi, A. Urich, A. Pospischil, G. Lilley, K. Unterrainer, H. Detz, P. Klang, A. M. Andrews, W. Schrenk, G. Strasser, and T. Mueller, *Nano Lett.*, **12**, 2773 (2012).
7. X. Gan, R.-J. Shiue, Y. Gao, I. Meric, T. F. Heinz, K. Shepard, J. Hone, S. Assefa, and D. Englund, *Nat. Photonics*, **7**, 883 (2013).

8. F. Bonaccorso, Z. Sun, T. Hasan, and A. C. Ferrari, *Nat. Photonics*, **4**, 611 (2010).
9. V. Ryzhii, N. Ryabova, M. Ryzhii, N. V. Baryshnikov, V. E. Karasik, V. Mitin, and T. Otsuji, *Opto-Electron. Rev.*, **20**, 15–25 (2012).
10. A. Tredicucci and M. S. Vitiello, *IEEE J. Sel. Top. Quantum Electron.*, **20**, 8500109 (2014).
11. L. Vicarelli, M. S. Vitiello, D. Coquillat, A. Lombardo, A. C. Ferrari, W. Knap, M. Polini, V. Pellegrini, and A. Tredicucci, *Nat. Mater.*, **11**, 865 (2012).
12. M. S. Vitiello, D. Coquillat, L. Viti, D. Ercolani, F. Teppe, A. Pitanti, F. Beltram, L. Sorba, W. Knap, and A. Tredicucci, *Nano Lett.*, **12**, 96 (2012).
13. A. Tomadin, A. Tredicucci, V. Pellegrini, M. S. Vitiello, and M. Polini, *Appl. Phys. Lett.*, **103**, 211120 (2013).
14. A. V. Muraviev, S. L. Rumyantsev, G. Liu, A. A. Balandin, W. Knap, and M. S. Shur, *Appl. Phys. Lett.*, **103**, 181114 (2013).
15. D. Spirito, D. Coquillat, S. L. De Bonis, A. Lombardo, M. Bruna, A. C. Ferrari, V. Pellegrini, A. Tredicucci, W. Knap, and M. S. Vitiello, *Appl. Phys. Lett.*, **104**, 061111 (2014).
16. L. Viti, D. Coquillat, D. Ercolani, L. Sorba, W. Knap, and M. S. Vitiello, *Opt. Express*, **22**, 8996 (2014).
17. C. O. Kim, S. Kim., D. H. Shin, S. S. Kang, J. M. Kim, C. W. Jang, S. S. Joo, J. S. Lee, J. H. Kim., S.-H. Choi, and E. Hwang, *Nat. Commun.*, **5**, 3249 (2014).
18. C.-H. Liu, Y.-C. Chang, T. B. Norris, and Z. Zhong, *Nat. Nanotechnol.*, **9**, 273 (2014).
19. A. K. Geim and I. V. Grigorieva, *Nature*, **499**, 419–425 (2013).
20. M. Liu, X. Yin, and X. Zhang, *Nano Lett.*, **12**, 1482–1485 (2012).
21. L. Britnell, R. V. Gorbachev, R. Jalil, B. D. Belle, F. Shedin, A. Mishenko, T. Georgiou, M. I. Katsnelson, L. Eaves, S. V. Morozov, N. M. R. Peres, J. Leist, A. K. Geim, K. S. Novoselov, and L. A. Ponomarenko, *Science*, **335**, 947–950 (2012).
22. V. Ryzhii, T. Otsuji, M. Ryzhii, N. Ryabova, S. O. Yurchenko, V. Mitin, and M. S. Shur, *J. Phys. D: Appl. Phys.*, **46**, 065102 (2013).
23. T. Georgiou, R. Jalil, B. D. Bellee, L. Britnell, R. V. Gorbachev, S. V. Morozov, Y.-J. Kim, A. Cholinia, S. J. Haigh, O. Makarovsky, L. Eaves, L. A. Ponomarenko, A. K. Geim, K. S. Nonoselov, and A. Mishchenko, *Nat. Nanotechnol.*, **8**, 100–103 (2013).
24. L. Britnell, R. V. Gorbachev, A. K. Geim, L. A. Ponomarenko, A. Mishchenko, M. T. Greenaway, T. M. Fromhold, K. S. Novoselov, and L. Eaves, *Nat. Commun.*, **4**, 1794–1799 (2013).

25. V. Ryzhii, T. Otsuji, M. Ryzhii, V. G. Leiman, S. O. Yurchenko, V. Mitin, and M. S. Shur, *J. Appl. Phys.*, **112**, 104507 (2012).
26. V. Ryzhii, T. Otsuji, M. Ryzhii, and M. S. Shur, *J. Phys. D: Appl. Phys.*, **45**, 302001 (2012).
27. V. Ryzhii, A. Satou, T. Otsuji, M. Ryzhii, V. Mitin, and M. S. Shur, *J. Phys. D: Appl. Phys.*, **46**, 315107 (2013).
28. V. Ryzhii, M. Ryzhii, V. Mitin, M. S. Shur, A. Satou, and T. Otsuji, *J. Appl. Phys.*, **113**, 174506 (2013).
29. V. Ryzhii, A. A. Dubinov, V. Y. Aleshkin, M. Ryzhii, and T. Otsuji, *Appl. Phys. Lett.*, **103**, 163507 (2013).
30. V. Ryzhii, A. A. Dubinov, T. Otsuji, V. Y. Aleshkin, M. Ryzhii, and M. S. Shur, *Opt. Express*, **21**, 31567 (2013).
31. V. Ryzhii, T. Otsuji, V. Y. Aleshkin, A. A. Dubinov, M. Ryzhii, V. Mitin, and M. S. Shur, *Appl. Phys. Lett.*, **104**, 163505 (2014).
32. V. Ryzhii, M. Ryzhii, V. Mitin, A. Satou, and T. Otsuji, Jpn. *J. Appl. Phys., Part I*, **50**, 094001 (2011).
33. H. C. Liu, *Appl. Phys. Lett.*, **60**, 1507 (1992).
34. V. Ryzhii, *J. Appl. Phys.*, **81**, 6442 (1997).
35. E. Rosencher, B. Vinter, F. Luc, L. Thibaudeau, P. Bois, and J. Nagle, *IEEE Trans. Quantum Electron.*, **30**, 2875 (1994).
36. K. K. Choi, *The Physics of Quantum Well Infrared Photodetectors* (World Scientific, Singapore, 1997).
37. S. Luryi, *Appl. Phys. Lett.*, **52**, 501 (1988).
38. R. M. Feenstra, D. Jena, and G. Gu, *J. Appl. Phys.*, **111**, 043711 (2012).
39. F. T. Vasko, *Phys. Rev. B*, **87**, 075424 (2013).
40. H. Shi, H. Pan, Y.-W. Zhang, and B. Yakobson, *Phys. Rev. B*, **87**, 155304 (2013).
41. M. Ryzhii and V. Ryzhii, *IEEE Trans. Electron Devices*, **47**, 1935 (2000).
42. M. Ryzhii, V. Ryzhii, R. Suris, and C. Hamaguchi, *Phys. Rev. B*, **61**, 2742 (2000).
43. V. V. Popov, O. V. Polischuk, T. V. Teperik, X. G. Peralta, S. J. Allen, N. J. M. Horing, and M. C. Wanke, *J. Appl. Phys.*, **94**, 3556 (2003).
44. S. A. Mikhailov and K. Ziegler, *J. Phys.: Condens. Matter*, **20**, 384204 (2008).
45. V. Ryzhii, A, Satou, and T. Otsuji, *J. Appl. Phys.*, **101**, 024509 (2007).
46. M. Graf, G. Scalari, D. Hofstetter, J. Faist, H. Beere, E. Linfeld, D. Ritchie, and G. Davies, *Appl. Phys. Lett.*, **84**, 475 (2004).

47. H. C. Liu, C. Y. Song, A. J. Spring Thorpe, and J. C. Cao, *Appl. Phys. Lett.*, **84**, 4068 (2004).

48. J. C. Cao and H. C. Liu, in *Advances in Infrared Photodetectors*, edited by S. D. Gunapala, D. R. Rhiger, and C. Jagadish (Academic Press, San Diego, 2011), p. 195.

49. J. Yu, G. Liu, A. V. Sumant, V. Goyal, and A. A. Balandin, *Nano Lett.*, **12**, 1603 (2012).

50. R. A. Suris and V. A. Fedirko, *Sov. Phys. Semicond.*, **12**, 629 (1978).

51. S. Barbieri, F. Mango, F. Beltram, M. Lazzarino, and L. Sorba, *Appl. Phys. Lett.*, **67**, 250 (1995).

52. J. K. Choi, V. Mitin, R. Ramaswamy, V. Pogrebnyak, M. Pakmehr, A. Muravjov, M. Shur, J. Gill, I. Medhi, B. Karasik, and A. Sergeev, *IEEE Sens. J.*, **13**, 80 (2013).

53. X. G. Peralta, S. J. Allen, M. C. Wanke, N. E. Harff, J. A. Simmons, M. P. Lilly, J. L. Reno, P. J. Burke, and J. P. Eisenstein, *Appl. Phys. Lett.*, **81**, 1627 (2002).

54. E. A. Shanner, M. Lee, M. C. Wanke, A. D. Grine, J. L. Reno, and S. J. Allen, *Appl. Phys. Lett.*, **87**, 193507 (2005).

55. V. Ryzhii, A. Satou, T. Otsuji, and M. S. Shur, *J. Appl. Phys.*, **103**, 014504 (2008).

56. S. Ghosh, I. Calizo, D. Teweldebrhan, E. P. Pokatilov, D. L. Nika, A. A. Balandin, W. Bao, F. Miao, and C. N. Lau, *Appl. Phys. Lett.*, **92**, 151911 (2008).

57. E. Pop, V. Varshney, and A. K. Roy, *MRS Bull.*, **37**, 1273 (2012).

58. H. Yan, X. Li, B. Chandra, G. Tulevski, Y. Wu, M. Freitag, W. Zhu, P. Avouris, and F. Xia, *Nat. Nanotechnol.*, **7**, 330 (2012).

59. S. J. Haigh, A. Gholinia, R. Jalil, S. Romani, L. Britnell, D. C. Elias, K. S. Novoselov, L. A. Ponomarenko, A. K. Geim, and R. Gorbachev, *Nat. Mater.*, **11**, 764 (2012).

60. M. Xu, T. Lian, M. Shi, and H. Chen, *Chem. Rev.*, **113**, 3766 (2013).

61. Q. H. Wang, K. Kalantar-Zadeh, A. Kis, J. N. Coleman, and M. S. Strano, *Nat. Nanotechnol.*, **7**, 699 (2012).

62. W. J. Yu, Y. Liu, H. Zhou, A. Yin, Z. Li, Y. Huang, and X. Duan, *Nat. Nanotechnol.*, **8**, 952 (2013).

Chapter 47

Electron Capture in van der Waals Graphene-Based Heterostructures with WS_2 Barrier Layers*

V. Ya. Aleshkin,[a] A. A. Dubinov,[a] M. Ryzhii,[b] V. Ryzhii,[c] and T. Otsuji[c]

[a]*Institute for Physics of Microstructures of the Russian Academy of Sciences and Lobachevsky State University of Nizhny Novgorod, Nizhny Novgorod 603950, Russia*

[b]*Department of Computer Science and Engineering, University of Aizu, Aizuwakamatsu, Fukushima 965-8580, Japan*

[c]*Institute of Electrical Communication, Tohoku University, Sendai 980-8577, Japan*

v-ryzhii@riec.tohoku.ac.jp

We consider the capture processes of hot electrons propagating across the van der Waals heterostructures with graphene layers (GLs) and WS_2 barrier layers. The capture probability of hot

*Reprinted with permission from V. Ya. Aleshkin, A. A. Dubinov, M. Ryzhii, V. Ryzhii, and T. Otsuji (2015). Electron capture in van der Waals graphene-based heterostructures with WS_2 barrier layers, *J. Phys. Soc. Jpn.*, **84**(9), 094703. Copyright © 2015 The Physical Society of Japan.

Graphene-Based Terahertz Electronics and Plasmonics: Detector and Emitter Concepts
Edited by Vladimir Mitin, Taiichi Otsuji, and Victor Ryzhii

ISBN 978-981-4800-75-4 (Hardcover), 978-0-429-32839-8 (eBook)
www.jennystanford.com

electrons into GLs as a function of the electron energy and the electric field as well as the capture probability averaged over the electron energy distribution are calculated accounting for their scattering on the optical phonons and on the electrons localized in GLs. Our calculations show that the total probability of these processes is rather low (less than 0.5%). Since the capture probability of hot electrons into GLs essentially determines the performance of the vertical hot-electron transistors with the GL base and the vertical hot-electron terahertz photodetectors, the obtained results support the prospects of these and other GL devices.

47.1 Introduction

The unique properties of graphene layers (GL) [1] enable their use in different electron and optoelectonic devices with enhanced performance and new functional abilities. Among the variety of such GL-based devices, there are devices in which the vertical transport of electrons across the GL is essential. These, for example, are the vertical graphene-base transistors (GBTs) in which the GL base is crossed by hot electrons propagating from the emitter to the collector [2-4], vertical graphene hot-electron detectors (GHEDs) of terahertz radiation with floating GLs [5], and terahertz lasers with the vertical injection from a wide gap photoexcited emitter [6]. As examples, Fig. 47.1 shows schematically the structures and the energy band diagrams of a GBT and a GHED, in which some portion of the electrons originating from the emitter and moving above the barriers created by the layers surrounding the GL is captured into the latter. The capture parameter p_c, which by analogy with quantum-well (QW) heterostructures can be defined as [7] $p_c = \upsilon_c/u$, where $\upsilon_c \propto \langle W_c \rangle$ is the capture rate (velocity), $\langle W_c \rangle$ is the average capture probability of the electrons crossing the GL and u is their average (drift) velocity, characterizes the electron flux into the GL and, hence, important characteristics of the devices in question. In the case of GBTs, the common emitter gain β, which is the ration of the collector and the base currents is equal to $\beta = (1 - p_c)/p_c$. The latter values becomes very large when $p_c \ll 1$. The photoelectric gain g and hence, the detector responsivity R and detectivity D^* in GL-CDs also dramatically increase with decreasing p_c: $g \propto 1/p_c$,

$R \propto 1/p_c$, and $D^* \propto 1/\sqrt{p_c}$. This is similar to the situation in QW infrared photodetectors, in which the electron capture into QWs plays a similar role (see, for instance, Refs. [7, 8]). By analogy with multiple-QW systems [9–14], the electric-field dependence of the capture probability can be the reason of the domain formation in GL structures [15]. In the model used recently for the calculation of the GHED characteristics, the electron capture into GLs was considered phenomenologically with p_c as a constant or as a given function of the electric field in the barriers [5, 15].

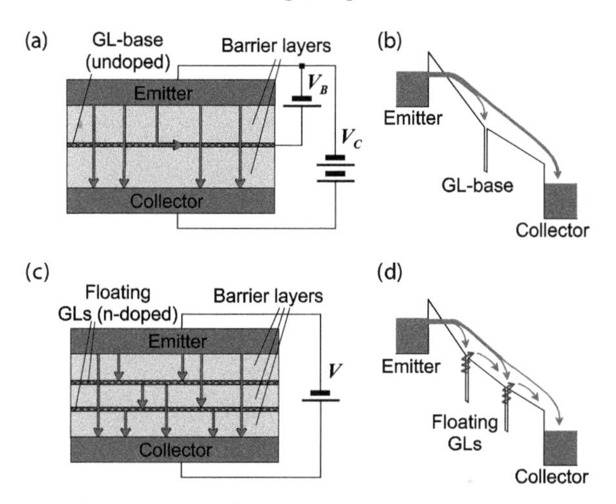

Figure 47.1 Schematic view of a GBT structure in common-emitter circuit and its band diagram, (a) and (b), as well a GHED structures with floating internal GLs and its band diagram, (c) and (d). Smooth arrows indicating electron transport over barriers and capture of some electrons into GLs. Wavy arrows correspond to emission of electrons from GLs in a GHED due to incident radiation.

In this chapter, we calculate the energy and electric-field dependences of the capture probability W_c and its average value in the van der Waals heterostructures with GLs [16]. For definiteness, we consider such heterostructures in which the GLs clad between the WS_2 inter GL barrier layers. The following scattering mechanisms resulting in the capture of electrons propagating above the WS_2 barriers and crossing the GL are taken into account: the interaction of electrons (i) with the optical phonons in the barrier material, i.e., WS_2, (ii) with the GL optical phonons, and (iii) with the equilibrium two-dimensional electron gas (2DEG) formed either by n-type doping or induced by the voltage applied to the outside contacts (in GHEDs). The scattering on the two-dimensional hole gas (2DHG) in the GBT base can be considered in a similar way. Our calculations show that the in the structure under consideration the capture parameter is rather low ($p_c \lesssim 0.5\%$). It can be markedly smaller than in the standard QW heterostructures. This promises the realization of GBTs and GHEDs with fairly higher performance. The obtained results can also be applied to different GL van der Waals heterostructures [16] (in particular with the hBN barrier layers) and other barrier materials.

47.2 Model and Electron Wave Functions

We consider a model, in which the GL constitutes a delta-layer sandwiched by the WS_2 barrier layers. The energy spectra of electrons in the GL sandwiched by the WS_2 layers is assumed to be linear [17] (with the characteristic velocity $v_W \simeq 10^8$ cm/s). This is confirmed by many experiments (see, for example, Ref. [18]). The electron energy dispersion in the barrier layers is assumed to be parabolic (with the effective mass m). As shown previously [19], the bottom of the conduction band in the bulk WS_2 is in the H-point. Since the energy difference between the minima in the K-point and H-point determined experimentally [19–21] is relatively small, i.e., it is less than the accuracy of the calculations, we assume that the electrons in the barriers are in the K-point.

The Schrodinger equation for the electron wave function in the initial state $\psi_i = \psi_i(z)$ dependent on the coordinate z (perpendicular to the GL plane) in the system under consideration can be presented as

$$-\frac{\hbar^2}{2m}\frac{d^2\psi_i}{dz^2} - eFz\psi_i - \alpha(k_{||})\delta(z)\psi_i = \varepsilon\psi_i, \qquad (47.1)$$

where ε is the electron energy associated with the motion along the z, $\hbar$ is the Plank constant, e is the electron charge, F is the electric field in the barrier(s), $\alpha(k_{||}) = \hbar\sqrt{-2(\varepsilon_f - \hbar k_{||}v_W + \hbar^2 k_{||}^2/2m)/m}$, ε_f is the electron potential energy in the GL (with respect to the bottom of the conduction band of WS_2), $k_{||}$ is the electron wave vector along the

GL, and the position of the GL corresponds to $z = 0$. The Dirac delta function in the last term in the left-side of Eq. (47.1) describes the GL potential confining electrons in it. Expression for $\alpha(k_{\parallel})$ corresponds to the above assumption of a linear dispersion of electrons between WS_2 layers and a quadratic dispersion of electrons in WS_2.

In the region $z > 0$, Eq. (47.1) yields

$$\frac{d^2\psi_i}{d\xi^2} + \xi\psi_i = 0, \quad \xi = \left(z + \frac{\varepsilon_i}{eF}\right)\left(\frac{2meF}{\hbar^2}\right)^{1/3}. \tag{47.2}$$

The general solution of this equation can be presented as follows:

$$\psi_i = a_1 Ai(-\xi) + a_2 Bi(-\xi), \tag{47.3}$$

where a_1 and a_2 are unknown constants and $Ai(\xi)$ and $Bi(\xi)$ are the Airy functions. Using the following asymptotic expansion for large negative values of the argument for the Airy functions [22]:

$$Ai(-\xi) \approx \pi^{-1/2}\xi^{-1/4}\sin\left(\frac{2}{3}\xi^{3/2} + \frac{\pi}{4}\right)$$

and

$$Bi(-\xi) \approx \pi^{-1/2}\xi^{-1/4}\cos\left(\frac{2}{3}\xi^{3/2} + \frac{\pi}{4}\right),$$

we find that $a_1 = ia$ and $a_2 = a$, so that the solution of Eq. (47.2) is given by

$$\psi_i(z) = a[iAi(-\zeta) + Bi(-\zeta)]. \tag{47.4}$$

In this case, a and the electron flux C_0 are related to each other as $C_0 = |a|^2(2meF/\hbar^2)^{1/3}$. In the following, we use the following normalization: $a = 1$.

The solution of Eq. (47.1) for $z < 0$ is presented as

$$\psi_i = a_3 Ai(-\xi) + a_4 Bi(-\xi). \tag{47.5}$$

Constants a_3 and a_4, which are found from the conditions of continuity of ψ_i and its for derivatives, are as follows:

$$a_3 = i + \pi\alpha\left(\frac{2m}{\hbar^2}\right)\left(\frac{\hbar^2}{2meF}\right)^{1/3}$$
$$\times [Bi^2(-\xi_0) + iAi(-\xi_0)Bi(-\xi_0)]$$

and

$$a_4 = 1 - \pi\alpha\left(\frac{2m}{\hbar^2}\right)\left(\frac{\hbar^2}{2meF}\right)^{1/3} \times [Bi(-\xi_0)Ai(-\xi_0) - iAi^2(-\xi_0)],$$

where $\xi_0 = e_i(2m/\hbar^2 e^2 F^2)^{1/3}$.

Deriving the wave function, ψ_f, of the final (localized) state, one can neglect the influence of the electric field [due to the fact that the energy localization $|\varepsilon_f| \gg eF/\kappa(0)$]. Taking this into account, we find

$$\psi_f(z,k_{||}) = \sqrt{\kappa(k_{||})\exp(-\kappa(k_{||}|z|)}, \qquad (47.6)$$

where $\kappa(k_{||}) = \sqrt{2m|\varepsilon_f| - \hbar k_{||} \upsilon_W + \hbar^2 k_{||}^2/2m)}/\hbar$.

47.3 Electron Capture Probability at Different Scattering Mechanisms

47.3.1 Capture due to the Emission of Barrier Optical Phonons

Optical phonons in the WS_2 barrier can be regarded as three-dimensional. Since WS_2 is a uniaxial crystal, the difference between the high and low frequency dielectric constants is small for polarization perpendicular to the *c*-axis (the axis *z*) [23]. Considering the results of Ref. [24] (case ii) for finding of the probability of the electron scattering with the emission of the polar optical phonons in uniaxial crystals, one can use the following expression for the electron flux into the GL:

$$C_B = \sum_{q,k_f} \frac{2\pi}{\hbar} \left|\langle f|H_{int}|i\rangle\right|^2 (N_q + 1)\delta(E_i - E_f - \hbar\Omega_z), \qquad (47.7)$$

where

$$\left|\langle f|H_{int}|i\rangle\right|^2 = \frac{2\pi e^2\hbar}{Vq^2}\frac{\omega_{zL}^2}{\Omega_z}\left(\frac{1}{\varepsilon_z^\infty} - \frac{1}{\varepsilon_z^0}\right) I^2(q_z, q_{||})\cos^2\theta,$$

$$I(q_z, q_{||}) = \int dz \exp(-iq_z z)\psi_f(z, q_{||})\psi_i(z),$$

$$\Omega_z^2 = \omega_z^2 \sin^2\theta + \omega_{zL}^2 \cos^2\theta,$$

$$E_i = \varepsilon_i + \hbar^2 k_i^2/2m, \quad E_f = \varepsilon_f + \hbar \upsilon_W k_f,$$

θ is the angle between wave vector q and the axis z, V is the crystal volume, ω_z and ω_{zL} are the transverse and longitudinal optical phonon frequency, respectively, N_q is the number of the optical phonons with wave vector q. To simplify the calculation, we first consider the case $k_i = 0$. In this case,

$$C_B = \frac{e^2\omega_{zL}^2}{\hbar\upsilon_W}\left(\frac{1}{\varepsilon_z^\infty} - \frac{1}{\varepsilon_z^0}\right)\int_0^\pi d\theta \frac{\cos^2\theta}{\Omega_z} | I(q_0 \cot\theta, q_0|^2$$
$$\times\left[\frac{1}{\exp(\hbar\Omega_z/T)-1}+1\right], \tag{47.8}$$

where $q_0 = (\varepsilon_i - \varepsilon_f - \hbar\Omega_z)/\hbar\upsilon_W$, T is the lattice temperature (in the energy units). The probability of electron capture in the GL due to the interaction with the barrier optical phonons is equal to

$$W_B(\varepsilon_i) = \frac{C_B}{C_0} = \frac{e^2\omega_{zL}m}{\hbar^2\upsilon_W}\left(\frac{\hbar^2}{2meF}\right)^{1/3}\left(\frac{1}{\varepsilon_z^\infty} - \frac{1}{\varepsilon_z^0}\right)$$
$$\times\int_0^\pi d\theta \frac{\omega_{zL}}{\Omega_z}\cos^2\theta \, | I(q_0\cot\theta, q_0)|^2 . \tag{47.9}$$

Equation (47.9) is valid when $k_i \ll (|\varepsilon_f| - \hbar\omega_{zL})/\hbar\upsilon_W$.

47.3.2 Capture due to the Emission of GL Optical Phonons

The matrix element of the Hamiltonian of the electron interaction with the graphene optical phonons can be presented as [25]:

$$\langle\Psi_i(k')|H_{int}|\Psi_f(k)\rangle$$
$$= \delta_{k+q,k'}\frac{1}{2}\frac{\partial t}{\partial b}\sum_m \frac{\mathbf{d}_m}{b}[\mathbf{u}_B(\mathbf{q})e^{iqd_m} - \mathbf{u}_A(\mathbf{q})]$$
$$\times(e^{-i\phi(k')-ik'd_m} + e^{i\phi(k)+ikd_m}), \tag{47.10}$$

where $\Psi_f(k)$ is the Bloch function, b is the length of the cells in the cell space, and $\partial t/\partial b \approx 45$ eV/nm [25]. Here, $\mathbf{d}_m$ are the position vectors from the carbon atom to its three nearest atoms, $\mathbf{u}_{A,B}$ are the displacements sublattices A and B due to the optical phonons. There are two types of the optical oscillations for $q = 0$ (Γ-point of the Brillouin zone) with the same energy (TO and LO). For each type of the lattice optical oscillations, the square of the matrix element modulus is independent of the wave vector (intravalley scattering) [25]:

$$\left|\left\langle \Psi_i(k')\middle| H_{\text{int}} \middle| \Psi_f(k)\right\rangle\right|^2 = \frac{\delta_{k+q,k'}}{S}\frac{9\hbar}{4\rho\omega}\left(\frac{\partial t}{\partial b}\right)^2, \tag{47.11}$$

where ρ is the density of graphene (7.6×10^{-7} kg/cm^2) and ω is the optical phonon frequency. Taking in to account the overlap of wave functions of the initial and final states, Eq. (47.11) can be written as follows:

$$\left|\left\langle \Psi_i(k')\middle| H_{\text{int}} \middle| \Psi_f(k)\right\rangle\right|^2 = \frac{\delta_{k+q,k'}}{S}\frac{9\hbar}{4\rho\omega}\left(\frac{\partial t}{\partial b}\right)^2 \times |\psi_f^*(0,q)\psi_i(0)|^2\, d^2, \tag{47.12}$$

where d is the GL thickness. In this case, the flux of the captured electrons is equal to

$$C_{\text{GL}} = \frac{9d^2}{8\pi\rho\omega}\left(\frac{\partial t}{\partial b}\right)^2 \int d^2q\,\delta(\varepsilon_i - \varepsilon_f - \hbar\omega - \hbar\upsilon_W q) \times |\psi_f^*(0,q)\psi_i(0)|^2. \tag{47.13}$$

Integrating Eq. (47.13), we obtain:

$$C_{\text{GL}} = \frac{9d^2}{4\hbar\upsilon_W\rho\omega}\left(\frac{\partial t}{\partial b}\right)^2 [Ai^2(-\xi_0) + Bi^2(-\xi_0)]Q\kappa(Q), \tag{47.14}$$

where $Q = (\varepsilon_i - \varepsilon_f - \hbar\omega)/\hbar\upsilon_W$, and the probability of capture of electrons in GL:

$$W_{\text{GL}}(\varepsilon_i, k_i = 0) = \frac{9d^2}{4\hbar^2\upsilon_W\rho\omega}\left(\frac{\partial t}{\partial b}\right)^2\left(\frac{\hbar^2}{2meF}\right)^{1/3} \times [Ai^2(-\xi_0) + Bi^2(-\xi_0)]Q\kappa(Q). \tag{47.15}$$

Note that deriving Eq. (47.15), we have assumed that the occupation numbers of optical phonons in the GL are very small because their energy is much larger than the thermal energy, i.e., this type of the electron capture is due to the spontaneous emission of optical phonons.

47.3.3 Capture Associated with the Electron-Electron Interaction

If we neglect the screening and the exchange term, then the interaction Hamiltonian and the scattering probability can be presented as follows:

$$W_{\mathrm{i,f}} = \frac{2\pi}{\hbar}\left|\langle f|H_{\mathrm{int}}|i\rangle\right|^2 \delta(E_{\mathrm{i}} - E_{\mathrm{f}}) \tag{47.16}$$

with the Hamiltonian

$$H_{\mathrm{int}} = -\frac{e^2}{\varepsilon_0 |\mathbf{r}_1 - \mathbf{r}_2|}.$$

In the case under consideration, the electron energy in GL depends on the longitudinal wave vector, so that the complete wave functions have the form:

$$\Psi_{\mathrm{i}}(\mathbf{r}_1, \mathbf{r}_2) = \psi_{\mathrm{i}}(z_1)\psi_{\mathrm{i}}(z_2, k_2)\exp(i\mathbf{k}_{\mathrm{i}}\mathbf{r}_1 + i\mathbf{k}_2\mathbf{r}_2)/S,$$

$$\Psi_{\mathrm{f}}(\mathbf{r}_1, \mathbf{r}_2) = \psi_{\mathrm{f}}(z_1, k'_1)\psi_{\mathrm{f}}(z_2, k'_2), \exp(i\mathbf{k}'_1\mathbf{r}_1 + i\mathbf{k}'_2\mathbf{r}_2)/S.$$

Integrating over the plane perpendicular to the axis z, it is convenient to transfer to the center of mass. Then we obtain the following expression for the matrix element of the ee-scattering:

$$\begin{aligned}\langle f|H_{\mathrm{int}}|i\rangle &= \frac{2e^2}{\varepsilon_0 S}\int dq_z \frac{I_1(q_z, k'_1)I_2(q_z, k_2, k'_2)}{(q_{||}^2 + q_z^2)}\delta_{k_i + k_2; k'_1 + k'_2} \\ &= \frac{M(k'_1, k_2, k'_2)}{S}\delta_{k_i + k_2; k'_1 + k'_2},\end{aligned} \tag{47.17}$$

where $I_1(q_z, k'_1) = \int dz \exp(iq_z z)\psi_{\mathrm{i}}(z)\ \psi^*_{\mathrm{f}}(z, k'_1)$, $I_2(q_z, k_2, k'_2) = \int dz \exp(-iq_z z)\ \psi_{\mathrm{f}}(z, k_2)\ \psi^*_{\mathrm{f}}(z, k'_2)$, and $q_{||} = k_{\mathrm{i}} - k'_1$. The expression for $M(k'_1, k_2, k'_2)$ can be written as:

$$M(k_1',k_2,k_2') = \frac{4\pi e^2}{\varepsilon_0} \frac{\sqrt{\kappa(k_2)\kappa(k_2')}(\kappa(k_2)+\kappa(k_2'))}{\beta^2 - k_1'^2}$$
$$\times \int dz \psi_\mathrm{i}(z) \psi_\mathrm{f}^*(z,k_1')$$
$$\times \left[\frac{\exp(-k_1'|z|)}{k_1'} - \frac{\exp(-\beta|z|)}{\beta} \right], \tag{47.18}$$

where $\beta = \kappa(k_2) + \kappa(k'_2)$. The electron flux as a result of electron-electron scattering in the GL can be expressed via $W_{\mathrm{i,f}}$ as follows:

$$C_\mathrm{ee}(\varepsilon_\mathrm{i},k_\mathrm{i}) = \Sigma_{k_2,k_1',k_2'} W_{k_\mathrm{i},k_2 \to k_1',k_2'}$$
$$\times f(k_2)[1-f(k_1')][1-f(k_2')], \tag{47.19}$$

where $f(k)$ is the distribution function of the electrons localized in the GL.

The initial energy is $\varepsilon_\mathrm{i} + \hbar^2 k_\mathrm{i}^2/2m + \varepsilon_\mathrm{f} + \hbar v_\mathrm{W} k_2$, and the final energy is $2\varepsilon_\mathrm{f} + \hbar v_\mathrm{W} k'_1 + \hbar v_\mathrm{W} k'_2 = 2\varepsilon_\mathrm{f} + \hbar v_\mathrm{W} k'_1 + \hbar v_\mathrm{W} |\mathbf{k}_\mathrm{i} + \mathbf{k}_2 - \mathbf{k}'_1|$. In the case $k_i = 0$, Eq. (47.19) can be rewritten as:

$$C_\mathrm{ee}(\varepsilon_\mathrm{i},k_\mathrm{i}=0)$$
$$= \frac{1}{(2\pi)^3 \hbar} \int d^2k_1' [1-f(k_1')]$$
$$\times \int d^2k_2 |M(k_1',k_2,k_2')|^2 f(k_2)[1-f(\mathbf{k}_2-\mathbf{k}_1')]$$
$$\times \delta[\varepsilon_\mathrm{i} - \varepsilon_\mathrm{f} + \hbar v_\mathrm{W}(k_2 - k_1' - |\mathbf{k}_2 - \mathbf{k}_1'|)]. \tag{47.20}$$

After the integration in Eq. (47.20) over the variable k_2, this equation transforms to the following:

$$C_\mathrm{ee} = \frac{4e^4}{\hbar^2 v_\mathrm{W} \varepsilon_0^2} \int_{k_0/2}^{\infty} dk_1' [1-f(k_1')]$$
$$\times \int_{k_3}^{k_\mathrm{f}} dk_2 f(k_2)[1-f(k_0+k_2-k_1')]$$
$$\times |I_3(k_1',k_2,k_0+k_2-k_1')|^2$$
$$\times \frac{2(k_0+k_2-k_1')}{\sqrt{1-\left[1+\frac{k_0}{k_1'} - \frac{k_0(2k_1'-k_0)}{2k_2k_1'}\right]^2}}, \tag{47.21}$$

where

$$k_0 = (\varepsilon_i - \varepsilon_f)/\hbar v_W,$$

$$k_f = k'_1 - k_0/2,$$

$$k_3 = \max\{k'_1 - k_0, (k_i - k_0/2)k_0/(2k_i + k_0)\},$$

$$I_3(k'_1, k_2, k'_2) = \frac{\sqrt{\kappa(k_2)\kappa(k'_2)}\beta}{\beta^2 - k_1'^2} \int dz \psi_i(z) \psi_f^*(k'_1) \times \left[\frac{\exp(-k'_1 |z|)}{k'_1} - \frac{\exp(-\beta |z|)}{\beta} \right].$$

As a result, for the probability of electron capture due to electron-electron scattering in the GL, we arrive at

$$W_{ee}(\varepsilon_i, k_i = 0) = \frac{4e^4 m}{\hbar^3 v_W \varepsilon_0^2} \left(\frac{\hbar^2}{2meF} \right)^{1/3} \times \int_{k_0/2}^{\infty} dk'_1 [1 - f(k'_1)] \int_{k_3}^{k_f} dk_2 f(k_2)[1 - f(k_0 - k_2 - k'_1)] \times \frac{2(k_0 + k_2 - k'_1) | I_3(k'_1, k_2, k_0 + k_2 - k'_1)|^2}{\sqrt{1 - \left[1 + \frac{k_0}{k'_1} - \frac{k_0(2k'_1 - k_0)}{2k_2 k'_1} \right]^2}}. \quad (47.22)$$

47.3.4 Results of Numerical Calculations

The obtained formulas for the probabilities of the capture associated with the pertinent scattering mechanisms, W_B, W_{GL}, and W_{ee}, were used for the calculations of these quantities as well as the net capture probability $W_c = W_B + W_{GL} + W_{ee}$ as functions of the energy of the electrons incident on the GL at given values of the electric fields in the barrier and of the barrier electric field at different electron energies.

We assume the following parameters related to the WS_2 barrier layer: $m = 0.27m_0$ for the effective electron mass [26] (m_0 is the free electron mass), $\hbar\omega_z = 54.57$ meV. $\hbar\omega_{zL} = 54.64$ meV for the energies of optical phonons [23], and ε_z^0 = 8.2 for the static dielectric constant [27]. The optical phonon energy in GLs and their thickness are set to be $\hbar\omega$ = 196 meV [28] and 0.35 nm [29], respectively. We believe that the electron gas in the GL is induced by the applied voltages, so that the contribution of the impurities to the scattering of the incident electrons and their capture in comparison with the electron-electron scattering is negligible.

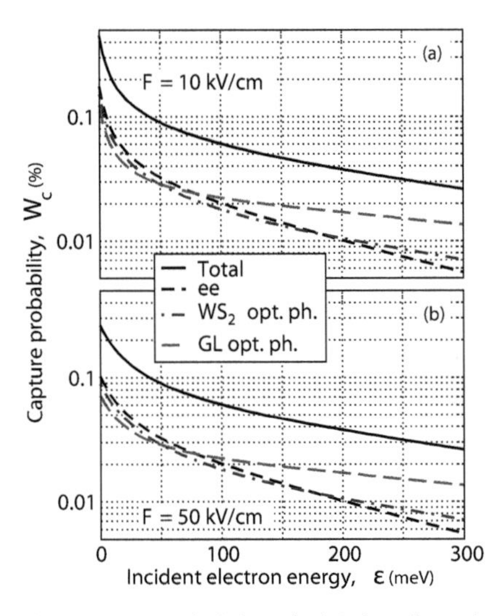

Figure 47.2 Total capture probability (solid lines) and contributions of different scattering mechanisms (broken lines) versus incident electron energy for (a) F = 10 kV/cm and (b) 50 kV/cm.

Figure 47.2 shows the total capture probability W_c as a function of the energy of electrons crossing the GL ε_i calculated for different values of the electric field in the barrier F (solid lines) as well as the contributions of the scattering on optical phonons in the barrier and the GL, W_B and W_{GL} and the electron-electron scattering W_{ee} (broken lines) at the temperature T = 300 K the concentration of electrons in the GL Σ_{GL} = 10^{12} cm^{-2}. As seen from Fig. 47.2, for the electrons with energies less than 75 meV, the electron-electron scattering mechanism provides a larger contribution. At higher electron energies, their capture is associated primarily with the emission of optical phonons in the GL. Figure 47.2 also indicates that the total capture probability at the electric fields under consideration does not exceed 5×10^{-3}. The smallness of the capture probability can be attributed to a relative weakness of the electron interaction with the polar optical phonons in the WS_2 barrier layer because of a small difference in the high-frequency and low-frequency dielectric constants of WS_2 [23]. Interaction with the optical phonons in the GL is weak as well due to a small overlap of the electron wave functions of the initial state with the GL, where this interaction takes place. The electron-electron interaction is also provides a moderate contribution because of a relatively weak overlap of the electron wave functions of the initial and final states, and the necessity of a large electron momentum transfer.

Figure 47.3 shows the electric-field dependences of the capture probability calculated for the following two electron energies: ε = 15 and 30 meV. It is instructive that these voltage dependences are rather weak. This is due to small electron energies ε_i in the examples under consideration in comparison with the GL ionization energy, i.e., with a big difference between the energy of the incident and localized electrons (so that their scattering is accompanied by a large momentum transfer), and the energy of optical phonons in the GL. Some decrease in W_c with increasing ε_i can be explained by the modification of the electron wave functions by the electric field.

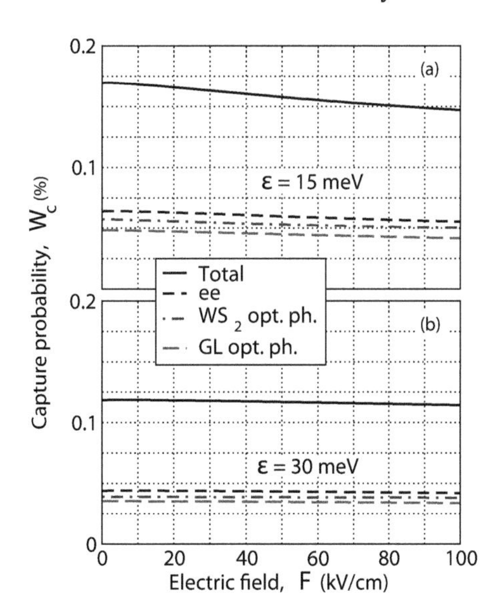

Figure 47.3 The same quantities as in Fig. 47.2 but versus electric field for different incident electron energies (a) ε = 15 meV and (b) ε = 30 meV.

Figure 47.4 shows the total capture probability of electrons with the energy ε = 10 meV as a function of the barrier electric field at different temperatures of the electrons localized in the GL (coinciding with the lattice temperatures *T*). The origin of such a dependence is associated with a larger contribution of low energy electrons in the GL (their fraction rises with decreasing temperature) to their scattering with the incident electrons.

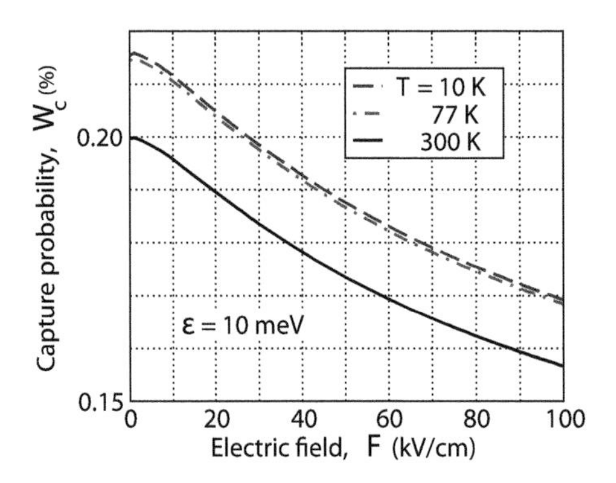

Figure 47.4 Total capture probability versus electric field at different temperatures.

47.4 Capture Parameter and Capture Velocity

The capture parameter p_c (or the capture velocity υ_c), which substantially determines the characteristics of the GL-based devices with the vertical electron transport are defined as (compare with Refs. [7, 15])

$$p_c = eC/j, \quad \upsilon_c = euC/j = p_c u, \tag{47.23}$$

where j is the density of the electron current from the emitter, $C = C_B + C_{GL} + C_{ee}$ is the net rate of the capture into a GL of the incident electrons, $u = b(F)F$ is the electron drift velocity, and $b(F)$ is the electron mobility in the barrier layer at the electric field F, which accounts for the scattering of electrons on impurities, phonons, localized electrons and the quantum-mechanical reflection from the GLs. The capture velocity by its meaning closely resembles the velocity (rate) of surface recombination at sample boundaries.

In the case of collision dominated electron transport across the barrier (drift transport) when the electron distribution function in the barriers is the Maxwellian function displaced by the drift velocity u with the effective electron temperature T_B [15],

$$C \propto \int_0^\infty dp_z \, \upsilon_z W_c(\varepsilon_z) \times \left\{ \exp\left[-\frac{(p_z - mu)^2}{2mT_B} \right] + \exp\left[-\frac{(p_z + mu)^2}{2mT_B} \right] \right\}$$

$$\simeq 2 \int_0^\infty d\varepsilon \, W_c(\varepsilon) \exp\left(-\frac{\varepsilon}{T_B} \right), \tag{47.24}$$

where $W_c(\varepsilon)$ is the net capture probability: $W_c(\varepsilon) = W_B^{ph}(\varepsilon) + W_{GL}^{ph}(\varepsilon) + W_{ee}(\varepsilon)$. In this case, the current density is given by

$$j \propto \int_0^\infty dp_z \, \upsilon_z \left\{ \exp\left[-\frac{(p_z - mu)^2}{2mT_B} \right] - \exp\left[-\frac{(p_z + mu)^2}{2mT_B} \right] \right\}$$

$$\simeq 2 \int_0^\infty d\varepsilon \, \exp\left(-\frac{\varepsilon}{T_B} \right) \left(\frac{\sqrt{2m\varepsilon} u}{T_B} \right). \tag{47.25}$$

Thus, at not too weak electric fields (when the drift of electrons prevails over their diffusion and $u \lesssim \upsilon_T$, where $\upsilon_T = \sqrt{T_B/2m}$ is the thermal velocity in the barrier layer), one can arrive at the following formulas for the capture parameter and capture velocity:

$$p_c = \frac{\int_0^\infty d\varepsilon \, W_c(\varepsilon) \exp\left(-\frac{\varepsilon}{T} \right)}{\int_0^\infty d\varepsilon \, \exp\left(-\frac{\varepsilon}{T_B} \right) \left(\frac{\sqrt{2m\varepsilon} u}{T} \right)}$$

$$= \frac{2\upsilon_T}{\sqrt{\pi} u} \int_0^\infty \frac{d\varepsilon \, W_c(\varepsilon)}{T} \exp\left(-\frac{\varepsilon}{T} \right). \tag{47.26}$$

Hence, p_c and υ_c can be presented as

$$p_c \simeq \frac{\langle W_c \rangle \sqrt{T_B/2m}}{b(F)F} \gtrsim \langle W_c \rangle,$$

$$\upsilon_c \simeq \langle W_c \rangle \sqrt{T_B/2m}, \tag{47.27}$$

where $\langle W_c \rangle$ is the average capture probability. The electron temperature in the barrier layers T_B can strongly depend on the electric field in these layers. Thus, the average capture probability is the field-dependent due to the following two effects: the effect of the electric field on the electron wave functions (see Figs. 47.3 and 47.4) and the electron heating by the electric field via the dependence of T_B on F. However the T_B–F relation depends on the thickness of the barriers, the net thickness of the heterostructure, and the features of the device design, because of possible electron heat removal by GLs and through the contacts. In this regard, both F and T_B can be considered to be relatively independent parameters determining the average capture probability and the capture velocity. Figure 47.5 shows the average capture probability $\langle W_c \rangle$ (equal to the capture parameters p_c and the capture velocity υ_c as functions of the electron temperature T_B in the barrier layers at different electric fields.

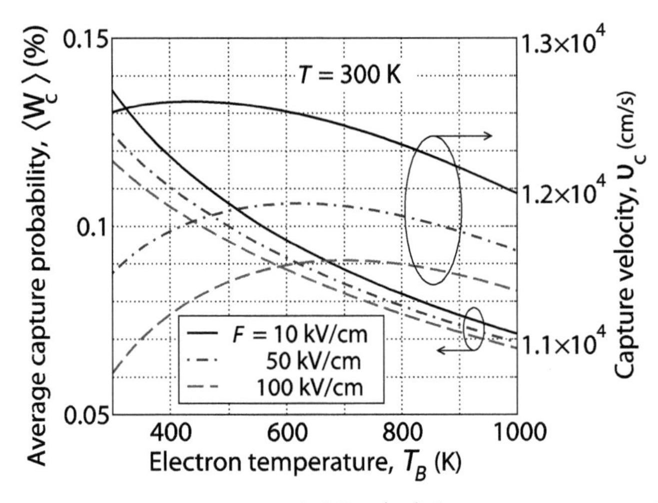

Figure 47.5 Average capture probability $\langle W_c \rangle$ (capture parameter p_c) and capture velocity υ_c versus electron temperature in the barrier.

47.5 Conclusion

We calculated the probability of electron capture in GLs W_c sandwiched between the WS_2 barrier layers and the pertinent capture parameter $p_c \simeq \langle W_c \rangle$ in presence of the external electric field associated with the following three scattering mechanisms: the emission of WS_2 optical phonons, the emission of GL optical phonons, and the interaction of the incident hot electrons with the electrons localized in GLs. The interaction of the hot electrons with the holes in the GLs of p-type (not considered here) provides a similar contribution to the hot-electron capture. The capture probability and the capture parameter as functions of the incident electron energy and the electric field in the barriers were found. It was shown that W_c and p_c do not exceed the value of 0.5% at room temperature. The obtained values of W_c and p_c are smaller than that in the QW-heterostructures based of the standard semiconductor materials [8, 30–32]. The smallness of the capture probability into GLs can be attributed to a relative weakness of the electron interaction with the polar optical phonons in the barrier layers (namely the WS_2 layers) because of a small difference in the high-frequency and low-frequency dielectric constants of WS_2 [23]. The electron interaction with optical phonons in the GL is weak as well due to a small overlap of the electron wave functions of the initial states with the GL, where this interaction takes place. The electron-electron interactions also provide a moderate contribution because of a relatively small overlap of the electron wave functions of the initial and final states and because of a large electron momentum transfer which accompanies the transitions associated with such interactions. This in contrast to the electron capture into QWs at least in the heterostructures based on A_3B_5 compounds, where the capture probability and the capture velocity are much higher. In particular, as shown previously [32], the capture velocity υ_c in the case of $Al_xGa_{1-x}As$-GaAs-$Al_xGa_{1-x}As$ heterostructure with the GaAs QW and $x = 0.33$ is about $\upsilon_c \simeq (5 - 7) \times 10^5$ cm/s, which is one order of magnitude or more larger than in the case of the GL structure with the WS_2 barrier layers (see Fig. 47.5). The same is related to the capture probability p_c, which, according to the data extracted from the experiments, is normally not smaller than $p_c \simeq 0.5$ (see Refs. [8, 33]).

Due to this, GBTs, GHEDs, and similar devices GL-based can surpass the pertinent QW-based devices. The developed theory can be applied to the analysis of the capture processes in different van der Waals GL-based heterostructures.

Acknowledgments

The authors are grateful to M. S. Shur and V. Mitin for numerous fruitful discussions in particular those related to the importance of weakeness of the capture processes in the van der Waals GL-base heterostructures. The work was supported by the Japan Society for promotion of Science (Grant-in-Aid for Specially Promoting Research 23000008), Japan, the Dynasty Foundation, Russia, and the Ministry of Education and Sciences (MES) of Russian Federation (agreement between MES and Lobachevsky State University of Nizhny Novgorod, 02.B.49.21.0003, August, 27 2013).

References

1. A. H. Castro Neto, F. Guinea, N. M. R. Peres, K. S. Novoselov, and A. K. Geim, *Rev. Mod. Phys.*, **81**, 109 (2009).
2. S. Vaziri, G. Lupina, C. Henkel, A. D. Smith, M. Ostling, J. Dabrowski, G. Lippert, W. Mehr, and M. C. Lemme, *Nano Lett.*, **13**, 1435 (2013).
3. W. Mehr, J. Ch. Scheytt, J. Dabrowski, G. Lippert, Y.-H. Xie, M. C. Lemme, M. Ostling, and G. Lupina, *IEEE Electron Device Lett.*, **33**, 691 (2012).
4. C. Zeng, E. B. Song, M. Wang, S. Lee, C. M. Torres, Jr., J. Tang, B. H. Weiler, and K. L. Wang, *Nano Lett.*, **13**, 2370 (2013).
5. V. Ryzhii, A. Satou, T. Otsuji, M. Ryzhii, V. Mitin, and M. S. Shur, *J. Appl. Phys.*, **116**, 114504 (2014).
6. A. R. Davoyan, M. Yu. Morozov, V. V. Popov, A. Satou, and T. Otsuji, *Appl. Phys. Lett.*, **103**, 251102 (2013).
7. E. Rosencher, B. Vinter, F. Luc, L. Thibaudeau, P. Bois, and J. Nagle, *IEEE Trans. Quantum Electron.*, **30**, 2875 (1994).
8. H. C. Schneider and H. C. Liu, *Quantum Well Infrared Photodetectors* (Springer, Berlin, 2007).
9. M. Ershov, V. Ryzhii, and C. Hamaguchi, *Appl. Phys. Lett.*, **67**, 3147 (1995).
10. L. Thibaudeau, P. Bois, and J. Y. Duboz, *J. Appl. Phys.*, **79**, 446 (1996).
11. V. Ryzhii and H. C. Liu, Jpn. *J. Appl. Phys.*, **38**, 5815 (1999).
12. V. Ryzhii, M. Ryzhii, and H. C. Liu, *J. Appl. Phys.*, **92**, 207 (2002).
13. M. Ryzhii, V. Ryzhii, R. Suris, and C. Hamaguchi, *Phys. Rev. B*, **61**, 2742 (2000).

14. V. Ryzhii, I. Khmyrova, M. Ryzhii, R. Suris, and C. Hamaguchi, *Phys. Rev. B*, **62**, 7268 (2000).

15. V. Ryzhii, T. Otsuji, M. Ryzhii, V. Ya. Aleshkin, A. A. Dubinov, V. Mitin, and M. S. Shur, *J. Appl. Phys.*, **117**, 154504 (2015).

16. A. K. Geim and I. V. Giigoiieva, *Nature*, **499**, 419 (2013).

17. S.-S. Li and C.-W. Zhang, *J. Appl. Phys.*, **114**, 183709 (2013).

18. L. Britnell, R. V. Gorbachev, A. K. Geim, L. A. Ponomarenko, A. Mishenko, M. T. Greenaway, T. M. Fromhold, K. S. Novoselov, and L. Eaves, *Nat. Commun.*, **4**, 1794 (2013).

19. A. Klein, S. Tiefenbacher, V. Eyert, C. Pettenkofer, and W. Jaegermann, *Phys. Rev. B*, **64**, 205416 (2001).

20. A. Kuc, N. Zibouche, and T. Heine, *Phys. Rev. B*, **83**, 245213 (2011).

21. W. S. Yun, S. W. Han, S. C. Hong, I. G. Kim, and J. D. Lee, *Phys. Rev. B*, **85**, 033305 (2012).

22. L. D. Landau and E. M. Lifshitz, *Quantum Mechanics: Non-Relativistic Theory* (Elsevier, Amsterdam, 1981).

23. A. Molina-Sanchez and L. Wirtz, *Phys. Rev. B*, **84**, 155413 (2011).

24. B. C. Lee, K. W. Kim, M. Dutta, and M. A. Stroscio, *Phys. Rev. B*, **56**, 997 (1997).

25. F. Rana, P. A. George, J. H. Strait, J. Dawlaty, S. Shivaraman, M. Chandrashekhar, and M. G. Spencer, *Phys. Rev. B*, **79**, 115447 (2009).

26. H. Shi, H. Pan, Y.-W. Zhang, and B. Yakobson, *Phys. Rev. B*, **87**, 155304 (2013).

27. A. Kumar and P. K. Ahluwalia, *Physica B*, **407**, 4627 (2012).

28. S. Piscanec, M. Lazzeri, F. Mauri, and A. C. Ferrari, *Eur. Phys. J.: Spec. Top.*, **148**, 159 (2007).

29. A. N. Sidorov, M. M. Yazdanpanah, R. Jalilian, P. J. Ouseph, R. W. Cohn, and G. U. Sumanasekera, *Nanotechnology*, **18**, 135301 (2007).

30. J. A. Brum and G. Bastard, *Phys. Rev. B*, **33**, 1420 (1986).

31. J. M. Gerard, B. Deveaud, and A. Regeny, *Appl. Phys. Lett.*, **63**, 240 (1993).

32. D. Bradt, Y. M. Sirenko, and V. Mitin, *Semicond. Sci. Technol.*, **10**, 260 (1995).

33. B. F. Levine, *J. Appl. Phys.*, **74**, R1 (1993).

Chapter 48

Resonant Plasmonic Terahertz Detection in Vertical Graphene-Base Hot-Electron Transistors*

V. Ryzhii,[a,b] T. Otsuji,[a] M. Ryzhii,[c] V. Mitin,[d] and M. S. Shur[e]
[a]*Research Institute of Electrical Communication, Tohoku University, Sendai 980-8577, Japan*
[b]*Center for Photonics and Infrared Engineering, Bauman Moscow State Technical University and Institute of Ultra High Frequency Semiconductor Electronics of RAS, Moscow 111005, Russia*
[c]*Department of Computer Science and Engineering, University of Aizu, Aizu-Wakamatsu 965-8580, Japan*
[d]*Department of Electrical Engineering, University at Buffalo, SUNY, Buffalo, New York 1460-1920, USA*
[e]*Department of Electrical, Computer, and System Engineering and Physics, Applied Physics, and Astronomy, Rensselaer Polytechnic Institute, Troy, New York 12180, USA*
v-ryzhii@riec.tohoku.ac.jp

*Reprinted with permission from V. Ryzhii, T. Otsuji, M. Ryzhii, V. Mitin, and M. S. Shur (2015). Resonant plasmonic terahertz detection in vertical graphene-base hot-electron transistor, *J. Appl. Phys.*, **118**, 204501. Copyright © 2015 AIP Publishing LLC.

Graphene-Based Terahertz Electronics and Plasmonics: Detector and Emitter Concepts
Edited by Vladimir Mitin, Taiichi Otsuji, and Victor Ryzhii

ISBN 978-981-4800-75-4 (Hardcover), 978-0-429-32839-8 (eBook)
www.jennystanford.com

We analyze dynamic properties of vertical graphene-base hot-electron transistors (GB-HETs) and consider their operation as detectors of terahertz (THz) radiation using the developed device model. The GB-HET model accounts for the tunneling electron injection from the emitter, electron propagation across the barrier layers with the partial capture into the GB, and the self-consistent oscillations of the electric potential and the hole density in the GB (plasma oscillations), as well as the quantum capacitance and the electron transit-time effects. Using the proposed device model, we calculate the responsivity of GB-HETs operating as THz detectors as a function of the signal frequency, applied bias voltages, and the structural parameters. The inclusion of the plasmonic effect leads to the possibility of the GB-HET operation at the frequencies significantly exceeding those limited by the characteristic RC-time. It is found that the responsivity of GB-HETs with a sufficiently perfect GB exhibits sharp resonant maxima in the THz range of frequencies associated with the excitation of plasma oscillations. The positions of these maxima are controlled by the applied bias voltages. The GB-HETs can compete with and even surpass other plasmonic THz detectors.

48.1 Introduction

Recently vertical hot-electron transistors (HETs) with the graphene base (GB) and the bulk emitter and collector separated from the base by the barrier layers—the hot-electron GB-HETs—made of SiO_2 and Al_2O_3 were fabricated and studied [1–4]. These HETs are fairly promising devices despite their modest characteristics at the present. Similar devices can be based on graphene layer (GL) heterostructures with the hBN, WS_2, and other barrier layers [5–8]. The history of different versions of HETs, including those with the thin metal base and the quantum-well (QW) base in which the carriers are generated from impurities or induced by the applied voltages, as well as HETs with resonant-tunneling emitter, is rather long (see, for example, Refs. [9–16]). Figure 48.1 shows schematically the GB-HET structure and its band diagrams. Depending on the GB doping, the base-emitter and collector-base voltages, V_B and V_C, and the thicknesses of barrier layers separating the base from the emitter and collector, W_E and W_C, respectively, the GB can be filled either with electrons or holes (compare Figs. 48.1b and c).

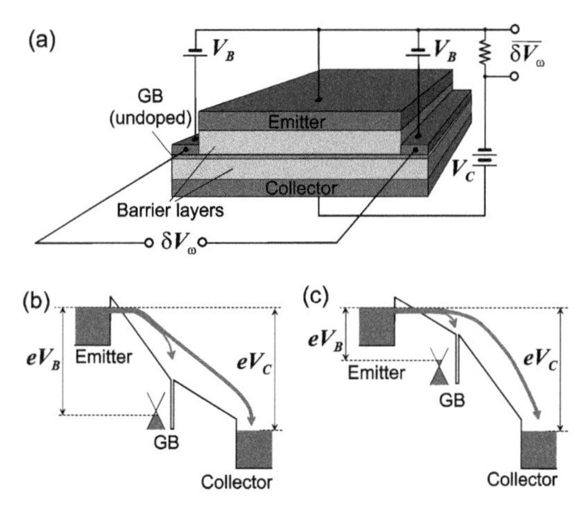

Figure 48.1 (a) Structure of a GB-HET with undoped GB and GB-HET band diagrams (potential profile in the direction perpendicular to the GB plane) at different relations between V_B and V_C: (b) with GB filled with holes and (c) with GB filled with electrons. Arrows show propagation of electrons across the barrier layers and capture of some portion of electrons into GB.

In GB-HETs (as in HETs), a significant fraction of the electrons injected from the top n-type emitter contact due to tunneling through the barrier top crosses the GB and reaches the collector, while the fraction of the electrons captured into the GB can be rather small. The tunneling electrons create the emitter-collector current J_C. The variations of the base-emitter voltage V_B result in the variation of J_C.

In the case of the GB with a two-dimensional electron gas (2DEG), the GB-HET can also be referred to as the N-n-N heterostructure HETs with the GB of n-type. In another case, when the GB comprises a two-dimensional hole gas (2DHG), the GB-HET can also be called the N-p-N heterostructure bipolar transistors (HBTs) with the GB of p-type.

The electrons captured into the GB support the emitter-base current J_B. The transistor gain in the common-emitter configuration is equal to $g_0 = J_C/J_B = (1 - p)/p$, where p is the average capture probability of the capture of electrons into the GB during their transit. One of the main potential advantages of GB-HETs is the high-speed operation associated with the combination of a short transit time (due to the vertical structure), a high gain g_0 (because of a low probability, p, of the hot electron capture into the GB), and a low GB resistance (owing to a high mobility of the carriers in graphene). As shown recently [17], p in the graphene heterostructures with proper inter-graphene barrier layers can be very small. Very small values of p in GB-HETs can lead to a fairly high gain. The use of the undoped base with the induced carriers leads to the exclusion of the scattering of the hot electrons crossing the GB with the impurities and to an increase of the mobility of the carriers localized in the GB.

By analogy with HETs with the QW-base made of the standard materials [18, 19], GB-HETs can exhibit resonant response to the incoming signals associated with the excitation of the plasma oscillations. The resonant plasma frequencies are determined by the characteristic plasma wave velocity s (which increases with the carrier density Σ_0) and the GB lateral sizes $2L$, while the quality factor of the plasma resonances is mainly limited by the carrier momentum relaxation time τ associated with the scattering on impurities, various imperfections, and phonons. Because of the specific features of the carrier statistics and dynamics in the graphene layers [20], the plasma velocity $s > v_W$, where $v_W \simeq 10^8$ cm/s is the characteristic velocity of the Dirac energy spectrum, and s can markedly exceed that in the QW heterostructures. This promotes the realization of the plasmonic resonances in the terahertz (THz) range even in the GB-HETs with fairly large lateral sizes. The possibility of achieving the elevated carrier mobilities in the GB can enable sharp plasmonic resonances in GB-HETs at room temperatures or above.

The application of the ac voltages $\pm\delta V_\omega/2$ at the signal frequency ω, associated with the incoming radiation (received by an antenna) to the side contacts to the GB, results in the variation of the local ac potential difference $\delta\varphi_\omega = \delta\varphi_\omega(x)$ (the axis x is directed in the GB plane) between the GB and the emitter contact. The emitter, base, and collector current densities j_E, j_B, and j_C include the dc components $j_{E,0}$, $j_{B,0}$, and $j_{C,0}$ (determined by the applied bias voltages V_B and V_C), the ac components $\delta j_{E,\omega}$, $\delta j_{B,\omega}$, and $\delta j_{C,\omega}$ (proportional to δV_ω, and (due to the nonlinear dependence of the tunneling injection current on the local potential difference between the GB and the emitter) the rectified dc components $\overline{\delta j_{E,\omega}}$, $\overline{\delta j_{B,\omega}}$, and $\overline{\delta j_C}$, (proportional to $|\delta V_\omega|^2$, i.e., to the intensity of the incoming radiation received by an antenna).

The net rectified current $\overline{\delta j_{C,\omega}}$ can serve as the output current in the GB-HETs radiation detectors. The detector responsivity $R_\omega \propto \overline{\delta j_{C,\omega}}/|\delta V_\omega|^2$. Due to the possibility of the plasmonic resonances, the rectified component (the detector output signal) can be resonantly large, similar to that in the HET detector [21] and other plasmonic THz detectors using different transistor structures, including those incorporating graphene [22–34].

In this chapter, we develop the GB-HET device model and evaluate the GB-HET characteristics as a radiation detector of radiation, in particular, in the THz range of signal frequencies.

48.2 Device Model and Related Equations

We consider a GB-HET with the highly conducting tunneling emitter and collector of the n-type, the undoped GB, and the barrier layers between the emitter and the GB and between the GB and the collector made of the materials like hBN and WS_2 (as shown in Fig. 48.1) with a 2DHG induced by the applied bias voltages. Such a GB-HET has the N-p-N structure with the band diagram shown in Fig. 48.1b. The GB-HET device model accounts for the tunneling injection of hot electrons from the emitter to the barrier layer (above the barrier top), their propagation across the barrier layers, partial capture of hot-electrons into the GB, and the excitation of the self-consistent oscillations of the electric potential and the hole density in the GB (plasma oscillations). The quantum capacitance and the electron transit-time effects are also taken into account.

We assume that the bias voltages V_B and V_C are applied between the GL base and the emitter and between the collector and the emitter contacts, respectively. In the framework of the gradual channel approximation [35], which is valid if W_E, $W_C \ll 2L$, the 2DHG density $\Sigma = \Sigma(x, t)$ in the GL base and its local potential $\varphi = \varphi(x, t)$ (counted from the potential of the emitter) are related to each other as

$$\Sigma = \frac{\kappa}{4\pi e}\left(\frac{\varphi - \mu/e - V_{bi}}{W_E} + \frac{\varphi - \mu/e - V_{bi} - V_C}{W_C}\right), \tag{48.1}$$

where κ is the dielectric constant of these layers, e is the hole charge, V_{bi} is the built-in voltage between the contact material and an undoped GL, and μ is the 2DHG Fermi energy in the GB. In a degenerate 2DHG, $\mu = \hbar v_W \sqrt{\pi\Sigma}$, where $\hbar$ is the Planck constant. The dependence of the right side of Eq. (48.1) on the hole Fermi energy is interpreted as the effect of quantum capacitance [36, 37].

The ac voltages $\pm\delta V_\omega/2$ and ω are applied between the side GB contacts connected with an antenna, so that the ac potential of the GB $\delta\varphi_\omega = \delta\varphi_\omega(x)$ obeys the following (asymmetric) conditions:

$$\delta\psi_\omega|_{x=\pm L} = \pm\delta V_\omega/2. \tag{48.2}$$

The side contacts can serve as the slot wave guide transforming the incoming THz radiation signals being received by an antenna into the ac voltage (see Fig. 48.1a). The ac component of the collector-emitter voltage $\delta V_{C,\omega}$ can also arise due to the ac potential drop across the load resistance. However, in the GB-HET with the wiring under consideration, $\delta V_{C,\omega}$ can be disregarded providing that the GB-HET structure is symmetrical (see following text). For GB-HETs with the degenerate 2DHG, the relation between the variations of the hole density and the potential, $\delta\Sigma_\omega$ and $\delta\varphi_\omega$ (the ac components), as follows from Eq. (48.1), can be expressed via the net capacitance per unit area $C = C_g C_{quant}/(C_{quant} + C_g)$, which accounts for the geometrical capacitance $C_g = (C_E + C_C) = (\kappa/4\pi)(W_E^{-1} + W_C^{-1})$ (with $C_E \propto W_E^{-1}$ and $C_C \propto W_C^{-1}$ being the geometrical emitter and collector capacitances, respectively) and the quantum capacitance [36, 37]. $C_{quant} = (2e^2\sqrt{\Sigma_0}/\sqrt{\pi}\hbar v_W) = (2e^2\mu_0/\pi\hbar^2 v_W^2)$, where Σ_0 and μ_0 are the pertinent dc values of the hole density and the Fermi energy

$$\varepsilon\delta\Sigma_\omega = C\delta\varphi_\omega, \tag{48.3}$$

Considering the electron tunneling from the emitter to the states above the top of the barrier (through the triangular barrier), the emitter electron tunneling current density j_E can be presented as

$$j_E = j_E^t \exp\left(-\frac{F}{F_E}\right). \tag{48.4}$$

Here $F = (a\sqrt{m}\Delta^{3/2}/e\hbar)$ is the characteristic tunneling field, Δ is the activation energy for electrons in the contact, m is the effective electron mass in the barrier material, $a \sim 1$ is a numerical coefficient, $F_E = (\varphi - V_{bi} - \mu/e)/W_E$ is the electric field in the emitter barrier, and j_E^t is the maximum current density which can be provided the emitter contact, As follows from Eqs. (48.2) and (48.3), the ac component and the rectified component of the emitter tunneling current density $\delta j_{\mathrm{E},\omega}$ and $\overline{\delta j_{\omega}}$ (for $F \gg F_0$) are, respectively, given by

$$\delta j_{\mathrm{E},\omega} = j_{\mathrm{E},0}\frac{F}{F_{\mathrm{E},0}^2}\delta F_{\mathrm{E},\omega} = \sigma_{\mathrm{E}}\delta F_{\mathrm{E},\omega}, \quad (48.5)$$

$$\overline{\delta j_{\mathrm{E},\omega}} \simeq \frac{j_{\mathrm{E},0}}{2}\left[\frac{1}{2}\left(\frac{F}{F_{\mathrm{E},0}}\right)^2 - \left(\frac{F}{F_{\mathrm{E},0}}\right)\right]\left|\frac{\delta F_{\mathrm{E},\omega}}{F_{\mathrm{E},0}}\right|^2$$

$$\simeq \frac{\sigma_{\mathrm{E}}F}{4}\left|\frac{\delta F_{\mathrm{E},\omega}}{F_{\mathrm{E},0}}\right|^2. \quad (48.6)$$

Here $j_{\mathrm{E},0} = j_{\mathrm{E}}^{\mathrm{t}}\exp(-F/F_{\mathrm{E},0})$, $F_{\mathrm{E},0} = (V_{\mathrm{E}} - V_{\mathrm{bi}} - \mu_0/e)W_{\mathrm{E}}$, and $\sigma_{\mathrm{E}} = j_0(F/F_{\mathrm{E},0}^2)$ are the emitter dc density (in the absence of the ac signals), the dc electric field in the emitter barrier, and the emitter differential conductance, respectively. The dc hole Fermi energy in the GB μ_0 obeys the following equation:

$$\mu_0^2 = \frac{\kappa\hbar^2 v_{\mathrm{W}}^2}{4e}\left[\left(V_{\mathrm{E}} - V_{\mathrm{bi}} - \frac{\mu_0}{e}\right)\left(\frac{1}{W_{\mathrm{E}}} + \frac{1}{W_{\mathrm{C}}}\right) - \frac{V_{\mathrm{C}}}{W_{\mathrm{C}}}\right]. \quad (48.7)$$

The ac electric field components perpendicular to the GB plane in the emitter barrier, as well as in the collector barrier, are, respectively, given by

$$\delta F_{\mathrm{E},\omega} = \frac{C_{\mathrm{quant}}}{(C_{\mathrm{quant}} + C_{\mathrm{g}})}\frac{\delta\varphi_{\omega}}{W_{\mathrm{E}}}, \quad (48.8)$$

$$\delta F_{\mathrm{C},\omega} = -\frac{C_{\mathrm{quant}}}{(C_{\mathrm{quant}} + C_{\mathrm{g}})}\frac{\delta\varphi_{\omega}}{W_{\mathrm{C}}}. \quad (48.9)$$

Using the Shockley-Ramo theorem, one can find the ac component the electron current densities coming to the base and the collector

$$\delta j_{\mathrm{EB},\omega} = \sigma_{\mathrm{EB},\omega}\delta F_{\mathrm{E},\omega},\ \delta j_{\mathrm{EC},\omega} = \sigma_{\mathrm{EC},\omega}\delta F_{\mathrm{E},\omega}. \quad (48.10)$$

Here

$$\sigma_{\mathrm{EB},\omega} = \sigma_{\mathrm{E}}\left[\frac{1}{W_{\mathrm{E}}}\int_0^{W_{\mathrm{E}}} dz e^{i\omega z/v} - \frac{(1-p)}{W_{\mathrm{C}}}\int_{W_{\mathrm{E}}}^{W_{\mathrm{E}}+W_{\mathrm{C}}} dz e^{i\omega z/v}\right]$$

$$= \frac{\sigma_{\mathrm{E}}}{i\omega}\left[\frac{e^{i\omega\tau_{\mathrm{E}}} - 1}{\tau_{\mathrm{E}}} - \frac{(1-p)e^{i\omega\tau_{\mathrm{E}}}(e^{i\omega\tau_{\mathrm{C}}} - 1)}{\tau_{\mathrm{C}}}\right], \quad (48.11)$$

$$\sigma_{\mathrm{EC},\omega} = \sigma_{\mathrm{E}} \frac{(1-p)}{W_{\mathrm{C}}} \int_{\mathrm{W_E}}^{\mathrm{W_E+W_C}} dz e^{i\omega z/\upsilon}$$

$$= \frac{\sigma_{\mathrm{E}}}{i\omega} \frac{(1-p)e^{\mathrm{i}\omega\tau_{\mathrm{E}}}(e^{\mathrm{i}\omega\tau_{\mathrm{C}}}-1)}{\tau_{\mathrm{C}}}, \quad (48.12)$$

where υ is the drift velocity of the hot electrons crossing the barriers above their tops (which is assumed to be constant) and $\tau_{\mathrm{E}} = W_{\mathrm{E}}/\upsilon$ and $\tau_{\mathrm{C}} = W_{\mathrm{C}}/\upsilon$ are the electron transit time across the emitter and collector barrier layers. The axis z is directed perpendicular to the GB plane. Because under the boundary conditions (48.2), $\delta\varphi_\omega(x) = -\delta\varphi_\omega(-x)$ (see following text) and, hence, $\delta F_{\mathrm{E},\omega}(x) = -\delta F_{\mathrm{E},\omega}(x)$, the net ac

$$\delta J_{\mathrm{EB},\omega} = \int_{-\mathrm{L}}^{\mathrm{L}} dx \delta j_{\mathrm{EB},\omega} = \sigma_{\mathrm{EB},\omega} \int_{-\mathrm{L}}^{\mathrm{L}} dx \delta F_{\mathrm{E},\omega} = 0, \quad (48.13)$$

$$\delta J_{\mathrm{EC},\omega} = \int_{-\mathrm{L}}^{\mathrm{L}} dx \delta j_{\mathrm{EC},\omega} = \sigma_{\mathrm{EC},\omega} \int_{-\mathrm{L}}^{\mathrm{L}} dx \delta F_{\mathrm{E},\omega} = 0. \quad (48.14)$$

Simultaneously for the rectified components of the dc densities from the emitter to the GB and from the emitter to the collector, one obtains

$$\overline{\delta j_{\mathrm{EB},\omega}} \simeq \frac{p\sigma_{\mathrm{E}} F}{4} \left| \frac{\delta F_{\mathrm{E},\omega}}{F_{\mathrm{E},0}} \right|^2 = p\Gamma |\delta\varphi_\omega|^2, \quad (48.15)$$

$$\overline{\delta j_{\mathrm{EC},\omega}} \simeq \frac{(1-p)\sigma_{\mathrm{E}} F}{4} \left| \frac{\delta F_{\mathrm{E},\omega}}{F_{\mathrm{E},0}} \right|^2 = (1-p)\Gamma |\delta\varphi_\omega|^2, \quad (48.16)$$

where

$$\Gamma = \frac{\sigma_{\mathrm{E}} F}{4F_{\mathrm{E},0}^2 W_{\mathrm{E}}^2} \frac{C_{\mathrm{quant}}^2}{(C_{\mathrm{quant}} + C_{\mathrm{g}})^2}. \quad (48.17)$$

Consequently, the net rectified components of the dc emitter-base and emitter-collector currents are given by

$$\overline{\delta J_{\mathrm{EB},\omega}} = \int_{-\mathrm{L}}^{\mathrm{L}} dx \overline{\delta j_{\mathrm{EB},\omega}} \simeq p\Gamma \int_{-\mathrm{L}}^{\mathrm{L}} dx |\delta\varphi_\omega|^2, \quad (48.18)$$

$$\overline{\delta J_{\mathrm{EC},\omega}} = \int_{-\mathrm{L}}^{\mathrm{L}} dx \overline{\delta j_{\mathrm{EC},\omega}} \simeq (1-p)\Gamma \int_{-\mathrm{L}}^{\mathrm{L}} dx\, |\delta\varphi_{\omega}|^2, \tag{48.19}$$

respectively.

The ac hole current along the GB is given by

$$\delta J_{\mathrm{BB},\omega} = -\sigma_{\mathrm{BB},\omega} \left. \frac{d\delta\varphi_{\omega}}{dx} \right|_{\mathrm{x=L}}, \tag{48.20}$$

where the lateral ac conductivity of the GB $\sigma_{\mathrm{BB},\omega}$ is given by (see, for example, Refs. [38–42])

$$\sigma_{\mathrm{BB},\omega} = \frac{ie^2 \mu_0}{\pi\hbar(\omega + i/\tau)}. \tag{48.21}$$

48.3 Plasma Oscillations and Rectified Current

To calculate the rectified current components using Eqs. (48.15) and (48.16), one needs to find the spatial distributions of the ac potentials in the GB $\delta\varphi_{\omega} = \delta\varphi_{\omega}(x)$.

For this purpose, we use the hydrodynamic equations for the hole transport along the GB [43, 44] (see also Refs. [21, 22]) coupled with the Poisson equation solved using the gradual channel approximation [i.e., using Eq. (48.1)]. Linearizing the hydrodynamic equations and Eq. (48.1) and taking into account that the ac component of the hole Fermi energy is expressed via the variation of their density and the potential, we arrive at the following equation for the ac component of the GB potential $\delta\varphi_{\omega}$ (compare with Refs. [16, 21, 22]), which should be solved with boundary conditions given by Eq. (48.2):

$$\frac{d^2\delta\varphi_{\omega}}{dx^2} + \frac{(\omega + i\nu)(\omega + i\bar{\nu})}{s^2}\delta\varphi_{\omega} = 0. \tag{48.22}$$

Here s is the characteristic velocity of the plasma waves in the gated graphene layers, which is given by $s = \sqrt{e^2\Sigma_0/mC} \propto \Sigma_0^{1/4}$ [18], $m = \mu_0/v_{\mathrm{W}}^2 \propto \sqrt{\Sigma_0}$ being the so-called "fictitious" effective hole (electron) mass in graphene layers [27], $\bar{\nu} = (\sigma_{\mathrm{B},\omega}/W_{\mathrm{E}}C_{\mathrm{g}})$ and $\nu = 1/\tau + \tilde{\nu}$, where τ is the hole momentum relaxation time in the 2DHG, and $\tilde{\nu} = \tilde{\nu}_{visc} + \tilde{\nu}_{\mathrm{rad}}$ is associated with the contribution of the 2DHG viscosity to the damping (see, for example, Ref. [21]) and with the radiation damping of the plasma oscillations. The latter mechanism is associated with the recoil that the holes in the GB feel emitting radiation (the pertinent term in the force acting on the holes is referred to as the Abraham-Lorentz force or the radiation reaction force [45, 46]). Taking into account that the viscosity damping rate is proportional to the second spatial derivative, for the gated plasmons with the acoustic-like spectrum $\tilde{\nu}_{\rm visc} = \xi\omega^2/s^2$ (ξ is the 2DHG viscosity) and considering that $\tilde{\nu}_{\rm rad} \propto (2e^2/3mc^3)\omega^2$ (Refs. [45–47]), we put $\nu = 1/\tau + \eta\omega^2$), where η is the pertinent damping parameter and c is the speed of light. If the viscosity damping surpasses the radiation damping, one can set $\eta \simeq \zeta/s^2$.

The characteristic plasma-wave velocity s is determined by Σ_0 (as well as the thicknesses of the emitter and collector barrier layers $W_{\rm E}$ and $W_{\rm C}$) [20] and, hence, can be changed by the variations of the bias voltages $V_{\rm E}$ and $V_{\rm B}$. Because of this the characteristic, plasmonic frequency $\Omega = \pi s/L$ can be effectively controlled by these voltages.

Equation (48.22) with Eq. (48.2) yields

$$\delta\varphi_\omega = \frac{\sin\left[\sqrt{(\omega+i\nu)(\omega+i\bar{\nu})}x/s\right]}{\sin\left[\sqrt{(\omega+i\nu)(\omega+i\bar{\nu})}L/s\right]}\frac{\delta V_\omega}{2}. \quad (48.23)$$

One can see from Eq. (48.23) that the spatial dependence of $\delta\varphi_\omega$ is rather complex. In particular, when ω approaches to $n\Omega$, where $n = 1, 2, 3, \ldots$, this distribution can be oscillatory with fairly high amplitude of the spatial oscillations when $\Omega \gg \nu$.

Substituting $\delta\varphi_\omega$ given by Eq. (48.23) into Eqs. (48.15) and (48.16), we arrive at the following:

$$\overline{\delta j_{{\rm EB},\omega}} \simeq p\Gamma \times \left|\frac{\sin\left[\sqrt{(\omega+i\nu)(\omega+i\bar{\nu})}x/s\right]}{\sin\left[\sqrt{(\omega+i\nu)(\omega+i\bar{\nu})}L/s\right]}\right|^2 \frac{|\delta V_\omega|^2}{4}, \quad (48.24)$$

$$\overline{\delta j_{{\rm EC},\omega}} \simeq (1-p)\Gamma \times \left|\frac{\sin\left[\sqrt{(\omega+i\nu)(\omega+i\bar{\nu})}x/s\right]}{\sin\left[\sqrt{(\omega+i\nu)(\omega+i\bar{\nu})}L/s\right]}\right|^2 \frac{|\delta V_\omega|^2}{4}, \quad (48.25)$$

$$\overline{\delta J_{{\rm EB},\omega}} \simeq p\Gamma \times \int_{-L}^{L} dx \left|\frac{\sin\left[\sqrt{(\omega+i\nu)(\omega+i\bar{\nu})}x/s\right]}{\sin\left[\sqrt{(\omega+i\nu)(\omega+i\bar{\nu})}L/s\right]}\right|^2 \frac{|\delta V_\omega|^2}{4}, \quad (48.26)$$

$$\overline{\delta J_{\mathrm{EC},\omega}} \simeq (1-p)\Gamma \times \int_{-L}^{L} dx \left| \frac{\sin\left[\sqrt{(\omega+i\nu)(\omega+i\bar{\nu})}x/s\right]}{\sin\left[\sqrt{(\omega+i\nu)(\omega+i\bar{\nu})}L/s\right]} \right|^2 \frac{|\delta V_\omega|^2}{4}, \quad (48.27)$$

As follows from Eqs. (48.24)–(48.27), $\overline{\delta j_{\mathrm{EC},\omega}}$ = [(1 - p)/p] $\overline{\delta j_{\mathrm{EB},\omega}}$ and $\overline{\delta J_{\mathrm{EC},\omega}}$ = [(1 - p)/p] $\overline{\delta J_{\mathrm{EB},\omega}}$ (as for the pertinent dc in the absence of the THz signals), so that $\overline{\delta J_{\mathrm{EC},\omega}} \gg \overline{\delta J_{\mathrm{EB},\omega}}$.

Using Eqs. (48.20) and (48.23) for the ac between the GB contacts, we obtain

$$\delta J_{\mathrm{BB},\omega} = -\frac{ie^2\mu_0 \cot\left[\sqrt{(\omega+i\nu)(\omega+i\bar{\nu})}L/s\right]}{\pi\hbar s} \times \sqrt{\frac{\omega+i\bar{\nu}}{w+i\nu}} \frac{\delta V_\omega}{2}. \quad (48.28)$$

Due to the symmetry of the GB-HET structure and the asymmetric spatial distribution of $\delta\varphi_\omega$, there is no rectified component of the lateral current in the GB-base, i.e., $\overline{\delta J_{\mathrm{BB},\omega}}$ = 0.

At very low signal frequencies $\omega \ll \nu$, $|\bar{\nu}| \simeq 4\pi p \sigma_{\mathrm{E}}/\kappa$, from Eqs. (48.23) and (48.27) we obtain

$$\delta\varphi_\omega \simeq \frac{\sinh\left(\sqrt{\nu\bar{\nu}}x/s\right)}{\sinh\left(\sqrt{\nu\bar{\nu}}L/s\right)} \frac{\delta V_\omega}{2}, \quad (48.29)$$

$$\overline{\delta J_{\mathrm{EC},\omega}} \simeq (1-p)\Gamma L \left[\frac{\dfrac{\sinh\left(2\sqrt{\nu\bar{\nu}}L/s\right)}{\left(2\sqrt{\nu\bar{\nu}}L/s\right)} - 1}{\sinh^2\left(\sqrt{\nu\bar{\nu}}L/s\right)} \right] \frac{|\delta V_\omega|^2}{4}$$

$$\simeq (1-p)\Gamma L \frac{|\delta V_\omega|^2}{6} \quad (48.30)$$

Because in reality $|\bar{\nu}| \ll \nu$, there is an intermediate range of frequencies $|\bar{\nu}| \ll \omega < \nu$. Assuming that $|\bar{\nu}| \ll \nu$, $\Omega^2/\omega < \nu$, from Eq. (48.27) we arrive at

$$\overline{\delta J_{\text{EC},\omega}} \simeq (1-p)\Gamma L\left(\frac{s}{L}\sqrt{\frac{2}{\nu\omega}}\right)\frac{|\delta V_\omega|^2}{4}$$

$$\propto (1-p)\sqrt{\frac{\Omega^2}{\nu\omega}}\,|\delta V_\omega|^2\,. \qquad (48.31)$$

If the characteristic frequency of the plasma oscillations $\Omega = \pi s/L \gg \nu$, the ac GB potential amplitude $|\delta\varphi_\omega|$ can markedly exceed the amplitude of the input ac signal δV_ω when the frequency is close to one of the resonant plasma frequencies $n\Omega$, where $n = 1, 2, 3, \ldots$ is the plasma resonance index. In this case, the rectified emitter-collector current is pronouncedly stratified, i.e., its density $\overline{\delta\, j_{\text{EC},\omega}}$ is a nearly periodic function of the coordinate x.

Figure 48.2 shows the spatial distributions of the rectified emitter-collector current density (corresponding to the current stratification) calculated for different frequencies using Eq. (48.25). It is assumed that $\tau = 1$ ps, $s = 2.5 \times 10^8$ cm/s, and $L = 1.5$ μm. As seen in Fig. 48.2, the spatial stratification of the current is rather pronounced when ω is close to the plasma resonant frequencies (the frequencies $\Omega/2\pi = 5/6 \simeq 0.83$ THz and $2\Omega/2\pi = 5/3 \simeq 1.66$ THz): the current exhibits two streams centered at $|x|/L \simeq 0.5$ when $\omega \simeq 0.83$ THz and four streams centered at $|x|/L \simeq 0.25$ when $\omega \simeq 1.66$ THz. At the frequencies far from the resonances, the spatial current distribution becomes weakly nonuniform.

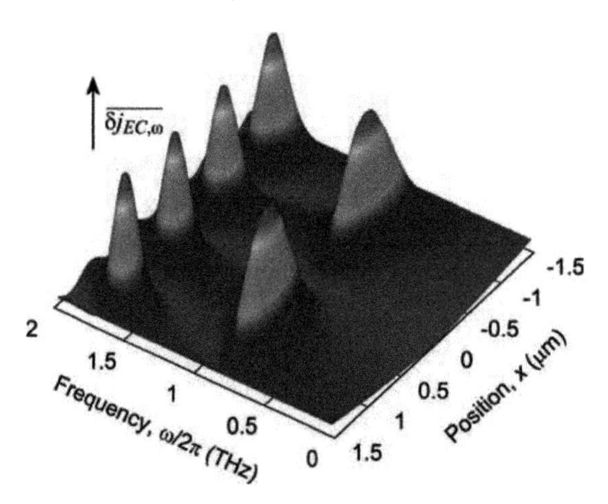

Figure 48.2 Spatial distributions of rectified component of the emitter-collector current density (in arbitrary units) $\overline{\delta j_{\text{EC},\omega}}$ at different frequencies ω ($\tau = 1$ ps, $s = 2.5 \times 10^8$ cm/s, and $L = 1.5$ μm).

48.4 GB-HET Detector Responsivity

48.4.1 Current Responsivity

The rectified current $\overline{\delta J_{\mathrm{EC},\omega}}$ can be considered as the output signal used for the detection of electromagnetic radiation (in particular, THz radiation). The current detector responsivity using this output signal is defined as

$$R_{\omega} = \frac{\overline{\delta J_{\mathrm{EC},\omega}} H}{SI_{\omega}}, \tag{48.32}$$

where H is the GB-HET lateral size in the direction along the contacts to the GB, I_{ω} is intensity of the incident radiation, and S is the antenna aperture. The latter is given by $S = \lambda_{\omega}^2 G/4\pi$ [47, 48], where G is the antenna gain, $\lambda_{\omega} = 2\pi c/\omega$ is the radiation wavelength, and c is the speed of light in vacuum. Taking into account that $I_{\omega} = c\mathcal{E}_{\omega}^2/8\pi$, where $\mathcal{E}_{\omega}$ is the radiation electric field in vacuum, and estimating δV_{ω} as $\delta V_{\omega} = \lambda_{\omega}\mathcal{E}_{\omega}/\pi$, one can arrive at

$$|\delta V_{\omega}|^2 = \frac{8\lambda_{\omega}^2 I_{\omega}}{\pi c}. \tag{48.33}$$

Considering Eqs. (48.27), (48.32), and (48.33), for the current responsivity (in A/W units), we find

$$R_{\omega} = \frac{R}{L}\int_{-L}^{L} dx \left| \frac{\sin\left[\sqrt{(\omega + i\nu)(\omega + i\bar{\nu})}x/s\right]}{\sin\left[\sqrt{(\omega + i\nu)(\omega + i\bar{\nu})}L/s\right]} \right|^2, \tag{48.34}$$

where

$$R \simeq \frac{8(1-p)\Gamma LH}{cG} = \rho\frac{LH}{W_{\mathrm{E}}^2}, \tag{48.35}$$

$$\rho = \frac{8(1-p)j_0}{cG}\left(\frac{F}{F_{\mathrm{E},0}^2}\right)^2. \tag{48.36}$$

As follows from Eq. (48.34), at relatively long hole momentum relaxation times τ, the GB-HET responsivity exhibits a series of very sharp and high peaks. It is instructive that in the GB-HETs under consideration, the height of the resonant peaks does not decrease with an increasing peak index (as in some other devices using the plasmonic resonances). Although some decrease in the peaks height with the increasing resonance index attributed to the effect of viscosity and radiative damping takes place, the pertinent effect is relatively weak (see following text).

In the limiting cases $\omega \ll \nu$, $|\bar{v}|$ and $|\bar{v}| \ll \nu$, $\Omega^2/\omega < \nu$, corresponding to Eqs. (48.25) and (48.26), one obtains

$$R_\omega = R_0 \simeq \frac{2}{3}R, \tag{48.37}$$

and

$$R_\omega \simeq R\left(\frac{s}{L}\sqrt{\frac{2}{\nu\omega}}\right) = R\sqrt{\frac{2\Omega^2}{\pi^2\nu\omega}} < R, \tag{48.38}$$

respectively. The quantities R and R_ω depend on the geometrical and quantum capacitances and, therefore, on the barrier layers thicknesses

$$R_\omega \propto R \propto \frac{1}{W_E^2}\frac{C_{quant}^2}{(C_{quant}+C_g)^2}$$

$$= \frac{1}{W_E^2}\left[1+\frac{\kappa}{4\pi\, C_{quant}(W_E^{-1}+W_C^{-1})}\right]^{-2}. \tag{48.39}$$

For the doping level of the emitter contact $N_D = 10^{18}$ cm^{-3}, assuming that the thermal electron velocity $\upsilon_T = 10^7$ cm/s, we obtain $j_E^t = 1.6 \times 10^6$ A/cm^2. Setting $\Delta = 0.2$ eV and the effective mass in the barrier layer $m \simeq 2.5 \times 10^{-28}$ g, we arrive at $F \sim 2 \times 10^6$ V/cm. Setting also $F_{E,0} = 5 \times 10^5$ V/cm (to provide the hole density in the GB about of $\Sigma = 10^{12}$ cm^{-2}), we find $j_{E,0} \simeq 2.9 \times 10^4$ A/cm^2 and $\sigma_E \simeq 0.23\ \Omega^{-1}$cm^{-1}. At these parameters, one obtains also $C_{quant} \simeq 2.6 \times 10^6$ cm^{-1}. Using these data and setting in addition $p \ll 1$, $W_E W_C/(W_E + W_C) = 5$ nm, $\kappa = 4$, and $G = 1.5$, from Eq. (48.36) we obtain $\rho \simeq 2 \times 10^{-4}$ A/W.

If $L = 1.5$ μm and $H = 10$ μm, Eqs. (48.35) and (48.36) yield $R \simeq 30$ A/W and $R_0 \simeq 20$ A/W. These parameters are also used in the estimates in the following text.

In the case of high quality factor of the plasma resonances Ω/ν, the quantities $\overline{\delta J_{EB,\omega}}$ and $\overline{\delta J_{EC,\omega}}$ and, hence, the responsivity as functions of the signal frequency ω described by Eqs. (48.26), (48.27), and (48.34) exhibit sharp peaks at $\omega \simeq n\Omega$, attributed to the resonant excitation of the plasma oscillations (standing plasma waves). The peak width is primarily determined by the frequency ν. Indeed, for $\Omega/\nu \gg 1$ Eq. (48.34) yields

$$\max R_\omega \simeq R_\Omega \simeq \frac{3R_0}{2}\left(\frac{2\Omega}{\pi\nu}\right)^2 \gg R_0. \tag{48.40}$$

Using the preceding estimate for R_0, the peak values of the responsivity at $\Omega = 5/3$ THz, and $\tau = 1$ ps is approximately equal to $\max R_\omega \simeq R_\Omega \simeq 1.33 \times 10^3$ A/W.

Figure 48.3 shows the frequency dependence of the normalized GB-HET current responsivity calculated using Eq. (48.34) for several sets of parameters. As seen, the positions of the responsivity peaks shift toward higher frequencies when the plasma frequency increases, i.e., when s increases and/or L decreases because $\Omega \propto s/L$ (compare the curves "1" and "2"). A shortening of the momentum relaxation time τ leads to a smearing of the peaks (compare the curve "3" corresponding to $\tau = 1$ ps and the curve "4" corresponding to $\tau = 0.5$ ps). Figure 48.4 shows the lowering and broadening of the resonance peaks of the GB-HET current responsivity with the decreasing momentum relaxation time τ described by Eq. (48.34). As seen, at $\tau < 0.3 - 0.4$ ps, the responsivity peaks vanish, while at $\tau =$ 1 ps they are fairly sharp and high.

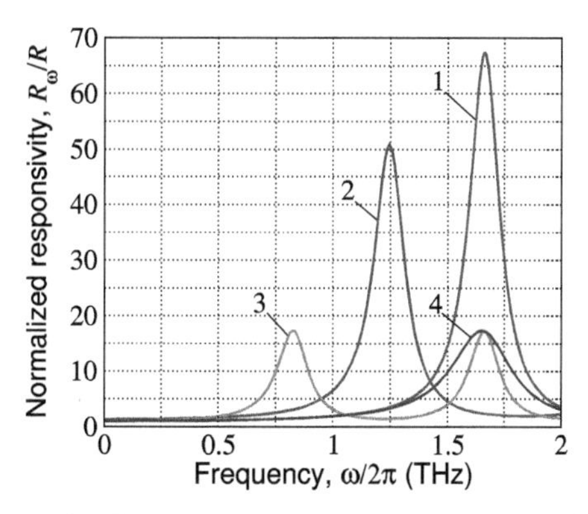

Figure 48.3 Normalized responsivity versus signal frequency for GB-HETs with different parameters: 1 – $\tau = 1$ ps, $s = 5 \times 10^8$ cm/s, $L = 1.5$ μm; 2 – $\tau = 1$ ps, $s = 5 \times 10^8$ cm/s, $L = 2.0$ μm; 3 – $\tau = 1$ ps, $s = 2.5 \times 10^8$ cm/s, $L = 1.5$ μm; 4 – $\tau = 0.5$ ps, $s = 5 \times 10^8$ cm/s, $L = 1.5$ μm. These parameters correspond to $\Omega/2\pi = 5/3$, 5/4, 5/6, and 5/3 THz, respectively.

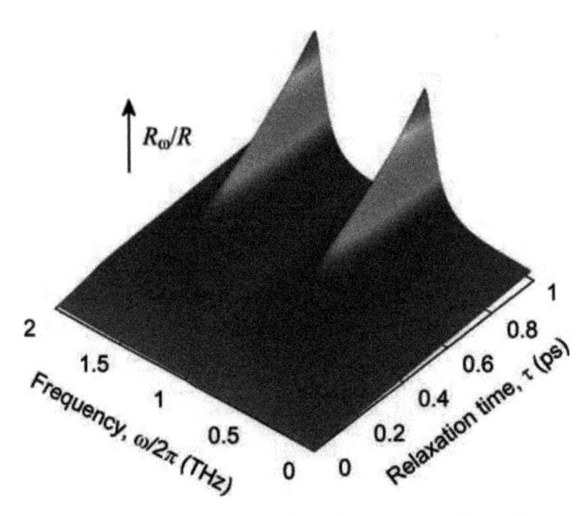

Figure 48.4 Normalized responsivity (in arbitrary units) as a function of signal frequency and momentum relaxation time ($s = 2.5 \times 10^8$ cm/s and $L = 1.5$ μm, i.e., $\Omega/2\pi = 5/6$ THz).

48.4.2 Voltage Responsivity

The variation of the dc component $\overline{\delta J_{\mathrm{EC},\omega}}$ caused by the ac signals results in a change of the voltage drop, $\overline{\delta V_\omega} = -\delta V_{\mathrm{C},0}$, across the load resistor in the collector circuit (see Fig. 48.1). Considering that this leads to an extra variation of the dc emitter and collector currents $\overline{\delta J_{\mathrm{E},0}} = [\sigma_\mathrm{E} C_\mathrm{g}/(C_{\mathrm{quant}} + C_\mathrm{g})(W_\mathrm{E} + W_\mathrm{C})]\delta V_{\mathrm{C},0}$ and $\overline{\delta J_{\mathrm{EC},0}} = [(1 - p)\sigma_\mathrm{E} C_\mathrm{g}/(C_{\mathrm{quant}} + C_\mathrm{g})]\delta V_{\mathrm{C},0}$ because of its dependence on the collector contact dc potential. The latter dependence is essentially associated with the effect of quantum capacitance. Taking this into account, for the voltage responsivity of the GB-HET under consideration $R_\omega^\mathrm{V} = \overline{\delta V_\omega} H/SI_\omega = \delta V_{\mathrm{C},0} H/SI_\omega$, we obtain

$$R_\omega^\mathrm{V} = \frac{R^\mathrm{V}}{L}\int_{-L}^{L} dx \left|\frac{\sin\left[\sqrt{(\omega + i\nu)(\omega + i\overline{\nu})}x/s\right]}{\sin\left[\sqrt{(\omega + i\nu)(\omega + i\overline{\nu})}L/s\right]}\right|^2. \tag{48.41}$$

Here

$$R^\mathrm{V} = R\frac{r_\mathrm{C}}{\left[1 + \frac{C_\mathrm{g}}{(C_{\mathrm{quant}} + C_\mathrm{g})}\frac{2LH}{(W_\mathrm{E} + W_\mathrm{C})}r_\mathrm{C}\sigma_\mathrm{E}\right]}, \tag{48.42}$$

where r_C is the load resistance. It is instructive that because of the absence of the ac collector current (i.e., the ac through the load resistor) associated with the GB-HET structure symmetry and the asymmetry of the applied ac signal voltage and the ac potential spatial distribution along the GB, the RC-factor of the voltage responsivity is independent of the signal frequency.

For $W_E = W_C = W$, Eqs. (48.41) and (48.42) yield

$$R_\omega^V = R_\omega \frac{r_C}{\left[1 + \dfrac{\sigma_E r_C (LH/W^2)W}{(1 + 2\pi\, C_{quant}\, W/\kappa)}\right]}. \quad (48.43)$$

At $r_C \ll (1 + 2\pi C_{quant} W/\kappa)(W/LH\sigma_E) = \bar{r}_C$, Eq. (48.43) yields the obvious formulas

$$R_0^V = R_0 r_C, \quad (48.44)$$

at low frequencies and

$$R_\Omega^V = \frac{3R_0}{2}\left(\frac{2\Omega}{\pi\nu}\right)^2 r_C, \quad (48.45)$$

at the plasma resonance $\omega = \Omega$.

At $r_C > (1 + 2\pi C_{quant} W/\kappa)(W/LH\sigma_E) = \bar{r}_C$, from Eq. (48.43) we obtain, respectively

$$R_\Omega^V \simeq R_0 \frac{(W^2/LH)}{\sigma_E W}\left(1 + \frac{2\pi\, C_{quant}\, W}{\kappa}\right), \quad (48.46)$$

and

$$R_\Omega^V \simeq \frac{3R_0}{2}\left(\frac{2\Omega}{\pi\nu}\right)^2 \frac{(W^2/LH)}{\sigma_E W}\left(1 + \frac{2\pi\, C_{quant}\, W}{\kappa}\right). \quad (48.47)$$

For $\tau = 1$ ps, $s = 5 \times 10^8$ cm/s, and $L = 1.5$ μm (as for the curve "1" in Fig. 48.3), so that $\Omega = 5/3$ THz, as well as $\sigma_E = 0.23$ A/V cm, $C_{quant} = 2.6 \times 10^6$ cm^{-1} (see the estimate in the previous subsection), and $\kappa = 4$ from Eq. (48.47) we obtain the following estimate: $R_\Omega^V \simeq 2 \times 10^5$ V/W. For the preceding parameters, one obtains $\bar{r}_C \simeq 150\ \Omega$. Even at smaller r_C, the voltage responsivity can be fairly large. Setting $r_C = 5 - 10\ \Omega$, we obtain $R_0^V \simeq 100 - 200$ V/W and $R_\Omega^V \simeq (6.6 - 13.2) \times 10^3$ V/W, respectively.

Considering Eqs. (48.42) and (48.39), we find the following dependences of the current and voltage responsivities R_ω and R_ω^V on the emitter and collector barriers thickness W (at not too large r_C):

$$R_{\omega}^{V} = R_{\omega} r_{C} \propto \frac{4\pi^2 C_{quant}^2}{\kappa^2 (1 + 2\pi\, C_{quant}\, W/\kappa)^2}. \tag{48.48}$$

According to Eq. (48.48), R_ω and R_{ω}^{V} markedly decrease in the range $W > \kappa/2\pi C_{quant} \simeq 10$ nm.

48.5 Discussion

As follows from Eqs. (48.34) and (48.40), an increase in ν with increasing frequency ω due to the reinforcement of the plasma oscillation damping associated with the viscosity and the radiative damping might lead to the gradual lowering of the resonant peaks with their index n. However, our estimates show that the contribution of these two mechanisms to the net damping is small compared to the damping associated with the hole momentum relaxation (collisional damping). Indeed, disregarding the radiative damping and assuming the 2DHG viscosity to be $\zeta = 10$ cm^2/s (i.e., smaller than in the standard 2DEG and 2DHG in the GaAs based heterostructures [21, 49, 50]) and $s = (2.5 - 5) \times 10^8$ cm/s, we obtain $\eta \simeq (4 - 16) \times 10^{-17}$ s. Hence in the frequency range $\omega/2\pi \leq 2$THz (as in Fig. 48.2), we find $\bar{\nu}_{visc} = \eta\omega^2 \leq (6.3 - 25.3) \times 10^9$ s$^{-1} \ll 1/\tau$. Therefore the heights of the responsivity peaks in Fig. 48.2 in the curve "3" at $\omega \simeq 0$ 8 and 1.6 THz are virtually equal. However, the peaks corresponding to higher resonances with the frequencies in the range 5–10 THz can be markedly lowered and smeared because in this range $\bar{\nu}_{visc}$ can become comparable with $1/\tau$. For example, for the same values of η, $\tau = 1$ ps, and $\omega/2\pi = 5$–10 THz, we obtain $\bar{\nu}_{visc}$ $\tau \simeq 0.16 - 0.64$.

Equation (48.42) describes the saturation of the voltage responsivity R_{Ω}^{V} peak value with increasing load resistance r_C. This is associated with the effect of the voltage drop across the load on the potential drop between the emitter and the base and, hence, the hole Fermi energy in the GB, which determines the injection current. Such an effect is due to the finite value of the GB quantum capacitance (see, Refs. [12, 13])—if C_{quant} tends to infinity, the emitter-base voltage becomes independent of r_C and the saturation of the R_{Ω}^{V} - r_C dependence vanishes.

The THz detectors using a similar operation principle and InP double HBTs (DHBTs) were recently fabricated and studied experimentally [51–53]. The estimated responsivities of the DHBTs in question for the non-resonant detection regime are somewhat smaller but of the same order of magnitude than those given by Eq. (48.43) and the pertinent estimates. However, the experimental values of the responsivity are much smaller than the values predicted in the preceding text for the resonant detection [see Eqs. (48.44) and (48.46) and the estimates based on these equations]. Apart from the parasitic effects and the absence of any spatial coupling antenna, this can be attributed to the doping of the base in the InP-DHBTs, which inevitably leads to relatively a shorter hole momentum relaxation time τ compared to that in the GB (where the 2DHG is induced by the applied voltages). Possibly, a higher probability of the hot electron capture into the InGaAs base (that had a relatively large thickness of 28 nm) in the DHBTs in comparison with the GB-HETs can be an additional factor.

The GB-HET current and voltage responsivities are determined by several characteristics: the characteristics of the tunneling emitter, geometrical characteristics of the GB-HET structure, materials of the emitter as well as the emitter and collector barrier layers, and applied bias voltages. The diversity of these factors enables the optimization of the GB-HETs operating as resonant plasmonic detectors, in particular, an increase in the responsivity in comparison with the values obtained in the preceding estimates. The resonant plasmonic THz detectors can based on not only the GB-HET structure shown in Fig. 48.1a (with an extra antenna connected to the GB side contacts) but also based on lateral structures with the GB contacts forming a periodic array.

The comparison of the GB-HETs [1–4] and InP-DHBTs [51–53] with the GB-HETs under consideration highlights the following advantages of the latter: (i) a longer momentum relaxation time of holes τ in the GB; (ii) a higher plasma-wave velocity s that enables higher resonant plasma frequencies; (iii) a smaller capture probability of hot electrons into the GB and, consequently, larger (or even much larger) fraction of the hot electrons reaching the collector; and (iv) coupling the incoming THz signal to the GB resulting in the absence of the ac in the emitter-collector circuit and preventing the RC effects usually hindering the high-frequency operation.

48.6 Conclusions

We developed an analytical model for vertical heterostructure HETs with the GB of the p-type sandwiched between the wide-gap emitter and collector layers and the N-type contacts. Using this model, we described the GB-HET dynamic properties and studied the GB-HET operation as detectors of THz radiation. The main features of the GB-HETs are high hole mobility in the GB, low probability the capture of the hot electrons injected from the emitter and crossing the GB, and the absence of the collector ac. These features enable pronounced voltage-controlled plasmonic response of the GB-HETs to the incoming THz radiation, high hot-electron injection efficiency, and the elimination of the RC-limitations leading to elevated GB-HET current and voltage responsivities in the THz range of frequencies, particularly at the plasmonic resonances at room temperature. This might provide the superiority of the GB-HET-based THz detectors over other plasmonic THz detectors based on the standard heterostructures. Thus the THz detectors based on the GB-HETs can be interesting for different applications.

Acknowledgments

The authors are grateful to D. Coquillat and F. Teppe for the information related to their experimental data on InP HBTs operating as THz detectors. The work was supported by the Japan Society for Promotion of Science (Grant-in-Aid for Specially Promoted Research 23000008) and by the Russian Scientific Foundation (Project No. 14-29-00277). The works at UB and RPI were supported by the U.S. Air Force Award No. FA9550-10-1-391 and by the U.S. Army Research Laboratory Cooperative Research Agreement, respectively.

References

1. W. Mehr, J. Ch. Scheytt, J. Dabrowski, G. Lippert, Y.-H. Xie, M. C. Lemme, M. Ostling, and G. Lupina, *IEEE Electron Device Lett.*, **33**, 691 (2012).
2. B. D. Kong, C. Zeng, D. K. Gaskill, K. L. Wang, and K. W. Kim, *Appl. Phys. Lett.*, **101**, 263112 (2012).

3. S. Vaziri, G. Lupina, C. Henkel, A. D. Smith, M. Ostling, J. Dabrowski, G. Lippert, W. Mehr, and M. C. Lemme, *Nano Lett.*, **13**, 1435 (2013).
4. C. Zeng, E. B. Song, M. Wang, S. Lee, C. M. Torres, Jr., J. Tang, B. H. Weiler, and K. L. Wang, *Nano Lett.*, **13**, 2370 (2013).
5. L. Britnel, R. V. Gorbachev, R. Jalil, B. D. Belle, F. Shedin, A. Mishenko, T. Georgiou, M. I. Katsnelson, L. Eaves, S. V. Morozov, N. M. R. Peres, J. Leist, A. K. Geim, K. S. Novoselov, and L. A. Ponomarenko, *Science*, **335**, 947 (2012).
6. T. Georgiou, R. Jalil, B. D. Bellee, L. Britnell, R. V. Gorbachev, S. V. Morozov, Y.-J. Kim, A. Cholinia, S. J. Haigh, O. Makarovsky, L. Eaves, L. A. Ponimarenko, A. K. Geim, K. S. Nonoselov, and A. Mishchenko, *Nat. Nanotechnol.*, **8**,100 (2013).
7. L. Britnel, R. V. Gorbachev, A. K. Geim, L. A. Ponomarenko, A. Mishchenko, M. T. Greenaway, T. M. Fromhold, K. S. Novoselov, and L. Eaves, *Nat. Commun.*, **4**,1794 (2013).
8. M. Liu, X. Yin, and X. Zhang, *Nano Lett.*, **12**, 1482 (2012).
9. C. A. Mead, *Proc. IRE*, **48**, 359 (1960).
10. J. M. Shannon, *IEE J. Solid-State Electron Devices*, **3**, 142 (1979).
11. M. Heiblum, D. C. Thomas, C. M. Knoedler, and M. I. Nathan, *Surf. Sci.*, **174**, 478 (1986).
12. S. Luryi, *IEEE Electron Device Lett.*, **6**, 178 (1985).
13. S. Luryi, in *High Speed Semiconductor Devices*, edited by S. M. Sze (Wiley, New York, 1990), p. 399.
14. J. Xu and M. S. Shur, Double base hot electron transistor, U.S. patent 4,901,122 (13 February 1990).
15. M. S. Shur, R. Gaska, A. Bykhovski, M. A. Khan, and J. W. Yang, *Appl. Phys. Lett.*, **76**, 3298 (2000).
16. M. Asada et al., *Jpn. J. Appl. Phys., Part 1*, **47**, 4375 (2008).
17. V. Ryzhii, T. Otsuji, M. Ryzhii, V. Ya. Aleshkin, A. A. Dubinov, V. Mitin, and M. S. Shur, *J. Appl. Phys.*, **117**, 154504 (2015).
18. V. Ryzhii, *Appl. Phys. Lett.*, **70**, 2532 (1997).
19. V. Ryzhii, Jpn. *J. Appl. Phys., Part 1*, **37**, 5937 (1998).
20. V. Ryzhii, A. Satou, and T. Otsuji, *J. Appl. Phys.*, **101**, 024509 (2007).
21. M. I. Dyakonov and M. S. Shur, *IEEE Trans. Electron Devices*, **43**, 1640 (1996).
22. W. Knap, Y. Deng, S. Rumyantsev, J.-Q. Lu, M. S. Shur, C. A. Saylor, and L. C. Brunel, *Appl. Phys. Lett.*, **80**, 3433 (2002).

23. X. G. Peralta, S. J. Allen, M. C. Wanke, N. E. Harff, J. A. Simmons, M. P. Lilly, J. L. Reno, P. J. Burke, and J. P. Eisenstein, *Appl. Phys. Lett.*, **81**, 1627 (2002).
24. T. Otsuji, M. Hanabe, and O. Ogawara, *Appl. Phys. Lett.*, **85**, 2119 (2004).
25. J. Lusakowski, W. Knap, N. Dyakonova, L. Varani, J. Mateos, T. Gonzales, Y. Roelens, S. Bullaert, A. Cappy, and K. Karpierz, *J. Appl. Phys.*, **97**, 064307 (2005).
26. F. Teppe, W. Knap, D. Veksler, M. S. Shur, A. P. Dmitriev, V. Yu. Kacharovskii, and S. Rumyantsev, *Appl. Phys. Lett.*, **87**, 052107 (2005).
27. V. Ryzhii, A. Satou, W. Knap, and M. S. Shur, *J. Appl. Phys.*, **99**, 084507 (2006).
28. A. El Fatimy, F. Teppe, N. Dyakonova, W. Knap, D. Seliuta, G. Valusis, A. Shcherepetov, Y. Roelens, S. Bollaert, A. Cappy, and S. Rumyantsev, *Appl. Phys. Lett.*, **89**, 131926 (2006).
29. J. Torres, P. Nouvel, A. Akwaoue-Ondo, L. Chusseau, F. Teppe, A. Shcherepetov, and S. Bollaert, *Appl. Phys. Lett.*, **89**, 201101 (2006).
30. S. A. Boubanga Tombet, Y. Tanimoto, A. Satou, T. Suemitsu, Y. Wang, H. Minamide, H. Ito, D. V. Fateev, V. V. Popov, and T. Otsuji, *Appl. Phys. Lett.*, **104**, 262104 (2014).
31. Y. Kurita, G. Ducournau, D. Coquillat, A. Satou, K. Kobayashi, S. A. Boubanga-Tombet, Y. M. Meziani, V. V. Popov, W. Knap, T. Suemitsu, and T. Otsuji, *Appl. Phys. Lett.*, **104**, 251114 (2014).
32. L. Vicarelli, M. S. Vitiello, D. Coquillat, A. Lombardo, A. C. Ferrari, W. Knap, M. Polini, V. Pellegrini, and A. Tredicucci, *Nat. Mater.*, **11**, 865 (2012).
33. V. Ryzhii, T. Otsuji, M. Ryzhii, and M. S. Shur, *J. Phys. D: Appl. Phys.*, **45**, 302001 (2012).
34. V. Ryzhii, A. Satou, T. Otsuji, M. Ryzhii, V. Mitin, and M. S. Shur, *J. Phys. D: Appl. Phys.*, **46**, 315107 (2013).
35. M. S. Shur, *Physics of Semiconductor Devices* (Prentice Hall, New Jersey, 1990).
36. S. Luryi, *Appl. Phys. Lett.*, **52**, 501 (1988).
37. T. Fang, A. Konar, H. Xing, and D. Jena, *Appl. Phys. Lett.*, **91**, 092109 (2007).
38. L. A. Falkovsky and A. A. Varlamov, *Eur. Phys. J. B*, **56**, 281 (2007).
39. V. P. Gusynin and S. G. Sharapov, *Phys. Rev. B*, **73**, 245411 (2006).
40. J. Cserti, *Phys. Rev. B*, **75**, 033405 (2007).

41. L. A. Falkovsky, *J. Phys.: Conf. Ser.*, **129**, 012004 (2008).
42. A. H. Castro Neto, F. Guinea, N. M. R. Peres, K. S. Novoselov, and A. K. Geim, *Rev. Mod. Phys.*, **81**, 109 (2009).
43. D. Svintsov, V. Vyurkov, S. Yurchenko, T. Otsuji, and V. Ryzhii, *J. Appl. Phys.*, **111**, 083715 (2012).
44. B. N. Narozhny, I. V. Gornyi, M. Titov, M. Schutt, and A. D. Mirlin, *Phys. Rev. B*, **91**, 035414 (2015).
45. M. A. Kats, N. Y. Geneved, Z. Gaburro, and F. Capasso, *Opt. Express*, **19**, 21748 (2011).
46. L. D. Landau and E. M. Lifshitz, *The Theory of Fields* (Pergamon, Oxford, 1971).
47. D. J. Griffiths, *Introduction to Electrodynamics* (Benjamin Cummings, 1999).
48. R. E. Colin, *Antennas and Radiowave Propagation* (McGraw-Hill, New York, 1985).
49. M. Müller, J. Schmalian, and L. Fritz, *Phys. Rev. Lett.*, **103**, 025301 (2009).
50. M. Mendoza, H. J. Herrmann, and S. Succi, *Phys. Rev. Lett.*, **106**, 156601 (2011).
51. D. Coquillat, V. Nodjiadjim, A. Konczykowska, M. Riet, N. Dyakonova, C. Consejo, F. Teppe, J. Godin, and W. Knap, in *IRMMW-THz: International Conference on Infrared, Millimeter, and Terahertz Waves*, Tucson, AZ, USA, 2014.
52. D. Coquillat, V. Nodjiadjim, A. Konczykowska, N. Dyakonova, C. Consejo, S. Ruffenach, F. Teppe, M. Riet, A. Muraviev, A. Gutin, M. Shur, J. Godin, and W. Knap, *J. Phys.: Conf. Ser.*, **647**, 012036 (2015).
53. D. Coquillat, V. Nodjiadjim, A. Konczykowska, N. Dyakonova, C. Consejo, S. Ruffenach, F. Teppe, M. Riet, A. Muraviev, A. Gutin, M. Shur, J. Godin, and W. Knap, in *Proceedings of the 40th IRMMW-THz: International Conference on Infrared, Millimeter, and Terahertz Waves*, Hong Kong, China, 2015.

Chapter 49

Nonlinear Response of Infrared Photodetectors Based on van der Waals Heterostructures with Graphene Layers*

V. Ryzhii,[a,b,c] M. Ryzhii,[d] D. Svintsov,[e] V. Leiman,[e] V. Mitin,[f] M. S. Shur,[g] and T. Otsuji[a]

[a]*Research Institute of Electrical Communication, Tohoku University, Sendai 980-8577, Japan*

[b]*Institute of Ultra High Frequency Semiconductor Electronics of RAS, Moscow 117105, Russia*

[c]*Center for Photonics and Infrared Engineering, Bauman Moscow State Technical University, Moscow 111005, Russia*

[d]*Department of Computer Science and Engineering, University of Aizu, Aizu-Wakamatsu 965-8580, Japan*

[e]*Laboratory of 2D Materials' Optoelectronics, Moscow Institute of Physics and Technology, Dolgoprudny 141700, Russia*

[f]*Department of Electrical Engineering, University at Buffalo, Buffalo, New York 1460-1920, USA*

[g]*Department of Electrical, Computer, and Systems Engineering and Department of Physics, Applied Physics, and Astronomy, Rensselaer Polytechnic Institute, Troy, New York 12180, USA*

v-ryzhii@riec.tohoku.ac.jp

---

*Reprinted with permission from V. Ryzhii, M. Ryzhii, D. Svintsov, V. Leiman, V. Mitin, M. S. Shur, and T. Otsuji (2017). Nonlinear response of infrared photodetectors based on van der Waals heterostructures with graphene layers, *Opt. Express*, **25**(5), 5536–5549. Copyright © 2017 Optical Society of America.

Graphene-Based Terahertz Electronics and Plasmonics: Detector and Emitter Concepts
Edited by Vladimir Mitin, Taiichi Otsuji, and Victor Ryzhii

ISBN 978-981-4800-75-4 (Hardcover), 978-0-429-32839-8 (eBook)
www.jennystanford.com

We report on the device model for the infrared photodetectors based on the van der Waals (vdW) heterostructures with the radiation absorbing graphene layers (GLs). These devices rely on the electron interband photoexcitation from the valence band of the GLs to the continuum states in the conduction band of the inter-GL barrier layers. We calculate the photocurrent and the GL infrared photodetector (GLIP) responsivity at weak and strong intensities of the incident radiation and conclude that the GLIPs can surpass or compete with the existing infrared and terahertz photodetectors. The obtained results can be useful for the GLIP design and optimization.

49.1 Introduction

Unique properties of graphene [1] have enabled its use in different detectors of terahertz (THz) and infrared (IR) radiation [2–10]. Progress in the fabrication of the van der Waals (vdW) heterostructures [11] incorporating the graphene layers (GLs) stimulated the development of new devices, which provide the advantages of combining GLs with such materials as hBN, WS_2, InSe, and similar materials [12–21]. Recently (see preprint [22]), we proposed and evaluated GL-vdW IR photodetectors using the interband transitions of the electrons in the GLs with their subsequent escape to the continuum states in the barrier layers. The features of these photodetectors (in the following referred to as GLIPs) are associated with the specifics of the interband absorption and slow capture of the photoexcited electrons propagating across the heterostructure in the continuum states above the barriers. This promises high values of the GLIP responsivity combined with a high speed due to the short electron transit times. The device model presented in [22] is limited to the GLIP operation at low and moderate IR radiation intensities. However, the applications of the GLIPs for the IR detection at the background limited conditions and for the generation of the THz radiation using the photomixing or the ultra-short pulsed excitation require the model accounting for the GLIP response to high power optical signals. In this chapter, we evaluated the GLIP response at elevated IR radiation intensities. This, in particular, can provide the insight on the GLIP operation in the background limited infrared performance (BLIP) regime and evaluate the dynamic range of the GLIP operation. Using the developed device model, we calculate the dependences of the photocurrent on the incident IR radiation intensity in GLIPs with different structural parameters (different conduction band offsets, numbers of the GLs, and electron capture efficiency).

49.2 Device Structure and Model

The GLIP under consideration consists of the GL-vdW heterostructure, which comprises the $N \geq 1$ undoped (inner) GLs clad by the barrier layers and two the doped contact GLs [so that the heterostructure includes $(N + 2)$ GLs in total and $(N + 1)$ barrier layers]. The inner GLs have the indices $n = 1, 2, 3, \ldots, N$. The index n marks the GL between the n-th and $(n + 1)$-th barrier layer. The barrier layers have the indices $n = 1$ (near-emitter barrier) and $n = 2, 3, \ldots, N + 1$ (other barriers, including that at the collector GL). It is assumed that the conduction band offset, Δ, between the GLs and barrier layers is smaller than the pertinent valence band offset. A sufficiently strong dc bias voltage V is applied between the contact GLs. The emitter and collector GLs are assumed to be doped either by donors or acceptors to provide their sufficiently high lateral conductivity.

The GLIP operation is associated with the following processes [22]: (1) the photogeneration of the electron-hole pairs in the GLs due to the interband radiative transitions; (2) the tunneling injection of the thermalized electrons from the ground states in the GLs and the escape of the photogenerated electrons from their excited states followed by the propagation across the barrier layers; (3) the electron capture from the continuum states above the inter-GL barriers into the inner GLs.

Figures 49.1a and b schematically show the band diagrams of a GLIP with the n-type emitter and collector GLs [22] without the IR irradiation at relatively low bias voltages (with the electric field, E_E, in the near-emitter barrier layer small in comparison with the electric field, E_B, in the heterostructure bulk, i.e., in other barrier layers) and under a strong intensity of the incident IR radiation (when E_E is larger than E_B), respectively. The band profile at a weak irradiation can also correspond to Fig. 49.1a.

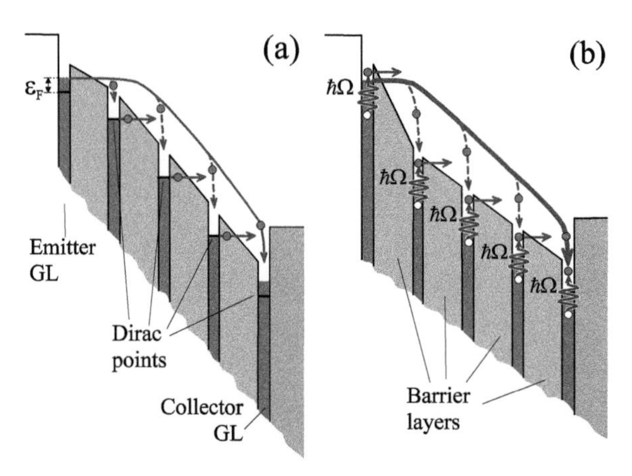

Figure 49.1 Band diagrams of a GLIP [22] (a) in dark at relatively low voltages when $E_E < E_B$ (weaker inclination of the band profile in the near-emitter barrier layer than in the barrier layers in the structure bulk) and (b) under strong irradiation when $E_E > E_B$ (the band profile in the near-barrier is steeper than in others). Solid and wavy arrows indicate generation of the electron-hole pair (electron photoexcitation due to the absorption of normally incident radiation) and different electron paths, respectively.

The consideration the GLIP operation at the elevated radiation intensities requires to account for the self-consistent potential redistribution across the heterostructure due to the competition between the processes of the electron escape from and capture into the GLs, the variation of the electron (hole) density in the emitter GL caused by the variation of the self-consistent electric field at the emitter GL, and, in principle, the saturation of the radiative interband transitions and the pertinent drop in the IR radiation absorption with the increasing intensity. Our calculations are based on the concept of the capture probability used in the theory of single and multiple quantum-well photodetectors (QWIPs) [23–27] and, more recently, in the theory of the electron transport in GL-vdW heterostructures [28]. The main difference between GLIPs and QWIPs is using the interband radiative transitions and the intraband (intersubband) transitions in GLIPs and QWIPs, respectively. Due to this, the GLIPs are sensitive to the normally incident IR radiation and their response is virtually independent of the GL doping. A small efficiency of the electron capture into GLs [29] promotes elevated GLIP responsivity due to the effect of the photoelectric gain.

49.3 Main Equations

To simplify the model, we neglect the thermionic emission of electrons and holes from the GLs in comparison with the tunneling and radiative processes due to relatively large ratios of the work functions under consideration and the thermal energy (see Appendix A) and disregard the local electric-field dependence of the capture efficiency [28, 34]. The balance between the electrons leaving the GLs due to their tunneling and the photoexcitation (direct or followed by tunneling) and the electrons captured into the GLs is governed by the following equations:

$$\frac{j}{e} = G_E + \beta\,\theta_E I, \quad \frac{jp_n}{e} = G_n + \beta\,\theta_n I. \tag{49.1}$$

Here j is the density of the electron current across GLs, e is the electron charge, p_n is the capture efficiency [22–27] for the electrons crossing the GL with the index $n = 1, 2, \ldots, N$, G_E, $R_E = \beta\theta_E I$, G_n, and $R_n = \beta\theta_n I_n$ are the rates of the injection and photoemission from the emitter GL and the inner GLs (with the index $n = 1, 2, \ldots, N$), respectively, $\beta = \pi\alpha/\sqrt{\kappa_\infty}$ is the probability of the interband photon absorption in GLs [1], $\alpha = e^2/\hbar c \simeq 1/137$ is the fine structure constant, c is the speed of light in vacuum, and $\sqrt{\kappa_\infty}$ is the barrier material refractive index, θ_n is the probability of the photoexcited electrons tunneling escape from the n-th GL to the continuum states above the inter-GL barriers, I is the IR radiation photon flux crossing the GLs. The quantity θ_n characterizes the escape of the photoexcited electrons from the GLs. The above value of β corresponds to the linear energy spectrum of carriers in GLs: $\varepsilon = v_W\,|\mathbf{p}|$, where $v_W \simeq 10^8$ cm/s and $\mathbf{p}$ is the carrier momentum.

At the sufficiently strong IR irradiation, the electric field, $E_E = E_1$, at the emitter GL can markedly exceed the electric fields, E_n (with $n = 2, 3, \ldots, N + 1$), in other inter-GL barriers. At such conditions, the injection of the thermalized electrons from the emitter GL to the heterostructure bulk is determined by the thermally activated tunneling, while such an injection from the inner GLs is small in comparison with the photoemission from these GLs. Hence, we set

$$G_E = \frac{j_m}{e}\exp\left(-\frac{\gamma_E^{3/2} E_{tunn}}{E_E}\right), \quad G_n = \frac{j_m}{e}\exp\left(-\frac{E_{tunn}}{E_n}\right) \tag{49.2}$$

Here $j_{\rm m}$ is the maximum current density which can be extracted from an *undoped* GL and $\gamma_{\rm E} = (\Delta - \varepsilon_{\rm F}\sqrt{1+E_{\rm F}/F})/\Delta$, where Δ is the conduction band offset between the GLs and the barrier layer material, $E_{\rm tunn} = 4\sqrt{2m}\,\Delta^{3/2}/3e\hbar$ is the field characterizing the tunneling through the triangular barrier top [30], $\varepsilon_{\rm F}$ is the Fermi energy of the thermalized electrons in the n-type emitter GL in the absence of the electric field at the emitter GL (determined by the emitter GL donor density Σ_0), $F = 4e\varepsilon_{\rm F}^2/\kappa\hbar^2 v_{\rm W}^2$, m and κ are the electron effective mass and the dielectric constant in the barrier layers. In the GLIPs with the emitter GL of p-type, one needs to replace $\varepsilon_{\rm F}$ and F by $-\varepsilon_{\rm F}$ and $-F$, respectively, although in the following we focus on the GLIPs with donor doped emitter GL. In contrast to our previous work [22], the quantity $\gamma_{\rm E}$ accounts for the variation of the Fermi energy in the emitter GL by the electric field. This effect can be pronounced in the case of elevated bias voltages considered below.

The quantities $\theta_{\rm E}$ and $\theta_{\rm n}$ are given by the following expressions:

$$\theta_{\rm E} = \frac{1}{1+\dfrac{\tau_{\rm esc}}{\tau_{\rm relax}}\exp\left(\dfrac{\eta_\Omega^{3/2}E_{\rm tunn}}{E_{\rm E}}\right)}, \quad \theta_{\rm n} = \frac{(1-\beta)^{\rm n}}{1+\dfrac{\tau_{\rm esc}}{\tau_{\rm relax}}\exp\left(\dfrac{\eta_\Omega^{3/2}E_{\rm tunn}}{E_{\rm n+1}}\right)}, \tag{49.3}$$

where $\eta_\Omega = [(\Delta - \hbar\Omega/2)/\Delta]$ if $\hbar\Omega/2 < \Delta$ and $\eta_\Omega = 0$ if $\hbar\Omega/2 \geq \Delta$. $\tau_{\rm esc}$ is the try-to-escape time, $\tau_{\rm relax}$ is the characteristic time of the photoexcited electrons energy relaxation, and $n = 1, 2, \ldots, N$.

Using Eq. (49.1) together with Eqs. (49.2) and (49.3), we arrive at the following set of $(N + 1)$ equations governing the electric fields, $E_{\rm E}$ and $E_{\rm n}$, in the barriers:

$$\frac{1}{1+\dfrac{\tau_{\rm esc}}{\tau_{\rm relax}}\exp\left(\dfrac{\eta_\Omega^{3/2}E_{\rm tunn}}{E_{\rm E}}\right)} - \frac{(1-\beta)^{\rm n}}{p}\,\frac{1}{1+\dfrac{\tau_{\rm esc}}{\tau_{\rm relax}}\exp\left(\dfrac{\eta_\Omega^{3/2}E_{\rm tunn}}{E_{\rm n+1}}\right)}$$

$$= \frac{j_{\rm m}}{e\beta I}\left[\frac{(1-\beta)^{\rm n}}{p}\exp\left(-\frac{E_{\rm tunn}}{E_{\rm n}}\right) - \exp\left(-\frac{\gamma_{\rm E}^{3/2}E_{\rm tunn}}{E_{\rm E}}\right)\right], \tag{49.4}$$

This set of equations should be supplemented by the following condition corresponding to the voltage applied between the emitter and collector equal to V: $\sum_{n=1}^{n+1} E_n = V/d$, where d is the thickness of the barrier layers.

In the GLIPs with not too large number of the GLs $N < \beta^{-1}$, taking into account that $(1 - \beta)^n \simeq \exp(-n\beta) \simeq 1 - n\beta \simeq 1$ (if $N\beta \ll 1$), the electric fields in the barriers in the heterostructure bulk are very close to each other, i.e., $E_n \simeq E_B$ (for $n = 2, 3, \ldots, N + 1$) with $E_B = (V/d - E_E)/N$. In this case, considering Eq. (49.4), we arrive at the following equations for E_E and E_B as functions of the IR radiation intensity I:

$$\frac{1}{1+\frac{\tau_{esc}}{\tau_{relax}}\exp\left(\frac{\eta_\Omega^{3/2}E_{tunn}}{E_E}\right)} - \frac{1}{p}\frac{1}{1+\frac{\tau_{esc}}{\tau_{relax}}\exp\left(\frac{N\eta_\Omega^{3/2}E_{tunn}}{V/d - E_E}\right)}$$

$$= \frac{j_m}{e\beta I}\left[\frac{1}{p}\exp\left(-\frac{NE_{tunn}}{V/d - E_E}\right) - \exp\left(-\frac{\gamma_E^{3/2}E_{tunn}}{E_E}\right)\right], \qquad (49.5)$$

$$E_B = \frac{(V/d - E_E)}{N}. \qquad (49.6)$$

49.4 Dark Characteristics

In the absence of irradiation at relatively low voltages ($V \ll dE_{tunn}$), Eq. (49.5) yields the following analytical expressions for the electric fields $E_E = E_E^{dark}$ and $E_B = E_B^{dark}$:

$$E_E^{dark} \simeq \frac{\gamma_0^{3/2}V}{(\gamma_0^{3/2} + N)d}\left[1 - \frac{NV\ln p}{(\gamma_0^{3/2} + N)^2 dE_{tunn}}\right]$$

$$\simeq \frac{\gamma_0^{3/2}V}{(\gamma_0^{3/2} + N)d}, \qquad (49.7)$$

$$E_B^{dark} = \frac{V}{(\gamma_0^{3/2} + N)d}\left[1 + \frac{\gamma^{3/2}V\ln p}{(\gamma_0^{3/2} + N)^2 dE_{tunn}}\right]$$

$$\simeq \frac{V}{(\gamma_0^{3/2} + N)d}, \qquad (49.8)$$

where $\gamma_0 = (\Delta - \varepsilon_F)/\Delta$. One can see that at low voltages, the ratio of the electric fields at the emitter and in the bulk $E_E/E_B \simeq \gamma_0^{3/2} < 1$. This case corresponds to the band diagram schematically shown in Fig. 49.1a. At elevated voltages V and sufficiently small capture efficiency p when $V > NdE_{tunn}/\ln(1/p)$, instead of Eqs. (49.7) and (49.8) we obtain

$$E_E^{dark} \simeq \frac{V}{d} - \frac{NE_{tunn}}{\ln(1/p)}, \quad E_B^{dark} \simeq \frac{E_{tunn}}{\ln(1/p)}. \tag{49.9}$$

In the latter case, as follows from Eqs. (49.9), $E_E > E_B$. The difference between the electric fields in the emitter barrier and in the bulk $E_E^{dark} - E_B^{dark}$ change the sign at the voltage V_0 when $E_E^{dark} = E_B^{dark} = V_0/d(N + 1)$, which satisfies the following equation:

$$\frac{V_0}{dE_{tunn}} = \frac{1}{(N+1)\ln(1/p)}\left[1 - \frac{\varepsilon_F}{\Delta}\sqrt{1 + \frac{V_0}{d(N+1)F}}\right]^{3/2}. \tag{49.10}$$

This implies that $V_0 < dE_{tunn}/(N + 1)\ln(1/p) \ll dE_{tunn}$. At $V < V_0$ and at $V > V_0$, one obtains $E_E^{dark} < E_B^{dark}$ and $E_E^{dark} > E_B^{dark}$ respectively.

49.5 Photoresponse at Low IR Radiation Intensities

At low intensities, the processes of the electron photoexcitation from the GLs can be considered as a perturbation in comparison with the tunneling processes of the thermalized electrons, so that the electric fields E_E and E_B are close to their values in the dark given by Eqs. (49.7)–(49.9). In this situation, for the range of low voltages, from Eqs. (49.1)–(49.3) one can find an analytic expression for the photocurrent density $j_{photo} = j - j_{dark}$ (where j_{dark} is the current density in the absence of irradiation, i.e., the dark current density):

$$j_{photo} \simeq \frac{\dfrac{e\beta I}{p}\dfrac{N}{(\gamma_0^{3/2} + N)}}{1 + \dfrac{\tau_{esc}}{\tau_{relax}}\exp\left[\dfrac{\eta^{3/2}(\gamma_0^{3/2} + N)dE_{tunn}}{V}\right]} \propto \frac{e\beta I}{p}. \tag{49.11}$$

As follows from Eq. (49.11) the photocurrent density can be markedly larger than the current density of the electrons photoexcited from all the GLs if the photoelectric gain $g = 1/p(\gamma_0^{3/2} + N) \gg 1$. This is associated with the amplification of the photoexcited electrons current by the extra injection from the emitter GL. A similar effect takes place in single and multiple-QW QWIPs [27].

Equation (49.11) leads to the expression for the GLIP responsivity $\mathcal{R} = j_{\text{photo}}/\hbar\Omega I_{\text{in}}$, where I_{in} is the intensity of the incoming radiation, which is weakly dependent on the number of GLs in the GLIP at the voltage $V \propto (N + 1)\, dE_{\text{tunn}}$ and markedly increases with a decrease in the capture efficiency p. Since the latter in the GLIPs can be very small [29], their responsivity should be fairly large. Using Eq. (49.11) and taking into account the IR radiation reflection from the device surface [that yields $I = 4I_{\text{in}}/(\sqrt{\kappa_\infty} + 1)^2$], for the characteristic GLIP responsivity (at low intensities) $R = \mathcal{R}|_{\hbar\Omega=\Delta/2}$ we obtain

$$\mathcal{R} \simeq \frac{2\beta e}{p\Delta(\sqrt{\kappa_\infty}+1)^2(1+\tau_{\text{esc}}/\tau_{\text{relax}})} \cdot \frac{N}{(\gamma_0^{3/2}+N)}. \qquad (49.12)$$

For $\kappa_\infty = 5$, $p = 0.01 - 0.02$, $\tau_{\text{esc}}/\tau_{\text{relax}} = 0.1$, $N = 1$, $\varepsilon_F = 0.1$ eV ($\gamma_0 = 0.75$), $\hbar\Omega = 0.8$ eV, $\Delta = 0.4$ eV, and, Eq. (49.12) yields $\mathcal{R} \simeq 0.135 - 0.270$ A/W.

Figure 49.2 shows the low intensity responsivity $\mathcal{R}$ as a function of the photon energy $\hbar\Omega$ calculated for different GL parameters (different values of Δ, N, p, and $\tau_{\text{esc}}/\tau_{\text{relax}}$) at different normalized voltages $U = V/dE_{\text{tunn}}$. Naturally, the responsivity $\mathcal{R}$ is somewhat smaller than the characteristic responsivity R [22] calculated without accounting for the reflection effect.

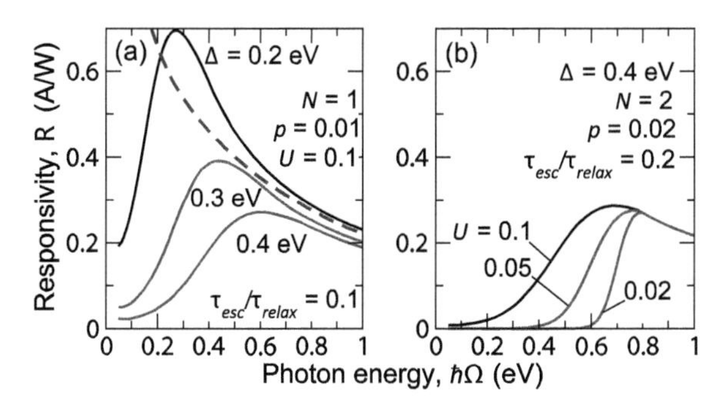

Figure 49.2 Spectral dependences of the responsivity $\mathcal{R}$ of GLIPs (a) with different conduction band offsets Δ, at a given normalized voltage $U = 0.2$ and (b) with $\Delta = 0.4$ eV at different normalized voltages U. Dashed line in the left panel corresponds to $\Delta = 0.4$ eV and $U = 0.25$.

It is worth noting that the GLIPs can exhibit good values of the responsivity even at $\hbar\Omega < 2\Delta$ if the bias voltage V is sufficiently high, see Fig. 49.2a. As seen from Fig. 49.2b, the GLIP responsivity in the range $\hbar\Omega < 2\Delta$ is very sensitive to the applied voltage. The latter is because an increase in the voltage leads to a substantial rise of the escape rate of the photoexcited electrons from the GLs. This can be used for an effective voltage control of the GLIP spectral characteristics.

An increase in V leads to a higher responsivity but also to a stronger dark current. The latter decreases the dark current limited detectivity.

The responsivity of the GLIPs intended for far-IR range of the spectrum with a smaller Δ can exhibit high values than those in the above estimate. Smaller values of the capture efficiency p [29] also promote an increase in the GLIP responsivity. Comparison of the GLIP responsivity $\mathcal{R}$ with the QWIP responsivity $\mathcal{R}_{QWIP}$ yields the following rough estimate for their ratio: $\mathcal{R}/\mathcal{R}_{QWIP} \sim (p_{QWIP}/p)(\beta/\beta_{QWIP})$. Since, $p_{QWIP} > p$) and $\beta > \beta_{QWIP}$, the responsivity ration can be large (about the order of magnitude or even more).

49.6 Nonlinear Response: High IR Radiation Intensities

At a strong IR irradiation, the photoexitation pronouncedly affects the electric fields in the emitter barrier, E_E, and in other barriers, E_B. When I varies from zero to a certain value I_V, the electric fields E_E and E_B vary from their values in dark to $E_E \simeq V/d$ and $E_B \simeq 0$, respectively.

Figure 49.3 shows the variation of E_E and E_B with increasing normalized intensity $S = e\beta I/j_m$ at different normalized voltages $U = V/dE_{tunn}$, found from the numerical solution of Eqs. (49.5) and (49.6), for a GLIP with $N = 1$, $\hbar\Omega = 0.6$ eV, and different values of the normalized voltage $U = V/dE_{tunn}$. Here and in the consequent figures we assume that $\Delta = 0.4$ eV, $\varepsilon_F = 0.1$ eV, $f = E_{tunn}/F = 30$, $\tau_{esc}/\tau_{relax} = 0.1$, and $p = 0.01$. One can see from Fig. 49.3 that at the chosen voltages, $E_E < E_B$ at relatively low S (and, hence, $E_E^{dark} < E_B^{dark}$). However, at certain values of S (different for different U), the quantity $E_E - E_B$ changes its sign. This is in line with Eqs. (49.9) and (49.10).

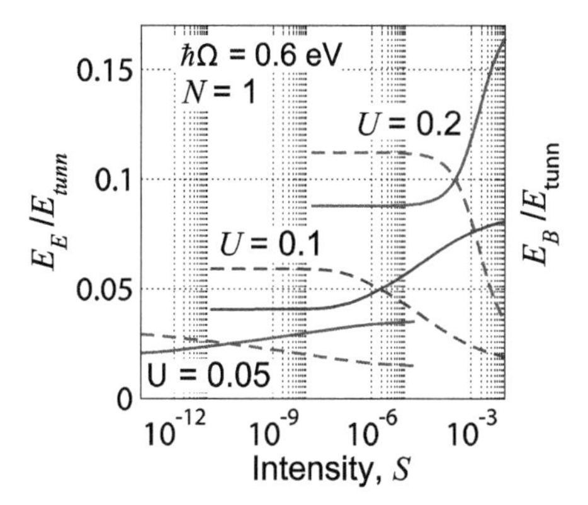

Figure 49.3 Normalized electric fields E_E/E_{tunn} (solid lines) and E_B/E_{tunn} (dashed lines) versus the normalized intensity S for a GL with $N = 1$ and $\hbar\Omega = 0.60$ eV at different normalized voltages U.

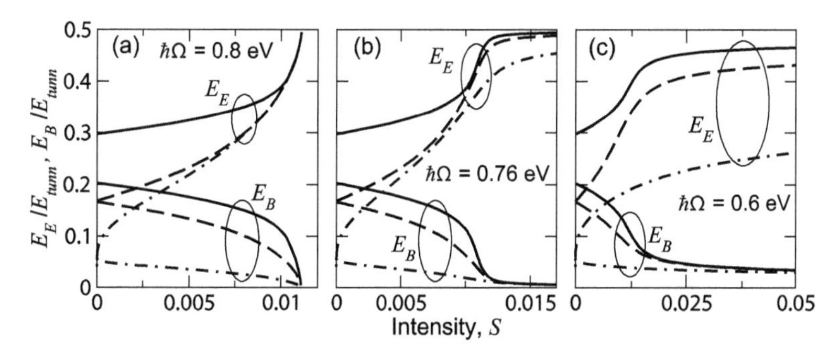

Figure 49.4 Normalized electric fields E_E/E_{tunn} and E_B/E_{tunn} versus the normalized intensity S for GLIPs with different number of GLs N and photon energies $\hbar\Omega$ at $U = 0.5$: solid lines correspond to $N = 1$, dashed lines - $N = 2$, and dashed-dotted lines - $N = 8$.

Figures 49.4 and 49.5 show the dependences of the electric fields E_E and E_B and the dependences of the normalized current $J = j_{photo}/j_m$ on the normalized intensity S calculated numerically using Eqs. (49.5) and (49.6) and Eq. (49.1), respectively, for GLIPs with $\Delta = 0.4$ eV, different number of the GLs N and different photon energies $\hbar\Omega$. As seen from Fig. 49.4, E_E tends to V/d (i.e., E_E/E_{tunn} tends to $U_V = 0.5$) and E_B tends to zero when the intensity approaches a certain value $S_V = e\beta\, I_V/j_m$. In particular, as follows from Eq. (49.5) at $\hbar\Omega \geq 2\Delta$ (when $\eta_\Omega = 0$), we can obtain

$$I_{\rm V} = \frac{j_{\rm m}p}{e\beta(1-p)}\left(1+\frac{\tau_{\rm esc}}{\tau_{\rm relax}}\right)\exp\left(-\gamma_{\rm V/d}^{3/2}\frac{dE_{\rm tunn}}{V}\right). \qquad (49.13)$$

At this intensity, $j \simeq j_{\rm photo} = j_{\rm V}$, where

$$j_{\rm V} = \frac{j_{\rm m}}{(1-p)}\exp\left(-\frac{\gamma_{\rm V/d}^{3/2}dE_{\rm tunn}}{V}\right) \simeq j_{\rm m}\exp\left(-\frac{\gamma_{\rm V/d}^{3/2}dE_{\rm tunn}}{V}\right). \qquad (49.14)$$

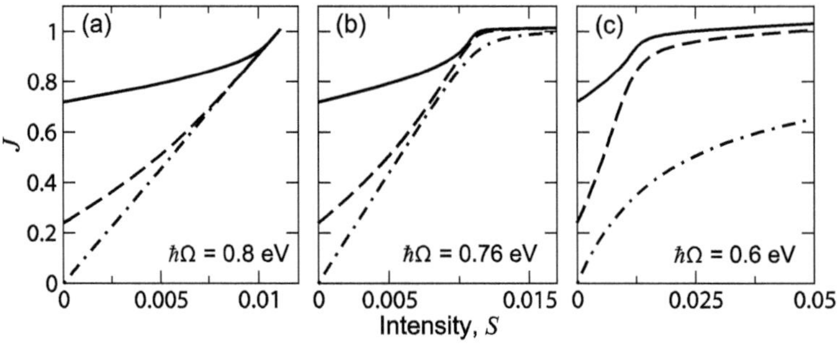

Figure 49.5 Normalized current density J versus normalized intensity S in GLIPs with different N and $\hbar\Omega$ at $U = 0.5$: as in Fig. 49.4, solid lines correspond to $N = 1$, dashed lines - $N = 2$, and dashed-dotted lines - $N = 8$.

As seen from Fig. 49.5, the normalized current J increases with the normalized intensity S (and, hence, $j_{\rm photo}$ increases with I). Thus, the current density $j_{\rm photo}$ increases with the increasing voltage V and intensity I. However, the $j_{\rm photo} - I$ dependence, being a linear one at low intensities, slows down at moderate and elevated intensities and tends to a saturation. This is attributed to the saturation of E_E as a function of I (as pointed above E_E tends to V/d), so that the exponents in the left parts of Eqs. (49.2) and (49.3) also approach to constants. Such a behavior is in line with the results on numerical calculations of $E_{\rm E}$ and $E_{\rm B}$ shown in Figs. 49.3 and 49.4.

Slowing down of the $J - S$ and, therefore, $j_{\rm photo} - I$ dependences results in a moderate rise of J with increasing S seen in the spectral characteristics shown in Fig. 49.6 for a GLIP with $N = 1$ at different S and U. Indeed, when S increases by two orders of magnitude, J increases only by a factor smaller than ten.

Figure 49.7 shows the spectral characteristics of the responsivity $\mathcal{R}$ for a GLIP with $N = 1$ at different S and U. The relatively slow $j_{\rm photo}$

$-I$ dependence implies that the GLIP responsivity $\mathcal{R}$ is a decreasing function of the intensity I at elevated values of the latter. The spectral characteristics of the photocurrent density and responsivity, j_{photo} versus $\hbar\Omega$ and R versus $\hbar\Omega$ at high radiation intensity and voltage are mainly determined not by a decrease in the effective activation energy of the photoexcited electrons with increasing $\hbar\Omega$ (i.e., the factor η_{ω}) but by a decreasing factor $1/\hbar\Omega$ in formula $\mathcal{R} = j_{\text{photo}}/\hbar\Omega I$.

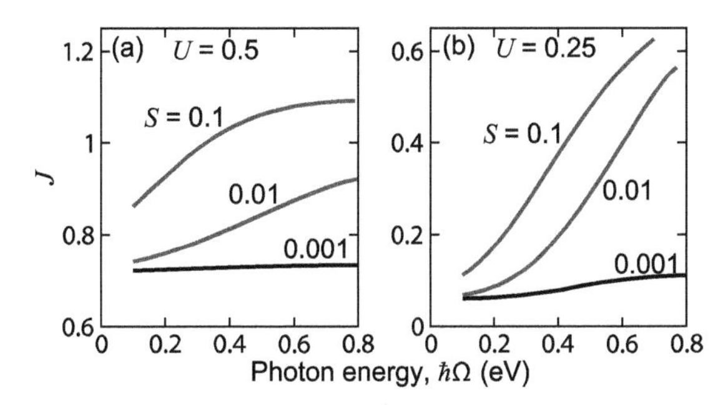

Figure 49.6 Spectral dependences of the normalized current density J at different normalized intensities S at (a) $U = 0.5$ and (b) $U = 0.25$.

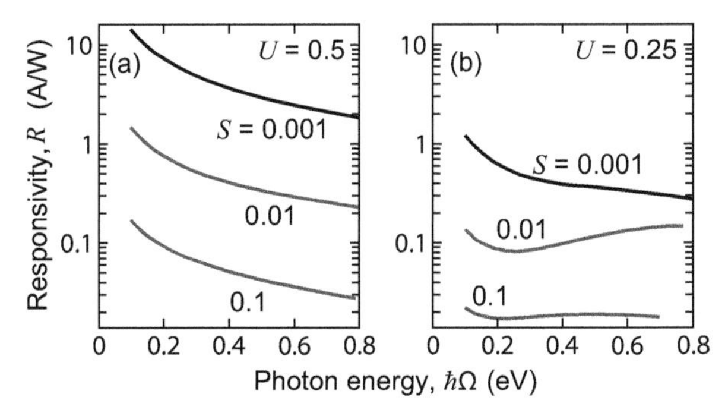

Figure 49.7 Spectral dependences of the responsivity, $\mathcal{R}$, of a GLIP with $N = 1$ at different normalized intensities S and voltages U.

One can see from Fig. 49.7 that the GLIP responsivity in the photon energy range $\hbar\Omega < 2\Delta$ at relatively high voltages can markedly exceed that at relatively low voltages shown in Figs. 49.2a and b. The dashed line in Fig. 49.2a, obtained considering the weak irradiation as a perturbation, and the dependence for $S = 0.001$ in

Fig. 49.7b are similar. Relatively high values of the GLIP responsivity in the range $\hbar\Omega < 2\Delta$ are attributed to a substantial increase in the barrier tunneling transparency at elevated voltages. However, the elevated values of the responsivity at high voltages are accompanied by rather high dark currents (corresponding to the values J at $S = 0$ in Fig. 49.5).

49.7 Comments

In our model, we neglected the dependence of the efficiency, p, of the electron capture to the n-GL on the electric fields in the barriers surrounding this GL, i.e., on E_n and E_{n+1}. The inclusion of this effect, can account for charging not only the GL adjacent to the emitter GL but also a few following GLs, although their charges would be relatively small (as was demonstrated considering the dark currents in the multiple-QW and -GL heterostructures [25, 26, 28, 32, 33]). Since the capture efficiency is much more sensitive to the average electric field $\bar{E} = V/d(N+1)$ in the heterostructure [34], our simplification appears to be justified. Nevertheless, the dependence of p versus $\bar{E}$ can somewhat affect the GLIP current-voltage characteristics both in dark conditions and under irradiation. When I exceeds I_V, Eq. (49.5) becomes invalid, because in this case it formally yields $E_B < 0$. At $I > I_V$, the electron transport in the GLIP bulk becomes diffusive, and one might expect that a further increase in I does not lead to a marked variation of the current. Thus, the quantity I_V is actually the IR radiation intensity corresponding to the photocurrent saturation. It defines the dynamic range of the GLIP normal operation ($I < I_V$). Estimating $j_m = e\Sigma_T/\tau_{esc}$, where $\Sigma_T = (\pi/6)(T/\hbar v_W)^2$ is the electron and hole density in the undoped GL [1], from Eqs. (49.11) and (49.12), we arrive at

$$I_V \simeq \frac{\pi p}{6\beta}\left(\frac{T}{\hbar v_W}\right)^2\left(\frac{1}{\tau_{esc}}+\frac{1}{\tau_{relax}}\right)\exp\left(-\gamma_{V/d}^{3/2}\frac{dE_{tunn}}{V}\right), \quad (49.15)$$

$$j_V \simeq \frac{\pi e}{\beta\tau_{esc}}\left(\frac{T}{\hbar v_W}\right)^2 \exp\left(-\gamma_{V/d}^{3/2}\frac{dE_{tunn}}{V}\right). \quad (49.16)$$

Assuming that $T = 100$–300 K, $\tau_{esc} = 10^{-13}$ s, $\tau_{esc}/\tau_{relax} = 0.1$, $\Delta = 0.4$ eV, $\varepsilon_F = 0.1$ eV, $p = 0.01$, $\kappa = \kappa_\infty = 5$, $m = 0.28\, m_0$ (m_0 is the electron mass in vacuum), $V/dE_{\text{tunn}} \sim 0.1$, we obtain $I_V \simeq 4 \times (10^{22} - 4 \times 10^{23})$ s^{-1}cm^{-2}. For $\hbar\Omega = 0.80$ eV, this photon flux corresponds to $P_V = \hbar\Omega I_V \simeq (5 - 50)$ kW/cm^2. This implies that at $I \sim I_V$, the power density absorbed by a GLIP with the N inner GLs is about of $P_V^{\text{abs}} \sim \beta N P_V \simeq (50 - 500)\, N$ Wcm^{-2}. Hence, the GLIP operation at such an intensity can be realized in the pulsed regime.

An increase in V leads to an increase in E_{E} and, consequently in the decrease of the factor $\gamma_{V/\text{d}}$ in Eqs. (49.15) and (49.16). At sufficiently large V, this factor can be close to zero. This implies that the device model under consideration is limited by the voltage range $U = V/dE_{\text{tunn}} \simeq U_{\text{max}} = V_{\text{max}}/dE_{\text{tunn}} \sim (\Delta/\varepsilon_{\text{F}})^2(F/E_{\text{tunn}})$. At the above parameters, this inequality yields $U \simeq U_{\text{max}} \sim 0.5$.

Above we also disregarded the following processes which, in principle, might affect the GLIP characteristics: the thermionic emission of electrons from the GLs, particularly, the hot electrons in the GLs heated by the absorbed IR radiation and the saturation of the radiation absorption. The former mechanism leads to an increase in the current emitted by the GLs due the thermionic emission contribution. However, as shown by our estimates, this contribution is relatively small in a wide range of the GLIP parameters, applied voltages, and radiation intensities corresponding to the normal device operation (except, possibly, the range of relatively low photon energies (see Appendix A).

The saturation of the radiation absorption in the GL due to the Pauli principle associated with the accumulation of the photoexcited electrons with the energy $\hbar\Omega/2$ and in the vicinity of this energy, should result in the saturation of the photocurrent as a function of the intensity. However, this can occur at sufficiently high intensities beyond real practical GLIP applications (see Appendix B).

49.8 Conclusions

We calculated the current in the GL-vdW heterostructures for GLIPs caused by the IR irradiation in a wide dynamic range—from low to high intensities. A strong IR radiation results in nonuniform electric-field distributions with a high-electric field domain near the emitter. The spatial redistribution of the electric-field leads to the photocurrent amplification (the effect of photoelectric gain) and to a slowing down of the photocurrent rise with increasing intensity. Since the capture efficiency in the GLIPs can be much smaller than in the standard QW-heterostructures, the GLIP responsivity can markedly exceed that of other QW-photodetectors, in particular, QWIPs [27] as well as the unitravelling-carrier photodiodes [35]. A substantial advantage of the GLIPs is associated with their sensitivity to the normally incident IR radiation. The saturation of the IR radiation absorption due to the Pauli principle can be an additional mechanism of the current-intensity characteristic nonlinearity but at relatively high IR radiation powers. The obtained results can be useful for the development and optimization of the GLIPs in starring arrays, optical communication devices, and terahertz photomixers.

Appendix A: Thermionic Electron Emission

The thermionic emission of the electrons thermalized in the GLs can contribute to the net current in the GLIPs [18]. If the effective temperature of these electrons is close to the lattice temperature T, the thermionic contribution is relatively weak if $\Delta/T > edE_{\rm tunn}N/V$. The latter takes place at not too low bias voltages V.

The heating of the thermalized electrons due to the radiation absorption leads to the variation of the net current j, which increases with the increasing I. Such a mechanism can be essential in the GLIPs with relatively small Δ at elevated temperatures. This might limit the applicability of the results obtained above.

The heating of the thermalized electron-hole system in the GLs is associated with a fraction of the kinetic energy of the photoexcited electrons and all of the energy of the photoexcited holes going to this system, particularly, in the emitter GL. This implies that the electron-hole system in the emitter GL approximately gets the power of $S \simeq \beta I\hbar\Omega(1/2 + w)$. Here w is the probability that a photoexcited electron does not escape from the GL: $w^{-1} = 1 + (\tau_{\rm relax}/\tau_{\rm esc})\exp(-\eta_{\Omega}^{3/2} E_{\rm tunn}/E_{\rm E})$. Depending on $\tau_{\rm esc}/\tau_{\rm relax}$ and $\hbar\Omega$, w varies in the range $0 < w < 1/2$. Neglecting the energy contribution of the Drude absorption (due to high photon energy in the spectral range under consideration) and considering that the specific heat of the degenerate electron system in the emitter GL is given by the standard formula $C = (2\pi\varepsilon_{\rm F}T/3\hbar^2 v_{\rm W}^2)$, and equalizing the energy received from the IR radiation and the energy transferring to phonons, one can obtain the following estimate for the effective temperature T_E:

$$\frac{T_E}{T} = \sqrt{1 + \frac{I}{I_T}}, \quad I_T = \frac{2(T/\hbar v_W)^2}{3\beta\tau_\varepsilon(1+2w)}\frac{\varepsilon_F}{\hbar\Omega}, \tag{49.A1}$$

where τ_ε is the electron energy relaxation time. Assuming the same parameters as for the estimate for I_V in Section 49.5 and setting $\tau_\varepsilon = (10^{-11} - 10^{-12})$ s, we obtain $I_T \simeq 3 \times (10^{23} - 10^{24})$ s^{-1}cm^{-2}. These values are of the same order of magnitude or larger than I_V.

At low intensities, $T_E/T \simeq 1 + I/2I_T$. In this case, the variation of the thermionic current density from the emitter GL due to the electron heating by the absorbed IR radiation is estimated as $j_m \exp\left(-\frac{\gamma_0\Delta}{T}\right)\left(\frac{\gamma_0\Delta}{2T}\right)\frac{I}{I_T}$. Comparing this value with the photocurrent density, associated with the photoemitted electrons (at $\hbar\Omega \simeq 2\Delta$ when $\eta_\Omega \simeq 0$), we find that the latter substantially exceeds the thermionic component if

$$\left(\frac{4\varepsilon_F}{\pi\Delta}\right)\left(\frac{\tau_{esc}}{\tau_\varepsilon}\right)\left(\frac{T}{\gamma_0\Delta}\right)\exp\left(\frac{\gamma_0\Delta}{T}\right) \gg 1. \tag{49.A2}$$

Inequality (49.18) is satisfied in a wide range of parameters and temperatures. Indeed, setting $\Delta = 0.4$ eV, $\Delta/\varepsilon_F = 4$, and $\tau_{esc}/\tau_\varepsilon = 0.1$, for $T = 0.025$ eV ($\simeq 300$ K) the term in the right-hand side of this inequality is approximately equal to 430. At $\Delta = 0.2$ eV and $T = 0.017$ eV ($\simeq 200$ K), this term is about of 30. When $\hbar\Omega$ is markedly smaller that 2Δ, the rate of the photoexcited electrons escape from the GLs can be decreased. At such photon energies, the heating mechanism can be crucial. However, the device model for the GLIPs operating in the range of relatively small photon energies should be generalized accordingly, that is out of the scope of the present work.

At elevated radiation intensities, considering Eq. (49.A1), the density of the thermionic current created by the hot electrons can be estimated as $J_T \simeq j_m \exp\left[-\gamma_{V/d}\frac{\Delta}{T}\left(1+\frac{I_T}{I_V}\right)^{-1}\right]$. The comparison of J_T and J_V given be Eq. (49.11) shows that $J_T \ll J_V$ when

$$\frac{I}{I_T} \ll \left(\frac{\Delta}{T}\right)^2\left(\frac{V}{dE_{tunn}}\right)^2. \tag{49.A3}$$

The term in the right-hand side of inequality (49.A3) is large except the cases of impractically small the band offsets and applied voltages. Since the validity of the formulas obtained in Sections 49.4 and 49.5 is limited by $I \leq I_V$, taking into account that $I_V \leq I_T$ and considering inequality (49.19), one might expect that the nonlinearity mechanism associated with the electric field distribution is more crucial than the heating mechanism.

Appendix B: Saturation of Absorption

At sufficiently strong irradiation, the distribution function of the photoexcited electrons at the energies close $\varepsilon \simeq \hbar\Omega/2$ can become close to 1/2. This can lead to the interband transitions saturation [31] and, hence, to the drop of the GLIP responsivity.

Taking into account the linearity of the GL energy spectrum and considering the Pauli principle, the interband photoexcitation rate with the estimated values of the electron distribution function at the energy $\varepsilon_\Omega = \hbar\Omega/2$ can be presented as

$$R = \beta\theta\, I(1 - 2f_\Omega), \quad f_\Omega \simeq \frac{\pi\hbar^2 v_W^2}{2\hbar^2\Omega\Delta\Omega}\Sigma. \qquad (49.B1)$$

Here θ is given by Eqs. (49.3), Σ is the density of the photoexcited electrons in the GL, and $\Delta\Omega$ is the incident IR radiation spectral width. Estimating Σ considering the balance between the electron photoexcitation and the escape and substituting the obtained value to Eq. (49.10), we arrive at

$$R = \frac{\beta\theta\, I}{1 + I/I_S}, \qquad (49.B2)$$

where for the most interesting situations when $\tau_{esc} \ll \tau_{relax}$ and $\hbar\Omega \sim 2\Delta$ one obtains $I_S \simeq (\Omega\Delta\Omega/\pi\beta v_W^2\, \tau_{esc})$.

As follows from Eq. (49.B1), the effect of the absorption saturation can be accounted for by replacing the factor $j_m/e\beta I$ in Eq. (49.5) by $j_m\,(1 + I/I_S)/e\beta I$. This leads to a saturation of j as a function of I and to decrease in the GLIP responsivity $\mathcal{R} \propto (\hbar\Omega I)^{-1}$ with increasing I. As a result, at large intensities, the modulation parameter $M = \delta j/\delta I$ should drop and tend to zero at very strong irradiation.

Setting $\hbar\Omega = 0.8$ eV, $\hbar\Delta\Omega \sim \hbar/\tau_{esc} \sim 0.02$ eV, we find $I_S \sim 2.6 \times 10^{27}$ s^{-1} cm^{-2} (or $S_S = \hbar\Omega I_S \sim 330$ MW/cm^2). The above results are in line with those predicted theoretically [31] and obtained experimentally [36]. Comparing the characteristic intensities I_V and I_S, one can conclude that the nonlinearity mechanism invoked in our model is more crucial at much lower intensities than the saturation mechanism.

Due to short tunneling delay times, the bandwidth of the GLIP response to the modulated IR radiation is primarily limited by the electron transit time $t_{trans} \propto (N + 1)d$ across the barrier layers. This time in the GLIPs with sufficiently thin inter-GL layers can be fairly short. In this regard, the GLIPs can surpass or, at least, compete with such IR detectors as unitravelling-carrier PDs [35]. The GLIP operation using the heating mechanism is much slow, because the pertinent speed is limited by the electron energy relaxation time $\tau_\varepsilon \gg \tau_{esc}, \tau_{trans}$.

Funding

The work at RIEC and UoA was supported by the Japan Society for Promotion of Science, KAKENHI Grant No. 16H06361. The work at RIEC with UB was supported by the RIEC Nation-Wide Cooperative Research Project. VR also acknowledges the support by the Russian Scientific Foundation, Grant No.14-29-00277. The work by VL and DS was supported by the Russian Foundation for Basic Research, Grant No. 16-37-60110. The work at RPI were supported by the US ARL Cooperative Research Agreement.

Acknowledgments

The authors are thankful to A. Satou, V. Y. Aleshkin, and A. A. Dubinov for useful discussions.

References

1. A. H. Castro Neto, F. Guinea, N. M. R. Peres, K. S. Novoselov, and A. K. Geim, The electronic properties of graphene, *Rev. Mod. Phys.*, **81**, 109–162 (2009).
2. T. Mueller, F. N. A. Xia, and P. Avouris, Graphene photodetectors for high speed optical communications, *Nat. Photonics*, **4**, 297–301 (2010).

3. V. Ryzhii, M. Ryzhii, V. Mitin, and T. Otsuji, Terahertz and infrared photodetection using p-i-n multiple-graphene layer structures, *J. Appl. Phys.*, **107**, 054512 (2010).

4. F. Bonaccorso, Z. Sun, T. Hasan, and A. C. Ferrari, Graphene photonics and optoelectronics, *Nat. Photonics*, **4**, 611622 (2010).

5. V. Ryzhii, N. Ryabova, M. Ryzhii, N. V. Baryshnikov, V. E. Karasik, V. Mitn, and T. Otsuji, Terahertz and infrared photodetectors based on multiple graphene layer and nanoribbon structures, *Opto-Electron. Rev.*, **20**, 15–25 (2012).

6. A. Tredicucci and M. S. Vitielo, Device concepts for graphene-based terahertz photonics, *IEEE J. Sel. Top. Quantum Electron.*, **20**, 8500109 (2014).

7. E. C. Peters, E. J. H. Lee, M. Burghard, and K. Kern, Gate dependent photocurrents at a graphene p-n junction, *Appl. Phys. Lett.*, **97**, 193102 (2010).

8. S. C. W. Song, M. S. Rudner, C. M. Marcus, and L. S. Levitov, Hot carrier transport and photocurrent response in graphene, *Nano Lett.*, **11**, 4688–4692 (2011).

9. C. H. Liu, Y.-C. Chang, T. B. Norris, and Z. Zhong, Photodetectors with ultra broadband and high responsivity at room temperature, *Nat. Nanotechnol.*, **9**, 273–278 (2014).

10. C. O. Kim, S. Kim, D. H. Shin, S. S. Kang, J. Min. Kim, Ch. W. Jang, S. S. Joo, J. S. Lee, Ju H. Kim, S.-Ho Choi, and E. Hwang, High photoresponsivity in an all-graphene p-n vertical junction photodetector, *Nat. Commun.*, **5**, 3249 (2014).

11. A. K. Geim and I. V. Grigorieva, Van der Waals heterostructures, *Nature*, **499**, 419–425 (2013).

12. F. Xia, H. Wang, Di Xiao, M. Dubey, and A. Ramasubramaniam, Two Dimensional Material Nanophotonics, *Nat. Photonics*, **8**, 899–907 (2014).

13. G. W. Mudd, S. A. Svatek, L. Hague, O. Makarovsky, Z. R. Kudrynsky, C. J.Mellor, P. H. Beton, L. Eaves, K. S. Novoselov, Z. D. Kovalyuk, E. E. Vdovin, A. J. Marsden, N. R. Wilson, and A. Patane, High broad-band photoresponsivity ofmechanically formed InSe graphene van der Waals heterostructures, *Adv. Mater.*, **27**, 37603766 (2015).

14. V. Ryzhii, A. Satou, T. Otsuji, M. Ryzhii, V. Ya. Aleshkin, A. A. Dubinov, V. Mitin, and M. S. Shur, Vertical hot electron terahertz detectors, *J. Appl. Phys.*, **116**, 114504 (2014).

15. V. Ryzhii, M. Ryzhii, V. Mitin, M. S. Shur, A. Satou, and T. Otsuji, Terahertz photomixing using plasma resonances in double-graphene layer structures, *J. Appl. Phys.*, **113**, 174505 (2013).
16. V. Ryzhii, T. Otsuji, V. Ya. Aleshkin, A. A. Dubinov, M. Ryzhii, V. Mitin, and M. S. Shur, Voltage-tunable terahertz and infrared photodetectors based on double-graphene-layer structures, *Appl. Phys. Lett.*, **104**, 163505 (2014).
17. V. Ryzhii V, T. Otsuji T, M. Ryzhii M, V. Ya. Aleshkin, A. A. Dubinov, D. Svintsov, V. Mitin, and M. S. Shur, Graphene vertical cascade interband terahertz and infrared photodetector, *2D Mater.*, **2**, 025002 (2015).
18. Q. Ma, T. I. Andersen, N. L. Nair, N. M. Gabor, M. Massicotte, C. H. Lui, A. F. Young, W. Fang, K. Watanabe, T. Taniguchi, J. Kong, N. Gedik, F. H. L. Koppens, and P. Jarillo-Herrero, Tuning ultrafast electron thermalization pathways in a van der Waals heterostructure, *Nat. Phys.*, **12**, 455–459 (2016).
19. M. Massicotte, P. Schmidt, F. Vialla, K. G. Schadler, A. Reserbat-Plantey, K. Watanabe, T. Taniguchi, K. J. Tielrooij, and F. H. L. Koppens, Picosecond photoresponse in van der Waals heterostructures, *Nat. Nanotechnol.*, **11**, 4246 (2016).
20. G. Gong, H. Zhang, W. Wang, L. Colombo, R. M. Wallace, and K. Cho, Band alignment of two-dimensional transition metal dichalcogenides: application in tunnel field effect transistors, *Appl. Phys. Lett.*, **103**, 053513 (2013).
21. I. Gierz, F. Calegari, S. Aeschlimann, M. Chavez Cervantes, C. Cacho, R. T. Chapman, E. Springate, S. Link, U. Starke, C. R. Ast, and A. Cavalleri, Tracking primary thermalization events in graphene with photoemission at extreme time scales, *Phys. Rev. Lett.*, **115**, 086803 (2015).
22. V. Ryzhii, M. Ryzhii, D. Svintsov, V. Leiman, V Mitin, M. S. Shur, and T. Otsuji, Infrared photodetectors based on graphene van der Waals heterostructures, *Infrared Phys. Technol.*, **84**, 72–81 (2017).
23. H. C. Liu, Photoconductive gain mechanism of quantum well intersubband infrared detectors, *Appl. Phys. Lett.*, **60**, 1507–1509(1992).
24. F. Rosencher, B. Vinter, F. Luc, L. Thibaudeau, P. Bois, and J. Nagle, Emission and capture of electrons in multiquantum-well structures, *IEEE J. Quantum Electron.*, **30**, 2875–2888 (1994).
25. L. Thibaudeau, P. Bois, and J. Y. Duboz, A self-consistent model for quantum well infrared photodetectors, *J. Appl. Phys.*, **79**, 446–452 (1996).

26. V. Ryzhii, Characteristics of quantum-well infrared photodetectors, *J. Appl. Phys.*, **81**, 6442–6448 (1997).
27. H. Schneider and H. C. Lui, *Quantum Well Infrared Photodetectors* (Springer, 2007).
28. V. Ryzhii, T. Otsuji, M. Ryzhii, V. Ya. Aleshkin, A. A. Dubinov, V. Mitin, and M. S. Shur, Vertical electron transport in van der Waals heterostructures with graphene layers, *J. Appl. Phys.*, **117**, 154504 (2015).
29. V. Ya. Aleshkin, A. A. Dubinov, M. Ryzhii, and V. Ryzhii, Electron capture in van der Waals graphene-based heterostructures with WS_2 barrier layers, *J. Phys. Soc. Jpn.*, **84**, 094703 (2015).
30. S. M. Sze S M 1999 *Physics of Semiconductor Devices* (Wiley, 1999), p. 103.
31. F. T. Vasko, Saturation of interband absorption in graphene, *Phys. Rev. B*, **82**, 345422 (2010).
32. M. Ershov, V. Ryzhii, and C. Hamaguchi, Contact and distributed effects in quantum well infrared photodetectors, *Appl. Phys. Lett.*, **67**, 3147–3149 (1995).
33. V. Ryzhii V, I. Khmyrova, M. Ryzhii, R. Suris, and C. Hamaguchi, Phenomenological theory of electric-field domains induced by infrared radiation in multiple quantum well structures, *Phys. Rev. B*, **62**, 7268–7274 (2000).
34. V. Ryzhii and R. Suris, Nonlocal hot-electron transport and capture model for multiple-quantum-well structures excited by infrared radiation, *Jpn. J. Appl. Phys.*, **40**, 513–517 (2001).
35. T. Ishibashi, Unitraveling-carrier photodiodes for terahertz applications, *IEEE J. Sel. Top.*, **20**, 3804210 (2014).
36. Z. Sun, T. Hasan, F. Torrisi, D. Popa, G. Privitera, F. Wang, F. Bonaccorso, D. M. Basko, and A. C. Ferrari, Graphene mode-locked ultrafast laser, *ACS Nano*, **4**(2), 803–810 (2010).

Chapter 50

Effect of Doping on the Characteristics of Infrared Photodetectors Based on van der Waals Heterostructures with Multiple Graphene Layers*

V. Ryzhii,[a,b,c] M. Ryzhii,[d] V. Leiman,[e] V. Mitin,[f] M. S. Shur,[g] and T. Otsuji[a]
[a]*Research Institute of Electrical Communication, Tohoku University, Sendai 980-8577, Japan*
[b]*Institute of Ultra High Frequency Semiconductor Electronics of RAS, Moscow 117105, Russia*
[c]*Center for Photonics and Infrared Engineering, Bauman Moscow State Technical University, Moscow 111005, Russia*
[d]*Department of Computer Science and Engineering, University of Aizu, Aizu-Wakamatsu 965-8580, Japan*
[e]*Laboratory of 2D Materials' Optoelectronics, Moscow Institute of Physics and Technology, Dolgoprudny 141700, Russia*
[f]*Department of Electrical Engineering, University at Buffalo, Buffalo, New York 1460-1920, USA*
[g]*Department of Electrical, Computer, and Systems Engineering and Department of Physics, Applied Physics, and Astronomy, Rensselaer Polytechnic Institute, Troy, New York 12180, USA*
v-ryzhii@riec.tohoku.ac.jp

Graphene-Based Terahertz Electronics and Plasmonics: Detector and Emitter Concepts
Edited by Vladimir Mitin, Taiichi Otsuji, and Victor Ryzhii

ISBN 978-981-4800-75-4 (Hardcover), 978-0-429-32839-8 (eBook)
www.jennystanford.com

We study the operation of infrared photodetectors based on van der Waals heterostructures with multiple graphene layers (GLs) and n-type emitter and collector contacts. The operation of such GL infrared photodetectors (GLIPs) is associated with the photoassisted escape of electrons from the GLs into the continuum states in the conduction band of the barrier layers due to the interband photon absorption, the propagation of these electrons, and the electrons injected from the emitter across the heterostructure and their collection by the collector contact. The space charge of the holes trapped in the GLs provides a relatively strong injection and large photoelectric gain. We calculate the GLIP responsivity and dark current detectivity as functions of the energy of incident infrared photons and the structural parameters. It is shown that both the periodic selective doping of the inter-GL barrier layers and the GL doping lead to a pronounced variation of the GLIP spectral characteristics, particularly near the interband absorption threshold, while the doping of GLs solely results in a substantial increase in the GLIP detectivity. The doping "engineering" opens wide opportunities for the optimization of GLIPs for operation in different parts of the radiation spectrum from near infrared to terahertz.

50.1 Introduction

The gapless energy spectrum of graphene layers (GLs) enables their use in the interband detectors of infrared radiation (see, for example, Refs. [1–5]). The incorporation of the GLs into the van der Waals (vdW) heterostructures based on such materials as hBN, WS_2, InSe, GaSe, and similar materials [6–15] can enable the creation of novel GL infrared photodetectors (GLIPs) with improved characteristics. As discussed in the recent review [16], mixed-dimensional van der Waals heterostructures have already shown considerable promise with expectations for more improvements in material quality and reproducibility in the near future. In particular, the feasibility of the exfoliation/transfer/sampling technique including doping techniques in vdW heterostructures was demonstrated [11–19]. Recently, we proposed and evaluated the IR detectors using the vdW heterostructures with the GLs clad by the widegap barrier layers—GL infrared photodetectors (GLIPs) [20, 21]. The GLs serve as photosensitive elements, in which electron-hole pairs are generated due to the interband absorption of IR radiation. The photogenerated electrons tunnel from the GLs through the barrier top to the continuum states in the barrier layers and support the terminal current. The photogenerated holes, which are confined in the GLs, form the space charge. The space charge is determined by the balance of the photogeneration and capture of the electrons propagating above the barriers. Figures 50.1 and 50.2 show the GLIP schematic view and the fragment of the device band diagram with the indicated main electron processes (the photoexcitation and the tunneling from and capture into the GLs). The space charge affects the electric field at the device emitter and, therefore, controls the injected electron current. In the devices based on the heterostructures with a low efficiency of the electron capture into the GLs, the injected current can markedly exceed the current created by the photoexcited electrons. This provides a relatively high photoelectric gain and detector responsivity. The rates of the escape of the photoexcited and thermalized electrons from the GLs and the capture of the electrons propagating across the barrier layers strongly depend of the potential profile near the GLs. The doping of the barrier layers, in particular, the selective doping using the delta layers of donors and acceptors as shown in Figs. 50.1 and 50.2 (which is called the "dipole" doping [22]) can markedly modify this profile resulting in the appearance of the "tooth" adjacent to each inner GL at the donor sheet side. The barrier doping was also effectively used in the unitravelling-carrier (UTC) photodiodes to reinforce the injection of the electrons photogenerated in the emitter of these devices [23]. The doping of the GLs can also lead to shift of the Fermi level in the GLs with respect to the Dirac level. The latter affects the spectrum of the electron photoexcitation, the escape rate of the thermalized electrons, and the capture processes.

The characteristics of the GLIPs considered previously [20, 21] are primarily predetermined by the electron affinities of GLs and barrier materials. In this chapter, we show that the proper doping of the barrier layers and GLs by acceptors and donors can pronouncedly modify the GLIP characteristics and result in an increase in the GLIP responsivity and detectivity, particularly, in the low-energy part of the infrared spectrum.

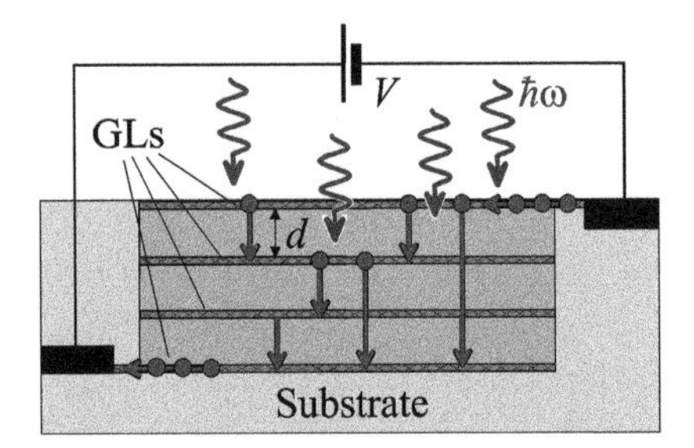

Figure 50.1 Schematic view of the GLIP heterostructure. Horizontal arrows correspond to electron flow from the emitter along the emitter GL and along the collector GL to the collector contact. Vertical arrows indicate flow of the electrons injected from the emitter GL and photoexcited from the inner GLs across one, two, or more inter-GL barriers before being captured.

50.2 Structure of GLIPs and Their Operation Principle

The GLIP under consideration consists of the GL-vdW heterostructure, which comprises $N = 1, 2, 3, \ldots$ GLs clad by the barrier layers and the two emitter and collector GLs (the top and bottom GLs, respectively). The latter GLs are doped by donors to provide their sufficiently high lateral conductivity. In contrast to our previously considered GLIPs [20, 21], we assume that the inter-GL barriers are selectively doped by acceptors and donors (as shown in Fig. 50.2) with equal densities Σ_B. The inner GLs can also be doped by acceptors with the density Σ_{GL}.

To provide the localization of the photoexcited holes, the valence band offset, Δ_V, between the GLs and barrier layers is larger than the conduction band offset Δ (i.e., $\Delta < \Delta_V$ or $\Delta \ll \Delta_V$). A sufficiently strong dc bias voltage V is applied between the contact GLs.

The GLIP operation is associated with the following processes [20, 21]: (1) the photoexcitation of the electron-hole pairs in the GLs due to the interband radiative transitions; (2) the tunneling injection of the thermalized electrons from the ground states in the GLs and the escape of the photoexcited electrons from their excited states followed by the propagation across the barrier layers; (3) the electron capture from the continuum states above the inter-GL barriers into the inner GLs.

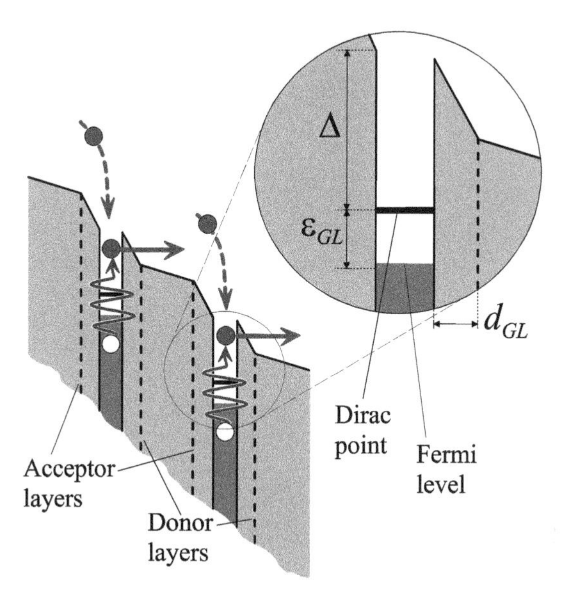

Figure 50.2 A fragment of the GLIP band diagram of a GLIP with barrier layers doped by acceptors and donors ("dipole" doped) and GLs doped by acceptors. Wavy, solid, and dashed arrows indicate the processes of the electron photoexcitation, tunneling, and capture, respectively. The inset shows the barrier "tooth" and its parameters.

50.3 Equations of the Model

Generalizing the results of the recent calculations [20, 21], the density of the current across the GLIP caused by the incident infrared radiation with the intensity I (inside the device) and photon energy $\hbar\omega$ can be presented as

$$j_{\text{photo}} = \frac{e\beta_\omega \theta_\omega I}{\left(\gamma_{\text{E}}^{3/2} + N\right)} \left[\gamma_{\text{E}}^{3/2} + \frac{(1-\beta_\omega)[1-(1-\beta_\omega)^N]}{p\beta_\omega} \right] \tag{50.1}$$

with

$$\beta_\omega = \frac{\beta_{\text{exp}}\left(\dfrac{\hbar\omega}{2T}\right)}{2\left[\cosh\left(\dfrac{\varepsilon_{\text{GL}}}{T}\right) + \cosh\left(\dfrac{\hbar\omega}{2T}\right)\right]}, \tag{50.2}$$

and

$$\theta_{\omega} = \frac{1}{1+\dfrac{\tau_{\rm esc}}{\tau_{\rm relax}}\exp\left(\dfrac{\eta_{\omega}^{3/2}E_{\rm tunn}}{E_{\rm GL}}\right)}. \tag{50.3}$$

Here, the factor depending on $\hbar\omega$ in Eq. (50.2) reflects the Pauli exclusion principle, the quantity θ_ω is the probability of the escape from the GL of the electrons photoexcited owing to the interband absorption of the photons with the energy $\hbar\omega$, $\beta = \pi\alpha/\sqrt{\kappa_{\rm B}}$, $\alpha = e^2/\hbar c$ is the fine structure constant, e and $\hbar$ are the electron charge and the Planck constant, $\sqrt{\kappa_{\rm B}}$ is the barrier material refractive index, T is the temperature, $\varepsilon_{\rm GL}$ is Fermi energy in the inner GLs counted from the Dirac point, p is the capture efficiency [24–29] [which in the heterostructures under consideration can be very small: $p \ll 1$ (Ref. [30])], N is the number of the inner GLs, and $\gamma_{\rm E} = (\Delta - \varepsilon_{\rm E})/\Delta < 1$, where $\varepsilon_{\rm E}$ is the electron Fermi energy in the emitter GL. The parameter $\gamma_{\rm E}$ plays the role of the emitter ideality parameter. It depends on the features of the electron injection from the emitter into the GLIP heterostructure bulk [26–31]. For the "ideal" emitter contact $\gamma_{\rm E} = 0$, and the electric field in the near-emitter barrier is close to zero. This corresponds to the situation when the emitter provides the injection of such an amount of electrons which is dictated by the conditions in the device bulk.

The probability, θ_ω, of the photoexcited electrons escape from the GLs is determined by the ratio of the try-to-escape time $\tau_{\rm esc}$ and the electron energy relaxation time $\tau_{\rm relax}$, and by the tunneling exponent. The latter depends on the energy (with respect to the barrier top) of the photoexcited electrons via the factor $\eta_\omega = (\Delta - \hbar\omega/2)/\Delta$ if $(\Delta - \Delta_{\rm GL}) \le \hbar\omega \le 2\Delta$ and $\eta_\omega = 0$ if $\hbar\omega > 2\Delta$, the characteristic "tunneling" field $E_{\rm tunn} = 4\sqrt{2m}\Delta^{3/2}/3e\hbar$ [32], (where m is the electron effective mass in the barrier layer), the height of the barrier "tooth" adjacent to the GL $\Delta_{\rm GL}$, which is determined by the real electric field in the barriers at the inner GLs $E_{\rm GL}$. Considering the doping of the barrier layers, we obtain the following formulas for the electric fields near the GLs $E_{\rm GL}$ and in the bulk of the barriers $E_{\rm B}$:

$$E_{\rm GL} = \frac{V}{\left(\gamma_{\rm E}^{3/2} + N\right)d} + \frac{V_{\rm B}}{d}\left(1 - \frac{2d_{\rm GL}}{d}\right), \tag{50.4}$$

$$E_{\rm B} = \frac{V}{\left(\gamma_{\rm E}^{3/2} + N\right)d} - \frac{V_{\rm B}}{d}\left(\frac{2d_{\rm GL}}{d}\right), \tag{50.5}$$

with $V_{\rm B} = 4\pi\, e\Sigma_{\rm B}d/\kappa_{\rm B}$ and $\Delta_{\rm GL} = E_{\rm GL}d_{\rm GL}$, where d is the barrier layer thickness and $d_{\rm GL}$ is the spacing between the GLs and the donor sheets (see the inset in Fig. 50.2). The barrier doping effectively increases the rate of the photoexcited electron tunneling rate when the "tooth" height $\Delta_{\rm GL}$ is sufficiently large, in particular, if $\Delta_{\rm GL} \simeq \Delta$. For the definiteness, the latter relationship is assumed in the following. The Fermi energy of holes in the inner GLs at not too high temperatures can be expressed via the acceptor and hole density $\Sigma_{\rm GL}$ as $\varepsilon_{\rm GL} = \hbar v_{\rm W}\sqrt{\pi\Sigma_{\rm GL}}$, where $v_{\rm W} \simeq 10^8$ cm/s,

50.4 GLIP Responsivity

Equation (50.1) for the GLIPs with not too large (realistic) number of the GLs in the GLIP part ($N \ll \beta_\omega^{-1}$), considered in the following, can be somewhat simplified leading to the following formula for the GLIP responsivity $R_\omega \propto j_{\rm photo}/\hbar\omega I$:

$$R_\omega \simeq R\frac{\left(\dfrac{\Delta}{2\hbar\omega}\right)\dfrac{N}{\left(\gamma_{\rm E}^{3/2}+N\right)}}{\left[1+\dfrac{\tau_{\rm esc}}{\tau_{\rm relax}}\exp\left(\dfrac{\eta_\omega^{3/2}E_{\rm tunn}}{E_{\rm GL}}\right)\right]} \times\frac{\exp\left(\dfrac{\hbar\omega}{2T}\right)}{\left[\cosh\left(\dfrac{\varepsilon_{\rm GL}}{T}\right)+\cosh\left(\dfrac{\hbar\omega}{2T}\right)\right]} \tag{50.6}$$

with

$$R = \frac{e\xi\beta}{p\Delta}. \tag{50.7}$$

Here, the factor ξ is determined by conditions of reflection of the incident radiation from the GLIP top interface [14]. Equation (50.6) turns to that derived and used previously for GLIPs with the undoped barriers and inner GLs [20, 21] at $\varepsilon_{\rm GL} = 0$ and $\Sigma_{\rm B} = 0$ when $E_{\rm GL}$ is replaced by $V/(\gamma_{\rm E}^{3/2} + N)d$. For example, setting $\xi = 1$, $p = 0.01$, $\kappa_{\rm B} =$

5, and Δ = 0.1 – 0.5 eV, for the characteristic responsivity we obtain $R \simeq$ (2 – 10) A/W.

The value of the GLIP responsivity given by Eqs. (50.5) and (50.6) corresponds to the photoconductive gain $g = [p(\gamma_E^{3/2} + N)]^{-1} \simeq (pN)^{-1}$ (compared with Refs. [24, 25, 29, 30]). The origin of this gain is associated with the accumulation of the charges formed in the GLs by the photogenerated holes. The latter is due to a much more effective confinement of the photogenerated holes than the photogenerated (photoexcited) electrons that stems from the condition $\Delta < \Delta_V$ accepted in Section 50.2. If this condition is violated, the escape probability of the photogenerated holes from the GLs can become rather high leading to vanishing of the photoconducting gain effect.

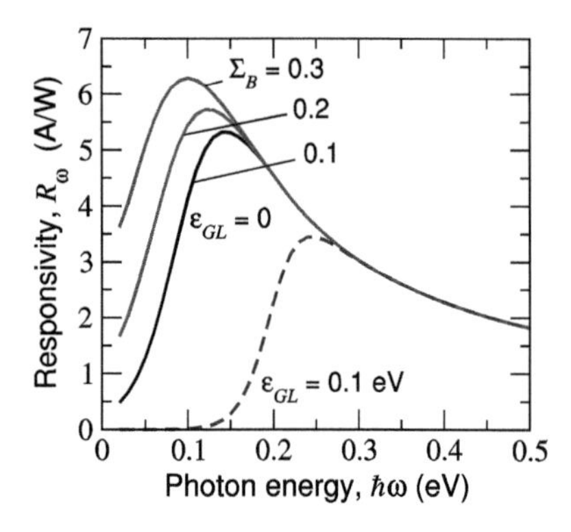

Figure 50.3 Spectral characteristics of GLIPs with Δ = 0.1 meV, five undoped GLs (Σ_{GL} = 0), and different donor and acceptor densities in the barriers Σ_B (in units 10^{12} cm^{-2}) at T = 100 K. The dashed line corresponds to GLIP with doped GLs (ε_{GL} = 0.1 eV, $\Sigma_{GL} = 0.8 \times 10^{12}$ cm^{-1}) – curves for different barrier doping are undistinguished.

Figures 50.3 and 50.4 show the responsivity calculated using Eq. (50.6) with Eqs. (50.4) and (50.5) as a function of the photon energy for the GLIPs with different barrier heights Δ and different doping levels of the inter-GL barriers and GLs. For the definiteness, the following general parameters are assumed: N = 5, ξ = 1, p = 0.01, $\gamma_E \ll 1$, d = 10 nm, d_{GL} = 2 nm, τ_{esc}/τ_{relax} = 0.1, T = 100 K, and $U = V/dE_{tunn}$ = 0.5. m = (0.14 – 0.28)m_0 (m_0 is the mass of bare electrons) and d = 10 nm. The value U = 0.5 corresponds to V/N = 0.026 – 0.038 V at Δ = 0.1 eV and V/N = 0.11 – 0.15 V at Δ = 0.25 eV.

One can see that an increase in the barrier doping level, which results in higher tunneling transparency of the barrier for the photoexcited electrons and, hence, higher probability of their escape, leads to a substantial increase in the responsivity at relatively low photon energies ($\hbar\omega < 2\Delta$). The responsivity of the GLIPs with smaller Δ is higher than that of the GLIPs with larger Δ in the low photon energy range (compare the curves in Figs. 50.3 and 50.4). An increase in the responsivity at relatively low photon energies exhibited by the curve for $\Sigma_B = 0.3 \times 10^{12}$ cm^{-2} in Fig. 50.4 is attributed to the factor $1/\hbar\omega$ in Eq. (50.6) (see also a comment in Section 50.6). Marked values of the responsivity in the range $\hbar\omega \lesssim 0.05$ eV (about several A/W, as seen from Figs. 50.3 and 50.4) imply that the GLIPs with properly doped barrier layers can operate not only in near- and mid-infrared spectral ranges but also in the terahertz range.

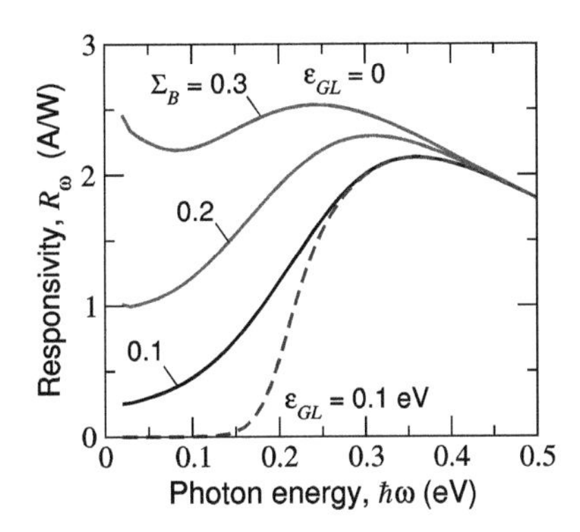

Figure 50.4 The same as in Fig. 50.3 but for GLIPs with $\Delta = 0.25$ eV.

The GL doping by acceptors also modifies the responsivity spectral dependence: its increase (and, therefore, increase in the Fermi energy ε_{GL}) gives rise to a marked shift of this dependence toward higher photon energies (compare the solid and dashed lines in Figs. 50.3 and 50.4). In the case of doped GLs, the barrier doping weakly affects the spectral dependence in question (the curves corresponding to different values of Σ_B and $\varepsilon_{GL} = 0.1$ eV are practically merged).

In principle, the temperature smearing of the electron energy distributions in the GLs somewhat affects the photon absorption probability at $\hbar\omega \simeq 2(\Delta + \varepsilon_{GL})$ due to the degeneracy of the electron system near the Fermi level. However, the variation of the temperature in the range $T = 50$–200 only slightly changes the above spectral dependences.

50.5 GLIP Detectivity

The dark current limited detectivity is usually determined as

$$D^*_\omega = \frac{R_\omega}{\sqrt{4egj_{\text{dark}}}}, \tag{50.8}$$

where g is the photoconductive gain which was introduced in Section 50.4. The dark current in the GLIPs is determined by the tunneling of the thermalized electrons from the GLs (amplified by the electron injection from the emitter GL). One can assume that the main contribution to this tunneling is provided by the electrons with the energies close to the Fermi level. Due to the specific features of the tunneling barrier shape, the tunneling exponent depends on the barrier parameters Δ_B and d. Considering this and generalizing the pertinent equations obtained for the GLIPs with the undoped barriers and inner GLs, one can use the following relationship for the dark current:

$$j_{\text{dark}} = \frac{j_{\max} f_E}{p} \exp\left[-\frac{(\eta^{3/2} + F)E_{\text{tunn}}}{E_{\text{GL}}}\right], \tag{50.9}$$

where for $\Delta_{\text{GL}} = \Delta$

$$\eta = 1 + \frac{\varepsilon_{\text{GL}}}{\Delta} = 1 + \frac{\hbar v_W \sqrt{\pi \Sigma_{\text{GL}}}}{\Delta}, \tag{50.10}$$

$$F = \left(\frac{E_{\text{GL}}}{E_B} - 1\right)\left(\frac{\varepsilon_{\text{GL}}}{\Delta}\right)^{3/2}$$

$$= \frac{(d/d_{\text{GL}})}{\dfrac{V}{(\gamma^{3/2} + N)V_B} - 1}\left(\frac{\hbar v_W \sqrt{\pi \Sigma_{\text{GL}}}}{\Delta}\right)^{3/2}, \tag{50.11}$$

$j_{\max}$ is the maximum current density which can be extracted from the emitter GL, and f_E is the pre-exponential factor, which depends on the emitter ideality factor γ_E. Disregarding for brevity the effects associated with the emitter nonideality (analyzed previously [20, 31]), we put in the following $\gamma_E = 0$ and $f_E = 1$.

At elevated temperatures, the thermionic escape of electrons from the GLs can also contribute to the dark current. The pertinent dark current density can be presented as

$$j_{\rm dark}^{\rm (therm)} = \frac{cj_{\rm max}}{p}\exp\left(-\frac{\Delta_{\rm therm}}{T}\right), \tag{50.12}$$

where the quantity $\Delta_{\rm therm}$ plays the role of the thermionic activation energy for the thermalized electrons in the GLs and $c \sim 1$. Generally, $\Delta - \Delta_{\rm GL} + \varepsilon_{\rm GL} \lesssim \Delta_{\rm therm} \lesssim \Delta + \varepsilon_{\rm GL}$.

Considering for simplicity the interpolation formula for the net dark current density in which the contributions given by Eqs. (50.9) and (50.12) are summarized, we arrive at the following equation for the dark current limited GLIP detectivity:

$$D_\omega^* = D^* \frac{\left(\dfrac{\Delta}{2\hbar\omega}\right)}{(1+\Theta_{\rm therm})^{1/2}} \frac{\exp\left[\dfrac{(\eta^{3/2}+F)E_{\rm tunn}}{2E_{\rm GL}}\right]}{\left[1+\dfrac{\tau_{\rm esc}}{\tau_{\rm relax}}\exp\left(\dfrac{\eta_\omega^{3/2}E_{\rm tunn}}{E_{\rm GL}}\right)\right]}$$

$$\times \frac{\exp\left(\dfrac{\hbar\omega}{2T}\right)}{\left[\cosh\left(\dfrac{\varepsilon_{\rm GL}}{T}\right)+\cosh\left(\dfrac{\hbar\omega}{2T}\right)\right]} \tag{50.13}$$

with

$$D^* = \frac{\sqrt{N}e\xi\beta}{\sqrt{4ej_{\rm max}\Delta}}. \tag{50.14}$$

Here

$$\Theta_{\rm therm} = c\exp\left[\frac{(\eta^{3/2}+F)dNE_{\rm tunn}}{E_{\rm GL}} - \frac{\Delta_{\rm therm}}{T}\right] \tag{50.15}$$

is the quantity characterizing the relative contribution of the tunneling and thermionic processes. For the GLIPs with undoped barrier setting $E_{\rm GL} = V/dN$, $F = 0$, and $\Delta_{\rm therm} \simeq \Delta + \varepsilon_{\rm GL}$, Eq. (50.15) yields

$$\Theta_{\rm therm} \simeq \exp\left(\frac{\eta^{3/2}NdE_{\rm tunn}}{V} - \frac{\Delta+\varepsilon_{\rm GL}}{T}\right). \tag{50.16}$$

Assuming $N = 5$ and $j_{\rm max} = 1.6 \times (10^5 - 10^6)$ A/cm^2, at the same other parameters as in the above estimate of the characteristic responsivity R, we obtain $D^* \simeq (0.5 - 7) \times 10^5$ cm$\sqrt{\rm Hz}$/W. Due to a large first exponential factor in Eq. (50.13), the real detectivity $D_{\omega}^{*} \gg D^{*}$. The GLIPs with a larger number of the inner GLs N can exhibit higher values of the dark current limited detectivity [because $D^{*} \propto \sqrt{N}$ (Ref. [33])]. The values of j_{max} used here correspond, in particular, to the electron density in the emitter GL $\Sigma_E = 10^{12}$cm^{-2} and the try-to-escape time $\tau_{esc} = 10^{-13} - 10^{-12}$ s.

According to Eq. (50.13), the spectral dependence of the detectivity repeats that of the responsivity (shown, in particular, in. Figs. 50.3 and 50.4). The dipole doping of the barrier layers leads to an increase in the GLIP responsivity (primarily in the range of relatively low photon energies) but simultaneously to an increase of the dark current and, hence, a drop of the detectivity. The doping of GLs by acceptors, which modifies the spectral characteristics, promotes the dark current lowering and a rise of the detectivity. In principle, carefully choosing the levels of both types of doping, one can expect the optimal relationship between the responsivity and the detectivity. However, taking into account the fact that realization of both types of doping in one device can markedly complicate its fabrication, we restrict ourselves by considering the detectivity of the GLIPs with the doping of the GLs only. Therefore, we focus on the GLIP detectivity as a function of the GL doping and the temperature assuming that the barrier layers are undoped.

Figure 50.5 shows the detectivity of the GLIPs with undoped barrier layers calculated using Eqs. (50.13) and (50.16) as a function the Fermi energy ε_{GL} (which is determined by the acceptor density in the GLs) for different temperatures. Figures 50.6 and 50.7 demonstrate how the detectivity of the GLIPs with different barrier heights Δ and Fermi energy ε_{GL} (i.e., different acceptor densities in the GLs Σ_{GL}) at different photon energy $\hbar\omega$ varies with increasing temperature T. We set $N = 5$, $\tau_{esc}/\tau_{relax} = 0.1$, and $U = 0.5$. The Fermi energy ε_{GL} changes from zero to 0.1 eV in the acceptor density range $\Sigma_{GL} = (0 - 8) \times 10^{12}$ cm^{-2}.

As seen from Fig. 50.5, the detectivity D_{ω}^{*} is a nonmonotonic function of the Fermi energy ε_{GL} and the acceptor density Σ_{GL} in the GLs with pronounced maxima at certain values ε_{GL} and Σ_{GL}. A pronounced increase in the detectivity is attributed to an increase in the barrier height for the thermalized electrons $\Delta + \varepsilon_{GL}$ with increasing doping level (see Fig. 50.2) leading to a diminishing of the tunneling and thermionic electron escape and, consequently, to a weaker dark current.

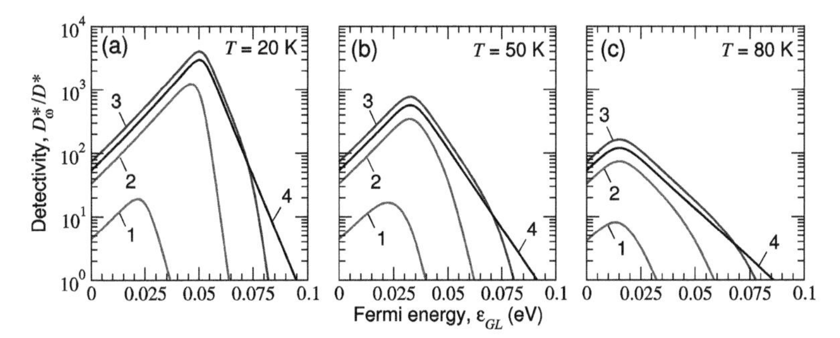

Figure 50.5 GLIP detectivity for Δ = 0.1 eV and undoped barriers (Σ_B = 0) as a function of the Fermi energy ε_{GL} (acceptor density in GLs Σ_{GL}) for different photon energies $\hbar\omega$ at (a) T = 20 K, (b) T = 50 K, and (c) T = 80 K : 1—$\hbar\omega$ = 0.05 eV, 2—0.1 eV, 3—$\hbar\omega$ = 0.15 eV, and 4—0.25 eV.

A steep detectivity roll-off at increased acceptor densities is associated with the Pauli principle leading to an abrupt drop of the photon absorption and, hence, the responsivity when E_{GL} becomes close to or larger than $\hbar\omega/2$. Some difference in the steepness of the detectivity roll-off seen in Figs. 50.5a–c is due to a stronger smearing of the Fermi-Dirac distribution in the GLs at higher temperatures. The dependences shown in Figs. 50.5–50.7 indicate a marked decrease in the detectivity maximum with increasing temperature. This is explained by an increase in the role of thermionic processes at elevated temperatures.

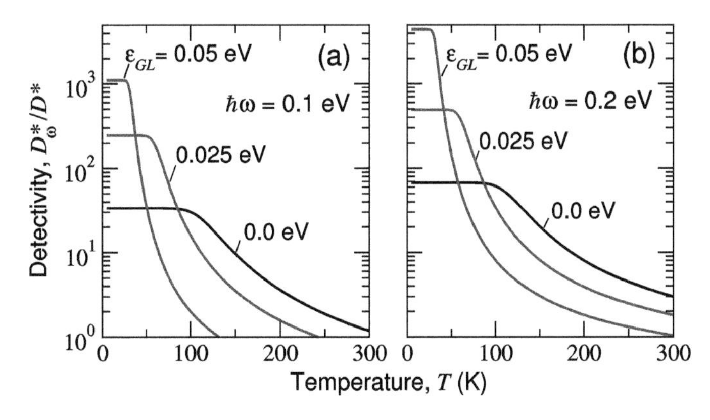

Figure 50.6 Temperature dependences of detectivity of GLIPs with Δ= 0.1 eV, undoped barriers and different Fermi energies ε_{GL} for (a) $\hbar\omega$ = 0.1 eV and (b) $\hbar\omega$ = 0.2 eV.

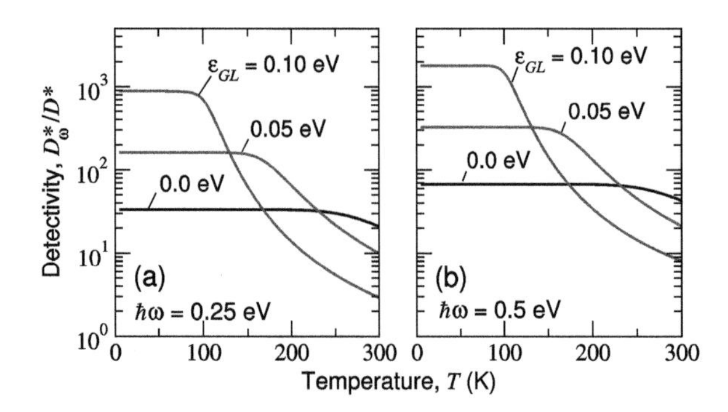

Figure 50.7 The same as in Fig. 50.6 but for $\Delta = 0.25$ eV and (a) $\hbar\omega = 0.25$ eV and (b) $\hbar\omega = 0.5$ eV.

One needs to point out that the values of the detectivity D^*_ω at certain values of the Fermi energies (acceptor densities) can be rather high. Taking into account the values of D^* obtained in the above estimate, for the D^*_ω maximum we find max $D^*_\omega > 10^9$ cm $\sqrt{\text{Hz}}/\text{W}$.

One can see from Figs. 50.6 and 50.7 that the detectivity being a flat function of the temperature steeply drops at T exceeding a certain temperature:

$$T_{\text{therm}} \sim \frac{\Delta^{3/2}}{(\Delta + \varepsilon_{\text{GL}})^{1/2}} \left(\frac{V}{NdE_{\text{tunn}}} \right). \quad (50.17)$$

This is associated with the inclusion of the thermionic contribution to the electron escape from the GLs. Although the enhancement of the GL doping results in a pronounced increase in the detectivity, it leads to a shrinking of the temperature range where D^*_ω [and crossings of the curves in Figs. 50.6 and 50.7]. Indeed, Eq. (50.17) yields $T_{\text{therm}} \propto \Delta^{3/2}/(\Delta + \varepsilon_{\text{GL}})^{1/2}$, i.e., a decreasing T_{therm} versus ε_{GL} relationship.

50.6 Discussion

As follows from the above results, the GLIPs under consideration can have high responsivity and detectivity in near and far IR spectral ranges. The GLIPs with $\Delta \sim 0.1$ eV can operate also in the terahertz range $\hbar\omega \sim 0.025 - 0.05$ eV($\hbar\omega/2\pi \sim 6 - 12$ THz), exhibiting reasonable values of the responsivity and detectivity.

In Section 50.4, we have omitted the analysis of the responsivity spectral behavior in the range of very low photon energies, because the interband absorption in this range can be complicated by the smearing of the carrier spectrum, fluctuations of the potential profile (existence of the electron-hole puddles), and formation of a narrow energy gap (due to specific doping or substrate properties) [34–39]—factors which are not described by Eqs. (50.2) and (50.3) and, hence, are beyond our device model.

The barrier height for the thermalized electrons in the GLIPs with doped GLs is equal to $\Delta_{therm} = \Delta^{(doped)} + \varepsilon_{GL}$. This height determines both the tunneling and thermionic dark currents. Comparing the dark currents in a GLIP with the acceptor doped GLs and with that in a GLIP with undoped GLs but with a higher barrier ($\Delta^{(undoped)} = \Delta_{therm} > \Delta^{(doped)}$), one can find that these currents are equal to each other or, at least, of the same order of magnitude. In contrast, taking into account that the barrier heights for the electrons photoexcited in the doped and undoped GLs are equal to $\Delta^{(doped)} - \hbar\omega/2$ and $\Delta^{(undoped)} - \hbar\omega/2$, respectively, we conclude that the escape rate of the photoexcited electrons and, hence, the GLIP responsivity in the former case is larger than that in the latter case (because $\Delta^{(doped)} < \Delta^{(undoped)}$). Thus, the GL doping by acceptors offers better GLIP performance in comparison with the GLIPs without doping of the GLs but at an elevated barrier height (larger difference in the GL and barrier material affinities).

Calculating both the photocurrent and the dark current in the GLIPs with the doped barriers, we have disregarded the effect of the donor and acceptor spatial fluctuation in the device plane. These fluctuations can lead to pronounced fluctuation of the electric field E_{GL} and, consequently, the tunneling current created by the photoexcited and thermalized electrons (see, for example, Refs. [40, 41]). As for the photocurrent and, hence, the GLIP responsivity the fluctuations in question promote an increase in these quantities. This implies that the values of the GLIP responsivity can somewhat exceed those obtained above. Since, considering, the detectivity focused on the GLIPs with undoped barrier layers, the problem of doping fluctuation is out of the scope of this work.

Since the photoexcited and injected electrons acquire kinetic energy propagating across the device under the electric field, they can be hot. If the electron energy relaxation length $L_\varepsilon = v_d\tau_\varepsilon$, where

v_d and τ_ε are the electron drift velocity and the energy relaxation time, respectively, exceeds the heterostructure period d, the electron effective temperature T_{eff} is mainly determined by the applied bias voltage V [42]. An increase in V leads to an increase of T_{eff} and to a drop of the capture efficiency p [30] (see also references therein). Since the responsivity $R_\omega \propto 1/p$ [see Eqs. (50.6) and (50.7)], the electron heating promotes higher values of the responsivity. As demonstrated by the particle Monte Carlo modeling of the electron capture into quantum wells (QWs) in heterostructures (albeit made of the standard material) with doped barriers [43], the doping affecting the potential profile in the barrier layers can result in a somewhat steeper drop of the capture efficiency with increasing voltage. The hot propagating electrons can provide a heating of the carriers localized in the GLs enhancing the electron thermionic escape. Apart from this, some fraction of the energy of the absorbed photons goes to the carrier heating [21]. This can lead to a decrease in the GLIP detectivity. However, one might expect that at a small capture efficiency and not too high radiation intensities, the negative impact of the heating is not too strong. The electric field across the GLs, modifying the wave functions of the photoexcited and thermalized electrons and, hence, the try-to-escape time, can lead to an increase in both the photocurrent and the dark current. This promotes higher values of the GLIP responsivity, but can add complexity to the voltage dependences of the GLIP detectivity. The consideration of the latter, as well as more rigorous treatment of the thermo-assisted tunneling, requires a generalization of the GLIP model that is beyond the scope of this work.

Generally, the selection of materials for the GLIPs from a wide variety of them is a matter of using of a proper band alignment (see, for example, Refs. [14, 44]). Several already fabricated and experimentally studied devices using the vdW heterostructures with the GLs, which can serve as the reference points for the GLIP realization, were reported recently [45–49]. The GLIPs with relatively low barriers able to operate in the THz range can be based on the Oxide family materials with the electron affinity close to that in GLs (for example, RuO_2 and TiO_2 [50]).

The comparison of the GLIP characteristics with the characteristics of photodetectors, using similar operation principles, namely with intersubband quantum well infrared photodetectors (QWIPs) [25, 29] shows the following advantages of the former:

(i) A higher responsivity due to both higher probability of the electron photoexcitation associated with the use of the interband transitions in the GLs and the intraband (intersubband) transitions in the QWs;

(ii) A higher detectivity associated with higher responsivity and weaker dark current (due to a larger activation energy);

(iii) Sensitivity to normally incident radiation (because of the use of the interband transition), so there is no need in special coupling structures;

(iv) No need in the GL doping, although such a doping, as shown above, provides an opportunity to vary the GLIP characteristics, in particular, enhancing the GLIP performance;

(v) Possibility of the GLIP operation in the range $\hbar\omega \simeq 0.025$ – 0.05 eV ($\omega/2\pi \simeq$ 6–12 THz), where using more conventional materials (e.g., III–V compounds) is hindered by optical phonon absorption;

(vi) Additional optimization is possible by changing the applied bias. As seen from Eqs. (50.4) and (50.5), such an optimization could be reached in the voltage range on the order of $V \sim 8\pi e\Sigma_B d_{GL}(\gamma_E^{3/2} + N)/\kappa_B$.

50.7 Conclusions

We studied the effect of the barrier layer and GL doping on the responsivity and detectivity of the GLIPs intended for the detection of infrared radiation. Using the developed device model, we demonstrated that the doping can result in a substantial modification of the spectral characteristics and enhancement of the GLIP responsivity and detectivity. The obtained results can be used for the characteristic optimization of the GLIP operating in different parts of the spectrum, including the terahertz range.

Acknowledgments

The authors are grateful to V. Ya. Aleshkin, A. A. Dubinov, A. Satou, and D. Svintsov for discussions and useful comments. The work at

RIEC and UoA was supported by the Japan Society for Promotion of Science, KAKENHI Grant No. 16H06361. The work at RIEC with UB was supported by the RIEC Nation-Wide Cooperative Research Project. VR also acknowledges the support by the Russian Scientific Foundation, Grant No. 14-29-00277. The work by V. L. was supported by the Russian Foundation for Basic Research, Grant No. 16-29-03402. The work at RPI was supported by the U.S. ARL Cooperative Research Agreement.

References

1. T. Mueller, F. N. A. Xia, and P. Avouris, *Nat. Photonics*, **4**, 297 (2010).
2. V. Ryzhii, M. Ryzhii, V. Mitin, and T. Otsuji, *J. Appl. Phys.*, **107**, 054512 (2010).
3. F. Bonaccorso, Z. Sun, T. Hasan, and A. C. Ferrari, *Nat. Photonics*, **4**, 611 (2010).
4. V. Ryzhii, N. Ryabova, M. Ryzhii, N. V. Baryshnikov, V. E. Karasik, V. Mitn, and T. Otsuji, *Opto-Electron. Rev.*, **20**, 15 (2012).
5. A. Tredicucci and M. S. Vitielo, *IEEE J. Sel. Top. Quantum Electron.*, **20**, 8500109 (2014).
6. A. K. Geim and I. V. Grigorieva, *Nature*, **499**, 419 (2013).
7. F. Xia, H. Wang, D. Xiao, M. Dubey, and A. Ramasubramaniam, *Nat. Photonics*, **8**, 899 (2014).
8. G. W. Mudd, S. A. Svatek, L. Hague, O. Makarovsky, Z. R. Kudrynsky, C. J. Mellor, P. H. Beton, L. Eaves, K. S. Novoselov, Z. D. Kovalyuk, E. E. Vdovin, A. J. Marsden, N. R. Wilson, and A. Patane, *Adv. Mater.*, **27**, 3760 (2015).
9. S. A. Svatek, G. W. Mudd, Z. R. Kudrynskyi, O. Makarovsky, Z. D. Kovalyuk, C. J. Mellor, L. Eaves, P. H. Beton, and A. Patane, *J. Phys.: Conf. Ser.*, **647**, 012001 (2015).
10. V. Ryzhii, T. Otsuji, V. Y. Aleshkin, A. A. Dubinov, M. Ryzhii, V. Mitin, and M. S. Shur, *Appl. Phys. Lett.*, **104**, 163505 (2014).
11. V. Ryzhii V, T. Otsuji, T. M. Ryzhii, M. V. Ya. Aleshkin, A. A. Dubinov, D. Svintsov, V. Mitin, and M. S. Shur, *2D Mater.*, **2**, 025002 (2015).
12. Q. Ma, T. I. Andersen, N. L. Nair, N. M. Gabor, M. Massicotte, C. H. Lui, A. F. Young, W. Fang, K. Watanabe, T. Taniguchi, J. Kong, N. Gedik, F. H. L. Koppens, and P. Jarillo-Herrero, *Nat. Phys.*, **12**, 455 (2016).
13. M. Massicotte, P. Schmidt, F. Vialla, K. G. Schadler, A. Reserbat-Plantey, K. Watanabe, T. Taniguchi, K. J. Tielrooij, and F. H. L. Koppens, *Nat. Nanotechnol.*, **11**, 42 (2016).

14. G. Gong, H. Zhang, W. Wang, L. Colombo, R. M. Wallace, and K. Cho, *Appl. Phys. Lett.*, **103**, 053513 (2013).
15. Z. Ben Aziza, H. Henck, D. Pierucci, M. G. Silly, E. Lhuillier, G. Patriarche, F. Sirotti, M. Eddrief, and A. Querghi, *ACS Nano*, **10**, 9679–9686 (2016).
16. D. Jariwala, T. J. Marks, and M. C. Hersam, *Nat. Mater.*, **16**, 170 (2017).
17. A. Castellanos-Gomez, M. Buscema, R. Molenar, D. Singh, L. Jansen, H. S. J. van der Zant, and G. A. Steel, *2D Mater.*, **1**, 011002 (2014).
18. M. Arai, R. Moriya, N. Yabuki, S. Masubuchi, K. Ueno, and T. Machida, *Appl. Phys. Lett.*, **107**, 103107 (2015).
19. Y. Sata, R. Moriya, S. Masabuchi, K. Watanabe, T. Taniguchi, and T. Machida, Jpn. *J. Appl. Phys., Part 1*, **56**, 04CK09 (2017).
20. V. Ryzhii, M. Ryzhii, D. Svintsov, V. Leiman, V. Mitin, M. S. Shur, and T. Otsuji, *Infrared Phys. Technol.*, **84**, 72 (2017).
21. V. Ryzhii, M. Ryzhii, D. Svintsov, V. Leiman, V. Mitin, M. S. Shur, and T. Otsuji, *Opt. Express*, **25**, 5536 (2017).
22. T. Akinwande, J. Zoum, M. S. Shur, and A. Gopinath, *IEEE Electron Device Lett.*, **11**, 332 (1990).
23. T. Ishibashi, *IEEE J. Sel. Top.*, **20**, 3804210 (2014).
24. H. C. Liu, *Appl. Phys. Lett.*, **60**, 1507 (1992).
25. F. Rosencher, B. Vinter, F. Luc, L. Thibaudeau, P. Bois, and J. Nagle, *IEEE J. Quantum Electron.*, **30**, 2875 (1994).
26. M. Ershov, V. Ryzhii, and C. Hamaguchi, *Appl. Phys. Lett.*, **67**, 3147 (1995).
27. L. Thibaudeau, P. Bois, and J. Y. Duboz, *J. Appl. Phys.*, **79**, 446 (1996).
28. V. Ryzhii, *J. Appl. Phys.*, **81**, 6442 (1997).
29. H. Schneider and H. C. Lui, *Quantum Well Infrared Photodetectors: Physics and Applications* (Springer, Berlin, 2007).
30. V. Ya. Aleshkin, A. A. Dubinov, M. Ryzhii, and V. Ryzhii, *J. Phys. Soc. Jpn.*, **84**, 094703 (2015).
31. V. Ryzhii, T. Otsuji, M. Ryzhii, V. Y. Aleshkin, A. A. Dubinov, V. Mitin, and M. S. Shur, Vertical electron transport in van der Waals heterostructures with graphene layers, *J. Appl. Phys.*, **117**, 154504 (2015).
32. S. M. Sze, *Physics of Semiconductor Devices* (Wiley, New York, 1999), p. 103.
33. K. K. Choi, *Physics of Quantum Well Infrared Photodetectors* (World Scientific, Singapore, 1997).

34. J. Martin, N. Akerman, G. Ulbricht, T. Lohmann, J. H. Smet, K. von Klitzing, and A. Yakoby, *Nat. Phys.*, **4**, 144 (2008).

35. Y. Zhang, V. Brar, C. Girit, A. Zett, and M. F. Cromme, *Nat. Phys.*, **5**, 722 (2009).

36. V. Ryzhii, M. Ryzhii, and T. Otsuji, *Appl. Phys. Lett.*, **99**, 173504 (2011).

37. F. T. Vasko, V. V. Mitin, V. Ryzhii, and T. Otsuji, *Phys. Rev. B*, **86**, 235424 (2012).

38. J. Jung, A. M. DaSilva, A. H. MacDonald, and S. Adam, *Nat. Commun.*, **6**, 6308 (2015).

39. M. Lundie, Z. Sljivancanin, and S. Tomic, *J. Phys.: Conf. Ser.*, **526**, 012003 (2014).

40. M. E. Raikh and I. M. Ruzhin, in *Mesoscopic Phenomena in Solids*, edited by B. L. Altshuler, P. A. Lee, and R. A. Webb (Elsevier, Amsterdam, 1991), p. 315.

41. A. Y. Shik, *Electronic Properties of Inhomogeneous Semiconductors* (Gordon and Breach, Luxemburg, 1995).

42. V. Ryzhii and R. Suris, Jpn. *J. Appl. Phys., Part 1*, **40**, 513 (2001).

43. M. Ryzhii, V. Ryzhii, and M. Willander, *Jpn. J. Appl. Phys., Part 1*, **38**, 6650 (1999).

44. J. Kang, S. Tongay, J. Zhou, J. B. Li, and J. Q. Wu, *Appl. Phys. Lett.*, **102**, 012111 (2013).

45. T. Georgiou, R. Jalil, B. D. Belle, L. Britnell, R. V. Gorbachev, S. V. Morozov, Y.-J. Kim, A. Gholinia, S. J. Haigh, O. Makarovsky, L. Eaves, L. A. Ponomarenko, A. K. Geim, K. S. Novoselov, and A. Mishchenko, *Nat. Nanotechnol.*, **7**, 100 (2013).

46. L. Britnell, R. V. Gorbachev, A. K. Geim, L. A. Ponomarenko, A. Mishchenko, M. T. Greenaway, T. M. Fromhold, K. S. Novoselov, and L. Eaves, *Nat. Commun.*, **4**, 1794 (2013).

47. A. Zubair, A. Nourbakhsh, J. Y. Hong, M. Qi, Y. Song, D. Jena, J. Kong, M. S. Dresselhaus, and T. Palacios, *Nano Lett.*, **17**, 3089 (2017).

48. X. Li, S. Lin, X. Lin, Z. Xu, P. Wang, S. Zhang, H. Zhong, W. Xu, Z. Wu, and W. Fang, *Opt. Express*, **24**, 134 (2016).

49. F. Withers, O. Del Pozo-Zamudio, A. Mishchenko, A. P. Rooney, A. Gholinia, K. Watanabe, T. Taniguchi, S. J. Haigh, A. K. Geim, A. I. Tartakovskii, and K. S. Novoselov, *Nat. Mater.*, **14**, 301–306 (2015).

50. M. Batmunkh, C. J. Shearer, M. J. Biggs, and J. G. Shapter, *J. Mater. Chem. A*, **3**, 9020 (2015).

Chapter 51

Real-Space-Transfer Mechanism of Negative Differential Conductivity in Gated Graphene-Phosphorene Hybrid Structures: Phenomenological Heating Model*

V. Ryzhii,[a,b,c] M. Ryzhii,[d] D. Svintsov,[e] V. Leiman,[e] P. P. Maltsev,[b] D. S. Ponomarev,[b] V. Mitin,[f] M. S. Shur,[g,h] and T. Otsuji[a]

[a]*Research Institute of Electrical Communication, Tohoku University, Sendai 980-8577, Japan*

[b]*Institute of Ultra High Frequency Semiconductor Electronics, RAS, Moscow 117105, Russia*

[c]*Center for Photonics and Infrared Engineering, Bauman, Moscow State Technical University, Moscow 111005, Russia*

[d]*Department of Computer Science and Engineering, University of Aizu, Aizu-Wakamatsu 965-8580, Japan*

[e]*Center for Photonics and 2D Material, Moscow Institute of Physics and Technology, Dolgoprudny 141700, Russia*

[f]*Department of Electrical Engineering, University at Buffalo, SUNY, Buffalo, New York 1460-1920, USA*

[g]*Departments of Electrical, Electronics, and Systems Engineering and Physics, Applied Physics, and Astronomy, Rensselaer Polytechnic Institute, Troy, New York 12180, USA*

[h]*Electronics of the Future, Inc., Vienna, Virginia 22181, USA*

v-ryzhii@riec.tohoku.ac.jp

---

*Reprinted with permission from V. Ryzhii, M. Ryzhii, D. Svintsov, V. Leiman, P. P. Maltsev, D. S. Ponomarev, V. Mitin, M. S. Shur, and T. Otsuji (2018). Real-space-transfer mechanism of negative differential conductivity in gated graphene-phosphorene hybrid structures: phenomenological heating model, *J. Appl. Phys.*, **124**, 114501. Copyright © 2018 AIP Publishing LLC.

Graphene-Based Terahertz Electronics and Plasmonics: Detector and Emitter Concepts
Edited by Vladimir Mitin, Taiichi Otsuji, and Victor Ryzhii

ISBN 978-981-4800-75-4 (Hardcover), 978-0-429-32839-8 (eBook)
www.jennystanford.com

We analyze the nonlinear carrier transport in the gated graphene-phosphorene (G-P) hybrid structures—the G-P field-effect transistors using a phenomenological model. This model assumes that due to high carrier densities in the G-P-channel, the carrier system, including the electrons and holes in both the G- and P-layers, is characterized by a single effective temperature. We demonstrate that a strong electric-field dependence of the G-P-channel conductivity and substantially nonlinear current-voltage characteristics, exhibiting a negative differential conductivity, are associated with the carrier heating and the real-space carrier transfer between the G- and P-layers. The predicted features of the G-P-systems can be used in the detectors and sources of electromagnetic radiation and in the logical circuits.

51.1 Introduction

Unique properties of Graphene (G) [1] and recent advances in technology of van der Waals materials [2, 3] present an excellent opportunity for developing effective electronic and optoelectronic devices [3–15]. Combining the G-layers with the gapless energy spectrum and enhanced electron (and hole) mobility and a few-layer black phosphorus layer or phosphorene (P) [16–23] exhibiting the flexibility of the band structure open up remarkable prospects for the creation of novel devices, in particular, photodetectors [24].

The G-P hybrid structures can be used in the real-space-transfer (RST) devices. The RST devices exhibiting the negative differential conductivity (NDC) have attracted a lot of attention since their proposal by Gribnikov in early 1970s [25] and further developments (for example, Refs. [26–31] see also Refs. [32, 33] and references therein). The RST devices exhibit interesting features including high speed operation. Most of the RST devices have been based on the A_3B_5 heterostructures. Their operation is associated with the electric-field heating of the electron gas in the narrow-gap channel (in particular, GaAs channel), resulting in the transfer of the hot relatively light electrons to an adjacent wide-gap layer with a higher electron effective mass (such as an $Al_xGa_{1-x}As$-layer). A decrease in the fraction of the light electrons accompanied by an increase in the heavy electrons with an increasing electric field results in the roll-off of the net electron system conductivity and could lead to the NDC.

In this chapter, we propose to use the effect of the RST in the gated graphene-phosphorene (G-P) hybrid sandwich-like structures, i.e., in the G-P-channel field-effect transistors (G-P-FETs) and evaluate the characteristics of such devices. In contrast to the effect of NDC in the standard semiconductors (in which the net electron or hole density does not markedly vary even at a strong heating), in the G-P-channels the net carrier density can be pronouncedly changed. This adds a substantial complexity to the operation of the G-P-channel devices.

51.2 Model

Figure 51.1 demonstrates a schematic view of the gated G-P structure (i.e., a FET with the G-P channel). It is assumed that the P-layer consisting of a few atomic layers is oriented in such a way that the direction from the FET source to its drain corresponds to the zigzag direction. The dynamics of electrons and holes in this direction is characterized by a large effective mass. As a result, the RST of the electrons and holes from the G-layer (where their mobility can be very high) to the P-layer (with relatively low mobility in the direction in question) can enable sharp current-voltage characteristics with an elevated peak-to-valley ratio (i.e., a large absolute value of the NDC). The selection of the P-layer for the hybrid structure under consideration is associated not only with a high electron and hole mass (and low mobility) in the electric-field direction, but also with a wide opportunity to provide a desirable height of the barrier between the P-layer conduction band bottom and the Dirac point in the G-layer. This can be realized by a proper choice of the number of atomic layers in the P-layer [16, 17].

For the sake of definiteness, we consider the P-layer consisting of a few atomic P-layers (N = 2–3), assuming that it is generally doped (the pristine P-layers are of p-type). The gate voltage V_G can substantially vary the carrier densities in the G-P-channel, so that the latter comprises the two-dimensional electron and hole gases in both the G- and P-layers.

The band gap Δ and the energy spacing Δ^e and Δ^h, between the Dirac point in the G-layer and edges of the conduction and valence bands, (determined by the pertinent work functions) depend on the number N. In the heterostructure under consideration, $\Delta^e \sim \Delta^h \simeq$ 0.4–0.45 eV ($N = 3$) [16].

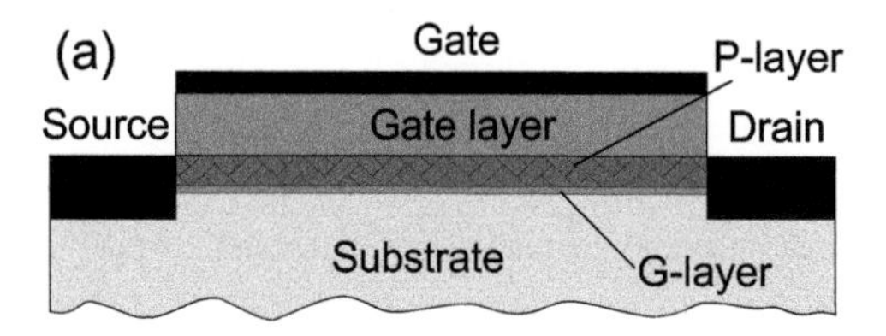

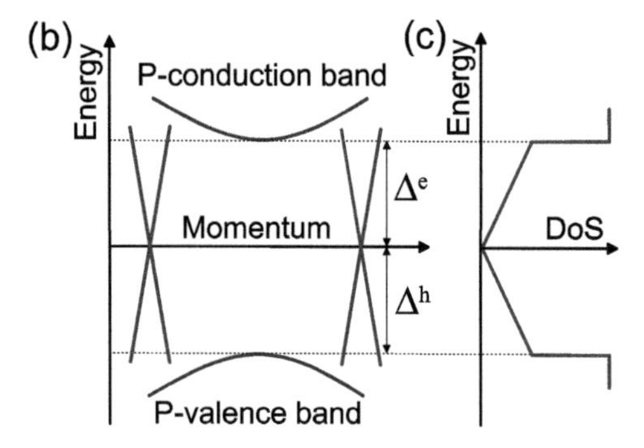

Figure 51.1 Schematic view of (a) the G-P FET structure, (b) energy spectrum, and (c) density of states (DoS). Cone-shaped parts of the energy spectrum correspond to the K- and K′-valleys in G-layers.

To calculate the G-P FET current-voltage characteristics and evaluate the NDC, as a first step, we use a simplified semi-classical phenomenological model for the electrons and holes in such a FET channel, although for more rigorous treatment of the RST, the quantum approach is needed [34–36].

We set the dispersion relations for the electrons and holes in the G- and P-layers as follows:

$$\varepsilon_G^e = \upsilon_W\sqrt{p_x^2 + p_y^2}, \quad \varepsilon_p^e = \Delta^e + \frac{p_x^2}{2m_{xx}} + \frac{p_y^2}{2m_{yy}}, \tag{51.1}$$

$$\varepsilon_G^h = -\upsilon_W\sqrt{p_x^2 + p_y^2}, \quad \varepsilon_p^h = -\Delta^h - \frac{p_x^2}{2m_{xx}} - \frac{p_y^2}{2m_{yy}}, \tag{51.2}$$

respectively. Here, $v_W \simeq 10^8$ cm/s is the characteristic velocity of electrons in the G-layers, $m_{xx} = M$ and $m_{yy} = m$ are the components of the effective mass tensor ($M \gg m$), and p_x and p_y are the carrier momenta in the source-drain direction and the perpendicular direction, respectively. The components of the effective mass tensor for both conduction and valence bands for $N = 2$–3 are approximately as follows: $m \simeq 0.06\ m_0$ and $M \simeq 1.35\ m_0$ ($N = 3$), where m_0 is the mass of bare electrons.

Figure 51.1 shows schematically the G-P-FET structure, the energy spectrum of electrons and holes, and the pertinent energy dependence of the density of states, corresponding to Eqs. (51.1) and (51.2).

At a relatively high frequency of the electron-electron, hole-hole, and electron-hole collisions, the electron-hole plasma in the G-P channel can be characterized by its common electron effective temperature T (in the energy units, equal for both the G- and P-layers) but generally different quasi-Fermi energies $|\mu^e| \neq |\mu^h|$, so that the electron and hole distribution functions are $f^e(\varepsilon) = \{\exp[(\varepsilon - \mu^e)/T] + 1\}^{-1}$ and $f^h(\varepsilon) = \{\exp[(\varepsilon - \mu^h)/T] + 1\}^{-1}$, respectively.

The model under consideration is based on the following assumptions:

(1) The sufficiently frequent inter-carrier collisions enable the establishment of the quasi-Fermi distributions with the common effective temperature, T for all electron and hole components in both G- and P-layers. Sufficiently strong interactions between the electrons (and holes) belonging to neighboring layers promote the inter-layer equilibrium [37–39].

(2) Due to heavy electron and hole effective masses M and $\sqrt{mM}$, the conductivity of the P-layer is relatively small because this layer mobility in the direction corresponding to the mass M is proportional to $1/\sqrt{mM}\ M$ [40]. Hence, we disregard the P-layer conductivity in comparison with the G-layer conductivity. The same assumption is valid if the P-layer is disconnected from the source and drain contacts. We also neglect the contribution of the heavy carriers to the energy balance.

(3) The momentum relaxation of the electrons and holes in the G-layer (light electrons and holes) is due to their scattering on defects, impurities, acoustic phonons, and the heavy electrons and holes in the P-layer. The latter and the charged

impurities are assumed to be screened (see Section 50.5). We believe that the energy relaxation at room temperature under consideration is associated with optical phonons in the G-layer, and the interband transitions assisted by the optical phonons are the main recombination-generation mechanisms, neglecting the Auger generation-recombination processes. Due to the prohibition of the Auger processes in the G-layers with the ideal linear gapless energy spectrum because of the energy and momentum conservation laws [41], even in non-ideal G-layers (see Ref. [42] and references therein), there is an ambiguity of the characteristic times ratio, $\tau_0^{\rm inter}/\tau_{\rm Auger}^{\rm inter}$, of the interband transitions mediated by the optical phonons and electron-hole (Auger) processes, particularly, at different carrier temperatures. The characteristic time $\tau_{\rm Auger}^{\rm inter}$ can be fairly different depending on the dielectric constant of the substrate κ and the spacing between the G-P-channel and the gate $W_{\rm g}$ [42]. The case of $\tau_0^{\rm inter} \ll \tau_{\rm Auger}^{\rm inter}$ which is under consideration in the following, can conditionally correspond to a large κ and a small $W_{\rm g}$.

(4) At relatively short characteristic times of the electron-hole generation-recombination associated with the optical phonons, the quasi-Fermi energies (counted from the Dirac point) of the electron and hole components can be generally different ($\mu^{\rm e} \neq -\mu^{\rm h}$).

51.3 Main Equations of the Model

51.3.1 Conductivity of the G-P Channel

Considering the above model, we use the following formula for the net conductivity of the G-P channel σ (see, in particular, Refs. [43–53])

$$\sigma = -\frac{e^2 T_0 \tau_0}{\pi \hbar^2}\left(\frac{T}{T_0}\right)^{l+1}\frac{\Sigma_{\rm I}}{(\Sigma_{\rm I}+\Sigma_{\rm P})}$$

$$\times\int_0^\infty d\xi\xi^{l+1}\frac{d}{d\xi}[f^{\rm e}(\xi)+f^{h}(\xi)]. \tag{51.3}$$

Here, $\tau(p) = \tau_0(pv_W/T_0)^l = \sigma_0(T/T_0)^l\xi^l$ is the momentum relaxation time, $\xi = pv_W/T_0$ is the normalized carrier energy in the G-layer, Σ_I is the density of the scatterers, $\Sigma_P(T)$ is the carrier density in the P-layer at the effective temperature T (which, due to their large effective mass, can also be considered as the effectively screened Coulomb scatterers for the light electron and holes in the G-layer), and $\hbar$ is the Planck constant. If the weakly screened Coulomb scattering prevails, $l = 1$ (for example, Refs. [43, 45, 46, 50, 52]). In the case of dominant scattering on neutral scatterers, $l = -1$.

Due to the above assumptions, in the following we set $l = -1$. In this case, Eq. (51.3) yields

$$\sigma = \sigma_0[f^e(0) + f^h(0)]\frac{\Sigma_I}{(\Sigma_I + \Sigma_P)}. \qquad (51.4)$$

Here, $\sigma_0 = (e^2T_0\tau_0/\pi\hbar^2)$ is the characteristic conductivity (it is equal to the low electric-field conductivity in the case of neutral scatterers).

51.3.2 Carrier Interband Balance

The carrier densities, Σ_G and Σ_P, in the G- and P-layer are, respectively, given by

$$\Sigma_G = \Sigma_0\left(\frac{T}{T_0}\right)^2\left[\mathcal{F}_1\left(\frac{\mu^e}{T}\right) + \mathcal{F}_1\left(\frac{\mu^h}{T}\right)\right], \qquad (51.5)$$

$$\Sigma_P = \Sigma_N\left(\frac{T}{T_0}\right) \times \ln\left\{\left[1 + \exp\left(\frac{\mu^e - \Delta^e}{T}\right)\right] \times \left[1 + \exp\left(\frac{\mu^h - \Delta^h}{T}\right)\right]\right\}, \qquad (51.6)$$

where

$$\mathcal{F}_1(a) = \int_0^\infty \frac{d\xi\xi}{\exp(\xi - a) + 1}$$

is the Fermi-Dirac integral [54], $\Sigma_0 = 2T_0^T/\pi\hbar^2 v_W^2$, and $\Sigma_N = NT_0\sqrt{mM}/\pi\hbar^2$. The factor N in the latter formula reflects the fact that the density of states in the few-layer P-layer scales roughly with the layer number N [38].

The net surface charge density in the G-P-channel $e\Sigma = \kappa\,|V_G - V_{CNP}\,|/4\pi W_g$ induced by the applied gate voltage V_G comprises the electron, Σ_G^e and Σ_P^e, and hole, Σ_G^h and Σ_P^h densities. Here, $V_g = V_G - V_{CNP}$ the gate-voltage swing, $V_{CNP} \propto \Sigma_I^{ch}$ is the voltage, which corresponds to the charge-neutrality point, Σ_I^{ch} is the density of non-compensated charged impurities, κ and W_g are the background dielectric constant and the thickness of the gate layer, respectively, and e is the electron charge. Considering the above, the gate voltage swing V_g and the quantities T, μ^e, and μ^h are related to each other as

$$\frac{V_g}{V_0} = \left(\frac{T}{T_0}\right)^2 \left[\mathcal{F}_1\left(\frac{\mu^e}{T}\right) - \mathcal{F}_1\left(\frac{\mu^h}{T}\right)\right]$$

$$+\gamma_0 \frac{T}{T_0} \ln \frac{\left[1 + \exp\left(\frac{\mu^e - \Delta^e}{T}\right)\right]}{\left[1 + \exp\left(\frac{\mu^h - \Delta^h}{T}\right)\right]}, \qquad (51.7)$$

Here, $\gamma_0 = N\sqrt{mM}v_W^2/2T_0$, $V_0 = 8eT_0^2\,W_g/\kappa\hbar^2\,v_W^2$ and N is the number of the monolayers in the P-layer (for a moderate N). The parameter γ_0 can be large due to a relatively high density of states in the P-layer.

In the limit $T = 0$ at not too high gate voltage swing $V_G - V_{Dirac}$ when the P-layer is empty, Eq. (51.7) yields the standard expression for the Fermi energy degenerate electron gas in the G-layer: $\mu = \hbar v_W \sqrt{\pi\Sigma}$. If $\Sigma_N = 0$ at very high electron and hole effective temperatures, μ tends to zero as $\mu \propto T^{-1}$.

In the following for definiteness and simplicity, we set $\Delta^e = \Delta^h = \Delta/2$ [this is approximately valid for $N = 2$ and $N = 3$ (Ref. [16])].

51.3.3 Generation-Recombination and Energy Balance Equations

The equation governing the interband balance of the carriers can be generally presented as

$$G_{Auger} + G_{op} + G_{ac} + G_{rad} = 0, \qquad (51.8)$$

where the terms in Eq. (51.8) correspond to different processes: the interband Auger generation-recombination processes and the processes associated with optical-phonon, acoustic-phonon, and radiative transitions (in particular, indirect transitions for which some selection restrictions are lifted). At sufficiently fast processes of the optical phonon decay into acoustic phonons followed by their effective removal, this is confirmed by high values of the G-layer thermal conductivity [55–58], the optical phonon system in the heterostructures under consideration is in equilibrium with the thermal bath with the temperature T_0.

For the optical-phonon term (under a rather natural assumption $\hbar\omega_0 > |\mu^e|, |\mu^h|, T$), we use the following expression [59, 60] (see also Ref. [61])

$$G_{op} = \frac{\Sigma_0}{\tau_0^{inter}}\left[\mathcal{N}_0 - (\mathcal{N}_0 + 1)\exp\left(\frac{\mu^e + \mu^h - \hbar\omega_0}{T}\right)\right]. \tag{51.9}$$

Here, $\mathcal{N}_0 = [\exp(\hbar\omega_0/T_0) - 1]^{-1} \simeq \exp(-\hbar\omega_0/T_0)$ is the equilibrium distribution function (the Planck distribution function) optical phonons in the G-layer with the energy $\hbar\omega_0$, τ_0^{inter} is the characteristic time of the spontaneous optical phonon emission accompanied by the interband transitions.

In the situation under consideration, we present the energy balance equation in the following form:

$$\frac{\hbar\omega_0\Sigma_0}{\tau_0^{inter}}\left[(\mathcal{N}_0 + 1)\exp\left(\frac{\mu^e + \mu^h - \hbar\omega_0}{T}\right) - \mathcal{N}_0\right]$$
$$+ \frac{\hbar\omega_0\Sigma_0}{\tau_0^{intra}}\left[(\mathcal{N}_0 + 1)\exp\left(-\frac{\hbar\omega_0}{T}\right) - \mathcal{N}_0\right] = \sigma E^2. \tag{51.10}$$

Here, τ_0^{intra} is the characteristic time of the spontaneous optical phonon emission accompanied by the intraband electron and hole transitions, $\Sigma_G \propto [\mathcal{F}_1(\mu^e/T) + \mathcal{F}_1(\mu^h/T)](T/T_0)^2$ is approximately equal to the net carrier density in the G-layer (a small deviation from the real carrier density is associated with the dependence of the optical phonon emission and absorption probability on the carrier energy, which, in turn, is due to the density of state linearity) and $E \simeq V_{sd}/L$ is the longitudinal source-to-drain electric field in the channel (V_{sd} is the voltage applied between the source and drain contacts and L is the length of the channel).

Due to relatively high values of $\hbar\omega_0$ compared to T_0, we set in the following $\mathcal{N}_0 \simeq \exp(-\hbar\omega_0/T_0) \ll 1$. Setting $\hbar\omega_0 \simeq 0.2$ eV and $T_0 = 0.025$ eV, we obtain the estimate $\mathcal{N}_0 \simeq 3.35 \times 10^{-4}$.

As assumed above, the electron-hole generation-recombination processes are associated primarily with the optical phonon spontaneous emission, (the pertinent characteristic time $\tau_0^{\rm inter}$ is much shorter than $\tau_{\rm Auger}^{\rm inter}$ associated with the Auger processes), so that the equation governing the electron and hole balance acquires the following form:

$$\exp\left(\frac{\mu^{\rm e} + \mu^{\rm h} - \hbar\omega_0}{T}\right) - \exp\left(-\frac{\hbar\omega_0}{T}\right) = 0. \qquad (51.11)$$

Equation (51.11) yields

$$\mu^{\rm e} + \mu^{\rm h} = \hbar\omega_0\left(1 - \frac{T}{T_0}\right). \qquad (51.12)$$

51.3.4 General Set of the Equations

Considering Eqs. (51.4)–(51.6) and (51.11), we arrive at the following set of the equations governing the carrier effective temperature, quasi-Fermi energy, conductivity, and the G-P channel current-voltage characteristics:

$$\frac{V_{\rm g}}{V_0} = \left[\mathcal{F}_1\left(\frac{\mu}{T}\right) - \mathcal{F}_1\left(-\frac{\mu}{T} - \hbar\omega_0\left(\frac{1}{T_0} - \frac{1}{T}\right)\right)\right]\frac{T^2}{T_0^2} + \gamma_0\left(\frac{T}{T_0}\right)$$
$$\times\left\{\exp\left(\frac{\mu - \Delta/2}{T}\right) - \exp\left[-\frac{\mu + \Delta/2}{T} - \hbar\omega_0\left(\frac{1}{T_0} - \frac{1}{T}\right)\right]\right\}, \qquad (51.13)$$

$$\frac{\sigma}{\sigma_0} = \frac{1}{1 + \dfrac{\Sigma_{\rm N}}{\Sigma_{\rm I}}\dfrac{T}{T_0}\exp\left(-\dfrac{\Delta}{2T}\right)\left\{\exp\left(\dfrac{\mu}{T}\right) + \exp\left[-\dfrac{\mu}{T} - \hbar\omega_0\left(\dfrac{1}{T_0} - \dfrac{1}{T}\right)\right]\right\}}$$
$$\times\left\{\frac{1}{\exp\left(-\dfrac{\mu}{T}\right) + 1} + \frac{1}{\exp\left[\dfrac{\mu}{T} + \hbar\omega_0\left(\dfrac{1}{T_0} - \dfrac{1}{T}\right)\right] + 1}\right\}, \qquad (51.14)$$

$$\left[\mathcal{F}_1\left(\frac{\mu}{T}\right)+\mathcal{F}_1\left(-\frac{\mu}{T}-\hbar\omega_0\left(\frac{1}{T_0}-\frac{1}{T}\right)\right)\right]$$
$$\times\left[\exp\left(-\frac{\hbar\omega_0}{T_0}\right)-\exp\left(-\frac{\hbar\omega_0}{T_0}\right)\right]\frac{T^2}{T_0^2}=\frac{\sigma}{\sigma_0}\frac{E^2}{E_0^2}, \quad (51.15)$$

$$j=\sigma(E)E. \quad (51.16)$$

Here,

$$E_0=\sqrt{\frac{2\hbar\omega_0}{T_0}}\left(\frac{T_0}{ev_W\sqrt{\tau_0\tau_0^{\mathrm{intra}}}}\right). \quad (51.17)$$

The channel current can be normalized by

$$j_0=\sigma_0E_0=\frac{eT_0^2}{\pi v_W\hbar^2}\sqrt{\frac{2\hbar\omega_0}{T_0}\frac{\tau_0}{\tau_0^{\mathrm{intra}}}}. \quad (51.18)$$

In the particular case when $V_g = 0$, i.e., it corresponds to the Dirac point ($V_G = V_{CNP}$), the above system of equations can be simplified. Indeed, in such a case, Eq. (51.13) results in

$$\frac{\mu}{T}=-\frac{\hbar\omega_0}{2}\left(\frac{1}{T_0}-\frac{1}{T}\right) \quad (51.19)$$

and, hence, Eq. (51.14) becomes as follows:

$$\frac{\sigma}{\sigma_0}=\frac{1}{1+\frac{2\Sigma_N}{\Sigma_I}\frac{T}{T_0}\exp\left(-\frac{\Delta}{2T}\right)\exp\left[-\frac{\hbar\omega_0}{2}\left(\frac{1}{T_0}-\frac{1}{T}\right)\right]}$$
$$\times\frac{2}{1+\exp\left[\frac{\hbar\omega_0}{2}\left(\frac{1}{T_0}-\frac{1}{T}\right)\right]}. \quad (51.20)$$

Equation (51.20) explicitly demonstrates an increase in the conductivity Σ with increasing effective temperature T.

51.4 Numerical Results

The set of Eqs. (51.13)–(51.16) was solved numerically, to obtain the effective-temperature and electric-field dependences. The pertinent calculation results are shown in Figs. 51.2–51.7. We set $N = 3$, Δ = 900 meV, $\sqrt{mM} = 0.29m_0 = 2.65 \times 10^{-28}$ g, $\hbar\omega_0$ = 200 meV, and T_0 = 25 meV, so that $\gamma_0 = 32$, $\Sigma_N = 1.05 \times 10^{13}$ cm^{-2}. The scatterer density was assumed to be in the range $\Sigma_I = 10^{10} - 10^{12}$ cm^{-2}. The relative gate voltage swing V_g/V_0 varied from zero to ten. The dependences on all plots below are normalized by j_0 and E_0 corresponding to $\Sigma_I = 10^{11}$ cm^{-2} assuming that $\tau_0 \propto \Sigma_I^{-1}$.

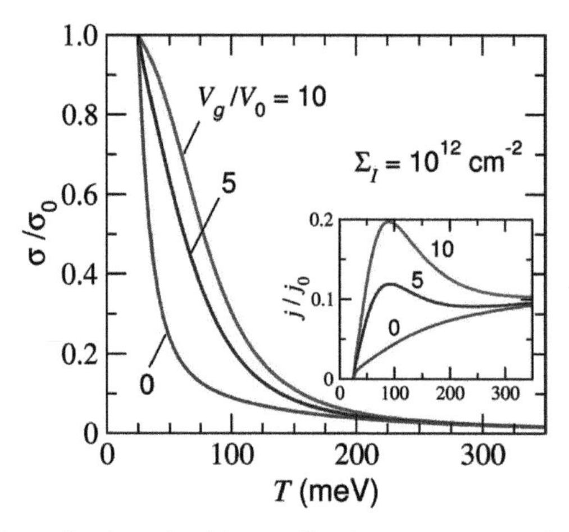

Figure 51.2 Normalized conductivity vs effective temperature. The inset shows the pertinent temperature dependence of the normalized current density.

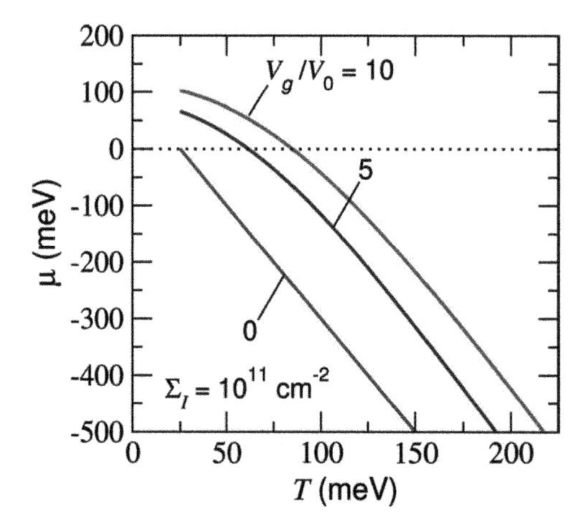

Figure 51.3 Quasi-Fermi energy μ vs effective temperature for $\Sigma_I = 10^{11}$ cm^{-2} at different normalized gate-voltage swings V_g/V_0.

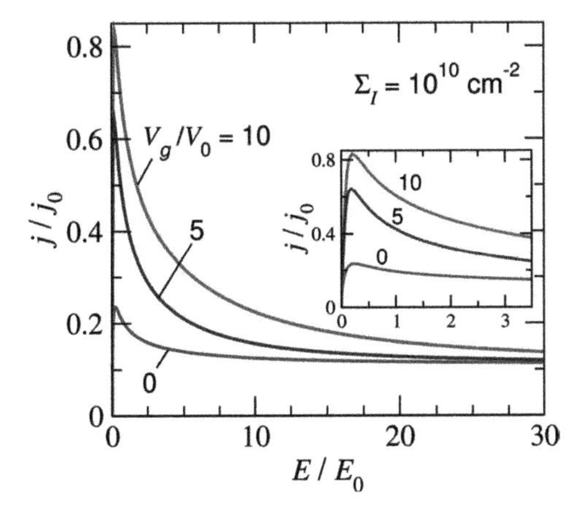

Figure 51.4 Current-voltage characteristics for Σ_I = 10^{10} cm^{-2} at different normalized gate-voltage swings V_g/V_0.

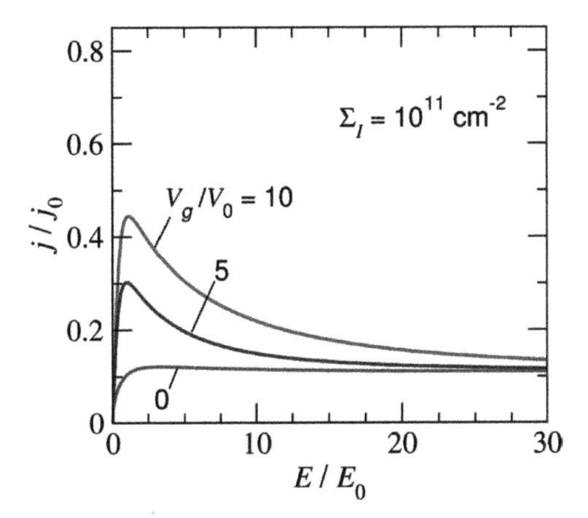

Figure 51.5 The same as in Fig. 51.4, but for Σ_I = 10^{11} cm^{-2}.

An increase in the electric field leads to a rise of the effective temperature. Figure 51.2 shows examples (for Σ_I = 10^{12} cm^{-2}) of the normalized G-P channel conductivity σ/σ_0 as a function of the effective temperature $T \geq T_0$ = 25 meV. As seen from the plots in Fig. 51.2, σ/σ_0 exhibits a steep drop with increasing T at all V_g/V_0. At smaller values of Σ_I, the σ/σ_0 versus T relation becomes even steeper. However, as shown in the inset in Fig. 51.2, the j/j_0 – T relation, found as an example for Σ_I = 10^{11} cm^{-2}, can be qualitatively different depending on V_g/V_0. Figure 51.3 shows that the quasi-Fermi μ also exhibits a steep drop when T increases (for $V_g > 0$). Moreover, μ changes its sign at certain values of T depending on V_g/V_0. This implies that the electron gas in the G-layer being degenerate at a moderate heating, i.e., at weak electric fields, becomes nondegenerate with increasing effective temperature. The variations of the effective temperature and the quasi-Fermi energy markedly affect the distribution of the carriers between the G- and P-layers and, hence, the current-voltage characteristics.

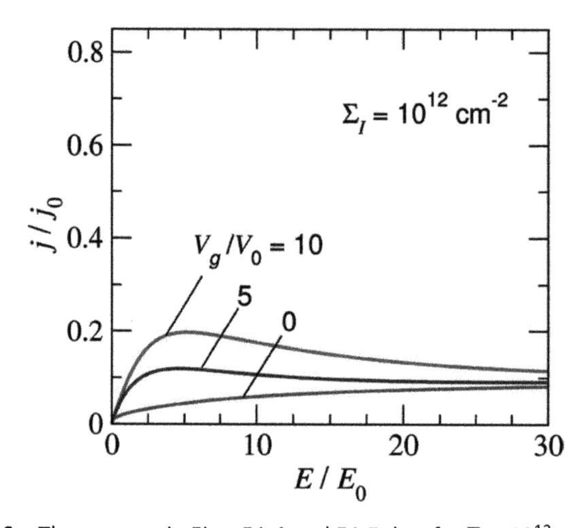

Figure 51.6 The same as in Figs. 51.4 and 51.5, but for $\Sigma_I = 10^{12}$ cm^{-2}.

Figures 51.4–51.6 demonstrate the current-voltage characteristics of the G-P channels with different scatterer density at different gate voltages. One can see that the shape of the current-voltage characteristics, in particular, the height of the current peaks and the peak-to-valley ratio, are different in the samples with different scatterer densities Σ_I. However, the NDC preserves when Σ_I varies in a rather wide range.

Setting $\tau_0 \simeq 1$ ps and $\tau_0^{\text{inter}} \simeq 0.1 - 1.0$ ps for room temperature, Eqs. (51.17) and (51.18) yield the following estimates: $E_0 \simeq 86 - 272$ V/cm and $j_0 \simeq 3.5 - 11.0$ A/cm. At $\kappa = 4$ and $W_g = 10^{-5}$–10^{-6} cm, the gate voltage is normalized by $V_0 \simeq 0.494 - 4.94$ V. Assuming that $\Sigma_I = 10^{10}$ cm^{-2} and using the peak values of the current density from Fig. 51.4, we obtain the following estimate for the FET transconductance $g = \Delta j/\Delta V_g$: $g \simeq 42 - 420$ mS/mm. In the FETs with higher background dielectric constant κ and thinner gate layer W_g, the transconductance can be markedly larger.

In Fig. 51.7, we compare the current-voltage characteristics of two G-P channels both with $\Sigma_I = 10^{11}$ cm^{-2} but with different energy gaps (Δ = 900 meV and Δ = 600 meV). As seen, the current-voltage characteristics maxima increase with increasing Δ and somewhat (weakly) shift toward higher electric fields. In the P-layers with the number of the atomic layers ($N > 3$), $\Delta^e \neq \Delta^h$ (actually $\Delta^e > \Delta^h$). The results obtained above are qualitatively valid in such cases as well, but in the above formulas one needs to replace $\Delta/2$ by Δ^h. One needs to note that the shape of the curves shown in Figs. 51.4–51.7 does not change with varying parameter τ_0^{intra} because they correspond to the normalized quantities j/j_0 and E/E_0 with j_0/E_0 independent of τ_0^{intra}.

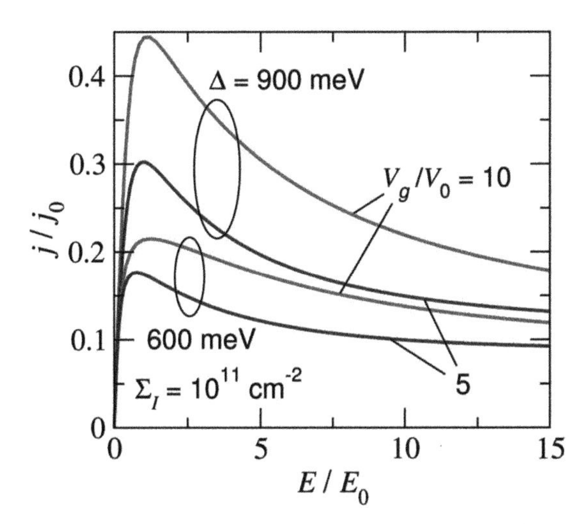

Figure 51.7 Current-voltage characteristics for Δ = 900 meV (as in Fig. 51.5) and Δ = 600 meV at different V_g/V_0 ($\Sigma_I = 10^{11}$ cm^{-2}).

51.5 Discussion

51.5.1 Screening in the G-P Channel

The screening length in a semiconductor with DoS(ε), shown in Fig. 51.1c, is given by

$$l_s^{-1} = -\frac{2\pi e^2}{\kappa}\int_0^{+\infty} \mathrm{DoS}(\varepsilon)\frac{d(f^e + f^h)}{d\varepsilon}d\varepsilon. \qquad (51.21)$$

As the G- and P-layers are located close to each other (at the distance below the Fermi wavelength), their inverse screening lengths are additive. A simple evaluation with linear DoS in graphene and constant DoS in phosphorene leads us to

$$l_s^{-1} = \frac{e^2}{\kappa\hbar^2}\left\{\frac{4T}{\upsilon_W^2}e\ln\left[(1+e^{\mu^e/T})(1+e^{\mu^h/T})\right] + 2N\sqrt{mM}\right.$$
$$\left.\times\left[\frac{1}{\exp\left(\frac{\Delta/2-\mu^e}{T}\right)+1} + \frac{1}{\exp\left(-\frac{\Delta/2+\mu^h}{T}\right)+1}\right]\right\}. \qquad (51.22)$$

The first term comes from screening in the G-layer, and the second one—from the P-layer. All the first factors in square brackets can be considered as an effective relativistic mass in G-layers. At $\mu^e = \mu^h = 0$, Eq. (51.22) yields

$$l_s^{-1} = \frac{e^2}{\kappa\hbar^2}\left[8\ln 2\frac{T}{\upsilon_W^2} + 2N\sqrt{mM}\right] > 8\alpha_g k_T \ln 2, \qquad (51.23)$$

where $\alpha_g = (e^2/\hbar\upsilon_W\,\kappa)$ is the coupling constant for G-layers and $k_T = T/\hbar\upsilon_W$ is the characteristic carrier wavenumber. From Eq. (51.23), we have the following estimate: $k_T l_s \lesssim (8\alpha_g \ln 2)^{-1}$, at $\kappa = 4$, one obtains $k_T l_s \simeq 137/600 \ln 2 \simeq 0.158 \ll 1$. The dependence of the screening length l_s on the quasi-Fermi energy $\mu = \mu^e$ (assuming that $\mu^h = \hbar\omega_0\,(1 - T/T_0) - \mu^e$) calculated using Eq. (51.23) is plotted in Fig. 51.8 ($\kappa = 4$ and $N = 3$). One can see that an increase in $|\mu|$ leads to a decrease in the screening length l_s. The temperature dependence for $\mu = 0$ shown in the inset in Fig. 51.8 indicates that l_s decreases with increasing T except a narrow region near T_0, i.e., at low electric fields. Relatively ineffective screening at such fields somewhat affects the low-field conductivity {due to a distinction between the momentum dependences for the Coulomb scattering and the scattering on the neutral disorder [see Eq. (51.3)]}, but is not important at high fields at which the NDC appears because of the screening reinforcement. One can see that an increase in T (the carrier heating) and μ leads to a decrease in l_s in comparison to the above estimate, so that both inverse characteristic wavenumbers $k_T^{-1} = \hbar v_W/T$ and $k_\mu^{-1} = \hbar v_W/\mu$ are markedly larger than l_s (see the dashed lines in Fig. 51.8). This implies that the assumption of the complete screening approximation [in particular, setting $l = -1$ in Eq. (51.3)] is well-justified, particularly taking into account the contribution of the heavy carriers in the P-layer. In the case of intermediate screening, one can set $l = 0$ (as it is done, for example, in Ref. [62], but contradict the above evaluation of the screening). This might somewhat slows down the NDC due to a weaker conductivity temperature dependence but does not change the qualitative picture of the effect under consideration.

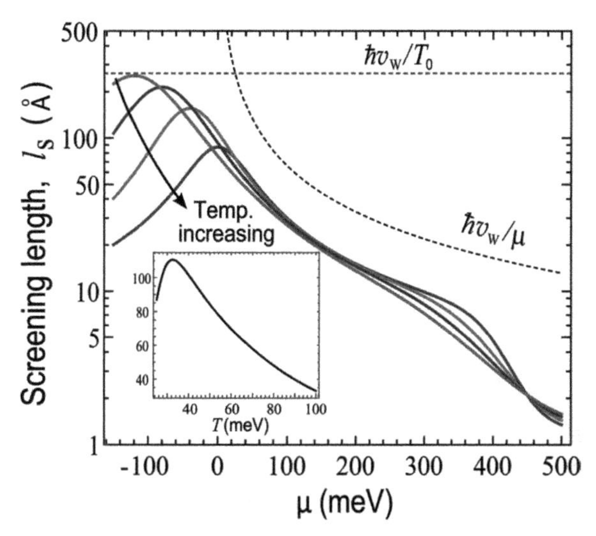

Figure 51.8 Dependence of screening length in the G-P channel as a function of Fermi energy at various carrier effective temperatures increasing from T = 25 meV with 10 meV step. Number of P-layers N = 3 and background dielectric constant κ = 4. Electrons and holes have different quasi-Fermi energies bound by $\mu^h = \hbar\omega_0(1 - T/T_0) - \mu^e$. Inset: detailed temperature dependence of the screening length at charge neutrality point μ = 0. Dashed lines correspond to $k_T^{-1}|_{T=T_0} = \hbar v_W/T_0$ and $k_\mu^{-1} = \hbar v_W/\mu$ versus μ.

51.5.2 Mutual Scattering of Electrons and Holes

At the gate voltages corresponding to the states close to the Dirac point, the mutual scattering of the electrons and holes in the G-layer can affect the conductivity of the latter. However, due to special features of the scattering of the carriers with the linear dispersion law [46, 47], such a scattering is similar to the scattering on uncharged and screened charged impurities, as well as acoustic phonons and defects. Hence, the inclusion of the inter-carrier scattering into the model should not markedly change the above results.

51.5.3 Optical Phonon Heating

In the case of relatively slow optical phonon decay processes, the heating of the optical phonon system can be substantial, so that its effective temperature T_0^{opt} can markedly exceed the thermal bath temperature T_0 being close to the lattice temperature in the G-P channel (the temperature of acoustic phonons in this channel) and the carrier effective temperatures T in the G-P channel. If such a decay is a "bottleneck," one can put $T_0^{opt} \simeq T$. In this case, both T and T_0^{opt} are determined by the lattice processes of the heat removal (characterized by their specific parameters and the device configuration). In this case, in Eqs. (51.3)–(51.7) one needs to replace T_0 by $T_0^{opt} \simeq T$, so that Eq. (51.11) yields $\mu^e + \mu^h \simeq 0$. One needs also to replace the left-hand side of Eq. (51.15) by the normalized value, Q, of the heat flow from the G-P channel to the substrate and the gate layer. For a simplified analysis, one can set $Q = K(T - T_0)(T/T_0)^q$, where K characterizes the heat conductivity of the interfaces between the G-P-channel and the surrounding layers the side contacts, as well as an efficiency of the heat removal via the side contacts and q is a number. Assume for the definiteness that $V_g = 0$ (i.e., $\mu^e = \mu^h = 0$) and set $q = 1$, we arrive at the following dependence of the channel current density on the common temperature of the carriers and the optical phonons in the G-P channel:

$$\frac{j}{j_0} \simeq \sqrt{\frac{K(T - T_0)}{1 + \frac{2\Sigma_N}{\Sigma_I}\frac{T}{T_0}\exp\left(-\frac{\Delta}{2T}\right)}}. \tag{51.24}$$

One can find that the latter $j - T$ relation exhibits a maximum at a certain temperature $T = T_{max} \sim \Delta/[2 \ln \Gamma(1 - \ln \Gamma/b)]$, where $\Gamma = (\Sigma_N\Delta/\Sigma_I T_0)$ and $b = \Delta/2T_0$. At the parameters used in the above calculations, the latter estimate gives $T_{max} \simeq 48 - 82$ meV. Due to a monotonic increase in T with increasing E, this implies that the pertinent current-voltage characteristic exhibits the NDC at sufficiently large E when $T > T_{\min}$ (at least when $q = 1$). The more detailed consideration of this case requires a more accurate model that is out of the scope of the present chapter.

51.5.4 Relaxation of Substrate Optical Phonons

If the electrons and holes effectively interact with the optical phonons of several types, say, with the G-layer optical phonons and the substrate optical phonons, Eqs. (51.9)–(51.11) should be properly generalized. In particular, considering both interband and intraband transition, instead of Eq. (51.11) governing the interband carrier balance one can arrive at

$$\frac{1}{\tau_0^{\text{inter}}}\exp\left(\frac{\mu^{\text{e}}+\mu^{\text{h}}-\hbar\omega_0}{T}\right)+\frac{1}{\tau_1^{\text{inter}}}\exp\left(\frac{\mu^{\text{e}}+\mu^{\text{h}}-\hbar\omega_1}{T}\right)$$
$$=\frac{1}{\tau_0^{\text{inter}}}\exp\left(-\frac{\hbar\omega_0}{T}\right)+\frac{1}{\tau_1^{\text{inter}}}\exp\left(-\frac{\hbar\omega_1}{T_0}\right). \quad (51.25)$$

Here, ω_1 and τ_1^{inter} are the pertinent parameters for the substrate optical phonons. As follows from Eqs. (51.25) and (51.12) for $\mu^{\text{e}} + \mu^{\text{h}}$ should be replaced by the following:

$$\frac{\mu^{\text{e}}+\mu^{\text{h}}}{T}=\ln\left[\frac{\tau_1^{\text{inter}}\exp(-\hbar\omega_0/T_0)+\tau_0^{\text{inter}}\exp(-\hbar\omega_1/T_0)}{\tau_1^{\text{inter}}\exp(-\hbar\omega_0/T)+\tau_0^{\text{inter}}\exp(-\hbar\omega_1/T)}\right]. \quad (51.26)$$

If $\tau_1^{\text{inter}} \gg \tau_0^{\text{inter}}$ and $\omega_1 \sim \omega_0$, Eq. (51.22) yields

$$\mu^{\text{e}}+\mu^{\text{h}}\simeq\hbar\omega_0\left(1-\frac{T}{T_0}\right)+T\left(\frac{\tau_0^{\text{inter}}}{\tau_1^{\text{inter}}}\right)$$
$$\times\left[\exp\left(\frac{\hbar\omega_0-\hbar\omega_1}{T}\right)-\exp\left(\frac{\hbar\omega_0-\hbar\omega_1}{T_0}\right)\right]$$
$$\simeq\hbar\omega_0\left(1-\frac{T}{T_0}\right)\left[1+\left(\frac{\tau_0^{\text{inter}}}{\tau_1^{\text{inter}}}\right)\frac{(\omega_0-\omega_1)}{\omega_0}\right]. \quad (51.27)$$

Equation (51.27) shows that the contribution of the substrate optical phonons can be roughly accounted for by a renormalization of the quantity $\hbar\omega_0$.

In the opposite case, $\tau_1^{\text{inter}} \ll \tau_0^{\text{inter}}$ and $\omega_1 \ll \omega_0$, from Eq. (51.26) we obtain

$$\mu^{\text{e}} + \mu^{\text{h}} \simeq \hbar\omega_1 \left(1 - \frac{T}{T_0}\right) \ll \hbar\omega_0 \left(1 - \frac{T}{T_0}\right), \tag{51.28}$$

i.e., $\mu^{\text{e}} + \mu^{\text{h}}$ can be close to zero in a wide range of the effective temperature T. Hence, marked modifications of the above results can occur only in the case of relatively strong interaction with the low energy substrate optical phonons.

51.5.5 Possible Applications of the G-P Devices

A steep decrease in the G-P channel conductivity with increasing effective temperature (see Fig. 51.2) can be used for detection of the incident radiation in a wide spectral range from terahertz (THz) to near infrared. The intraband and interband transitions caused by the absorbing photons result in the carrier heating and their redistribution between the G- and P-layers and, hence, in a decrease in the G-P-channel conductivity. This heating effect (providing the negative photoconductivity) can have substantial stronger influence on the conductivity than that associated with the photogeneration of the extra carriers. The RST can markedly affect the response of the electron-hole plasma generated in the G-P channel by ultrashort optical pulses increasing the efficiency of the photoconducting antennas comprising the G-P structures or their arrays.

NDC can lead to the instability of the electron-hole plasma in the G-P channel. This instability can be used for the generation of high-frequency oscillations in the device and the output microwave or THz radiation (like in the Gunn diodes). The possibility of THz operation can be limited by the speed of the carrier exchange between the G- and P-layers. Due to an effective coupling of these layers and strong overlap of the pertinent wave function, one might expect that the inverse times of the G-P carrier exchange fall into the terahertz range.

Depending on the contact properties, the RST device with NDC could be used as a switch between low and high voltage states or a tunable current limiter. In a regime when the NDC leads to the formation of propagating high field domains, profiling the P-layer enables the applications for logical circuits (including the non-

Boolean logic circuits) and functional generators. Inserting a number of RST devices into a THz waveguide will enable an operation in a highly efficient hybrid mode of quenching the domain formation for a high power integrated THz source. However, an effective operation of the G-P RST devices in the THz range necessitates a sufficiently short time of the carrier transfer between the G- and P-layers and back. This issue requires a detailed consideration which is beyond our model and the scope of this work.

51.6 Conclusions

We proposed the FETs with the G-P channel and calculated their characteristics using the developed device model. We demonstrated that the carrier heating, in particular, by the source-to-drain electric field leads to a substantial RST of the carriers between the G- and P-layers. As a result, the population of the heavy carriers in the P-layer strongly increases that results in a pronounced scattering reinforcement of the light carriers in the G-layer and, consequently, the drop of the G-P channel conductivity could lead to NDC. The FETs under consideration can be used for detection and generation of electro-magnetic radiation and exhibit nontrivial characteristics useful for the logical circuits and functional generators.

Acknowledgments

The authors are grateful to Professor V. Vyurkov for valuable comments. This work was supported by Japan Society for Promotion of Science (Grant Nos. 16H06361 and 16K14243), Russian Science Foundation (Grant No. 14-2900277), and Russian Foundation for Basic Research (Grant Nos. 16-37-60110 and 18-07-01379). It was also partially supported by the RIEC Nation-Wide Collaborative Research Project, Japan. The work at RPI was supported by the Office of Naval Research (Project Monitor Dr. Paul Maki).

References

1. A. H. Castro Neto, F. Guinea, N. M. R. Peres, K. S. Novoselov, and A. K. Geim, The electronic properties of graphene, *Rev. Mod. Phys.*, **81**, 109 (2009).

2. A. Geim and I. V. Grigorieva, Van der Waals heterostructures, *Nature*, **499**, 419 (2013).
3. M. Chhowalla, D. Jena, and H. Zhang, Two-dimensional semiconductors for transistors, *Nat. Rev. Mater.*, **1**, 16052 (2016).
4. A. Tredicucci and M. Vitiello, Device concepts for graphene-based terahertz photonics, *J. Sel. Top. Quantum*, **20**, 130 (2014).
5. Q. Bao and K. P. Loh, Graphene photonics, plasmonics, and broadband optoelectronic devices, *Nano*, **6**, 3677 (2012).
6. F. Bonaccorso, Z. Sun, T. Hasan, and A. Ferrari, Graphene photonics and optoelectronics, *Nat. Photonics*, **4**, 611 (2010).
7. V. Ryzhii, M. Ryzhii, and T. Otsuji, Negative dynamic conductivity of graphene with optical pumping, *J. Appl. Phys.*, **101**, 083114 (2007).
8. V. Ryzhii, M. Ryzhii, A. Satou, T. Otsuji, A. A. Dubinov, and V. Y. Aleshkin, Feasibility of terahertz lasing in optically pumped epitaxial multiple graphene layer structures, *J. Appl. Phys.*, **106**, 084507 (2009).
9. V. Ryzhii, M. Ryzhii, V. Mitin, and T. Otsuji, Toward the creation of terahertz graphene injection laser, *J. Appl. Phys.*, **110**, 094503 (2011).
10. V. Ryzhii, A. A. Dubinov, V. Y. Aleshkin, M. Ryzhii, and T. Otsuji, Injection terahertz laser using the resonant inter-layer radiative transitions in double-graphene-layer structure, *Appl. Phys. Lett.*, **103**, 163507 (2013).
11. S. Boubanga-Tombet, S. Chan, T. Watanabe, A. Satou, V. Ryzhii, and T. Otsuji, Ultrafast carrier dynamics and terahertz emission in optically pumped graphene at room temperature, *Phys. Rev. B*, **85**, 035443 (2012).
12. T. Li, L. Luo, M. Hupalo, J. Zhang, M. C. Tringides, J. Schmalian, and J. Wang, Femtosecond population inversion and stimulated emission of dense Dirac fermions in graphene, *Phys. Rev. Lett.*, **108**, 167401 (2012).
13. I. Gierz, J. C. Petersen, M. Mitrano, C. Cacho, I. E. Turcu, E. Springate, A. Stohr, A. Kohler, U. Starke, and A. Cavalleri, Snapshots of nonequilibrium Dirac carrier distributions in graphene, *Nat. Mater.*, **12**, 1119 (2013).
14. E. Gruber, R. A. Wilhelm, R. Petuya, V. Smejkal, R. Kozubek, A. Hierzenberger, B. C. Bayer, I. Aldazabal, A. K. Kazansky, F. Libish, A. V. Krasheninnikov, M. Schleberger, S. Facsko, A. G. Borisov, A. Arnau, and F. Aumayr, Ultrafast electronic response of graphene to a strong and localized electric field, *Nat. Commun.*, **7**, 13948 (2016).
15. D. Yadav, G. Tamamushi, T. Watanabe, J. Mitsushio, Y. Tobah, K. Sugawara, A. A. Dubinov, M. Ryzhii, V. Ryzhii, and T. Otsuji, Terahertz

light-emitting graphene-channel transistor toward single-mode lasing, *Nanophotonics*, **7**, 741–752 (2018).

16. Y. Cai, G. Zhang, and Y.-W. Zhang, Layer-dependent band alignment and work function of few-layer phosphorene, *Sci. Rep.*, **4**, 6677 (2014).
17. X. Ling, H. WAng, S. Huang, F. Xia, and M. S. Dresselhaus, The renaissance of black phosphorus, *PNAS*, **112**, 4523 (2015).
18. E. Leong, R. J. Suess, A. B. Sushkov, H. D. Drew, T. E. Murphy, and M. Mittendorff, Terahertz photoresponse of black phosphorus, *Opt. Express*, **25**(11), 12666 (2017).
19. M. Buscema, D. J. Groenendijk, S. I. Blanter, G. A. Steele, H. S. J. van der Zant, and A. Castellanos-Gomez, Fast and broadband photoresponse of few-layer black phosphorus field-effect transistors, *Nano Lett.*, **14**, 3347 (2014).
20. Y. Deng, Z. Luo, N. J. Conrad, H. Liu, Y. Gong, S. Najmaei, P. M. Ajayan, J. Lou, X. Xu, and P. D. Ye, Black phosphorus-monolayer MoS2 van der Waals heterojunction P-N diode, *ACS Nano*, **8**, 8292 (2014).
21. M. Engel, M. Steiner, and Ph. Avouris, A black phosphorus photodetector for multispectral high-resolution imaging, *Nano Lett.*, **14**, 6414 (2014).
22. Z.-P. Ling, J.-T. Zhu, X. Liu, and K.-W. Ang, Interface engineering for the enhancement of carrier transport in black phosphorus transistor with ultra-thin high-κ gate dielectric, *Sci. Rep.*, **6**, 26609 (2016).
23. F. Ahmed, Y. D. Kim, M. S. Choi, X. Liu, D. Qu, Z. Yang, J. Hu, I. P. Herman, J. Hone, and W. J. Yoo, High electric field carrier transport and power dissipation in multilayer black phosphorus field effect transistor with dielectric engineering, *Adv. Funct. Mater.*, **27**, 1604025 (2017).
24. F. H. L. Koppens, T. Mueller, Ph. Avouris, A. C. Ferrari, M. S. Vitiello, and M. Polini, Photodetectors based on graphene, other two-dimensional materials and hybrid systems, *Nat. Nanotechnol.*, **9**, 780 (2014).
25. Z. S. Gribnikov, *Sov. Phys. Semicond.*, **6**, 1204 (1973).
26. K. Hess, H. Morkoc, H. Shichijo, and B. G. Streetman, Negative differential resistance through real-space electron transfer, *Appl. Phys. Lett.*, **35**, 469 (1979).
27. A. Kastalsky and S. Luryi, Novel real-space hot-electron transfer devices, *IEEE Electron Device Lett.*, **4**, 334 (1983).
28. A. Kastalsky, S. Luryi, A. C. Gossard, and R. Hendel, A field-effect transistor with a negative differential resistance, *IEEE Electron Device Lett.*, **5**, 57 (1984).

29. N. Sawaki and I. Akasaki, Scattering and real space transfer in multi-quantum well structures, *Physica B+C*, **134**, 494 (1985).
30. I. C. Kizilyalli and K. Hess, Physics of real-space transfer transistors, *J. Appl. Phys.*, **65**, 2005 (1989).
31. S. Luryi and M. R. Pinto, Symmetry of the real-space transfer and collector-controlled states in charge injection transistors, *Semicond. Sci. Technol.*, **7**(3B), B520 (1999).
32. Z. S. Gribnikov, K. Hess, and G. A. Kosinovsky, Nonlocal and nonlinear transport in semiconductors: real-space transfer effects, *J. Appl. Phys.*, **77**, 1337 (1995).
33. E. Sermuksnis, J. Liberis, A. Matulionis, V. Avrutin, R. Ferreyra, U. Ozgur, and H. Morkoc, Hot-electron real-space transfer and longitudinal transport in dual AlGaN/AlN/{AlGaN/GaN} channels, *Semicond. Sci. Technol.*, **30**(3), 035003 (2015).
34. R. Q. Yang, Quantum real-space transfer in semiconductor heterostructures, *Appl. Phys. Lett.*, **73**, 3265 (1998).
35. Z. S. Gribnikov, N. Z. Vagidov, R. R. Bashirov, V. V. Mitin, and G. I. Haddad, Quantum real-space transfer in a heterostructure overgrown on the cleaved edge of a superlattice, *J. Appl. Phys.*, **93**, 330 (2003).
36. C. Jin, Z. Chen, and J. Chen, Novel quantum real-space transfer in semiconductor heterostructures, *Proc. SPIE*, **8419**, 84191Y (2012).
37. G. Zhang, A. Chaves, S. Huang, F. Wang, Q. Xing, T. Low, and H. Yan, Determination of layer-dependent exciton binding energies in few-layer black phosphorus, *Sci. Adv.*, **4**(3), eaap9977 (2018).
38. T. Low, R. Roldan, H. Wang, F. Xia, P. Avouris, L. M. Moreno, and F. Guinea, Plasmons and screening in monolayer and multilayer black phosphorus, *Phys. Rev. Lett.*, **113**, 106802 (2014).
39. S. Yuan, A. N. Rudenko, and M. I. Katsnelson, Transport and optical properties of single- and bilayer black phosphorus with defects, *Phys. Rev. B*, **91**,115436(2015).
40. J. Xi, M. Long, D. Wang, and Z. Shuai, First principles prediction of charge mobility in carbon and organic nanomaterials, *Nanoscale*, **4**, 4348 (2012).
41. M. S. Foster and I. L. Aleiner, Slow imbalance relaxation and thermoelectric transport in graphene, *Phys. Rev. B*, **79**, 085415 (2009).
42. G. Alymov, V. Vyurkov, V. Ryzhii, A. Satou, and D. Svintsov, Auger recombination in Dirac materials: a tangle of many-body effects, *Phys. Rev. B*, **97**, 205411 (2018).

43. T. Ando, Screening effect and impurity scattering in monolayer graphene, *J. Phys. Soc. Jpn.*, **75**, 074716 (2006).

44. L. A. Falkovsky and A. A. Varlamov, Space-time dispersion of graphene conductivity, *Eur. Phys. J. B*, **56**, 281 (2007).

45. E. H. Hwang, S. Adam, and S. D. Sarma, Carrier transport in two-dimensional graphene layers, *Phys. Rev. Lett.*, **98**, 186806 (2007).

46. F. T. Vasko and V. Ryzhii, Voltage and temperature dependencies of conductivity in gated graphene, *Phys. Rev. B*, **76**, 233404 (2007).

47. V. Vyurkov and V. Ryzhii, Effect of Coulomb scattering on graphene conductivity, *JETP Lett.*, **88**, 370 (2008).

48. A. Kashuba, Conductivity of defectless graphene, *Phys. Rev. B*, **78**, 085415 (2008).

49. E. H. Hwang and S. Das Sarma, Acoustic phonon scattering limited carrier mobility in two-dimensional extrinsic graphene, *Phys. Rev. B*, **77**, 115449 (2008).

50. E. H. Hwang and S. Das Sarma, Screening induced temperature dependent transport in 2D graphene, *Phys. Rev. B*, **79**, 165404 (2009).

51. S. V. Morozov, K. S. Novoselov, M. I. Katsnelson, F. Schedin, D. C. Elias, J. A. Jaszczak, and A. K. Geim, Giant intrinsic carrier mobilities in graphene and its bilayer, *Phys. Rev. Lett.*, **100**, 016602 (2008).

52. J.-H. Chen, C. Jang, S. Adam, M. Fuhrer, E. Williams, and M. Ishigami, Charged-impurity scattering in graphene, *Nat. Phys.*, **4**, 377 (2008).

53. H. Hirai, H. Tsuchiya1, Y. Kamakura, N. Mori, and M. Ogawa, Electron mobility calculation for graphene on substrates, *J. Appl. Phys.*, **116**, 083703 (2014).

54. J. S. Blakemore, *Semiconductor Statistics* (Dover, 1987).

55. A. A. Balandin, S. Ghosh, W. Bao, I. Calizo, D. Teweldebrhan, F. Miao, and C. N. Lau, Superior thermal conductivity of single-layer graphene, *Nano Lett.*, **8**, 902 (2008).

56. S. Ghosh, I. Calizo, D. Teweldebrhan, E. P. Pokatilov, D. L. Nika, A. A. Balandin, W. Bao, F. Miao, and C. N. Lau, Extremely high thermal conductivity of graphene: prospects for thermal management applications in nanoelectronic circuits, *Appl. Phys. Lett.*, **92**, 151911 (2008).

57. A. A. Balandin, Thermal properties of graphene and nanostructured carbon materials, *Nat. Mater.*, **10**, 569 (2011).

58. E. Pop, V. Varshney, and A. K. Roy, Thermal properties of graphene: fundamentals and applications, *MRS Bull.*, **37**, 1273 (2012).

59. V. Ryzhii, M. Ryzhii, V. Mitin, A. Satou, and T. Otsuji, Effect of heating and cooling of photogenerated electron-hole plasma in optically pumped graphene on population inversion, *Jpn. J. Appl. Phys.*, **50**, 094001 (2011).
60. V. Ryzhii, T. Otsuji, M. Ryzhii, N. Ryabova, S. O. Yurchenko, V. Mitin, and M. S. Shur, Graphene terahertz uncooled bolometers, *J. Phys. D: Appl. Phys.*, **46**, 065102 (2013).
61. F. Rana, P. A. George, J. H. Strait, S. Sharavaraman, M. Charasheyhar, and M. G. Spencer, Carrier recombination and generation rates for intravalley and intervalley phonon scattering in graphene, *Phys. Rev. B*, **79**, 115447 (2009).
62. J. N. Heyman, J. D. Stein, Z. S. Kaminski, A. R. Banman, A. M. Massari, and J. T. Robinson, Carrier heating and negative photoconductivity in graphene, *J. Appl. Phys.*, **117**, 015101 (2015).

Chapter 52

Negative Photoconductivity and Hot-Carrier Bolometric Detection of Terahertz Radiation in Graphene-Phosphorene Hybrid Structures*

V. Ryzhii,[a,b,c,d] M. Ryzhii,[e] D. S. Ponomarev,[b,c] V. G. Leiman,[c] V. Mitin,[a,f] M. S. Shur,[g,h] and T. Otsuji[a]

[a]*Research Institute of Electrical Communication, Tohoku University, Sendai 980-8577, Japan*
[b]*Institute of Ultra High Frequency Semiconductor Electronics of RAS, Moscow 117105, Russia*
[c]*Center of Photonics and Two-Dimensional Materials, Moscow Institute of Physics and Technology, Dolgoprudny 141 700, Russia*
[d]*Center for Photonics and Infrared Engineering, Bauman Moscow State Technical University, Moscow 111005, Russia*
[e]*Department of Computer Science and Engineering, University of Aizu, Aizu-Wakamatsu 965-8580, Japan*
[f]*Department of Electrical Engineering, University at Buffalo, Buffalo, New York 1460-192*
[g]*Department of Electrical, Computer, and Systems Engineering and Department of Physics, Applied Physics, and Astronomy, Rensselaer Polytechnic Institute, Troy, New York 12180, USA*
[h]*Electronics of the Future, Inc., Vienna, VA 22181, USA*
v-ryzhii@riec.tohoku.ac.jp

Graphene-Based Terahertz Electronics and Plasmonics: Detector and Emitter Concepts
Edited by Vladimir Mitin, Taiichi Otsuji, and Victor Ryzhii

ISBN 978-981-4800-75-4 (Hardcover), 978-0-429-32839-8 (eBook)
www.jennystanford.com

We consider the effect of terahertz (THz) radiation on the conductivity of the ungated and gated graphene (G)-phosphorene (P) hybrid structures and propose and evaluated the hot-carrier uncooled bolometric photodetectors based on the GP-lateral diodes (GP-LDs) and GP-field-effect transistors (GP-FETs) with the GP channel. The operation of the GP-LDs and GP-FET photodetectors is associated with the carrier heating by the incident radiation absorbed in the G-layer due to the intraband transitions. The carrier heating leads to the relocation of a significant fraction of the carriers into the P-layer. Due to a relatively low mobility of the carriers in the P-layer, their main role is associated with a substantial reinforcement of the scattering of the carriers. The GP-FET bolometric photodetector characteristics are effectively controlled by the gate voltage. A strong negative conductivity of the GP-channel can provide much higher responsivity of the THz hot-carriers GP-LD and GP-FET bolometric photodetectors in comparison with the bolometers with solely the G-channels.

52.1 Introduction

Unique energy spectra, of graphene (G) [1] and a few-layer black phosphorus layer or phosphorene (P) [2], their optical and electric properties, and recent advances in technology open remarkable prospects for the creation of novel devices using G-layers [3–6], the P-layers [2, 7–10], and different hybrid structures including the G-P hybrid structures [11–13]. In particular, the GP hybrid systems can be used for the improvement of various devices. The possibility of the layer-dependent alignment work function control [2, 14] provides substantial flexibility in the device design. In this chapter, we propose and evaluate the detector of the terahertz (THz) radiation based on a lateral diode (LD) and a field-effect transistor (FET) with the GP channel, GP-LD and GP-FET, respectively. The operation of the GP-LD and GP-FET photodetectors with the GP-channel is associated with the carrier heating by the incident radiation absorbed in the G-layer leading to a variation of the channel conductivity [15, 16]. This principle is used in the hot-carrier bolometers based on the G-channel exhibiting the negative or positive photoconductivity (see, for example [17–21] and references therein). However, a major disadvantage of using G-layers in the bolometric photodetectors is that the conductivity of pristine G-layers is weakly dependent on the carrier temperature. This can be overcome by the introduction of the barrier regions (by partitioning of the channel into nanoribbons in which the energy gap is opened [17] or using disordered G-layers [20]). In the G-P bolometers under consideration, the carrier heating caused by the absorbed radiation leads to the transfer of a significant portion of the carriers into the P-layer. This results in a decrease of the density of the highly mobile carriers in the G-layer and in a reinforcement of the scattering of these carriers on the carriers residing in the P-layer. Due to a high effective mass of the carriers in the P-layer and, hence, a relatively low mobility, their main role is associated with a substantial reinforcement of the scattering of the carriers in the G-layer. As a result, the conductivity of the GP-channel can markedly drop with the carrier heating. We demonstrate that the effect of the negative THz photoconductivity in the G-P channels can be much stronger than that for the G-channels, particularly at room temperature. Therefore, the GP-LDs and GP-FETs could effectively operate as the uncooled hot-carrier THz bolometers with an elevated responsivity.

52.2 Model

Figure 52.1 demonstrates a schematic view of the ungated and gated G-P structures (i.e., a GP-LD and a GP-FET with the GP-channels). It is assumed that the P-layer consisting of a few atomic layers, is oriented in such a way that the direction from the source to its drain corresponds to the zigzag direction. The dynamics of electrons and holes in this direction is characterized by a huge effective mass. As a result, a substantial amount of the electrons and holes can relocate from the G-layer (where their mobility could be very high) to the P-layer (with a low mobility).

For the sake of definiteness, we consider the P-layer consisting of several atomic P-layers ($N = 4 - 5$), assuming that in both GP-LD and GP-FET structures it is p-doped (the pristine P-layers are of p-type). At the incident THz photon energies $\hbar\Omega < 2\mu$, where $|\mu|$ is the Fermi energy of the main carriers (holes), the carrier heating is associated with its Intraband (Drude) absorption in the G-layer.

The band gap Δ_G and the energy spacing Δ^e and Δ^h, between the Dirac point in the G-layer and the edges of the conduction and valence bands, (determined by the pertinent work functions) depend on the number N. In the G-P channel under consideration (with $N = 4$ or 5), the band structure is asymmetric: $\Delta^e > \Delta^h = \Delta$ [14].

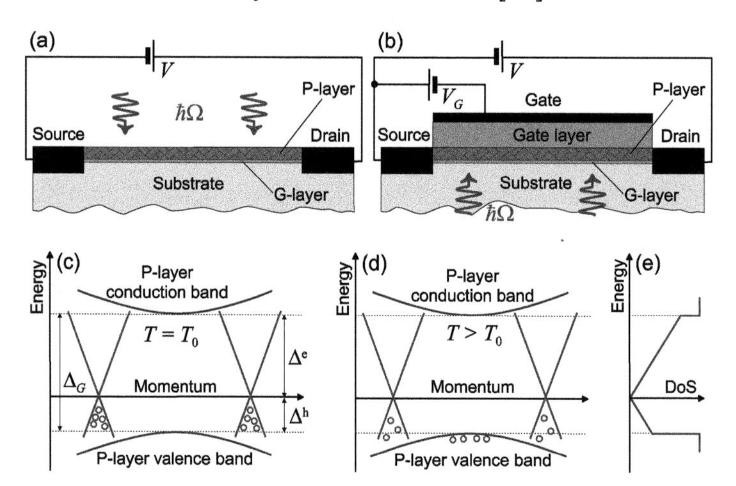

Figure 52.1 The structures of (a) the GP-LD and (b) the GP-FET (b), their asymmetric with respect to the Dirac point ($\Delta^e > \Delta^h$) energy band diagrams with the G-Dirac cones and the parabolic extrema corresponding to the P-layer at (c) $T = T_0$ and (d) $T > T_0$, and (e) the energy dependence of the density of state (DoS). Open circles correspond to the holes in the valence bands of G- and P-layers.

The dispersion relation for the holes in the G- and P-layers can be presented as

$$\varepsilon_G^e = v_W\sqrt{p_x^2 + p_y^2}, \quad \varepsilon_P^e = \Delta^e + \frac{p_x^2}{2m_{xx}} + \frac{p_y^2}{2m_{yy}}, \tag{52.1}$$

$$\varepsilon_G^h = -v_W\sqrt{p_x^2 + p_y^2}, \quad \varepsilon_P^e = -\Delta^h - \frac{p_x^2}{2m_{xx}} - \frac{p_y^2}{2m_{yy}}, \tag{52.2}$$

respectively. Here $v_W \simeq 10^8$ cm/s is the characteristic velocity of electrons in the G-layers, $m_{xx} = M$ and $m_{yy} = m$ are the components of the effective mass tensor ($M \gg m$), p_x and p_y are the carrier momenta in the source-drain direction and the perpendicular direction, respectively. The components of the effective mass tensor for the valence bands ($N = 4 - 5$) are approximately equal to $m \simeq 0.04m_0$ and $M \simeq 1.01m_0$, where m_0 is the mass of a free electron.

The model under consideration is based on the following assumptions:

(1) The sufficiently frequent electron-electron, electron-hole, and hole-hole collisions enable the establishment of the distinct quasi-Fermi electron and hole distributions with the common effective temperature T in both G- and P-layers and split quasi-Fermi levels due to the carrier-phonon interband scattering. This is consistent with the numerous experimental studies in which the G-layer was excited with an optical or infrared pump pulse and probed with photoelectron or optical spectroscopy at different photon energies (see, for example, [18] and the references therein). Sufficiently strong interactions between the electrons and holes belonging to both G- and P-layers promote the inter-layer equilibrium [22–24]. Hence, the electron and hole distribution functions are the following functions of the carrier energy ε: $f^{e,h} = [\exp(\varepsilon - \mu^{e,h})/T + 1]^{-1}$ (where μ^e and $\mu^h = \mu$ are the quasi-Fermi energies counted from the Dirac point).

(2) Due to heavy electron and hole effective masses M and $\sqrt{mM}$, the conductivity of the P-layer is relatively small because this layer mobility in the direction corresponding to the mass M is proportional to $1/\sqrt{mM}\ M$ [25], so that the P-layer conductivity could be neglected in comparison with the G-layer conductivity. Thus, the main role of the carrier relocation from the G-layer into the P-layer is associated with an intensification of the carriers (in the G-layer) scattering on the carriers (in the P-layer) when the concentration of the latter increases with the carrier system heating.

(3) The momentum relaxation of the electrons and holes in the G-layer (which we refer to as the "light" electrons and holes) is due to their scattering on acoustic phonons, neutral defects, and heavy particles in the P-layer. In contrast to the G-channel-based THz bolometers intended for the operation at very low temperatures at which the carrier energy relaxation is due to the interaction with acoustic phonons, the energy relaxation in the uncooled bolometric detectors under consideration is associated with the optical phonons in the G-layer. The interband transitions assisted by the optical phonons [26, 27]

and with the Auger processes (see [28, 29] and the discussion therein) are assumed to be the main recombination-generation mechanisms. We characterize the relative role of these processes by the parameter $\eta = \tau_{\text{Auger}}/(\tau_{\text{Auger}} + \tau_0^{\text{inter}})$, where τ_{Auger} and τ_0^{inter} are the times characterizing the pertinent interband transitions (we call this parameter as the Auger parameter). When $\eta \simeq 1$, the quasi-Fermi energies can be markedly different ($\mu^{\text{e}} \neq -\mu^{\text{h}}$).

52.3 Conductivity of the G-P-Channel

The net surface charge density in the GP-channel, which comprises the electron and hole charges in both the G- and P-layers, induced by the acceptors and the gate voltage V_{G} is equal to $e\Sigma = \kappa\,|V_{\text{G}} - V_{\text{A}}|/4\pi W_{\text{g}}$ (where $V_{\text{G}} = V_{\text{A}} > 0$, which is proportional to the acceptor density, corresponds to the charge neutrality point, κ and W_g are the background dielectric constant and the thickness of the gate layer, and e is the electron charge). By introducing the voltage (gate) swing $V_{\text{g}} = V_{\text{G}} - V_{\text{A}}$, we unify the consideration of the GP-LDs and GP-FETs. In particular, the case of GP-LDs corresponds to $V_{\text{G}} = 0$, so that $V_g = -V_{\text{A}} < 0$, while in the GP-FETs V_{g} can be both negative and positive.

The gate voltage swing $V_{\text{g}} = V_{\text{G}} - V_{\text{A}}$ or its dimensional value $U_{\text{g}} = V_{\text{g}}/V_0$ and the quantities T, μ^{e}, and μ^{h} are related to each other as

$$U_{\text{g}} = \left(\frac{T}{T_0}\right)^2 \left[\mathcal{F}_1\left(\frac{\mu^{\text{e}}}{T}\right) - \mathcal{F}_1\left(\frac{\mu^{\text{h}}}{T}\right)\right]$$
$$-\gamma_{\text{N}}\frac{T}{T_0}\ln\left[1+\exp\left(\frac{\mu^h - \Delta}{T}\right)\right]. \tag{52.3}$$

Here $\mathcal{F}_1(\text{a}) = \int_0^\infty d\xi\xi\,[\exp(\xi - a) + 1]^{-1}$ is the Fermi-Dirac integral [30], $U_{\text{g}} = (V_{\text{G}} - V_{\text{A}})/V_0 = V_{\text{G}}/V_0 - \pi\Sigma_{\text{A}}\hbar^2 v_{\text{W}}^2/2T_0^2$, $V_0 = 8eT_0^2\,W_{\text{g}}/\kappa\hbar^2 v_{\text{W}}^2$, $\gamma_{\text{N}} = N\sqrt{mM}v_{\text{W}}^2/2T_0$, and $\hbar$ is the Planck constant. For $N = 5$ and therefore setting, ($\sqrt{mM} \simeq 0.2m_0$ and $M/m \simeq 25$), $W_{\text{g}} = 10 - 1000$ nm and $\kappa = 4$, and $T = 25$ meV ($\simeq 300$ K), one can obtain $\gamma_{\text{N}} \simeq 110$ and $V_0 \simeq 0.05 - 4.94$ V. A large value of γ_{N} is due to a relatively high density of states in the P-layer.

In equilibrium at sufficiently low temperatures, when the P-layer is empty (the second term in the right-hand side of Eq. (52.3) is negligible), Eq. (52.3) yields $\mu^e = -\mu^h = -\mu \simeq -\hbar v_W \sqrt{\kappa |V_g| / 4eW_g}$ when $|V_g|$ is relatively large.

When the electron-hole system in the G-P channel is heated by the source-to-drain DC voltage or by the incident radiation, the electron and hole quasi-Fermi levels can be split: $\mu^e \neq -\mu^h$. Accounting for the competition between the optical phonon mediated and the Auger generation-recombination processes, the equation governing the carrier interband balance can result in the following equation relating μ^e and μ^h at an arbitrary effective temperature T:

$$\mu^e + \mu^h = \eta \hbar \omega_0 \left(1 - \frac{T}{T_0}\right). \tag{52.4}$$

where $\hbar\omega_0$ is the optical phonon energy. Equation (52.4) generalizes that obtained previously [13] for the case of the dominant optical phonon generation-recombination processes by the introduction of a phenomenological factor $\eta = \tau_{Auger}/(\tau_{Auger} + \tau_{Opt})$.

Considering Eq. (52.4), we rewrite Eq. (52.3) as

$$U_g = \left(\frac{T}{T_0}\right)^2 \left[\mathcal{F}_1\left(-\frac{\mu}{T} - \eta\hbar\omega_0\left(\frac{1}{T_0} - \frac{1}{T}\right)\right) - \mathcal{F}_1\left(\frac{\mu}{T}\right)\right]$$
$$-\gamma_N \frac{T}{T_0} \ln\left[1 + \exp\left(\frac{\mu - \Delta}{T}\right)\right]. \tag{52.5}$$

In particular, using Eq. (52.5), one can obtain immediately the dependence of the hole Fermi energy $\mu_0 = \mu|_{T=T_0}$ on the voltage swing U_g.

Focusing on the GP-channels with dominant carrier scattering on acoustic phonons, neutral defects, on each other, and on the short-range screened heavy carriers, the momentum relaxation time $\tau(p)$ as a function of the carrier momentum p can be set as $\tau_p = \tau_0(T_0/pv_W)$ $[\Sigma_G/(\Sigma_G + \Sigma_P)]$, where $\tau_0 \propto \Sigma_G^{-1}$ is the momentum relaxation time in the G-layer with the effective scatterer density Σ_G at $T = T_0$ and $(\Sigma_G + \Sigma_P)$ is the net scatterer density, which accounts for the density, Σ_P, of the heavy carriers in the P-layer. In this case, the GP-channel conductivity could be presented as (in line with [15, 16, 31–36]):

$$\sigma_{\rm GP} = -\frac{\sigma_0 \Sigma_{\rm G}}{(\Sigma_{\rm G} + \Sigma_{\rm P})} \int_0^{\infty} d\xi \frac{d(f^{\rm e} + f^{\rm h})}{d\xi} \quad (52.6)$$

with $\sigma_0 = (e^2 T_0 \tau_0 / \pi \hbar^2)$ being the G-layer low electric-field conductivity. Using Eqs. (52.4) and (52.6), the GP-channel conductivity can be expressed via the G-layer conductivity $\sigma_{\rm G}$ (without the P-layer conductivity) with the latter expressed via the effective temperature T and the hole quasi-Fermi energy $\mu = \mu^{\rm h}$:

$$\sigma_{\rm GP} = \frac{\sigma_{\rm G} \Sigma_{\rm G}}{(\Sigma_{\rm G} + \Sigma_{\rm P})}, \quad (52.7)$$

$$\sigma_{\rm G} = \sigma_0 \left[\frac{1}{\exp\left(\frac{\mu}{T} + \eta \hbar \omega_0 \left(\frac{1}{T_0} - \frac{1}{T}\right)\right) + 1} + \frac{1}{\exp\left(-\frac{\mu}{T}\right) + 1} \right]. \quad (52.8)$$

The density of scatterers (heavy holes), $\Sigma_{\rm P}$, in the P-layer, which exponentially increases with increasing $|\mu|$ and T, can also be expressed via these quantities:

$$\Sigma_{\rm P} = \Sigma_{\rm N} \frac{T}{T_0} \ln\left[1 + \exp\left(\frac{\mu - \Delta}{T}\right)\right], \quad (52.9)$$

where $\Sigma_{\rm N} = N T_0 \sqrt{mM} / \pi \hbar^2 \propto \gamma_{\rm N}$. The factor N in the latter formula reflects the fact that the density of states in the few-layer P-layer increases with the layer number N [15].

The second factor in the right-hand side of Eq. (52.6) reflects an increase in the scatterer density associated with the inclusion of the scattering on the heavy holes in the P-layer. It is instructive that at $\eta = 0$ when $\mu^{\rm e} = -\mu^{\rm h} = -\mu$, the G-channel conductivity $\sigma_{\rm G} = \sigma_0$ is independent of T. This is because of the specific of the carrier scattering in the system under consideration (scattering on acoustic phonons, neutral defects and effectively screened charged scatterers) [35–38].

Using Eqs. (52.7)–(52.9), we obtain

$$\sigma_{\mathrm{GP}} = \frac{\sigma_{\mathrm{G}}}{1+\dfrac{\Sigma_{\mathrm{N}}}{\Sigma_{\mathrm{G}}}\dfrac{T}{T_0}\ln\left[1+\exp\left(\dfrac{\mu-\Delta}{T}\right)\right]}$$

$$= \frac{\sigma_0}{1+\dfrac{\Sigma_{\mathrm{N}}}{\Sigma_{\mathrm{G}}}\dfrac{T}{T_0}\ln\left[1+\exp\left(\dfrac{\mu-\Delta}{T}\right)\right]}$$

$$\times\left[\frac{1}{\exp\left(\dfrac{\mu}{T}+\eta\hbar\omega_0\left(\dfrac{1}{T_0}-\dfrac{1}{T}\right)\right)+1}+\frac{1}{\exp\left(-\dfrac{\mu}{T}\right)+1}\right]. \quad (52.10)$$

Equations (52.5) and (52.10) describe the dependences of the quasi-Fermi energy μ and the G-P-channel conductivity σ_{GP} on the effective temperature T and the voltage swing U_g. Solving these equations, one can obtain the characteristics of the GP-channel in wide ranges of the normalized voltage swing U_g, carrier effective temperature T, and the density Σ_{G}.

52.4 Negative Photoconductivity in the G-P Channels

The variation of the current density, $J - J_0$, in the GP-channel for small effective carrier temperature variations is given by

$$J - J_0 \simeq E_{\mathrm{SD}} \left.\frac{d\sigma_{\mathrm{GP}}}{dT}\right|_{T=T_0} \cdot (T - T_0), \quad (52.11)$$

where J_0 is the linear density of the source-drain dc current in the absence of the irradiation (i.e., the dark current density, $E_{\mathrm{SD}} = V_{\mathrm{SD}}/L$ and V_{SD} are the source-drain electric field and voltage, respectively, L is the length of the GP-channel, and σ_{GP} is given by Eq. (52.10). An increase in the carrier effective temperature T (the carrier heating) caused by the irradiation corresponds to the negative photoconductivity temperature when the conductivity derivative $(d\sigma_{\mathrm{GP}}/dT)|_{T=T_0} < 0$. The latter is in line with the experimental observations [18, 19, 37, 38]. Figure 52.2 shows $\Lambda_{\mathrm{GP}} = \left.\dfrac{1}{\sigma_0}\dfrac{d\sigma_{\mathrm{GP}}}{d\ln T}\right|_{T=T_0}$, and $\Lambda_G = \dfrac{1}{\sigma_0} \dfrac{d\sigma_G}{d \ln T}\Big|_{T=T_0}$ (i.e., with $N = 0$) found as functions of U_g using Eqs. (52.5) and (52.10) for different values of the scatterer density Σ_G in the G-layer. One can see from Fig. 52.2 that both the quantities Λ_{GP} and Λ_G are negative. Here and in the following, we assume that $\hbar\omega_0 = 200$ meV, $\Delta = 200$ meV, $\Sigma_N = 1.2 \times 10^{13}$ cm^{-2}, $\Sigma_G = 5 \times (10^{10} - 10^{11})$ cm^{-2} corresponding to $\tau_0 = (0.24 - 2.4)$ ps, $\eta = 0.1 - 0.9$, $T_0 = 25$ meV ($\simeq 300$ K), and $\gamma_N = 110$. The scatterer densities range $\Sigma_G = 5 \times (10^{10} - 10^{11})$ cm^{-2} at $\mu_0 = 75$ meV, corresponds to the rather practical values of the carrier mobility in G-layers $b_G \simeq (30 - 300) \times 10^3$ cm^2/s V [39–42].

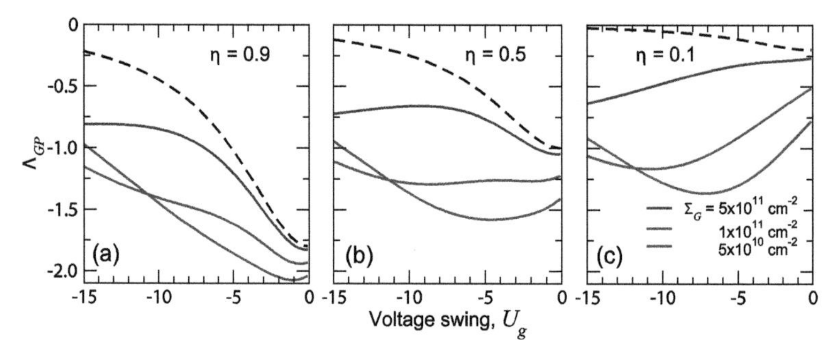

Figure 52.2 The temperature derivative of the GP-channel conductivity, Λ_{GP}, versus U_g for $\Delta = 200$ meV and different scatterer densities Σ_G at (a) $\eta = 0.9$, (b) $\eta = 0.5$, and (c) $\eta = 0.1$. Dashed line corresponds to Λ_G ($N = 0$, i.e., $\Sigma_N = 0$) and the same values of η.

At small $|U_g|$, the absolute values of these quantities $|\Lambda_{GP}| \gtrsim |\Lambda_G|$. However, at sufficiently large $|U_g|$, $|\Lambda_{GP}|$ can be substantially larger than $|\Lambda_G|$, particularly, at $\eta \ll 1$. This is attributed to a steeper effective temperature dependence of the GP-channel conductivity due to an increasing hole population in the P-layer. The latter implies that the GP-channel can exhibit a stronger temperature dependence and, hence, a stronger effect of the negative photoconductivity than the G-channel. The comparison of the plots in Figs. 52.2a–c demonstrates that the relative intensification of the Auger processes (a decrease in the parameter η) leads to a marked decrease in $|\Lambda_G|$, diminishing the temperature dependence of the G-channel, while these processes weakly affect the temperature dependence of the GP-channel and, hence, the quantities $|\Lambda_{GP}|$ and $J - J_0$.

52.5 Responsivity of the GP-Photodetectors

We limit our following consideration by the GP photodetectors operating as hot-carrier bolometers, so that the incident radiation does not produce a marked amount of the extra electrons and holes and the variation of the carrier density is associated primarily with the heating processes. This happens when $|U_g|$ is sufficiently large (to provide large hole Fermi energy μ), the photon energy $\hbar\Omega$ is not too large ($\hbar\Omega < 2\mu_0$), and the carrier momentum relaxation time τ_0 is not too short, so that the intraband absorption dominates the interband absorption (see Appendix A).

Under irradiation, the carrier effective temperature varies. Its value can be found considering the balance of the power, S_{abs}, receiving by the carriers due to the absorption of the incident radiation with the photon energy $\hbar\Omega$ and the power, $S_{lattice}$, which the carriers transfer to the lattice. As assumed above, the latter is associated with the interband transitions accompanied by the emission and absorption of the G-channel optical phonons having the energy $\hbar\omega_0$. The power received by the carrier system is given by

$$S_{abs} = \frac{4\pi\sigma_{GP}}{c\sqrt{\kappa}}\hbar\Omega I_\Omega = \frac{4\pi}{c\sqrt{\kappa}}\frac{\sigma_{GP}}{(1+\Omega^2\tau^2)}\hbar\Omega I_\Omega. \qquad (52.12)$$

where c is the speed of light, I_Ω is the radiation photon flux, $\sigma_{GP,\Omega} = \sigma_{GP}/(1 + \Omega^2\tau^2)$ is the high-frequency G-P channel conductivity, τ is the average hole momentum relaxation time in the GP-channel, which considering that $\tau_p \propto 1/p$ can be estimated as

$$\tau \simeq \frac{1}{[2[\mathcal{F}_1(\mu_0/T_0)+\mathcal{F}_1(-\mu_0/T_0)]^{1/2}}\frac{\tau_0}{(1+P_N)}. \qquad (52.13)$$

Here $P_N = (\Sigma_N/\Sigma_G)\ln\{1 + \exp[-(\Delta - \mu_0)/T_0]\}$. At $\mu_0 \lesssim T_0$ and $\mu_0 \gg T_0$, Eq. (52.13) yields $\tau \simeq [\sqrt{3}\tau_0/\pi\ (1 + P_N)]$ and $\tau \simeq [\tau_0/(I + P_N)](T_0/\mu_0) \simeq \tau_0/\sqrt{2U_g}$, respectively.

Taking into account Eqs. (52.7), (52.10), and (52.12) at $T = T_0$, we obtain

$$S_{abs} = \left(\frac{4\pi\sigma_0}{c\sqrt{\kappa}}\right)\frac{\hbar\Omega I_\Omega}{(1+P_N)}\frac{1}{(1+\Omega^2\tau^2)}, \qquad (52.14)$$

As previously [17, 27], we set

$$S_{\text{lattice}} = \hbar\omega_0 \frac{\Sigma_0}{\tau_0^{\text{intra}}}\left[(\mathcal{N}_0+1)\exp\left(-\frac{\hbar\omega_0}{T}\right)-\mathcal{N}_0\right] \tag{52.15}$$

where $\mathcal{N}_0 \simeq \exp(-\hbar\omega_0/T_0)$ is the equilibrium number of optical phonons, τ_0^{intra} is the characteristic time of the spontaneous emission of optical phonons at the intraband transitions, and $\Sigma_0 \simeq \pi T_0^2/3\hbar^2 v_W^2$ (at $\mu_0 \lesssim T_0$) and $\Sigma_0 \simeq \mu_0^2/\pi\hbar^2 v_W^2$ (at $\mu_0 \gg T_0$) are the carrier densities in the G-channel at $T = T_0$. For simplicity, below we use the following interpolation formulas:

$$\tau = \frac{\tau_0}{(1+P_N)}\sqrt{\frac{3}{\pi^2\left(1+\frac{6|U_g|}{\pi^2}\right)}}, \tag{52.16}$$

$$\Sigma_0 = \frac{\pi T_0^2}{3\hbar^2 v_W^2}\left(1+\frac{6|U_g|}{\pi^2}\right). \tag{52.17}$$

Equalizing S_{abs} and S_{lattice} given by Eqs. (52.14) and (52.15), and taking into account Eq. (52.16), we arrive at the following equation which relates the variation of the carrier effective temperature $T - T_0$ and the photon flux I_Ω:

$$\frac{T-T_0}{T_0} = \left(\frac{12\alpha\hbar v_W^2 \tau_0 \tau_0^\varepsilon}{\pi\sqrt{\kappa T_0^2}}\right) \times \frac{\hbar\Omega I_\Omega}{(1+P_N)(1+6|U_g|/\pi^2)(1+\Omega^2\tau^2)}, \tag{52.18}$$

where $\alpha = e^2/c\hbar \simeq 1/137$ is the fine structure constant. The latter formula corresponds to the hole energy relaxation time $\tau_0^\varepsilon = \tau_0^{\text{intra}} (T_0/\hbar\omega_0)^2 \exp(\hbar\omega_0/T_0)$. Setting $\tau_0^{\text{intra}} = 0.7$ ps (for example, [43]), one obtains $\tau_0^\varepsilon \simeq 32.6$ ps.

Taking into account the variation of the current density, $J - J_0$, in the GP-channel caused by the irradiation, the GP bolometer intrinsic current responsivity R_{GP} can be presented by the following expression:

$$R_{GP} = \frac{(J-J_0)H}{\hbar\Omega I_\Omega A}. \tag{52.19}$$

Here A = LH is the GP-channel area, L is the channel length (the spacing between the source and the drain), and H is the channel width, i.e., its size in the direction perpendicular to the current direction. Using Eqs. (52.11) and (52.18), we arrive at the following:

$$R_{\mathrm{GP}} = R_0 \frac{|\Lambda_{\mathrm{GP}}|}{[(1+P_{\mathrm{N}})(1+6|U_{\mathrm{g}}|/\pi^2)(1+\Omega^2\tau^2)]} \tag{52.20}$$

Here

$$R_0 = \frac{12\alpha}{\pi^2\sqrt{\kappa}} \frac{e}{T_0} \frac{ev_{\mathrm{W}}^2\tau_0^2\tau_0^{\varepsilon}E_{\mathrm{SD}}}{\hbar L} \propto \tau_0^2. \tag{52.21}$$

In the above calculations we have not accounted for the carrier heating by the source-drain electric field assuming that it is weak, so that T is very close to T_0. The pertinent condition is as follows:

$$E_{\mathrm{SD}} \ll \overline{E}_{\mathrm{SD}} = \frac{\pi T_0}{ev_{\mathrm{W}}} \sqrt{\frac{(1+P_{\mathrm{N}})(1+6|U_{\mathrm{g}}|/\pi^2)}{3\tau_0\tau_0^{\varepsilon}}}$$

$$= \overline{E^*}_{\mathrm{SD}} \sqrt{\frac{(1+P_{\mathrm{N}})(1+6|U_{\mathrm{g}}|/\pi^2)}{3}}. \tag{52.22}$$

For $E_{\mathrm{SD}} = \overline{E^*}_{\mathrm{SD}} = (\pi T_0/ev_{\mathrm{W}}\sqrt{\tau_0\tau_0^{\varepsilon}}$, the quantity R_0, given by Eq. (52.21), is equal to

$$\max R_0 = \frac{12\alpha}{\pi\sqrt{\kappa}} \frac{v_{\mathrm{W}}\tau_0^{3/2}}{\hbar L} \propto \tau_0^{3/2}. \tag{52.23}$$

At $\tau_0 = (0.24 - 2.4)$ ps, $\tau_0^{\varepsilon} = 32.6$ ps, and $L = 10^{-3}$ cm, for the quantities $\overline{E^*}_{\mathrm{SD}}$ and max R_0 one can find $\overline{E^*}_{\mathrm{SD}} \simeq (92 - 290)$ V/cm and max $R_0 \simeq (1.5 - 43.3)$ A/W, respectively

The voltage responsivity $R_{\mathrm{GP}}^{\mathrm{V}} = Rr_{\mathrm{L}}$, where r_{L} is the load resistance. Setting r_{L} equal to the GP-channel resistance, i.e., $r_{\mathrm{L}} = L(1 + P_{\mathrm{N}})/H\sigma_0$. In this case, for $R_{\mathrm{GP}}^{\mathrm{V}}$ one obtains

$$R_{\mathrm{GP}}^{\mathrm{V}} = R_0^{V} \frac{|\Lambda_{\mathrm{GP}}|}{[(1+6|U_{\mathrm{g}}|/\pi^2)(1+\Omega^2\tau^2)]} \tag{52.24}$$

with

$$R_0^{\mathrm{V}} = \frac{12\alpha}{\pi\sqrt{\kappa}} \frac{\hbar v_{\mathrm{W}}^2\tau_0\tau_0^{\varepsilon}E_{\mathrm{SD}}}{T_0^2 H} \propto \tau_0. \tag{52.25}$$

Setting $E_{\mathrm{SD}} = \overline{E^*}_{\mathrm{SD}}$, we arrive at the following expression for the characteristic value of the GP bolometer voltage responsivity:

$$\max R_0^{\rm V} = \frac{12\alpha}{\sqrt{\kappa}}\sqrt{\tau_0\tau_0^{\varepsilon}}\left(\frac{\hbar v_{\rm W}}{eT_0H}\right). \tag{52.26}$$

For τ_0 = 0.24–2.4 ps, τ_0^{ε} = 32.6 ps, and $H = 10^{-2}$ cm, we obtain max $R_0^{\rm V} \simeq (1.85 - 5.85) \times 10^2$ V/W. Naturally, at weaker source-drain electric fields $E_{\rm SD} < \overline{E}^*_{\rm SD} \sim \overline{E}_{\rm SD}$, the quantities R_0 and $R_0^{\rm V}$ are smaller than max R_0 and max $R_0^{\rm V}$.

Using Eqs. (52.20) and (52.24) at relatively low frequencies Ω, the current and voltage responsivities can be presented as

$$R_{\rm GP} = \frac{R_0|\Lambda_{\rm GP}|}{(1+P_{\rm N})(1+6|U_{\rm g}|/\pi^2)}, \quad R_{\rm GP}^{\rm V} = \frac{R_0^{\rm V}|\Lambda_{\rm GP}|}{(1+6|U_{\rm g}|/\pi^2)}. \tag{52.27}$$

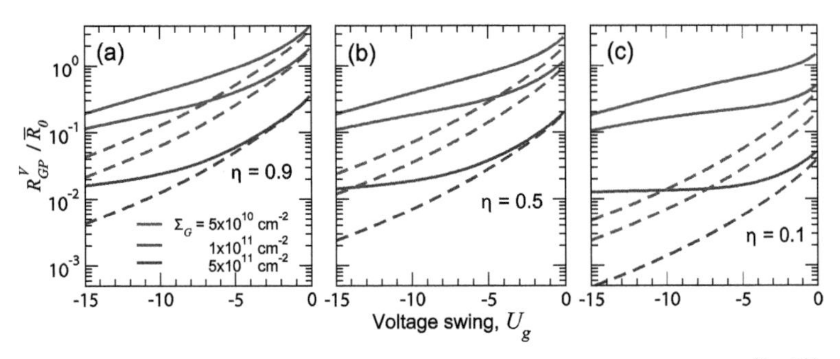

Figure 52.3 The normalized GP-bolometer voltage responsivity $R_{\rm GP}^{\rm V}/\overline{R}_0^{\rm V}$ versus the voltage swing $U_{\rm g}$ for the same parameters as in Fig. 52.2: (a) η = 0.9, (b) η = 0.5, and (b) η = 0.1. Dashed lines correspond to the G-bolometers with the same parameters of the G-layer.

Figures 52.3 and 52.4 show the GP-bolometers voltage low-frequency responsivity $R_{\rm GP}^{\rm V}$ normalized by the quantity $\overline{R}_0^{\rm V} = R_{\rm GP}^{\rm V}\big|_{\tau_0=1.2\rm ps}$ as a function of the voltage swing $U_{\rm g}$ calculated using Eqs. (52.10) and (52.27) for the same structural parameters as for Fig. 52.2 (given in Section 52.4). The normalized responsivity of the G-detectors (with the G-channel) is also shown by the dashed lines. First, as seen from Figs. 52.3 and 52.4, the responsivity sharply decreases with an increase in the scatterer density $\Sigma_{\rm G}$ and the voltage swing $U_{\rm g}$. This is attributed to a weaker carrier heating at their stronger scattering and their larger density. The latter markedly rises with increasing $U_{\rm g}$. Second, the GP-bolometer responsivity, moderately exceeding that of the G-bolometer responsivity at small

$|U_g|$, becomes orders of magnitudes larger at elevated values of $|U_g|$ (compare the solid and dashed lines in Figs. 52.3 and 52.4). The difference in the GP- and G-bolometers responsivities becomes fairly pronounced at smaller Auger parameter η (at stronger Auger generation-recombination processes). This correlates with a drop of Λ_G clearly seen in Fig. 52.2.

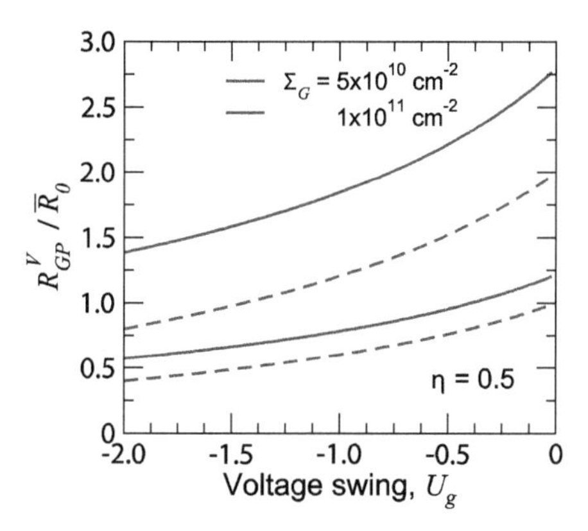

Figure 52.4 Zoom of the same plots as in Fig. 52.3b.

52.6 Bandwidth and Gain-Bandwidth Product

As follows from Eqs. (52.20) and (52.24), the GP-bolometer responsivity decreases when the photon frequency $\Omega > 2\pi f_{GP}$, where the cut-off frequency is given by

$$f_{GP} = \frac{(1+P_N)}{2\sqrt{3}\tau_0}\sqrt{\left(1+\frac{6|U_g|}{\pi^2}\right)}$$
$$\simeq \frac{1}{2\sqrt{3}\tau_0}\left[1+\left(\frac{\Sigma_N}{\Sigma_G}\right)\exp\left(\sqrt{2U_g}-\frac{\Delta}{T_0}\right)\right]$$
$$\times\sqrt{\left(1+\frac{6|U_g|}{\pi^2}\right)}. \qquad (52.28)$$

At small $|U_g|$, Eq. (52.27) yields $f_{GP} \simeq (1/2\sqrt{3}\pi\tau_0)$, so that for the values of τ_0 used above, one obtains $f_{GP} \simeq 0.12 - 1.2$ THz. At relatively large $|U_g|$, the frequency $f_{GP} \simeq (1 + P_N)\sqrt{2|U_g|}/2\pi$ can be much higher than that at $U_g \simeq 0$. This is seen from Fig. 52.5, which shows the cut-off frequency f_{GP} as a function of the normalized voltage swing calculated using Eq. (52.28). The cut-off frequency f_{GP} is larger than the pertinent frequency for the G-bolometers f_G by a factor $(1 + P_N)$. At large values of Σ_N/Σ_G and U_g, this factor can be larger than that at $U_g \simeq 0$.

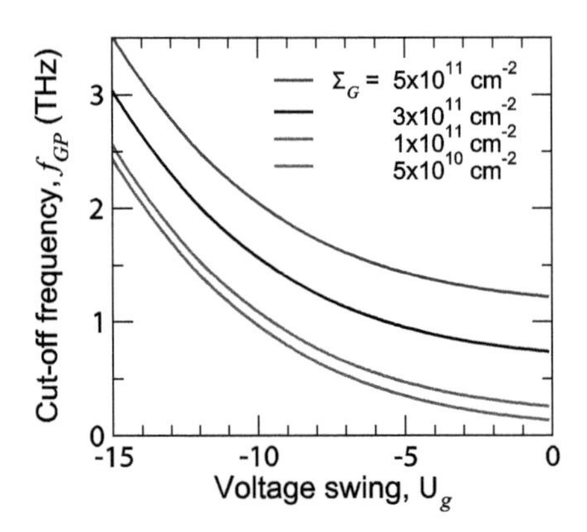

Figure 52.5 The cut-off frequency f_{GP} versus the voltage swing U_g for different scatterer densities Σ_G.

The comparison of the gain-bandwidth products of the GP- and G-bolometers, defined as max $R^V_{GP} f_{GP}$ and $R^V_G f_G$, yields the following estimate for these factors ratio $K \simeq \Lambda_{GP}\Lambda_G$. As seen from Fig. 52.3, K markedly exceeds unity, particularly at $\eta \ll 1$.

52.7 Detectivity of the GP-Bolometers

The dark-current-limited detectivity of the GP-bolometers D^*_{GP} can be evaluated as (see, for example), [44]:

$$D^*_{GP} = \frac{R_{GP}}{\sqrt{4eJ_0H/A}}. \tag{52.29}$$

where J_0H is the net source-drain dc current

Equations (52.18) and (52.24) for relatively low frequencies Ω yield

$$D^*_{GP} = D^*_0 \frac{|\Lambda_{GP}|}{\sqrt{(1+P_N)(1+6|U_g|/\pi^2)}} \tag{52.30}$$

with

$$D^*_0 = \frac{6\alpha}{\pi^{3/2}\sqrt{\kappa}} \frac{v_W^2 \tau_0^{3/2} \tau_0^{\varepsilon}}{T_0} \sqrt{\frac{eE_{SD}}{LT_0}} \propto \tau_0^{3/2}. \tag{52.31}$$

for an arbitrary E_{SD}, and

$$\max D_0^* = \frac{6\alpha}{\pi\sqrt{\kappa}} \frac{v_W^{3/2}\tau_0^{5/4}(\tau_0^{\varepsilon})^{3/4}}{T_0 L^{1/2}} \tag{52.32}$$

for $E_{SD} = \overline{E^*}_{SD}$ Using the same parameters as for the above estimates of max R_0^V and assuming that $L = 10^{-3}$ cm, at $E_{SD} = \overline{E^*}_{GP}$, one obtains $D_0^* \simeq (0.122 - 2.17) \times 10^9$ cm $\sqrt{\text{Hz}}$ /W.

The GP-bolometer dark-current-limited detectivity D_{GP}^*, given by Eq. (52.30), exhibits a fairly steep drop with increasing U_g resembling that of the responsivity R_{GP}^V shown in Fig. 52.3. One should note that the difference in the detectivities for a smaller Σ_G and those corresponding to a larger Σ_G is more pronounced that the pertinent difference in the responsivities.

52.8 Discussion

52.8.1 General Comments

As seen from Eq. (52.11), the quantity $\Lambda_{GB} = \left.\frac{1}{\sigma_0}\frac{d\Sigma_{GP}}{d\ln T}\right|_{T=T_0}$ determines the variation of the GP-channel conductivity due to the carrier heating. Its absolute value $|\Lambda_{GP}|$ exhibits a maximum at a certain value of U_g, which depends on Σ_G. This is seen from Fig. 52.3. Considering only the variation of the channel conductivity associated with the carrier transfer to the P-layer, from Eq. (52.10) we find for Λ_{GP} and the maximum of its modulus $|\Lambda_{GP}|^{max}$

$$\Lambda_{GP} \simeq -\frac{P_N}{(1+P_N)^2}\left(1+\frac{\Delta}{T_0}\right), \tag{52.33}$$

$$|\Lambda_{GP}|^{max} \simeq \frac{1}{4}\left(1+\frac{\Delta}{T_0}\right), \tag{52.34}$$

respectively. This maximum is reached at $P_N = 1$. i.e., at relatively moderate population of the P-layer ($\Sigma_P = \Sigma_G$) that corresponds to $\mu_0 \simeq \Delta - T_0 \ln(\Sigma_N/\Sigma_G)$ or $U_g = -[\Delta/T_0 - \ln(\Sigma_N/\Sigma_G)]^2 = U_g^{max}$. For $\Sigma_G = 5 \times 10^{10}$, 1×10^{11}, and 5×10^{11} cm^{-2}, the pertinent values of U_g are approximately equal to −6, −10, and −23, respectively. The latter is in line with plots in Fig. 52.2. It is interesting that max $|\Lambda_{GP}|^{max}$ and $|U_g|$ are a linear and a quadratic functions of Δ, respectively.

The obtained results show that the responsivity and detectivity of the GP-bolometers steeply decrease with increasing voltage swing U_g (see Fig. 52.3). Thus, it is practical to use the range of relatively small $|U_g|$, although the bandwidth of the GP-bolometers extends with increasing $|U_g|$ as seen from Fig. 52.5. Thus, there is an opportunity of the voltage control of the cut-off frequency. In the GP-LDs, the minimum value of $|U_g|$ is determined by the acceptor density in the GP-channel, so that this density should be minimized to achieve acceptable characteristics. Apart from easier fabrication, the GP-LDs can exhibit the enhance performance of the whole bolometric photodetector due to a more effective THz radiation input.

As demonstrated above, the characteristics of the GL-LD and GL-FET bolometers under consideration can be markedly different depending on the Auger parameter η. This parameter depends on the substrate material, particularly, on its dielectric constant κ. The calculations [29] predicted the optical phonon recombination time in the G-layers from less than a picosecond to several picoseconds at the carrier densities under consideration above and room temperature. The experimental results of the carrier recombination dynamics in G-layer [45] were interpreted assuming that the interband relaxation is associated with the optical phonon processes rather than the carrier-carrier processes, so that $\tau_{Opt} < \tau_{Auger}$ (and η is close to unity). The recent calculations [29] (as well as the previous one's [46]) showed that an increase in κ leads to a virtually linear increase in τ_{Auger} and, hence, in an increase in η. Although, this increase in τ_{Auger} is not too pronounced—the change in κ from 5 to 25 results in a fourfold rise of τ_{Auger} at room temperature [29]. For example, in the case of GP-LDs with SiO_2 and hBN substrates, in which $\kappa \sim 4 - 5$, $\tau_{Auger} \lesssim 1$ ps, whereas bin the case of the HfO_2 substrate, $\tau_{Auger} \gtrsim 2$ ps. In the GP-FETs, the screening of the carrier interaction by a highly conducting gate can substantially suppress the Auger processes. Indeed, using the data obtained recently [29], one can find that τ_{Auger} being $\tau_{Auger} \simeq 1$ ps at $\kappa = 5$ and the gate layer thickness $W_g = 10 - 15$ nm, becomes $\tau_{Auger} \simeq 6$ ps at $W_g = 2$ nm. One needs to point out that in the case of high-κ substrates, additional recombination channel associated with the substrate polar phonons [47] can promote father increase in η. Setting $\tau_{Opt} = (1 - 3)$ ps, we find that the latter values of τ_{Auger} correspond to $\eta = 0.25 - 0.85$.

Hence, the range of the Auger parameter η variations assumed in the above calculations appear to be reasonable.

The values of the GP-detector responsivity demonstrated in Fig. 52.3 are of the same order of magnitude or can exceed the room temperature responsivity of the proposed and realized THz photodetectors based on different heterostructures [21, 48–59], including those based on the P-channel [7, 60, 61] (although in G-based devices at very low temperatures much higher responsivities have been achieved [20]).

52.8.2 Assumptions

The main assumptions of our device model are fairly natural and practical. We disregarded the contribution of the carriers in the P-layer to the net conductivity of the G-P-channel. The pertinent condition can be presented as $\sigma_0 \gtrsim eb_P\Sigma_N \exp[(\Delta - \mu_0)/T_0]$ or $\tau_0 \gtrsim \pi\hbar^2\Sigma_N \exp[(\Delta - \mu_0)/T_0]b_P/eT_0$ where b_P is the carrier mobility in the P-layer. Assuming that b_P = 330 – 540 cm^2/s V [60] (see also [2]) Σ_N = 1.2 × 10^{13} cm^{-2}, and μ_0 < 100 meV (i.e., $|U_g|$ < 25, see Fig. 52.2), the above inequalities are valid if $\tau_0 \gg 0.005$ ps. The values of τ_0 assumed in our calculations well satisfy this requirement.

(i) Above we estimated the scattering time τ_p and, hence, τ_0 as in [15]: $\tau_p^{-1} = \upsilon_{SP}/\hbar$, so that $\tau_0^{-1} = (\upsilon_S/\upsilon_W)(T_0/\hbar)$, where $\upsilon_S = (\pi^2 U_S^2 l_S^2 \Sigma_G/4\hbar^2\upsilon_W)$, $U_S = e^2/\kappa l_S$ is the characteristic potential of the scatter, and l_S is the screening length. Setting l_S = 5 nm or smaller (see also the estimate for l_S at μ_0 = 75 meV [13] and below), we find $\upsilon_S \simeq 2 \times 10^7$ cm/s. At $\Sigma_G = 10^{12}$ cm^{-2} and T_0 = 25 meV, the latter yields $\tau_0 \simeq 0.12$ ps. For the scattering on the acoustic phonons due to the deformation potential interaction with the longitudinal vibrations at T_0 = 25 meV, one obtains $\upsilon_S(ac) \simeq 8 \times 10^5$ cm/s and $\tau_0^{ac} \simeq 3$ ps. The contribution of the hole-hole scattering in the G-layer to its dc and ac conductivity are small (despite substantially non-parabolic hole spectrum [59]). The role of the hole-electron scattering is also small due low electron densities, particularly, at high gate voltages.

(ii) The interband absorption of the incident radiation with the photons with the energies $\hbar\Omega \leq 2\pi\hbar f_{GP} < 2\mu_0$ in the G-layer disregarded in our model, is practically prohibited due to the Pauli blocking. At small values of μ_0 (i.e., small $|U_g|$), this absorption is weak in comparison with the intraband (Drude) absorption if

$$\frac{\pi\alpha}{4} \ll \frac{4\pi\Sigma_{GP}}{c(1+\Omega^2\tau^2)}. \qquad (52.35)$$

At $\Omega < \tau^{-1} = 2\pi f_{GD}$, inequality (52.31) implies $\tau_0 > (\pi\hbar/16T_0) \simeq$ 0.005 ps.

(iii) Considering the features of the DoS (see Fig. 52.1e), the screening length, l_S, of the charges in the GP-channel at low and relatively high voltage swing U_g and $T = T_0$ is given by

$$l_S \lesssim \frac{\kappa\hbar^2 v_W^2}{8\ln 2e^2 T_0}, \quad l_S \lesssim \frac{\kappa\hbar^2 v_W^2}{4e^2\mu_0}, \qquad (52.36)$$

respectively. Assuming $\kappa = 4$ and setting $\mu_0 < T_0 = 25$ meV and $\mu_0 = 60$ meV, from Eq. (52.32) we obtain $l_S \simeq 3.4 - 7.5$ nm. The products of the characteristic carrier wavenumbers $k_T = T_0/\hbar v_w$ and $k_{\mu_0} = \mu_0/4\hbar v_w$ and the pertinent values of l_S are $k_T l_S \lesssim$ 0.31 and $k_\mu l_S \lesssim 0.43$. The latter indicates a rather short range interaction (an effective screening of the charged impurities and the heavy carriers in the P-layer) in the device under consideration at its working conditions.

52.9 Conclusions

We studied the effect of THz photoconductivity of the G- and GP-channels and showed that their conductivity decreases under the THz irradiation (the effect of negative conductivity). It was revealed that this effect in G-channels is determined by the competition of the interband transitions associated with optical phonons and the Auger generation-recombination processes and vanishes when the latter processes prevail. However, the negative conductivity in the GP-channels is weakly sensitive to the relative roles of the latter process. The negative photoconductivity of the ungated and gated GP-channels (GP-LDs and GP-FETs) under the THz irradiation, enables using these devices as bolometric THz photodetectors. We evaluated the responsivity, bandwidth, and detectivity characteristics of such THz bolometers and demonstrated that an effective transfer of the carriers from the G-layer into the P-layer, caused by their heating due to the intraband absorption of the THz radiation, leads to the substantial decrease in the G-P-channel conductivity. This effect of the negative THz photoconductivity is associated primarily with the intensification of the light carrier scattering in the G-layer on the heavy carriers in the P-layer. Using the developed device model for the GP-LD and GP-FET bolometers, we demonstrated that these photodetectors can exhibit a fairly high responsivity in a wide range of the THz frequencies at the room temperature. The main requirement to achieve the elevated photodetector performance is having sufficiently high values of the G-layer mobility. The main characteristics of the GP-FET bolometers are effectively controlled by the gate voltage. The GP-LD and GP-FET THz bolometric photodetectors can substantially surpass the THz bolometers with the G-channel and compete and even outperform the existing devices. Further enhancement of the GP-LD and GP-FET THz bolometer can be realized using the GP-GP-...-GP superlattice heterostructures as the channel, integrating the GP-LDs and GP-FETs with THz microcavities or waveguides, and implementing different schemes of the plasmonic enhancement of the THz absorption.

Appendix A: Short-Range versus Long-Range Scattering

As seen from Fig. 52.2, in the G-channels the quantity $\Lambda_G < 0$. This implies that the carrier heating in the G-layers by the absorbed radiation leads to a decrease in the conductivity, i.e. to the negative photoconductivity. This phenomenon was observed in the experiments (see, for example, [18, 19, 37, 38]). As shown above, the G-layer negative photoconductivity at the room temperatures can appear when the short-range scattering dominates and the Auger generation-recombination processes are weaker than those associated with the optical phonons. In the case of the dominant long-range scattering, the G-layer conductivity rises with increasing carrier effective temperature [15]. In the model [18], the G-layer conductivity was considered assuming that $\tau_p = \tau_0$ is independent of carrier momentum p. In such a model,

$$\sigma_{\rm G} \propto \int_0^\infty d\xi \xi \frac{d(f^{\rm e}+f^{\rm h})}{d\xi}$$
$$= \ln\left[(1+e^{\mu^{\rm e}/T})(1+e^{\mu^{\rm h}/T})\right]. \quad (52.A1)$$

At small and high ratios $\mu^{\rm e}/T$ and $\mu^{\rm h}/T$, Eq. (52.A1) yields $\sigma_{\rm G} \propto [4 + (\mu^{\rm e} + \mu^{\rm h})/2T]$ and $\sigma_{\rm G} \propto (\mu^{\rm e} + \mu^{\rm h})/T]$, respectively. This results (accounting for Eq. (52.4)) in $\sigma_{\rm G} \propto [4 + \eta\hbar\omega(1/T - 1/T_0]$ and $\sigma_{\rm G} \propto \eta\hbar\omega(1/T - 1/T_0)$. One can see that at $\tau_{\rm p} = const$, as in the case $\tau_{\rm p} \propto 1/p$ considered by us, the G-layer conductivity decreases with increasing carrier temperature. However, this effect vanishes (the conductivity becomes insensitive to the carrier temperature variation and, hence, to the irradiation) when the Auger parameter η tends to zero. Thus, even in this case, the carrier temperature dependence of the GP-channel conductivity $\sigma_{\rm GP}$, related to $\sigma_{\rm G}$ according to Eq. (52.10), corresponds to the negative photoconductivity with the main contribution of the carrier transfer to the P-layer.

On the contrary, if the long-range scattering with $\tau_{\rm p} \propto p$ would dominate, $\sigma_{\rm G}$ could be a rising function of the carrier temperature leading to the positive G-layer photoconductivity. This can surpass the effect of the G-to-P carrier transitions. Both types of the G-layer photoconductivity (negative and positive) depending on the photon energy and the enviromental gases have been observed, for example, in [38].

Appendix B: Frequency Dependence of the GP- and G-Channel Conductivity

Following the standard procedures (see, for example, [32]), the ac conductivity can be presented as (compare with Eq. (52.6))

$$\sigma_{\rm GP,\Omega} = -\frac{\sigma_0\Sigma_{\rm G}}{\Sigma_{\rm G}+\Sigma_{\rm P}} \int_0^\infty \frac{d\xi[d(f^{\rm e}+f^{\rm h})/d\xi]}{1+\dfrac{\Omega^2\tau_0^2\Sigma_{\rm G}^2}{(\Sigma_{\rm G}+\Sigma_{\rm P})^2\xi^2}}. \quad (52.B1)$$

At low and high frequencies frequencies, Eq. (52.B1) can be rewritten as

$$\sigma_{\rm GP,\Omega} \simeq \sigma_{\rm GP} \simeq \frac{\sigma_0\Sigma_{\rm G}}{(\Sigma_{\rm G}+\Sigma_{\rm P})}, \quad (52.B2)$$

$$\sigma_{\mathrm{GP},\Omega} \simeq \frac{\sigma_{\mathrm{GP}}}{\Omega^2\tau^2} \simeq \frac{\sigma_0\Sigma_{\mathrm{G}}}{(\Sigma_{\mathrm{G}}+\Sigma_{\mathrm{P}})} \frac{1}{\Omega^2\tau^2}, \tag{52.B3}$$

respectively, where

$$\frac{1}{\tau} = \frac{2}{\tau_0} \frac{(\Sigma_{\mathrm{G}}+\Sigma_{\mathrm{P}})}{\Sigma_{\mathrm{G}}} \left[\mathcal{F}_1\left(-\frac{\mu_0}{T_0}\right) + \mathcal{F}_1\left(\frac{\mu_0}{T_0}\right)\right]^{1/2}. \tag{52.B4}$$

For $U_g \simeq 0$ and $U_g \ggg 1$, Eq. (52.B4) yields

$$\tau \simeq \frac{\sqrt{3}\tau_0}{\pi(1+P_{\mathrm{N}})} \qquad \tau \simeq \frac{1}{(1+P_{\mathrm{N}})\sqrt{2|U_g|}}, \tag{52.B5}$$

so that τ as a function of $|U_g|$ can, for example, be interpolated by Eq. (52.16). In the case of the dominating long-range scattering, for the cut-off frequency one obtains $f_{\mathrm{GP}} \simeq 1/2\pi\tau_0$.

Acknowledgments

The authors are grateful to P. P. Maltsev, A. Satou, D. Svintsov, and V. Vyurkov for useful discussions. VR is also thankful to N. Ryabova for assistance. The work was supported by Japan Society for Promotion of Science, KAKENHI Grant No. 16H06361, the Russian Science Foundation (Grant No.14-29-00277), Russian Foundation for Basic Research (Grant No. 18-07-01145), RIEC Nation-Wide Collaborative Research Project, and by Office of Naval Research (Project Monitor Dr. Paul Maki).

References

1. A. H. Castro Neto, F. Guinea, N. M. R. Peres, K. S. Novoselov, and A. K. Geim, The electronic properties of graphene, *Rev. Mod. Phys.*, **81**, 109–162 (2009).
2. X. Ling, H. Wang, S. Huang, F. Xia, and M. S. Dresselhaus, The renaissance of black phosphorus, *Proc. Natl. Acad. Sci. U.S.A.*, **112**, 4523–4530 (2015).
3. F. Bonaccorso, Z. Sun, T. Hasan, and A. Ferrari, Graphene photonics and optoelectronics, *Nat. Photonics*, **4**, 611–622 (2010).
4. V. Ryzhii, M. Ryzhii, V. Mitin, and T. Otsuji, Toward the creation of terahertz graphene injection laser, *J. Appl. Phys.*, **110**, 094503 (2011).

5. Q. Bao and K. P. Loh, Graphene photonics, plasmonics, and broadband optoelectronic devices, *ACS Nano*, **6**, 3677–3677 (2012).

6. A. Tredicucci and M. Vitiello, Device concepts for graphene-based terahertz photonics, *J. Sel. Top. Quant.*, **20**, 130–138 (2014).

7. M. Buscema, D. J. Groenendijk, S. I. Blanter, G. A. Steele, H. S. J. van der Zant, and A. Castellanos-Gomez, Fast and broadband photoresponse of few-layer black phosphorus field-effect transistors, *Nano Lett.*, **14**, 3347–3352 (2014).

8. M. Engel, M. Steiner, and Ph. Avouris, A black phosphorus photo-detector for multispectral high-resolution imaging, *Nano Lett.*, **14**, 6414–6417 (2014).

9. E. Leong, R. J. Suess, A. B. Sushkov, H. D. Drew, T. E. Murphy, and M. Mittendorff, Terahertz photoresponse of black phopsporus, *Opt. Express*, **25**(11), 12666–12674 (2017).

10. F. Ahmed, Y. D. Kim, M. S. Choi, X. Liu, D. Qu, Z. Yang, J. Hu, I. P. Herman, J. Hone, W. J. Yoo, High electric field carrier transport and power dissipation in multilayer black phosphorus field effect transistor with dielectric engineering, *Adv. Funct. Mater.*, **27**, 1604025 (2017).

11. Y. Deng, Z. Luo, N. J. Conrad, H. Liu, Y. Gong, S. Najmaei, P. M. Ajayan, J. Lou, X. Xu, P. D. Ye, Black phosphorus-monolayer MoS_2 van der Waals heterojunction p-n diode, *ACS Nano*, **8**, 8292–8299 (2014).

12. F. H. L. Koppens, T. Mueller, Ph. Avouris, A, C. Ferrari, M. S. Vitiello, and M. Polini, Photodetectors based on graphene, other two-dimensional materials and hybrid systems, *Nat. Nanotechnol.*, **9**, 780–793 (2014).

13. V. Ryzhii, M. Ryzhii, D. Svitsov, V. Leiman, P. P. Maltsev, D. S. Ponomarev, V. Mitin, M. S. Shur, and T. Otsuji, Real-space-transfer mechanism of negative differential conductivity in gated graphene-phosphorene hybrid structures: phenomenological heating model, *J. Appl. Phys.*, **124**, 114501 (2018).

14. Y. Cai, G. Zhang, and Y.-W. Zhang, Layer-dependent band alignment and work function of few-layer phosphorene, *Sci. Rep.*, **4**, 6677 (2014).

15. F. T. Vasko and V. Ryzhii, Voltage and temperature dependencies of conductivity in gated graphene, *Phys. Rev. B*, **76**, 233404 (2007).

16. O. G. Balev, F. T. Vasko, and V. Ryzhii, Carrier heating in intrinsic graphene by a strong dc electric field, *Phys. Rev. B*, **79**, 165432 (2009).

17. V. Ryzhii, T. Otsuji, M. Ryzhii, N. Ryabova, S. O. Yurchenko, V. Mitin, and M. S. Shur, Graphene terahertz uncooled bolometers, *J. Phys. D: Appl. Phys.*, **46**, 065102 (2013).

18. J. N. Heyman, J. D. Stein, Z. S. Kaminski, A. R. Banman, A. M. Massari, and J. T. Robinson, Carrier heating and negative photoconductivity in graphene, *J. Appl. Phys.*, **117**, 015101 (2015).

19. Xu Du, D. E. Prober, H. Vora, and C. Mckitterick, Graphene-based bolometers, *Graphene 2D Mater.*, **1**, 1–22 (2014).

20. Q. Han, T. Gao, R. Zhang, Yi Chen, J. Chen, G. Liu, Y. Zhang, Z. Liu, X. Wu, and D. Yu, Highly sensitive hot electron bolometer based on disordered graphene, *Sci. Rep.*, **3**, 3533 (2013).

21. G. Skoblin, J. Sun, and A. Yurgens, Graphene bolometer with thermoelectric readout and capacitive coupling to an antenna, *Appl. Phys. Lett.*, **112**, 063501 (2018)

22. G. Zhang, A. Chaves, S. Huang F. Wang, Q. Xing, T. Low, and H. Yan, Determination of layer-dependent exciton binding energies in few-layer black phosphorus, *Sci. Adv.*, **4**(3), eaap9977 (2018).

23. T. Low, R. Roldn, H. Wang, F. Xia, P. Avouris, L. M. Moreno, and F. Guinea, Plasmons and screening in monolayer and multilayer black phosphorus, *Phys. Rev. Lett.*, **113**, 106802 (2014).

24. S. Yuan, A. N. Rudenko, and M. I. Katsnelson, Transport and optical properties of single- and bilayer black phosphorus with defects, *Phys. Rev. B*, **91**, 115436 (2015).

25. J. Xi, M. Long, D. Wang, and Z. Shuai, First principles prediction of charge mobility in carbon and organic nanomaterials, *Nanoscale*, **4**, 4348–4369 (2012).

26. F. Rana, P. A. George, J. H. Strait, S. Sharavaraman, M. Charasheyhar, and M. G. Spencer, Carrier recombination and generation rates for intravalley and intervalley phonon scattering in graphene, *Phys. Rev. B*, **79**, 115447 (2009).

27. V. Ryzhii, M. Ryzhii, V. Mitin, A. Satou, and T. Otsuji, Effect of heating and cooling of photogenerated electron-hole plasma in optically pumped graphene on population inversion, *Jpn. J. Appl. Phys.*, **50**, 094001 (2011).

28. M. S. Foster and I. L. Aleiner, Slow imbalance relaxation and thermoelectric transport in graphene, *Phys. Rev. B*, **79**, 085415 (2009).

29. G. Alymov, V. Vyurkov, V. Ryzhii, A. Satou, and D. Svintsov, Auger recombination in Dirac materials: a tangle of many-body effects, *Phys. Rev. B*, **97**, 205411 (2018).

30. J. S. Blakemore, *Semiconductor Statistics* (Dover, 1987).

31. T. Ando, Screening Effect and impurity scattering in monolayer graphene, *J. Phys, Soc. Jpn.*, **75**, 074716 (2006)

32. L. A. Falkovsky and A. A. Varlamov, Space-time dispersion of graphene conductivity, *Eur. Phys. J. B*, **56**, 281–284 (2007).
33. E. H. Hwang, S. Adam, and S. D. Sarma, Carrier transport in two-dimensional graphene layers, *Phys. Rev. Lett.*, **98**, 186806 (2007).
34. V. Vyurkov and V. Ryzhii, Effect of Coulomb scattering on graphene conductivity, *JETP Lett.*, **88**, 370–372 (2008).
35. E. H. Hwang and S. Das Sarma, Acoustic phonon scattering limited carrier mobility in two-dimensional extrinsic graphene, *Phys. Rev. B*, **77**, 115449 (2008).
36. E. H. Hwang and S. Das Sarma, Screening induced temperature dependent transport in 2D graphene, *Phys. Rev. B*, **79**, 165404 (2009).
37. G. Jnawali, Y. Rao, H. G. Yan, and T. F. Heinz, Observation of a transient decrease in terahertz conductivity of single-layer graphene induced by ultrafast optical excitation, *Nano Lett.*, **13**, 524–530 (2013).
38. C. J. Docherty, C. T. Lin, H. J. Joyce, R. J. Nicholas, L. M. Hertz, L. J. Li, and M. B. Johnston, Extreme sensitivity of graphene photoconductivity to environmental gases, *Nat. Commun.*, **3**, 1228 (2012).
39. S. V. Morozov, K. S. Novoselov, M. I. Katsnelson, F. Schedin, D. C. Elias, J. A. Jaszczak, and A. K. Geim, Giant intrinsic carrier mobilities in graphene and its bilayer, *Phys. Rev. Lett.*, **100**, 016602 (2008).
40. H. Hirai, H. Tsuchiya, Y. Kamakura, N. Mori, and M. Ogawa, Electron mobility calculation for graphene on substrates, *J. Appl. Phys.*, **116**, 083703 (2014).
41. L. Banszerus, M. Schmitz, S. Engels, J. Dauber, M. Oellers, F. Haupt, K. Watanabe, T. Taniguchi, B. Beschoten, and C. Stampfer, Ultrahigh-mobility graphene devices from chemical vapor deposition on reusable copper, *Sci. Adv.*, **1**(6), e1500222 (2015).
42. L. Wang, I. Meric, P. Y. Huang, Q. Gao, Y. Gao, H. Tran, T. Taniguchi, K. Watanabe, L. M. Campos, D. A. Muller, J. Guo, P. Kim, J. Hone, K. L. Shepard, and C. R. Dean, One-dimensional electrical contact to a two-dimensional material, *Science*, **342**, 614–617 (2013).
43. K. J. Tielrooij, J. C. W. Song, S. A. Jensen, A. Centeno, A. Pesquaera, A. Z. Elorza, M. Bonn, L. S. Levitov, and F. H. L. Koppens, Photoexcitation cascade and multiple hot-carrier generation in graphene, *Nat. Phys.*, **9**, 248–252 (2013).
44. H. Schneider and H,C, Liu, *Quantum Well Infrared Photodetectors: Physics and Applications* (Springer, NY, 2007).
45. J. M. Dawlaty, S. Shivaraman, M. Chandrashekhar, F. Rana, and M. G. Spencer, Measurement of ultrafast carrier dynamics in epitaxial graphene, *Appl. Phys. Lett.*, **92**, 042116 (2008).

46. F. Rana, Electron-hole generation and recombination rates for coulomb scattering in graphene, *Phys. Rev. B*, **76**, 155431 (2007).

47. F. Rana, J. H. Strait, H, Wang, and C, Manolatou, Ultrafast carrier recombination and generation rates for plasmon emission and absorption in graphene, *Phys. Rev. B*, **84**, 045437 (2011).

48. S. D. Gunapala, S. V. Bandara, J. K. Liu, J. M. Mumolo, S. B. Rafol, D. Z. Ting, A. Soibel, and C. Hill, Quantum well infrared photodetector technology and applications, *IEEE J. Sel. Top. Quantam Electron.*, **20**(6), 154–165 (2014).

49. V. Ryzhii, T. Otsuji, V. E. Karasik, M.Ryzhii, V. Leiman, V. Mitin, and M. S. Shur, Comparison of intersubband quantum-well and interband graphene layer infrared photodetectors, *IEEE J. Quantum Electron.* **54**(2), 1–8 (2018).

50. V. Ryzhii, M. Ryzhii, M. S. Shur, V. Mitin, A. Satou, and T. Otsuji, Resonant plasmonic terahertz detection in graphene split-gate field-effect transistors with lateral pn junctions, *J. Phys. D: Appl. Phys.*, **49**, 315103 (2016).

51. M. Mittendorff, S. Winnerl, J. Kamann, J. Eroms, D. Weiss, H. Schneider, and M. Helm, Ultrafast graphene-based broadband THz detector, *Appl. Phys. Lett.*, **103**, 021113 (2013).

52. V. Ryzhii, M. Ryzhii, D. Svintsov, V. Leiman, V. Mitin, M. S. Shur, and T. Otsuji, Nonlinear response of infrared photodetectors based on van der Waals heterostructures with graphene layers, *Opt. Express*, **25**, 5536–5549 (2017).

53. V. Ryzhii, M. Ryzhii, V. Leiman, V. Mitin, M. S. Shur, and T. Otsuji, Effect of doping on the characteristics of infrared photodetectors based on van der Waals heterostructures with multiple graphene layers, *J. Appl. Phys.*, **122**, 054505 (2017).

54. V. Ya. Aleshkin, A. A. Dubinov, S. V. Morozov, M. Ryzhii, T. Otsuji, V. Mitin, M. S. Shur, and V. Ryzhii, Interband infrared photodetectors based on HgTe-CdHgTe quantum-well heterostructures, *Opt. Mater. Express*, **8**, 1349 (2018).

55. V. Ryzhii, M. Ryzhii, V. Mitin, and T. Otsuji, Terahertz and infrared photodetection using p-i-n multiple-graphene-layer structures, *J. Appl. Phys.*, **107**, 054512 (2010).

56. A. V. Muraviev, S. L. Rumyantsev, G. Liu, A. A. Balandin, W. Knap, and M. S. Shur, Plasmonic and bolometric terahertz detection by graphene field-effect transistor, *Appl. Phys. Lett.*, **103**, 181114 (2013).

57. Y. Wang, W. Yin, Q. Han, X. Yang, H. Ye, Q. Lv, and D. Yin, Bolometric effect in a waveguide-integrated graphene photodetector, *Chin. Phys. B*, **25**, 118103 (2016).

58. D. A. Bandurin, D. Svintsov, I. Gayduchenko, S. G. Xu, A. Principi, M. Moskotin, I. Tretyakov, D. Yagodkin, S. Zhukov, T. Taniguchi, K. Watanabe, I. V. Grigorieva, M. Polini, G. Goltsman, A. K. Geim, and G. Fedorov, Resonant terahertz detection using graphene plasmons, *Nat. Commun.*, **9**, 5392 (2018).

59. D. S. Ponomarev, D. V. Lavrukhin, A. E. Yachmenev, R. A. Khabibullin, I. E. Semenikhin, V. V. Vyurkov, M. Ryzhii, T. Otsuji, and V. Ryzhii, Lateral terahertz hot-electron bolometer based on an array of Sn nanothreads in GaAs, *J. Phys. D: Appl. Phys.*, **51**, 135101 (2018).

60. L. Viti, J. Hu, D. Coquillat, A. Politano, W. Knap, and M. S. Vitiello, Efficient terahertz detection in black-phosphorus nano-transistors with selective and controllable plasma-wave, bolometric and thermoelectric response, *Sci. Rep.*, **6**, 20474 (2016).

61. E. Leong, R. J. Suess, A. B. Sushkov, H. D. Drew, T. E. Murphy, and M. Mittendorff, Terahertz photoresponse of black phosphorus, *Opt. Express*, **25**, 12666–12674 (2017).

62. D. Svintsov, V. Ryzhii, A. Satou, T. Otsuji, and V. Vyurkov, Carrier-carrier scattering and negative dynamic conductivity in pumped graphene, *Opt. Express*, **22**, 19873–19886 (2014).

Index